国外优秀数学著作
原 版 系 列

[美] 迈克尔·弗里德(Michael D.Fried)
[美] 摩西·贾登(Moshe Jarden) 著

算术域

Field Arithmetic (Third Edition)

哈尔滨工业大学出版社
HARBIN INSTITUTE OF TECHNOLOGY PRESS

黑版贸审字 08—2017—069 号

Reprint from the English language edition:
Field Arithmetic
by Michael D. Fried and Moshe Jarden
Copyright © Springer-Verlag Berlin Heidelberg 2008
This work is published by Springer Nature
The registered company is Springer-Verlag GmbH
All Rights Reserved
This reprint has been authorised by Springer Nature for distribution in China Mainland.

图书在版编目(CIP)数据

算术域:第 3 版:英文/(美)迈克尔·弗里德,(美)摩西·贾登著. —哈尔滨:哈尔滨工业大学出版社,2018.1
书名原文:Field Arithmetic,Third Edition
ISBN 978-7-5603-6776-7

Ⅰ.①算… Ⅱ.①迈…②摩… Ⅲ.①算术-研究-英文 Ⅳ.①O121

中国版本图书馆 CIP 数据核字(2017)第 174105 号

策划编辑	刘培杰
责任编辑	张永芹　杜莹雪
封面设计	孙茵艾
出版发行	哈尔滨工业大学出版社
社　　址	哈尔滨市南岗区复华四道街 10 号　邮编 150006
传　　真	0451-86414749
网　　址	http://hitpress.hit.edu.cn
印　　刷	哈尔滨市工大节能印刷厂
开　　本	787mm×1092mm　1/16　印张 52.5　字数 968 千字
版　　次	2018 年 1 月第 1 版　2018 年 1 月第 1 次印刷
书　　号	ISBN 978-7-5603-6776-7
定　　价	158.00 元

(如因印装质量问题影响阅读,我社负责调换)

To those precious colleagues who can appreciate the goals of and connections to other areas. To those who acknowledge the depth of what we already know from the absorbed contribution of previous generations before we address our papers. To those who can transcend the hubris of today's mathematical community.

לְרִנָּה בְּאַהֲבָה
עַל תְּמִיכָה וְאַהֲדָה

To those precious colleagues who can appreciate
the goals of and connections to other areas. To
those who acknowledge the depth of what we already
know from time absorbed contribution of
previous generations before we address our Tc
part. To those who can transcend the limits of
today's mathematical community.

Table of Contents

Chapter 1. Infinite Galois Theory and Profinite Groups 1
  1.1 Inverse Limits 1
  1.2 Profinite Groups 4
  1.3 Infinite Galois Theory 9
  1.4 The p-adic Integers and the Prüfer Group 12
  1.5 The Absolute Galois Group of a Finite Field 15
  Exercises 16
  Notes 18

Chapter 2. Valuations and Linear Disjointness 19
  2.1 Valuations, Places, and Valuation Rings 19
  2.2 Discrete Valuations 21
  2.3 Extensions of Valuations and Places. 24
  2.4 Integral Extensions and Dedekind Domains 30
  2.5 Linear Disjointness of Fields 34
  2.6 Separable, Regular, and Primary Extensions 38
  2.7 The Imperfect Degree of a Field 44
  2.8 Derivatives 48
  Exercises 50
  Notes 51

Chapter 3. Algebraic Function Fields of One Variable 52
  3.1 Function Fields of One Variable 52
  3.2 The Riemann-Roch Theorem 54
  3.3 Holomorphy Rings 56
  3.4 Extensions of Function Fields 59
  3.5 Completions 61
  3.6 The Different 67
  3.7 Hyperelliptic Fields 70
  3.8 Hyperelliptic Fields with a Rational quadratic Subfield . . 73
  Exercises 75
  Notes 76

Chapter 4. The Riemann Hypothesis for Function Fields 77
  4.1 Class Numbers 77
  4.2 Zeta Functions 79
  4.3 Zeta Functions under Constant Field Extensions 81
  4.4 The Functional Equation 82
  4.5 The Riemann Hypothesis and Degree 1 Prime Divisors . . 84
  4.6 Reduction Steps 86
  4.7 An Upper Bound 87
  4.8 A Lower Bound 89

Exercises 91
Notes 93

Chapter 5. Plane Curves 95
5.1 Affine and Projective Plane Curves 95
5.2 Points and prime divisors 97
5.3 The Genus of a Plane Curve 99
5.4 Points on a Curve over a Finite Field 104
Exercises 105
Notes 106

Chapter 6. The Chebotarev Density Theorem 107
6.1 Decomposition Groups 107
6.2 The Artin Symbol over Global Fields 111
6.3 Dirichlet Density 113
6.4 Function Fields 115
6.5 Number Fields 121
Exercises 129
Notes 130

Chapter 7. Ultraproducts 132
7.1 First Order Predicate Calculus 132
7.2 Structures 134
7.3 Models 135
7.4 Elementary Substructures 137
7.5 Ultrafilters 138
7.6 Regular Ultrafilters 139
7.7 Ultraproducts 141
7.8 Regular Ultraproducts 145
7.9 Nonprincipal Ultraproducts of Finite Fields 147
Exercises 147
Notes 148

Chapter 8. Decision Procedures 149
8.1 Deduction Theory 149
8.2 Gödel's Completeness Theorem 152
8.3 Primitive Recursive Functions 154
8.4 Primitive Recursive Relations 156
8.5 Recursive Functions 157
8.6 Recursive and Primitive Recursive Procedures 159
8.7 A Reduction Step in Decidability Procedures 160
Exercises 161
Notes 162

Chapter 9. Algebraically Closed Fields 163
9.1 Elimination of Quantifiers 163
9.2 A Quantifiers Elimination Procedure 165
9.3 Effectiveness 168
9.4 Applications 169
Exercises 170
Notes 170

Chapter 10. Elements of Algebraic Geometry 172
10.1 Algebraic Sets 172
10.2 Varieties 175
10.3 Substitutions in Irreducible Polynomials 176
10.4 Rational Maps 178
10.5 Hyperplane Sections 180
10.6 Descent 182
10.7 Projective Varieties 185
10.8 About the Language of Algebraic Geometry 187
Exercises 190
Notes 191

Chapter 11. Pseudo Algebraically Closed Fields 192
11.1 PAC Fields 192
11.2 Reduction to Plane Curves 193
11.3 The PAC Property is an Elementary Statement 199
11.4 PAC Fields of Positive Characteristic 201
11.5 PAC Fields with Valuations 203
11.6 The Absolute Galois Group of a PAC Field 207
11.7 A non-PAC Field K with K_{ins} PAC 211
Exercises 217
Notes 218

Chapter 12. Hilbertian Fields 219
12.1 Hilbert Sets and Reduction Lemmas 219
12.2 Hilbert Sets under Separable Algebraic Extensions . . . 223
12.3 Purely Inseparable Extensions 224
12.4 Imperfect fields 228
Exercises 229
Notes 230

Chapter 13. The Classical Hilbertian Fields 231
13.1 Further Reduction 231
13.2 Function Fields over Infinite Fields 236
13.3 Global Fields 237
13.4 Hilbertian Rings 241
13.5 Hilbertianity via Coverings 244

13.6 Non-Hilbertian g-Hilbertian Fields	248
13.7 Twisted Wreath Products	252
13.8 The Diamond Theorem	258
13.9 Weissauer's Theorem	262
Exercises	264
Notes	266

Chapter 14. Nonstandard Structures 267

14.1 Higher Order Predicate Calculus	267
14.2 Enlargements	268
14.3 Concurrent Relations	270
14.4 The Existence of Enlargements	272
14.5 Examples	274
Exercises	275
Notes	276

Chapter 15. Nonstandard Approach
to Hilbert's Irreducibility Theorem 277

15.1 Criteria for Hilbertianity	277
15.2 Arithmetical Primes Versus Functional Primes	279
15.3 Fields with the Product Formula	281
15.4 Generalized Krull Domains	283
15.5 Examples	286
Exercises	289
Notes	290

Chapter 16. Galois Groups over Hilbertian Fields 291

16.1 Galois Groups of Polynomials	291
16.2 Stable Polynomials	294
16.3 Regular Realization of Finite Abelian Groups	298
16.4 Split Embedding Problems with Abelian Kernels	302
16.5 Embedding Quadratic Extensions in $\mathbb{Z}/2^n\mathbb{Z}$-extensions	306
16.6 $\mathbb{Z}_p$-Extensions of Hilbertian Fields	308
16.7 Symmetric and Alternating Groups over Hilbertian Fields	315
16.8 GAR-Realizations	321
16.9 Embedding Problems over Hilbertian Fields	325
16.10 Finitely Generated Profinite Groups	328
16.11 Abelian Extensions of Hilbertian Fields	332
16.12 Regularity of Finite Groups over Complete Discrete Valued Fields	334
Exercises	335
Notes	336

Chapter 17. Free Profinite Groups 338

17.1 The Rank of a Profinite Group	338

Table of Contents   v

   17.2 Profinite Completions of Groups 340
   17.3 Formations of Finite Groups 344
   17.4 Free pro-$\mathcal{C}$ Groups 346
   17.5 Subgroups of Free Discrete Groups 350
   17.6 Open Subgroups of Free Profinite Groups 358
   17.7 An Embedding Property 360
   Exercises 361
   Notes 362

Chapter 18. The Haar Measure 363
   18.1 The Haar Measure of a Profinite Group 363
   18.2 Existence of the Haar Measure 366
   18.3 Independence 370
   18.4 Cartesian Product of Haar Measures 376
   18.5 The Haar Measure of the Absolute Galois Group . . . 378
   18.6 The PAC Nullstellensatz 380
   18.7 The Bottom Theorem 382
   18.8 PAC Fields over Uncountable Hilbertian Fields 386
   18.9 On the Stability of Fields 390
   18.10 PAC Galois Extensions of Hilbertian Fields 394
   18.11 Algebraic Groups 397
   Exercises 400
   Notes 401

Chapter 19. Effective Field Theory and Algebraic Geometry . . . 403
   19.1 Presented Rings and Fields 403
   19.2 Extensions of Presented Fields 406
   19.3 Galois Extensions of Presented Fields 411
   19.4 The Algebraic and Separable Closures of Presented Fields . 412
   19.5 Constructive Algebraic Geometry 413
   19.6 Presented Rings and Constructible Sets 422
   19.7 Basic Normal Stratification 425
   Exercises 427
   Notes 428

Chapter 20. The Elementary Theory of e-Free PAC Fields . . . 429
   20.1 $\aleph_1$-Saturated PAC Fields 429
   20.2 The Elementary Equivalence Theorem
                 of $\aleph_1$-Saturated PAC Fields 430
   20.3 Elementary Equivalence of PAC Fields 433
   20.4 On e-Free PAC Fields 436
   20.5 The Elementary Theory of Perfect e-Free PAC Fields . 438
   20.6 The Probable Truth of a Sentence 440
   20.7 Change of Base Field 442
   20.8 The Fields $K_s(\sigma_1,\ldots,\sigma_e)$ 444

20.9 The Transfer Theorem 446
20.10 The Elementary Theory of Finite Fields 448
Exercises 451
Notes 453

Chapter 21. Problems of Arithmetical Geometry 454
21.1 The Decomposition-Intersection Procedure 454
21.2 C_i-Fields and Weakly C_i-Fields 455
21.3 Perfect PAC Fields which are C_i 460
21.4 The Existential Theory of PAC Fields 462
21.5 Kronecker Classes of Number Fields 463
21.6 Davenport's Problem 467
21.7 On permutation Groups 472
21.8 Schur's Conjecture 479
21.9 Generalized Carlitz's Conjecture 489
Exercises 493
Notes 495

Chapter 22. Projective Groups and Frattini Covers 497
22.1 The Frattini Groups of a Profinite Group 497
22.2 Cartesian Squares 499
22.3 On $\mathcal{C}$-Projective Groups 502
22.4 Projective Groups 506
22.5 Frattini Covers 508
22.6 The Universal Frattini Cover 513
22.7 Projective Pro-p-Groups 515
22.8 Supernatural Numbers 520
22.9 The Sylow Theorems 522
22.10 On Complements of Normal Subgroups 524
22.11 The Universal Frattini p-Cover 528
22.12 Examples of Universal Frattini p-Covers 532
22.13 The Special Linear Group $SL(2, \mathbb{Z}_p)$ 534
22.14 The General Linear Group $GL(2, \mathbb{Z}_p)$ 537
Exercises 539
Notes 542

Chapter 23. PAC Fields and Projective Absolute Galois Groups . . 544
23.1 Projective Groups as Absolute Galois Groups 544
23.2 Countably Generated Projective Groups 546
23.3 Perfect PAC Fields of Bounded Corank 549
23.4 Basic Elementary Statements 550
23.5 Reduction Steps 554
23.6 Application of Ultraproducts 558
Exercises 561
Notes 561

Chapter 24. Frobenius Fields ... 562
24.1 The Field Crossing Argument ... 562
24.2 The Beckmann-Black Problem ... 565
24.3 The Embedding Property and Maximal Frattini Covers ... 567
24.4 The Smallest Embedding Cover of a Profinite Group ... 569
24.5 A Decision Procedure ... 574
24.6 Examples ... 576
24.7 Non-projective Smallest Embedding Cover ... 579
24.8 A Theorem of Iwasawa ... 581
24.9 Free Profinite Groups of at most Countable Rank ... 583
24.10 Application of the Nielsen-Schreier Formula ... 586
Exercises ... 591
Notes ... 592

Chapter 25. Free Profinite Groups of Infinite Rank ... 594
25.1 Characterization of Free Profinite Groups
                by Embedding Problems ... 595
25.2 Applications of Theorem 25.1.7 ... 601
25.3 The Pro-$\mathcal{C}$ Completion of a Free Discrete Group ... 604
25.4 The Group Theoretic Diamond Theorem ... 606
25.5 The Melnikov Group of a Profinite Group ... 613
25.6 Homogeneous Pro-$\mathcal{C}$ Groups ... 615
25.7 The S-rank of Closed Normal Subgroups ... 620
25.8 Closed Normal Subgroups with a Basis Element ... 623
25.9 Accessible Subgroups ... 625
Notes ... 633

Chapter 26. Random Elements in Free Profinite Groups ... 635
26.1 Random Elements in a Free Profinite Group ... 635
26.2 Random Elements in Free pro-p Groups ... 640
26.3 Random e-tuples in $\hat{\mathbb{Z}}^n$ ... 642
26.4 On the Index of Normal Subgroups
              Generated by Random Elements ... 646
26.5 Freeness of Normal Subgroups
              Generated by Random Elements ... 651
Notes ... 654

Chapter 27. Omega-Free PAC Fields ... 655
27.1 Model Companions ... 655
27.2 The Model Companion in an Augmented Theory of Fields ... 659
27.3 New Non-Classical Hilbertian Fields ... 664
27.4 An abundance of ω-Free PAC Fields ... 667
Notes ... 670

Chapter 28. Undecidability 671
28.1 Turing Machines 671
28.2 Computation of Functions by Turing Machines 672
28.3 Recursive Inseparability of Sets of Turing Machines 676
28.4 The Predicate Calculus 679
28.5 Undecidability in the Theory of Graphs 682
28.6 Assigning Graphs to Profinite Groups 687
28.7 The Graph Conditions 688
28.8 Assigning Profinite Groups to Graphs 690
28.9 Assigning Fields to Graphs 694
28.10 Interpretation of the Theory of Graphs
                               in the Theory of Fields . . . 694
Exercises 697
Notes 697

Chapter 29. Algebraically Closed Fields with
                               Distinguished Automorphisms . . 698
29.1 The Base Field K 698
29.2 Coding in PAC Fields with Monadic Quantifiers 700
29.3 The Theory of Almost all $\langle \tilde{K}, \sigma_1, \ldots, \sigma_e \rangle$'s 704
29.4 The Probability of Truth Sentences 706

Chapter 30. Galois Stratification 708
30.1 The Artin Symbol 708
30.2 Conjugacy Domains under Projection 710
30.3 Normal Stratification 715
30.4 Elimination of One Variable 717
30.5 The Complete Elimination Procedure 720
30.6 Model-Theoretic Applications 722
30.7 A Limit of Theories 725
Exercises 726
Notes 729

Chapter 31. Galois Stratification over Finite Fields 730
31.1 The Elementary Theory of Frobenius Fields 730
31.2 The Elementary Theory of Finite Fields 735
31.3 Near Rationality of the Zeta Function of a Galois Formula . 739
Exercises 748
Notes 750

Chapter 32. Problems of Field Arithmetic 751
32.1 Open Problems of the First Edition 751
32.2 Open Problems of the Second Edition 754
32.3 Open problems 758

Table of Contents	ix
References	761
Index	780

Introduction to the Third Edition

The third edition of "Field Arithmetic" improves the second edition in two ways. First it removes many typos and mathematical inaccuracies that occur in the second edition. In particular, it fills out a big gap in the References of the second edition, where unfortunately all references between "Gilmore and Robinson" and "Kantor and Lubotzky" are missing. Secondly, the third edition reports on five open problems of the second edition that were solved since that edition appeared in 2005.

János Kollár solved Problem 2 by proving that if each projective plane curve defined over a field K has a K-rational point, then K is PAC.

János Kollár also solved Problem 3 and proved that if K is a PAC field, w is a valuation of $\tilde{K}$, and V is a variety defined over K, then $V(K)$ is w-dense in $V(\tilde{K})$.

János Kollár partially settled Problem 21. He proved that every PAC field of characteristic 0 is C_1.

Problem 31 was affirmatively solved by Lior Bary-Soroker by establishing an analog of the diamond theorem for the finitely generated non-Abelian free profinite groups.

Finally, Eric Rosen suggested to reorganize Corollary 28.5.3 of the second edition that led to an affirmative solution of Problem 33.

Unfortunately, a full account of the first four solutions is out of the scope of the present volume.

Much of the improvment made in the present edition is due to Arno Fehm and Dan Haran. I am really indebted to them for their contribution.

Tel Aviv, Autumn 2007   Moshe Jarden

Introduction to the Second Edition

The first edition of "Field Arithmetic" appeared in 1986. At the end of that edition we gave a list of twenty-two open problems. It is remarkable that since then fifteen of them were partially or fully solved. Parallel to this, Field Arithmetic has developed in many directions establishing itself as an independent branch of Algebra and Number Theory. Some of these developments have been documented in books. We mention here "Groups as Galois groups" [Völklein] on consequences of the Riemann existence theorem, "Inverse Galois Groups" [Malle-Matzat] with a comprehensive report on finite Galois groups over number fields, "Profinite groups" [Ribes-Zalesskii] including the cohomology of profinite groups, "Analytic pro-p Groups" [Dixon-du.Sautoy-Mann-Segal] on closed subgroups of $GL(n, \mathbb{Z}_p)$, "Subgroup Growth" [Lubotzky-Segal] on counting the number of subgroups of finitely generating groups, and "Multi-Valued Fields" [Ershov7] on the model theory of fields with several valuations. This led to an official recognition of Field Arithmetic by the Mathematical Reviews in the form of MSC number 12E30.

Introduction to the Second Edition

The extent which Field Arithmetic has reached makes it impossible for us to report in one extended volume about all exciting results which have been achieved. We have therefore made several choices which best suit the spirit of this book but do not extend beyond the scope of one volume.

The new results and additional topics have made it necessary to reorganize and to enlarge the sections dealing with background material. Of course, we took the opportunity afforded by editing a second edition to correct flaws and mistakes which occurred in the first edition and to add more details to proofs wherever it seemed useful.

We list the major changes and additions we made in the book:

Chapter 2 has been reorganized. Sections 2.5–2.9 of the first edition, which survey the theory of algebraic function fields of one variable, were moved to Chapter 3. Sections 2.5–2.8 dealing with linear disjointness, regular extensions, and separability appeared in the first edition as sections 9.1–9.3. A nice application of linear disjointness is Leptin's construction (which preceded that of Warehouse) of a Galois group isomorphic to a given profinite group (Proposition 2.6.12).

In addition to the introductory material about the theory of algebraic function fields of one variable, Chapter 3 now includes a proof of the Riemann-Hurwitz formula and a discussion of hyperelliptic curves.

The proof of Theorem 4.9 of the first edition, estimating the number of zeros of an absolutely irreducible polynomial over a finite field, had a flaw. This has been fixed in the proof of Theorem 5.4.1.

Likewise, the inequality given by [Fried-Jarden3, Prop. 5.16] is inaccurate. This inaccuracy is fixed in Proposition 6.4.8.

We find it more convenient to use the language of algebraic sets as introduced in [Weil5] for model theoretic applications. Section 10.8 translates the basic concepts of that language to the now more commonly used language of schemes.

Theorem 10.14 of the first edition (due to Frey-Prestel) says that the Henselian hull of a PAC field K is K_s. Proposition 11.5.3 (due to Prestel) strengthens this theorem. It says that K is w-dense in $\tilde{K}$ for every valuation w of $\tilde{K}$.

What we called "a separably Hilbertian field" in the first edition, is now called "a Hilbertian field" (Section 12.1). This agrees with the common usage and seems more appropriate for applications.

Section 13.5 gives an alternative definition of Hilbertianity via coverings leading to the notion of "g-Hilbertianity". This sets the stage for a generalization of a theorem of Zannier: Every global field has an infinite normal extension N which is g-Hilbertian but not Hilbertian (Theorem 13.6.2). Moreover, there is a unique factorization subring R of N with infinitely many irreducible elements (Example 15.5.8). This answers negatively Problems 14.20 and 14.21 of the first edition.

Chapter 13 includes now one of the major results of Field Arithmetic which we call "Haran's diamond theorem": Let M_1 and M_2 be Galois extensions of a Hilbertian field K and M a field between K and $M_1 M_2$ not contained in M_1 nor in M_2. Then M is Hilbertian (Theorem 13.8.3). In particular, if N is a Galois extension of K, then N is not the compositum of two Galois extensions of K neither of which is contained in the other. This settles Problems 12.18 and 12.19 of the first edition.

The immediate goal of Hilbert's irreducibility theorem was to realize the groups S_n and A_n as Galois groups over $\mathbb{Q}$. Chapter 16 is dedicated to realizations of Galois groups over arbitrary Hilbertian fields. One of the most important of these results is due to Harbater (Proposition 16.12.1): Let K be a complete valued field, t an indeterminate, and G a finite group. Then G is **regular** over K, that is, $K(t)$ has a finite Galois extension F, regular over K, with $\mathrm{Gal}(F/K(t)) \cong G$. Unfortunately, none of the three proofs of this theorem fits into the scope of this book.

Section 16.6 proves a theorem of Whaples: Let K be a field and p a prime number. Suppose $\mathbb{Z}/p\mathbb{Z}$ (resp. $\mathbb{Z}/4\mathbb{Z}$ if $p=2$) occurs as a Galois group over K. Then $\mathbb{Z}_p$ is realizable over K.

Section 16.7 generalizes a theorem of Hilbert: Let K be a field and $n \geq 2$ an integer with $\mathrm{char}(K) \nmid (n-1)n$. Then A_n is regular over K.

One of the most far-reaching attempts to realize arbitrary finite Galois groups over Hilbertian fields uses Matzat's notion of GAR realization of simple finite groups: Let K be a Hilbertian field and $\alpha\colon G \to \mathrm{Gal}(L/K)$ a finite embedding problem over K. Suppose every composition factor of $\mathrm{Ker}(\alpha)$ has a GAR realization over K. Then the embedding problem is solvable. This leads in particular to the realization of many finite groups over $\mathbb{Q}$ (Remark 16.9.5).

Chapter 17 deals mainly with **Melnikov's formations** $\mathcal{C}$ (i.e. sets consisting of all finite groups whose composition factors belong to a given set of finite simple groups). We prove that every free abstract group F is residually-$\mathcal{C}$. Thus, if the free pro-$\mathcal{C}$ group with a given rank m exists, then the canonical injection of F into $\hat{F}_m(\mathcal{C})$ is injective (Proposition 17.5.11 – Ribes-Zalesskii).

Konrad Neumann improved former results of Fried-Geyer-Jarden and proved that every field is stable (Theorem 18.9.3). This allows the construction of PAC Hilbertian Galois extensions of arbitrary countable Hilbertian fields (Theorems 18.10.2 and 18.10.3). We survey Neumann's proof in Section 18.9. The full proof unfortunately falls outside the scope of this book.

It seemed to be well known that the concept of absolute irreducibility of a variety is elementary. Unfortunately, we could find no solid proof for it in the literature. Proposition 19.5.9 fills in the gap by proving that result.

Section 21.2 includes now the classical results about C_i-fields and not only the corresponding results about weakly C_i-fields as was the case in Section 19.2 of the first edition.

Sections 21.8 gives a complete proof of Schur's Conjecture: If $f(X)$ is a polynomial with coefficients in a global field K with $\mathrm{char}(K) \nmid \deg(f)$ and f permutes $O_K/\mathfrak{p}$ for infinitely many primes $\mathfrak{p}$ of K, then each composition factor of f is linearly related over K to a Dickson polynomial of a prime degree. Section 21.7 proves all lemmas about permutation groups which are used in the proof of Schur's Conjecture (Theorem 21.8.13). This includes the classification of subgroups of $\mathrm{AGL}(1, \mathbb{F}_l)$ (Lemma 21.7.2), and the theorems of Schur (Proposition 21.7.7) and Burnside (Proposition 21.7.8) about doubly transitive permutation groups.

Section 21.9 contains the Fried-Cohen version of Lenstra's proof of the generalized Carlitz's Conjecture: Let p be a prime number, q a power of p, and $f \in \mathbb{F}_q[X]$ a polynomial of degree $n > 1$ which is not a power of p. Suppose f permutes infinitely many finite extensions of $\mathbb{F}_q$. Then $\gcd(n, q-1) = 1$.

The universal Frattini p-cover of a finite group plays a central role in Fried's theory of modular towers. Section 22.11 introduces the former concept and proves its basic properties. Corollary 22.13.4 shows then that $\mathrm{PSL}(2, \mathbb{Z}_p)$ is a p-Frattini cover of $\mathrm{PSL}(2, \mathbb{F}_p)$ although it is not the universal p-Frattini cover.

Chapter 23 puts together material on PAC fields which appeared in Section 20.5 and Chapter 21 of the first edition.

The Beckmann-Black Problem is a refinement of the inverse problem of Galois Theory. Débes proved that the problem has an affirmative solution over PAC fields (Theorem 24.2.2).

Chapter 25 substantially extends the study of free profinite groups F of infinite rank which appeared in Section 24.4 of the first edition. Most of the material goes back to Melnikov. We characterize closed normal subgroups of F by their S-ranks, and prove that a closed subgroup of F is accessible if and only if it is homogeneous.

The first part of Chapter 25 reproduces the group theoretic version of Haran's diamond theorem.

Chapter 26 is completely new. It describes the properties of the closed subgroup $\langle \mathbf{x} \rangle$ and the closed normal subgroup $[\mathbf{x}]$ generated by a random e-tuple $\mathbf{x} = (x_1, \ldots, x_e)$ of elements of a finitely generated free profinite group F of finite rank $n \geq 2$. For example, with probability 1, $\langle \mathbf{x} \rangle \cong \hat{F}_e$ (Proposition 26.1.7). This solves Problem 16.16 of the first edition. In addition, with a positive probability, $[\mathbf{x}]$ has infinite rank and is isomorphic to $\hat{F}_\omega$ (Theorem 26.4.5 and Corollary 26.5.7). The latter result is based on the Golod-Shafarevich Inequality.

Chapter 28 considers an infinite field K which is finitely generated over its base field. It proves that for $e \geq 2$ the theory of all sentences θ which hold in almost all structures $\langle \tilde{K}, \sigma_1, \ldots, \sigma_e \rangle$ with $(\sigma_1, \ldots, \sigma_e) \in \mathrm{Gal}(K)^e$ is undecidable. Moreover, the probability that a sentence θ hold in $\langle \tilde{K}, \sigma_1, \ldots, \sigma_e \rangle$ is in general a nonrational number.

Perhaps the most significant achievement of Field Arithmetic since the first edition appeared is the solution of Problem 24.41 of that edition: The absolute Galois group of a countable PAC Hilbertian field is free of rank $\aleph_0$. It was originally proved in characteristic 0 with complex analysis by Fried-Völklein. Then it was proved in the general case by Pop using rigid geometry and by Haran-Jarden-Völklein using "algebraic patching". The two latter methods also lead to the proof that $\mathrm{Gal}(C(t))$ is a free profinite group if C is an arbitrary algebraically closed field (Harbater, Pop, Haran-Jarden). The method of Fried-Völklein led to the theory of modular towers of Fried.

A remote goal in Galois theory is the classification of absolute Galois groups among all profinite groups. In this framework, one tries to construct new absolute Galois groups out of existing ones. For example, for all fields $K_1, \ldots, K_n$ there exists a field K with $\mathrm{Gal}(K)$ isomorphic to the free product of $\mathrm{Gal}(K_1), \ldots, \mathrm{Gal}(K_n)$ (Pop, Melnikov, Ershov, Koenigsmann). Generalization of this result to infinite families of closed subgroups generalize the concepts "projective groups" and "PRC fields" or "PpC fields" to "relatively projective groups" and "pseudo closed fields" (Haran-Jarden-Pop). They generalize the classification of projective groups as those profinite groups appearing as absolute Galois groups of PAC fields.

All of the exciting material mentioned in the preceding two paragraphs lie unfortunately outside the scope of this volume.

It is my pleasure to thank colleagues and friends who critically read parts of the manuscript of the present edition of "Field Arithmetic": Michael Bensimhoun, David Brink, Gregory Cherlin, Michael Fried, Wulf-Dieter Geyer, Peter Müller, Dan Haran, Wolfgang Herfort, Alexander Lubotzky, Nikolay Nikolov, Dan Segal, Aharon Razon, and Irene Zimmermann.

Tel Aviv, Spring 2004   Moshe Jarden

Introduction to the First Edition

Our topic is the use of algebraic tools — coming mainly from algebraic geometry, number theory, and the theory of profinite groups — in the study of the elementary properties of classes of fields, and related algorithmic problems. (We take the precise definition of "elementary" from first order logic.) This subject has its more distant roots in Tarski's observation that, as a consequence of elimination theory, the full elementary theory of the class of all algebraically closed fields is decidable; this relies on the Euclid algorithm of finding the greatest common divisor of two polynomials in one variable over a field. In its first phase this line of thought led to similar results on real closed fields and p-adic fields.

The subject took a new turn with the work of James Ax [Ax2] on the elementary theory of the class of finite fields, which represents a radical departure in terms of the algebraic methods used. The analysis is based entirely on three properties of a finite field K:

(1a) K is perfect.

(1b) K has a unique extension of each degree.

(1c) There is an explicitly computable function $q(d, m)$ such that any absolutely irreducible variety V defined over K will have a K-rational point if $|K| > q(\dim(V), \deg(V))$.

The validity of the third condition for finite fields is a consequence of Riemann's hypothesis for curves over finite fields. Methods of logic, specifically ultraproducts, led Ax to consider this condition for infinite fields as well, in which case the lower bound afforded by the function q is vacuous, and the condition becomes:

(2) Every absolutely irreducible variety over K has a K-rational point.

Fields satisfying (2) are said to be pseudo algebraically closed, or PAC.

The second condition may be interpreted as a description of the absolute Galois group $\mathrm{Gal}(K)$ as a profinite group: $\mathrm{Gal}(K)$ is the free profinite group on one generator. In Ax' approach it was convenient to have an Abelian absolute Galois group, but a strong trend in later work has been the systematic analysis of situations involving progressively more general Galois groups. One of our central goals here is the presentation of the general theory of PAC fields in its modern form, and its connections with other branches of algebra. From what we have said so far, some connections with algebraic geometry and profinite groups are visible; a number theoretic connection will appear shortly.

One important feature of PAC fields is that they occur in profusion in nature and are in fact typical in the following sense. Since the absolute Galois group $\mathrm{Gal}(\mathbb{Q})$ of the rationals is a compact topological group, it carries a canonical invariant probability measure, the Haar measure. We can therefore ask for the probability that the fixed field $\tilde{\mathbb{Q}}(\boldsymbol{\sigma})$ of a sequence $\boldsymbol{\sigma} = (\sigma_1, ..., \sigma_e)$ of automorphisms of $\tilde{\mathbb{Q}}$ will be PAC; and we find that this occurs with probability 1. In addition, the absolute Galois group of $\tilde{\mathbb{Q}}(\boldsymbol{\sigma})$ is free on the e generators $\sigma_1, ..., \sigma_e$, again with probability 1. These facts are consequences of Hilbert's irreducibility theorem for $\mathbb{Q}$ (Chapter 13), at least in the context of countable fields. We will develop other connections between the PAC property and Hilbertianity.

There are also remarkable connections with number theory via the Chebotarev density theorem (Chapters 6, 13, 16, 20, 21, 31). For example, the probability that a given elementary statement ψ holds for the field $\tilde{\mathbb{Q}}(\sigma)$ coincides with the Dirichlet density of the set of primes for which it holds for the field $\mathbb{F}_p$, and this density is rational. Thus, the "probability 1" theory of the fixed fields $\tilde{\mathbb{Q}}(\sigma)$ coincides with the theory of "all sufficiently large" finite fields, which by Ax' work is an algorithmically decidable theory.

Ax' results extend to the "probability 1" theory of the fields $\tilde{\mathbb{Q}}(\boldsymbol{\sigma})$ for $\boldsymbol{\sigma}$ of length $e > 1$, by somewhat different methods (Chapter 20), although the connection with finite fields is lost. The elementary theories of such fields are largely determined by three properties: PAC, characteristic zero, and having an absolute Galois group which is free on e generators. To determine the full elementary theory of one such field K, it is also necessary to describe the intersection $K \cap \tilde{\mathbb{Q}}$.

Although the absolute Galois group of a PAC field need not be free, it can be shown to be projective in a natural sense, and conversely any projective profinite group occurs as the Galois group of some PAC field. In extending the theory from PAC fields with free Galois group to the general (projective) case, certain obstacles arise: for example, the algorithmic results do not extend. There is nonetheless a quite general theory, which enables us to identify some broad classes of projective profinite groups for which the associated classes of profinite groups behave well, and also to pinpoint unruly behavior in other case.

One approach to the algorithmic problems associated with PAC fields leads to the study of profinite groups G with the embedding property (the terminology reflects a preoccupation with the corresponding fields): for each pair of continuous epimorphisms $\varphi\colon G \to A$, $\alpha\colon B \to A$, where B is a finite quotient of G, we require that φ should factor through α. A perfect PAC field whose absolute Galois group is a group with the embedding property is called a Frobenius field. The elementary theory of all Frobenius fields can be computed quite explicitly. The algorithm has some relationship with elimination theory as used by Tarski. We associate to each elementary statement in the language of PAC fields a stratification of affine space into basic normal locally closed algebraic sets, each equipped with a Galois extension of its function field, and the given statement is reinterpreted as a statement about conjugacy classes of subgroups of the specified Galois groups. When the initial statement has no quantifiers this is a fairly trivial procedure, but addition of quantifiers corresponds to a special kind of "projection" of these Galois stratifications.

This procedure has not yet been closely examined from the point of view of computational complexity. Like most procedures which operate by tracing through a series of projections, it is effective but hopelessly inefficient in its present form. It is not yet clear whether it is substantially less efficient than Tarski's procedure for algebraically closed fields, nor whether, like that procedure, it can significantly reorganized and sped up.

The Galois stratification algorithm relies on techniques of effective algebraic geometry, and also involves substantial algorithmic problems of a new type connected with the theory of profinite groups. Specifically, it is necessary to determine, given two collections $A_1, ..., A_m$ and $B_1, ..., B_n$ of finite groups, whether or not there is a projective group with the embedding property which has each A_i as (continuous) image, but none of the groups B_j. The solution to this problem depends on recent work on projective covers (Chapter 22) and embedding covers (Chapter 24). Ultimately our decision problem reduces to the determination of the finite quotients of the projective cover of the embedding cover of a single finite group.

The theory of projective covers leads also to the undecidability results alluded to earlier. A fairly natural encoding of graphs into profinite groups is lifted by this theory into the class of projective profinite groups, and then by looking at the corresponding PAC fields we see that their elementary theories encode algorithmically undecidable problems (the analogous results for graphs are well known).

In the final chapter we return to our point of departure, the theory of finite fields. The zeta function of a Galois formula over a finite field is defined, and using a result of Dwork and Bombieri we show that some integral power of each such function is a product of an exponential and a rational function over $\mathbb{Q}$.

One of the goals of this book is to serve as a bridge between algebraists and logicians. For the algebraist there is a self contained introduction to the logic and model theory background for PAC fields (Chapter 7). Chapter 14 gives the "nonstandard" framework that suffices for Weissauer's proof of Hilbert's irreducibility theorem (Chapter 15), and Chapters 8 and 28 include basic recursion theory. On the other hand, for logicians with basic algebraic background (e.g. Lang's book "Algebra") Chapter 4 has the Stepanov-Bombieri elementary proof of the Riemann hypotheses for curves, and Chapter 6 gives an elementary proof of the Chebotarev density theorem. Both groups of readers may find the extensive treatment of profinite groups (Chapters 1, 17, 18, 22, 24, 25 and 26) and of Hilbertian fields (Chapters 12, 13, 15, and 16) valuable.

Although PAC fields arise over arithmetically rich fields, they themselves lack properties that we associate with the arithmetic, say, of the rationals. For example, a PAC field F admits no orderings and all Henselizations of F are separably closed (Section 11.5). Many PAC field results generalize to pseudo real closed (PRC) fields.

A field F is **PRC** if each absolutely irreducible variety defined over F has an F-rational point provided it has a nonsingular $\hat{F}$-rational point in each real closure $\hat{F}$ of F. Thus, a PRC field without orderings is PAC. This, and the development of the theory of **pseudo p-adically closed PpC fields** are outside the scope of this book. We refer to [Prestel1], to [Jarden12], [Haran-Jarden2], and to [Haran-Jarden3] for literature about PRC fields and to [Haran-Jarden4] for PpC fields. Similarly, we give no account of the theories of real closed fields and p-adically closed fields that preceded the development of the theory of PAC fields. In particular, for Hilbert's 17th problem and the Ax-Kochen-Ershov p-adic theory, we refer the reader to [Prestel2], [Ax-Kochen1, Ax-Kochen2, and Ax-Kochen3], and [Prestel-Roquette].

ACKNOWLEDGEMENT: We are indebted to several colleagues who corrected errors in the process of critically reading the manuscript. In particular, Wulf-Dieter Geyer, Gregory L. Cherlin, and Dan Haran made crucial contributions.

Michael D. Fried, Gainesville, Florida
Moshe Jarden, Tel Aviv, Israel
Summer 1986

Notation and Convention

$\mathbb{Z}$ = the ring of rational integers.
$\mathbb{Z}_p$ = the ring of p-adic integers.
$\mathbb{Q}$ = the field of rational numbers.
$\mathbb{R}$ = the field of real numbers.
$\mathbb{C}$ = the field of complex numbers.
$\mathbb{F}_q$ = the field with q elements.
K_s = the separable closure of a field K.
K_{ins} = the maximal purely inseparable extension of a field K.
$\tilde{K}$ = the algebraic closure of a field K.
$\text{Gal}(L/K)$ = the Galois group of a Galois extension L/K.
We call a polynomial $f \in K[X]$ **separable** if f has no multiple root in $\tilde{K}$.
$\text{Gal}(f, K)$ = the Galois group of a separable polynomial $f \in K[X]$ over a field K viewed as a permutation group of the roots of f.
$\text{Gal}(K) = \text{Gal}(K_s/K)$ = the absolute Galois group of a field K.
$\text{irr}(x, K)$ = the monic irreducible polynomial of an algebraic element x over a field K.
Whenever we form the compositum EF of field extensions of a field K we tacitly assume that E and F are contained in a common field.
$|A| = \#A$ = the cardinality of a set A.
$R^\times$ = the group of invertible elements of a ring R.
$\text{Quot}(R)$ = the quotient field of an integral domain R.
$A \subset B$ means "the set A is properly contained in the set B".
$a^x = x^{-1}ax$, for elements a and x of a group G.
$H^x = \{h^x \mid h \in H\}$, for a subgroup H of G.
Given subgroups A, B of a group G, we use "$A \leq B$" for "A is a subgroup of B" and "$A < B$" for "A is a proper subgroup of B".
Given an Abelian (additive) group A and a positive integer n, we write A_n for the subgroup $\{a \in A \mid na = 0\}$. For a prime number p we let $A_{p^\infty} = \bigcup_{i=1}^\infty A_{p^i}$.
For a group B that acts on a group A from the right, we use $B \ltimes A$ to denote the semidirect product of A and B.
Bold face letters stand for n-tuples, e.g. $\mathbf{x} = (x_1, \ldots, x_n)$.
$\text{ord}(x)$ is the order of an element x in a group G.
For a positive integer n and an integer a with $\gcd(a, n) = 1$, we use $\text{ord}_n a$ to denote the **order of a modulo n**. Thus, $\text{ord}_n a$ is the minimal positive integer d with $a^d \equiv 1 \mod n$.
In the context of groups, S_n (resp. A_n) stands for the full permutation group (resp. alternative group) of $\{1, \ldots, n\}$.
In the context of groups, C_n stands for the cyclic multiplicative group of order n. Likewise we use $\mathbb{Z}/n\mathbb{Z}$ for the additive multiplicative group of order n.
In the context of fields, ζ_n stands for a primitive root of unity of order n.
$\dot{\bigcup}_{i \in I} B_i$ is the disjoint union of sets B_i, $i \in I$.

Chapter 1.
Infinite Galois Theory and Profinite Groups

The usual Galois correspondence between subgroups of Galois groups of finite Galois extensions and intermediate fields is not valid for infinite Galois extensions. The Krull topology restores this correspondence for closed subgroups (Proposition 1.3.1).

Since Galois groups are inverse limits of finite groups, they are profinite. Conversely, we define profinite groups, independently of Galois theoretic properties. Each profinite group actually appears as a Galois group (Corollary 1.3.4). In particular, we study the procyclic groups $\mathbb{Z}_p$ and $\hat{\mathbb{Z}}$ and prove that every finite field has the latter as its absolute Galois group.

1.1 Inverse Limits

Our interest in inverse limits comes from infinite Galois theory: Infinite Galois groups are inverse limits of finite Galois groups. As a preparation to the study of "profinite groups" we define in this section inverse limits of topological spaces and characterize inverse limits of finite topological spaces.

Let I be a set with a **partial ordering** $\leq$; that is, $\leq$ is a binary relation which is reflexive, transitive, and $a \leq b$ and $b \leq a$ imply $a = b$. We call $(I, \leq)$ a **directed partially ordered set** if in addition
(1) for all $i, j \in I$ there exists $k \in I$ with $i \leq k$ and $j \leq k$.

An **inverse system** (also called a **projective system**) over a directed partially ordered set $(I, \leq)$ is a data $(S_i, \pi_{ji})_{i,j \in I}$ where S_i is a set and $\pi_{ji} \colon S_j \to S_i$ is a map for all $i, j \in I$ with $i \leq j$ satisfying the following rules:
(2a) π_{ii} the identity map for each $i \in I$.
(2b) $\pi_{ki} = \pi_{ji} \circ \pi_{kj}$ if $i \leq j \leq k$.

Let S be the subset of the cartesian product $\prod_{i \in I} S_i$ consisting of all elements $s = (s_i)_{i \in I}$ with $\pi_{ji}(s_j) = s_i$ for all $i \leq j$. Note: S may be empty. Let $\mathrm{pr}_i \colon \prod_{j \in I} S_j \to S_i$ be the projection on the ith coordinate. Denote the restriction of pr_i to S by π_i. Then $\pi_i = \pi_{ji} \circ \pi_j$ for every $i \leq j$. We say $(S, \pi_i)_{i \in I}$ is the **inverse** (or **projective**) **limit** of the family $(S_i)_{i \in I}$ with respect to the maps π_{ji}. Denote S by $\varprojlim S_i$.

Let $(S_i', \pi_{ji}')_{i,j \in I}$ be another inverse system over I. Suppose for each $i \in I$ we are given a map $\theta_i \colon S_i \to S_i'$ with $\pi_{ji}' \circ \theta_j = \theta_i \circ \pi_{ji}$ for all $i \leq j$. (We say that the maps θ_i, $i \in I$, are **compatible**.) Then there exists a unique map $\theta \colon \varprojlim S_i \to \varprojlim S_i'$ satisfying $\pi_i' \circ \theta = \theta_i \circ \pi_i$ for each $i \in I$: θ maps $s = (s_i) \in \varprojlim S_i$ onto $\theta(s)$ with $\theta(s)_i = \theta_i(s_i)$. Denote θ by $\varprojlim \theta_i$.

Similarly, let X be a set and for each $i \in I$ let $\theta_i \colon X \to S_i$ be a map satisfying $\pi_{ji} \circ \theta_j = \theta_i$ whenever $i \leq j$ (Again, we say that the maps θ_i,

$i \in I$, are **compatible**.) Then there exists a unique map $\theta: X \to \varprojlim S_i$ with $\pi_i \circ \theta = \theta_i$ for each $i \in I$.

When S_i are topological spaces, we assume π_{ji} are continuous. Then we equip $\varprojlim S_i$ with the topology induced from the product topology of $\prod_{i \in I} S_i$. Recall that the product topology on $\prod_{i \in I} S_i$ has a basis consisting of the sets $\prod_{i \in I} U_i$, with U_i open in S_i for each $i \in I$, and $U_i = S_i$ for all but finitely many $i \in I$. Since pr_i is continuous, so is π_i, $i \in I$. If $\theta_i: S_i \to S_i'$ are continuous, then $\theta: \varprojlim S_i \to \varprojlim S_i'$ is also continuous.

LEMMA 1.1.1: *The collection of all subsets of $S = \varprojlim S_i$ of the form $\pi_i^{-1}(U_i)$ with U_i open in S_i is a basis for the topology of S.*

Proof: Let $s \in S$. By definition, each basic open neighborhood of s has the form $V = S \cap \left(\prod_{j \in J} V_j \times \prod_{i \in I \smallsetminus J} S_j \right)$, where J is a finite subset of I and V_j is an open subset of S_j, $j \in J$. Take $k \in I$ with $k \geq j$ for all $j \in J$. Then $U_k = \bigcap_{j \in J} \pi_{kj}^{-1}(V_j)$ is an open subset of S_k and $\pi_k^{-1}(U_k)$ is an open neighborhood of s in V. Therefore, the collection $\pi_i^{-1}(U_i)$, $i \in I$, is a basis for the topology of S. $\square$

LEMMA 1.1.2: *In the notation above, if each S_i, $i \in I$, is a Hausdorff space, then $\varprojlim S_i$ is a closed subset of $\prod S_i$.*

Proof: Suppose $s = (s_i) \in \prod S_i$ does not belong to $\varprojlim S_i$. Then there are $i, j \in I$ with $i \leq j$ and $\pi_{ji}(s_j) \neq s_i$. Take open disjoint neighborhoods U_i and U_i' of s_i and $\pi_{ji}(s_j)$, respectively. Then $U_i \times \pi_{ji}^{-1}(U_i') \times \prod_{k \neq i,j} S_k$ is an open neighborhood of s in $\prod S_i$ that does not intersect $\varprojlim S_i$. $\square$

If, in addition, each S_i is compact, then Tychonoff's theorem implies $\prod S_i$ is also compact. Thus, $\varprojlim S_i$ with the induced topology is compact.

LEMMA 1.1.3: *The inverse limit S of an inverse system of nonempty compact Hausdorff spaces S_i, $i \in I$, is a nonempty compact Hausdorff space.*

Proof: We only need to prove that S is nonempty. Indeed, $S = \bigcap_{k \geq j} R_{kj}$, where $R_{kj} = \{s \in \prod S_i \mid \pi_{kj}(s_k) = s_j\}$. The natural map $\mathrm{pr}_k \times \mathrm{pr}_j: \prod S_i \to S_k \times S_j$ is continuous. The Hausdorff property of S_j implies $T = \{(s_k, s_j) \in S_k \times S_j \mid \pi_{kj}(s_k) = s_j\}$ is a closed subset of $S_k \times S_j$. Hence, $R_{kj} = (\mathrm{pr}_k \times \mathrm{pr}_j)^{-1}(T)$ is a closed subset of $\prod S_i$. Since $\prod S_i$ is compact, we only need to show that the intersection of finitely many of the R_{kj} is nonempty. Indeed, let $j_1 \leq k_1, \ldots, j_n \leq k_n$ be n pairs in I. Choose $l \in I$ with $k_i \leq l$, $i = 1, \ldots, n$, and choose $s_l \in S_l$. Define $s_{j_i} = \pi_{l,j_i}(s_l)$ and $s_{k_i} = \pi_{l,k_i}(s_l)$, for $i = 1, \ldots, n$. For each $r \in I \smallsetminus \{j_1, \ldots, j_n, k_1, \ldots, k_n\}$ let s_r an arbitrary element of S_r. Then $s = (s_i) \in \bigcap_{i=1}^n R_{k_i, j_i}$. $\square$

1.1 Inverse Limits

COROLLARY 1.1.4: *The inverse limit of an inverse system of nonempty finite sets is nonempty.*

Proof: Equip each of the finite sets with the discrete topology. □

COROLLARY 1.1.5: *Let $(S_i, \pi_{ji})_{i,j \in I}$ and $(S'_i, \pi'_{ji})_{i,j \in I}$ be inverse systems of compact Hausdorff spaces. Let $\theta_i \colon S_i \to S'_i$ be a compatible system of surjective continuous maps. Put $S = \varprojlim S_i$, $S' = \varprojlim S'_i$, and $\theta = \varprojlim \theta_i$. Then $\theta \colon S \to S'$ is surjective.*

Proof: Let $s' = (s'_i)$ be an element of S'. Then $(\theta_i^{-1}(s'_i), \pi_{ji})_{i,j \in I}$ is an inverse system of nonempty compact Hausdorff spaces. By Lemma 1.1.3, the inverse limit of $\theta_i^{-1}(s'_i)$ is nonempty. Each element in the inverse limit is mapped by θ onto s'. □

COROLLARY 1.1.6: *Let X be a compact space, $(S_i, \pi_{ji})_{i,j \in I}$ an inverse system of Hausdorff spaces, and $\theta_i \colon X \to S_i$ a compatible system of continuous surjective maps. Put $\theta = \varprojlim \theta_i$. Then $\theta \colon X \to \varprojlim S_i$ is surjective.*

Proof: Consider $s = (s_i)_{i \in I} \in \varprojlim S_i$. Then $\theta_i^{-1}(s_i)$ is a closed nonempty subset of X. For $i_1, \ldots, i_n \in I$ there is a $j \in I$ with $i_1, \ldots, i_n \leq j$. Then $\theta_j^{-1}(s_j) \subseteq \theta_{i_1}^{-1}(s_{i_1}) \cap \cdots \cap \theta_{i_n}^{-1}(s_{i_n})$, so $\theta_{i_1}^{-1}(s_{i_1}) \cap \cdots \cap \theta_{i_n}^{-1}(s_{i_n})$ is nonempty. Since X is compact, there is an $x \in X$ which belongs to $\theta_i^{-1}(s_i)$ for every $i \in I$. It satisfies, $\theta(x) = s$. Thus, θ is surjective. □

A **profinite space** is an inverse limit of an inverse system of finite discrete spaces.

LEMMA 1.1.7: *A compact Hausdorff space S is profinite if and only if its topology has a basis consisting of open-closed sets.*

Proof: Suppose first $S = \varprojlim S_i$ is an inverse limit of finite discrete spaces S_i. Let $\pi_i \colon S \to S_i$ be the projection on the ith coordinate, $i \in I$. By Lemma 1.1.1, the sets $\pi_i^{-1}(U_i)$, where $i \in I$ and U_i is a subset of S_i, form a basis of the topology of S. Since S_i is discrete, U_i is open-closed. Hence, so is $\pi_i^{-1}(U_i)$.

Now suppose the topology of S has a basis consisting of open-closed sets. An **open-closed partition** of S is a finite set $\mathbf{A}$ of nonempty open-closed disjoint subsets of S whose union is S. Denote the set of all open-closed partitions of S by $\mathcal{A}$. Let $\mathbf{A}, \mathbf{A'}, \mathbf{B} \in \mathcal{A}$. Write $\mathbf{A} \leq \mathbf{B}$ if for each $B \in \mathbf{B}$ there is an $A \in \mathbf{A}$ with $B \subseteq A$. Then A is unique. Next note that $\mathbf{C} = \{A \cap A' \mid A \in \mathbf{A}, A' \in \mathbf{A'}\}$ belongs to $\mathcal{A}$ and satisfies $\mathbf{A}, \mathbf{A'} \leq \mathbf{C}$. Thus, $(\mathcal{A}, \leq)$ is a directed partially ordered set.

When $\mathbf{A} \leq \mathbf{B}$, define a map $\pi_{\mathbf{B},\mathbf{A}}$ from $\mathbf{B}$ to $\mathbf{A}$ by $\pi_{\mathbf{B},\mathbf{A}}(B) = A$, where A is the unique element of $\mathbf{A}$ containing B. Equip each $\mathbf{A} \in \mathcal{A}$ with the discrete topology. Then $(\mathbf{A}, \pi_{\mathbf{B},\mathbf{A}})_{\mathbf{A},\mathbf{B} \in \mathcal{A}}$ is an inverse system of finite discrete spaces.

Its limit $S' = \varprojlim \mathbf{A}$ is a profinite space. We construct a homeomorphism of S onto S'.

Let $s \in S$ and $\mathbf{A} \in \mathcal{A}$. Define $\theta_\mathbf{A}(s)$ to be the unique A in $\mathbf{A}$ which contains s. Then $\theta_\mathbf{A} \colon S \to \mathbf{A}$ is a continuous surjective map. If $\mathbf{A} \leq \mathbf{B}$, then $\pi_{\mathbf{B},\mathbf{A}} \circ \theta_\mathbf{B} = \theta_\mathbf{A}$. Hence, by Corollary 1.1.6, there is a continuous surjective map $\theta \colon S \to S'$ satisfying $\pi_\mathbf{A} \circ \theta = \theta_\mathbf{A}$ for each $\mathbf{A} \in \mathcal{A}$.

Suppose s, s' are distinct elements of S. Since S is Hausdorff, there are disjoint open-closed subsets A and A' of S with $s \in A$ and $s' \in A'$. Put $A'' = S \smallsetminus (A \cup A')$. Then $\mathbf{A} = \{A, A', A''\}$ is an open-closed partition of S, $\theta_\mathbf{A}(s) = A$, and $\theta_\mathbf{A}(s') = A'$. Thus, $\theta_\mathbf{A}(s) \neq \theta_\mathbf{A}(s')$, so $\theta(s) \neq \theta(s')$. Therefore, θ is bijective. Since S is compact and S' is Hausdorff, θ is a homeomorphism. Consequently, S is a profinite space. □

Remark 1.1.8: Totally disconnected spaces. Let S be a compact Hausdorff space. Suppose S has a basis for its topology consisting of open-closed sets. It is not difficult to see that S is **totally disconnected**; that is each s in S is its own connected component. Conversely, if S is totally disconnected, then the topology of S has a basis consisting of open-closed subsets [Ribes-Zalesski, Thm. 1.1.12]. □

1.2 Profinite Groups

We survey here the basic properties of compact groups. In particular, this will apply to profinite groups.

TOPOLOGICAL GROUPS. A **topological group** is a group G equipped with a topology in which the product $(x, y) \mapsto xy$ and the inverse map $x \mapsto x^{-1}$ are continuous. It follows that for each $a \in G$ the maps $x \mapsto ax$, $x \mapsto xa$, and $x \mapsto x^{-1}$ are homeomorphisms. We always assume $\{1\}$ is a closed subset of G. Consequently, $\{a\}$ is a closed subset of G for each $a \in G$.

It follows that G is a Hausdorff space. Indeed, let a, b be distinct elements of G. The identity $1 \cdot 1^{-1} = 1$ and the continuity of the group operations give open neighborhoods U and W of 1 with $UW^{-1} \subseteq G \smallsetminus \{a^{-1}b\}$. Thus, aU and bW are disjoint open neighborhoods of a and b, respectively, as needed.

In addition, each closed subgroup H of G of finite index is open. Indeed, there are $a_i \in G$, $i \in I$, with I finite and $G = \bigcup_{i \in I} a_i H$. Each of the sets $a_i H$ is closed and there is a $j \in I$ with $a_j H = H$. Therefore, $H = G \smallsetminus \bigcup_{i \neq j} a_i H$ is open.

Conversely, if G is compact, then every open subgroup H of G is of finite index. Otherwise, G would be a disjoint union of infinitely many cosets of H, each of which is open.

Let N be a closed normal subgroup of G and $\pi \colon G \to G/N$ the quotient map. Equip G/N with the quotient topology. Thus, a subset $\bar{U}$ of G/N is open if and only if $\pi^{-1}(\bar{U})$ is open. It follows that the group operations of G/N are continuous. In addition, a subset $\bar{C}$ of G/N is closed if and only if $\pi^{-1}(\bar{C})$ is closed. In particular, since $N = \pi^{-1}(1)$ is closed, $\{1\}$ is a closed subset of G/N. Consequently, G/N is a topological group and $\pi\colon G \to G/N$ is a continuous map. Moreover, if U is an open subset of G, then $\pi^{-1}(\pi(U)) = \bigcup_{n \in N} nU$, so $\pi(U)$ is open in G/N. Therefore, π is an open map.

Suppose again G is compact. Let θ be a continuous homomorphism of G into a topological group H. Since H is Hausdorff, θ is a closed map. Suppose in addition, θ is surjective. Put $N = \mathrm{Ker}(\theta)$. Let $\pi\colon G \to G/N$ be the quotient map and $\bar{\theta}\colon G/N \to H$ the map induced by θ. Then $\bar{\theta}$ is an isomorphism of abstract groups. In addition, $\bar{\theta}$ is a continuous bijective map of the compact group G/N onto the Hausdorff group H. Hence, $\bar{\theta}$ is an isomorphism of topological groups and θ is an open map. This is the **first isomorphism theorem for compact groups**.

Let H be a closed subgroup and N a closed normal subgroup of G. Then $HN = \{hn \mid h \in H,\ n \in N\}$ is the image of the compact group $H \times N$ under the continuous map $(h, n) \mapsto hn$. Hence, HN is a closed subgroup of G. As in the preceding paragraph, the group theoretic isomorphism $\theta\colon HN/N \to H/H \cap N$ defined by $\theta(hN) = h(H \cap N)$, $h \in H$, is a homeomorphism, so θ is an isomorphism of topological groups.

Similarly, if M and N are closed normal subgroups of G with $N \leq M$, then M/N is a closed normal subgroup of G/N and the map $G/M \to (G/N)/(M/N)$ given by $gM \mapsto (gN)M/N$, $g \in G$, is an isomorphism of topological groups.

Here is one way to equip an abstract group G with a topology. Let $\mathcal{N}$ be a family of normal subgroups of G, closed under finite intersections, such that the intersection of all $N \in \mathcal{N}$ is 1. Take $\mathcal{N}$ to be a basis for the open neighborhoods of 1. A basis for the open neighborhoods of $a \in G$ is the family $\mathcal{N}_a = \{aN \mid N \in \mathcal{N}\}$. The union of the $\mathcal{N}_a$ is then a basis for a group topology of G. For example, the identity $xN \cdot yN = xyN$ for normal subgroups N implies multiplication is continuous.

Let $(G_i, \pi_{ji})_{i,j \in I}$ be an inverse system of topological groups and continuous homomorphisms $\pi_{ji}\colon G_j \to G_i$, for each $i, j \in I$ with $j \geq i$. Then $G = \varprojlim G_i$ is a topological group and the projections $\pi_i\colon G \to G_i$ are continuous homomorphisms. Let $\langle G'_i, \pi'_{ji} \rangle_{i,j \in I}$ be another system of topological groups with $G' = \varprojlim G'_i$. Suppose $\theta_i\colon G_i \to G'_i$, $i \in I$, is a compatible system of continuous homomorphisms. Then the corresponding map $\theta\colon G \to G'$ is a continuous homomorphism.

PROFINITE GROUPS. In this book, we primarily consider an inverse system of finite groups $(G_i, \pi_{ji})_{i,j \in I}$, each equipped with the discrete topology. We call the inverse limit $G = \varprojlim G_i$ a **profinite group**. By Lemma 1.1.3, G is a compact group. By Lemma 1.1.1, the open-closed sets $\pi_i^{-1}(g_i)$, $g_i \in G_i$, $i \in I$, form a basis for the topology on G. In particular, the open normal subgroups of G form a basis for the open neighborhoods of 1.

Remark 1.2.1: Basic rules. Here are some basic rules for profinite groups G and H which we use in this book without explicit reference:

(a) A subgroup H of G is open if and only if H is closed of a finite index. The intersection of all open normal subgroups of G is 1. Every open subset of G is a union of cosets $g_i N_i$ with N_i open normal and $g_i \in G$.

(b) Every profinite group is compact, Hausdorff, and has a basis for its topology consisting of open-closed sets (Lemmas 1.1.3 and 1.1.7).

(c) A subset C of a profinite group is closed if and only if C is compact (use (b)).

(d) A subset B of a profinite group is open-closed if it is a union of finitely many cosets $g_i N$ with N open normal and $g_i \in G$ (use (a) and the compactness of B).

(e) Every homomorphism $\varphi \colon G \to H$ is tacitly assumed to be continuous. In particular, φ maps compact subsets of G onto compact subsets of H. Hence, φ maps closed subsets of G onto closed subsets of H (use (c)).

(f) By the first isomorphism theorem for compact groups, every epimorphism $\varphi \colon G \to H$ of profinite groups is an open map. In particular, φ maps open subgroups of G onto open subgroups of H. □

We list below special properties of profinite groups not shared by all compact groups:

LEMMA 1.2.2: *Let $\{H_i \mid i \in I\}$ be a **directed** family of closed subset of a profinite group G; that is:*

(1) *for every finite subset J of I there is an $i \in I$ with $H_i \leq \bigcap_{j \in J} H_j$.*

Put $H = \bigcap_{i \in I} H_i$. Then:

(a) *For every open subgroup U of G containing H, there is an $i \in I$ with $H_i \leq U$.*

(b) *$\bigcap_{i \in I} K H_i = KH$ and $\bigcap_{i \in I} H_i K = HK$ for every closed subset K of G.*

(c) *Let $\varphi \colon G \to \bar{G}$ be an epimorphism. Then $\varphi(\bigcap_{i \in I} H_i) = \bigcap_{i \in I} \varphi(H_i)$.*

(d) *Let K and K' be closed subgroups of G. Then the set $S = \{\sigma \in G \mid K^\sigma = K'\}$ is closed.*

(e) *Let K and K' be closed subgroups of G which contain H. Suppose each H_i is normal and KH_i and $K'H_i$ are conjugate. Then K and K' are conjugate.*

(f) *Let $G = \varprojlim G_i$ be an inverse limit of finite groups, $\pi_i \colon G \to G_i$ the quotient maps, and K, K' closed subgroups of G. Suppose $\pi_i(K)$ and $\pi_i(K')$ are conjugate in G_i for each i. Then K and K' are conjugates.*

Proof of (a): The set $H_i \cap (G \smallsetminus U)$ is closed for every $i \in I$ and $\bigcap_{i \in I} H_i \cap (G \smallsetminus U) = H \cap (G \smallsetminus U) = \emptyset$. Since G is compact, there exists a finite subset J of I such that $\bigcap_{j \in J} H_j \cap (G \smallsetminus U) = \emptyset$. By (1) there exists $i \in I$ such that $H_i \leq \bigcap_{j \in J} H_j$. Hence, $H_i \leq U$.

Proof of (b): Let $g \in \bigcap_{i \in I} K H_i$. Then, for each $i \in I$ there are $k_i \in K$ and $h_i \in H_i$ with $g = k_i h_i$. Hence, the closed subset $H_i \cap g^{-1} K$ of G is nonempty.

1.2 Profinite Groups

By (1), the intersection of finitely many of the sets $H_i \cap g^{-1}K$ is nonempty. Since G is compact, there exists $h \in \bigcap_{i \in I} H_i \cap g^{-1}K$. It satisfies $h \in H$ and $h = g^{-1}k$ for some $k \in K$. Therefore, $g = kh^{-1} \in KH$.

Proof of (c): Let $K = \text{Ker}(\varphi)$. Then φ induces a bijection between the set of closed subgroups lying between K and G and the set of closed subgroups of $\bar{G}$. Hence, by (b), $\varphi(H) = \varphi(KH) = \varphi(\bigcap_{i \in I} KH_i) = \bigcap_{i \in I} \varphi(KH_i) = \bigcap_{i \in I} \varphi(H_i)$.

Proof of (d): Denote the set of all open normal subgroups of G by $\mathcal{N}$. For each $N \in \mathcal{N}$, let S_N be the inverse image under the quotient map $G \to G/N$ of the finite set
$$\{s \in G/N \mid (KN/N)^s = K'N/N\}.$$
Then, S_N is closed. If $s \in S_N$ for each $N \in \mathcal{N}$, then, by (b), $K^s = \bigcap_{N \in \mathcal{N}} K^s N = \bigcap_{N \in \mathcal{N}} K'N = K'$. Hence, $S = \bigcap_{N \in \mathcal{N}} S_N$. Therefore, S is closed.

Proof of (e): For each $i \in I$, $S_i = \{\sigma \in G \mid K^\sigma H_i = K' H_i\}$ is a nonempty subset of G. By (d), S_i is closed. In the notation of (1) we have $S_i \subseteq \bigcap_{j \in J} S_j$. Hence, by compactness, there is a $\sigma \in \bigcap_{i \in I} S_i$. Thus, $K^\sigma H_i = K' H_i$ for each $i \in I$. We conclude from (b) that $K^\sigma = K'$.

Proof of (f): Apply (e) to $H_i = \text{Ker}(\pi_i)$. □

LEMMA 1.2.3: *Each closed subgroup H of a profinite group G is the intersection of open subgroups.*

Proof: Let $\mathcal{N}$ be the set of all open normal subgroups of G. By Lemma 1.2.2(b), $\bigcap_{N \in \mathcal{N}} NH = H$. □

The following result gives a sufficient condition for a compact group to be profinite. We use it below to prove that the category of profinite groups is closed under various natural operations.

LEMMA 1.2.4: *Let G be a compact group and $\{N_i \mid i \in I\}$ a directed family of closed normal subgroups of G of finite index satisfying*
(2) $\bigcap_{i \in I} N_i = 1$.
Then $G = \varprojlim G/N_i$ and G is a profinite group.

Proof: By definition, $\varprojlim G/N_i$ is a profinite group. We only need to prove the isomorphism. The quotient maps $G \to G/N_i$ define a continuous embedding θ of G into $\varprojlim G/N_i$. If $(g_i N_i)_{i \in I}$ is an element of the latter group, then the closed subsets $g_i N_i$ of G have the finite intersection property. Since G is compact, there exists $g \in \bigcap_{i \in I} g_i N_i$. Then, $\theta(g) = (g_i N_i)_{i \in I}$. Thus, θ is bijective. Compactness of G and the Hausdorff property of $\varprojlim G/N_i$ imply that θ is an isomorphism of topological groups. □

LEMMA 1.2.5: *The following statements hold for each closed subgroup H of a profinite group G:*
(a) *The group H is profinite. Moreover, for each open normal subgroup M of H there exists an open normal subgroup N of G with $H \cap N \leq M$. If M is normal in G, then N can be chosen such that $H \cap N = M$.*
(b) *Every open subgroup H_0 of H is the intersection of an open subgroup of G with H.*
(c) *Every homomorphism φ_0 of H into a finite group A extends to a homomorphism $\varphi \colon H' \to A$, where H' is an open subgroup of G containing H. If H is normal in G, then H' can be chosen to be normal.*

Proof of (a): If N is an open normal subgroup of G, then $H \cap N$ is an open normal subgroup of H. The family of all groups $H \cap N$ is directed and its intersection is trivial. Hence, $H = \varprojlim H/H \cap N$ is a profinite group.

Let M be an open normal subgroup of H. By Lemma 1.2.2(a), $H \cap N \leq M$ for some open normal subgroup N of G. If $M \triangleleft G$, then $MN \triangleleft G$ and $H \cap MN = M$.

Proof of (b): The intersection of all conjugates of H_0 in H is an open normal subgroup of H. Hence, by the first statement of the lemma, there is an open normal subgroup N of G such that $M = H \cap N \leq H_0$. Then $G_0 = H_0 N$ is an open subgroup of G and $H \cap G_0 = H_0$.

Proof of (c): By (a), G has an open normal subgroup N with $H \cap N \leq \mathrm{Ker}(\varphi_0)$. Put $H' = HN$ and define $\varphi \colon H' \to A$ by $\varphi(hn) = \varphi_0(h)$. Then φ is a well defined homomorphism which extends φ_0. □

LEMMA 1.2.6: *These statements hold:*
(a) *If N is a closed normal subgroup of a profinite group G, then G/N is profinite.*
(b) *The cartesian product, $G = \prod_{i \in I} G_i$, of profinite groups is profinite.*
(c) *Every inverse limit, $G = \varprojlim G_i$, of profinite groups is profinite.*

Proof of (a): A closed normal group N is an intersection of open subgroups (Lemma 1.2.3) and therefore of open normal subgroups N_i, $i \in I$, of G. Now apply Lemma 1.2.4 to G/N and $\{N_i/N \mid i \in I\}$.

Proof of (b): Consider a finite subset J of I. For each $j \in J$ let N_j be an open normal subgroup of G_j. Then $N = \prod_{j \in J} N_j \times \prod_{i \in I \smallsetminus J} G_i$ is an open normal subgroup of G. The family of all these normal subgroups is directed and its intersection is 1. By Lemma 1.2.4, $G = \varprojlim G/N$ is a profinite group.

Proof of (c): Observe that G is a closed subgroup of the cartesian product $\prod_{i \in I} G_i$ (Lemma 1.1.2). By (b) this product is profinite. Now apply Lemma 1.2.5. □

1.3 Infinite Galois Theory

LEMMA 1.2.7: *Each epimorphism $\varphi \colon G \to A$ of profinite groups has a **continuous set theoretic section** $\varphi' \colon A \to G$. That is, φ' is a continuous map satisfying $\varphi \circ \varphi' = \mathrm{id}_A$.*

Proof: Let $K = \mathrm{Ker}(\varphi)$. We split the proof into two parts.

PART A: *K is finite.* Lemma 1.2.5 gives an open subgroup H of G with $K \cap H = 1$. Thus, φ maps H bijectively onto $B = \varphi(H)$. As both H and B are Hausdorff and compact, $\varphi|_H$ is a homeomorphism. Set $\beta = (\varphi|_H)^{-1}$.

Next let $(A : B) = n$, choose $a_1, \ldots, a_n \in A$ and $g_1, \ldots, g_n \in G$ with $A = \bigcup_{i=1}^n a_i B$ and $\varphi(g_i) = a_i$, $i = 1, \ldots, n$. Define $\varphi' \colon A \to G$ by $\varphi'(a_i b) = g_i \beta(b)$, $b \in B$. Then φ' is a continuous set theoretic section of φ.

PART B: *K is arbitrary.* Denote the set of all closed normal subgroups of G which are contained in K by $\mathcal{L}$. For each $L \in \mathcal{L}$, let $\varphi_L \colon G/L \to A$ be the epimorphism induced by φ. For each $L' \in \mathcal{L}$ with $L' \leq L$ let $\varphi_{L',L} \colon G/L' \to G/L$ be the quotient map.

Let Φ' be the set of all pairs (L, φ'_L) where $L \in \mathcal{L}$ and $\varphi'_L \colon A \to G/L$ is a set theoretic section of φ_L. Since $\varphi_K \colon G/K \to A$ is an isomorphism, it has an inverse φ'_K. Hence, (K, φ'_K) is in Φ'.

Define a partial ordering on Φ' as follows: $(L', \varphi'_{L'}) \leq (L, \varphi'_L)$ if $L' \leq L$ and $\varphi_{L',L} \circ \varphi'_{L'} = \varphi'_L$. Suppose $\Phi'' = \{(L_i, \varphi'_i) \mid i \in I\}$ is a descending chain in Φ'; that is, every two elements of Φ'' are comparable. Put $L'' = \bigcap_{i \in I} L_i$. Then, by Lemmas 1.2.6(a) and 1.2.4, $G/L'' = \varprojlim G/L_i$. Thus, the compatible maps $\varphi'_i \colon A \to G/L_i$ give a section $\varphi'_{L''} \colon A \to G/L''$ to $\varphi_{L''}$. Thus, $(L'', \varphi'_{L''})$ is a lower bound for Φ''.

Zorn's lemma gives a minimal element (L, φ'_L) of Φ'. Assume $L \neq 1$. Then L has a proper open subgroup L' which is normal in G. Consider the epimorphism $\varphi_{L',L} \colon G/L' \to G/L$ with the finite kernel L/L'. Part A gives a set theoretic section $\varphi'_{L,L'}$ to $\varphi_{L',L}$. Set $\varphi'_{L'} = \varphi'_{L,L'} \circ \varphi'_L$. Then $(L', \varphi'_{L'})$ is an element of Φ' which is smaller than (L, φ'_L). This contradiction to the minimality of (L, φ'_L) proves that $L = 1$. Put $\varphi' = \varphi'_L$. Then φ' is a continuous set theoretic section of φ. $\square$

1.3 Infinite Galois Theory

Let N be a Galois extension of a field K. The Galois group $\mathrm{Gal}(N/K)$ associated with N/K consists of all automorphisms of N that fix each element of K. If N/K is a finite extension and H_1, H_2 are subgroups of $\mathrm{Gal}(N/K)$ with the same fixed fields in N, then $H_1 = H_2$. This is not the case any more if N/K is infinite.

Consider for example the case where $K = \mathbb{F}_p$ for some prime number p and $N = \bigcup_{i=1}^\infty \mathbb{F}_{p^i}$. Let φ be the **Frobenius automorphism** of $N/\mathbb{F}_p$. It is defined by the rule $\varphi x = x^p$ for each $x \in N$. Let G_0 be the discrete subgroup of $\mathrm{Gal}(N/\mathbb{F}_p)$ generated by φ. It is a countable group and $\mathbb{F}_p$ is its fixed field in N. On the other hand, each element of $\mathrm{Gal}(\mathbb{F}_{p^{2i}}/\mathbb{F}_p)$ has exactly two extensions to $\mathbb{F}_{p^{2^{i+1}}}$. Hence, there are $2^{\aleph_0}$ sequences $(\sigma_1, \sigma_2, \sigma_3, \ldots)$ with $\sigma_i \in \mathrm{Gal}(F_{p^{2^i}}/\mathbb{F}_p)$ such that the restriction of σ_{i+1} to $\mathbb{F}_{p^{2^i}}$ is σ_i for $i = 1, 2, 3, \ldots$. Each such sequence defines a unique $\sigma \in \mathrm{Gal}(N/\mathbb{F}_p)$ whose restriction to $\mathbb{F}_{p^{2^i}}$ is σ_i, $i = 1, 2, 3, \ldots$. It follows that the cardinality of $\mathrm{Gal}(N/\mathbb{F}_p)$ is $2^{\aleph_0}$. In particular, $\mathrm{Gal}(N/\mathbb{F}_p)$ is different from G_0 but has the same fixed field, namely $\mathbb{F}_p$.

The Galois correspondence is restored for closed subgroups of $\mathrm{Gal}(N/K)$ in the "Krull topology" which we now introduce. Denote the set of all intermediate fields $K \subseteq L \subseteq N$, with L/K finite and Galois, by $\mathcal{L}$. To each $L \in \mathcal{L}$ associate the (finite) Galois group $\mathrm{Gal}(L/K)$. If $L' \in \mathcal{L}$ and $L \subseteq L'$, then $\mathrm{res}_L : \mathrm{Gal}(L'/K) \to \mathrm{Gal}(L/K)$ is an epimorphism. Consider the inverse limit $\varprojlim \mathrm{Gal}(L/K)$, with L ranging over $\mathcal{L}$. Every $\sigma \in \mathrm{Gal}(N/K)$ defines a unique element $(\mathrm{res}_L \sigma)_{L \in \mathcal{L}}$ of $\varprojlim \mathrm{Gal}(L/K)$. Conversely, every $(\sigma_L)_{L \in \mathcal{L}} \in \varprojlim \mathrm{Gal}(L/K)$ defines a unique $\sigma \in \mathrm{Gal}(N/K)$ with $\mathrm{res}_L \sigma = \sigma_L$ for each $L \in \mathcal{L}$. Thus, $\sigma \mapsto (\mathrm{res}_L \sigma)_{L \in \mathcal{L}}$ is an isomorphism $\mathrm{Gal}(N/K) \cong \varprojlim \mathrm{Gal}(L/K)$. This isomorphism induces a topology on $\mathrm{Gal}(N/K)$ through the topology on $\varprojlim \mathrm{Gal}(L/K)$: the **Krull topology**. Thus, under the Krull topology, $\mathrm{Gal}(N/K)$ becomes a profinite group and the family $\mathcal{N} = \{\mathrm{Gal}(N/L) \mid L \in \mathcal{L}\}$ is a basis for the open neighborhoods of 1. If N/K is a finite extension, then the Krull topology is discrete.

Suppose L is a finite extension of K contained in N. Then its **Galois closure** $\hat{L}$ is the smallest Galois extension of K that contains L. It is finite over K and is contained in N. Write $\mathrm{Gal}(N/L)$ as a union of right cosets of $\mathrm{Gal}(N/\hat{L})$ to see that $\mathrm{Gal}(N/L)$ is an open closed subgroup of $\mathrm{Gal}(N/K)$.

Suppose L is an arbitrary extension of K in N. Then L is the union of a family $\{L_i \mid i \in I\}$ of finite extensions of K. Hence, $\mathrm{Gal}(N/L) = \bigcap_{i \in I} \mathrm{Gal}(N/L_i)$. Therefore, $\mathrm{Gal}(N/L)$ is a closed subgroup of $\mathrm{Gal}(N/K)$.

If S is a set of automorphisms of N, then

$$N(S) = \{x \in N \mid \sigma x = x \text{ for every } \sigma \in S\}$$

is the fixed field of S in N. $N(S)$ is also the fixed field in N of the closed subgroup $\langle S \rangle$ of $\mathrm{Gal}(N/K)$ generated by S. If $S = \{\sigma_1, \ldots, \sigma_e\}$ is a finite set, replace $N(S)$ by $N(\sigma_1, \ldots, \sigma_e)$.

Let M be a Galois extension of K in N. Denote by res_M (or res) the homomorphism from $\mathrm{Gal}(N/K)$ into $\mathrm{Gal}(M/K)$ that maps $\sigma \in \mathrm{Gal}(N/K)$ onto its restriction $\mathrm{res}_M \sigma$ to M. It is a continuous surjective map.

PROPOSITION 1.3.1: *Let N be a Galois extension of a field K. Then $L \mapsto \mathrm{Gal}(N/L)$ is a bijection from the family of fields L lying between K and N onto the family of closed subgroups of $G = \mathrm{Gal}(N/K)$. The inverse map is $H \mapsto N(H)$.*

Proof: Consider a field extension L of K in N. Put $H = \mathrm{Gal}(N/L)$. Then $L \subseteq N(H)$. Each $x \in N(H)$ is contained in a finite Galois extension $M \subseteq N$

1.3 Infinite Galois Theory

of L. Since the map res: $\mathrm{Gal}(N/L) \to \mathrm{Gal}(M/L)$ is surjective, $\sigma x = x$ for every $\sigma \in \mathrm{Gal}(M/L)$. By finite Galois theory, $x \in L$. Hence, $N(\mathrm{Gal}(N/L)) = L$.

Conversely, let H be a closed subgroup of G and put $L = N(H)$. Then $H \leq \mathrm{Gal}(N/L)$. Consider $\sigma \in \mathrm{Gal}(N/L)$. In order to prove $\sigma \in H$ it suffices to show that σ is in the closure of H. Indeed, let $M \subseteq N$ be a finite Galois extension of K. Then $M \cap L = M(\mathrm{res}_M H)$. Finite Galois theory shows $\mathrm{res}_M \sigma \in \mathrm{Gal}(M/M \cap L) = \mathrm{res}_M H$. Hence, $H \cap \sigma \mathrm{Gal}(N/M)$ is nonempty. It follows that $\mathrm{Gal}(N/N(H)) = H$. □

As in finite Galois theory [Lang7, pp. 192–199], Proposition 1.3.1 gives the following rules for the Galois correspondence:
(1a) $L_1 \subseteq L_2 \iff \mathrm{Gal}(N/L_2) \leq \mathrm{Gal}(N/L_1)$.
(1b) $H_1 \leq H_2 \iff N(H_2) \subseteq N(H_1)$.
(1c) $N(H_1) \cap N(H_2) = N(\langle H_1, H_2 \rangle)$, where $\langle H_1, H_2 \rangle$ is the closed subgroup of G generated by the closed subgroups H_1 and H_2.
(1d) $\mathrm{Gal}(N/L_1 \cap L_2) = \langle \mathrm{Gal}(N/L_1), \mathrm{Gal}(N/L_2) \rangle$.
(1e) $\mathrm{Gal}(N/L_1 L_2) = \mathrm{Gal}(N/L_1) \cap \mathrm{Gal}(N/L_2)$.
(1f) $N(H_1 \cap H_2) = N(H_1)N(H_2)$.

Since Galois groups are compact, their images under restriction are closed. As in the finite case this produces other theorems of infinite Galois theory.
(2a) $N(\sigma H \sigma^{-1}) = \sigma N(H)$ and
(2b) $\mathrm{Gal}(N/\sigma L) = \sigma \mathrm{Gal}(N/L) \sigma^{-1}$, for every $\sigma \in G$.
(2c) A closed subgroup H of G is normal if and only if $L = N(H)$ is a Galois extension of K.
(2d) If L is a Galois extension of K and $L \subseteq N$, then res: $\mathrm{Gal}(N/K) \to \mathrm{Gal}(L/K)$ is a continuous open epimorphism with kernel $\mathrm{Gal}(N/L)$ and $\mathrm{Gal}(L/K) \cong \mathrm{Gal}(N/K)/\mathrm{Gal}(N/L)$.
(2e) res: $\mathrm{Gal}(LM/M) \to \mathrm{Gal}(L/L \cap M)$ is an isomorphism for every Galois extension L of K and every extension M of K.
(2f) If, in (2e), M is also a Galois extension of K, then $\sigma \mapsto (\mathrm{res}_L \sigma, \mathrm{res}_M \sigma)$ is an isomorphism

$$\mathrm{Gal}(LM/L \cap M) \cong \mathrm{Gal}(L/L \cap M) \times \mathrm{Gal}(M/L \cap M)$$

and

$$\mathrm{Gal}(LM/K) \cong \{(\sigma, \tau) \in \mathrm{Gal}(L/K) \times \mathrm{Gal}(M/K) \mid \mathrm{res}_{L \cap M} \sigma = \mathrm{res}_{L \cap M} \tau\}.$$

In both cases we use the product topology on products of groups.

The first isomorphism of (2f) is a special case of the second one. To prove the second isomorphism, note first that the map $\sigma \mapsto (\mathrm{res}_L \sigma, \mathrm{res}_M \sigma)$ is a continuous injective map of the left hand side onto the right hand side. Hence, it suffices to prove surjectivity. Thus, consider $\rho \in \mathrm{Gal}(L/K)$ and $\tau \in \mathrm{Gal}(M/K)$ with $\mathrm{res}_{L \cap M} \rho = \mathrm{res}_{L \cap M} \tau$. Extend ρ to an automorphism ρ_1

of LM and let $\rho_0 = \mathrm{res}_M \rho_1$. Then $\rho_0^{-1}\tau \in \mathrm{Gal}(M/L \cap M)$. By (2e) there is a $\lambda \in \mathrm{Gal}(LM/L)$ with $\mathrm{res}_M \lambda = \rho_0^{-1}\tau$. The element $\sigma = \rho_1 \lambda$ of $\mathrm{Gal}(LM/K)$ satisfies $\mathrm{res}_L \sigma = \rho$ and $\mathrm{res}_M \sigma = \tau$, as desired.

These statements are useful when N is the separable closure K_s of K. Denote $\mathrm{Gal}(K_s/K)$ by $\mathrm{Gal}(K)$, the **absolute Galois group** of K.

Next we show that profinite groups are Galois groups.

LEMMA 1.3.2: *Suppose a profinite group G acts faithfully as automorphisms of a field F. Suppose for each $x \in F$, the stabilizer $S(x) = \{\sigma \in G \mid \sigma x = x\}$, is an open subgroup of G. Then F is a Galois extension of the fixed field $K = F(G)$ and $G = \mathrm{Gal}(F/K)$.*

Proof: If G is a finite group, this is a result of Artin [Lang7, p. 264]. In general the group $H = S(x_1) \cap \cdots \cap S(x_n)$ is open in G for every $x_1, \ldots, x_n \in F$. Therefore, so is the intersection N of all the conjugates of H (Exercise 4). The finite quotient group G/N acts faithfully on the field $L = K(Gx_1, \ldots, Gx_n)$. It has K as its fixed field. Thus, L is a finite Galois extension of K and $G/N \cong \mathrm{Gal}(L/K)$.

The field F is the union of all above L and 1 is the intersection of all the above N. Hence, F is a Galois extension of K and $\mathrm{Gal}(F/K) \cong \varprojlim \mathrm{Gal}(L/K) \cong \varprojlim G/N \cong G$. □

PROPOSITION 1.3.3: *Let L/K be a Galois extension and $\alpha \colon G \to \mathrm{Gal}(L/K)$ an epimorphism of profinite groups. Then there is a Galois extension F/E and an isomorphism $\varphi \colon \mathrm{Gal}(F/E) \to G$ such that F is a purely transcendental extension of L, $L \cap E = K$, and $\alpha \circ \varphi = \mathrm{res}_L$.*

Proof: Let X be the disjoint union of all quotient groups G/N, where N ranges over all open normal subgroups of G. Consider the elements of X as independent over L and set $F = L(X)$. Define an action of each σ of G on F by $\sigma(\tau N) = \sigma \tau N$, for $\tau N \in X$, and $\sigma(a) = \alpha(\sigma)(a)$ for $a \in L$. This action of G is faithful. We have $S(\tau N) = N$ and $S(a) = \alpha^{-1}(\mathrm{Gal}(L/K(a)))$. Any $u \in F$ is a rational function with integral coefficients in $a_1, \ldots, a_m \in L$ and $x_1, \ldots, x_n \in X$. The stabilizer $S(u)$ of u contains the open subgroup $S(a_1) \cap \cdots \cap S(a_m) \cap S(x_1) \cap \cdots \cap S(x_n)$. Hence, $S(u)$ is an open subgroup. Let E be the fixed field of G in F. By Lemma 1.3.2, $G = \mathrm{Gal}(F/E)$ and the conclusion of the proposition follows from the definitions. □

COROLLARY 1.3.4 (Leptin): *Every profinite group is isomorphic to a Galois group of some Galois extension.*

1.4 The p-adic Integers and the Prüfer Group

The first examples of profinite groups are the group $\mathbb{Z}_p$ of p-adic numbers and the **Prüfer group** $\hat{\mathbb{Z}} = \varprojlim \mathbb{Z}/n\mathbb{Z}$.

1.4 The p-adic Integers and the Prüfer Group

THE p-ADIC GROUP $\mathbb{Z}_p$. Let p be a rational prime. Consider the quotient rings $\mathbb{Z}/p^i\mathbb{Z}$ with their canonical homomorphisms $\mathbb{Z}/p^j\mathbb{Z} \to \mathbb{Z}/p^i\mathbb{Z}$ given for $j \geq i$ by $x + p^j\mathbb{Z} \mapsto x + p^i\mathbb{Z}$. The inverse limit $\mathbb{Z}_p = \varprojlim \mathbb{Z}/p^i\mathbb{Z}$ is the **ring of p-adic integers**. It is a **profinite ring**. Each $x \in \mathbb{Z}_p$ is a sequence $(x_i + p^i\mathbb{Z})_{i \in \mathbb{N}}$ where $x_i \in \mathbb{Z}$ and $x_j \equiv x_i \mod p^i\mathbb{Z}$ for $j \geq i$. Each integer $m \geq 0$ corresponds to a basic neighborhood of x consisting of all elements $y = (y_i + p^i\mathbb{Z})_{i \in \mathbb{N}}$ with $y_m \equiv x_m \mod p^m\mathbb{Z}$.

The map $a \mapsto (a + p^i\mathbb{Z})_{i \in \mathbb{N}}$ is an embedding of $\mathbb{Z}$ into $\mathbb{Z}_p$. Identify $\mathbb{Z}$ with its image in $\mathbb{Z}_p$. The sequence $(x_i)_{i \in \mathbb{N}}$ converges to $x = (x_i + p^i\mathbb{Z})_{i \in \mathbb{N}}$ in the p-adic topology. Hence, $\mathbb{Z}$ is dense in $\mathbb{Z}_p$. Yet, $\mathbb{Z}$ is not equal to $\mathbb{Z}_p$. For example, if $p \neq 2$, then $(\sum_{i=0}^{n-1} p^i + p^n\mathbb{Z})_{n \in \mathbb{N}}$ belongs to $\mathbb{Z}_p$ but not to $\mathbb{Z}$. For $p = 2$, $(\sum_{i=0}^{n-1} 4^i + 2^n\mathbb{Z})_{n \in \mathbb{N}}$ belongs to $\mathbb{Z}_2$ but not to $\mathbb{Z}$ (see also Exercise 15).

LEMMA 1.4.1: *The ring $\mathbb{Z}_p$ has the following properties:*
(a) *An element $x = (x_i + p^i\mathbb{Z})_{i \in \mathbb{N}}$ is invertible if and only if $p \nmid x_1$.*
(b) *$\mathbb{Z}_p$ is an integral domain.*

Proof of (a): Suppose $x' = (x_i' + p^i\mathbb{Z})_{i \in \mathbb{N}}$ is an inverse of x. Then $x_1'x_1 \equiv 1 \mod p$, hence $p \nmid x_1$.

Conversely, suppose $p \nmid x_1$. Then for each i, $p \nmid x_i$. Hence, there exists $x_i' \in \mathbb{Z}$ which is unique modulo p^i with $x_i'x_i \equiv 1 \mod p^i$. Thus, $x' = (x_i' + p^i\mathbb{Z})_{i \in \mathbb{N}}$ is in $\mathbb{Z}_p$ and $x'x = 1$.

Proof of (b): Let $x = (x_i + p^i)_{i \in \mathbb{N}}$ and $y = (y_i + p^i)_{i \in \mathbb{N}}$ be nonzero elements of $\mathbb{Z}_p$. Then there are $m, n \in \mathbb{N}$ with $x_m \not\equiv 0 \mod p^m$ and $y_n \not\equiv 0 \mod p^n$. Hence, $x_{m+n}y_{m+n} \not\equiv 0 \mod p^{m+n}$. Therefore, $xy \neq 0$. Consequently, $\mathbb{Z}_p$ is an integral domain. □

LEMMA 1.4.2:
(a) *For each i, $p^i\mathbb{Z}_p$ is the kernel of the projection $\pi_i \colon \mathbb{Z}_p \to \mathbb{Z}/p^i\mathbb{Z}$. Thus, $p^i\mathbb{Z}_p$ is an open subgroup of $\mathbb{Z}_p$ of index p^i.*
(b) *If H is a subgroup of $\mathbb{Z}_p$ of a finite index, then $H = p^i\mathbb{Z}_p$ for some $i \in \mathbb{N}$.*
(c) *0 is the only closed subgroup of $\mathbb{Z}_p$ of infinite index.*
(d) *$p\mathbb{Z}_p$ is the unique closed maximal subgroup of $\mathbb{Z}_p$.*
(e) *All nonzero closed subgroups of $\mathbb{Z}_p$ are isomorphic to $\mathbb{Z}_p$.*

Proof of (a): Suppose $x = (x_j + p^j\mathbb{Z})_{j \in \mathbb{N}}$ belongs to $\mathrm{Ker}(\pi_i)$. Then $x_i \equiv 0 \mod p^i$. Hence, $x_j \equiv 0 \mod p^i$ for each $j \geq i$. Write $x_j = p^i y_j$ for $j \geq i$ and $y_j = y_i$ for $j < i$. Let $z = (y_{j+i} + p^j\mathbb{Z})_{j \in \mathbb{N}}$. Then $z \in \mathbb{Z}_p$ and $p^i z = x$. Indeed, $p^i z_j = p^i y_{j+i} = x_{j+i} \equiv x_j \mod p^j$ for every positive integer j.

Proof of (b): Conversely, let H be a subgroup of $\mathbb{Z}_p$ of index n. Suppose $n = kp^i$ where $p \nmid k$. By Lemma 1.4.1(a), k is invertible in the ring $\mathbb{Z}_p$, so $n\mathbb{Z}_p = p^i\mathbb{Z}_p$. Thus, $p^i\mathbb{Z}_p = n\mathbb{Z}_p \leq H$. It follows, $p^i = (\mathbb{Z}_p : p^i\mathbb{Z}_p) \geq (\mathbb{Z}_p : H) = kp^i$. Therefore, $k = 1$ and $H = p^i\mathbb{Z}_p$.

Proof of (c): A closed subgroup J of $\mathbb{Z}_p$ of infinite index is the intersection of infinitely many open groups (Lemma 1.2.3). Hence, by (a), all subgroups $p^i\mathbb{Z}_p$ contain J. Consequently $J = 0$.

Proof of (d): By (a), (b), and (c), $p\mathbb{Z}_p$ is the unique closed maximal subgroup of $\mathbb{Z}_p$.

Proof of (e): By Lemma 1.4.1(b), the map $x \mapsto p^i x$ is an isomorphism of $\mathbb{Z}_p$ onto $p^i \mathbb{Z}_p$. □

Every element $x = (x_i + p^i\mathbb{Z})_{i\in\mathbb{N}}$ has a unique representation as a formal power series $\sum_{i=0}^{\infty} a_i p^i$, with $0 \le a_i < p$ for all i. Indeed, $x_n \equiv \sum_{i=0}^{n-1} a_i p^i \mod p^n$, for every $n \in \mathbb{N}$.

LEMMA 1.4.3: *Let $\alpha\colon \mathbb{Z}_p \to \mathbb{Z}/p^n\mathbb{Z}$ be an epimorphism with $n \ge 1$ and H a closed subgroup of $\mathbb{Z}_p$. Suppose $\alpha(H) = \mathbb{Z}/p^n\mathbb{Z}$. Then $H = \mathbb{Z}_p$.*

Proof: By Lemma 1.4.2(a), $\operatorname{Ker}(\alpha) = p^n \mathbb{Z}_p$. Thus, by assumption, $H + p^n \mathbb{Z}_p = \mathbb{Z}_p$. Assume $H \ne \mathbb{Z}_p$. Then, by Lemma 1.4.2(d), $H \le p\mathbb{Z}_p$. Therefore, $\mathbb{Z}_p = H + p^n \mathbb{Z}_p \le p\mathbb{Z}_p < \mathbb{Z}_p$. It follows from this contradiction that $H = \mathbb{Z}_p$. □

In the terminology of Section 22.5, Lemma 1.4.3 says that $\alpha\colon \mathbb{Z}_p \to \mathbb{Z}/p^n\mathbb{Z}$ is a Frattini cover.

THE PRÜFER GROUP. For each $n \in \mathbb{N}$ consider the quotient group $\mathbb{Z}/n\mathbb{Z}$ and the canonical homomorphisms $\mathbb{Z}/n\mathbb{Z} \to \mathbb{Z}/m\mathbb{Z}$ defined for $m|n$ by $x + n\mathbb{Z} \mapsto x + m\mathbb{Z}$. The inverse limit $\hat{\mathbb{Z}} = \varprojlim \mathbb{Z}/n\mathbb{Z}$ is the **Prüfer group**. Like with $\mathbb{Z}_p$, embed $\mathbb{Z}$ as a dense subgroup of $\hat{\mathbb{Z}}$ by $x \mapsto (x + n\mathbb{Z})_{n\in\mathbb{N}}$. Thus, $\hat{\mathbb{Z}}$ is the closure of the subgroup generated by 1. Write $\hat{\mathbb{Z}} = \langle 1 \rangle$ and say that 1 **generates** $\hat{\mathbb{Z}}$. Also, the subgroups $n\mathbb{Z}$ of $\mathbb{Z}$ form a basis for the neighborhoods of 0 in the induced topology.

LEMMA 1.4.4: *For each $n \in \mathbb{N}$, $n\hat{\mathbb{Z}}$ is an open subgroup of $\hat{\mathbb{Z}}$ of index n and $n\hat{\mathbb{Z}} \cong \hat{\mathbb{Z}}$. If H is a subgroup of $\hat{\mathbb{Z}}$ of index n, then $H = n\hat{\mathbb{Z}}$.*

Proof of: Suppose $x = (x_k + k\mathbb{Z})_{k\in\mathbb{Z}}$ lies in the kernel Z_n of the projection $\hat{\mathbb{Z}} \to \mathbb{Z}/n\mathbb{Z}$. Then $x_n \equiv 0 \mod n$. Hence, for each $r \in \mathbb{N}$ we have $x_{rn} \equiv x_n \equiv 0 \mod n$, so $x_{rn} = ny_{rn}$ for some $y_{rn} \in \mathbb{Z}$. Let $z = (y_{rn} + r\mathbb{Z})_{r\in\mathbb{Z}}$. If r' is a multiple of r, then $ny_{r'n} \equiv ny_{rn} \mod rn$, so $y_{r'n} \equiv y_{rn} \mod r$. Therefore, $z \in \hat{\mathbb{Z}}$. Moreover, $x_r \equiv x_{rn} \equiv ny_{rn} = nz_r \mod r$. Hence, $x = nz$. Consequently, $n\hat{\mathbb{Z}}$ is an open subgroup of $\hat{\mathbb{Z}}$ of index n.

Next note that the map $x \mapsto nx$ is an isomorphism of $\hat{\mathbb{Z}}$ onto $n\hat{\mathbb{Z}}$. Indeed, if $nx = 0$, then $nx_{rn} \equiv 0 \mod rn$, so $x_r \equiv x_{rn} \equiv 0 \mod r$ for each $r \in \mathbb{N}$. Hence, $x = 0$.

Finally, if H is a subgroup of $\hat{\mathbb{Z}}$ of index n, then $n\hat{\mathbb{Z}}$ is contained in H and has the same index. Therefore, $H = n\hat{\mathbb{Z}}$. □

We conclude by relating $\hat{\mathbb{Z}}$ to the groups $\mathbb{Z}_p$.

LEMMA 1.4.5: *The group $\hat{\mathbb{Z}}$ is topologically isomorphic to the cartesian product $\prod \mathbb{Z}_p$ where p ranges over all primes numbers.*

Proof: Let $n = \prod p^{k_p}$ be the decomposition of a positive integer n into a product of prime powers. The Chinese remainder theorem gives a canonical isomorphism $\prod \mathbb{Z}/p^{k_p}\mathbb{Z} \to \mathbb{Z}/n\mathbb{Z}$. Combine this with the projection $\prod \mathbb{Z}_p \to \prod \mathbb{Z}/p^{k_p}\mathbb{Z}$ to obtain a continuous epimorphism $f_n \colon \prod \mathbb{Z}_p \to \mathbb{Z}/n\mathbb{Z}$. The maps f_n form a compatible system, so they give a continuous homomorphism $f \colon \prod \mathbb{Z}_p \to \hat{\mathbb{Z}}$. Since $\prod \mathbb{Z}_p$ is compact and $\hat{\mathbb{Z}}$ is Hausdorff, $\mathrm{Im}(f)$ is a closed subgroup of $\hat{\mathbb{Z}}$. Moreover, $\mathbb{Z}$ embeds diagonally in $\prod \mathbb{Z}_p$ and $f(m) = m$ for each $m \in \mathbb{Z}$. Thus, $\mathbb{Z} \subseteq \mathrm{Im}(f)$. Since $\mathbb{Z}$ is dense in $\hat{\mathbb{Z}}$, we have $\mathrm{Im}(f) = \hat{\mathbb{Z}}$, so f is surjective. The kernel of f is $\bigcap \mathrm{Ker}(f_n) = 0$. Hence, f is also injective. The compactness and Hausdorff properties imply that f is a topological isomorphism. $\square$

As a consequence of Lemma 1.4.5, $\mathbb{Z}_p$ is both a closed subgroup and a quotient of $\hat{\mathbb{Z}}$, for each prime p.

1.5 The Absolute Galois Group of a Finite Field

For every prime power q there exists a field $\mathbb{F}_q$ (unique up to isomorphism) with q elements. It is characterized within its algebraic closure $\tilde{F}_q$ by

$$\mathbb{F}_q = \{x \in \tilde{F}_q \mid x^q = x\}.$$

The field $\mathbb{F}_q$ has, for each $n \in \mathbb{N}$, exactly one extension, $\mathbb{F}_{q^n}$, of degree n. It is Galois with a cyclic group generated by the **Frobenius automorphism** $\pi_{q,n}$ defined by $\pi_{q,n}(x) = x^q$ for $x \in \mathbb{F}_{q^n}$. The map $a + n\mathbb{Z} \mapsto \pi_{q,n}^a$ is an isomorphism of $\mathbb{Z}/n\mathbb{Z}$ onto $\mathrm{Gal}(\mathbb{F}_{q^n}/\mathbb{F}_q)$. If $m|n$, there is a canonical commutative diagram

$$\begin{array}{ccc} \mathbb{Z}/n\mathbb{Z} & \longrightarrow & \mathbb{Z}/m\mathbb{Z} \\ \downarrow & & \downarrow \\ \mathrm{Gal}(\mathbb{F}_{q^n}/\mathbb{F}_q) & \xrightarrow{\mathrm{res}} & \mathrm{Gal}(\mathbb{F}_{q^m}/\mathbb{F}_q). \end{array}$$

Take the inverse limits to obtain an isomorphism $\hat{\mathbb{Z}} \cong \mathrm{Gal}(\mathbb{F}_q)$ mapping the identity element 1 of $\hat{\mathbb{Z}}$ to the **Frobenius automorphism** π_q, defined on all of $\tilde{\mathbb{F}}_q$ by $\pi_q(x) = x^q$.

Let l be a prime number and $\mathbb{F}_q^{(l)} = \bigcup_{i=1}^{\infty} \mathbb{F}_{q^{l^i}}$. Then the projection $\hat{\mathbb{Z}} \mapsto \mathbb{Z}_l$ corresponds to $\mathrm{res} \colon \mathrm{Gal}(\mathbb{F}_q) \to \mathrm{Gal}(\mathbb{F}_q^{(l)}/\mathbb{F}_q)$. By Lemma 1.4.5, $\mathrm{Gal}(\mathbb{F}_q) \cong \prod \mathrm{Gal}(\mathbb{F}_q^{(l)}/\mathbb{F}_q)$. On the other hand, let N_l be the fixed field of $\mathbb{Z}_l$ in $\tilde{\mathbb{F}}_q$. Then $\mathrm{Gal}(N_l) = \mathbb{Z}_l$, $\mathbb{F}_q^{(l)} N_l = \tilde{\mathbb{F}}_q$, and $\mathbb{F}_q^{(l)} \cap N_l = \mathbb{F}_q$. It follows,

$$\mathrm{Gal}(\mathbb{F}_q) = \mathrm{Gal}(\mathbb{F}_q^{(l)}) \times \mathrm{Gal}(N_l) \cong \prod_{l' \neq l} \mathbb{Z}_{l'} \times \mathbb{Z}_l.$$

Exercises

1. Let (S_i, π_{ji}) be an inverse system of finite sets with all π_{ji} surjective and let $S = \varprojlim S_i$. Use Lemma 1.1.3 to prove that all maps $\pi_i \colon S \to S_i$ determined by the π_{ji}'s are surjective.

2. Let $H_1, \ldots, H_r$ be closed subgroups of a profinite group G. Prove that

$$H_1 \cap \cdots \cap H_r = \bigcap_N (H_1 N \cap \cdots \cap H_r N),$$

where N ranges over all open normal subgroups of G. Hint: Use Lemma 1.2.2(b).

3. Suppose H is a closed subgroup of a profinite group G. Prove: If $HN/N = G/N$ for every open normal subgroup N of G, then $H = G$.

4. Let H be an open subgroup of index n of a profinite group G. Denote the intersection of all conjugates of H in G by N. Note: Multiplication of G on the left cosets of H induces a homomorphism of G into the symmetric group S_n with kernel N. Conclude that G/N is isomorphic to a subgroup of S_n and $(G : N) \leq n!$.

5. Let G be a compact group and H an open subgroup. Suppose H is profinite. Prove: G is profinite. Hint: Use Lemma 1.2.4.

6. Let S be a set of rational primes. Consider the profinite group $\mathbb{Z}_S = \varprojlim \mathbb{Z}/n\mathbb{Z}$, with n running over all positive integers with prime factors in S.
   (a) Prove: The finite homomorphic images of $\mathbb{Z}_S$ are exactly the groups $\mathbb{Z}/n\mathbb{Z}$, where the prime factors of n belong to S.
   (b) Embed $\mathbb{Z}$ in $\mathbb{Z}_S$ and determine the topology on $\mathbb{Z}$ induced by that of $\mathbb{Z}_S$.
   (c) Prove that $\mathbb{Z}_S$ is **procyclic** (i.e. $\mathbb{Z}_S$ is the closure of a group generated by one element).
   (d) Follow the proof of Lemma 1.4.5 to prove that $\mathbb{Z}_S \cong \prod_{p \in S} \mathbb{Z}_p$.

7. Let G be a procyclic group (Exercise 6). Use that G is a homomorphic image of $\hat{\mathbb{Z}}$ to show there exists a set S of primes with $G = \prod_{p \in S} G_p$, where for each $p \in S$ either $G_p \cong \mathbb{Z}/p^{i_p}\mathbb{Z}$ for some $i_p \in \mathbb{N}$ or $G_p \cong \mathbb{Z}_p$. In particular, if G is torsion free, then $G \cong \prod_{p \in S} \mathbb{Z}_p$.

8. Let G be a closed subgroup of $\hat{\mathbb{Z}}$. Prove that every finite quotient of G is a cyclic group. Conclude that G is procyclic and therefore that there exists a set S of prime numbers with $G \cong \prod_{l \in S} \mathbb{Z}_l$.

9. Let G be a profinite group. Prove that each of the following statements is equivalent to $G \cong \hat{\mathbb{Z}}$.
   (a) G has exactly one open subgroup of each index n.
   (b) G is procyclic and there is an epimorphism $\pi \colon G \to \hat{\mathbb{Z}}$.

10. Let G be a procyclic group. Use Exercise 8 to prove that each epimorphism $\pi\colon G \to \hat{\mathbb{Z}}$ is an isomorphism.

11. Let G be a profinite group with at most one open subgroup of every index n.
(a) Prove that every open subgroup is normal.
(b) Observe that (a) holds for every finite homomorphic image $\bar{G}$ of G. Conclude that $\bar{G}$ is nilpotent.
(c) Let P be a finite p-group with the above properties. Prove that every element x of P of maximal order generates P.
(d) Conclude that G is a procyclic group.

12. Define powers in a profinite group G with exponents in $\hat{\mathbb{Z}}$ in the following way: Let $g \in G$ and $\nu \in \hat{\mathbb{Z}}$. Then there exists a sequence $\{\nu_1, \nu_2, \nu_3, \ldots\}$ of elements of $\mathbb{Z}$ that converges to ν. By compactness there is a subsequence of $\{g^{\nu_1}, g^{\nu_2}, g^{\nu_3}, \ldots\}$ that converges to an element h of G.
(a) Prove h does not depend on the sequence $\{\nu_1, \nu_2, \nu_3, \ldots\}$. So we may denote h by g^ν. Hint: If N is a normal subgroup of G of index n, then $x^n \in N$ for every $x \in G$.
(b) Prove the usual rules for the power operations. For example,
$$g^\mu g^\nu = g^{\mu+\nu}, \qquad (g^\mu)^\nu = g^{\mu\nu}, \qquad \text{and} \qquad g^\nu h^\nu = (gh)^\nu \text{ if } gh = hg.$$
(c) Prove that the map $(g, \nu) \mapsto g^\nu$ of $G \times \hat{\mathbb{Z}}$ into G is continuous.

13. Multiplication in the groups $\mathbb{Z}/p^i\mathbb{Z}$ is compatible with the canonical maps $\mathbb{Z}/p^{i+1}\mathbb{Z} \to \mathbb{Z}/p^i\mathbb{Z}$. Therefore, it defines a multiplication in the additive group $\mathbb{Z}_p = \varprojlim \mathbb{Z}/p^i\mathbb{Z}$.
(a) Prove: $\mathbb{Z}_p$ is an integral domain (the quotient field of which, $\mathbb{Q}_p$, is the **field of p-adic numbers**).
(b) Show: Every closed subgroup of $\mathbb{Z}_p$ is an ideal of $\mathbb{Z}_p$.
(c) Show: $p\mathbb{Z}_p$ is the unique maximal ideal of $\mathbb{Z}_p$; observe that $\mathbb{Z}_p/p\mathbb{Z}_p \cong \mathbb{F}_p$.
(d) Deduce: $\alpha \in \mathbb{Z}_p$ is a unit (i.e. invertible in this ring) if and only if α is congruent modulo $p\mathbb{Z}_p$ to one of the numbers $1, 2, \ldots, p-1$. Hint: Form the inverse of $1 + \beta p$, $\beta \in \mathbb{Z}_p$ using the geometric series for $1/(1+x)$.

14. Define multiplication in the additive group $\hat{\mathbb{Z}}$ in a manner analogous to the definition of multiplication in $\mathbb{Z}_p$. Prove that this makes $\hat{\mathbb{Z}}$ a commutative topological ring with zero divisors.
(a) Prove that the isomorphism of additive groups $\hat{\mathbb{Z}} \cong \prod \mathbb{Z}_p$ established in Lemma 1.4.5 is an isomorphism of rings.
(b) Prove that every closed subgroup of $\hat{\mathbb{Z}}$ is also an ideal.

15. Use the power series representation of the elements of $\mathbb{Z}_p$ (Section 1.4) to show $|\mathbb{Z}_p| = 2^{\aleph_0}$. Conclude that $\mathbb{Q}_p$ has elements that are transcendental over $\mathbb{Q}$.

16. For a prime number p, let K_p be $\mathbb{Q}(\zeta_p)$ if $p \neq 2$ and $\mathbb{Q}(\sqrt{-1})$ if $p = 2$. Also, let $L_p = \bigcup_{i=1}^{\infty} \mathbb{Q}(\zeta_{p^i})$.
   (a) Prove that if K' is a field such that $K_p \subseteq K' \subset L_p$, then $\mathrm{Gal}(L_p/K') \cong \mathbb{Z}_p$ and $[K' : \mathbb{Q}] < \infty$.
   (b) Prove that $\mathrm{Gal}(L_p/\mathbb{Q})$ is isomorphic to $\mathbb{Z}_p \times \mathbb{Z}/(p-1)\mathbb{Z}$ if $p \neq 2$ and to $\mathbb{Z}_2 \times \mathbb{Z}/2\mathbb{Z}$ if $p = 2$.

Notes

More about topological groups can be found in [Pontryagin].

A detailed exposition on Galois theory of finite extensions appears in [Lang7, Chapter VI, Section 1]. For finite fields see [Lang1, Chapter V, Section 5].

Leptin's proof of Corollary 1.3.4 uses linear disjointness of fields [Leptin]. We reproduce it in Proposition 2.6.12. The proofs of Lemma 1.3.2 and Proposition 1.3.3 appear in [Waterhouse].

This chapter overlaps with [Ribes, Chapter 1].

Chapter 2.
Valuations and Linear Disjointness

Sections 2.1–2.4 introduce the basic elements of the theory of valuations, especially discrete valuations, and of Dedekind domains. These sections are primarily a survey. We prove that an overring of a Dedekind domain is again a Dedekind domain (Proposition 2.4.7).

The rest of the chapter centers around the notion of linear disjointness of fields. We use this notion to define separable, regular, and primary extensions of fields. In particular, we prove that an extension F/K with a K-rational place is regular. Section 2.8 gives a useful criterion for separability with derivatives.

2.1 Valuations, Places, and Valuation Rings

The literature treats arithmetic theory of fields through three intimately connected classes of objects: valuations, places, and valuation rings. We briefly review the basic definitions.

Call an Abelian (additive) group Γ with a binary relation $<$ an **ordered group** if the following statements hold for all $\alpha, \beta, \gamma \in \Gamma$.
(1a) Either $\alpha < \beta$, or $\alpha = \beta$, or $\beta < \alpha$.
(1b) If $\alpha < \beta$ and $\beta < \gamma$, then $\alpha < \gamma$.
(1c) If $\alpha < \beta$, then $\alpha + \gamma < \beta + \gamma$.

Some examples of ordered groups are the additive groups $\mathbb{Z}$, $\mathbb{R}$, and $\mathbb{Z} \oplus \mathbb{Z}$ with the order $(m,n) < (m',n')$ if either $m < m'$ or $m = m'$ and $n < n'$ (the **lexicographic order**).

A **valuation** v of a field F is a map of F into a set $\Gamma \cup \{\infty\}$, where Γ is an ordered group, with these properties:
(2a) $v(ab) = v(a) + v(b)$.
(2b) $v(a+b) \geq \min(v(a), v(b))$.
(2c) $v(a) = \infty$ if and only if $a = 0$.
(2d) There exists $a \in F^\times$ with $v(a) \neq 0$.

By definition the symbol ∞ satisfies these rules:
(3a) $\infty + \infty = \alpha + \infty = \infty + \alpha = \infty$; and
(3b) $\alpha < \infty$ for each $\alpha \in \Gamma$.

Condition (2) implies several more properties of v:
(4a) $v(1) = 0$, $v(-a) = v(a)$.
(4b) If $v(a) < v(b)$, then $v(a+b) = v(a)$ (Use the identity $a = (a+b) - b$ and (2b));
(4c) If $\sum_{i=1}^n a_i = 0$, then there exist $i \neq j$ such that $v(a_i) = v(a_j)$ and $v(a_i) = \min(v(a_1), \ldots, v(a_n))$ (Use (2b) and (4b)).

We refer to the pair (F, v) as a **valued field**.

The subgroup $\Gamma_v = v(F^\times)$ of Γ is the **value group** of v. The set $O_v = \{a \in F \mid v(a) \geq 0\}$ is the **valuation ring** of v. It has a unique maximal ideal $\mathfrak{m}_v = \{a \in F \mid v(a) > 0\}$. Refer to the residue field $\bar{F}_v = O_v/\mathfrak{m}_v$ as the **residue field** of F at v. Likewise, whenever there is no ambiguity, we denote the coset $a + \mathfrak{m}_v$ by $\bar{a}$ and call it the **residue** of a at v.

Two valuations v_1, v_2 of a field F with value groups Γ_1, Γ_2 are **equivalent** if there exists an isomorphism $f \colon \Gamma_1 \to \Gamma_2$ with $v_2 = f \circ v_1$. Starting from Section 2.2, we abuse our language and say that v_1 and v_2 are **distinct** if they are inequivalent.

A **place** of a field F is a map φ of F into a set $M \cup \{\infty\}$, where M is a field, with these properties:

(5a) $\varphi(a+b) = \varphi(a) + \varphi(b)$.
(5b) $\varphi(ab) = \varphi(a)\varphi(b)$.
(5c) There exist $a, b \in F$ with $\varphi(a) = \infty$ and $\varphi(b) \neq 0, \infty$.

By definition the symbol ∞ satisfies the following rules:

(6a) $x + \infty = \infty + x = \infty$ for each $x \in M$.
(6b) $x \cdot \infty = \infty \cdot x = \infty \cdot \infty = \infty$ for each $x \in M^\times$.
(6c) Neither $\infty + \infty$, nor $0 \cdot \infty$ are defined.

It is understood that (5a) and (5b) hold whenever the right hand side is defined. These conditions imply that $\varphi(1) = 1$, $\varphi(0) = 0$ and $\varphi(x^{-1}) = \varphi(x)^{-1}$. In particular, if $x \neq 0$, then $\varphi(x) = 0$ if and only if $\varphi(x^{-1}) = \infty$.

We call an element $x \in F$ with $\varphi(x) \neq \infty$ **finite** at φ, and say that φ is finite at x. The subring of all elements finite at φ, $O_\varphi = \{a \in F \mid \varphi(a) \neq \infty\}$, is the **valuation ring** of φ. It has a unique maximal ideal $\mathfrak{m}_\varphi = \{a \in F \mid \varphi(a) = 0\}$. The quotient ring $O_\varphi/\mathfrak{m}_\varphi$ is a field which is canonically isomorphic to the **residue field** $\bar{F}_\varphi = \{\varphi(a) \mid a \in O_\varphi\}$ of F at φ. The latter is a subfield of M. Call φ a **K-place** if K is a subfield of F and $\varphi(a) = a$ for each $a \in K$.

Two places φ_1 and φ_2 of a field F with residue fields M_1 and M_2 are **equivalent** if there exists an isomorphism $\lambda \colon M_1 \to M_2$ with $\varphi_2 = \lambda \circ \varphi_1$.

A **valuation ring** of a field F is a proper subring O of F such that if $x \in F^\times$, then $x \in O$ or $x^{-1} \in O$. The subset $\mathfrak{m} = \{x \in O \mid x^{-1} \notin O\}$ is the unique maximal ideal of O (Exercise 1). The map $\varphi \colon F \to O/\mathfrak{m} \cup \{\infty\}$ which maps $x \in O$ onto its residue class modulo $\mathfrak{m}$ and maps $x \in F \smallsetminus O$ onto ∞ is a place of F with valuation ring O. Denote the units of O by $U = \{x \in O \mid x^{-1} \in O\}$. Then $F^\times/U$ is a multiplicative group ordered by the rule $xU \leq yU \iff yx^{-1} \in O$. The map $x \mapsto xU$ defines a valuation of F with O being its valuation ring.

These definitions easily give a bijective correspondence between the valuation classes, the place classes and the valuation rings of a field F.

An isomorphism $\sigma \colon F \to F'$ of fields induces a bijective map of the valuations and places of F onto those of F' according to the following rule: If v is a valuation of F, then $\sigma(v)$ is defined by $\sigma(v)(x) = v(\sigma^{-1}x)$ for every $x \in F'$. If φ is a place of F, then $\sigma(\varphi)(x) = \varphi(\sigma^{-1}x)$. In particular, σ

induces an isomorphism $\bar{F}_\varphi \cong \bar{F}'_{\sigma(\varphi)}$ of residue fields. It is also clear that if φ corresponds to v, then $\sigma(\varphi)$ corresponds to $\sigma(v)$.

A valuation v of a field F is **real** (or of **rank** 1) if Γ_v is isomorphic to a subgroup of $\mathbb{R}$. Real valuations satisfy the so called **weak approximation theorem**, a generalization of the Chinese remainder theorem [Cassels-Fröhlich, p. 48]:

PROPOSITION 2.1.1: *Consider the following objects: inequivalent real valuations $v_1, \ldots, v_n$ of a field F, elements $x_1, \ldots, x_n$ of F, and real numbers $\gamma_1, \ldots, \gamma_n$. Then there exists $x \in F$ with $v_i(x - x_i) \geq \gamma_i$, $i = 1, \ldots, n$.*

2.2 Discrete Valuations

A valuation v of a field F is **discrete** if $v(F^\times) \cong \mathbb{Z}$. In this case we normalize v by replacing it with an equivalent valuation such that $v(F^\times) = \mathbb{Z}$. Each element $\pi \in F$ with $v(\pi) = 1$ is a **prime element** of O_v.

Prime elements of a unique factorization domain R produce discrete valuations of $F = \mathrm{Quot}(R)$. If p is a prime element of R, then every element x of $F^\times$ has a unique representation as $x = up^m$, where u is relatively prime to p and $m \in \mathbb{Z}$. Define $v_p(x)$ to be m. Then v_p is a discrete valuation of F. Suppose p' is another prime element of R. Then $v_{p'}$ is equivalent to v_p if and only if $p'R = pR$, that is if $p' = up$ with $u \in R^\times$.

Example 2.2.1: Basic examples of discrete valuations.

(a) The ring of integers $\mathbb{Z}$ is a unique factorization domain. For each prime number p the residue field of $\mathbb{Q}$ at v_p is $\mathbb{F}_p$. When p ranges over all prime numbers, v_p ranges over all valuations of $\mathbb{Q}$ (Exercise 3).

(b) Let $R = K[t]$ be the ring of polynomials in an indeterminate t over a field K. Then R is a unique factorization domain. Then prime elements of R are the irreducible polynomials p over K. Units of R are the elements u of $K^\times$, so $v_p(u) = 0$ and we say v_p is **trivial** on K. The residue field of $K(t)$ at v_p is isomorphic to the field $K(a)$, where a is a root of p.

There is one additional valuation, v_∞, of $K(t)$ which is trivial on K. It is defined for a quotient $\frac{f}{g}$ of elements of $K[t]$ by the formula $v_\infty(\frac{f}{g}) = \deg(g) - \deg(f)$. The set of v_p's and v_∞ give all valuation of $K(t)$ trivial on K. Thus, all valuations of $K(t)$ which are trivial on K are discrete (Exercise 4).

An arbitrary irreducible polynomial p may have several roots $a \in \tilde{K}$. Each of them defines a place $\varphi_a \colon K(t) \to \tilde{K} \cup \{\infty\}$ by $\varphi_a(t) = a$ and $\varphi_a(c) = c$ for each $c \in K$. These places are equivalent. If $p(t) = t - a$, then φ_a is the unique place of $K(t)$ corresponding to v_p. Similarly, there is a unique place φ_∞ corresponding to v_∞. It is defined by $\varphi_\infty(t) = \infty$.

We may view each $f(t) \in K(t)$ as a function from $K \cup \{\infty\}$ into itself: $f(a) = \varphi_a(f(t))$. Explicitly, write $f(t) = \frac{g(t)}{h(t)}$ with $g, h \in K[X]$ and $\gcd(g, h) = 1$. Let $a \in K$. Then $f(a) = \frac{g(a)}{h(a)}$ if $h(a) \neq 0$ and $f(a) = \infty$

if $h(a) = 0$. To compute $f(\infty)$ let $u = t^{-1}$ and write $f(t) = \frac{g_1(u)}{h_1(u)}$ with $g_1, h_1 \in K[X]$ and $\gcd(g_1, h_1) = 1$. Then $f(\infty) = \frac{g_1(0)}{h_1(0)}$ if $h_1(0) \neq 0$ and $f(\infty) = \infty$ if $h_1(0) = 0$.

Suppose for example $f(t) \in K[t]$ and $f \neq 0$. Then f maps K into itself and $f(\infty) = \infty$. Now suppose $f(t) = \frac{at+b}{ct+d}$ with $ad - bc \neq 0$ and $c \neq 0$, then $f(\infty) = \frac{a}{c}$.

When K is algebraically closed, each irreducible polynomial is linear. Hence, each valuation of $K(t)$ which is trivial over K is either v_{t-a} for some $a \in K$ or v_∞. □

More examples of discrete valuations arise through extensions of the basic examples (Section 2.3).

LEMMA 2.2.2: *Every discrete valuation ring R is a principal ideal domain.*

Proof: Let v be the valuation of $K = \text{Quot}(R)$ with $O_v = R$ and $v(K^\times) = \mathbb{Z}$. Choose a prime element π of R. Now consider a nonzero ideal $\mathfrak{a}$ of R. Then the minimal integer m with $\pi^m \in \mathfrak{a}$ is positive. It satisfies, $\mathfrak{a} = \pi^m R$. □

As a consequence of Lemma 2.2.2, finitely generated modules over R have a simple structure.

PROPOSITION 2.2.3: *Let R be a discrete valuation ring, p a prime element of R, $K = \text{Quot}(R)$, and M a finitely generated R-module. Put $\bar{K} = R/pR$. Let $r = \dim_K M \otimes_R K$, $n = \dim_{\bar{K}} M/pM$, and $m = n - r$. Then there is a unique m-tuple of positive integers $(k_1, k_2, \ldots, k_m)$ with $k_1 \leq k_2 \leq \cdots \leq k_m$ and $M \cong R/p^{k_m}R \oplus \cdots \oplus R/p^{k_1}R \oplus R^r$. Moreover, r is the maximal number of elements of M which are linearly independent over R and n is the minimal number of generators of M.*

Proof: By Lemma 2.2.2, R is a principal ideal domain, so $M = M_{\text{tor}} \oplus N$, where $M_{\text{tor}} = \{m \in M \mid rm = 0 \text{ for some } r \in R,\ r \neq 0\}$ and N is a free R-module [Lang7, p. 147, Thm. 7.3]. Both M_{tor} and N are finitely generated [Lang7, p. 147, Cor. 7.2]. In particular, $N \cong R^s$ for some integer $s \geq 0$. Suppose $m \in M_{\text{tor}}$ and $am = 0$ with $a \in R$, $a \neq 0$. Then, $m \otimes 1 = am \otimes \frac{1}{a} = 0$. Hence, $M_{\text{tor}} \otimes_R K = 0$ and $M \otimes_R K \cong K^s$. Therefore, $s = r$.

By [Lang7, p. 151, Thm. 7.7], $M_{\text{tor}} \cong R/q_{m'}R \oplus \cdots \oplus R/q_1R$ where $q_1, \ldots, q_{m'}$ are elements of R which are neither zero nor units and $q_i | q_{i+1}$, $i = 1, \ldots, m' - 1$. Multiplying each q_i by a unit, we may assume $q_i = p^{k_i}$ with k_i an integer and $1 \leq k_1 \leq k_2 \leq \cdots \leq k_{m'}$. Moreover, the above cited theorem assures $Rq_1, \ldots, Rq_{m'}$ are uniquely determined by the above conditions. Hence, $k_1, \ldots, k_{m'}$ are also uniquely determined.

Combining the first two paragraphs gives:

$$M \cong R/p^{k_{m'}}R \oplus \cdots \oplus R/p^{k_1}R \oplus R^r.$$

Hence, $M/pM = (R/pR)^{m'+r} \cong \bar{K}^{m'+r}$, so $n = m' + r$ and $m' = m$.

2.2 Discrete Valuations

Now recall that elements $v_1, \ldots, v_s$ of M are **linearly independent** over R if $\sum_{i=1}^{s} a_i v_i = 0$ with $a_1, \ldots, a_s \in R$ implies $a_1 = \cdots = a_s = 0$. Alternatively, $v_1 \otimes 1, \ldots, v_s \otimes 1$ are linearly independent over K. Thus, r is the maximal number of R-linearly independent elements of M.

Finally, by Nakayama's lemma [Lang7, p. 425, Lemma 4.3], n is the minimal number of generators of M. □

Definition 2.2.4: Let R be an integral domain with quotient field F. An **overring** of R is a ring $R \subseteq R' \subset F$. It is said to be **proper** if $R \neq R'$. □

LEMMA 2.2.5: *A discrete valuation ring O has no proper overrings.*

Proof: Let R be an overring of O. Assume there exists $x \in R \smallsetminus O$. Then x^{-1} is a nonunit of O. Choose a prime element π for O. Then $x = u\pi^{-m}$ for some $u \in O^\times$ and a positive integer m. Hence, $\pi^{-1} = u^{-1}\pi^{m-1}x \in R$. Therefore, $u'\pi^k \in R$ for all $u' \in O^\times$ and $k \in \mathbb{Z}$. We conclude that $R = \text{Quot}(O)$. □

Composita of places attached to discrete valuations of rational function fields of one variable give rise to useful places of rational function fields of several variables.

Construction 2.2.6: Composition of places. Suppose ψ is a place of a field K with residue field L and φ is a place of L with residue field M. Then $\psi^{-1}(O_\varphi)$ is a valuation ring of K with maximal ideal $\psi^{-1}(\mathfrak{m}_\varphi)$ and residue field $\psi^{-1}(O_\varphi)/\psi^{-1}(\mathfrak{m}_\varphi) \cong O_\varphi/\mathfrak{m}_\varphi \cong M$. Define a map $\varphi \circ \psi \colon K \to M \cup \{\infty\}$ as follows: $\varphi \circ \psi(x) = \varphi(\psi(x))$ if $\psi(x) \neq \infty$ and $\varphi \circ \psi(x) = \infty$ if $\psi(x) = \infty$. Then $\varphi \circ \psi$ is a homomorphism on $\psi^{-1}(O_\varphi)$ and $\{x \in K \mid \varphi \circ \psi(x) = \infty\} = K \smallsetminus \psi^{-1}(O_\varphi)$. Therefore, $\varphi \circ \psi$ is a place of K, called the **compositum** of ψ and φ, $O_{\varphi \circ \psi} = \psi^{-1}(O_\varphi)$, and $\mathfrak{m}_{\varphi \circ \psi} = \psi^{-1}(\mathfrak{m}_\varphi)$.

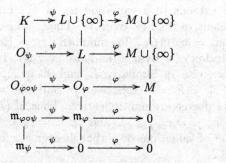

In addition, $L = \bar{K}_\psi$ and $M = \bar{L}_\varphi = \bar{K}_{\varphi \circ \psi}$. □

LEMMA 2.2.7: *Let K be a field, $a_1, \ldots, a_r$ elements of $\tilde{K}$, $t_1, \ldots, t_r$ indeterminates, and L a finite extension of K. Then there exists a K-place $\varphi \colon K(\mathbf{t}) \to K(\mathbf{a}) \cup \{\infty\}$ such that $\varphi(t_i) = a_i$, $i = 1, \ldots, r$. Moreover, every extension of φ to an L-place of $L(\mathbf{t})$ maps $L(\mathbf{t})$ onto $L(\mathbf{a}) \cup \{\infty\}$.*

Proof: For each i there is a $K(a_1, \ldots, a_{i-1}, t_{i+1}, \ldots, t_r)$-place

$$\varphi_i \colon K(a_1, \ldots, a_{i-1}, t_i, t_{i+1} \ldots, t_r) \to K(a_1, \ldots, a_{i-1}, a_i, t_{i+1}, \ldots, t_r)$$

with $\varphi_i(t_i) = a_i$ (Example 2.2.1). The compositum $\varphi = \varphi_r \circ \cdots \circ \varphi_1$ is a K-place of $K(t_1,\ldots,t_r)$ with residue field $K(a_1,\ldots,a_r)$ and $\varphi(t_i) = a_i$, $i = 1,\ldots,r$.

Let now φ be an extension of φ to an L-place of $L(\mathbf{t})$. Choose a basis $b_1,\ldots,b_n$ for L/K. Then $b_1,\ldots,b_n$ is also a basis for $L(\mathbf{t})/K(\mathbf{t})$. Hence, each $f \in L(\mathbf{t})$ has a presentation $f = \sum_{i=1}^{n} b_i f_i$ with $f_i \in K(\mathbf{t})$. Assume without loss that $\frac{f_i}{f_1}$ is finite under φ for $i = 1,\ldots,n$. Then $f = f_1 \sum_{i=1}^{n} \frac{f_i}{f_1} b_i$ and $\varphi(f) \in L(a_1,\ldots,a_r) \cup \{\infty\}$. Thus, $\varphi(L(\mathbf{t})) = L(\mathbf{a}) \cup \{\infty\}$. □

2.3 Extensions of Valuations and Places

The examples of Section 2.2 and the following extension results give a handle on describing valuations of function fields in one variable.

PROPOSITION 2.3.1 (Chevalley [Lang4, p. 8, Thm. 1]): *Let φ_0 be a homomorphism of an integral domain R into an algebraically closed field M and let F be a field containing R. Then φ_0 extends either to an embedding φ of F into M or to a place φ of F into $M \cup \{\infty\}$.*

When F is algebraic over R, the proposition has a more precise form:

Let $f \in R[X]$ be an irreducible polynomial over $E = \text{Quot}(R)$ and $\bar{f} \in M[X]$ the result of applying φ_0 to the coefficients of f. Suppose $\bar{f}$ is not identically zero. Assume x and $\bar{x}$ are roots of f and $\bar{f}$ in $\tilde{E}$ and M, respectively. Then φ_0 extends to a place φ of $E(x)$ into $M \cup \{\infty\}$ with $\varphi(x) = \bar{x}$ [Lang4, p. 10, Thm. 2]. Moreover, if φ_0 is injective, so is φ [Lang4, p. 8, Prop. 2].

In particular, suppose v is a valuation of a field E and F is an extension of E. Then v extends to a valuation w_0 of F. Each valuation w of F which is equivalent to w_0 **lies over** v. Thus, w lies over v if and only if $O_v \subseteq O_w$ and $\mathfrak{m}_v = \mathfrak{m}_w \cap O_v$. The number $e_{w/v} = (w(F^\times) : w(E^\times))$ is the **ramification index** of w over v (and also over E). The field degree $[F : E]$ bounds $e_{w/v}$ (Exercise 5). Similarly, $\bar{E}_v$ embeds in $\bar{F}_w$ to give the inequality $f_{w/v} = [\bar{F}_w : \bar{E}_v] \leq [F : E]$ (Exercise 7). Both the ramification index and the residue field degree are multiplicative. Thus, if (F', w') is an extension of (F, w), then $e_{w'/v} = e_{w'/w} e_{w/v}$ and $f_{w'/v} = f_{w'/w} f_{w/v}$. If $[F : E] < \infty$, then the number of valuations of F that lie over v is finite (a consequence of Proposition 2.3.2).

PROPOSITION 2.3.2: *Let F/E be a finite extension of fields and v a valuation of E. Let $w_1,\ldots,w_g$ be all inequivalent extensions of v to F. Then*

$$(1) \qquad \sum_{i=1}^{g} e_{w_i/v} f_{w_i/v} \leq [F : E]$$

[Bourbaki2, p. 420, Thm. 1]. *If, in addition, v is discrete and F/E is separa-*

2.3 Extensions of Valuations and Places

ble, then each w_i is discrete and (see [Bourbaki2, p. 425, Cor. 1])

$$\text{(2)} \qquad \sum_{i=1}^{g} e_{w_i/v} f_{w_i/v} = [F:E].$$

Suppose $(F,w)/(E,v)$ is an extension of discrete valued fields. In particular, $w(a) = v(a)$ for each $a \in E$. By definition, $e_{w/v} = (w(F^\times) : v(E^\times))$. However, as in Section 2.2, it is customary to replace v and w by equivalent valuations with $v(E^\times) = w(F^\times) = \mathbb{Z}$. The new valuations satisfy

$$w(a) = e_{w/v} v(a) \quad \text{for each} \quad a \in E.$$

Whenever we speak about an extension of discrete valuations, we mean they are normalized and satisfy the latter relation.

Suppose F is a finite Galois extension of E with a Galois group G. Let w be a discrete valuation of F and let $\sigma \in G$. Then, $\sigma(w)$ is a valuation of F (Section 2.1), both w and $\sigma(w)$ lie over the same valuation v of E, and

$$e_{w/v} = e_{\sigma(w)/v} \quad \text{and} \quad f_{w/v} = f_{\sigma(w)/v}.$$

Conversely, suppose w and w' are two discrete valuations of F over the same valuation v of E. Then there exists $\sigma \in G$ such that $\sigma(w) = w'$ (Exercise 9). Thus, if $w_1, \ldots, w_g$ are all distinct valuations of F that lie over v, then they all have the same residue degree f and ramification index e over v. In this case formula (2) simplifies to

$$\text{(3)} \qquad efg = [F:E].$$

The subgroups

$$D_w = D_{w/v} = \{\sigma \in G \mid \sigma O_w = O_w\}$$
$$I_w = I_{w/v} = \{\sigma \in G \mid w(x - \sigma x) > 0 \text{ for all } x \in O_w\}$$

are the **decomposition group** and the **inertia group**, respectively, of w over E. Obviously $I_w \triangleleft D_w$. If $\bar{F}_w/\bar{E}_v$ is separable, then [Serre3, p. 33]

$$\text{(4)} \qquad |I_w| = e_{w/v} \quad \text{and} \quad |D_w| = e_{w/v} f_{w/v}.$$

Section 2.6 generalizes the notion of separable algebraic extension of fields to arbitrary extensions of fields. In particular, purely transcendental extensions of fields are separable. We use this notion in the following definition. Suppose $(F,w)/(E,v)$ is an arbitrary extension of valued fields. We say w is **unramified** (resp. **tamely ramified**) over v (or also over E) if $\bar{F}_w/\bar{E}_v$ is a separable extension and $e_{w/v} = 1$ (resp. $\text{char}(\bar{E}_v) \nmid e_{w/v}$). We say v is **unramified** (resp. **tamely ramified**) in F if each extension of v to F is unramified (resp. tamely ramified) over v.

Example 2.3.3: Purely transcendental extensions. Let (E, v) be a valued field. Consider a transcendental element t over E. Extend v to a valuation v' of $E(t)$ as follows.

First define v' on $E[t]$ by the following rule:

$$(5) \qquad v'\Big(\sum_{i=0}^m a_i t^i\Big) = \min\big(v(a_0), \ldots, v(a_m)\big)$$

for $a_0, \ldots, a_m \in E$. The same argument used to prove Gauss' Lemma proves that $v'(fg) = v'(f) + v'(g)$ for all $f, g \in E[t]$.

Indeed, let $f(t) = \sum_{i=0}^m a_i t^i$ and $g(t) = \sum_{j=0}^n b_j t^j$. Let r be the minimal integer with $v(a_r) = \min\big(v(a_0), \ldots, v(a_m)\big)$ and let s be the minimal integer with $v(b_s) = \min\big(v(b_0), \ldots, v(b_n)\big)$. If $i + j = r + s$ and $(i, j) \neq (r, s)$, then either $i < r$ or $j < s$. In both cases $v(a_r) + v(b_s) < v(a_i) + v(b_j)$. Hence

$$v'\Big(\sum_{i=0}^m a_i t^i\Big) + v'\Big(\sum_{j=0}^n b_j t^j\Big) = v(a_r) + v(b_s)$$

$$= \min\Big(\sum_{i+j=k} v(a_i b_j) \;\Big|\; k = 0, \ldots, m+n\Big)$$

$$= v'\Big(\sum_{i=0}^m a_i t^i \cdot \sum_{j=0}^n b_j t^j\Big),$$

as claimed.

We extend v' to $E(t)$ by the rule $v'(\frac{f}{g}) = v'(f) - v'(g)$. Then we prove $v'(u_1 + u_2) \geq \min\big(v'(u_1), v'(u_2)\big)$ first for $u_1, u_2 \in E[t]$ and then for $u_1, u_2 \in E(t)$. Thus, v' is a valuation of $E(t)$. Note that the residue of t at v' is transcendental over $\bar{E}_v$. Indeed, suppose $\sum_{i=0}^n \bar{a}_i \bar{t}^i = 0$ for some $a_0, \ldots, a_n \in O_v$. Then $\min\big(v(a_0), \ldots, v(a_n)\big) = v'\big(\sum_{i=0}^n a_i t^i\big) > 0$. Hence, $\bar{a}_i = 0$, $i = 0, \ldots, n$.

It follows that, $\overline{E(t)}_{v'} = \bar{E}_v(\bar{t})$ is a rational function field over $\bar{E}_v$. By definition, $\Gamma_{v'} = \Gamma_v$. In particular, if v is discrete, then so is v' and $e_{v'/v} = 1$.

Suppose v'' is another extension of v to $E(t)$ with the residue of t at v'' transcendental over $\bar{E}_v$. We show that $v'' = v'$. Indeed, for $a_0, \ldots, a_n \in E$, not all zero, choose j between 0 and n with $v(a_j) = \min\big(v(a_0), \ldots, v(a_n)\big)$. Then $\sum_{i=0}^n \overline{a_i/a_j} \bar{t}^i \neq 0$. Therefore,

$$v''\Big(\sum_{i=0}^n a_i t^i\Big) = v(a_j) + v''\Big(\sum_{i=0}^n (a_i/a_j) t^i\Big)$$

$$= \min\big(v(a_0), \ldots, v(a_n)\big) = v'\Big(\sum_{i=0}^n a_i t^i\Big),$$

as claimed. □

2.3 Extensions of Valuations and Places

LEMMA 2.3.4: *Let v be a discrete valuation of a field E, $h \in O_v[X]$ a monic irreducible polynomial of degree n, x a root of $h(X)$ in $\tilde{E}$, and $F = E(x)$. Suppose the residue polynomial $\bar{h}(X)$ is separable. Then v is unramified in F.*

Proof: By assumption, $\bar{h}(X) = \prod_{i=1}^{r} h_i(X)$, where $h_i \in \bar{E}_v[X]$ are distinct monic irreducible polynomials. For each i between 1 and r choose a root a_i of $h_i(X)$ in $(\bar{E}_v)_s$. Use Proposition 2.3.1 to extend the residue map $O_v \to \bar{E}_v$ to a place φ_i of F with $\varphi_i(x) = a_i$. Denote the corresponding valuation by w_i. Then $\bar{E}_v(a_i) \subseteq \bar{F}_{w_i}$. Since $h_i(X)$ and $h_j(X)$ have no common root for $i \neq j$, the valuations $w_1, \ldots, w_r$ are mutually inequivalent extensions of v. Label any further extensions of v to valuations of F as $w_{r+1}, \ldots, w_g$. By (1)

$$n = \sum_{i=1}^{r} \deg(h_i) = \sum_{i=1}^{r} [\bar{E}_v(a_i) : \bar{E}_v] \leq \sum_{i=1}^{g} e_{w_i/v} f_{w_i/v} \leq n.$$

Hence, $e_{w_i/v} = 1$ and $\bar{E}_v(a_i) = \bar{F}_{w_i}$ for $i = 1, \ldots, r$. Moreover, $w_1, \ldots, w_r$ are all extensions of v to F and each of them is unramified over E. Therefore, v is unramified in F. □

The converse of Lemma 2.3.4 requires $\bar{E}_v$ to be infinite.

LEMMA 2.3.5: *Let v be a discrete valuation of a field E. Let F be a separable extension of E of degree n. Suppose v is unramified in F and $\bar{E}_v$ is an infinite field. Then F/E has a primitive element x with $\mathrm{irr}(x, E) \in O_v[X]$ and the residue of $\mathrm{irr}(x, E)$ at v is a separable polynomial.*

Proof: Let $w_1, \ldots, w_g$ be all extensions of v to F. By (2), $[F : E] = \sum_{i=1}^{g} [\bar{F}_{w_i} : \bar{E}_v]$. Moreover, for each i the extension $\bar{F}_{w_i}/\bar{E}_v$ is finite and separable. Hence, we may choose c_i in F with $w_i(c_i) = 0$ and the residue $\bar{c}_i$ of c_i at w_i is a primitive element of $\bar{F}_{w_i}/\bar{E}_v$. Let $h_i = \mathrm{irr}(\bar{c}_i, \bar{E}_v)$. Since $\bar{E}_v$ is infinite, we may choose $c_1, \ldots, c_g$ such that $\bar{c}_1, \ldots, \bar{c}_g$ are mutually nonconjugate over $\bar{E}_v$. Thus, $h_1, \ldots, h_g$ are relatively prime.

Use Proposition 2.1.1 to find $x \in F$ with $w_i(x - c_i) > 0$, $i = 1, \ldots, g$. Then, $w_i(x) = 0$, $i = 1, \ldots, g$. Extend each w_i to the Galois closure of F/E. Then all E-conjugates of x have nonnegative values under each extended valuation. Hence, the elementary symmetric polynomials in the E-conjugates of x belong to O_v. Therefore, $f(X) = \mathrm{irr}(x, E) \in O_v[X]$.

Let $\bar{f}$ be the residue of f at v. By construction, $\bar{f}(\bar{c}_i) = 0$, therefore $h_i | \bar{f}$, $i = 1, \ldots, g$. Since $h_1, \ldots, h_g$ are relatively prime, $\prod_{i=1}^{g} h_i | \bar{f}$. Hence,

$$[F : E] = \sum_{i=1}^{g} [\bar{F}_{w_i} : \bar{E}_v] = \sum_{i=1}^{g} \deg(h_i)$$
$$\leq \deg(\bar{f}) = \deg(f) = [E(x) : E] \leq [F : E].$$

Consequently, $E(x) = F$, as desired. □

Example 3.5.4 shows the assumption on $\bar{E}_v$ to be infinite is necessary for Lemma 2.3.5 to hold.

The next lemma says that arbitrary change of the base field preserves unramified discrete valuations.

LEMMA 2.3.6: *Let (E,v) be a discrete valued field. Consider a separable algebraic extension F of E and a discrete valued field (E_1, v_1) which extends (E,v). Suppose v is unramified in F. Then v_1 is unramified in FE_1.*

Proof: Suppose without loss that $[F:E] < \infty$. Let $F_1 = FE_1$. Suppose first that $\bar{E}_v$ is infinite. Choose x as in Lemma 2.3.5 and let $f(X) = \mathrm{irr}(x, E)$. Then $F = E(x)$ and $\bar{f}(X)$ is separable. Hence, $F_1 = E_1(x)$ and $\bar{f}(X)$ is still separable. By Lemma 2.3.4, v_1 is unramified in F_1.

In the general case we consider an extension w_1 of v_1 to a valuation of F_1. Denote the restriction of w_1 to F by w. Let t be transcendental over F_1. Example 2.3.3 extends v (resp. w, v_1, w_1) in a canonical way to a discrete valuation v' (resp. w', v'_1, w'_1) of $E(t)$ (resp. $F(t)$, $E_1(t)$, $F_1(t)$). Further, $e_{v'/v} = 1$ (resp. $e_{w'/w} = 1$, $e_{v'_1/v_1} = 1$, $e_{w'_1/w_1} = 1$) and $\overline{E(t)}_{v'} = \bar{E}_v(\bar{t})$ (resp. $\overline{F(t)}_{w'} = \bar{F}_w(\bar{t})$, $\overline{E_1(t)}_{v'_1} = \bar{E}_{1,v_1}(\bar{t})$, $\overline{F_1(t)}_{w'_1} = \bar{F}_{1,w_1}(\bar{t})$), where $\bar{t}$ is transcendental over $\bar{F}_{1,w_1}$. Moreover, w'_1 extends w' and v'_1 extends v' giving this diagram:

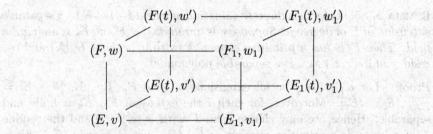

We claim v' is unramified in $F(t)$. Indeed, $\overline{F(t)}_{w'} = \bar{F}_w \cdot \bar{E}_v(\bar{t})$ is a separable extension of $\overline{E(t)}_{v'}$. Also, $e_{w'/v'} = e_{w'/v'}e_{v'/v} = e_{w'/v} = e_{w'/w}e_{w/v} = 1$. Hence, w' is unramified over v'. If u^* is an arbitrary extension of v' to $F(t)$ and u is its restriction to F, then the residue of t at u^* is $\bar{t}$, which is transcendental over $\bar{F}_u$. Thus, by uniqueness of the construction in Example 2.3.3, $u^* = u'$, where u' is the canonical extension of u to $F(t)$. By the above, u^* is unramified over v'.

Since $\overline{E(t)}_{v'}$ is infinite, the first paragraph of the proof implies v'_1 is unramified in $F_1(t)$. Thus, $\bar{F}_{1,w_1}(\bar{t})/\bar{E}_{1,v_1}(\bar{t})$ is a separable extension and $e_{w'_1/v'_1} = 1$. Therefore, $\bar{F}_{1,w_1}/\bar{E}_{1,v_1}$ is a separable extension and

$$e_{w_1/v_1} = e_{w_1/v_1}e_{w'_1/w_1} = e_{w'_1/v_1} = e_{w'_1/v'_1}e_{v'_1/v_1} = 1.$$

Consequently, v_1 is unramified in F_1. □

Combine the multiplicativity of the ramification index and the residue field degree with Lemma 2.3.6 to prove:

2.3 Extensions of Valuations and Places

COROLLARY 2.3.7: *Let $(E,v) \subseteq (E',v') \subseteq (E'',v'')$ be a tower of discrete valued fields. The following hold:*
(a) *v''/v is unramified if and only if v''/v' and v'/v are unramified.*
(b) *v is unramified in E'' if and only if v is unramified in E' and each extension of v to E' is unramified in E''.*
(c) *Let F_1 and F_2 be field extensions of E which are contained in a common field. Suppose F_1/E is separable algebraic and v is unramified in F_1 and in F_2. Then v is unramified in $F_1 F_2$.*

Example 2.3.8: Radical extensions. Let (E,v) be a discrete valued field and n a positive integer with $\mathrm{char}(\bar{E}_v) \nmid n$. Consider an extension $F = E(x)$ of degree n of E where $x^n = a$ is in E. Let w be an extension of v to a valuation of F and let $e = e_{w/v}$. Assume both v and w are normalized. Then

(6) $$nw(x) = ev(a) \quad \text{and} \quad e \leq n.$$

There are three cases to consider:

CASE A: $\gcd(n, v(a)) = 1$. By (6), $n|e$, so $n = e$. By (2), w is the unique extension of v to F. Therefore, v **totally ramifies** in F.

CASE B: $n \nmid v(a)$. By (6), $e \neq 1$. Hence, w ramifies over E.

CASE C: $n|v(a)$. Choose $\pi \in E$ with $v(\pi) = 1$. Write $a = b\pi^{kn}$ with $k \in \mathbb{Z}$ and $b \in E$ such that $v(b) = 0$. Then $y = x\pi^{-k}$ satisfies $y^n = b$ and $F = E(y)$. Moreover, $Y^n - \bar{b}$ decomposes over $(\bar{E}_v)_s$ into distinct linear factors. Therefore, by Lemma 2.3.4, v is unramified in F. $\square$

Example 2.3.9: Artin-Schreier Extensions. Let (E,v) be a discrete valued field of positive characteristic p. An **Artin-Schreier extension** F of degree p has the form $E(x)$ where $x^p - x = a$ with $a \in E$. We consider two cases:

CASE A: $v(a) < 0$ and $p \nmid v(a)$. Let w be an extension of v to F. Then $w(x)$ must be negative and $w(x^p) < w(x)$. Hence, $pw(x) = ev(a)$, where $e = e_{w/v}$. Hence, $p = e$ and $w(x) = v(a)$. Thus, v totally ramifies in F.

CASE B: $v(a) \geq 0$. Then $X^p - X - \bar{a}$ is a separable polynomial. By Lemma 2.3.4, v is unramified in F.

In particular, if $v(a) > 0$, then $X^p - X = \prod_{i=0}^{p-1}(X - i)$ in $\bar{E}_v$. Hence, by Proposition 2.3.2, v has exactly p extensions to F. Label them $v_0, \ldots, v_{p-1}$ with $v_i(x - i) > 0$, $i = 0, \ldots, p-1$. Since $v_i(x-i) < v_i((x-i)^p)$, we conclude from $(x-i)^p - (x-i) = a$ that $v_i(x-i) = v(a)$. $\square$

LEMMA 2.3.10 (Eisenstein's Criterion): *Let R be a unique factorization domain, p a prime element of R, and $f(X) = a_n X^n + a_{n-1} X^{n-1} + \cdots + a_0$ a polynomial with coefficients $a_i \in R$. Then each of the following conditions suffices for f to be irreducible over $\mathrm{Quot}(R)$:*
(a) *$p \nmid a_n$, p divides $a_0, \ldots, a_{n-1}$, and $p^2 \nmid a_0$.*

(b) $p \nmid a_0$, p divides $a_1, \ldots, a_n$, and $p^2 \nmid a_n$.

Proof of (a): See [Lang7, p. 183].

Proof of (b): By (a), the polynomial $X^n f(X^{-1}) = a_n + a_{n-1}X + \cdots + a_0 X^n$ is irreducible over K. Therefore, $f(X)$ is irreducible. $\square$

Example 2.3.11: Ramification at infinity. Let K be a field, t an indeterminate, and $f(X) = a_n X^n + \cdots + a_0 \in K[X]$ with $a_n \neq 0$. By Eisenstein criterion, $f(X) - t$ is irreducible over $\tilde{K}(t)$. Choose a root x of $f(X) = t$ in $\widetilde{K(t)}$. Let $v = v_\infty$ be the valuation of $K(t)$ with $v(t) = -1$ which is trivial on K and let w be a valuation of $K(x)$ lying over v. The relation $a_n x^n + \cdots + a_0 = t$ implies $w(x) < 0$. Hence, $-e_{w/v} = w(t) = w(f(x)) = nw(x)$. Since $e_{w/v} \leq [K(x) : K(t)] \leq n$, this implies $e_{w/v} = [K(x) : K(t)] = n$ and $w(x) = -1$. Hence, v is totally ramified in $K(x)$. In particular, w is the unique valuation of $K(x)$ lying over $K(t)$. $\square$

2.4 Integral Extensions and Dedekind Domains

Integral extensions of $\mathbb{Z}$ in number fields are Dedekind domains. Although they are in general not unique factorization domain, their ideals uniquely factor as products of prime ideals. In this section we survey the concepts of integral extensions of rings and of Dedekind domains and prove that every overring of a Dedekind domain is again a Dedekind domain.

Let F be a field containing an integral domain R. An element $x \in F$ is **integral over** R if it satisfies an equation of the form $x^n + a_{n-1}x^{n-1} + \cdots + a_0 = 0$ with $a_1, \ldots, a_n \in R$. The set of all elements of F which are integral over R form a ring (e.g. by Proposition 2.4.1 below), the **integral closure** of R in F. Call R **integrally closed** if R coincides with its integral closure in Quot(R). For example, every valuation ring O of F is integrally closed. Indeed, assume $x \in F \smallsetminus O$ and x is integral over O. Then $x^n + a_{n-1}x^{n-1} + \cdots + a_0 = 0$ for some $a_0, \ldots, a_{n-1} \in O$. Then x^{-1} is in the maximal ideal $\mathfrak{m}$ of O and $1 + a_{n-1}x^{-1} + \cdots + a_0 x^{-n} = 0$. Thus, $1 \in \mathfrak{m}$, a contradiction.

PROPOSITION 2.4.1 ([Lang4, p. 12]): *An element x of F is integral over R if and only if every place of F finite on R is finite at x. Thus, the integral closure of R in F is the intersection of all valuation rings of F which contain R. In particular, every valuation ring of F is integrally closed.*

Suppose φ is a place of a field F and K is a subfield of F. We say that φ is **trivial** on K, or also that φ is a place of F/K, if $\varphi(x) \neq \infty$ for all $x \in K$. Then $\varphi(y) \neq 0$ for all $y \in K^\times$. Thus, φ maps K isomorphically onto $\varphi(K)$.

LEMMA 2.4.2: *Let $K \subseteq L \subseteq F$ be a tower of fields and φ a place of F. Suppose φ is trivial on K and L is algebraic over K. Then φ is trivial on L.*

Proof: Each $x \in L$ is integral over K, so by Proposition 2.4.1, $\varphi(x) \neq \infty$. Thus, φ is also trivial on L. $\square$

2.4 Integral Extensions and Dedekind Domains

Let S be a subring of F containing R. Call S **integral over** R if every element of S is integral over R. If $S = R[x_1, \ldots, x_m]$ and S is integral over R, then S is a finitely generated R-module. Indeed, every element of S is a linear combination with coefficients in R of the set of monomials $x_1^{\alpha_1} x_2^{\alpha_2} \cdots x_m^{\alpha_m}$, where $0 \le \alpha_i < \deg\bigl(\mathrm{irr}(x_i, \mathrm{Quot}(R))\bigr)$. Propositions 2.3.1 and 2.4.1 give the following:

PROPOSITION 2.4.3: *Let $R \subseteq S$ be integral domains with S finitely generated as an R-algebra. Suppose S is integral over R. Then the following hold:*
(a) *S is finitely generated as an R-module.*
(b) *Let $\varphi \colon R \to M$ be a homomorphism into an algebraically closed field M. Then the set of all homomorphisms $\psi \colon S \to M$ that extend φ is finite and nonempty.*

Suppose $R_1 \subseteq R_2 \subseteq R_3$ are integral domains. Proposition 2.4.1 implies that R_3 is integral over R_1 if and only if R_2 is integral over R_1 and R_3 is integral over R_2.

Call an integral domain R **Noetherian** if every ideal of R is finitely generated. For example, since a discrete valuation ring O is a principal ideal domain, it is integrally closed and Noetherian.

If R is an integral domain and $\mathfrak{p}$ is a prime ideal of R, then

$$R_{\mathfrak{p}} = \{\tfrac{a}{b} \mid a \in R \quad \text{and} \quad b \in R \smallsetminus \mathfrak{p}\}$$

is the **local ring** of R at $\mathfrak{p}$. It has a unique maximal ideal, $\mathfrak{p} R_{\mathfrak{p}}$. If R is a Noetherian domain, then $R_{\mathfrak{p}}$ is also Noetherian. If R is integrally closed, then so is $R_{\mathfrak{p}}$.

LEMMA 2.4.4: *Suppose R is an integral domain. Then $R = \bigcap R_{\mathfrak{m}}$, where $\mathfrak{m}$ ranges over all maximal ideals of R. More generally, $\mathfrak{a} = \bigcap \mathfrak{a} R_{\mathfrak{m}}$ for each ideal $\mathfrak{a}$ of R.*

Proof: Suppose x belongs to each $\mathfrak{a} R_{\mathfrak{m}}$. For each $\mathfrak{m}$, $x = a_{\mathfrak{m}}/b_{\mathfrak{m}}$, with $a_{\mathfrak{m}} \in \mathfrak{a}$ and $b_{\mathfrak{m}} \in R \smallsetminus \mathfrak{m}$. Denote the ideal generated by all the $b_{\mathfrak{m}}$'s by $\mathfrak{b}$. If $\mathfrak{b} \ne R$, then $\mathfrak{b}$ is contained in a maximal ideal $\mathfrak{m}$. Hence, $b_{\mathfrak{m}} \in \mathfrak{m}$, a contradiction. Hence, $\mathfrak{b} = R$. In particular, $1 = \sum_{\mathfrak{m} \in M} b_{\mathfrak{m}} c_{\mathfrak{m}}$ where M is a finite set of maximal ideals, and $c_{\mathfrak{m}} \in R$ for each $\mathfrak{m} \in M$. Therefore $x = \sum_{\mathfrak{m} \in M} x b_{\mathfrak{m}} c_{\mathfrak{m}} = \sum_{\mathfrak{m} \in M} a_{\mathfrak{m}} c_{\mathfrak{m}} \in \mathfrak{a}$. $\square$

Let R be an integral domain with the quotient field F. A nonzero R-submodule $\mathfrak{a}$ of F is said to be a **fractional ideal** of R if there exists a nonzero $x \in R$ with $x\mathfrak{a} \subseteq R$. In particular, every ideal of R is a fractional ideal. Define the **product**, $\mathfrak{ab}$, of two fractional ideals $\mathfrak{a}$ and $\mathfrak{b}$ to be the R-submodule generated by the products ab, with $a \in \mathfrak{a}$ and $b \in \mathfrak{b}$. Define the **inverse** of a fractional ideal $\mathfrak{a}$ as $\mathfrak{a}^{-1} = \{x \in F \mid x\mathfrak{a} \subseteq R\}$. If $a \in \mathfrak{a}$, then $a\mathfrak{a}^{-1} \subseteq R$. Therefore, both $\mathfrak{ab}$ and $\mathfrak{a}^{-1}$ are fractional ideals.

PROPOSITION 2.4.5 ([Cassels-Fröhlich, p. 6]): *The following conditions on an integral domain R are equivalent:*
(a) *R is Noetherian, integrally closed, and its nonzero prime ideals are maximal.*
(b) *R is Noetherian and the local ring, $R_{\mathfrak{p}}$, of every nonzero prime ideal $\mathfrak{p}$ is a discrete valuation ring.*
(c) *Every fractional ideal $\mathfrak{a}$ is **invertible** (i.e. $\mathfrak{a}\mathfrak{a}^{-1} = R$).*

When these conditions hold, R is called a **Dedekind domain**.

By Proposition 2.4.5, the set of all fractional ideals of a Dedekind domain R forms an Abelian group, with R being the unit. One proves that this group is free and the maximal ideals of R are free generators of this group. Thus, every ideal $\mathfrak{a}$ of R has a unique presentation $\mathfrak{a} = \mathfrak{p}_1^{m_1}\mathfrak{p}_2^{m_2}\cdots\mathfrak{p}_r^{m_r}$, as the product of powers of maximal ideals with positive exponents [Cassels-Fröhlich, p. 8].

Every principal ideal domain is a Dedekind domain. Thus, $\mathbb{Z}$ and $K[x]$, where x is a transcendental element over a field K, are Dedekind domains. By the same reason, every discrete valuation ring is a Dedekind domain.

In the notation of Proposition 2.4.5(b), $R_{\mathfrak{p}}$ is the valuation ring of a discrete valuation $v_{\mathfrak{p}}$ of $K = \text{Quot}(R)$. The corresponding place $\varphi_{\mathfrak{p}}$ is finite on R. Conversely, if φ is such a place, then $\mathfrak{p} = \{x \in R \mid \varphi(x) = 0\}$ is a nonzero prime ideal of R. Since $R_{\mathfrak{p}} \subseteq O_{\varphi}$, Lemma 2.2.5 implies that $R_{\mathfrak{p}} = O_{\varphi}$. This establishes a bijection between the nonzero prime ideals of R and the equivalence classes of places of K finite on R.

PROPOSITION 2.4.6 ([Cassels-Fröhlich, p. 13]): *Let S be the integral closure of a Dedekind domain R in a finite algebraic extension of $\text{Quot}(R)$. Then S is also a Dedekind domain.*

Let $\mathfrak{p}$ be a prime ideal of R. Then $\mathfrak{p}S = \mathfrak{P}_1^{e_1}\mathfrak{P}_2^{e_2}\cdots\mathfrak{P}_r^{e_r}$, where $\mathfrak{P}_1, \mathfrak{P}_2, \ldots, \mathfrak{P}_r$ are the distinct prime ideals of S that lie over $\mathfrak{p}$; that is, $\mathfrak{P}_i \cap R = \mathfrak{p}$, $i = 1, \ldots, r$. For each i we have $\mathfrak{p}S_{\mathfrak{P}_i} = \mathfrak{P}_i^{e_i} S_{\mathfrak{P}_i}$. Hence, e_i is the ramification index of $v_{\mathfrak{P}_i}$ over $v_{\mathfrak{p}}$. We say $\mathfrak{P}_i$ is **unramified** over K if $v_{\mathfrak{P}_i}/v_{\mathfrak{p}}$ is unramified; that is, $e_i = 1$ and $S/\mathfrak{P}_i$ is a separable extension of $R/\mathfrak{p}$. The prime ideal $\mathfrak{p}$ is **unramified** in L if each $\mathfrak{P}_i$ is unramified over K.

By Proposition 2.4.6, the integral closure of $\mathbb{Z}$ in a finite extension L of $\mathbb{Q}$ is a Dedekind domain, O_L, called the **ring of integers** of L.

PROPOSITION 2.4.7 (Noether-Grell): *Every overring R' of a Dedekind domain R is a Dedekind domain.*

Proof: We show that R' satisfies Condition (b) of Proposition 2.4.5.

PART A: *An injective map.* If $\mathfrak{p}'$ is a nonzero prime ideal of R', then $\mathfrak{p} = R \cap \mathfrak{p}'$ is a nonzero prime ideal of R. Indeed, for $0 \neq x \in \mathfrak{p}'$, write $x = \frac{a}{b}$, where $a, b \in R$. Thus, $0 \neq a = bx \in R \cap \mathfrak{p}' = \mathfrak{p}$. Since $R_{\mathfrak{p}} \subseteq R'_{\mathfrak{p}'}$ and $R_{\mathfrak{p}}$ is a discrete valuation ring, Lemma 2.2.5 implies that $R_{\mathfrak{p}} = R'_{\mathfrak{p}'}$. Hence,

(1) $$\mathfrak{p}R_{\mathfrak{p}} = \mathfrak{p}'R'_{\mathfrak{p}'}.$$

2.4 Integral Extensions and Dedekind Domains

In addition, $\mathfrak{p}'R'_{\mathfrak{p}'} \cap R' = \mathfrak{p}'$. Therefore, the map $\mathfrak{p}' \mapsto R \cap \mathfrak{p}'$ from the set of nonzero prime ideals of R' into the set of nonzero prime ideals of R is injective.

PART B: *A finiteness condition.* Let x be a nonzero element of R', $\mathfrak{p}'$ a prime ideal of R' which contains x, and $\mathfrak{p} = R \cap \mathfrak{p}'$. Then $R_\mathfrak{p} = R'_{\mathfrak{p}'}$. Hence, $v_\mathfrak{p}(x) > 0$, where $v_\mathfrak{p}$ is the valuation of Quot(R) corresponding to $\mathfrak{p}$. But this relation holds only for the finitely many prime ideals of R that appear with positive exponents in the factorization of the fractional ideal xR. Hence, by Part A, x belongs to only finitely many prime ideals of R'.

PART C: *The ring R' is Noetherian.* Let $\mathfrak{a}$ be a nonzero ideal of R'. Choose a nonzero element $x \in \mathfrak{a}$ and denote the finite set of prime ideals of R' that contain x by P. For each $\mathfrak{p} \in P$ the local ring $R'_\mathfrak{p}$ is a discrete valuation domain. Hence, there exists $a_\mathfrak{p} \in \mathfrak{a}$ such that $\mathfrak{a}R'_\mathfrak{p} = a_\mathfrak{p} R'_\mathfrak{p}$. Denote the ideal of R' generated by x and by all $a_\mathfrak{p}$, for $\mathfrak{p} \in P$, by $\mathfrak{a}_0$. It is contained in $\mathfrak{a}$. To show that $\mathfrak{a}$ is finitely generated, we need only prove that $\mathfrak{a} \subseteq \mathfrak{a}_0$.

Indeed, consider a prime ideal $\mathfrak{q}$ of R' not in P. Then $x \notin \mathfrak{q}$, so $\mathfrak{a}_0 \not\subseteq \mathfrak{q}$. Hence, $\mathfrak{a}_0 R'_\mathfrak{q} = R'_\mathfrak{q}$. It follows from Lemma 2.4.4 that $\mathfrak{a}_0 = \bigcap_{\mathfrak{p} \in P} \mathfrak{a}_0 R'_\mathfrak{p}$. Therefore, $\mathfrak{a} \subseteq \bigcap_{\mathfrak{p} \in P} \mathfrak{a} R'_\mathfrak{p} = \bigcap_{\mathfrak{p} \in P} a_\mathfrak{p} R'_\mathfrak{p} \subseteq \bigcap_{\mathfrak{p} \in P} \mathfrak{a}_0 R'_\mathfrak{p} = \mathfrak{a}_0$, as desired. $\square$

LEMMA 2.4.8: *Let (E,v) be a discrete valued field, F_1, F_2, F finite separable extensions of E with $F = F_1 F_2$, and w an extension of v to F. Suppose v is unramified in F_1. Then the residue fields with respect to w satisfy $\bar{F} = \bar{F}_1 \bar{F}_2$.*

Proof: Choose a finite Galois extension N of E which contains F and an extension w' of w to N. Denote the decomposition groups of w' over E, F_1, F_2, F by $D_E, D_{F_1}, D_{F_2}, D_F$, respectively. Let E', F'_1, F'_2, F' be the fixed fields in N of $D_E, D_{F_1}, D_{F_2}, D_F$, respectively. Let $v' = w'|_{E'}$. Since all valuations of N lying over v' are conjugate over E', the definition of E' as the fixed field of D_E implies that w' is the unique extension of v' to N. Also, $D_{F_1} = \text{Gal}(N/F_1) \cap D_E$, so $F_1 E' = F'_1$. By Lemma 2.3.6, v' is unramified in F'_1. Finally, by [Serre3, p. 32, Prop. 21(c)], the residue fields of E, F_1, F_2, F at w coincide with the residue fields of E', F'_1, F'_2, F' at w', respectively.

We may therefore replace E, F_1, F_2, F, respectively, by E', F'_1, F'_2, F', if necessary, to assume that $w|_{F_1}$ is the unique extension of v to F_1. Now put $w_i = w|_{F_i}$, $i = 1, 2$. By Proposition 2.4.1, O_{w_1} is the integral closure of O_v in F_1. Since v is unramified in F_1, Proposition 2.3.2 implies $[F_1 : E] = [\bar{F}_1 : \bar{E}]$, where the bar denotes reduction modulo w.

Choose $x \in O_{w_1}$ such that $\bar{x}$ is a primitive element for the separable extension $\bar{F}_1/\bar{E}$. Let $f = \text{irr}(x, E)$ and $p = \text{irr}(\bar{x}, \bar{E})$. Then $f \in O_v[X]$ and $f(x) = 0$. Hence, $\bar{f}(\bar{x}) = 0$ and $p | \bar{f}$. Therefore,

$$[F_1 : E] \geq \deg(f) \geq \deg(\bar{p}) = [\bar{F}_1 : \bar{E}] = [F_1 : E].$$

Consequently, $p = \bar{f}$, $F_1 = E(x)$, and $\bar{F}_1 = \bar{E}(\bar{x})$.

By Lemma 2.3.6, w_2 is unramified in F. Thus, we may apply the result of the preceding paragraph to F/F_2 and conclude that $\bar{F} = \bar{F}_2(\bar{x})$. Consequently, $\bar{F}_1 \bar{F}_2 = \bar{E}(\bar{x})\bar{F}_2 = \bar{F}_2(\bar{x}) = \bar{F}$. □

2.5 Linear Disjointness of Fields

Central to field theory is the concept "linear disjointness of fields", an analog of linear independence of vectors.

We repeat the convention made in "Notation and Convention" that whenever we form the compositum of fields, we tacitly assume they are contained in a common field.

LEMMA 2.5.1: *Let E and F be extensions of a field K. The following conditions are equivalent:*
(a) *Each m-tuple $(x_1, \ldots, x_m)$ of elements of E which is linearly independent over K is also linearly independent over F.*
(b) *Each n-tuple $(y_1, \ldots, y_n)$ of elements of F which is linearly independent over K is also linearly independent over E.*

Proof: It suffices to prove that (a) implies (b). Let $y_1, \ldots, y_n$ be elements of F for which there exist $a_1, \ldots, a_n \in E$ with $a_1 y_1 + \cdots + a_n y_n = 0$. Let $\{x_j \mid j \in J\}$ be a linear basis for E over K and write $a_i = \sum_{j \in J} a_{ij} x_j$ with a_{ij} elements of K, only finitely many different from 0. Then

$$\sum_{j \in J} \Big(\sum_{i=1}^n a_{ij} y_i \Big) x_j = 0.$$

By (a), $\{x_j \mid j \in J\}$ is linearly independent over F. Hence, $\sum a_{ij} y_i = 0$ for every j. If $y_1, \ldots, y_m$ are linearly independent over K, then $a_{ij} = 0$ for every i and j, so $a_i = 0$, $i = 1, \ldots, m$. Thus, $y_1, \ldots, y_m$ are linearly independent over E. This proves (b). □

Definition: With E and F field extensions of a field K, refer to E and F as **linearly disjoint over** K if (a) (or (b)) of Lemma 2.5.1 holds. □

COROLLARY 2.5.2: *Let E and F be extensions of a field K such that $[E : K] < \infty$. Then E and F are linearly disjoint over K if and only if $[E : K] = [EF : F]$. If in addition $[F : K] < \infty$, then this is equivalent to $[EF : K] = [E : K][F : K]$.*

Proof: If E and F are linearly disjoint over K and $w_1, \ldots, w_n$ is a basis for E/K, then $w_1, \ldots, w_n$ is also a basis for EF over F. Hence, $[EF : F] = n = [E : K]$. Conversely, suppose $[E : K] = [EF : F]$ and let $x_1, \ldots, x_m \in E$ be linearly independent over K. Extend $\{x_1, \ldots, x_m\}$ to a basis $\{x_1, \ldots, x_n\}$ of E/K. Since $\{x_1, \ldots, x_n\}$ generates EF over F and $n = [EF : F]$, $\{x_1, \ldots, x_n\}$ is a basis of EF/F. In particular, $x_1, \ldots, x_m$ are linearly independent over F. □

2.5 Linear Disjointness of Fields

Let E/K be a finite Galois extension. If $E \cap F = K$, then, by Corollary 2.5.2, E and F are linearly disjoint over K. The condition, $E \cap F = K$ is equivalent to "res: $\mathrm{Gal}(EF/F) \to \mathrm{Gal}(E/K)$ is an isomorphism" and also to "res: $\mathrm{Gal}(F) \to \mathrm{Gal}(E/K)$ is surjective." For arbitrary extensions this condition is clearly necessary, but not sufficient. Let L be a degree $n > 1$ extension of K for which L' is conjugate to L over K and $L' \cap L = K$. Then $[LL' : K] \leq n(n-1)$. Thus, according to Corollary 2.5.2, L and L' are not linearly disjoint over K. For example, $\mathbb{Q}(\sqrt[3]{2})$ is not linearly disjoint from $\mathbb{Q}(\zeta_3 \sqrt[3]{2})$ over $\mathbb{Q}$ although their intersection is $\mathbb{Q}$.

LEMMA 2.5.3 (Tower Property): *Let $K \subseteq E$ and $K \subseteq L \subseteq F$ be four fields. Then E is linearly disjoint from F over K if and only if E is linearly disjoint from L over K and EL is linearly disjoint from F over L.*

Proof: The only nontrivial part is to show that if E and F are linearly disjoint over K, then EL and F are linearly disjoint over L.

Apply Lemma 2.5.1. Suppose that $y_1, \ldots, y_m$ are elements of F which are linearly independent over L, but $a_1, \ldots, a_m$ are elements of EL such that $\sum_{i=1}^m a_i y_i = 0$. Clear denominators to assume that $a_i \in L[E]$, so that $a_i = \sum a_{ij} x_j$ with $a_{ij} \in L$, where $\{x_j \mid j \in J\}$ is a linear basis for E over K. Then $\sum_j (\sum_i a_{ij} y_i) x_j = 0$. By assumption, the x_j are linearly independent over F. Hence, $\sum_j a_{ij} y_i = 0$, so $a_{ij} = 0$ for all i and j. Consequently, $a_i = 0$, $i = 1, \ldots, m$. $\square$

LEMMA 2.5.4: *Let L be a separable algebraic extension of a field K and let M be a purely inseparable extension of K. Then L and M are linearly disjoint over K.*

Proof: Let $\hat{L}$ be the Galois closure of L/K. Then $\hat{L} \cap M = K$. Hence, $\hat{L}$ and M are linearly disjoint over K. Therefore, by Lemma 2.5.3, L and M are linearly disjoint over K. $\square$

Let $E_1, \ldots, E_n$ be n extensions of a field K. We say that $E_1, \ldots, E_n$ are **linearly disjoint** over K if $E_1 \cdots E_{m-1}$ and E_m are linearly disjoint over K for $m = 2, \ldots, n$. Induction on n shows that this is the case if and only if the following condition holds: If w_{i,j_i}, $j_i \in J_i$, are elements of E_i which are linearly independent over K, $i = 1, \ldots, n$, then $\prod_{i=1}^n w_{i,j_i}$, $(j_1, \ldots, j_n) \in J_1 \times \cdots \times J_n$, are linearly independent over K.

It follows that $E_1, \ldots, E_n$ are linearly disjoint over K if and only if the canonical homomorphism of $E_1 \otimes_K \cdots \otimes_K E_n$ into $E_1 \cdots E_n$ that maps $x_1 \otimes \cdots \otimes x_n$ onto $x_1 \cdots x_n$ is injective. It also follows that if $E_1, \ldots, E_n$ are linearly disjoint over K, then $E_{\pi(1)}, \ldots, E_{\pi(n)}$ are linearly disjoint over K for every permutation π of $\{1, \ldots, n\}$.

The application of tensor products makes the following lemma an easy observation.

LEMMA 2.5.5: *Let $E_1, \ldots, E_n$ (resp. $F_1, \ldots, F_n$) be linearly disjoint field extensions of K (resp. L). For each i between 1 and n let $\varphi_i \colon E_i \to F_i \cup \{\infty\}$,* be either a place or an embedding. Suppose $\varphi_1, \ldots, \varphi_n$ coincide on K and $\varphi_i(K) = L$, $i = 1, \ldots, n$. Let $E = E_1 \cdots E_n$ and $F = F_1 \cdots F_n$. Then there exists a place $\varphi \colon E \to \tilde{F} \cup \{\infty\}$ that extends each of the φ_i's. If each φ_i is an isomorphism of E_i onto F_i, then φ is an isomorphism of E onto F.

Proof: Let O_i be the valuation ring of φ_i if φ_i is a place and E_i if φ_i is an isomorphism. By assumption, the map $x_1 \cdots x_n \to x_1 \otimes \cdots \otimes x_n$ is an isomorphism $O_1 \cdots O_n \cong O_1 \otimes_K \cdots \otimes_K O_n$ of rings. Hence, there exists a ring homomorphism $\varphi_0 \colon O_1 \cdots O_n \to F$ such that $\varphi_0(x) = \varphi_i(x)$ for each $x \in O_i$, $i = 1, \ldots, n$. Extend φ_0 to a place $\varphi \colon E \to \tilde{F} \cup \{\infty\}$ (Proposition 2.3.1). If $x \in E_i \smallsetminus O_i$, then $\varphi(x^{-1}) = \varphi_i(x^{-1}) = 0$, so $\varphi(x) = \varphi_i(x) = \infty$. We conclude that φ coincides with φ_i on E_i. $\square$

Finally, define a family $\{E_i \mid i \in I\}$ of field extensions of K to be **linearly disjoint over** K if every finite subfamily is linearly disjoint over K. It follows from the discussion preceding Lemma 2.5.5 that a sequence $(E_1, E_2, E_3, \ldots)$ of fields extensions of K is linearly disjoint over K if E_n is linearly disjoint from $E_1 \cdots E_{n-1}$ for $2, 3, 4, \ldots$. Then, $E_{\pi(1)}, E_{\pi(2)}, E_{\pi(3)}, \ldots$ are linearly disjoint for every permutation π of $\mathbb{N}$.

LEMMA 2.5.6: *Let $\{L_i \mid i \in I\}$ be a linearly disjoint family of Galois extensions of a field K. Then $\mathrm{Gal}(\prod_{i \in I} L_i / K) \cong \prod_{i \in I} \mathrm{Gal}(L_i / K)$.*

Proof: Since $\prod_{i \in I} \mathrm{Gal}(L_i/K) \cong \varprojlim \prod_{i \in I_0} \mathrm{Gal}(L_i/K)$, we may assume I is finite. In this case, the embedding $\mathrm{Gal}(\prod_{i \in I} L_i/K) \to \prod_{i \in I} \mathrm{Gal}(L_i/K)$ given by $\sigma \mapsto (\sigma|_{L_i})_{i \in I}$ is surjective (Lemma 2.5.5). Therefore, it is an isomorphism. $\square$

LEMMA 2.5.7: *Let K be a field, $K_1, K_2, K_3, \ldots$ a linearly disjoint sequence of extensions of K, and L a finite separable extension of K. Then there exists a positive integer n such that $L, K_n, K_{n+1}, K_{n+2}, \ldots$ are linearly disjoint over K.*

Proof: Replace L by its Galois closure over K, if necessary, to assume L is Galois over K. Assume for each positive integer n the field L is not linearly disjoint from $K_n K_{n+1} K_{n+2} \cdots$ over K. Then $L_n = L \cap K_n K_{n+1} K_{n+2} \cdots$ is a proper extension of K. Since L has only finitely many extensions that contain K and since $L_n \supseteq L_{n+1} \supseteq L_{n+2} \supseteq \cdots$, there is an m such that $L_n = L_m$ for all $n \geq m$. Since L_m is a finite extension of K, there is an $n > m$ with $L_m \subseteq K_m \cdots K_{n-1}$. Similarly, there exists $r > n$ with $L_m \subseteq K_n \cdots K_{r-1}$. By assumption, $K_m \cdots K_{n-1}$ and $K_n \cdots K_{r-1}$ are linearly disjoint over K. In particular, their intersection is K. Therefore, $L_m = K$. This contradiction proves there exists n such that $L, K_n, K_{n+1}, K_{n+2}, \ldots$ are linearly disjoint over K. $\square$

LEMMA 2.5.8: *Let v be a discrete valuation of a field K and L, M finite extensions of K. Suppose v is unramified in L but totally ramified in M. Then L and M are linearly disjoint over K.*

2.5 Linear Disjointness of Fields

Proof: Let L_0 be the maximal separable extension of K in L and v_0 an extension of v to L_0. Then L/L_0 is purely inseparable. Hence, v_0 is ramified in L. Therefore, $L = L_0$ and L/K is separable.

Since v is unramified in each of the conjugates of L over K, it is unramified in their compositum (Corollary 2.3.7). We may therefore replace L by the Galois closure of L/K, if necessary, to assume L/K is Galois.

Let $m = [L \cap M : K]$. Choose an extension w of v to $L \cap M$. Then $e(w/v) = 1$ on one hand and $e(w/v) = m$ on the other hand. Thus, $L \cap M = K$. Therefore, L is linearly disjoint from M over K. $\square$

Example 2.5.9: Roots of unity. For each n consider the Galois extension $\mathbb{Q}(\zeta_n)$ of $\mathbb{Q}$ obtained by adjoining a primitive root of unity of order n. It is well known that $\varphi(n) = [\mathbb{Q}(\zeta_n) : \mathbb{Q}]$ is the number of integers between 1 and n which are relatively prime to n [Lang7, p. 278, Thm. 3.1]. If m is relatively prime to n, then $\varphi(mn) = \varphi(m)\varphi(n)$ [LeVeque, p. 28, Thm. 3-7]. In addition, $\mathbb{Q}(\zeta_m, \zeta_n) = \mathbb{Q}(\zeta_{mn})$. Hence, $[\mathbb{Q}(\zeta_m, \zeta_n) : \mathbb{Q}] = [\mathbb{Q}(\zeta_{mn}) : \mathbb{Q}] = \varphi(mn) = \varphi(m)\varphi(n) = [\mathbb{Q}(\zeta_m) : \mathbb{Q}][\mathbb{Q}(\zeta_n) : \mathbb{Q}]$. It follows from Corollary 2.5.2 that $\mathbb{Q}(\zeta_m)$ and $\mathbb{Q}(\zeta_n)$ are linearly disjoint over $\mathbb{Q}$. $\square$

Here is an application of linear disjointness to integral closures of domains.

LEMMA 2.5.10: *Let K be a field, L a separable algebraic extension of K, and R an integrally closed integral domain containing K. Let $E = \operatorname{Quot}(R)$, $F = EL$, and S the integral closure of R in F. Suppose E and L are linearly disjoint over K. Then $S = RL \cong R \otimes_K L$.*

Proof: Assume without loss L/K is finite. Choose a basis $w_1, \ldots, w_n$ for L/K. Let $\sigma_1, \ldots, \sigma_n$ be the distinct K-embeddings of L into K_s. Then $\det(\sigma_i w_j) \neq 0$.

Each element of L is integral over K, hence over R, so $RL \subseteq S$. Conversely, let $x \in S$. By the linear disjointness, $w_1, \ldots, w_n$ form a basis for F/E. Hence, $x = \sum_{j=1}^n e_j w_j$ with $e_j \in E$, $j = 1, \ldots, n$. Also, each σ_i extends to an E-embedding of F into E_s (Lemma 2.5.5). Thus, $\sigma_i x = \sum_{j=1}^n e_j \sigma_i w_j$, $i = 1, \ldots, n$. Apply Kramer's law to present each e_k as a polynomial in $\sigma_i x, \sigma_i w_j$, with $i, j = 1, \ldots, n$, divided by $\det(\sigma_i w_j)$. Thus, e_k is an element of E which is integral over R. Since R is integrally closed, $e_k \in R$, $k = 1, \ldots, n$. Consequently, $x \in RL$, as needed. $\square$

We generalize the tower property to families of field extensions:

LEMMA 2.5.11: *Let K be a field and I a set. For each $i \in I$ let F_i/E_i be a field extension with $K \subseteq E_i$. Suppose $\{F_i \mid i \in I\}$ is linearly disjoint over K. Denote the compositum of all E_i's by E. Then the set $\{F_i E \mid i \in I\}$ is linearly disjoint over E. Moreover, for each $i \in I$, the field F_i is linearly disjoint from E over E_i.*

Proof: It suffices to consider the case where $I = \{1, 2, \ldots, n\}$. By induction suppose $F_i E_1 \cdots E_{n-1}$, $i = 1, \ldots, n-1$, are linearly disjoint over $E_1 \cdots E_n$.

By assumption, $F_1 \cdots F_{n-1}$ is linearly disjoint from F_n over K. Hence, by the tower property, $F_1 \cdots F_{n-1}$ is linearly disjoint from E over $E_1, \ldots, E_{n-1}$, so $F_i E$, $i = 1, \ldots, n-1$, are linearly disjoint over E.

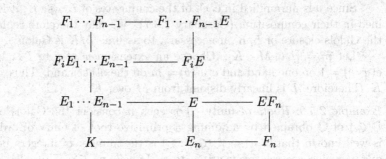

Moreover, $F_1 \cdots F_{n-1} E$ is linearly disjoint from EF_n over E. Consequently, E is linearly disjoint from F_n over E_n and $F_i E$, $i = 1, \ldots, n$ are linearly disjoint over E, as claimed. $\square$

2.6 Separable, Regular, and Primary Extensions

Based on the notion of linear disjointness we define here three type of field extensions. We say that a field extension F/K is **separable** (resp. **regular, primary**) if F is linearly disjoint from K_{ins} (resp. $\tilde{K}$, K_s) over K.

SEPARABLE EXTENSIONS. We generalize the notion of "separable algebraic extension" to arbitrary field extensions.

Let K be a field of positive characteristic p. The field generated over K by the pth roots of all elements of K is denoted $K^{1/p}$. We denote the maximal purely inseparable extension of K by K_{ins} (or K^{1/p^∞}). Let F be a finitely generated extension of K. A collection $t_1, \ldots, t_r \in F$ of elements algebraically independent over K is a **separating transcendence basis** if $F/K(t_1, \ldots, t_r)$ is a finite separable extension.

LEMMA 2.6.1: *An extension F of a field K is separable if it satisfies one of the following equivalent conditions:*
(a) *F is linearly disjoint from K_{ins} over K.*
(b) *F is linearly disjoint from $K^{1/p}$ over K.*
(c) *Every finitely generated extension E of K which is contained in F has a separating transcendence basis.*

Moreover, a separating transcendence basis can be selected from a given set of generators for F/K.

Proof: The implications "(a) => (b)" and "(c) => (a)" are immediate consequences of the tower property (Lemma 2.5.3). For "(b) => (c)" see [Lang 4, p. 54]. Lemma 19.2.4 gives a constructive proof. $\square$

In particular, every separable algebraic extension satisfies conditions (a), (b), and (c) of Lemma 2.6.1. Now apply the rules of linear disjointness.

2.6 Separable, Regular, and Primary Extensions

COROLLARY 2.6.2:
(a) If E/K and F/E are separable extensions, then F/K is also separable.
(b) If F/K is a separable extension, then E/K is separable for every field $K \subseteq E \subseteq F$.
(c) Every extension of a perfect field is separable.
(d) If E/K is a purely inseparable extension and F/K is a separable extension, then E and F are linearly disjoint over K.

Example 2.6.3: A separable tower does not imply separable steps. Consider the tower of fields $\mathbb{F}_p \subset \mathbb{F}_p(t^p) \subset \mathbb{F}_p(t)$, where t is transcendental over $\mathbb{F}_p$. The extension $\mathbb{F}_p(t)/\mathbb{F}_p$ is separable, but $\mathbb{F}_p(t)/\mathbb{F}_p(t^p)$ is not. ☐

REGULAR EXTENSIONS. Finitely generated regular extensions characterize absolutely irreducible varieties (Section 10.2)

LEMMA 2.6.4: *A field extension F/K is regular if it satisfies one of the following equivalent conditions:*
(a) F/K is separable and K is algebraically closed in F.
(b) F is linearly disjoint from $\tilde{K}$ over K.

Proof: The implication "(b) => (a)" is immediate.

To prove "(a) => (b)", it suffices to assume that F/K is finitely generated. Then F/K has a separating transcendence basis, $t_1, \ldots, t_r$, which is also a separating transcendence basis for the extension FK_s/K_s. Since $\tilde{K} = (K_s)_{\text{ins}}$, Lemma 2.6.1 implies that FK_s is linearly disjoint from $\tilde{K}$ over K_s. Also, K_s/K is a Galois extension and $F \cap K_s = K$. Hence, F is linearly disjoint from K_s over K. Therefore, by Lemma 2.5.3, F is linearly disjoint from $\tilde{K}$ over K. ☐

COROLLARY 2.6.5:
(a) If E/K and F/E are regular extensions, then F/K is regular.
(b) If F/K is a regular extension, then E/K is regular for every field E lying between K and F.
(c) Every extension of an algebraically closed field is regular.
(d) Let m be a cardinal number and K_α, $\alpha \leq m$, an ascending transfinite sequence of fields such that $K_\gamma = \bigcup_{\alpha < \gamma} K_\alpha$ for each limit ordinal number $\gamma \leq m$. Suppose $K_{\gamma+1}$ is a regular extension of K_γ for all $\gamma < m$. Then K_m is a regular extension of each K_β with $\beta < m$.

Proof of (d): Let $\delta \leq m$ be a transfinite number. By transfinite induction assume K_γ is regular extension of K_β for all $\beta \leq \gamma < \delta$. Now distinguish between two cases:

CASE A: δ *is a limit number.* Consider $\beta < \delta$, elements $a_1, \ldots, a_r \in \tilde{K}_\beta$ linearly independent over K_β, and elements $u_1, \ldots, u_r \in K_\delta$ satisfying $\sum_{i=1}^r a_i u_i = 0$. Then there exists an ordinal number γ with $\beta \leq \gamma < \delta$ and $u_1, \ldots, u_r \in K_\gamma$. Since K_γ/K_β is regular, $\tilde{K}_\beta$ is linearly disjoint from K_γ

over K_β, so $a_1 = \cdots = a_r = 0$. Therefore, K_δ is linearly disjoint from $\tilde{K}_\beta$ over K_β.

CASE B: $\delta = \gamma + 1$ *is a successor number*. By assumption, both K_γ/K_β and $K_{\gamma+1}/K_\gamma$ are regular extensions. Hence, by (a), K_δ/K_β is a regular extension. □

The next lemma gives a criterion for a regular extension F/K to be linearly disjoint from another extension of K in terms of "algebraic independence". To define this notion consider an arbitrary field extension F/K and a subset T of F. We say that T is **algebraically independent** over K if $f(t_1, \ldots, t_n) \neq 0$ for all $t_1, \ldots, t_n \in T$ and for each nonzero $f \in K[X_1, \ldots, X_n]$. If in addition $F/K(T)$ is an algebraic extension, then T is a **transcendence base** of F/K. The cardinality of T depends only on F/K. It is the **transcendence degree** of F/K. We denote it by $\text{trans.deg}(F/K)$. For example, $\text{trans.deg}(F/K) = 0$ if and only if F/K is an algebraic extension. If S is a subset of F such that $F/K(S)$ is algebraic, then S contains a transcendence base for F/K [Lang7, p. 356, Thm. 1.1]. In particular, if F/K is finitely generated, then $\text{trans.deg}(F/K) < \infty$. The converse is false. For example, $\text{trans.deg}(\tilde{\mathbb{Q}}/\mathbb{Q}) = 0$ although $\tilde{\mathbb{Q}}/\mathbb{Q}$ is not finitely generated.

If T_0 is a subset of F which is algebraically independent over K, choose a transcendence base T_1 for $F/K(T_0)$. Then $T_0 \cap T_1 = \emptyset$ and $T_0 \cup T_1$ is a transcendence base for F/K. This argument also gives the additivity of the transcendence degree for a tower $K \subseteq E \subseteq F$ of fields:

(1) $\qquad \text{trans.deg}(F/K) = \text{trans.deg}(E/K) + \text{trans.deg}(F/E)$.

Now consider two extensions E and F of a field K. We say that E and F are **algebraically independent** over K if
(2) every m-tuple $(t_1, \ldots, t_m)$ of elements of E which is algebraically independent over K is also algebraically independent over F.

It follows that E and F are algebraically independent over K if and only if E_0 and F are algebraically independent over K for every subfield E_0 of E which is finitely generated over K. Hence, in order to prove that algebraic independence is a symmetric relation, we may consider finitely generated extensions E and F of K, assume that (2) holds, and prove condition (2) with the roles of E and F exchanged. Indeed, let $u_1, \ldots, u_n$ be elements of F which are algebraically independent over K. Enlarge n, if necessary, to assume that $u_1, \ldots, u_n$ form a transcendence base of F/K. Then $F/K(\mathbf{u})$ is algebraic and therefore so is $EF/E(\mathbf{u})$. After reordering the u_i, we may assume $u_1, \ldots, u_m$ form a transcendence base for EF/E. Assumption (2) implies that $\text{trans.deg}(E/K) = \text{trans.deg}(EF/F)$. Hence, by (1), $m = \text{trans.deg}(EF/E) = \text{trans.deg}(EF/K) - \text{trans.deg}(E/K) = \text{trans.deg}(EF/K) - \text{trans.deg}(EF/F) = \text{trans.deg}(F/K) = n$. Therefore, $u_1, \ldots, u_n$ are algebraically independent over E, as desired.

Like linear disjointness, algebraic independence has the tower property: Let $K \subseteq L \subseteq M$ and $K' \subseteq L' \subseteq M'$ be fields with $K \subseteq K'$, $L' = K'L$ and

2.6 Separable, Regular, and Primary Extensions

$M' = L'M$. Then trans.deg$(M/K) =$ trans.deg$(L/K)+$trans.deg(M/L) and trans.deg$(M'/K') =$ trans.deg$(L'/K') +$ trans.deg(M'/L'). Also,

$$\text{trans.deg}(L'/K') \leq \text{trans.deg}(L/K), \text{ trans.deg}(M'/L') \leq \text{trans.deg}(M/L).$$

Hence, M is algebraically independent from K' over K if and only if L is algebraically independent from K' over K and M is algebraically independent from L' over L.

By considering monomials in elements $x_1, \ldots, x_n$ of E, it is clear that if E and F are linearly disjoint over K, then they are also algebraically independent over K. The converse, however, is false: Any two extensions of K one of which is algebraic are algebraically independent over K. Lemma 2.6.7 below gives a partial converse.

LEMMA 2.6.6: *Let F and $\bar{F}$ be fields, T (resp. $\bar{T}$) an algebraically independent set over F (resp. over $\bar{F}$), $\varphi_0: F \to \bar{F} \cup \{\infty\}$ a place, and $\varphi_1: T \to \bar{T}$ a bijective map. Then there exists a place $\varphi: F(T) \to \bar{F}(\bar{T}) \cup \{\infty\}$ extending both φ_0 and φ_1.*

Proof: The case where T consists of one element t is covered by Example 2.3.3. In the general case well order T and apply transfinite induction. □

LEMMA 2.6.7: *Let E be a regular extension of a field K and let F be an extension of K. If E and F are algebraically independent over K, then E and F are linearly disjoint over K.*

Proof (Artin): Let $x_1, \ldots, x_n$ be elements of E for which there exist $a_1, \ldots, a_n \in F$, not all zero, such that $\sum a_i x_i = 0$. Use Proposition 2.3.1 to choose a K-place φ of F into $\tilde{K} \cup \{\infty\}$. Let T be a transcendence base for E over K. Then the elements of T are algebraically independent over F. Hence, by Lemma 2.6.6, φ extends to a $K(T)$-place of $F(T)$. Since E is an algebraic extension of $K(T)$, φ extends to an E-place of EF into $\tilde{E} \cup \{\infty\}$ (Lemma 2.4.2).

With no loss we may divide $a_1, \ldots, a_n$ by, say a_1, to assume that $a_1 = 1$ and that all the a_i are finite under φ. Thus, $\sum \varphi(a_i) x_i = 0$ is a nontrivial linear combination of the x_i over $\tilde{K}$. But E is linearly disjoint from $\tilde{K}$ over K. Hence, $x_1, \ldots, x_n$ are also linearly dependent over K. □

COROLLARY 2.6.8:
(a) *Let E be a regular extension of a field K, algebraically independent from an extension F of K. Then EF is a regular extension of F.*
(b) *If two regular extensions E and F of K are algebraically independent, then EF/K is regular.*

Proof: For (a) note that E is also algebraically independent from $\tilde{F}$ over K. By Lemma 2.6.7, E is linearly disjoint from $\tilde{F}$ over K. Hence, by Lemma 2.5.3, EF is linearly disjoint from $\tilde{F}$ over F. Therefore, EF/F is regular.

For (b) use (a) and Corollary 2.6.5(a). □

LEMMA 2.6.9: *Each of the following conditions on a field extension F/K implies that F/K is regular:*
(a) *For all $u_1,\ldots,u_n \in F^\times$, there exists a K-place $\varphi\colon F \to \tilde{K} \cup \{\infty\}$ with $\varphi(u_1),\ldots,\varphi(u_n) \in K^\times$.*
(b) *There exists a K-place $\varphi\colon F \to K \cup \{\infty\}$.*

Proof: We prove that F/K satisfies Condition (b) of Lemma 2.6.4. Consider $w_1,\ldots,w_n$ in $\tilde{K}$ which are linearly independent over K. Assume there exist $u_1,\ldots,u_n$ in F, not all zero, such that $\sum_{i=1}^n u_i w_i = 0$. Omitting the terms with $u_i = 0$, we may assume $u_i \neq 0$ for all i. In Case (a) choose a K-place $\varphi\colon F \to \tilde{K} \cup \{\infty\}$ with $\varphi(u_i) \in K^\times$ for each i such that $u_i \neq 0$. In Case (b) divide $u_1,\ldots,u_n$ with one of them, say with u_1, to assume that $u_1 = 1$ and that $\varphi(u_i) \in K$ for $i = 1,\ldots,n$. Now apply Proposition 2.3.1 and extend φ to a place $\tilde{\varphi}\colon F\tilde{K} \to \tilde{K} \cup \{\infty\}$. By Lemma 2.4.2, $\tilde{\varphi}$ is trivial on $\tilde{K}$. In other words, the restriction of $\tilde{\varphi}$ to $\tilde{K}$ is an automorphism.

In particular, $\tilde{\varphi}(w_1),\ldots,\tilde{\varphi}(w_n)$ are linearly independent over K. Since $\sum_{i=1}^n \varphi(u_i)\tilde{\varphi}(w_i) = 0$, this implies $\varphi(u_i) = 0$ for $i = 1,\ldots,n$. This contradiction proves that $w_1,\ldots,w_n$ are linearly independent over F. We conclude that F is linearly disjoint from K over $\tilde{K}$. □

Example 2.6.10: Purely transcendental extensions. Let $t_1,\ldots,t_n$ be algebraically independent elements over a field K. For each i between 1 and n the map $t_i \to 0$ extends to a $K(t_1,\ldots,t_{i-1})$-place φ_i of $K(t_1,\ldots,t_i)$ onto $K(t_1,\ldots,t_{i-1}) \cup \{\infty\}$. Hence, by Lemma 2.6.9(b),

$$K(t_1,\ldots,t_i)/K(t_1,\ldots,t_{i-1})$$

is a regular extension. Therefore, by Corollary 2.6.9(a), $K(\mathbf{t})/K$ is a regular extension.

Of course, we can also prove the latter result directly: Let $f_1,\ldots,f_m$ be elements of $K(\mathbf{t})$ which are linearly dependent over $\tilde{K}$. Thus, there are $\tilde{c}_1,\ldots,\tilde{c}_m \in \tilde{K}$ not all zero with $\sum_{i=0}^m \tilde{c}_i f_i = 0$. Clearing denominators, we may assume all $f_i \in K[\mathbf{t}]$. Write $f_i(\mathbf{t}) = \sum_{\mathbf{j}} a_{i\mathbf{j}} t_1^{j_1} \cdots t_n^{j_n}$. Then $\sum_{\mathbf{j}} \left(\sum_{i=1}^m \tilde{c}_i a_{i\mathbf{j}}\right) t_1^{j_1} \cdots t_n^{j_n} = \sum_{i=1}^m \tilde{c}_i f_i(\mathbf{t}) = 0$. Hence, $\sum_{i=1}^m \tilde{c}_i a_{i\mathbf{j}} = 0$ for all $\mathbf{j}$. Thus, the homogeneous linear system of equations $\sum_{i=1}^m X_i a_{i\mathbf{j}} = 0$ with coefficients $\mathbf{a}_{i\mathbf{j}} \in K$ has a nonzero solution in $\tilde{K}^n$. Therefore, it has a nonzero solution in K^n. In other words, there are $c_1,\ldots,c_n \in K$ not all zero with $\sum_{i=1}^m c_i a_{i\mathbf{j}} = 0$ for all $\mathbf{j}$. They satisfy $\sum_{i=1}^m c_i f_i = 0$. Hence, $f_1,\ldots,f_m$ are linearly dependent over K. This completes the direct proof that $K(\mathbf{t})/K$ is a regular extension.

We have not defined composition of places. But if we had, we could compose the places $\varphi_i\colon K(t_1,\ldots,t_i) \to K(t_1,\ldots,t_{i-1})$, $i = n, n-1,\ldots,1$, of the first paragraph to a K-place $\varphi\colon K(\mathbf{t}) \to K$ satisfying $\varphi(t_i) = 0$, $i = 1,\ldots,n$. Again, by Lemma 2.6.9(b), this would prove that $K(\mathbf{t})/K$ is regular. Also, given $a_1,\ldots,a_n \in K$, we can replace t_i by $t_i - a_i$ to produce a K-place $\psi\colon K(\mathbf{t}) \to K$ with $\psi(t_i) = a_i$, $i = 1,\ldots,n$.

2.6 Separable, Regular, and Primary Extensions

Let now T be an arbitrary set of algebraically independent elements over K and $\varphi_0 \colon T \to K$ a map. Then every finitely generated subextension of $K(T)/K$ is regular. Therefore, $K(T)/K$ is regular. Moreover, using transfinite induction, it is possible to construct a K-place $\varphi \colon K(T) \to K$ which extends φ_0. $\square$

Example 2.6.11: Absolutely irreducible polynomials. Now consider a polynomial $f \in K[T_1, \ldots, T_n, X]$. Suppose f is **absolutely irreducible**; that is, f is irreducible in $\tilde{K}[T_1, \ldots, T_n, X]$. Let x be a root of $f(\mathbf{t}, X)$ in $\widetilde{K(\mathbf{t})}$. Then $[K(\mathbf{t}, x) : K(\mathbf{t})] = \deg_X f = [\tilde{K}(\mathbf{t}, x) : \tilde{K}(\mathbf{t})]$. By Corollary 2.5.2, $K(\mathbf{t}, x)$ is linearly disjoint from $\tilde{K}(\mathbf{t})$ over $K(\mathbf{t})$. By Example 2.6.10, $K(\mathbf{t})$ is linearly disjoint from $\tilde{K}$ over K. Hence, by the tower property (Lemma 2.5.3), $K(\mathbf{t}, x)$ is linearly disjoint from $\tilde{K}$ over K. Therefore, $K(\mathbf{t}, x)/K$ is a regular extension.

Conversely, suppose f is irreducible in $K[\mathbf{T}, X]$ and $K(\mathbf{t}, x)/K$ is a regular extension. Reversing the above arguments shows that f is absolutely irreducible. $\square$

As an application we rephrase Corollary 1.3.4 and supply a new proof. It is a simplified version of [Leptin].

PROPOSITION 2.6.12: *Let G be a profinite group and K a field. Then there is a Galois extension F/E with $K \subseteq E$ and $\mathrm{Gal}(F/E) \cong G$.*

Proof: Write G as a projective limit $\varprojlim G_i$ of finite groups G_i with i ranging over a directed set I. By definition, G is a closed subgroup of $\prod_{i \in I} G_i$. Suppose we have constructed an algebraic extension F/E with $K \subseteq E$ and $\mathrm{Gal}(F/E) \cong \prod_{i \in I} G_i$. Let E' be the fixed field of G in F. Then $\mathrm{Gal}(F/E') \cong G$.

In order to construct F/E with $\mathrm{Gal}(F/E) \cong \prod_{i \in I} G_i$, we choose a family $(x_i^\sigma)_{i \in I, \sigma \in G_i}$ of algebraically independent elements over K. For each $i \in I$ let $F_i = K(x_i^\sigma \mid \sigma \in G_i)$. The group G_i acts on F_i by the rule $(x_i^\sigma)^\tau = x_i^{\sigma\tau}$ and $a^\tau = a$ for $a \in K$. Let E_i be the fixed field. Then $K \subseteq E_i$ and F_i/E_i is a Galois extension with Galois group G_i [Lang7, p. 264].

Denote the compositum of all E_i's by E and the compositum of all F_i's by F. By Example 2.6.10, each F_i is a regular extension of K. By construction, the set $\{F_i \mid i \in I\}$ is algebraically independent over K. Hence, by Lemma 2.6.7, the set $\{F_i \mid i \in I\}$ is linearly disjoint over K. It follows from Lemma 2.5.11, that the set $\{EF_i \mid i \in I\}$ is linearly disjoint over E. Moreover, E is linearly disjoint from F_i over E_i. Therefore, $\mathrm{Gal}(EF_i/E) \cong \mathrm{Gal}(F_i/E_i) \cong G_i$. It follows from Lemma 2.5.6 that $\mathrm{Gal}(F/E) \cong \prod_{i \in I} G_i$, as desired. $\square$

PRIMARY EXTENSIONS. We use primary extensions in the study of C_i-fields (Section 21.2).

LEMMA 2.6.13: *A field extension F/K is primary if it satisfies one of the following equivalent conditions:*

(a) $F \cap \tilde{K}/K$ is a purely inseparable extension.
(b) The field F is linearly disjoint from K_s over K.

Proof: Clearly "(b) => (a)." The implication "(a) => (b)" holds since $F \cap K_s = K$ and K_s/K is a Galois extension. □

COROLLARY 2.6.14:
(a) If E/K and F/E are primary extensions, then so is F/K.
(b) If F/K is a primary extension, then E/K is primary, for every field $K \subseteq E \subseteq F$.
(c) Every extension of a separably closed field is primary.
(d) An extension F/K is regular if and only if it is separable and primary.

LEMMA 2.6.15:
(a) Let E be a primary extension of a field K which is algebraically independent from an extension F of K. Then EF is a primary extension of F.
(b) If two primary extensions E and F of K are algebraically independent, then EF/K is primary.

Proof: Assertion (b) follows from (a) and from Corollary 2.6.14(a). To prove (a), choose a transcendence base T for E/K and let M be the maximal separable extension of $K(T)$ in E. Then M is a separable and primary extension of K. Hence, by Lemma 2.6.14(d), it is regular. Also, M is algebraically independent from F_s over K. By Lemma 2.6.7, MF is linearly disjoint from F_s over F. Since EF is a purely inseparable extension of MF, it is linearly disjoint from MF_s. It follows that EF is linearly disjoint from F_s over F; that is, EF is a primary extension of F. □

2.7 The Imperfect Degree of a Field

We classify fields of positive characteristic by their imperfect degree and characterize those fields for which every finite extension has a primitive element as fields of imperfect degree 1.

Let F be a field of positive characteristic p. Consider a subfield F_0 of F that contains the field F^p of all pth powers in F. Observe that for $x_1, \ldots, x_n \in F$, the set of monomials

(1) $$x_1^{i_1} \cdots x_n^{i_n}, \quad 0 \leq i_1, \ldots, i_n \leq p-1,$$

generates $F_0(\mathbf{x})$ over F_0. Hence, $[F_0(\mathbf{x}) : F_0] \leq p^n$. If $[F_0(\mathbf{x}) : F_0] = p^n$, then $x_1, \ldots, x_n$ are said to be p-**independent over** F_0. Equivalently, each of the fields $F_0(x_1), \ldots, F_0(x_n)$ has degree p over F_0 and they are linearly disjoint over F_0. This means that the set of monomials (1) is linearly independent over F_0. A subset B of F is p-**independent over** F_0, if every finite subset of B is p-independent over F_0. If in addition $F_0(B) = F$, then B is said to be a p-**basis** for F over F_0. As in the theory of vector spaces, each maximal p-independent subset of F over F_0 is a p-basis for F over F_0.

2.7 The Imperfect Degree of a Field

If $x_1, \ldots, x_n \in F$ are p-independent over F^p, we call them p-**independent** elements of F. The p-power $p^n = [F : F^p]$ is the **imperfect degree** of F and n is the **imperfect exponent** of F. We say that F is n-**imperfect**. Thus, a perfect field has imperfect exponent 0. Both quantities are **infinite** if $[F : F^p] = \infty$. In this case F is ∞-**imperfect**.

LEMMA 2.7.1 (Exchange Principle): *Let F_0 be a subfield of F which contains F^p.*
(a) *Let $x_1, \ldots, x_m, y_1, \ldots, y_n \in F$ be such that $x_1, \ldots, x_m$ are p-independent over F_0 and $x_1, \ldots, x_m \in F_0(y_1, \ldots, y_n)$. Then $m \leq n$, and there is a reordering of $y_1, \ldots, y_n$ so that $y_1, \ldots, y_m \in F_0(x_1, \ldots, x_m, y_{m+1}, \ldots, y_n)$.*
(b) *Every subset of F which is p-independent over F_0 extends to a p-basis for F over F_0.*

Proof: We use induction on m. Assume the lemma is true for $m = k$. Thus, for $m = k + 1$ we may assume that

$$x_{k+1} \in F_0(x_1, \ldots, x_k, y_{k+1}, \ldots, y_n) = F_1.$$

Then $[F_1 : F_0] \leq p^n$ and there exists l between $k+1$ and n such that

$$y_l \in F_0(x_1, \ldots, x_{k+1}, y_{k+1}, \ldots, y_{l-1}),$$

since otherwise $[F_1 : F_0] \geq p^{n+1}$, a contradiction. Thus, y_l can be exchanged for x_{k+1}. This proves the first part of the lemma for $m = k + 1$.

For the last part start from a subset A of K which is p-independent over F_0. Use Zorn's lemma to prove the existence of a maximal subset B of F which contains A and which is p-independent over F_0. Then B is a p-basis of F over F_0. $\square$

LEMMA 2.7.2: *Suppose F is a finitely generated extension of transcendence degree n of a perfect field K of positive characteristic p. Then the imperfect exponent of F is n.*

Proof: Choose a separating transcendence basis $t_1, \ldots, t_n$ for F/K. Then $K(\mathbf{t})^p = K(\mathbf{t}^p)$ and $t_1, \ldots, t_n$ is a p-basis for $K(\mathbf{t})/K(\mathbf{t}^p)$; that is, $[K(\mathbf{t}) : K(\mathbf{t}^p)] = p^n$. Since $K(\mathbf{t})$ is a purely inseparable extension of $K(\mathbf{t}^p)$ and F^p is a separable extension of $K(\mathbf{t}^p)$, these extensions of $K(\mathbf{t}^p)$ are linearly disjoint. Also, F is both a separable extension and a purely inseparable extension of $K(\mathbf{t})F^p$. Hence, $F = K(\mathbf{t})F^p$. Consequently, $[F : F^p] = [K(\mathbf{t}) : K(\mathbf{t}^p)] = p^n$, as claimed. $\square$

LEMMA 2.7.3: *Let B a subset of F which is p-independent over F^p and F' a separable extension of F. Then B is p-independent over $(F')^p$. If, in addition, F' is separable algebraic over F, then the imperfect degree of F' is equal to that of F.*

Proof: Assume without loss that B consists of n elements. Then $[(F')^p(B) : (F')^p] = [F^p(B) : F^p] = p^n$. Hence, B is p-independent over $(F')^p$.

Suppose now F'/F is separably algebraic. Then F' is both separably and purely inseparable over $F(F')^p$, so, $F' = F(F')^p$. Hence, $[F' : (F')^p] = [F : F^p]$. Therefore, the imperfect degree of F' is equal to that of F. □

LEMMA 2.7.4: *Let K be a field of positive characteristic p, let $a, b_1, \ldots, b_m$ be p-independent elements of K, and let $x_1, \ldots, x_m$ be algebraically independent over K. Suppose $y_1, \ldots, y_m$ satisfy*

(2) $$ax_i^p + b_i y_i^p = 1, \quad i = 1, \ldots, m.$$

Then K is algebraically closed in $K(\mathbf{x}, \mathbf{y}) = K_m$.

Proof: We use induction on m.

PART A: $m = 1$. Let $x = x_1$, $y = y_1$, and $b = b_1$ and assume that u is a nonzero element of K_1 which is algebraic over K. Then u is also algebraic over $K(a^{1/p}, b^{1/p})$. But $K(x, y, a^{1/p}, b^{1/p}) = K(x, a^{1/p}, b^{1/p})$ is a purely transcendental extension of $K(a^{1/p}, b^{1/p})$. Hence, $u \in K(a^{1/p}, b^{1/p})$ and therefore $u^p \in K$. Write

(3) $$u = \frac{h_0(x)}{h(x)} + \frac{h_1(x)}{h(x)} y + \cdots + \frac{h_k(x)}{h(x)} y^k$$

with $k \leq p-1$, $h(x), h_0(x), \ldots, h_k(x) \in K[x]$ and $h(x), h_k(x) \neq 0$. With no loss we may assume that x does not divide the greatest common divisor of $h(x), h_0(x), \ldots, h_k(x)$. Raise (3) to the pth power, multiply it by $h(x)^p$ and substitute $y^p = (1 - ax^p)b^{-1}$ to obtain:

(4) $(h(x)u)^p = h_0(x)^p + h_1(x)^p(1-ax^p)b^{-1} + \cdots + h_k(x)^p(1-ax^p)^k b^{-k}$.

If $h(0) = 0$, then the substitution $x = 0$ in (4) gives

$$0 = h_0(0)^p + h_1(0)^p b^{-1} + \cdots + h_k(0)^p b^{-k},$$

Therefore, $h_0(0) = h_1(0) = \cdots = h_k(0) = 0$, contrary to assumption. Thus, we may assume $h(0) \neq 0$. Then the substitution $x = 0$ in (4) shows that $u \in K(b^{1/p})$. Similarly, $u \in K(a^{1/p})$. Since a and b are p-independent in K, $u \in K(a^{1/p}) \cap K(b^{1/p}) = K$.

Thus, K is algebraically closed in $K(x, y)$.

PART B: *Induction*. Assume the Lemma is true for $m - 1$. Then K is algebraically closed in $K_{m-1} = K(x_1, \ldots, x_{m-1}; y_1, \ldots, y_{m-1})$. If we prove that a and b_m are p-independent in K_{m-1}, then with K_{m-1} replacing K in Part A, K_{m-1} is algebraically closed in K_m, so K is algebraically closed in K_m.

Since $x_1, \ldots, x_m$ are algebraically independent over K, the field $K(a^{1/p}, b_1^{1/p}, \ldots, b_m^{1/p})$ is linearly disjoint from $E_{m-1} = K(x_1, \ldots, x_{m-1})$ over K. Thus,

(5) $$[E_{m-1}(a^{1/p}, b_1^{1/p}, \ldots, b_m^{1/p}) : E_{m-1}] = p^{m+1}.$$

2.7 The Imperfect Degree of a Field

Also, from (2)

$$K_{m-1} = E_{m-1}(y_1, \ldots, y_{m-1}) \quad \text{and}$$
$$K_{m-1}(a^{1/p}, b_m^{1/p}) = E_{m-1}(a^{1/p}, b_1^{1/p}, \ldots, b_m^{1/p}).$$

Thus,

(6) $\quad [K_{m-1} : E_{m-1}] \leq p^{m-1} \quad \text{and} \quad [K_{m-1}(a^{1/p}, b_m^{1/p}) : K_{m-1}] \leq p^2.$

Combine (5) and (6) to conclude that (6) consists of equalities. In particular, a and b_m are p-independent in K_{m-1}. $\square$

LEMMA 2.7.5: *The following conditions on a field K of positive characteristic p are equivalent:*
(a) *The imperfect exponent of K is at most 1.*
(b) *Every finite extension of K has a primitive element.*
(c) *If K is algebraically closed in a field extension F, then F is regular over K.*

Proof: If K is perfect, then (a), (b), and (c) are true. Therefore, we may assume $\text{char}(K) = p > 0$ and K is imperfect.

Proof of "(a) $\Longrightarrow$ (b)": By assumption, $[K^{1/p} : K] = [K : K^p] = p$. Hence, $K_1 = K^{1/p}$ is the unique purely inseparable extension of K of degree p. Moreover, $K_1 = K(a^{1/p})$ for some $a \in K$, so $K_n = K(a^{1/p^n})$ is a purely inseparable extension of K of degree p^n.

Assume that for each $m \leq n$, K_m is the unique purely inseparable extension of K of degree p^m. Let L be a purely inseparable extension of K of degree p^{n+1}. If we prove that $L = K_{n+1}$, then we may conclude by induction that each finite purely inseparable extension of K has a primitive element.

To this end choose $x \in L \smallsetminus K_n$. Let m be the smallest positive integer with $x^{p^m} \in K$. Then $K(x)$ is a purely inseparable extension of K of degree p^m. If $m \leq n$, then by the induction hypothesis $K(x) = K_m \subseteq K_n$, so $x \in K_n$. This contradiction proves that $m = n+1$ and $L = K(x)$.

The same argument implies that $x^p \in K_n$. Hence, with $q = p^n$, we have $x^p = \sum_{i=0}^{q-1} c_i a^{i/p^n}$ for some $c_0, \ldots, c_{q-1} \in K$. Therefore,

$$x = \sum_{i=0}^{q-1} c_i^{1/p} a^{i/p^{n+1}} \in K_1(a^{1/p^{n+1}}) = K_{n+1}.$$

It follows that $L \subseteq K_{n+1}$. As both fields have degree p^{n+1} over K, they coincide, as desired.

Now let E be a finite extension of K. Denote the maximal separable extension of K in E by E_0. By the primitive element theorem, $E_0 = K(x)$. Since E_0 is both separable and purely inseparable over KE_0^p we have $E_0 = KE_0^p$. Therefore $[E_0 : E_0^p] = [K : K^p] = p$. Apply the first part of the proof to E_0 and conclude that $E = E_0(y)$, for some element y. Thus, $E = K(x,y)$ with x separable over K. By [Waerden3, §6.10], E/K has a primitive element

Proof of "(b) $\Longrightarrow$ (c)": Let $K(x)$ be a finite extension of K and let $f = \mathrm{irr}(x, K)$. If K is algebraically closed in F, then f remains irreducible over F. Otherwise, its factors would have coefficients algebraic over K and in F, and therefore in K. Thus, F is linearly disjoint from $K(x)$ over K. Hence, (b) implies that F is regular over K.

Proof of "(c) $\Longrightarrow$ (a)": Assume a and b are p-independent elements of K. Then $[K(a^{1/p}, b^{1/p}) : K] = p^2$. Let x and y be transcendental elements over K with $ax^p + by^p = 1$. Put $F = K(x, y)$. By Lemma 2.7.4, K is algebraically closed in F. Hence, by (c), F is regular over K. Therefore, $[F(a^{1/p}, b^{1/p}) : F] = [K(a^{1/p}, b^{1/p}) : K] = p^2$. On the other hand, , $F(a^{1/p}) = F(b^{1/p})$, so $[F(a^{1/p}, b^{1/p}) : F] \le p$. This contradiction proves that the imperfect exponent of K is at most 1. $\square$

Remark 2.7.6: *Relative algebraic closedness does not imply regularity.* Let K be a field of positive characteristic p. Suppose K has p-independent elements a, b (e.g. $K = \mathbb{F}_p(t, u)$ where t, u are algebraically independent over $\mathbb{F}_p$). Let x, y be transcendental elements over K with $ax^p + by^p = 1$. Put $F = K(x, y)$. The proof of "(c) $\Longrightarrow$ (a)" then shows that K is algebraically closed in F but F is not linearly disjoint from $K^{1/p}$ over K. Thus, F is not a separable extension of K. A fortiori, F/K is not regular. $\square$

2.8 Derivatives

We develop a criterion for a finitely generated field extension of positive characteristic p to be separable in terms of derivatives..

Definition 2.8.1: A map $D \colon F \to F$ is called a **derivation** of the field F if $D(x + y) = D(x) + D(y)$ and $D(xy) = D(x)y + xD(y)$ for all $x, y \in F$.

If D vanishes on a subfield K of F, then D is a derivation of F **over** K (or a K-**derivation**). $\square$

Let $F(x)$ be a field extension of F and $f \in F[X]$. Suppose D extends to $F(x)$. Then D satisfies the classical chain rule:

$$(1) \qquad D(f(x)) = f^D(x) + f'(x)D(x),$$

where f^D is the polynomial obtained by applying D to the coefficients of f and f' is the usual derivative of f. There are three cases:

CASE 1: *x is separably algebraic over F.* Then, with $f = \mathrm{irr}(x, F)$, $f'(x) \ne 0$. By (1), $0 = f^D(x) + f'(x)D(x)$. Thus, D extends uniquely to $F(x)$.

CASE 2: *x is transcendental.* Then D extends to $F(x)$ by rule (1) and $D(x)$ may be chosen arbitrarily.

2.8 Derivatives

CASE 3: x satisfies $x^{p^m} = a \in F$, for some m. Then D extends to $F(x)$ if and only if $D(a) = 0$. In this case $D(x)$ may be chosen arbitrarily.

LEMMA 2.8.2: *A necessary and sufficient condition for a finitely generated extension F/K to be separably algebraic is that 0 is the only K-derivation of F.*

Proof: Necessity follows from Case 1.

Now suppose F/K is not separably algebraic. Then we may write $F = K(x_1, \ldots, x_n)$ such that x_i is transcendental over $K(x_1, \ldots, x_{i-1})$ for $i = 1, \ldots, k$, x_i is separably algebraic over $K(x_1, \ldots, x_{i-1})$ for $i = k+1, \ldots, l$, and x_i is purely inseparable over $K(x_1, \ldots, x_{i-1})$ for $i = l+1, \ldots, n$. Moreover, either $n > l$ or $n = l$ and $k > 0$. If $n > l$, then Case 1 allows us to extend the zero derivation of $K(x_1, \ldots, x_{n-1})$ to a nonzero derivation of F. If $n = l$ and $k > 0$, then by Case 2, the zero derivation of $K(x_1, \ldots, x_{k-1})$ extends to a nonzero derivation D of $K(x_1, \ldots, x_k)$. Applying Case 3 several times, we may then extend D to a derivation of F. □

LEMMA 2.8.3: *Let F/K be a finitely generated extension of positive characteristic p and transcendence degree n. Then F/K is separable if and only if $[F : KF^p] = p^n$. In this case $t_1, \ldots, t_n$ form a p-basis for F over KF^p if and only they form a separating transcendence basis for F/K.*

Proof: Suppose first $[F : KF^p] = p^n$. Let $t_1, \ldots, t_n$ be a p-basis for F/KF^p. Every derivation D of F vanishes on F^p. If D vanishes on $K(\mathbf{t})$, it vanishes on $F = K(\mathbf{t}) \cdot F^p$. By Lemma 2.8.2, $F/K(\mathbf{t})$ is separably algebraic and $t_1, \ldots, t_n$ is a separating transcendence basis for F/K.

Conversely, suppose F/K is separable. Let $t_1, \ldots, t_n$ be a separating transcendence basis for F/K. The extension $F/K(\mathbf{t}) \cdot KF^p$ is both separable and purely inseparable. Hence, $F = K(\mathbf{t}) \cdot KF^p$. Since $F^p/K(\mathbf{t})^p$ is separably algebraic and since $K(\mathbf{t}^p)F^p = KF^p$, we conclude that $KF^p/K(\mathbf{t}^p)$ is separably algebraic.

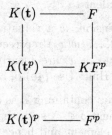

Therefore, KF^p is linearly disjoint from $K(\mathbf{t})$ over $K(\mathbf{t}^p)$, and $[F : KF^p] = [K(\mathbf{t}) : K(\mathbf{t}^p)] = p^n$. Moreover, $\mathbf{t}$ is a p-basis for F/KF^p. □

COROLLARY 2.8.4: *Let F/K be a finitely generated separable extension of positive characteristic p and let $t \in F$.*
(a) *If there exists a derivation D of F/K such that $D(t) \neq 0$, then F is a separable extension of $K(t)$.*

(b) *If t is transcendental over K and $F/K(t)$ is separable, then there exists a derivation D of F/K such that $D(t) \neq 0$.*

Proof of (a): By assumption, $t \notin KF^p$. Let $n = \text{trans.deg}(F/K)$. By Lemma 2.8.3, $[F : KF^p] = p^n$. Hence, t can be extended to a p-basis $t, t_2, \ldots, t_n$ for F/KF^p. Again, by Lemma 2.8.3, $t, t_2, \ldots, t_n$ is a separating transcendence basis for F/K. Therefore, F is a separable extension of $K(t)$.

Proof of (b): Let $t_2, \ldots, t_n$ be a separating transcendence basis for $F/K(t)$. By Case 2, there exists a derivation D_0 of $K(t, t_2, \ldots, t_n)/K$ such that $D_0(t) = 1$, $D_0(t_2) = 0$, ..., $D_0(t_n) = 0$. By Case 1, D_0 extends to a derivation D of F/K. □

Exercises

1. Let O be a valuation ring of a field F and consider the subset $\mathfrak{m} = \{x \in O \mid x^{-1} \notin O\}$. Show that if $x \in \mathfrak{m}$ and $a \in O$, then $ax \in \mathfrak{m}$. Prove that $\mathfrak{m}$ is closed under addition. Hint: Use the identity $x + y = (1 + xy^{-1})y$ for $y \neq 0$. Show that $\mathfrak{m}$ is the unique maximal ideal of O.

2. Use Exercise 1 to prove that every valuation ring is integrally closed.

3. Let v be a valuation of $\mathbb{Q}$. Observe that $v(n) \geq v(1) = 0$, for each $n \in \mathbb{N}$. Hence, there exists a smallest $p \in \mathbb{N}$ such that $v(p) > 0$. Prove that p is a prime element of O_v and v is equivalent to v_p. Hint: If a positive integer m is relatively prime to p, then there exist $x, y \in \mathbb{Z}$ such that $xp + ym = 1$.

4. Let v be a valuation of the rational function field $F = K(t)$ which is trivial on K. Suppose there exists $p \in K[t]$ with $v(p) > 0$. Now suppose p has smallest degree with this property. Show that v is equivalent to v_p. Otherwise, there exists $f \in K[t]$ such that $v(f(t)) < 0$. Conclude that $v(t) < 0$, and that v is equivalent to v_∞.

5. Let F/E be a field extension, w a valuation of F, and $x_1, \ldots, x_e$ elements of F such that $w(x_1), \ldots, w(x_e)$ represent distinct classes of $w(F^\times)$ modulo $w(E^\times)$. Show that $x_1, \ldots, x_e$ are linearly independent over E. Thus, $(w(F^\times) : w(E^\times)) \leq [F : E]$. Hint: Use (4b) of Section 2.1.

6. Let Δ be an ordered group containing $\mathbb{Z}$ as a subgroup of index e. Show there exists no positive element $\delta \in \Delta$ such that $e\delta < 1$. Conclude that Δ contains a smallest positive element and hence that $\Delta \cong \mathbb{Z}$. Combine this with Exercise 5 to prove that if the restriction of w to E is discrete, then w is discrete.

7. In the notation of Exercise 5, let v be the restriction of w to E. Let $y_1, \ldots, y_f$ be elements of F with $w(y_1), \ldots, w(y_f) \geq 0$ with residue classes $\bar{y}_1, \ldots, \bar{y}_f$ linearly independent over $\bar{E}_v$. Show that $y_1, \ldots, y_f$ are linearly independent over E. Conclude that $[\bar{F}_w : \bar{E}_v] \leq [F : E]$. Hint: If $a_1, \ldots, a_f \in$

F are not all zero, then there exists j, $1 \leq j \leq f$ such that $v\left(\frac{a_1}{a_j}\right), \ldots, v\left(\frac{a_f}{a_j}\right) \geq 0$.

8. Let v be a discrete valuation of a field K and let w be an extension of v to a finite Galois extension L of K. Assume that w' is also an extension of v to L such that $w' \neq \sigma(w)$ for all $\sigma \in \text{Gal}(L/K)$. Combine Exercise 7 with Proposition 2.1.1 to produce $x \in L$ such that $w'(x) > 0$ and $w(\sigma x - 1) > 0$ for all $\sigma \in \text{Gal}(L/K)$. With $y = N_{L/K}(x)$, conclude that the former condition gives $v(y) > 0$, while the latter implies $v(y-1) > 0$. Use this contradiction to prove that $\text{Gal}(L/K)$ acts transitively on the extensions of v to L.

9. Let $L, K_1, \ldots, K_n$ be extensions of a field K. Let $L_i = K_i L$, $i = 1, \ldots, n$. Suppose K_i is linearly disjoint from L over K for $i = 1, \ldots, n$ and $L_1, \ldots, L_n$ are linearly disjoint over L. Prove that $K_1, \ldots, K_n$ are linearly disjoint over K.

10. Let v be a discrete valuation of a field K and let L and M be two finite extensions of K such that v is unramified in L and totally ramified in M. Prove that L and M are linearly disjoint over K. Hint: Consider the Galois hull $\hat{L}$ of L/K.

11. Let E be a regular extension of a perfect field K and let F be a purely inseparable extension of E. Prove that F/K is a regular extension.

12. Let K be a field algebraically closed in an extension F. Prove that $K(x)$ is linearly disjoint from F for every $x \in \tilde{K}$. Hint: Check the irreducibility of $\text{irr}(x, K)$ over F.

13. Prove that a field extension F/K is primary if and only if $FK_{\text{ins}} \cap \tilde{K} = K_{\text{ins}}$. Use this criterion to give another proof to Lemma 2.6.14(a).

14. Let F/K be a finitely generated field extension of characteristic $p > 0$ and of transcendence degree 1. Prove that for each positive integer n, KF^{p^n} is the unique subfield E of F which contains K such that F/E is a purely inseparable extension of degree p^n.

15. (Geyer) The following example shows that Lemma 2.4.8 is false for arbitrary real valuations. Consider the field $\mathbb{Q}_2$ of 2-adic numbers. Show that the field $K = \mathbb{Q}_2(\sqrt[n]{2} \mid n \in \mathbb{N})$ is a totally ramified extension of $\mathbb{Q}_2$ with value group $\mathbb{Q}$. Hence, each extension of K is unramified. Prove that the residue field of both $K(\sqrt{3})$ and $K(\sqrt{-1})$ is $\mathbb{F}_2$. However, their compositum contains $K(\sqrt{-3})$ and therefore has $\mathbb{F}_4$ as its residue field.

Notes

The terminology "algebraic independence" for field extensions replaces "freeness" which we used in [Fried-Jarden3].

Corollary 4 of [Lang4, p. 61] proves Lemma 2.6.15(a) only under the condition (our notation) that E is a separable extension of K.

Chapter 3.
Algebraic Function Fields of One Variable

Sections 3.1–3.4 survey the theory of functions of one variable; the Riemann-Roch Theorem; properties of holomorphy rings of function fields; and extensions of the field of constants.

Sections 3.5–3.6 include a proof of the Riemann-Hurwitz formula.

The rest of the chapter applies these concepts and results to hyperelliptic curves.

3.1 Function Fields of One Variable

Call a field extension F/K an **algebraic function field of one variable** (briefly a **function field**) if these conditions hold:

(1a) The transcendence degree of F/K is 1.
(1b) F/K is finitely generated and regular.

In this case there exists $t \in F$, transcendental over K, with $F/K(t)$ a finite separable extension. All valuations of $K(t)$ trivial on K are discrete (Example 2.2.1), so their extensions to F are also discrete (Proposition 2.3.2). Also, since the residue fields of the valuations of $K(t)$ are finite extensions of K, so are the residue fields of the valuations of F.

We define a **prime divisor** of F/K as an equivalence class of K-places of F. For $\mathfrak{p}$ a prime divisor of F/K, choose a place $\varphi_\mathfrak{p}$ in $\mathfrak{p}$. Then $\varphi_\mathfrak{p}$ fixes the elements of K and maps F into $\tilde{K} \cup \{\infty\}$. Denote its residue field by $\bar{F}_\mathfrak{p}$. As mentioned above, $\bar{F}_\mathfrak{p}$ is a finite extension of K of degree $\deg(\mathfrak{p}) = [\bar{F}_\mathfrak{p} : K]$ which we call the **degree** of $\mathfrak{p}$. Also, choose a valuation $v_\mathfrak{p}$ corresponding to $\mathfrak{p}$ and normalize it so that $v_\mathfrak{p}(F^\times) = \mathbb{Z}$. Each element π of F with $v_\mathfrak{p}(\pi) = 1$ is a **local parameter** of F at $\mathfrak{p}$.

Denote the free Abelian group that the prime divisors of F/K generate by $\mathrm{Div}(F/K)$. Each element $\mathfrak{a}$ of $\mathrm{Div}(F/K)$ is a **divisor** of F/K. It has the form $\mathfrak{a} = \sum \alpha_\mathfrak{p} \mathfrak{p}$, where $\mathfrak{p}$ runs over the prime divisors of F/K, the $\alpha_\mathfrak{p}$ are integers and all but finitely many of them are zero. Define a homomorphism $v_\mathfrak{p} \colon \mathrm{Div}(F/K) \to \mathbb{Z}$ by $v_\mathfrak{p}(\mathfrak{a}) = \alpha_\mathfrak{p}$.

The symbol $v_\mathfrak{p}$ appears for two distinct functions. We show these uses are compatible as follows. Introduce the **divisor** of a nonzero element x of F as $\mathrm{div}(x) = \sum v_\mathfrak{p}(x)\mathfrak{p}$. Since $v_\mathfrak{p}(x) = 0$ for all but finitely many $\mathfrak{p}$ the right hand side is well defined. If x is a **constant** (i.e. $x \in K$), then $\mathrm{div}(x) = 0$. If $x \in F \smallsetminus K$, then the place of $K(x)$ taking x to 0 (resp. to ∞) has finitely many extensions to the field F (Proposition 2.3.1). Equivalence classes of these extensions are the **zeros** (resp. **poles**) of x. Thus, $\mathrm{div}(x)$ is not zero if x is not a constant. Define the **divisor of zeros** and the **divisor of poles**

3.2 The Riemann-Roch Theorem

of x as follows:

$$\mathrm{div}_0(x) = \sum_{v_\mathfrak{p}(x)>0} v_\mathfrak{p}(x)\mathfrak{p}, \qquad \mathrm{div}_\infty(x) = -\sum_{v_\mathfrak{p}(x)<0} v_\mathfrak{p}(x)\mathfrak{p}.$$

Then $\mathrm{div}(x) = \mathrm{div}_0(x) - \mathrm{div}_\infty(x)$. In particular, $v_\mathfrak{p}(x)$ (using $v_\mathfrak{p}$ as a valuation) is the same as $v_\mathfrak{p}$ applied to the divisor $\mathrm{div}(x)$.

Define the **degree** $\deg(\mathfrak{a})$ of a divisor $\mathfrak{a}$ to be $\sum v_\mathfrak{p}(\mathfrak{a}) \deg(\mathfrak{p})$. Then $\deg \colon \mathrm{Div}(F/K) \to \mathbb{Z}$ is a homomorphism. Since

$$\deg(\mathrm{div}_0(x)) = \deg(\mathrm{div}_\infty(x)) = [F : K(x)]$$

for $x \in F \smallsetminus K$ ([Deuring3, p. 27] or Proposition 2.3.2), $\deg(\mathrm{div}(x)) = 0$.

To each divisor $\mathfrak{a}$ attach a vector space $\mathcal{L}(\mathfrak{a})$ over K:

$$\mathcal{L}(\mathfrak{a}) = \{x \in F^\times \mid \mathrm{div}(x) + \mathfrak{a} \geq 0\}.$$

The phrase '$\mathfrak{b} \geq 0$' for a divisor $\mathfrak{b}$ means that $v_\mathfrak{p}(\mathfrak{b}) \geq 0$ for every prime divisor $\mathfrak{p}$. The K-vector space $\mathcal{L}(\mathfrak{a})$ is finite dimensional [Deuring3, p. 23] (Note: Deuring uses the notation $L(\mathfrak{a})$ for $\mathcal{L}(-\mathfrak{a})$). Denote the nonnegative integer $\dim_K \mathcal{L}(\mathfrak{a})$ by $\dim(\mathfrak{a})$.

Extend the notation $0 \leq \mathfrak{b}$ to a partial ordering on $\mathrm{Div}(F/K)$ by writing $\mathfrak{a} \leq \mathfrak{b}$ if $v_\mathfrak{p}(\mathfrak{a}) \leq v_\mathfrak{p}(\mathfrak{b})$ for every prime divisor $\mathfrak{p}$ of F/K. This implies $\mathcal{L}(\mathfrak{a}) \subseteq \mathcal{L}(\mathfrak{b})$. In this case, $\mathcal{L}(\mathfrak{a}) \subset \mathcal{L}(\mathfrak{b})$ is equivalent to $\dim(\mathfrak{a}) < \dim(\mathfrak{b})$.

For divisors $\mathfrak{a}_1, \ldots, \mathfrak{a}_n$ of F/K, write

$$\min(\mathfrak{a}_1, \ldots, \mathfrak{a}_n) = \sum_\mathfrak{p} \min\bigl(v_\mathfrak{p}(\mathfrak{a}_1), \ldots, v_\mathfrak{p}(a_n)\bigr)\mathfrak{p}.$$

This is the maximal divisor of F/K less than or equal to $\mathfrak{a}_i$ for $i = 1, \ldots, n$. Similarly, define $\max(\mathfrak{a}_1, \ldots, \mathfrak{a}_n)$.

Call a divisor of the form $\mathrm{div}(x)$, where $x \in F^\times$, a **principal divisor**. Since $\mathrm{div}(xy) = \mathrm{div}(x) + \mathrm{div}(y)$, the set of principal divisors of F/K is a subgroup of $\mathrm{Div}(F/K)$. The quotient group is called the **group of divisor classes** of F/K. Denote it by $\mathcal{C}$. Every element of $\mathcal{C}$ is a **class of divisors**. Call two divisors **linearly equivalent** if they differ by a principal divisor. Two linearly equivalent divisors have the same degree and the same dimension. This defines the degree and the dimension of a class of divisors.

LEMMA 3.1.1: *Let F/K be a function field and $x_1, \ldots, x_m \in F^\times$. Put $\mathfrak{a} = \sum_{i=1}^m \mathrm{div}_\infty(x_i)$ and $R = K[x_1, \ldots, x_m]$. Then $\bigcup_{k=1}^\infty \mathcal{L}(k\mathfrak{a})$ is the integral closure of R in F.*

Proof: Denote the integral closure of R in F by S and the set of poles of $x_1, \ldots, x_m$ by P. First consider a nonzero $z \in \mathcal{L}(k\mathfrak{a})$. Then $\mathrm{div}(z) + k\mathfrak{a} \geq 0$. Hence, $v_\mathfrak{p}(z) \geq 0$ for each prime divisor $\mathfrak{p}$ of F/K which is not in P; that is for each $\mathfrak{p}$ with $\varphi_\mathfrak{p}$ finite on R. By Proposition 2.4.1, $z \in S$.

Conversely, let $z \in S$. By Proposition 2.4.1, $v_{\mathfrak{p}}(z) \geq 0$ for each $\mathfrak{p} \notin P$. For each $\mathfrak{p} \in P$ the set $I_{\mathfrak{p}} = \{1 \leq i \leq m \mid v_{\mathfrak{p}}(x_i) < 0\}$ is not empty. Choose a positive integer k with $v_{\mathfrak{p}}(z) - k \sum_{i \in I_{\mathfrak{p}}} v_{\mathfrak{p}}(x_i) \geq 0$ for all $\mathfrak{p} \in P$. Then $z \in \mathcal{L}(k\mathfrak{a})$. □

3.2 The Riemann-Roch Theorem

The degree of a divisor $\mathfrak{a}$ of a function field F/K can be often read out of its definition. It is more difficult to compute the dimension of $\mathfrak{a}$ directly from the definition. The Riemann-Roch theorem allows us in many cases to compute $\dim(\mathfrak{a})$ from $\deg(\mathfrak{a})$ and a constant of F/K called the "genus". In this section we present the Riemann-Roch theorem in three forms and explain how to apply it in the computation of $\dim(\mathfrak{a})$.

Let F/K be a function field. A **repartition** of F/K is a function α from prime divisors of F/K into F (denote the image of $\mathfrak{p}$ by $\alpha_{\mathfrak{p}}$) with $v_{\mathfrak{p}}(\alpha_{\mathfrak{p}}) \geq 0$ for all but finitely many $\mathfrak{p}$. Denote the set of all repartitions of F/K by $\mathbb{A}$. The following definitions turn $\mathbb{A}$ into an F-algebra: For $\alpha, \beta \in \mathbb{A}$ and $x \in F$,

$$(\alpha + \beta)_{\mathfrak{p}} = \alpha_{\mathfrak{p}} + \beta_{\mathfrak{p}}, \quad (\alpha\beta)_{\mathfrak{p}} = \alpha_{\mathfrak{p}}\beta_{\mathfrak{p}} \quad \text{and} \quad (x\alpha)_{\mathfrak{p}} = x \cdot \alpha_{\mathfrak{p}}.$$

Extend $v_{\mathfrak{p}}$ to $\mathbb{A}$ by $v_{\mathfrak{p}}(\alpha) = v_{\mathfrak{p}}(\alpha_{\mathfrak{p}})$. If $\mathfrak{a}$ is a divisor of F/K, then

$$\Lambda(\mathfrak{a}) = \{\alpha \in \mathbb{A} \mid v_{\mathfrak{p}}(\alpha) + v_{\mathfrak{p}}(\mathfrak{a}) \geq 0 \text{ for every } \mathfrak{p}\}$$

is a vector space over K. Identify F as a subset of $\mathbb{A}$ by the diagonal mapping. Then $\Lambda(\mathfrak{a})$ contains $\mathcal{L}(\mathfrak{a})$ and $\mathbb{A}/(\Lambda(\mathfrak{a}) + F)$ is a finite dimensional K-vector space. Moreover, there exists a nonnegative integer g, called the **genus** of F/K, which is independent of $\mathfrak{a}$ such that

$$(1) \qquad \dim(\mathfrak{a}) - \deg(\mathfrak{a}) + g - 1 = \dim_K \frac{\mathbb{A}}{\Lambda(\mathfrak{a}) + F}.$$

This is the first form of the Riemann-Roch theorem [Deuring3, p. 34].

The second form of the theorem interprets the right hand side of (1) as the dimension of a space of differentials. A **differential** (also called **Weil differential**) of the function field F/K is a K-linear map of $\mathbb{A}$ into K that vanishes on some subspace of the form $\Lambda(\mathfrak{a}) + F$, where $\mathfrak{a}$ is a divisor of F/K. The set of all differentials that vanish on $\Lambda(\mathfrak{a}) + F$, denoted $\Omega(\mathfrak{a})$, is the dual space of $\mathbb{A}/(\Lambda(\mathfrak{a}) + F)$. We denote its dimension by $\delta(\mathfrak{a})$ and rewrite (1) as

$$(2) \qquad \dim(\mathfrak{a}) = \deg(\mathfrak{a}) - g + 1 + \delta(\mathfrak{a}).$$

The third form of the theorem identifies $\delta(\mathfrak{a})$ with the dimension of a divisor of F/K related to $\mathfrak{a}$. To each nonzero differential ω of F/K there corresponds a unique divisor of F/K, denoted $\mathrm{div}(\omega)$, with $\mathrm{div}(\omega) \geq \mathfrak{a}$ if and only if $\omega \in \Omega(\mathfrak{a})$ [Deuring3, p. 39].

3.2 The Riemann-Roch Theorem

THEOREM 3.2.1: *Let ω be a nonzero differential of an algebraic function field F/K of one variable of genus g. Put $\mathfrak{w} = \text{div}(\omega)$. Then, for every divisor $\mathfrak{a}$ of F/K we have*

(3) $$\dim(\mathfrak{a}) = \deg(\mathfrak{a}) - g + 1 + \dim(\mathfrak{w} - \mathfrak{a}).$$

Proof: For each $x \in F^\times$ define a K-linear map $x\omega \colon \mathbb{A} \to K$ by $(x\omega)(\alpha) = \omega(x\alpha)$. If ω vanishes on $\Lambda(\mathfrak{a})$, then $x\omega$ vanishes on $\Lambda(\text{div}(x) + \mathfrak{a})$. Hence, $x\omega$ is a differential and $\text{div}(x\omega) = \text{div}(x) + \text{div}(\omega)$. The map $x \mapsto x\omega$ gives an isomorphism of $\mathcal{L}(\text{div}(\omega) - \mathfrak{a})$ onto $\Omega(\mathfrak{a})$. Hence, $\dim(\text{div}(\omega) - \mathfrak{a}) = \dim \Omega(\mathfrak{a}) = \delta(\mathfrak{a})$. Thus, (3) is a reformulation of (2). □

If ω' is another nonzero differential, then there exists $x \in F^\times$ with $\omega' = x\omega$. Thus, $\text{div}(\omega') = \text{div}(x) + \text{div}(\omega)$ [Deuring3, p. 37] and the **divisor of a differential** is determined up to a principal divisor. The class of these divisors is the **canonical class** of F/K. Each of its elements is a **canonical divisor**.

A few additional observations help us apply the third form of the Riemann-Roch theorem in (3):

LEMMA 3.2.2: *The following holds for each function field F/K of genus g:*
(a) $\dim(0) = 1$, and if $x \in F^\times$, then $\deg(\text{div}(x)) = 0$ and $\dim(\text{div}(x)) = 1$.
(b) *If $\mathfrak{w}$ is a canonical divisor, then* $\dim(\mathfrak{w}) = g$ *and* $\deg(\mathfrak{w}) = 2g - 2$ *Hence,*
$\dim_K(\mathbb{A}/\Lambda(0) + F) = \dim(\Omega(0)) = \dim(\mathcal{L}(\mathfrak{w})) = g$.
(c) $\deg(\mathfrak{a}) < 0$ *implies* $\dim(\mathfrak{a}) = 0$.
(d) $\deg(\mathfrak{a}) > 2g - 2$ *implies* $\delta(\mathfrak{a}) = 0$ *and* $\dim(\mathfrak{a}) = \deg(\mathfrak{a}) - g + 1$.

Proof of (a): $\mathcal{L}(0) = K$ because each $x \in F \smallsetminus K$ has a pole. Similarly, $\mathcal{L}(\text{div}(x)) = Kx^{-1}$.

Proof of (b): Take $\mathfrak{a} = 0$ and then $\mathfrak{a} = \mathfrak{w}$ in (3).

Proof of (c): Use that $\deg(\text{div}(x)) = 0$.

Proof of (d): By (b) $\deg(\mathfrak{w} - \mathfrak{a}) < 0$. Hence, by (c), $\delta(\mathfrak{a}) = \dim(\mathfrak{w} - \mathfrak{a}) = 0$. Thus, (3) simplifies to $\dim(\mathfrak{a}) = \deg(\mathfrak{a}) + 1 - g$. □

Here is our first application of Lemma 3.2.2(d):

LEMMA 3.2.3: *Let F/K be a function field of genus g, $\mathfrak{p}$ a prime divisor of F/K, and n a positive integer satisfying $(n-1)\deg(\mathfrak{p}) > 2g - 2$. Then there exists $x \in F^\times$ with $\text{div}_\infty(x) = n\mathfrak{p}$.*

Proof: By Lemma 3.2.2(d), $\dim((n-1)\mathfrak{p}) = (n-1)\deg(\mathfrak{p}) + 1 - g$ and $\dim(n\mathfrak{p}) = n\deg(\mathfrak{p}) + 1 - g$. Hence, $\dim((n-1)\mathfrak{p}) < \dim(n\mathfrak{p})$, so $\mathcal{L}((n-1)\mathfrak{p}) \subset \mathcal{L}(n\mathfrak{p})$. Every element $x \in \mathcal{L}(n\mathfrak{p}) \smallsetminus \mathcal{L}((n-1)\mathfrak{p})$ will satisfy $\text{div}_\infty(x) = n\mathfrak{p}$.
□

Example 3.2.4: Rational function field. Let $F = K(t)$, where t is a transcendental element over K. Denote the prime divisor corresponding to the valuation v_∞ (Section 2.2) by $\mathfrak{p}_\infty$. We determine the linear space $\mathcal{L}(n\mathfrak{p}_\infty)$, where n is a positive integer.

If $u \in \mathcal{L}(n\mathfrak{p}_\infty)$, then $v_\mathfrak{p}(u) \geq 0$ for every prime divisor $\mathfrak{p} \neq \mathfrak{p}_\infty$. This means $u \in K[t]$. Also $v_{\mathfrak{p}_\infty}(u) \geq -n$, so the degree of u as a polynomial in t is bounded by n. Thus, $\mathcal{L}(n\mathfrak{p}_\infty) = \{u \in K[t] \mid \deg(u) \leq n\}$ and $\dim(n\mathfrak{p}_\infty) = n+1$. For $n > 2g-2$, Lemma 3.2.2(d) gives $n+1 = n-g+1$. Hence, $g = 0$.

Conversely, let F/K be a function field of one variable of genus 0. Assume F/K has a prime divisor $\mathfrak{p}$ of degree 1 (e.g. if K is algebraically closed). Lemma 3.2.3 gives $t \in F^\times$ with $\mathrm{div}_\infty(t) = \mathfrak{p}$. By Section 3.1, $[F : K(t)] = \deg(\mathrm{div}_\infty(t)) = 1$. Consequently, $F = K(t)$.

Consider now a polynomial $f(t) = \sum_{i=0}^n a_i t^i$ with $a_i \in K$ and $a_n \neq 0$. For each $i < n$ we have $v_\infty(a_n t^n) < v_\infty(a_i t^i)$, so $v_\infty(f(t)) = -n$. For $\mathfrak{p} \neq \mathfrak{p}_\infty$ we have $v_\mathfrak{p}(t) \geq 0$, hence $v_\mathfrak{p}(f(t)) \geq 0$. Thus, $\mathrm{div}_\infty(f(t)) = \deg(f)\mathfrak{p}_\infty$.

Suppose $g(t) \in K[t]$ is relatively prime to $f(t)$. Put $u = \frac{f(t)}{g(t)}$. Then t is a root of the polynomial $h(u,T) = u \cdot g(T) - f(T)$. Since $h(u,T)$ is linear in u and $\gcd(f,g) = 1$, $h(u,T)$ is irreducible over $K(u)$. In addition, $\deg_T(h) = \max(\deg(f), \deg(g))$. Therefore,

(4) $$[K(t) : K(u)] = \max\big(\deg(f), \deg(g)\big).$$

In particular, suppose $K(u) = K(t)$. Then, $f(t)$ and $g(t)$ are linear and relatively prime. This means, $u = \frac{at+b}{ct+d}$ with $a,b,c,d \in K$ and $ad - bc \neq 0$. □

3.3 Holomorphy Rings

Let F/K be a function field of genus g. Denote the set of prime divisors of F/K by $\mathcal{R}$. For each $\mathfrak{p} \in \mathcal{R}$ let $O_\mathfrak{p} = \{x \in F \mid v_\mathfrak{p}(x) \geq 0\}$ be the corresponding valuation ring. To every subset S of $\mathcal{R}$ we attach the **holomorphy ring** $O_S = \bigcap_{\mathfrak{p} \in S} O_\mathfrak{p}$.

By definition, $K \subseteq O_S$. If S is empty, then, by definition, $O_S = F$. If $S = \mathcal{R}$, then the elements of O_S have no poles. They are therefore constants. Thus, $O_\mathcal{R} = K$.

The case where S is a nonempty proper subset of $\mathcal{R}$ requires a strengthening of the weak approximation theorem (Proposition 2.1.1).

PROPOSITION 3.3.1 (Strong Approximation Theorem): *Let S be a finite subset of $\mathcal{R}$. Consider $\mathfrak{q} \in \mathcal{R} \smallsetminus S$ and let $S' = S \cup \{\mathfrak{q}\}$. Suppose for each $\mathfrak{p} \in S$ we have an $x_\mathfrak{p} \in F$ and a positive integer $m_\mathfrak{p}$. Then there exists $x \in F$ with*

(1) $\quad v_\mathfrak{p}(x - x_\mathfrak{p}) = m_\mathfrak{p}$ *for each $\mathfrak{p} \in S$ and $v_\mathfrak{p}(x) \geq 0$ for each $\mathfrak{p} \in \mathcal{R} \smallsetminus S'$.*

3.3 Holomorphy Rings

Moreover, if m is an integer with

(2) $$m \cdot \deg(\mathfrak{q}) > 2g - 2 + \sum_{\mathfrak{p} \in S}(m_\mathfrak{p} + 1)\deg(\mathfrak{p}),$$

then x can be chosen such that, in addition to (1), it satisfies $v_\mathfrak{q}(x) \geq -m$.

Proof: Let m be a positive integer satisfying (2). Consider the divisor $\mathfrak{a} = m\mathfrak{q} - \sum_{\mathfrak{p} \in S} m_\mathfrak{p} \mathfrak{p}$. Then $\deg(\mathfrak{a}) > 2g - 2$. By Lemma 3.2.2(d), $\delta(\mathfrak{a}) = 0$, so $\mathbb{A} = F + \Lambda(\mathfrak{a})$. Define $\xi \in \mathbb{A}$ by $\xi_\mathfrak{p} = x_\mathfrak{p}$ for $\mathfrak{p} \in S$ and $\xi_\mathfrak{p} = 0$ for $\mathfrak{p} \in \mathcal{R} \smallsetminus S$. Then there exists $y \in F$ such that $y - \xi \in \Lambda(\mathfrak{a})$:

(3) $v_\mathfrak{p}(y - x_\mathfrak{p}) \geq m_\mathfrak{p}$ for $\mathfrak{p} \in S$, $v_\mathfrak{q}(y) \geq -m$, and $v_\mathfrak{p}(y) \geq 0$ for $\mathfrak{p} \in \mathcal{R} \smallsetminus S'$.

Now consider the divisor $\mathfrak{b} = m\mathfrak{q} - \sum_{\mathfrak{p} \in S}(m_\mathfrak{p} + 1)\mathfrak{p}$. For each $\mathfrak{p} \in S$ we have $\deg(\mathfrak{b} + \mathfrak{p}) > \deg(\mathfrak{b}) > 2g - 2$. By Lemma 3.2.2(d),

$$\dim(\mathfrak{b} + \mathfrak{p}) = \deg(\mathfrak{b} + \mathfrak{p}) - g + 1 > \deg(\mathfrak{b}) - g + 1 = \dim(\mathfrak{b}).$$

Hence, $\mathcal{L}(\mathfrak{b}) \subset \mathcal{L}(\mathfrak{b} + \mathfrak{p})$. Choose $z_\mathfrak{p} \in \mathcal{L}(\mathfrak{b} + \mathfrak{p}) \smallsetminus \mathcal{L}(\mathfrak{b})$. Then $v_\mathfrak{p}(z_\mathfrak{p}) = m_\mathfrak{p}$. Also $v_{\mathfrak{p}'}(z_\mathfrak{p}) \geq m_{\mathfrak{p}'} + 1$ if $\mathfrak{p}' \in S \smallsetminus \{\mathfrak{p}\}$, $v_\mathfrak{q}(z_\mathfrak{p}) \geq -m$, and $v_{\mathfrak{p}'}(z_\mathfrak{p}) \geq 0$ for $\mathfrak{p}' \in \mathcal{R} \smallsetminus S'$.

Let $P = \{\mathfrak{p} \in S \mid v_\mathfrak{p}(y - x_\mathfrak{p}) > m_\mathfrak{p}\}$ and let $Q = S \smallsetminus P$. Then $z = \sum_{\mathfrak{p} \in P} z_\mathfrak{p}$ has the following property:
(4) $v_\mathfrak{p}(z) = m_\mathfrak{p}$ if $\mathfrak{p} \in P$, $v_\mathfrak{p}(z) \geq m_\mathfrak{p} + 1$ if $\mathfrak{p} \in Q$, $v_\mathfrak{q}(z) \geq -m$, and $v_\mathfrak{p}(z) \geq 0$ for $\mathfrak{p} \in \mathcal{R} \smallsetminus S'$.

Combine (3) and (4) to see that $x = z + y$ satisfies $v_\mathfrak{p}(x - x_\mathfrak{p}) = m_\mathfrak{p}$, for $\mathfrak{p} \in S$, $v_\mathfrak{q}(x) \geq -m$, and $v_\mathfrak{p}(x) \geq 0$ for $\mathfrak{p} \in \mathcal{R} \smallsetminus S'$. $\square$

If $\mathfrak{p}$ belongs to a subset S of $\mathcal{R}$, then $O_S \subseteq O_\mathfrak{p}$. Also, $P = \{x \in O_S \mid v_\mathfrak{p}(x) > 0\}$ is a prime ideal of O_S, the **center** of $\mathfrak{p}$ at O_S. Denote the local ring of O_S at P by $O_{S,P}$.

PROPOSITION 3.3.2 (Holomorphy Ring Theorem): *Let S be a nonempty proper subset of $\mathcal{R}$. Then S has these properties:*
(a) $\mathrm{Quot}(O_S) = F$.
(b) *If $\mathfrak{p} \in S$ and P is the center of $\mathfrak{p}$ at O_S, then $O_\mathfrak{p} = O_{S,P}$.*
(c) *If $\mathfrak{q} \in \mathcal{R} \smallsetminus S$, then $O_S \not\subseteq O_\mathfrak{q}$.*
(d) *Every nonzero prime ideal of O_S is the center of a prime $\mathfrak{p} \in S$.*
(e) *Distinct primes in S have distinct centers at O_S, and the center of each $\mathfrak{p} \in S$ is a maximal ideal of O_S.*
(f) O_S *is a Dedekind domain.*

Proof of (a): Consider $z \in F \smallsetminus K$. Since $S \subset \mathcal{R}$, there is a $\mathfrak{q} \in \mathcal{R} \smallsetminus S$. There are only finitely many $\mathfrak{p} \in S$ with $v_\mathfrak{p}(z) < 0$. Hence, Proposition 3.3.1 gives $y \in F$ such that the following holds for each $\mathfrak{p} \in S$: If $v_\mathfrak{p}(z) < 0$, then $v_\mathfrak{p}(y - z^{-1}) = v_\mathfrak{p}(z^{-1}) + 1$, so $v_\mathfrak{p}(y) = -v_\mathfrak{p}(z)$; while if $v_\mathfrak{p}(z) \geq 0$, then

$v_{\mathfrak{p}}(y) \geq 0$. Let $x = yz$. Then both x and y belong to O_S. If $z \in O_S$, then $z \in \mathrm{Quot}(O_S)$. If $z \notin O_S$, then there is a $\mathfrak{p} \in S$ with $v_{\mathfrak{p}}(z) < 0$. Hence, $v_{\mathfrak{p}}(y) = v_{\mathfrak{p}}(z^{-1}) \neq \infty$, so $y \neq 0$. Therefore, $z = xy^{-1} \in \mathrm{Quot}(O_S)$.

Proof of (b): Let $z \in O_{\mathfrak{p}} \smallsetminus K$. As in the proof of (a), there exists $y \in F$ such that $v_{\mathfrak{p}}(y) = 0$, $v_{\mathfrak{p}'}(y) = -v_{\mathfrak{p}'}(z)$ if $\mathfrak{p}' \in S \smallsetminus \{\mathfrak{p}\}$ and $v_{\mathfrak{p}'}(z) < 0$, while $v_{\mathfrak{p}'}(y) \geq 0$ if $\mathfrak{p}' \in S \smallsetminus \{\mathfrak{p}\}$ and $v_{\mathfrak{p}'}(z) \geq 0$. Therefore, $x = yz$ is in O_S and $z \in O_{S,P}$. Since the inclusion $O_{S,P} \subseteq O_{\mathfrak{p}}$ is clear, $O_{\mathfrak{p}} = O_{S,P}$.

Proof of (c): If $S \cup \{\mathfrak{q}\} = \mathcal{R}$ and $O_S \subseteq O_{\mathfrak{q}}$, then $O_S = O_{\mathcal{R}} = K$, a contradiction to (a). Therefore, assume that $S \cup \{\mathfrak{q}\}$ is a proper subset of $\mathcal{R}$. By Proposition 3.3.1, there exists $x \in F$ such that $v_{\mathfrak{q}}(x) = -1$ and $v_{\mathfrak{p}}(x) \geq 0$ for each $\mathfrak{p} \in S$. This element belongs to O_S but not to $O_{\mathfrak{q}}$.

Proof of (d): Let P be a nonzero prime ideal of O_S. Proposition 2.3.1 extends the quotient map $O_S \to O_S/P$ to a place φ of F trivial on K. Let $\mathfrak{p}$ be the prime divisor of F/K which is defined by φ. Then $O_S \subseteq O_{\mathfrak{p}}$ and $P = \{x \in O_S \mid v_{\mathfrak{p}}(x) > 0\}$. By (c), $\mathfrak{p} \in S$.

Proof of (e): Let $\mathfrak{p}$ and $\mathfrak{p}'$ be two distinct prime divisors in S. By the strong approximation theorem (Proposition 3.3.1), there exists $x \in O_S$ with $v_{\mathfrak{p}}(x) > 0$ and $v_{\mathfrak{p}'}(x) = 0$. This means the center P of $\mathfrak{p}$ is not contained in the center of $\mathfrak{p}'$. The maximality of P now follows from (d).

Proof of (f): Let $\mathfrak{q} \in \mathcal{R} \smallsetminus S$, let $S' = \mathcal{R} \smallsetminus \{\mathfrak{q}\}$ and choose $\mathfrak{p}' \in S'$. Since $O_{S'}$ is an overring of $O_{S'}$ it suffices (Proposition 2.4.7) to prove that $O_{S'}$ is a Dedekind domain. The strong approximation theorem gives $x \in O_{S'}$ with $v_{\mathfrak{p}'}(x) > 0$. Since x must have a pole, it must be $\mathfrak{q}$. Thus, $K[x] \subseteq O_{\mathfrak{p}}$ if and only if $\mathfrak{p} \in S'$. Therefore, by Proposition 2.4.1, $O_{S'} = \bigcap_{\mathfrak{p} \in S'} O_{\mathfrak{p}}$ is the integral closure of $K[x]$ in F. Since $K[x]$ is a Dedekind domain, $O_{S'}$ is also a Dedekind domain (Proposition 2.4.6). □

COROLLARY 3.3.3: *The following conditions are equivalent for a nonempty subset S of $\mathcal{R}$:* (a) $S = \mathcal{R}$; (b) $O_S = K$; *and* (c) O_S *is a field.*

Proof: Every nonconstant element of F has a pole. Therefore, (a) implies (b). The implication "(b) => (c)" is trivial. To prove that (c) implies (a), note that $O_S \neq F$, since S is nonempty. If $S \neq \mathcal{R}$, then, by Proposition 3.3.2(a), the quotient field of O_S is F. Hence, O_S is not a field. □

The following converse to Proposition 3.3.2 is useful:

PROPOSITION 3.3.4: *Let F/K be a function field and R a proper subring of F containing K. Suppose R is integrally closed and $\mathrm{Quot}(R) = F$. Then there is a nonempty subset S of $\mathcal{R}$ with $R = O_S$. Thus, R is a Dedekind domain.*

Proof: Let S be the set of all prime divisors of F/K which are finite on R. By Proposition 2.4.1, $O_S = R$. As $R \subset F$, S is nonempty. By Corollary 3.3.3, $S \neq \mathcal{R}$. Hence, by Proposition 3.3.2(f), R is a Dedekind domain. □

3.4 Extensions of Function Fields

Let E/K and F/L be algebraic function fields of one variable. We say that F/L is an **extension** of E/K if $E \subseteq F$, $K \subseteq L$, and $L \cap E = K$. We call F/L a **constant field extension** of E/K if E is linearly disjoint from L over K and $F = EL$. Thus, in this case $[F:E] = [L:K]$. Recall that E/K is a regular extension ((1b) of Section 3.1). Hence, the linear disjointness of E and L over K is automatic if L is an algebraic extension of K. Likewise, E is linearly disjoint from L over K if $L = K(x_1, \ldots, x_n)$ and $x_1, \ldots, x_n$ are algebraically independent over E (Lemma 2.6.7).

Let F/L be an extension of E/K, $\mathfrak{p}$ be a prime divisor of E/K, and $\mathfrak{P}$ be a prime divisor of F/L that **lies over** $\mathfrak{p}$, that is $v_\mathfrak{P}$ lies over $v_\mathfrak{p}$. Denote the ramification index of $v_\mathfrak{P}$ over $v_\mathfrak{p}$ by $e_{\mathfrak{P}/\mathfrak{p}}$. Since both $v_\mathfrak{P}$ and $v_\mathfrak{p}$ are discrete and normalized, $v_\mathfrak{P}(x) = e_{\mathfrak{P}/\mathfrak{p}} v_\mathfrak{p}(x)$ for each $x \in E$. Refer to $\mathfrak{P}$ as **unramified** over E if $v_\mathfrak{P}$ is unramified over E. If F is separable algebraic over E, then only finitely many prime divisors of F/L are ramified over E [Deuring3, p. 111].

Over every prime divisor of E/K there lie only finitely many prime divisors of F/L [Deuring3, p. 96]. Use this result to embed the group of divisors $\mathrm{Div}(E/K)$ of E/K into $\mathrm{Div}(F/L)$ as follows: For $\mathfrak{p}$ a prime divisor of E/K and $\mathfrak{P}_1, \ldots, \mathfrak{P}_d$ the prime divisors of F/L lying over $\mathfrak{p}$, map $\mathfrak{p}$ to the divisor $\sum_{i=1}^d e_{\mathfrak{P}_i/\mathfrak{p}} \mathfrak{P}_i$ of F/L. Extend this map to $\mathrm{Div}(E/K)$ by linearity. The principal divisor of x (in $\mathrm{Div}(E/K)$) maps to the principal divisor of x in $\mathrm{Div}(F/L)$, so there is no ambiguity using $\mathrm{div}(x)$ for that divisor. In particular, for every divisor $\mathfrak{a}$ of E/K we have $\mathcal{L}_E(\mathfrak{a}) = E \cap \mathcal{L}_F(\mathfrak{a})$.

Suppose $[F:E] < \infty$. If F/E is separable, apply (2) of Section 2.3 to conclude:

$$(1) \qquad \sum_{i=1}^d e_{\mathfrak{P}_i/\mathfrak{p}} [\bar{F}_{\mathfrak{P}_i} : \bar{E}_\mathfrak{p}] = [F:E].$$

Even if F/E is not separable, the following argument shows (1) still holds: By Lemma 3.2.3, there are an integer m and $x \in E$ with $\mathrm{div}_\infty(x) = m\mathfrak{p}$. Apply the rules $\deg_F(\mathrm{div}_\infty(x)) = [F:L(x)]$ and $\deg_E(\mathrm{div}_\infty(x)) = [E:K(x)]$ to conclude that (1) is true in general [Deuring3, p. 97].

Next consider a function field F/K and let E be a proper extension of K contained in F. Since F is linearly disjoint from $\tilde{K}$ over K, so is E. In particular, there is a transcendental element x in E. Then $[F:K(x)] = \deg_F(\mathrm{div}_\infty(x)) < \infty$. Thus, $[E:K(x)] < \infty$, E/K is a function field of one variable, and $[F:E] < \infty$.

LEMMA 3.4.1: *Let E/K and F/K be algebraic function fields of one variable with $E \subseteq F$. Then, $\deg_F(\mathfrak{a}) = [F:E]\deg_E(\mathfrak{a})$ for each $\mathfrak{a} \in \mathrm{Div}(E/K)$.*

Proof: Assume by linearity that $\mathfrak{a} = \mathfrak{p}$ is a prime divisor. Let $\mathfrak{P}_1, \ldots, \mathfrak{P}_d$ be the prime divisors of F/K which lie over $\mathfrak{p}$. Then $K \subseteq \bar{E}_\mathfrak{p} \subseteq \bar{F}_{\mathfrak{P}_i}$, so

$\deg_F \mathfrak{P}_i = f_{\mathfrak{P}_i/\mathfrak{p}} \deg_E(\mathfrak{p})$, $i = 1, \ldots, d$. By (1),

$$\deg_F(\mathfrak{p}) = \sum_{i=1}^{d} e_{\mathfrak{P}_i/\mathfrak{p}} \deg_F(\mathfrak{P}_i)$$

$$= \sum_{i=1}^{d} e_{\mathfrak{P}_i/\mathfrak{p}} f_{\mathfrak{P}_i/\mathfrak{p}} \deg_E(\mathfrak{p}) = [F:E] \deg_E(\mathfrak{p}). \quad \square$$

PROPOSITION 3.4.2: *Let F/L be a constant field extension of an algebraic function field E/K of one variable with L/K separable. Then:*
(a) *For a divisor $\mathfrak{a}$ of E/K, $\deg_E(\mathfrak{a}) = \deg_F(\mathfrak{a})$, $\dim_E(\mathfrak{a}) = \dim_F(\mathfrak{a})$, and $\mathcal{L}_E(\mathfrak{a})L = \mathcal{L}_F(\mathfrak{a})$.*
(b) *$\mathrm{genus}(E/K) = \mathrm{genus}(F/L)$.*
(c) *If $\mathfrak{p}$ and $\mathfrak{P}$ are respective prime divisors of E/K and F/L, with $\mathfrak{P}$ lying over $\mathfrak{p}$, then $\bar{F}_\mathfrak{P} = L\bar{E}_\mathfrak{p}$ and $\mathfrak{P}$ is unramified over $\mathfrak{p}$.*
(d) *Let R be an integrally closed subring of E containing K. Then RL is the integral closure of R in F.*
(e) *Let $\mathfrak{p}$ be a prime divisor of E/K of degree 1. Then there is a unique prime divisor $\mathfrak{P}$ of F/L lying over $\mathfrak{p}$ and $\deg(\mathfrak{P}) = 1$.*
(f) *Let $x_1, \ldots, x_m$ be elements of E. Denote the integral closure of $K[\mathbf{x}]$ in E by S. Then SL is the integral closure of $L[\mathbf{x}]$ in F.*

Proof of (a): [Deuring3, p. 126] shows $\deg_E(\mathfrak{a}) = \deg_F(\mathfrak{a})$. We show $\dim_E(\mathfrak{a}) = \dim_F(\mathfrak{a})$. Let B be a basis of $\mathcal{L}_E(\mathfrak{a})$ over K. Then B is contained in $\mathcal{L}_F(\mathfrak{a})$ and is linearly independent over L. Therefore, $\dim_E(\mathfrak{a}) \leq \dim_F(\mathfrak{a})$.

Conversely, let B' be a basis of $\mathcal{L}_F(\mathfrak{a})$. Then B' is contained in EL_0 for some finitely generated extension L_0/K of L/K. By [Deuring3, p. 132], $\dim_{EL_0}(\mathfrak{a}) = \dim_E(\mathfrak{a})$. Hence, $\dim_F(\mathfrak{a}) = |B'| \leq \dim_{EL_0}(\mathfrak{a}) = \dim_E(\mathfrak{a})$. Therefore, $\dim_E(\mathfrak{a}) = \dim_F(\mathfrak{a})$.

It follows that $\mathcal{L}_E(\mathfrak{a})L \subseteq \mathcal{L}_F(\mathfrak{a})$ have the same dimension over L, so $\mathcal{L}_E(\mathfrak{a})L = \mathcal{L}_F(\mathfrak{a})$.

Proof of (b): Let $g_E = \mathrm{genus}(E/K)$ and $g_F = \mathrm{genus}(F/L)$. Choose a divisor $\mathfrak{a}$ of E/K with $\deg_E(\mathfrak{a}) > \max(2g_E - 2, 2g_F - 2)$. By Lemma 3.2.2(d), $\dim_E(\mathfrak{a}) = \deg_E(\mathfrak{a}) + 1 - g_E$ and $\dim_F(\mathfrak{a}) = \deg_F(\mathfrak{a}) + 1 - g_F$. Therefore, by (a), $g_E = g_F$.

Proof of (c): The general case reduces to the case where L/K is finitely generated. In this case [Deuring3, p. 128] proves that $\bar{F}_\mathfrak{P} = L\bar{E}_\mathfrak{p}$.

Now choose a primitive element c of L/K and let $f = \mathrm{irr}(c, K)$. By the preceding paragraph, $\bar{F}_\mathfrak{P} = \bar{E}(c)$. The reduction of f modulo $\mathfrak{P}$ is f itself. Since L/K is separable, f is separable. Hence, by Lemma 2.3.4, $\mathfrak{P}/\mathfrak{p}$ is unramified.

Proof of (d): Assume without loss that $[L:K] < \infty$. Choose a basis $w_1, \ldots, w_n$ of L/K. Let $\sigma_1, \ldots, \sigma_n$ be the K-embeddings of L into K_s. Since

E is linearly disjoint from L over K each σ_i extends uniquely to an E-embedding of F into E_s.

Now consider $z \in F$ which is integral over R. Write $z = \sum_{j=1}^{n} a_j w_j$ with $a_1, \ldots, a_n \in E$. Then $\sigma_i z = \sum_{j=1}^{n} a_j \sigma_i w_j$, $i = 1, \ldots, n$. By Cramer's rule, $a_j = \frac{b_j}{\Delta} = \frac{b_j \Delta}{\Delta^2}$, where b_j is in the ring generated over R by the a_k's and the $\sigma_k w_l$, and $\Delta = \det(\sigma_k w_l)_{1 \le k,l \le n} \in K^\times$. By assumption, each b_j is integral over R. Since L/K is separable, $\Delta^2 \in K^\times$ [Lang7, p. 286, Cor. 5.4]. Hence, each a_j is integral over R. Since R is integrally closed, $a_j \in R$, $j = 1, \ldots, n$. Consequently, $z \in RL$.

Proof of (e): To prove (e), it suffices to consider two cases. In one case, $L = K(u)$ with u purely transcendental over K. [Deuring3, pp. 128–129] handles this case. The other case is when L/K is separable and finite. Let in this case $\mathfrak{P}_1, \ldots, \mathfrak{P}_d$ be the prime divisors of F/L lying over $\mathfrak{p}$. By (c), $e_{\mathfrak{P}_i/\mathfrak{p}} = 1$, $i = 1, \ldots, m$. By assumption, $\deg(\mathfrak{p}) = 1$, so $\bar{E}_\mathfrak{p} = K$. Hence, by (c), $\bar{F}_{\mathfrak{P}_i} = L$, $i = 1, \ldots, d$. Since F/L is a constant field extension of E/K, we have $[F : E] = [L : K]$. Therefore, by (1), $d = 1$ and $\bar{F}_{\mathfrak{P}_1} = L$. Thus, $\deg(\mathfrak{P}_1) = 1$.

Proof of (f): Put $\mathfrak{a} = \sum_{i=1}^{m} \mathrm{div}_{\infty,E}(x_i)$. By Lemma 3.1.1, $S = \bigcup_{k=1}^{\infty} \mathcal{L}_E(k\mathfrak{a})$. By (a), $SL = \bigcup_{k=1}^{\infty} \mathcal{L}_E(k\mathfrak{a})L = \bigcup_{k=1}^{\infty} \mathcal{L}_F(k\mathfrak{a})$. Hence, by Lemma 3.1.1, SL is the integral closure of $L[\mathbf{x}]$ in F. □

3.5 Completions

The completion of a function field F/K at a prime divisor $\mathfrak{p}$ gives a powerful tool to investigate the behavior of F at $\mathfrak{p}$. For example, it allows us to determine the decomposition of $\mathfrak{p}$ to prime divisors in finite extensions of F (Proposition 3.5.3). We also use completions to define the 'different' of an extension. This notion plays a central role in the Riemann-Hurwitz genus formula, to be introduced in the next section.

Let v be a rank-1 valuation of a field F. Then v induces a topology on F. Two elements x, y of F are 'close' in this topology if $v(x - y)$ is 'large'. A sequence $\{x_i\}_{i=1}^{\infty}$ of elements of F is a **Cauchy sequence** if for every integer m there exists k such that $i, j \ge k$ implies $v(x_i - x_j) \ge m$. If every Cauchy sequence converges, F is **complete**. Each F embeds as a dense subfield in a complete field $\hat{F}_v$ with a valuation v extending the valuation of F [Borevich-Shafarevich, Chap. 1, Sec. 4.1]. In particular, $\hat{F}_v$ has the same residue field and value group at v as F. We call $(\hat{F}_v, v)$ (or also just $\hat{F}_v$) the **completion** of (F, v). We also say that $\hat{F}_v$ is the **completion** of F at v. The completion $(\hat{F}_v, v)$ of (F, v) is unique up to an F-automorphism.

Example 3.5.1: $\mathbb{Q}_p$ and $K((t))$. The completion of $\mathbb{Q}$ at the p-adic valuation v_p (Section 2.2) is the field $\mathbb{Q}_p$ of p-**adic numbers**. Every element x of $\mathbb{Q}_p^\times$ has a unique presentation as a convergent (in the v_p-topology) power series

$\sum_{n=m}^{\infty} a_n p^n$ with $m = v_p(x)$, $a_n \in \mathbb{Z}$, $0 \leq a_n \leq p - 1$, and $a_m \neq 0$. The valuation ring $\mathbb{Z}_p$ of $\mathbb{Q}_p$ consists of all $x \in \mathbb{Q}_p$ with $m \geq 0$.

Next consider a field K and a transcendental element t over K. Let v be the unique valuation of $K(t)$ with $v(t) = 1$ (Section 2.2). The completion of $K(t)$ at v is the field $K((t))$ of **formal power series** in t with coefficients in K. Each nonzero element f of $K((t))$ has a unique presentation $f = \sum_{n=m}^{\infty} a_n t^n$ with $m = v(f)$ and $a_n \in K$. The valuation ring of $K((t))$ is the ring $K[[t]]$ of formal power series in t with coefficients in K. The residue field of $K((t))$ under v is K. By Lemma 2.6.9(b), $K((t))$ is a regular extension of K.

Consider now a finite extension L of K with a basis $w_1, \ldots, w_d$. Then $w_1, \ldots, w_d$ are linearly independent over $K((t))$. Let $x = \sum_{n=m}^{\infty} a_n t^n$ with $a_n \in L$ be an element of $L((t))$. For each n write $a_n = \sum_{i=1}^{d} a_{ni} w_i$ with $a_{ni} \in K$. Then, $x = \sum_{i=1}^{d} (\sum_{n=m}^{\infty} a_{ni} t^n) w_i$. Therefore, $L((t)) = K((t))L$ and $w_1, \ldots, w_d$ form a basis for $L((t))/K((t))$. $\square$

The next result contains various versions of Hensel's Lemma:

PROPOSITION 3.5.2: *Let (F, v) be a complete discrete valued field.*
(a) *Let $f \in O_v[X]$ and $a \in O_v$ with $v(f(a)) > 2v(f'(a))$. Then there is a unique $x \in F$ with $f(x) = 0$ and $v(x - a) \geq v(f(a)) - v(f'(a))$ [Cassels-Fröhlich, p. 83].*
(b) *Let $f \in O_v[X]$ be a monic polynomial. Denote reduction at v by a bar. Suppose $\bar{f}(X) = \zeta(X)\eta(X)$ with $\zeta, \eta \in \bar{F}_v[X]$ monic and relatively prime. Then there are monic polynomials $g, h \in O_v[X]$ with $\bar{g} = \zeta$, $\bar{h} = \eta$, and $f(X) = g(X)h(X)$ [Zariski-Samuel2, p. 279, Thm. 17].*
(c) *Let F' be a finite algebraic extension of F. Then v has a unique extension v' to F' and F' is complete under v' [Cassels-Fröhlich, p. 56, Thm.].*

Here is a global application of completions:

PROPOSITION 3.5.3: *Let (E, v) be a discrete valued field. Denote its completion by $(\hat{E}, \hat{v})$. Consider a finite separable extension F of E. Let z be a primitive element for F/E which is integral over O_v. Put $h = \mathrm{irr}(z, E)$. Let $h = h_1 \cdots h_r$ be the decomposition of h into a product of irreducible polynomials over $\hat{E}$. For each i let z_i be a root of h_i in $\hat{E}_s$. Denote the unique extension of $\hat{v}$ to $\hat{F}_i = \hat{E}(z_i)$ by $\hat{v}_i$. The following holds:*
(a) *The map $z \mapsto z_i$ extends to a E-embedding of F into $\hat{F}_i$.*
(b) *$\hat{F}_i$ is the completion of F at the restriction v_i of $\hat{v}_i$ to F.*
(c) *The valuations $v_1, \ldots, v_r$ are mutually nonequivalent. Every extension of v to F coincides with one of the v_i's.*
(d) *$e_{\hat{v}_i/\hat{v}} = e_{v_i/v}$, $f_{\hat{v}_i/\hat{v}} = f_{v_i/v}$, and $\bar{h}_i = h_{i0}^e$ with $h_{i0} \in \bar{E}[X]$ irreducible and $e \in \mathbb{N}$.*
(e) *The map $g(z) \mapsto (g(z_1), \ldots, g(z_r))$ for $g \in \hat{E}[X]$ is an $\hat{E}$-isomorphism of $\hat{E} \otimes_E F$ onto $\bigoplus_{i=1}^{r} \hat{F}_i$.*

3.5 Completions

(f) Each z in F satisfies $\operatorname{trace}_{F/E} z = \sum_{i=1}^{r} \operatorname{trace}_{\hat{F}_i/\hat{E}} z_i$ and $\operatorname{norm}_{F/E} z = \prod_{i=1}^{r} \operatorname{norm}_{\hat{F}_i/\hat{E}} z_i$.

Proof: See [Cassels-Fröhlich, p. 57, Thm.]. The last statement of (d) follows from Proposition 3.5.2(b). □

Example 3.5.4: Dedekind. We apply Proposition 3.5.3 to show the necessity of the assumption "$\bar{E}_v$ is infinite" in Lemma 2.3.5.

Consider the polynomial $f(X) = X^3 - X^2 - 2X - 8$. Observe that $f(X)$ has no root modulo 3, hence no root in $\mathbb{Q}$. It is therefore irreducible. Let z be a root of $f(X)$ in $\tilde{\mathbb{Q}}$ and $F = \mathbb{Q}(z)$. Then $[F : \mathbb{Q}] = 3$. Next observe that $v_2(f(0)) = 3$, $v_2(f'(0)) = 1$, $v_2(f(1)) = 1$, $v_2(f'(1)) = 0$, $v_2(f(2)) = 3$, and $v_2(f'(2)) = 1$. Hence, by Proposition 3.5.2(a), f has three roots x_1, x_2, x_3 in $\mathbb{Q}_2$ with $v_2(x_1) \geq 2$, $v_2(x_2 - 1) \geq 1$, and $v_2(x_3 - 2) \geq 2$. Thus, $f(X) = (X - x_1)(X - x_2)(X - x_3)$ is a decomposition of $f(X)$ over $\mathbb{Q}_2$ into three distinct irreducible polynomials. By Proposition 3.5.3, v_2 has three distinct extensions to F: w, w', and w''. By Proposition 2.3.2, $\bar{F}_w = \mathbb{F}_2$.

Now assume $F/\mathbb{Q}$ has a primitive element z with $g = \operatorname{irr}(z, \mathbb{Q}) \in O_{v_2}[X]$ such that $\bar{g}$ is separable. Then $\deg(g) = 3$. By the preceding paragraph, $\bar{F}_w = \mathbb{F}_2$. Hence, all three distinct roots of $\bar{g}$ belong to $\mathbb{F}_2$. But $\mathbb{F}_2$ has only two elements, so z does not exist. □

LEMMA 3.5.5: *Let (F, v) be a discrete valued field and I a nonzero O_v submodule of F which is not F. Then I is a fractional ideal of O_v and there exists $m \in \mathbb{Z}$ with $I = \mathfrak{m}_v^{-m}$. If F is a finite extension of a field E, and O_v is the unique valuation ring of F lying over $O_v \cap E$ (e.g. (E, v) is complete), then $\operatorname{trace}_{F/E} I$ is a fractional ideal of $O_v \cap E$.*

Proof: Choose $\pi \in F$ with $v(\pi) = 1$. Let x be a nonzero element of I. If $x' \in F$ and $v(x') \geq v(x)$, then $x' = \frac{x'}{x} x \in I$. Since $I \neq F$, this implies $v(I)$ is bounded from below. Hence, $-m = \inf(v(x) \mid x \in I)$ is an integer. Therefore, $I = \pi^{-m} O_v$.

Now suppose O_v is the unique valuation ring of F over $O = O_v \cap E$. By Proposition 2.4.1, O_v is the integral closure of O in F. Let a be an element of A with $v(a)$ sufficiently large. By the preceding paragraph, $aI \subseteq O_v$. Then $a \cdot \operatorname{trace}_{F/E}(I) = \operatorname{trace}_{F/E}(aI) \subseteq \operatorname{trace}_{F/E}(O_v) \subseteq A$. Thus, $\operatorname{trace}_{F/E}(I)$ is a fractional ideal of A. □

Using completions, we generalize the notions of repartition and differential (Section 3.2) of a function field.

Let $\mathfrak{p}$ be a prime divisor of F/K. Suppose the residue field $\bar{F}_\mathfrak{p}$ of F at $\mathfrak{p}$ is separable over K. By Hensel's Lemma, $\bar{F}_\mathfrak{p}$ embeds into $\hat{F}_\mathfrak{p}$. Indeed, choose $a \in F$ with $\bar{F}_\mathfrak{p} = K(\bar{a})$. Put $f = \operatorname{irr}(\bar{a}, K)$. Then $v_\mathfrak{p}(f(a)) > 0$ and $v_\mathfrak{p}(f'(a)) = 0$. By Proposition 3.5.2(a), there is an $x' \in \hat{F}_\mathfrak{p}$ with $f(x') = 0$. The map $\bar{a} \mapsto x'$ extends to a K-embedding of $\bar{F}_\mathfrak{p}$ into $\hat{F}_\mathfrak{p}$.

Let π be an element of $\hat{F}_\mathfrak{p}$ with $v_\mathfrak{p}(\pi) = 1$. Then the completion $\hat{F}_\mathfrak{p}$ is isomorphic to the field $\bar{F}_\mathfrak{p}((\pi))$ of formal power series in π over $\bar{F}_\mathfrak{p}$. Every element of this field has the form $\alpha = \sum_{i=m}^{\infty} a_i \pi^i$, where m is an integer and $a_i \in \bar{F}_{\mathfrak{p}}$ [Chevalley2, p. 46]. If $a_m \neq 0$, then $v_{\mathfrak{p}}(\alpha) = m$.

Consider the cartesian product $\prod \hat{F}_{\mathfrak{p}}$ with $\mathfrak{p}$ ranging over all prime divisors of F/K. An $\alpha \in \prod \hat{F}_{\mathfrak{p}}$ is an **adele** if $v_{\mathfrak{p}}(\alpha_{\mathfrak{p}}) \geq 0$ for all but finitely many $\mathfrak{p}$. In particular, each repartition of F/K is an adele. Denote the set of adeles by $\hat{\mathbb{A}}$ (or by $\hat{\mathbb{A}}_F$, if F is not clear from the context). It is an F-subalgebra of $\prod \hat{F}_{\mathfrak{p}}$ which contains the algebra $\mathbb{A}$ of repartitions. For each prime divisor $\mathfrak{p}$ of F/K embed $\hat{F}_{\mathfrak{p}}$ in $\hat{\mathbb{A}}$ as follows. Identify $x \in \hat{F}_{\mathfrak{p}}$ with the adele ξ having $\xi_{\mathfrak{p}} = x$ and $\xi_{\mathfrak{p}'} = 0$ if $\mathfrak{p}' \neq \mathfrak{p}$.

For each $\mathfrak{a} \in \mathrm{Div}(F/K)$ consider the K-vector space

$$\hat{\Lambda}(\mathfrak{a}) = \{\alpha \in \hat{\mathbb{A}} \mid v_{\mathfrak{p}}(\alpha) + v_{\mathfrak{p}}(\mathfrak{a}) \geq 0 \text{ for each } \mathfrak{p}\},$$

where we have abbreviated $v_{\mathfrak{p}}(\alpha_{\mathfrak{p}})$ by $v_{\mathfrak{p}}(\alpha)$. Then $\mathbb{A} \cap \hat{\Lambda}(\mathfrak{a}) = \Lambda(\mathfrak{a})$ and $\mathbb{A} + \hat{\Lambda}(\mathfrak{a}) = \hat{\mathbb{A}}$. Indeed, let $\hat{\alpha} \in \hat{\mathbb{A}}$. For each $\mathfrak{p}$ choose $\alpha_{\mathfrak{p}} \in F$ with $v_{\mathfrak{p}}(\alpha_{\mathfrak{p}} - \hat{\alpha}_{\mathfrak{p}}) \geq -v_{\mathfrak{p}}(\mathfrak{a})$. Then $\alpha = (\alpha_{\mathfrak{p}})$ belongs to $\mathbb{A}$, $\alpha - \hat{\alpha}$ is in $\hat{\Lambda}(\mathfrak{a})$, and $\hat{\alpha} = \alpha - (\alpha - \hat{\alpha}) \in \mathbb{A} + \hat{\Lambda}(\mathfrak{a})$. Therefore

(1) $\qquad \mathbb{A} \cap (\hat{\Lambda}(\mathfrak{a}) + F) = \Lambda(\mathfrak{a}) + F$ and $\mathbb{A} + (\hat{\Lambda}(\mathfrak{a}) + F) = \hat{\mathbb{A}}$.

Hence, $\hat{\mathbb{A}}/(\hat{\Lambda}(\mathfrak{a}) + F) \cong \mathbb{A}/(\Lambda(\mathfrak{a}) + F)$. Thus, in the notation of Section 3.2,

(2) $\qquad \dim_K \left(\hat{\mathbb{A}}/(\hat{\Lambda}(\mathfrak{a}) + F) \right) = \delta(\mathfrak{a}) = \dim(\mathfrak{w} - \mathfrak{a}),$

where $\mathfrak{w}$ is a canonical divisor of F/K.

Recall that a differential of F/K is a K-linear map, $\omega \colon \mathbb{A} \to K$, which vanishes on a subspace of the form $\Lambda(\mathfrak{a}) + F$. By (1) we can extend ω uniquely to a K-linear map $\hat{\omega} \colon \hat{\mathbb{A}} \to K$ which vanishes on $\hat{\Lambda}(\mathfrak{a}) + F$. So, from now on, a **differential** of F/K is a K-linear map $\hat{\omega} \colon \hat{\mathbb{A}} \to K$ which vanishes on $\hat{\Lambda}(\mathfrak{a}) + F$ for some $\mathfrak{a} \in \mathrm{Div}(F/K)$. The restriction of ω to $\mathbb{A}$ is a differential in the old sense. The **divisor** of $\hat{\omega}$ is the maximum of all $\mathfrak{a} \in \mathrm{Div}(F/K)$ such that $\hat{\omega}$ vanishes on $\hat{\Lambda}(\mathfrak{a})$. Denote it by $\mathrm{div}(\hat{\omega})$. By the above, $\mathrm{div}(\hat{\omega}) = \mathrm{div}(\omega)$.

Denote the K-vector space of all differentials that vanish on $\hat{\Lambda}(\mathfrak{a}) + F$ by $\hat{\Omega}(\mathfrak{a})$. The natural map $\omega \mapsto \hat{\omega}$ is an isomorphism of $\Omega(\mathfrak{a})$ onto $\hat{\Omega}(\mathfrak{a})$. In particular, these spaces have the same dimension $\delta(\mathfrak{a})$.

Suppose F/E is a finite separable extension of function fields of one variable over K. Let $\alpha \in \hat{\mathbb{A}}_E$. Given a prime divisor $\mathfrak{P}$ of F/K, denote its restriction to E by $\mathfrak{p}$ and let $\alpha_{\mathfrak{P}} = \alpha_{\mathfrak{p}}$. This identifies $\hat{\mathbb{A}}_E$ as a subalgebra of $\hat{\mathbb{A}}_F$. The isomorphisms $\hat{E}_{\mathfrak{p}} \otimes_E F \cong \bigoplus_{\mathfrak{P} \mid \mathfrak{p}} \hat{F}_{\mathfrak{P}}$ of Proposition 3.5.3(e) combine to an isomorphism

(3) $\qquad \hat{\mathbb{A}}_E \otimes_E F \cong \hat{\mathbb{A}}_F$

3.5 Completions

[Cassels-Fröhlich, p. 64 or Artin3, p. 244, Thm. 2]. In addition, define a trace function $\text{trace}_{F/E}\colon \hat{\mathbb{A}}_F \to \hat{\mathbb{A}}_E$:

$$\text{trace}_{F/E}(\beta)_{\mathfrak{p}} = \sum_{\mathfrak{P}|\mathfrak{p}} \text{trace}_{\hat{F}_{\mathfrak{P}}/\hat{E}_{\mathfrak{p}}}(\beta_{\mathfrak{P}}).$$

PROPOSITION 3.5.6: *Let F/E be a finite separable extension of function fields of one variable over a field K. The following holds for all $\alpha, \beta \in \hat{\mathbb{A}}_F$:*
(a) $\text{trace}_{F/E}(\alpha + \beta) = \text{trace}_{F/E}(\alpha) + \text{trace}_{F/E}(\beta)$.
(b) $\text{trace}_{F/E}(\alpha\beta) = \alpha \cdot \text{trace}_{F/E}(\beta)$ *if* $\alpha \in \hat{\mathbb{A}}_E$.
(c) *The trace of an element x of F coincides with the trace of x as an adele.*
(d) *Let $\mathfrak{P}$ be a prime divisor of F/K and $\mathfrak{p} = \mathfrak{P}|_E$. Then, the trace of an element x of $\hat{F}_{\mathfrak{P}}$ to $\hat{E}_{\mathfrak{p}}$ coincides with the trace of x as an adele.*
(e) *There exists $\alpha \in \hat{\mathbb{A}}_F$ such that $\text{trace}_{F/E}(\alpha) \neq 0$.*

Proof: Statements (a) and (b) follow from the corresponding properties of the trace function on fields. Statement (c) follows from Proposition 3.5.3(f). Statement (d) follows from the definition. Finally, use (c) and the corresponding fact for the trace of fields [Lang7, p. 286, Thm. 5.2] to prove (e). □

Remark 3.5.7: Complementary modules.

(a) Let (E, v) be a complete discrete valued field. Denote the valuation ring of (E, v) by O_E. Let F be a finite separable extension of E. Denote the unique extension of v to F by v. Choose a generator π_F of the maximal ideal of O_F. The **complementary module** of O_F over E is

$$O'_{F/E} = \{x \in F \mid \text{trace}_{F/E}(xO_F) \subseteq O_E\}.$$

It is a fractional ideal of O_F which contains O_F [Lang5, p. 58, Cor.]. By Lemma 3.5.5, $O'_{F/E} = \pi_F^{-d_{F/E}} O_F$ for some nonnegative integer $d_{F/E}$. Call $d_{F/E}$ the **different exponent** of F/E. It is known [Lang5, p. 62, Prop. 8] that $d_{F/E} > 0$ if and only if F/E is ramified. Moreover, $d_{F/E} \geq e_{F/E} - 1$. If the residue field extension is separable, then equality holds if and only if F/E is tamely ramified [Serre4, p. 67, Prop. 13].

(b) Let F/E be a finite separable extension of function fields of one variable over a field K. Consider prime divisors $\mathfrak{p}$ and $\mathfrak{P}$ of E/K and F/K, respectively, with $\mathfrak{P}$ lying over $\mathfrak{p}$. Choose completions $\hat{E}_{\mathfrak{p}}$ and $\hat{F}_{\mathfrak{P}}$ with $\hat{E}_{\mathfrak{p}} \subseteq \hat{F}_{\mathfrak{P}}$. Denote the valuation rings of $\hat{E}_{\mathfrak{p}}$ and $\hat{F}_{\mathfrak{P}}$, by $\hat{O}_{\mathfrak{p}}$ and $\hat{O}_{\mathfrak{P}}$, respectively. Occasionally write $d_{\hat{F}_{\mathfrak{P}}/\hat{E}_{\mathfrak{p}}}$ as $d_{\mathfrak{P}/\mathfrak{p}}$ or $d_{\mathfrak{P}/E}$ and call $d_{\mathfrak{P}/\mathfrak{p}}$ the **different exponent** of $\mathfrak{P}$ over $\mathfrak{p}$ (or over E). By (a), $d_{\mathfrak{P}/\mathfrak{p}} > 0$ if and only if $\mathfrak{P}/\mathfrak{p}$ is ramified. This happens for only finitely many $\mathfrak{P}$'s. □

Chapter 3. Algebraic Function Fields of One Variable

LEMMA 3.5.8: *Let $E \subseteq F \subseteq F'$ be function fields of one variable over a field K with F'/E separable. Consider prime divisors $\mathfrak{p}$, $\mathfrak{P}$, $\mathfrak{P}'$ of E/K, F/K, F'/K, respectively, with $\mathfrak{P}$ lying over $\mathfrak{p}$ and $\mathfrak{P}'$ lying over $\mathfrak{P}$. Then $d_{\mathfrak{P}'/\mathfrak{p}} = e_{\mathfrak{P}'/\mathfrak{P}} d_{\mathfrak{P}/\mathfrak{p}} + d_{\mathfrak{P}'/\mathfrak{P}}$. In particular, if $\mathfrak{P}'/\mathfrak{P}$ is unramified, then $d_{\mathfrak{P}'/\mathfrak{p}} = d_{\mathfrak{P}/\mathfrak{p}}$. If $\mathfrak{P}/\mathfrak{p}$ is unramified, then $d_{\mathfrak{P}'/\mathfrak{p}} = d_{\mathfrak{P}'/\mathfrak{P}}$.*

Proof: Suppose the formula holds. If $\mathfrak{P}'/\mathfrak{P}$ is unramified, then $d_{\mathfrak{P}'/\mathfrak{P}} = 0$ and $e_{\mathfrak{P}'/\mathfrak{P}} = 1$. Hence, $d_{\mathfrak{P}'/\mathfrak{p}} = d_{\mathfrak{P}/\mathfrak{p}}$. If $\mathfrak{P}/\mathfrak{p}$ is unramified, then $d_{\mathfrak{P}/\mathfrak{p}} = 0$, so $d_{\mathfrak{P}'/\mathfrak{p}} = d_{\mathfrak{P}'/\mathfrak{P}}$.

To prove the formula, assume without loss, E, F, and F' are complete with respective valuation rings O_E, O_F, and $O_{F'}$. Choose prime elements π_E, π_F, and $\pi_{F'}$ for the respective maximal ideals. Put $d = d_{F/E}$, $d' = d_{F'/F}$, and $e' = e_{F'/F}$. We have to prove that $e'd + d' = d_{F'/E}$.

First note that $\pi_{F'}^{e'} O_{F'} = \pi_F O_{F'}$. Hence,

$$\text{trace}_{F'/E}(\pi_{F'}^{-e'd-d'} O_{F'}) = \text{trace}_{F/E}(\text{trace}_{F'/F}(\pi_F^{-d} \pi_{F'}^{-d'} O_{F'}))$$
$$= \text{trace}_{F/E}(\pi_F^{-d} \text{trace}_{F'/F}(\pi_{F'}^{-d'} O_{F'}))$$
$$\subseteq \text{trace}_{F/E}(\pi_F^{-d} O_F) \subseteq O_E.$$

Thus,

(4) $$\pi_{F'}^{-e'd-d'} O_{F'} \subseteq O'_{F'/E}.$$

By definition, $\pi_F^{-d-1} \notin O'_{F/E}$. Hence, $\text{trace}_{F/E}(\pi_F^{-d-1} O_F) \not\subseteq O_E$. By the second part of Lemma 3.5.5, $\text{trace}_{F/E}(\pi_F^{-d-1} O_F)$ is a fractional ideal of O_E. Hence, by the first part of Lemma 3.5.5, $\text{trace}_{F/E}(\pi_F^{-d-1} O_F) = \pi_E^{-m} O_E$ for some positive integer m. Therefore, $\pi_E^{-1} O_E \subseteq \text{trace}_{F/E}(\pi_F^{-d-1} O_F)$. Similarly, $\pi_F^{-1} O_F \subseteq \text{trace}_{F'/F}(\pi_{F'}^{-d'-1} O_{F'})$.

Assume $\pi_{F'}^{-e'd-d'} O_{F'} \neq O'_{F'/E}$. Then, by (4), $\pi_{F'}^{-e'd-d'-1} O_{F'} \subseteq O'_{F'/E}$. Therefore, by the preceding paragraph,

$$O_E \supseteq \text{trace}_{F'/E}(\pi_{F'}^{-e'd-d'-1} O_{F'})$$
$$= \text{trace}_{F/E}(\pi_F^{-d} \text{trace}_{F'/F}(\pi_{F'}^{-d'-1} O_{F'}))$$
$$\supseteq \text{trace}_{F/E}(\pi_F^{-d-1} O_F) \supseteq \pi_E^{-1} O_E,$$

which is a contradiction. Therefore, $\pi_{F'}^{-e'd-d'} O_{F'} = O'_{F'/E} = \pi_{F'}^{-d_{F'/E}} O_{F'}$. Consequently, $e'd + d' = d_{F'/E}$, as claimed. $\square$

3.6 The Different

The Riemann-Hurwitz genus formula enables us to compute the genus of a function field of one variable F over a field K from the genus of a function subfield E in terms of the 'different' of the extension F/E. This formula is in particular useful when the genus of E is known, e.g. when E is the field of rational functions over K.

Let F/E be a finite separable extension of function fields of one variable over a field K. The **different** of F/E is a divisor of F/K:

$$\text{Diff}(F/E) = \sum d_{\mathfrak{P}/E}\mathfrak{P},$$

where $\mathfrak{P}$ ranges over all prime divisors of F/K and $d_{\mathfrak{P}/E}$ is the different exponents of $\mathfrak{P}/E$. By Remark 3.5.7(a), $d_{\mathfrak{P}/E} \geq 0$ for all $\mathfrak{P}$. Moreover, $d_{\mathfrak{P}/E} > 0$ if and only if $\mathfrak{P}$ is ramified over E. Since only finitely many $\mathfrak{P}$ ramify over E, $\text{Diff}(F/E)$ is well defined, $\text{Diff}(F/E) \geq 0$, and $\deg(\text{Diff}(F/E)) \geq 0$.

THEOREM 3.6.1 (Riemann-Hurwitz Genus Formula): *Let F/E be a finite separable extension of function fields of one variable over a field K. Put $g_E = \text{genus}(E/K)$ and $g_F = \text{genus}(F/K)$. Then*

(1) $\qquad 2g_F - 2 = [F:E](2g_E - 2) + \deg(\text{Diff}(F/E)).$

Proof: Let $\mathfrak{p}_i$, $i \in I$, be the prime divisors of E/K. Choose a nonzero differential ω of E/K. Write $\text{div}(\omega) = \sum_{i \in I} k_i \mathfrak{p}_i$ with $k_i \in \mathbb{Z}$ and $k_i = 0$ for all but finitely many $i \in I$. By definition,
(2) ω vanishes on $\hat{\Lambda}_E(\sum_{i \in I} k_i \mathfrak{p}_i)$ but not on $\hat{\Lambda}_E((k_r + 1)\mathfrak{p}_r + \sum_{i \neq r} k_i \mathfrak{p}_i)$ for any $r \in I$.

Define a map $\Omega \colon \hat{\mathbb{A}}_F \to K$:

$$\Omega(\alpha) = \omega(\text{trace}_{F/E}(\alpha)).$$

It is K-linear and vanishes on F (Proposition 3.5.6).

For each $i \in I$ let $\mathfrak{p}_i = \sum_{j \in J_i} e_{ij} \mathfrak{p}_{ij}$, where the $\mathfrak{p}_{ij}$ are the distinct prime divisors of F/K lying over $\mathfrak{p}_i$ and the e_{ij} are the corresponding ramification indices. We simplify notation and for all $i \in I$ and $j \in J_i$ let $E_i = \hat{E}_{\mathfrak{p}_i}$, $O_i = \hat{O}_{\mathfrak{p}_i}$, $v_i = v_{\mathfrak{p}_i}$, $F_{ij} = \hat{F}_{\mathfrak{p}_{ij}}$, $O_{ij} = \hat{O}_{\mathfrak{p}_{ij}}$, $v_{ij} = v_{\mathfrak{p}_{ij}}$, and

(3) $\qquad d_{ij} = v_{ij}(\text{Diff}(F/E)) = -v_{ij}(O'_{ij}),$

where O'_{ij} is the complementary module of O_{ij} with respect to $\text{trace}_{F/E}$ (Remark 3.5.7). Also, for all $\alpha \in \hat{\mathbb{A}}_E$ and $\beta \in \hat{\mathbb{A}}_F$ let $\alpha_i = \alpha_{\mathfrak{p}_i}$ and $\beta_{ij} = \beta_{\mathfrak{p}_{ij}}$. We prove that
(4a) Ω vanishes on $\hat{\Lambda}_F(\sum_{i,j}(e_{ij}k_i + d_{ij})\mathfrak{p}_{ij})$ and
(4b) for each $r \in I$ and each $s \in J_r$ the differential Ω does not vanish on $\hat{\Lambda}_F((e_{rs}k_r + d_{rs} + 1)\mathfrak{p}_{rs} + \sum_{(i,j) \neq (r,s)}(e_{ij}k_i + d_{ij})\mathfrak{p}_{ij}).$

This will imply that

$$\mathrm{div}_F(\Omega) = \sum_{i,j}(e_{ij}k_i + d_{ij})\mathfrak{p}_{ij} = \mathrm{div}_E(\omega) + \mathrm{Diff}(F/E).$$

The formulas $\deg(\mathrm{div}_F(\Omega)) = 2g_F - 2$ and $\deg(\mathrm{div}_E(\omega)) = [F:E](2g_E - 2)$ (Lemmas 3.2.2(b) and 3.4.1) will give (1).

For all i,j choose $\pi_i \in E$ and $\pi_{ij} \in F$ with $v_i(\pi_i) = 1$ and $v_{ij}(\pi_{ij}) = 1$.

PROOF OF (4a): Consider $\alpha \in \hat{\Lambda}_F(\sum_{ij}(e_{ij}k_i + d_{ij})\mathfrak{p}_{ij})$. Then $v_{ij}(\alpha) \geq -e_{ij}k_i - d_{ij}$. Hence, $v_{ij}(\pi_i^{k_i}\alpha) \geq -d_{ij}$. By (3), $(\pi_i^{k_i}\alpha)_{ij} \in O'_{ij}$, so $\mathrm{trace}_{F_{ij}/E_i}(\pi_i^{k_i}\alpha)_{ij} \in O_i$. Thus,

$$(\mathrm{trace}_{F/E}(\pi_i^{k_i}\alpha))_i = \sum_{j \in J_i} \mathrm{trace}_{F_{ij}/E_i}(\pi_i^{k_i}\alpha) \in O_i,$$

hence $v_i(\mathrm{trace}_{F/E}(\alpha)) + k_i \geq 0$ for each $i \in I$. Therefore, $\mathrm{trace}_{F/E}(\alpha) \in \hat{\Lambda}_E(\sum_{i\in I} k_i\mathfrak{p}_i)$. Consequently, by (2), $\Omega(\alpha) = \omega(\mathrm{trace}_{F/E}(\alpha)) = 0$.

PROOF OF (4b): Assume there exist $r \in I$ and $s \in J_r$ such that Ω vanishes on

$$V = \hat{\Lambda}_F\big((e_{rs}k_r + d_{rs} + 1)\mathfrak{p}_{rs} + \sum_{(i,j)\neq(r,s)}(e_{ij}k_i + d_{ij})\mathfrak{p}_{ij}\big).$$

By Lemma 3.5.5, there is an $m \in \mathbb{Z}$ with

$$I_{rs} = \mathrm{trace}_{F_{rs}/E_r}(\pi_r^{-k_r}\pi_{rs}^{-d_{rs}-1}O_{rs}) = \pi_r^{-m}O_r$$

We distinguish between two cases:

CASE A: $m \geq k_r + 1$. Consider $\beta \in \hat{\Lambda}_E((k_r + 1)\mathfrak{p}_r + \sum_{i\neq r} k_i\mathfrak{p}_i)$. Then $v_r(\beta) \geq -k_r - 1$. If $v_r(\beta) \geq -k_r$, then $\beta \in \hat{\Lambda}_E(\sum_{i\in I} k_i\mathfrak{p}_i)$, so $\omega(\beta) = 0$ (by (2)). Otherwise, $v_r(\beta) = -k_r - 1$. In this case write $\beta = \alpha + \gamma$, where $\alpha, \gamma \in \hat{\mathbb{A}}_E$ satisfy $\alpha_r = (1 - \pi_r)\beta_r$, $\gamma_r = \pi_r\beta_r$, $\alpha_i = 0$, and $\gamma_i = \beta_i$ for $i \neq r$. Then, $\alpha_r \in \pi_r^{-k_r-1}O_r \subseteq \pi_r^{-m}O_r = I_{rs}$, so there is a $\delta_{rs} \in \pi_r^{-k_r}\pi_{rs}^{-d_{rs}-1}O_{rs}$ with $\mathrm{trace}_{F_{rs}/E_r}\delta_{rs} = \alpha_r$. For $(i,j) \neq (r,s)$ let $\delta_{ij} = 0$. Then $\delta \in V$, $\alpha = \mathrm{trace}_{F/E}\delta$, and $\omega(\alpha) = \omega(\mathrm{trace}(\delta)) = \Omega(\delta) = 0$. Also, $\gamma \in \hat{\Lambda}_E(\sum_{i\in I} k_i\mathfrak{p}_i)$, so $\omega(\gamma) = 0$ (by (2)). It follows that $\omega(\beta) = 0$. Thus, ω vanishes on $\hat{\Lambda}_E((k_r + 1)\mathfrak{p}_r + \sum_{i\neq r} k_i\mathfrak{p}_i)$, in contradiction to (2).

CASE B: $m \leq k_r$. Then,

$$\mathrm{trace}_{F_{rs}/E_r}(\pi_r^{-k_r}\pi_{rs}^{-d_{rs}-1}O_{rs}) = \pi_r^{-m}O_r \subseteq \pi_r^{-k_r}O_r.$$

Hence, $\mathrm{trace}_{F_{rs}/E_r}(\pi_{rs}^{-d_{rs}-1}O_{rs}) \subseteq O_r$. Therefore, $\pi_{rs}^{-d_{rs}-1}O_{rs} \subseteq O'_{rs}$, so $v_{rs}(O'_{rs}) \leq -d_{rs} - 1$. This contradiction to (3) completes the proof of Case B and the proof of the whole theorem. $\square$

Remark 3.6.2: *Applications of the Riemann-Hurwitz formula.* Let F/E be a finite separable extension of function fields of one variable of a field K. We say that F/E is **unramified** (resp. **tamely ramified**) if each prime divisor of F/K is unramified (resp. tamely ramified) over E. Let $g_E = \text{genus}(E/K)$ and $g_F = \text{genus}(F/K)$. Suppose $[F:E] \geq 2$.

(a) Comparison of genera: We have mentioned at the beginning of this section that $\deg(\text{Div}(F/E)) \geq 0$. Hence, by (1), $g_F \geq g_E$. Both g_E and g_F have the same value if and only if $g_E = 1$ and F/E is unramified, or $g_E = 0$ and $\deg(\text{Diff}(F/E)) = 2([F:E] - 1)$.

In particular, if $F = K(t)$, then $g_F = 0$ (Example 3.2.4). Hence, $g_E = 0$. Each prime divisor of F/K of degree 1 induces a prime divisor of E/K of degree 1. We conclude from Example 3.2.4 that $E = K(u)$ is also a rational function field. This is **Lüroth's theorem**. This theorem actually holds for arbitrary algebraic extension $K(t)/E$ and not only for separable extensions and may be proved by elementary arguments on polynomials [Waerden1, p. 218].

(b) An analog of a theorem of Minkowski: Suppose F/E is unramified. In this case the Riemann-Hurwitz formula simplifies to

$$g_F - 1 = [F:E](g_E - 1).$$

Hence, $g_E > 0$. In other words, a function field E/K of genus 0 has no proper finite unramified extension F which is regular over K. In particular, $K(t)$ has no finite proper separable unramified extension F which is regular over K. This is an analog of a theorem of Minkowski saying that $\mathbb{Q}$ has no proper unramified extensions [Janusz, p. 57, Cor. 11.11].

(c) The Hurwitz-Riemann formula for tamely ramified extensions: Suppose F/E is tamely ramified. By Remark 3.5.7, the Riemann-Hurwitz formula simplifies to

$$2g_F - 2 = [F:E](2g_E - 2) + \sum_{\mathfrak{p}} \sum_{\mathfrak{P}|\mathfrak{p}} (e_{\mathfrak{P}/\mathfrak{p}} - 1) \deg(\mathfrak{P}).$$

(d) An analog of Minkowski's theorem in the tamely ramified case: Suppose K is algebraically closed and $g_E = 0$. Then $E = K(t)$ and the degree of each prime divisor is 1. Suppose F/E is a proper tamely ramified extension. Then E has at least two prime divisors that ramify in F.

Indeed, assume E has only one prime divisor $\mathfrak{p}$ that ramifies in F. Let $\mathfrak{P}_1, \ldots, \mathfrak{P}_r$ be the prime divisors of $\mathfrak{p}$ in F. Then

$$-2 \leq 2g_F - 2 = -2[F:E] + \sum_{i=1}^{r} (e_{\mathfrak{P}_i/\mathfrak{p}} - 1)$$
$$= -2[F:E] + [F:E] - r = -[F:E] - r$$

Hence, $3 \leq [F:E] + r \leq 2$, a contradiction.

In particular, if K is algebraically closed and $\operatorname{char}(K) = 0$, then every proper extension of $K(t)$ is ramified over at least two prime divisors.

(e) Generation of Galois groups by inertia groups: Let K be an algebraically closed field, $E = K(t)$, and F a finite Galois extension of E. Denote the prime divisors of F/K which ramify over E by $\mathfrak{P}_1, \ldots, \mathfrak{P}_r$. For each i let D_i be the decomposition group of $\mathfrak{P}_i$ over E. Since K is algebraically closed, D_i is also the inertia group of $\mathfrak{P}_i$ over E. Let E_i be the fixed field of D_i in F. Then $\mathfrak{P}_i|_{E_i}$ is unramified over E. Hence, $E_0 = E_1 \cap \cdots \cap E_r$ is unramified over E. By (b), $E_0 = E$. In other words, the inertia groups of the prime divisors of F/K which ramify over E generate $\operatorname{Gal}(F/E)$.

(f) Quasi-p groups: Let K be an algebraically closed field of positive characteristic p. Consider the rational field $E = K(t)$, a prime divisor $\mathfrak{p}$ of E/K, and a finite Galois extension F of E which is ramified only over $\mathfrak{p}$. Denote the fixed field in F of all p-Sylow subgroups of $\operatorname{Gal}(F/E)$ by E_p. Then E_p/E is a Galois extension of degree relatively prime to p. Hence, E_p/E is tamely ramified. The only prime divisor of E/K which is possibly ramified in E_p is $\mathfrak{p}$. It follows from (d) that $E_p = E$. In other words, $\operatorname{Gal}(F/E)$ is generated by its p-Sylow subgroups. One says that $\operatorname{Gal}(F/E)$ is **quasi-p**.

(g) Abhyankar's conjecture: Let K be an algebraically closed field of positive characteristic p. Let $E = K(t)$ and $\mathfrak{p}$ a prime divisor of E/K. In 1957, Abhyankar conjectured that for each finite quasi-p group G there exists a finite Galois extension F of E which is unramified outside $\mathfrak{p}$ and $\operatorname{Gal}(F/E) \cong G$ [Abhyankar]. Serre proved the conjecture for solvable G in 1990 [Serre9]. Raynaud treated all other cases in 1991 [Raynaud].

(h) The generalized Abhyankar's conjecture: Let again K be an algebraically closed field of positive characteristic p. Put $E = K(t)$. Consider a set $S = \{\mathfrak{p}_1, \ldots, \mathfrak{p}_r\}$ of prime divisors of E/K and a finite Galois extension F of E which is unramified over E outside S. Let $G = \operatorname{Gal}(F/E)$. Denote the subgroup of G generated by all p-Sylow subgroups of G by $G(p)$. Let E_p be the fixed field of $G(p)$ in F. Then E_p is a Galois extension of E which is tamely ramified over E and unramified outside S. By [Grothendieck, XIII, Cor. 2.12], $\operatorname{Gal}(E_p/E)$ is generated by $r - 1$ elements.

This led Abhyankar to conjecture that for every finite group G such that $G/G(p)$ is generated by $r - 1$ elements there is a Galois extension F of E which is unramified outside S and $\operatorname{Gal}(F/E) \cong G$. Harbater proved this conjecture in [Harbater2] by reducing it to the special case $r = 1$ proved by Raynaud. □

3.7 Hyperelliptic Fields

We demonstrate the concepts and results of this chapter to a study of a special kind of algebraic function fields of one variable which we now introduce. A function field F/K is **hyperelliptic** if its genus is at least 2 and if it is a quadratic extension of a function field E/K of genus 0. We then say E is a **quadratic subfield** of F. It turns out that E is uniquely determined.

3.7 Hyperelliptic Fields

Moreover, we will be able to identify E from the arithmetic of F.

LEMMA 3.7.1: *Let F/K be a function field of genus g. Let $\mathfrak{a}$, $\mathfrak{b}$ be nonnegative divisors of F/K and let $\mathfrak{w}$ be a canonical divisor of F/K.*
(a) *If $\mathcal{L}(\mathfrak{b}-\mathfrak{a}) = \mathcal{L}(\mathfrak{b})$, and $\dim(\mathfrak{b}) \geq 1$, then $\dim(\mathfrak{a}) = 1$.*
(b) *If $g \geq 1$ and $\mathfrak{a} > 0$, then $\dim(\mathfrak{w}-\mathfrak{a}) < \dim(\mathfrak{w})$.*
(c) *If $g \geq 1$ and $x_1, \ldots, x_g$ is a basis for $\mathcal{L}(\mathfrak{w})$, then*

$$-\mathfrak{w} = \min\bigl(\operatorname{div}(x_1), \ldots, \operatorname{div}(x_g)\bigr).$$

Proof of (a): Since $\mathfrak{a} \geq 0$, we have $K \subseteq \mathcal{L}(\mathfrak{a})$. Conversely, let $x \in \mathcal{L}(\mathfrak{a})$. Then $\operatorname{div}(x) + \mathfrak{a} \geq 0$. Consider $y \in \mathcal{L}(\mathfrak{b})$. By assumption, $y \in \mathcal{L}(\mathfrak{b}-\mathfrak{a})$, hence $\operatorname{div}(y) + \mathfrak{b} - \mathfrak{a} \geq 0$. Therefore, $\operatorname{div}(xy) + \mathfrak{b} \geq 0$, so $xy \in \mathcal{L}(\mathfrak{b})$.

Apply this result to a basis $y_1, \ldots, y_n$ of $\mathcal{L}(\mathfrak{b})$. Find $a_{ij} \in K$ such that $xy_i = \sum_{j=1}^n a_{ij} y_j$, $i = 1, \ldots, n$. Hence, $\det(xI - A) = 0$, where $A = (a_{ij})_{1 \leq i,j \leq n}$. Therefore, x satisfies a monic equation with coefficients in K. Since K is algebraically closed in F, we have $x \in K$. Consequently, $\dim(\mathfrak{a}) = 1$.

Proof of (b): Assume $\dim(\mathfrak{w}-\mathfrak{a}) = \dim(\mathfrak{w})$. Then $\mathcal{L}(\mathfrak{w}-\mathfrak{a}) = \mathcal{L}(\mathfrak{w})$. By Lemma 3.2.2(b), $\dim(\mathfrak{w}-\mathfrak{a}) = \dim(\mathfrak{w}) = g$. So, by (a), $\dim(\mathfrak{a}) = 1$. By Theorem 3.2.1, $\dim(\mathfrak{a}) = \deg(\mathfrak{a}) + 1 - g + \dim(\mathfrak{w}-\mathfrak{a})$. Hence, $\deg(\mathfrak{a}) = 0$, which is a contradiction. Therefore, $\dim(\mathfrak{w}-\mathfrak{a}) < \dim(\mathfrak{w})$.

Proof of (c): Denote $\min\bigl(\operatorname{div}(x_1), \ldots, \operatorname{div}(x_g)\bigr)$ by $\mathfrak{m}$. Since $\operatorname{div}(x_i) + \mathfrak{w} \geq 0$ we have $\mathfrak{m} \geq -\mathfrak{w}$.

If $\mathfrak{m} > -\mathfrak{w}$, then there exists a prime divisor $\mathfrak{p}$ of F/K with $\mathfrak{m}-\mathfrak{p} \geq -\mathfrak{w}$. Hence, $\operatorname{div}(x_i) + \mathfrak{w} - \mathfrak{p} \geq 0$, so $x_i \in \mathcal{L}(\mathfrak{w} - \mathfrak{p})$ for $i = 1, \ldots, g$. Therefore, $\mathcal{L}(\mathfrak{w} - \mathfrak{p}) = \mathcal{L}(\mathfrak{w})$, contradicting (b). Consequently, $\mathfrak{m} = -\mathfrak{w}$. □

PROPOSITION 3.7.2: *Let F/K be a function field of genus $g \geq 2$. Consider a canonical divisor $\mathfrak{w}$ of F/K. Let $x_1, \ldots, x_g$ be a basis of $\mathcal{L}(\mathfrak{w})$. If $E = K\bigl(\frac{x_2}{x_1}, \ldots, \frac{x_g}{x_1}\bigr)$ is a proper subfield of F, then $\operatorname{genus}(E/K) = 0$ and $[F : E] = 2$. Thus, F/K is a hyperelliptic field.*

Proof: Let $\mathfrak{w}' = \operatorname{div}(x_1) + \mathfrak{w}$. Then $\mathfrak{w}'$ is also a canonical divisor and $1, \frac{x_2}{x_1}, \ldots, \frac{x_g}{x_1}$ is a basis of $\mathcal{L}(\mathfrak{w}')$. Replace $\mathfrak{w}$ by $\mathfrak{w}'$, if necessary, to assume that $x_1 = 1$ and $E = K(x_2, \ldots, x_g)$.

Since $g \geq 2$, the element x_2 is transcendental over K. Hence, E is also an algebraic function field of one variable over K and $d = [F : E] < \infty$ (Section 3.4). Denote $\operatorname{genus}(E/K)$ by g_E and let $\mathfrak{w}_E$ be a canonical divisor of E/K. By Lemma 3.7.1(c), $\mathfrak{w} = -\min\bigl(0, \operatorname{div}(x_2), \ldots, \operatorname{div}(x_g)\bigr)$. In particular, $\mathfrak{w} \geq 0$. By Lemma 3.2.2(b), $\deg(\mathfrak{w}) = 2g - 2 \geq 2$. Therefore,

(1) $\qquad\qquad\qquad\qquad \mathfrak{w} > 0.$

Observe that $\operatorname{div}(x_i) \in \operatorname{Div}(E/K)$, $i = 2, \ldots, g$, so $\mathfrak{w} \in \operatorname{Div}(E/K)$. We may therefore apply Riemann-Roch to E/K and $\mathfrak{w}$:

(2) $\qquad \dim\bigl(\mathcal{L}_E(\mathfrak{w})\bigr) = \deg_E(\mathfrak{w}) + 1 - g_E + \dim\bigl(\mathcal{L}_E(\mathfrak{w}_E - \mathfrak{w})\bigr).$

Since $1, x_1, \ldots, x_g$ are in E and generate $\mathcal{L}_F(\mathfrak{w})$, we have $\mathcal{L}_E(\mathfrak{w}) = \mathcal{L}_F(\mathfrak{w})$. Hence, by Lemma 3.2.2(b), $\dim(\mathcal{L}_E(\mathfrak{w})) = g$. Applying Lemma 3.2.2(b) again and Lemma 3.4.1, we have $2g-2 = \deg_F(\mathfrak{w}) = d \cdot \deg_E(\mathfrak{w})$. Substituting this in (2) gives

$$(3) \qquad g = \frac{2g-2}{d} + 1 - g_E + \dim\big(\mathcal{L}_E(\mathfrak{w}_E - \mathfrak{w})\big),$$

which may be rewritten as

$$(4) \qquad d\big(\dim\big(\mathcal{L}_E(\mathfrak{w}_E - \mathfrak{w})\big) - g_E\big) = (g-1)(d-2)$$

By assumption, both g and d are at least 2. Hence, the right hand side of (4) is at least 0. On the other hand, by Lemma 3.2.2(b) and by (1),

$$(5) \qquad \dim\big(\mathcal{L}_E(\mathfrak{w}_E - \mathfrak{w})\big) \leq \dim\big(\mathcal{L}_E(\mathfrak{w}_E)\big) = g_E.$$

Hence, both sides of (4) are 0. Since $g \geq 2$, this gives $d = 2$.

Finally, by (3), $g_E = \dim(\mathcal{L}_E(\mathfrak{w}_E - \mathfrak{w}))$. Hence, by (5), $\dim(\mathcal{L}_E(\mathfrak{w}_E - \mathfrak{w})) = \dim(\mathcal{L}_E(\mathfrak{w}_E))$. We conclude from Lemma 3.7.1(b) that $g_E = 0$. $\square$

LEMMA 3.7.3: *Let F/K be a function field of genus $g \geq 1$, $\mathfrak{w}$ a canonical divisor of F/K, and $x_1, \ldots, x_g$ a basis of $\mathcal{L}_F(\mathfrak{w})$. Let E/K be a subfield of genus 0 of F/K. Then $\sum_{i=1}^g Ex_i \neq F$.*

Proof: For each i we have $\hat{\Lambda}_E(0)x_i \subseteq \hat{\Lambda}_F(\mathfrak{w})$. Hence,

$$(6) \qquad \sum_{i=1}^g \big(\hat{\Lambda}_E(0) + E\big)x_i \subseteq \hat{\Lambda}_F(\mathfrak{w}) + F.$$

Choose a canonical divisor $\mathfrak{w}_E$ of E/K. By (2) of Section 3.5,

$$\dim_K\big(\hat{\mathbb{A}}_E/(\hat{\Lambda}_E(0) + E)\big) = \dim(\mathfrak{w}_E) = \text{genus}(E/K) = 0.$$

Hence,

$$(7) \qquad \hat{\mathbb{A}}_E = \hat{\Lambda}_E(0) + E.$$

On the other hand,

$$\dim_K\big(\hat{\mathbb{A}}_F/(\hat{\Lambda}_F(\mathfrak{w}) + F)\big) = \dim(0) = 1.$$

Hence, $\hat{\Lambda}_F(\mathfrak{w}) + F \subset \hat{\mathbb{A}}_F$, so by (6) and (7)

$$(8) \qquad \sum_{i=1}^g \hat{\mathbb{A}}_E x_i \subset \hat{\mathbb{A}}_F.$$

If $\sum_{i=1}^g Ex_i = F$, then by (3) of Section 3.5, $\hat{\mathbb{A}}_F = \sum_{i=1}^g \hat{\mathbb{A}}_E x_i$, contradicting (8). Consequently, $\sum_{i=1}^g Ex_i \neq F$. $\square$

3.8 Hyperelliptic Fields with a Rational Quadratic Subfield

PROPOSITION 3.7.4: *Let F/K be a hyperelliptic function field, $\mathfrak{w}$ a canonical divisor of F/K, and $x_1,\ldots,x_g$ a basis of $\mathcal{L}_F(\mathfrak{w})$. Then $E = K\left(\frac{x_2}{x_1},\ldots,\frac{x_g}{x_1}\right)$ is the only quadratic subfield of F.*

Proof: By definition, F has quadratic subfields of genus 0. Let E' be one of them. By Lemma 3.7.3, $\sum_{i=1}^{g} E'x_i \neq F$. Since $[F : E'] = 2$, this implies $x_i \in E'$, $i = 1,\ldots,g$. It follows that $E \subseteq E'$. In particular, $E \subset F$. By Proposition 3.7.2, $[F : E] = 2$. Comparing degrees implies $E' = E$. □

3.8 Hyperelliptic Fields with a Rational Quadratic Subfield

A hyperelliptic field with a rational quadratic subfield is generated by two generators which satisfy an equation of a special type. This situation arises, for example, when the hyperelliptic field has a prime divisor of degree 1 (Example 3.2.4).

PROPOSITION 3.8.1: *Let F/K be a hyperelliptic field of genus g. Suppose the quadratic subfield of F is $K(x)$ with x indeterminate. Then $F = K(x,y)$, where y satisfies a relation $y^2 + h_1(x)y + h_2(x) = 0$ with $h_1, h_2 \in K[X]$, $\deg(h_1) \leq g+1$, and $\deg(h_2) \leq 2g+2$. If $\mathrm{char}(K) \neq 2$, we may choose y with $y^2 = f(x)$, where $f \in K[X]$ is a polynomial with no multiple root and $\deg(f(x)) \leq 2g+2$.*

Proof: Let $\mathfrak{p}_\infty$ be the pole of x in $K(x)/K$. By Lemma 3.4.1, $\deg_F(n\mathfrak{p}_\infty) = 2n$ for each positive integer n. If $n \geq g$, then $2n > 2g-2$. By Lemma 3.2.2(d), $\dim(\mathcal{L}_F(n\mathfrak{p}_\infty)) = 2n + 1 - g$. In particular, for $n = g$ the elements $1, x, \ldots, x^g$ form a basis for $\mathcal{L}_F(g\mathfrak{p}_\infty)$.

For $n = g+1$ we have $\dim\bigl(\mathcal{L}_F((g+1)\mathfrak{p}_\infty)\bigr) = g+3$. Hence, two more elements are needed to complete $1, x, \ldots, x^g$ to a basis of $\mathcal{L}_F((g+1)\mathfrak{p}_\infty)$. We take one of them as x^{g+1} and denote the other one by y. By Riemann-Roch, $\dim(\mathcal{L}_{K(x)}((g+1)\mathfrak{p}_\infty)) = g+2$. So, $1, x, \ldots, x^{g+1}$ form a basis of $\mathcal{L}_{K(x)}((g+1)\mathfrak{p}_\infty)$. If y is in $K(x)$, then y is also in $\mathcal{L}_{K(x)}((g+1)\mathfrak{p}_\infty)$. This implies that $1, x, \ldots, x^{g+1}, y$ are linearly dependent over K. We conclude from this contradiction that $y \notin K(x)$, so $F = K(x,y)$.

Next note that $\dim\bigl(\mathcal{L}_F((2g+2)\mathfrak{p}_\infty)\bigr) = 3g+5$. All $3g+6$ elements $1, x, \ldots, x^{2g+2}, y, yx, \ldots, yx^{g+1}, y^2$ belong to $\mathcal{L}_F((2g+2)\mathfrak{p}_\infty)$. Therefore, there are $a_i, b_j, c \in K$, not all zero, with $\sum_{i=0}^{2g+2} a_i x^i + \sum_{j=0}^{g+1} b_j y x^j + cy^2 = 0$. Since $y \notin K(x)$, we have $c \neq 0$. Let $h_1(X) = \sum_{j=0}^{g+1} c^{-1} b_j X^j$, $h_2(X) = \sum_{i=0}^{2g+2} c^{-1} a_i X^i$. Then

(1) $$y^2 + h_1(x)y + h_2(x) = 0.$$

If $\mathrm{char}(K) \neq 2$, replace y by $2y + h_1(x)$, if necessary, to assume (1) has the form $y^2 = f(x)$. Here $f \in K[X]$ has degree at most $2g+2$. Finally, if f has multiple roots, then $f = g^2 h$ with h having no multiple roots. Replace y by $yg(x)^{-1}$ to assume f has no multiple roots. □

Our first task is to compute the genus of the hyperelliptic field in characteristic $\neq 2$ from $\deg(f)$.

PROPOSITION 3.8.2: *Let K be a field of characteristic $\neq 2$ and $f \in K[X]$ a polynomial of degree $d \geq 1$ with no multiple roots. Let $F = K(x,y)$, with x transcendental over K and $y^2 = f(x)$. Then F/K is an algebraic function field of one variable of genus $\frac{d-1}{2}$ if d is odd and of genus $\frac{d-2}{2}$ if d is even. In particular, if $d \geq 5$, then F/K is hyperelliptic.*

Proof: Since $f(x)$ is not a square in $\tilde{K}(x)$,

$$[K(x,y) : K(x)] = [\tilde{K}(x,y) : \tilde{K}(x)] = 2.$$

Hence, $K(x,y)$ is linearly disjoint from $\tilde{K}(x)$ over $K(x)$. Since x is transcendental over K, $K(x)$ is linearly disjoint form $\tilde{K}$ over K. Hence, by the tower property of linear disjointness (Lemma 2.5.3), F is linearly disjoint from $\tilde{K}$ over K. It follows that F/K is an algebraic function field of one variable.

Since $\text{char}(K) \neq 2$, $F/K(x)$ is tamely ramified. Since the genus of $K(x)$ is 0, the Riemann-Hurwitz formula reduces to

$$(2) \qquad 2g - 2 = -4 + \sum_{\mathfrak{p}} \sum_{\mathfrak{P}|\mathfrak{p}} (e_{\mathfrak{P}/\mathfrak{p}} - 1) \deg(\mathfrak{P})$$

(Remark 3.6.2(c)). If $\mathfrak{p}$ is a prime divisor of $K(x)$ which ramifies in F, then $\mathfrak{p}$ has only one extension $\mathfrak{P}$ to F, the ramification index of $\mathfrak{P}$ is 2, and its residue degree is 1. Hence, (2) simplifies to

$$(3) \qquad 2g = -2 + \sum \deg(\mathfrak{p}),$$

where $\mathfrak{p}$ ranges over all prime divisors of $K(x)/K$ which ramify in F.

Let $f(x) = p_1(x) \cdots p_r(x)$ be the decomposition of $f(x)$ into a product of distinct irreducible polynomials in $K[x]$. To each p_i there corresponds a prime divisor $\mathfrak{p}_i$ of $K(x)/K$ of degree $\deg(p_i)$ and a valuation v_i such that $v_i(f(x)) = v_i(p_i(x)) = 1$ (Example 2.2.1(b)). By Example 2.3.8, each $\mathfrak{p}_i$ ramifies in F. In addition, let v_∞ be the valuation of $K(x)/K$ with $v_\infty(x) = -1$ and let $\mathfrak{p}_\infty$ be the corresponding prime divisor; its degree is 1. Since $v_\infty(f(x)) = -d$, the prime divisor $\mathfrak{p}_\infty$ is ramified in F if d is odd and unramified if d is even (Example 2.3.8). All other prime divisors of $K(x)$ are unramified in F. The sum of the degrees of the ramified prime divisors is $\delta + \sum_{i=1}^r \deg(p_i) = \delta + d$, where $\delta = 1$ if d is odd and $\delta = 0$ if d is even. It follows from (3) that $g = \frac{d-1}{2}$ if d is odd and $g = \frac{d-2}{2}$ if d is even. □

In characteristic 2 we compute the genus of only a special type of a hyperelliptic field.

Exercises

LEMMA 3.8.3: *Let (K,v) be a valued field of characteristic 2 and t an element of K with $v(t) = 1$. Consider an Artin-Schreier extension $L = K(x)$ with $x^2 + x = \frac{1}{t}$. Denote the unique extension of v to L by w. Let $y = tx$. Then $w(y) = 1$, $O_w = O_v[y]$ is the integral closure of O_v in L, and $O'_w = y^{-2}O_w$.*

Proof: By Example 2.3.11, v has a unique extension w to L. It totally ramifies over K and satisfies $w(x) = -1$, $w(t) = 2$, and $w(y) = 1$. In particular, O_w is the integral closure of O_v in L (Proposition 2.4.1). Since $\bar{K}_v = \bar{L}_w$, each $z \in O_w$ can be written as $z = a + by$ with $a \in O_v$ and $b \in O_w$. Hence, $O_w = O_v[y]$ [Lang5, p. 26, Prop. 23].

To compute $O'_w = \{z \in L \mid \text{trace}_{L/K}(zO_w) \subseteq O_v\}$ observe that $g(X) = \text{irr}(y, K) = Y^2 + tY + t$ and $g'(X) = t$. Therefore, $O'_w = t^{-1}O_w = y^{-2}O_w$ [Lang5, p. 59, Cor.]. Of course, one may also compute O'_w directly. □

PROPOSITION 3.8.4: *Let K be a field of characteristic 2 and $a_1, \ldots, a_d$ distinct elements of K. Let $F = K(x,y)$ with x transcendental over K, $h \in K[X]$ with $\deg(h) \leq d$, and $y^2 + y = \frac{h(x)}{(x-a_1)\cdots(x-a_d)}$. Then F/K is an algebraic function field of one variable of genus $d - 1$. In particular, if $d \geq 3$, then F/K is hyperelliptic.*

Proof: For each i between 1 and d let v_i be the discrete normalized valuation of $K(x)/K$ with $v_i(x - a_i) = 1$. Let $t = (x - a_1) \cdots (x - a_d)$. Then $v_i(t) = 1$. Extend v_i to a valuation $\tilde{v}_i$ of $\tilde{K}(x)/\tilde{K}$. If y is in $\tilde{K}(x)$, then $2\tilde{v}_i(y) = -1$, which is a contradiction. Hence, y is not in $\tilde{K}(x)$, so $[K(x,y) : K(x)] = [\tilde{K}(x,y) : \tilde{K}(x)] = 2$. It follows that F is linearly disjoint from $\tilde{K}$ over K, so F/K is an algebraic function field of one variable.

By Example 2.3.9, $v_1, \ldots, v_d$ are the only valuations of $K(x)/K$ that ramify in F. Hence, only $v_1, \ldots, v_d$ contribute to the different of $F/K(x)$. Let w_i be the unique extension of v_i to F. Then $\hat{F}_{w_i} = \widehat{K(x)}_{v_i}(y)$. By Lemma 3.8.3, the contribution of w_i to $\deg(\text{Diff}(F/K(x)))$ is 2. Hence, $\deg(\text{Diff}(F/K(x))) = 2d$. Let $g = \text{genus}(F/K)$. By Riemann-Hurwitz (Theorem 3.6.1), $2g - 2 = -4 + 2d$. Hence, $g = d - 1$, as claimed. □

Exercises

In the following exercises F is a function field of one variable over a field K and t is a transcendental element over K.

1. Let $\mathfrak{a}$ be a divisor of F/K. Note: If $\mathfrak{a} \geq 0$ and $\deg(\mathfrak{a}) = 0$, then $\mathfrak{a} = 0$.
   (a) Prove that if $\deg(\mathfrak{a}) = 0$ and $\mathfrak{a}$ is not a principal divisor, then $\dim(\mathfrak{a}) = 0$.
   (b) Let g be the genus of F. Suppose $\mathfrak{a}$ is a noncanonical divisor with $\deg(\mathfrak{a}) = 2g - 2$. Show that $\dim(\mathfrak{a}) = g - 1$. Hint: Use Riemann-Roch and (a).

2. Let $\mathfrak{a}$ and $\mathfrak{b}$ be divisors of F/K with $\mathfrak{b} \geq 0$. Use Riemann-Roch to prove that
   (a) $\dim(\mathfrak{a}) \leq \dim(\mathfrak{a} + \mathfrak{b}) \leq \dim(\mathfrak{a}) + \deg(\mathfrak{b})$; and

(b) $\dim(\mathfrak{a}) \leq \max(0, \deg(\mathfrak{a}) + 1)$.

Hint: Write $\mathfrak{a}$ as the difference of its "positive" and "negative" parts.

3. Let $\mathfrak{p}_1, \ldots, \mathfrak{p}_n$ be distinct prime divisors of F/K and let $m_1, \ldots, m_n$ be positive integers such that $\sum_{i=1}^n m_i \deg(\mathfrak{p}_i) > 2g - 1$. If K is infinite, prove that there exists $x \in F^\times$ such that $\mathrm{div}_\infty(x) = \sum_{i=1}^n m_i \mathfrak{p}_i$. In particular, $[F : K(x)] = \sum_{i=1}^n m_i \deg(\mathfrak{p}_i)$. Hint: Consider $\mathfrak{a} = \sum_{i=1}^n m_i \mathfrak{p}_i$ and $\mathfrak{a}_j = \mathfrak{a} - \mathfrak{p}_j$ for $j = 1, \ldots, n$. For each j, $1 \leq j \leq n$, distinguish between the cases $\deg(\mathfrak{a}_j) \leq 2g - 2$ and $\deg(\mathfrak{a}_j) > 2g - 2$, and prove that $\mathcal{L}(\mathfrak{a}_j) \subset \mathcal{L}(\mathfrak{a})$.

4. Suppose $F = K(t)$. Prove, in the notation of Example 3.2.4, that $-2\mathfrak{p}_\infty$ is a canonical divisor of F/K.

5. Suppose the genus of F/K is 0. Prove that every divisor $\mathfrak{a}$ with $\deg(\mathfrak{a}) = 0$ is principal. Hint: Compute $\dim(\mathfrak{a})$ and apply Exercise 1(a).

6. In the notation of Proposition 3.3.2 let $\mathfrak{p}_1, \ldots, \mathfrak{p}_r \in S$, $k_1, \ldots, k_r \in \mathbb{N}$, and P_i be the center of $\mathfrak{p}_i$ at O_S, $i = 1, \ldots, r$. Consider the ideal $A = \prod_{i=1}^n P_i^{k_i}$ of O_S. Prove that the divisor $\mathfrak{a} = \sum_{i=1}^r k_i \mathfrak{p}_i$ of F/K satisfies $\deg(\mathfrak{a}) = (O_S : A)$. Hint: Prove that $O_S/A \cong \prod_{i=1}^r O_S/P_i^{k_i}$ and that $O_S/P_i \cong P_i^k/P_i^{k+1}$, as groups, for each nonnegative integer k.

7. Prove that every algebraic function field F/K of genus 2 is hyperelliptic with a rational quadratic subfield. Hint: Choose a positive canonical divisor $\mathfrak{w}$ of F/K. Then choose a nonconstant x in $\mathcal{L}(\mathfrak{w})$.

8. Let K be a field of characteristic that does not divide n. Consider the function field $F = K(x, y)$ over K with x, y satisfying $x^n + y^n = 1$. Prove that $\mathrm{genus}(F) = \frac{(n-1)(n-2)}{2}$. Hint: Use Example 2.3.8 and the Riemann-Hurwitz formula (Remark 3.6.2(c)).

Notes

In addition to [Chevalley2] one may find a proof of the Riemann-Roch theorem in [Lang4, Chapter 10, Section 2], [Deuring3, Section 15], and [Stichtenoth, Section I.5].

The content of Section 3.7 is borrowed from [Artin3]. One may also find it in [Stichtenoth, Section VI.2] along with computations of genera of various function fields using the Riemann-Hurwitz genus formula. However, in contrast to our exposition, [Stichtenoth] assumes the fields of constants to be perfect.

Chapter 4.
The Riemann Hypothesis for Function Fields

In this chapter K is a finite field of characteristic p with q elements. Let F be an algebraic function field of one variable over K and g the genus of F/K. Denote the group of divisors and the group of divisor classes of F/K by $\mathcal{D}$ and $\mathcal{C}$, respectively.

The series $\zeta(s) = \sum_{n=1}^{\infty} n^{-s}$ defines the classical Riemann zeta function of a complex variable. It converges absolutely for $\mathrm{Re}(s) > 1$. Hence, $\zeta(s)$ is an analytic function in this domain. The series diverges for $s = 1$. The function, however, can be analytically continued to a function meromorphic on the whole s-plane. The production of this requires two stages: an analytic continuation of $\zeta(s) + \frac{1}{1-s}$ to the half plane $\mathrm{Re}(s) > 0$ by a rearrangement of the series via Abel summation; and then, (the difficult part) a demonstration that $\frac{\zeta(s)}{\zeta(1-s)}$ is the product of the classical Gamma function and an elementary function in the domain $0 < \mathrm{Re}(s) < 1$. The expression that results from the last stage is called the functional equation for $\zeta(s)$. The resulting analytic continuation of $\zeta(s)$ yields a function with a simple pole at $s = 1$ with residue 1 and zeros at the points $-2, -4, -6, -8, \ldots$. There are no other zeros in the domains $\mathrm{Re}(s) \geq 1$ and $\mathrm{Re}(s) \leq 0$ [Titchmarsh, p. 30]. The classical Riemann hypothesis is still unproven. It states that the only zeros of $\zeta(s)$ in the strip $0 < \mathrm{Re}(s) < 1$ lie on the line $\mathrm{Re}(s) = \frac{1}{2}$; its applications are legion.

There is an analog for F/K of the Riemann zeta function (Section 4.2). It satisfies a functional equation (Proposition 4.4.1). Our main goal is the proof of an analog to the Riemann hypothesis (Theorem 4.5.1). In Chapter 5 we extract from this an explicit estimate for the number of points on any curve over a finite field.

4.1 Class Numbers

The assumption that K is finite results in the finiteness of other sets connected to F/K. For example, F/K has only finitely many ideal classes of degree 0 and only finitely many nonnegative divisors of a given degree n. The main result of this section (Lemma 4.1.4) computes the latter number in terms of the former one.

Let $\mathfrak{p}$ be a prime divisor of F/K. Its residue field $\bar{F}_{\mathfrak{p}}$ is a finite extension of K of degree $\deg(\mathfrak{p})$. Thus, $\bar{F}_{\mathfrak{p}}$ is a finite field whose order we indicate by $N\mathfrak{p} = q^{\deg(\mathfrak{p})}$, the **norm** of $\mathfrak{p}$. Extend the definition of the norm to arbitrary divisors by the formula $N\mathfrak{a} = q^{\deg(\mathfrak{a})}$. Then $N(\mathfrak{a} + \mathfrak{b}) = N\mathfrak{a} \cdot N\mathfrak{b}$.

Definition 4.1.1: Denote the number of divisor classes of F/K of degree zero by h; the **class number** of F/K. □

If $g = 0$, then every divisor of degree 0 of F/K is principal (Exercise 5 of Chapter 6). Therefore, $h = 1$ if F is $K(x)$ or any genus 0 function field.

LEMMA 4.1.2: *Only finitely many nonnegative divisors of F/K have degree equal to a given integer m. In addition, the class number of F/K is finite.*

Proof: Let $x \in F$ be transcendental element over K and let $E = K(x)$. Denote the collection of all prime divisors of E/K of degree $\leq m$ by S_0. Each element of S_0, except possibly $\mathfrak{p}_\infty$, corresponds to a monic irreducible polynomial in $K[x]$ of degree $\leq m$. Hence, S_0 is a finite set. Only finitely many prime divisors of F/K lie over a given prime divisor $\mathfrak{p}_0$ of E/K. Each of them has degree at least as large as $\deg(\mathfrak{p}_0)$. Thus, there are only finitely many prime divisors of F/K of degree $\leq m$. Therefore, the set $\mathcal{A}_m$ of all nonnegative divisors of F/K of degree m is finite.

For the second part of the lemma choose a nonnegative divisor $\mathfrak{m}$ of degree $m \geq g$. Denote the set of divisor classes of degree m by $\mathcal{C}_m$. For each $\mathfrak{b} \in \mathcal{C}_m$ Riemann-Roch (Theorem 3.2.1) implies that $\dim(\mathfrak{b}) \geq m - g + 1 \geq 1$. Hence, there is an $x \in F^\times$ with $\mathrm{div}(x) + \mathfrak{b} \geq 0$, so the class of $\mathfrak{b}$ contains a nonnegative divisor. It follows that the map $\mathcal{A}_m \to \mathcal{C}_m$ mapping each $\mathfrak{a} \in \mathcal{A}_m$ onto each class is surjective. Therefore, by the preceding paragraph, $\mathcal{C}_m$ is finite.

Finally, the map $\mathfrak{a} \mapsto \mathfrak{a} + \mathfrak{m}$ induces a bijective map of $\mathcal{C}_0$ onto $\mathcal{C}_m$. Consequently, $\mathcal{C}_0$ is a finite group. $\square$

LEMMA 4.1.3: *The number of nonnegative divisors in a given class of divisors C of F/K is $\frac{q^{\dim(C)}-1}{q-1}$.*

Proof: If $\mathfrak{a} \geq 0$, then $\mathrm{div}(a) + \mathfrak{a} \geq 0$ for each $a \in K^\times$, so $K \subseteq \mathcal{L}(\mathfrak{a})$. Hence, $\dim(\mathfrak{a}) \geq 1$. This gives the formula if $\dim(C) = 0$. Suppose $\dim(C) = n > 0$. Let $\mathfrak{c}$ be a divisor in C. The number of nonnegative divisors in C is equal to the number of principal divisors $\mathrm{div}(x)$ with $x \in \mathcal{L}(\mathfrak{c})$. Let $x_1, \ldots, x_n$ be a basis for $\mathcal{L}(\mathfrak{c})$ over K. The number of elements of $F^\times$ of the form $\sum_{i=1}^n a_i x_i$, with $a_1, \ldots, a_n \in K$ is equal to $q^n - 1$. Since $\mathrm{div}(x) = \mathrm{div}(x')$ if and only if there exists an $a \in K^\times$ such that $x' = ax$, the formula follows. $\square$

Denote the greatest common divisor of the degrees of the divisors of F/K by δ. Eventually we prove that $\delta = 1$. In the meantime notice that a positive integer n is a multiple of δ if and only if there exists a divisor of degree n. Indeed, there are divisors $\mathfrak{a}_1, \ldots, \mathfrak{a}_r$ with $\delta = \gcd(\deg(\mathfrak{a}_1), \ldots, \deg(\mathfrak{a}_r))$. For each multiple n of δ there are $a_1, \ldots, a_r \in \mathbb{Z}$ with $\sum_{i=1}^r a_i \deg(\mathfrak{a}_i) = n$. The divisor $\mathfrak{a} = \sum_{i=1}^r a_i \mathfrak{a}_i$ satisfies $\deg(\mathfrak{a}) = n$. In particular, since the degree of the canonical divisor is $2g - 2$ (Lemma 3.2.2(b)) δ divides $2g - 2$.

LEMMA 4.1.4: *Let A_n be the number of nonnegative divisors of F/K of degree n. If $n \geq 0$ is a multiple of δ larger than $2g - 2$, then $A_n = h \frac{q^{n-g+1}-1}{q-1}$.*

Proof: Let n be a multiple of δ with $n > 2g - 2$. Choose a divisor $\mathfrak{c}$ of degree n. By Riemann-Roch, $\dim(\mathfrak{c}) = n - g + 1$. Moreover, the map $\mathfrak{a} \mapsto \mathfrak{a} + \mathfrak{c}$ defines a bijection of the set of divisor classes of degree 0 with the set of

4.2 Zeta Functions

divisor classes of degree m. Hence, there are h divisor classes of degree n. It follows that, $A_n = h \frac{q^{n-g+1}-1}{q-1}$. □

4.2 Zeta Functions

The zeta function of a function field over K is a rational function with coefficients in $\mathbb{Q}$ and with simple poles at several points including 1.

We define the **zeta function** of the function field F/K to be the Dirichlet series

(1) $$\zeta(s) = \zeta_{F/K}(s) = \sum_{\mathfrak{a} \geq 0}(N\mathfrak{a})^{-s}$$

where $\mathfrak{a}$ runs over the nonnegative divisors of F/K. Check the domain of convergence of the series (1) using the substitution $t = q^{-s}$ and the identity $N\mathfrak{a} = q^{\deg(\mathfrak{a})}$. We obtain a power series for $\zeta(s)$ in terms of t:

(2) $$Z(t) = \sum_{\mathfrak{a} \geq 0} t^{\deg(\mathfrak{a})} = \sum_{n=0}^{\infty} A_n t^n,$$

where A_n is the number of nonnegative divisors of degree n. By Lemma 4.1.4

$$Z(t) = \sum_{n=0}^{2g-2} A_n t^n + h \sum_{m=d}^{\infty} \frac{q^{m\delta-g+1}-1}{q-1} t^{m\delta}$$

where $d = \frac{2g-2+\delta}{\delta}$. The right hand side converges for $|t| < q^{-1}$; (i.e. for $\mathrm{Re}(s) > 1$) and

(3) $$Z(t) = \Phi(t) + \frac{hq^{g-1+\delta}}{q-1} \cdot \frac{t^{2g-2+\delta}}{1-(qt)^\delta} - \frac{h}{q-1} \cdot \frac{t^{2g-2+\delta}}{1-t^\delta},$$

where

$$\Phi(t) = \sum_{n=0}^{2g-2} A_n t^n$$

is a polynomial of degree $\leq 2g - 2$. We summarize:

PROPOSITION 4.2.1: *The power series $Z(t)$ in (2) converges in the circle $|t| < q^{-1}$. Formula (3) continues $Z(t)$ to a meromorphic function on the whole plane. The only poles of $Z(t)$ occur for values of t with $t^\delta = 1$ or $t^\delta = q^{-\delta}$, and they are simple.*

The Dirichlet series for $\zeta(s)$ in (1) converges in the right half plane $\mathrm{Re}(s) > 1$. The substitution $t = q^{-s}$ in (3) continues $\zeta(s)$ to a meromorphic function in the whole plane.

Like the Riemann zeta function, $\zeta_{F/K}(s)$ has a multiplicative presentation. Let $a_1, a_2, a_3, \ldots$ be a sequence of complex numbers of absolute value less than 1. We say that the infinite product $\prod_{i=1}^{\infty} \frac{1}{1-a_i}$ **converges** (resp. **absolutely converges**) if the limit

$$\lim_{n\to\infty} \prod_{i=1}^{n} \frac{1}{1-a_i} \quad (\text{resp. } \lim_{n\to\infty} \prod_{i=1}^{n} \frac{1}{1-|a_i|})$$

converges to a nonzero complex number.

PROPOSITION 4.2.2: *If* $\operatorname{Re}(s) > 1$ *and* $|t| < q^{-1}$, *then*

(4) $$\zeta(s) = \prod_{\mathfrak{p}} \frac{1}{1-(N\mathfrak{p})^{-s}} = \prod_{\mathfrak{p}} \frac{1}{1-t^{\deg(\mathfrak{p})}},$$

where $\mathfrak{p}$ *runs over the prime divisors of* F/K. *The product converges absolutely. Therefore, it is independent of the order of the factors. In particular, if* $\operatorname{Re}(s) > 1$, *then* $\zeta(s) \neq 0$.

Proof: The prime divisors are free generators of the group of divisors. Thus, for every positive integer m, if $\operatorname{Re}(s) > 1$, then

$$\prod_{N\mathfrak{p}\leq m} \frac{1}{1-(N\mathfrak{p})^{-s}} = \prod_{N\mathfrak{p}\leq m} \sum_{k=0}^{\infty} (N\mathfrak{p})^{-sk} = \sum_{\substack{\mathfrak{a}\geq 0 \\ N\mathfrak{a}\leq m}} (N\mathfrak{a})^{-s} + \sideset{}{'}\sum_{\substack{\mathfrak{a}\geq 0 \\ N\mathfrak{a}> m}} (N\mathfrak{a})^{-s},$$

where the prime in the second sum means that $\mathfrak{a}$ runs over all nonnegative divisors with norm exceeding m whose prime divisors $\mathfrak{p}$ satisfy $N\mathfrak{p} \leq m$. It follows that

$$\left| \zeta(s) - \prod_{N\mathfrak{p}\leq m} \frac{1}{1-(N\mathfrak{p})^{-s}} \right| \leq \sum_{\substack{\mathfrak{a}\geq 0 \\ N\mathfrak{a}> m}} |(N\mathfrak{a})^{-s}|$$

and the right hand side converges to zero as m tends to infinity.

Now we prove that $\zeta(s) \neq 0$ when $\operatorname{Re}(s) > 1$. Indeed, consider a positive integer m. Then,

$$\left| \zeta(s) \prod_{N\mathfrak{p}\leq m} 1-(N\mathfrak{p})^{-s} \right| = \left| \prod_{N\mathfrak{p}>m} \frac{1}{1-(N\mathfrak{p})^{-s}} \right|$$

$$= \left| 1 + \sideset{}{''}\sum (N\mathfrak{a})^{-s} \right| \geq 1 - \sum_{\substack{\mathfrak{a}>0 \\ N\mathfrak{a}>m}} (N\mathfrak{a})^{\operatorname{Re}(s)} > 0,$$

(where the double primes mean summation over all positive divisors $\mathfrak{a}$ with only prime divisors $\mathfrak{p}$ satisfying $N\mathfrak{p} > m$) if m is large enough. Therefore, $\zeta(s) \neq 0$.

In particular, $\prod_{\mathfrak{p}} \frac{1}{1-(N\mathfrak{p})^{-\operatorname{Re}(s)}}$ converges. Since, $|N\mathfrak{p}^{-s}| = N\mathfrak{p}^{-\operatorname{Re}(s)}$, this means that $\prod_{\mathfrak{p}} \frac{1}{1-(N\mathfrak{p})^{-s}}$ absolutely converges. By Exercise 3, the value of the product is independent of the order of the factors. $\square$

4.3 Zeta Functions under Constant Field Extensions

The analytic properties of the zeta function of a function field F/K proved in Proposition 4.2.2 result in the conclusion that $\delta = 1$ (Corollary 4.3.3).

Denote the unique extension of K of degree r by K_r. Then $F_r = FK_r$ is a function field of one variable over K_r. Use r as a subscript to denote the "extension" of objects of F to F_r. For example, if $\mathfrak{a}$ is a divisor of F/K, then $\deg_r \mathfrak{a}$ denotes the degree of $\mathfrak{a}$ as a divisor of F_r/K_r. We have already noted that $\deg_r \mathfrak{a} = \deg(\mathfrak{a})$, $\dim_r \mathfrak{a} = \dim(\mathfrak{a})$ and $g_r = g$ (Proposition 3.4.2).

LEMMA 4.3.1: *Let $\mathfrak{p}$ be a prime divisor of F/K. Then $\mathfrak{p}$ decomposes in F_r as $\mathfrak{p} = \mathfrak{P}_1 + \mathfrak{P}_2 + \cdots + \mathfrak{P}_d$, where $\mathfrak{P}_1, \mathfrak{P}_2, \ldots, \mathfrak{P}_d$ are distinct prime divisors of F_r/K_r, $\deg(\mathfrak{P}_i) = r^{-1} \cdot \mathrm{lcm}(r, \deg(\mathfrak{p}))$ and $d = \gcd(r, \deg(\mathfrak{p}))$.*

Proof: Put $m = \deg(\mathfrak{p})$. Since $\mathfrak{p}$ is unramified in F_r (Proposition 3.4.2), $\mathfrak{P}_1, \ldots, \mathfrak{P}_d$ are distinct. Moreover, $\overline{(F_r)}_{\mathfrak{P}_i} = K_r \bar{F}_{\mathfrak{p}}$. Hence, $[\overline{(F_r)}_{\mathfrak{P}_i} : K] = \mathrm{lcm}(r, m)$, and thus $\deg(\mathfrak{P}_i) = [\overline{(F_r)}_{\mathfrak{P}_i} : K_r] = r^{-1} \cdot \mathrm{lcm}(r, m)$, $i = 1, \ldots, d$. Also, by Propositions 2.6 and 3.4.2, $r = [K_r : K] = [F_r : F] = d \cdot [\bar{F}_{r, \mathfrak{P}_i} : \bar{F}_{\mathfrak{p}}] = d \cdot \mathrm{lcm}(r, m) m^{-1}$. Therefore, $d = \gcd(r, m)$. □

PROPOSITION 4.3.2: *For every complex number t,*

$$(1) \qquad Z_r(t^r) = \prod_{\xi^r = 1} Z(\xi t),$$

where ξ runs over the rth roots of unity.

Proof: Since both sides of (1) are rational functions of t (by (3) of Section 4.2), it suffices to prove (1) for $|t| < q^{-r}$. First apply the product formula (4) of Section 4.2, then (Lemma 4.3.1):

$$(2) \qquad Z_r(t^r)^{-1} = \prod_{\mathfrak{p}} \prod_{\mathfrak{P} | \mathfrak{p}} (1 - t^{r \cdot \deg(\mathfrak{P})}) = \prod_{\mathfrak{p}} (1 - t^{\mathrm{lcm}(r, \deg(\mathfrak{p}))})^{\gcd(r, \deg(\mathfrak{p}))}$$

$$(3) \qquad \prod_{\xi^r = 1} Z(\xi t)^{-1} = \prod_{\mathfrak{p}} \prod_{\xi^r = 1} (1 - (\xi t)^{\deg(\mathfrak{p})}).$$

Thus, (1) follows if we show equality of the corresponding factors on the right hand sides of (2) and (3). Indeed, for a fixed $\mathfrak{p}$ let $m = \deg(\mathfrak{p})$ and $d = \gcd(r, m)$. We must show that

$$(4) \qquad (1 - t^{rm/d})^d = \prod_{\xi^r = 1} (1 - (\xi t)^m).$$

Substitute $t^m = x^{-1}$ in (4) to rewrite it as

$$(5) \qquad (x^{r/d} - 1)^d = \prod_{\xi^r = 1} (x - \xi^m).$$

Both monic polynomials in (5) have each (r/d)th root of unity as a zero of multiplicity d. Indeed, if ζ_r is a primitive root of unity of order r, then ζ_r^m is a primitive root of unity of order r/d and each power of ζ_r^m appears d times among $1, \zeta_r^m, \zeta_r^{2m}, \ldots, \zeta_r^{(r-1)m}$. Therefore, the polynomials are equal. $\square$

COROLLARY 4.3.3 (F. K. Schmidt): $\delta = 1$.

Proof: By (2) of Section 4.2, $Z(t) = \sum_{m=0}^{\infty} A_{m\delta} t^{m\delta}$. Hence, if $\xi^\delta = 1$, then $Z(\xi t) = Z(t)$. From (1), $Z_\delta(t^\delta) = Z(t)^\delta$. However, by (3) of Section 4.2 (applied to $Z_\delta(t^\delta)$ instead of to $Z(t)$), $Z_\delta(t^\delta)$ has a simple pole at $t = 1$, while $Z(t)^\delta$ has a pole of order δ at $t = 1$. Consequently, $\delta = 1$. $\square$

COROLLARY 4.3.4: *For every integer n there are exactly h divisor classes of F/K of degree n.*

Proof: By Corollary 4.3.3 there is a divisor $\mathfrak{c}$ of degree n. The map $\mathfrak{a} \to \mathfrak{a} + \mathfrak{c}$ induces a bijection of the set of divisor classes of F/K of degree 0 onto the set of divisor classes of degree m. Hence, the number of elements in the latter set equals the number of elements in the former set, namely h. $\square$

Example 4.3.5: By definition, A_1 is the number of prime divisors of F/K of degree 1. If $g = 0$, then $h = 1$ and $A_1 = q+1$ (Lemma 4.1.4). Thus, F/K has prime divisors of degree 1. By Example 3.2.4, $F = K(x)$ is a rational function field. If $\delta = 1$ (Corollary 4.3.3) and $g = 1$, then $A_1 = h$ (Lemma 4.1.4); in other words the class number is equal to the number of prime divisors of degree 1. Since this is the order of a group (Definition 4.1.1), it is again positive. By Lemma 4.1.4 and by (2) of Section 4.1.2, $Z(t) = \frac{1}{(1-t)(1-qt)}$. $\square$

4.4 The Functional Equation

Like the Riemann zeta functions, $Z(t)$ satisfies a functional equation relating its values in t and $\frac{1}{qt}$. The main tool in the proof is the Riemann-Roch theorem.

PROPOSITION 4.4.1: *$Z(t)$ satisfies the functional equation*

$$(\sqrt{q}t)^{1-g} Z(t) = (\sqrt{q}t)^{g-1} Z\left(\frac{1}{qt}\right).$$

Proof: If $g = 0$, the result follows from the explicit presentation $Z(t) = \frac{1}{(1-t)(1-qt)}$ in Example 4.3.5. Therefore, assume $g > 0$.

The basic idea is to split $Z(t)$ into the sum of a polynomial $P(t)$ and an infinite series $Q(t)$, each of which satisfies the same functional equation in the statement of the proposition.

Apply Lemmas 4.1.3 and 4.1.4 with $\delta = 1$ (Corollary 4.3.3) to obtain:

$$Z(t) = \sum_{\mathfrak{a} \geq 0} t^{\deg(\mathfrak{a})} = \sum_{\deg(C)=0}^{2g-2} \sum_{\substack{\mathfrak{a} \geq 0 \\ \mathfrak{a} \in C}} t^{\deg(\mathfrak{a})} + \sum_{n=2g-1}^{\infty} A_n t^n$$

4.4 The Functional Equation

$$= \sum_{\deg(C)=0}^{2g-2} \frac{q^{\dim(C)}-1}{q-1} t^{\deg(C)} + \sum_{n=2g-1}^{\infty} h \frac{q^{n-g+1}-1}{q-1} t^n$$

$$= \left[\frac{1}{q-1} \sum_{\deg(C)=0}^{2g-2} q^{\dim(C)} t^{\deg(C)}\right]$$

$$+ \left[\frac{h}{q-1} \sum_{n=2g-1}^{\infty} q^{n-g+1} t^n - \frac{h}{q-1} \sum_{n=0}^{\infty} t^n\right]$$

$$= P(t) + Q(t),$$

where $P(t)$ (resp. $Q(t)$) is the expression in the first (resp. second) brackets. We have also used that F/K has exactly h divisor classes of each degree (Corollary 4.3.4).

First we analyze $P(t)$. The Riemann-Roch theorem relates $\dim(C)$ to $\dim(W-C)$, where W is the canonical class. Recall that $\deg(W) = 2g-2$ (Lemma 3.2.2(b)). Hence

$$\dim(C) - \frac{1}{2}\deg(C) = \frac{1}{2}\deg(C) + 1 - g + \dim(W-C)$$

$$= \dim(W-C) - \frac{1}{2}\deg(W-C).$$

As C varies over all divisor classes of degree between 0 and $2g-2$ so does $W-C$. Hence

$$(\sqrt{q}t)^{2-2g} P(t) = \frac{(\sqrt{q}t)^{2-2g}}{q-1} \sum_{\deg(C)=0}^{2g-2} q^{\dim(C)} t^{\deg(C)}$$

$$= \frac{1}{q-1} \sum_{\deg(C)=0}^{2g-2} q^{\dim(C)-\frac{1}{2}\deg(C)} (\sqrt{q}t)^{2-2g+\deg(C)}$$

$$= \frac{1}{q-1} \sum_{\deg(C)=0}^{2g-2} q^{\dim(W-C)-\frac{1}{2}\deg(W-C)} (\sqrt{q}t)^{-\deg(W-C)}$$

$$= \frac{1}{q-1} \sum_{\deg(C')=0}^{2g-2} q^{\dim(C')} (qt)^{-\deg(C')} = P\left(\frac{1}{qt}\right).$$

Now evaluate the geometric series involved in the expression for $Q(t)$:

$$Q(t) = \frac{h}{q-1} \left[\sum_{n=2g-1}^{\infty} q^{n-g+1} t^n - \sum_{n=0}^{\infty} t^n\right] = \frac{h}{q-1}\left[\frac{q^g t^{2g-1}}{1-qt} - \frac{1}{1-t}\right]$$

A direct computation shows that

$$(\sqrt{q}t)^{2-2g} Q(t) = Q\left(\frac{1}{qt}\right).$$

4.5 The Riemann Hypothesis and Degree 1 Prime Divisors

We reformulate here the Riemann hypothesis for a function field F of one variable over a finite field and draw an estimate for the number of prime divisors of F of degree 1.

Rewrite formula (3) of Section 4.2 with $\delta = 1$ (Corollary 4.3.3) as

$$Z(t) = \Phi(t) + \frac{hq^g}{q-1} \cdot \frac{t^{2g-1}}{1-qt} - \frac{h}{q-1} \cdot \frac{t^{2g-1}}{1-t}$$

where $\Phi(t)$ a polynomial of degree $\leq 2g - 2$. Hence

$$Z(t) = \frac{L(t)}{(1-t)(1-qt)}$$

with $L(t) = a_0 + a_1 t + \cdots + a_{2g} t^{2g}$ a polynomial with rational coefficients. We determine some of these coefficients:

First: $a_0 = L(0) = Z(0) = A_0 = 1$, since the zero divisor is the only nonnegative divisor of degree 0. Second: A_1 is equal to the number of prime divisors of F/K of degree 1. Write $N = A_1$, so that

$$L(t) = (1-t)(1-qt) \sum_{n=0}^{\infty} A_n t^n \equiv 1 + (N - (q+1))t \mod t^2.$$

Therefore,

(1) $\qquad a_1 = N - (q+1)$

Now let $x \in F$ be transcendental over K and write $F_0 = K(x)$. The Zeta function of F_0/K is $Z_0(t) = \frac{1}{(1-t)(1-qt)}$ (Example 4.3.5). By Proposition 4.4.1

(2) $\qquad L(t) = \frac{Z(t)}{Z_0(t)} = \frac{q^{g-1} t^{2g-2} Z\left(\frac{1}{qt}\right)}{q^{-1} t^{-2} Z_0\left(\frac{1}{qt}\right)} = q^g t^{2g} L\left(\frac{1}{qt}\right).$

This functional equation for $L(t)$, written explicitly, has the form

$$\sum_{i=0}^{2g} a_i t^i = \sum_{i=0}^{2g} a_{2g-i} q^{i-g} t^i.$$

Equivalently, $a_i = q^{i-g} a_{2g-i}$. In particular, deduce for $i = 0$ and $i = 1$ that

(3) $\qquad a_{2g} = q^g \quad \text{and} \quad a_{2g-1} = q^{g-1}(N - (q+1)).$

4.5 The Riemann Hypothesis and Degree 1 Prime Divisors

These formulas imply that $\deg(L(t)) = 2g$. Decompose $L(t)$ over $\mathbb{C}$ as

(4) $$L(t) = \prod_{i=1}^{2g}(1 - \omega_i t)$$

where the ω_i^{-1}'s are the zeros of $L(t)$. Formulas (1), (3), and (4) give

(5) $$q^g = \prod_{i=1}^{2g} \omega_i \quad \text{and} \quad N - (q+1) = -\sum_{i=1}^{2g} \omega_i.$$

Moreover, the functional equation (2) for $L(t)$ implies that

$$L\left(\frac{1}{\omega_i}\right) = 0 \text{ if and only if } L\left(\frac{\omega_i}{q}\right) = 0.$$

Rename the roots $\omega_1, \ldots, \omega_{2g}$ as $\omega_1, \omega_1', \ldots, \omega_f, \omega_f', \sqrt{q}, \ldots, \sqrt{q}, -\sqrt{q}, \ldots, -\sqrt{q}$ with $f \leq g$ such that $\omega_i \omega_i' = q$, $i = 1, \ldots, f$, and $\sqrt{q}$ (resp. $-\sqrt{q}$) appear k (resp. l) times. Then $2f + k + l = 2g$ shows that if k is odd, then l is odd. In this case (5) gives $q^g = q^f q^{k/2} (-1)^l q^{l/2} = -q^g$, a contradiction. Hence, both k and l are even and we may take $f = g$. Thus,

$$L(t) = \prod_{i=1}^{g}(1 - \omega_i t)(1 - \omega_i' t),$$

with $\omega_i \omega_i' = q$ for $i = 1, \ldots, g$.

Here is a reformulation of the Riemann hypothesis for the function field F/K. Sections 4.6 - 4.8 complete the proof.

THEOREM 4.5.1:
(a) The zeros of the function $\zeta_{F/K}(s)$ lie on the line $\mathrm{Re}(s) = \frac{1}{2}$.
(b) The zeros of the function $Z_{F/K}(t)$ lie on the circle $|t| = q^{-\frac{1}{2}}$.
(c) If the ω_i are the inverses of the zeros of the polynomial $L_{F/K}(t)$, then

$$|\omega_i| = \sqrt{q}, \quad i = 1, 2, \ldots, 2g.$$

Note that (c) is equivalent to (b), since the poles of $Z(t)$ are $t = 1$ and $t = q^{-1}$. Theorem 4.5.1 with (5) provides an estimate on the number of prime divisors of degree 1:

THEOREM 4.5.2: Let F be a function field of one variable over a finite field K of q elements. Denote the genus of F/K by g and let N be the number of prime divisors of F/K of degree 1. Then $|N - (q+1)| \leq 2g\sqrt{q}$.

4.6 Reduction Steps

Theorem 4.5.2 is a consequence of the Riemann hypothesis. As a first step, this section shows that an appropriate version of Theorem 4.5.2 implies the Riemann hypothesis. As in Section 4.3, denote the unique extension of K of degree r by K_r.

LEMMA 4.6.1: *The Riemann hypothesis holds for the function field F/K if and only if it holds for the function field F_r/K_r.*

Proof: Use (2) and (4) of Section 4.5 to express $L(t)$. Then apply Proposition 4.3.2 to compute $L_r(t^r)$:

$$L_r(t^r) = \frac{Z_r(t^r)}{Z_{r,0}(t^r)} = \prod_{\xi^r = 1} \frac{Z(\xi t)}{Z_0(\xi t)} = \prod_{\xi^r = 1} L(\xi t)$$

$$= \prod_{\xi^r = 1} \prod_{i=1}^{2g}(1 - \omega_i \xi t) = \prod_{i=1}^{2g}(1 - \omega_i^r t^r).$$

Hence, $L_r(t) = \prod_{i=1}^{2g}(1 - \omega_i^r t)$. Thus,

(1) $\omega_1^r, \ldots, \omega_{2g}^r$ are the inverses of the zeros of L_r.

Since $|\omega_i| = \sqrt{q}$ if and only if $|\omega_i^r| = \sqrt{q^r}$, the lemma follows. $\square$

Denote the number of prime divisors of F_r/K_r of degree 1 by N_r.

LEMMA 4.6.2: *Let F be a function field of one variable over a field K of q elements. If there exists a constant c such that $|N_r - (q^r + 1)| \leq cq^{r/2}$ for every positive integer r, then the Riemann hypothesis holds for F/K.*

Proof: Apply the differential operator $D = -t\frac{d \log}{dt}$ to both sides of the formula $L(t) = \prod_{i=1}^{2g}(1 - \omega_i t)$:

(2) $$D(L(t)) = \sum_{i=1}^{2g} \frac{\omega_i t}{1 - \omega_i t} = \sum_{n=1}^{\infty} \Big(\sum_{r=1}^{2g} \omega_i^r\Big) t^r$$

Combine (1) with (5) of Section 4.5 to obtain $-\sum_{i=1}^{2g} \omega_i^r = N_r - (q^r + 1)$. The hypothesis of the lemma thus implies $|\sum_{i=1}^{2g} \omega_i^r| \leq cq^{r/2}$. Therefore the radius of convergence R of the right hand side of (2) satisfies $R \geq q^{-\frac{1}{2}}$. But (2) implies that the ω_i^{-1} are the only singularities of $D(L(t))$. Hence, $R = \min_{1 \leq i \leq 2g} |\omega_i^{-1}|$. Therefore, $|\omega_i| \leq \sqrt{q}$ for $i = 1, \ldots, 2g$. This together with the equality $q^g = \prod_{i=1}^{2g} \omega_i$ ((5) of Section 4.5), implies that $|\omega_i| = \sqrt{q}$ for $i = 1, \ldots, 2g$. $\square$

4.7 An Upper Bound

Assume, by extension of constants if necessary, that K and F satisfy these conditions:
(1a) $q = a^2$ is a square;
(1b) $q > (g+1)^4$; and
(1c) F has a prime divisor $\mathfrak{o}$ of degree 1.

By Lemma 4.6.1, a proof of the Riemann hypothesis under these conditions suffices for the general case. We prove a result that has, as a special case, the inequality

$$(2) \qquad N - (q+1) < (2g+1)\sqrt{q}.$$

Let σ be an automorphism of F over K. It induces a permutation of the prime divisors of F/K. If $\mathfrak{p}$ is a prime divisor of F/K, the $\mathfrak{p}^\sigma$ is the prime divisor corresponding to the place

$$\varphi_\mathfrak{p}^\sigma(x) = \varphi_\mathfrak{p}(\sigma x) \text{ for } x \in F$$

(where, as in Section 2.1, $\varphi_\mathfrak{p}^\sigma = \sigma^{-1}\varphi_\mathfrak{p}$). Also, recall that the map $x \mapsto x^q$ is an automorphism $\tilde{F}$ over K. Hence, the formula

$$\varphi_\mathfrak{p}^q(x) = \varphi_\mathfrak{p}(x)^q$$

defines a place $\varphi_\mathfrak{p}^q$ of F/K. Although $\varphi_\mathfrak{p}^q$ is equivalent to $\varphi_\mathfrak{p}$, it is convenient to use $\varphi_\mathfrak{p}^q$, because $\varphi_\mathfrak{p} = \varphi_\mathfrak{p}^q$ if and only if $\deg(\mathfrak{p}) = 1$.

The remainder of this section investigates the expression

$$N^{(\sigma)} = \sum_{\varphi_\mathfrak{p}^\sigma = \varphi_\mathfrak{p}^q} \deg(\mathfrak{p})$$

in order to show that

$$(3) \qquad N^{(\sigma)} - (q+1) < (2g+1)\sqrt{q}.$$

If σ is the identity automorphism, then $N^{(\sigma)} = N$ and (3) becomes (2).

With the notation $m = a - 1$, $n = a + 2g$, and $r = m + an$, rewrite (3) as

$$(4) \qquad N^{(\sigma)} - 1 \leq r.$$

With $\mathfrak{o}$ as in (1c), consider the ascending sequence of K-vector spaces $\mathcal{L}(\mathfrak{o}) \subseteq \mathcal{L}(2\mathfrak{o}) \subseteq \mathcal{L}(3\mathfrak{o}) \subseteq \cdots$. By Exercise 2 of Chapter 3,

$$(5) \qquad \dim(\mathcal{L}(i\mathfrak{o})) - \dim(\mathcal{L}((i-1))\mathfrak{o}) \leq 1.$$

88   Chapter 4. The Riemann Hypothesis for Function Fields

With k a positive integer, let I_k be the set of i, $1 \leq i \leq k$ for which equality holds in (5). For each $i \in I_k$, choose $u_i \in \mathcal{L}(i\mathfrak{o}) - \mathcal{L}((i-1)\mathfrak{o})$. Then $\text{div}_\infty(u_i) = i\mathfrak{o}$ and the system $\{u_i \mid i \in I_k\}$ is a basis for $\mathcal{L}(k\mathfrak{o})$. In particular, this holds for $k = m$. Since $a^2 = q$, a is a power of $\text{char}(K)$. Thus, the set $\mathcal{L}(n\mathfrak{o})^a = \{y^a \mid y \in \mathcal{L}(n\mathfrak{o})\}$, consisting of elements in the field F^a is a K-vector space with basis $\{u_j^a \mid j \in I_n\}$ and the same dimension as $\mathcal{L}(n\mathfrak{o})$. Therefore,

$$\mathcal{L} = \Big\{ \sum_{i \in I_m} u_i y_i^a \mid y_i \in \mathcal{L}(n\mathfrak{o}) \Big\}$$

is a K-vector space generated by the set $U = \{u_i u_j^a \mid i \in I_m \text{ and } j \in I_n\}$.

LEMMA 4.7.1: *The set U is linearly independent over K.*

Proof: It suffices to prove that $\{u_i \mid i \in I_m\}$ is linearly independent over the field F^a. Indeed, assume that there exist $y_i \in F$, $i \in I_m$, not all zero, such that $\sum_{i \in I_m} u_i y_i^a = 0$. Then there exist distinct $i, j \in I_m$ such that $v_\mathfrak{o}(u_i y_i^a) = v_\mathfrak{o}(u_j y_j^a)$, $y_i \neq 0$, $y_j \neq 0$ ((2) of Section 2.1). Thus, $-i + av_\mathfrak{o}(y_i) = -j + av_\mathfrak{o}(y_j)$, so $i \cong j \mod a$. Since $1 \leq i, j \leq m < a$, this is a contradiction. □

By Lemma 4.7.1, $\dim(\mathcal{L}) = \dim(\mathcal{L}(m\mathfrak{o})) \cdot \dim(\mathcal{L}(n\mathfrak{o}))$. Apply Riemann-Roch to the right hand side terms:

(6) $\dim(\mathcal{L}) \geq (m - g + 1)(n - g + 1) = q + \sqrt{q} - g(g+1)$.

Now consider the K-vector space

$$\mathcal{L}' = \Big\{ \sum_{i \in I_m} (\sigma^{-1} u_i)^a y_i \mid y_i \in \mathcal{L}(n\mathfrak{o}) \Big\}.$$

Check that $\mathcal{L}' \subseteq \mathcal{L}(ma\mathfrak{o}^\sigma + n\mathfrak{o})$ and $\deg(ma\mathfrak{o}^\sigma + n\mathfrak{o}) = q + 2g > 2g - 2$. By Riemann-Roch,

(7) $\dim(\mathcal{L}') \leq \dim(ma\mathfrak{o}^\sigma + n\mathfrak{o}) = \deg(ma\mathfrak{o}^\sigma + n\mathfrak{o}) - g + 1 = q + g + 1$.

By (1b), $\sqrt{q} - g(g+1) > g + 1$. Thus, the right side of (6) is greater than the right side of (7). Therefore,

(8) $\dim(\mathcal{L}') < \dim(\mathcal{L})$.

Finally, define an additive map σ^* from $\mathcal{L}$ into $\mathcal{L}'$ by

$$\sigma^*\Big(\sum_{i \in I_m} u_i y_i^a \Big) = \sum_{i \in I_m} (\sigma^{-1} u_i)^a y_i$$

By (8), the kernel of σ^* is nontrivial. Hence, there exist $y_i \in \mathcal{L}(n\mathfrak{o})$, $i \in I_m$, not all zero, with

$$\sum_{i \in I_m} (\sigma^{-1} u_i)^a y_i = 0. \tag{9}$$

In particular, $u = \sum_{i \in I_m} u_i y_i^a$ is a nonzero element of $\mathcal{L}(r\mathfrak{o})$. If $\mathfrak{p}$ is a prime divisor of F/K and $\mathfrak{p} \neq \mathfrak{o}$, then $\varphi_{\mathfrak{p}}(y_i) \neq \infty$ and $\varphi_{\mathfrak{p}}(u_i) \neq \infty$. If in addition $\varphi_{\mathfrak{p}}^\sigma = \varphi_{\mathfrak{p}}^q$, then (9) implies

$$\varphi_{\mathfrak{p}}(u) = \sum_{i \in I_m} \varphi_{\mathfrak{p}}(u_i) \varphi_{\mathfrak{p}}(y_i)^a = \sum_{i \in I_m} \varphi_{\mathfrak{p}}(\sigma^{-1} u_i)^q \varphi_{\mathfrak{p}}(y_i)^a$$

$$= \varphi_{\mathfrak{p}} \Big(\sum_{i \in I_m} (\sigma^{-1} u_i)^a y_i \Big)^a = 0$$

Hence, $\mathfrak{p}$ occurs in the divisor of zeros of u. In other words,

$$\sum_{\substack{\mathfrak{p} \neq \mathfrak{o} \\ \mathfrak{p}^\sigma = \mathfrak{p}^q}} \mathfrak{p} \leq \mathrm{div}_0(u).$$

Thus, $N^{(\sigma)} - 1 \leq \deg(\mathrm{div}_0(u)) = \deg(\mathrm{div}_\infty(u)) \leq r$ and the proof of (4) is complete.

4.8 A Lower Bound

In the notation of Section 4.7, we establish a lower bound inequality of the form $N^{(\sigma)} - (q+1) > c'\sqrt{q}$ where c' is an explicit constant depending only on F. This will complete the construction of Lemma 4.6.2 toward the proof of the Riemann hypothesis for F/K.

LEMMA 4.8.1: *Let F be a function field of one variable over a field K with q elements. Let F' be a finite Galois extension of F with a Galois group G such that K is also algebraically closed in F' and let $\sigma \in G$. Then $N^{(\sigma)}(F) = [F':F]^{-1} \sum_{\tau \in G} N^{(\sigma\tau)}(F')$.*

Proof: Let $\mathfrak{p}'$ be a prime divisor of F'/K and let $\mathfrak{p}$ be its restriction to F. Then $\varphi_{\mathfrak{p}}^\sigma = \varphi_{\mathfrak{p}}^q$ if and only if there exists $\tau \in G$ such that $\varphi_{\mathfrak{p}'}^{\sigma\tau} = \varphi_{\mathfrak{p}'}^q$. The number of such τ is the ramification index $e_{\mathfrak{p}}$ of $\mathfrak{p}'$ over $\mathfrak{p}$ (Section 2.3). Put $f_{\mathfrak{p}'/\mathfrak{p}} = [\bar{F}'_{\mathfrak{p}'} : \bar{F}_{\mathfrak{p}}]$ and denote the number of prime divisors of F' lying over $\mathfrak{p}$ by $g_{\mathfrak{p}'/\mathfrak{p}}$. Then:

$$\sum_{\tau \in G} N^{(\sigma\tau)}(F') = \sum_{\tau \in G} \sum_{\varphi_{\mathfrak{p}'}^{\sigma\tau} = \varphi_{\mathfrak{p}'}^q} \deg(\mathfrak{p}') = \sum_{\varphi_{\mathfrak{p}}^\sigma = \varphi_{\mathfrak{p}}^q} \sum_{\mathfrak{p}'|\mathfrak{p}} e_{\mathfrak{p}'/\mathfrak{p}} \deg(\mathfrak{p}')$$

$$= \sum_{\varphi_{\mathfrak{p}}^\sigma = \varphi_{\mathfrak{p}}^q} e_{\mathfrak{p}'/\mathfrak{p}} f_{\mathfrak{p}'/\mathfrak{p}} g_{\mathfrak{p}'/\mathfrak{p}} \deg(\mathfrak{p})$$

$$= [F':F] \sum_{\varphi_{\mathfrak{p}}^\sigma = \varphi_{\mathfrak{p}}^q} \deg(\mathfrak{p}) = [F':F] N^{(\sigma)}(F). \quad \square$$

90   Chapter 4. The Riemann Hypothesis for Function Fields

Let F be a function field of one variable over a field K of q elements and let σ be an automorphism of F over K of finite order. Denote the fixed field of σ in F by E. Then F is a finite Galois extension of E. As a finite field, K is perfect. Hence, there exists $x \in E$, transcendental over K, such that E is a finite separable extension of $K(x)$. Let $\hat{F}$ be the Galois closure of $F/K(x)$ and let $\hat{K}$ be the algebraic closure of K in $\hat{F}$. Then $\hat{F}$, as well as $F\hat{K}$, are function fields of one variable over $\hat{K}$ and σ extends to an automorphism of $\hat{F}$ over $\hat{K}(x)$. After an additional finite extension of $\hat{K}$ (if necessary), assume that Condition (1) of Section 4.7 holds for $\hat{F}/\hat{K}$, and therefore for $F\hat{K}/\hat{K}$ (use Proposition 3.4.2).

Starting with a given F we have extended the field of constants so as to assume these conditions:

(1a) F/K has a separating transcendence element x; the field F has a finite extension $\hat{F}$, which is Galois over $K(x)$, and K is algebraically closed in $\hat{F}$;
(1b) q is a square larger than $(\hat{g}+1)^4$ where $\hat{g}$ is the genus of $\hat{F}/\hat{K}$; and
(1c) $\hat{F}/\hat{K}$ has a prime divisor of degree 1.

LEMMA 4.8.2: *Under these conditions*

(2) $$N^{(\sigma)}(F) - (q+1) \geq -\frac{n-m}{m}(2\hat{g}+1)\sqrt{q},$$

where $m = [\hat{F} : F]$ *and* $n = [\hat{F} : K(x)]$.

Proof: Let $H = \mathrm{Gal}(\hat{F}/F)$ and $G = \mathrm{Gal}(\hat{F}/K(x))$. From Lemma 4.8.1

(3) $$N^{(\sigma)}(F) = \frac{1}{m}\sum_{\tau \in H} N^{(\sigma\tau)}(\hat{F}) \text{ and } q+1 = N(K(x)) = \frac{1}{n}\sum_{\theta \in G} N^{(\theta)}(\hat{F}).$$

Apply inequality (3) of Section 4.7:

$$\sum_{\theta \in G} N^{(\theta)}(\hat{F}) = \sum_{\tau \in H} N^{(\sigma\tau)}(\hat{F}) + \sum_{\theta \in G \smallsetminus \sigma H} N^{(\theta)}(\hat{F})$$
$$\leq \sum_{\tau \in H} N^{(\sigma\tau)}(\hat{F}) + \sum_{\theta \in G \smallsetminus \sigma H} (q+1+(2\hat{g}+1)\sqrt{q})$$
$$= \sum_{\tau \in H} N^{(\sigma\tau)}(\hat{F}) + (n-m)(q+1+(2\hat{g}+1)\sqrt{q}).$$

From the second half of (3),

$$\sum_{\tau \in H} N^{(\sigma\tau)}(\hat{F}) \geq n(q+1) - (n-m)(q+1+(2\hat{g}+1)\sqrt{q})$$
$$= m(q+1) - (n-m)(2\hat{g}+1)\sqrt{q}.$$

Thus, the first half of (3) implies that
$$N^{(\sigma)}(F) \geq (q+1) - \frac{n-m}{m}(2\hat{g}+1)\sqrt{q}. \quad \square$$

Note that the condition (1), as well as the numbers $\hat{g}, m, n$, are independent of extension of the field of constants. We may therefore combine Lemma 4.8.2 with the results of Section 4.7 to conclude:

PROPOSITION 4.8.3: *Let F be a function field of one variable over a finite field K and let σ be an automorphism of F over K of finite order. Then K has a finite extension K' with q' elements and there exists a positive constant c such that for every positive integer r we have*
$$|N^{(\sigma)}(F'_r) - ((q')^r - 1)| \leq c(q')^{r/2},$$
where K'_r is the unique extension of K' of degree r, $F'_r = F'K'_r$, and σ extends to an automorphism, also denoted by σ, of F'_r over K'_r.

In particular, Proposition 4.8.3 is valid in the case $\sigma = 1$. By Lemma 4.6.2, the Riemann hypothesis is true for the function field F'/K'. It follows from Lemma 4.6.1 that it is also true for F/K.

Exercises

1. For real valued functions f, g write
$$f(x) = O(g(x)) \text{ as } x \to a$$
if there exists a positive constant c such that $|f(x)| \leq c|g(x)|$ for all values of x in a neighborhood of a.

    Let F be a function field of one variable over a field K of q elements. Denote the set of prime divisors of F/K of degree r by $P_r(F/K)$. Follow these instructions to prove that
$$|P_r(F/K)| = \frac{1}{r}q^r + O(q^{r/2})$$

    (a) Use Theorem 4.5.2 to prove that $|P_1(F_r/K_r)| = q^r + O(q^{r/2})$.
    (b) Observe that if $\mathfrak{P} \in P_1(F_r/K_r)$ and if $\mathfrak{p}$ is the prime divisor of F/K that lies below $\mathfrak{P}$, then $\deg(\mathfrak{p})|r$.
    (c) Deduce from Lemma 4.3.1 that if $d|r$, then over each $\mathfrak{p} \in P_d(F/K)$ there lie exactly d elements of $P_1(F_r/K_r)$. Thus,
$$|P_1(F_r/K_r)| = \sum_{d|r} d|P_d(F/K)|.$$
    (d) Use the estimates $|P_d(F/K)| = O(q^d)$ and $d \leq \frac{r}{2}$ for proper divisors d of r and the inequality $q + q^2 + \cdots + q^{\frac{r}{2}} \leq 2q^{\frac{r}{2}}$ to complete the proof.

2. Let F be a function field of one variable over a field K of q elements and genus g. Let σ be an automorphism of F over K of finite order and let $N^{(\sigma)}$ be as in Section 4.7. Prove that $|N^{(\sigma)} - (q+1)| \leq 2g\sqrt{q}$. Hint: Extend σ to an automorphism $\tilde{\sigma}$ of $F\tilde{K}/\tilde{K}$ by $\tilde{\sigma}x = x^q$ for each $x \in \tilde{K}$. Then let E be the fixed field of $\tilde{\sigma}$ in $F\tilde{K}$. Show that E is a function field of one variable over K and that $E\tilde{K} = F\tilde{K}$.

3. This exercise establishes basic facts about infinite products which lie behind the multiplicative presentation of the zeta functions. Consider a sequence $\{z_i\}_{i=1}^\infty$ of nonzero complex numbers. If the sequence of the **partial products** $\prod_{i=1}^n z_i$ converges to a nonzero complex number z, we say that the infinite product $\prod_{i=1}^\infty z_i$ converges and that z is its **value**. We say that $\prod_{i=1}^n (1 + z_i)$ **absolutely converges** if $\prod_{i=1}^\infty (1 + |z_i|)$ converges.

The logarithm function makes a connection between the theory of infinite products and the theory of infinite series:

PROPOSITION ([Knopp, p. 434]): Let $a_i \neq -1$, $i = 1, 2, 3, \ldots$ be complex numbers. Then the product $\prod_{i=1}^\infty (1 + a_i)$ converges if and only if, the series $\sum_{i=1}^\infty \log(1+a_i)$ whose terms are the principal values of $\log(1+a_i)$ converges. If l is the sum of this series, then $\prod_{i=1}^\infty (1 + a_i) = e^l$.

(a) Prove that if $|z| \leq \frac{1}{2}$, then $|z| \leq 2|\log(1+z)| \leq 2|z|$.

(b) Suppose $0 < |a_i| \leq \frac{1}{2}$ for $i = 1, 2, 3, \ldots$ and consider the following series:

$$\sum_{i=1}^\infty |\log(1+a_i)|, \quad \sum_{i=1}^\infty |a_i|, \quad \sum_{i=1}^\infty \log(1+|a_i|).$$

Use (a) to prove that the convergence of each these series implies the convergence of the two others.

(c) Suppose $0 < |a_i| \leq \frac{1}{2}$ for $i = 1, 2, 3, \ldots$ and $\prod_{i=1}^\infty (1 - |a_i|)^{-1}$ converges. Then for each permutation π of $\mathbb{N}$, the product $\prod_{i=1}^\infty (1 - a_{\pi(i)})^{-1}$ converges and its value is independent of π.

4. Let F be a function field of one variable over a field K with q elements. Consider $\sigma \in \text{Aut}(F/K)$. As in Section 4.7, define $N^{(\sigma)}$ to be the sum of all $\deg(\mathfrak{p})$, where $\mathfrak{p}$ ranges over all prime divisors of F/K for which $\varphi_\mathfrak{p}^\sigma = \varphi_\mathfrak{p}^q$. Extend σ to an automorphism of $F\tilde{K}$ by the formula $\tilde{\sigma}x = x^q$ for each $x \in \tilde{K}$. Use Lemma 4.3.1 for r a multiple of $\deg(\mathfrak{p})$ to prove that $N^{(\sigma)}$ is also the number of prime divisors $\mathfrak{P}$ of $F\tilde{K}/\tilde{K}$ for which $\varphi_\mathfrak{P}^{\tilde{\sigma}} = \varphi_\mathfrak{P}^q$.

5. Let F be a function field of one variable over a field with q elements and of genus g. Let σ be an element of $\text{Aut}(F/K)$ of order n. Define $N^{(\sigma)}$ as in Exercise 4. Prove that

$$|N^{(\sigma)} - (q+1)| \leq 2g\sqrt{q}.$$

Hint: Let $\tilde{\sigma}$ be as in Exercise 4. Denote the fixed field of $\tilde{\sigma}$ in $F\tilde{K}$ by F'. Prove that F' is also the fixed field of $\tilde{\sigma}$ in FK_r for each multiple r of n. Deduce that F' is a function field of one variable over K such that $F\tilde{K} = F'\tilde{K}$. Use Exercise 4 to prove that $N^{(\sigma)}$ is the number of prime divisors of F'/K of degree 1. Then apply Theorem 4.5.2.

6. Let q be a power of an odd prime. Prove that $ax^2 + by^2 = c$ has a solution $(x,y) \in \mathbb{F}_q^2$ if $a,b,c \in \mathbb{F}_q^\times$. Let (x_0, y_0) be one of these solutions. Use the substitution $x = x_0 + tz$, $y = y_0 + t$ and solve for t in terms of z to show that the function field of $aX^2 + bY^2 - c = 0$ is isomorphic to $\mathbb{F}_q(z)$. Conclude that the function field has exactly $q+1$ degree 1 places. Hint: Assume $a = 1$ and observe that $c - by^2$ takes on $(q+1)/2$ values in $\mathbb{F}_q$. On the other hand there are $\frac{q+1}{2}$ squares in $\mathbb{F}_q$.

7. Let q be a prime power $\equiv 1 \bmod 3$ and let α be a generator of $\mathbb{F}_q^\times$. Follow these instructions to prove that $\alpha X^3 + \alpha^2 Y^3 = 1$ has a solution in $\mathbb{F}_q$:

Let $A_i = \{\alpha^i x^3 \mid x \in \mathbb{F}_q^\times\}$, $i = 0, 1, 2$ and note that $\alpha A_0 = A_1$, $\alpha A_1 = A_2$ and $\alpha A_2 = A_0$. Let $A_i + A_j = \{x + y \mid x \in A_i \wedge y \in A_j\}$. Now assume that $A_1 + A_2 \subseteq A_1 \cup A_2$ to obtain a contradiction. For this use the notation $A_1 + A_2 = m_1 A_1 \cup m_2 A_2$ to indicate that for each $a_i \in A_i$ there are m_i pairs $(x,y) \in A_1 \times A_2$ such that $x + y = a_i$ (independent of $a_i \in A_i$), $i = 1, 2$. Multiply this "equality" by α (resp. α^2) to compute $A_2 + A_0$ (resp. $A_0 + A_1$). For $a_0 \in A_0$, use this to compute the number of triples $(x, y, z) \in A_0 \times A_1 \times A_2$ such that $x + y + z = a_0$ in two ways: first compute $(A_0 + A_1) + A_2$ and then $A_0 + (A_1 + A_2)$. The resulting expressions for m_1 and m_2 will lead to the contradiction $m_1 = m_2 = 0$.

The attractiveness of the Riemann hypothesis for curves over finite fields has resulted in extensive lists of problems in a number of books treating the combinatorics of finite fields. For the sake of completeness we reference two such problem sources: [Ireland-Rosen, Chap. 8, pp. 105–107; Chap. 11, pp. 169–171] [Lidl-Niederreiter, see notes, Chap. 6, pp. 339–346]. In those chapters that use the Riemann hypothesis, our material and problems will tend to concentrate on the connections between the Riemann hypothesis and arithmetic properties of fields that are special to this book (e.g. an explicit form of Hilbert's irreducibility theorem for global fields that follows from the Riemann hypothesis - Theorem 13.3.4).

Notes

An extensive survey of the literature giving estimates on the number of points on an affine variety V appears in [Lidl-Niederreiter, pp. 317–339].

Although considerable literature on the Riemann hypothesis for curves over finite fields (Theorem 4.5.1) existed long before the two proofs given by Weil [Weil2], subsequent concerns included two sophisticated — and interrelated — developments. Both of Weil's proofs employed elements of the theory of algebraic geometry outside the domain of algebraic curves. Indeed, Weil's far reaching generalization of Theorem 4.5.1 was suggested to him by the latter proof. This generalization, now called the Riemann hypothesis for nonsingular projective varieties over finite fields (proved by Deligne in [Deligne]). Since, however, Weil's theorem had so many applications to apparently elementary results about finite fields, many practitioners were anxious for a more accessible proof.

Stepanov [Stepanov1] was the first to make serious progress on an *elementary proof* (e.g. not applying the theory of algebraic surfaces) of Weil's result. He introduced elements of diophantine approximation to the problem in the case of hyperelliptic curves, in a style suggested by the Thue-Siegel-Roth theorem. To do this he constructed an auxiliary nonvanishing function on the hyperelliptic curve with a "lot" of prescribed zeros of "high" multiplicity. Eventually [Stepanov2] realized Theorem 4.5.1 for all hyperelliptic curves over finite fields. Continuing with the purely diophantine approximation approach, Stepanov [Stepanov3], for prime fields (and some extra conditions on the equations) and W. M. Schmidt [W.M.Schmidt1 and W.M.Schmidt2] were able to prove Theorem 4.5.1 for all curves. Indeed, Schmidt's method were even applicable to prove results like those of Deligne even for some complete intersections.

Our proof, however, of Theorem 4.5.1 follows [Bombieri1]. In this proof the Riemann hypotheses replaces diophantine approximation to give the construction of the auxiliary functions that appear in Stepanov's proof (the function u at the end of Section 4.7).

Voloch [Voloch] has an elementary proof of the Riemann hypothesis for function fields that sometimes gives a better bound than Weil's estimate (see also [Stöhr-Voloch]).

Chapter 5. Plane Curves

The estimate on the number of prime divisors of degree 1 of a function field F over $\mathbb{F}_q$ (Theorem 4.5.2) leads in this chapter to an estimate on the number N of K-rational zeros of an absolutely irreducible polynomial $f \in \mathbb{F}_q[X,Y]$. We prove (Theorem 5.4.1) that $|N + (q+1)| \leq (d-1)(d-2)\sqrt{q} + d$, where $d = \deg(f)$.

5.1 Affine and Projective Plane Curves

Let K be a field and let Ω be an algebraically closed extension of K of infinite transcendence degree over K. We denote by $\mathbb{A}^2$ the **affine plane**: all pairs $(x,y) \in \Omega^2$. We denote by $\mathbb{P}^2$ the **projective plane**: all nonzero triples $\mathbf{x} = (x_0, x_1, x_2) \in \Omega^3$ modulo the equivalence relation $\mathbf{x} \sim \mathbf{x}'$ if and only if $\mathbf{x}' = c\mathbf{x}$ for some $c \in \Omega$. We denote the equivalence class of (x_0, x_1, x_2) by $(x_0{:}x_1{:}x_2)$. Embed the affine plane $\mathbb{A}^2$ in $\mathbb{P}^2$ by the map $(x,y) \to (1{:}x{:}y)$. With this understood, the points of $\mathbb{A}^2$ are then referred to as the **finite points** on $\mathbb{P}^2$, whereas the points of the form $(0{:}x_1{:}x_2)$ are the **points at infinity** on $\mathbb{P}^2$.

An **affine plane curve defined over K** is a set

(1) $$\Gamma = \{(x,y) \in \mathbb{A}^2 \mid f(x,y) = 0\},$$

where $f \in K[X,Y]$ is a nonconstant absolutely irreducible polynomial. Write it in the form

(2) $$f(X,Y) = f_d(X,Y) + f_{d-1}(X,Y) + \cdots + f_0(X,Y)$$

where $f_k(X,Y)$ is a homogeneous polynomial of degree k, for $k = 0, \ldots, d$, and $f_d(X,Y) \neq 0$. Then d is the **degree** of Γ. Attach to f the homogeneous polynomial $f^*(X_0, X_1, X_2)$ of degree d:

(3) $$f^*(X_0, X_1, X_2) = f_d(X_1, X_2) + X_0 f_{d-1}(X_1, X_2) + \cdots + X_0^d f_0(X_1, X_2)$$

and let
$$\Gamma^* = \{(x_0{:}x_1{:}x_2) \in \mathbb{P}^2 \mid f^*(x_0, x_1, x_2) = 0\}$$

Then Γ^* is the **projective plane curve corresponding** to Γ. It is also called the **projective completion** of Γ. We have $\Gamma = \Gamma^* \cap \mathbb{A}^2$, and $\Gamma^* \smallsetminus \Gamma$ is a finite set corresponding to the points of $f_d(X_1, X_2) = 0$ in $\mathbb{P}^1$. The infinite points of Γ^* are sometimes referred to as the **points at infinity** on Γ.

If $f(X,Y) = a + bX + cY$ with $a, b, c \in K$ and $b \neq 0$ or $c \neq 0$ (i.e. $\deg(f) = 1$), then Γ is a **line**. The points on the corresponding projective line satisfy $aX_0 + bX_1 + cX_2 = 0$. The **line at infinity** is given by $X_0 = 0$.

Call a point (x,y) of an affine curve Γ **generic** if $\mathrm{trans.deg}_K K(x,y) = 1$. Because f is absolutely irreducible, the field $F = K(x,y)$ is a regular extension of K. Thus, F is a function field of one variable which we call the **function field** of Γ over K. Up to K-isomorphism it is independent of the choice of the generic point.

The map $(X,Y) \mapsto (x,y)$ extends to a K-epimorphism of rings $K[X,Y] \to K[x,y]$ with $f(X,Y)K[X,Y]$ as the kernel (Gauss' lemma). Thus, a polynomial $g \in K[X,Y]$ vanishes on Γ if and only if $g(x,y) = 0$, or, equivalently, g is a multiple of f.

Define the **genus** of Γ (and of Γ^*) to be the genus of F.

The **coordinate ring** of Γ over K is $R = K[x,y]$. There is a bijective correspondence between the set of all maximal ideals $\mathfrak{m}$ of R with quotient field $R/\mathfrak{m} = K$ and the set $\Gamma(K)$ of all K-rational points of Γ. If $(a,b) \in \Gamma(K)$, then the corresponding maximal ideal of R is $\mathfrak{m} = \{g(x,y) \in R \mid g(a,b) = 0\}$. If $\mathfrak{p}$ is a prime ideal of R and $\mathfrak{p} \subseteq \mathfrak{m}$, then the transcendence degree of the quotient field of $R/\mathfrak{p}$ is either 0 or 1. In the latter case $\mathfrak{p} = 0$, and in the former case $R/\mathfrak{p}$ is already a field so that $\mathfrak{p} = \mathfrak{m}$.

The **local ring** of Γ at $\mathbf{a} = (a,b)$ over K is the local ring of R at $\mathfrak{m}$:

$$O_{\Gamma,(a,b),K} = R_{\mathfrak{m}} = \left\{ \frac{g(x,y)}{h(x,y)} \mid g(x,y), h(x,y) \in R \text{ and } h(a,b) \neq 0 \right\}.$$

The unique nonzero prime ideal of $R_{\mathfrak{m}}$ is generated by the elements of $\mathfrak{m}$. As a local ring of a Noetherian domain, $R_{\mathfrak{m}}$ is itself a Noetherian domain.

Similarly, a point $(x_0{:}x_1{:}x_2)$ of Γ^* with $x_0 \neq 0$ is said to be **generic** over K, if $\left(\frac{x_1}{x_0}, \frac{x_2}{x_0}\right)$ is a generic point of Γ. Define the **local ring** of a point $\mathbf{a} = (a_0{:}a_1{:}a_2)$ of Γ^* over K as

$$O_{\Gamma^*,\mathbf{a},K} = \left\{ \frac{g(x_0,x_1,x_2)}{h(x_0,x_1,x_2)} \middle| \begin{array}{l} g,h \text{ are homogeneous polynomials of} \\ \text{the same degree and } h(a_0,a_1,a_2) \neq 0 \end{array} \right\}$$

If $a_0 \neq 0$, then this ring coincides with the local ring of the corresponding point $\left(\frac{a_1}{a_0}, \frac{a_2}{a_0}\right)$ of Γ. If $a_1 \neq 0$, then regard $\mathbf{a}$ as a point on the affine curve Γ_1, defined by the equation $f^*(X_0,1,X_2) = 0$. Then the projective completion of Γ_1 is also Γ^*, but the line $X_1 = 0$ is taken as "the line at infinity". This shows that the local ring of a projective plane curve at each point is equal to the local ring of some affine curve at a point. It is therefore a Noetherian domain.

In any case, call a K-rational point of Γ (or of Γ^*) K-**normal** if its local ring over K is integrally closed.

Two affine plane curves Γ_1 and Γ_2 are K-**isomorphic** if their coordinate rings are K-isomorphic. In this case generators of the coordinate ring of Γ_2 (resp. Γ_1) can be expressed as polynomials in the generators of the coordinate ring of Γ_1 (resp. Γ_2), and the composition of these two polynomial maps is the identity when applied to generators of the coordinate ring of Γ_1 (resp. Γ_2).

Two projective plane curves Γ_1^* and Γ_2^* are **K-isomorphic** if
(4a) for each $\mathbf{x} \in \Gamma_1^*$ there exist homogeneous polynomials g_0, g_1, g_2 in $K[X_0, X_1, X_2]$ of the same degree such that $y_i = g_i(\mathbf{x}) \neq 0$ for at least one i, $0 \leq i \leq 2$, $\mathbf{y} \in \Gamma_2^*$, and there exist homogeneous polynomials h_0, h_1, h_2 in $K[X_0, X_1, X_2]$ of the same degree such that $h_i(\mathbf{y}) = x_i$, $i = 0, 1, 2$; and
(4b) the same condition with the roles of Γ_1^* and Γ_2^* exchanged.

For example, if g_0, g_1, g_2 are linear polynomials with a nonsingular coefficient matrix, then (g_0, g_1, g_2) is called **nonsingular homogeneous linear transformation**.

The function fields of two isomorphic plane curves are K-isomorphic. So are the local rings of corresponding points. It follows that the genera of isomorphic plane curves are the same; and if $\mathbf{p}_1$ and $\mathbf{p}_2$ are corresponding points of the curves, then $\mathbf{p}_1$ is K-normal if and only if $\mathbf{p}_2$ is K-normal. In particular, both curves have the same number of nonnormal points.

If Γ is a plane curve defined over a field K and L is an algebraic extension of K, then Γ is also defined over L. The function field F_L of Γ over L is the extension of F by the field of constants L. By Proposition 3.4.2, the genus of Γ remains unchanged if L is separable over K.

5.2 Points and Prime Divisors

Let Γ be an affine plane curve of degree d defined by an absolutely irreducible equation $f(X, Y) = 0$ over a field K. We establish a bijective correspondence between the K-rational points of f and the prime divisors of the function field of f of degree 1 (Lemma 5.2.2). We prove that a K-rational of Γ is normal if and only if it is simple (Lemma 5.2.3).

Let (x, y) be a generic point of Γ over K. Denote the coordinate ring and the function field, respectively, of Γ over K by $R = K[x, y]$ and $F = K(x, y)$. Consider a K-rational point (a, b) of Γ. Then the map $(x, y) \mapsto (a, b)$ uniquely extends to a K-homomorphism φ of the local ring $O_{(a,b)}$ into K. The homomorphism φ extends further (not necessarily uniquely) to a $\tilde{K}$-valued place φ' of F. Call (a, b) the **center** of the corresponding prime divisor $\mathfrak{p}$ of F/K. Then $\mathfrak{p}$ lies over the unique prime divisor $\mathfrak{p}_0$ of $K(x)/K$ determined by the map $x \mapsto a$. Since $[F : K(x)] \leq d$, there exist at most d prime divisors $\mathfrak{p}$ of F/K with the point (a, b) as a center on Γ. Since each point of Γ^* is a finite point of some affine representative of Γ^* (Section 5.1) this holds for each K-rational point of the projective completion Γ^* of Γ.

If a point $\mathbf{p}$ of $\Gamma^*(K)$ is K-normal, then its local ring $O_\mathbf{p}$ is a Noetherian integrally closed domain. By Section 5.1, each nonzero prime ideal of $O_\mathbf{p}$ is maximal. Hence, by Proposition 2.4.5, $O_\mathbf{p}$ is a discrete valuation domain. Therefore, there exists a unique prime divisor $\mathfrak{p}$ of F/K with $\mathbf{p}$ as a center on Γ. The degree of $\mathfrak{p}$ is 1.

Conversely, consider a prime divisor $\mathfrak{p}$ of F/K of degree 1. Suppose that $\varphi_\mathfrak{p}$ is finite on R. Then $(a, b) = \varphi_\mathfrak{p}(x, y)$ is a K-rational point of Γ and it is the center of $\mathfrak{p}$ on Γ.

The following result shows that the number of prime divisors $\mathfrak{p}$ of F/K which are not finite on R does not exceed d.

PROPOSITION 5.2.1 (Noether's Normalization Theorem): *Let $K[x_1,\ldots,x_n]$ be a finitely generated integral domain over a field K with a quotient field F. If the transcendence degree of F over K is r, then there exist elements $t_1,\ldots,t_r$ in $K[\mathbf{x}]$ such that $K[\mathbf{x}] = K[t_1,\ldots,t_r,x_1,\ldots,x_{n-r}]$ (after rearranging the x_i's) and $K[\mathbf{x}]$ is integral over $K[\mathbf{t}]$* [Lang4, p. 22].

If K is an infinite field, then $t_1,\ldots,t_r$ can be chosen to be linear combinations of $x_1,\ldots,x_r$ with coefficients in K [Zariski-Samuel1, p. 266].

Lemma 19.5.1 gives a constructive version of Noether's normalization theorem.

Return now to the plane curve Γ. Noether's normalization theorem, in its linear form, allows us to replace x by a linear combination t of x and y such that R is integral over $K[t]$ and $[F:K(t)] \leq d$. If K is finite, replace K by a suitable finite extension to achieve the linear dependence of t on x and y. If a place $\varphi_\mathfrak{p}$ of F is not finite on R, then it is infinite at t (Lemma 2.4.1). Hence, $\mathfrak{p}$ lies over the infinite prime divisor $\mathfrak{p}_\infty$ of $K(t)/K$. There are at most d prime divisors of F/K that lie over $\mathfrak{p}_\infty$. This proves our contention. Now we summarize.

LEMMA 5.2.2: *Let Γ be an affine plane curve of degree d defined over a field K. Denote the coordinate ring and the function field, respectively, of Γ over K by R and F. Then:*
(a) *For each K-normal point $\mathbf{p} \in \Gamma(K)$ there exists exactly one prime divisor $\mathfrak{p}$ of F/K with center $\mathbf{p}$ on Γ; the degree of $\mathfrak{p}$ is 1.*
(b) *There are at most d prime divisors of F/K whose centers on Γ are a given point $\mathbf{p}$.*
(c) *If a prime divisor $\mathfrak{p}$ of F/K is of degree 1 and if $\varphi_\mathfrak{p}$ is finite on R, then its center is in $\Gamma(K)$.*
(d) *There are at most d prime divisors of F/K which are not finite on R.*

Finally, we point out that the K-normal points and simple points on Γ are the same. Here, a point (a,b) on Γ is **simple** if $\frac{\partial f}{\partial X}(a,b) \neq 0$ or $\frac{\partial f}{\partial Y}(a,b) \neq 0$.

LEMMA 5.2.3: *A K-rational point (a,b) of Γ is normal if and only if it is simple. Moreover, if $\frac{\partial f}{\partial X}(a,b) \neq 0$, then F has a discrete normalized valuation v with $v(y-b) = 1$.*

Proof: Suppose first that (a,b) is normal. Then, its local ring $R = O_{(a,b)}$ over K is a discrete valuation domain. Hence, $R(x-a) \subseteq R(y-b)$ or $R(y-b) \subseteq R(x-a)$.

Suppose for example that $R(x-a) \subseteq R(y-b)$. Then there exist $g,h,q \in K[X,Y]$ such that $h(a,b) \neq 0$ and
$$h(X,Y)(X-a) = g(X,Y)(Y-b) + q(X,Y)f(X,Y).$$

Apply $\frac{\partial}{\partial X}$ to both sides and substitute (a,b) for (X,Y):

$$h(a,b) = q(a,b)\frac{\partial f}{\partial X}(a,b).$$

Hence, $\frac{\partial f}{\partial X}(a,b) \neq 0$ and therefore (a,b) is simple.

Conversely, suppose that $\frac{\partial f}{\partial X}(a,b) \neq 0$. We show that $y - b$ generates the maximal ideal $\mathfrak{m}$ of R. Indeed,

$$f(X,Y) = \frac{\partial f}{\partial X}(a,b)(X-a) + \frac{\partial f}{\partial Y}(a,b)(Y-b) + \text{ higher terms,}$$

and

$$0 = f(x,y) = \frac{\partial f}{\partial X}(a,b)(x-a) + \frac{\partial f}{\partial Y}(a,b)(y-b) + (x-a)u + (y-b)v,$$

with $u, v \in \mathfrak{m}$. Hence,

$$0 = \left(\frac{\partial f}{\partial X}(a,b) + u\right)(x-a) + \left(\frac{\partial f}{\partial Y}(a,b) + v\right)(y-b).$$

But, since $\frac{\partial f}{\partial X}(a,b)$ is a nonzero constant and $u \in \mathfrak{m}$, the coefficient of $x - a$ is a unit of R. Therefore, $x - a \in R(y-b)$, so $\mathfrak{m} = R(x-a, y-b) = R(y-b)$.

Since R is Noetherian, the ideal $\mathfrak{a} = \bigcap_{n=1}^{\infty} \mathfrak{m}^n$ is a finitely generated R-module. Since $\mathfrak{m}\mathfrak{a} = \mathfrak{a}$, Nakayama's Lemma [Lang4, p. 195, Prop. 1] implies that $\mathfrak{a} = 0$ (alternatively, use Krull's intersection theorem [Eisenbud, p. 152]). Hence, each $z \in R$ has a unique representation $z = w(y-b)^n$, with w a unit of R and $n \geq 0$. Thus, R is a discrete valuation domain. Therefore, R is integrally closed (Exercise 3 of Chapter 2).

Let v be the normalized discrete valuation of F corresponding to R. Since $y - b$ generates $\mathfrak{m}$, we have $v(y-b) = 1$. $\square$

For a projective curve Γ^* defined by $f^*(X_0, X_1, X_2) = 0$ as in (3) of Section 5.1, the point $(a_0{:}a_1{:}a_2)$ is **simple** if $\frac{\partial f}{\partial X_i}(a_0, a_1, a_2) \neq 0$, for some i, $i = 0, 1$ or 2. Otherwise, $(a_0{:}a_1{:}a_2)$ is **singular**.

5.3 The Genus of a Plane Curve

Here we bound the genus and the number of nonnormal points of a plane curve by a function of its degree. We first prove a finiteness result for the coordinate ring of a plane curve.

LEMMA 5.3.1: *Let $R = K[x,y]$ be an integral domain with quotient field F of transcendence degree 1 over K. Let S be the integral closure of R in F. Then S/R is a finitely generated K-vector space.*

Proof: It is well known that S is a finitely generated R-module ([Lang4, p. 120] or [Zariski-Samuel1, p. 267]). Hence, there exists a nonzero element

$z \in R$ such that $zS \subseteq R$. Thus, $S/R \cong zS/zR \subseteq R/zR$. It suffices to prove that $\dim_K R/zR < \infty$.

By Noether's normalization theorem (Proposition 5.2.1), we may assume without loss that R is integral over $K[x]$. The element z satisfies an equation of the form $z^m + a(x)z^{m-1} + \cdots + g(x) = 0$, with $0 \neq g(x) \in K[x]$ of degree, say, k. Since $g(x)$ belongs to zR, every power of x is a linear combination modulo zR of $1, x, \ldots, x^{k-1}$ with coefficients in K.

If d is the degree of a monic equation for y over $K[x]$, then each element of R can be written as $h(x, y)$, with $h \in K[X, Y]$ is of degree at most $d - 1$ in Y. Our lemma follows. □

In the notation of the proof of Lemma 5.3.1, S is the integral closure in F of the Dedekind domain $K[x]$. Hence, S is a Dedekind domain and therefore every nonzero prime ideal of S is maximal. By Section 5.2, every nonzero prime ideal of R is maximal.

Consider a nonzero prime ideal $\mathfrak{p}$ of R. Let $R_\mathfrak{p}$ be the local ring of R at $\mathfrak{p}$. Then the integral closure of $R_\mathfrak{p}$ in F is $S_\mathfrak{p} = \{\frac{s}{b} \mid s \in S \text{ and } b \in R \smallsetminus \mathfrak{p}\}$.

LEMMA 5.3.2:
(a) The quotient S/R is isomorphic to the direct sum $\sum S_\mathfrak{p}/R_\mathfrak{p}$, where $\mathfrak{p}$ runs over the nonzero prime ideals of R.
(b) Let $\mathfrak{p}_1, \ldots, \mathfrak{p}_k$ be the prime ideals of S that lie over a prime ideal $\mathfrak{p}$ of R. Then $\dim_K S_\mathfrak{p}/R_\mathfrak{p} \geq k - 1$.

Proof of (a): Denote the **conductor** of S over R by $\mathfrak{c} = \{c \in R \mid cS \subseteq R\}$. It is a nonzero ideal of both R and S. The local ring $R_\mathfrak{p}$ is integrally closed, and therefore equal to $S_\mathfrak{p}$, if and only if $\mathfrak{c} \not\subseteq \mathfrak{p}$ [Zariski-Samuel1, p. 269]. Let $\mathfrak{c} = \prod_{i=1}^{k} \prod_{j=1}^{l(i)} \mathfrak{p}_{ij}^{e_{i,j}}$ be a factorization of $\mathfrak{c}$ into the product of prime ideal powers of S where $\mathfrak{p}_{i1}, \ldots, \mathfrak{p}_{i,l(i)}$ all lie over the same prime ideal $\mathfrak{p}_i$ of R, $i = 1, \ldots, k$. Then $S_{\mathfrak{p}_{ij}} \subseteq R_{\mathfrak{p}_i}$, $i = 1, \ldots, k$. If $\mathfrak{p}$ is a nonzero prime ideal of R not in the set $\{\mathfrak{p}_1, \ldots, \mathfrak{p}_k\}$, then $R_\mathfrak{p} = S_\mathfrak{p}$. Hence, with $R_i = R_{\mathfrak{p}_i}$ and $S_i = S_{\mathfrak{p}_i}$, $i = 1, \ldots, k$, Lemma 2.4.4 gives $S \cap R_1 \cap \cdots \cap R_k = \bigcap R_\mathfrak{p} = R$, where $\mathfrak{p}$ runs over the nonzero prime ideals of R. Therefore, the map $s + R \mapsto (s + R_1, \ldots, s + R_k)$ for $s \in S$, is an injective homomorphism into $\sum_{i=1}^{k} S_i/R_i$. The proof of the lemma is complete if we show that the map is surjective.

Let $s_1, \ldots, s_k$ be elements of S and let $a_1, \ldots, a_k \in R$ such that $a_i \notin \mathfrak{p}_i$. The ideals $\mathfrak{q}_i = \prod_{j=1}^{l(i)} \mathfrak{p}_{ij}^{e_{i,j}}$ of S are pairwise relatively prime, $i = 1, \ldots, k$. Hence, by the weak approximation theorem (Proposition 2.1.1), there exist $s \in S$, $q_i \in \mathfrak{q}_i$, and $a'_i \in R \smallsetminus \mathfrak{p}_i$ such that

$$s - \frac{s_i}{a_i} = \frac{q_i}{a'_i} \qquad i = 1, \ldots, k.$$

For each i choose $b_i \in (R \smallsetminus \mathfrak{p}_i) \cap \prod_{j \neq i} \mathfrak{q}_j$. Then $c_i = b_i q_i \in \mathfrak{c} \subseteq R$ and $s = \frac{s_i}{a_i} + \frac{c_i}{b_i a'_i}$. Thus, $s + R_i = \frac{s_i}{a_i} + R_i$ for $i = 1, \ldots, k$.

Proof of (b): Now let $\mathfrak{p}_1, \ldots, \mathfrak{p}_k$ be the prime ideals of S that lie over $\mathfrak{p}$. For $i = 1, \ldots, k - 1$ use the weak approximation theorem to find $s_i \in S$

5.3 The Genus of a Plane Curve

such that $s_i \equiv 1 \mod \mathfrak{p}_i$ and $s_i \equiv 0 \mod \mathfrak{p}_j$, for $j = 1, \ldots, k$ and $j \neq i$. We need only show that $s_1, \ldots, s_{k-1}$ are K-linearly independent modulo $R_{\mathfrak{p}}$. Indeed, if $r = \sum_{i=1}^{k-1} a_i s_i \in R_{\mathfrak{p}}$ with $a_i \in K$ (and $s_i \in \mathfrak{p}_k$), then, r belongs to $\mathfrak{p}R_{\mathfrak{p}}$. Since $a_j \equiv \sum_{i=1}^{k-1} a_i s_i \equiv 0 \mod \mathfrak{p}_k$, this gives $a_j = 0$, $j = 1, \ldots, k-1$. Consequently, $\dim_K S_{\mathfrak{p}}/R_{\mathfrak{p}} \geq k - 1$. ☐

LEMMA 5.3.3: *Let Γ be an affine plane curve defined over a field K with a generic point $\mathbf{x} = (x_1, x_2)$ and let $\mathbf{a} \in \Gamma(K)$. Consider a separable algebraic extension L of K and let $O_{\mathbf{a},K}$ (resp. $O'_{\mathbf{a},K}$) be the local ring of Γ at $\mathbf{a}$ over K (resp. its integral closure). Then*
(a) $O_{\mathbf{a},L} \cap K(\mathbf{x}) = O_{\mathbf{a},K}$; *and*
(b) $\dim_K O'_{\mathbf{a},K}/O_{\mathbf{a},K} \leq \dim_L O'_{\mathbf{a},L}/O_{\mathbf{a},L}$.

Proof: Without loss assume that L/K is finite. Let $w_1, \ldots, w_n$ be a linear basis for L/K. Since Γ is defined by an absolutely irreducible polynomial, $K(\mathbf{x})/K$ is a regular extension of K. Hence, $K(\mathbf{x})$ and L are linearly disjoint over K. Therefore, $w_1, \ldots, w_n$ is also a linear basis for $L(\mathbf{x})/K(\mathbf{x})$.

Consider $u \in O_{\mathbf{a},L} \cap K(\mathbf{x})$. Then $u = f(\mathbf{x})/g(\mathbf{x})$ with $f(\mathbf{x}), g(\mathbf{x}) \in L[\mathbf{x}]$ and $g(\mathbf{a}) \neq 0$. Write $f(\mathbf{x}) = \sum f_i(\mathbf{x})w_i$ and $g(\mathbf{x}) = \sum g_i(\mathbf{x})w_i$, with $f_i(\mathbf{x}), g_i(\mathbf{x}) \in K[\mathbf{x}]$, $i = 1, \ldots, n$. From $\sum u g_i(\mathbf{x})w_i = \sum f_i(\mathbf{x})w_i$ deduce that $u g_i(\mathbf{x}) = f_i(\mathbf{x})$, $i = 1, \ldots, n$. Since there is an i with $g_i(\mathbf{a}) \neq 0$, we have $u \in O_{\mathbf{a},K}$. This proves (a).

To prove (b), consider $u_1, \ldots, u_m \in O'_{\mathbf{a},K}$ linearly dependent over L modulo $O_{\mathbf{a},L}$. We show that they are also linearly dependent over K modulo $O_{\mathbf{a},K}$. Indeed, there exist $b_1, \ldots, b_m \in L$ not all zero and $v \in O_{\mathbf{a},L}$ such that $\sum b_i u_i = v$. Write $b_i = \sum_{j=1}^{n} a_{ij} w_j$ with $a_{ij} \in K$ and let $v_j = \sum_{i=1}^{m} a_{ij} u_i$. Then $v_j \in K(\mathbf{x})$ and $\sum v_j w_j = v$. For each $K(\mathbf{x})$-embedding σ of $L(\mathbf{x})$ into the algebraic closure of $K(\mathbf{x})$ we have $\sum v_j w_j^\sigma = v^\sigma$. Since $L(\mathbf{x})/K(\mathbf{x})$ is separable, $\det(w_j^\sigma) \neq 0$ [Lang7, p. 286, Cor. 5.4]. Apply Cramer's rule to solve the system $\sum v_j w_j^\sigma = v^\sigma$, all σ, and write v_j as a linear combination of the v^σ's with coefficients in the Galois closure $\hat{L}$ of L/K. Since $\mathbf{a}$ is K-rational, each v^σ is in $O_{\mathbf{a},\hat{L}}$. Thus, $v_j \in O_{\mathbf{a},\hat{L}} \cap K(\mathbf{x})$. Hence, by (a), $v_j \in O_{\mathbf{a},K}$. Therefore, the u_i's are linearly dependent over K modulo $O_{\mathbf{a},K}$. ☐

PROPOSITION 5.3.4: *Let K be an algebraically closed field and Γ^* a projective plane curve of degree d and genus g defined over K. Then*

(1) $$g = \frac{1}{2}(d-1)(d-2) - \sum \dim_K O'_{\mathbf{p}}/O_{\mathbf{p}},$$

where $\mathbf{p}$ runs over the points of $\Gamma^*(K)$, $O_{\mathbf{p}}$ denotes the local ring of Γ^* at $\mathbf{p}$ and $O'_{\mathbf{p}}$ is the integral closure of $O_{\mathbf{p}}$ in the function field of Γ^* over K.

Proof: The five parts of the proof draw information from the form of Γ^* after a change of variables that puts d distinct points at infinity.

PART A: *A linear transformation.* Choose a line L over K passing through no singular point of Γ^* and is not tangent to Γ^*. Then L cuts Γ^* in d distinct simple points $\mathfrak{p}_1,\ldots,\mathfrak{p}_d$ of Γ^* [Seidenberg, p. 37]. By Lemma 5.2.3, they are normal. After a suitable linear homogeneous transformation [Seidenberg, Chapter 5] we may assume that L is the line at infinity and that the infinite points on the X-axis and the Y-axis, $(0{:}1{:}0)$ and $(0{:}0{:}1)$ do not belong to the set $\{\mathfrak{p}_1,\ldots,\mathfrak{p}_d\}$. Such a transformation does not change the genus or the degree of the curve. Moreover, corresponding points have the same local rings. View Γ^* as the projective completion of an affine curve Γ defined by an equation $f(X,Y) = 0$, as given by (1) of Section 5.1. Let (x,y) be a generic point of Γ over K, let $R = K[x,y]$ be the coordinate ring of Γ, and let F be its function field.

PART B: *The divisor at ∞.* By Part A, $\mathfrak{p}_i = (0{:}a_i{:}1)$, with $a_i \in K^\times$ for $i = 1,\ldots,d$, and $a_1,\ldots,a_d$ distinct. In the notation of (2) of Section 5.1, obtain the factorization

$$(2) \qquad f_d(X,Y) = c\prod_{i=1}^{d}(X - a_iY), \text{ with } 0 \neq c \in K.$$

Therefore, $f(x,Y)$ is an irreducible polynomial of degree d over $K[x]$. Thus, R is integral over $K[x]$ and each element of R can be uniquely expressed as a polynomial $h(x,y)$ with coefficients in $K[x]$ such that $\deg_Y(h) \leq d - 1$. Similarly R is integral over $K[y]$.

Over the infinite prime divisor $\mathfrak{p}_\infty$ of $K(x)$ there lie exactly d distinct prime divisors $\mathfrak{p}_1,\ldots,\mathfrak{p}_d$ of F/K, with $\mathfrak{p}_i$ being the unique prime divisor with center $\mathfrak{p}_i$. In particular, $\mathfrak{p}_\infty$ is unramified in F (Proposition 2.3.2) and we may therefore normalize $v_{\mathfrak{p}_i}$ such that $v_{\mathfrak{p}_i}(x) = -1$. Since R is integral over $K[x]$, this implies $v_{\mathfrak{p}_i}(y) < 0$, so $v_{\mathfrak{p}_i}\left(\frac{x}{y}\right) \geq 0$. by (2) of Section 5.1, $c\prod_{i=1}^{d}\left(\frac{x}{y} - a_i\right) + f_{d-1}\left(\frac{x}{y},1\right)y^{-1} + \cdots + f_0\left(\frac{x}{y},1\right)y^{-d} = 0$. Hence, $\frac{x}{y}$ has residue a_i at $\mathfrak{p}_i$, so $v_{\mathfrak{p}_i}(y) = -1$. Let $\mathfrak{o} = \mathfrak{p}_1 + \cdots + \mathfrak{p}_d$. Then the pole divisors $\mathrm{div}_\infty(x)$ and $\mathrm{div}_\infty(y)$ of x and y in F are both $\mathfrak{o}$.

PART C: *The integral closure of R in F.* Let S be the integral closure of R in F. By Lemma 3.1.1, $S = \bigcup_{n=1}^{\infty} S_n$, where $S_n = \mathcal{L}(n\mathfrak{o})$. Let $R_n = R \cap S_n$, $n = 1, 2, \ldots$. By Lemma 5.3.1, $\dim_K S/R < \infty$. Hence, for n sufficiently large $R + S_n = R + S_{n+1} = R + S_{n+2} = \ldots$. Therefore, $S = R + S_n$, so $S_n/R_n \cong S/R$. For $n > 2g - 2$, Lemma 3.2.2(d) implies that $\dim_K S_n = nd - g + 1$. It follows that

$$(3) \qquad nd - g + 1 = \dim_K R_n + \dim_K S/R.$$

PART D: For $n > d$, we have

$$(4) \qquad R_n = \{h(x,y) \in R \mid \deg(h) \leq n \text{ and } \deg_Y(h) \leq d - 1\}.$$

5.3 The Genus of a Plane Curve

Indeed, let $k \geq 0$ be an integer and $h_k \in K[X,Y]$ a homogeneous polynomial of degree k with $\deg_Y(h_k) \leq d-1$. If $i+j=k$, then $\operatorname{div}_\infty(x^i y^j) = k\mathbf{o}$ (Part B). Hence, $h_k(x,y) \in S_k$.

We prove that $h_k(x,y) \notin S_{k-1}$: Since K is algebraically closed, we may factor $h_k(x,y)$ as

$$h_k(x,y) = \prod_{j=1}^{d-1}(x - b_j y) \cdot ax^{k-d+1}$$

with $b_j \in K$ and $a \in K^\times$. Then

$$v_{\mathfrak{p}_i}(h_k(x,y)) = \sum_{j=1}^{d-1} v_{\mathfrak{p}_i}(x - b_j y) - (k - d + 1).$$

If $h_k(x,y) \in S_{k-1}$, then for each i we have $\sum_{j=1}^{d-1} v_{\mathfrak{p}_i}(x - b_j y) - (k-d+1) + k - 1 \geq 0$. Hence, there exists $1 \leq j \leq d-1$ with $v_{\mathfrak{p}_i}(x - b_j y) \geq 0$. Therefore, $b_j = a_i$. But this implies that two of the a_i's are equal, a contradiction.

To complete Part D, write each $h(x,y)$ in $K[x,y]$ with $\deg_Y(h) \leq d-1$ as $h(x,y) = \sum_{k=1}^{m} h_k(x,y)$ with h_k as above and $h_m(x,y) \neq 0$. If $m \leq n$, then $h_k(x,y) \in S_k \subseteq S_n$, $k = 0, \ldots, m$, and therefore $h(x,y) \in R_n$. If $m > n$, then $h(x,y) \in S_m \smallsetminus S_{m-1}$ and therefore $h(x,y) \notin R_n$.

PART E: *Computation of the genus.* By Part D, the set $\bigcup_{j=0}^{d-1}\{x^i y^j \mid i = 0, \ldots, n-j\}$ is a basis of R_n. Hence

$$\dim_K R_n = \sum_{j=0}^{d-1}(n - j + 1) = nd + 1 - \frac{1}{2}(d-1)(d-2).$$

Substitute this in (3) to conclude that

(5) $$g = \frac{1}{2}(d-1)(d-2) - \dim_K S/R.$$

By Lemma 5.3.2(a), $S/R = \bigoplus S_P/R_P$, where P ranges over all nonzero prime ideals of R. Since K is algebraically closed, there is a bijective correspondence between the nonzero prime ideals P of R and the finite points $\mathbf{p}$ of Γ^* and we have $R_P = O_\mathbf{p}$. Hence, $S/R = \bigoplus_{\mathbf{p} \in \Gamma^*(K)} O'_\mathbf{p}/O_\mathbf{p}$. Substituting this expression for S/R in (5), we get (1). $\square$

COROLLARY 5.3.5: *Let Γ^* be a projective plane curve of degree d defined over a perfect field K. Then* $\operatorname{genus}(\Gamma^*) \leq \frac{1}{2}(d-1)(d-2)$.

Proof: The genus of Γ^* does not change by going from K to its algebraic closure (Proposition 3.4.2). Now apply Lemma 5.3.3 and Proposition 5.3.4. $\square$

COROLLARY 5.3.6: *Let Γ^* be a projective plane curve of degree d defined over an algebraically closed field K. Then genus$(\Gamma^*) = \frac{1}{2}(d-1)(d-2)$ if and only if Γ^* is* **smooth**; *that is, all K-rational points of Γ^* are simple.*

Proof: The condition "Γ^* is smooth" is equivalent to $O_{\mathbf{p}} = O'_{\mathbf{p}}$ for all K-rational points $\mathbf{p}$ of Γ^*. Now apply Proposition 5.3.4. □

5.4 Points on a Curve over a Finite Field

Using the formula for the genus of an absolutely irreducible curve Γ over $\mathbb{F}_q$ (Proposition 5.3.4), we translate the estimate of the number of prime divisors of degree 1 of the function field of Γ to an estimate on $|\Gamma(\mathbb{F}_q)|$ which involves only q and the degree of Γ.

THEOREM 5.4.1: *Let $f \in \mathbb{F}_q[X,Y]$ be an absolutely irreducible polynomial of degree d. Denote the affine curve defined by the equation $f(X,Y) = 0$ by Γ. Then*

$$q + 1 - (d-1)(d-2)\sqrt{q} - d \leq |\Gamma(\mathbb{F}_q)| \leq q + 1 + (d-1)(d-2)\sqrt{q}.$$

Proof: Put $K = \mathbb{F}_q$. Denote the function field of Γ over K by F, and denote the number of prime divisors of degree 1 of F/K by N. By Theorem 4.5.2,

$$(1) \qquad q + 1 - 2g\sqrt{q} \leq N \leq q + 1 + 2g\sqrt{q},$$

where $g = \text{genus}(F/K)$. Let Γ^* be the projective completion of Γ. For each $\mathbf{p} \in \Gamma^*(K)$ let $k(\mathbf{p})$ (resp. $k_1(\mathbf{p})$) be the number of prime divisors (resp. prime divisors of degree 1) of F/K with center at $\mathbf{p}$. Then

$$(2) \qquad N - |\Gamma^*(K)| = \sum_{\mathbf{p} \in \Gamma^*(K)} (k_1(\mathbf{p}) - 1) \leq \sum_{\mathbf{p} \in \Gamma^*(K)} (k(\mathbf{p}) - 1).$$

Let

$$(3) \qquad \delta = \sum_{\mathbf{p} \in \Gamma^*(K)} \dim_K O'_{\mathbf{p},K}/O_{\mathbf{p},K} \text{ and } \tilde{\delta} = \sum_{\mathbf{p} \in \Gamma^*(K)} \dim_{\tilde{K}} O'_{\mathbf{p},\tilde{K}}/O_{\mathbf{p},\tilde{K}}.$$

Consider $\mathbf{p} \in \Gamma^*(K)$. If $\mathbf{p}$ is K-normal, then $\dim_K O'_{\mathbf{p},K}/O_{\mathbf{p},K} = 0$ and $k_1(\mathbf{p}) = k(\mathbf{p}) = 1$ (Lemma 5.2.2(a)). If $\mathbf{p}$ is not K-normal, then $\dim_K O'_{\mathbf{p},K}/O_{\mathbf{p},K} \geq 1$. Hence, by (3), and by Lemma 5.3.3,

$$\sum_{\mathbf{p} \in \Gamma^*(K)} (1 - k_1(\mathbf{p})) \leq \delta \leq \tilde{\delta}.$$

We conclude from (2), (1), and Proposition 5.3.4 that

$$|\Gamma(K)| \leq |\Gamma^*(K)| = N + \sum_{\mathbf{p} \in \Gamma^*(K)} (1 - k_1(\mathbf{p})) \leq q + 1 + 2g\sqrt{q} + \tilde{\delta}$$

$$= q + 1 + (d-1)(d-2)\sqrt{q} - 2\tilde{\delta}\sqrt{q} + \tilde{\delta} \leq q + 1 + (d-1)(d-2)\sqrt{q}.$$

Exercises

Next we give a lower bound for $|\Gamma(K)|$. Note first that Γ^* has at most d points at infinity [Seidenberg, p. 37]. Hence, $|\Gamma(K)| \geq |\Gamma^*(K)| - d$. Therefore, by (2), Lemma 5.3.2(b), (1), and Proposition 5.3.4

$$|\Gamma(K)| \geq |\Gamma^*(K)| - d \geq N - \sum_{\mathbf{p} \in \Gamma^*(K)} (k(\mathbf{p}) - 1) - d$$

$$\geq N - \sum_{\mathbf{p} \in \Gamma^*(K)} \dim_K O'_{K,\mathbf{p}}/O_{K,\mathbf{p}} - d \geq q + 1 - 2g\sqrt{q} - \tilde{\delta} - d$$

$$= q + 1 - (d-1)(d-2)\sqrt{q} + 2\tilde{\delta}\sqrt{q} - \tilde{\delta} - d$$

$$\geq q + 1 - (d-1)(d-2)\sqrt{q} - d.$$

This completes the proof of the theorem. $\square$

COROLLARY 5.4.2: *The curve Γ of Theorem 5.4.1 satisfies the following conditions:*
(a) *For each m there exists q_0 such that $|\Gamma(\mathbb{F}_q)| \geq m$ for all $q \geq q_0$.*
(b) *If $q > (d-1)^4$, then $\Gamma(K)$ is not empty.*

Proof of (b): By Theorem 5.4.1, $|\Gamma(\mathbb{F}_q)| \geq q + 1 - (d-1)(d-2)\sqrt{q} - d = \sqrt{q}(\sqrt{q} - (d-1)(d-2)) - (d-1) > \sqrt{q} - (d-1) > 0.$ $\square$

Exercises

1. Use Proposition 5.3.4 to prove that any projective plane curve of degree 2 is smooth.

2. Let Γ^* be a projective plane curve of degree 3. Use Proposition 5.3.4 to show that either Γ^* is smooth, in which case it has genus 1, or Γ^* has exactly one singular point, in which case it has genus 0 and therefore its function field is rational (Example 3.2.4).

3. Consider the affine plane curve Γ defined by the equation $Y^2 - 2X^2 - X^3 = 0$ over a field K of characteristic $\neq 2$ that does not contain $\sqrt{2}$. Note that $(0,0)$ is a singular point of Γ. Let (x,y) be a generic point of Γ over K. Show that the map $(x,y) \to (0,0)$ does not extend to a K-rational (rather than $\tilde{K}$-rational) place of $K(x,y)$ (i.e. the singular point of Γ is not the center of a K-rational place of the function field of Γ). Hint: Consider the element $\frac{x}{y}$ of $K(x,y)$.

4. Prove directly that the function field of the curve $Y^2 - 2X^2 - X^3 = 0$ is rational.

5. Consider the projective plane curve Γ^* defined for $d \geq 2$ by $X_0 X_1^{d-1} - X_0 X_2^{d-1} - X_2^d = 0$ over an algebraically closed field K.
(a) Use Lemma 5.2.3 to show that the only nonnormal point of Γ^* is $(1:0:0)$.

(b) Let (x,y) be a generic point of the affine part Γ of Γ^* defined by $X^{d-1} - Y^{d-1} - Y^d = 0$ and let $z = \frac{x}{y}$. Use Lemma 2.4.4 to conclude that $K[y,z]$ is the integral closure of $K[x,y]$ in $K(x,y)$. Conclude that the genus of Γ is 0.

(c) Prove that $S_0 = \{\frac{f(y,z)}{g(x,y)} \mid f \in K[Y,Z],\ g \in K[X,Y],\ g(0,0) \neq 0\}$ is the integral closure of the local ring R_0 of Γ at $(0,0)$.

(d) Use Proposition 5.3.4 to conclude that the set $\{x^i z^j \mid j = 1, \ldots, d-2;\ i = 0, \ldots, j-1\}$ gives a basis for S_0/R_0 over K.

6. Count the number of points on the projective plane curve $X_0^3 + X_1^3 + X_2^3 = 0$ over the finite field $\mathbb{F}_q$, where q is a prime power such that $\gcd(q-1, 3) = 1$. Hint: The map $x \mapsto x^3$ from $\mathbb{F}_q^\times$ into itself is bijective.

7. Let $f \in \mathbb{F}_q[X]$ be a polynomial of degree d such that $g(X,Y) = \frac{f(Y)-f(X)}{Y-X}$ is absolutely irreducible. Suppose that either q is not a power of 2 or $d \geq 3$. Use Theorem 5.4.1 to prove that if $q > (d-1)^4$, then there exist distinct $x, y \in \mathbb{F}_q$ such that $f(y) = f(x)$. Hint: Observe that $g(X,X)$, the derivative of $f(X)$, has at most $d-1$ zeros.

Notes

The proof of Proposition 5.3.4 is an elaboration on [Samuel, p. 52].

Denote the maximum number of $\mathbb{F}_q$-points on a curve of genus g which is defined over $\mathbb{F}_q$ by $N_q(g)$. For fixed q, put $A(q) = \limsup_{g \to \infty} \frac{1}{g} N_q(g)$. Weil's estimate $N_q(g) \leq q + 1 + 2g\sqrt{q}$ (Theorem 4.5.2 and the proof of Theorem 5.4.1) implies that $A(q) \leq 2q^{\frac{1}{2}}$. [Serre6] improves Weil's estimate (via interpretation of the Frobenius as an endomorphism on Jacobians) to give the bound $N_q(g) \leq q + 1 + g[2\sqrt{q}]$ (where $[x]$ is the greatest integer not exceeding x). Thus, $A(q) \leq [2\sqrt{q}]$. But [Drinfeld-Vlâdut] obtains the much improved estimate $A(q) \leq \sqrt{q} - 1$. When q is a square [Ihara] and [Tsfasman-Vlâdut-Zink] have shown this bound to be exact. As for a lower estimate, [Serre6] proves the existence of $c > 0$ such that $A(q) \geq c \log(q)$ (e.g. $A(2) \geq \frac{8}{39}$). The exact lower and upper bound for $A(q)$ for general q have yet to be found. The case $q = 2$ has application to coding theory as first noted by [Goppa] (or p. 530 of [Lidl-Niederreiter] for a survey of work in this direction).

Chapter 6. The Chebotarev Density Theorem

The major connection between the theory of finite fields and the arithmetic of number fields and function fields is the Chebotarev density theorem. Explicit decision procedures and transfer principles of Chapters 20 and 31 depend on the theorem or some analogs. In the function field case our proof, using the Riemann hypothesis for curves, is complete and elementary. In particular, we make no use of the theory of analytic functions. The number field case, however, uses an asymptotic formula for the number of ideals in an ideal class, and only simple properties of analytic functions. In particular, we do not use Artin's reciprocity law (or any equivalent formulation of class field theory). This proof is close to Chebotarev's original field crossing argument, which gave a proof of a piece of Artin's reciprocity law for cyclotomic extensions.

6.1 Decomposition Groups

Let R be an integrally closed domain with quotient field K. Consider a finite Galois extension L of K with Galois group G, and denote the integral closure of R in L by S. Suppose $\mathfrak{p}$ is a prime ideal of R. By Chevalley's theorem (Proposition 2.3.1), there exists a prime ideal $\mathfrak{P}$ of S **lying over** $\mathfrak{p}$ (i.e. $\mathfrak{p} = R \cap \mathfrak{P}$). Denote the quotient fields of $R/\mathfrak{p}$ and $S/\mathfrak{P}$, respectively, by $\bar{K}$ and by $\bar{L}$. The set of all $\sigma \in G$ satisfying $\sigma(\mathfrak{P}) = \mathfrak{P}$ is a group $D_\mathfrak{P}$ called the **decomposition group** of $\mathfrak{P}$ over K. Its fixed field in L is the **decomposition field** of $\mathfrak{P}$ over K.

For $x \in S$ denote the equivalence class of x modulo $\mathfrak{P}$ by $\bar{x}$. Each $\sigma \in D_\mathfrak{P}$ induces a unique automorphism $\bar{\sigma}$ of $\bar{L}$ over $\bar{K}$ satisfying $\bar{\sigma}\bar{x} = \overline{\sigma x}$ for each $x \in S$. The map $\sigma \mapsto \bar{\sigma}$ is a homomorphism of $D_\mathfrak{P}$ into $\mathrm{Aut}(\bar{L}/\bar{K})$. Its kernel is the **inertia group** $I_\mathfrak{P}$ of $\mathfrak{P}$ over K,

$$I_\mathfrak{P} = \{\sigma \in G \mid \sigma x \in x + \mathfrak{P} \text{ for every } x \in S\}.$$

The fixed field of $I_\mathfrak{P}$ in L is the **inertia field** of $\mathfrak{P}$ over K. If $\sigma \in G$, then $\sigma S = S$ and $\sigma \mathfrak{P}$ is another prime ideal of S that lies over $\mathfrak{p}$. In this case $D_{\sigma\mathfrak{P}} = \sigma \cdot D_\mathfrak{P} \cdot \sigma^{-1}$ and $I_{\sigma\mathfrak{P}} = \sigma \cdot I_\mathfrak{P} \cdot \sigma^{-1}$. Conversely, any two prime ideals of S lying over the same prime ideal of R are conjugate over K [Lang7, p. 340].

In the proof of Lemma 6.1.1 we use the expression "to **localize R and S at $\mathfrak{p}$**". This means that we replace R and $R_\mathfrak{p}$, $\mathfrak{p}$ by $\mathfrak{p}R_\mathfrak{p}$, S by $S_\mathfrak{p} = \{\frac{s}{a} \mid s \in S \text{ and } a \in R \smallsetminus \mathfrak{p}\}$, and $\mathfrak{P}$ by $\mathfrak{P}S_\mathfrak{p}$. The local ring $R_\mathfrak{p}$ is integrally closed, $\mathfrak{p}R_\mathfrak{p}$ is its maximal ideal, and $\bar{K} = R_\mathfrak{p}/\mathfrak{p}R_\mathfrak{p}$. In addition, $S_\mathfrak{p}$ is the integral closure of $R_\mathfrak{p}$ in L [Lang7, p. 338, Prop. 1.8 and 1.9], and $\mathfrak{P}S_\mathfrak{p}$ is a prime ideal that lies over $\mathfrak{p}R_\mathfrak{p}$. Thus, $S_\mathfrak{p}/\mathfrak{P}S_\mathfrak{p}$ is a domain which is integral over the field $\bar{K}$. Hence, $S_\mathfrak{p}/\mathfrak{P}S_\mathfrak{p} = \bar{L}$ is a field [Lang7, p. 339, Prop. 1.11] and $\mathfrak{P}S_\mathfrak{p}$ is a maximal ideal. Moreover, $D_{\mathfrak{P}S_\mathfrak{p}} = D_\mathfrak{P}$ and $I_{\mathfrak{P}S_\mathfrak{p}} = I_\mathfrak{P}$.

LEMMA 6.1.1:
(a) The field extension $\bar{L}/\bar{K}$ is normal, and the map $\sigma \mapsto \bar{\sigma}$ from $D_{\mathfrak{P}}$ into $\mathrm{Aut}(\bar{L}/\bar{K})$ is surjective.
(b) Suppose $\bar{L}/\bar{K}$ is separable and $[\bar{L}:\bar{K}] = [L:K]$. Then $\bar{L}/\bar{K}$ is Galois, $D_{\mathfrak{P}} = \mathrm{Gal}(L/K)$, and the map $\sigma \mapsto \bar{\sigma}$ is an isomorphism of $\mathrm{Gal}(L/K)$ onto $\mathrm{Gal}(\bar{L}/\bar{K})$.

Proof of (a): Denote the decomposition field of $\mathfrak{P}$ by L_0, let $S_0 = S \cap L_o$, and let $\mathfrak{P}_0 = \mathfrak{P} \cap L_0$. We prove $S_0/\mathfrak{P}_0 = R/\mathfrak{p}$.

Suppose $x \in S_0$. We need only to find $a \in R$ such that $x \equiv a \bmod \mathfrak{P}_0$. For each $\sigma \in G \smallsetminus D_{\mathfrak{P}}$ we have $\sigma^{-1}\mathfrak{P} \neq \mathfrak{P}$. If $\sigma^{-1}\mathfrak{P} \cap L_0 = \mathfrak{P}_0$, then there exists $\tau \in \mathrm{Gal}(L/L_0) = D_{\mathfrak{P}}$ such that $\tau\sigma^{-1}\mathfrak{P} = \mathfrak{P}$. Therefore, $\sigma \in D_{\mathfrak{P}}$, a contradiction. Thus, $\sigma^{-1}\mathfrak{P} \cap L_0 \neq \mathfrak{P}_0$. Localize at $\mathfrak{p}$, to assume that $\mathfrak{p}$ is a maximal ideal of R. Then $\mathfrak{P}_0$ and $\sigma^{-1}\mathfrak{P} \cap L_0$ are maximal ideals of S_0. Hence, $\mathfrak{P}_0 + \sigma^{-1}\mathfrak{P}_0 \cap L_0 = S_0$. By the Chinese remainder theorem [Lang7, p. 94], there exists $y \in S_0$ with $y \equiv x \bmod \mathfrak{P}_0$ and $y \equiv 1 \bmod \sigma^{-1}\mathfrak{P} \cap L_0$ for every $\sigma \in G \smallsetminus D_{\mathfrak{P}}$. Thus, $y \equiv x \bmod \mathfrak{P}$ and $\sigma y \equiv 1 \bmod \mathfrak{P}$ for every $\sigma \in G \smallsetminus D_{\mathfrak{P}}$. Since L_0/K is a separable extension, the element $a = \mathrm{norm}_{L_0/K}(y)$ of R is a product of y and elements σy with σ running over nonidentity coset representatives of $D_{\mathfrak{P}}$ in G. Consequently $a \equiv x \bmod \mathfrak{P}_0$, as desired.

To continue localize S_0 and S at $\mathfrak{P}_0$ to assume $\bar{K} = S_0/\mathfrak{P}_0$ and $\bar{L} = S/\mathfrak{P}$. Write each element of $\bar{L}$ as $\bar{x}$, with $x \in S$, and let $f = \mathrm{irr}(x, L_0)$. Then $f(X) = \prod_{i=1}^m (X - x_i)$, where $x_i \in S$, $i = 1, \ldots, m$, are the conjugates of x over L_0. The coefficients of $f(X)$ belong to $L_0 \cap S = S_0$. Hence, $\bar{f}(X) \in \bar{K}[X]$. In addition, $\bar{x}$ is a root of the polynomial $\bar{f}(X) = \prod_{i=1}^m (X - \bar{x}_i)$ with roots in $\bar{L}$. Hence, $\bar{L}$ is a normal extension of $\bar{K}$. Moreover, $[\bar{K}(\bar{x}) : \bar{K}] \leq \deg(f) \leq [L:L_0]$.

Separable extensions have primitive generators [Lang7, p. 243]. Hence, the maximal subfield E of $\bar{L}$ which is separable over $\bar{K}$ satisfies $[E:\bar{K}] \leq [L:L_0]$. Thus, $E = \bar{K}(\bar{x})$ for some $x \in S$. In the above notation, if $\tau \in \mathrm{Aut}(\bar{L}/\bar{K}) \cong \mathrm{Gal}(E/\bar{K})$, we have $\tau\bar{x} = \bar{x}_j$ for some j, $1 \leq j \leq m$. The map $x \mapsto x_j$ extends to a field automorphism $\sigma \in \mathrm{Gal}(L/L_0)$ with $\bar{\sigma} = \tau$. Consequently, The map $\sigma \mapsto \bar{\sigma}$ from $D_{\mathfrak{P}}$ into $\mathrm{Aut}(\bar{L}/\bar{K})$ is surjective.

Proof of (b): By assumption, $[L:K] = [\bar{L}:\bar{K}] = |\mathrm{Gal}(\bar{L}/\bar{K})| \leq |D_{\mathfrak{P}}| \leq [L:K]$. Hence, $|D_{\mathfrak{P}}| = |\mathrm{Gal}(L/K)| = |\mathrm{Gal}(\bar{L}/\bar{K})|$. Therefore, $D_{\mathfrak{P}} = \mathrm{Gal}(L/K)$ and the map $\mathrm{Gal}(L/K) \to \mathrm{Gal}(\bar{L}/\bar{K})$ is an isomorphism. $\square$

The **discriminant** of $f \in R[X]$, $\mathrm{disc}(f)$, gives information about ramification. Assume f is monic and $\prod_{i=1}^n (X - x_i)$ is the factorization of f into linear factors. Then

$$(1) \qquad \mathrm{disc}(f) = (-1)^{\frac{n(n-1)}{2}} \prod_{i \neq j}(x_i - x_j) = (-1)^{\frac{n(n-1)}{2}} \prod_{j=1}^n f'(x_j).$$

This is an element of R, and $\mathrm{disc}(f) \neq 0$ if and only if the x_i's are distinct.

6.1 Decomposition Groups

Assume f is irreducible. Then (1) implies

$$\mathrm{disc}(f) = (-1)^{\frac{n(n-1)}{2}} \mathrm{norm}_{K(x_1)/K}(f'(x_1)).$$

We call $\mathrm{norm}_{K(x_1)/K}(f'(x_1))$ the **discriminant** of x_1 over K.

LEMMA 6.1.2: *Let R be an integrally closed domain with quotient field K. Let S be the integral closure of R in a finite separable extension L of K. Assume $L = K(z)$ with z integral over R and let $f = \mathrm{irr}(z, K)$. Suppose $d = \mathrm{disc}(f)$ is a unit of R. Then $S = R[z]$.*

Proof: Let $n = [L : K]$. Assume $y = a_0 + a_1 z + \cdots + a_{n-1} z^{n-1}$, with $a_i \in K$, is element of S. We must prove $a_i \in R$, $i = 0, \ldots, n-1$. To this end, let $\sigma_1, \ldots, \sigma_n$ be the isomorphisms of L over K into $\tilde{K}$. Then each of $\sigma_i y = \sum_{j=0}^{n-1} a_j \sigma_i z^j$, $i = 1, \ldots, n$, is integral over R.

Solve for the a_j's by Cramer's rule. This gives $a_j = \frac{b_j}{\Delta} = \frac{\Delta b_j}{\Delta^2}$, $j = 0, \ldots, n-1$, where $\Delta = \det(\sigma_i z^j)$ with b_j integral over R, $j = 0, \ldots, n-1$. But Δ is a Vandermonde determinant:

$$\Delta^2 = \pm \prod_{i \neq j} (\sigma_i z - \sigma_j z) = \pm \mathrm{norm}_{L/K} f'(z) = \pm d.$$

Since d is a unit of R, a_j is integral over R, $j = 0, \ldots, n-1$. Since R is integrally closed, all a_j are in R. □

Definition 6.1.3: Ring covers. As in the preceding lemmas, consider two integrally closed integral domains $R \subseteq S$ with $K \subseteq L$ their respective quotient fields such that L/K is finite and separable. Suppose $S = R[z]$, where z is integral over R and the discriminant of z over K is a unit of R. In this set up we say S/R is a **ring cover** and L/K is the corresponding **field cover**. In this case, Lemma 6.1.2 implies that S is the integral closure of R in L. Call the element z a **primitive element for the cover**. If in addition L/K is Galois, call S/R is a **Galois ring cover**. □

We summarize the previous results for ring covers:

LEMMA 6.1.4: *Let S/R be a Galois ring cover with L/K the corresponding field cover. Then for every prime ideal $\mathfrak{p}$ of R and for every prime ideal $\mathfrak{P}$ of S lying over $\mathfrak{p}$ the following holds. The quotient field, $\bar{L}$, of $S/\mathfrak{P}$ is a Galois extension of the quotient field, $\bar{K}$, of $R/\mathfrak{p}$. The map $\sigma \mapsto \bar{\sigma}$ of $D_{\mathfrak{P}}$ into $\mathrm{Gal}(\bar{L}/\bar{K})$ given by $\bar{\sigma}\bar{x} = \overline{\sigma x}$ for $x \in S$ is an isomorphism.*

Remark 6.1.5: Creating ring covers. Let $K = K_0(x_1, \ldots, x_n)$ be a finitely generated extension of a field K_0. The subring $R = K_0[x_1, \ldots, x_n]$ of K is not necessarily integrally closed. But, there exists a nonzero $x_{n+1} \in K$ with $R' = K_0[x_1, \ldots, x_{n+1}]$ integrally closed ([Lang4, p. 120]; a constructive proof of this fact appears in Section 19.7). Suppose z is a primitive generator for L/K, $f \in R[Z]$ is irreducible polynomial over K, and $f(z) = 0$. Multiply x_{n+1} by the inverse of the product of the leading coefficient and the discriminant of f. Then $S' = R'[z]$ is a ring cover of R' with z a primitive element. □

Remark 6.1.6: Decomposition groups of places. Suppose L/K is a finite Galois extension and φ is a place of L with a valuation ring O. Then $\bar{L} = \varphi(O)$ is the residue field of L under φ. Also, $R = O \cap K$ is the valuation ring of the restriction of φ to K and $\bar{K} = \varphi(R)$ is its residue field. By Proposition 2.4.1, O contains the integral closure S of R in L. Let $\mathfrak{m}$ be the maximal ideal of O, $\mathfrak{P} = S \cap \mathfrak{m}$, and $\mathfrak{p} = \mathfrak{P} \cap R$. Then $O = S_\mathfrak{P}$ [Lang4, p. 18, Thm. 4] and $\mathfrak{P}$ is maximal [Lang7, p. 339, Prop 1.11]. Hence, $\bar{L} = S/\mathfrak{P}$ and $R/\mathfrak{p} \cong \bar{K}$. We call $D_\varphi = D_\mathfrak{P}$ and $I_\varphi = I_\mathfrak{P}$ the **decomposition group** and **inertia group**, respectively, of φ over K. The fixed fields of D_φ and I_φ in L are the **decomposition field** and the **inertia field**, respectively, of φ over K. By lemma 6.1.1, $\bar{L}/\bar{K}$ is a normal extension and the map $\sigma \mapsto \bar{\sigma}$ from $D_\mathfrak{P}$ into $\text{Aut}(\bar{L}/\bar{K})$ is surjective.

Suppose now $\bar{L}/\bar{K}$ is separable and $I_\varphi = 1$ (This holds if S/R is a ring-cover.) By Lemma 6.1.1, $\bar{L}/\bar{K}$ is Galois and D_φ is isomorphic to $\text{Gal}(\bar{L}/\bar{K})$ under the map $\sigma \mapsto \bar{\sigma}$. Denote the decomposition field of φ over K by L_0. For each $x \in O \cap L_0$ and each $\sigma \in D_\varphi$ we have $\bar{\sigma}\bar{x} = \overline{\sigma x} = \bar{x}$. Hence, $\bar{x} \in \bar{K}$. Therefore, $\varphi(L_0) = \bar{K} \cup \{\infty\}$.

Let $\mathfrak{P}_0 = \mathfrak{P} \cap L_0$. Then each prime ideal of S that lies over $\mathfrak{P}_0$ is conjugate to $\mathfrak{P}$ by an element of $D_\mathfrak{P}$. Hence, $\mathfrak{P}$ is the only prime ideal of S lying over $\mathfrak{P}_0$. By Proposition 2.3.2, $e(\mathfrak{P}/\mathfrak{P}_0)[\bar{L} : \bar{L}_0] \leq [L : L_0]$. Since $[\bar{L} : \bar{L}_0] = [\bar{L} : \bar{K}] = [L : L_0]$, we have $e(\mathfrak{P}/\mathfrak{P}_0) = 1$. In particular, if O is discrete, $\mathfrak{p}$ is unramified in L. $\square$

Remark 6.1.7: Ring covers under change of base ring. Consider an integrally closed domain R with quotient field K. Let L be a finite separable extension of K, S the integral closure of R in L, z an element of S with $L = K(z)$, and $f = \text{irr}(z, K)$. Then $\text{norm}_{L/K}(f'(z)) = \prod f'(z)^\sigma$, where σ ranges over all K-embeddings of L into K_s. Each $f'(z)^\sigma$ is integral over R. Hence, $\text{norm}_{L/K}(f'(z))$ is a unit of R if and only if $f'(z)$ is a unit of S.

Suppose $f'(z)$ is a unit of S. By Lemma 6.1.2, $S = R[z]$ and z is a primitive element of the ring-cover S/R. Let φ be a homomorphism of R into an integrally closed domain $\bar{R}$. Put $\bar{K} = \text{Quot}(\bar{R})$. Then φ extends to a homomorphism ψ of S into the algebraic closure of $\bar{K}$ (Proposition 2.4.1). Put $\bar{z} = \psi(z)$, $\bar{S} = \bar{R}[\bar{z}]$, $\bar{L} = \text{Quot}(\bar{S})$, $\bar{f} = \varphi(f)$, and $g = \text{irr}(\bar{z}, \bar{K})$. Then $\bar{f}(\bar{z}) = 0$, $\bar{f}'(\bar{z})$ is a unit of $\bar{R}[\bar{z}]$, and there is a monic polynomial $h \in \bar{K}[X]$ with $\bar{f}(X) = g(X)h(X)$. The coefficients of h are polynomials in the roots of $\bar{f}$. Since the latter are integral over $\bar{R}$, we have $h \in \bar{R}[X]$. Deduce from $g'(\bar{z})h(\bar{z}) = \bar{f}'(\bar{z})$ that $g'(\bar{z})$ is a unit of $\bar{R}[\bar{z}]$. Hence, by Lemma 6.1.2, $\bar{S}/\bar{R}$ is a ring-cover with $\bar{z}$ as a primitive element. In particular, $\bar{S}$ is the integral closure of $\bar{R}$ in $\bar{L}$. Thus, if $\bar{R} = \bar{K}$, then $\bar{S} = \bar{L}$.

As an example, let L/K be a finite separable extension, z a primitive element for L/K and $f = \text{irr}(z, K)$. Then $f'(z) \neq 0$. Hence, $L = K[z]$ is a ring-cover of K. Therefore, $R[z]/R$ is a ring-cover whenever R is an integrally closed ring containing K.

For another example suppose in the notation of the first two paragraphs

6.2 The Artin Symbol over Global Fields

that R is a valuation ring and $\mathrm{Ker}(\varphi)$ is the maximal ideal of R. Then $\bar{R} = \bar{K}$ is a field and $\bar{S} = \bar{K}[\bar{z}]$ is an integral extension of $\bar{K}$. Hence, $\bar{S} = \bar{L}$ is a field. In addition, the local ring of S at $\mathrm{Ker}(\psi)$ is a valuation ring lying over R [Lang4, p. 18, Thm. 4.7]. Hence, φ and ψ extend uniquely to places of K and L with residue fields $\bar{K}$ and $\bar{L}$, respectively. □

The next result is another consequence of Lemma 6.1.2 which may be applied to covers.

LEMMA 6.1.8:
(a) *Let R be an integral domain with quotient field K, L a finite Galois extension of K, and S the integral closure of R in L. Consider a monic polynomial $f \in R[X]$ having all of its roots in L, a prime ideal $\mathfrak{p}$ of R, and a prime ideal $\mathfrak{P}$ of S lying over $\mathfrak{p}$. Assume $\mathrm{disc}(f) \notin \mathfrak{p}$ and denote reduction modulo $\mathfrak{P}$ by a bar. Also let $\sigma \in D_{\mathfrak{P}}$ and F a field containing $\bar{K}$ such that $\bar{L} \cap F = \bar{L}(\bar{\sigma})$. Then the number of the roots of f in $L(\sigma)$ is equal to the number of the roots $\bar{f}$ in F.*
(b) *Suppose $L = K(z)$ with z integral over R, $f = \mathrm{irr}(z, K)$, and $\mathrm{disc}(f) \notin \mathfrak{p}$. Then $\bar{L}/\bar{K}$ is separable, $I_{\mathfrak{P}} = 1$, and $\mathfrak{p}$ is unramified in L.*

Proof of (a): The roots of $\bar{f}$ are distinct, because $\mathrm{disc}(\bar{f}) \neq 0$. In addition $\deg(f) = \deg(\bar{f})$. Thus, $x \mapsto \bar{x}$ maps the roots of f bijectively onto the roots of $\bar{f}$. For x a root of f, $\sigma x = x$ if and only if $\bar{\sigma}\bar{x} = \bar{x}$. Moreover, since all of the roots of $\bar{f}$ belong to $\bar{L}$, we have $\bar{x} \in \bar{L}(\bar{\sigma})$ if and only if $\bar{x} \in F$. Consequently, the number of the roots of f in $L(\sigma)$ is equal to the number of the roots of $\bar{f}$ in in F.

Proof of (b): Replace R by $R_{\mathfrak{p}}$, if necessary, to assume R is a local ring and $\mathfrak{p}$ is its maximal ideal. By Lemma 6.1.2, $S = R[z]$ is the integral closure of R in L. Under the assumptions of (b), $\bar{L} = \bar{K}(\bar{z})$. From (a), $\bar{L}$ is a separable extension of $\bar{K}$. Also, if $\bar{\sigma} = 1$ for some $\sigma \in D_{\mathfrak{P}}$, then $\bar{\sigma}\bar{z} = \bar{z}$. Hence, $\sigma z = z$, so $\sigma = 1$. Hence, $I_{\mathfrak{P}} = 1$. By (4) of Section 2.3, $\mathfrak{p}$ is unramified in L. □

6.2 The Artin Symbol over Global Fields

The Artin symbol over number fields is a generalization of the Legendre symbol for quadratic residues. Here we define the Artin symbol over global fields and state some of its basic properties.

Let R be a Dedekind domain with quotient field K. Consider a finite separable extension L of K. Let S be the integral closure of R in L. Take $z \in S$ with $L = K(z)$. If $f = \mathrm{irr}(z, K)$, then $d = \mathrm{disc}(f) \in R$ and $d \neq 0$. Consider $R_1 = R[d^{-1}]$ and $S_1 = S[d^{-1}]$. Then d is a unit in R_1 and $S_1 = R_1[z]$ (Lemma 6.1.2). Thus, adjoining d^{-1} gives a ring cover S_1/R_1 for L/K. Maximal ideals $\mathfrak{P}$ for which $\mathfrak{P}S_1 = S_1$ are exactly those containing d. For all others $S/\mathfrak{P} \cong S_1/\mathfrak{P}S_1$ and $S/\mathfrak{P} = (R/\mathfrak{p})[\bar{z}]$.

If in addition L/K is a Galois extension, then extending $\mathfrak{P}$ to S_1 leaves the decomposition group and the inertia group unchanged. Hence, if $\mathfrak{P}$ does not contain d, then $\mathfrak{P}S_1$, and therefore also $\mathfrak{P}$, is unramified over K (Lemma 6.1.8). Thus, a prime ideal $\mathfrak{p}$ of R not containing d, does not ramify in L. Since only finitely many prime ideals of R contain d, only finitely many prime ideals ramify in L.

Denote the greatest common divisor of all the principal ideals $f'(z)S$ with $z \in S$ by $\mathrm{Diff}(S/R)$ and call it the **different** of S over R. Then

$$(\mathrm{Diff}(S/R))^{-1} = \{x \in L \mid \mathrm{trace}_{L/K}(xS) \subseteq R\}.$$

A prime ideal $\mathfrak{P}$ of S ramifies over K if and only if it divides $\mathrm{Diff}(S/R)$ [Lang5, p. 62]. Hence, the prime divisors of the **discriminant** $D_{S/R} = \mathrm{norm}_{L/K}\mathrm{Diff}(S/R)$ of S over R (an ideal of R), are exactly those primes that ramify in L.

Call K a **global field** if K is either a finite extension of $\mathbb{Q}$ (K is a **number field**) or K is a function field of one variable over a finite field. In the number field case denote the integral closure of $\mathbb{Z}$ in K by O_K. In the function field case K is a finite separable extension of $\mathbb{F}_p(t)$, where $p = \mathrm{char}(K)$ and t is transcendental over $\mathbb{F}_p$. Denote the integral closure of $\mathbb{F}_p[t]$ in K by O_K. With the understanding it depends on t, call O_K the **ring of integers of** K. The local ring of O_K at a prime ideal $\mathfrak{p}$ is a valuation ring. Denote its residue class field by $\bar{K}_\mathfrak{p}$. Note that $\bar{K}_\mathfrak{p}$ is a finite field. We call $N\mathfrak{p} = |\bar{K}_\mathfrak{p}|$ the **absolute norm** of $\mathfrak{p}$.

Let L be a finite Galois extension of K. Suppose $\mathfrak{p}$ is unramified in L. If $\mathfrak{P}$ is a prime ideal of O_L over $\mathfrak{p}$, then reduction modulo $\mathfrak{P}$ gives a canonical isomorphism of the decomposition group $D_\mathfrak{P}$ and $\mathrm{Gal}(\bar{L}_\mathfrak{P}/\bar{K}_\mathfrak{p})$ (Lemma 6.1.4). The latter group is cyclic. It contains a canonical generator Frob, the **Frobenius automorphism**. It acts on $\bar{L}_\mathfrak{P}$ by this rule:

(1) $$\mathrm{Frob}(x) = x^{N\mathfrak{p}} \quad \text{for } x \in \bar{L}_\mathfrak{P}.$$

Call the element of $D_\mathfrak{P}$ that corresponds to Frob the **Frobenius automorphism** at $\mathfrak{P}$ and denote it by $\left[\frac{L/K}{\mathfrak{P}}\right]$. It is uniquely determined in $\mathrm{Gal}(L/K)$ by the condition

(2) $$\left[\frac{L/K}{\mathfrak{P}}\right] x \equiv x^{N\mathfrak{p}} \mod \mathfrak{P} \quad \text{for all } x \in O_L.$$

Let $\mathbb{F}_{q^n}$ be the algebraic closure of $\mathbb{F}_q$ in L and let $N\mathfrak{p} = q^k$. By (2), the restriction of $\left[\frac{L/K}{\mathfrak{P}}\right]$ to $\mathbb{F}_{q^n}$ is Frob_q^k.

If $K \subseteq K' \subseteq L$ and K'/K is a Galois extension, this immediately implies

(3) $$\mathrm{res}_{K'}\left[\frac{L/K}{\mathfrak{P}}\right] = \left[\frac{K'/K}{\mathfrak{P} \cap K'}\right].$$

If $\sigma \in \mathrm{Gal}(L/K)$, then $\left[\frac{L/K}{\sigma\mathfrak{P}}\right] = \sigma\left[\frac{L/K}{\mathfrak{P}}\right]\sigma^{-1}$. Therefore, as $\mathfrak{P}$ ranges over prime ideals of O_L lying over $\mathfrak{p}$, the Frobenius automorphism ranges over

6.3 Dirichlet Density

some conjugacy class in $\text{Gal}(L/K)$ that depends on $\mathfrak{p}$. This conjugacy class is the **Artin symbol**, $\left(\frac{L/K}{\mathfrak{p}}\right)$. It is tacit in this symbol that $\mathfrak{p}$ is unramified in L. If L/K is Abelian, then $\left(\frac{L/K}{\mathfrak{p}}\right)$ is one element, $\left[\frac{L/K}{\mathfrak{P}}\right]$. In this case write $\left(\frac{L/K}{\mathfrak{p}}\right)$ for $\left[\frac{L/K}{\mathfrak{P}}\right]$.

In defining the Frobenius automorphism and the Artin symbol, replacing O_K by the local ring $O_{K,\mathfrak{p}}$ does not change these objects. If K is a function field over $\mathbb{F}_q$, the local rings $O_{K,\mathfrak{p}}$ bijectively correspond to prime divisors $\mathfrak{p}'$ of $K/\mathbb{F}_q$ finite at t. We use $\left(\frac{L/K}{\mathfrak{p}'}\right)$ as a substitute for $\left(\frac{L/K}{\mathfrak{p}}\right)$. Since each prime divisor $\mathfrak{p}'$ of $K/\mathbb{F}_q$ is either finite at t or at t^{-1}, the symbol $\left(\frac{L/K}{\mathfrak{p}'}\right)$ is well defined if $\mathfrak{p}'$ is unramified in L.

Example 6.2.1: *Quadratic extensions of* $\mathbb{Q}$. Let a be a nonsquare integer and p an odd prime number not dividing a. Put $L = \mathbb{Q}(\sqrt{a})$. Then p is unramified in L (Example 2.3.8). Let $\mathfrak{p}$ be a prime divisor of L lying over p. By elementary number theory [LeVeque, p. 46],

$$\left(\frac{L/\mathbb{Q}}{p}\right)\sqrt{a} \equiv (\sqrt{a})^p = a^{\frac{p-1}{2}}\sqrt{a} \equiv \left(\frac{a}{p}\right)\sqrt{a} \mod \mathfrak{p}.$$

Thus, the Frobenius symbol $\left(\frac{L/\mathbb{Q}}{p}\right)$ acts $\sqrt{a}$ as the Legendre symbol $\left(\frac{a}{p}\right)$. □

6.3 Dirichlet Density

For K a global field denote the set of all prime ideals of O_K by $P(K)$. If A is a subset of $P(K)$, then the **Dirichlet density**, $\delta(A)$, of A is the limit

$$\delta(A) = \lim_{s \to 1^+} \frac{\sum_{\mathfrak{p} \in A}(N\mathfrak{p})^{-s}}{\sum_{\mathfrak{p} \in P(K)}(N\mathfrak{p})^{-s}},$$

if it exists. The Dirichlet density is a quantitative measure on subsets of $P(K)$. We apply it to test if specific subsets are infinite. Clearly $\delta(A)$ is a real number between 0 and 1. For example, $\delta(P(K)) = 1$. If K is a number field, then

$$\lim_{s \to 1^+} \sum_{\mathfrak{p} \in P(K)} (N\mathfrak{p})^{-s} = \infty$$

[Lang5, p. 162]. Relation (19) of Section 6.4 implies that the same holds if K is a function field. Hence, in both cases, $\delta(A) = 0$ if $\sum_{\mathfrak{p} \in A}(N\mathfrak{p})^{-1}$ is finite. In particular, $\delta(A) = 0$ if A is finite. If A and B are disjoint subsets of $P(K)$ having a density, then $\delta(A \cup B) = \delta(A) + \delta(B)$.

Here is the main result:

THEOREM 6.3.1 (Chebotarev Density Theorem): *Let L/K be a finite Galois extension of global fields and let $\mathcal{C}$ be a conjugacy class in $\mathrm{Gal}(L/K)$. Then the Dirichlet density of $\{\mathfrak{p} \in P(K) \mid \left(\frac{L/K}{\mathfrak{p}}\right) = \mathcal{C}\}$ exists and is equal to $\frac{|\mathcal{C}|}{[L:K]}$.*

Section 6.4 proves Theorem 6.3.1 for function fields and Section 6.5 proves the theorem for number fields. A non obvious special case is Dirichlet's theorem showing the arithmetic progression $\{a, a+n, a+2n, \ldots\}$ has infinitely many primes when $\gcd(a,n) = 1$.

COROLLARY 6.3.2 (Dirichlet): *Suppose a and n are relatively prime positive integers. Then the Dirichlet density of $\{p \in P(\mathbb{Q}) \mid p \equiv a \mod n\}$ is $\frac{1}{\varphi(n)}$, where $\varphi(n)$ is the Euler totient function.*

Proof: Denote a primitive nth root of unity by ζ_n and let $L = \mathbb{Q}(\zeta_n)$. Then $\mathrm{Gal}(L/K)$ is isomorphic to $(\mathbb{Z}/n\mathbb{Z})^\times$. If $\sigma \in \mathrm{Gal}(L/K)$ and $\sigma\zeta_n = \zeta_n^a$, then this isomorphism maps σ to $a \mod n$. Also, for a, b relatively prime to n, we have $\zeta_n^a \equiv \zeta_n^b \mod p$ if and only if $\zeta_n^a = \zeta_n^b$. Thus, for $p \nmid n$, $p \equiv a \mod n$ if and only if $\left(\frac{L/\mathbb{Q}}{p}\right)(\zeta_n) = \zeta_n^a$. Now apply Theorem 6.3.1. $\square$

Example 6.3.3: Let $f \in \mathbb{Z}[X]$ be a monic polynomial. Write

$$f(X) = \prod_{i=1}^{r} f_i(X)$$

with $f_1(X), \ldots, f_r(X)$ monic and irreducible and let d be the product of the discriminants of $f_1, \ldots, f_r$ ((1) of Section 6.1). Consider the following hypotheses:
(1) $f(X) \equiv 0 \mod p$ has a solution for all but finitely many primes p.
(2) $f(X) \equiv 0 \mod p$ has a solution for all primes $p \nmid d$.
Let L be the splitting field of f over $\mathbb{Q}$. According to Theorem 6.3.1, each element of $\mathrm{Gal}(L/\mathbb{Q})$ has the form $\left[\frac{L/\mathbb{Q}}{\mathfrak{p}}\right]$ for infinitely many prime ideals $\mathfrak{p}$ of O_L. Therefore, Lemma 6.1.8(a) implies each of (1) and (2) is equivalent to the following.
(3) Each $\sigma \in \mathrm{Gal}(L/\mathbb{Q})$ fixes a root of $f(X)$.
In particular, (1) and (2) are equivalent. $\square$

Example 6.3.4: Let K be a number field and B the set of all prime ideals of O_K whose absolute norm is not a prime number. Then $\sum_{\mathfrak{p} \in B} \frac{1}{N\mathfrak{p}} \leq \sum_p \frac{[K:\mathbb{Q}]}{p^2} < \infty$. Hence, $\delta(B) = 0$.

Suppose L is a finite Galois extension of K and $\mathcal{C}$ a conjugacy class in $\mathrm{Gal}(L/K)$. Then, in view of the preceding paragraph, the Chebotarev density theorem gives infinitely many prime ideals $\mathfrak{p}$ of O_K such that $\left(\frac{L/K}{\mathfrak{p}}\right) = \mathcal{C}$ and $N\mathfrak{p}$ is a prime number. $\square$

6.4 Function Fields

This section contains the proof of the Chebotarev density theorem in the function field case. Apart from elementary algebraic manipulations it depends only on the Riemann hypothesis for curves.

To fix notation, let q be a power of a prime number. Consider a function field K over $\mathbb{F}_q$, a finite Galois extension L of K, and a conjugacy class $\mathcal{C}$ of $\text{Gal}(L/K)$ with c elements. Let $\mathbb{F}_{q^n}$ be the algebraic closure of $\mathbb{F}_q$ in L and fix a separating transcendence element t for $K/\mathbb{F}_q$. Denote the Frobenius element of $\mathbb{F}_{q^n}/\mathbb{F}_q$ by Frob_q.

As in Section 6.2, let O_K be the integral closure of $\mathbb{F}_q[t]$ in L and let $P(K)$ be the set of all nonzero prime ideals of O_K. Denote the set of prime divisors of $K/\mathbb{F}_q$ by $\mathbb{P}(K)$. Identify $P(K)$ with a cofinite subset of $\mathbb{P}(K)$. Thus, $C = \{\mathfrak{p} \in \mathbb{P}(K) \mid \left(\frac{L/K}{\mathfrak{p}}\right) = \mathcal{C}\}$ and $\{\mathfrak{p} \in P(K) \mid \left(\frac{L/K}{\mathfrak{p}}\right) = \mathcal{C}\}$ differ by finitely many elements. It suffices therefore to prove that $\delta(C) = \frac{c}{[L:K]}$.

In addition to $n = [\mathbb{F}_{q^n} : \mathbb{F}_q]$, two more degrees enter the proof: $d = [K : \mathbb{F}_q(t)]$ and $m = [L : K\mathbb{F}_{q^n}]$ as in the following diagram.

$$
\begin{array}{ccccc}
K & \xrightarrow{n} & K \cdot \mathbb{F}_{q^n} & \xrightarrow{m} & L \\
{\scriptstyle d}\big\downarrow & & \big\downarrow & & \\
\mathbb{F}_q(t) & \xrightarrow{n} & \mathbb{F}_{q^n}(t) & &
\end{array}
$$

Fix the following notation:

$\mathbb{P}'(K) = \{\mathfrak{p} \in \mathbb{P}(K) \mid \mathfrak{p} \text{ is unramified over } \mathbb{F}_q(t) \text{ and in } L\}$
$\mathbb{P}_k(K) = \{\mathfrak{p} \in \mathbb{P}'(K) \mid \deg(\mathfrak{p}) = k\}$
$\mathbb{P}'_k(K) = \{\mathfrak{p} \in \mathbb{P}'(K) \mid \deg(\mathfrak{p}) = k\}$
$C_k(L/K, \mathcal{C}) = \{\mathfrak{p} \in \mathbb{P}'_k(K) \mid \left(\frac{L/K}{\mathfrak{p}}\right) = \mathcal{C}\}$
$D_k(L/K, \tau) = \{\mathfrak{P} \in \mathbb{P}(L) \mid \mathfrak{P} \cap K \in \mathbb{P}_k(K) \text{ and } \left[\frac{L/K}{\mathfrak{P}}\right] = \tau\}$,
$\qquad\qquad\qquad\qquad\qquad\qquad\qquad\qquad$ for $\tau \in \text{Gal}(L/K)$

$C' = \bigcup_{k=1}^{\infty} C_k(L/K, \mathcal{C})$
$g_K = $ the genus of K
$\text{Frob}_q \;=\;$ the Frobenius automorphism of $\text{Gal}(\mathbb{F}_q)$ and also of $\text{Gal}(\mathbb{F}_{q^n}/\mathbb{F}_k)$ for each k.

The sets C' and C differ by only finitely many elements. Hence, they have the same Dirichlet density. To compute this density, we compute the cardinality of each finite set $C_k(L/K, \mathcal{C})$. This is also of independent interest, especially when $k = 1$.

LEMMA 6.4.1: *Let k be a positive integer, $\mathfrak{p} \in C_k(L/K, \mathcal{C})$, and $\tau \in \mathcal{C}$.*
(a) *There are exactly $[L : K]/\text{ord}(\tau)$ primes of $\mathbb{P}(L)$ over $\mathfrak{p}$.*
(b) *If C_k is a subset of $C_k(L/K, \mathcal{C})$ and*

$$D_k(\tau) = \{\mathfrak{P} \in D_k(L/K, \tau) \mid \mathfrak{P} \cap K \in C_k\},$$

then $|C_k| = |\mathcal{C}| \cdot \text{ord}(\tau) \cdot |D_k(\tau)| \cdot [L:K]^{-1}$.

Proof of (a): Suppose $\mathfrak{P} \in \mathbb{P}(L)$ lies over $\mathfrak{p}$. Then, by (2) of Section 2.3,

$$[L:K] = e_{\mathfrak{P}/\mathfrak{p}} f_{\mathfrak{P}/\mathfrak{p}} g_{\mathfrak{P}/\mathfrak{p}}$$

where $f_{\mathfrak{P}/\mathfrak{p}}$ is the order of the decomposition group $\langle [\frac{L/K}{\mathfrak{P}}] \rangle$. In our case $e_{\mathfrak{P}/\mathfrak{p}} = 1$ and τ is conjugate to $[\frac{L/K}{\mathfrak{P}}]$. Thus, $f_{\mathfrak{P}/\mathfrak{p}} = \text{ord}(\tau)$ and (a) holds.

Proof of (b): For each $\sigma \in \text{Gal}(L/K)$, $D_k(\sigma\tau\sigma^{-1}) = \sigma D_k(\tau)$. If $\tau' \in \mathcal{C}$ and $\tau' \neq \tau$, then $D_k(\tau')$ and $D_k(\tau)$ are disjoint. Thus, $\bigcup_{\tau \in \mathcal{C}} D_k(\tau)$ is the set of primes of $\mathbb{P}(L)$ lying over the elements of C_k. By (a),

$$\frac{|C_k| \cdot [L:K]}{\text{ord}(\tau)} = \sum_{\tau \in \mathcal{C}} |D_k(\tau)| = |\mathcal{C}| \cdot |D_k(\tau)|,$$

and the formula follows. $\square$

LEMMA 6.4.2: *Let $K \subseteq K' \subseteq L$ and $\tau \in \text{Gal}(L/K')$. Denote the algebraic closure of $\mathbb{F}_q$ in K' by $\mathbb{F}_{q^r}$. Suppose $r|k$. Then*

$$D_k(L/K, \tau) = D_{k/r}(L/K', \tau) \cap \{\mathfrak{P} \in \mathbb{P}(L) \mid \deg(\mathfrak{P} \cap K) = k\}.$$

Proof: Let $\mathfrak{P} \in \mathbb{P}(L)$. Suppose $\mathfrak{p} = \mathfrak{P} \cap K \in \mathbb{P}_k(K)$ and $\mathfrak{p}' = \mathfrak{P} \cap K' \in \mathbb{P}(K')$. Then, $N\mathfrak{p} = \bar{K}_{\mathfrak{p}} = q^k$ and $N\mathfrak{p}' = \bar{K}'_{\mathfrak{p}'} = q^{rl}$, where $l = \deg(\mathfrak{p}') = [\bar{K}'_{\mathfrak{p}'} : K']$. By (2) of Section 6.2,

(1) $\qquad \left[\frac{L/K}{\mathfrak{P}}\right] = \tau \iff \tau x \equiv x^{q^k} \mod \mathfrak{P}$ for every $x \in O_L$; and

(2) $\qquad \left[\frac{L/K'}{\mathfrak{P}}\right] = \tau \iff \tau x \equiv x^{q^{rl}} \mod \mathfrak{P}$ for every $x \in O_L$.

Thus, it suffices to show $[\frac{L/K}{\mathfrak{P}}] = \tau$ implies $rl = k$.

Since $\tau \in \text{Gal}(L/K')$, (1) implies $x \equiv x^{q^k} \mod \mathfrak{P}$ for every $x \in O_{K'}$. Hence, $\overline{K'}_{\mathfrak{p}'} \subseteq \mathbb{F}_{q^k}$. On the other hand, $\overline{K'}_{\mathfrak{p}'} \supseteq \bar{K}_{\mathfrak{p}} = \mathbb{F}_{q^k}$. Therefore, $\mathbb{F}_{q^{rl}} = \overline{K'}_{\mathfrak{p}'} = \mathbb{F}_{q^k}$. Consequently, $rl = k$. $\square$

COROLLARY 6.4.3: *With the hypotheses of Lemma 6.4.2, let $\mathcal{C}$ and $\mathcal{C}'$ be the respective conjugacy classes of τ in $\text{Gal}(L/K)$ and in $\text{Gal}(L/K')$ and $C'_{k/r} = C_{k/r}(L/K', \mathcal{C}') \smallsetminus \{\mathfrak{p}' \in P(K') \mid \deg(\mathfrak{p}' \cap K) \leq \frac{k}{2}\}$. Then*

$$|C_k(L/K, \mathcal{C})| = \frac{|\mathcal{C}||C'_{k/r}|}{|\mathcal{C}'|[K':K]}.$$

6.4 Function Fields

Proof: Let $l = \frac{k}{r}$. Then

$$D'_k(\tau) = D_l(L/K',\tau) \cap \{\mathfrak{P} \in \mathbb{P}(L) \mid \deg(\mathfrak{P} \cap K) = k\}$$

is the set of primes in $D_l(L/K',\tau)$ lying over

$$C''_k = C_l(L/K',\mathcal{C}') \cap \{\mathfrak{p}' \in \mathbb{P}(K') \mid \deg(\mathfrak{p}' \cap K) = k\}.$$

$D'_k(\tau)$ is also the set of primes in $D_l(L/K,\tau)$ over $C_k(L/K,\mathcal{C})$. By Lemma 6.4.2, $D'_k(\tau) = D_k(L/K,\tau)$.

Applying Lemma 6.4.1 twice gives a chain of equalities:

$$\frac{[L:K]}{|\mathcal{C}| \cdot \text{ord}(\tau)}|C_k(L/K,\mathcal{C})| = |D_k(L/K,\tau)| = |D'_k(\tau)| = \frac{[L:K']}{|\mathcal{C}'| \cdot \text{ord}(\tau)}|C''_k|.$$

Thus, it suffices to show that $C''_k = C'_l$. Indeed, if $\mathfrak{p}' \in \mathbb{P}(K')$ is of degree l and $\mathfrak{p} = \mathfrak{p}' \cap K$, then $\mathbb{F}_q \subseteq \bar{K}_\mathfrak{p} \subseteq \bar{K}'_{\mathfrak{p}'} = \mathbb{F}_{q^{rl}} = \mathbb{F}_{q^k}$. Hence, $\deg(\mathfrak{p})|k$. Thus, either $\deg(\mathfrak{p}) = k$ or $\deg(\mathfrak{p}) \leq \frac{k}{2}$. Therefore, $C''_k = C'_l$. □

LEMMA 6.4.4: Let k be a positive integer such that

$$(3) \qquad \text{res}_{\mathbb{F}_{q^n}}(\tau) = \text{res}_{\mathbb{F}_{q^n}}(\text{Frob}_q^k)$$

for every $\tau \in \mathcal{C}$. Let n' be a multiple of n and $L' = L\mathbb{F}_{q^{n'}}$. Then L'/K is a Galois extension, $\mathbb{F}_{q^{n'}}$ is the algebraic closure of $\mathbb{F}_q$ in L' and $g_L = g_{L'}$ (Proposition 3.4.2). Moreover, for each $\tau \in \mathcal{C}$ there exists a unique $\tau' \in \text{Gal}(L'/K)$ with $\text{res}_L \tau' = \tau$ and $\text{res}_{\mathbb{F}_{q^{n'}}}(\tau') = \text{res}_{\mathbb{F}_{q^{n'}}}(\text{Frob}_q^k)$. Furthermore:

(a) $\text{ord}(\tau') = \text{lcm}(\text{ord}(\tau), [\mathbb{F}_{q^{n'}} : \mathbb{F}_{q^{n'}} \cap \mathbb{F}_{q^k}])$;
(b) $\mathcal{C}' = \{\tau' \mid \tau \in \mathcal{C}\}$ is a conjugacy class in $\text{Gal}(L'/K)$; and
(c) $C_k(L'/K,\mathcal{C}) = C_k(L/K,\mathcal{C}')$.

Proof: Given $\tau \in \mathcal{C}$, the existence of τ' follows from (3), because $K\mathbb{F}_{q^{n'}} \cap L = K\mathbb{F}_{q^n}$. Uniqueness follows from $L' = \mathbb{F}_{q^{n'}} L$. To prove (a), note that

$$\text{ord}(\tau') = \text{lcm}\left(\text{ord}(\text{res}_L\tau'), \text{ord}(\text{res}_{\mathbb{F}_{q^{n'}}}\tau')\right)$$
$$= \text{lcm}\left(\text{ord}(\tau), [\mathbb{F}_{q^{n'}} : \mathbb{F}_{q^{n'}} \cap \mathbb{F}_{q^k}]\right).$$

Since $\mathbb{F}_{q^{n'}}/\mathbb{F}_q$ is an Abelian extension, assertion (b) follows from the uniqueness of τ'.

To prove (c), we show that $\mathfrak{P} \in \mathbb{P}(L')$ and $\mathfrak{p} = \mathfrak{P} \cap K \in \mathbb{P}_k(K)$ imply $\left[\frac{L/K}{\mathfrak{P} \cap L}\right] = \tau \iff \left[\frac{L'/K}{\mathfrak{P}}\right] = \tau'$. Indeed, if $\left[\frac{L/K}{\mathfrak{P} \cap L}\right] = \tau$, then $\tau x \equiv x^{q^k} \mod \mathfrak{P} \cap L$ for each $x \in O_L$. By definition, $\text{Frob}_q^k x = x^{q^k}$ for each $x \in \mathbb{F}_{q^{n'}}$. Since $O_{L'} = \mathbb{F}_{q^n} O_L$ (Proposition 3.4.2(d)), $\tau' x \equiv x^{q^k} \mod \mathfrak{P}$ for each $x \in O_{L'}$. Hence, $\left[\frac{L'/K}{\mathfrak{P}}\right] = \tau'$. The converse is a special case of (3) of Section 6.2. □

COROLLARY 6.4.5: *If* $L = \mathbb{F}_{q^n}K$ *and* $\tau \in \mathrm{Gal}(L/K)$ *satisfies (3), then*
$$C_k(K/K, 1) = C_k(L/K, \{\tau\}).$$

Now we give estimates for the key sets.

LEMMA 6.4.6: *Suppose* $L = K\mathbb{F}_{q^n}$, $\mathcal{C} = \{\tau\}$, *and* $\tau|_{\mathbb{F}_{q^n}} = \mathrm{Frob}_q$. *Then*

(4) $\quad\quad |\#C_1(L/\mathbb{F}_q, \mathcal{C}) - q| < 2(g_L\sqrt{q} + g_L + d).$

Proof: First note that $C_1(L/\mathbb{F}_q, \mathcal{C}) = \mathbb{P}'_1(K)$ and each $\mathfrak{p} \in \mathbb{P}(\mathbb{F}_q)$ is unramified in L. Thus, $\mathbb{P}_1(\mathbb{F}_q) \smallsetminus C_1(L/K, \mathcal{C})$ consists exactly of all prime divisors of $\mathrm{Diff}(K/\mathbb{F}_q(t))$. By Riemann-Hurwitz (Theorem 3.6.1), $\deg(\mathrm{Diff}(K/\mathbb{F}_q(t))) = 2(g_K + d - 1)$. By Theorem 4.5.2,
$$|\#\mathbb{P}_1(K) - (q+1)| \leq 2g_K\sqrt{q}.$$
Hence, $|\#C_1(L/K, \mathcal{C}) - q| \leq 2g_K\sqrt{q} + 1 + 2(g_K + d - 1)$. Since $g_K = g_L$, this proves (4). □

Denote the situation when a divides b and $a < b$ by a pd b.

LEMMA 6.4.7: *Let* K' *be a degree* km *extension of* K *containing* $\mathbb{F}_{q^k}$. *Then*

(5) $\quad \#\{\mathfrak{q} \in \mathbb{P}(K') \mid \deg(\mathfrak{q} \cap K) \text{ pd } k\} \leq 2m(q^{k/2} + (2g_K + 1)q^{k/4}).$

Proof: If $j|k$ and $\mathfrak{p} \in \mathbb{P}_j(K)$, then $\mathbb{F}_{q^j} \subseteq \mathbb{F}_{q^k}$. By Lemma 4.3.1, $\mathfrak{p}$ decomposes in $K\mathbb{F}_{q^j}$ into j prime divisors of degree 1. Each has exactly one extension to $K\mathbb{F}_{q^k}$. The latter decomposes in K' into at most m prime divisors. Hence,

$$\#\{\mathfrak{q} \in P(K') \mid \deg(\mathfrak{q} \cap K) \text{ pd } k\} = \sum_{\substack{j|k \\ j \leq k/2}} \{\mathfrak{q} \in P(K') \mid \deg(\mathfrak{q} \cap K) = j\}$$

(6)
$$\leq m \sum_{\substack{j|k \\ j \leq k/2}} \{\mathfrak{q} \in P(K\mathbb{F}_{q^k}) \mid \deg(\mathfrak{q} \cap K) = j\}$$

$$= m \sum_{\substack{j|k \\ j \leq k/2}} \{\mathfrak{q} \in P(K\mathbb{F}_{q^j}) \mid \deg(\mathfrak{q} \cap K) = j\}$$

$$\leq m \sum_{\substack{j|k \\ j \leq k/2}} |\mathbb{P}_1(K\mathbb{F}_{q^j})|$$

By Theorem 4.5.2, $|\mathbb{P}_1(KF_{q^j})| \leq q^j + 2g_K q^{j/2} + 1$. Now verify the inequalities $\sum_{j \leq k/2} q^j \leq 2q^{k/2}$ and $\frac{k}{2} \leq 2q^{k/4}$ and use them to deduce (5) from (6). □

We now come to the key result of this Section, Proposition 6.4.8. It is the main result from which the Chebotarev density theorem follows. It is also the main ingredient in an arithmetic proof of Hilbert irreducibility theorem (Lemma 13.3.3). A variant of it (Lemma 31.2.1) is crucial to the Galois stratification procedure for the elementary theory of finite fields.

Recall: $\mathbb{F}_{q^n}$ is the algebraic closure of $\mathbb{F}_q$ in L.

6.4 Function Fields

PROPOSITION 6.4.8: *Let a be a positive integer with $\operatorname{res}_{\mathbb{F}_{q^n}} \tau = \operatorname{res}_{\mathbb{F}_{q^n}} \operatorname{Frob}_q^a$ for each $\tau \in \mathcal{C}$. Let k be a positive integer. If $k \not\equiv a \mod n$, then $C_k(L/K, \mathcal{C})$ is empty. If $k \equiv a \mod n$, then*

$$(7) \quad \left| \#C_k(L/K, \mathcal{C}) - \frac{c}{km} q^k \right|$$
$$< \frac{2c}{km} \left[(m + g_L) q^{k/2} + m(2g_K + 1) q^{k/4} + g_L + dm \right].$$

Proof: Suppose $\mathcal{P} \in \mathbb{P}(L)$ lies over $\mathfrak{p} \in C_k(L/K, \mathcal{C})$. Then, $\left[\frac{L/K}{\mathcal{P}}\right] \in \mathcal{C}$, so

$$\operatorname{res}_{\mathbb{F}_{q^n}}(\operatorname{Frob}_q^a) = \operatorname{res}_{\mathbb{F}_{q^n}} \left[\frac{L/K}{\mathcal{P}}\right] = \operatorname{res}_{\mathbb{F}_{q^n}}(\operatorname{Frob}_q^k).$$

Hence, $k \equiv a \mod n$.

Conversely, suppose $k \equiv a \mod n$. Let $\tau \in \mathcal{C}$ and $n' = nk \cdot \operatorname{ord}(\tau)$. Extend L to $L' = L\mathbb{F}_{q^{n'}}$. Then

$$[L' : K\mathbb{F}_{q^{n'}}] = [L : K\mathbb{F}_{q^n}] = m.$$

Since $k \equiv a \mod n$, there exists $\tau' \in \operatorname{Gal}(L'/K)$ with $\tau'|_L = \tau$ and $\tau'|_{\mathbb{F}_{q^{n'}}} = \operatorname{Frob}_q^k$. Thus,

$$\operatorname{ord}(\tau') = \operatorname{lcm}(\operatorname{ord}(\tau), \operatorname{ord}(\operatorname{Frob}_q^k)) = \operatorname{lcm}(\operatorname{ord}(\tau), [\mathbb{F}_{q^{n'}} : \mathbb{F}_{q^k}])$$
$$= \operatorname{lcm}(\operatorname{ord}(\tau), n \cdot \operatorname{ord}(\tau)) = n \cdot \operatorname{ord}(\tau).$$

Denote the conjugacy class of τ' in $\operatorname{Gal}(L'/K)$ by $\mathcal{C}'$. By Lemma 6.4.4(c), $C_k(L'/K, \mathcal{C}') = C_k(L/K, \mathcal{C})$.

Denote the fixed field of τ' in L' by K'. Then $K' \cap \mathbb{F}_{q^{n'}} = K' \cap \tilde{\mathbb{F}}_q = \mathbb{F}_{q^k}$ and $K'\mathbb{F}_{q^{n'}} = L'$.

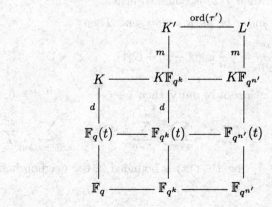

Thus, $[K':K\mathbb{F}_{q^k}] = [L':K\mathbb{F}_{q^{n'}}] = [L:K\mathbb{F}_{q^n}] = m$. Hence, $[K':\mathbb{F}_{q^k}(t)] = dm$. Applying Lemma 6.4.3 with L', K, $\mathcal{C}'$, $\{\tau'\}$, k replacing F, E, $\mathcal{C}$, $\mathcal{C}'$, r we conclude that

$$|C_k(L'/K,\mathcal{C}')| = \frac{c}{[K':K]}|C_1(L'/K',\{\tau'\})\smallsetminus\{\mathfrak{q}\in\mathbb{P}(K')|\ \deg(\mathfrak{q}\cap K)\ \text{pd}\ k\}|.$$

Since $[K':K] = km$, Lemma 6.4.7 implies

(8) $\qquad |\#C_k(L'/K,\mathcal{C}') - \dfrac{c}{km}\#C_1(L'/K',\{\tau'\})|$

$$\leq \frac{c}{km}\#\{\mathfrak{q}\in\mathbb{P}(K')|\ \deg(\mathfrak{q}\cap K)\ \text{pd}\ k\}$$

$$\leq \frac{c}{km}\cdot 2m(q^{k/2} + (2g_K+1)q^{k/4})$$

By Lemma 6.4.6, now with K', L', n', τ', q^k replacing K, L, n, τ, q,

(9) $\qquad |\#C_1(L'/K',\{\tau'\}) - q^k| < 2(g_{L'}q^{k/2} + g_{L'} + dm).$

Now we combine (8) and (9) using the equalities $g_L = g_{L'}$ and $C_k(L/K,\mathcal{C}) = C_k(L'/K,\mathcal{C}')$ to prove (7):

$$|\#C_k(L/K,\mathcal{C}) - \frac{c}{km}q^k| = |\#C_k(L'/K,\mathcal{C}') - \frac{c}{km}q^k|$$

$$\leq |\#C_k(L'/K,\mathcal{C}') - \frac{c}{km}\#C_1(L'/K',\{\tau'\})| + \frac{c}{km}|\#C_1(L'/K',\{\tau'\}) - q^k|$$

$$\leq \frac{c}{km}\cdot 2m(q^{k/2} + (2g_K+1)^{k/4}) + \frac{2c}{km}(g_L q^{k/2} + g_L + dm)$$

$$= \frac{2c}{km}((m+g_L)q^{k/2} + (2g_K+1)q^{k/4} + g_L + dm). \qquad \square$$

We deduce the function field case of Theorem 6.3.1 by summing over k in the conclusion of Proposition 6.4.8. We use the big O notation, i.e. for real valued functions $f(x)$ and $g(x)$ write

$$f(x) = O(g(x)), \qquad x\to a$$

to mean there exists a constant c with $|f(x)| \leq c|g(x)|$ for all x close to a. In particular, if $g(x) = 1$, then $f(x)$ is bounded near a.

LEMMA 6.4.9: *Let a and n be positive integers. Then*

$$\sum_{j=0}^{\infty}\frac{x^{a+jn}}{a+jn} = -\frac{1}{n}\log(1-x) + O(1), \qquad x\to 1^-.$$

Proof: If $\zeta \neq 1$ is an nth root of unity, then $1 + \zeta + \cdots + \zeta^{n-1} = 0$. Hence

$$-\frac{1}{n}\sum_{i=0}^{n-1}\log(1-\zeta^i x)\zeta^{-ia} = \frac{1}{n}\sum_{k=1}^{\infty}\frac{x^k}{k}\sum_{i=0}^{n-1}\zeta^{i(k-a)} = \sum_{k\equiv a\bmod n}\frac{x^k}{k}.$$

Since, for $1 \leq i \leq n-1$, $\log(1-\zeta^i x)$ is bounded in the neighborhood of 1, the result follows. $\qquad\square$

6.5 Number Fields

LEMMA 6.4.10: *Suppose $0 < a \leq n$ is an integer with $\mathrm{res}_{\mathbb{F}_{q^n}} \tau = \mathrm{res}_{\mathbb{F}_{q^n}} \mathrm{Frob}_q^a$ for each $\tau \in \mathcal{C}$. Then,*

(10) $$\sum_{\mathfrak{p} \in C} (N\mathfrak{p})^{-s} = -\frac{c}{[L:K]} \log(1 - q^{1-s}) + O(1), \qquad s \to 1^+.$$

Proof: The set $C' = \bigcup_{k=1}^{\infty} C_k(L/K, \mathcal{C})$ differs from C by only finitely many elements. We apply Proposition 6.4.8 and Lemma 6.4.9 for $x = q^{1-s}$ to compute:

$$\sum_{\mathfrak{p} \in C} (N\mathfrak{p})^{-s} = \sum_{j=0}^{\infty} \sum_{\mathfrak{p} \in C_{a+jn}(L/K, \mathcal{C})} (N\mathfrak{p})^{-s}$$

$$= \sum_{j=0}^{\infty} \left(\frac{c}{m(a+jn)} q^{a+jn} + O(q^{\frac{1}{2}(a+jn)}) \right) q^{-(a+jn)s}$$

$$= \frac{c}{m} \sum_{j=0}^{\infty} \frac{q^{(1-s)(a+jn)}}{a+jn} + O\left(q^{(\frac{1}{2}-s)a} \sum_{j=0}^{\infty} q^{(\frac{1}{2}-s)jn} \right)$$

$$= -\frac{c}{mn} \log(1 - q^{1-s}) + O(1) + O\left(\frac{q^{(\frac{1}{2}-s)a}}{1 - q^{(\frac{1}{2}-s)n}} \right)$$

$$= -\frac{c}{[L:K]} \log(1 - q^{1-s}) + O(1), \qquad s \to 1^+. \qquad \square$$

When $L = K$, Lemma 6.4.10 simplifies to

(19) $$\sum_{\mathfrak{p} \in P(E)} N\mathfrak{p}^{-s} = -\log(1 - q^{1-s}) + O(1), \qquad s \to 1^+.$$

Dividing (18) by (19) and taking the limit as $s \to 1^+$ gives the Dirichlet density of C:

$$\delta(C) = \lim_{s \to 1^+} \frac{\sum_{\mathfrak{p} \in C} (N\mathfrak{p})^{-s}}{\sum_{\mathfrak{p} \in P(E)} (N\mathfrak{p})^{-s}} = \frac{c}{[L:K]}$$

This concludes the Chebotarev density theorem for function fields.

6.5 Number Fields

Let L/K be a finite Galois extension of number fields. The proof of the Chebotarev density theorem for L/K splits into eight parts. It uses the asymptotic formula (2) for counting ideals with bounded norm in a given class (which we quote without proof). We say L/K is **cyclotomic** if $L \subseteq K(\zeta)$ with ζ a root of 1. The case that L/K is cyclotomic produces the general case from an easy reduction to L/K cyclic.

PART A: *Ideals with a bounded norm in a given class.* Let $\mathfrak{c}$ be a nonzero ideal of O_K. Denote the group of all fractional ideals of K relatively prime to $\mathfrak{c}$ by $J(\mathfrak{c})$. Let $P(\mathfrak{c})$ be the subgroup of all principal fractional ideals xO_K, where x satisfies the following conditions.

(1a) If $\mathfrak{p}$ is a prime ideal of O_K that divides $\mathfrak{c}$, then x lies in the local ring $O_\mathfrak{p}$ of O_K at $\mathfrak{p}$ and $x \equiv 1 \bmod \mathfrak{c}O_\mathfrak{p}$.

(1b) x is **totally positive**: σx is positive for each embedding $\sigma\colon K \to \mathbb{R}$.

The factor group $G(\mathfrak{c}) = J(\mathfrak{c})/P(\mathfrak{c})$ is finite [Lang5, p. 127]. Denote the order of $G(\mathfrak{c})$ by $h_\mathfrak{c}$.

Extend the absolute norm $N\mathfrak{p}$ of prime ideals multiplicatively to all fractional ideals. Consider a class $\mathcal{K}$ of $J(\mathfrak{c})$ modulo $P(\mathfrak{c})$. Denote the number of ideals $\mathfrak{a} \in \mathcal{K}$ (of O_K) with $N\mathfrak{a} \leq n$ by $j(\mathcal{K}, n)$ The key asymptotic formula is:

(2) $$j(\mathcal{K}, n) = \rho_\mathfrak{c} n + O(n^{1-[K:\mathbb{Q}]^{-1}}), \quad n \to \infty,$$

where $\rho_\mathfrak{c}$ is a positive constant dependent on $\mathfrak{c}$ and K but not on $\mathcal{K}$ [Lang5, p. 132].

PART B: *Abelian characters.* A **character** of a finite Abelian group G is a homomorphism $\chi\colon G \to \mathbb{C}^\times$. Define multiplication of characters by $(\chi_1\chi_2)(\sigma) = \chi_1(\sigma)\chi_2(\sigma)$. The set $\hat{G}$ of characters of G forms a group isomorphic to G. Here are the standard character formulas.

(3a) $\sum_{\sigma \in G} \chi_1(\sigma^{-1})\chi_2(\sigma) = |G|$ if $\chi_1 = \chi_2$ and 0 otherwise.
(3b) $\sum_{\chi \in \hat{G}} \chi(\sigma^{-1})\chi(\tau) = |G|$ if $\sigma = \tau$ and 0 otherwise.
(3c) $\prod_{\chi \in \hat{G}}(1 - \chi(\sigma)X) = (1 - X^f)^{|G|/f}$ if $f = \mathrm{ord}(\sigma)$.

Formulas (3a) and (3b) are known as the **orthogonality relations** [Goldstein, p. 113]. Formula (3c) follows in the case where $G = \langle\sigma\rangle$ by observing that the zeros of both sides are the roots of 1 of order f and from the relation $|\hat{G}| = |G|$. The general case follows from the cyclic case and the canonical isomorphism $\widehat{\langle\sigma\rangle} \cong \hat{G}/\langle\sigma\rangle^\perp$, where $\langle\sigma\rangle^\perp = \{\chi \in \hat{G} \mid \chi(\sigma) = 1\}$ [Goldstein, p. 112].

PART C: *L-series.* For a given ideal $\mathfrak{c}$ of O_K and a character χ of $G(\mathfrak{c})$ consider the Dirichlet series

(4) $$L_\mathfrak{c}(s, \chi) = \sum_{\gcd(\mathfrak{a},\mathfrak{c})=1} \frac{\chi(\mathfrak{a})}{(N\mathfrak{a})^s}, \quad \mathrm{Re}(s) > 1,$$

where $\mathfrak{a}$ ranges over all ideals of O_K relatively prime to $\mathfrak{c}$, and $\chi(\mathfrak{a}) = \chi(\mathcal{K})$ if $\mathcal{K}$ is the class of $\mathfrak{a}$. Call $L_\mathfrak{c}(s, \chi)$ an **L-series**. The function $\chi(\mathfrak{a})$ is multiplicative on $J(\mathfrak{c})$. Therefore, $L_\mathfrak{c}(s, \chi)$ satisfies the Euler identity

(5) $$L_\mathfrak{c}(s, \chi) = \prod_{\mathfrak{p} \nmid \mathfrak{c}} \left(1 - \frac{\chi(\mathfrak{p})}{(N\mathfrak{p})^s}\right)^{-1} \quad \text{for } \mathrm{Re}(s) > 1$$

We quote the following result of complex analysis:

6.5 Number Fields

LEMMA 6.5.1 ([Lang5, p. 158]): *Let $\{a_i\}_{i=1}^{\infty}$ be a sequence of complex numbers, for which there is a $0 \leq \sigma < 1$ and a complex number ρ with*

$$\sum_{i=1}^{n} a_i = \rho n + O(n^{\sigma}) \quad \text{as} \quad n \to \infty.$$

Then $f(s) = \sum_{n=1}^{\infty} a_n n^{-s}$ for $\operatorname{Re}(s) > 1$ analytically continues to $\operatorname{Re}(s) > \sigma$, except for a simple pole with residue ρ at $s = 1$.

LEMMA 6.5.2: *The function $L_{\mathfrak{c}}(s,\chi)$ has an analytic continuation to the half plane $\operatorname{Re}(s) > 1 - \frac{1}{[K:\mathbb{Q}]}$. If $\chi = 1$, then it has a simple pole at $s = 1$ with residue $h_{\mathfrak{c}}\rho_{\mathfrak{c}}$. If $\chi \neq 1$, then $L_{\mathfrak{c}}(s,\chi)$ is analytic in the entire half plane.*

Proof: We use (2) to substitute for $j(\mathcal{K},t)$ and use the orthogonality relation (3a) for the finite group $G(\mathfrak{c})$, to conclude that

$$\sum_{\substack{(\mathfrak{a},\mathfrak{c})=1 \\ N\mathfrak{a} \leq n}} \chi(\mathfrak{a}) = \sum_{\mathcal{K} \in G(\mathfrak{c})} \sum_{\substack{\mathfrak{a} \in \mathcal{K} \\ N\mathfrak{a} \leq n}} \chi(\mathfrak{a}) = \sum_{\mathcal{K} \in G(\mathfrak{c})} \chi(\mathcal{K})\left(\rho_{\mathfrak{c}} n + O(n^{1-[K:\mathbb{Q}]^{-1}})\right)$$

$$= \begin{cases} h_{\mathfrak{c}}\rho_{\mathfrak{c}} n + O(n^{1-[K:\mathbb{Q}]^{-1}}) & \text{if } \chi = 1 \\ O(n^{1-[K:\mathbb{Q}]^{-1}}) & \text{if } \chi \neq 1, \end{cases} \quad n \to \infty.$$

Thus, our Lemma is a special case of Lemma 6.5.1. $\square$

PART D: *Special case of Artin's reciprocity law.* This law is the central result of class field theory. Consider a finite Abelian extension L/K of number fields. Let $\mathfrak{c}$ be an ideal of O_K divisible by all prime ideals ramifying in L (we say that $\mathfrak{c}$ is **admissible**). If a prime ideal $\mathfrak{p}$ does not divide $\mathfrak{c}$, then $\left(\frac{L/K}{\mathfrak{p}}\right)$ defines a unique element of $\operatorname{Gal}(L/K)$. The map $\mathfrak{p} \mapsto \left(\frac{L/K}{\mathfrak{p}}\right)$ extends to a homomorphism $\omega_{\mathfrak{c}}\colon J(\mathfrak{c}) \to \operatorname{Gal}(L/K)$ called the **reciprocity map**. When referring to the extension L/K we will denote $\omega_{\mathfrak{c}}$ by $\omega_{L/K,\mathfrak{c}}$. Let L' be any Abelian extension of K containing L. Suppose each prime ideal of O_K ramified in L' divides $\mathfrak{c}$. Then, $\operatorname{res}_L(\omega_{L'/K,\mathfrak{c}}(\mathfrak{p})) = \omega_{L/K,\mathfrak{c}}(\mathfrak{p})$ for each prime $\mathfrak{p} \in J(\mathfrak{c})$ ((3) of Section 6.2). So,

$$\text{(6)} \qquad \operatorname{res}_L(\omega_{L'/K,\mathfrak{c}}(\mathfrak{a})) = \omega_{L/K,\mathfrak{c}}(\mathfrak{a}) \qquad \text{for each } \mathfrak{a} \in J(\mathfrak{c})$$

Class field theory proves that $\omega_{\mathfrak{c}}$ is surjective [Lang5, p. 199, Thm. 1]. In order to describe its kernel, let $\operatorname{Norm} = \operatorname{Norm}_{L/K}$ be the norm map of fractional ideals of O_L onto fractional ideals of O_K. If $\mathfrak{A}$ is an ideal of O_L, then $\operatorname{Norm}(\mathfrak{A})$ is the ideal of O_K generated by $\operatorname{norm}_{L/K}(a)$ for all $a \in \mathfrak{A}$. If $\mathfrak{B}$ is another ideal, then $\operatorname{Norm}(\mathfrak{A}\mathfrak{B}) = \operatorname{Norm}(\mathfrak{A})\operatorname{Norm}(\mathfrak{B})$, so Norm extends multiplicatively to the group of all fractional ideals of O_L. If $\mathfrak{P}$ is a prime ideal of O_L, $\mathfrak{p} = \mathfrak{P} \cap O_K$, and $f = f_{\mathfrak{P}/\mathfrak{p}}$, then $\operatorname{Norm}(\mathfrak{P}) = \mathfrak{p}^f$. Finally, $\operatorname{Norm}_{L/\mathbb{Q}}$ and the absolute norm of ideals relate to each other by the formula $\operatorname{Norm}_{L/\mathbb{Q}}(\mathfrak{A}) = N(\mathfrak{A})\mathbb{Z}$. [Janusz, pp. 35–37].

Artin reciprocity law gives an admissible ideal $\mathfrak{c}$ of O_K such that $\omega_{\mathfrak{c}}: J(\mathfrak{c}) \to \text{Gal}(L/K)$ is surjective and $\text{Ker}(\omega_{\mathfrak{c}}) = \text{Norm}(\mathfrak{c})P(\mathfrak{c})$ [Lang5, p. 205, Thm. 3]. Here we prove one part of Artin reciprocity law for cyclotomic extensions of K.

LEMMA 6.5.3: *Let ζ be a primitive mth root of 1, L a subfield of $K(\zeta)$ containing K, and $\mathfrak{c}$ an ideal of O_K divisible by m. Then $P(\mathfrak{c}) \subseteq \text{Ker}(\omega_{\mathfrak{c}})$.*

Proof: Each prime ideal of O_K which ramifies in $K(\zeta)$ divides m [Goldstein, p. 98]. This defines $\omega_{\mathfrak{c}} = \omega_{L/K,\mathfrak{c}}$. Use (6) to assume that $L = K(\zeta)$.

Consider the natural embedding $i: \text{Gal}(L/K) \to (\mathbb{Z}/m\mathbb{Z})^{\times}$ determined by $\sigma(\zeta) = \zeta^{i(\sigma)}$, $\sigma \in \text{Gal}(L/K)$. If $\mathfrak{p} \in J(\mathfrak{c})$ is prime and $\mathfrak{P}$ is a prime ideal of O_L lying over $\mathfrak{p}$, then $\omega_{\mathfrak{c}}(\mathfrak{p})(\zeta) \equiv \zeta^{N\mathfrak{p}} \mod \mathfrak{P}$. Since reduction modulo $\mathfrak{P}$ is injective on $\{1, \zeta, \ldots, \zeta^{m-1}\}$ [Goldstein, p. 97, Prop. 6-2-2], $\omega_{\mathfrak{c}}(\mathfrak{p})(\zeta) = \zeta^{N\mathfrak{p}}$. Hence, $i(\omega_{\mathfrak{c}}(\mathfrak{p})) \equiv N\mathfrak{p} \mod m$. Therefore,

(7) $\qquad i(\omega_{\mathfrak{c}}(\mathfrak{a})) \equiv N\mathfrak{a} \mod m$ for each $\mathfrak{a} \in J(\mathfrak{c})$.

Now let $xO_K \in P(\mathfrak{c})$, with $x \in K^{\times}$. Since x is totally positive (by (1b)), $N_{K/\mathbb{Q}}(x)$ is a positive rational number congruent by (1a) to 1 mod m. It generates the same fractional $\mathbb{Z}$-ideal as $N(xO_K)$ [Janusz, p. 37]. Since $N(xO_K)$ is also a positive rational number, $N(xO_K) = N_{K/\mathbb{Q}}(x) \equiv 1 \mod m$. We conclude from (7) that $\omega_{\mathfrak{c}}(xO_K) = 1$. $\square$

PART E: *Cyclotomic L-series and the Dedekind zeta function.* Let L and $\mathfrak{c}$ be as in Lemma 6.5.3. Then $\omega_{\mathfrak{c}}$ induces a homomorphism $\bar{\omega}_{\mathfrak{c}}$ from $G(\mathfrak{c}) = J(\mathfrak{c})/P(\mathfrak{c})$ onto a subgroup G of $\text{Gal}(L/K)$. (In Corollary 6.5.5(c), we prove that $G = \text{Gal}(L/K)$.) Thus, for each $\chi \in \hat{G}$, $\chi \circ \bar{\omega}_{\mathfrak{c}}$ is a character of $G(\mathfrak{c})$.

In the notation of Part C, the L-series $L_{\mathfrak{c}}(s, 1)$ of the trivial character is the **Dedekind zeta function** of K with respect to $\mathfrak{c}$.

$$\zeta_{\mathfrak{c}}(s, K) = \sum_{\gcd(\mathfrak{a},\mathfrak{c})=1} \frac{1}{(N\mathfrak{a})^s} = \prod_{\mathfrak{p} \nmid \mathfrak{c}} \left(1 - \frac{1}{(N\mathfrak{p})^s}\right)^{-1}, \qquad \text{Re}(s) > 1$$

LEMMA 6.5.4: *In the above notation let $\mathfrak{C} = \mathfrak{c}O_L$ and let $n = (\text{Gal}(L/K) : G)$. Then*

(8) $\qquad \zeta_{\mathfrak{C}}(s, L) = \prod_{\chi \in \hat{G}} L_{\mathfrak{c}}(s, \chi \circ \bar{\omega}_{\mathfrak{c}})^n.$

Proof: Let $\mathfrak{p}$ be a prime ideal of O_K with $\mathfrak{p} \nmid \mathfrak{c}$. Suppose $\mathfrak{p}$ factors in L into a product of g prime ideals $\mathfrak{P}$, each of degree f. Since $\mathfrak{p}$ is unramified in L, we have $fg = [L:K] = n|G|$ and $\omega_{\mathfrak{c}}(\mathfrak{p}) = \left(\frac{L/K}{\mathfrak{p}}\right)$ is of order f. By (3c) and the relation $N\mathfrak{P} = N\mathfrak{p}^f$,

$$\prod_{\chi \in \hat{G}} \left(1 - \frac{\chi \circ \bar{\omega}_{\mathfrak{c}}(\mathfrak{p})}{(N\mathfrak{p})^s}\right)^n = \left(1 - \frac{1}{(N\mathfrak{p})^{sf}}\right)^{n|G|/f} = \prod_{\mathfrak{P}|\mathfrak{p}} \left(1 - \frac{1}{(N\mathfrak{P})^s}\right).$$

Applying the product over all $\mathfrak{p} \nmid \mathfrak{c}$, we conclude (8). $\square$

6.5 Number Fields

COROLLARY 6.5.5: *Let χ_1 be a nontrivial character of G. Then:*
(a) $L_{\mathfrak{c}}(1, \chi_1 \circ \bar{\omega}_{\mathfrak{c}}) \neq 0$;
(b) $\log \zeta_{\mathfrak{c}}(s, K) = -\log(s-1) + O(1)$, $s \to 1^+$; and
(c) $G = \text{Gal}(L/K)$.

Proof: Since $\bar{\omega}_{\mathfrak{c}} \colon G(\mathfrak{c}) \to G$ is surjective, $\chi \circ \bar{\omega}_{\mathfrak{c}}$ is a nontrivial character of $G(\mathfrak{c})$ for each nontrivial character of G. By Lemma 6.5.2, $\zeta_{\mathfrak{c}}(s, K)$ has a simple pole at $s = 1$. All other factors of the right hand side of (8) are regular at $s = 1$. Assume $L_{\mathfrak{c}}(s, \chi_1 \circ \bar{\omega}_{\mathfrak{c}})$ has a zero at $s = 1$, then the zero of $L_{\mathfrak{c}}(s, \chi \circ \bar{\omega}_{\mathfrak{c}})^n$ in (8) at $s = 1$ cancels the pole of $\zeta_{\mathfrak{c}}(s, K)^n$. Hence, $\zeta_{\mathfrak{c}}(s, L)$ is analytic at $s = 1$, contradicting Lemma 6.5.1. This proves (a). Formula (b) follows from Lemma 6.5.2.

Finally, the left side of (8) has a pole of order 1 at $s = 1$. By (a), the right side has pole of order n. Therefore, $n = 1$ and $G = \text{Gal}(L/K)$. □

LEMMA 6.5.6: *If χ is a character of G, then*

$$\log L_{\mathfrak{c}}(s, \chi \circ \bar{\omega}_{\mathfrak{c}}) = \sum_{\mathfrak{p} \nmid \mathfrak{c}} \frac{\chi \circ \bar{\omega}_{\mathfrak{c}}(\mathfrak{p})}{(N\mathfrak{p})^s} + O(1), \quad s \to 1^+.$$

Proof: We apply the Euler identity (5) for $\text{Re}(s) > 1$ to obtain:

$$\log L_{\mathfrak{c}}(s, \chi \circ \bar{\omega}_{\mathfrak{c}}) = -\sum_{\mathfrak{p} \nmid \mathfrak{c}} \log\left(1 - \frac{\chi \circ \bar{\omega}_{\mathfrak{c}}(\mathfrak{p})}{(N\mathfrak{p})^s}\right) = \sum_{\mathfrak{p} \nmid \mathfrak{c}} \sum_{k=1}^{\infty} \frac{\chi \circ \bar{\omega}_{\mathfrak{c}}(\mathfrak{p})^k}{k \cdot (N\mathfrak{p})^{ks}}.$$

Next let $\sigma = \text{Re}(s)$. Then

$$\left|\sum_{\mathfrak{p} \nmid \mathfrak{c}} \sum_{k=2}^{\infty} \frac{|\chi \circ \bar{\omega}_{\mathfrak{c}}(\mathfrak{p})|^k}{k(N\mathfrak{p})^{ks}}\right| \leq \sum_{p} \sum_{\mathfrak{p}|p} \sum_{k=2}^{\infty} \frac{1}{p^{f_{\mathfrak{p}/p} k \sigma}}$$

$$\leq [K : \mathbb{Q}] \sum_{p} \sum_{k=2}^{\infty} \frac{1}{p^{k\sigma}} = [K : \mathbb{Q}] \sum_{p} \frac{1}{p^{2\sigma}} \left(\frac{1}{1 - p^{-\sigma}}\right)$$

$$\leq 2[K : \mathbb{Q}] \sum_{p} \frac{1}{p^2} < \infty$$

and the lemma follows. □

In particular,

$$\log \zeta_{\mathfrak{c}}(s, K) = \sum_{\mathfrak{p}} \frac{1}{(N\mathfrak{p})^s} + O(1), \quad s \to 1^+.$$

Combining this formula with Corollary 6.5.5(b) gives an alternative expression for the Dirichlet density of a set A of primes of K:

(9) $$\delta(A) = \lim_{s \to 1^+} \frac{\sum_{\mathfrak{p} \in A} (N\mathfrak{p})^{-s}}{-\log(s-1)}.$$

PART F: *The Chebotarev density theorem for cyclotomic extensions.* In the notation of Part E, let $\sigma \in G = \text{Gal}(L/K)$. To compute the Dirichlet density of the set $C = \{\mathfrak{p} \in P(K) \mid \left(\frac{L/K}{\mathfrak{p}}\right) = \sigma\}$, or equivalently of $C' = \{\mathfrak{p} \in P(K) \mid \mathfrak{p} \nmid c,\ \omega_c(\mathfrak{p}) = \sigma\}$, we apply the Dirichlet character argument. This singles out elements of C' from all primes by evaluating

$$f(s) = \sum_{\chi \in \hat{G}} \chi(\sigma^{-1}) \log(L_c(s, \chi \circ \bar{\omega}_c))$$

in two ways. On one hand, by Corollary 6.5.5,

$$f(s) = -\log(s-1) + O(1), \qquad s \to 1^+.$$

On the other hand, by Lemma 6.5.6,

$$f(s) = \sum_{\mathfrak{p} \nmid c} \left(\sum_{\chi \in \hat{G}} \chi(\sigma^{-1}) \chi(\omega_c(\mathfrak{p})) \right) \frac{1}{(N\mathfrak{p})^s} + O(1), \qquad s \to 1^+.$$

Orthogonality relation (3b) shows

$$f(s) = [L:K] \sum_{\mathfrak{p} \in C'} \frac{1}{(N\mathfrak{p})^s} + O(1), \qquad s \to 1^+.$$

Hence, $\sum_{\mathfrak{p} \in C} (N\mathfrak{p})^{-s} = -[L:K]^{-1} \log(s-1) + O(1)$. Therefore, by (9), $\delta(C) = [L:K]^{-1}$. This proves the Chebotarev density theorem for cyclotomic extensions.

PART G: *The Abelian case.* Let L be a finite Abelian extension of K and let $\sigma \in \text{Gal}(L/K)$. Suppose M is a finite cyclotomic extension of K with $L \cap M = K$. Choose $\tau \in \text{Gal}(M/K)$ with $\text{ord}(\sigma)|\text{ord}(\tau)$. Let $F = LM$. Consider the unique ρ of $\text{Gal}(F/K)$ satisfying $\text{res}_L \rho = \sigma$ and $\text{res}_M \rho = \tau$, and let $E = F(\rho)$. Then, $[M : E \cap M] = \text{ord}(\tau) = \text{ord}(\rho) = [F : E]$. Hence, $F = ME$ and F/E is a cyclotomic extension.

By Part F, the Dirichlet density of $A'_\tau = \{\mathfrak{q} \in P(E) \mid \left(\frac{F/E}{\mathfrak{q}}\right) = \rho\}$ is $\frac{1}{[F:E]}$. Since

$$\sum_{N(\mathfrak{q} \cap K) < N\mathfrak{q}} \frac{1}{(N\mathfrak{q})^s} \leq \sum_{N(\mathfrak{q} \cap \mathbb{Q}) < N\mathfrak{q}} \frac{1}{(N\mathfrak{q})^s} \leq [E:\mathbb{Q}] \sum_{q \text{ prime}} \frac{1}{q^{2s}} < \infty,$$

$\delta(A'_\tau)$ is the Dirichlet density of

$$A''_\tau = \left\{ \mathfrak{q} \in P(E) \mid \left(\frac{F/E}{\mathfrak{q}}\right) = \rho,\ \mathfrak{q} \text{ unramifies over } K, \text{ and } N(\mathfrak{q} \cap K) = N\mathfrak{q} \right\}.$$

6.5 Number Fields

Thus, if $\mathfrak{p} \in P(K)$ lies under $\mathfrak{q} \in A''_\tau$, then $\mathfrak{p}$ belongs to

$$A_\tau = \{\mathfrak{p} \in P(K) \mid \left(\frac{F/K}{\mathfrak{p}}\right) = \rho\}.$$

Conversely, if $\mathfrak{q} \in P(E)$ lies over $\mathfrak{p} \in A_\tau$, then $\left(\frac{E/K}{\mathfrak{p}}\right) = \mathrm{res}_E \rho = 1$. Hence, $\mathfrak{p}$ splits in E, $N\mathfrak{p} = N\mathfrak{q}$, and $\mathfrak{q} \in A''_\tau$. In particular, exactly $[E:K]$ primes $\mathfrak{q}$ of $P(E)$ lie over $\mathfrak{p}$. By Part F, $\delta(A''_\tau) = \frac{1}{[F:E]}$, so

(10) $$\delta(A_\tau) = \frac{\delta(A''_\tau)}{[E:K]} = \frac{1}{[F:K]}.$$

For $T = T(M/K) = \{\tau \in \mathrm{Gal}(M/K) \mid \mathrm{ord}(\sigma)|\mathrm{ord}(\tau)\}$, (10) yields

(11) $$\delta\left(\bigcup_{\tau \in T} A_\tau\right) = \sum_{\tau \in T} \delta(A_\tau) = \frac{1}{[L:K]} \cdot \frac{|T(M/K)|}{[M:K]}.$$

By (3) of Section 6.2,

(12) $$\bigcup_{\tau \in T} A_\tau \subseteq \left\{\mathfrak{p} \in P(K) \mid \left(\frac{L/K}{\mathfrak{p}}\right) = \sigma\right\} = C_\sigma.$$

LEMMA 6.5.7: *Let L/K be a finite extension of number fields and let $m \in \mathbb{N}$. Then there exists a cyclic cyclotomic extension M/K of degree m such that $M \cap L = K$.*

Proof: It suffices to find M such that $M \cap L = \mathbb{Q}$ and $[M:\mathbb{Q}] = m$. For this we may replace L by the maximal cyclotomic extension of $\mathbb{Q}$ in L to assume that $L \subseteq \mathbb{Q}(\zeta_n)$ for some $n \in \mathbb{N}$. Use Corollary 6.3.2, which follows just from Part F. Thus, there exists a prime $q > n$ with $m|q - 1$. Since $\mathbb{Q}(\zeta_q)/\mathbb{Q}$ is cyclic of degree $q - 1$, it contains a cyclic extension M of $\mathbb{Q}$ of order m. Since $\gcd(n,q) = 1$, $L \cap M \subseteq \mathbb{Q}(\zeta_n) \cap \mathbb{Q}(\zeta_q) = \mathbb{Q}$. □

Let $\mathrm{ord}(\sigma) = p_1^{a_1} \cdots p_r^{a_r}$ be the factorization of $\mathrm{ord}(\sigma)$ into a product of distinct prime powers. Lemma 6.5.7 gives a cyclic cyclotomic extension M/K of degree $p_1^{b_1} \cdots p_r^{b_r}$ with $b_1 \geq a_1, \ldots, b_r \geq a_r$ and $L \cap M = K$. Then

$$|T(M/K)| = \prod_{i=1}^{r}(p_i^{b_i} - p_i^{a_i-1}) = [M:K]\prod_{i=1}^{r}(1 - p_i^{a_i-1-b_i}).$$

The product on the right approaches 1 as $b_1, \ldots, b_r$ tend to ∞. Hence, given $\varepsilon > 0$, there exists M as above with

$$\left|\frac{\#T(M/K)}{[M:K]} - 1\right| < \varepsilon.$$

By (11) and (12) there exists $s_0 > 1$ with

$$(13) \qquad \frac{\sum_{\mathfrak{p} \in C_\sigma}(N\mathfrak{p})^{-s}}{\sum_{\mathfrak{p} \in P(K)}(N\mathfrak{p})^{-s}} > \frac{1}{[L:K]} - \frac{\varepsilon}{[L:K]}, \qquad 1 < s < s_0.$$

As σ ranges on $\mathrm{Gal}(L/K)$, the sum of the left hand sided of (13) is 1. Hence, by (13)

$$(14) \qquad \frac{\sum_{\mathfrak{p} \in C_\sigma}(N\mathfrak{p})^{-s}}{\sum_{\mathfrak{p} \in P(K)}(N\mathfrak{p})^{-s}} < \frac{1}{[L:K]} + \varepsilon \frac{[L:K]-1}{[L:K]}, \qquad 1 < s < s_0.$$

It follows from (13) and (14) that each C_σ has Dirichlet density $[L:K]^{-1}$. This proves the Chebotarev density theorem for Abelian extensions.

PART H: *Reduction to the cyclic case.* We now reduce the Chebotarev density theorem for an arbitrary finite Galois extension L/K of number fields to the case that L/K is cyclic.

Let $\mathcal{C}$ be a conjugacy class in $\mathrm{Gal}(L/K) = G$ and let

$$C = \{\mathfrak{p} \in P(K) \mid \left(\frac{L/K}{\mathfrak{p}}\right) = \mathcal{C}\}.$$

Choose $\tau \in \mathcal{C}$ and let $K' = L(\tau)$ be its fixed field. Denote the set of primes $\mathfrak{q} \in P(K')$ which are unramified over K, have a relative degree 1 over K, and satisfy $\left(\frac{L/K'}{\mathfrak{q}}\right) = \{\tau\}$ by D'. If $\mathfrak{q} \in D'$ and $\mathfrak{p} = \mathfrak{q} \cap K$, then $N\mathfrak{p} = N\mathfrak{q}$. Hence, if $\mathfrak{P} \in P(L)$ lies over $\mathfrak{q}$, then $\left[\frac{L/K}{\mathfrak{P}}\right] = \left[\frac{L/K'}{\mathfrak{P}}\right] = \tau$, so $\mathfrak{p} \in C$. Moreover, since $\mathrm{ord}(\tau) = [L:K']$, $\mathfrak{P}$ is the unique element of $P(L)$ over $\mathfrak{q}$. Conversely, if $\mathfrak{p} \in C$, then there exists a prime $\mathfrak{P} \in P(L)$ that lies over $\mathfrak{p}$ with $\left[\frac{L/K}{\mathfrak{P}}\right] = \tau$. Then $\mathfrak{q} = \mathfrak{P} \cap K'$ belongs to D' and lies over $\mathfrak{p}$.

Let $C_G(\tau)$ be the centralizer of τ in G. The map $\sigma \mapsto \sigma\tau\sigma^{-1}$ from G onto $\mathcal{C}$ has fibers whose order is $|C_G(\tau)|$, so $|G| = |\mathcal{C}| \cdot |C_G(\tau)|$. All primes $\mathfrak{P}' \in P(L)$ over $\mathfrak{p}$ that satisfy $\left[\frac{L/K}{\mathfrak{P}'}\right] = \tau$ are conjugate to $\mathfrak{P}$ by some element $\sigma \in C_G(\tau)$. Hence, their number is $\frac{|C_G(\tau)|}{|D_\mathfrak{P}|} = \frac{|G|}{[L:K'] \cdot |\mathcal{C}|}$.

This is therefore the number of $\mathfrak{q} \in D'$ that lie over $\mathfrak{p}$. Hence, for $s > 1$

$$(15) \qquad \sum_{\mathfrak{p} \in C} \frac{1}{(N\mathfrak{p})^s} = \frac{|\mathcal{C}|[L:K']}{[L:K]} \sum_{\mathfrak{q} \in D'} \frac{1}{(N\mathfrak{q})^s}.$$

The set D' differs from $D = \{\mathfrak{q} \in P(K') \mid \left(\frac{L/K'}{\mathfrak{q}}\right) = \{\tau\}\}$ by only primes of relative degree at least 2 over K. For these primes we have

$$\sum_{\deg(\mathfrak{q}) \geq 2} \frac{1}{(N\mathfrak{q})^s} \leq [K':\mathbb{Q}] \sum_q \frac{1}{q^2} < \infty.$$

Exercises

(Note: This elimination of the set of absolute degree 2 primes won't work in the function field case, for there are only finitely many primes of degree 1.) Hence, by Part G for cyclic extensions

$$\sum_{\mathfrak{q}\in D'}(N\mathfrak{q})^{-s} = \sum_{\mathfrak{q}\in D}(N\mathfrak{q})^{-s} + O(1) = -\frac{1}{[L:K']}\log(s-1) + O(1), \quad s\to 1^+.$$

Combining this with (15) gives

$$\sum_{\mathfrak{p}\in C}(N\mathfrak{p})^{-s} = -\frac{|C|}{[L:K]}\log(s-1) + O(1), \quad s\to 1^+.$$

Finally, by (9), $\delta(C) = \frac{|C|}{[L:K]}$, as stated. $\square$

Exercises

1. Consider the ring $R = \mathbb{Z}[\zeta_p][x]$, where p is a prime number and x is an indeterminate. Let K be the quotient field of R and $L = K(x^{1/p})$. Give an example, in Lemma 6.1, where the residue class field extension $\bar{L}/\bar{K}$ is not Galois.

2. Let S/R be a ring cover. Consider prime ideals $\mathfrak{p}$ of R and $\mathfrak{P}$ of S with $\mathfrak{P}\cap R = \mathfrak{p}$. Prove that $\mathfrak{p}S_\mathfrak{P} = \mathfrak{P}S_\mathfrak{P}$ (This, together with the separability of $S_\mathfrak{P}/\mathfrak{P}S_\mathfrak{P}$ over $R_\mathfrak{p}/\mathfrak{p}R_\mathfrak{p}$ means that S/R is an **unramified extension of rings**).

    Hint: Assume without loss that R and S are local rings with maximal ideals $\mathfrak{p}$ and $\mathfrak{P}$, respectively. Let z be a primitive element for S/R and $f = \mathrm{irr}(z, \mathrm{Quot}(R))$. Denote the reduction modulo $\mathfrak{p}S$ with a bar. Observe that $\bar{S}$ is a local ring with maximal ideal $\bar{\mathfrak{P}}$. Prove that $\bar{S} \cong \bar{R}[X]/\bar{f}(X)\bar{R}[X]$ and the right hand side is a direct sum of fields corresponding to the irreducible factors of $\bar{f}(X)$. Conclude that there is only one such factor.

3. Let L_1, L_2 be finite Galois extensions of a global field K and let $L = L_1L_2$. Consider a prime ideal $\mathfrak{P}$ of O_L and put $\mathfrak{P}_i = \mathfrak{P}\cap L_i$, $i = 1, 2$. Prove that $\left[\frac{L/K}{\mathfrak{P}}\right]$ is the unique element of $\mathrm{Gal}(L/K)$ whose restriction to L_i is $\left[\frac{L_i/K}{\mathfrak{P}_i}\right]$, $i = 1, 2$.

4. Show that if $f(X) \in \mathbb{Z}[X]$ is irreducible, then there exists infinitely many primes p for which $f(X) \equiv 0 \mod p$ has no solution. Hint: Use the equivalence of (1) and (3) of Section 6.3 and Lemma 13.3.2.

5. Let L/K be a finite Galois extension of global fields. Denote the set of prime ideals $\mathfrak{p}$ of O_K that split completely in L ($\mathfrak{p}O_L = \mathfrak{P}_1\cdots\mathfrak{P}_n$, where $n = [L:K]$) by $\mathrm{Splt}(L/K)$.

    (a) Show that a prime $\mathfrak{p} \in P(K)$, unramified in L, belongs to $\mathrm{Splt}(L/K)$ if and only if $\left(\frac{L/K}{\mathfrak{p}}\right) = 1$.

(b) (Bauer) Suppose L and L' are finite Galois extensions of a global field K such that $\mathrm{Splt}(L/K)$ and $\mathrm{Splt}(L'/K)$ differ by a finite set., Prove that $L = L'$. Hint: Apply the Chebotarev density theorem to the field LL'.

6. Let K and L be number fields with equal zeta functions (Section 6.5, Part E). For each positive integer n prove that the number of ideals of O_K with absolute norm n is the number of ideals of O_L with absolute norm n. Apply this to any a prime p unramified in KL. Prove that p splits completely in K if and only if p splits completely in L. Conclude from Exercise 5 that if K and L are Galois over $\mathbb{Q}$, then $K = L$

7. Let K be a global field. Denote the set of all prime ideals $\mathfrak{p}$ of O_K whose absolute degree is at least 2 by $P'(K)$. That is, $N\mathfrak{p} = p^d$ where $d \geq 2$. When K is a number field prove that the Dirichlet density as well as the natural density of $P'(K)$ is 0. If, however K is a function field show that almost all primes of $P(K)$ belong to $P'(K)$. Thus, $\delta(P'(K)) = 1$.

Notes

Frobenius [Frobenius] conjectured what we now call the Chebotarev density theorem for finite Galois extension L/K of number fields. His result replaced the conjugacy class appearing in the conjecture by the union of all conjugates of σ^i, where σ is a given element of $\mathrm{Gal}(L/K)$ and i ranges over all integers relatively prime to $\mathrm{ord}(\sigma)$. A fine account of the **Frobenius density theorem** appears in [Janusz].

Chebotarev [Tschebotarev] used cyclotomic fields to prove the Frobenius conjecture via a more difficult version of the field crossing argument of Part G of Section 6.5. Artin [Artin1] introduced his L-series; then he proved his reciprocity law and applied it to reprove the conjecture [Artin2]. Our proof is a mixture of both methods, with the addition of Deuring's reduction to the cyclic case [Deuring1]. It was elaborated for this book by Haran. Deuring's reduction was reproduced in [MacCluer].

For the function field case note that Reichardt proved Proposition 6.19 when $a = k = 1$ and K is algebraically closed in F, (i.e. $m = 1$) [Reichardt]. The restriction $m = 1$ does not appear explicitly in [Reichardt]. Without it, however, the result as well as its proof would be false. It is also interesting to note that [Reichardt] appeared before Weil proved the Riemann hypothesis for curves. Thus, Reichardt's (analytic) proof uses only that the maximum of the real parts of the zeros of the Zeta function is less than 1.

Serre [Serre2] gives a unified approach to the number field and function field case. He considers a scheme X of finite type over $\mathbb{Z}$, takes an étale Galois covering of X and attaches an L-series $L(X, \chi; s)$ to the cover. He says that an induction on $\dim(X)$ shows if $\chi \neq 1$, then $L(X, \chi; s)$ is holomorphic and $\neq 0$ at the point $s = \dim(X)$. This implies the Chebotarev density theorem by the classical Dirichlet argument (e.g. as in Part F of Section 6.5).

Serre's program for the function field case appears in [Fried9]. An early version of our proof for the function field case appears in [Jarden10].

Notes

The proof of [Fried-Jarden3, Prop. 5.16] applies [Fried-Jarden3, Lemma 5.14] in a faulty way. Indeed, d on [Fried-Jarden3, p. 63, line -3] should be replaced by md. This version of Field Arithmetic follows [Geyer-Jarden4, Appendix] and corrects this mistake also improving the estimate of [Fried-Jarden3, Prop. 5.16].

There are several effective versions of the Chebotarev density theorem. One of the most valuable for the problems in this book is [Lagarias-Montegomery-Odlyzko, p. 416, Theorem]. This isolates the contribution of the absolute discriminant d_L of a number field L over $\mathbb{Q}$ to the error term. It proves that there is an effectively computable constant A with the following property: For each Galois extension L/K of $\mathrm{Gal}(L/K)$ of number fields and each conjugacy class $\mathcal{C}$ of $\mathrm{Gal}(L/K)$ there is a prime $\mathfrak{p}$ of K, unramified in L with $\left(\frac{L/K}{\mathfrak{p}}\right) = \mathcal{C}$, $p = N_{L/\mathbb{Q}}\mathfrak{p}$ is a rational prime and $p \leq 2d_L^A$. This result is independent of the generalized Riemann hypothesis.

Chapter 7. Ultraproducts

We develop the basic concepts of logic and model theory required for applications to field theory. These include the Skolem-Löwenheim theorem, Loš theorem and an $\aleph_1$-saturation property for ultraproducts. Finally, we apply regular ultraproducts of families of models to the theory of finite fields.

7.1 First Order Predicate Calculus

There is no general test to decide whether a given polynomial $f(X_1, \ldots, X_n)$ with integral coefficients has a zero in $\mathbb{Z}^n$; this is the negative solution to Hilbert's 10th problem. A partial decision test must satisfy two criteria: it must be conclusive for a significant body of polynomials; and it must be effective in concrete situations. The simplest, and most famous, such test is the **congruence test**, whereby we test the congruence $f(X_1, \ldots, X_n) \equiv 0 \bmod p$ for solutions for all primes p. Regard the coefficients of f as elements of $\mathbb{F}_p$ to see that the above congruence is equivalent to solving the equation $f(X_1, \ldots, X_n) = 0$ in $\mathbb{F}_p$. If $f(X_1, \ldots, X_n) = 0$ has no solution in $\mathbb{F}_p$ for one p, then $f(X_1, \ldots, X_n) = 0$ has no solution in $\mathbb{Z}$.

This chapter develops language and technique for the formulation of analogs of the diophantine problem and of the corresponding congruence test over general rings and fields. We start with the introduction of a first order language, the concept of a theory in the first order language, and a model for this theory.

A **language** (more precisely, **first order language**) consists of **letters**, rules for combining letters into **meaningful words**, and, finally, an interpretation of the meaningful words. The language we now describe depends on functions μ and ν from sets I and J to $\mathbb{N}$ and on a set K. It is denoted $\mathcal{L}(\mu, \nu, K)$. Here are its letters:
(1a) Countably many **variable symbols**: $X_1, X_2, X_3, \ldots$;
(1b) **constant symbols** c_k, one for each $k \in K$;
(1c) a $\mu(i)$-**ary relation symbol**, R_i, one for each $i \in I$;
(1d) the **equality symbol** $=$;
(1e) a $\nu(i)$-**ary function symbol**, F_j, one for each $j \in J$;
(1f) the **negation symbol** $\neg$, and the disjunction symbol $\vee$;
(1g) the **existential symbol** $\exists$; and
(1h) parentheses ( ) and brackets [ ].

A finite sequence of letters of $\mathcal{L}(\mu, \nu, K)$ is a **string**. Among the strings of $\mathcal{L}(\mu, \nu, K)$ the (**meaningful**) **words** include, "terms", "formulas" and "sentences"; which we now define.

The collection of **terms** of $\mathcal{L}(\mu, \nu, K)$ is the smallest collection of strings that contains all of the following:
(2a) all the variable symbols X_i;

7.1 First Order Predicate Calculus

(2b) all the constant symbols c_k; and

(2c) all the strings $F_j(t_1, \ldots, t_{\nu(j)})$ where $j \in J$ and $(t_1, \ldots, t_{\nu(j)})$ is a $\nu(j)$-tuple of previously defined terms.

We list the **atomic formulas**:

(3a) $t = t'$ for each pair of terms (t, t'); and

(3b) $R_i(t_1, \ldots, t_{\mu(i)})$, for all $i \in I$ and all $\mu(i)$-tuples $(t_1, \ldots, t_{\mu(i)})$ of terms.

The set of **formulas** is the smallest collection of strings containing all atomic formulas and satisfying the following:

(4a) $\neg[\varphi]$ is a formula, if φ is a formula;

(4b) $\varphi_1 \vee \varphi_2$ is a formula if φ_1 and φ_2 are formulas; and

(4c) $(\exists X_l)[\varphi]$ is a formula, if φ is a formula and $l \in \mathbb{N}$.

This definition allows us to prove a property of formulas or to make definitions depending on formulas by an **induction on structure**. We first prove the property for atomic formulas. Then, assuming its validity for φ, φ_1 and φ_2 we prove it for $\neg[\varphi]$, $\varphi_1 \vee \varphi_2$ and $(\exists X_l)[\varphi]$.

As a first example we define the notion of **free occurrence of a variable in a formula** by an induction on structure: Any occurrence of X in an atomic formula φ is **free**. If an occurrence of X in a formula φ is free, and ψ is an arbitrary formula, then this occurrence is free in $\neg\varphi$, $\varphi \vee \psi$ and $(\exists Y)[\varphi]$, for Y distinct from X. Any occurrence of X which is not free is **bounded**.

Any variable X which has a free occurrence in a formula φ is said to be a **free** variable of φ. Frequently we write $\varphi(X_1, \ldots, X_n)$ (or $t(X_1, \ldots, X_n)$) to indicate that $X_1, \ldots, X_n$ include all the free variables of φ (or t).

Example: In the formula

$$(\exists X)[\overset{1}{X} = Y \vee (\exists X)[\overset{2}{X} = c]] \vee \neg R(\overset{3}{X}, Y)$$

occurrences 1 and 2 are bounded, while occurrence 3 is free; both occurrences of Y are free. Hence, X and Y are free variables of the formula. □

A formula without free variables is a **sentence**. Some abbreviations simplify this language:

(5a) $\varphi \wedge \psi$ for $\neg[\neg\varphi \vee \neg\psi]$ ($\wedge$ is the **conjunction symbol**);

(5b) $\varphi \rightarrow \psi$ for $\neg\varphi \vee \psi$ ($\rightarrow$ is the **implication symbol**);

(5c) $\varphi \leftrightarrow \psi$ for $[\varphi \rightarrow \psi] \wedge [\psi \rightarrow \varphi]$ ($\leftrightarrow$ is the **double implication symbol**);

(5d) $(\forall X_l)[\varphi]$ for $\neg(\exists X_l)[\neg\varphi]$ ($\forall$ is the **universal quantifier**);

(5e) $\bigwedge_{i=1}^{n} \varphi_i$ for $\varphi_1 \wedge \varphi_2 \wedge \cdots \wedge \varphi_n$; and

(5f) $\bigvee_{i=1}^{n} \varphi_i$ for $\varphi_1 \vee \varphi_2 \vee \cdots \vee \varphi_n$.

7.2 Structures

The sentences of a first order language are interpreted in "structures" of this language. In each of these structures, they are either true or false.

A **structure** for the language $\mathcal{L}(\mu,\nu,K)$ is a system

$$\mathcal{A} = \langle A; \bar{R}_i, \bar{F}_j, \bar{c}_k \rangle_{i \in I, j \in J, k \in K}$$

where A is a nonvoid set, called the **domain** of $\mathcal{A}$, $\bar{R}_i$ is a $\mu(i)$-ary relation of A (i.e. a subset of $A^{\mu(i)}$), $\bar{F}_j \colon A^{\nu(j)} \to A$ is a $\nu(j)$-ary function on A, and $\bar{c}_k$ is an element of A, called a **constant**.

Sometimes we use the same letter for the logical symbol and its interpretation in the structure. Also, for well known binary relations and binary functions we write the relation and function symbols as usual, between the argument (e.g. "$a \leq b$" for "a less than or equal to b").

Occasionally we add to $\mathcal{L} = \mathcal{L}(\mu,\nu,K)$ a new constant symbol $\tilde{a}$ for each $a \in A$. This gives an extended structure $\mathcal{L}(A) = \mathcal{L}(\mu,\nu,K \cup \{\tilde{a} \mid a \in A\})$.

A **substitution** into A is a function, $f(X_i) = x_i$, from the set of variables into A. The following recursive rules extend this uniquely to a function from the set of terms into A:

(1a) $f(c_k) = \bar{c}_k$, and

(1b) $f(F_j(t_1,\ldots,t_{\nu(j)})) = \bar{F}_j(f(t_1),\ldots,f(t_{\nu(j)}))$, where $t_1,\ldots,t_{\nu(j)}$, are terms for which f has already been defined.

Define the **truth value** of a formula φ under a substitution f (either "true" or "false") by induction on structure:

(2a) $t = t'$ is true if $f(t) = f(t')$; and

(2b) $R_i(t_1,\ldots,t_{\mu(i)})$ is true if $(f(t_1),\ldots,f(t_{\mu(i)})) \in \bar{R}_i$.

Continue by assuming that the truth values of φ, φ_1, and φ_2 have been defined for all possible substitutions. Then

(3a) $\neg \varphi$ is true if φ is false;

(3b) $\varphi_1 \vee \varphi_2$ is true if φ_1 is true or if φ_2 is true (so if both φ_1 and φ_2 are true, then $\varphi_1 \vee \varphi_2$ is also true); and

(3c) $(\exists X_l)[\varphi]$ is true if there exists an x in A such that φ is true under the substitution g defined by: $g(X_l) = x$ and $g(X_m) = f(X_m)$ if $m \neq l$.

The truth values of the additional logical symbols introduced above are as follows:

(4a) $\varphi_1 \wedge \varphi_2$ is true if both φ_1 and φ_2 are true;

(4b) $\varphi \to \psi$ is true if the truth of φ implies the truth of ψ (i.e. either φ is false, or both φ and ψ are true);

(4c) $\varphi \leftrightarrow \psi$ is true if both φ and ψ are true or both φ and ψ are false; and

(4d) $(\forall X_l)[\varphi]$ is true if for each x in A, φ is true under the substitution g defined by $g(X_l) = x$ and $g(X_m) = f(X_m)$ if $m \neq l$.

By an easy induction on structure one observes that the truth value of a formula $\varphi(X_1,\ldots,X_n)$ under a substitution f depends only on $f(X_1) = x_1,\ldots,f(X_n) = x_n$. If φ is true under f, write $\mathcal{A} \models \varphi(x_1,\ldots,x_n)$. In particular, for φ a sentence, the truth value of φ is independent of f. It is either true in $\mathcal{A}$, or false in $\mathcal{A}$. In the former case write $\mathcal{A} \models \varphi$, and in the latter, $\mathcal{A} \not\models \varphi$.

7.3 Models

Models of a first order language $\mathcal{L}$ are generalization of groups, rings, fields, and ordered sets. They are structures where a given set of sentences, called axioms, is true.

A **theory** in a first order language $\mathcal{L} = \mathcal{L}(\mu, \nu, K)$ is a set of sentences T of $\mathcal{L}$. A structure $\mathcal{A}$ for $\mathcal{L}$ is called a **model** of T if $\mathcal{A} \models \theta$ for every $\theta \in T$. In this case write $\mathcal{A} \models T$. If T' is another theory in $\mathcal{L}$ for which every model of T' is also a model of T, write $T' \models T$. If Π is a theory of $\mathcal{L}$ and T is the theory of all sentences θ of $\mathcal{L}$ such that $\Pi \models \theta$, then Π is said to be a **set of axioms** for T.

Denote the class of all models of a theory T by $\mathrm{Mod}(T)$. If $\mathcal{A}$ is a structure for a language $\mathcal{L}$, then $\mathrm{Th}(\mathcal{A}, \mathcal{L})$ is the set of all sentences of $\mathcal{L}$ which are true in $\mathcal{A}$.

Example 7.3.1: The theory of fields. Denote the first order language that contains the two binary functions symbols + (addition) and · (multiplication), and two constant symbols 0 and 1 by $\mathcal{L}(\mathrm{ring})$. For each integral domain R let $\mathcal{L}(\mathrm{ring}, R)$ be the language $\mathcal{L}(\mathrm{ring})$ extended by all elements of R as constant symbols. Denote the usual axioms for the theory of fields by Π:

$$(\forall X)(\forall Y)(\forall Z)[(X+Y)+Z = X+(Y+Z)];$$
$$(\forall X)(\forall Y)[X+Y = Y+X];$$
$$(\forall X)[X+0 = X];$$
$$(\forall X)(\exists Y)[X+Y = 0];$$
$$(\forall X)(\forall Y)(\forall Z)[(XY)Z = X(YZ)];$$
$$(\forall X)(\forall Y)[XY = YX];$$
$$(\forall X)[1 \cdot X = X];$$
$$(\forall X)[X \neq 0 \rightarrow (\exists Y)[XY = 1]];$$
$$1 \neq 0; \text{ and}$$
$$(\forall X)(\forall Y)(\forall Z)[X(Y+Z) = XY + XZ].$$

Every model of Π is a field. Extend Π by all equalities — the **positive diagram** of R —

(1) $\qquad a_1 + b_1 = c_1$ and $a_2 b_2 = c_2$, for $a_i, b_i, c_i \in R$

that are true in R. Denote the set obtained by $\Pi(R)$. A model of $\Pi(R)$ is a field that contains a subset $\bar{R} = \{\bar{a} \mid a \in R\}$ whose elements satisfy the equalities

$$\bar{a}_1 + \bar{b}_1 = \bar{c}_1 \quad \text{and} \quad \bar{a}_2 \bar{b}_2 = \bar{c}_2$$

whenever the corresponding equalities of (1) are true in R. That is, $\bar{R}$ is a homomorphic image of R.

If $R = K$ is a field, then $\bar{K}$ is an isomorphic copy of K. Thus, a model of $\Pi(K)$ is, (up to an isomorphism) a field containing K. □

Example 7.3.2: Irreducible Polynomials. Let R be an integral domain. An **elementary statement** about models of $\Pi(R)$ is a mathematical statement that applies to each member of $\text{Mod}(\Pi(R))$ and for which there exists a sentence θ of $\mathcal{L}(\text{ring}, R)$ which is true in any given model F if and only if the statement is true.

Consider, for example, a polynomial $f(X_1, \ldots, X_n)$ of degree d with coefficients in R. Then "$f(\mathbf{X})$ is irreducible" is an elementary statement about models of $\Pi(R)$. Indeed, it is equivalent to the conjunction of the statements "there exist no polynomials g, h of degree d_1 and d_2 respectively such that $f(\mathbf{X}) = g(\mathbf{X})h(\mathbf{X})$," where (d_1, d_2) runs over all pairs of positive integers with $d_1 + d_2 = d$. Rewrite the phrase "there exists no polynomial $g(X_1, \ldots, X_n)$ of degree d_1", as "$\neg(\exists u_1) \cdots (\exists u_k)$" where $u_1, \ldots, u_k$ are variables for the coefficients of $g(\mathbf{X})$. A system of equalities between corresponding coefficients on both sides of "=" replaces "$f(\mathbf{X}) = g(\mathbf{X})h(\mathbf{X})$."

Similarly, we may consider a polynomial

$$f(\mathbf{u}, X_1, \ldots, X_n) = \sum u_\mathbf{i} X_1^{i_1} \cdots X_n^{i_n}$$

with intermediate coefficients $u_\mathbf{i}$. The same argument as above gives a formula $\varphi(\mathbf{u})$ in $\mathcal{L}(\text{ring})$ such that for each field K and all tuples $\mathbf{a}$ with entries in K, the polynomial $f(\mathbf{a}, \mathbf{X})$ is irreducible in $K[\mathbf{X}]$ if and only if $\varphi(\mathbf{a})$ is true in K. □

Two structures $\mathcal{A} = \langle A, R_i, F_j, c_k \rangle$ and $\mathcal{B} = \langle B, S_i, G_j, d_k \rangle$ of the language $\mathcal{L} = \mathcal{L}(\mu, \nu, K)$ are **isomorphic**, if there exists a bijective function $f \colon A \to B$ such that
(2a) $(a_1, \ldots, a_{\mu(i)}) \in R_i \iff (f(a_1), \ldots, f(a_{\mu(i)})) \in S_i$, for each $i \in I$;
(2b) $f(F_j(a_1, \ldots, a_{\nu(j)})) = G_j(f(a_1), \ldots, f(a_{\nu(j)}))$, for each $j \in J$; and
(2c) $f(c_k) = d_k$, for each $k \in K$.
In this case write $\mathcal{A} \cong \mathcal{B}$.

The structures $\mathcal{A}$ and $\mathcal{B}$ are **elementarily equivalent** if $\mathcal{A} \models \theta \iff \mathcal{B} \models \theta$ for every sentence θ of $\mathcal{L}$. If this is the case, we write $\mathcal{A} \equiv \mathcal{B}$. Clearly, if $\mathcal{A} \cong \mathcal{B}$, then $\mathcal{A} \equiv \mathcal{B}$. But we will have many examples that show the converse is false.

Two fields L and L' that contain a field K are isomorphic as models of $\Pi(K)$ if and only if there exists a field isomorphism of L onto L' that fixes every element of K: write $L \cong_K L'$. If, however, they are elementarily equivalent as models of $\Pi(K)$, we write $L \equiv_K L'$.

Call $\mathcal{A}$ a **substructure** of $\mathcal{B}$, ($\mathcal{A} \subseteq \mathcal{B}$, and $\mathcal{B}$ is an **extension** of $\mathcal{A}$) if $A \subseteq B$, $R_i = A^{\mu(i)} \cap S_i$ for each $i \in I$, $F_j(a_1, \ldots, a_{\nu(j)}) = G_j(a_1, \ldots, a_{\nu(j)})$ for each $j \in J$ and all $a_1, \ldots, a_{\nu(j)} \in A$; and $c_k = d_k$ for each $k \in K$.

7.4 Elementary Substructures

More generally, an **embedding** of $\mathcal{A}$ into $\mathcal{B}$ is an injective map $f\colon A \to B$ that satisfies Condition (2). Note that an arbitrary map $f\colon A \to B$ is an embedding of $\mathcal{A}$ into $\mathcal{B}$ if and only if for each quantifier free formula $\varphi(X_1, \ldots, X_n)$ of $\mathcal{L}$ and for all $a_1, \ldots, a_n \in A$, the condition $\mathcal{A} \models \varphi(a_1, \ldots, a_n)$ implies $\mathcal{B} \models \varphi(f(a_1), \ldots, f(a_n))$. Indeed, an application of the latter condition to the formula $X_1 \neq X_2$ implies that f is injective.

Suppose now that $\mathcal{A} \subseteq \mathcal{B}$. We say $\mathcal{A}$ is **existentially closed** in $\mathcal{B}$ if for each quantifier free formula $\varphi(X_1, \ldots, X_n)$ and for all $b_1, \ldots, b_n \in B$ with $\mathcal{B} \models \varphi(b_1, \ldots, b_n)$ there exist $a_1, \ldots, a_n \in A$ with $\mathcal{A} \models \varphi(a_1, \ldots, a_n)$.

Call $\mathcal{A}$ an **elementary substructure** of $\mathcal{B}$ and $\mathcal{B}$ an **elementary extension** of $\mathcal{A}$ (in symbols $\mathcal{A} \prec \mathcal{B}$) if $\mathcal{A} \subseteq \mathcal{B}$ and if for each formula $\varphi(X_1, \ldots, X_n)$ of $\mathcal{L}$ and for every $a_1, \ldots, a_n$ in A, the truth of $\varphi(a_1, \ldots, a_n)$ in $\mathcal{A}$ is equivalent to its truth in $\mathcal{B}$. It follows, in particular, that a sentence θ of $\mathcal{L}$ is true in $\mathcal{A}$ if and only if it is true in $\mathcal{B}$, (i.e. $\mathcal{A} \equiv \mathcal{B}$). The converse is false (Example 7.3.3). If, however, $\mathcal{A} \subseteq \mathcal{B}$ then "$\mathcal{A} \equiv \mathcal{B}$ as models of $\mathcal{L}(A)$" is equivalent to "$\mathcal{A} \prec \mathcal{B}$ as models of $\mathcal{L}$". Transitivity of elementarily equivalence follows immediately: $\mathcal{A} \prec \mathcal{B}$, $\mathcal{B} \prec \mathcal{C}$ implies $\mathcal{A} \prec \mathcal{C}$. In addition, $\mathcal{A} \subseteq \mathcal{B} \prec \mathcal{C}$ and $\mathcal{A} \prec \mathcal{C}$ imply $\mathcal{A} \prec \mathcal{B}$.

Example 7.3.3: Elementary subfields. If a field K is an elementary subfield of a field F, then F is a regular extension of K. In other words, K is algebraically closed in F and F/K is separable (Lemma 2.6.4).

First of all let $x \in \tilde{K} \cap F$ and $f = \mathrm{irr}(x, K)$. Then the sentence $(\exists X)[f(X) = 0]$ holds in F, so also in K. Therefore, $\deg(f) = 1$, hence $x \in K$. Consequently, $\tilde{K} \cap F = K$.

Let now $u_1, \ldots, u_n \in K$ and $b_1, \ldots, b_n \in F$ with $\sum_{i=1}^n b_i u_i^{1/p} = 0$. Then, $\sum_{i=1}^n b_i^p u_i = 0$. Hence, $(\exists X_1) \cdots (\exists X_n) \sum_{i=1}^n X_i^p u_i$ is true in F, so also in K. In other words, there exist $a_1, \ldots, a_n \in K$ with $\sum_{i=1}^n a_i u_i^{1/p} = 0$. Therefore, F is linearly disjoint from $K^{1/p}$ over K. Consequently, F/K is separable.

For example, let x be an indeterminate. Then, $\mathbb{Q}(x^2) \cong \mathbb{Q}(x)$. Hence, $\mathbb{Q}(x^2) \equiv \mathbb{Q}(x)$. But $\mathbb{Q}(x)$ is a proper algebraic extension of $\mathbb{Q}(x^2)$. Therefore, $\mathbb{Q}(x^2)$ is not an elementary subfield of $\mathbb{Q}(x)$. $\square$

7.4 Elementary Substructures

We develop criteria for one structure to be an elementary substructure of another.

Let m be a cardinal number. Consider a transfinite sequence $\{\mathcal{A}_\alpha \mid \alpha < m\}$ of structures for a language $\mathcal{L} = \mathcal{L}(\mu, \nu, K)$ with $\mathcal{A}_\alpha = \langle A_\alpha, R_{\alpha i}, F_{\alpha j}, c_{\alpha k} \rangle$. Suppose $\mathcal{A}_\alpha \subseteq \mathcal{A}_\beta$ for each $\alpha \leq \beta < m$ and define the **union** $\bigcup_{\alpha < m} \mathcal{A}_\alpha$ to be the structure $\mathcal{A}_m = \langle A_m, R_{mi}, F_{mj}, c_{mk} \rangle$ with $A_m = \bigcup_{\alpha < m} A_\alpha$, $R_{mi} = \bigcup_{\alpha < m} R_{mi}$, $F_{mj}(x_1, \ldots, x_{\nu(j)}) = F_{\alpha j}(x_1, \ldots, x_{\nu(j)})$ if $x_1, \ldots, x_{\nu(j)} \in A_\alpha$, and $c_{mk} = c_{0k}$. Then $\mathcal{A}_\alpha \subseteq \mathcal{A}_m$ for each $\alpha < m$.

LEMMA 7.4.1:
(a) If $\mathcal{A}_\alpha \prec \mathcal{A}_\beta$ for each $\alpha \leq \beta < m$, then $\mathcal{A}_\alpha \prec \mathcal{A}_m$ for each $\alpha < m$.
(b) If $\mathcal{B}$ is another structure of $\mathcal{L}$ such that $\mathcal{A}_\alpha \prec \mathcal{B}$ for each $\alpha < m$, then $\mathcal{A}_m \prec \mathcal{B}$.

Proof of (a): Use an induction on structure to prove that for each formula $\varphi(X_1, \ldots, X_n)$ of $\mathcal{L}$, for each $\alpha < m$, and for every $x_1, \ldots, x_n \in A_\alpha$

(1) $$\mathcal{A}_\alpha \models \varphi(\mathbf{x}) \iff \mathcal{A}_m \models \varphi(\mathbf{x}).$$

Proof of (b): Let $\varphi(X_1, \ldots, X_n)$ be a formula of $\mathcal{L}$ and let $x_1, \ldots, x_n \in A_m$. Then there exists $\alpha < m$ such that $x_1, \ldots, x_n \in A_\alpha$. It follows from (1) that

$$\mathcal{A}_m \models \varphi(\mathbf{x}) \iff \mathcal{A}_\alpha \models \varphi(\mathbf{x}) \iff \mathcal{B} \models \varphi(\mathbf{x}). \qquad \square$$

PROPOSITION 7.4.2 (Skolem-Löwenheim): Let $\mathcal{L} = \mathcal{L}(\mu, \nu, K)$ be a countable language, let $\mathcal{B} = \langle B, S_i, G_j, d_k \rangle$ be a structure of $\mathcal{L}$ and let A_0 be a countable subset of B. Then $\mathcal{B}$ has a countable elementary substructure $\mathcal{A} = \langle A, R_i, F_j, c_k \rangle$ such that $A_0 \subseteq A$.

Proof: We construct an ascending chain of countable sets $A_0 \subseteq A_1 \subseteq A_2 \subseteq \cdots \subseteq B$. Suppose that A_n has already been constructed, then A_{n+1} consists of all d_k's with $k \in K$; the $G_j(y_1, \ldots, y_{\nu(j)})$ for all $j \in J$ and $\nu(j)$-tuple $(y_1, \ldots, y_{\nu(j)})$ of elements of A_n; and an element $x_m \in B$ such that $\mathcal{B} \models \varphi(x_1, \ldots, x_m)$ for each formula $\varphi(X_1, \ldots, X_m)$ and for every $x_1, \ldots, x_{m-1} \in A_n$ satisfying $\mathcal{B} \models (\exists X_m)[\varphi(x_1, \ldots, x_{m-1}, X_m)]$.

Define $\mathcal{A}$ as follows: $A = \bigcup_{n=1}^{\infty} A_n$, $R_i = A^{\mu(i)} \cap S_i$, $F_j(y_1, \ldots, y_{\mu(j)}) = G_j(y_1, \ldots, y_{\mu(j)})$ for $y_1, \ldots, y_{\mu(j)} \in A$, and $c_k = d_k$. By the choice of the x_n's above, the function F_j is well defined. Hence, $\mathcal{A}$ is a countable substructure of $\mathcal{B}$. Now use an induction on structure to prove that for each formula $\varphi(X_1, \ldots, X_n)$ of $\mathcal{L}$ and for each $\mathbf{x} \in A^n$, $\mathcal{A} \models \varphi(\mathbf{x})$ if and only if $\mathcal{B} \models \varphi(\mathbf{x})$. This proves that $\mathcal{A} \prec \mathcal{B}$. $\qquad \square$

7.5 Ultrafilters

A **filter** on a set S is a nonempty family $\mathcal{D}$ of subsets of S t which are "big" in a sense made precise by the following condition:
(1a) $\emptyset \notin \mathcal{D}$.
(1b) If $A, B \in \mathcal{D}$, then $A \cap B \in \mathcal{D}$.
(1c) If $A \in \mathcal{D}$ and $A \subseteq B \subseteq S$, then $B \in \mathcal{D}$.

If, in addition,
(2) For each $A \subseteq S$ either $A \in \mathcal{D}$ or $S \smallsetminus A \in \mathcal{D}$,
then $\mathcal{D}$ is an **ultrafilter**. In this case $\mathcal{D}$ also satisfies
(3) $A \cup B \in \mathcal{D}$ implies $A \in \mathcal{D}$ or $B \in \mathcal{D}$.

Example 7.5.1:

(a) The family of all **cofinite** subsets of S (i.e. those subsets whose complements are finite) is a filter of S.

(b) The family $\mathcal{D}_a$ of all subsets of S that contain a given element a of S is an ultrafilter on S, called a **principal ultrafilter**. From (3), an ultrafilter $\mathcal{D}$ is principal if and only if it contains a finite set.

A family $\mathcal{D}_0$ of subsets of S satisfies the **finite intersection property** if $A_1, \ldots, A_n \in \mathcal{D}_0$ implies $A_1 \cap \cdots \cap A_n \neq \emptyset$. If one adds to $\mathcal{D}_0$ all the sets $B \subseteq S$ that contain finite intersections $A_1 \cap \cdots \cap A_n$ of elements of $\mathcal{D}_0$, then one obtains a filter $\mathcal{D}_1$. By Zorn's Lemma there exists a maximal filter $\mathcal{D}$ of S that contains $\mathcal{D}_1$. □

Our next Lemma says that $\mathcal{D}$ is an ultrafilter.

LEMMA 7.5.2: *A filter $\mathcal{D}$ on a set S is an ultrafilter if and only if it is maximal.*

Proof: Suppose $\mathcal{D}$ is maximal and let $A \subseteq S$. Assume that $S \smallsetminus A \notin \mathcal{D}$. Then $\mathcal{D} \cup \{A\}$ has the finite intersection property. Indeed, for $D_1, \ldots, D_n \in \mathcal{D}$ let $D = D_1 \cap \cdots \cap D_n$. If $D \cap A = \emptyset$, then $D \subseteq S \smallsetminus A$. Hence, $S \smallsetminus A \in \mathcal{D}$, a contradiction. By the comment above there exists a filter $\mathcal{D}'$ on S containing $\mathcal{D} \cup \{A\}$. By the maximality of $\mathcal{D}$, $A \in \mathcal{D}$. Thus, $\mathcal{D}$ is an ultrafilter. The converse is clear. □

COROLLARY 7.5.3: *Every family $\mathcal{D}_0$ of subsets of S that satisfies the finite intersection property is contained in an ultrafilter.*

A somewhat stronger assumption implies the existence of nonprincipal ultrafilters.

LEMMA 7.5.4: *Let $\mathcal{D}_0$ be a family of subsets of a set S that has the following property: If $A_1, \ldots, A_n \in \mathcal{D}_0$, then $A_1 \cap \cdots \cap A_n$ is an infinite set. Then there exists a nonprincipal ultrafilter $\mathcal{D}$ on S that contains $\mathcal{D}_0$.*

Proof: The family $\mathcal{D}_1$ that consists of all subsets of S that contain a set in $\mathcal{D}_0$ and all cofinite subsets of S has the finite intersection property. Choose an ultrafilter $\mathcal{D}$ that contains $\mathcal{D}_1$; it is a nonprincipal ultrafilter. □

7.6 Regular Ultrafilters

The family $\mathcal{F}$ of all finite subsets of an infinite set S has properties dual to those of a filter:
(1a) $S \notin \mathcal{F}$;
(1b) $A, B \in \mathcal{F}$ implies $A \cup B \in \mathcal{F}$; and
(1c) $B \in \mathcal{F}$ and $A \subseteq B$ imply $A \in \mathcal{F}$.
Later we will work with another family that satisfies the same conditions - the family of zero sets of a measure space S. Therefore we give both cases a unified treatment. Let $\mathcal{F}$ be a nonempty family of subsets of S that satisfies (1). Call the elements of $\mathcal{F}$ **small sets**.

Define **Boolean polynomials** in $Z_1, \ldots, Z_m$ recursively: The variables $Z_1, \ldots, Z_m$ are Boolean polynomials, and if U, U_1, U_2 are Boolean polynomials, then U', $U_1 \cup U_2$, and $U_1 \cap U_2$ are Boolean polynomials.

Evaluate a Boolean polynomial $P(Z_1, \ldots, Z_m)$ at subsets $A_1, \ldots, A_m$ of S by interpreting the symbols $\cup$, $\cap$, and $'$ as union, intersection, and taking the complement, respectively.

With addition and multiplication given by

$$A + B = (A \smallsetminus B) \cup (B \smallsetminus A), \quad A \cdot B = A \cap B, \quad 0 = \emptyset, \quad 1 = S,$$

the family of all subsets of S becomes an algebra over the field $\mathbb{F}_2$, in which each element except 1 is a zero divisor. The family $\mathcal{F}$ of small sets is an ideal of this algebra.

Two subsets A and B of S are congruent modulo $\mathcal{F}$ in this algebra if and only if they differ from each other by a small set, (i.e. $(A \smallsetminus B) \cup (B \smallsetminus A) \in \mathcal{F}$). We then say that A and B are **almost equal**, and write $A \approx B$.

Clearly, if $A_1 \approx B_1$ and $A_2 \approx B_2$, then $S \smallsetminus A_1 \approx S \smallsetminus B_1$, $A_1 \cup A_2 \approx B_1 \cup B_2$, and $A_1 \cap A_2 \approx B_1 \cap B_2$. Thus, in general, if $P(Z_1, \ldots, Z_m)$ is a Boolean polynomial in $Z_1, \ldots, Z_m$ and $A_i \approx B_i$, $i = 1, \ldots, m$, are subsets of S, then $P(A_1, \ldots, A_m) \approx P(B_1, \ldots, B_m)$.

If the difference $A \smallsetminus B$ of two subsets A, B of S is a small set, then we say that A is **almost contained** in B.

A family of subsets of S which is closed under unions, intersections and taking complements is called a **Boolean algebra of sets**.

The **Boolean algebra generated by a family** $\mathcal{A}_0$ of subsets of S is the intersection of all Boolean algebras of families of S that contain S. It consists of all expressions $P(A_1, \ldots, A_m)$, where $P(Z_1, \ldots, Z_m)$ is a Boolean polynomial and $A_1, \ldots, A_m$ belong to $\mathcal{A}_0$. Denote this family by $\mathcal{A}$. The family $\mathcal{A}'$ of all subsets of S which are almost equal to a set in $\mathcal{A}$ is a Boolean algebra that contains both $\mathcal{A}$ and $\mathcal{F}$. $\mathcal{A}'$ is the Boolean algebra generated by $\mathcal{A}_0$ and $\mathcal{F}$.

Call an ultrafilter $\mathcal{D}$ on S **regular** (with respect to $\mathcal{F}$) if it contains no small set. In particular, if $A \in \mathcal{D}$ and $B \approx A$, then $B \in \mathcal{D}$. For example, nonprincipal ultrafilters on S are regular with respect to the family of finite subsets of S.

LEMMA 7.6.1: *Let S be a set and let $\mathcal{F}$ be a family of small subsets of S. Suppose that a family $\mathcal{D}_0$ of subsets of S satisfies*
(2) $A_1, \ldots, A_n \in \mathcal{D}_0$ *implies* $A_1 \cap \cdots \cap A_n$ *is not a small set.*
Then there exists a regular ultrafilter $\mathcal{D}$ on S that contains $\mathcal{D}_0$.

Proof: Repeat the proof of Lemma 7.5.4. ☐

We will apply the next result to model theoretic results for families of fields.

7.7 Ultraproducts

PROPOSITION 7.6.2 ([Ax2, p. 265]): *Let S be a set, $\mathcal{F}$ a family of small subsets of S, $\mathcal{A}$ a Boolean algebra of subsets of S that contains $\mathcal{F}$, and C a subset of S. Suppose $C \notin \mathcal{A}$. Then there exist two regular ultrafilters $\mathcal{D}$ and $\mathcal{D}'$ such that $\mathcal{D} \cap \mathcal{A} = \mathcal{D}' \cap \mathcal{A}$ but $C \in \mathcal{D}$ and $C \notin \mathcal{D}'$.*

Proof: Denote the collection of all $A \in \mathcal{A}$ that almost contain C or almost contain $S \smallsetminus C$ by $\mathcal{A}_0$. Suppose $A_i \in \mathcal{A}_0$ almost contains C for $i = 1, \ldots, m$, and $B_j \in \mathcal{A}_0$ almost contains $S \smallsetminus C$, $j = 1, \ldots, n$, and $A_1 \cap \cdots \cap A_m \cap B_1 \cap \cdots \cap B_n \approx \emptyset$. Take complements to deduce that

$$(S \smallsetminus A_1) \cup \cdots \cup (S \smallsetminus A_m) \cup (S \smallsetminus B_1) \cup \cdots \cup (S \smallsetminus B_n) \approx S$$

Each $S \smallsetminus A_i$ is almost contained in $S \smallsetminus C$ and each $S \smallsetminus B_i$ is almost contained in C. Hence $C \approx (S \smallsetminus B_1) \cup \cdots \cup (S \smallsetminus B_n)$, and therefore $C \in \mathcal{A}$, a contradiction. Thus, $\mathcal{A}_0$ satisfies (2).

By Zorn's Lemma, $\mathcal{A}$ has a maximal subcollection $\mathcal{A}_1$ that contains $\mathcal{A}_0$ and has property (2). In particular, $\mathcal{A}_1$ is closed under finite intersections.

CLAIM: *The family $\mathcal{A}_1 \cup \{C\}$ satisfies (2).* Indeed, if $A_1, \ldots, A_m \in \mathcal{A}_1$ and $A_1 \cap \cdots \cap A_m \cap C \approx \emptyset$, then C is almost contained in $S \smallsetminus A$, where $A = A_1 \cap \cdots \cap A_m \in \mathcal{A}$. Hence, $S \smallsetminus A \in \mathcal{A}_0 \subseteq \mathcal{A}_1$. But this is a contradiction, because $A \in \mathcal{A}_1$.

By Lemma 7.6.1 there exists a regular ultrafilter $\mathcal{D}$ on S that contains $\mathcal{A}_1 \cup \{C\}$. Obviously $\mathcal{D} \cap \mathcal{A}$ contains $\mathcal{A}_1$ and satisfies (2). The maximality of $\mathcal{A}_1$ implies that $\mathcal{D} \cap \mathcal{A} = \mathcal{A}_1$. Similarly, there exists a regular ultrafilter $\mathcal{D}'$ on S that contains $\mathcal{A}_1 \cup \{S \smallsetminus C\}$. It also satisfies $\mathcal{D}' \cap \mathcal{A} = \mathcal{A}_1$. Hence, $\mathcal{D}' \cap \mathcal{A} = \mathcal{D} \cap \mathcal{A}$. $\square$

7.7 Ultraproducts

From a given family of structures, ultraproducts allow us to create new structures which retain, sometimes in a particularly useful form, those elementary properties that hold for almost all structures in the family. Furthermore, those elementary properties that hold only for a small subfamily no longer hold in the new structures. In many cases we are able to establish simple criteria under which two such new models are elementarily equivalent (e.g. Lemma 20.3.3). This has been a successful route to the investigation of the elementary theory of many algebraic structures.

To be more explicit, consider a language $\mathcal{L} = \mathcal{L}(\mu, \nu, K)$ and a set S together with an ultrafilter $\mathcal{D}$ on S. Suppose that for each $s \in S$ we are given a structure $\mathcal{A}_s = \langle A_s, R_{is}, F_{js}, c_{ks} \rangle$ for $\mathcal{L}$. We construct the **ultraproduct** of $\mathcal{A}_s$, $s \in S$, **modulo** $\mathcal{D}$, denoted $\prod \mathcal{A}_s / \mathcal{D} = \mathcal{A} = \langle A, R_i, F_j, c_k \rangle$, in the following way.

Define an equivalence relation on the cartesian product $\prod_{s \in S} A_s$ by

$$a \sim b \iff \{s \in S \mid a_s = b_s\} \in \mathcal{D}.$$

The domain A of $\mathcal{A}$ is the elements of $\prod A_s$ modulo this relation. For simplicity use representatives of the equivalence classes instead of the classes themselves. With this convention define R_i by

$$(a_1,\ldots,a_{\mu(i)}) \in R_i \iff \{s \in S \mid (a_{1s},\ldots,a_{\mu(i),s}) \in R_{is}\} \in \mathcal{D}.$$

Define F_j by the rule: $F_j(b_1,\ldots,b_{\nu(j)})$ is the equivalence class of the function $s \mapsto F_{js}(b_{1s},\ldots,b_{\nu(j),s})$. Similarly c_k is the equivalence class of the function $s \mapsto c_{ks}$. Check that R_i, F_j, and c_k are well defined. An induction on the structure of terms shows that if $t(X_1,\ldots,X_n)$ is a term of $\mathcal{L}$ and $x_1,\ldots,x_n \in A$, then

(1) $$\{s \in S \mid t(x_1,\ldots,x_n)_s = t(x_{1s},\ldots,x_{ns})\} \in \mathcal{D}.$$

This gives rise to the fundamental property of ultraproducts:

PROPOSITION 7.7.1 (Łoś): *Let $\mathcal{D}$ be an ultrafilter on S. Suppose for each $s \in S$ that $\mathcal{A}_s = \langle A_s, R_{is}, G_{js}, c_{ks}\rangle$ is a structure for a language $\mathcal{L}(\mu,\nu,K)$. Consider the ultraproduct $\mathcal{A} = \langle A, R_i, G_j, c_k\rangle = \prod \mathcal{A}_s/\mathcal{D}$. Then, for every formula $\varphi(X_1,\ldots,X_n)$ and for every $x_1,\ldots,x_n \in A$:*

(2) $$\mathcal{A} \models \varphi(x_1,\ldots,x_n) \iff \{s \in S \mid \mathcal{A}_s \models \varphi(x_{1s},\ldots,x_{ns})\} \in \mathcal{D}$$

Proof: Do an induction on structure: Statement (2) follows easily for atomic formulas from (1) and the induction steps present no difficulty. We demonstrate the induction step for the existential quantifier.

Suppose that (2) is true for φ and for each n-tuple of elements of A. Let $x_1,\ldots,x_{n-1} \in A$ and assume that

$$S_0 = \{s \in S \mid \mathcal{A}_s \models (\exists X_n)[\varphi(x_{1s},\ldots,x_{n-1,s},X_n)]\} \in \mathcal{D}.$$

For each $s \in S_0$ there exists $x_{ns} \in A_s$ such that $\mathcal{A}_s \models \varphi(x_{1s},\ldots,x_{ns})$. If $s \notin S_0$, let $x_{ns} \in A_s$ be arbitrary. Let x_n be the equivalence class defined by the x_{ns}'s. Then the set $\{s \in S \mid \mathcal{A}_s \models \varphi(x_{1s},\ldots,x_{ns})\}$ contains S_0, hence belongs to $\mathcal{D}$. By the induction assumption, $\mathcal{A} \models \varphi(x_1,\ldots,x_n)$. Therefore, $\mathcal{A} \models (\exists X_n)[\varphi(x_1,\ldots,x_{n-1},X_n)]$. $\square$

COROLLARY 7.7.2: *If θ is a sentence of $\mathcal{L}$, then*

$$\mathcal{A} \models \theta \iff \{s \in S \mid \mathcal{A}_s \models \theta\} \in \mathcal{D}.$$

Example 7.7.3: Ultraproducts of fields. If the structures $\mathcal{A}_s$ is a field for each $s \in S$, then so is $\mathcal{A}$. If each of them is algebraically closed, then so is $\mathcal{A}$. But even if each of them is algebraic over a given field K, $\mathcal{A}$ may be not.

For example, let $S = \mathbb{N}$ and let $\mathcal{D}$ be a nonprincipal ultrafilter on $\mathbb{N}$. For each $n \in \mathbb{N}$ choose an element $x_n \in \tilde{\mathbb{Q}}$ of degree greater than n over $\mathbb{Q}$. Then let $K_n = \mathbb{Q}(x_n)$, $K = \prod K_n/\mathcal{D}$, and x be the equivalence class of

7.7 Ultraproducts

$(x_1, x_2, x_3, \ldots)$. For each n, almost all x_i satisfy no equation of degree n over $\mathbb{Q}$. Hence, by Corollary 7.7.2, so does x. Consequently, x is transcendental over $\mathbb{Q}$. □

Ultraproducts also satisfy a saturation property. Let $\mathcal{A}$ be a structure with a domain A for a language $\mathcal{L}$. Extend $\mathcal{L}$ to a language $\mathcal{L}(A)$ by adding a new constant symbol for each element of A. We say that $\mathcal{A}$ is $\aleph_1$-**saturated** if the following holds.

If $r(1) < r(2) < r(3) < \cdots$ is an increasing sequence of positive integers and for each $n \in \mathbb{N}$, $\varphi_n(X_1, \ldots, X_{r(n)})$ is a formula of $\mathcal{L}(A)$ such that

$$(3) \qquad \mathcal{A} \models (\exists X_1) \cdots (\exists X_{r(n)}) \bigwedge_{t=1}^{n} \varphi_t(X_1, \ldots, X_{r(t)}),$$

then there exist $x_1, x_2, x_3, \ldots$ in A such that $\mathcal{A} \models \varphi_n(x_1, \ldots, x_{r(n)})$ for each $n \in \mathbb{N}$.

LEMMA 7.7.4: *Let $\mathcal{D}$ be a nonprincipal ultrafilter on $\mathbb{N}$. Suppose for each $n \in \mathbb{N}$ that $\mathcal{A}_n$ is a structure with a domain A_n for a language $\mathcal{L}$. Then the ultraproduct $\mathcal{A} = \prod \mathcal{A}_n / \mathcal{D}$ is $\aleph_1$-saturated.*

Proof: To simplify notation assume that $r_n = n$ in the preceding definition. Suppose (3) holds for each $n \in \mathbb{N}$. Then

$$D_n = \{s \in \mathbb{N} \mid \mathcal{A}_s \models (\exists X_1) \cdots (\exists X_n) \bigwedge_{t=1}^{n} \varphi_t(X_1, \ldots, X_t)\} \in \mathcal{D}$$

for each $n \in \mathbb{N}$. Clearly $D_1 \supseteq D_2 \supseteq D_3 \supseteq \cdots$. Since $\mathcal{D}$ is nonprincipal, $D'_n = D_n \smallsetminus \{1, 2, \ldots, n\} \in \mathcal{D}$. Also $D'_1, D'_2, D'_3, \ldots$ is a decreasing sequence with empty intersection.

Now define $x_1, x_2, x_3, \ldots$ in A as follows. If $s \in D'_n \smallsetminus D'_{n-1}$, choose $x_{1s}, \ldots, x_{ns}$ in A_s such that $\mathcal{A}_s \models \bigwedge_{t=1}^{n} \varphi_t(x_{1s}, \ldots, x_{ts})$. Thus, for each $n \in \mathbb{N}$, x_{ns} is well defined for all $s \in D'_n$. For $s \in \mathbb{N} \smallsetminus D'_n$ choose $x_{ns} \in A_s$ arbitrarily. From this definition, for each $n \in \mathbb{N}$, the set $\{s \in \mathbb{N} \mid \mathcal{A}_s \models \varphi_n(x_{1s}, \ldots, x_{ns})\}$ contains the union $\bigcup_{p=n}^{\infty}(D'_p \smallsetminus D'_{p-1}) = D'_n$. Therefore Proposition 7.7.1 gives $\mathcal{A} \models \varphi_n(x_1, \ldots, x_n)$. □

If all structures $\mathcal{A}_s$ are the same, say $\mathcal{A}_s = \mathcal{A}$, then the ultraproduct $\prod \mathcal{A}_s / \mathcal{D}$, denoted by $\mathcal{A}^S / \mathcal{D}$, is called the **ultrapower** of $\mathcal{A}$ to S modulo $\mathcal{D}$. Denote the domain of $\mathcal{A}$ by A. Consider the diagonal embedding of A into A^S. That is, map $a \in A$ onto the constant function $a_s = a$. This gives a canonical injective map of A into $A^S / \mathcal{D}$. Indeed, if the images of two elements a and b of A are equal, then the set $\{s \in S \mid a_s = b_s\}$ belongs to $\mathcal{D}$ and is therefore nonempty. It follows that $a = b$. We identify A with its image to conclude from Proposition 7.7.1 the following result:

PROPOSITION 7.7.5: *If $\mathcal{D}$ is an ultrafilter on a set S and $\mathcal{A}$ is a structure for a language $\mathcal{L}$, then $\mathcal{A}$ is an elementary substructure of $\mathcal{A}^S/\mathcal{D}$.*

The compactness theorem of model theory is now an easy corollary.

PROPOSITION 7.7.6 (The Compactness Theorem): *Let T be a set of sentences in a first order language $\mathcal{L}$. If each finite subset of T has a model, then T has a model.*

Proof: Denote the collection of all finite subsets of T by F. For each $\Phi \in F$ let $D_\Phi = \{\Phi' \in F \mid \Phi \subseteq \Phi'\}$. Then, $D_\Phi \cap D_{\Phi'} = D_{\Phi \cup \Phi'}$. Hence, the family $\mathcal{D}_0 = \{D_\Phi \mid \Phi \in F\}$ has the finite intersection property. By Corollary 7.5.3, there exists an ultrafilter $\mathcal{D}$ on F that contains $\mathcal{D}_0$. Choose a model, M_Φ, for each $\Phi \in F$. Then, $M = \prod M_\Phi/\mathcal{D}$ is a model of T, because if $\theta \in T$, then $D_{\{\theta\}} \in \mathcal{D}$. □

A similar construction characterizes existential closedness by ultraproducts:

PROPOSITION 7.7.7: *Let $\mathcal{A} \subseteq \mathcal{B}$ be structures of a language $\mathcal{L}$ with domains $A \subseteq B$. Then $\mathcal{A}$ is existentially closed in $\mathcal{B}$ if and only if there exists an ultrapower $\mathcal{A}^*$ of $\mathcal{A}$ and there exists an embedding $\beta\colon \mathcal{B} \to \mathcal{A}^*$ whose restriction to $\mathcal{A}$ is the canonical embedding $\mathcal{A} \to \mathcal{A}^*$.*

Proof: Suppose first that there exist $\mathcal{A}^*$ and β as above. Let $\varphi(X_1,\ldots,X_n)$ be a quantifier free formula in $\mathcal{L}(A)$ and let $b_1,\ldots,b_n$ be elements of B such that $\mathcal{B} \models \varphi(b_1,\ldots,b_n)$. Then, $\mathcal{A}^* \models \varphi(\beta(b_1),\ldots,\beta(b_n))$. Hence, $\mathcal{A}^* \models (\exists X_1)\cdots(\exists X_n)[\varphi(X_1,\ldots,X_n)]$. We conclude from Proposition 7.7.5 that there exist $a_1,\ldots,a_n \in A$ with $\mathcal{A} \models \varphi(a_1,\ldots,a_n)$.

Conversely, suppose that $\mathcal{A}$ is existentially closed in $\mathcal{B}$. Denote the constant symbol of $\mathcal{L}(B)$ associated with an element $b \in B$ by $\tilde{b}$. Let T be the set of sentences $\varphi(\tilde{b}_1,\ldots,\tilde{b}_n)$ of $\mathcal{L}(B)$ such that $\varphi(X_1,\ldots,X_n)$ is a quantifier free formula of $\mathcal{L}(A)$ and $b_1,\ldots,b_n$ are elements of B with $\mathcal{B} \models \varphi(b_1,\ldots,b_n)$. Denote the collection of all finite subsets of T by F. For each $\Phi \in F$ let $D_\Phi = \{\Phi' \in F \mid \Phi \subseteq \Phi'\}$. Then $D_\Phi \cap D_{\Phi'} = D_{\Phi \cup \Phi'}$, hence the family $\mathcal{D}_0 = \{D_\Phi \mid \Phi \in F\}$ has the finite intersection property. By Corollary 7.5.3 there exists an ultrafilter $\mathcal{D}$ on F that contains $\mathcal{D}_0$.

Consider an element Φ in F. List the elements of Φ as $\varphi_i(\tilde{b}_{i1},\ldots,\tilde{b}_{in})$, where $\varphi_i(X_1,\ldots,X_n)$ is a quantifier free formula of $\mathcal{L}(A)$ and $b_{i1},\ldots,b_{in}$ are elements of B such that $\mathcal{B} \models \varphi_i(b_{i1},\ldots,b_{in})$, $i = 1,\ldots,m$. For each i choose new variable symbols $X_{i1},\ldots,X_{in}$ and consider the quantifier free formula

$$\bigwedge_{i=1}^m \varphi_i(X_{i1},\ldots,X_{in}) \wedge \bigwedge [X_{ij} = X_{kl}]$$

of $\mathcal{L}(A)$, where in the second conjunct, i,j,k,l range over all indices such that $b_{ij} = b_{kl}$. Since $\mathcal{B} \models \bigwedge_{i=1}^m \varphi_i(b_{i1},\ldots,b_{in}) \wedge \bigwedge [b_{ij} = b_{kl}]$ there exist $b_{ij}^\Phi \in A$ such that $\mathcal{A} \models \bigwedge_{i=1}^m \varphi_i(b_{i1}^\Phi,\ldots,b_{in}^\Phi) \wedge \bigwedge [b_{ij}^\Phi = b_{kl}^\Phi]$. Extend $\mathcal{A}$ to a structure

$\mathcal{A}_\Phi$ of $\mathcal{L}(B)$ by choosing b_{ij}^Φ for the constant symbol $\tilde{b}_{ij}$, for all i and j. If b is an element of B which is different from all b_{ij}, choose an arbitrary $a \in A$ for $\tilde{b}$. Then $\mathcal{A}_\Phi$ is well defined and each of the sentences $\varphi_i(\tilde{b}_{i1},\ldots,\tilde{b}_{in})$ is valid in $\mathcal{A}_\Phi$.

Finally, consider the ultraproduct $\prod \mathcal{A}_\Phi/\mathcal{D}$ with domain $A^* = A^F/\mathcal{D}$. For each $b \in B$ let b^* be the element of A^* that corresponds to the constant symbol $\tilde{b}$. Put $\mathcal{A}^* = \mathcal{A}^F/\mathcal{D}$. Then the map $\beta\colon B \to A^*$ that maps b onto b^* is an embedding of $\mathcal{B}$ into $\mathcal{A}^*$ whose restriction to $\mathcal{A}$ is the canonical map $\mathcal{A} \to \mathcal{A}^*$. Indeed, consider a quantifier free formula $\varphi(X_1,\ldots,X_n)$ of $\mathcal{L}(A)$ and elements $b_1,\ldots,b_n \in B$ with $\mathcal{B} \models \varphi(b_1,\ldots,b_n)$. Then $\mathcal{A}_\Phi \models \varphi(b_1^\Phi,\ldots,b_n^\Phi)$ for each Φ in F that contains φ. Hence, by Corollary 7.7.2, $\prod \mathcal{A}_\Phi/\mathcal{D} \models \varphi(b_1^*,\ldots,b_n^*)$. Consequently, $\mathcal{A}^* \models \varphi(b_1^*,\ldots,b_n^*)$. □

A variation of the proof of Proposition 7.7.7 proves the following characterization of elementary embedding:

PROPOSITION 7.7.8 (Scott): *An embedding $\alpha\colon \mathcal{A} \to \mathcal{B}$ of structures is elementary if and only if there exists an ultrapower $\mathcal{A}^*$ of $\mathcal{A}$ and an elementary embedding $\beta\colon \mathcal{B} \to \mathcal{A}^*$ such that $\beta \circ \alpha\colon \mathcal{A} \to \mathcal{A}^*$ is the canonical embedding.*

Likewise a lemma of Frane [Bell-Slomson, p. 161] says that $\mathcal{A} \equiv \mathcal{B}$ if and only if $\mathcal{B}$ is elementarily embeddable in an ultrapower of $\mathcal{A}$. A theorem of Shelah supersedes Frane's lemma: $\mathcal{A} \equiv \mathcal{B}$ if and only if there exists a set I and an ultrafilter $\mathcal{D}$ on I such that $\mathcal{A}^I/\mathcal{D} \cong \mathcal{B}^I/\mathcal{D}$ [Shelah].

7.8 Regular Ultraproducts

Regular ultraproducts generalize non-principal ultraproducts. They appear in Chapter 20 as ultraproducts of fields $\tilde{\mathbb{Q}}(\mathbf{s})$ with sets of measure 0 replacing finite sets of σ's. The general model theoretic notions that we develop in this section will be applied in Chapter 20 to investigate the model theory of the $\tilde{\mathbb{Q}}(\sigma)$'s.

Let S be a set equipped with a family $\mathcal{F}$ of small subsets. Suppose that for each $s \in S$, $\mathcal{A}_s$ is a structure for a fixed language $\mathcal{L}$. The **truth set** of a sentence θ of $\mathcal{L}$ is defined to be the following subset of S:

$$A(\theta) = \{s \in S \mid \mathcal{A}_s \models \theta\}.$$

The map $\theta \mapsto A(\theta)$ preserves the Boolean operations:

$$A(\theta_1 \vee \theta_2) = A(\theta_1) \cup A(\theta_2),\ A(\theta_1 \wedge \theta_2) = A(\theta_1) \cap A(\theta_2),\ A(\neg\theta) = S \smallsetminus A(\theta).$$

More generally, if $P(Z_1,\ldots,Z_m)$ is a Boolean polynomial, then

(1) $$A(P(\theta_1,\ldots,\theta_m)) = P(A(\theta_1),\ldots,A(\theta_m)),$$

where $P(\theta_1,\ldots,\theta_m)$ is obtained from $P(Z_1,\ldots,Z_m)$ by first replacing $\cup$, $\cap$, and $'$ by $\vee$, $\wedge$, and $\neg$, respectively, and then substituting $\theta_1,\ldots,\theta_m$ for the $Z_1,\ldots,Z_m$.

Let T be the theory of all sentences θ of $\mathcal{L}$ that are true in $\mathcal{A}_s$ for **almost all** $s \in S$ (i.e. for all $s \in S$ excluding a small subset).

If $\mathcal{D}$ is a regular ultrafilter, we say that $\prod \mathcal{A}_s/\mathcal{D}$ is a **regular ultraproduct**.

PROPOSITION 7.8.1:
(a) *A sentence θ of $\mathcal{L}$ is in T if and only if it is true in every regular ultraproduct of the $\mathcal{A}_s$'.*
(b) *Every model of T is elementarily equivalent to a regular ultraproduct of the $\mathcal{A}_s$'.*

Proof of (a): If θ belongs to T, then, by Corollary 7.7.2, θ is true in every regular ultraproduct of the $\mathcal{A}_s$'.

Conversely, if $\theta \notin T$, then $A(\neg\theta)$ is not small. Hence, by Lemma 7.6.1, there exists a regular ultrafilter $\mathcal{D}$ on S which contains $A(\neg\theta)$. Therefore, by Corollary 7.7.2, θ is false in $\prod \mathcal{A}_s/\mathcal{D}$.

Proof of (b): Let $\mathcal{A}$ be a model of T. Then, by (1), $\mathcal{D}_0 = \{A(\theta) \mid \mathcal{A} \models \theta\}$ is closed under finite intersections. Observe that no $A(\theta) \in \mathcal{D}_0$ is small, since this would mean then $\neg\theta \in T$, so $\mathcal{A} \models \neg\theta$. By Lemma 7.6.1, there exists a regular ultrafilter $\mathcal{D}$ on S which contains $\mathcal{D}_0$. By Corollary 7.7.2, $\mathcal{A} \equiv \prod \mathcal{A}_s/\mathcal{D}$. □

In concrete situations we often seek, in addition to the above data, a special set Λ_0 of sentences of $\mathcal{L}$ with this property:
(2) If $\mathcal{A}$ and $\mathcal{A}'$ are models of T, then $\mathcal{A} \equiv \mathcal{A}'$ if and only if $\mathcal{A}$ and $\mathcal{A}'$ satisfy the same sentences of Λ_0.

Call Λ_0 a **set of basic test sentences**. Every Boolean combination of basic test sentences is called a **test sentence**.

Denote the Boolean algebra generated by $\{A(\lambda) \mid \lambda \in \Lambda_0\} \cup \mathcal{F}$ by Λ. From (2), if $\mathcal{D}$ and $\mathcal{D}'$ are two regular ultrafilters on S, then

(3) $$\prod \mathcal{A}_s/\mathcal{D} \equiv \prod \mathcal{A}_s/\mathcal{D}' \iff \mathcal{D} \cap \Lambda = \mathcal{D}' \cap \Lambda.$$

PROPOSITION 7.8.2: *Suppose that Λ_0 is a set of basic test sentences. Then for each sentence θ of $\mathcal{L}$ there exists a test sentence λ such that*
(a) $A(\theta) \approx A(\lambda)$, *and*
(b) $\theta \leftrightarrow \lambda$ *belongs to T.*

Proof: We have only to prove that $A(\theta) \in \Lambda$. Assume $A(\theta)$ does not belong to Λ. By Proposition 7.6.2, there exist regular ultrafilters $\mathcal{D}$ and $\mathcal{D}'$ of S such that $\mathcal{D} \cap \Lambda = \mathcal{D}' \cap \Lambda$, but $A(\theta) \in \mathcal{D}$ and $A(\theta) \notin \mathcal{D}'$. This contradicts (3), since $\prod \mathcal{A}_s/\mathcal{D} \equiv \prod \mathcal{A}_s/\mathcal{D}'$. □

7.9 Nonprincipal Ultraproducts of Finite Fields

We return to the starting point and the motivation of this book, the theory of finite fields.

Let S be a countable set. Define small sets as the finite subsets of S. For each $s \in S$ let F_s be a finite field. Suppose
(1) for each $n \in \mathbb{N}$ there are only finitely many $s \in S$ such that $|F_s| \leq n$.
A concrete example arises by letting S be the collection of prime divisors of a global field K. The finite fields are the corresponding residue fields.

Every finite field is perfect. That is, for every prime p, every F_s satisfies the following sentence of $\mathcal{L}(\text{ring})$:

(2) $$p = 0 \rightarrow (\forall X)(\exists Y)[Y^p = X].$$

By Section 1.5:

(3) $$\text{Gal}(F_s) \cong \hat{\mathbb{Z}}$$

for each $s \in S$. Also, by Theorem 5.4.2(a), and (1), given positive integers d and n, the following statement is true for almost all $s \in S$:
(4) For every absolutely irreducible polynomial $f \in F_s[X,Y]$ of degree d there are n distinct points $(x_i, y_i) \in F_s \times F_s$ with $f(x_i, y_i) = 0$, $i = 1, \ldots, n$.

In Proposition 20.4.4 we will display a sequence Π_1 of sentences of $\mathcal{L}(\text{ring})$ such that a field F is a model of Π_1, if and only if $\text{Gal}(F) \cong \hat{\mathbb{Z}}$. Also, there are sentences $\pi_{d,n}$ such that $F \models \pi_{d,n}$ if and only if F satisfies (4) (Proposition 11.3.2). We will be able then to prove the following result:

PROPOSITION 7.9.1: *Suppose S satisfies (1) and $F = \prod F_s / \mathcal{D}$ is a nonprincipal ultraproduct. Then F is a perfect field, $\text{Gal}(F) \cong \hat{\mathbb{Z}}$ and for every nonconstant absolutely irreducible polynomial $f \in F(X,Y)$ there exist infinitely many points $(x,y) \in F \times F$ with $f(x,y) = 0$.* □

Exercises

1. Consider the language $\mathcal{L}$ which has only one relation symbol, $<$, but neither function symbols nor constant symbols. Let $\mathcal{A}$ and $\mathcal{B}$ be the sets $\{1, 2, 3, \ldots\}$ and $\{0, 1, 2, 3, \ldots\}$ with the usual ordering. Then $\mathcal{A}$ and $\mathcal{B}$ are models for $\mathcal{L}$ and $\mathcal{A} \subseteq \mathcal{B}$. Observe that $\mathcal{A} \cong \mathcal{B}$, hence $\mathcal{A} \equiv \mathcal{B}$, but $\mathcal{A}$ is not an elementary submodel of $\mathcal{B}$.

2. Extend a given language $\mathcal{L}$ to a language $\mathcal{L}'$ by adding a new unary predicate symbol P. Let $\mathcal{A}'$ be a structure for $\mathcal{L}'$ with a domain A' such that the subset A of A' corresponding to P is a domain of a structure for $\mathcal{L}$. To each formula φ of $\mathcal{L}$ assign a formula φ' of $\mathcal{L}'$ by induction: $\varphi' = \varphi$ if φ is an atomic formula; $(\neg \varphi)' = \neg(\varphi')$, $(\varphi \vee \psi)' = \varphi' \vee \psi'$, and $((\exists X)\varphi)' = (\exists X)[P(X) \wedge \varphi']$. Refer to this as **relativizing** the variables of $\mathcal{L}$ to P.

(a) Prove that if $\varphi(X_1, \ldots, X_n)$ is a formula of $\mathcal{L}$ and $a_1, \ldots, a_n \in A$, then

$$\mathcal{A} \models \varphi(\mathbf{a}) \iff \mathcal{A}' \models \varphi'(\mathbf{a}).$$

(b) Observe that $((\forall X)\varphi)' = (\forall X)[P(X) \to \varphi']$.

3. Use transfinite induction to generalize the Skolem-Löwenheim theorem (Proposition 7.4.2) to uncountable languages.

4. Let $\mathcal{L}$ be a first order language and let S be a set of representatives of all elementary equivalence classes of structures for $\mathcal{L}$. For every sentence θ of $\mathcal{L}$ define $D(\theta)$ to be $\{M \in S \mid M \models \theta\}$. Show that the family of all $D(\theta)$ is a basis for a totally disconnected, Hausdorff and compact topology on S.

Hint: From Exercise 3, S is a set. To prove the compactness of S, note that if $\{\theta_i \mid i \in I\}$ is a set of sentences of $\mathcal{L}$ such that $\bigcap_{i \in J} D(\theta_i) \neq \emptyset$ for every finite subset J of I, then, Proposition 7.7.6 (the compactness theorem) asserts that $\bigcap_{i \in I} D(\theta_i) \neq \emptyset$.

5. For each prime p let K_p be a field such that $[L : K_p]$ is a p-power for every finite extension L of K_p. Let $\mathcal{D}$ be a nonprincipal ultrafilter of the set of primes. Prove that the ultraproduct $K = \prod K_p / \mathcal{D}$ is an algebraically closed field. Hint: Show that every polynomial $f \in K[X]$ of degree ≥ 2 is reducible.

Notes

For more about models and ultraproducts see [Bell-Slomson].

In [Ax1] and [Ax2] Ax uses the technique of nonprincipal ultraproducts to investigate the theory of finite fields. This theory is then modified in [Jarden-Kiehne] and [Jarden5] to deal with families of fields indexed by measure spaces. In this context, regular ultrafilters and regular ultraproducts replace the nonprincipal ultrafilters and ultraproducts of Ax. This chapter generalizes these to the abstract framework of model theory.

Chapter 8. Decision Procedures

The later chapters of this book give decision procedures for various theories of fields. If T denotes such a theory, we might give a naive description of a decision procedure as a system of instructions for deciding in a finite number of steps whether or not a given sentence θ belongs to T. We elaborate, more precisely, on two types of procedures, model theoretic and algebraic.

Model theoretic procedures require deduction theory (Section 8.1) and the Gödel completeness theorem (Section 8.2). With recursion theory (Sections 8.3 - 8.6) we are able to give precise definitions for our two types of procedures: model theoretic procedures are recursive while algebraic procedures are primitive recursive.

Model theoretic applications appear in Chapter 20, while algebraic applications appear in Chapter 9 for the theory of algebraically closed fields, and in greater diversity in Chapters 30 and 31.

8.1 Deduction Theory

We study the notion of a "formal proof" of a sentence in a first order language $\mathcal{L}$. All proofs we consider are carried out within a fixed "deduction system".

A "deduction system" is a finite set $\mathcal{D}$ of decidable "inference rules" and a set of formulas called "logical axioms".

An **inference rule** is a relation $R(\chi_1, \ldots, \chi_k, \psi)$ between formulas of $\mathcal{L}$. We say that a formula ψ is a **consequence** of a set F of formulas if there are $\chi_1, \ldots, \chi_k$ in F such that $R(\chi_1, \ldots, \chi_k, \psi)$ holds.

Every logical axiom is a **logically valid formula**, that is a formula $\varphi(x_1, \ldots, x_n)$ which is true in each structure $\mathcal{A}$ for $\mathcal{L}$ and under every substitution. For example, "$X = X$" is a logically valid formula. Tautologies (e.g. $\theta \vee \neg \theta$ with θ a sentence of $\mathcal{L}$) give others. We give a formal definition of tautology.

Consider a map f from a set $\{Z_1, \ldots, Z_m\}$ of variables into the set $\{\text{true, false}\}$. Extend f to Boolean polynomials $P(Z_1, \ldots, Z_m)$ by the following recursive rules:

(1a) $\qquad f(\neg P) = \begin{cases} \text{true} & \text{if } f(P) = \text{false} \\ \text{false} & \text{otherwise,} \end{cases}$

(1b) $\qquad f(P_1 \vee P_2) = \begin{cases} \text{true} & \text{if } f(P_1) = \text{true or } f(P_2) = \text{true} \\ \text{false} & \text{otherwise, and} \end{cases}$

(1c) $\qquad f(P_1 \wedge P_2) = \begin{cases} \text{true} & \text{if } f(P_1) = \text{true and } f(P_2) = \text{true} \\ \text{false} & \text{otherwise.} \end{cases}$

If $f(P(Z_1, \ldots, Z_m)) = \text{true}$ for each map f, we call the Boolean polynomial $P(Z_1, \ldots, Z_m)$ a **tautology**.

Since there are only 2^m maps

$$f\colon \{Z_1, \ldots, Z_m\} \to \{\text{true, false}\},$$

we can effectively determine if a Boolean polynomial is a tautology. Thus, if $P(Z_1, \ldots, Z_m)$ is a tautology and $\varphi_1, \ldots, \varphi_m$ are formulas of $\mathcal{L}$, then $P(\varphi_1, \ldots, \varphi_m)$ is a logically valid formula. Sometimes we abuse our language and refer to $P(\varphi_1, \ldots, \varphi_m)$ as a tautology.

A **formal proof from a set T of sentences of $\mathcal{L}$** is a finite sequence $(\varphi_1, \ldots, \varphi_n)$ of formulas of $\mathcal{L}$ such that each φ_m is either a logical axiom or an element of T or it is it is a consequence of $\{\varphi_1, \ldots, \varphi_{m-1}\}$ by one of the inference rules of $\mathcal{D}$. In this case call $(\varphi_1, \ldots, \varphi_n)$ a **formal proof** of φ_n from T. Its **length** is n. A formula φ is said to be **logically deducible** from T if there exists a formal proof of φ from T: write $T \vdash \varphi$. If $T \vdash \theta$ implies that $\theta \in T$ for every sentence θ of $\mathcal{L}$, then T is said to be **deductively closed**. Two formulas φ and ψ are said to be **equivalent** modulo T if $T \vdash [\varphi \leftrightarrow \psi]$. Note that if $(\varphi_1, \ldots, \varphi_n)$ is a formal proof from T, then $(\varphi_1, \ldots, \varphi_m)$ is also a formal proof from T, for each $m \leq n$. This allows induction arguments on the length of the proofs.

Let us now specify a specific deduction theory $\mathcal{D}$ which we use throughout the chapter and the rest of this book. $\mathcal{D}$ has two rules of inference:
(2a) **Modus Ponens**: ψ is a consequence of φ and $\varphi \to \psi$.
(2b) **Rule of Generalization**: $(\forall X)\varphi$ is a consequence of φ.

There are three types of logical axioms in $\mathcal{D}$:
(3a) **Sentential axioms**: If $P(Z_1, \ldots, Z_m)$ is a tautology and $\varphi_1, \ldots, \varphi_m$ are formulas of $\mathcal{L}$, then $P(\varphi_1, \ldots, \varphi_m)$ is a logical axiom.
(3b) **Quantifier axioms**:
(3b1) If φ and ψ are formulas and X is not a free variable of φ, then

$$(\forall X)[\varphi \to \psi] \to [\varphi \to (\forall X)[\psi]] \text{ and } (\forall X)[\varphi \wedge \psi] \to [\varphi \wedge (\forall X)[\psi]]$$

are logical axioms.
(3b2) Let φ be a formula and t a term. Suppose that no interval of the form $(\exists Y)\psi$, where ψ is a formula and Y occurs in t, contains occurencies of X which are free in φ. Then the formula $\varphi(t) \to (\exists X)[\varphi(X)]$ is a logical axiom.

Here we write $\varphi(t)$ for the formula obtained from φ by replacing all free occurences of X by t. Of course the notation $\varphi(t)$ may be ambiguous if φ has several free variables, but we use it only when it is clear from the context which variable is to be replaced.

Note that the restriction we put on t is necessary. Indeed, if $\varphi(X, Z)$ is the formula $(\exists Y)[Y \neq X] \to [Z \neq Z]$ and if we take t to be Y, then we get the formula

$$[(\exists Y)[Y \neq Y] \to [Z \neq Z]] \to (\exists X)[(\exists Y)[Y \neq X] \to [Z \neq Z]]$$

8.1 Deduction Theory

which is not logically valid. Indeed, since $(\exists Y)[Y \neq Y]$ is false, $(\exists Y)[Y \neq Y] \to [Z \neq Z]$ is logically true. However, if we choose a structure with a domain $A = \{x, y, z, \ldots\}$ such that $x \neq y$ and we substitute z for Z, then under the substitution $X = x$, $(\exists Y)[Y \neq x]$ is true but $z \neq z$ is false, so $(\exists X)\big[(\exists Y)[Y \neq X] \to [z \neq z]\big]$ is false.

(3c) **Identity axioms**: If Y, Z are variables, $t(X_1, \ldots, X_n)$ is a term and $\varphi(X_1, \ldots, X_n)$ is an atomic formula, then the formulas

$$Y = Y$$
$$Y = Z \to [t(X_1, \ldots, X_{i-1}, Y, X_{i+1}, \ldots, X_n)$$
$$= t(X_1, \ldots, X_{i-1}, Z, X_{i+1}, \ldots, X_n)]$$
$$Y = Z \to [\varphi(X_1, \ldots, X_{i-1}, Y, X_{i+1}, \ldots, X_n)$$
$$\to \varphi(X_1, \ldots, X_{i-1}, Z, X_{i+1}, \ldots, X_n)]$$

are logical axioms.

The proof of the following result contains our first examples of formal proofs.

LEMMA 8.1.1 (Deduction Theorem): *If T is a theory of $\mathcal{L}$, if θ is a sentence of $\mathcal{L}$, and if φ is a formula of $\mathcal{L}$, then $T \cup \{\theta\} \vdash \varphi$ implies $T \vdash [\theta \to \varphi]$.*

Proof: We use induction on the length of the proof of φ from $T \cup \{\theta\}$. There are four cases:

CASE 1: φ is a logical axiom or φ belongs to T. Then

$$(\varphi, \varphi \to [\theta \to \varphi], \theta \to \varphi)$$

is a proof of $\theta \to \varphi$ from T.

CASE 2: If φ equals θ, use the tautology $\theta \to \theta$.

CASE 3: The formulas ψ and $\psi \to \varphi$ appear in a proof of φ of length n from $T \cup \{\theta\}$, and φ is a consequence of $\varphi_1, \ldots, \varphi_{n-1}$ by Modus Ponens. Then both ψ and $\psi \to \varphi$ have proofs from $T \cup \{\theta\}$ of length less than n. By induction, $T \vdash [\theta \to \psi]$ and $T \vdash [\theta \to (\psi \to \varphi)]$. Furthermore, the formula

$$[\theta \to [\psi \to \varphi]] \to [[\theta \to \psi] \to [\theta \to \varphi]]$$

is a tautology. Hence, $T \vdash [\theta \to \varphi]$.

CASE 4: φ is of the form $(\forall X)\psi$, the formula ψ appears in a proof of φ from $T \cup \{\theta\}$, and φ is inferred from ψ by the Rule of Generalization. We may use the quantifier axiom $(\forall X)[\theta \to \psi] \to [\theta \to (\forall X)[\psi]]$ and the induction hypothesis $T \vdash [\theta \to \psi]$, together with the two rules of inference to conclude that $T \vdash [\theta \to \varphi]$. □

LEMMA 8.1.2: *Suppose T is a set of sentences, φ is a formula with a free variable X, and c is a constant that appears neither in T nor in φ. Then $T \vdash \varphi(c)$ implies $T \vdash \varphi$.*

Proof: Note that $\varphi(c)$ does not belong to T. If $\varphi(c)$ is a logical axiom, so is φ. Otherwise, in a proof of $\varphi(c)$ from T, the formula $\varphi(c)$ is inferred from previous formulas only by Modus Ponens or by Generalization. Now use induction on the length of the proof. $\square$

8.2 Gödel's Completeness Theorem

Consider a theory T of a language $\mathcal{L}$. We prove in this section that a sentence θ of $\mathcal{L}$ is logically deducible from T if and only if θ holds in each model of T.

Call T **consistent** if there exists a sentence θ of $\mathcal{L}$ which is not deducible from T. Since $\tau \wedge \neg \tau \to \theta$ is a tautology for each formula τ, this is equivalent to saying that the **contradiction** $\tau \wedge \neg \tau$ is not deducible from T.

LEMMA 8.2.1: *Let T be a theory and let θ be a sentence of a language $\mathcal{L}$.*
(a) *If T is consistent and $T \vdash \theta$, then $T \cup \{\theta\}$ is consistent.*
(b) *If $T \cup \{\theta\}$ is inconsistent, then $T \vdash \neg \theta$.*
(c) *If T is consistent and $T \cup \{\theta\}$ is inconsistent, then $T \cup \{\neg \theta\}$ is consistent.*

Proof of (a): If $T \cup \{\theta\}$ were inconsistent, then $T \cup \{\theta\} \vdash \tau \wedge \neg \tau$ for each formula τ. By the deduction theorem, $T \vdash \theta \to [\tau \wedge \neg \tau]$. Therefore, $T \vdash \tau \wedge \neg \tau$, in contradiction to the assumption on T.

Proof of (b): Our assumption implies that $T \cup \{\theta\} \vdash \theta \wedge \neg \theta$. Hence, by Lemma 8.1.1, $T \vdash [\theta \to \theta \wedge \neg \theta]$. Now use the tautology $[\neg \theta \vee (\theta \wedge \neg \theta)] \to \neg \theta$ to deduce that. $T \vdash \neg \theta$.

Proof of (c): Combine (a) and (b). $\square$

LEMMA 8.2.2: *Let T be a consistent theory of a language $\mathcal{L}$, φ a formula with at most one free variable X and c a new constant symbol. Extend T to a theory T' in the language $\mathcal{L}(c)$ by the sentence $(\exists X)[\varphi] \to \varphi(c)$. Then T' is consistent.*

Proof: If T' is inconsistent, Lemma 8.2.1 gives $T \vdash \neg[(\exists X)[\varphi] \to \varphi(c)]$. Thus, $T \vdash [(\exists X)[\varphi] \wedge \neg \varphi(c)]$. Hence, by Lemma 8.1.2, $T \vdash (\exists X)[\varphi] \wedge \neg \varphi$. It follows that $T \vdash [(\exists X)[\varphi] \wedge (\forall X)[\neg \varphi]]$, so $T \vdash [(\exists X)[\varphi] \wedge \neg(\exists X)[\varphi]]$. Therefore, contrary to assumption, T is inconsistent. $\square$

A theory T of a language $\mathcal{L}$ is said to be **complete** if for each sentence θ of $\mathcal{L}$ either $\theta \in T$ or $\neg \theta \in T$.

A **witness** with respect to T for a formula φ with a free variable X is a constant c for which $T \vdash [(\exists X)[\varphi] \to \varphi(c)]$.

8.2 Gödel's Completeness Theorem

LEMMA 8.2.3: *Let T be a consistent theory of a countable language $\mathcal{L}$. Then one can extend $\mathcal{L}$ and T, respectively, to a new countable language $\mathcal{L}'$ and a theory T' of $\mathcal{L}'$ such that*
(a) *T' is consistent and complete; and*
(b) *every formula $\varphi(X)$ of $\mathcal{L}'$ has a witness c with respect to T'.*

Proof: Consider a sequence $c_1, c_2, c_3, \ldots$ of distinct new constant symbols and let $\mathcal{L}' = \mathcal{L}(c_1, c_2, c_3, \ldots)$. Order all formulas of $\mathcal{L}'$ with at most one free variable X in a sequence $\varphi_1, \varphi_2, \varphi_3, \ldots$. Define an increasing sequence $T = T_0 \subset T_1 \subset T_2 \subset \ldots$ of theories of $\mathcal{L}'$ inductively as follows: Let $c = c_k$ be the first constant symbol that does not appear in $T_n \cup \{\varphi_n\}$ and define $T_{n+1} = T_n \cup \{(\exists X)[\varphi_n] \to \varphi_n(c)\}$. By Lemma 8.2.2, and induction on n, T_n is a consistent theory for every n. Since each proof from $T_\infty = \bigcup_{n=0}^\infty T_n$ is a proof from T_n for some n, T_∞ is also a consistent theory. By construction, each formula $\varphi(X)$ of $\mathcal{L}'$ has a witness c with respect to T_∞.

Now order all sentences of $\mathcal{L}'$ in a sequence, $\theta_1, \theta_2, \theta_3, \ldots$. Inductively define a second increasing sequence $T_\infty = T_0' \subseteq T_1' \subseteq T_2' \subseteq \ldots$ of consistent theories of $\mathcal{L}'$ as follows: If $T_n' \cup \{\theta_{n+1}\}$ is consistent, then $T_{n+1}' = T_n' \cup \{\theta_{n+1}\}$; otherwise $T_n' \cup \{\neg\theta_{n+1}\}$ is consistent (Lemma 8.2.1). Define T_{n+1}' to be $T_n' \cup \{\neg\theta_{n+1}\}$. The theory $T' = \bigcup_{n=1}^\infty T_n'$ has the desired properties. □

PROPOSITION 8.2.4: *Let T' be a consistent and complete theory of a countable language $\mathcal{L}'$. Suppose that each formula $\varphi(X)$ of $\mathcal{L}'$ has a witness with respect to T'. Then T' has a model.*

Proof: Define an equivalence relation on the set of all **constant terms** t of $\mathcal{L}'$ (i.e. terms without a variable) as follows:

$$t_1 \sim t_2 \iff T' \vdash [t_1 = t_2].$$

Denote the equivalence class of t by $\bar{t}$ and let M be the set of all equivalence classes: M is domain of the model $\mathcal{M}$ for T'.

For R an m-ary relation symbol of $\mathcal{L}'$, define the corresponding relation $\bar{R}$ on M by

$$(\bar{t}_1, \ldots, \bar{t}_m) \in \bar{R} \iff T' \vdash R(t_1, \ldots, t_m).$$

For F an n-ary function symbol of $\mathcal{L}'$, define the corresponding function $\bar{F}$ on M by

$$\bar{F}(\bar{t}_1, \ldots, \bar{t}_n) = \overline{F(t_1, \ldots, t_n)}.$$

Finally, if c is a constant symbol of $\mathcal{L}'$, then $\bar{c}$ is taken to be the corresponding constant of $\mathcal{M}$. The validity of these definitions follows from the identity axioms.

The definition of $\bar{F}$ implies, by an induction on structure, that if $s(X_1, \ldots, X_n)$ is a term with at most n variables $X_1, \ldots, X_n$ and if $t_1, \ldots, t_n$ are constant terms, then $\overline{s(t_1, \ldots, t_n)} = s(\bar{t}_1, \ldots, \bar{t}_n)$. Therefore if $\psi(X_1, \ldots, X_n)$ is an atomic formula and $t_1, \ldots, t_n$ are constant terms, then

(1) $$\mathcal{M} \models \psi(\bar{t}_1, \ldots, \bar{t}_n) \iff T' \vdash \psi(t_1, \ldots, t_n)$$

We claim that (1) also holds for an arbitrary formula ψ. Since T' is complete, if (1) holds for a given formula ψ, then it also holds for $\neg\psi$. If (1) holds for ψ_1, ψ_2, then it also holds for $\psi_1 \vee \psi_2$. Finally, suppose that (1) holds for $\psi(X_1, \ldots, X_n)$. We prove that (1) holds for $(\exists X_1)\psi$.

Indeed, suppose $T' \vdash (\exists X_1)\psi(X_1, t_2, \ldots, t_n)$ for some constant terms $t_2, \ldots, t_n$. Let c be a witness for the formula $\psi(X_1, t_2, \ldots, t_n)$. Then

$$T' \vdash (\exists X_1)\psi(X_1, t_2, \ldots, t_n) \to \psi(c, t_2, \ldots, t_n).$$

Hence $T' \vdash \psi(c, t_2, \ldots, t_n)$, so, $\mathcal{M} \models \psi(\bar{c}, \bar{t}_2, \ldots, \bar{t}_n)$, by the induction hypothesis. Therefore, $\mathcal{M} \models (\exists X_1)\psi(X_1, \bar{t}_2, \ldots, \bar{t}_n)$.

Conversely, suppose $\mathcal{M} \models (\exists X_1)\psi(X_1, \bar{t}_2, \ldots, \bar{t}_n)$. Then there exists a constant term t_1 with $\mathcal{M} \models \psi(\bar{t}_1, \bar{t}_2, \ldots, \bar{t}_n)$. By (1), $T' \models \psi(t_1, t_2, \ldots, t_n)$. By the Quantifier Axiom (3b2) of Section 8.1, $T' \vdash \psi(t_1, t_2, \ldots, t_n) \to (\exists X_1)\psi(X_1, t_2, \ldots, t_n)$. Consequently, $T' \vdash (\exists X_1)\psi(X_1, t_2, \ldots, t_n)$.

It follows that (1) holds for every formula ψ, and thus for every sentence ψ of $\mathcal{L}'$. Consequently, $\mathcal{M}$ is a model of T'. $\square$

The combination of Lemma 8.2.3 with Proposition 8.2.4 gives the following basic result:

THEOREM 8.2.5 (Gödel's Completeness Theorem): *If T is a consistent theory of a language $\mathcal{L}$, then T has a model.*

Proof: By the compactness theorem, we may assume that T is finite. Hence, we may assume that $\mathcal{L}$ is countable. Lemma 8.2.3 extends $\mathcal{L}$ and T, respectively, to a countable language $\mathcal{L}'$ and a theory T' of $\mathcal{L}'$ such that T' is consistent and complete. By Proposition 8.2.4, T' has a model $\mathcal{M}$. Then $\mathcal{M}$ is also a model of T. $\square$

COROLLARY 8.2.6: *Let T be a theory and let θ be a sentence of a countable language $\mathcal{L}$. Then*

$$T \models \theta \iff T \vdash \theta.$$

Proof: Since all logical axioms are logically valid, we conclude "$\Longleftarrow$" by induction on the length of the proof of θ from T.

If $T \vdash \theta$ does not hold, then the theory $T' = T \cup \{\neg\theta\}$ is consistent (Lemma 8.2.1). From Theorem 8.2.5, the theory T' has a model, and therefore $T \models \theta$ does not hold. $\square$

8.3 Primitive Recursive Functions

Consider the family $\mathcal{F}_n$ of all functions from N^n to N, where N is the set of nonnegative integers and let $\mathcal{F} = \bigcup_{n=1}^{\infty} \mathcal{F}_n$. Among the functions of $\mathcal{F}$, those that suit the recursive operation of "everyday" mathematics are called primitive recursive functions (Definition 8.3.1). Recursive functions include further, less "computable", functions (Definition 8.5.1). All our decidability and undecidability results are expressed in terms of these functions.

8.3 Primitive Recursive Functions

Consider the following basic functions:
(1a) The identically zero function: $f(x) = 0$.
(1b) The successor function: $f(x) = x + 1$.
(1c) The projection functions: $f(x_1, \ldots, x_n) = x_i$, for $n \in \mathbb{N}$ and $1 \leq i \leq n$.

Definition 8.3.1: The set of **primitive recursive functions** is the smallest subset of $\mathcal{F}$ which contains the functions (1) and is closed under the following operations:

(2a) Composition: If $g \in \mathcal{F}_m$ and $h_1, \ldots, h_m \in \mathcal{F}_n$ are primitive recursive functions, then the function

$$f(x_1, \ldots, x_n) = g(h_1(x_1, \ldots, x_n), \ldots, h_m(x_1, \ldots, x_n))$$

is also primitive recursive.

(2b) Inductive definition: If $f_0 \in \mathcal{F}_n$ and $g \in \mathcal{F}_{n+2}$ are primitive recursive functions, then the function $f \in \mathcal{F}_{n+1}$, which is defined by the following induction,

$$f(x_1, \ldots, x_n, 0) = f_0(x_1, \ldots, x_n)$$
$$f(x_1, \ldots, x_n, y+1) = g(x_1, \ldots, x_n, y, f(x_1, \ldots, x_n, y))$$

is also primitive recursive. □

We list some standard primitive recursive functions. In each case the definition is based on previously defined functions.
(3a) The constant functions, $f(x) = k$ (apply (2a), (1b), (1a) and induction on k).
(3b) The identity function, $f(x) = x$ (use (1c) with $n = 1$).
(3c) The addition function, $f(x, y) = x + y$ (use induction on y).
(3d) multiplication function, $f(x, y) = xy$.
(3e) The exponential functions, $f(x, y) = x^y$ with $f(x, 0) = 1$.
(3f) The factorial function, $f(x) = x!$ with $f(0) = 1$.
(3g) The predecessor function, $\mathrm{pd}(0) = 0$ and $\mathrm{pd}(x+1) = x$.
(3h) The sign function, $\mathrm{sgn}(0) = 0$ and $\mathrm{sgn}(x+1) = 1$.
(3i) Subtraction up to 0, $f(x, y) = x - y$ if $x \geq y$ and zero otherwise, denoted $f(x, y) = x \dotdiv y$ ($x \dotdiv (y+1) = \mathrm{pd}(x \dotdiv y)$).
(3j) The $\overline{\mathrm{sgn}}$ function, $\overline{\mathrm{sgn}}(x) = 1 \dotdiv x$. (i.e. $\overline{\mathrm{sgn}}(0) = 1$ and $\overline{\mathrm{sgn}}(x) = 0$ if $x \geq 1$).
(3k) The minimum function, $\min(x, y) = y \dotdiv (y \dotdiv x)$.
(3l) The maximum function, $\max(x, y) = x + y \dotdiv \min(x, y)$.
(3m) The absolute value, $|x - y| = (x \dotdiv y) + (y \dotdiv x)$.
(3n) The remainder of division of x by y function, $\mathrm{rm}(0, y) = 0$ and

$$\mathrm{rm}(x+1, y) = (\mathrm{rm}(x, y) + 1)\mathrm{sgn}(y \dotdiv (\mathrm{rm}(x, y) + 1)).$$

(3o) The integral value function $\left[\frac{x}{y}\right]$,

$$\left[\frac{0}{y}\right] = 0 \quad \text{and} \quad \left[\frac{x+1}{y}\right] = \left[\frac{x}{y}\right] + \mathrm{sgn}(y \dotdiv (\mathrm{rm}(x, y) + 1))$$

Thus, $x = \left[\frac{x}{y}\right]y + \mathrm{rm}(x,y)$ and $0 \le \mathrm{rm}(x,y) < y$ for all $x \ge 0$ and all $y > 0$.

Finally, certain **long summations** and **long multiplications** are primitive recursive. If $g(x_1, \ldots, x_n, y)$ is a primitive recursive function, then both of the functions $\sum_{y<z} g(x_1, \ldots, x_n, y)$ and $\prod_{y<z} g(x_1, \ldots, x_n, y)$ are primitive recursive. Indeed, define long summations inductively:

(4a) $\quad \displaystyle\sum_{y<0} g(\mathbf{x}, y) = 0 \quad$ and $\quad \displaystyle\sum_{y<z+1} g(\mathbf{x}, y) = \sum_{y<z} g(\mathbf{x}, y) + g(\mathbf{x}, z),$

(4b) $\quad \displaystyle\sum_{y \le z} g(\mathbf{x}, y) = \sum_{y<z+1} g(\mathbf{x}, y),$

(4c) $\quad \displaystyle\sum_{w<y\le z} g(\mathbf{x}, y) = \sum_{y<z \dot- w} g(\mathbf{x}, y+w+1).$

Long multiplications can similarly be defined inductively.

8.4 Primitive Recursive Relations

The definition of primitive recursive relations depends on their characteristic functions. Recall that the **characteristic function** $\chi_R(x_1, \ldots, x_n)$ of an n-ary relation $R(x_1, \ldots, x_n)$ on N has value 1 if $R(x_1, \ldots, x_n)$ is true and value 0 if $R(x_1, \ldots, x_n)$ is false. The relation R is said to be **primitive recursive** if χ_R is primitive recursive.

If $g_1, \ldots, g_n$ are m-ary primitive recursive functions and R is an n-ary primitive recursive relation, then $S(\mathbf{x}) = R(g_1(\mathbf{x}), \ldots, g_n(\mathbf{x}))$ is an m-ary relation with the characteristic function $\chi_R(g_1(\mathbf{x}), \ldots, g_n(\mathbf{x}))$. Therefore S is also a primitive recursive relation.

If Q and R are primitive recursive n-ary relations, then $Q \cap R$, $Q \cup R$ and Q' (= complement of Q) are primitive recursive relations with respective characteristic functions $\chi_Q \chi_R$, $\max(\chi_Q, \chi_R)$ and $1 \dot- \chi_Q$.

We may use primitive recursive relations to get new primitive recursive functions. Here is an example: Let $R_1, \ldots, R_m$ be disjoint n-ary primitive recursive relations whose union is N^n and let $g_1, \ldots, g_m$ be arbitrary primitive recursive n-ary functions. Let $f(\mathbf{x}) = g_i(\mathbf{x})$ if $R_i(\mathbf{x})$ is true, $i = 1, \ldots, m$. Then

$$f(\mathbf{x}) = \sum_{i=1}^{m} \chi_{R_i}(\mathbf{x}) g_i(\mathbf{x}),$$

and hence f is primitive recursive.

We now define an operator that interprets the existential quantifier as a relation.

8.5 Recursive Functions

For an $(n+1)$-ary relation $R(x_1,\ldots,x_n,y)$, and $a,b \in N$ with $a < b$ define an $(n+2)$-ary relation $(Ey)_{a<y<b}R(\mathbf{x},y)$ by the condition that it is true if there exists y with $a < y < b$ such that $R(\mathbf{x},y)$ is true. The characteristic function of the new relation is $\min(1, \sum_{a<y<b} \chi_R(\mathbf{x},y))$. Therefore, it is primitive recursive if R is primitive recursive.

The **bounded minimum operator** $(\mu y)_{a<y<z}$ turns a relation into a function. Here is its definition: Let $R(x_1,\ldots,x_n,y)$ be an $(n+1)$-ary relation. Then $(\mu y)_{a<y<z} R(\mathbf{x},y)$ is the smallest y, with $a < y < z$, such that $R(\mathbf{x},y)$ is true, if there exists any, and z otherwise. The identity

$$(\mu y)_{a<y<z} R(\mathbf{x},y) = a + \sum_{a<s\leq z} \prod_{a<t<s} (1 \dotdiv \chi_R(\mathbf{x},t))$$

shows that if R is a primitive recursive relation, then $(\mu y)_{a<y<z} R(\mathbf{x},y)$ is a primitive recursive function.

The following are consequences of these definitions:

(1a) The relation "$x < y$" is primitive recursive, since its characteristic function is $\operatorname{sgn}(y \dotdiv x)$.

(1b) The relation "$y > 0$ and y divides x," denoted by $y|x$, is primitive recursive, since its characteristic function is $\operatorname{sgn}(y)(1 \dotdiv \operatorname{rm}(x,y))$.

(1c) The relation "x is a prime," denoted $\operatorname{pr}(x)$, is primitive recursive, since it is defined by the condition $x > 1 \land \neg(Ey)_{1<y<x} y|x$.

(1d) The function p_i, the ith prime, can be defined by induction:

$$p_1 = 2 \quad \text{and} \quad p_{i+1} = (\mu x)_{p_i < x < p_i!+1} \operatorname{pr}(x).$$

8.5 Recursive Functions

The family of **recursive functions** is the smallest subset of $\mathcal{F}$ which contains all primitive recursive functions and is closed under composition, inductive definition (Definition 8.3.1) and the **minimum operator** which we now define.

Definition 8.5.1: Let $R(\mathbf{x},y)$ be an $(n+1)$-ary relation on N such that for each $\mathbf{x}$ there exists y for which $R(\mathbf{x},y)$ is true. Then $(\mu y)R(\mathbf{x},y)$ is the smallest y for which $R(\mathbf{x},y)$ is true. □

Remark 8.5.2: If we can decide for each $(\mathbf{x},y)$ whether or not $R(\mathbf{x},y)$ is true, then we can also compute $(\mu y)R(\mathbf{x},y)$ by a check of the validity of the relations $R(\mathbf{x},0), R(\mathbf{x},1), R(\mathbf{x},2),\ldots$ in order. By assumption, we can discover the first y for which $R(\mathbf{x},y)$ is true in a finite number of steps. There is not, however, a bound on the number of steps in terms of R and $\mathbf{x}$. □

In contrast there is a bound for the number of steps needed to compute primitive recursive functions at given arguments.

Following Ackermann, we use induction on m to define a sequence $\varphi_m(y)$, of primitive recursive functions:

$$\varphi_0(y) = y + 1;$$

and with φ_m defined, induct on m to define $\varphi_{m+1}(y)$ by

$$\varphi_{m+1}(0) = \varphi_m(1) \quad \text{and} \quad \varphi_{m+1}(y+1) = \varphi_m(\varphi_{m+1}(y)).$$

These functions grow very fast with m. Indeed, $\varphi_1(y) = y + 2$, $\varphi_2(y) = 2y + 3$, $\varphi_3(y)$ grows with y as $5 \cdot 2^y$, $\varphi_4(y)$ grows as an exponential tower of y stories, $\varphi_5(y)$ is the iteration y times of $\varphi_4(1)$, and so on. In particular, every primitive recursive function is bounded by one of the φ_n's. More precisely, for every primitive recursive function $g(x_1, \ldots, x_n)$ there exists an m such that for all $x_1, \ldots, x_n$

(1) $$g(x_1, \ldots, x_n) < \varphi_m(x_1 + \cdots + x_n).$$

An examination of the proof of this statement [Hermes, p. 84] shows that m can be computed by induction on structure as follows.

For the identically zero function, the successor function, and the projection functions, take for m the values 0, 1 and 0, respectively.

If $g(\mathbf{x}) = h_0(h_1(\mathbf{x}), \ldots, h_k(\mathbf{x}))$ and the index that corresponds to h_i is m_i, $i = 0, \ldots, k$, then g satisfies (1) with $m = m_0 + \max(m_1 + \cdots + m_k) + 6$.

Finally, for the induction operation, if $g(\mathbf{x}, 0) = h_0(\mathbf{x})$ and $g(\mathbf{x}, y+1) = h_1(\mathbf{x}, y, g(\mathbf{x}, y))$, and if the indices that correspond to h_0, h_1 are m_0, m_1, respectively, then g satisfies (1) with $m = \max(m_0, m_1) + 5$.

Define the **Ackermann function** to be $\varphi(x, y) = \varphi_x(y)$. Although this definition is inductive it does not use the primitive recursive induction principle (2b) of Section 8.3. Indeed, if φ were primitive recursive, then $\psi(x) = \varphi(x, x)$ would also be. From (1) there would exist m such that $\psi(x) < \varphi(m, x)$ for all x. For $m = x$ this would give the contradiction $\varphi(m, m) = \psi(m) < \varphi(m, m)$. Hence, φ is not primitive recursive. However, the Ackermann function is recursive [Hermes, p. 90]. Thus, *the set of primitive recursive functions is a proper subset of the set of recursive functions.*

Define the number of steps (abbr. ns) needed to compute the basic primitive recursive functions, (1) of Section 8.3, to be 1. Also, if $f(\mathbf{x}) = g(h_1(\mathbf{x}), \ldots, h_m(\mathbf{x}))$ then

$$\text{ns}(f; \mathbf{x}) = \text{ns}(g; h_1(\mathbf{x}), \ldots, h_m(\mathbf{x})) + \sum_{i=1}^{n} \text{ns}(h_i : \mathbf{x}).$$

Finally, if $f(\mathbf{x}, y)$ is defined by induction on y as in (2b) of Section 8.3, then

$$\text{ns}(f; \mathbf{x}, y+1) = \text{ns}(f_0; \mathbf{x}) + \sum_{t=1}^{y} \text{ns}(g; \mathbf{x}, t, f(\mathbf{x}, t)).$$

If $f(\mathbf{x})$ is a primitive recursive function, then so is the function $\mathrm{ns}(f;\mathbf{x})$. Hence, if the definition of f is explicitly given, then one can compute an m, according to the above rules, such that $\mathrm{ns}(f;\mathbf{x}) < \varphi_m(x_1+\cdots+x_n)$, for all $\mathbf{x}$. This gives us the required bound on the number of steps needed to compute f at $\mathbf{x}$.

8.6 Recursive and Primitive Recursive Procedures

We need a way to label, with integers, the elements of a given language in order to apply the concepts of recursive and primitive recursive functions to algorithms. Gödel numbering (below) gives such a labeling. As an illustration we give an explicit Gödel numbering for $\mathcal{L}(\mathrm{ring})$. The numbering is an injective map ν from the set of all terms and all formulas of $\mathcal{L}(\mathrm{ring})$ to $\mathbb{N}$. Define ν by induction, first on terms and then on formulas: $\nu(0) = 2$; $\nu(1) = 2\cdot 3$; $\nu(X_i) = 2^2\cdot 3^i$, $i = 1, 2, \ldots$; $\nu(t+u) = 2^3 3^{\nu(t)} 5^{\nu(u)}$; $\nu(tu) = 2^4 3^{\nu(t)} 5^{\nu(u)}$; $\nu(t{=}u) = 2^5 3^{\nu(t)} 5^{\nu(u)}$ for terms t, u; $\nu(\neg\varphi) = 2^6 3^{\nu(\varphi)}$; $\nu(\varphi\vee\psi) = 2^7 3^{\nu(\varphi)} 5^{\nu(\psi)}$; and $\nu((\exists X_i)[\varphi]) = 2^8 3^i 5^{\nu(\varphi)}$ for formulas φ, ψ.

Call $\nu(t)$ (resp. $\nu(\varphi)$) the **Gödel number** of t (resp. φ). An analysis following that of Section 8.3 shows that the set of all Gödel numbers is primitive recursive. Moreover, from the Gödel number of a term or of a formula one may effectively reconstruct the corresponding term or formula. Thus, we call a theory T of $\mathcal{L}(\mathrm{ring})$ **recursive** (resp. **primitive recursive**) if its image by the Gödel numbering, $\nu(T)$, is a recursive (resp. primitive recursive) set. If, in this case, the actual recursive definition of the characteristic function $\chi_{\nu(T)}$, is given, then we may translate it back into terms of the language $\mathcal{L}(\mathrm{ring})$. From this we obtain a finite set of instructions, which applied to a sentence θ, decides for us, whether θ belongs to T. This is what we call a **decision procedure**, recursive or primitive recursive as T is recursive or primitive recursive. We then say that T is a **decidable theory**.

In practice this is turned about. It is usual to start from a set of instructions for $\mathcal{L}(\mathrm{ring})$, which apply to a sentence θ and from which we decide whether θ belongs to T or not. These instructions arise from certain basic operations by compositions, inductions, and minimalizations. The Gödel numbering ν translates these basic operations into recursive operations on numbers. In principle, a proof of recursiveness (or primitive recursiveness) follows from inspecting these operations on numbers. In practice we rely on a direct analysis of the original set of instructions for $\mathcal{L}(\mathrm{ring})$.

The distinction between recursive procedures and primitive recursive procedures lies in the use of the minimum operator. Procedures that use only the bounded minimum operator are primitive recursive.

Experience shows that the technical distinction between primitive recursive and recursive procedures corresponds to two distinct areas of mathematics. In showing that a specific procedure is recursive, Zorn's lemma (e.g. via ultraproducts) is a common tool. The procedure itself may use enumeration of all proofs of $\mathcal{L}(\mathrm{ring})$, and the Gödel completeness theorem. This is the sort of procedure that most model theorists prefer. On the other hand, primitive recursive procedures involve constructive field theory and algebraic geometry and rarely use model theoretic tools. Since algebraists tend to single out particular statements for intense analysis, this appeals to their desire to know the relation of these statements to the underlying theory.

In this book we present both methods. Thus, we acquaint the reader with model theory as well as algebraic geometry. Section 8.7 presents the basics that will be applied in the recursive procedures of Chapters 19 and 23. On the other hand, the elimination of quantifiers procedure of Chapter 9 is applied in Chapters 9, 30 and 31 to give primitive recursive procedures.

In both cases we extend the language $\mathcal{L}(\text{ring})$ to the language $\mathcal{L}(\text{ring}, K)$ where K is "an explicitly given field with elimination theory" (Chapter 19). Since we often avoid details steps that show our procedures to be recursive (or primitive recursive), we justify our precise definitions (rather than the use of the naive concepts of "effective," "computable," "decidable," etc) on two counts. First: naive definition of "decidable" is unsatisfactory for a proof that a specific theory is undecidable. For example, Chapter 28 will use the diagonal method — a concept based on recursive functions — to demonstrate our main undecidability results. Second: Several of the decidability results that we do give can be interpreted as a (albeit, sophisticated) collation of a small number of operations that can be programmed in a computer. For example, factorization of polynomials in $\mathbb{Z}[X]$ appears in almost all our procedures. Many of these operations are the object of considerable research at the time of this book. It is conceivable that the number of steps in many of our procedures can be bounded by a specific Ackermann function φ_m (Section 8.5).

8.7 A Reduction Step in Decidability Procedures

Suppose Π is a set of sentences in a language $\mathcal{L}$ which is equipped with a Gödel numbering ν. Then ν extends to a Gödel numbering of all proofs from Π. For example, if $(\theta_1, \ldots, \theta_n)$ is a formal proof in $\mathcal{L}(\text{ring})$, then

$$\nu(\theta_1, \ldots, \theta_n) = 2^9 3^{\nu(\theta_1)} 5^{\nu(\theta_2)} \ldots p_{n+1}^{\nu(\theta_n)},$$

extends the Gödel numbering defined in Section 8.6, and it induces an ordering on all proofs from Π.

It can be shown that the set of logical axioms (Section 8.1) is primitive recursive and that if Π is a recursive set, then the set of all proofs from Π is recursive. Let f be the function with $f(0) = 0$ and $f(n) = \nu(\theta)$ where θ is the consequence of the nth proof from Π. Then f is a recursive function from N onto the set of Gödel numbers of all logical consequences of Π (see also Exercise 4). This does not, however, mean that the set of logical consequences of Π must be recursive, since the image of a recursive function need not be recursive.

This concept sometimes allows us to reduce the decidability of a theory to the decidability of a simpler theory.

LEMMA 8.7.1: *Let T be a deductively closed theory of a given countable language $\mathcal{L}$. Suppose T has a recursive sequence $\Pi = \{\pi_1, \pi_2, \pi_3, \ldots\}$ of axioms (Section 7.3). Suppose also that Λ is a decidable theory of $\mathcal{L}$ and that for each sentence θ of $\mathcal{L}$ there exists $\lambda \in \Lambda$ such that $\theta \leftrightarrow \lambda$ belongs to T. Then there exists a recursive procedure which when applied to a given sentence θ of $\mathcal{L}$ finds a sentence λ of Λ with the property that $\theta \leftrightarrow \lambda$ belongs to T. If in addition $\Lambda \cap T$ is decidable, then T is also decidable.*

Proof: Let θ be a sentence of $\mathcal{L}$. There exists a sentence $\lambda \in \Lambda$ such that $\theta \leftrightarrow \lambda$ belongs to T. By Corollary 8.2.6, $T \vdash \theta \leftrightarrow \lambda$.

To find such a λ explicitly, we make a list of all formal proofs from Π and examine them one by one (i.e. according to their Gödel numbers). After a check of finitely many proofs we must come to a proof of a sentence of the form $\theta \leftrightarrow \lambda$ with $\lambda \in \Lambda$. Since Λ is decidable, we can recognize this proof. Applying Corollary 8.2.6, once again, we conclude that $\theta \leftrightarrow \lambda$ belongs to T. Thus, θ is in T if and only if λ is in T. Therefore, if $\Lambda \cap T$ is decidable, we can decide whether or not λ, and hence θ, belongs to T. □

The assumptions of Lemma 8.7.1 hold in the situation described in Section 7.8. Suppose S is a set equipped with a family of small subsets, as in Section 7.8. For every $s \in S$, $\mathcal{A}_s$ is a structure for a fixed countable language $\mathcal{L}$. Denote the theory of all sentences θ of $\mathcal{L}$ that are true in almost all $\mathcal{A}_s$ by T. Assume Λ_0 is a decidable theory of $\mathcal{L}$ having this property:
(1) If $\mathcal{A}$ and $\mathcal{A}'$ are models of T, then $\mathcal{A} \equiv \mathcal{A}'$ if and only if $\mathcal{A}$ and $\mathcal{A}'$ satisfy the same sentences of Λ_0.

Then the Boolean algebra Λ generated by Λ_0 is also decidable. It follows from Proposition 7.8.2 that for each sentence θ of $\mathcal{L}$ there exists a sentence λ of Λ such that $\theta \leftrightarrow \lambda$ belongs to T. We conclude:

PROPOSITION 8.7.2: *In the situation above, if T has a decidable set of axioms Π and if $T \cap \Lambda$ is decidable, then T is decidable. Moreover, for each sentence θ of $\mathcal{L}$ one can find a sentence λ of Λ such that $\theta \leftrightarrow \lambda$ belongs to T.*

Note that this decidability procedure gives no bound on the number of steps necessary to decide if a given sentence θ belongs to T. For example, it gives no bound on the lengths of the proofs of $\theta \leftrightarrow \lambda$ from T; nor does it limit the number of axioms from Π that appear in the proof.

Exercises

1. Give an example of formulas φ and ψ for which

$$(\forall X)[\varphi \to \psi] \to [\varphi \to (\forall X)[\psi]]$$

is not a logically valid formula.

2. Prove by induction on structure that if X is a free variable in a formula φ and t is a term none of his variables other that X appear in φ, then $\varphi(t) \to (\exists X)\varphi$ is a logically valid formula.

3. Prove that the functions $\text{rm}(x, y)$ and $\left[\frac{x}{y}\right]$ in (3n) and (3o) of Section 8.3 are indeed the remainder and quotient, respectively, of division of x by y.

4. For a positive integer x, let $p(x)$ be the largest prime divisor of x, and let $e(x)$ be the exponent of the maximal power of $p(x)$ that divides x. Prove that both $p(x)$ and $e(x)$ are primitive recursive functions. Hint: Both $p(x)$ and $e(x)$ are bounded by x. Use the bounded minimum operator.

5. Prove (1) of Section 8.5 using the statement following it.

6. Show that the set of all Gödel's numbers of elements of $\mathcal{L}(\text{ring})$ (in Section 8.6) is primitive recursive.

Notes

Gödel proved the completeness theorem in 1930 [Gödel]. Our proof of the theorem appears in [Henkin].

The reduction step described in Section 8.7, generalizes a well known procedure [Ax2, p. 265].

Chapter 9. Algebraically Closed Fields

We establish a simple algebraic elimination of quantifiers procedure for the theory of algebraically closed fields. This theory is model complete (Corollary 9.3.2). Among the applications are Hilbert's Nullstellensatz and the Bertini-Noether theorem.

9.1 Elimination of Quantifiers

A theory T of a language $\mathcal{L}$ admits an **elimination of quantifiers** if every formula of $\mathcal{L}$ is equivalent modulo T to a formula without quantifiers. Suppose that T contains its logical consequences, has a recursive set of axioms and there exists a decision procedure for the quantifier free sentences of T. Then, T is decidable (Proposition 8.7.2). Moreover, if T admits elimination of quantifiers, then T is "model complete" (Lemma 9.1.2)

Definition: A formula $\varphi(X_1, \ldots, X_n)$ is in **prenex normal form** if it has the form

$$(1) \qquad (Q_1 Y_1) \cdots (Q_m Y_m) \psi(X_1, \ldots, X_n, Y_1, \ldots, Y_m),$$

where each of the Q_i is either the quantifier $\exists$ or the quantifier $\forall$, and $\psi(X_1, \ldots, X_n, Y_1, \ldots, Y_m)$ is a formula without quantifiers.

Call two formulas φ and φ' of the same language **logically equivalent** if $\varphi \leftrightarrow \varphi'$ is a logically valid formula. □

LEMMA 9.1.1: *Every formula $\varphi(X_1, \ldots, X_n)$ of a language $\mathcal{L}$ is logically equivalent to a formula $\varphi^*(X_1, \ldots, X_n)$ in prenex normal form.*

Proof: Use induction on the structure of formulas as follows: Since atomic formulas have no variables, they are already in a prenex normal form. Assume that φ has form (1). Then $\neg \varphi$ has the form $(Q'_1 Y_1) \cdots (Q'_m Y_m) \neg \psi(\mathbf{X}, \mathbf{Y})$, where $Q'_i = \exists$ if $Q_i = \forall$ and $Q'_i = \forall$ if $Q_i = \exists$ and $\mathbf{X} = (X_1, \ldots, X_n)$ and $\mathbf{Y} = (Y_1, \ldots, Y_m)$. If φ has form (1) and φ' has the form

$$(Q'_1 Y'_1) \cdots (Q'_{m'} Y'_{m'}) \psi'(\mathbf{X}, \mathbf{Y}'),$$

then after replacing the Y'_i's, if necessary, by new variables different from the Y_i's and the X_i's, the formula $\varphi \vee \varphi'$ is logically equivalent to

$$(Q_1 Y_1) \cdots (Q_m Y_m)(Q'_1 Y'_1) \cdots (Q'_{m'} Y'_{m'})[\psi(\mathbf{X}, \mathbf{Y}) \vee \psi'(\mathbf{X}, \mathbf{Y}')].$$

Finally, if φ has form (1), then both $(\exists Z)\varphi$ and $(\forall Z)\varphi$ have form (1). □

The proof of Lemma 9.1.1 gives φ^* effectively from φ. The number of letters in φ is a bound on the number of steps in this procedure. In any decision procedure we therefore assume that the sentences (or if necessary, the formulas) are in prenex normal form.

A **quantifier elimination procedure** for a theory T of a language $\mathcal{L}$ is a procedure that constructs, from a given formula $\psi(X, Z_1, \ldots, Z_n)$, a formula $\psi'(Z_1, \ldots, Z_n)$ for which $T \models (\exists X)[\psi(X, \mathbf{Z})] \leftrightarrow \psi'(\mathbf{Z})$. This means that for each model $\mathcal{A}$ of T with a domain A and each n-tuple $(a_1, \ldots, a_n)$ of A, $\mathcal{A} \models (\exists X)\psi(X, \mathbf{a})$ if and only if $\mathcal{A} \models \psi'(\mathbf{a})$. That is, the quantifier $\exists$ has been **eliminated**.

Elimination of $\exists$ also gives elimination of $\forall$: If $T \models (\exists X)[\neg\psi(X, \mathbf{Z})] \leftrightarrow \psi''(\mathbf{Z})$, then $T \models (\forall X)[\psi(X, \mathbf{Z})] \leftrightarrow \neg\psi''(\mathbf{Z})$.

Starting from a formula $\varphi(X_1, \ldots, X_n)$ of form (1), one finds a formula $\psi_1(\mathbf{X}, Y_1, \ldots, Y_{m-1})$ such that

$$T \models (Q_m Y_m)[\psi(\mathbf{X}, Y_1, \ldots, Y_m)] \leftrightarrow \psi_1(\mathbf{X}, Y_1, \ldots, Y_{m-1}).$$

Then

$$T \models (Q_1 Y_1) \cdots (Q_m Y_m)[\psi(\mathbf{X}, Y_1, \ldots, Y_m)]$$
$$\leftrightarrow (Q_1 Y_1) \cdots (Q_{m-1} Y_{m-1})[\psi_1(\mathbf{X}, Y_1, \ldots, Y_{m-1})].$$

By induction, this eliminates the quantifiers $Q_m, Q_{m-1}, \ldots, Q_1$ one at a time to derive a quantifier free formula $\psi_m(\mathbf{X})$ which is equivalent to $\varphi(\mathbf{X})$ modulo T. If $\mathcal{A} \subseteq \mathcal{B}$ are two models of T and if $a_1, \ldots, a_n$ are elements of the domain of $\mathcal{A}$, then $\mathcal{A} \models \psi_m(\mathbf{a})$ if and only if $\mathcal{B} \models \psi_m(\mathbf{a})$. Therefore, $\mathcal{A} \models \varphi(\mathbf{a})$ if and only if $\mathcal{B} \models \varphi(\mathbf{a})$. Since φ is arbitrary this means that $\mathcal{A} \prec \mathcal{B}$.

We call a theory T **model complete** if for any two models $\mathcal{A} \subseteq \mathcal{B}$ of T, we have $\mathcal{A} \prec \mathcal{B}$.

LEMMA 9.1.2: *If a theory T admits an elimination of quantifiers, then T is model complete.*

An application of a quantifier elimination procedure to a sentence θ produces a sentence θ' without quantifiers such that $T \models \theta \leftrightarrow \theta'$. If, in addition, we can decide if a sentence θ' without quantifiers belongs to T, and if T is deductively closed, then T is decidable.

A quantifier free formula $\psi(X_1, \ldots, X_n)$ is derived from Boolean polynomials $P(Z_1, \ldots, Z_m)$ by substituting an atomic formula $\tau_i(X_1, \ldots, X_n)$ for Z_i, $i = 1, \ldots, m$. To simplify ψ, put $P(Z_1, \ldots, Z_m)$ into disjunctive-conjunctive form

$$P'(Z_1, \ldots, Z_m) = \bigvee_{i \in I} \left(\bigwedge_{j \in J} Z_{ij} \wedge \bigwedge_{j \in J'} \neg Z_{ij} \right)$$

where each Z_{ij} is in $\{Z_1, \ldots, Z_m\}$. Thus, $P(Z_1, \ldots, Z_m)$ and $P'(Z_1, \ldots, Z_m)$ have the same truth value for each function $f: \{Z_1, \ldots, Z_m\} \to \{\text{true, false}\}$. Apply **de-Morgans's laws** (e.g. $X \wedge (Y \vee Z) \equiv (X \wedge Y) \vee (X \wedge Z)$ or $\neg(X \vee Y) \equiv (\neg X) \wedge (\neg Y)$) to go from P to P'. It is clear that ψ is logically equivalent to the formula $P'(\tau_1, \ldots, \tau_m)$.

For additional simplification we use the logical equivalence of $(\exists X)[\psi \vee \psi']$ and $(\exists X)[\psi] \vee (\exists X)[\psi']$. Thus, we are reduced to considering a formula of the form $(\exists Y)[\bigwedge_{j \in J} \tau_j(Y, X_1, \ldots, X_n)]$, with $\tau_j(Y, \mathbf{X})$ either an atomic formula or the negation of an atomic formula, $j \in J$. We only need to find a formula $\psi'(X_1, \ldots, X_n)$ equivalent to this modulo T.

9.2 A Quantifiers Elimination Procedure

The key to this elementary procedure is a division algorithm for polynomials.

We consider an integral domain R and use the language $\mathcal{L}(R, \text{ring})$. The models for the axioms $\Pi(R)$ (Example 7.3.1), are exactly the fields F that contain a homomorphic image of R. Extend $\Pi(R)$ to a new set of axioms, $\tilde{\Pi}(R)$, by addition of the sentences

$$(\forall Z_0) \cdots (\forall Z_{n-1})(\exists X)[X^n + Z_{n-1}X^{n-1} + \cdots + Z_0 = 0], \quad n = 1, 2, \ldots.$$

A model F of $\Pi(R)$ is also a model of $\tilde{\Pi}(R)$ if and only if F is algebraically closed. Our ultimate goal in the following discussion is to introduce a quantifier elimination procedure for $\tilde{\Pi}(R)$.

Each term of $\mathcal{L}(\text{ring}, R)$ is equivalent modulo $\Pi(R)$ to a polynomial $f(X_1, \ldots, X_n)$ with coefficients in R. Hence, every atomic formula is equivalent modulo $\Pi(R)$ to an equation $f(X_1, \ldots, X_n) = 0$. A conjunction

$$\bigwedge_{i=1}^{m} g_i(\mathbf{X}) \neq 0$$

of inequalities is equivalent to the inequality $g_1(\mathbf{X}) \cdots g_m(\mathbf{X}) \neq 0$. Thus, by Section 9.1, for a quantifier elimination procedure for $\tilde{\Pi}(R)$ we only need to eliminate Y from formulas of the form

(1) $\quad (\exists Y)[f_1(\mathbf{X}, Y) = 0 \wedge \cdots \wedge f_m(\mathbf{X}, Y) = 0 \wedge g(\mathbf{X}, Y) \neq 0]$

with $f_1, \ldots, f_m, g \in R[X_1, \ldots, X_n, Y]$.

THEOREM 9.2.1: Let R be an integral domain and $\varphi(X_1, \ldots, X_n)$ a formula of $\mathcal{L}(\text{ring}, R)$. Then $\varphi(X_1, \ldots, X_n)$ is equivalent modulo $\tilde{\Pi}(R)$ to a quantifier free formula $\psi(X_1, \ldots, X_r)$. If φ is a sentence, then there exists a nonzero $c \in R$ such that either φ is true in every model of $\tilde{\Pi}(R) \cup \{c \neq 0\}$ or φ is false in every model of $\tilde{\Pi}(R) \cup \{c \neq 0\}$.

Proof: We may assume φ is given by (1). The proof consists of three parts:

PART A: *Reduction to the case that only one of $f_1, \ldots, f_m$ involves Y.* Move each of the conjuncts that appears in (1) and that does not involve Y to the left of $(\exists Y)$ according to the rule "$(\exists Y)[\varphi \wedge \psi] \equiv \varphi \wedge (\exists Y)[\psi]$ if Y does not appear in φ." Thus, we assume $\deg_Y(f_i(\mathbf{X}, Y)) \geq 1$, $i = 1, \ldots, m$ and $m \geq 2$.

We now perform an induction on $\sum \deg_Y(f_i(\mathbf{X}, Y))$:

Let $p(\mathbf{X}, Y)$ and $q(\mathbf{X}, Y)$ be polynomials with coefficients in R such that $0 \leq \deg_Y(p(\mathbf{X}, Y)) \leq \deg_Y(q(\mathbf{X}, Y)) = d$. Write $p(\mathbf{X}, Y)$ in the form

(2) $$p(\mathbf{X}, Y) = a_k(\mathbf{X})Y^k + a_{k-1}(\mathbf{X})Y^{k-1} + \cdots + a_0(\mathbf{X})$$

with $a_j \in R[\mathbf{X}]$. For each j with $0 \leq j \leq k$ let

$$p_j(\mathbf{X}, Y) = a_j(\mathbf{X})Y^j + a_{j-1}(\mathbf{X})Y^{j-1} + \cdots + a_0(\mathbf{X}).$$

If $a_j(\mathbf{X})$ is not identically zero, division of $q(\mathbf{X}, Y)$ by $p_j(\mathbf{X}, Y)$ produces $q_j(\mathbf{X}, Y)$ and $r_j(\mathbf{X}, Y)$ in $R[\mathbf{X}, Y]$ for which

(3) $$a_j(\mathbf{X})^d q(\mathbf{X}, Y) = q_j(\mathbf{X}, Y) p_j(\mathbf{X}, Y) + r_j(\mathbf{X}, Y),$$

and $\deg_Y(r_j) < \deg_Y(p_j) \leq d$.

Let F be a model of $\Pi(R)$. If $x_1, \ldots, x_n, y$ are elements of F such that $a_k(\mathbf{x}) = \cdots = a_{j+1}(\mathbf{x}) = 0$ and $a_j(\mathbf{x}) \neq 0$, then $[p(\mathbf{x}, y) = 0 \wedge q(\mathbf{x}, y) = 0]$ is equivalent in F to $[p_j(\mathbf{x}, y) = 0 \wedge r_j(\mathbf{x}, y) = 0]$. Therefore, the formula $[p(\mathbf{X}, Y) = 0 \wedge q(\mathbf{X}, Y) = 0]$ is equivalent modulo $\Pi(R)$ to the formula

(4) $$\bigvee_{j=0}^{k} [a_k(\mathbf{X}) = 0 \wedge \cdots \wedge a_{j+1}(\mathbf{X}) = 0$$

$$\wedge\ a_j(\mathbf{X}) \neq 0 \wedge p_j(\mathbf{X}, Y) = 0 \wedge r_j(\mathbf{X}, Y) = 0]$$
$$\vee\ [a_k(\mathbf{X}) = 0 \wedge \cdots \wedge a_0(\mathbf{X}) = 0 \wedge q(\mathbf{X}, Y) = 0].$$

Also, for each j between 0 and k, $\deg_Y(p_j(\mathbf{X}, Y)) \leq \deg_Y(p(\mathbf{X}, Y))$ and $\deg_Y(r_j(\mathbf{X}, Y)) < \deg_Y(p(\mathbf{X}, Y))$, and the last disjunct has only one polynomial, $q(\mathbf{X}, Y)$, that involves Y.

Apply the outcome of (4) to $f_1(\mathbf{X}, Y)$ and $f_m(\mathbf{X}, Y)$ (of (1)). With the rule "$(\exists Y)[\varphi \vee \psi] \equiv (\exists Y)[\varphi] \vee (\exists Y)[\psi]$" we have replaced (1) by disjunctions of statements of form (1) in each of which the sum corresponding to $\sum \deg_Y(f_i(\mathbf{X}, Y))$ is smaller. Using the induction assumption, we conclude that m may be taken to be at most 1.

PART B: *Reduction to the case that $m = 0$.* Continue the notation of Part A which left us at the point of considering how to eliminate Y from $p(\mathbf{X}, Y)$ in

(5) $$(\exists Y)[p(\mathbf{X}, Y) = 0 \wedge g(\mathbf{X}, Y) \neq 0].$$

Consider a model $\tilde{F}$ of $\tilde{\Pi}(R)$ and elements $x_1, \ldots, x_n$ in $\tilde{F}$. If $p(\mathbf{x}, Y)$ is not identically zero, then (since $\tilde{F}$ is algebraically closed) the statement

$$\tilde{F} \models (\exists Y)[p(\mathbf{x}, Y) = 0 \wedge g(\mathbf{x}, Y) \neq 0]$$

9.2 A Quantifiers Elimination Procedure

is equivalent to the statement

$$p(\mathbf{x}, Y) \text{ does not divide } g(\mathbf{x}, Y)^k \text{ in } \tilde{F}[Y].$$

Therefore, with $q(\mathbf{X}, Y) = g(\mathbf{X}, Y)^k$ and in the notation of (2) and (3), formula (5) is equivalent modulo $\tilde{\Pi}(R)$ to the formula

$$\bigvee_{j=0}^{k} [a_k(\mathbf{X}) = 0 \wedge \cdots \wedge a_{j+1}(\mathbf{X}) = 0 \wedge a_j(\mathbf{X}) \neq 0 \wedge (\exists Y)[r_j(\mathbf{X}, Y) \neq 0]]$$

$$\vee [a_k(\mathbf{X}) = 0 \wedge \cdots \wedge a_0(\mathbf{X}) = 0 \wedge (\exists Y)[g(\mathbf{X}, Y) \neq 0]],$$

a disjunction of statements of form (1) with $m = 0$.

PART C: *Completion of the proof.* By Part B we are at the point of removing Y from a statement of the form

$$(\exists Y)[a_l(\mathbf{X})Y^l + a_{l-1}(\mathbf{X})Y^{l-1} + \cdots + a_0(\mathbf{X}) \neq 0].$$

Since models of $\tilde{\Pi}(R)$ are infinite fields, this formula is equivalent modulo $\tilde{\Pi}(R)$ to

$$a_l(\mathbf{X}) \neq 0 \vee a_{l-1}(\mathbf{X}) \neq 0 \vee \cdots \vee a_0(\mathbf{X}) \neq 0.$$

We have completed the procedure to transform a given formula $\varphi(X_1, \ldots, X_n)$ of $\mathcal{L}(\text{ring}, R)$ modulo $\tilde{\Pi}(R)$ to an equivalent quantifier free formula $\psi(X_1, \ldots, X_n)$. In particular, if φ is a sentence, then ψ can be taken to be of the form

(6) $$\bigvee_{i \in I} \bigwedge_{j \in J} [a_{ij} = 0 \wedge c_i \neq 0].$$

with $a_{ij}, c_i \in R$. Let c be the product of all nonzero c_i's and all the nonzero a_{ij}'s (or $c = 1$ if there are none). If $\tilde{F}$ is a model of $\tilde{\Pi}(R) \cup \{c \neq 0\}$, then (6) is true in $\tilde{F}$ if and only if it is true in R. $\square$

Here is a useful reformulation of the second part of Theorem 9.2.1:

COROLLARY 9.2.2: *Let R be an integral domain with $K = \text{Quot}(R)$ and let θ be a sentence of $\mathcal{L}(\text{ring}, R)$. Then there exists a nonzero c in R with the following properties:*
(a) *If $\tilde{K} \models \theta$, then $\tilde{F} \models \theta$ for each algebraically closed field $\tilde{F}$ which contains a homomorphic image $\bar{R}$ of R in which the image $\bar{c}$ of c is nonzero.*
(b) *If there exists an algebraically closed field $\tilde{F}$ which contains a homomorphic image $\bar{R}$ of R such that $\bar{c} \neq 0$ and $\tilde{F} \models \theta$, then $\tilde{K} \models \theta$.*

9.3 Effectiveness

Section 19.1 gives a precise definition for "presented ring." Intuitively these are rings in which we explicitly recognize the elements and in which we can explicitly carry out addition and multiplication, say in the manner of the elementary arithmetic of $\mathbb{Z}$ and $\mathbb{Q}$. In such a ring the proof of Theorem 9.2.1 allows an explicit computation of formula ψ (and of c if φ is a sentence) from φ. Furthermore, it produces an explicit bound for the number of basic operations required to produce ψ and c. Thus, the proof gives a primitive recursive elimination of quantifiers.

In particular, consider the case $R = \mathbb{Z}$. Models of $\tilde{\Pi}(\mathbb{Z})$ are classified according to their characteristic. If p is a prime, define $\tilde{\Pi}_p(\mathbb{Z})$ to be $\tilde{\Pi}(\mathbb{Z}) \cup \{p = 0\}$. For characteristic zero we define $\tilde{\Pi}_0(\mathbb{Z}) = \tilde{\Pi}(\mathbb{Z}) \cup \{p \neq 0 |\ p \text{ is a prime}\}$. The element c in Theorem 9.2.1 has finitely many prime divisors. For each prime divisor p of c one can decide statement (6) of Section 9.2. If it is true, then $\tilde{\Pi}_p(\mathbb{Z}) \models \psi$. Otherwise, $\tilde{\Pi}_p(\mathbb{Z}) \models \neg\psi$. Running over all prime divisors of c, this gives a procedure for deciding if ψ is true in all algebraically closed fields.

THEOREM 9.3.1:
(a) If R is a presented ring, we can explicitly compute $\psi(X_1, \ldots, X_n)$ of Theorem 9.2.1 from a given formula $\varphi(X_1, \ldots, X_n)$. If φ is a sentence, we can explicitly compute c from Theorem 9.2.1 and Corollary 9.2.2.
(b) The theory of algebraically closed fields of a given characteristic is primitive recursive.
(c) The theory of algebraically closed fields is primitive recursive.

Note that $\tilde{\mathbb{Q}}$ (resp. $\tilde{\mathbb{F}}_p$) is an elementary subfield of every algebraically closed field of characteristic 0 (resp. p). Therefore, from Lemma 9.1.2 we conclude the following:

COROLLARY 9.3.2: *The theory of algebraically closed fields is model complete. Moreover, the theory of algebraically closed fields of a given characteristic is complete.*

For $R = \mathbb{Z}$ Corollary 9.2.2 takes the following form:

COROLLARY 9.3.3: *Let θ be a sentence of $\mathcal{L}(\text{ring}, \mathbb{Z})$. Then we can explicitly compute $c \in \mathbb{N}$ with the following properties:*
(a) *If $\tilde{\mathbb{Q}} \models \theta$, then $\tilde{F} \models \theta$ for each algebraically closed field $\tilde{F}$ of characteristic p with $p \nmid c$.*
(b) *If there exists a prime p with $p \nmid c$ and $\tilde{\mathbb{F}}_p \models \theta$, then $\tilde{\mathbb{Q}} \models \theta$.*

COROLLARY 9.3.4: *Let θ be a sentence of $\mathcal{L}(\text{ring}, \mathbb{Z})$. Then θ is true in each algebraically closed field of characteristic 0 if and only if for infinitely many primes p, θ is true in each algebraically closed field of characteristic p.*

9.4 Applications

Model completeness of the theory of algebraically closed fields provides alternative proofs to some basic results of algebraic geometry. One of them is the weak version of Hilbert's Nullstellensatz:

PROPOSITION 9.4.1: *Let K be a field, $f_1(\mathbf{X}), \ldots, f_m(\mathbf{X}) \in K[X_1, \ldots, X_n]$ and $\mathfrak{a}$ be the ideal generated by the f_i's. Suppose $\mathfrak{a} \neq K[\mathbf{X}]$. Then there exist $x_1, \ldots, x_n$ in $\tilde{K}$ such that $f_i(\mathbf{x}) = 0$, $i = 1, \ldots, m$.*

Proof: By Zorn's Lemma, there exists a maximal ideal $\mathfrak{m}$ of $K[\mathbf{X}]$ that contains $\mathfrak{a}$. The quotient ring $F = K[\mathbf{X}]/\mathfrak{m}$ is a field containing K and the n-tuple $(X_1 + \mathfrak{m}, \ldots, X_m + \mathfrak{m})$ annihilates every polynomial in $\mathfrak{m}$. Thus, the sentence

$$(1) \qquad (\exists X_1) \ldots (\exists X_m)[\bigwedge_{i=1}^{m} f_i(X_1, \ldots, X_m) = 0]$$

of $\mathcal{L}(\text{ring}, K)$ is true in $\tilde{F}$. By Corollary 9.3.2, $\tilde{K} \prec \tilde{F}$. Hence, (1) is also true in $\tilde{K}$. $\square$

We deduce a stronger version of the Nullstellensatz from the weaker version:

PROPOSITION 9.4.2: *Let $f_1, \ldots, f_m, g$ be polynomials in $K[X_1, \ldots, X_n]$. If g vanishes at each common zero in $\tilde{K}^n$ of $f_1, \ldots, f_m$, then some power of g belongs to the ideal $\mathfrak{a}$ of $K[\mathbf{X}]$ generated by $f_1, \ldots, f_m$.*

Proof: Let Y be an additional variable. Assume without loss that $g \neq 0$. Consider the polynomial $1 - Yg(\mathbf{X})$. Then the $m + 1$ polynomials

$$f_1(\mathbf{X}), \ldots, f_m(\mathbf{X}), \ 1 - Yg(\mathbf{X})$$

of $K[\mathbf{X}, Y]$ have no common zero in $\tilde{K}^{n+1}$. By Proposition 9.4.1, they therefore generate the whole ring. In particular, there are polynomials $a_1, \ldots, a_m, b$ in $K[\mathbf{X}, Y]$ such that

$$(2) \qquad 1 = \sum_{i=1}^{m} a_i(\mathbf{X}, Y) f_i(\mathbf{X}) + b(\mathbf{X}, Y)(1 - Yg(\mathbf{X}))$$

Substitute $Y = g(\mathbf{X})^{-1}$ in (2) to get $1 = \sum_{i=1}^{m} a_i(\mathbf{X}, g(\mathbf{X})^{-1}) f_i(\mathbf{X})$. With $r \geq \max_{1 \leq i \leq m} \deg_Y a_i(\mathbf{X}, Y)$, we have

$$g(\mathbf{X})^r = \sum_{i=1}^{m} g(\mathbf{X})^r a_i(\mathbf{X}, g(\mathbf{X})^{-1}) f_i(\mathbf{X})$$

is in $\mathfrak{a}$. $\square$

A version of the Bertini-Noether theorem follows from a special case of the quantifier elimination procedure:

PROPOSITION 9.4.3: *Let $f(X_1,\ldots,X_n)$ be a polynomial with coefficients in an integral domain R. Then there exists a nonzero element c of R such that if $f(\mathbf{X})$ is absolutely irreducible, then $f(\mathbf{X})$ remains absolutely irreducible under each homomorphism φ of R into a field satisfying $\varphi(c) \neq 0$.*

If, in addition, R is a presented ring, then c can be explicitly computed from the coefficients of $f(\mathbf{X})$.

Proof: Let $\tilde{K}$ be the algebraic closure of $\operatorname{Quot}(R)$. The discussion following Example 7.3.1 gives a sentence θ of $\mathcal{L}(\text{ring}, R)$ such that $\tilde{K} \models \theta$ if and only if $f(\mathbf{X})$ is irreducible in $\tilde{K}[\mathbf{X}]$, i.e. $f(\mathbf{X})$ is absolutely irreducible. The proposition is thus a special case of Theorems 9.2.1 and 9.3.1. $\square$

Exercises

1. Find the remainder of division of $f_2(X) = X^{100}$ by $f_1(X) = X^2 - 3X + 2$.

2. For each of the following pairs of polynomials, g and h, check if g divides h.

    (a) $g(X_1, X_2) = p(X_1) - p(X_2)$, $h(X_1, X_2) = q(p(X_1)) - q(p(X_2))$ with $p, q \in \mathbb{Q}[X]$.

    (b) $g(X_1, X_2) = X_1^2 + X_2^3$, $h(X_1, X_2) = \sum_{i=0}^{n} a_i X_1^i X_2^{n-i}$ with $a_i \in \mathbb{Q}$, $i = 0, \ldots, n$.

    (c) $g(X_1, X_2) = X_1^2 + 2X_1 X_2 + 2X_2^2 + 1$, $h(X_1, X_2) = X_1^4 + 2X_1^2 + 4X_2^4 + 4X_2^2 + 1$.

3. Let h and g be as in Exercise 2(c). Use the proof of Theorem 9.2.1 to eliminate X_2 from the formula $(\exists X_2)[h = 0 \wedge g \neq 0]$.

4. Strengthen Hilbert's Nullstellensatz (Proposition 9.4.1) as follows: Let m, n, d be positive integers. Show that there exists a positive integer $\nu = \nu(m, n, d)$ with the following property: Let K be a field and $g, f_1, \ldots, f_m \in K[X_1, \ldots, X_n]$ polynomials of degree at most d such that g vanishes at every common zero, $\mathbf{x} \in \check{K}^n$, of $f_1, \ldots, f_m$. Then there exists $\nu \geq r$ and polynomials $c_1, \ldots, c_m \in K[\mathbf{X}]$ of degree at most ν satisfying $g(\mathbf{X})^r = \sum_{i=1}^{m} c_i(\mathbf{X}) f_i(\mathbf{X})$.

    Hint: Observe that it suffices to find the c_i's in $\tilde{K}[\mathbf{X}]$ (Use a linear basis for $\tilde{K}/K$ that contains 1.) If the theorem were not true for some m, n, d, then for each ν there would be an algebraically closed field K_ν for which the above statement were false. Use ultraproducts of these fields to deduce a contradiction to Proposition 9.3.3 (A. Robinson). Alternatively, use the compactness theorem 7.7.6.

Notes

The elimination of quantifiers procedure of this chapter (i.e. without using the full development of resultants as they appear in Chapter 19) seems to be folklore.

It is interesting to note that the converse of this result is also true: If K is an infinite field whose theory in $\mathcal{L}(\text{ring})$ admits an elimination of quantifiers, then K is algebraically closed [Macintyre-McKenna-v.d.Dries].

It is well known [Sacks, p. 54], that if T is a theory in a language $\mathcal{L}$, then $\mathcal{L}$ can be extended to a language $\mathcal{L}'$ in which T admits an elimination of quantifiers. Indeed, the proof of Proposition 7.4.2 (Skolem-Löwenheim) does this. The resulting elimination of quantifiers, however, gives no decision procedure since quantifier free sentences of $\mathcal{L}'$ contain new function symbols with no natural algebraic interpretation. They therefore cannot be computed.

Tarski extends $\mathcal{L}(\text{ring})$ by the relation symbol $<$ for the ordering, and establishes an algebraic quantifier elimination procedure for the theory of real closed fields which yields an algebraic decision procedure for this theory [Tarski].

Macintyre extends $\mathcal{L}(\text{ring})$ by unary relations V and P_n, one for each positive integer n. This generalizes the model of p-adically closed fields, where V extends the valuation ring property and P_n generalizes nth powers [Macintyre]. He shows that this theory of p-adically closed fields admits an elimination of quantifiers and a model theoretic decision procedure.

The standard algebraic proof of the Hilbert Nullstellensatz [Lang4, p. 32], uses Chevalley's extension of places theorem (Proposition 2.3.1).

The Bertini-Noether Theorem 9.4.3 may be deduced algebraically from the Hilbert Nullstellensatz 9.4.2 [Lang3, p. 157].

The work [Kollár1] gives an optimal effective version of Exercise 4, independent of the field K.

Chapter 10. Elements of Algebraic Geometry

Here we present the algebraic geometry background for the study of PAC fields. The central result is a descent argument which associates to each variety V defined over a finite extension L of a field K a variety W defined over K. Throughout this chapter and subsequent chapters we make the following convention: Whenever we are given a collection of field extensions of a given field K we assume that all of them are contained in a common field.

10.1 Algebraic Sets

This section and Section 10.2 give a basic exposition on affine algebraic geometry.

We start with a field K, the **basic field**, and an algebraically closed extension Ω of K, the **universal domain**, which has infinite transcendence degree over K. **Affine n-space** $\mathbb{A}^n$ is the set of all **points** $\mathbf{x} = (x_1, \ldots, x_n)$, with coordinates x_i in Ω. To each subset $\mathfrak{a}_0$ of $K[\mathbf{X}] = K[X_1, \ldots, X_n]$ we attach the **algebraic set**

$$V(\mathfrak{a}_0) = \{\mathbf{x} \in \mathbb{A}^n \mid f(\mathbf{x}) = 0 \text{ for every } f \in \mathfrak{a}_0\}.$$

If $\mathfrak{a}$ is the ideal generated by $\mathfrak{a}_0$, then $V(\mathfrak{a}) = V(\mathfrak{a}_0)$.

PROPOSITION 10.1.1 (Hilbert's Basis Theorem [Lang 8, p. 186]): *The polynomial ring $K[\mathbf{X}]$ in n-variables over a field K is **Noetherian** — the following equivalent conditions hold:*
(a) *Every ideal in $K[\mathbf{X}]$ is finitely generated.*
(b) *Every ascending sequence of ideals of $K[\mathbf{X}]$ is eventually stationary.*

If $f_1, \ldots, f_m$ are generators of the ideal $\mathfrak{a}$, then $V(\mathfrak{a}) = V(f_1, \ldots, f_m)$. The empty set is $V(1)$ and $\mathbb{A}^n$ is $V(0)$.

If $\mathfrak{a}$ and $\mathfrak{b}$ are subsets of $K[\mathbf{X}]$, then $V(\mathfrak{a}) \cup V(\mathfrak{b}) = V(\mathfrak{ab})$, where $\mathfrak{ab} = \{\sum_{i=1}^n f_i g_i \mid f_i \in \mathfrak{a} \text{ and } g_i \in \mathfrak{b}\}$. For a given family, $\{\mathfrak{a}_i \mid i \in I\}$ of subsets of $K[\mathbf{X}]$, we have $\bigcap_{i \in I} V(\mathfrak{a}_i) = V(\bigcup_{i \in I} \mathfrak{a}_i)$. Thus, both the union of two algebraic sets and the intersection of an arbitrary family of algebraic sets are again algebraic sets. Therefore, the family of all complements of algebraic sets is the collection of open sets for a topology, the **Zariski K-topology**, on $\mathbb{A}^n$. Algebraic sets of $\mathbb{A}^n$ are the **Zariski K-closed subsets** (which we abbreviate in this chapter to K-closed subsets).

The V correspondence has an inverse: Attach to each subset A of $\mathbb{A}^n$ the following ideal of $K[\mathbf{X}]$:

$$I(A) = I_K(A) = \{f \in K[\mathbf{X}] \mid f(\mathbf{x}) = 0 \text{ for every } \mathbf{x} \in A\}.$$

10.1 Algebraic Sets

For an ideal $\mathfrak{a}$ of $K[\mathbf{X}]$, $\sqrt{\mathfrak{a}} = \{f \in K[\mathbf{X}] \mid \exists r \in \mathbb{N} \text{ with } f^r \in \mathfrak{a}\}$ is the **radical** of $\mathfrak{a}$. The I and V functions have the following properties:
(1a) $\mathfrak{a}_1 \subseteq \mathfrak{a}_2 \subseteq K[\mathbf{X}] \Longrightarrow V(\mathfrak{a}_2) \subseteq V(\mathfrak{a}_1)$.
(1b) $A_1 \subseteq A_2 \subseteq \mathbb{A}^n \Longrightarrow I(A_2) \subseteq I(A_1)$.
(1c) $A_1, A_2 \subseteq \mathbb{A}^n \Longrightarrow I(A_1 \cup A_2) = I(A_1) \cap I(A_2)$.
(1d) If $\mathfrak{a}$ is an ideal of $K[\mathbf{X}]$, then, since $\mathfrak{a}$ is finitely generated (Proposition 10.1.1), the Hilbert's Nullstellensatz (Proposition 9.4.2) implies that $I(V(\mathfrak{a})) = \sqrt{\mathfrak{a}}$.
(1e) If $\mathfrak{p}$ is a prime ideal of $K[\mathbf{X}]$, then $I(V(\mathfrak{p})) = \mathfrak{p}$.
(1f) $A \subseteq \mathbb{A}^n \Longrightarrow V(I(A)) = \bar{A}$, the closure of A.

Call a K-closed set V **reducible** if it is a union $V = V_1 \cup V_2$ of two proper K-closed subsets. Otherwise, it is **irreducible** and V is called a K-**variety**. From the above, V is irreducible if and only if $\mathfrak{p} = I(V)$ is a prime ideal of $K[\mathbf{X}]$ (i.e. $K[\mathbf{X}]/\mathfrak{p}$ is an integral domain). In this case there is a K-embedding of $K[\mathbf{X}]/\mathfrak{p}$ in Ω. Let x_i be the image of $X_i + \mathfrak{p}$ in this embedding, $i = 1, \ldots, n$. Then $\mathbf{x} = (x_1, \ldots, x_n)$ is a **generic point** of V over K, $K[\mathbf{x}]$ is the **coordinate ring** of V, $K(\mathbf{x})$ is the **function field** of V, and there is a canonical short exact sequence

$$0 \to \mathfrak{p} \to K[\mathbf{X}] \to K[\mathbf{x}] \to 0 \,.$$

For $\mathbf{x}'$ a point of V, the map $\mathbf{x} \mapsto \mathbf{x}'$ extends to a K-homomorphism of rings $K[\mathbf{x}] \to K[\mathbf{x}']$. We say that $\mathbf{x}'$ is a K-**specialization** of $\mathbf{x}$.

Conversely, if a K-closed set V has a point $\mathbf{x}$ such that all points of V are K-specialization of $\mathbf{x}$, then $I(V) = \{f \in K[\mathbf{X}] \mid f(\mathbf{x}) = 0\}$. Hence, $I(V)$ is a prime ideal of $K[\mathbf{X}]$, V is a K-variety and $\mathbf{x}$ is a generic point of V over K.

Although $\mathbf{x}$ is not uniquely determined, if $\mathbf{x}'$ is another a generic point of V, then the above K-homomorphism $K[\mathbf{x}] \to K[\mathbf{x}']$ is an isomorphism. In particular, the transcendence degree of $K(\mathbf{x})$ over K (denoted $\dim_K(\mathbf{x})$) is an invariant of V called the **dimension** of V and denoted $\dim(V)$. Clearly, $\dim(\mathbb{A}^n) = n$, and if $V_1 \subseteq V_2$, then $\dim(V_1) \leq \dim(V_2)$. Conversely,
(2) If $V_1 \subseteq V_2$ are K-varieties of the same dimension, then $V_1 = V_2$ [Lang 4, p. 29].

The **local ring** of V at a point $\mathbf{a}$ is

$$O_{V,\mathbf{a}} = \Big\{ \frac{f(\mathbf{x})}{g(\mathbf{x})} \mid f, g \in K[\mathbf{X}] \text{ and } g(\mathbf{a}) \neq 0 \Big\}.$$

It is determined by $\mathbf{x}$ up to a K-isomorphism. Suppose an element y of $\widetilde{K(\mathbf{x})}$ is integral over $O_{V,\mathbf{a}}$. Then every K-place φ of $K(\mathbf{x},y)$ with $\varphi(\mathbf{x}) = \mathbf{a}$ is finite at y (Proposition 2.4.1). Moreover, $\varphi(y)$ is algebraic over $K(\mathbf{a})$.

A K-variety of dimension 0 is the set of all K-conjugates of a point $(a_1, \ldots, a_n) \in \tilde{K}^n$. A K-variety of dimension 1 is called a K-**curve**. Note that a K-variety need not be absolutely irreducible (e.g. $V(X^2 + Y^2)$ over $\mathbb{Q}$). Therefore, a K-curve may not be a curve defined over K as in Section 5.1. A K-variety of dimension $n - 1$ in $\mathbb{A}^n$ is called a K-**hypersurface**. Each hypersurface is of the form $V(f)$, where $f \in K[\mathbf{X}]$ is a nonconstant irreducible polynomial (Exercise 9). If $f(\mathbf{X}) = a_0 + a_1 X_1 + \cdots + a_n X_n$ is a linear polynomial, then $V(f)$ is called a **hyperplane**.

The study of arbitrary K-closed sets can be reduced to the study of K-varieties. Indeed, by Hilbert's basis theorem (Proposition 10.1.1) every strictly descending sequence of algebraic sets in $\mathbb{A}^n$ is finite. Hence, every algebraic set A is a finite union $A = \bigcup V_i$, of K-varieties. Omit V_i if it is contained in some V_j, for $j \neq i$, to obtain a representation of A as a union that is unique, up to the order of the V_i. These K-varieties are called the K-**components** of A. The **dimension** of A is defined as the maximal dimension of its components, or equivalently, the maximal transcendence degree over K of the points of A.

LEMMA 10.1.2 (Dimension Theorem): *If V and W are K-varieties neither of which contains the other, then $\dim(V \cap W) < \min\{\dim(V), \dim(W)\}$.*

Proof: If, say, $\dim(V \cap W) = \dim(V)$, then there exists a point $\mathbf{x} \in V \cap W$ whose transcendence degree over K is $\dim(V)$. Hence, $\mathbf{x}$ is a generic point for V. Therefore, $V \subseteq V \cap W$, contradiction the assumption on V and W. $\square$

There is also a lower bound on the dimension of intersections [Lang4, p. 36]:

LEMMA 10.1.3 (Dimension Theorem): *If V is a K-variety and H is an irreducible hypersurface that neither contains V nor is disjoint from V, then the dimension of each K-component of $H \cap V$ is $\dim(V) - 1$.*

COROLLARY 10.1.4: *If $f_1, \ldots, f_m$ are forms in $K[X_0, \ldots, X_n]$ and $m \leq n$, then $f_1, \ldots, f_m$ have a nontrivial common zero in $\tilde{K}$.*

COROLLARY 10.1.5: *Let $n \geq 2$ and $f \in K[X_0, \ldots, X_n]$ nonzero form. Suppose there exists no $\mathbf{a} \in \tilde{K}^{n+1}$ with $\mathbf{a} \neq 0$, $f(\mathbf{a}) = 0$, and $\frac{\partial f}{\partial X_i}(\mathbf{a}) = 0$, $i = 0, \ldots, n$. Then f is absolutely irreducible.*

Proof: Assume $f = gh$ with $g, h \in \tilde{K}[\mathbf{X}]$ of degree at least 1. Write $g = \sum_{i=k}^{l} g_i$ and $h = \sum_{j=r}^{s} h_j$, where each g_i is a form in $\tilde{K}[\mathbf{X}]$ of degree i, each h_j is a form in $\tilde{K}[\mathbf{X}]$ of degree j, and $g_k, g_l, h_r, h_s \neq 0$. Let $d = \deg(f)$. Then $g_k h_r \neq 0$. Hence, $k + r = d$. Similarly $l + s = d$. It follows from $k \leq l$ and $r \leq s$ that $k = l$ and $r = s$. Thus, both g and h are forms.

By Corollary 10.1.4, there exists $\mathbf{a} \in \tilde{K}^{n+1}$ with $g(\mathbf{a}) = h(\mathbf{a}) = 0$ and $\mathbf{a} \neq 0$. Hence, $f(\mathbf{a}) = 0$ and $\frac{\partial f}{\partial X_i}(\mathbf{a}) = \frac{\partial g}{\partial X_i}(\mathbf{a}) h(\mathbf{a}) + g(\mathbf{a}) \frac{\partial h}{\partial X_i}(\mathbf{a}) = 0$ for each i. This contradiction proves that f is irreducible over $\tilde{K}$. $\square$

Remark 10.1.6: In geometric terms, the assumption made on f in Corollary 10.1.5 means that the closed subset of $\mathbb{P}^n$ defined by f (Section 10.7) has no singular points. ☐

10.2 Varieties

Consider a K-variety $V = V(f_1, \ldots, f_m)$ in $\mathbb{A}^n$, with $f_i \in K[\mathbf{X}], i = 1, \ldots, m$. If L is a field extension of K, then $f_i \in L[\mathbf{X}]$. Therefore, V can be considered also as an L-algebraic set. It may happen, however, that V does not remain irreducible over L. For example, an irreducible polynomial $f \in K[X]$ in one variable and of degree > 1, decomposes into linear factors over $\tilde{K}$. Those K-varieties V that remain irreducible over every extension of K are called **absolutely irreducible**. Following [Weil5] and [Lang4] we call them **varieties**.

Suppose that a K-variety V is irreducible over an extension L of K. It is possible that the L-ideal $L \cdot I_K(V)$ is a proper subideal of $I_L(V)$. For example, if $\mathrm{char}(K) = p$ and a is an element of K whose pth root α does not belong to K, then the K-variety V of dimension 0 defined by $X^p - a = 0$ consists of a unique point, namely α. Therefore, it is absolutely irreducible. But $\tilde{K} \cdot I_K(V) = \tilde{K}[X](X^p - a)$ is a proper subideal of $I_{\tilde{K}}(V) = \tilde{K}[X](X - \alpha)$. To consider this we introduce an additional concept.

An absolutely irreducible K-variety V is said to be **defined** over K if $I_{\tilde{K}}(V)$ is generated by polynomials in $K[\mathbf{X}]$ (i.e. $I_{\tilde{K}}(V) = \tilde{K} \cdot I_K(V)$).

LEMMA 10.2.1 (Weil [Lang4, p. 66]): *For a point $\mathbf{x} \in \mathbb{A}^n$ and two fields $K \subseteq L$, let $\mathfrak{p}_K$ (resp. $\mathfrak{p}_L$) be the prime ideal of $K[\mathbf{X}]$ (resp. $L[\mathbf{X}]$) that consists of all polynomials which vanish at $\mathbf{x}$. Then $\mathfrak{p}_L = L \cdot \mathfrak{p}_K$ if and only if L is linearly disjoint from $K(\mathbf{x})$ over K.*

COROLLARY 10.2.2:
(a) *Let V be a K-variety with a generic point $\mathbf{x}$. Then V is a variety defined over K if and only if $K(\mathbf{x})/K$ is a regular extension.*
(b) *Let $f \in K[X_1, \ldots, X_n]$ be an irreducible polynomial. Choose $x_1, \ldots, x_n$ with $f(x_1, \ldots, x_n) = 0$ and $\mathrm{trans.deg}(K(\mathbf{x})/K) = n - 1$. Then f is absolutely irreducible if and only if $K(\mathbf{x})/K$ is regular.*
(c) *Let F be a regular extension of K and $x_1, \ldots, x_n$ elements of F. Then the set of all K-specializations of $\mathbf{x}$ is a variety in $\mathbb{A}^n$ defined over K.*

Proof of (a): Suppose $K(\mathbf{x})/K$ is regular and let L be an extension of K. To prove that V is L-irreducible, we may assume that $\tilde{K} \subseteq L$ and that $K(\mathbf{x})$ and L are algebraically independent over K (otherwise replace $\mathbf{x}$ by another generic point). By Lemma 2.6.7, L is linearly disjoint from $K(\mathbf{x})$ over K. Let $\mathfrak{p}_K$ and $\mathfrak{p}_L$ be as in Lemma 10.2.1. By Lemma 10.2.1, $I_L(V) \subseteq \mathfrak{p}_L = L \cdot \mathfrak{p}_K = L \cdot I_K(V) \subseteq I_L(V)$. Hence, $I_L(V) = \mathfrak{p}_L$ is a prime ideal. That is, V is L-irreducible. Moreover, $I_L(V) = L \cdot I_K(v)$. Thus, V is a variety defined over K.

For the converse apply Lemma 10.2.1.

Proof of (b): Apply (a) to the hypersurface $V(f)$.

Proof of (c): By Section 10.1, $\mathbf{x}$ is a generic point of a K-variety V in $\mathbb{A}^n$. By Corollary 2.6.5(b), $K(\mathbf{x})/K$ is a regular extension. Hence, by (a), V is a variety defined over K. □

LEMMA 10.2.3 (Weil [Lang 4, p. 74]): *Every absolutely irreducible K-variety V has a smallest field of definition L, which is a finite purely inseparable extension of K.*

Consider the $\tilde{K}$-components, $V_1, \ldots, V_m$ of a K-algebraic set A. The uniqueness of the decomposition $A = \bigcup_{i=1}^m V_i$ implies that the Galois group $\mathrm{Gal}(K)$ permutes the set $\{V_1, \ldots, V_m\}$. In particular, if A is a K-variety, then the action of $\mathrm{Gal}(K)$ on $\{V_1, \ldots, V_m\}$ is transitive. Thus, all V_i have the same dimension. Conversely, if A is a $\tilde{K}$-closed set and $A^\sigma = A$ for all $\sigma \in \mathrm{Gal}(K)$, then A is a K-closed set [Lang 4, p. 74].

LEMMA 10.2.4: *Let K be a separably closed field. Then every K-variety is absolutely irreducible.*

Proof: Let V be a K-variety. Then V has only one $\tilde{K}$-component because $\mathrm{Gal}(K)$ is trivial. Thus, V is a variety. □

Note that if $K \subseteq L$, then every K-closed (resp. open) subset of $\mathbb{A}^n$ is also L-closed (resp. open). We call a subset A of $\mathbb{A}^n$ **Zariski-closed** (resp. **open**) if A is L-closed (resp. open) for some subfield L of Ω. The collection of all Zariski-open subsets of $\mathbb{A}^n$ forms the **Zariski topology** of $\mathbb{A}^n$.

LEMMA 10.2.5: *The Zariski topology of K^n coincides with its Zariski K-topology.*

Proof: Let L be an extension of K in Ω and $g \in L[\mathbf{X}]$ a nonzero polynomial. Choose a basis $\{w_i \mid i \in I\}$ for L/K and write $g = \sum_{i \in I} w_i g_i$ where $g_i \in K[\mathbf{X}]$ and all but finitely many g_i are zero. Then $\{\mathbf{x} \in K^r \mid g(\mathbf{x}) = 0\} = \bigcap_{i \in I} \{\mathbf{x} \in K^r \mid g_i(\mathbf{x}) = 0\}$. Hence, each Zariski-closed subset of K^r is also Zariski K-closed. □

10.3 Substitutions in Irreducible Polynomials

The substitution of a variable in an irreducible polynomial by another polynomial need not give an irreducible polynomial. For example $X - T^2$ is irreducible, but $Y^2 - T^2$ is not. The following lemma gives sufficient conditions for this to happen.

LEMMA 10.3.1 (Geyer): *Let $f \in K[T, X_1, \ldots, X_n]$ be an irreducible polynomial with $\partial f/\partial T \neq 0$. Let $g \in K[Y_1, \ldots, Y_m]$ be a nonconstant polynomial such that $g(\mathbf{Y}) - c$ is absolutely irreducible for each $c \in \tilde{K}$. Then the polynomial $f(g(\mathbf{Y}), \mathbf{X})$ is irreducible in $K[\mathbf{X}, \mathbf{Y}]$.*

Proof: Let V be the K-closed subset of $\mathbb{A}^{1+n+m}$ defined by the equations $f(T, \mathbf{X}) = 0$ and $g(\mathbf{Y}) = T$. We choose $\mathbf{a} \in (\tilde{K})^n$ with $\frac{\partial f}{\partial T}(T, \mathbf{a}) \neq 0$. Then

10.3 Substitutions in Irreducible Polynomials

we choose $c \in \tilde{K}$ with $f(c, \mathbf{a}) = 0$. Finally, we choose $\mathbf{b}, \mathbf{b}' \in (\tilde{K})^m$ with $g(\mathbf{b}) = c$ and $g(\mathbf{b}') \neq c$. Thus, V is nonempty and $g(\mathbf{Y}) - T$ does not vanish on $V(f)$. Hence, by the dimension theorem (Lemma 10.1.3) the dimension of each K-component of V is $n + m - 1$.

Consider two points $(t, \mathbf{x}, \mathbf{y})$ and $(t', \mathbf{x}', \mathbf{y}')$ of V of dimension $n + m - 1$ over K. Since T occurs in $f(T, \mathbf{X})$, t is algebraic over $K(\mathbf{x})$. Since $g(\mathbf{y}) = t$, $y_1, \ldots, y_m$ are algebraically dependent over $K(t)$. If $\dim_K \mathbf{x}$ were less than n, then $\dim_K(t, \mathbf{x}, \mathbf{y}) \leq n - 1 + m - 1 < n + m - 1$, contradicting our assumption. Hence, $\dim_K \mathbf{x} = n$. Similarly, $\dim_K \mathbf{x}' = n$. Hence, since $f(T, \mathbf{X})$ is irreducible, the map $(t, \mathbf{x}) \to (t', \mathbf{x}')$ extends to a K-isomorphism $\theta_0 \colon K(t, \mathbf{x}) \to K(t', \mathbf{x}')$. By Corollary 9.3.2, $g(\mathbf{Y}) - t$ is irreducible over the algebraic closure of $K(\mathbf{x})$, and therefore also over $K(t, \mathbf{x})$. Since $\dim_{K(t,\mathbf{x})}(\mathbf{y}) = \dim_{K(t',\mathbf{x}')}(\mathbf{y}') = m - 1$, θ_0 extends to an isomorphism $\theta \colon K(t, \mathbf{x}, \mathbf{y}) \to K(t', \mathbf{x}', \mathbf{y}')$ with $\theta(y_j) = y'_j$, $j = 1, \ldots, m$. This means that V has only one K-component. That is, V is irreducible over K.

Now consider a generic point $(t, \mathbf{x}, \mathbf{y})$ of V over K. Then $(\mathbf{x}, \mathbf{y})$ is a generic point over K of the projection V' of V on the affine space $\mathbb{A}^{n+m}$ in the variables $(\mathbf{X}, \mathbf{Y})$. Moreover, $\dim(V') = \dim(V) = n + m - 1$. Hence, V' is a K-hypersurface in $\mathbb{A}^{n+m}$. That is, $V' = V(h)$, with $h \in K[\mathbf{X}, \mathbf{Y}]$ an irreducible polynomial (Section 10.1). By definition, h vanishes on the K-algebraic set defined by $f(g(\mathbf{Y}), \mathbf{X}) = 0$. Hence, by Hilbert's Nullstellensatz (Proposition 9.4.2), $h(\mathbf{X}, \mathbf{Y})^r = f(g(\mathbf{Y}), \mathbf{X}) g_1(\mathbf{X}, \mathbf{Y})$, for some $r \in \mathbb{N}$ and $g_1 \in K[\mathbf{X}, \mathbf{Y}]$. Since $h(\mathbf{X}, \mathbf{Y})$ is irreducible, there is an s, $1 \leq s \leq r$, with $f(g(\mathbf{Y}), \mathbf{X}) = h(\mathbf{X}, \mathbf{Y})^s$. If $s = 1$, we are done. So, assume $s > 1$. Then

$$(1) \quad \frac{\partial f}{\partial T}(t, \mathbf{x}) \frac{\partial g}{\partial Y_j}(\mathbf{y}) = s \cdot h(\mathbf{x}, \mathbf{y})^{s-1} \frac{\partial h}{\partial Y_j}(\mathbf{x}, \mathbf{y}) = 0, \qquad j = 1, \ldots, m.$$

However, since $\dim_K(\mathbf{x}) = n$ and $\frac{\partial f}{\partial T} \neq 0$ we have $\frac{\partial f}{\partial T}(t, \mathbf{x}) \neq 0$. In addition, since $\dim_K(\mathbf{y}) = m$ and $g(\mathbf{Y})$ is absolutely irreducible, there exists j, $1 \leq j \leq m$, such that $\frac{\partial g}{\partial Y_j}(\mathbf{y}) \neq 0$. This contradicts (1). □

COROLLARY 10.3.2: *Let $f \in K(T_1, \ldots, T_r)[X_1, \ldots, X_n]$ be an irreducible polynomial. Then $f(\mathbf{T}, \mathbf{X}) = h_0(\mathbf{T})^{-1} h_1(\mathbf{T}) f_1(\mathbf{T}, \mathbf{X})$ with $h_0, h_1 \in K[\mathbf{T}]$ nonzero polynomials and $f_1 \in K[\mathbf{T}, \mathbf{X}]$ irreducible. Suppose $\frac{\partial f}{\partial T_i} \neq 0$, $i = 1, \ldots, r$. Let $g_i \in K[Y_{i1}, \ldots, Y_{im}]$, $i = 1, \ldots, r$, be nonconstant polynomials such that $g_i(\mathbf{Y}_i) + c$ is absolutely irreducible for every $c \in \tilde{K}$. Then the polynomial $f(\mathbf{g}(\mathbf{Y}), \mathbf{X}) = f(g_1(\mathbf{Y}_1), \ldots, g_r(\mathbf{Y}_r), X_1, \ldots, X_n)$ is defined and irreducible in $K(\mathbf{Y})[\mathbf{X}]$.*

Proof: By the description of f, $h_0(\mathbf{g}(\mathbf{Y})) \neq 0$. Hence, $f(\mathbf{g}(\mathbf{Y}), \mathbf{X})$ is defined. To prove its irreducibility, successively substitute $T_r = g_r(\mathbf{Y}_r)$, $T_{r-1} = g_{r-1}(\mathbf{Y}_{r-1}), \ldots, T_1 = g_1(\mathbf{Y}_1)$ in $f_1(\mathbf{T}, \mathbf{X})$ as an induction application of Lemma 10.3.1. □

Remark 10.3.3: If $\operatorname{char}(K) = 0$ and if T_i appears in f, then the condition $\frac{\partial f}{\partial T_i} \neq 0$, $i = 1, \ldots, r$, is automatic. □

10.4 Rational Maps

Let V and W be K-varieties in $\mathbb{A}^n$ and $\mathbb{A}^m$ with generic points $\mathbf{x}$ and $\mathbf{y}$, respectively. Assume with $\mathbf{y} = (y_1, \ldots, y_m)$ that $y_j \in K(\mathbf{x})$, $j = 1, \ldots, m$. Then $I = \{g \in K[\mathbf{X}] \mid g(\mathbf{x})y_j \in K[\mathbf{x}], \ j = 1, \ldots, m\}$ is a nonzero ideal of $K[\mathbf{X}]$ and $V(I)$ is a proper K-closed subset of V. For each $\mathbf{a} \in V_0 = V \smallsetminus V(I)$ there exist $f_1, \ldots, f_m, g \in K[\mathbf{X}]$ such that $g(\mathbf{a}) \neq 0$ and $g(\mathbf{x})y_j = f_j(\mathbf{x})$, $j = 1, \ldots, m$. Hence $\mathbf{b} = \left(\frac{f_1(\mathbf{a})}{g(\mathbf{a})}, \ldots, \frac{f_m(\mathbf{a})}{g(\mathbf{a})}\right)$ is a well defined point of W which depends on $\mathbf{a}$ but not on $f_1, \ldots, f_m, g$. Let $\mathbf{b} = \varphi(\mathbf{a})$ and call the map $\varphi: V_0 \to W$ obtained in this way a K-**rational map** of V to W. The K-open subset V_0 of V is the **domain** of **definition** of φ.

In particular, if $\mathbf{x}$ is a generic point of V, then $\mathbf{x}$ belongs to V_0 and $\mathbf{y} = \varphi(\mathbf{x})$.

If the coordinates of a point $\mathbf{a}$ of V belong to K, we say that $\mathbf{a}$ is K-**rational**. Denote the set of all K-rational points of V by $V(K)$. From the definition, the restriction of φ to $V_0(K)$ maps this set into $W(K)$.

Suppose $\mathbf{y}$ is a generic point of W. Choose $g \in K[Y_1, \ldots, Y_m]$ such that $g(\mathbf{y}) \neq 0$ and x_i is integral over $K[\mathbf{y}, g(\mathbf{y})^{-1}]$, $i = 1, \ldots, n$. Then $W_0 = \{\mathbf{b} \in W \mid g(\mathbf{b}) \neq 0\}$ is a nonempty K-open subset of W and $W_0 \subseteq \varphi(V_0)$. We say in this case that φ is **dominant**. We also call $[K(\mathbf{x}) : K(\mathbf{y})]$ the **degree** of φ. If $\dim(V) = \dim(W)$, then $\deg(\varphi) < \infty$.

If $K(\mathbf{x})/K(\mathbf{y})$ is a separable extension, we say that the rational map φ is **separable**.

A K-rational map $\varphi: V \to W$ is said to be K-**birational** if there exists a K-rational map $\psi: W \to V$ and open subsets V_0 and W_0 of the respective domains of definitions of φ and ψ such that $\varphi \circ \psi$ and $\psi \circ \varphi$ are the identity maps on W_0 and V_0, respectively. Then φ (and ψ) is a K-**birational equivalence** and V and W are K-**birationally equivalent**.

If $\mathbf{x}$ is a generic point of V, then $\mathbf{x} \in V_0$, so $\mathbf{y}' = \varphi(\mathbf{x}) \in W_0$. It follows that $\psi(\mathbf{y}') = \mathbf{x}$. Thus, $K(\mathbf{y}') = K(\mathbf{x})$, hence $\dim(V) \leq \dim(W)$. By symmetry, $\dim(V) = \dim(W)$ and $\mathbf{y}'$ is a generic point of W. In particular, (Corollary 10.2.2) V is a variety defined over K if and only if W is one.

Conversely, if $\mathbf{y}$ is a generic point of W and $K(\mathbf{x}) = K(\mathbf{y})$, then V and W are K-birationally equivalent.

Use of birational equivalence reduces many questions about algebraic sets to questions about hypersurfaces.

LEMMA 10.4.1: *Every variety V defined over K is K-birationally equivalent to a hypersurface W.*

Proof: Let $\mathbf{x} = (x_1, \ldots, x_n)$ be a generic point of V over K. Then $K(\mathbf{x})/K$ is a regular extension. It therefore has a separating transcendence basis $\mathbf{t} =$

10.4 Rational Maps

$(t_1, \ldots, t_r)$. In particular, $K(\mathbf{x})/K(\mathbf{t})$ is a finite separable extension. By the primitive element theorem, there exists $y \in K(\mathbf{x})$ such that $K(\mathbf{x}) = K(\mathbf{t}, y)$. Let $f = \mathrm{irr}(y, K(\mathbf{t}))$. Multiply f by a suitable element of $K(\mathbf{t})$ and replace y if necessary to assume that the coefficients of f belong to $K[\mathbf{t}]$ and that their greatest common divisor is 1 (i.e. $f(\mathbf{t}, Y)$ is a **primitive polynomial**). The field $K(\mathbf{t}, y)$ is linearly disjoint from $\tilde{K}(\mathbf{t})$ over $K(\mathbf{t})$ (Corollary 10.2.2). Hence, $f(\mathbf{t}, Y)$ is irreducible over $\tilde{K}(\mathbf{t})$. By Gauss' Lemma, $f(\mathbf{T}, Y)$ is an absolutely irreducible polynomial. The corresponding hypersurface $W = V(f)$ is K-birationally equivalent to V. □

The next proposition generalizes Proposition 9.4.3:

PROPOSITION 10.4.2 (Bertini-Noether): *Let R be an integral domain and $f_1, \ldots, f_k \in R[X_1, \ldots, X_n]$ polynomials such that $V = V(f_1, \ldots, f_k)$ is a variety defined over $K = \mathrm{Quot}(R)$. Then there exists a nonzero $c \in R$ such that if $r \mapsto \bar{r}$ is a homomorphism of R into a field $\bar{K}$ with $\bar{c} \neq 0$, then $\bar{V} = V(\bar{f}_1, \ldots, \bar{f}_k)$ is a variety defined over $\bar{K}$ and $\dim(\bar{V}) = \dim(V)$.*

Proof: Apply Lemma 10.4.1 to find a K-birational map φ from V into a hypersurface $V(g)$ in $\mathbb{A}^{1+\dim(V)}$, where g is an absolutely irreducible polynomial with coefficients in K. Let ψ be the inverse map of φ. The statement "g is an absolutely irreducible polynomial and φ is a birational map from $V(f_1, \ldots, f_k)$ into $V(g)$ with ψ as its inverse" is equivalent modulo $\tilde{\Pi}(R)$ (notation of Section 9.2) to a sentence θ in $\mathcal{L}(\mathrm{ring}, R)$ (Example 7.3.2). Denote the product of all the elements of R that appear in the denominators of any polynomial in θ (including those elements involved in the open sets where $\varphi \circ \psi$ and $\psi \circ \varphi$ are the identity maps) by c_0. Add the conjunct $[c_0 \neq 0]$ to θ. By Theorem 9.2.1, there exists a nonzero element $c \in R$ such that $\tilde{\Pi}(R) \models [c \neq 0 \to \theta \wedge [c_0 \neq 0]]$. Therefore, if $r \mapsto \bar{r}$ is a homomorphism as in the statement of this proposition, then $V(\bar{f}_1, \ldots, \bar{f}_k)$ is birationally equivalent to $V(\bar{g})$ and $\bar{g}$ is absolutely irreducible. Hence, $\bar{V} = V(\bar{f}_1, \ldots, \bar{f}_k)$ is an absolutely irreducible variety defined over $\bar{K}$ of dimension equal to $\dim(V)$. □

Remark: If R is an explicitly given integral domain, then g, φ, ψ, and c can be explicitly computed from the coefficients of $f_1, \ldots, f_k$. □

Two special cases of Proposition 10.4.2 are of particular interest:

COROLLARY 10.4.3:
(a) Let R be the ring of integers of a global field K. Let $f_1, \ldots, f_k \in R[X_1, \ldots, X_n]$ be polynomials such that $V = V(f_1, \ldots, f_k)$ is a variety defined over K. Then, for all but finitely many prime ideals $\mathfrak{p}$ of R, the reduced algebraic set $\bar{V}_{\mathfrak{p}} = V(\bar{f}_1, \ldots, \bar{f}_k)$ modulo $\mathfrak{p}$ is a variety defined over $\bar{K}_{\mathfrak{p}}$ and $\dim(\bar{V}_{\mathfrak{p}}) = \dim(V)$.
(b) Let K_0 be a field and let $R = K_0[\mathbf{t}]$ be the coordinate ring of a K_0-variety U with generic point $\mathbf{t} = (t_1, \ldots, t_m)$. Let $f_1, \ldots, f_k$ be polynomials in $K_0[T_1, \ldots, T_m, X_1, \ldots, X_n]$ such that $V = V(f_1(\mathbf{t}, \mathbf{X}), \ldots, f_k(\mathbf{t}, \mathbf{X}))$ is a

variety defined over $K = \text{Quot}(R)$. Then U has a nonempty Zariski K-open subset U_0 such that for each $\mathbf{a} \in U_0$ the algebraic set $\bar{V}_{\mathbf{a}} = V(f_1(\mathbf{a}, X), \ldots, f_k(\mathbf{a}, \mathbf{X}))$ is a variety defined over $K_0(\mathbf{a})$ and $\dim(\bar{V}_{\mathbf{a}}) = \dim(V)$.

Call a K-rational map $\varphi\colon V \to W$ of K-varieties a K-**morphism** if V is the definition domain of φ. In this case $K[\varphi(\mathbf{x})] \subseteq K[\mathbf{x}]$ for each generic point $\mathbf{x}$ of V (Lemma 2.4.4). Thus $\varphi(\mathbf{x}) = (f_1(\mathbf{x}), \ldots, f_m(\mathbf{x}))$ with $f_1, \ldots, f_m \in K[\mathbf{X}]$. If $\mathbf{y}$ is a generic point of W and $\mathbf{y}' = \varphi(\mathbf{x})$, then the K-specialization $\mathbf{y} \to \mathbf{y}'$ extends to a K-homomorphism of $K[\mathbf{y}]$ into $K[\mathbf{x}]$.

The K-morphism φ is **finite** if $\mathbf{y}'$ is a generic point of W and if $K[\mathbf{x}]$ is an integral extension of $K[\mathbf{y}']$. In this case, by Proposition 2.4.3, for each $\mathbf{b} \in W$ there exists $\mathbf{a} \in V$ with $\varphi(\mathbf{a}) = \mathbf{b}$. In addition, $\varphi^{-1}(\mathbf{b})$ is a finite subset of V (a consequence of Proposition 2.4.3(b)).

The K-morphism is a K-**isomorphism** if $K[\mathbf{x}] \cong K[\mathbf{y}]$, equivalently, if there exists a K-morphism $\psi\colon W \to V$ such that both $\varphi \circ \psi$ and $\psi \circ \varphi$ are the identity maps. In particular, isomorphic K-varieties are K-birationally equivalent.

The projection $\pi\colon \mathbb{A}^n \to \mathbb{A}^{n-1}$ defined by $\pi(x_1, \ldots, x_n) = (x_1, \ldots, x_{n-1})$ is the crucial K-morphism in the stratification procedure of Section 30.2.

10.5 Hyperplane Sections

We show that the "general" intersection of a variety with a hyperplane is a variety. This often reduces questions about arbitrary varieties to questions about curves.

LEMMA 10.5.1: *Let F be a finitely generated extension of a field K and let E a subfield of F which contains K. Then E is also finitely generated extension of K.*

Proof: Let $t_1, \ldots, t_m$ be a transcendence basis for E/K. Extend it to a transcendence basis, $t_1, \ldots, t_n$, for F/K. Then $F_0 = K(t_1, \ldots, t_n)$ is linearly disjoint from E over $E_0 = K(t_1, \ldots, t_m)$, because E/E_0 is algebraic. Since F/F_0 is both algebraic and finitely generated, $[E : E_0] \leq [F : F_0] < \infty$. Consequently, E/K is finitely generated. $\square$

PROPOSITION 10.5.2 (Matsusaka-Zariski): *Let F be a regular finitely generated extension of a field K and y, z elements of F algebraically independent over K. Suppose that there exists a derivation D of F over K with $D(z) \neq 0$. Then there exists a finite subset C of K such that F is a regular extension of $K(y + cz)$ for each $c \in K \smallsetminus C$.*

Proof: With one possible exception, every element c of K satisfies $D(y + cz) \neq 0$. Hence, (Corollary 2.8.4), F is a separable extension of $K(y + cz)$.

For c_1 and c_2 distinct elements of K such that F is a separable extension of $K(y + c_i z)$, denote the algebraic closure of $K(y + c_i z)$ in F by E_i, $i = 1, 2$.

10.5 Hyperplane Sections

Observe that $K(y,z) = K(y+c_1z, y+c_2z)$ has transcendence degree 2 over K. Hence, E_1 and E_2 are algebraically independent over K. Since they are subextensions of the regular extension F/K, they are themselves regular over K. Thus (Lemma 2.6.7), E_1 and E_2 are linearly disjoint over K. By Lemma 2.5.3, $E_1(y,z)$ and $E_2(y,z)$ are linearly disjoint over $K(y,z)$. In particular, they are either distinct or they are both equal to $K(y,z)$.

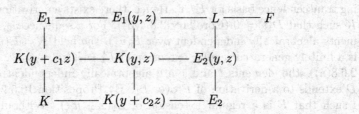

Both fields are contained in the maximal separable algebraic extension L of $K(y,z)$ in F. By Lemma 10.5.1, L is finitely generated over K. Hence, L is a finite separable extension of $K(y,z)$. Thus, there are only finitely many intermediate fields between $K(y,z)$ and L.

It follows that there are only finitely many $c_1 \in K$ for which $E_1(y,z)$ is a proper extension of $K(y,z)$. For c_2 not one of these elements, $E_2(y,z) = K(y,z)$ and $K(y+c_2z) = E_2$ is algebraically closed in F. If, in addition, c_2 is also chosen so that $F/K(y+c_2z)$ is separable, we conclude that $F/K(y+c_2z)$ is regular. □

COROLLARY 10.5.3: *Every variety V of dimension $r \geq 1$ defined over an infinite field K contains an absolutely irreducible curve C defined over K.*

Proof: We may assume that $r \geq 2$. Let $\mathbf{x} = (x_1, \ldots, x_n)$ be a generic point of V over K. Then $F = K(\mathbf{x})$ is a regular extension of K. By Lemma 2.6.1, a separating transcendence basis for F/K can be selected from $x_1, \ldots, x_n$. In particular, one of the coordinates of $\mathbf{x}$, say x_1, is transcendental over K and $F/K(x_1)$ is separable. By Corollary 2.8.4(b), there exists a derivation D of F/K with $D(x_1) \neq 0$ and there exists a second coordinate, say x_n, such that x_1 and x_n are algebraically independent over K. By Proposition 10.5.2, there exists $c \in K$ such that F is a regular extension of $K(t)$, where $t = x_1 + cx_n$. Let H_t be the hyperplane in $\mathbb{A}^n$ defined over $K(t)$ by the equation $X_1 + cX_n - t = 0$. Consider the algebraic set $V \cap H_t$ over $K(t)$. The point $\mathbf{x}$ belongs to $V \cap H_t$. If $\mathbf{x}' \in V \cap H_t$, then $t = x'_1 + cx'_n$. Hence, $\mathbf{x}'$ is a $K(t)$-specialization of $\mathbf{x}$. It follows that $V \cap H_t$ is a variety defined over $K(t)$ with $\mathbf{x}$ a generic point. Its dimension is the transcendence degree of $F/K(t)$, which is $r - 1$. By Corollary 10.4.3(b), there exists $\bar{t} \in K$ such that $V \cap H_{\bar{t}}$ is a variety defined over K of dimension $r - 1$.

By induction on $\dim(V)$, we conclude that $V \cap H_{\bar{t}}$, and hence also V, contain a curve defined over K. □

Here is another application of proposition 10.5.2:

PROPOSITION 10.5.4: *For an arbitrary field K let $f \in K[T, X, Y_1, \ldots, Y_m]$ be an absolutely irreducible polynomial with $\partial f/\partial Y_1 \neq 0$. Let u_0 and u_1 be algebraically independent over K. Then with $L = K(u_0, u_1)$, the polynomial $f(T, u_0 + u_1 T, \mathbf{Y})$ is irreducible in $\tilde{L}[T, \mathbf{Y}]$.*

Proof: Let $(t, x, \mathbf{y})$ be a generic point of the hypersurface of $\mathbb{A}^{2+m}$ defined by $f(T, X, \mathbf{Y}) = 0$ and let $E = K(t, x, \mathbf{y})$. Then $t, x, y_2, \ldots, y_m$ form a separating transcendence basis for E/K. Hence, there exists a derivation D of E over K such that $D(t) \neq 0$ (Corollary 2.8.4(b)). For a sequence $u_2, u_3, \ldots$ of elements algebraically independent over $E(u_1)$, the field $F = E(u_1, u_2, u_3, \ldots)$ is a finitely generated regular extension of $L_1 = K(u_1, u_2, u_3, \ldots)$ (Corollary 2.6.8(a)), the elements t and x are algebraically independent over L_1, and D extends to a derivation of F over L_1. By Proposition 10.5.2, there exists i such that F is a regular extension of $L_1(x - u_i t)$. Without loss assume that $i = 1$. Then F is a regular extension of $K(u_1, x - u_1 t)$. Therefore, $E(u_1)/K(x - u_1 t, u_1)$ is regular (Corollary 2.6.5(b)). Since t and x are algebraically independent over $K(u_1)$, the element $x - u_1 t$ is transcendental over $K(u_1)$. Hence, $x - u_1 t$ and u_1 are algebraically independent over K, so we may assume that $u_0 = x - u_1 t$.

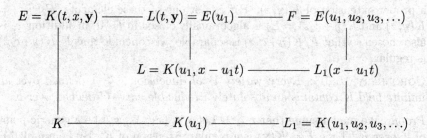

Consider now the polynomial $g(T, \mathbf{Y}) = f(T, u_0 + u_1 T, Y)$ over L. Observe that $(t, \mathbf{y})$ is a generic point of $V(g)$ over L, so $V(g)$ is L-irreducible. By the preceding paragraph, $L(t, \mathbf{y}) = E(u_1)$ is a regular extension of L. Hence, by Corollary 10.2.2, $g(T, \mathbf{Y})$ is absolutely irreducible. □

10.6 Descent

If a variety V is defined over a field K and L is an extension of K, then V is also defined over L. In this section we associate to a variety V, defined over L, a variety W which is defined over K, such that each K-rational point of W induces an L-rational point of V.

Let V (resp. W) be a variety defined over K in $\mathbb{A}^n$ (resp. $\mathbb{A}^m$) with generic point $\mathbf{x}$ (resp. $\mathbf{y}$). Replace $\mathbf{y}$ by another generic point of W, if necessary, to assume that $K(\mathbf{y})$ and $K(\mathbf{x})$ are algebraically independent over K. By Corollary 2.6.8(b) $K(\mathbf{x}, \mathbf{y})$ is a regular extension of K. The variety generated by $(\mathbf{x}, \mathbf{y})$ over K is the **cartesian product** $V \times W = \{(\mathbf{a}, \mathbf{b}) \mid \mathbf{a} \in V \text{ and } \mathbf{b} \in W\}$. If $V = V(f_1, \ldots, f_r)$ and $W = V(g_1, \ldots, g_s)$

10.6 Descent

with $f_1, \ldots, f_r \in K[X_1, \ldots, X_n]$ and $g_1, \ldots, g_s \in K[Y_1, \ldots, Y_m]$, then $V \times W$ is the set of zeros of $f_1, \ldots, f_r, g_1, \ldots, g_s$. Similar definitions and statements are valid for the cartesian product of several varieties.

Let V be a variety in $\mathbb{A}^n$ defined over a field L, $I = I_L(V)$, and σ an isomorphism of L onto a field L'. Then, $I' = \sigma I$ is an ideal of $L'[X_1, \ldots, X_n]$. The algebraic set $V' = V(I')$ is a variety defined over L', which we denote by σV. Assume that the transcendence degree of the universal domain Ω over L is the same as over L'. Then extend σ to an automorphism of Ω such that if $\mathbf{x}$ and $\mathbf{x}'$ are, respectively, generic points of V and V' over L and L', then $\sigma \mathbf{x} = \mathbf{x}'$. It follows that if $\mathbf{a} \in V$, then $\sigma \mathbf{a} \in V'$. For $\varphi \colon W \to V$ an L-rational map of varieties defined over L, apply σ to the coefficients of the polynomials appearing in φ to obtain a rational map $\sigma(\varphi) \colon \sigma W \to \sigma V$.

Construction 10.6.1: Separable descent [Weil3, Weil4]. Let L be a separable extension of degree d of a field K. Denote the d distinct K-embeddings of L into $\tilde{K}$ by $\sigma_1, \ldots, \sigma_d$ with $\sigma_1 = 1$. Choose a basis $w_1, \ldots, w_d$ for L/K. For each i between 1 and d define a linear map $\lambda_i \colon \mathbb{A}^{nd} \to \mathbb{A}^n$ at a point $\mathbf{y} = (y_{jk} \mid 1 \leq j \leq d,\ 1 \leq k \leq n)$ by $\lambda_i(\mathbf{y}) = \mathbf{x}_i$ with

$$(1) \qquad x_{ik} = \sum_{j=1}^{d} (\sigma_i w_j) y_{jk}.$$

The $d \times d$ matrix $(\sigma_i w_j)$ is nonsingular [Lang7, p. 286]. Hence, the linear map $\Lambda = (\lambda_1, \ldots, \lambda_d)$ from $\mathbb{A}^{nd}$ into $\mathbb{A}^n \times \cdots \times \mathbb{A}^n$ (d factors) is a linear isomorphism. Moreover, λ_1 is defined over L and $\lambda_i = \sigma_i \lambda_1$, $i = 1, \ldots, d$. $\square$

PROPOSITION 10.6.2: *In the notation of Construction 10.6.1 let V be a variety defined over L in $\mathbb{A}^n$. Let $\mathbf{x}_1, \ldots, \mathbf{x}_d$ generic points of $\sigma_1 V, \ldots, \sigma_d V$, respectively, such that $\tilde{K}(\mathbf{x}_1), \ldots, \tilde{K}(\mathbf{x}_d)$ are algebraically independent over $\tilde{K}$. Define $\mathbf{y}$ by (1). Then $\mathbf{y}$ generates a variety W over K, the restriction of Λ to W is an isomorphism onto $\sigma_1 V \times \cdots \times \sigma_d V$, and the restriction of λ_1 to W maps W onto V.*

Proof: Consider a finite Galois extension M of K that contains L and extend $\sigma_1, \ldots, \sigma_d$ to elements of $\mathrm{Gal}(M/K)$. By Lemma 2.6.7, $M(\mathbf{x}_1), \ldots, M(\mathbf{x}_d)$ are linearly disjoint extensions of M. From the discussion preceding Construction 10.6.1, each $\sigma \in \mathrm{Gal}(M/K)$ extends to an automorphism of $F = M(\mathbf{x}_1, \ldots, \mathbf{x}_d)$ such that for $1 \leq i, i' \leq d$,

$$\sigma|_L \cdot \sigma_i = \sigma_{i'} \iff \sigma \mathbf{x}_i = \mathbf{x}_{i'}.$$

For $\mathbf{y} = \Lambda^{-1}(\mathbf{x})$ and for each k, $1 \leq k \leq n$, the elements σy_{jk} satisfy the same system of linear equations

$$\sigma x_{ik} = \sum_{j=1}^{d} (\sigma \sigma_i w_j)(\sigma y_{jk}), \qquad i = 1, \ldots, d$$

as the elements y_{jk}, $j = 1, \ldots, d$. Hence, $\sigma \mathbf{y} = \mathbf{y}$. Therefore, $M \cap K(\mathbf{y}) = K$. Since the matrix $(\sigma_i w_j)_{1 \leq i,j \leq d}$ is invertible, $M \cdot K(\mathbf{y}) = F$. Since F is a regular extension of M, $K(\mathbf{y})$ is also linearly disjoint from $\tilde{K}$ over K. This means that the K-algebraic set $W = \Lambda^{-1}(\sigma_1 V \times \cdots \times \sigma_d V)$, generated by $\mathbf{y}$, is a variety defined over K. □

Unlike separable descent, purely inseparable descent is not linear:

PROPOSITION 10.6.3 ([Roquette1]): *Let K be a field of characteristic $p > 0$ and let L be a purely inseparable extension of K. Then, for every variety V defined over L there exists a variety W defined over K and an L-birational morphism $\pi \colon W \to V$.*

Proof: A variety V which is defined over L is already defined over a subfield of L of finite degree, say p^k, over K. Use induction on k to assume that $[L : K] = p$ and hence that $L^p \subseteq K$. Let $\mathbf{x} = (x_1, \ldots, x_n)$ be a generic point of V over L. Then $F = L(\mathbf{x})$ is a regular extension of L of transcendence degree $r = \dim(V)$. Apply Lemma 2.8.3 to find $z_1, \ldots, z_r$, a p-basis for F over $F^p L$. Then $N = F(z_1^{1/p}, \ldots, z_r^{1/p})$ has degree p^r over F. Furthermore, $E = N^p K$ is a subfield of F and

(2) $$EL = N^p L = F^p L(z_1, \ldots, z_r) = F.$$

By Lemma 10.5.1, E is finitely generated over K.

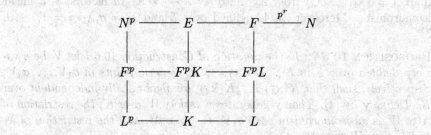

Write $E = K(y_1, \ldots, y_m)$. Then $x_i = \frac{f_i(\mathbf{y})}{g(\mathbf{y})}$ for some $f_i, g \in L[Y_1, \ldots, Y_m]$, $i = 1, \ldots, n$. Since $L^p \subseteq K$, we have $g(\mathbf{y})^p \in K[\mathbf{y}]$. Multiply the numerator and denominator of $x_1, \ldots, x_n$ by $g(\mathbf{y})^{p-1}$ to assume that $g \in K[Y_1, \ldots, Y_m]$. Let $y_{m+1} = g(\mathbf{y})^{-1}$ and denote the K-variety with generic point $(y_1, \ldots, y_{m+1})$ by W. Then the map

$$(b_1, \ldots, b_{m+1}) \mapsto (f_1(b_1, \ldots, b_m)b_{m+1}, \ldots, f_n(b_1, \ldots, b_m)b_{m+1})$$

for $\mathbf{b} \in W$, is an L-birational morphism $\pi \colon W \to V$.

To conclude the proof we still must show that E is a regular extension of K (i.e. that W is a variety defined over K). This follows if we prove that N^p is a regular extension of L^p, or equivalently that N is a regular extension of L.

By (2), $[N : N^p L] = p^r$. Hence, by Lemma 2.8.3, N is a separable extension of L. It remains to show that L is algebraically closed in N. Indeed, if $a \in \tilde{L} \cap N$, then $a^p \in \tilde{L} \cap F = L$. Since N is separable over L, $a \in L$. □

If L is an arbitrary finite extension of a field K, then the number of distinct embeddings of L into $\tilde{K}$ is equal to the degree over K of the maximal separable extension L_0 of K in L. Combine Propositions 10.6.2 and 10.6.3:

PROPOSITION 10.6.4: *Let V be a variety defined over a finite extension L of a field K and let $\sigma_1, \ldots, \sigma_d$ be the distinct K-embeddings of L into $\tilde{K}$. Then there exists a variety W defined over K and an L-morphism $\pi \colon W \to V$ such that the map $(\sigma_1(\pi), \ldots, \sigma_d(\pi)) \colon W \to \sigma_1 V \times \cdots \times \sigma_d V$ is a birational morphism.*

10.7 Projective Varieties

In this section we give the basic definitions of objects associated with projective varieties which are used in this book.

Introduce an equivalence relation on the affine space $\mathbb{A}^{n+1}$ without the origin 0. Call points $(a_0, \ldots, a_n)$ and $(b_0, \ldots, b_n)$ of $\mathbb{A}^{n+1} \smallsetminus \{0\}$ **equivalent** if there exists $\lambda \in \Omega$ such that $(b_0, \ldots, b_n) = (\lambda a_0, \ldots, \lambda a_n)$. Denote the equivalence class of $(a_0, \ldots, a_n)$ by $(a_0 : \cdots : a_n)$. The quotient set of $\mathbb{A}^{n+1} \smallsetminus \{0\}$ under this equivalence relation, denoted $\mathbb{P}^n$, is the n-**dimensional projective space**. If a form $f \in K[X_0, \ldots, X_n]$ vanishes at a point $(a_0, \ldots, a_n) \in \mathbb{A}^{n+1} \smallsetminus \{0\}$, then f vanishes at all points equivalent to $(a_0, \ldots, a_n)$. Thus, we say that f vanishes at the point $(a_0 : \cdots : a_n)$ of $\mathbb{P}^n$. If $\mathfrak{a}_0$ is a set of forms in $K[\mathbf{X}]$, then

(1) $\qquad V(\mathfrak{a}_0) = \{\mathbf{a} \in \mathbb{P}^n \mid f(\mathbf{a}) = 0 \text{ for each } f \in \mathfrak{a}_0\}$

is the K-**algebraic set** defined by $\mathfrak{a}_0$. In Section 10.1 we used the same notation to denote the corresponding K-algebraic set in $\mathbb{A}^{n+1}$. The projective set (1) is obtained from the affine set by deleting the origin and dividing out by the above equivalence relation. Thus, unless it is clear from the context, we must indicate whether our set is affine or projective.

The union of finitely many projective K-algebraic sets is a K-algebraic set. The intersection of any family of K-algebraic sets is a K-algebraic set. The empty set and $\mathbb{P}^n$ are K-algebraic sets. Thus, the K-algebraic sets are the **Zariski K-closed sets** of a topology on $\mathbb{P}^n$ called the **Zariski K-topology**. A K-**variety** is a Zariski K-closed set which is **irreducible** over K (i.e. not the union of two proper Zariski K-closed sets). Every Zariski K-closed set V decomposes as a finite union $V = \bigcup_{i=1}^m V_i$, where the V_i are K-varieties and $V_i \not\subseteq V_j$ for $i \neq j$. The V_i's are said to be the K-**components** of V. An absolutely irreducible projective K-variety is also called a **projective variety**.

The projective space $\mathbb{P}^n$ can be covered by a union of $n+1$ K-open sets
$$U_i = \{\mathbf{a} \in \mathbb{P}^n \mid a_i \neq 0\}, \quad i = 0, \ldots, n.$$
The map $\varphi_i \colon U_i \to \mathbb{A}^n$ defined by
$$\varphi_i(a_0:\cdots:a_n) = \left(\frac{a_0}{a_i}, \ldots, \frac{a_{i-1}}{a_i}, \frac{a_{i+1}}{a_i}, \ldots, \frac{a_n}{a_i}\right)$$
is a homeomorphism of U_i with its induced K-topology onto $\mathbb{A}^n$ with its K-topology. Therefore, $V = \bigcup_{i=0}^n U_i \cap V$ is a covering of V by open subsets and φ_i maps $U_i \cap V$ homeomorphically onto the affine Zariski K-closed set $\varphi_i(U_i \cap V)$. We call $V = \bigcup_{i=0}^n U_i \cap V$ the **standard decomposition** of V into affine Zariski K-closed subset of $\mathbb{A}^n$.

The maps $\varphi_i \colon U_i \to \mathbb{A}^n$ have an inverse: For simplicity we consider only the case where $i = 0$. For each $f \in K[Y_1, \ldots, Y_n]$ of degree d we attach the form $f^*(X_0, \ldots, X_n) = X_0^d f\left(\frac{X_1}{X_0}, \ldots, \frac{X_n}{X_0}\right)$ of degree d and note that $f(Y_1, \ldots, Y_n) = f^*(1, Y_0, \ldots, Y_n)$. If $A = V(f_1, \ldots, f_m)$ is an affine Zariski K-closed subset of $\mathbb{A}^n$, then the projective Zariski K-closed subset $A^* = V(f_1^*, \ldots, f_m^*)$ of $\mathbb{P}^n$ is the closure of A in $\mathbb{P}^n$ and $\varphi_0(A^* \cap U_0) = A$. In particular, if A is a K-variety, then so is A^*.

Suppose V is a nonempty K-variety in $\mathbb{P}^n$. Then $I = \{i \mid V \cap U_i \neq \emptyset\}$ is nonempty. In this case all $\varphi_i(V \cap U_i)$ with $i \in I$ are mutually K-birationally equivalent. We can therefore define the **function field** of V to be the function field of $\varphi_i(V \cap U_i)$, $i \in I$. Specifically, there exists a point $\mathbf{x} = (x_0:\cdots:x_n)$ in V such that $I = \{i \mid x_i \neq 0\}$. Moreover, $\mathbf{x}_i = \varphi_i(\mathbf{x})$ is a generic point of $\varphi_i(V \cap U_i)$ and the function field of V over K is $F = K(\mathbf{x}_i)$, $i \in I$. We define $\dim(V)$ as $\mathrm{trans.deg}(F/K)$, that is, the dimension of $\varphi_i(V \cap U_i)$ for each $i \in I$.

The dimension theorems 10.1.2 and 10.1.3 carry over to projective K-varieties: If V and W are K-varieties, neither of which contains the other, then $\dim(V \cap W) < \min(\dim(V), \dim(W))$. If $h \in K[X_0, \ldots, X_n]$ is a form that does not vanish on V and $r = \dim(V)$, then each K-component of $V \cap V(h)$ has dimension $r - 1$.

If ψ is a K-place of F, then there exists an $i \in I$ such that ψ is finite at $\frac{x_0}{x_i}, \ldots, \frac{x_n}{x_i}$. Then $\mathbf{a}_i = \psi\left(\frac{x_0}{x_i}, \ldots, \frac{x_{i-1}}{x_i}, \frac{x_{i+1}}{x_i}, \ldots, \frac{x_n}{x_i}\right)$ is a point of $\varphi_i(U_i \cap V)$, a K-specialization of $\mathbf{x}_i$. It defines a point $\mathbf{a} \in V$ independent of i such that $\varphi_i(\mathbf{a}) = \mathbf{a}_i$. Call $\mathbf{a}$ a K-specialization of $\mathbf{x}$. Thus, each K-place of F defines a point of V. Refer to this property of projective K-varieties as **completeness**.

A **K-rational map** $\pi \colon V \to W$ between two projective varieties V in $\mathbb{P}^n$ and W in $\mathbb{P}^m$ is a map given by $\pi(\mathbf{a}) = (f_0(\mathbf{a}):\cdots:f_m(\mathbf{a}))$, where $f_0, \ldots, f_m \in K[X_0, \ldots, X_n]$ are forms of the same degree not vanishing on all V. If $g_0, \ldots, g_m$ are such forms, with $(\mathbf{f}(\mathbf{a})) = (\mathbf{g}(\mathbf{a}))$ on a nonempty open set of V, then we identify the two rational maps given by $\mathbf{f}$ and $\mathbf{g}$. If for each point $\mathbf{a} \in V$, π has a representation $\mathbf{f}$ such that $\mathbf{f}(\mathbf{a}) \neq 0$, we call π a **K-morphism**.

Example 10.7.1: *A morphism between projective curves.* Consider the parabola $P = V(X_1^2 - X_0 X_2)$ in $\mathbb{P}^2$ over an algebraically closed field K. Then $\pi_1 \colon P \to \mathbb{P}^1$ defined by $\pi_1(\mathbf{x}) = (x_1 \colon x_2)$ is a rational map with domain of definition including all points of P except $(1{:}0{:}0)$. Similarly, $\pi_2 \colon P \to \mathbb{P}^1$ defined by $\pi_2(\mathbf{x}) = (x_0 \colon x_1)$ is a rational map with domain of definition including all points of P except $(0{:}0{:}1)$. These maps agree on their common domain of definition (i.e. $(x_1{:}x_2) = \frac{x_1}{x_0}(x_0{:}x_1)$ if $(x_0{:}x_1{:}x_2) \in P$ and $x_0 x_1 \ne 0$). Hence, they define a morphism $\pi \colon P \to \mathbb{P}^1$. It is not difficult to prove that there exist no forms $g, h \in K[X_0, X_1, X_2]$ of the same degree with $\pi(\mathbf{x}) = (g(\mathbf{x}) : h(\mathbf{x}))$ for all $\mathbf{x} \in V(K)$ (Exercise 9). This shows the need to vary the forms in the definition of morphism of projective varieties in the neighborhoods of distinct points of P. □

10.8 About the Language of Algebraic Geometry

A central task of Field Arithmetic is the study of solutions of systems of polynomial equations over fields which vary in certain distinguished classes. Methods of Algebraic Geometry are indispensable in this study. Likewise, it is both useful and attractive to use methods of Logic and Model Theory, like ultraproducts, to move around in those classes. Moreover, Logic and Model Theory form a natural framework for decidability problems of first order theories in the language of ring theory. The basic objects in all these problems are tuples of elements in fields. It is therefore natural to use the language of varieties which [Weil5] establishes and which we present in this chapter.

We are aware of the fact that most algebraic geometers today use the language of schemes as established by Serre and Grothendieck. This language is more powerful than the languages of varieties. Unfortunately, the terminology of the two languages is not always compatible.

In this section, we assume that the reader is familiar with the language of schemes, say like in [Mumford2] or in [Hartshorne], and establish a lexicon which translates the concepts of the language of varieties as introduced in this chapter into concepts of the language of schemes.

Consider a basic field K and a universal extension Ω of K. With each K-closed set A we associate a reduced affine scheme of finite type over K in the following way. A has the form $V(\mathfrak{a})$, where $\mathfrak{a}$ is an ideal of $K[\mathbf{X}]$, where $\mathbf{X} = (X_1, \ldots, X_n)$ for some positive integer n. The radical $\sqrt{\mathfrak{a}}$ of $\mathfrak{a}$ satisfies $A = V(\sqrt{\mathfrak{a}})$ and it is uniquely determined by A (Hilbert's Nullstellensatz, (1d) of Section 10.1). We may therefore consider the ring $R = K[\mathbf{X}]/\sqrt{\mathfrak{a}}$ which is finitely generated over K and is **reduced** (i.e. it has no nilpotent elements other than 0). Let A' be $\mathrm{Spec}(R)$ together with its structure sheaf. The natural embedding $K \hookrightarrow R$ induces a morphism $A' \to \mathrm{Spec}(K)$. This is the reduced affine scheme of finite type over K that we associate with A. To shorten notation, we will use A' instead of $A' \to \mathrm{Spec}(K)$.

Conversely, each reduced affine scheme Y of finite type over K is isomorphic to $\mathrm{Spec}(R)$, where $R = K[X_1,\ldots,X_n]/\mathfrak{a}$ is reduced. Thus, $Y = A'$ where $A = V(\mathfrak{a})$. If Y is isomorphic to the spectrum of another reduced ring $S = K[X_1,\ldots,X_m]/\mathfrak{b}$, then there is an isomorphism $R \cong_K S$. This gives rise to an isomorphism of A and $V(\mathfrak{b})$.

More generally, each morphism $Z \to Y$ of reduced affine schemes of finite type over K with $Z = \mathrm{Spec}(S)$ and $Y = \mathrm{Spec}(R)$ corresponds to a K-homomorphism $R \to S$ of rings. The latter gives rise to a morphism $B \to A$ of K-algebraic sets.

In fact, we have a bijective functorial correspondence of $\mathrm{Mor}(B,A)$ and $\mathrm{Mor}(B',A')$ which factors through $\mathrm{Hom}(R,S)$ [Mumford2, p. 113, Cor. 1]. Thus, we have an equivalence of the category of K-algebraic sets and the category of reduced schemes of finite type over K [Cohn, p. 19, Prop. 3.1].

Each point $\mathbf{a} = (a_1,\ldots,a_n)$ of A uniquely corresponds to a K-homomorphism $h_\mathbf{a} \colon R \to \Omega$ which maps the coset $X_i + \sqrt{\mathfrak{a}}$ onto a_i, $i=1,\ldots,n$. The homomorphism $h_\mathbf{a}$ uniquely corresponds to $\mathfrak{p} = \mathrm{Ker}(h_\mathbf{a})$, which is a point of $A' = \mathrm{Spec}(R)$. If $\sigma \in \mathrm{Aut}(\Omega/K)$ and $\mathbf{b} = \sigma(\mathbf{a})$, then $\mathfrak{p} = \mathrm{Ker}(h_\mathbf{b})$. Hence, K-isomorphism classes of points of A bijectively correspond to points of A'.

The scheme which corresponds to $\mathbb{A}^n$ is $\mathrm{Spec}(K[\mathbf{X}])$, which is more precisely denoted by $\mathbb{A}^n_K$. The inclusion $A \subseteq \mathbb{A}^n$ induces a closed immersion $A' \hookrightarrow \mathbb{A}^n_K$. In this sense, the correspondence $A \rightsquigarrow A'$ respects inclusions, unions, and finite intersections. We may therefore extend this correspondence to complements of algebraic sets and establish a homeomorphism of the Zariski K-topology of $\mathbb{A}^n$ onto the Zariski topology of $\mathbb{A}^n_K$, hence also of the Zariski topologies of A and A'.

Beside the prime ideals of R, which constitute the points of A', there is in the language of schemes another type of points, namely **geometric points**. For each field L which contains K, an L**-rational point** of A' is a morphism $\mathbf{p}\colon \mathrm{Spec}(L) \to A'$ of schemes over K. It corresponds to a K-homomorphism $h_\mathbf{a}\colon R \to L$, where $\mathbf{a} = (a_1,\ldots,a_n)$ has coordinates in L. The set of L-rational points of A' is denoted by $A'(L)$. Likewise, we have denoted the set of all L-rational points of A by $A(L)$. The map $\mathbf{a} \mapsto \mathrm{Ker}(h_\mathbf{a})$ is a Zariski homeomorphism of $A(L)$ onto $A'(L)$. Thus, we may identify the two sets. In particular, A itself in the language of this book is $A(\Omega)$.

One may show that A is K-irreducible if and only if A' is irreducible. In this case A has a generic point $\mathbf{x}$ over K. Every other point of A is a K-specialization of $\mathbf{x}$. In other words, A is the closure of $\mathbf{x}$ in the Zariski topology. The corresponding point of A' is the zero ideal of R. It has the same property as $\mathbf{x}$, namely, A' is the closure of the zero ideal in the Zariski topology. The only difference between the two languages here is that A' has only one generic point, whereas A has many, namely all K-conjugates of $\mathbf{x}$. Nevertheless, by looking at $A(\Omega)$, the scheme theorist can discover all generic points of A'. In particular, one may also speak, for example, about 'generic points of A' in Ω which are algebraically independent over K'.

Another way to discover generic points in the scheme theoretic language is by looking at the function field F of A'. Each element of F is locally defined

10.8 About the Language of Algebraic Geometry

as a function from A' into K induced by the quotient of two polynomials in $K[\mathbf{X}]$. Since A' is irreducible, this function uniquely corresponds to an element of $K(\mathbf{x})$, where $\mathbf{x}$ is a generic point of A over K. Thus, F is K-isomorphic to $K(\mathbf{x})$.

Suppose now that K' is an extension of K which is contained in Ω such that Ω is still a universal extension of K'. This happens, for example, if the transcendence degree of K' over K is finite. Then A is also a K'-algebraic set but A' is not a scheme over K'. The language of schemes replaces A' by the fiber product $A' \times_{\mathrm{Spec}(K)} \mathrm{Spec}(K')$, which one usually shortens to $A' \times_K K'$ or even to $A'_{K'}$. The properties of fiber products imply that for each extension L of K', it is possible to identify the sets $A'(L)$ and $A'_{K'}(L)$. Thus, we may safely use the notation $A(L)$ as before.

Assume further that A is K-irreducible. Then A is absolutely irreducible if and only if A is K'-irreducible for each extension K' of K or, equivalently, for $K' = \tilde{K}$. This happens exactly when $A(\tilde{K})$ is irreducible in the Zariski topology over $\tilde{K}$. Thus, the concept of absolute irreducibility in both languages is the same.

If A is absolutely irreducible, then A is defined over K if and only if $K(\mathbf{x})/K$ is a regular extension (Corollary 10.2.2). If we choose $\mathbf{x}$ to be algebraically independent from K' over K, then this happens if and only if for all K', the fields $K(\mathbf{x})$ and K' are linearly disjoint over K. In other words, the canonical map $K(\mathbf{x}) \otimes_K K' \to K'(\mathbf{x})$ is injective. In particular, $K[\mathbf{x}] \otimes_K K'$ is an integral domain. Since $A'_{K'}$ is defined as $\mathrm{Spec}(K[\mathbf{x}] \otimes_K K')$, this means that $A'_{K'}$ is irreducible. If we start from an absolutely irreducible reduced scheme B of finite type over K', we may say that it is **defined over** K if there exists a K-irreducible algebraic set A such that $A'_{K'} \cong B$.

Finally, although it plays only a minor role in our book, we briefly mention how to interpret Weil's "abstract varieties" in the language of schemes. Weil starts in [Weil5, §VII,3] from a system $\mathcal{V} = \langle V_i, \varphi_{ij} \rangle_{i,j \in I}$, where I is a finite set, V_i is an absolutely irreducible variety defined over K, φ_{ii} is the identity map, and $\varphi_{ij}: V_i \to V_j$ is a birational map defined over K with $\varphi_{jk} \circ \varphi_{ij} = \varphi_{ik}$ for all $i, j, k \in I$. For $i \in I$ choose a generic point $\mathbf{x}_i$ such that $\varphi_{ij}(\mathbf{x}_i) = \mathbf{x}_j$ for all $i, j \in I$. Assume that if $(\mathbf{a}_i, \mathbf{a}_j)$ is a specialization of $(\mathbf{x}_i, \mathbf{x}_j)$, then φ_{ij} is defined at $\mathbf{a}_i$, so $\varphi_{ij}(\mathbf{a}_i) = \mathbf{a}_j$ and we call $\mathbf{a}_i$ and $\mathbf{a}_j$ **equivalent**. The **abstract variety** V determined by $\mathcal{V}$ is the union $\bigcup_{i \in I} V_i$ modulo this equivalence relation. Weil defines a topology on V such that for each $i \in I$, the image of V_i in V is open [Weil5, Prop. VII,10]. Since the map of V_i into V is injective, this covers V with a union of affine varieties. In the above notation, if $\mathbf{a}_i$ and $\mathbf{a}_j$ are equivalent, then there are open affine subsets U_i of V_i and U_j of V_j which contain $\mathbf{a}_i$ and $\mathbf{a}_j$, respectively, such that φ_{ij} maps U_i isomorphically onto U_j. Moreover, if $\mathbf{x}$ is a generic point of V, then the diagonal of $V \times V$ consists of exactly of all specializations of $(\mathbf{x}, \mathbf{x})$, so it therefore closed. Hence, V is a scheme (which is reduced and absolutely irreducible). In particular, this procedure works for projective varieties. Thus, the concept of projective variety in both languages is the same.

Exercises

1. Let $f \in K[X_1, \ldots, X_n]$ be an irreducible polynomial with $\deg_{X_n}(f) \geq 1$ and let $(x_1, \ldots, x_n)$ be a zero of f for which $x_1, \ldots, x_{n-1}$ are algebraically independent over K. Then $F = K(\mathbf{x})$ is a regular extension of K if and only if the polynomial f is absolutely irreducible. Hint: Use the Gauss Lemma.

2. Use Exercise 1 to prove that every K-hypersurface in $\mathbb{A}^n$ has the form $V(f)$, where $f \in K[X_1, \ldots, X_n]$ is irreducible.

3. Let $f \in K[X, Y_1, \ldots, Y_m, Z]$, with $\partial f/\partial Y_1 \neq 0$, be an absolutely irreducible polynomial over a field K. Then $f(X, \mathbf{Y}, T_0 + T_1 X + \cdots + T_n X^n)$ is an absolutely irreducible polynomial in the variables $X, \mathbf{Y}, \mathbf{T}$, for every positive integer n. Hint: Deduce the statement for $n = 1$ from Proposition 10.5.4. Now use induction.

4. For a field K let $f_1, \ldots, f_r \in K[X]$ be irreducible polynomials, none of which is a multiple of the other. Denote by V the algebraic set defined by the system
$$f_i(X) - Y_i^n = 0, \quad i = 1, \ldots, r.$$
Prove that if $\text{char}(K) \nmid n$, then V is a variety defined over K [Duret1].

   Hint: Choose a transcendental element x over K and let $y_1, \ldots, y_r$ be elements such that $f_i(x) = y_i^n$, $i = 1, \ldots, r$. Use Kummer's theory [Lang7, p. 295, Thm. 8.2] to show that $[\tilde{K}(x, \mathbf{y}) : \tilde{K}(x)] = [K(x, \mathbf{y}) : K(x)] = n^r$. Deduce that $(x, \mathbf{y})$ is a generic point for V over $\tilde{K}$. Now apply Corollary 10.2.2.

5. For a field K of characteristic p let $f_1, \ldots, f_r \in K[X]$ be irreducible polynomials of degree less than p. Suppose that $f_1(X), \ldots, f_r(X)$ are linearly independent over $\mathbb{F}_p$. Prove that the algebraic set defined by the system
$$f_i(X) = Y_i^p - Y_i, \quad i = 1, \ldots, r$$
is a variety defined over K [Duret1].

   Hint: Follow the hint to Exercise 4 and replace Kummer's theory by Artin-Schreier theory [Lang7, p. 296, Thm. 8.3].

6. Use notation of Section 10.7 and show that the K-open sets in $\mathbb{P}^n$ are exactly the sets whose intersection with U_i is K-open, $i = 1, \ldots, n+1$.

7. Let $L = \mathbb{Q}(\sqrt{a})$ with $a \in \mathbb{Q}$ a nonsquare. Let V be the variety defined over L by the equation $f(X_1, X_2) = f_1(X_1, X_2) + \sqrt{a} f_2(X_1, X_2) = 0$ in $\mathbb{A}^2$ with $f_1, f_2 \in \mathbb{Q}[X_1, X_2]$. Let $Y_{11}, Y_{21}, Y_{12}, Y_{22}$ be variables representing coordinates of $\mathbb{A}^4$. Show that the variety W that satisfies the conclusion of Proposition 10.6.2 is defined by the equations $g_1(\mathbf{Y}) = 0 = g_2(\mathbf{Y})$ where $g_1, g_2 \in \mathbb{Q}[Y_{11}, Y_{21}, Y_{12}, Y_{22}]$ and $g_1(\mathbf{Y}) + \sqrt{a} g_2(\mathbf{Y}) = f(Y_{11} + \sqrt{a} Y_{21}, Y_{12} + \sqrt{a} Y_{22})$.

8. Let K be an algebraically closed field and let $\pi\colon V \to V$ be a K-morphism of a projective variety V. Prove that if the map $\pi\colon V(K) \to V(K)$ is injective, then it is also surjective [Ax3, p. 190].

Hint: First consider the case where $K = \tilde{F}_p$ and check the restriction of π to $V(K_0)$, with K_0 a finite subfield of K such that V (resp. π) is K_0-closed (resp. K_0-morphism). Then use Corollary 9.3.2 to extend the result for an arbitrary algebraically closed field K of characteristic p. Finally, apply Corollary 9.3.3 to deduce the result for the case when $\text{char}(K) = 0$.

9. Let K, P, and π be as in Example 10.7.1. Prove there exist no forms $g, h \in K[X_0, X_1, X_2]$ of the same degree d such that $\pi(\mathbf{a}) = (g(\mathbf{a}) : h(\mathbf{a}))$ for all $\mathbf{a} \in P$. Hint: Assume g and h exist. Let $\mathbf{x} = (x_0{:}x_1{:}x_2)$ be a generic point for P over K. Then

$$g(\mathbf{x}) = g_0(x_0, x_2) + g_1(x_0, x_2)x_1 \quad \text{and} \quad h(\mathbf{x}) = h_0(x_0, x_2) + h_1(x_0, x_2)x_1,$$

where $g_0, h_0 \in K[X_0, X_2]$ are forms of degree d and $g_1, h_1 \in K[X_0, X_2]$ forms of degree $d - 1$. Distinguish between two cases: $X_0 | g_1$ and $X_0 \nmid g_1$. In the first case prove that $g(0{:}0{:}1) = h(0{:}0{:}1) = 0$. In the second case find $t \in K$ with $g(1{:}t{:}t^2) = h(1{:}t{:}t^2) = 0$. This will give a contradiction.

10. Define the genus of a curve Γ over a field K to be the genus of its function field. Use the Weil estimate (Theorem 4.5.2) to prove that every projective curve Γ over a finite field has a K-rational point.

Notes

Most of Sections 10.1–10.3 is in [Lang4]. Our approach, however, to rational maps is different. While Lang (who follows [Weil5]) prefers to identify a rational map $\varphi\colon V \to W$ with the closure (in the Zariski topology) of its graph, we have chosen to define it as an actual map from an open subset of V into W given by rational functions.

[Inaba, p. 16] proves Proposition 10.5.4 in a somewhat different manner.

Chapter 11.
Pseudo Algebraically Closed Fields

Let $\tilde{K}$ be an algebraically closed field and consider the ideal I generated by polynomials $f_1, \ldots, f_m$ in $\tilde{K}[X_1, \ldots, X_n]$. The Hilbert Nullstellensatz asserts that if I is not the whole ring, then $f_1, \ldots, f_m$ have a common $\tilde{K}$-zero. Conversely, this property is a sufficient condition for a field K to be algebraically closed. Proposition 7.9.1 is a motivation for a weaker Nullstellensatz:

Suppose K is a nonprincipal ultraproduct of finite fields. If $f(X,Y) \in K[X,Y]$ is an absolutely irreducible polynomial, then $f(X,Y)$ has a K-rational zero.

Theorem 11.2.3 establishes that a field K with this property has a stronger property:

(*) Each nonempty variety defined over K has a K-rational point.

Fields with property (*), a central topic of this book, are called **pseudo algebraically closed** (**PAC**).

If $\mathrm{char}(K) > 0$, then we show that the K-rational point in (*) can be chosen to preserve p-independence of given elements in the function field of V (Proposition 11.4.1). Algebraic extensions of PAC fields are PAC (Corollary 11.2.5). If K is PAC, then K has no orderings (Theorem 11.5.1). If w is a valuation of K_s, then K is w-dense in K_s (Proposition 11.5.3) and the Henselian closure of K at $w|_K$ is K_s itself (Corollary 11.5.5). Finally, we prove that $\mathrm{Gal}(K)$ is projective (Theorem 11.6.2).

11.1 PAC Fields

We start by proving few basic properties of PAC fields.

PROPOSITION 11.1.1: *Let K be a PAC field and V a variety defined over K. Then the set $V(K)$ is dense in V in the Zariski K-topology. In particular, K is infinite.*

Proof: Let $\mathbf{x} = (x_1, \ldots, x_n)$ be a generic point of V over K and $W = V(g_1, \ldots, g_m)$ a K-algebraic set that does not contain V. Then one of the g_i, say g_1, does not vanish at $\mathbf{x}$. Write $y \cdot g_1(\mathbf{x}) = 1$ and consider the algebraic set V' in $\mathbb{A}^{n+1}$ with generic point $(\mathbf{x}, y)$. The function field of V' over K is equal to $K(\mathbf{x}, y) = K(\mathbf{x})$, the function field of V over K. Since $K(\mathbf{x}, y)$ is therefore a regular extension of K, V' is a variety defined over K (Corollary 10.2.2). By assumption, V' has a K-rational point $(\mathbf{x}', y')$ with $y' \cdot g_1(\mathbf{x}') = 1$. In particular $g_1(\mathbf{x}') \neq 0$. But since $\mathbf{x}'$ is a K-specialization of $\mathbf{x}$, it is a point of V. The inequality $g_1(\mathbf{x}') \neq 0$ implies that $\mathbf{x}' \notin W$. Thus, $V(K)$ contains a point outside of $W(K)$. Since W is arbitrary, $V(K)$ is dense in V relative to the Zariski K-topology.

Finally, if $a_1, \ldots, a_m$ are elements of K, then $K = \mathbb{A}^1(K)$ has an element a_{m+1} outside the set $V\bigl(\prod_{i=1}^{m}(X - a_i)\bigr)$. Consequently, K is infinite. □

By Hilbert Nullstellensatz, algebraically closed fields are PAC. So are separably closed fields [Lang4, p. 76]. Both statements are also immediate consequences of Theorem 11.2.3 below. We shall see many nonseparably closed PAC fields.

For PAC fields there is an analog of the strong form of the Hilbert Nullstellensatz:

COROLLARY 11.1.2: *If a polynomial $g \in K[\mathbf{X}]$ vanishes on $V(K)$, then $g \in I_K(V)$.*

Proof: By Proposition 11.1.1, $g(\mathbf{x}) = 0$ for each generic point $\mathbf{x}$ of V. Hence, g vanishes on V, so $g \in I_K(V)$. □

We reformulate Proposition 11.1.1 in the language of homomorphisms:

PROPOSITION 11.1.3: *Let K be a PAC field, $F = K(x_1, \ldots, x_n)$ a finitely generated regular extension of K, and y a nonzero element of $K[\mathbf{x}]$. Then there exists a K-homomorphism $\varphi \colon K[\mathbf{x}] \to K$ such that $\varphi(y) \neq 0$.*

Proof: Let V be the variety generated by $\mathbf{x}$ over K (Corollary 10.2.2). Let $g \in K[\mathbf{X}]$ with $y = g(\mathbf{x})$. Then $U = \{\mathbf{a} \in V \mid g(\mathbf{a}) \neq 0\}$ is a nonempty Zariski open subset of V. Proposition 11.1.1 gives $\mathbf{a} \in U(K)$. Then specialization $\mathbf{x} \to \mathbf{a}$ extends to a K-homomorphism $\varphi \colon K[\mathbf{x}] \to K$ with $\varphi(y) \neq 0$. □

Remark 11.1.4: *Places into PAC fields.* Under the assumptions of Proposition 11.1.3, it is possible to extend φ to a K-place $\varphi \colon F \to K \cup \{\infty\}$ with $\varphi(y) \neq 0$. The proof uses the theory of simple points of varieties which we have not developed in this book.

Indeed, let V, U and $\mathbf{a}$ be as in the proof of Proposition 11.1.3. Following [Jarden-Roquette, Lemma A1], we may assume $\mathbf{a} = (0, \ldots, 0)$ and make U of the proof of Proposition 11.1.3 smaller such that $\mathbf{a}$ is a simple point of V. Then the local ring $O = O_{V,\mathbf{a}}$ is a regular ring of dimension $r = \dim(V)$. Thus, the maximal ideal M of O is generated by r elements $t_1, \ldots, t_r$. The degree function with respect to $t_1, \ldots, t_r$ is a valuation of F. Let φ_0 be the corresponding place. Put $u_i = \frac{t_i}{t_r}$ and $\bar{u}_i = \varphi_0(u_i)$, $i = 0, \ldots, r-1$. Then $\bar{u}_1, \ldots, \bar{u}_{r-1}$ are algebraically independent over K. By Lemma 2.2.7, there is a K-place φ_1 of $K(\bar{u}_1, \ldots, \bar{u}_{r-1})$ which maps each $\bar{u}_i$ onto 0. The composition $\varphi_1 \circ \varphi_0$ is a K-place of F into $K \cup \{\infty\}$ extending the homomorphism φ of Proposition 11.1.3. □

11.2 Reduction to Plane Curves

The PAC property of a field K can be tested on special types of varieties: open subsets of hypersurfaces (the primitive element theorem), open subsets of plane curves (intersecting with hypersurfaces), and finally on just affine hypersurfaces. We also prove that every algebraic extension of a PAC field is PAC.

LEMMA 11.2.1 (Frey-Geyer): *Let L be an algebraic extension of an infinite field K and $f \in K[X,Y]$ a nonconstant absolutely irreducible polynomial with only finitely many zeros in L. Then there exists a polynomial $g \in K[X, Z]$ which is monic in Z such that $f^*(X, Z) = f(X, g(X, Z))$ is absolutely irreducible and has no zero in L.*

Proof: If Y does not occur in f, then f is a nonconstant linear polynomial in X, so it has infinitely many zeors in L. We deduce from this contradiction that $\deg_Y f \geq 1$.

Let $(\xi_1, \eta_1), \ldots, (\xi_m, \eta_m)$ be a set of representatives of the conjugacy classes of the L-rational zeros of f over K. If $\text{char}(K) = p \neq 0$, then $f_1(X^p, Y^p)$ is reducible over $\tilde{K}$ for each $f_1 \in K[X, Y]$, so f is not of this form. Switch X and Y if necessary to assume that $\frac{\partial f}{\partial X}$ is not identically zero. Apply a nonsingular linear transformation of the variables over K to assure that $\xi_1, \ldots, \xi_m$ are distinct and nonconjugate over K. Let $b \in K[Y]$ be the leading coefficient of $f(X, Y)$ as a polynomial in X. Finally, let t be an indeterminate. Since $\deg_Y f \geq 1$, all roots of $f(X,t)$ are transcendental over K, hence different from all conjugates of ξ_i over K. Thus, $f(X,t)$ is relatively prime to $f(X, \eta_i)$ in $\tilde{K}(t)[X]$. Therefore, there are $p_{i1}, p_{i2} \in \tilde{K}[t, X]$ and a nonzero $q \in \tilde{K}[t]$ which is independent of i such that $p_{i1}(t, X)f(X, \eta_i) + p_{i2}(t, X)f(X, t) = q(t)$. If $\eta \in \tilde{K}$ is not a zero of $q(t)$, then $f(X, \eta_i)$ and $f(X, \eta)$ have no common zero. Consequently, each zero of $f(X, \eta)$ is K-conjugate to none of the ξ_i's, $i = 1, \ldots, m$.

It follows that there exists an $\eta_0 \in K$ and a $\xi_0 \in \tilde{K}$ such that

(1a) $\qquad f(\xi_0, \eta_0) = 0, \quad b(\eta_0) \neq 0, \quad \frac{\partial f}{\partial X}(\xi_0, \eta_0) \neq 0, \quad \text{and}$

(1b) $\qquad \eta_0 \neq \eta_i, \quad i = 1, \ldots, m.$

In particular, ξ_0 is K-conjugate to none of the $\xi_1, \ldots, \xi_m$.

By Hilbert's Nullstellensatz L is not algebraically closed. Therefore, there exists a polynomial $h(Z) = Z^n + \alpha_1 Z^{n-1} + \cdots + \alpha_n$, with $n > 1$ and coefficients in K, that has no root in L. The remaining parts of the proof construct $g(X, Z)$ and establish properties of $f^*(X, Z)$.

PART A: *Construction of g.* The polynomials $g_i(X) = \text{irr}(\xi_i, K)$, $i = 0, \ldots, m$ are distinct. Apply the Chinese remainder theorem to find for each j between 1 and $n - 1$ a polynomial $a_j(X)$ in $K[X]$ such that
(2a) $a_j(X) \equiv 0 \mod g_0(X)$; and
(2b) $a_j(X) \equiv \alpha_j \mod g_i(X)$, $i = 1, \ldots, m$.

Similarly, there exists an $a_n \in K[X]$ such that
(3a) $a_n(X) \equiv \eta_0 \mod g_0(X)^2$; and
(3b) $a_n(X) \equiv \alpha_n + \eta_i \mod g_i(X)$, $i = 1, \ldots, m$.

Let $g(X, Z) = Z^n + a_1(X)Z^{n-1} + \cdots + a_n(X)$ and $f^*(X, Z) = f(X, g(X, Z))$.

11.2 Reduction to Plane Curves

PART B: $f^*(X, Z)$ is absolutely irreducible. Let x and z be transcendental elements over K such that $f^*(x, z) = 0$. Let $y = g(x, z)$. Then $K(x, y)$, the function field of $f(X, Y) = 0$, is a regular extension of K. Moreover,

$$[\tilde{K}(x, z) : \tilde{K}(x, y)] \leq [K(x, z) : K(x, y)] \leq \deg_Z g \leq n.$$

Therefore, it suffices to prove that the degree of z over $\tilde{K}(x, y)$ is n. But, the specialization $(x, y) \to (\xi_0, \eta_0)$ defines a discrete valuation v of $\tilde{K}(x, y)$ which is trivial on $\tilde{K}$ (Section 5.2). From (2a), $v(a_j(x)) > 0$, $j = 1, \ldots, n-1$, and from (3a), $v(a_n(x) - \eta_0) \geq 2$. By (1) and Lemma 6.1.8(b), v is unramified over $\tilde{K}(y)$. Hence, $v(\eta_0 - y) = 1$. Therefore, $v(a_n(x) - y) = v((a_n(x) - \eta_0) + (\eta_0 - y)) = 1$. Consequently, $g(x, Z) - y$ is an Eisenstein polynomial for z over $\tilde{K}(x, y)$ (with respect to v). It follows that $g(x, Z) - y$ is irreducible and $[\tilde{K}(x, z) : \tilde{K}(x, y)] = n$.

PART C: $f^*(X, Z) = 0$ has no solution in L. Suppose (ξ, ζ) is a zero of f^* for some $\xi, \zeta \in L$. Then $f(\xi, g(\xi, \zeta)) = 0$. Therefore, there exists i with $1 \leq i \leq m$ and an automorphism $\sigma \in \text{Gal}(K)$ such that $\sigma(\xi) = \xi_i$ and $\eta_i = g(\xi_i, \sigma(\zeta))$. By (2b) and (3b)

$$\sigma(h(\zeta)) = h(\sigma(\zeta)) = \sigma(\zeta)^n + \alpha_1 \sigma(\zeta)^{n-1} + \cdots + \alpha_n$$
$$= \sigma(\zeta)^n + a_1(\xi_i)\sigma(\zeta)^{n-1} + \cdots + a_{n-1}(\xi_i)\sigma(\zeta) + a_n(\xi_i) - \eta_i$$
$$= g(\xi_i, \sigma(\zeta)) - \eta_i = 0.$$

Hence, $h(\zeta) = 0$, contrary to the choice of h. □

Here is a reformulation of Lemma 11.2.1 in geometric terms:

LEMMA 11.2.2: *Let L be an algebraic extension of an infinite field K and Γ an affine plane curve defined over K such that $\Gamma(L)$ is finite. Then, there is an affine plane curve Γ^* defined over K and there is a finite K-morphism $\pi \colon \Gamma^* \to \Gamma$ such that $\Gamma^*(L)$ is empty.*

THEOREM 11.2.3: *Let L be an algebraic extension of an infinite field K. Suppose every plane curve defined over K has an L-rational point. Then L is PAC.*

Proof: First of all note that, by Lemma 11.2.2, $\Gamma(L)$ is infinite for every affine plane curve Γ defined over K. Now consider a variety V defined over L. Suppose without loss that $\dim(V) \geq 1$. Then V is defined over a subfield L_0 of L of finite degree over K. Proposition 10.6.4 provides a variety W defined over K with $\dim(W) \geq 1$ and an L_0-morphism $\pi \colon W \to V$. By Corollary 10.5.3, W contains a curve C defined over K. From Lemma 10.4.1, C is K-birationally equivalent to a plane curve Γ. Then $\Gamma(L)$ is infinite, hence $C(L)$ is nonempty, so $V(L)$ is nonempty. Consequently, L is a PAC field. □

COROLLARY 11.2.4 ([Ershov1]): *Infinite algebraic extensions of finite fields are PAC fields.*

Proof: Let K be an infinite algebraic extension of $\mathbb{F}_p$. Consider an absolutely irreducible plane curve Γ of degree d defined by $f(X,Y) = 0$ with coefficients in K. For every q_0 there exists a p-power $q \geq q_0$ such that $\mathbb{F}_q \subseteq K$ and $f(X,Y) \in \mathbb{F}_q[X,Y]$. By Theorem 5.4.1,

$$\Gamma(\mathbb{F}_q) \geq (q+1) - (d-1)(d-2)\sqrt{q} - d.$$

Hence, $\Gamma(K)$ is an infinite set. By Theorem 11.2.3, K is a PAC field. □

COROLLARY 11.2.5 (Ax-Roquette): *Every algebraic extension of a PAC field is a PAC field.*

Example 11.2.6: A minimal PAC field. Problem 16.48 of [Fried-Jarden3] asks for a PAC field N all of whose proper subfields are non-PAC. We call N a **minimal PAC field**. The results of this chapter allow us to give a quick example of a minimal PAC field.

Start from a finite prime field $K = \mathbb{F}_p$. By Section 1.5, $\text{Gal}(K) \cong \hat{\mathbb{Z}}$, and for each prime number l, K has a unique extension K_{l^∞} such that $\text{Gal}(K_{l^\infty}/K) \cong \mathbb{Z}_l$. By Corollary 11.2.4, K_{l^∞} is a PAC field. Each proper subfield of K_{l^∞} is a finite field (Lemma 1.4.2(c)), so by Proposition 11.1.1, is not PAC. Thus, K_{l^∞} is a minimal PAC field. □

PROBLEM 11.2.7: *Does there exists a minimal PAC field which is not an algebraic extension of a finite field?*

Remark 11.2.8: Criterion for PAC with projective curves. Let K be a field. If each affine curve defined over K has a K-rational point, then each projective curve over K has a K-rational point. We sketch a proof that the converse is also true. The proof is based on Lemma 2 of [Frey]:

(4) Let F be a function field of one variable over K. Suppose F/K has only finitely many prime divisors of degree 1. Then F has a finite separable extension F' which is regular over K and has no prime divisors of degree 1.

The proof of (4) goes along the same lines as the proof of Lemma 11.2.1 except that it is not necessary to make a linear transformation as in the beginning of that proof. Instead of (η_0, ξ_0) one chooses a prime divisor $\mathfrak{p}_0$ of F/K of degree greater than 1. In particular, the proof works even if K is finite.

Assume now that K is not PAC. Then, by Theorem 11.2.3, there exists an affine curve C over K with no K-rational points. Let F be the function field of C over K. Then F/K has only finitely many prime divisors of degree 1 (Section 5.2). Let F' be as in (4). Choose a projective K-normal model Γ for F'/K [Lang4, p. 134]. By Section 5.2, each K-rational point of Γ gives rise to a prime divisor of F'/K of degree 1. Hence, Γ has no K-rational point. This proves the projective criterion for PAC fields. □

11.2 Reduction to Plane Curves

Example 11.2.9: Plane projective curves over a finite field without rational points. Since finite fields are non-PAC (Proposition 11.1.1), Remark 11.2.8 implies that for each prime power q there exists a projective curve Γ defined over $\mathbb{F}_q$ with no K-rational points. Here we give for each prime power $q = p^k$ an explicit projective plane curve Γ defined over $\mathbb{F}_q$ without $\mathbb{F}_q$-rational points. The curve Γ is defined by a homogeneous absolutely irreducible equation $f(X_0, X_1, X_2) = 0$. We distinguish between several cases.

CASE A: For $p > 3$ take $f(X_0, X_1, X_2) = X_0^{q-1} + X_1^{q-1} + X_2^{q-1}$. The absolute irreducibility of f reduces to the absolute irreducibility of the polynomial $1 + X^{q-1} + Y^{q-1}$. For the latter observe that $1 + X^{q-1}$ has simple roots in $\tilde{\mathbb{F}}_p$ and apply Eisensteins's criterion (Lemma 2.3.10) over the ring $\tilde{\mathbb{F}}_p[X]$. Alternatively, a singular point $(x_0{:}x_1{:}x_1)$ of Γ must satisfy $-x_0^{q-2} = 0$, $-x_1^{q-2} = 0$, and $-x_2^{q-2} = 0$, so $x_0 = x_1 = x_2 = 0$, which is a contradiction. Thus, Γ has no singular points. By Corollary 10.1.5, f is absolutely irreducible. Next we use the relation $x^{q-1} = 1$ for $x \in \mathbb{F}_q^\times$ to conclude that if $(x_0, x_1, x_2) \in \mathbb{F}_q^3 \smallsetminus \{(0,0,0)\}$, then $x_0^{q-1} + x_1^{q-1} + x_2^{q-1}$ is 1, 2, or 3. Since $p > 3$, this value is not 0 in $\mathbb{F}_q$.

CASE B: For $q = 3^k$ and $k \geq 2$ choose $\alpha \in \mathbb{F}_q \smallsetminus \mathbb{F}_3$ and let $f(X_0, X_1, X_2) = \alpha X_0^{q-1} + X_1^{q-1} + X_2^{q-1}$. The values of f on $\mathbb{F}_q^3 \smallsetminus \{(0,0,0)\}$ are 1, 2, α, $\alpha + 1$, or $\alpha + 2$. None of them is 0. The absolute irreducibility of f follows as in Case A.

CASE C: For $q = 3$ let $f(X_0, X_1, X_2) = X_0^6 + X_1^6 + X_2^6 + X_0^2 X_1^2 X_2^2$. Then use that $x^2 = 1$ for each $x \in \mathbb{F}_3^\times$ to observe that f takes only the nonzero values 1 and 2 on $\mathbb{F}_3^3 \smallsetminus \{(0,0,0)\}$.

Next we prove that f is irreducible over $\tilde{\mathbb{F}}_3$. Since f is homogeneous, it suffices to prove that the dehomogenized polynomial $g(X, Y) = 1 + X^6 + Y^6 + X^2 Y^2$ is irreducible in $\tilde{\mathbb{F}}_3[X, Y]$. To this end choose an indeterminate x and let y be a root of the equation $g(x, Y) = 0$. We have to prove that $[\tilde{\mathbb{F}}_3(x, y) : \tilde{\mathbb{F}}_3(x)] = 6$.

To this end let $t = x^2$, $u = y^2$, and $h(T, U) = 1 + T^3 + U^3 + TU$. Then $h(t, u) = 0$. If $h(t, U)$ is reducible in $\tilde{\mathbb{F}}_3(t)[U]$, then there are $a, b \in \tilde{\mathbb{F}}_3$ such that $1 + T^3 + (aT + b)^3 + T(aT + b) = 0$. This leads to the contradiction $a + 1 = 0$ and $a = 0$. Thus, $h(t, U)$ is irreducible, so $[\tilde{\mathbb{F}}_3(t, u) : \tilde{\mathbb{F}}_3(t)] = 3$. The same argument, with x replacing t implies that $[\tilde{\mathbb{F}}_3(x, u) : \tilde{\mathbb{F}}_3(x)] = 3$. Also, $[\tilde{\mathbb{F}}_3(x) : \tilde{\mathbb{F}}_3(t)] = 2$, so $[\tilde{\mathbb{F}}_3(x, u) : \tilde{\mathbb{F}}_3(t, u)] = 2$. By symmetry, $[\tilde{\mathbb{F}}_3(x, y) : \tilde{\mathbb{F}}_3(t, u)] = 2$. Therefore, we are reduced to proving that $\tilde{\mathbb{F}}_3(x, u) \cap \tilde{\mathbb{F}}_3(x, y) = \tilde{\mathbb{F}}_3(t, u)$.

$$\begin{array}{ccccc}
\tilde{\mathbb{F}}_3(x) & \xrightarrow{\;3\;} & \tilde{\mathbb{F}}_3(x, u) & \xrightarrow{\;2\;} & \tilde{\mathbb{F}}_3(x, y) \\
{\scriptstyle 2}\big\downarrow & & {\scriptstyle 2}\big\downarrow & & {\scriptstyle 2}\big\downarrow \\
\tilde{\mathbb{F}}_3(t) & \xrightarrow{\;3\;} & \tilde{\mathbb{F}}_3(t, u) & \xrightarrow{\;2\;} & \tilde{\mathbb{F}}_3(t, y)
\end{array}$$

Next observe that $(-1,0)$ is a zero of the affine plane curve defined by $h(T,U) = 0$ satisfying $\frac{\partial h}{\partial U}(-1,0) = -1$. Hence, by Lemma 5.2.3, $\tilde{\mathbb{F}}_3(t,u)$ has a discrete valuation v with $v(t+1) = 1$ and $v(u) > 0$. In particular, $v(t) = 0$. Hence, by Case C of Example 2.3.8, v is unramified in $\tilde{\mathbb{F}}_3(x,u)$. Assume $v(u) = 2k$ for some $k \in \mathbb{N}$. Then $v((1+t)^3) = 3$, $v(u^3) = 6k$, and $v(tu) = 2k$ are distinct integers. On the other hand, $(1+t)^3 + u^3 + tu = 0$. This contradiction to Rule (4c) of Section 2.1 implies that $2 \nmid v(u)$. It follows from Case B of Example 2.3.8 that v ramifies in $\tilde{\mathbb{F}}_3(t,y)$. Consequently, $\tilde{\mathbb{F}}_3(x,u) \cap \tilde{\mathbb{F}}_3(t,y) = \tilde{\mathbb{F}}_3(t,u)$, as claimed.

Exercise 5 gives a smooth plane projective curve over $\mathbb{F}_3$ without $\mathbb{F}_3$-rational points. That example stands in analogy to the curve appearing in Case E below.

CASE D: For $q = 2^k$ and $k \geq 3$ choose $\alpha \in \mathbb{F}_q \smallsetminus \mathbb{F}_4$ and let $f(X_0, X_1, X_2) = \alpha^2 X_0^{q-1} + \alpha X_1^{q-1} + X_2^{q-1}$. Then α is a root of no quadratic polynomial with coefficients in $\mathbb{F}_2$. Hence, $\alpha^2 X_0^{q-1} + \alpha X_1^{q-1} + X_2^{q-1} = 0$ has no nontrivial solution in $\mathbb{F}_q$.

CASE E: For $q = 4$ and $q = 2$ let

$$f(X_0, X_1, X_2) = X_0^6 + X_1^6 + X_2^6 + X_0^3 X_1^3 + X_0^3 X_2^3 + X_1^3 X_2^3 + X_0^2 X_1^2 X_2^2$$

Using that $x^3 = 1$ for each $x \in \mathbb{F}_4^\times$, we may check that $f(x_0, x_1, x_2) = 1$ if $(x_0, x_1, x_2) \in \mathbb{F}_4^3 \smallsetminus \{(0,0,0)\}$ and one of the coordinates is 0. Otherwise, $f(x_0, x_1, x_2) = 6 \cdot 1 + x_0^2 x_1^2 x_2^2 \neq 0$. It follows that $\Gamma(\mathbb{F}_4)$ is empty.

In order to prove that Γ is absolutely irreducible it suffices to prove that Γ is smooth (Corollary 10.1.5). Indeed, let $(x_0:x_1:x_2)$ be a singular point of Γ. Taking partial derivatives, we find that $f(x_0, x_1, x_2) = 0$ and

(5a) $\qquad x_0^2 x_1^3 + x_0^2 x_2^3 = 0$

(5b) $\qquad x_0^3 x_1^2 + x_1^2 x_2^3 = 0$

(5c) $\qquad x_0^3 x_2^2 + x_1^3 x_2^2 = 0$

If $x_0 = 0$, then by (5c), $x_1 = 0$ or $x_2 = 0$. In both cases $f(x_0, x_1, x_2) = 0$ implies that $x_1 = 0$ and $x_2 = 0$, which is a contradiction. Thus, $x_0 \neq 0$. Similarly, $x_1, x_2 \neq 0$.

It follows from (5) that $x_1 = \omega x_0$ and $x_2 = \omega' x_0$ where $\omega^3 = (\omega')^3 = 1$. Hence, $f(x_0, x_1, x_0) = \omega^2 (\omega')^2 x_0^6 \neq 0$, which is a contradiction. Thus, Γ has no singular point. $\square$

11.3 The PAC Property is an Elementary Statement

Remark 11.2.10: Criterion for PAC with projective plane curves. Sharpening Remark 11.2.8 and generalizing Example 11.2.9, János Kollár proves in [Kollár2, Theorem 1] that for every non-PAC field K there exists a projective plane curve without K-rational points. ◻

11.3 The PAC Property is an Elementary Statement

Many model theoretic results require the equivalence of the statement "K is a PAC field" with a conjunction of a sequence of elementary statements on K.

Let R be an integral domain with $K = \text{Quot}(R)$. Denote the set of all polynomials $f \in R[X_1, \ldots, X_n]$ with $\deg_{X_j}(f) < d$, $j = 1, \ldots, n$, by $S_R(n, d)$. Define **Kronecker substitution** to be the map $S_d \colon S_R(n, d) \to S_R(1, d^n)$ deriving from the substitution of $Y^{d^{j-1}}$ for X_j, $j = 1, \ldots, n$: If

$$f(X_1, \ldots, X_n) = \sum a_i X_1^{i_1} X_2^{i_2} \ldots X_n^{i_n}, \qquad \text{where } a_i \in R,$$

then

$$S_d(f)(Y) = \sum_i a_i Y^{i_1 + i_2 d + \cdots + i_n d^{n-1}}.$$

Note that the set of coefficients of f coincides with the set of coefficients of $S_d(f)$. Also,

$$S_d(fg) = S_d(f) \cdot S_d(g) \qquad \text{if} \qquad fg \in S_R(n, d).$$

Since the map $(i_1, i_2 \ldots, i_n) \mapsto i_1 + i_2 d + \cdots + i_n d^{n-1}$ is a bijection from the set of all n-tuples of integers between 0 and $d-1$ onto the set of integers between 0 and $d^n - 1$, the map S_d is a bijection.

LEMMA 11.3.1: *If a polynomial $f \in S_R(n, d)$ factors in $\tilde{K}$, then it factors in a finite extension of K of degree at most $(d^n - 1)!$.*

Proof: If $f = gh$ with $g, h \in \tilde{K}[X_1, \ldots, X_n]$, then $S_d(f) = S_d(g) \cdot S_d(h)$. But $S_d(f)$ is a polynomial in one variable of degree at most $d^n - 1$ with coefficients in K. Hence, the coefficients of $S_d(g)$ and of $S_d(h)$ lie in a finite extension of K of degree at most $(d^n - 1)!$. The same holds for the coefficients of g and h. ◻

It follows that a polynomial $f \in S_K(n, d)$ is absolutely irreducible if and only if it is irreducible over each separable extension of K of degree at most $(d^n - 1)!$ and if it is not a pth power of a polynomial over $\tilde{K}$ (see also Lemma 12.3.1(a)). We use the primitive element theorem to rephrase the first condition. The second condition is equivalent to f having at least one nonzero partial derivative. Thus, the absolute irreducibility of f is equivalent to the following condition:

(1) For every irreducible polynomial $h \in K[T]$ of degree at most $(d^n - 1)!$, there exist no polynomials $g_1, g_2, g_3 \in K[T, \mathbf{X}]$ such that
  (a) $\deg_T(g_i) < (d^n - 1)!$ and $\deg_{X_j}(g_i) < d$, $i = 1, 2$ and $j = 1, \ldots, n$;
  (b) $\deg_T(g_3) < 2((d^n - 1)!)$ and $\deg_{X_j}(g_3) < d$, $j = 1, \ldots, n$;
  (c) $f(\mathbf{X}) = g_1(T, \mathbf{X})g_2(T, \mathbf{X}) + g_3(T, \mathbf{X})h(T)$; and
  (d) $\frac{\partial f}{\partial X_j} \neq 0$ for some j, $1 \leq j \leq n$.

Condition (1) is equivalent to a sentence in $\mathcal{L}(\text{ring})$. Theorem 11.2.3 implies that K is PAC if and only if for $d = 1, 2, 3, \ldots$ and for every absolutely irreducible polynomial $f \in K[X, Y]$ in $S_K(2, d)$ there is a point $(a, b) \in K^2$ with $f(a, b) = 0$. Thus, the PAC property of a field is equivalent to a conjunction of sentences of $\mathcal{L}(\text{ring})$:

PROPOSITION 11.3.2: *There exist explicit sentences $\theta_1, \theta_2, \theta_3, \ldots$ in $\mathcal{L}(\text{ring})$ such that a field K is PAC if and only if it satisfies each of the θ_i's.*

COROLLARY 11.3.3: *Every ultraproduct of PAC fields is a PAC field.*

As we have already observed in the proof of Corollary 11.2.4, for each d there exists q_0 such that if $q \geq q_0$, then $\Gamma(\mathbb{F}_q) \neq \emptyset$ for each absolutely irreducible polynomial $f \in S_{\mathbb{F}_q}(2, d)$. The combination of this observation with (1) yields an analog of Corollary 11.3.3:

COROLLARY 11.3.4 ([Ax2, Prop. 10]): *Every nonprincipal ultraproduct of distinct finite fields is a PAC field.*

The following result characterizes PAC fields in terms of existential closedness.

PROPOSITION 11.3.5: *A field K is PAC if and only if K is existentially closed in every regular extension.*

Proof: Suppose first that K is PAC and let F be a regular extension of K. Each quantifier free formula $\varphi(X_1, \ldots, X_n)$ in $\mathcal{L}(\text{ring}, K)$ has, up to equivalence, the form

(2) $$\bigvee_{i \in I} \bigwedge_{j \in J_i} [f_{ij}(\mathbf{X}) = 0 \wedge g_i(\mathbf{X}) \neq 0],$$

where f_{ij} and g_i are polynomials in $K[\mathbf{X}]$. Suppose that there exists $\mathbf{x}$ in F^n that satisfies (2). By Corollary 10.2.2(c), the set of all K-specializations of $\mathbf{x}$ is an absolutely irreducible variety which is defined over K. Since $g_i(\mathbf{x}) \neq 0$ for all $i \in I$, the polynomial $\prod_{i \in I} g_i(\mathbf{X})$ does not vanish on V. It follows from Proposition 11.1.1 that V has a K-rational point $\mathbf{a}$ such that $g_i(\mathbf{a}) \neq 0$ for all i. From $f_{ij}(\mathbf{x}) = 0$ we conclude that $f_{ij}(\mathbf{a}) = 0$ for all i and j. Hence, $K \models \varphi(\mathbf{a})$. This proves that K is existentially closed in F.

Conversely, suppose that K is existentially closed in every regular extension. Let V be an absolutely irreducible variety defined over K in $\mathbb{A}^n$. Let $\mathbf{x}$ be a generic point of V over K and let $F = K(\mathbf{x})$. By Corollary 10.2.2, F

is a regular extension of K. Hence, K is existentially closed in F. Choose generators $f_1,\ldots,f_m$ for $I_K(V)$. Since $f_i(\mathbf{x}) = 0$, $i = 1,\ldots,m$, there exists $\mathbf{a} \in K^n$ with $f_i(\mathbf{a}) = 0$, $i = 1,\ldots,m$. Thus, $\mathbf{a} \in V(K)$. Consequently, K is PAC. $\square$

11.4 PAC Fields of Positive Characteristic

We have observed in Proposition 11.1.3 that a field K is PAC if and only if it has the following property:

(1) For every integral domain $R = K[x_1,\ldots,x_n]$ with quotient field $F = K(x_1,\ldots,x_n)$ regular over K there exists a K-homomorphism $\varphi\colon R \to K$.

If $\mathrm{char}(K) = p > 0$, we explain how to choose φ so as to preserve p-independence.

PROPOSITION 11.4.1 ([Tamagawa]): *Let K be a PAC field of characteristic $p > 0$ with imperfect exponent s (possibly infinite — Section 2.7). Consider an integral domain R, finitely generated over K, with quotient field F regular over K. For $m \leq s$ and $y_1,\ldots,y_m \in R$, p-independent, there exists a K-homomorphism $\varphi\colon R \to K$ such that $\varphi(y_1),\ldots,\varphi(y_m)$ are p-independent.*

Proof: Put $E = F^p K(y_1,\ldots,y_m)$ and $p^k = [E : F^p K]$. Relabel the y_i's so that $y_1,\ldots,y_k$ are p-independent over $F^p K$. Then $E = F^p K(y_1,\ldots,y_k)$ and there exist p-independent elements $a_1,\ldots,a_n \in K$, such that $y_{k+1},\ldots,y_m \in F^p(a_1,\ldots,a_n,y_1,\ldots,y_k)$. Increase (if necessary) the number of a_i's, so that $m \leq n$. This is permissible since $m \leq s$. By the exchange principle (Lemma 2.7.1), we may reorder the a_i's so that

(2) $\qquad a_{n-m+k+1},\ldots,a_n \in F^p(a_1,\ldots,a_{n-m+k},y_1,\ldots,y_m).$

With $\eta_i = (a_{n-m+i} y_i)^{1/p}$, $i = 1,\ldots,k$, define F' to be $F(\eta_1,\ldots,\eta_k)$.

$$\begin{array}{ccccccc} F^p K & \xrightarrow{p^k} & (F')^p K = E & \xrightarrow{} & F & \xrightarrow{p^k} & F' \\ \big| & & \big| & & & & \\ F^p & \xrightarrow{p^k} & (F')^p & & & & \end{array}$$

The rest of the proof divides into parts according to properties of F'.

PART A: *Separability of the extension F'/K.* Note that

$$(F')^p K = F^p K(a_{n-m+1}y_1,\ldots,a_{n-m+k}y_k) = F^p K(y_1,\ldots,y_k) = E.$$

Hence $[(F')^p K : F^p K] = [E : F^p K] = p^k$. Therefore, $\eta_1^p,\ldots,\eta_k^p$ are p-independent over $F^p K$, hence over F^p. It follows that $\eta_1,\ldots,\eta_k$ are p-independent over F. Thus, $[F' : F] = p^k$, so $[F : F^p K] = [F : E][E : F^p K] = [F : E][F' : F] = [F' : (F')^p K]$. Let r be the transcendence degree of F over K. Since F/K is separable, Lemma 2.8.3 asserts that $[F : F^p K] = p^r$. Hence, $[F' : (F')^p K] = p^r$. Another application of Lemma 2.8.3 shows that F'/K is separable.

PART B: *K is algebraically closed in F'.* Consider $\theta \in F'$ algebraic over K. Since $\theta^p \in F$ is algebraic over K, we have $\theta^p \in K$. Write θ as

(3) $$\sum u_{\mathbf{i}} \eta_1^{i_1} \ldots \eta_k^{i_k}, \quad 0 \le i_1, \ldots, i_k < p \quad \text{and} \quad u_{\mathbf{i}} \in F.$$

The expression

$$\theta^p = \sum u_{\mathbf{i}}^p (a_{n-m+1} y_1)^{i_1} \cdots (a_{n-m+k} y_k)^{i_k}$$

gives a p-dependence relation for $a_{n-m+1} y_1, \ldots, a_{n-m+k} y_k$ over $F^p K$. By Part A, these elements are p-independent over $F^p K$. Thus, for each $\mathbf{i} \ne 0$, we have $u_{\mathbf{i}} = 0$. It follows that $\theta = u_0 \in F$. Therefore, since K is algebraically closed in F, $\theta \in K$.

PART C: *The p-independence covering property for F'/K.* We construct the morphism $\varphi \colon R \to K$ asserted by the theorem. From (2) there exist elements $v_{ij} \in F$ such that for $i = n - m + k + 1, \ldots, n$

(4) $$a_i = \sum_j v_{ij}^p a_1^{j_1} \cdots a_{n-m+k}^{j_{n-m+k}} y_1^{j_{n-m+k+1}} \cdots y_m^{j_{n+k}}.$$

Consider the ring $S = R[\eta_1, \ldots, \eta_k, y_1^{-1}, \ldots, y_m^{-1}, v_{ij}]_{i,j}$ whose quotient field is F'. From Parts A and B, F' is a regular extension of the PAC field K. Hence, there exists a K-homomorphism $\psi \colon S \to K$. It satisfies

(5) $$\psi(\eta_i)^p = a_{n-m+i} \psi(y_i) \text{ and } \psi(y_i) \ne 0, \quad i = 1, \ldots, k.$$

Apply ψ to (4):

(6) $$a_i = \sum_j \psi(v_{ij})^p a_1^{j_1} \cdots a_{n-m+k}^{j_{n-m+k}} \psi(y_1)^{j_{n-m+k+1}} \cdots \psi(y_m)^{j_{n+k}},$$

$i = n - m + k + 1, \ldots, n$. By (5) and (6),

$$a_1, \ldots, a_n \in K^p(a_1, \ldots, a_{n-m}, \psi(y_1), \ldots, \psi(y_m)).$$

Thus, since $a_1, \ldots, a_n$ are p-independent over K^p, so are

$$a_1, \ldots, a_{n-m}, \psi(y_1), \ldots, \psi(y_m).$$

The restriction φ of ψ to R is the desired homomorphism. □

11.5 PAC Fields with Valuations

Like algebraically closed fields, pseudo algebraically closed fields carry little arithmetical structure. For example, a PAC field cannot be ordered.

THEOREM 11.5.1: *Let K be a PAC field. If $\text{char}(K) \neq 2$, then every element of K is the sum of two squares. In particular, the field K is not formally real.*

Proof: Assume $a \in K^\times$. Consider the polynomial $X^2 + Y^2 - a$ as a polynomial in X over the unique factorization domain $\tilde{K}[Y]$. The prime element $Y - \sqrt{a}$ divides the constant term $Y^2 - a$ to the first degree. Hence, by the Eisenstein criterion $X^2 + Y^2 - a$ is absolutely irreducible. Thus, there exist $x, y \in K$ with $x^2 + y^2 = a$. Taking $a = -1$, we conclude that K is not formally real. $\square$

An analogous result holds for Henselian fields. Recall that a field K is said to be **Henselian** with respect to a valuation v if v has a unique extension (also denoted v) to every algebraic extension of K. In particular $v(\sigma(x)) = v(x)$ for each $x \in K_s$ and $\sigma \in \text{Gal}(K)$. It follows that every algebraic extension of an Henselian field K is Henselian.

It is well known that the following property of a valued field (K, v) is equivalent to the Henselian property [Ribenboim, p. 186]:

(1) For each monic polynomial $f \in K[X]$ with coefficients in the valuation ring of v and for each element $x \in K$ such that $v(f(x)) > 0$ and $v(f'(x)) = 0$ there exists $y \in K$ such that $f(y) = 0$ and $v(y - x) > 0$.

In particular, completions of fields with respect to rank-1 valuations are Henselian (Proposition 3.5.2). For example each of the p-adic fields $\mathbb{Q}_p$ and every field of power series $K((t))$ in one variable is Henselian.

The characterization (1) of Henselian fields has the consequence that if a field K_0 is algebraically closed in a Henselian field K, then K_0 is also Henselian.

Every valued field (K, v) has a minimal separable algebraic extension (K_v, v) which is Henselian. The valued field (K_v, v) is unique up to a K-isomorphism and is called the **Henselian closure** of (K, v) [Ribenboim, p. 176]. If (K_s, w) is an extension of (K_v, v), then K_v is the decomposition field of w over K. In other words, K_v is the fixed field in K_s of the group $\{\sigma \in \text{Gal}(K) \mid \sigma(w) = w\}$. The residue field and the value group of (K_v, v) are the same as those of (K, v). Thus, (K_v, v) is an **immediate extension** of (K, v) [Ribenboim, p. 184, Cor. 1]. In addition, suppose (K', v') is a separable extension of (K, v). Let w be an extension of v'. Then each $\sigma \in \text{Gal}(K')$ satisfies $\sigma \in \text{Gal}(K_v K')$ if and only if $\sigma w = w$. Hence, $K_v K' = K'_{v'}$.

For example, the algebraic part, $\mathbb{Q}_{p,\text{alg}} = \tilde{\mathbb{Q}} \cap \mathbb{Q}_p$ of $\mathbb{Q}_p$, is the Henselian closure of $\mathbb{Q}$ with respect to the p-adic valuation.

LEMMA 11.5.2: *Let (K, v) be a valued field and w an extension of v to a valuation of $\tilde{K}$. Denote the w-closure of K in $\tilde{K}$ by E. Then the following hold:*

(a) *For each $x \in \tilde{K}^\times$ there exists $a \in K^\times$ such that $v(a) > w(x)$.*

(b) E is a field.
(c) $E \cap K_s$ is contained in the Henselian closure K_v of (K,v) which is defined as the fixed field of the decomposition group of $w|_{K_s}$.
(d) K_s is w-dense in $\tilde{K}$.

Proof of (a): Assume without loss that $w(x) > 0$. Let $\sum_{i \in I} a_i x^i = 0$ be an equation for x with nonzero coefficients a_i in K. By (4c) of Section 2.1, there are $i, j \in I$ with $i < j$ and $w(a_i x^i) = w(a_j x^j)$. So, $v\left(\frac{a_i}{a_j}\right) = (j-i)w(x) \geq w(x)$. Thus, $a = \left(\frac{a_i}{a_j}\right)^2$ is in K and satisfies $v(a) > w(x)$.

Proof of (b): We prove only that if $x \in E$ and $x \neq 0$, then $x^{-1} \in E$. The proof that E is closed under addition and multiplication is left to the reader.

Indeed, let γ be an element of $w(\tilde{K}^\times)$. Choose $\delta \in w(\tilde{K}^\times)$ such that $\delta > \max(w(x), \gamma - 2w(x))$. By assumption, there exists $a \in K$ such that $w(a - x) > \delta$. Then $w(a) = w(x)$ and $w(x^{-1} - a^{-1}) = w(a - x) + 2w(x) > \gamma$. Consequently, $x^{-1} \in E$.

Proof of (c): Let $x \in E \cap K_s$, let $\sigma \in \mathrm{Gal}(K_v)$, and let $\gamma \in w(\tilde{K}^\times)$. Choose $a \in K$ with $w(a - x) > \gamma$. Since $\sigma(w|_{K_s}) = w|_{K_s}$, this implies that $w(a - \sigma x) > \gamma$. It follows that $w(x - \sigma x) > \gamma$. Since γ is arbitrary, $x = \sigma x$. Consequently, $x \in K_v$.

Proof of (d): Assume without loss that $\mathrm{char}(K) = p > 0$. Let $x \in \tilde{K}$ and let $\gamma \in w(\tilde{K}^\times)$. Then there are a power q of p and an $a \in K_s$ with $x^q = a$. Use (a) to choose $b \in K_s^\times$ with $v(b) > \max(q\gamma - w(x), v(a))$. Then $X^q - bX - a = \prod_{i=1}^q (X - x_i)$ with $x_1, \ldots, x_q \in K_s$. It follows that $\sum_{i=1}^q w(x - x_i) = w(x^q - bx - a) = w(bx) > q\gamma$. Hence, there exists i between 1 and q with $w(x - x_i) > \gamma$. Consequently, K_s is w-dense in $\tilde{K}$. □

PROPOSITION 11.5.3 (Prestel): *Let K be a PAC field and let w be a valuation of $\tilde{K}$. Then K is w-dense in $\tilde{K}$.*

Proof: By Lemma 11.5.2(b) and Corollary 11.2.5 the w-closure of K in $\tilde{K}$ is a PAC field. Thus, we may assume that K is w-closed in $\tilde{K}$ and prove that $K = \tilde{K}$.

To this end, let $f \in K[X]$ be a separable irreducible polynomial of degree $n \geq 1$ and let $f(X) = \prod_{i=1}^n (X - x_i)$ be its factorization in $\tilde{K}[X]$. Consider $\gamma \in \Gamma = v(\tilde{K}^\times)$ and choose $c \in K^\times$ such that $w(c) \geq n\gamma$ (Lemma 11.5.2(a)). Apply Lemma 2.3.10(b) (Eisenstein's criterion) over the ring $\tilde{K}[Y]$ to a linear factor of $f(Y)$ to deduce that $f(X)f(Y) - c^2$ is absolutely irreducible. Since K is PAC, there exist $x, y \in K$ such that $f(x)f(y) = c^2$. It follows from $w(f(x)) + w(f(y)) = 2w(c)$ that $w(f(x)) \geq n\gamma$ or $w(f(y)) \geq n\gamma$. Suppose for example that the first possibility occurs. Then $\sum_{i=1}^n w(x - x_i) \geq n\gamma$. Hence, there is an i with $w(x - x_i) \geq \gamma$.

Since $\{x_1, \ldots, x_n\}$ is a finite set, the latter conclusion implies that there exists i such that for each $\gamma \in \Gamma$ there exists an $x \in K$ with $w(x - x_i) \geq \gamma$. Thus, x_i belongs to the w-closure of K, which is, by assumption, K itself.

11.5 PAC Fields with Valuations

Since f is irreducible, $n = 1$. Therefore, $K = K_s$. By Lemma 11.5.2(d), K is w-dense in $\tilde{K}$. Consequently, $K = \tilde{K}$. □

Remark 11.5.4: Density. Exploiting Prestel's trick in the proof of Proposition 11.5.3, János Kollár proves in [Kollár2, Thm. 2] that if K is a PAC field, w is a valuation of $\tilde{K}$, and V is a variety over K, Then $V(K)$ is w-dense in $V(\tilde{K})$. □

COROLLARY 11.5.5 (Frey-Prestel): *The Henselian closure K_v of a PAC field K with respect to a valuation v is K_s. Consequently, $\bar{K}_v$ is separably closed and $v(K^\times)$ is a divisible group.*

Proof: Let w be an extension of v to K_s. By Proposition 11.5.3, K is w-dense in K_s. By Lemma 11.5.2(c), the w-closure of K in K_s is contained in K_v. Hence, $K_v = K_s$.

Now recall that K and K_v have the same residue field and the same value group. Consider a monic separable polynomial $\bar{f} \in \bar{K}_v[X]$ of positive degree. Lift $\bar{f}$ to a monic polynomial $f \in O_v[X]$. Then f is separable, so has a root in K_v. The residue of this root is a root of $\bar{f}$ in $\bar{K}_v$.

Next consider $a \in K^\times$ and a positive integer n which is not divisible by char(K). Choose $x \in K_v$ with $x^n = a$. Then $\frac{1}{n}v(a) \in v((K_v)^\times) = v(K^\times)$.

Finally, suppose char$(K) = p > 0$ and $v(a) < 0$. Then there is an $x \in K_v$ with $x^p - x - a = 0$. Hence, $\frac{1}{p}v(a) = v(x) \in v(K^\times)$. Consequently, $v(K^\times)$ is divisible. □

COROLLARY 11.5.6: *Let K be a PAC field which is not separably closed. Then K is Henselian with respect to no valuation.*

Denote the maximal Abelian, nilpotent, and solvable extensions of $\mathbb{Q}$ by $\mathbb{Q}_{ab}$, $\mathbb{Q}_{nil}$ and $\mathbb{Q}_{solv}$, respectively.

COROLLARY 11.5.7 ([Frey]): *The fields $\mathbb{Q}_{ab}$ and $\mathbb{Q}_{nil}$ are not PAC fields.*

Proof: By Corollary 11.2.5 we have only to prove the statement for $\mathbb{Q}_{nil}$. For each p, $\mathbb{Q}_{p,\text{alg}} = \tilde{\mathbb{Q}} \cap \mathbb{Q}_p$ is Henselian. Assume $\mathbb{Q}_{nil}$ is PAC. Then $\mathbb{Q}_{nil}\mathbb{Q}_{p,\text{alg}}$ is PAC (Corollary 11.2.5) and Henselian. By Corollary 11.5.5, $\mathbb{Q}_{nil}\mathbb{Q}_{p,\text{alg}} = \tilde{\mathbb{Q}}$. Hence, Gal$(\mathbb{Q}_{p,\text{alg}}) \cong$ Gal$(\mathbb{Q}_{nil} \cap \mathbb{Q}_{nil}\mathbb{Q}_{p,\text{alg}})$ is **pronilpotent**. That is, the Galois group of every irreducible separable polynomial over $\mathbb{Q}_{p,\text{alg}}$ is nilpotent.

In particular, this is true for $p = 5$. However, $X^3 + 5$ is irreducible over $\mathbb{Q}_5$ (e.g. by Eisenstein's criterion) and its discriminant $-27 \cdot 5^2$ [Lang7, p. 270] is not a square in $\mathbb{Q}_{5,\text{alg}}$ (e.g. -27 is a quadratic non-residue modulo 5). Hence, $X^3 + 5$ is irreducible over $\mathbb{Q}_{5,\text{alg}}$ and its discriminant is not a square in $\mathbb{Q}_{5,\text{alg}}$. Therefore, Gal$(X^3 + 5, \mathbb{Q}_{5,\text{alg}})$ is S_3 which is not nilpotent. We conclude from this contradiction to the preceding paragraph that $\mathbb{Q}_{nil}$ is not a PAC field. □

In contrast to $\mathbb{Q}_{ab}$ and $\mathbb{Q}_{nil}$, we have $\mathbb{Q}_{solv}\mathbb{Q}_{p,\text{alg}} = \tilde{\mathbb{Q}}$ for each prime number p.

PROPOSITION 11.5.8: *The Henselian closure of each valuation of $\mathbb{Q}_{\text{solv}}$ is $\tilde{\mathbb{Q}}$.*

Proof: Let v be a valuation of $\mathbb{Q}_{\text{solv}}$. Replacing v by an equivalent valuation, we may assume that $v|_{\mathbb{Q}} = v_p$ for some prime number p. Thus, $\mathbb{Q}_{\text{solv}}$ can be embedded in $\tilde{\mathbb{Q}}_p$ such that v is the restriction to $\mathbb{Q}_{\text{solv}}$ of the unique extension of v_p to $\tilde{\mathbb{Q}}_p$. Let $\mathbb{Q}_{p,\text{alg}} = \tilde{\mathbb{Q}} \cap \tilde{\mathbb{Q}}_p$ and $M = \mathbb{Q}_{\text{solv}}\mathbb{Q}_{p,\text{alg}}$. Then M is a Henselian closure of M at v. Assume $M \neq \tilde{\mathbb{Q}}$.

Since $\text{Gal}(\mathbb{Q}_p)$ is prosolvable [Cassels-Fröhlich, p. 31, Cor. 1], so is $\text{Gal}(M)$. Since M contains all roots of unity, there exist $b \in M$ and $n \in \mathbb{N}$ such that $\sqrt[n]{b} \notin M$. If $a \in M$ is sufficiently v_p-close to M, then by Krasner, $M(\sqrt[n]{b}) \subseteq M(\sqrt[n]{a})$ [Jarden14, Lemma 12.1]. Choose $w_1, \ldots, w_m \in \mathbb{Q}_{\text{solv}}$ and $b_1, \ldots, b_m \in \mathbb{Q}_{p,\text{alg}}$ such that $b = \sum_{i=1}^{m} b_i w_i$. Since $\mathbb{Q}$ is v_p-dense in $\mathbb{Q}_p$, we may choose $a_1, \ldots, a_p$ which are v_p-close to $b_1, \ldots, b_p$. Then $a = \sum_{i=1}^{m} a_i w_i$ is v-close to b and lies in $\mathbb{Q}_{\text{solv}}$. It follows that $\sqrt[n]{a} \in \mathbb{Q}_{\text{solv}} \subseteq M$, so $\sqrt[n]{b} \in M$. We conclude from this contradiction that $M = \tilde{\mathbb{Q}}$. □

Thus, Corollary 11.5.5 fails to solve the following problem.

PROBLEM 11.5.9:
(a) *Is $\mathbb{Q}_{\text{solv}}$ a PAC field?*
(b) *Does there exist an infinite non-PAC field K of a finite transcendence degree over its prime field such that K is not formally real and all of its Henselian closures are separably closed?*

Remark 11.5.10: On Problem 11.5.9(a). Problem 11.5.9(a) leads to other problems with respectable classical connections. Here is one example: By Theorem 11.2.3 we have only to check if each absolutely irreducible curve, $f(X, Y) = 0$, with $f \in \mathbb{Q}[X, Y]$ has a point with coordinates in $\mathbb{Q}_{\text{solv}}$. Let $E = \mathbb{Q}(x, y)$ be the function field of this curve. Suppose there exists $t \in E$ such that $E/\mathbb{Q}(t)$ is an extension whose Galois closure, $\hat{E}$, is solvable over $\mathbb{Q}(t)$. Then, each specialization $t \to t_0$ such that $t_0 \in \mathbb{Q}$ and x and y are integral over the corresponding local ring extends to a specialization $(x, y) \to (x_0, y_0)$ with $x_0, y_0 \in \mathbb{Q}_{\text{solv}}$. Since there are infinitely many such specializations, $f(X, Y) = 0$ certainly has infinitely many $\mathbb{Q}_{\text{solv}}$-rational points. This idea fails, however, if the function field $F = \mathbb{C}(x, y)$ over $\mathbb{C}$ for a "general" curve, $f(X, Y) = 0$, has no subfield $\mathbb{C}(t)$ with $\text{Gal}(\hat{F}/\mathbb{C}(t))$ solvable.

The proper subfields of a "general" curve of genus $g > 1$ are of genus 0 [Fried7, p. 26-27]. Therefore, with no loss, assume that there are no proper fields between $\mathbb{C}(t)$ and F; that is, $\text{Gal}(\hat{F}/\mathbb{C}(t))$ is a primitive solvable group. A theorem of Galois implies that $[F : \mathbb{C}(t)] = p^r$ for some prime p [Burnside2, p. 202]. Combining [Fried7, p. 26] and [Ritt] one may prove that for F "general" of genus suitably large (> 6) that if $r = 1$ then it is impossible for $\text{Gal}(\hat{F}/\mathbb{C}(t))$ to be solvable. But higher values of r have not yet been excluded. □

Remark 11.5.11: On Problem 11.5.9(b). Theorem D of [Geyer-Jarden5] gives for each characteristic p (including $p = 0$) an example of an infinite non-PAC field K of characteristic p which is not formally real and all of its Henselian closures of K are separably closed. This gives an affirmative answer to Problem 11.5.1(b) of [Fried-Jarden3]. The proof is based on results and ideas of [Efrat]. It uses Galois cohomology, valuations of higher rank, and the Jacobian varieties of curves.

Problem 11.5.9(b) is a reformulation of the older problem. □

11.6 The Absolute Galois Group of a PAC Field

We show that the absolute Galois groups of PAC fields are "projective" (Theorem 11.6.2). All decidability and undecidability results on PAC fields depend on this result.

LEMMA 11.6.1: *Let L/K be a finite Galois extension, B a finite group, and $\alpha\colon B \to \mathrm{Gal}(L/K)$ an epimorphism. Then there exists a finite Galois extension F/E with $\mathrm{Gal}(F/E) = B$ such that E is a regular finitely generated extension of K, F is a purely transcendental extension of L with $\mathrm{trans.deg}(F/L) = |B|$, and $\alpha = \mathrm{res}_{F/L}$.*

Proof: Let $\{y^\beta \mid \beta \in B\}$ be a set of indeterminates of cardinality $|B|$. Put $F = L(y^\beta \mid \beta \in B)$. Define an action of B on F by $(y^\beta)^{\beta'} = y^{\beta\beta'}$ and $a^{\beta'} = a^{\alpha(\beta')}$ for $\beta, \beta' \in B$ and $a \in L$. Then let E be the fixed field of B in F. By Galois theory, F is a Galois extension of E with Galois group B [Lang7, p. 264, Thm. 1.8] and $\alpha(\beta) = \mathrm{res}_{F/L}(\beta)$ for each $\beta \in B$. Moreover, F is a finitely generated extension of K. Hence, by Lemma 10.5.1, E is a finitely generated extension of K. By construction, $E \cap L = K$ and F is a purely transcendental extension of L, so F is linearly disjoint from $\tilde{K}$ over L. Hence, EL is linearly disjoint from $\tilde{K}$ over L. It follows from Lemma 2.5.3, that E is linearly disjoint from $\tilde{K}$ over K; that is E is a regular extension of K. □

Let E be a finitely generated extension of a field K and F a finite Galois extension of E. By Remark 6.1.5, F/E has a Galois ring cover S/R. Thus, $R = K[x_1, \ldots, x_m]$ is an integrally closed domain with quotient field E; S is the integral closure of R in F, and $S = R[z]$ where, if $f = \mathrm{irr}(z, E)$, then $f'(z)$ is a unit of S (Definition 6.1.3). Every homomorphism φ_0 of R onto a field $\bar{E}$ extends to a homomorphism φ of S onto a Galois extension $\bar{F}$ of $\bar{E}$. The map φ induces an embedding $\varphi^*\colon \mathrm{Gal}(\bar{F}/\bar{E}) \to \mathrm{Gal}(F/E)$ such that $\varphi(\varphi^*(\sigma)(x)) = \sigma(\varphi(x))$ for each $\sigma \in \mathrm{Gal}(\bar{F}/\bar{E})$ and $x \in S$ (Lemma 6.1.4).

THEOREM 11.6.2 ([Ax2, p. 269]): *Let K be a PAC field, A and B finite groups, and $\rho\colon \mathrm{Gal}(K) \to A$ and $\alpha\colon B \to A$ epimorphisms. Then there exists a homomorphism $\gamma\colon \mathrm{Gal}(K) \to B$ such that $\rho = \alpha \circ \gamma$ (i.e. $\mathrm{Gal}(K)$ is* **projective**).

Proof (Haran): Denote the fixed field of $\mathrm{Ker}(\rho)$ in K_s by L. Then L is a finite Galois extension of K and ρ defines an isomorphism $\mathrm{Gal}(L/K) \to A$. Thus, we may identify A with $\mathrm{Gal}(L/K)$ and ρ with the restriction map.

Let E and F be as in Lemma 11.6.1. Let S/R be a Galois ring cover for F/E. Since K is a PAC field, there exists a K-homomorphism $\varphi_0 \colon R \to K$ (Proposition 11.1.3). Let φ be an extension of φ_0 to S which is the identity on L. Then $M = \varphi(S)$ is a Galois extension of K which contains L and φ induces an embedding $\varphi^* \colon \mathrm{Gal}(M/K) \to \mathrm{Gal}(F/E)$ such that $\mathrm{res}_{F/L} \circ \varphi^* = \mathrm{res}_{M/L}$. Compose φ^* with the map $\mathrm{res}_M \colon \mathrm{Gal}(K) \to \mathrm{Gal}(M/K)$ to obtain the desired homomorphism $\gamma \colon \mathrm{Gal}(K) \to B$ with $\rho = \alpha \circ \gamma$. $\square$

There are non PAC fields K with $\mathrm{Gal}(K)$ projective (e.g. K is finite or $K = \mathbb{C}(t)$). On the other hand, if G is a projective group, then there exists some PAC field K such that $G \cong \mathrm{Gal}(K)$ (Corollary 23.1.2).

The projectivity of the absolute Galois group of a field K is closely related to the vanishing of the Brauer group $\mathrm{Br}(K)$ of K, although it is not equivalent to it. We survey the concept of the Brauer group and prove that $\mathrm{Br}(K) = 0$ if K is PAC.

A **central simple** K-algebra is a K-algebra A whose center is K and which has no two sided ideals except 0 and A. In particular, if D is a division ring with center K, then the ring $M_n(D)$ of all $n \times n$ matrices with entries in D is a central simple K-algebra for each positive integer n [Huppert, p. 472]. Conversely, if A is a finite dimensional central simple K-algebra, then, by a theorem of Wedderburn, there exists a unique division ring D with center K and a positive integer n such that $A \cong_K M_n(D)$ [Huppert, p. 472].

Suppose A' is another finite dimensional central simple K-algebra. Then A' is **equivalent** to A if there exists a positive integer n' such that $A' \cong M_{n'}(D)$. In particular, A is equivalent to D. We denote the equivalence class of A by $[A]$ and let $\mathrm{Br}(K)$ be the set of all equivalence classes of finite dimensional central simple K-algebras.

The tensor product of two finite dimensional central simple K-algebras is again a finite dimensional central simple K-algebra [Weil6, p. 166]. Moreover, the tensor product respects the equivalence relation between finite dimensional central simple K-algebras. Hence, $[A] \cdot [B] = [A \otimes_K B]$ is an associative multiplication rule on $\mathrm{Br}(K)$. Since $A \otimes_K B \cong B \otimes_K A$, multiplication in $\mathrm{Br}(K)$ is commutative. Further, the equivalence class $[K]$ is a unit in $\mathrm{Br}(K)$, because $A \otimes_K K \cong A$. Finally, let A° be the **opposite** algebra of A. It consists of all elements a°, with $a \in A$. Addition and multiplication are defined by the rules $a^\circ + b^\circ = (a+b)^\circ$ and $a^\circ b^\circ = (ba)^\circ$. One proves that $A \otimes_K A^\circ \cong M_n(K)$, where $n = \dim_K(A)$. Thus, $[A^\circ] = [A]^{-1}$. Therefore, $\mathrm{Br}(K)$ is an Abelian group.

For each field extension L of K the map $A \mapsto A \otimes_K L$ induces a homomorphism $\mathrm{res}_{L/K} \colon \mathrm{Br}(K) \to \mathrm{Br}(L)$. The kernel of $\mathrm{res}_{L/K}$ consists of all $[A]$ such that $A \otimes_K L \cong_L M_n(L)$ for some positive integer n. If A satisfies the latter relation, then A is said to **split** over L. If L' contains L, then A also splits over L'. It is known that each A splits over K_s [Weil6, p. 167]. Thus, $\mathrm{Br}(K_s)$ is trivial.

11.6 The Absolute Galois Group of a PAC Field

Construction 11.6.3: The reduced norm of a central simple algebra. Let A be a finite dimensional central simple algebra A over a field K. Choose a K_s-isomorphism $\alpha \colon A \otimes_K K_s \to M_n(K_s)$ for some positive integer n. In particular, $\dim_K A = \dim_{K_s}(M_n(K_s)) = n^2$. Let $\{e_{ij} \mid 1 \leq i,j \leq n\}$ be a basis of A over K. Then $\mathbf{e}_{ij} = \alpha(e_{ij} \otimes 1)$, $1 \leq i,j \leq n$, form a basis of $M_n(K_s)$ over K_s. Each $a \in A$ has a unique presentation as $a = \sum_{i,j=1}^n a_{ij} e_{ij}$ with $a_{ij} \in K$. The matrix $\mathbf{a} = (a_{ij})_{1 \leq i,j \leq n}$ satisfies

$$\alpha(a \otimes 1) = \sum_{i,j}^n a_{ij} \mathbf{e}_{ij} = \big(\lambda_{kl}(\mathbf{a})\big)_{1 \leq k,l \leq n},$$

where λ_{kl} are linear forms over K_s in the n^2 variables X_{ij}. Indeed, if $\mathbf{e}_{ij} = (\varepsilon_{ij,kl})_{1 \leq k,l \leq n}$, then $\lambda_{kl}(\mathbf{X}) = \sum_{ij=1}^n \varepsilon_{ij,kl} X_{ij}$, where $\mathbf{X} = (X_{ij})_{1 \leq i,j \leq n}$.

The **reduced norm** of a is defined by

(1) $\qquad\qquad\qquad \text{red.norm}(a) = \det(\alpha(a \otimes 1)).$

If $\alpha' \colon A \otimes_K K_s \to M_n(K_s)$ is another K_s-isomorphism, then $\alpha' \circ \alpha^{-1}$ is a K_s-isomorphism of $M_n(K_s)$. By Skolem-Noether [Weil5, p. 166, Prop. 4], $\alpha' \circ \alpha^{-1}$ is a conjugation by an invertible matrix of $M_n(K_s)$. Hence, $\det(\alpha'(a \otimes 1)) = \det(\alpha(a \otimes 1))$. Thus, red.norm$(a)$ is independent of the particular choice of α. Moreover, each $\sigma \in \text{Gal}(K)$ fixes red.norm(a). Therefore, red.norm$(a) \in K$.

Indeed, σ induces an isomorphism $1 \otimes \sigma^{-1}$ of $A \otimes_K K_s$ and an isomorphism σ_n of $M_n(K_s)$. Then $\alpha' = \sigma_n \circ \alpha \circ 1 \otimes \sigma^{-1} \colon A \otimes_K K_s \to M_n(K_s)$ is a K_s-isomorphism satisfying $\alpha'(a \otimes 1) = \sigma_n(\alpha(a \otimes 1))$ for each $a \in K$. It follows that $\sigma(\text{red.norm}(a)) = \sigma(\det(\alpha(a \otimes 1))) = \det(\sigma_n(\alpha(a \otimes 1))) = \det(\alpha'(a \otimes 1)) = \text{red.norm}(a)$, as claimed.

Now let $p(\mathbf{X}) = \det(\lambda_{kl}(\mathbf{X}))$. It is a homogeneous polynomial of degree n over K_s such that $p(\mathbf{a}) \in K$ for each $\mathbf{a} \in M_n(K)$. It follows that the coefficients of p belong to K (Exercise 9).

Next observe that the linear forms λ_{kl} are linearly independent over K_s because the $n^2 \times n^2$ matrix $(\mathbf{e}_{ij})_{1 \leq i,j \leq n}$ is nonsingular. We may therefore form a change of variables $Y_{kl} = \lambda_{kl}(\mathbf{X})$. It maps $p(\mathbf{X})$ onto $\det(\mathbf{Y})$, which is an absolutely irreducible polynomial. Hence, $p(\mathbf{X})$ is also absolutely irreducible. We call $p(\mathbf{X})$ the **reduced form** of A. □

THEOREM 11.6.4: *Let K be a PAC field. Then its Brauer group $\text{Br}(K)$ is trivial.*

Proof ([Ax2, p. 269]): Assume $\text{Br}(K)$ is nontrivial. Then there exists a division ring D with center K such that $\dim_K(D) = n^2$ and $n > 1$. Let $p(\mathbf{X})$ be the associated reduced form. Since $p(\mathbf{X})$ is an absolutely irreducible polynomial (Construction 11.6.3), $p(\mathbf{X})$ has a nontrivial zero $\mathbf{a} \in M_n(K)$ (Proposition 11.1.1). In the notation of Construction 11.6.3 (with $A = D$), let $a = \sum_{i,j=1}^n a_{ij} e_{ij}$. By (1), red.norm$(a) = p(\mathbf{a}) = 0$. On the other hand, a is a nonzero element of D, hence invertible. Therefore, $\alpha(a \otimes 1)$ is a regular matrix, so red.norm$(a) = \det(\alpha(a \otimes 1)) \neq 0$. This contradiction proves that $\mathrm{Br}(K)$ is trivial. □

Remark 11.6.5: Varieties of Severi-Brauer. An alternative proof of Theorem 11.6.4 uses **varieties of Severi-Brauer**. They are varieties V which are defined over a field K and are isomorphic over K_s to $\mathbb{P}^n$ for some positive integer n. There is a bijective correspondence between K-isomorphism classes of varieties V of Severi-Brauer and equivalence classes of finite dimensional central simple K-algebras A. If V has a K-rational point, then A splits over K [Jacobson, p. 113]. In particular, if K is PAC, this implies that $\mathrm{Br}(K) = 0$. □

The connection between the projectivity of the absolute Galois group of a field K and its Brauer group is based on the canonical isomorphism

$$(2) \qquad H^2(\mathrm{Gal}(K), K_s^\times) \cong \mathrm{Br}(K)$$

[Deuring2, p. 56, Satz 1 or Serre4, §X5]. Here we assume that the reader is familiar with Galois cohomology, e.g. as presented in [Ribes1] or in [Serre9]. In particular, it follows from (2) that
(3) every element of $\mathrm{Br}(K)$ has a finite order [Ribes1, p. 138, Cor. 6.7].

For each prime number p and a profinite group G the notation $\mathrm{cd}_p(G)$ stands for the **pth cohomological dimension** of G. It is the maximal positive integer n such that $H^n(G, A)_{p^\infty} = 0$ for each torsion G-module A. Finally, $\mathrm{cd}(G) = \sup_p (\mathrm{cd}_p(G))$ is the **cohomological dimension** of G.

PROPOSITION 11.6.6: *The following conditions on a field K are equivalent:*
(a) *$\mathrm{Gal}(K)$ is projective.*
(b) *$\mathrm{cd}(\mathrm{Gal}(K)) \leq 1$.*
(c) *For each prime number $p \neq \mathrm{char}(K)$ and for each finite separable extension L of K, $\mathrm{Br}(L)_{p^\infty}$ is trivial.*

Proof of "(a) $\iff$ (b)": Let p be a prime number. By [Ribes, p. 211, Prop. 3.1], $\mathrm{cd}_p(\mathrm{Gal}(K)) \leq 1$ if and only if for every finite Galois extension L of K and for every short exact sequence

$$0 \longrightarrow (\mathbb{Z}/p\mathbb{Z})^m \longrightarrow B \xrightarrow{\alpha} \mathrm{Gal}(L/K) \longrightarrow 1$$

there exists a homomorphism $\beta \colon \mathrm{Gal}(K) \to B$ such that $\alpha \circ \beta = \mathrm{res}_{K_s/L}$. By a theorem of Gruenberg (Corollary 22.4.3), the latter condition holds for all p if and only if $\mathrm{Gal}(K)$ is projective. Consequently, (a) and (b) are equivalent.

Proof of "(b) $\iff$ (c)": Let p be again a prime number. First suppose that $p \neq \mathrm{char}(K)$. Then $\mathrm{cd}_p(\mathrm{Gal}(K)) \leq 1$ if and only if $\mathrm{Br}(L)_{p^\infty}$ is trivial for every finite separable extension L of K [Ribes, p. 261, Cor. 3.7]. If $p = \mathrm{char}(K)$, then $\mathrm{cd}_p(\mathrm{Gal}(L)) \leq 1$ for every field L of characteristic p [Ribes, p. 256, Thm. 3.3]. Since the Brauer group of each field is torsion (by (3)), this establishes the equivalence of (b) and (c). □

11.7 A non-PAC Field K with K_{ins} PAC

PROPOSITION 11.6.7 ([Ribes, p. 264, Prop. 3.10]): *The following conditions on a field K are equivalent.*
(a) $\text{Br}(L)$ *is trivial for every finite separable extension L of K.*
(b) *The norm map*, $\text{norm}: N^\times \to L^\times$, *is surjective for every finite separable extension L of K and for every finite Galois extension N of L.*

We summarize consequences of the previous results for PAC fields:

COROLLARY 11.6.8: *The following statements hold for every PAC field K:*
(a) $\text{Gal}(K)$ *is projective.*
(b) $\text{Br}(K)$ *is trivial.*
(c) $\text{cd}(\text{Gal}(K)) \leq 1$.
(d) *The map* $\text{norm}: N^\times \to K^\times$ *is surjective for each finite Galois extension N of K.*

Proof: Let L be a finite separable extension of K. By Corollary 11.2.5, L is PAC. Hence, by Theorem 11.6.4, $\text{Br}(L)$ is trivial. Therefore, $\text{cd}(\text{Gal}(K)) \leq 1$ (Proposition 11.6.6) and the norm map $N^\times \to K^\times$ is surjective for each finite Galois extension N of K. □

Example 11.6.9 (Geyer): We construct an example of a field K with $\text{Gal}(K)$ projective, a finite Galois extension K' of K, and an element u of K which is not a norm of an element of K'. By Proposition 11.6.7, K has a finite separable extension L such that $\text{Br}(L) \neq 0$. This will show that it is impossible to omit the condition "$p \neq \text{char}(K)$" in Condition (c) of Proposition 11.6.6.

We start from a transcendental element u over $\mathbb{F}_2$ and let $K_0 = \mathbb{F}_2(u)_s$. Then choose an transcendental element t over K_0 and let $K = K_0(t)$. By Tsen's Theorem, $\text{Gal}(K)$ is projective [Ribes2, p. 276 or Jarden17, Thm. 1.1]. Consider the Artin-Schreier extension $K' = K(x)$ of K with $x^2 + x + t = 0$. Each element y of K' has the form $y = v + wx$ with $v, w \in K$ and $\text{norm}_{K'/K}(y) = (v + wx)(v + w(1 + x)) = v^2 + vw + w^2 t$. Write $v = \frac{f(t)}{h(t)}$ and $w = \frac{g(t)}{h(t)}$, where $f, g, h \in K_0[t]$ and $h \neq 0$. Let a (resp. b, c) be the leading coefficient of f (resp. g, h). If $\text{norm}_{K'/K} y = u$, then $f(t)^2 + f(t)g(t) + g(t)^2 t = h(t)^2 u$. Compare the leading coefficients of both sides of this equality. If $\deg(f) > \deg(g)$, then $a^2 = c^2 u$. If $\deg(f) \leq \deg(g)$, then $b^2 = c^2 u$. In both cases we find that u is a square in $\mathbb{F}_2(u)_s$, which is not the case. This contradiction proves that u is not a norm of an element of K'. □

11.7 A non-PAC Field K with K_{ins} PAC

Let L/K be a purely inseparable extension of fields. If K is PAC, then so is L (Corollary 11.2.5). Problem 12.4 of [Geyer-Jarden3] asks whether the converse is true. An example of Hrushovski shows that this is not the case. The main ingredient of this example is the analog of Mordell conjecture for function fields:

PROPOSITION 11.7.1 (Grauert-Manin [Samuel, pp. 107 and 118]): *Let K be a finitely generated regular extension of a field K_0 and C a nonconstant curve over K/K_0. Suppose the genus of C over $\tilde{K}$ is at least 2. Then $C(K)$ is a finite set.*

Here we say that C is a **nonconstant curve** over K/K_0 if C is defined over K and if C is not birationally equivalent over $\tilde{K}$ to a curve C_0 defined over $\tilde{K}_0$.

LEMMA 11.7.2: *Let $F = K(x_1, \ldots, x_n)$ be a finitely generated extension of a field K of positive characteristic p. Suppose K is algebraically closed in F. Then*

(1) $$\bigcap_{k=1}^{\infty} K(x_1^{p^k}, \ldots, x_n^{p^k}) = K.$$

Proof: Denote the left hand side of (1) by F_0. First suppose that K is perfect. Thus, $K^p = K$, so $F_0 = F_0^p$ is also perfect. In addition F_0, as a subfield of F, is finitely generated over K (Lemma 10.5.1). Assume F_0 is transcendental over K. Choose a transcendental basis $t_1, \ldots, t_r$ with $r \geq 1$. Then F_0 has a finite degree over $E = K(t_1, \ldots, t_r)$. On the other hand, since F_0 is perfect, $E(t_1^{1/p^m})$ is contained in F_0 and has degree p^m over E for each positive integer m. This contradiction proves that F_0 is algebraic over K. Since K is algebraically closed in F, we conclude that $F_0 = K$.

In the general case K_{ins} is a perfect field. Hence, by the preceding paragraph,

$$F_0 \subseteq F \cap \bigcap_{k=1}^{\infty} K_{\text{ins}}(x_1^{p^k}, \ldots, x_n^{p^k}) = F \cap K_{\text{ins}} = K.$$

Therefore, $F_0 = K$. □

LEMMA 11.7.3: *Let K be a finitely generated regular transcendental extension of a field K_0 of positive characteristic p. Let C be a curve which is defined over K and whose genus over $K\tilde{K}_0$ is at least 2. Let F be a finitely generated regular extension of K. Suppose C is a nonconstant curve over F/K_0. Then K has a finitely generated extension E which is contained in F such that F/E is a finite purely inseparable extension and $C(K) = C(E)$.*

Proof: Let $F = K(x_1, \ldots, x_n)$, and for each k write $F_k = K(x_1^{p^k}, \ldots, x_n^{p^k})$. By Lemma 11.7.2, the intersection of all F_k is K. Since $F\tilde{K}_0/K\tilde{K}_0$ is a regular extension, the genus of C over $F\tilde{K}_0$ is the same as the genus of C over $K\tilde{K}_0$ (Proposition 3.4.2(b)), so at least 2. By Proposition 11.7.1, $C(F)$ is a finite set. Hence, there exists a positive integer k such that $C(F_k) = C(K)$, so $E = F_k$ satisfies the assertion of the Lemma. □

11.7 A non-PAC Field K with K_{ins} PAC

Remark 11.7.4: *On Möbius transformations.* Let K be a field and x an indeterminate. To each matrix $A = \begin{pmatrix} a & b \\ c & d \end{pmatrix}$ in $\text{GL}(2,K)$ we associate a **Möbius transformation** τ_A (also called a **linear fractional transformation**). It is the K-isomorphism of $K(x)$ into $K(x)$ defined by the following rule:

$$\tau_A(x) = \frac{ax+b}{cx+d}. \tag{3}$$

If B is another matrix in $\text{GL}(2,K)$, then $\tau_A \circ \tau_B = \tau_{BA}$. If I is the unit matrix, then τ_I is the identity map of $K(x)$. In particular, $\tau_{A^{-1}} \circ \tau_A = \text{id}$, so τ_A is an automorphism of $K(x)/K$.

If $k \in K^\times$, then $\tau_{kI} = \text{id}$. Conversely, if $\tau_A = \text{id}$, then, by (3), $c(x')^2 + (d-a)(x') - b = 0$ for all $x' \in K$. Hence, $c = b = 0$ and $d = a$, so $A = aI$. Therefore, the kernel of the map $A \mapsto \tau_A$ consists of the group of scalar matrices. If τ is arbitrary element of $\text{Aut}(K(x)/K)$, then $K(x) = K(\tau(x))$. Hence, by Example 3.2.4, there exists $A \in \text{GL}(2,K)$ such that $\tau(x) = \tau_A(x)$. Thus, the map $A \mapsto \tau_A$ defines an isomorphism

$$\text{PGL}(2,K) \cong \text{Aut}(K(x)/K). \tag{4}$$

Substituting elements of $K \cup \{\infty\}$ in (3), we may also view τ_A as a bijective map of $K \cup \{\infty\}$ onto itself. For example, if $c \neq 0$, then $\frac{a \cdot \infty + b}{c \cdot \infty + d} = \frac{a}{c}$. Note that since the pairs (a,b) and (c,d) are linearly independent over K, no $x' \in K$ satisfies both $ax' + b = 0$ and $cx' + d = 0$. Hence, the value of $\tau_A(x')$ is well defined. The arithmetic with ∞ becomes clearer if we substitute $x = \frac{x_1}{x_0}$ in (3) and view τ_A as a bijective map of $\mathbb{P}^1(K)$ onto itself:

$$\tau_A(x_0{:}x_1) = (cx_1 + dx_0 : ax_1 + bx_0). \tag{5}$$

□

The map $x' \mapsto \tau_A(x')$ of $K \cup \{\infty\}$ onto itself defines determines τ_A. This is one of the consequences of Lemma 11.7.5:

LEMMA 11.7.5: Let K/K_0 be an extension of fields. Let (x_1,x_2,x_3) and (y_1,y_2,y_3) be triples of distinct elements of $K \cup \{\infty\}$. Then there is a unique Möbius transformation τ over K such that $\tau(x_i) = y_i$, $i = 1,2,3$. Moreover, τ can be presented as τ_A, where the entries of A belong to the field $K_0(x_1,x_2,x_3,y_1,y_2,y_3)$.

Proof of uniqueness: If $\tau_i(x_j) = y_j$ for $i = 1,2$ and $j = 1,2,3$, then $\tau = \tau_1^{-1}\tau_2$ satisfies $\tau(x_j) = x_j$ for $j = 1,2,3$. Suppose first that none of the x_j is ∞ and $\tau = \tau_A$ with $A = \begin{pmatrix} a & b \\ c & d \end{pmatrix}$. Then $cx_j^2 + (d-a)x_j - b = 0$ for $j = 1,2,3$. Hence, $b = c = 0$ and $a = d$.

Now assume that $x_1 = \infty$; that is, x_1 is $(0{:}1)$ in homogeneous coordinates. By (5), $c = 0$. Then we may assume that $d = 1$ and conclude from $ax_j + b = x_j$, $j = 2,3$, that $a = 1$ and $b = 0$.

In both cases $\tau = \text{id}$ and $\tau_1 = \tau_2$.

214   Chapter 11. Pseudo Algebraically Closed Fields

Proof of existence: The Möbius transformation $\tau(x) = \frac{1}{x-x'}$ maps the element x' of K onto ∞. Likewise, $\tau(x) = \frac{1}{x}$ exchanges 0 and ∞. Hence, we may assume that $x_1 = \infty$ and $y_1 = \infty$. Since $x_2 \neq x_3$,

$$\tau(x) = \frac{y_2 - y_3}{x_2 - x_3}x + \frac{x_2 y_3 - x_3 y_2}{x_2 - x_3}$$

maps x_i onto y_i, $i = 1, 2, 3$, as desired. $\square$

Remark 11.7.6: *Conservation of branch points.* Let K be an algebraically closed field, x an indeterminate, and F a finite separable extension of $K(x)$. Then F/K is a function field of one variable. For each $a \in K \cup \{\infty\}$ let $\varphi_a \colon K(x) \to K \cup \{\infty\}$ be the K-place of $K(x)$ with $\varphi_a(x) = a$. Denote the corresponding prime divisor of $K(x)/K$ by $\mathfrak{p}_a$. We say that a is a **branch point** of $F/K(x)$ (with respect to x) if $\mathfrak{p}_a$ ramifies in F. There are only finitely many prime divisors of $K(x)/K$ which ramify in F (Section 3.4), so $F/K(x)$ has only finitely many branch points.

If τ is a Möbius transformation of $K(x)$ and $\tau(x) = x'$, then τ maps the set of branch points of $F/K(x)$ with respect to x onto the set of branch points of $F/K(x)$ with respect to x'. By Lemma 11.7.5, there exists a Möbius transformation τ of $K(x)$ mapping ∞ onto a finite nonbranch point of $F/K(x)$. Replacing x by $\tau(x)$, if necessary, we may assume that ∞ is not a branch point.

Let $S = K[y_1, \ldots, y_n]$ be the integral closure of $K[x]$ in F. Then S is a Dedekind domain (Proposition 2.4.6). Denote the curve generated in $\mathbb{A}^n$ over K by $\mathbf{y}$ by C. The local ring $O_{C,\mathbf{b}}$ of C at each $\mathbf{b} \in C(K)$ is the local ring of S at the kernel of the K-homomorphism mapping $\mathbf{y}$ onto $\mathbf{b}$. This kernel is a maximal ideal of S. Hence, $O_{C,\mathbf{b}}$ is a valuation ring. The corresponding prime divisor $\mathfrak{q}_\mathbf{b}$ of F/K is uniquely determined by $\mathbf{b}$. Conversely, each prime divisor $\mathfrak{q}$ of F/K with $\varphi_\mathfrak{q}$ finite at x is also finite on S and $\varphi_\mathfrak{q}(\mathbf{y}) \in C(K)$. Thus, the map $\mathbf{b} \mapsto \mathfrak{q}_\mathbf{b}$ is a bijective correspondence between $C(K)$ and the set of prime divisors of F/K which are finite at x.

Choose a polynomial $f \in K[Y_1, \ldots, Y_n]$ such that $x = f(\mathbf{y})$. Then f defines a morphism $\pi \colon C \to \mathbb{A}^1$ by $\pi(\mathbf{b}) = f(\mathbf{b})$ for each $\mathbf{b} \in C(K)$. The prime divisor $\mathfrak{q}_\mathbf{b}$ of F/K lies over $\mathfrak{p}_a$ if and only if $\pi(\mathbf{b}) = a$.

Now let $\mathfrak{p}$ be a prime divisor of $K(x)/K$ which is finite at x. Denote the prime divisors of F/K lying over $\mathfrak{p}$ by $\mathfrak{q}_1, \ldots, \mathfrak{q}_r$ and the corresponding ramification indices by $e_1, \ldots, e_r$. Since K is algebraically closed, $\sum_{i=1}^r e_i = [F : K(x)]$ (Proposition 2.3.2). Hence, $\mathfrak{p}$ is ramified in F if and only if $r < [F : K(x)]$. Therefore, by the preceding two paragraphs, an element a of K is a branch point of $F/K(x)$ if and only if there are less than $[F : K(x)]$ points of $C(K)$ lying over a.

Next consider an algebraically closed subfield K_0 of K. Suppose there is a function field F_0 over K_0 which contains $x, y_1, \ldots, y_n$ such that $F_0 K = F$ and f has coefficients in K_0. Then $[F_0 : K_0(x)] = [F : K(x)]$ and C is already defined over K_0. By the preceding paragraph, "**a** is a branch point

11.7 A non-PAC Field K with K_{ins} PAC

with respect to x" is an elementary statement in the language $\mathcal{L}(\text{ring}, K_0)$. Let $a_1, \ldots, a_m$ be all branch points of $F_0/K_0(x)$. Then "a is a branch point with respect to x if and only if $a = a_i$ for some i between 1 and m" is an elementary statement which holds over K_0. Since K_0 is an elementary subfield of K (Corollary 9.3.2), the same statement holds over K. It follows that each branch point of $F/K(x)$ with respect to x belongs to K_0. □

REMARK 11.7.7: *Construction of nonconstant curves.* Let K_0 be a field. Choose five distinct elements $t_1, \ldots, t_5$ in some regular extension of K_0 such that $t_4 \notin \tilde{K}_0(t_1, t_2, t_3)$. Consider a regular field extension K of K_0 containing t_1, t_2, t_3, t_4, t_5. Put $f(X) = \prod_{i=1}^{5}(X - t_i)$ and define a curve C over K by the equation $Y^2 = f(X)$ if $\text{char}(K) \neq 2$ and $Y^2 + Y = \frac{1}{f(X)}$ if $\text{char}(K) = 2$. By Propositions 3.8.2 and 3.8.4, C is a hyperelliptic curve. More precisely, the genus g of C is 2 if $\text{char}(K) \neq 2$ and 4 if $\text{char}(K) = 2$.

CLAIM: *C is a nonconstant curve over K/K_0.*

Proof: Choose a generic point (x, y) for C over K and let $F = \tilde{K}(x, y)$. Then $\tilde{K}(x)$ is a quadratic subfield of F. Assume that C is a constant curve over K/K_0. Then there exists a function field of one variable F_0 over $\tilde{K}_0$ such that $F_0 \tilde{K} = F$. Since $F_0/\tilde{K}_0$ is a regular extension, F_0 is linearly disjoint from $\tilde{K}$ over $\tilde{K}_0$ (Lemma 2.6.7). In particular, $g = \text{genus}(F_0/\tilde{K}_0) = \text{genus}(F/\tilde{K})$ (Proposition 3.4.2). Denote the canonical divisor of $F_0/\tilde{K}_0$ by $\mathfrak{w}_0$ and denote the image of $\mathfrak{w}_0$ in $\text{Div}(F/\tilde{K})$ by $\mathfrak{w}$. By Proposition 3.4.2 and Lemma 3.2.2(b), $\deg(\mathfrak{w}) = \deg(\mathfrak{w}_0)$ and $\dim(\mathfrak{w}) = \dim(\mathfrak{w}_0)$. Hence, $\mathfrak{w}$ is the canonical divisor of $F/\tilde{K}$ (Exercise 1(b) of Chapter 3). Choose a basis $z_1, z_2, \ldots, z_g$ for $\mathcal{L}(\mathfrak{w}_0)$ over $\tilde{K}_0$. Then, by the linear disjointness, $z_1, z_2, \ldots, z_g$ form a basis for $\mathcal{L}(\mathfrak{w})$ over $\tilde{K}$. By Proposition 3.7.4, $\tilde{K}(\frac{z_2}{z_1}, \ldots, \frac{z_g}{z_1})$ is the unique quadratic subfield of F, so $\tilde{K}(\frac{z_2}{z_1}, \ldots, \frac{z_g}{z_1}) = \tilde{K}(x)$. In particular, $\tilde{K}(\frac{z_2}{z_1}, \ldots, \frac{z_g}{z_1})$ is a function field of genus 0 over $\tilde{K}$. Hence, $\tilde{K}_0(\frac{z_2}{z_1}, \ldots, \frac{z_g}{z_1})$ is a function field of genus 0 over $\tilde{K}$. By Example 3.2.4, there is an x_0 with $\tilde{K}_0(\frac{z_2}{z_1}, \ldots, \frac{z_g}{z_1}) = \tilde{K}_0(x_0)$. It satisfies $\tilde{K}(x_0) = \tilde{K}(x)$. By Lemma 11.7.5, there exists a Möbius transformation τ over $\tilde{K}$ such that $\tau(x) = x_0$.

For each i between 1 and 5 let $\varphi_i \colon \tilde{K}(x) \to \tilde{K} \cup \{\infty\}$ be the $\tilde{K}$-place with $\varphi_i(x) = t_i$. By Examples 2.3.8 and 2.3.9, t_i is a branch point of $F/\tilde{K}(x)$ with respect to x. Put $a_i = \tau(t_i)$. Then $\varphi_i(x_0) = \varphi_i(\tau(x)) = \tau(\varphi_i(x)) = \tau(t_i) = a_i$. Thus, a_i is a branch point of $F/\tilde{K}(x_0)$ with respect to x_0. It follows from Remark 11.7.6 that $a_i \in \tilde{K}_0$.

By Lemma 11.7.5, τ is already defined over $\tilde{K}_0(t_1, t_2, t_3)$. Hence, $t_4 = \tau^{-1}(a_4) \in \tilde{K}_0(t_1, t_2, t_3)$. This contradiction to the assumption we made above proves that C is a nonconstant curve over K/K_0. □

THEOREM 11.7.8 ([Hrushovski, Cor. 5]): *For each prime number p there exists a countable non-PAC field E of characteristic p such that E_{ins} is PAC.*

Proof: Choose algebraically independent elements t_1, t_2, t_3, t_4, t_5 over $\mathbb{F}_p$.

Let C be the curve defined over $K = \mathbb{F}_p(t_1, t_2, t_3, t_4, t_5)$ in Remark 11.7.7. Then, C is a nonconstant curve over $F/\mathbb{F}_p$ for every regular field extension F of $\mathbb{F}_p$ that contains K. By Proposition 11.7.1, $C(K)$ is a finite set.

By induction we construct two ascending towers of fields $K = E_1 \subseteq E_2 \subseteq E_3 \subseteq \cdots$ and $F_1 \subseteq F_2 \subseteq F_3 \subseteq \cdots$ and for each positive integer m we enumerate the absolutely irreducible varieties which are defined over E_m in a sequence, $V_{m1}, V_{m2}, V_{m3}, \ldots$ such that

(6a) E_m and F_m are finitely generated regular extensions of K,
(6b) F_m is a finite purely inseparable extension of E_m,
(6c) $C(E_m) = C(K)$, and
(6d) $V_{ij}(F_m) \neq \emptyset$ for $i, j = 1, \ldots, m-1$.

Indeed, suppose $E_1, \ldots, E_{m-1}$, $F_1, \ldots, F_{m-1}$, and V_{ij} for $i < m$ and all j have been defined such that they satisfy (6). Let V be the direct product of V_{ij} for $i, j = 1, \ldots, m-1$. It is an absolutely irreducible variety defined over E_{m-1}. Let $\mathbf{x}$ be a generic point of V over E_{m-1}. Then $E'_m = E_{m-1}(\mathbf{x})$ is a finitely generated regular extension of E_{m-1} and therefore also of K. Apply Lemma 11.7.3 to E_{m-1}, E'_m (instead K, F) and construct an extension E_m of E_{m-1} which is contained in E'_m such that E'_m/E_m is a finite purely inseparable extension and $C(E_m) = C(E_{m-1})$. By (6c) for $m-1$ we have $C(E_m) = C(K)$. Since E'_m is a regular extension of E_{m-1}, it is linearly disjoint from $F_{m-1}\tilde{K}$ over E_{m-1}. Hence, $F_m = E'_m F_{m-1}$ is linearly disjoint from $F_{m-1}\tilde{K}$ over F_{m-1}.

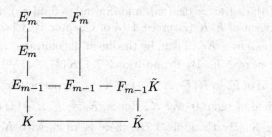

Since F_{m-1} is linearly disjoint from $\tilde{K}$ over K, F_m is linearly disjoint from $\tilde{K}$ over K. Thus, F_m is a regular extension of K. By construction, F_m is a finite purely inseparable extension of E_m and $V_{ij}(F_m) \neq \emptyset$ for $i, j = 1 \ldots, m-1$.

Let $E = \bigcup_{m=1}^{\infty} E_m$ and $F = \bigcup_{m=1}^{\infty} F_m$. Then E and F are countable regular extensions of K and F is purely inseparable over E. Hence, in order to prove that E_{ins} is PAC it suffices, by Theorem 11.2.3, to prove that each absolutely irreducible variety V defined over E has an F-rational point.

Indeed, if V is such a variety, then $V = V_{ij}$ for some i and j. Let $m = \max\{i, j\} + 1$. By (6d), V has an F_m-rational point, which is, of course, an F-rational point.

Finally, each point of $C(E)$ belongs to $C(E_m)$ for some m and therefore, by (6c), to $C(K)$. Thus, $C(E) = C(K)$ is a finite set. By Proposition 11.1.1, E is not PAC. □

PROBLEM 11.7.9: *Does there exists a finitely generated field extension E of $\mathbb{F}_p$ such that E is not PAC but E_{ins} is PAC?*

Exercises

1. Let V be an absolutely irreducible variety which is defined over a PAC field K. Suppose that A and B are Zariski K-closed subset of V such that $V(K) = A(K) \cup B(K)$. Use Proposition 11.1.1 to prove that $V = A$ or $V = B$. Conclude that $V(K)$ is connected in the Zariski-topology.

2. By the Artin-Schreier theorem, the only torsion that occurs in the absolute Galois group of a field K comes from real closed fields. Prove that if K is PAC, then $\mathrm{Gal}(K)$ is torsion free.

3. A more elementary characterization of PAC fields, just as useful as Theorem 11.2.3, can be derived from Lemma 10.4.1. Prove that a field K is PAC if and only if for each nonconstant absolutely irreducible polynomial $f \in K[T_1, \ldots, T_r, X]$, separable in X, and for every nonzero polynomial $g \in K[T_1, \ldots, T_r]$, there exist $a_1, \ldots, a_r, b \in K$ such that $f(\mathbf{a}, b) = 0$ and $g(\mathbf{a}) \neq 0$.

4. Prove the following assertion without using descent: If K is a PAC field of characteristic $p > 0$, then K_{ins} is also a PAC field.

5. Let Γ be the plane curve defined over $\mathbb{F}_3$ by the equation $X_0^4 + X_1^4 - X_2^4 - X_0^2 X_2^2 + X_1^3 X_2 = 0$. Prove that Γ is absolutely irreducible and has no $\mathbb{F}_3$-rational point. Hint: Show that Γ is smooth.

6. (a) Let $f \in L[X_1, \ldots, X_n]$ be a nonconstant homogeneous polynomial with at least one absolutely irreducible factor of multiplicity 1. Prove that if $a \in L^\times$, then $f(X_1, \ldots, X_n) - a$ is absolutely irreducible.

   (b) Let M/L be a finite Galois extension with basis $w_1, \ldots, w_n$. Describe the norm map $\mathrm{norm}_{M/L} \colon M \to L$ as a homogeneous polynomial of degree n in $X_1, \ldots, X_n$.

   (c) Combine (a) and (b) to show that if L is a finite separable extension of a PAC field K, then $\mathrm{norm}_{M/L}$ is surjective. Thus, show directly property (b) of Lemma 11.6.7.

7. This exercise suggest an alternative way to achieve some consequences of Corollary 11.5.5.

   (a) Use Eisenstein's criterion to prove that the polynomial $f(X, Y) = (X^q - X)(Y^q - Y) + 1$ defined over the field $\mathbb{F}_q$ (q a prime power) is absolutely irreducible and has no $\mathbb{F}_q$-rational zeros. Deduce that a PAC field is infinite.

   (b) (Ax) Let K be a field with a valuation v whose residue class field has q elements. Let $\pi \in K$ with $v(\pi) > 0$ and show that $(X^q - X - 1)(Y^q - Y - 1) - \pi = 0$ defines an absolutely irreducible variety with no K-rational point. Thus, show directly that K is not a PAC field. Conclude, in particular, that fields finitely generated over their prime fields are not PAC fields.

8. Use the idea included in the proof of Corollary 11.5.7 to show that $\mathrm{Gal}(\mathbb{Q}_p)$ is pronilpotent for no prime p.

9. Let K be a field, L an infinite extension of K, and $f \in L[X_1, \ldots, X_n]$. Prove that if $f(\mathbf{a}) \in K$ for each $\mathbf{a} \in L^n$, then the coefficients of f belong to K. Hint: For $n = 1$ use Cramer's rule to compute the coefficients of f. Then continue by induction on n.

Notes

Apparently PAC fields appear for the first time in J. Ax's papers [Ax1] and [Ax2]. Ax observes that the Riemann Hypothesis for curves implies that nonprincipal ultraproducts of finite fields are PAC fields. From this he establishes a recursive decision procedure for the theory of finite fields. Following a suggestion of the second author, Frey introduces in [Frey] the name PAC for fields over which each variety has rational points. He also proves a place theoretic predecessor to Lemma 11.2.1. Remark 11.2.1, due to János Kollár, settles Problem 11.2.10 of the second edition.

Our proof of Proposition 11.4.1 appears to be simpler than Tamagawa's.

In [Frey] Frey uses the theory of homogeneous spaces attached to elliptic curves and Tate's theory of bad reduction in order to prove Corollary 11.5.5 for real valuations. Prestel showed the authors the elementary direct proof of the Corollary for arbitrary valuations (private communication). A variation of that proof serves also for Proposition 11.5.3.

Chapter 12. Hilbertian Fields

David Hilbert proved his celebrated irreducibility theorem during his attempt to solve a central problem of Galois theory: Is every finite group realizable over $\mathbb{Q}$? He proved that a general specialization of the coefficients of the general polynomial of degree n to elements of $\mathbb{Q}$ gives a polynomial whose Galois group is S_n. Further, if $f \in \mathbb{Q}[T_1, \ldots, T_r, X]$ is an irreducible polynomial, then there exist $a_1, \ldots, a_r \in \mathbb{Q}$ such that $f(\mathbf{a}, X)$ remains irreducible. This result is now known as **Hilbert's irreducibility theorem**. Since then, many more finite groups have been realized over $\mathbb{Q}$. Most of those have been realized via Hilbert's theorem. This has brought the theorem to the center of the theory of fields.

Various alternative proofs of the irreducibility theorem apply to other fields (including all infinite finitely generated fields). We call them **Hilbertian fields**. We give several reductions of the irreducibility theorem, and one especially valuable result (Corollary 12.2.3):

Let K be a Hilbertian field, L a finite separable extension of K, and $f \in L[T_1, \ldots, T_r, X]$ an irreducible polynomial. Then there exist $a_1, \ldots, a_r \in K$ such that $f(\mathbf{a}, X)$ is irreducible in $L[X]$.

Chapters 13 and 15 give a proof of the Hilbertian property for most known Hilbertian fields. This lays the foundation to a subject central to the book, the model theory of PAC fields.

12.1 Hilbert Sets and Reduction Lemmas

Consider a field K and two sets $T_1, \ldots, T_r$ and $X_1, \ldots, X_n$ of variables. Let $f_1(\mathbf{T}, \mathbf{X}), \ldots, f_m(\mathbf{T}, \mathbf{X})$ be polynomials in $X_1, \ldots, X_n$ with coefficients in $K(\mathbf{T})$. Assume these are irreducible in the ring $K(\mathbf{T})[\mathbf{X}]$. For $g \in K[\mathbf{T}]$ a nonzero polynomial, denote the set of all r-tuples $(a_1, \ldots, a_r) \in K^r$ with $g(\mathbf{a}) \neq 0$ and $f_1(\mathbf{a}, \mathbf{X}), \ldots, f_m(\mathbf{a}, \mathbf{X})$ defined and irreducible in $K[\mathbf{X}]$ by $H_K(f_1, \ldots, f_m; g)$. Call $H_K(f_1, \ldots, f_m; g)$ a **Hilbert subset** of K^r. If in addition $n = 1$ and each f_i is separable in X, call $H_K(f_1, \ldots, f_m; g)$ a **separable Hilbert subset** of K^r. A **Hilbert set** (resp. separable Hilbert set) of K is a Hilbert subset (resp. separable Hilbert subset) of K^r for some positive integer r.

The intersection of finitely many Hilbert subsets of K^r is again a Hilbert subset of K^r. Hence, if each Hilbert subset of K^r is nonempty, then each Hilbert subset of K^r is Zariski K-dense in K^r. By Lemma 10.2.5, each Hilbert subset of K^r is Zariski dense in K^r. The same holds for separable Hilbert sets.

We say that K is **Hilbertian** if each separable Hilbert set of K is nonempty. In particular, a Hilbertian field must be infinite.

The next lemmas reduce the infiniteness of arbitrary Hilbert sets (resp. separable Hilbert sets) to the infiniteness of special Hilbert sets (resp. separable Hilbert sets).

LEMMA 12.1.1: *Each Hilbert subset (resp. separable Hilbert subset) $H_K(f_1, \ldots, f_m; g)$ of K^r contains a Hilbert subset (resp. separable Hilbert subset) $H_K(f_1', \ldots, f_m'; g')$ of K^r with f_i' irreducible in $K[\mathbf{T}, \mathbf{X}]$ and $f_i' \notin K[\mathbf{T}]$, $i = 1, \ldots, m$.*

Proof: By assumption, f_i is irreducible in $K(\mathbf{T})[\mathbf{X}]$. Hence, at least one X_j occurs in f_i. Multiply $f_i(\mathbf{T}, \mathbf{X})$ by a polynomial, $g_i(\mathbf{T})$, to ensure that its coefficients lie in $K[\mathbf{T}]$. Then divide the resulting polynomial by the greatest common divisor, $d_i(\mathbf{T})$, of its coefficients. Since $K[\mathbf{T}, \mathbf{X}]$ has unique factorization, we obtain an irreducible polynomial $f_i' \in K[\mathbf{T}, \mathbf{X}]$. Now put $g' = g \cdot g_1 d_1 \ldots g_m d_m$ and conclude that $H_K(f_1', \ldots, f_m'; g') \subseteq H_K(f_1, \ldots, f_m; g)$. Finally, suppose $n = 1$ and each $f_i(\mathbf{T}, X)$ is separable in X. Then so is each $f_i'(\mathbf{T}, X)$. □

LEMMA 12.1.2: *Suppose every Hilbert set of K of the form $H_K(f_1, \ldots, f_m; g)$ with f_i irreducible in $K[T, X_1, \ldots, X_n]$, $i = 1, \ldots, m$, is nonempty. Then every Hilbert set of K is nonempty.*

Proof: Start with irreducible polynomials $f_i \in K[T_1, \ldots, T_r, X_1, \ldots, X_n]$, $i = 1, \ldots, m$, and a nonzero polynomial $g \in K[T_1, \ldots, T_r]$. By assumption, there exists $a_1 \in K$ with $f_i(a_1, T_2, \ldots, T_r, \mathbf{X})$ irreducible, $i = 1, \ldots, m$, and $g(a_1, T_2, \ldots, T_r) \neq 0$. Repeat this procedure r times to find $a_1, \ldots, a_r \in K$ with $f_i(\mathbf{a}, \mathbf{X})$ irreducible and $g(\mathbf{a}) \neq 0$, $i = 1, \ldots, m$. By Lemma 12.1.1, every Hilbert set of K is nonempty. □

LEMMA 12.1.3: *Every Hilbert subset of K contains a Hilbert set of the form $H(f_1, \ldots, f_m; g)$, where f_i is an irreducible polynomial in $K[T, X]$ with $\deg_X(f_i) \geq 1$, $i = 1, \ldots, m$.*

Proof: Let $f \in K[T, X_1, \ldots, X_n]$ be an irreducible polynomial with $f \notin K[T]$ and $0 \neq g_0 \in K[T]$. By Lemma 12.1.1, it suffices to find irreducible polynomials $h_1, \ldots, h_r \in K[T, Y] \setminus K[T]$ and $0 \neq g \in K[T]$ with $H_K(h_1, \ldots, h_r; g) \subseteq H_K(f; g_0)$.

Indeed, let $d > \max_{1 \leq j \leq n} \deg_{X_j}(f)$. Apply the Kronecker substitution $S_d \colon S_{K[T]}(n, d) \to S_{K[T]}(1, d^n)$ on f (Section 11.3). Consider the factorization of $S_d(f)$ into irreducible factors of $K[T, Y]$:

(1) $$S_d(f)(T, Y) = \prod_{i \in I} h_i(T, Y).$$

The polynomials $S_d(f)(T, Y)$ and $f(T, X_1, \ldots, X_n)$ have the same coefficients in $K[T]$. Since f is irreducible in $K[T, X_1, \ldots, X_n]$, the greatest common divisor of its coefficients in $K[T]$ is 1, so none of the h_i is in $K[T]$.

Let $I = J \cup J'$ be a nontrivial partition of I. The exponent to which Y appears in each of the polynomials $\prod_{i \in J} h_i(T, Y)$ and $\prod_{i \in J'} h_i(T, Y)$ does not

12.1 Hilbert Sets and Reduction Lemmas

exceed $d^n - 1$. Since S_d is bijective on $S_{K[T]}(n, d)$ (Section 11.3), there exist polynomials $p_J, p_{J'} \in K[T, \mathbf{X}]$ with $\deg_{X_j}(p_J), \deg_{X_j}(p_{J'}) < d$, $j = 1, \ldots, n$, and

(2) $\quad S_d(p_J)(T, Y) = \prod_{i \in J} h_i(T, Y)$ and $S_d(p_{J'})(T, Y) = \prod_{i \in J'} h_i(T, Y)$.

Note: The product $p_J(T, \mathbf{X}) p_{J'}(T, \mathbf{X})$ contains a monomial of the form $g_J(T) X_1^{\nu_1} \cdots X_n^{\nu_n}$ in which at least one of the ν_j exceeds $d - 1$. Otherwise, the relation $S_d(f) = S_d(p_J) \cdot S_d(p_{J'}) = S_d(p_J p_{J'})$ would imply a nontrivial factorization of the irreducible polynomial f into $p_J p_{J'}$.

Let g be the product of g_0 with all g_J (one for each nontrivial partition). Let a be an element of K with each $h_i(a, Y)$ irreducible in $K[Y]$ and $g(a) \neq 0$. We show $f(a, \mathbf{X})$ is also irreducible.

Indeed, assume $f(a, \mathbf{X}) = q(\mathbf{X}) q'(\mathbf{X})$ is a nontrivial factorization of $f(a, \mathbf{X})$ in $K[\mathbf{X}]$, then (1) implies

$$\prod_{i \in I} h_i(a, Y) = S_d(f)(a, Y) = S_d(q)(Y) \cdot S_d(q')(Y).$$

Hence, (2) implies a nontrivial partition $I = J \cup J'$ of I with

$$S_d(q)(Y) = \prod_{i \in J} h_i(a, Y) = S_d(p_J)(a, Y) \quad \text{and}$$

$$S_d(q')(Y) = \prod_{i \in J'} h_i(a, Y) = S_d(p_{J'})(a, Y).$$

Hence, $q(\mathbf{X}) = p_J(a, \mathbf{X})$ and $q'(\mathbf{X}) = p_{J'}(a, \mathbf{X})$. Thus, $p_J(a, \mathbf{X}) p_{J'}(a, \mathbf{X}) = f(a, \mathbf{X})$. The left hand side contains the monomial $g_J(a) X_1^{\nu_1} \cdots X_n^{\nu_n}$ in which at least one ν_j exceeds $d - 1$, while $f(a, \mathbf{X})$ contains no such monomial. Therefore, $f(a, \mathbf{X})$ is irreducible. $\square$

LEMMA 12.1.4: Let $H = H_K(g_1, \ldots, g_m; h)$ be a Hilbert subset of K^r with $g_i \in K[T_1, \ldots, T_r, X]$ irreducible and $\deg_X(g_i) \geq 1$, $i = 1, \ldots, m$. Then H contains a Hilbert set of the form $H_K(f_1, \ldots, f_m)$ in which f_i is a monic irreducible polynomial in $K[\mathbf{T}, X]$ of degree at least 2 in X, $i = 1, \ldots, m$. If $g_1, \ldots, g_m$ are separable in X, then so are $f_1, \ldots, f_m$.

Moreover, suppose $\mathbf{a} \in K^r$ and none of the polynomials $f_i(\mathbf{a}, X)$ has a root in K. Then none of the polynomials $g_i(\mathbf{a}, X)$ with $\deg_X(g_i) \geq 2$ has a root in K.

Proof: Let c_i be the leading coefficient of g_i as a polynomial in X, $n_i = \deg_X(g_i)$, and $q = h c_1 \cdots c_m$. Choose a prime number $l \neq \text{char}(K)$ with q not an lth power in $K[\mathbf{T}]$. By assumption, $n_i \geq 1$. If $m = 0$ or $n_i = 1$, let $f_i(\mathbf{T}, X) = X^l - q(\mathbf{T})$. If $n_i \geq 2$, let

$$f_i(\mathbf{T}, X) = q(\mathbf{T})^{n_i} c_i(\mathbf{T})^{-1} g_i(\mathbf{T}, q(\mathbf{T})^{-1} X).$$

Then

$$f_i(\mathbf{T}, X) = X^{n_i} + b_{i,n_i-1}(\mathbf{T})X^{n_i-1} + q(\mathbf{T})\sum_{j=0}^{n_i-2} b_{ij}(\mathbf{T})X^j.$$

with $b_{ij} \in K[\mathbf{T}]$. In each case $f_i(\mathbf{T}, X)$ is monic in X and irreducible in $K[\mathbf{T}, X]$ (see [Lang7, p. 297, Lemma 9.1] for the case $m = 0$ or $n_i = 1$). We prove that $H_K(f_1, \ldots, f_m) \subseteq H$.

Let $\mathbf{a}$ be in K^r with $f_1(\mathbf{a}, X), \ldots, f_m(\mathbf{a}, X)$ irreducible in $K[X]$. Consider an i between 1 and m.

Suppose first $n_i = 1$. Then $X^l - q(\mathbf{a})$ is irreducible. Hence, $q(\mathbf{a}) \neq 0$. Therefore, $h(\mathbf{a}) \neq 0$, $c_i(\mathbf{a}) \neq 0$, and $g_i(\mathbf{a}, X) = c_i(\mathbf{a})X + b_i$ for some $b_i \in K$. Thus, $g_i(\mathbf{a}, X)$ is irreducible.

Now suppose $n_i \geq 2$. Then $q(\mathbf{a}) \neq 0$ (otherwise $f_i(\mathbf{a}, X) = X^{n_i} + b_{i,n_i-1}(\mathbf{a})X^{n_i}$ is reducible) and $g_i(\mathbf{a}, q(\mathbf{a})^{-1}X) = q(\mathbf{a})^{-n_i}c_i(\mathbf{a})f_i(\mathbf{a}, X)$ is irreducible. Hence, so is $g_i(\mathbf{a}, X)$. Similarly, if $n_i \geq 2$ and $f_i(\mathbf{a}, X)$ has no root in K, neither has $g_i(\mathbf{a}, X)$. □

COROLLARY 12.1.5: *Suppose each Hilbert set of the form* $H_K(f_1, \ldots, f_m)$ *with* $f_i \in K[T, X]$ *monic and of degree at least 2 in X is nonempty. Then every Hilbert set of K is nonempty.*

Proof: By Lemma 12.1.4 for $r = 1$, each Hilbert subset of K of the form $H_K(g_1, \ldots, g_m; h)$ with $g_i \in K[T, X]$ irreducible and $\deg_X(g_i) \geq 1$, $i = 1, \ldots, m$, is nonempty. Hence, by Lemma 12.1.3, each Hilbert subset of K is nonempty. Consequently, by Lemma 12.1.2, every Hilbert set of K is nonempty. □

The corresponding result for separable Hilbert sets is also true. We postpone its proof to Section 13.2. Here we show that separable Hilbert subsets of K^r contain especially simple separable Hilbert subsets of K^r.

LEMMA 12.1.6: *Let H be a separable Hilbert subset of K^r. Then there exists an irreducible polynomial $f \in K[T_1, \ldots, T_r, X]$, separable, monic, and of degree at least 2 in X with $H_K(f) \subseteq H$.*

Proof: Use Lemma 12.1.1 to assume that $H = H_K(f_1, \ldots, f_m; g)$ with $f_1, \ldots, f_m \in K[\mathbf{T}, X] \smallsetminus K[\mathbf{T}]$ irreducible and separable in X and $0 \neq g \in K[\mathbf{T}]$. For each i, $1 \leq i \leq m$, let x_i be a root of $f_i(\mathbf{T}, X)$ in $K(\mathbf{T})_s$. Then the finite separable extension $K(\mathbf{T}, x_1, \ldots, x_m)$ of $K(\mathbf{T})$ is contained in a separable extension $K(\mathbf{T}, y)$ of $K(\mathbf{T})$ of degree at least 2. Assume without loss, y is integral over $K[\mathbf{T}]$. Let $h_1 = \mathrm{irr}(y, K(\mathbf{T}))$. Then $h_1(T, X)$ is separable, monic, and of degree at least 2 in X. Rewrite x_i as $\frac{p_i(\mathbf{T}, y)}{p_0(\mathbf{T})}$ with $p_1, \ldots, p_m \in K[\mathbf{T}, X]$ and $0 \neq p_0 \in K[\mathbf{T}]$. Let $u_i(\mathbf{T}, x_i)$ be the leading coefficient of $\mathrm{irr}(y, K(\mathbf{T}, x_i))$. Write $\mathrm{norm}_{K(\mathbf{T}, x_i)/K(\mathbf{T})}(u_i(\mathbf{T}, x_i))$ as $\frac{q_i(\mathbf{T})}{q_0(\mathbf{T})}$ with $q_0, q_1, \ldots, q_m \in K[\mathbf{T}]$. Finally, let $r_i(\mathbf{T})$ be the leading coefficient of $f_i(\mathbf{T}, X)$,

$i = 1, \ldots, m$. Define g_1 to be the product $g \cdot p_0 \cdot q_0 \cdots q_m \cdot r_1 \cdots r_m$. We show that $H_K(h_1; g_1) \subseteq H_K(f_1, \ldots, f_m; g)$.

Indeed, if $\mathbf{a} \in H_K(h_1; g_1)$ and c is a root of $h_1(\mathbf{a}, X)$, then

(3) $$[K(c) : K] = [K(\mathbf{T}, y) : K(\mathbf{T})].$$

Moreover, with $b_i = \frac{p_i(\mathbf{a}, c)}{p_0(\mathbf{a})}$, the nonvanishing of $g_1(\mathbf{a})$ implies

$$[K(b_i) : K] \leq [K(\mathbf{T}, x_i) : K(\mathbf{T})] \text{ and } [K(c) : K(b_i)] \leq [K(\mathbf{T}, y) : K(\mathbf{T}, x_i)].$$

Multiply the terms in these two inequalities and apply (3) to conclude that they are equalities. Therefore, $f_1(\mathbf{a}, X), \ldots, f_m(\mathbf{a}, X)$ are irreducible over K.

By Lemma 12.1.4, there exists an irreducible polynomial $f \in K[\mathbf{T}, X]$, separable, monic, and of degree at least 2 in X, with $H_K(f) \subseteq H_K(h_1; g_1)$. □

12.2 Hilbert Sets under Separable Algebraic Extensions

Let L/K be a finite separable extension. We prove every Hilbert subset of L^r contains a Hilbert subset of K^r.

LEMMA 12.2.1: *Let L be a separable extension of degree d of an infinite field K and $\sigma_0, \ldots, \sigma_{d-1}$ distinct representatives of the cosets of $\mathrm{Gal}(L)$ in $\mathrm{Gal}(K)$. Suppose $f \in L[X_1, \ldots, X_n]$ is a nonconstant polynomial. Then there exist $c_1, \ldots, c_n \in L$ with $f(X_1 + c_1, \ldots, X_n + c_n)^{\sigma_i}$, $i = 0, \ldots, d-1$, pairwise relatively prime in $\tilde{K}[X_1, \ldots, X_n]$.*

Proof: Let θ be a primitive element for L/K. Choose algebraically independent elements t_{ik}, $i = 0, \ldots, d-1$ and $k = 1, \ldots, n$, over K. For every i and k consider

$$u_{ik} = \sum_{j=0}^{d-1} (\theta^{\sigma_i})^j t_{jk}.$$

Write $\mathbf{u}_i = (u_{i1}, \ldots, u_{in})$ and $\mathbf{u} = (\mathbf{u}_0, \ldots, \mathbf{u}_{d-1})$. The linear transformation $\mathbf{t} \to \mathbf{u}$ has as determinant the nth power of $\prod_{i<j}(\theta^{\sigma_j} - \theta^{\sigma_i}) \neq 0$. Thus, the t_{ik} are linear combinations of the u_{ik} with coefficients in $\tilde{K}$; so the u_{ik} are algebraically independent over K.

Write $f = f_1 \cdots f_m$, a product of $\tilde{K}$-irreducible factors. Since f_μ is nonconstant and the u_{ik} are algebraically independent, $f_\mu^{\sigma_i}(\mathbf{u}_i) \neq f_\nu^{\sigma_j}(\mathbf{u}_j)$ for $i \neq j$ and each μ and ν, $1 \leq \mu, \nu \leq m$. Therefore

$$h(\mathbf{t}) = \prod_{i<j} \prod_{\mu,\nu} (f_\mu^{\sigma_i}(\mathbf{u}_i) - f_\nu^{\sigma_j}(\mathbf{u}_j)) \neq 0,$$

so there exist $a_{ik} \in K$ with $h(\mathbf{a}) \neq 0$. Let

$$c_k = \sum_{j=0}^{d-1} a_{jk} \theta^j, \qquad k = 1, \ldots, n.$$

The $\tilde{K}$-specialization $\mathbf{t} \to \mathbf{a}$ maps u_{ik} to $c_k^{\sigma_i}$. Hence $f_\mu(\mathbf{c})^{\sigma_i} \neq f_\nu(\mathbf{c})^{\sigma_j}$ for $i \neq j$. Therefore, $f_\mu(\mathbf{X}+\mathbf{c})^{\sigma_i} \neq f_\nu(\mathbf{X}+\mathbf{c})^{\sigma_j}$ for all i, j, μ, ν with $0 \leq i < j \leq d-1$ and $1 \leq \mu, \nu \leq m$. Since $f_1(\mathbf{X}+\mathbf{c})^{\sigma_i}, \ldots, f_m(\mathbf{X}+\mathbf{c})^{\sigma_i}$ are exactly the irreducible factors of $f(\mathbf{X}+\mathbf{c})^{\sigma_i}$ in $\tilde{K}[\mathbf{X}]$, $f(\mathbf{X}+\mathbf{c})^{\sigma_0}, \ldots, f(\mathbf{X}+\mathbf{c})^{\sigma_{d-1}}$ are relatively prime in pairs. □

LEMMA 12.2.2: *Let L be a finite separable extension of a field K and $f \in L(T_1, \ldots, T_r)[X_1, \ldots, X_n]$ irreducible. Then, there exists an irreducible $p \in K(\mathbf{T})[\mathbf{X}]$ with $H_K(p) \subseteq H_L(f)$. If $n = 1$ and f is separable in X, then p is separable in X.*

Proof: Let S be a set of representatives of the left cosets of $\text{Gal}(L)$ in $\text{Gal}(K)$. Denote the algebraic closure of $K(\mathbf{T})$ by F. First consider the case where the f^σ, with $\sigma \in S$, are pairwise relatively prime in $F[\mathbf{X}]$. Let $p = \prod_{\sigma \in S} f^\sigma$. Since $\text{Gal}(K)$ fixes the coefficients of p, $p \in K(\mathbf{T})[\mathbf{X}]$. Moreover, if q is an irreducible factor of p in $K(\mathbf{T})[\mathbf{X}]$, then one of the f^σ's (and therefore all) divides q in $\tilde{K}(\mathbf{T})[\mathbf{X}]$. Since the f^σ are pairwise relatively prime, p divides q. Hence, p is irreducible in $K(\mathbf{T})[\mathbf{X}]$. In addition, if $n = 1$ and f is separable in X, then p is separable in X.

Let $a_1, \ldots, a_r$ be elements of K such that $p(\mathbf{a}, \mathbf{X})$ is defined and irreducible in $K[\mathbf{X}]$. If $f(\mathbf{a}, \mathbf{X}) = g(\mathbf{X})h(\mathbf{X})$ were a nontrivial decomposition in $L[\mathbf{X}]$, then $p(\mathbf{a}, \mathbf{X}) = (\prod_{\sigma \in S} g^\sigma(\mathbf{X}))(\prod_{\sigma \in S} h^\sigma(\mathbf{X}))$ would be a nontrivial decomposition in $K[\mathbf{X}]$, which is a contradiction. Therefore, $f(\mathbf{a}, \mathbf{X})$ is irreducible in $L[\mathbf{X}]$ and $H_K(p) \subseteq H_L(f)$.

In the general case apply Lemma 12.2.1 to find $c_1, \ldots, c_r \in L(\mathbf{T})$ with $f(\mathbf{T}, \mathbf{X} + \mathbf{c})^\sigma$ pairwise relatively prime in $F[\mathbf{X}]$, σ running over S. Let $g(\mathbf{T}, \mathbf{X}) = f(\mathbf{T}, \mathbf{X} + \mathbf{c})$. Irreducibility of $g(\mathbf{a}, \mathbf{X})$ is equivalent to that of $f(\mathbf{a}, \mathbf{X})$ for each $\mathbf{a} \in K^r$ such that both polynomials are defined. Thus, the lemma follows from the first part of the proof. □

COROLLARY 12.2.3: *Let L be a finite separable extension of a field K. Then every Hilbert subset (resp. separable Hilbert subset) of L^r contains a Hilbert subset (resp. separable Hilbert subset) of K^r. In particular, if K is Hilbertian, so is L.*

Remark 12.2.4: The converse of Corollary 12.2.3 is false: Example 13.9.5 gives a field K with an empty Hilbert set and a finite separable extension L whose Hilbert sets are all nonempty. □

12.3 Purely Inseparable Extensions

Corollary 12.2.3 is false if L/K is inseparable. For example, let $K = \mathbb{F}_p(t)$, with t transcendental over $\mathbb{F}_p$, and let $L = K(t^{1/p})$. Then $X^p - T$ is irreducible over L. Yet every $a \in K$ is a pth power in L, so $X^p - a$ is reducible over L. Hence, $H_L(X^p - T)$ contains no elements of K. Still, $X^p - t^{1/p}$ is irreducible over L, so $H_L(X^p - T)$ is nonempty. The following results generalize this

12.3 Purely Inseparable Extensions

observation. They assert: if every Hilbert set of K is nonempty, then every Hilbert set of L is nonempty.

First we list simple properties of purely inseparable extensions.

LEMMA 12.3.1: *Let L/K be a purely inseparable extension of fields of characteristic p.*
(a) *For each irreducible $g \in K[\mathbf{X}]$ there exist an irreducible $f \in L[\mathbf{X}]$ and $e \geq 0$ with $g = f^{p^e}$.*
(b) *Consider $f \in L[\mathbf{X}]$. Suppose $L^{p^e} \subseteq K$ and $g = f^{p^e}$ is irreducible in $K[\mathbf{X}]$. Then f is irreducible in $L[\mathbf{X}]$.*
(c) *Let $f \in L[\mathbf{X}]$ be a non pth power with $g = f^{p^e}$ irreducible in $K[\mathbf{X}]$. Then f is irreducible in $L[\mathbf{X}]$.*
(d) *If $f \in L[\mathbf{X}]$ is irreducible and $e \geq 0$ is the least nonnegative integer with $g = f^{p^e} \in K[\mathbf{X}]$, then g is irreducible in $K[\mathbf{X}]$.*

Proof of (a): Let $f_1 \cdots f_m$ be a factorization of g into irreducible factors in $L[\mathbf{X}]$. Choose $k \geq 0$ such that $f_i^{p^k} \in K[\mathbf{X}]$, $i = 1, \ldots, m$. The relation $g^{p^k} = f_1^{p^k} \cdots f_m^{p^k}$ and the unique factorization in $K[\mathbf{X}]$ imply that $f_1^{p^k} = g^r$ for some positive integer r. Unique factorization over $L[\mathbf{X}]$ then gives a positive integer s with $g = f_1^s$. Then $f_1^{p^k} = f_1^{rs}$ implies that $s = p^e$, so $g = f_1^{p^e}$ for some positive integer e.

Proof of (b): Let $f_1 \cdots f_m$ be a factorization of f into irreducible factors in $L[\mathbf{X}]$. Then $f_1^{p^e} \cdots f_m^{p^e}$ is a factorization of g in $K[\mathbf{X}]$. Since g is irreducible over K, we have $m = 1$. Therefore, f is irreducible over L.

Proof of (c): By (a), $g = h^{p^{e'}}$ for some irreducible $h \in L[\mathbf{X}]$ and $e' \geq 0$. Hence, $h^{p^{e'}} = f^{p^e}$, so $e' \geq e$. Therefore, $h^{p^{e'-e}} = f$. The assumption on f implies that $e' = e$ and $f = h$. Thus, f is irreducible in $L[\mathbf{X}]$.

Proof of (d): Let $g_1 \cdots g_m$ be a factorization of g into irreducible factors in $K[\mathbf{X}]$. By (a), $g_i = f_i^{p^{e_i}}$ for some irreducible polynomial f_i in $L[\mathbf{X}]$ and $e_i \geq 0$. Hence, $f^{p^e} = f_1^{p^{e_1}} \cdots f_m^{p^{e_m}}$. Therefore, $f_i = f$ and $p^{e_1} + \cdots + p^{e_m} = p^e$. In addition, $f^{p^{e_i}} = g_i \in K[X]$, so by assumption, $e_i \geq e$. It follows that $m = 1$ and g is irreducible in $K[\mathbf{X}]$. $\square$

LEMMA 12.3.2: *Let K be a field with $\mathrm{char}(K) = p > 0$, L a purely inseparable extension of K, and f an irreducible polynomial in $L[T_1, \ldots, T_r, X]$ with $\deg_X f \geq 1$.*
(a) *If f is separable in X, then $H_L(f)$ contains a separable Hilbert subset of K^r.*
(b) *If all occurrences of $T_1, \ldots, T_r, X$ in f are powers of p, then $H_L(f)$ contains a Hilbert subset of K^r.*

Proof: Let e be the least nonnegative integer with $g = f^{p^e} \in K[\mathbf{T}, X]$. By Lemma 12.3.1(d), g is irreducible in $K[\mathbf{T}, X]$. Distinguish between two cases to find a Hilbert subset of K contained in $H_L(f)$.

CASE A: *f is separable with respect to X.* Consider $\mathbf{a} \in K$ with $g(\mathbf{a}, X)$ irreducible in $K[X]$ and $f(\mathbf{a}, X)$ separable. Then, $f(\mathbf{a}, X)$ is not a pth power in $L[X]$. By Lemma 12.3.1(c), $f(\mathbf{a}, X)$ is irreducible in $L[X]$.

CASE B: *All occurrences of $T_1, \ldots, T_r, X$ in f are powers of p.* By assumption, f is not a pth power in $L[\mathbf{T}, X]$, so at least one coefficient $c(\mathbf{T})$ of $f(\mathbf{T}, X)$ is not a pth power in $L[\mathbf{T}]$. In particular, L is an infinite field. Hence, K is also infinite.

CLAIM: *There is a nonempty Zariski L-open subset U of $\mathbb{A}^r$ such that $c(\mathbf{a}) \notin L^p$ for all $\mathbf{a} \in U(L)$.* To prove the claim, write $c(\mathbf{T}) = \sum c_\mathbf{i} T_1^{i_1 p} \cdots T_r^{i_r p}$ with $\mathbf{i} = (i_1, \ldots, i_r)$, $0 \leq i_1 \leq d_1, \ldots, 0 \leq i_r \leq d_r$, and $c_\mathbf{i} \in L^r$. Let $d = (d_1 + 1) \cdots (d_r + 1)$. Assume there are d r-tuples

(1) $$(b_{k_1}, \ldots, b_{k_r})$$

with $0 \leq k_j \leq d_1$ and $b_{k_j} \neq b_{k_{j'}}$, if $j \neq j'$ and $1 \leq j, j' \leq r$ such that $c(b_{k_1}, \ldots, b_{k_r}) = \lambda^p$ with $\lambda \in L$. Thus, there is a $\lambda_{k_1, \ldots, k_r} \in L$ with

(2) $$\sum_\mathbf{i} c_\mathbf{i} b_{k_1}^{i_1 p} \cdots b_{k_r}^{i_r p} = \lambda_{k_1, \ldots, k_r}^p$$

Consider (2) as a system of d linear equations in the $c_\mathbf{i}$'s. The coefficient matrix of (2) is a Kronecker product of r Vandermonde matrices. The determinant of this matrix is

$$D = \det(b_{k_1}^{i_1 p} \cdots b_{k_r}^{i_r p}) = \prod_{j=1}^r \det(b_{k_j}^{i_j})^{d'_j p}$$

$$= \prod_{j=1}^r \prod_{j < j'} (b_{k_{j'}} - b_{k_j})^{d'_j p}$$

where $d'_j = (d_j + 1)^{-1}(d_1 + 1) \cdots (d_r + 1)$, $j = 1, \ldots, r$ (See [Bourbki, Algebra, p. 534] for the determinant of the Kronecker product of two matrices, from which the determinant of the Kronecker product of r matrices can be derived.) Thus, D is a pth power of a nonzero element of L. Applying Cramer's rule to the system (2), we conclude that $c_\mathbf{i} \in L^p$ for each $\mathbf{i}$. This contradiction to the assumption we made on $c(\mathbf{T})$ proves that there are at most $d - 1$ r-tuples (1). Let B be the set of all these r-tuples. Define,

$$U = \{(x_1, \ldots, x_r) \in \mathbb{A}^r \mid (x_1, \ldots, x_r) \notin B \text{ and } x_j \neq x_{j'} \text{ if } j \neq j'\}.$$

Then U is a Zariski L-open subset of $\mathbb{A}^r$ satisfying our claim.

Now we may use the claim to choose $\mathbf{a} \in K^r$ with $c(\mathbf{a}) \notin L^p$ (therefore $f(\mathbf{a}, X)$ is not a pth power in $L[X]$) and $g(\mathbf{a}, X)$ irreducible in $K[X]$. By Lemma 12.3.1(c), $f(\mathbf{a}, X)$ is irreducible in $L[X]$. □

12.3 Purely Inseparable Extensions

PROPOSITION 12.3.3: *Let L/K be an algebraic extension of fields of a finite separable degree. Then each separable Hilbert subset of L^r contains a separable Hilbert subset of K^r. Thus, if K is Hilbertian, then so is L.*

Proof: By assumption, L is a purely inseparable extension of a finite separable extension of K. Apply Corollary 12.2.3 to assume L/K is purely inseparable.

Let H be a separable Hilbert subset of L^r. By Lemma 12.1.6, H contains $H_L(f)$ with irreducible $f \in L[T, X]$ separable in X. From Lemma 12.3.2, $H_L(f)$ contains a separable Hilbert subset of K^r. □

LEMMA 12.3.4: *Let K be a field of positive characteristic p, L an extension of K with $L^{p^r} \subseteq K$ for some $r \geq 0$, and $f_1, \ldots, f_m$ irreducible polynomials in $L[T, X]$. Suppose $\deg_X f_i \geq 1$ and f_i is separable in T. Suppose in addition that each Hilbert subset of K is nonempty. Then there exists $\gamma \in L$ such that $H_L(f_i(T + \gamma, X))$ contains a Hilbert subset of K, $i = 1, \ldots, m$.*

Proof: Let i be between 1 and m. At least one of the coefficients $c_i(T)$ of $f_i(T, X)$ has nonzero derivative $c_i'(T)$. Choose a nonzero coefficient γ_i of $c_i'(T)$. Let e be the least nonnegative integer with $L^{p^e} \subseteq K$. Then $L^{p^{e-1}} \not\subseteq K$, so $L^{p^{e-1}} \not\subseteq \gamma_i^{p^{e-1}} K$, $i = 1, \ldots, m$. Since K is infinite, $L^{p^{e-1}} \not\subseteq \bigcup_{i=1}^m \gamma_i^{-p^{e-1}} K$. Hence, there exists $\lambda \in L$ independent of i with $(\gamma_i \lambda)^{p^{e-1}} \notin K$. This implies $(c_i'(T) \lambda)^{p^{e-1}} \notin K[T]$.

The Taylor expansion $c_i(T + \lambda Z) = c_i(T) + c_i'(T) \cdot \lambda Z + \cdots$ shows that $c_i(T + \lambda Z)^{p^{e-1}} \notin K[T, Z]$. Hence, $h_i(T, Z) = c_i(T + \lambda Z)^{p^e}$ is not a pth power of a polynomial belonging to $K[T, Z]$. Therefore, $X^p - h_i(T, Z)$ is irreducible over K. By assumption, there exists $b \in K$, independent of i, with $X^p - h_i(T, b)$ irreducible over K. Thus, e is the least nonnegative integer for which $c_i(T + \lambda b)^{p^e}$ is in $K[T]$. Hence, e is the least nonnegative integer for which $g_i(T, X) = f_i(T + \lambda b, X)^{p^e} \in K[T, X]$. By Lemma 12.3.1(d), $g_i(T, X)$ is irreducible in $K[T, X]$.

If for $a \in K$, the polynomial $g_i(a, X)$ is irreducible in $K[X]$, then from Lemma 12.3.1(b), $f_i(a + \lambda b, X)$ is irreducible in $L[X]$. Thus, $H_K(g_i) \subseteq H_L(f_i(T + \lambda b, X))$. □

PROPOSITION 12.3.5: *Let K be a field with all Hilbert subsets nonempty and L an algebraic extension of K. Then in each of the following cases all Hilbert sets of L are nonempty.*
(a) *$\mathrm{char}(K) = p > 0$ and $L^{p^r} \subseteq K$ for some $r \geq 0$.*
(b) *L is a finite extension of K.*

Proof of (a): Consider a Hilbert subset $H_L(f_1, \ldots, f_n)$ of L with $f_i \in L[T, X]$ irreducible and $\deg_X f_i \geq 2$, $i = 1, \ldots, n$. Reorder the f_i's, if necessary, to assume the following: For $i = 1, \ldots, m$, f_i is separable in X or inseparable in both T and X. For $i = m+1, \ldots, n$, f_i is separable in T. Lemma 12.3.4 gives γ in L such that $H_L(f_i(T + \gamma, X))$ contains a Hilbert subset H_i of K.

For each i between 1 and m, the polynomial $f_i(T+\gamma, X)$ is either separable in X or inseparable in both T and X. By Lemma 12.3.2, $H_L(f_i(T+\gamma, X))$ contains a Hilbert subset H_i of K, so $H = H_1 \cap \cdots H_m$ is a Hilbert subset of K which is contained in $H_L(f_1, \ldots, f_n)$. By Corollary 12.1.5, every Hilbert set of L is nonempty.

Proof of (b): Combine (a) with Lemma 12.2.2. □

12.4 Imperfect fields

Suppose $\text{char}(K) = p \neq 0$ and every Hilbert set of K is nonempty. Then there exists $a \in K$ with $X^p - a$ is irreducible. Thus, K is imperfect. We show that, conversely, every Hilbert set of an imperfect Hilbertian field is nonempty.

LEMMA 12.4.1: *Let $\text{char}(K) = p > 0$ and let $f \in K[X] \smallsetminus K^p[X]$ be irreducible and monic. Then $f(X^{p^m})$ is irreducible in $K[X]$ for each $m \geq 0$.*

Proof: By induction it suffices to prove the Lemma for $m = 1$. If x is a root of $f(X^p)$, then $y = x^p$ is a root of $f(X)$. We show $[K(x) : K(y)] = p$.

If not, then $K(x) = K(y)$ and $\text{irr}(x, K) = X^n + b_{n-1}X^{n-1} + \cdots + b_0$, where $n = [K(x) : K] = [K(y) : K] = \deg(f)$. Therefore, y is a root of $g(X) = X^n + b_{n-1}^p X^{n-1} + \cdots + b_0^p$. Hence, $f = g$ and $f \in K^p[X]$, contrary to our hypotheses.

It follows, $[K(x) : K] = p \cdot \deg(f) = \deg(f(X^p))$. Consequently, $f(X^p)$ is irreducible. □

LEMMA 12.4.2: *Let K be an imperfect field of characteristic p. Suppose $h \in K[T]$ is not a pth power. Then:*
(a) *The additive group K has infinitely many congruence classes modulo K^p.*
(b) *All $a \in K$ with $h(T + a) \in K^p[T]$ are congruent modulo K^p.*
(c) *For each $b \in K$ with $h(T + b) \notin K^p[T]$ only finitely many $c \in K$ satisfy $h(c^p + b) \in K^p$.*

Proof of (a): Since K is imperfect, K/K^p is a nonzero vector space over the infinite field K^p. Hence, K/K^p is infinite.

Proof of (b): Let $a, b \in K$ with $h(T+a), h(T+b) \in K^p[T]$. Since $h(T+a)$ is not a pth power, $h(T+a) = g(T^p) + f(T)$ where $g, f \in K^p[T]$ but $p \nmid \deg(f)$. Then, with $c = b - a$, we have $h(T + b) = g(T^p + c^p) + f(T + c)$. Let $f(T) = d_m^p T^m + d_{m-1}^p T^{m-1} + \cdots + d_0^p$ with $d_m \neq 0$. Then, $p \nmid m$ and the coefficient of T^{m-1} in $f(T+c)$ is $mcd_m^p + d_{m-1}^p$. Since $h(T+b)$ and $g(T^p + c^p)$ are in $K^p[T]$, this element belongs to K^p. Finally, $m^p = m \neq 0$. Therefore, c is a pth power.

Proof of (c): Write $h(T + b) \notin K^p[T]$ in the form

$$h(T + b) = f_1(T) + f_2(T)u_2 + \cdots + f_r(T)u_r,$$

where $r \geq 2$, $u_2, \ldots, u_r \in K$, $1, u_2, \ldots, u_r$ linearly independent over K^p, $f_1, \ldots, f_r \in K^p[T]$, and $f_2, \ldots, f_r \neq 0$. If $c \in K$ and $h(c^p + b) \in K^p$, then $f_1(c^p) + f_2(c^p)u_2 + \cdots + f_r(c^p)u_r \in K^p$. Therefore, $f_2(c^p) = \cdots = f_r(c^p) = 0$. The number of such c's is finite. □

PROPOSITION 12.4.3 ([Uchida]): *Let K be a field satisfying these conditions:*
(1a) *Every separable Hilbert subset of K is nonempty.*
(1b) *If $\mathrm{char}(K) > 0$, then K is imperfect.*

Then every Hilbert set of K is nonempty. In particular, K is Hilbertian.

Proof: If $\mathrm{char}(K) = 0$, then each Hilbert set of K is separable. Corollary 12.1.5 settles this case, so suppose $\mathrm{char}(K) = p > 0$ and K is imperfect. In this case, Corollary 12.1.5 shows it suffices to prove that each Hilbert set of the form $H_K(f_1, \ldots, f_m)$ with $f_1, \ldots, f_m \in K[T, X]$ irreducible, monic, and of degree at least 2 in X, is nonempty. Assume $f_1, \ldots, f_l$ are not separable in X and $f_{l+1}, \ldots, f_m$ are separable in X.

For each i, $1 \leq i \leq l$ there exists $g_i \in K[T, X]$ irreducible, separable and monic in X and q_i, a power of p different from 1, with $f_i(T, X) = g_i(T, X^{q_i})$. Since $f_i(T, X)$ is irreducible, g_i has a coefficient $h_i \in K[T]$ which is not a pth power. Choose $a_i \in K$ with $h_i(T + a_i) \in K^p[T]$ if there exists any, otherwise let $a_i = 0$. By (1b), K is infinite. Hence, by Lemma 12.4.2(a), there exists $b \in K \smallsetminus \bigcup_{i=1}^l (a_i + K^p)$. By Lemma 12.4.2(b), $h_i(T+b) \notin K^p[T]$, so $g_i(T+b, X) \in K(X)[T] \smallsetminus K(X)^p[T]$, $i = 1, \ldots, l$. Thus (Lemma 12.4.2(c)), the set C of all elements $c \in K$ with $h_i(c^p + b) \in K^p$ for some i, $1 \leq i \leq l$, is finite. Let $d(T) = \prod_{c \in C}(T - c)$ and let $g_i = f_i$, $i = l + 1, \ldots, m$. By Lemma 12.4.1, all $g_i(T^p + b, X)$ are irreducible, monic, and separable in X.

By (1a), there exists $a \in K$ such that $d(a) \neq 0$ and $g_i(a^p + b, X)$ is separable and irreducible in $K[X]$, $i = 1, \ldots, m$. Thus, $f_i(a^p + b, X)$ is irreducible for $i = l + 1, \ldots, m$. Now consider i between 1 and l. Since $d(a) \neq 0$, we have $a \notin C$, so $h_i(c^p + b) \notin K^p$. Hence, $g_i(a^p + b, X) \notin K^p[X]$. By Lemma 12.4.1, $f_i(a^p + b, X) = g_i(a^p + b, X^{q_i})$ is irreducible over K. It follows that $a^p + b \in H_K(f_1, \ldots, f_m)$. □

Exercises

1. Let K be a Hilbertian field and let $f \in K[T_1, \ldots, T_n, X]$ be a polynomial such that for every $\mathbf{t} \in K^n$ there exists a $x \in K$ with $f(\mathbf{t}, y) = 0$. Prove that $f(\mathbf{T}, X)$ has a factor of degree 1 in X.

2. Let $\{K_\alpha \mid \alpha < m\}$ be a transfinite ascending tower of Hilbertian fields. Suppose $K_{\alpha+1}$ is a regular extension of K_α for each $\alpha < m$. Prove that the union $L = \bigcup_{\alpha < m} K_\alpha$ is a Hilbertian field.

Notes

A field K is defined to be Hilbertian in [Fried-Jarden3] if each Hilbert set of K is nonempty. Most applications however use only separable Hilbertian sets. We have therefore decided to follow the convention of other authors and call a field Hilbertian if each separable Hilbert set of K is nonempty. Of course, if $\text{char}(K) = 0$, there is no difference between the two notions. When $\text{char}(K) > 0$ there is a simple relation between them. By Uchida's theorem (Proposition 12.4.3), Hilbertian fields in the old sense are imperfect Hilbertian fields in the new sense.

A great part of Section 12.1 occurs in Hilbert's original paper [Hilbert]. [Lang3, Chapter VIII] reproduces it, though its version of our Lemma 12.2.2 contains a flaw. Following Inaba [Lang3, p. 151, Proposition 3] claims that if L is a finite separable extension of a field K of degree d and if an irreducible polynomial $f \in L[X]$ has d distinct conjugates $f^{\sigma_1}, \ldots, f^{\sigma_d}$ over K, then $g = f^{\sigma_1} \cdots f^{\sigma_d}$ is irreducible in $K[X]$. Here is a counter-example.

Let $\gamma = \sqrt[3]{2}$ and ω be a primitive 3rd root of unity. Then the polynomial $f(X) = X^2 + \gamma X + \gamma^2 = (X - \gamma\omega)(X - \gamma\omega^2)$ is irreducible over $L = \mathbb{Q}(\gamma)$. It has distinct conjugates over $\mathbb{Q}$:

$$f^{\sigma_2}(X) = X^2 + \gamma\omega X + \gamma^2\omega^2 = (X - \gamma\omega^2)(X - \gamma) \quad \text{and}$$
$$f^{\sigma_3}(X) = X^2 + \gamma\omega^2 X + \gamma^2\omega = (X - \gamma)(X - \gamma\omega).$$

Yet the product $f(X)f^{\sigma_2}(X)f^{\sigma_3}(X) = (X^3 - 2)^2$ is reducible in $\mathbb{Q}[X]$.

Inaba treats purely inseparable extensions, though [Inaba, p. 12] claims that if L is a finite purely inseparable extension of a field K, then every Hilbert subset of L contains a Hilbert subset of K. This contradicts the example appearing in the first paragraph of Section 12.3.

Finally, the results of Sections 12.3 and 12.4 are due mainly to [Uchida].

Chapter 13. The Classical Hilbertian Fields

Global fields and functions fields of several variables have been known to be Hilbertian for three quarters of a century. These are the "classical Hilbertian fields".

The Hilbert property for rational function fields of one variable $K = K_0(t)$ over infinite fields K_0 is basically a combination of the Matsusaka-Zariski theorem (Proposition 10.5.2) with the Bertini-Noether theorem (Proposition 10.4.2). We show that every separable Hilbert subset H of K contains all elements of the form $a + bt$ with (a,b) in a nonempty Zariski K_0-open subset of $\mathbb{A}^2$ (Proposition 13.2.1).

Our unified approach to the proof of the irreducibility theorem for both number fields and function fields over finite fields uses Proposition 6.4.8, the main ingredient in the proof of the Chebotarev density theorem for function fields, to show that every separable Hilbert subset of a global field contains an arithmetic progression. Thus, we display an intimate connection between the irreducibility theorem and the Riemann hypothesis for curves.

Our proof of the irreducibility theorem for global fields specializes the parameters of the irreducible polynomials to integral elements. This leads to the notion of Hilbertian rings. Thus, finitely generated integral domains over $\mathbb{Z}$ and finitely generated transcendental ring extensions of an arbitrary field K_0 are Hilbertian (Proposition 13.4.1).

Section 13.5 describes the connection between the classical definition and the geometric definition of the Hilbert property due to Colliot-Thélène. This leads to the notion of a g-Hilbertian field. We prove that for each $g \geq 0$ each global field has an infinite normal extension which is g-Hilbertian but not Hilbertian (Theorem 13.6.2).

Each finite proper separable extension of a Galois extension of a Hilbertian field K is Hilbertian. This is a theorem of Weissauer. Moreover, if L_1 and L_2 are Galois extensions of K and neither of them contains the other, then $L_1 L_2$ is Hilbertian. The diamond theorem says even more: each extension M of K in $L_1 L_2$ which is contained in neither L_1 nor in L_2 is Hilbertian (Theorem 13.8.3). An essential tool in the proof is the twisted wreath product (Section 13.7).

13.1 Further Reduction

The main application of Hilbertianity of a field K is to reduce Galois extensions of $K(t_1, \ldots, t_r)$ to Galois extensions of K with the same Galois group.

LEMMA 13.1.1 (Hilbert): *Let K be a field, $t_1, \ldots, t_r$ indeterminates, and $F_j = K(\mathbf{t}, x_j)$ a finite separable extension of $K(\mathbf{t})$ with $x_j \neq 0$, $j = 1, \ldots, m$. Then:*
(a) $\mathbb{A}^r$ has a nonempty Zariski K-open subset U such that each every K-place

$\varphi \colon F_j \to \tilde{K} \cup \{\infty\}$ with $\mathbf{a} = \varphi(\mathbf{t}) \in U(\tilde{K})$ and $\varphi(K(\mathbf{t})) = K(\mathbf{a}) \cup \{\infty\}$ satisfies this:

(a1) $b_j = \varphi(x_j) \in K_s^\times$ and $\varphi(F_j) = K(\mathbf{a}, b_j) \cup \{\infty\}$, $j = 1, \ldots, m$.

(a2) For each j with $F_j/K(\mathbf{t})$ Galois, $K(\mathbf{a}, b_j)/K(\mathbf{a})$ is Galois. Moreover, the map $\sigma \mapsto \bar{\sigma}$ defined by $\bar{\sigma}(\varphi(y)) = \varphi(\sigma y)$ for $\sigma \in D_\varphi$ and $y \in F_j$ with $\varphi(y) \ne \infty$ is an isomorphism of the decomposition group D_φ onto $\mathrm{Gal}(K(\mathbf{a}, b_j)/K)$.

(b) K^r has a separable Hilbert subset H satisfying this: For each $\mathbf{a} \in H \cap U(K)$ and for each K-place $\varphi \colon F_j \to \tilde{K} \cup \{\infty\}$ satisfying (a1) and (a2), the map $\sigma \mapsto \bar{\sigma}$ defined in (a2) is an isomorphism $\mathrm{Gal}(F_j/K(\mathbf{t})) \cong \mathrm{Gal}(K(b_j)/K)$, $j = 1, \ldots, m$.

Proof: For each j between 1 and m let $f_j \in K[\mathbf{T}, X]$ be an irreducible polynomial with $f_j(\mathbf{t}, x_j) = 0$. Let $g_{j0} \in K[\mathbf{T}]$ be the leading coefficient of f_j viewed as a polynomial in X. Let $g_{j1} \in K[\mathbf{T}]$ be a nonzero polynomial and k a positive integer such that $\frac{g_{j1}(\mathbf{t})}{g_{j0}(\mathbf{t})^k}$ is the discriminant of $f_j(\mathbf{t}, X)$. Write $g = \prod_{j=1}^s g_{j0} g_{j1}$, $R = K[\mathbf{t}, g(\mathbf{t})^{-1}]$, and $S_j = R[x_j]$. By Lemma 6.1.2, S_j/R is a Galois ring-cover with x_j a primitive element. Define U as the set of all $\mathbf{a} \in \mathbb{A}^r$ with $g(\mathbf{a}) \ne 0$.

Let $\varphi \colon F_j \to \tilde{K} \cup \{\infty\}$ be a K-place such that $\mathbf{a} = \varphi(\mathbf{t}) \in U(\tilde{K})$ and $\varphi(K(\mathbf{t})) = K(\mathbf{a}) \cup \{\infty\}$ (By Lemma 2.2.7, there exists φ with these properties for each $\mathbf{a} \in U(\tilde{K})$.) Put $R = O_\varphi \cap K(\mathbf{t})$. By Remark 6.1.7, $R[x_j]/R$ is a ring-cover and $\varphi(F_j) = K(\mathbf{a}, b_j) \cup \{\infty\}$. This proves (a1). Assertion (a2) then follows from Lemma 6.1.4.

The set $H = H_K(f_1, \ldots, f_m)$ satisfies the requirements of (b). $\square$

Consider an arbitrary field K. Let $h_i \in K[T, X]$ be irreducible with $\deg_X(h_i) \ge 2$, $i = 1, \ldots, m$, and let $0 \ne g \in K[T]$. Put

$$H'_K(h_1, \ldots, h_m; g) = \{a \in K \mid g(a) \ne 0 \text{ and } \prod_{i=1}^m h_i(a, b) \ne 0 \text{ for each } b \in K\}.$$

If K is Hilbertian, then each set $H'_K(h_1, \ldots, h_m; g)$ is nonempty. The next lemma shows the converse:

LEMMA 13.1.2: *Let $f \in K[T_1, \ldots, T_r, X]$ be an irreducible polynomial, monic and separable in X with $\deg_X(f) \ge 2$. Then there exist absolutely irreducible polynomials $h_1, \ldots, h_m \in K[T_1, \ldots, T_r, X]$, monic and separable in X, with $\deg_X(h_i) \ge 2$, $i = 1, \ldots, m$, such that $H'_K(h_1, \ldots, h_m) \subseteq H_K(f)$.*

Proof: We prove the lemma in two ways. The first proof uses decomposition groups. The second one uses the beginning of what we later call "the intersection decomposition procedure" (Section 21.1.1). In both proofs we start from algebraically independent elements $t_1, \ldots, t_r$ over K and write $E = K(\mathbf{t})$.

Proof A: Let x be a root of $f(\mathbf{t}, X)$ in E_s, $F = E(x)$, and $\hat{F}$ the splitting field of $f(\mathbf{t}, X)$ over E. Choose a primitive element z for $\hat{F}/E$. List all proper

13.1 Further Reduction

extensions of E in $\hat{F}$ which are regular over K as $E_1, \ldots, E_m$. For each i between 1 and m choose a primitive element y_i for E_i/E which is integral over $K[\mathbf{t}]$. Thus, there is an irreducible polynomial $h_i \in K[\mathbf{T}, X]$ monic and separable in X with $h_i(\mathbf{t}, y_i) = 0$ and $\deg_X(h_i) = [E_i : E] \geq 2$. By Example 2.6.11, h_i is absolutely irreducible.

Lemma 13.1.1(a) gives a nonzero polynomial $g \in K[\mathbf{T}]$ such that for each $\mathbf{a} \in K^r$ with $g(\mathbf{a}) \neq 0$ each K-place $\varphi \colon \hat{F} \to \tilde{K} \cup \{\infty\}$ with $\varphi(\mathbf{t}) = \mathbf{a}$ has this property:

(2a) $\varphi(F) = K(\varphi(x)) \cup \{\infty\}$.
(2b) Let N be the residue field of $\hat{F}$ under φ. Then the decomposition group of φ is isomorphic to $\mathrm{Gal}(N/K)$.

CLAIM: $H'_K(h_1, \ldots, h_m; g) \subseteq H_K(f)$. Indeed, let $\mathbf{a} \in H'_K(h_1, \ldots, h_m; g)$. Use Lemma 2.2.7 to construct a K-place $\varphi \colon \hat{F} \to \tilde{K} \cup \{\infty\}$ with $\varphi(\mathbf{t}) = \mathbf{a}$ and $\varphi(K(\mathbf{t})) = K \cup \{\infty\}$. Since $f(\mathbf{a}, X)$ is monic, $b = \varphi(x)$ is in $\tilde{K}$. Denote the decomposition field of φ in $\hat{F}$ by E'. The restriction of φ to E' maps E' onto $K \cup \{\infty\}$ (Remark 6.1.6). Hence, by Lemma 2.6.9(b), E'/K is regular.

Assume $E' \neq E$. Then $E' = E_i$ for some i between 1 and m. Hence, $c = \varphi(y_i) \in K$ and $h_i(\mathbf{a}, c) = 0$. This contradiction to the choice of $\mathbf{a}$ proves that $E' = E$.

Denote the residue field of $\hat{F}$ under φ by N. By the preceding paragraph and (2b), $[N : K] = [\hat{F} : E]$. By (2a), the residue field of F under φ is $K(b)$. Then $[K(b) : K] \leq [F : E]$ and $[N : K(b)] \leq [\hat{F} : F]$. Hence, $[K(b) : K] = [F : E]$. Therefore, $f(\mathbf{a}, X)$ is irreducible over K, as claimed.

Finally, use Lemma 12.1.4 to eliminate g.

PROOF B: Let $f(\mathbf{t}, X) = \prod_{i=1}^n (X - x_i)$ be the factorization of $f(\mathbf{t}, X)$ in $E_s[X]$. Consider a nonempty proper subset I of $\{1, \ldots, n\}$ and let $f_I(X) = \prod_{i \in I}(X - x_i)$. Since f is irreducible, $f_I \notin E[X]$. Hence, $f_I(X)$ has a coefficient $y_I \notin E$. Thus, there is an irreducible polynomial $g_I \in K[\mathbf{T}, X]$, monic and separable in X with $\deg_X(g_I) \geq 2$, such that $g_I(\mathbf{t}, X) = \mathrm{irr}(y_I, E) \in K[\mathbf{t}, X]$.

Suppose g_I factors nontrivially over $\tilde{K}$: $g_I = g_{I,1} \cdots g_{I,k}$, where each $g_{I,j} \in \tilde{K}[\mathbf{T}, X]$ is irreducible and $k \geq 2$. Since $g_I(\mathbf{t}, X)$ is monic and separable in X, the factors $g_{I,j}$ are relatively prime. By Hilbert Nullstellensatz (Proposition 9.4.2), $V(g_{I,i}) \not\subseteq V(g_{I,j})$ for $i \neq j$. Thus (Lemma 10.1.2), $W_I = V(g_{I,1}, \ldots, g_{I,k})$ is a Zariski K-closed subset of $\mathbb{A}^{r+1}$ of dimension at most $r - 1$.

Denote the union of all W_I such that g_I factors nontrivially over $\tilde{K}$ by W. Let A be the K-Zariski closure of the projection of W on the first r coordinates. Then $\dim(A) \leq r - 1$. Hence, there exists a nonzero polynomial $q \in K[\mathbf{T}]$ which vanishes on A.

Now list all g_I which are absolutely irreducible as $h_1, \ldots, h_m$. Then

(3) $\qquad H'_K(h_1, \ldots, h_m; q) \subseteq H'_K(g_I \mid I \neq \emptyset,\ I \subset \{1, \ldots, n\})$.

Indeed, suppose $\mathbf{a}$ is in the left hand side of (3). Then $g_I(\mathbf{a}, b) \neq 0$ for each I with g_I absolutely irreducible and each $b \in K$. Assume there is an I with g_I reducible over $\tilde{K}$ and there is a $b \in K$ with $g_I(\mathbf{a}, b) = 0$. Then, in the above notation, there exists j between 1 and k with $g_{I,j}(\mathbf{a}, b) = 0$. As g_I is irreducible over K, all $g_{I,i}$ are K-conjugate to $g_{I,j}$. Hence, $g_{I,i}(\mathbf{a}, b) = 0$ for each i. Thus, $(\mathbf{a}, b) \in W(K)$ and so $\mathbf{a} \in A(K)$. Therefore, $q(\mathbf{a}) = 0$, in contradiction to the assumption on $\mathbf{a}$.

Next we prove that the right hand side of (3) is contained in $H_K(f)$. Assume for $\mathbf{a} \in K^r$ that $f(\mathbf{a}, X) = p(X)q(X)$ factors nontrivially in $K[X]$ with p and q monic. Extend the K-specialization $\mathbf{t} \to \mathbf{a}$ to a K-specialization $(\mathbf{t}, x_1, \ldots, x_n) \to (\mathbf{a}, c_1, \ldots, c_n)$ with $c_1, \ldots, c_n \in \tilde{K}$ (Propositions 2.3.1 and 2.3.3). Then, $f(\mathbf{a}, X) = \prod_{i=1}^n (X - c_i)$. For some nonempty proper subset I of $\{1, \ldots, n\}$, $p(X) = \prod_{i \in I}(X - c_i)$, the polynomial $f_I(X)$ maps onto $p(X)$, and y_I maps onto a coefficient b of $p(X)$. Then b lies in K and satisfies $g_I(\mathbf{a}, b) = 0$. Thus, $\mathbf{a}$ does not belong to the right hand side of (3).

Finally, use Lemma 12.1.4 to eliminate q. □

The following result is one ingredient of the proof of Lemma 13.1.4. That lemma will be used in the proof of the diamond theorem (Theorem 13.8.3).

LEMMA 13.1.3: *Let K be an infinite field, $t_1, \ldots, t_r$ algebraically independent elements over K, $E_0 = K(\mathbf{t})$, and E a finite separable extension of E_0. Then there are fields L, F_0, and F with these properties:*
(a) *L/K is a finite Galois extension.*
(b) *$F = \hat{E}L$, where $\hat{E}$ is the Galois closure of E/K.*
(c) *$F_0 L = F$ and $F_0 \cap L = K$.*
(d) *$\mathrm{Gal}(F/E_0) = \mathrm{Gal}(F/L(\mathbf{t})) \ltimes \mathrm{Gal}(F/F_0)$ (Definition 13.7.1).*
(e) *There is an L-place $\varphi \colon F \to L \cup \{\infty\}$ with $\varphi(F_0) = K \cup \{\infty\}$.*
(f) *Both F_0/K and F/L are regular extensions.*

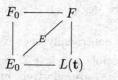

Proof: Let $\hat{E}$ be the Galois closure of E/E_0. Choose a primitive element x for $\hat{E}/E_0$. Remark 6.1.5 gives a nonzero polynomial $q \in K[\mathbf{T}]$ such that $K[\mathbf{t}, q(\mathbf{t})^{-1}, x]/K[\mathbf{t}, q(\mathbf{t})^{-1}]$ is a Galois ring-cover with x being a primitive element.

Choose $\mathbf{a} \in K^r$ with $q(\mathbf{a}) \neq 0$. Extend the specialization $\mathbf{t} \to \mathbf{a}$ to a K-place $\psi_0 \colon E_0 \to K \cup \{\infty\}$ (Lemma 2.2.7). Then extend ψ_0 to a place ψ of $\hat{E}$ into $\tilde{K} \cup \{\infty\}$. By Remark 6.1.7, $\psi(\hat{E}) = L \cup \{\infty\}$ where $L = K(\psi(x))$ is a finite Galois extension of K. Let $F = \hat{E}L$. By Remark 6.1.7, $L[\mathbf{t}, q(\mathbf{t})^{-1}, x]/L[\mathbf{t}, q(\mathbf{t})^{-1}]$ is a ring-cover and ψ extends to an L-place $\varphi \colon F \to L \cup \{\infty\}$. Hence, by Lemma 2.6.9(b), F/L is regular.

13.1 Further Reduction

Denote the decomposition field of ψ over E_0 by F_0. By Remark 6.1.6, $\psi(E_0) = K \cup \{\infty\}$. Hence, by Lemma 2.6.9(b), F_0/K is regular. In particular, $F_0 \cap L = K$. By Remark 6.1.7, $L[\mathbf{t}, q(\mathbf{t})^{-1}]/K[\mathbf{t}, q(\mathbf{t})^{-1}]$ is a ring-cover. Hence, $L[\mathbf{t}, q(\mathbf{t})^{-1}, x]$ is the integral closure of $K[\mathbf{t}, q(\mathbf{t})^{-1}]$ in F. Suppose σ is in the inertia group of φ over E_0; that is, $\sigma \in \mathrm{Gal}(F/F_0)$. Let $a \in L$. Then $\psi(\sigma a) = \bar\sigma\psi(a) = \psi(a)$. Since ψ is an isomorphism on L (see discussion after Proposition 2.3.1), $\sigma a = a$. Thus, σ lies in the inertia group of ψ over $L(\mathbf{t})$. Since $L[\mathbf{t}, q(\mathbf{t})^{-1}, x]/L[\mathbf{t}, q(\mathbf{t})^{-1}]$ is a ring-cover, the latter is trivial. Therefore, $\sigma = 1$. Thus, by Remark 6.1.6, $\mathrm{Gal}(F/F_0) \cong \mathrm{Gal}(L/K)$ and therefore $F_0 L = F$. Finally, (d) is a reinterpretation of (c). □

Let $f \in K[X]$ be a polynomial. We say f is **Galois** over K if f is irreducible, separable, and $K(x)$ is the splitting field of f over K for every root x of f.

If L' is an extension of L and f is irreducible over L', then L' is linearly disjoint from $L(x)$ over L. Hence, f is Galois over L' and $\mathrm{Gal}(L'(x)/L') \cong \mathrm{Gal}(L(x)/L)$.

LEMMA 13.1.4: *Suppose K is an infinite field. Let $f \in K[T_1, \ldots, T_r, X]$ be an irreducible polynomial which is monic and separable in X. Then there are a finite Galois extension L of K and an absolutely irreducible polynomial $g \in K[\mathbf{T}, X]$ which as a polynomial in X is monic, separable, and Galois over $L(\mathbf{T})$ such that $K^r \cap H_L(g) \subseteq H_K(f)$.*

Proof: Let $t_1, \ldots, t_r$ be algebraically independent elements over K, $E_0 = K(\mathbf{t})$, $x \in E_{0,s}$ a root of $f(\mathbf{t}, X)$, and $E = E_0(x)$. Let L, F_0 and F be the fields given by 13.1.3. Choose a primitive element y for F_0/E_0 which is integral over $K[\mathbf{t}]$. Then there is an irreducible polynomial $g \in K[\mathbf{T}, X]$, monic and separable in X such that $g(\mathbf{t}, y) = 0$. By Example 2.6.11, g is absolutely irreducible. Since $F = L(\mathbf{t}, y)$, the polynomial $g(\mathbf{t}, X)$ is Galois over $L(\mathbf{t})$.

Lemma 13.1.1(a) gives $q \in K[\mathbf{T}]$, $q \neq 0$ such that for each K-place $\varphi\colon F \to \tilde K \cup \{\infty\}$ with $\mathbf{a} = \varphi(\mathbf{t}) \neq \infty$, $\varphi(K(\mathbf{t})) = K \cup \{\infty\}$, and $q(\mathbf{a}) \neq 0$ this holds:

(4) Let $b = \varphi(x)$ and $c = \varphi(y)$. Then $\varphi(E) = K(b) \cup \{\infty\}$, $\varphi(F_0) = K(c) \cup \{\infty\}$, and $\varphi(F) = L(c) \cup \{\infty\}$.

Suppose $\mathbf{a} \in K^r \cap H_L(g; q)$. Then $g(\mathbf{a}, Y)$ is irreducible over L, $q(\mathbf{a}) \neq 0$,, $q(\mathbf{a}) \neq 0$, and $g(\mathbf{a}, c) = 0$. Use Lemma 2.2.7 to construct a K-place $\varphi\colon F \to \tilde K \cup \{\infty\}$ with $\varphi(\mathbf{t}) = \mathbf{a}$ and $\varphi(K(\mathbf{t})) = K \cup \{\infty\}$. By Lemma 13.1.3, $\deg_Y(g) = [F_0 : E_0] = [F : L(\mathbf{t})]$. Hence,

$$[L(c) : K] = [L(c) : L][L : K] = \deg_Y(g)[L : K]$$
$$= [F : L(\mathbf{t})][L(\mathbf{t} : K(\mathbf{t})] = [F : E_0].$$

In addition, $[K(b) : K] \leq [E : E_0]$ and $[L(c) : K(b)] \leq [F : E]$. Hence, $[K(b) : K] = [E : E_0]$. Since $f(\mathbf{a}, b) = 0$, the polynomial $f(\mathbf{a}, X)$ is irreducible over K.

Finally, use Lemma 12.1.4 to eliminate q. □

PROBLEM 13.1.5 ([Dèbes-Haran, p. 284]): *Let K be an infinite field and H a separable Hilbertian subset of K^r. Does there exist an absolutely irreducible polynomial $f \in K[T_1, \ldots, T_r, X]$ which is monic and separable in X such that $H_K(f) \subseteq H$?*

13.2 Function Fields over Infinite Fields

Function fields of several variables over infinite fields are Hilbertian. This is a consequence of the following result.

PROPOSITION 13.2.1: *Let K_0 be an infinite field, t an indeterminate, and $K = K_0(t)$. Then every separable Hilbert subset H of K contains a set $\{a + bt \mid (a, b) \in U(K_0)\}$ for some nonempty Zariski K_0-open subset U of $\mathbb{A}^2$. Thus, every Hilbert set of K is nonempty and K is Hilbertian.*

Proof: The first statement implies that every separable Hilbert subset of K is nonempty. If $\operatorname{char}(K) > 0$, then K is imperfect. Hence, by Uchida (Proposition 12.4.3), every Hilbert set of K is nonempty and K is Hilbertian.

Lemmas 12.1.6 and 13.1.2 reduce the proof of the first statement to the following one: For an absolutely irreducible polynomial $f \in K[X, Y]$, monic and separable in Y, with $\deg_Y(f) > 1$, there exists a nonempty Zariski K_0-open subset U of $\mathbb{A}^2$ such that $f(a + bt, Y)$ has no zeros in K for each $(a, b) \in U(K_0)$.

If necessary, multiply f by a suitable element of K and make a linear change in the variable Y to assume that $f(X, Y) = g(t, X, Y) \in K_0[t, X, Y]$ is an absolutely irreducible polynomial and $\frac{\partial g}{\partial Y} \neq 0$. Then $g(t, Z_0 + Z_1 t, Y)$ is an irreducible polynomial in $\tilde{L}[t, Y]$, where $L = K_0(Z_0, Z_1)$ (Proposition 10.5.4). By Proposition 9.4.3, there exists a nonzero polynomial $h \in K_0[Z_0, Z_1]$ such that $g(t, a + bt, Y)$ is irreducible in $\tilde{K}_0[t, Y]$ for every $a, b \in K_0$ such that $h(a, b) \neq 0$. In particular, $f(a + bt, Y)$ has no zeros in K if $h(a, b) \neq 0$. □

Proposition 13.2.1 implies the analog to Corollary 12.1.5 for separable Hilbert sets:

PROPOSITION 13.2.2: *Let K be a field. Suppose each separable Hilbert subset of K of the form $H_K(f)$ with irreducible $f \in K[T, X]$, separable, monic, and of degree at least 2 in X, is nonempty. Then K is Hilbertian.*

Proof: By Lemma 12.1.6, it suffices to consider a separable irreducible polynomial $f \in K[T_1, \ldots, T_r, X]$ and to prove that $H_K(f) \neq \emptyset$. The case $r = 1$ is covered by the assumption of the Proposition. Now, suppose $r \geq 2$ and the statement holds for $r - 1$. The assumption of the Proposition implies K is infinite. Hence, by Proposition 13.2.1, $f(T_1, \ldots, T_{r-1}, g(T_1, \ldots, T_{r-1}), X)$ is irreducible and separable in $K(T_1, \ldots, T_{r-1})[X]$ for some $g \in K(T_1, \ldots, T_{r-1})$. Write $g(t_1, \ldots, T_{r-1}) = \frac{g_1(T_1, \ldots, T_{r-1})}{g_0(T_1, \ldots, T_{r-1})}$ with $g_0, g_1 \in K[T_1, \ldots, T_{r-1}]$. Lemma

12.1.6 gives an irreducible polynomial $h \in K[T_1,\ldots,T_{r-1},X]$, separable, monic, and of degree at least 2 in X, such that

$$H_K(h) \subseteq H_K\bigl(f(T_1,\ldots,T_{r-1},g(T_1,\ldots,T_{r-1},X));g_0(T_1,\ldots,T_{r-1})\bigr).$$

The induction hypotheses gives $a_1,\ldots,a_{r-1} \in K$ such that $h(a_1,\ldots,a_{r-1},X)$ is irreducible and separable in $K[X]$. Let $a_r = g(a_1,\ldots,a_{r-1})$. Then $f(a_1,\ldots,a_r,X)$ is well defined and irreducible in $K[X]$. □

13.3 Global Fields

For K a global field it is easy to choose the h_i's in Lemma 13.1.2 with coefficients in the ring of integers O_K. If, in particular, K is a number field, $r = 1$, and each of the curves $h_i = 0$ has positive genus, a celebrated theorem of Siegel implies that each of the h_i's has only finitely many zeros $(a,b) \in O_K^2$ ([Lang3, p. 121] or [Robinson-Roquette]). In this case $H'_K(h_1,\ldots,h_m)$ is clearly infinite. If, however, the curve $h_i = 0$ is of genus zero, we may use Riemann-Hurwitz to replace $h_i(T,x)$ by $g_i(T,X) = h_i(m(T),X)$ for some $m(T) \in O_K[T]$, so that $g_i(T,X) = 0$ has positive genus. Thus, Siegel's theorem gives Hilbert's theorem effortlessly. Of course, Siegel's theorem, a deep result in arithmetic, applies only to number fields (more generally, to fields which are finitely generated over $\mathbb{Q}$ [Lang3, p. 127, Thm. 4]). Besides, its power masks subtle connections between the irreducibility theorem and other arithmetic results.

Our approach to the Hilbert irreducibility theorem for global fields is based on the Chebotarev density theorem for function fields over finite fields. More accurately, we use a special case of Proposition 6.4.8. As a bonus we will prove that each Hilbert set over $\mathbb{Q}$ contains arithmetic progressions.

We start the proof with a weak consequence of Bauer's theorem. In keeping with our elementary treatment we use only Euclid's argument for proving the infinitude of primes:

LEMMA 13.3.1: *Let L/K be a finite separable extension of global fields. Then there exist infinitely many primes $\mathfrak{p}$ of K such that $\bar{L}_\mathfrak{P} = \bar{K}_\mathfrak{p}$ for every prime $\mathfrak{P}$ of L lying over $\mathfrak{p}$.*

Proof: Assume, without loss, that $K = \mathbb{Q}$ if $\mathrm{char}(K) = 0$ and $K = \mathbb{F}_p(t)$ if $\mathrm{char}(K) = p$. In particular, O_K is a principal ideal domain with only finitely many units. Replace L by the Galois hull of L/K to assume that L/K is Galois.

Consider a primitive element $z \in O_L$ for the extension L/K with discriminant $d \in O_K$ (Section 6.1). Suppose $\mathfrak{p}_1,\ldots,\mathfrak{p}_k$ are prime ideals of O_K satisfying the conclusion of the lemma. For each i between 1 and k choose a nonzero $\pi_i \in \mathfrak{p}_i$ and let $\pi = d \cdot \pi_1 \cdots \pi_k$. Consider also $f(X) = \mathrm{irr}(z,K) = X^n + c_{n-1}X^{n-1} + \cdots + c_0$. Since $c_0 \neq 0$ and O_K has only finitely many units, there exists a positive integer m such that

$$c_0^{-1}f(c_0\pi^m) = c_0^{n-1}\pi^{mn} + c_{n-1}c_0^{n-2}\pi^{m(n-1)} + \cdots + 1$$

is a nonunit of O_K. Therefore, this element has a prime divisor $\mathfrak{p} = \mathfrak{p}_{k+1}$ different from $\mathfrak{p}_1, \ldots, \mathfrak{p}_k$ and relatively prime to d. Let $\mathfrak{P}$ be a prime ideal of O_L lying over $\mathfrak{p}_{k+1}$. Denote the reduction modulo $\mathfrak{P}$ by a bar. Then $\bar{f}(X)$ has a root in $\bar{K}_\mathfrak{p}$, namely $c_0 \pi^m$ modulo $\mathfrak{p}$, which we may assume to be $\bar{z}$. By Remark 6.1.7, $\bar{L}_\mathfrak{P} = \bar{K}_\mathfrak{p}(\bar{z}) = \bar{K}_\mathfrak{p}$. □

LEMMA 13.3.2: For H a proper subgroup of a finite group G, the set $G \smallsetminus \bigcup_{\sigma \in G} \sigma^{-1} H \sigma$ is nonempty.

For $f \in K[X]$ an irreducible separable polynomial with roots $x_1, \ldots, x_n$ and splitting field $N = K(x_1, \ldots, x_n)$, there exists $\sigma \in \mathrm{Gal}(N/K)$ such that $\sigma x_i \neq x_i$, for each $i = 1, \ldots, n$.

Proof: Let $H_1, \ldots, H_m$ be the distinct subgroups of G conjugate to H. Assume without loss that $m \geq 2$. Then m is equal to the index of the normalizer of H. Thus, $m \leq (G : H)$. The intersection $H_1 \cap \cdots \cap H_m$ contains 1, so $|\bigcup_{i=1}^m H_i| < \sum_{i=1}^m |H_i| = m \cdot |H| \leq |G|$.

With $\mathrm{Gal}(N/K) = G$ and $\mathrm{Gal}(N/K(x_1)) = H$ the conjugates of H are $\mathrm{Gal}(N/K(x_1)), \ldots, \mathrm{Gal}(N/K(x_m))$. The last statement of the lemma follows by choosing $\sigma \in G \smallsetminus \bigcup_{\tau \in G} \tau^{-1} H \tau$. □

LEMMA 13.3.3: Let q be a prime power, $\bar{K} = \mathbb{F}_q$, t an indeterminate, and $\bar{E} = \bar{K}(t)$. Consider a Galois extension $\bar{F} = \bar{K}(t, z)$ where $\bar{g}(t, z) = 0$ with $g \in \bar{K}[T, Z]$ an irreducible polynomial. Suppose $\bar{K}$ is algebraically closed in $\bar{F}$. Let $d = \deg(\bar{g})$, $m = \deg_Z(\bar{g})$, and $\mathcal{C}$ be a conjugacy class in $\mathrm{Gal}(\bar{F}/\bar{E})$. Then the number N of primes of $\bar{E}/\bar{K}$ of degree 1, unramified in $\bar{F}$, with Artin symbol equal to $\mathcal{C}$ satisfies

(1) $$\left| N - \frac{|\mathcal{C}|}{m} q \right| < 10 d^2 |\mathcal{C}| \sqrt{q}.$$

Proof: Write $g_{\bar{F}} = \mathrm{genus}(\bar{F})$. Since $\bar{K}$ is algebraically closed in $\bar{F}$, $\bar{g}$ is absolutely irreducible (Corollary 10.2.2). Hence, $g(T, Z) = 0$ defines a curve of degree d over $\bar{K}$. Recalling that $\bar{E}$ has genus 0, Proposition 6.4.8, with $k = 1$, gives the inequality

$$\left| N - \frac{|\mathcal{C}|}{m} q \right| < \frac{2|\mathcal{C}|}{m} \left[(m + g_{\bar{F}}) q^{\frac{1}{2}} + m q^{\frac{1}{4}} + g_{\bar{F}} + m \right].$$

Combining this with the inequality $g_{\bar{F}} \leq \frac{1}{2}(d-1)(d-2)$ for the genus of $\bar{F}$ (Corollary 5.3.5) we obtain (1). □

An **absolute value** of a field K is a map $|\ |: K \to \mathbb{R}$ which satisfies the following conditions for all $x, y \in K$:
(2a) $|x| \geq 0$ and $|x| = 0$ if and only if $x = 0$.
(2b) $|xy| = |x||y|$.
(2c) $|x + y| \leq |x| + |y|$.
(2d) There is an $a \in K$ with $|a| \neq 1$.

13.3 Global Fields

If, instead of (2c), $|\ |$ satisfies the stronger condition

(2c') $|x+y| \leq \max(|x|,|y|)$,

it is **ultrametric** (or **non-archimedean**), otherwise, it is **metric** (or **archimedean**). An absolute value $|\ |'$ is **equivalent** to $|\ |$ if there exists $c > 0$ such that $|x|' = |x|^c$ for all $x \in K$.

If $|\ |$ is ultrametric, then $O_{|\ |} = \{x \in K \mid |x| \leq 1\}$ is a valuation ring of K. In particular, $|n| \leq 1$ for each positive integer n. Conversely, suppose $|n| \leq 1$ for each positive integer n. Let z be an element of K with $|z| \leq 1$. Then $|1+z|^k = \left|\sum_{i=0}^{k}\binom{k}{i}z^i\right| \leq k+1$. Hence, $|1+z| \leq (k+1)^{1/k}$. Letting k go to infinity, we find that $|1+z| \leq 1$. Next let $x, y \in K$ with $|x| \leq |y|$ and $y \neq 0$. Then $|x+y| = |1+\frac{x}{y}||y| \leq |y| = \max(|x|,|y|)$. Consequently, $|\ |$ is ultrametric.

In particular, suppose $\text{char}(K) = p > 0$. Then $n^{p-1} = 1$ for each positive integer n not divisible by p. Hence, $|n| = 1$. Therefore, $|\ |$ is ultrametric.

If K is a global field, the map $|\ | \to O_{|\ |}$ is a bijection between the equivalence classes of ultrametric absolute values of K and the valuation rings of K. In particular, if K is a function field of one variable over a finite field K_0, then the equivalence classes of absolute values of K bijectively correspond to the prime divisors of K/K_0. If K is a number field, then the equivalence classes of ultrametric absolute values bijectively correspond to the prime ideals of O_K. In addition, each embedding $\sigma: K \to \mathbb{C}$ defines a metric absolute value $|\ |_\sigma$ of K: $|x|_\sigma = |\sigma x|$, where $|\ |$ is here the usual absolute value of $\mathbb{C}$. Each metric absolute value of K is equivalent to some $|\ |_\sigma$ and there are at most $[K:\mathbb{Q}]$ such equivalence classes [Cassels-Fröhlich, p. 57, Thm.]. In each case we call an equivalence class of an absolute value a **prime** of K. For each prime $\mathfrak{p}$ of K we choose an absolute value $|\ |_\mathfrak{p}$ which represents it.

The set of all primes of a global field K satisfies the **strong approximation theorem**: Let $\mathfrak{p}_0, \mathfrak{p}_1, \ldots, \mathfrak{p}_n$ be distinct primes of K. For each i between 1 and n consider an element a_i of K and let $\varepsilon > 0$. Then there exists $x \in K$ such that $|x - a_i|_{\mathfrak{p}_i} < \varepsilon$, $i = 1, \ldots, n$, and $|x|_\mathfrak{p} \leq 1$ for each prime $\mathfrak{p}$ not in $\{\mathfrak{p}_0, \mathfrak{p}_1, \ldots, \mathfrak{p}_n\}$ [Cassels-Fröhlich, p. 67]. In the function field case, this theorem is also a consequence of the strong approximation theorem (Proposition 3.3.1).

Generalizations of the following lemma appear as ingredients in the Galois stratification procedure of Chapter 30.

LEMMA 13.3.4: *Let K be a global field and $f \in K[T,X]$ an absolutely irreducible polynomial, monic and separable in X, with $\deg_X(f) > 1$. Then K has infinitely many ultrametric primes $\mathfrak{p}$ for which there is an $a_\mathfrak{p} \in O_K$ with this property: if $a \in O_K$ satisfies $a \equiv a_\mathfrak{p} \mod \mathfrak{p}$, then $f(a,b) \neq 0$ for every $b \in K$.*

Proof: Assume without loss that $f \in O_K[T,X]$. Let $E = K(t)$, with t an indeterminate, and let $x_1, \ldots, x_n$ be the roots of $f(t,X)$ in E_s. Denote the algebraic closure of K in the Galois extension $F = E(x_1, \ldots, x_n)$ of E by L.

Then F is a regular extension of L; write it as $F = L(t,z)$ with z integral over $O_L[t]$. Thus, there is an absolutely irreducible polynomial $g \in L[T,X]$ with $g(t,X) = \mathrm{irr}(z, L(t))$. By Lemma 13.3.1, the set of ultrametric primes $\mathfrak{p}$ of K for which $\bar{L}_\mathfrak{P} = \bar{K}_\mathfrak{p}$ for each prime $\mathfrak{P}$ of L over $\mathfrak{p}$, is infinite; call it A_1. The rest of the proof derives the result from the case where L replaces K and g replaces f.

PART A: *Geometrically exceptional primes.* Let A_2 be the restriction to K of the set of ultrametric primes $\mathfrak{P}$ of L for which both $f \bmod \mathfrak{P}$ and $g \bmod \mathfrak{P}$ are defined, separable in X, and absolutely irreducible. By Bertini-Noether (Proposition 9.4.3), A_2 is a cofinite set.

Express z as a polynomial in $x_1, \ldots, x_n$ (with coefficients that are ratios of elements of $O_K[t]$) and express each x_i as a polynomial in z (with coefficients that are ratios of elements of $O_L[t]$). Denote the cofinite set of ultrametric primes of K which divide none of the denominators of these coefficients by A_3.

Finally, consider $d(g;t) = \mathrm{norm}_{F/LE}(\frac{\partial g}{\partial X}(t,z)) \in O_L[t]$. Denote the cofinite set consisting of the restriction to K of the ultrametric primes $\mathfrak{P}$ of L for which $\deg(d(g;t)) = \deg(d(g;t) \bmod \mathfrak{P})$ by A_4. The set $A = A_1 \cap A_2 \cap A_3 \cap A_4$ is infinite.

PART B: *Reduction modulo primes of A.* For $\mathfrak{p} \in A$ and $\mathfrak{P}$ a prime of L over $\mathfrak{p}$, extend the residue map $O_L \to \bar{L}_\mathfrak{P} = \bar{K}_\mathfrak{p}$ to a place φ of F by choosing $\bar{t} = \varphi(t)$ as a transcendental element over $\bar{K}_\mathfrak{p}$. Denote $\varphi(x)$ by $\bar{x}$ for every $x \in F$ finite under φ. Let $\bar{E} = \bar{K}_\mathfrak{p}(\bar{t})$ and $\bar{F} = \bar{E}(\bar{z}) = \bar{E}(\bar{x}_1, \ldots, \bar{x}_n)$, where $\bar{x}_1, \ldots, \bar{x}_n$ are the distinct roots of the irreducible polynomial $\bar{f}(\bar{t}, X)$.

By the choice of $\mathfrak{p}$, the polynomial $\bar{f}(\bar{t}, X)$ is irreducible over $\bar{E}$. Hence, by Lemma 13.3.2, there exists $\sigma \in \mathrm{Gal}(\bar{F}/\bar{E})$ such that $\sigma \bar{x}_i \neq \bar{x}_i$, $i = 1, \ldots, n$. Denote the conjugacy class of σ in $\mathrm{Gal}(\bar{F}/\bar{E})$ by $\mathrm{Con}(\sigma)$. Next note that since $\bar{g}$ is absolutely irreducible and $\bar{g}(\bar{t}, \bar{z}) = 0$, the field $\bar{K}_\mathfrak{p}$ is algebraically closed in $\bar{F}$. In addition, the polynomial $d(\bar{g}; \bar{t}) = N_{\bar{F}/\bar{E}}(\frac{\partial \bar{g}}{\partial X}(\bar{t}, \bar{z}))$ with coefficients in $\bar{K}_\mathfrak{p}$ has the same degree as $d(g;t)$. If $\bar{a} \in \bar{K}_\mathfrak{p}$ is not a root of $d(\bar{g}; \bar{t})$, then the prime $\mathfrak{p}_{\bar{a}}$ corresponding to the specialization $\bar{t} \to \bar{a}$ of $\bar{E}$ is unramified in $\bar{F}$ (Lemma 6.1.8). The number of primes of degree 1 of $\bar{E}$ which ramify in $\bar{F}$ is therefore bounded by $1 + \deg(d(g;t))$.

Let A' be the set of all $\mathfrak{p} \in A$ finite at t for which there exists $\bar{a}_\mathfrak{p} \in \bar{K}_\mathfrak{p}$ with $d(\bar{f}; \bar{a}_\mathfrak{p}) \neq 0$ such that the Artin symbol $\left(\frac{\bar{F}/\bar{E}}{\mathfrak{p}_{\bar{a}}}\right)$ (here $\bar{a} = \bar{a}_\mathfrak{p}$) corresponding to the prime $\mathfrak{p}_{\bar{a}}$ equals $\mathrm{Con}(\sigma)$. By Lemma 13.3.3, A' is cofinite in A. For each $\mathfrak{p} \in A'$ there exists a prime Q of $\bar{F}$ lying over $\mathfrak{p}_{\bar{a}}$ with these properties: for $q = |\bar{K}_\mathfrak{p}|$

(3a) $\sigma x \equiv x^q \bmod Q$ for every $x \in \bar{F}$ integral with respect to Q;
and, since $d(\bar{f}; \bar{t}) = \prod_{i \neq j}(\bar{x}_i - \bar{x}_j)$,
(3b) $\bar{x}_i \neq \bar{x}_j \bmod Q$ for every $i \neq j$.

PART C: *Finding $a_\mathfrak{p}$.* For each i between 1 and n let j be an integer with $\sigma \bar{x}_i = \bar{x}_j$. Then $i \neq j$. Hence, by (3a) and (3b), $\bar{x}_i \neq \bar{x}_i^q \bmod Q$. That is, the polynomial $\bar{f}(\bar{a}_\mathfrak{p}, X)$ has no roots in $\bar{K}_\mathfrak{p}$.

Let $a_\mathfrak{p}$ be an element of O_k which is mapped to $\bar{a}_\mathfrak{p}$ by φ. If $a \in O_K$ satisfies $a \equiv a_\mathfrak{p} \mod \mathfrak{p}$, then $\bar{a} = \bar{a}_\mathfrak{p}$, and therefore $f(a,b) \neq 0$ for each $b \in O_K$. Since $f(a, X)$ is monic, this proves $f(a, X)$ has no zero in K. □

Let K be a global field. An **arithmetic progression** of O_K is a set of the form $a + \mathfrak{a}$, where $a \in O_K$ and $\mathfrak{a}$ is an ideal of O_K.

THEOREM 13.3.5: *Let K be a global field and H a separable Hilbert subset of K. Then:*
(a) *H contains an arithmetic progression of O_K.*
(b) *Let $\mathfrak{q}_0, \mathfrak{q}_1, \ldots, \mathfrak{q}_n$ be distinct primes of K, $b_1, \ldots, b_n$ elements of K, and $\varepsilon > 0$. Then there exists $x \in H$ with $|x - b_i|_{\mathfrak{q}_i} < \varepsilon$ for $i = 1, \ldots, n$ and $|x|_\mathfrak{q} \leq 1$ for each prime of K not in $\{\mathfrak{q}_0, \mathfrak{q}_1, \ldots, \mathfrak{q}_n\}$.*
(c) *The intersection of H with each arithmetical progression of O_K is nonempty.*

Proof: By Lemma 13.1.2 there exist absolutely irreducible polynomials $h_1, \ldots, h_m \in O_K[T, X]$, monic and separable in X, with $\deg_X(h_i) > 1$, $i = 1, \ldots, m$, and $H'_K(h_1, \ldots, h_m) \subseteq H$. Apply Lemma 13.3.4 to find distinct prime ideals $\mathfrak{p}_1, \ldots, \mathfrak{p}_m$ of O_K and elements $a_1, \ldots, a_m \in O_K$ such that $x \in H'_K(h_i)$ for each $x \in K$ with $|x - a_i|_{\mathfrak{p}_i} < 1$, $i = 1, \ldots, m$. The Chinese remainder theorem [Lang7, p. 94] produces an element $a \in O_K$ with $a + \mathfrak{p}_i = a_i + \mathfrak{p}_i$ for $i = 1, \ldots, m$. Thus, with $\mathfrak{a} = \mathfrak{p}_1 \ldots \mathfrak{p}_m$, we have $a + \mathfrak{a} \subseteq H$. This proves (a).

For the proof of (b) choose $\mathfrak{p}_1, \ldots, \mathfrak{p}_m$ above not in $\{\mathfrak{q}_0, \mathfrak{q}_1, \ldots, \mathfrak{q}_n\}$. Then use the strong approximation theorem for global fields [Cassels-Fröhlich, p. 67] to find $x \in K$ with $|x - a_i|_{\mathfrak{p}_i} < 1$, $i = 1, \ldots, m$, $|x - b_j|_{\mathfrak{q}_j} < \varepsilon$, $j = 1, \ldots, n$, and $|x|_\mathfrak{q} \leq 1$ for each prime $\mathfrak{q}$ not in $\{\mathfrak{p}_1, \ldots, \mathfrak{p}_m, \mathfrak{q}_0, \ldots, \mathfrak{q}_n\}$. Then $x \in H$ and $|x|_{\mathfrak{p}_i} \leq 1$ for $i = 1, \ldots, m$, as desired.

Each arithmetical progression in O_K has the form $A = \{x \in O_K \mid v_{\mathfrak{q}_i}(x - a) \geq k_i, \ i = 1, \ldots, r\}$ for some $a \in O_K$, ultrametric primes $\mathfrak{q}_1, \ldots, \mathfrak{q}_r$ which are finite on O_K, and positive integers $k_1, \ldots, k_r$. If K is a number field, choose $\mathfrak{q}_0$ to be a metric prime. If K is a function field of one variable over a finite field, choose $\mathfrak{q}_0$ to be a prime which is not finite on O_K. Statement (b) gives $x \in A$ with $v_\mathfrak{q}(x) \geq 0$ for all ultrametric primes $\mathfrak{q}$ of K which are finite on O_K. Thus, $x \in O_K$. □

13.4 Hilbertian Rings

We call an integral domain R with quotient field K **Hilbertian** if every separable Hilbert subset of K^r contains elements all of whose coordinates are in R. In this case, each overring of R is Hilbertian. In particular, K is Hilbertian.

PROPOSITION 13.4.1: *Let R be an integral domain with quotient field K. Suppose either R is finitely generated over $\mathbb{Z}$ or R is finitely generated over a field K_0 and K/K_0 is transcendental. Then R is Hilbertian.*

Proof: Let R_0 be either $\mathbb{Z}$ or K_0. In the former case let $K_0 = \mathbb{Q}$. By assumption, $R = R_0[u_1, \ldots, u_n]$. Assume without loss $u_1, \ldots, u_m$ with $m \leq n$ form a transcendental base for K/K_0. By Proposition 12.3.3, every separable Hilbertian subset of K^r contains a separable Hilbertian subset of $K_0(u_1, \ldots, u_m)$. We may therefore assume that $u_1, \ldots, u_n$ are algebraically independent over K_0.

Consider an irreducible polynomial $f \in K[T_1, \ldots, T_r, X]$ which is separable in X. We have to prove that $H_K(f) \cap R^r \neq \emptyset$ (Lemma 12.1.6). There are several cases to consider.

CASE 1: $r = 1$ and $n = 0$. Then $R = R_0 = \mathbb{Z}$. By Theorem 13.3.5 there exists $a \in R$ such that $f(a, X)$ is irreducible.

CASE 2: $r = 1$, $n = 1$, and $R_0 = K_0$. If K_0 is finite, then K is global and $R = O_K$. Thus, Theorem 13.3.5 gives $a \in R$ with $f(a, X)$ irreducible. If K_0 is infinite, then Proposition 13.2.1 gives $a, b \in K_0$ such that $f(a + bu_1, X)$ is irreducible.

CASE 3: $r = 1$, $n = 1$, and $R_0 = \mathbb{Z}$. Proposition 13.2.1 gives a nonempty Zariski $\mathbb{Q}$-open subset U of $\mathbb{A}^2$ such that $f(a + bu_1, X)$ is irreducible for all $(a, b) \in U(\mathbb{Q})$. Since $\mathbb{Z}$ is infinite, we may choose (a, b) in $U(\mathbb{Z})$.

CASE 4: $r = 1$ and $n \geq 2$. Then $R_0[u_1, \ldots, u_{n-1}]$ is an infinite ring. Hence, by Proposition 13.2.1, it has elements a, b such that $f(a + bu_1, X)$ is irreducible.

CASE 5: $r \geq 2$. Consider f as a polynomial in T_r, X with coefficients in the infinite ring $R_0[\mathbf{u}, T_1, \ldots, T_{r-1}]$. As such, f is irreducible. Replace R_0 and K_0 in Cases 2, 3, and 4 by $R_0[\mathbf{u}, T_1, \ldots, T_{r-1}]$ and $K_0(\mathbf{u}, T_1, \ldots, T_{r-1})$ to find $g \in R_0[\mathbf{u}, T_1, \ldots, T_{r-1}]$ such that $f(T_1, \ldots, T_{r-1}, g(T_1, \ldots, T_{r-1}), X)$ is irreducible. Now use an induction hypothesis on r to find $a_1, \ldots, a_{r-1} \in R$ with $f(a_1, \ldots, a_{r-1}, g(a_1, \ldots, a_{r-1}), X)$ irreducible. Then $a_r = g(a_1, \ldots, a_{r-1})$ is in R and $f(\mathbf{a}, X)$ is irreducible. ∎

The quotient fields of the rings mentioned in Proposition 13.4.1 are the **classical Hilbertian fields**.

THEOREM 13.4.2: *Suppose K is a global field or a finitely generated transcendental extension of an arbitrary field K_0. Then K is Hilbertian. Moreover, each Hilbert set of K is nonempty.*

Proof: As a consequence of Proposition 13.4.1, K is Hilbertian. To prove the second statement, assume $\text{char}(K) = p > 0$. Then K is imperfect. Hence, by Uchida (Proposition 12.4.3), every Hilbertian set of K is nonempty. ∎

The proof of Proposition 13.4.1 gives another useful version of Theorem 13.4.2:

13.4 Hilbertian Rings

PROPOSITION 13.4.3: *Let K be a finitely generated separable extension of a field K_0 with $m = \text{trans.deg}(K/K_0) > 0$. Let $r \leq m$ and H a separable Hilbert subset of K^r. Then H contains a point $(u_1, \ldots, u_r)$ with $u_1, \ldots, u_r$ algebraically independent over K_0.*

Proof: Choose a separating transcendence base $t_1, \ldots, t_m$ for K/K_0. Then $K/K_0(\mathbf{t})$ is a finite separable extension. By Corollary 12.2.3, H contains a separable Hilbert subset of $K_0(\mathbf{t})^r$. We may therefore assume $K = K_0(\mathbf{t})$. By Lemma 12.1.6, we may assume $H = H_K(f)$ with an irreducible $f \in K[T_1, \ldots, T_r, X]$, monic and separable in X.

For $m = 1$ and K_0 finite, Theorem 13.3.5 gives $u \in H$ transcendental over K_0. For $m = 1$ and K_0 infinite, Proposition 13.2.1 does the job. Assume $m \geq 2$. Then $K' = K_0(t_2, \ldots, t_m, T_2, \ldots, T_r)$ is an infinite field, t_1 is transcendental over K', and $K'(t_1) = K(T_2, \ldots, T_r)$. Consider f as an irreducible polynomial in T_1, X over $K'(t_1)$. Proposition 13.2.1 gives nonzero $g, h \in K_0[t_2, \ldots, t_m, T_2, \ldots, T_r]$ such that

$$f_1(T_2, \ldots, T_r, X) = f\bigl(g(T_2, \ldots, T_r) + h(T_2, \ldots, T_r)t_1, T_2, \ldots, T_r, X\bigr)$$

is irreducible over K.

Let $K_1 = K_0(t_1)$. Apply the induction hypothesis to K_1 and f_1 instead of to K and f and find $u_2, \ldots, u_r$ in K algebraically independent over K_1 such that $f(g(u_2, \ldots, u_r) + h(u_2, \ldots, u_r)t_1, u_2, \ldots, u_r, X)$ is irreducible over K. Put $u_1 = g(u_2, \ldots, u_r) + h(u_2, \ldots, u_r)t_1$. Then $f(\mathbf{u}, X)$ is irreducible over K and $K_0(u_1, u_2, \ldots, u_r) = K_0(t_1, u_2, \ldots, u_r)$. Hence, $u_1, \ldots, u_r$ are algebraically independent over K_0, as desired. □

Remark 13.4.4: More Hilbertian rings. Exercise 4 proves that each valuation ring of a Hilbertian field is Hilbertian.

Geyer extends this result to several valuations of rank 1. He considers a Hilbertian field K, nonequivalent absolute values $|\ |_1, \ldots, |\ |_n$ of K, and a separable Hilbert subset H of K^r. Let $\mathbf{a}_1, \ldots, \mathbf{a}_r$ be tuples in K^r and let $\varepsilon > 0$. Then there exists $\mathbf{x} \in H$ such that $|\mathbf{x} - \mathbf{a}_i|_i < \varepsilon$, $i = 1, \ldots, n$ [Geyer1, Lemma 3.4]. In particular, if $v_1, \ldots, v_n$ are valuations of rank 1, then their holomorphy ring, $R = \{a \in K \mid v_i(a) \geq 0,\ i = 1, \ldots, n\}$ is Hilbertian.

Proposition 19.7 of [Jarden14] generalizes Geyer's result. Here one considers valuations $v_1, \ldots, v_m$ and orderings $<_1, \ldots, <_n$ of a Hilbertian field K. Suppose the valuations and the orderings are **independent**. That is, they define distinct topologies on K. Let H be a separable Hilbert subset of K^r, $\mathbf{a}_1, \ldots, \mathbf{a}_m, \mathbf{b}_n, \ldots, \mathbf{b}_n \in K^r$, α_i an element of the valuation group of v_i, $i = 1, \ldots, m$, and $c_j \in K$ with $c_j >_j 0$ for $j = 1, \ldots, n$. Then there exist $\mathbf{x} \in H$ with $v_i(\mathbf{x} - \mathbf{a}_i) > \alpha_i$, $i = 1, \ldots, m$, and $-c_j <_j \mathbf{x} - \mathbf{b}_j <_j c_j$, $j = 1, \ldots, n$. Moreover, the holomorphy ring of arbitrary finitely many valuations of K is Hilbertian [Jarden14, Proposition 19.6].

In contrast, holomorphy ring of infinitely many valuations of a field K may not be Hilbertian even if K is Hilbertian. For example, if K_0 is a finite field, the holomorphy ring of all valuations of $K_0(t)$ is K_0. The latter field is not Hilbertian, although $K_0(t)$ is Hilbertian. More interesting examples can be found in Example 15.5.6. □

The following Lemma improves Exercise 2 of Section 12:

LEMMA 13.4.5: *Let m be a cardinal number and $\{K_\alpha \mid \alpha < m\}$ a transfinite sequence of fields. Suppose that for each $\alpha < m$ the field $K_{\alpha+1}$ is a proper finitely generated regular extension of K_α. Then $K = \bigcup_{\alpha < m} K_\alpha$ is a Hilbertian field.*

Proof: Let $f \in K(T_1, \ldots, T_r)[X]$ be a monic separable irreducible polynomial. By Lemma 12.1.6, it suffices to prove that $H_K(f) \neq \emptyset$. Indeed, the coefficients of f belong to K_α for some $\alpha < m$. By Theorem 13.4.2, $K_{\alpha+1}$ is Hilbertian. Find $a_1, \ldots, a_r \in K_{\alpha+1}$ such that $f(\mathbf{a}, X)$ is separable and irreducible in $K_{\alpha+1}[X]$. Choose a zero b of $f(\mathbf{a}, X)$ in the separable closure of $K_{\alpha+1}$. By Corollary 2.6.5(d), K is a regular extension of $K_{\alpha+1}$. In particular, K is linearly disjoint from $K_{\alpha+1}(b)$ over $K_{\alpha+1}$. Therefore, $f(\mathbf{a}, X)$ is irreducible in $K[X]$. □

The PAC and the Hilbert properties are in a sense opposing. Nevertheless there are fields that are both PAC and Hilbertian:

PROPOSITION 13.4.6 ([Fried3]): *Every field K has a regular extension F which is PAC and Hilbertian.*

Proof: Wellorder the varieties of dimension at least 1 that are defined over K in a transfinite sequence $\{V_\alpha \mid \alpha < m\}$. Use transfinite induction to define, for each $\beta < m$, a function field F_β for V_β which is algebraically independent from $\prod_{\alpha < \beta} F_\alpha$ (the composite of the F_α's with $\alpha < \beta$) over K. Since F_β is a regular extension of K, the field $\prod_{\alpha \leq \beta} F_\alpha$ is a proper finitely generated regular extension of $\prod_{\alpha < \beta} F_\alpha$ (Corollary 2.6.8(a)). It follows that $K_1 = \prod_{\alpha < m} F_\alpha$ is Hilbertian (Lemma 13.4.5) and regular over K (Lemma 2.6.5(d)). Moreover, every variety defined over K has a K_1-rational point.

Iterate this construction to obtain a countable tower $K \subset K_1 \subset K_2 \subset \cdots$ of fields such that K_i is Hilbertian, regular over K_{i-1} and every variety defined over K_i has a K_{i+1}-rational point. The union $F = \bigcup_{i=1}^{\infty} K_i$ is the desired field (Exercise 2 of Chapter 12). □

13.5 Hilbertianity via Coverings

Colliot-Thélène and Sansuc redefine Hilbertianity of a field by geometric terms [Serre8]. The new definition naturally leads to a weakening of Hilbertianity to g-Hilbertianity, where g is a nonnegative integer.

Definition 13.5.1: Thin sets. Let K be a field, V a K-variety, $\mathbf{x} = (x_1, \ldots, x_n)$ a generic point of V over K, and $F = K(\mathbf{x})$. A subset

13.5 Hilbertianity via Coverings

A of $V(K)$ is **thin** if

(1) $$A \subseteq V_0(K) \cup \bigcup_{i=1}^{m} \varphi_i(W_i(K)),$$

where V_0 is a Zariski K-closed subset of V with $\dim(V_0) < \dim(V)$, W_i is a K-variety with $\dim(W_i) = \dim(V)$, and $\varphi_i\colon W_i \to V$ is a dominant separable rational map with $\deg(\varphi_i) \geq 2$, $i = 1, \ldots, m$. □

Remark 13.5.2: *Simplification of (1)*. Let V be a K-variety and $\mathbf{x} = (x_1, \ldots, x_n)$ a generic point over K.

(a) Suppose V' is a K-variety which is birationally equivalent to V. Then, $V(K)$ is thin if and only if $V'(K)$ is thin. Moreover, enlarging V_0 is necessary, we may replace each W_i in (1) by a birationally equivalent K-variety. In addition, we may compose φ_i with any K-birational map.

(b) If $V(K)$ is not thin, then $V(K)$ is Zariski K-dense in V. Otherwise, V would have a Zariski K-closed subset V_0 of lower dimension such that $V(K) = V_0(K)$.

The converse is not true. For example, take K to be a field with absolute Galois group isomorphic to $\mathbb{Z}_p$ for some prime number p (Section 1.5). Choose a second prime number $l \neq p, \mathrm{char}(K)$. Let $\varphi\colon \mathbb{A}^1 \to \mathbb{A}^1$ be the morphism of degree l given by $\varphi(x) = x^l$. Then, $\varphi(K) = K$. Otherwise, there is an $a \in K$ which is not an l-power in K. Then $[K(\sqrt[l]{a}) : K] = l$ [Lang7, p. 297]. This contradicts the choice of K and l. It follows that K is thin.

(c) Suppose $V(K)$ is Zariski K-dense in V. Then, V is absolutely irreducible and defined over K.

To prove this statement, we choose a generic point $\mathbf{x}$ of V over K. By Proposition 10.2.2, it suffices to prove that $K(\mathbf{x})/K$ is regular. This will follow from the following two claims:

CLAIM A: *K is algebraically closed in $K(\mathbf{x})$.* Let $c \in K(\mathbf{x}) \cap \tilde{k}$. Write $c = \frac{f(\mathbf{x})}{g(\mathbf{x})}$ with $f, g \in K[\mathbf{X}]$ and $g(\mathbf{x}) \neq 0$. By assumption, there is a point $\mathbf{a} \in V(K)$ with $g(\mathbf{x}) \neq 0$. The K-specialization $\mathbf{x} \to \mathbf{a}$ extends to a K-homomorhphism $\varphi\colon K[\mathbf{x}] \to K$. In particular, $\varphi(c) = \frac{f(\mathbf{a})}{g(\mathbf{a})} \in K$. Since the restriction of φ to $K(c)$ is an isomorphism, $c \in K$.

CLAIM B: *$K(\mathbf{x})/K$ is a separable extension.* To prove this claim, we may assume $p = \mathrm{char}(K) > 0$. Now assume there are $c_1, \ldots, c_m \in K_{\mathrm{ins}}$ which are linearly independent over K and polynomials $f_1, \ldots, f_m \in K[\mathbf{X}]$ with $m \geq 1$, $f_i(\mathbf{x}) \neq 0$, $i = 1, \ldots, m$, and $\sum_{i=1}^{m} c_i f_i(\mathbf{x}) = 0$. Choose a power q of p with $c_i^q \in K$. Then $\sum_{i=1}^{m} c_i^q f_i^q(\mathbf{x}^q) = 0$. By assumption, there exists $\mathbf{a} \in V(K)$ with $f_1(\mathbf{a}) \neq 0$. Substituting $\mathbf{a}$ for $\mathbf{x}$ in the latter equality gives $\sum_{i=1}^{m} c_i^q f_i^q(\mathbf{a}^q) = 0$, so $\sum_{i=1}^{m} c_i f_i(\mathbf{a}) = 0$. The linear independence of $c_1, \ldots, c_m$ implies that $f_1(\mathbf{a}) = \cdots = f_m(\mathbf{a}) = 0$, which is a contradiction.

(d) Suppose $V(K)$ is nonthin. By (b), $V(K)$ is Zariski K-closed in V. Hence, by (c), V is a variety defined over K.

(e) Suppose V is a variety defined over K but W_i in (1) is not absolutely irreducible or W_i is absolutely irreducible but not defined over K. By (c), $W_i(K)$ is not Zariski K-closed in W_i. Thus, there exists a K-closed subset W_{i0} of W_i with $\dim(W_{i0}) < \dim(W_i)$ and $W_i(K) = W_{i0}(K)$. Then $\varphi_i(W_i(K)) = \varphi(W_{i0})(K)$ and $\dim(\varphi(W_{i0})) < \dim(W_i) = \dim(V)$. Increasing V_0 in (1), we may omit W_i from (1). Thus, without loss, we may assume that all W_i in (1) are varieties over K.

(f) Suppose, in the notation of (1), that V is a variety over K and V_0 is a K-variety. Then there exists a variety W_0 over K and a dominant separable map $\varphi_0\colon W_0 \to V$ with $\deg(\varphi_0) \geq 2$ and $V_0(K) \subseteq \varphi(W_0(K))$.

To this end, choose generic points $\mathbf{x}$ and $\mathbf{x}_0$ for V and V_0, respectively, over K. Let $\alpha\colon K[\mathbf{x}] \to K[\mathbf{x}_0]$ be the K-homomorphism with $\alpha(\mathbf{x}) = \mathbf{x}_0$. Since $\dim(V_0) < \dim(V)$, there is a nonzero u in $\mathrm{Ker}(\alpha)$. In particular, u is not in K. Since K is algebraically closed in $K(\mathbf{x})$ (because V is a variety over K), u is transcendental over K. Choose a transcendence base $t_1,\ldots,t_r$ for $K(\mathbf{x})/K(u)$, and a prime number $l > \max\bigl(\mathrm{char}(K), [K(\mathbf{x}) : K(u,\mathbf{t})]\bigr)$. Put $y = u^{1/l}$. Then, $[K(\mathbf{x},y) : K(\mathbf{x})] = [\tilde{K}(\mathbf{x},y) : \tilde{K}(\mathbf{x})] = l$. Hence, $K(\mathbf{x},y)/K$ is a regular extension. By Corollary 10.2.2(a), $(\mathbf{x},y)$ generate a variety W_0 over K. Let $\varphi_0\colon W_0 \to V$ be the projection $\varphi_0(\mathbf{x},y) = \mathbf{x}$. Then φ_0 is a dominant separable rational map of degree l.

By construction, $(\mathbf{x}_0, 0) \in W$. If $\mathbf{a} \in V_0(K)$, then the specialization $\mathbf{x}_0 \to \mathbf{a}$ extends to a specialization $(\mathbf{x}_0, 0) \to (\mathbf{a}, 0)$. Hence, $(\mathbf{a}, 0) \in W_0(K)$ and $\varphi_0(\mathbf{a}, 0) = \mathbf{a}$. Thus, $\varphi_0(W_0(K)) = V_0(K)$.

(g) Suppose V is a variety over K. For each i between 1 and m let $h_i \in K[X_1,\ldots,X_n,Y]$ be a polynomial which is monic and of degree at least 2 in Y. Let y_i be a root of $h_i(\mathbf{x}, Y)$ in $K(\mathbf{x})_s$. Suppose $K(\mathbf{x}, y_i)$ is a regular extension of K. Let $g \in K[\mathbf{X}]$ be a polynomial which does not vanish on V. Generalize the notation of Section 13.1 and define

$$H'_{K,V} = \{\mathbf{a} \in V(K) \mid g(\mathbf{a}) \neq 0 \text{ and } \prod_{i=1}^{m} h_i(\mathbf{a}, b) \neq 0 \text{ for all } b \in K\}.$$

If $V(K)$ is nonthin, then $H'_{K,V} \neq 0$. Indeed, $W_i = V \cap V(h_i)$ is a variety in $\mathbb{A}^{n+1}$ with generic point $(\mathbf{x}, y_i)$ over K. Let $\varphi_i\colon W_i \to V$ be the projection on the first n coordinates and let $V_0 = V(g)$. Then $V(K) \not\subseteq V_0(K) \cup \bigcup_{i=1}^{m} \varphi_i(W_i(K))$ (because $V(K)$ is nonthin), so there is a $\mathbf{a} \in V(K) \smallsetminus V_0(K) \cup \bigcup_{i=1}^{m} \varphi_i(W_i(K)) = H'_{K,V}$, as claimed.

Conversely, suppose $H'_{K,V} \neq \emptyset$ for all $H'_{K,V}$ as above such that $K(\mathbf{x}, y_i)$ is a regular extension of K, $i = 1,\ldots,m$. Giving φ_i and W_i as in (1), we may assume that W_i is absolutely irreducible (by (e)). Let F_i be the function field of W_i over K. Then F_i can be chosen to be a finite separable extension of $K(\mathbf{x})$. Choose a primitive element y_i for $F_i/K(\mathbf{x})$ which is integral over $K[\mathbf{x}]$. Let $h_i \in K[\mathbf{X}, Y]$ a polynomial such that $h_i(\mathbf{x}, Y) = \mathrm{irr}(y_i, K(\mathbf{x}))$. By (a), we may assume that $(\mathbf{x}, y_i)$ is a generic point of W_i over K and $\varphi_i\colon W_i \to V$

13.5 Hilbertianity via Coverings

is the projection on the first n coordinates. Let $V_0 = V(g)$. Then $H'_{K,V} \neq \emptyset$ is equivalent to $V(K) \not\subseteq V_0(K) \cup \bigcup_{i=1}^{m} \varphi_i(W_i(K))$.

(h) $\mathbb{A}^n(K)$ is nonthin if and only if for all absolutely irreducible polynomials $h_1, \ldots, h_m \in K[X_1, \ldots, X_n, Y]$ which are separable and monic in Y of degree at least 2 in Y and for all nonzero $g \in K[X_1, \ldots, X_n]$, the set $H'_{K,\mathbb{A}^n}$ is nonempty.

Indeed, a generic point $\mathbf{x} = (x_1, \ldots, x_n)$ for $\mathbb{A}^n$ over K consists of algebraically independent elements $x_1, \ldots, x_n$ over K. If $y \in K(\mathbf{x})_s$ is integral over K and $h \in K[\mathbf{X}, Y]$ is a monic polynomial in Y such that $h(\mathbf{x}, Y) = \mathrm{irr}(y, K(\mathbf{x}))$, then h is absolutely irreducible if and only if $K(\mathbf{x}, y)$ is a regular extension of K (Corollary 10.2.2). Thus, our statement is a special case of (g).

(i) Suppose $\mathbb{A}^1(K)$ is a thin set. Then so is $V(K)$ for every variety V over K.

Indeed, suppose $\dim(V) = r > 0$. Let $\mathbf{x}$ be a generic point of V over K. Then $K(\mathbf{x})$ is a regular extension of K of transcendence degree r. By (e) and (f), there exist varieties $W_1, \ldots, W_n$ over K and dominant separable maps $\varphi_i \colon W_i \to \mathbb{A}^1$, $i = 1, \ldots, m$ such that $K = \bigcup_{i=1}^m \varphi_i(W_i(K))$. Use (a) to assume that W_i is a curve defined by $g_i(T, Y) = 0$, where $g_i \in K[T, Y]$ is absolutely irreducible and is monic and separable in Y. Further, assume φ_i is the projection on the first coordinate.

Proposition 13.4.3 gives a transcendental element $t \in K(\mathbf{x})$ such that $g_i(t, Y)$ is irreducible over $K(\mathbf{x})$, $i = 1, \ldots, m$. Choose $y_i \in K(t)_s$ with $g_i(t, y_i) = 0$. Then $K(\mathbf{x})$ and $K(t, y_i)$ are linearly disjoint over $K(t)$. Let $\psi \colon V \to \mathbb{A}^1$ be the K-rational map defined by $\psi(\mathbf{x}) = t$. Then ψ is defined outside a Zariski K-closed subset V_0 of V of dimension less than r. Let W'_i be the K-variety generated by $(\mathbf{x}, y_i)$. Let $\varphi'_i \colon W'_i \to V$ be the projection $\varphi'_i(\mathbf{x}, y_i) = \mathbf{x}$. We prove that $V(K) = V_0(K) \cup \bigcup_{i=1}^m \varphi'_i(W'_i(K))$.

$$\begin{array}{ccc} K(\mathbf{x}) \relbar\joinrel\relbar K(\mathbf{x}, y_i) & \quad & V \xleftarrow{\varphi'_i} W'_i \\ \big| \qquad\qquad \big| & \quad & \psi \Big\downarrow \qquad \Big\downarrow \psi'_i \\ K(t) \relbar\joinrel\relbar K(t, y_i) & \quad & \mathbb{A}^1 \xleftarrow{\varphi_i} W_i \end{array}$$

Indeed, let $\mathbf{b} \in V(K) \smallsetminus V_0(K)$. Then $a = \psi(\mathbf{b}) \in K$. By assumption, there exist i between 1 and m and $\mathbf{c}_i \in W_i(K)$ with $\varphi_i(\mathbf{c}_i) = a$. By linear disjointness, $K[\mathbf{x}, y_i] = K[\mathbf{x}] \otimes_{K[t]} K[y_i]$. Hence, $(\mathbf{x}, y_i) \to (\mathbf{b}, c_i)$ is a K-specialization. In other words, $(\mathbf{b}, c_i) \in W'_i(K)$ and $\varphi'_i(\mathbf{b}, c_i) = \mathbf{b}$, as desired.

In the language of schemes, W'_i is the **fiber product** of $\varphi \colon V \to \mathbb{A}^1$ and $\varphi_i \colon W_i \to \mathbb{A}^1$. $\square$

All of this gives a new characterization of Hilbertian field.

PROPOSITION 13.5.3: *The following conditions on a field K are equivalent:*
(a) *K is Hilbertian.*

(b) $\mathbb{A}^1(K)$ is nonthin.
(c) There is a variety V over K such that $V(K)$ is nonthin.

Proof: By Lemma 13.1.2, K is Hilbertian if and only if the following condition holds: $H'_{K,\mathbb{A}^1}(h_1,\ldots,h_m;g) \neq \emptyset$ for all absolutely irreducible polynomials $h_i \in K[X,Y]$ which are separable and of degree at least 2 in Y, and for all nonzero $g \in K[X]$. By Remark 13.5.2(h), the latter condition is equivalent to $\mathbb{A}^1(K)$ being nonthin. Thus, (a) is equivalent to (b).

Remark 13.5.2(i) asserts that (b) is equivalent to (c). □

Remark 13.5.4: Examples of thin sets. Even if $\mathbb{A}^1(K)$ is nonthin, there may be varieties V over K such that $V(K)$ is thin. For example, each finitely generated extension K of $\mathbb{Q}$ is Hilbertian (Theorem 13.4.2). Hence, by Proposition 13.5.3, $\mathbb{A}^1(K)$ is nonthin. However, by Faltings, $C(K)$ is finite for each curve C over K of genus at least 2 [Faltings-Wüstholz, p. 205]. In particular, $C(K)$ is thin.

Moreover, let A be an Abelian variety over K of dimension d. For each integer $n \geq 2$, multiplication by n is a dominant rational map of degree n^{2d} of A onto itself [Mumford1, pp. 42 and 64]. By Mordell-Weil, $A(K)/nA(K)$ is a finite group [Lang3, p. 71]. Let $\mathbf{a}_1,\ldots,\mathbf{a}_m$ be representatives for the cosets of $A(K)$ modulo $nA(K)$. For each i, the map $\varphi_i(\mathbf{x}) = \mathbf{a}_i + n\mathbf{x}$ gives a morphism of degree n^{2d} of A onto A. Also, $A(K) = \bigcup_{i=1}^m \varphi_i(A(K))$. Therefore, $A(K)$ is thin.

For example, let E be an elliptic curve over K with $E(K)$ infinite. Then $E(K)$ is Zariski K-dense in E but $E(K)$ is thin. Thus, the converse of Remark 13.5.2(b) is not true.

Proposition 27.3.5 characterizes fields K for which $V(K)$ is nonthin for every variety V over K as 'ω-free PAC fields'. □

13.6 Non-Hilbertian g-Hilbertian Fields

In case where $A = \mathbb{A}^1$, one may try to bound the genera of the curves W_i appearing in (1) of Section 13.5. This leads to a weaker version of Hilbertianity.

Definition 13.6.1: g-Hilbertian fields. Let K be a field and g a nonnegative integer. We say that K is g**-Hilbertian** if K is not a finite union of sets $\varphi(C(K))$ with C a curve over K of genus at most g and $\varphi\colon C \to \mathbb{A}^1$ a dominant separable rational map of degree at least 2. □

By Proposition 13.5.3, K is Hilbertian if and only if K is g-Hilbertian for each $g \geq 0$. One may ask whether being g-Hilbertian for large g suffices for K to be Hilbertian. We show here that this is not the case.

THEOREM 13.6.2 (Zannier): *Let K be a global field and g a nonnegative integer. Then K has an infinite normal extension N which is g-Hilbertian but not Hilbertian.*

13.6 Non-Hilbertian g-Hilbertian Fields

Proof: The construction of N starts with a prime number $l \neq \mathrm{char}(K)$ and another prime number q with

(2) $$q > \max(\mathrm{char}(K), 2g - 2 + 2l).$$

Extend K, if necessary, to assume $\zeta_l, \zeta_q \in K$. Then use the Hilbertianity of K (Theorem 13.4.2) to choose $u \in K$ which is not a qth power. Thus, $\sqrt[q]{u} \notin K_{\mathrm{ins}}$. By Eisenstein's criterion (Lemma 2.3.10), $f(X,Y) = Y^l - X^{lq} + u$ is an absolutely irreducible polynomial.

Now define an ascending sequence of Galois extensions K_n of K inductively: $K_0 = K$ and $K_{n+1} = K_n(\sqrt[l]{a^{lq} - u} \mid a \in K_n)$. Then K_n/K is a **pro-l-extension**. That is, K_n/K is Galois and $[L : K]$ is an l-power for each finite subextension L/K of K_n/K.

Let $M(K) = \bigcup_{n=0}^{\infty} K_n$. This is an infinite pro-l extension of K. Let $N = N(K) = \bigcup M(K')$ with K' ranging over all finite purely inseparable extensions of K. Then N is a perfect field which is infinite and normal over K. For each $a \in N$ there exists $b \in N$ such that $b^l - a^{lq} + u = 0$. Hence, N is non-Hilbertian. We prove however that N is g-Hilbertian.

Note: We may replace K in the proof by any finite extension L in N. Indeed, then $M(L)$ is contained in $M(K')$ for some finite purely inseparable extension K' of K. Hence, $N = N(L)$.

CLAIM A: *Suppose M is a finite Galois extension of K in N. Then $[M : K]$ is a power of l and $M(\sqrt[q]{u})$ is a Galois extension of M of degree q.*

Indeed, there is a finite purely inseparable extension K' of K with $M \subseteq M(K')$, so $[M : K] = [MK' : K']$ is a power of q. By the choice of u, $K(\sqrt[q]{u})/K$ is Galois of degree 1, so $M(\sqrt[q]{u})/M$ is a Galois extension of degree q.

Let P be the set of all ultrametric prime divisors $\mathfrak{p}$ of K with $v_{\mathfrak{p}}(u) = 0$, $v_{\mathfrak{p}}(l) = 0$, $v_{\mathfrak{p}}(q) = 0$, and $\left(\frac{K(\sqrt[q]{u}/K)}{\mathfrak{p}}\right) \neq 1$.

CLAIM B: *Each $\mathfrak{p}$ in P is unramified in $M(K)$.*

Indeed, it suffices to consider a finite extension L of K in $M(K)$ in which $\mathfrak{p}$ is unramified, to take a prime divisor $\mathfrak{q}$ of L over $\mathfrak{p}$ and an element $a \in L$, and to prove that $\mathfrak{q}$ is unramified in $L(\sqrt[l]{a^{lq} - u})$. By Example 2.3.8, it suffices to prove that l divides $v_{\mathfrak{q}}(a^{lq} - u)$.

Suppose first $v_{\mathfrak{q}}(a) < 0$. Then $v_{\mathfrak{q}}(a^{lq} - u) = v_{\mathfrak{q}}(a^{lq}) = lqv_{\mathfrak{q}}(a)$. Now suppose $v_{\mathfrak{q}}(a) \geq 0$. Denote reduction modulo $\mathfrak{q}$ by a bar. The assumptions on $\mathfrak{q}$ imply $O_{\mathfrak{p}}[\sqrt[q]{u}]/O_{\mathfrak{p}}$ is a ring-cover. By Remark 6.1.7, $\bar{K}(\sqrt[q]{\bar{u}}) = \overline{K(\sqrt[q]{u})}$. Since $\left(\frac{K(\sqrt[q]{u}/K)}{\mathfrak{p}}\right) \neq 1$, this implies $[\bar{K}(\sqrt[q]{\bar{u}}) : \bar{K}] = q$. Next, let M be the Galois closure of L/K. Then $M \subseteq M(K)$. Hence, by Claim A, $[\bar{M} : \bar{K}]$ is a power of l. Therefore, $[\bar{L} : \bar{K}]$ is also a power of L. Thus, $\bar{K}(\sqrt[q]{\bar{u}}) \not\subseteq \bar{L}$, so $\bar{u}$ is not a qth power in $\bar{L}$. In particular, $\bar{a}^{lq} \neq \bar{u}$. Consequently, $v_{\mathfrak{q}}(a^{lq} - u) = 0$.

CLAIM C: *Let L be a finite extension of K in $M(K)$. Consider nonconstant polynomials $h_1,\ldots,h_r \in L[X]$. Suppose h_i has an irreducible L-factor of f_i which is separable, of multiplicity not divisible by l; suppose further the degree of each irreducible L-factor of h_i is less than q, $i = 1,\ldots,r$. Then O_K has an arithmetical progression A with $\sqrt[l]{h_i(a)} \notin N$ for all $a \in A$ and $i = 1,\ldots,r$.*

Indeed, write $h_i = f_i^k g_i$ with $f_i, g_i \in L[X]$, f_i irreducible, $l \nmid k$, and g_i relatively prime to f_i. Choose $k', l' \in \mathbb{Z}$ with $k' \geq 1$ and $kk' + ll' = 1$. Then $h_i^{k'} = f_i^{kk'} g_i^{k'} = f_i g_i^{k'} f_i^{-ll'}$. For every $a \in K$, $\sqrt[l]{h_i(a)} \notin N$ if and only if $\sqrt[l]{f_i(a)g_i(a)^{k'}} \notin N$. Each root of f_i in K_s is a simple root of $f_i g_i^{k'}$. Therefore, we may replace h_i by $f_i g_i^{k'}$, if necessary, to assume h_i has a simple root a_i in K_s.

Now extend L, if necessary, to assume L is Galois over K with Galois group G. By Claim A, $L(\sqrt[q]{u})/L$ is a Galois extension of degree q. Let M be the splitting field of $\prod_{i=1}^r \prod_{\sigma \in G} h_i^\sigma(X)$ over L. Since the degree of each irreducible L-factor of $h_1,\ldots,h_r$ is less than q, M is a finite Galois extension of K whose degree is divisible only by prime numbers smaller than q. Hence, M is a proper subfield of $M(\sqrt[q]{u})$. Denote the set of all ultrametric primes $\mathfrak{p}$ of K which unramify in $M(\sqrt[q]{u})$ with $v_\mathfrak{p}(u) = 0$, $v_\mathfrak{p}(l) = 0$, $v_\mathfrak{p}(q) = 0$, and $\left(\frac{M(\sqrt[q]{u})/K}{\mathfrak{p}}\right) \subseteq \mathrm{Gal}(M(\sqrt[q]{u})/M) \smallsetminus \{1\}$ by P_0. Then P_0 is a subset of P. By Chebotarev (Theorem 6.3.1), P_0 is infinite.

Omit finitely many elements from P_0 to assume that for each $\mathfrak{p} \in P_0$ all nonzero coefficients of the h_i are $\mathfrak{p}$-units and a_i is a simple root of h_i modulo $\mathfrak{p}$. Now choose distinct primes $\mathfrak{p}_1,\ldots,\mathfrak{p}_r$ in P_0.

Consider i between 1 and r. Let $\mathfrak{p} = \mathfrak{p}_i$. Choose a prime $\mathfrak{q} = \mathfrak{q}_i$ of L over $\mathfrak{p}$, a prime $\mathfrak{q}'$ of M over $\mathfrak{q}$, and denote reduction modulo $\mathfrak{q}'$ by a bar. Then $\bar{M} = \bar{K}$, so $\bar{h}_i$ is a polynomial with coefficients in $\bar{K}$ which decomposes into linear factors. In particular, $\bar{a}_i \in \bar{K}$, $\bar{h}_i(\bar{a}_i) = 0$, and $\bar{h}_i'(\bar{a}_i) \neq 0$.

Choose $b_i \in K$ with $v_{\mathfrak{q}'}(b_i - a_i) > 0$. Then, $v_\mathfrak{q}(h_i(b_i)) \geq 1$ and $v_\mathfrak{q}(h_i'(b_i)) = 0$. If $v_\mathfrak{q}(h_i(b_i)) \geq 2$, choose $\pi_i \in K$ with $v_\mathfrak{p}(\pi_i) = 1$. Since $\mathfrak{p}$ is unramified in M, $v_\mathfrak{q}(\pi_i) = 1$. Then $v_\mathfrak{q}(h_i(b_i + \pi_i) - h_i(b_i) - h_i'(b_i)\pi_i) \geq 2$. Hence, $v_\mathfrak{q}(h_i(b_i + \pi_i)) = 1$. Thus, replacing b_i by $b_i + \pi_i$, if necessary, we may assume that in any case $v_\mathfrak{q}(h_i(b_i)) = 1$.

The weak approximation theorem (Proposition 2.1.1) gives $b, \pi \in K$ with $v_{\mathfrak{p}_i}(b - b_i) \geq 2$ and $v_{\mathfrak{p}_i}(\pi - \pi_i) \geq 2$, $i = 1,\ldots,r$. Let $A = b + \pi^2 O_K$. Then $v_{\mathfrak{p}_i}(h_i(a)) = 1$ for each i and all $a \in A$.

Consider an arbitrary finite purely inseparable extension K' of K. Denote the unique extension of $\mathfrak{p}_i$ to K' by $\mathfrak{p}'_i$. Then $v_{\mathfrak{p}'_i}(h_i(a)) = 1$ if $\mathrm{char}(K) = 0$ and $v_{\mathfrak{p}'_i}(h_i(a))$ is a power of $\mathrm{char}(K)$ if $\mathrm{char}(K) > 0$. In each case, $l \nmid v_{\mathfrak{p}'_i}(h_i(a))$. By Claim B, applied to K' instead of to K, $\mathfrak{p}'_i$ is unramified in $M(K')$. Hence, $l \nmid v_i(h_i(a))$ for each extension v_i of $v_{\mathfrak{p}'_i}$ to $M(K')$, so $\sqrt[l]{h_i(a)} \notin M(K')$. Consequently, $\sqrt[l]{h_i(a)} \notin N$, as claimed.

CLAIM D: *Let L be a finite extension of K and C a curve of genus at most*

13.6 Non-Hilbertian g-Hilbertian Fields

g over L. Suppose C is defined by $Y^l = c\prod_{j=1}^{s} f_j(X)^{m_j}$, where $c \in K^\times$, $f_1, \ldots, f_s$ are distinct monic irreducible polynomials in $L[X]$, and $l \nmid m_j$, $j = 1, \ldots, s$. Then $\deg(f_j) < q$, $j = 1, \ldots, s$.

Indeed, replace m_j by its residue modulo l, if necessary, to assume $1 \leq m_j \leq l - 1$. Choose a transcendental element x over L and $y \in L(x)_s$ with $y^l = c\prod_{j=1}^{s} f_j(x)^{m_j}$. Then $F = L(x, y)$ is the function field of C over L. Denote the prime divisor of $L(x)/L$ corresponding to f_j by P_j (Section 2.2). Then $\deg(P_j) = \deg(f_j)$, $v_{P_j}(f_j(x)) = 1$, and $v_{P_j}(f_i(x)) = 0$ for $j \neq i$. Hence, $v_{P_j}(c\prod_{i=1}^{s} f_i(x)^{m_i}) = m_j$ is not divisible by l. By Example 2.3.8, P_j totally ramifies in F. Thus, F has a unique prime divisor Q_j lying over P_j, $e_{Q_j/P_j} = l$, and $\deg(Q_j) = \deg(P_j) = \deg(f_j)$. Since $l \neq \mathrm{char}(L)$, the ramification is tame. Hence, by Riemann-Hurwitz (Remark 3.6.2(c)), $2g - 2 \geq 2 \cdot \mathrm{genus}(F) - 2 \geq -2l + (l-1)\deg(f_j)$. Therefore, by (2), $\deg(f_j) \leq 2g - 2 + 2l < q$, $j = 1, \ldots, s$, as claimed.

CLAIM E: N is g-Hilbertian.

Indeed, assume $N = \bigcup_{i=1}^{m} \varphi_i(C_i(N))$ with C_i a curve over N of genus at most g and $\varphi_i \colon C_i \to \mathbb{A}^1$ a dominant separable rational map over N of degree at least 2. Replace K by a finite extension in N, if necessary, to assume that C_i and φ_i are defined over K and the genus of C_i over K is the same as the genus of C_i over N. Choose a transcendental element x over K and a generic point $\mathbf{z}_i$ for C_i over K with $\varphi_i(\mathbf{z}_i) = x$, $i = 1, \ldots, m$. Then $E_i = K(\mathbf{z}_i)$ is a finite separable extension of $K(x)$ of degree at least 2. Denote the Galois closure of E_i by F_i.

Next, list the cyclic extensions of $K(x)$ of degree l which are contained in one of the fields E_i as $D_1, \ldots, D_r$. For each j between 1 and r choose a primitive element y_j for $D_j/K(x)$ such that $y_j^l = h_j(x)$ with $h_j \in K[X]$. Then $h_j(x)$ is not an lth power in $K(x)$. Since each E_i is a regular extension of K, so is D_j/K. Hence, $\deg(h_j) \geq 1$. Since N is perfect, we may again replace K by a finite purely inseparable extension in N to assume that $h_j = c_j \prod_{k=1}^{s_j} h_{jk}^{m_{jk}}$ such that $c_j \in K^\times$, h_{jk} are distinct monic separable irreducible polynomials of positive degrees, and m_{jk} are positive integers. If $l | m_{jk}$ for some j, k, replace y_j by $y_j h_{jk}(x)^{-m_{jk}/l}$. Thus, assume without loss, $l \nmid m_{jk}$ for all j and k.

By Claim D, $\deg(h_{jk}) < q$. Hence, by Claim C, there exists an arithmetical progression A of O_K such that $\sqrt[l]{h_j(a)} \notin N$ for each $a \in A$, $j = 1, \ldots, r$. Let $F = \prod_{i=1}^{m} F_i$. Let H be the set of all $a \in K$ such that each K-place $\psi \colon F \to \tilde{K} \cup \{\infty\}$ with $\psi(x) = a$ maps F_i onto $\bar{F}_i \cup \{\infty\}$ and E_i onto $\bar{E}_i \cup \{\infty\}$ with these properties:

(3a) $\bar{F}_i/K$ is a Galois extension and $\mathrm{Gal}(\bar{F}_i/K) \cong \mathrm{Gal}(F_i/K(x))$.
(3b) ψ is finite at $\mathbf{z}_i$ and $\mathbf{c}_i = \psi(\mathbf{z}_i)$ satisfies $\bar{E}_i = K(\mathbf{c}_i)$ and $[K(\mathbf{c}_i) : K] = [E_i : K(x)]$.

By Lemma 13.1.1, H contains a separable Hilbert subset. Theorem 13.3.5(c) gives $a \in A \cap H$. By assumption, $a = \varphi_i(\mathbf{c})$ with $\mathbf{c} \in C_i(N)$ for some i between 1 and m. Extend the K-specialization $(x, \mathbf{z}_i) \to (a, \mathbf{c})$ to a place $\psi\colon F \to \tilde{K} \cup \{\infty\}$. Then (3) is true with $\mathbf{c}_i = \mathbf{c}$. Since F_i is the Galois closure of $E_i/K(\mathbf{x})$, $\bar{F}_i$ is the Galois closure of $K(\mathbf{c})/K$. Since $K(\mathbf{c}) \subseteq N$ and N/K is normal, $\bar{F}_i \subseteq N$. By Claim A, $\mathrm{Gal}(\bar{F}_i/K)$ is an l-group. By (3b), $[\bar{E}_i : K] = [E_i : K(x)] > 1$. Thus, $\mathrm{Gal}(\bar{F}_i/\bar{E}_i)$ is a proper subgroup of $\mathrm{Gal}(\bar{F}_i/K)$. Hence, $\mathrm{Gal}(\bar{F}_i/E_i)$ is contained in a normal subgroup of index l of $\mathrm{Gal}(\bar{F}_i/K)$ [Hall, p. 45, Cor. 4.2.2]. By (3a), $\mathrm{Gal}(F_i/E_i)$ is contained in a normal subgroup of $\mathrm{Gal}(F_i/K(x))$ of index l, so $K(x)$ has a cyclic extension of degree l in E_i. It is D_j for some j between 1 and r. Then $b = \psi(y_j)$ satisfies $b^l = h_j(a)$ and $b \in \bar{E}_i \subseteq N$. This contradicts the choice of a in A. Thus, our initial assumption $N \subseteq \bigcup_{i=1}^r \varphi_i(C_i(N))$ is false, so N is g-Hilbertian. □

13.7 Twisted Wreath Products

Given finite groups A and G, there are several ways of constructing a new group out of them. We describe here two of them: the semidirect product and the wreath product.

Definition 13.7.1: Semidirect products. Let A and G be profinite groups. Suppose G acts on A continuously from the right. That is, there is a continuous map $G \times A \to A$ mapping (σ, a) onto a^σ and satisfying these rules: $(ab)^\sigma = a^\sigma b^\sigma$, $a^1 = a$, and $(a^\sigma)^\tau = a^{\sigma\tau}$ for $a, b \in A$ and $\sigma, \tau \in G$. The **semidirect product** $G \ltimes A$ consists of all pairs $(\sigma, a) \in G \times A$ with the product rule $(\sigma, a)(\tau, b) = (\sigma\tau, a^\tau b)$. This makes $G \ltimes A$ a profinite group with unit element $(1,1)$ and $(\sigma, a)^{-1} = (\sigma^{-1}, a^{-\sigma^{-1}})$.

Identify each $\sigma \in G$ (resp. $a \in A$) with the pair $(\sigma, 1)$ (resp. $(1, a)$). This embeds G and A into $G \ltimes A$ such that
(1) A is normal, $G \cap A = 1$, and $GA = G \ltimes A$.

Each element of $G \ltimes A$ has a unique presentation as σa with $\sigma \in G$ and $a \in A$. The product rule becomes $(\sigma a)(\tau b) = (\sigma\tau)(a^\tau b)$ and the action of G on A coincides with conjugation: $a^\sigma = \sigma^{-1} a \sigma$.

The projection $\sigma a \mapsto \sigma$ of $G \ltimes A$ on G is an epimorphism with kernel A.

Conversely, suppose H is a profinite group and A, G are closed subgroups satisfying (1). Then H is the semidirect product $G \ltimes A$.

Likewise, consider a short exact sequence

$$1 \longrightarrow A \longrightarrow H \xrightarrow{\alpha} G \longrightarrow 1$$

Suppose the sequence **splits**. That is, there exists a homomorphism $\alpha'\colon G \to H$ satisfying $\alpha(\alpha'(g)) = g$ for each $g \in G$. Then α' is an embedding which we call a **group theoretic section** of α. Identifying G with $\alpha'(G)$, we have $H = G \ltimes A$.

Suppose $\varphi_1\colon G \to E$ and $\varphi_2\colon A \to E$ are homomorphisms of profinite groups and $\varphi_2(a^\sigma) = \varphi_2(a)^{\varphi_1(\sigma)}$ for all $a \in A$ and $\sigma \in G$. Then $\varphi(\sigma a) = \varphi_1(\sigma)\varphi_2(a)$ is a homomorphism $\varphi\colon G \ltimes A \to E$.

Here is a Galois theoretic interpretation of semidirect products: Let K, L, E, F be fields, with L/K Galois, F/K Galois, $EL = F$, and $E \cap L = K$. Then $\mathrm{Gal}(F/K) = \mathrm{Gal}(F/E) \ltimes \mathrm{Gal}(F/L)$. □

13.7 Twisted Wreath Products

Definition 13.7.2: *Twisted wreath products.* Let A and G be finite groups, G_0 a subgroup of G, and Σ is a system of representatives for the right cosets $G_0\sigma$, $\sigma \in G$. Thus,

$$G = \bigcup_{\sigma \in \Sigma} G_0\sigma = \bigcup_{\sigma \in \Sigma} \sigma^{-1}G_0.$$

Suppose G_0 acts on A from the right. Let

$$\operatorname{Ind}_{G_0}^G(A) = \{f \colon G \to A \mid f(\sigma\sigma_0) = f(\sigma)^{\sigma_0} \text{ for all } \sigma \in G \text{ and } \sigma_0 \in G_0\}.$$

Make $\operatorname{Ind}_{G_0}^G(A)$ a group by the rule $(fg)(\sigma) = f(\sigma)g(\sigma)$. Then $f^g(\sigma) = f(\sigma)^{g(\sigma)}$, where f^g denotes conjugation in $\operatorname{Ind}_{G_0}^G(A)$ and the right hand side is conjugation in A.

The group G acts on $\operatorname{Ind}_{G_0}^G(A)$ by $f^\sigma(\tau) = f(\sigma\tau)$. This gives rise to the semidirect product $G \ltimes \operatorname{Ind}_{G_0}^G(A)$, which we also denote by $A \operatorname{wr}_{G_0} G$. Each element of this group has a unique presentation as a product σf with $\sigma \in G$ and $f \in \operatorname{Ind}_{G_0}^G(A)$. The product and the inverse operation in $A \operatorname{wr}_{G_0} G$ are given by $(\sigma f)(\tau g) = \sigma\tau f^\tau g$ and $(\sigma f)^{-1} = \sigma^{-1} f^{-\sigma^{-1}}$. The identification $\sigma = \sigma \cdot 1$ and $f = 1 \cdot \sigma$ identifies G and $\operatorname{Ind}_{G_0}^G(A)$ as subgroups of $A \operatorname{wr}_{G_0} G$. Under this identification, $\operatorname{Ind}_{G_0}^G(A)$ is normal, $G \cap \operatorname{Ind}_{G_0}^G(A) = 1$, and
$G \cdot \operatorname{Ind}_{G_0}^G(A) = A \operatorname{wr}_{G_0} G$. Since $(\sigma \cdot 1)^{-1}(1 \cdot f)(\sigma \cdot 1) = 1 \cdot f^\sigma$, conjugation of f by σ in $A \operatorname{wr}_{G_0} G$ coincides with the action of σ on f. The map $\sigma f \mapsto \sigma$ is an epimorphism $\pi \colon A \operatorname{wr}_{G_0} G \to G$ with $\operatorname{Ker}(\pi) = \operatorname{Ind}_{G_0}^G(A)$. We call $A \operatorname{wr}_{G_0} G$ the **twisted wreath product** of A and G with respect to G_0. □

Twisted wreath products are usually non-Abelian:

LEMMA 13.7.3: *Let A be a nontrivial finite group, G a finite group, and G_0 a proper subgroup of G. Then $A \operatorname{wr}_{G_0} G$ is not commutative.*

Proof: Choose $a \in A$, $a \neq 1$. Define a function $f \in \operatorname{Ind}_{G_0}^G(A)$ by $f(\sigma) = a^\sigma$ for $\sigma \in G_0$ and $f(\sigma) = 1$ for $\sigma \in G \smallsetminus G_0$. Choose $\sigma \in G \smallsetminus G_0$. Then $f^\sigma(1) = f(\sigma) = 1 \neq a = f(1)$. Hence, $f^\sigma \neq f$. Consequently, $A \operatorname{wr}_{G_0} G$ is not commutative. □

The next result shows that not only twisted wreath products are in general non-Abelian but their centers are small:

LEMMA 13.7.4: *Let $\pi \colon A \operatorname{wr}_{G_0} G \to G$ be a twisted wreath product of finite groups, $H_1 \triangleleft A \operatorname{wr}_{G_0} G$, and $h_2 \in A \operatorname{wr}_{G_0} G$. Put $I = \operatorname{Ind}_{G_0}^G(A) = \operatorname{Ker}(\pi)$ and $G_1 = \pi(H_1)$. Suppose $A \neq 1$.*
(a) *Suppose $\pi(h_2) \notin G_0$ and $(G_1 G_0 : G_0) > 2$. Then there is an $h_1 \in H_1 \cap I$ with $h_1 h_2 \neq h_2 h_1$.*

(b) *Suppose $G_1 \not\leq G_0$ and $\pi(h_2) \notin G_1 G_0$. Then there is an $h_1 \in H_1 \cap I$ with $h_1^{h_2} \notin \langle h_1 \rangle^{h'}$ for all $h' \in \pi^{-1}(G_1 G_0)$. In particular, $h_1 h_2 \neq h_2 h_1$.*

Proof: Put $\sigma_2 = \pi(h_2)$. Consider $\sigma_1 \in G_1$ and $g \in I$. By definition, there are $f_1, f_2 \in I$ with $\sigma_1 f_1 \in H_1$ and $h_2 = \sigma_2 f_2$. Put $h_1 = g^{\sigma_1 f_1} g^{-1}$. Then $h_1 = [\sigma_1 f_1, g^{-1}] \in [H_1, I] \leq H_1 \cap I$. For each $\tau \in G$
$$h_1(\tau) = ((g^{\sigma_1})^{f_1})(\tau) g(\tau)^{-1} = g(\sigma_1 \tau)^{f_1(\tau)} g(\tau)^{-1}.$$

Hence, for all $\tau \in G$ and $f' \in I$ we have:

(2a) $h_1^{h_2}(1) = h_1^{\sigma_2 f_2}(1) = h_1(\sigma_2)^{f_2(1)} = g(\sigma_1 \sigma_2)^{f_1(\sigma_2) f_2(1)} g(\sigma_2)^{-f_2(1)}$,

(2b) $h_1^{\tau f'}(1) = h_1(\tau)^{f'(1)} = g(\sigma_1 \tau)^{f_1(\tau) f'(1)} g(\tau)^{-f'(1)}$, and

(2c) $h_1(1) = g(\sigma_1)^{f_1(1)} g(1)^{-1}$.

We apply (2) in the proofs of (a) and (b) to special values σ_1 and g. Choose $a \in A$, $a \neq 1$.

Proof of (a): Since $(G_1 G_0 : G_0) > 2$, there is a $\sigma_1 \in G_1$ with distinct cosets $\sigma_1^{-1} G_0, \sigma_2 G_0, G_0$. Thus, none of the cosets $\sigma_1 G_0, \sigma_2 G_0, \sigma_1 \sigma_2 G_0$ is G_0. Therefore, by definition of I, there is a $g \in I$ with $g(\sigma_1) = g(\sigma_2) = g(\sigma_1 \sigma_2) = 1$ and $g(1) = a$. By (2a), $h_1^{h_2}(1) = 1$. By (2c), $h_1(1) \neq 1$. Consequently, $h_1^{h_2} \neq h_1$, as desired.

Proof of (b): Since $G_1^{\sigma_2} = G_1 \not\leq G_0$, we have $G_1 \not\leq G_0^{\sigma_2^{-1}}$. Hence, $G_1 \cap G_0$ and $G_1 \cap G_0^{\sigma_2^{-1}}$ are proper subgroups of G_1. Their union is a proper subset of G_1. Thus, there is an element $\sigma_1 \in G_1 \smallsetminus G_0 \cup G_0^{\sigma_2^{-1}}$. It follows that $\sigma_2 \notin \sigma_1 \sigma_2 G_0$. By assumption, $\sigma_2 \notin G_1 G_0$. Therefore, there is a $g \in I$ with $g(G_1 G_0) = 1$, $g(\sigma_1 \sigma_2) = 1$, and $g(\sigma_2) = a^{-1}$.

Consider $\tau \in G_1 G_0$ and $f' \in I$. By (2a), $h_1^{h_2}(1) = a^{f_2(1)} \neq 1$. By (2b), $h_1^{\tau f'}(1) = 1$. Hence, $(h_1^k)^{\tau f'}(1) = 1$ for all integers k. It follows that $h_1^{h_2} \notin \langle h_1 \rangle^{h'}$ for all $h' \in \pi^{-1}(G_1 G_0)$. □

Remark 13.7.5: *Decomposition of $\mathrm{Ind}_{G_0}^G(A)$ into a direct product.* To each $a \in A$ associate a function $f_a: G \to A$:
$$f_a(\sigma) = \begin{cases} a^\sigma & \text{if } \sigma \in G_0 \\ 1 & \text{if } \sigma \notin G_0. \end{cases}$$

These functions satisfy the following rules: $f_a f_b = f_{ab}$ and $g^{-1} f_a g = f_{ag(1)}$ for all $a, b \in A$ and $g \in \mathrm{Ind}_{G_0}^G(A)$. Thus, the map $a \mapsto f_a$ identifies A with the normal subgroup $\{f \in \mathrm{Ind}_{G_0}^G(A) \mid f(G \smallsetminus G_0) = 1\}$ of $\mathrm{Ind}_{G_0}^G(A)$.

Applying $\sigma \in G$ on A gives the following normal subgroup of $\mathrm{Ind}_{G_0}^G(A)$:
$$A^\sigma = \{f \in \mathrm{Ind}_{G_0}^G(A) \mid f(G \smallsetminus \sigma^{-1} G_0) = 1\}.$$

13.7 Twisted Wreath Products

An arbitrary element $f \in \mathrm{Ind}_{G_0}^G(A)$ has a unique presentation $f = \prod_{\sigma \in \Sigma} f_\sigma$, with $f_\sigma \in A^\sigma$. Specifically, $f_\sigma(G \smallsetminus \sigma^{-1} G_0) = 1$ and $f_\sigma(\sigma^{-1} \sigma_0) = f(\sigma^{-1} \sigma_0)$ for all $\sigma \in \Sigma$ and $\sigma_0 \in G_0$. It follows that $\mathrm{Ind}_{G_0}^G(A) = \prod_{\sigma \in \Sigma} A^\sigma$.

The latter relation allows us to present A as a quotient of $\mathrm{Ind}_{G_0}^G(A)$ in various ways: Let $N = \{f \in \mathrm{Ind}_{G_0}^G(A) \mid f(1) = 1\}$. For each $\sigma \in G$ let $N^\sigma = \{f^\sigma \mid f \in N\} = \{f \in \mathrm{Ind}_{G_0}^G(A) \mid f(\sigma^{-1}) = 1\}$. Then the map $f \mapsto f(\sigma^{-1})$ gives rise to a short exact sequence $1 \to N^\sigma \to \mathrm{Ind}_{G_0}^G(A) \to A \to 1$. Like in the preceding paragraph, we find that

(3) $$N^\sigma = \{f \in \mathrm{Ind}_{G_0}^G(A) \mid f(\sigma^{-1}) = 1\} = \prod_{\substack{\tau \in \Sigma \\ G_0 \tau \neq G_0 \sigma}} A^\tau.$$

Note that G_0 leaves $N = N^1$ invariant, so $N \triangleleft \mathrm{Ind}_{G_0}^G(A) G_0$. We summarize some of the groups mentioned above in the following diagram:

(4)

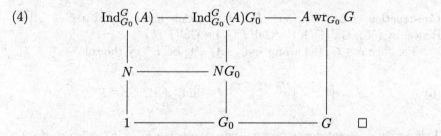

Remark 13.7.6: *Interpretation of twisted wreath products in Galois theory.*
(a) Let $\hat{F}/K$ be a finite Galois extension with $\mathrm{Gal}(\hat{F}/K) \cong A \, \mathrm{wr}_{G_0} G$ and $A \neq 1$. View $\mathrm{Ind}_{G_0}^G(A)$ and G under this isomorphism as subgroups of $\mathrm{Gal}(\hat{F}/K)$. Let F, L, L_0, and $\hat{K}$ be the fixed fields in $\hat{F}$, respectively, of the subgroups N, $\mathrm{Ind}_{G_0}^G(A)$, $\mathrm{Ind}_{G_0}^G(A) G_0$, and G. Galois theory interprets the various relations among the subgroups of $A \, \mathrm{wr}_{G_0} G$ as relations among fields:
(5a) $K \subseteq L_0 \subseteq L \subset F \subseteq \hat{F}$.
(5b) $L \cap \hat{K} = K$ and $L \hat{K} = \hat{F}$.
(5c) L/K, F/L_0, and $\hat{F}/K$ are finite Galois extensions.
(5d) The fields F^σ with $\sigma \in \Sigma$ are linearly disjoint over L. Moreover, $\hat{F} = \prod_{\sigma \in \Sigma} F^\sigma$.
(5e) There is a field F_0 with $L \cap F_0 = L_0$ and $F = L F_0$.

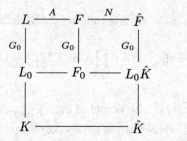

The assertion "F/L_0 is Galois" follows from "$N \triangleleft \mathrm{Ind}_{G_0}^G(A)G_0$". Thus, Condition (5e) follows from (5a)-(5d) by taking $F_0 = F \cap L_0\hat{K}$.

(b) Conversely, consider fields $K, L_0, L, F, \hat{F}, \hat{K}$ satisfying Conditions (5a)-(5d). Let

$$F_0 = F \cap (L_0\hat{K}), \quad G = \mathrm{Gal}(\hat{F}/\hat{K}) \cong \mathrm{Gal}(L/K),$$
$$G_0 = \mathrm{Gal}(\hat{F}/L_0\hat{K})) \cong \mathrm{Gal}(F/F_0) \cong \mathrm{Gal}(L/L_0),$$
$$A = \mathrm{Gal}(F/L) \cong \mathrm{Gal}(F_0/L_0).$$

In particular, Σ is a subset of $\mathrm{Gal}(\hat{F}/K)$. Suppose also that (5d) holds. Then $A \triangleleft \mathrm{Gal}(F/L_0)$. Thus, G_0 viewed as a subgroup of $\mathrm{Gal}(F/L_0)$ acts on A by conjugation. We construct an isomorphism $\varphi \colon A\,\mathrm{wr}_{G_0} G \to \mathrm{Gal}(\hat{F}/K)$ which is the identity on G and maps $\mathrm{Ind}_{G_0}^G(A)$ onto $\mathrm{Gal}(\hat{F}/L)$.

Construction of φ: By (5d), $\hat{F}/L$ is a Galois extension of degree $|A|^{|\Sigma|}$. Hence, by (5b), $\mathrm{Gal}(\hat{F}/K) = \mathrm{Gal}(\hat{F}/\hat{K}) \ltimes \mathrm{Gal}(\hat{F}/L)$.

For each $\sigma \in G$, the group $\mathrm{Ind}_{G_0}^G(A)$ acts on F^σ by the rule

(6) $$z^f = \left((z^{\sigma^{-1}})^{f(\sigma^{-1})}\right)^\sigma, \quad f \in \mathrm{Ind}_{G_0}^G(A), \quad z \in F^\sigma.$$

This action does not depend on σ. Indeed, assume $F^\sigma = F^{\sigma'}$ with $\sigma, \sigma' \in G$. Write $\sigma = \sigma_0\tau$ and $\sigma' = \sigma_0'\tau'$ with $\sigma_0, \sigma_0' \in G_0$ and $\tau, \tau' \in \Sigma$. Since F/F_0 is Galois, $F^\tau = F^{\tau'}$, so by (5a) and (5d), $\tau = \tau'$. Thus, $\sigma' = \rho\sigma$ with $\rho = \lambda'\lambda^{-1} \in G_0$. Moreover,

$$\rho^{-1}f(\sigma^{-1}\rho^{-1})\rho = f(\sigma^{-1}\rho^{-1})^\rho = f(\sigma^{-1}\rho^{-1}\rho) = f(\sigma^{-1}).$$

Hence, $\left((z^{(\rho\sigma)^{-1}})^{f((\rho\sigma)^{-1})}\right)^{\rho\sigma} = \left((z^{\sigma^{-1}})^{f(\sigma^{-1})}\right)^\sigma$.

If $z \in L$, then $z^f = z$, because $f(\sigma^{-1})$ as an element of A fixes L. Thus, the action (6) defines a homomorphism $\varphi_\sigma \colon \mathrm{Ind}_{G_0}^G(A) \to \mathrm{Gal}(F^\sigma/L)$. If $\varphi_\sigma(f) = 1$, then $(z^{\sigma^{-1}})^{f(\sigma^{-1})} = z^{\sigma^{-1}}$ for each $z \in F^\sigma$. Hence, $z^{f(\sigma^{-1})} = z$ for each $z \in F$. Therefore, $f(\sigma^{-1}) = 1$. It follows that $\mathrm{Ker}(\varphi_\sigma) = N^\sigma$, with N^σ as in (3).

Using (5d), the φ_σ's, with σ ranging on Σ, define a homomorphism

$$\varphi_0 \colon \mathrm{Ind}_{G_0}^G(A) \to \prod_{\sigma \in \Sigma} \mathrm{Gal}(F^\sigma/L) = \mathrm{Gal}(\hat{F}/L)$$

by $\varphi_0(f) = \prod_{\sigma \in \Sigma} \varphi_\sigma(f)$. The kernel of φ_0 is $\bigcap_{\sigma \in \Sigma} N^\sigma = 1$ and $|\mathrm{Ind}_{G_0}^G(A)| = |A|^{|\Sigma|} = |\mathrm{Gal}(\hat{F}/L)|$. Hence, φ_0 is an isomorphism.

13.7 Twisted Wreath Products

Next we show that φ_0 is compatible with the action of G. To this end let $z \in F^\sigma$ and $\tau \in G$. Then $z^{\tau^{-1}} \in F^{\sigma\tau^{-1}}$ and

$$\left(\left(z^{\tau^{-1}}\right)^f\right)^\tau = \left(\left(\left(z^{\tau^{-1}}\right)^{(\tau\sigma^{-1})}\right)^{f(\tau\sigma^{-1})}\right)^{(\sigma\tau^{-1})}\right)^\tau$$

$$= \left(\left(z^{\sigma^{-1}}\right)^{f(\tau\sigma^{-1})}\right)^\sigma = \left(\left(z^{\sigma^{-1}}\right)^{f^\tau(\sigma^{-1})}\right)^\sigma = z^{f^\tau}.$$

Hence, $\tau^{-1}\varphi_0(f)\tau = \varphi_0(f^\tau)$, as desired. This allows us to combine φ_0 with the identity map of G and define the isomorphism $\varphi \colon A\operatorname{wr}_{G_0} G \to \operatorname{Gal}(\hat{F}/K)$.

Having established φ, we say that the fields $K, \hat{K}, L_0, L, F, \hat{F}$ **realize the twisted wreath product** $A \operatorname{wr}_{G_0} G$. We say that the fields $K, L_0, L, F, \hat{F}$ **realize the twisted wreath product** if there exists a field $\hat{K}$ such that $K, \hat{K}, L_0, L, F, \hat{F}$ realize the twisted wreath product.

(c) Consider fields $K, \hat{K}, L_0, L, F, \hat{F}$ that realize $A \operatorname{wr}_{G_0} G$. Suppose E is a Galois extension of L_0 with $L \subseteq E \subseteq F$. Let $\bar{A} = \operatorname{Gal}(E/L)$ and $\hat{E} = \prod_{\sigma \in \Sigma} E^\sigma$. Then $\hat{E}$ is a Galois extension of K in $\hat{F}$. Let $J = \hat{E} \cap \hat{K}$. Then $\hat{E}\hat{K} = \hat{F}$. Hence, $\operatorname{Gal}(\hat{E}/J) \cong \operatorname{Gal}(\hat{F}/\hat{K}) = G$ and $\operatorname{Gal}(\hat{E}/L_0 J) \cong \operatorname{Gal}(\hat{F}/L_0\hat{K}) = G_0$. Moreover, $L \cap J = K$ and $LJ = \hat{E}$.

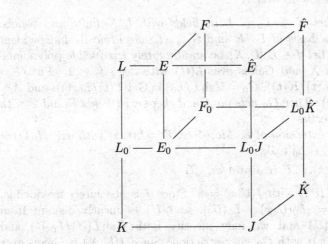

By (b), $K, J, L_0, L, E, \hat{E}$ realize $\bar{A} \operatorname{wr}_{G_0} G$. □

Remark 13.7.7: Wreath products. Suppose G_0 is the trivial subgroup of G. Then the twisted wreath product $A \operatorname{wr}_{G_0} G$ simplifies to the (usual) **wreath product** $A \operatorname{wr} G$. In this case $\operatorname{Ind}_{G_0}^G(A)$ is just the group A^G of all functions $f \colon G \to A$. Multiplication is carried out componentwise. Again, G acts on A^G by the formula $f^\sigma(\tau) = f(\sigma\tau)$. Thus, $A \operatorname{wr} G$ is the semidirect product $G \ltimes A^G$. Each element of $G \ltimes A^G$ has a unique representation as a product σf with $\sigma \in G$ and $f \in A^G$. The multiplication rule is $\sigma f \cdot \tau g = (\sigma\tau)(f^\tau g)$. Identify each $a \in A$ with the function $f_a \colon G \to A$ given by $f_a(1) = a$ and $f_a(\sigma) = 1$ for $\sigma \neq 1$. Then $\operatorname{Ind}_{G_0}^G(A) = \prod_{\sigma \in G} A^\sigma$.

Now Suppose $K \subseteq L \subseteq F \subseteq \hat{F}$ is a tower of fields and $\hat{K}$ is a field satisfying these conditions:
(7a) L/K, F/L, and $\hat{F}/K$ are finite Galois extensions.
(7b) $L \cap \hat{K} = K$ and $L\hat{K} = \hat{F}$. Write $G = \mathrm{Gal}(\hat{F}/\hat{K})$.
(7c) The fields F^σ with $\sigma \in G$ are linearly disjoint over L and $\hat{F} = \prod_{\sigma \in G} F^\sigma$.

By Remark 13.7.6(b), there is an isomorphism $\varphi\colon A\operatorname{wr} G \to \mathrm{Gal}(\hat{F}/K)$ which is the identity on G, maps A^G onto $\mathrm{Gal}(\hat{F}/L)$, and $\prod_{\sigma \neq 1} A^\sigma$ onto $\mathrm{Gal}(\hat{F}/F)$. By (7b), restriction to L maps G isomorphically onto $\mathrm{Gal}(L/K)$. We say that the fields $K, \hat{K}, L, F, \hat{F}$ **realize** the wreath product $A\operatorname{wr} G$. $\square$

13.8 The Diamond Theorem

The diamond theorem proved in this section says all fields 'captured' between two Galois extensions of a Hilbertian field are Hilbertian. In particular, this theorem implies that a non-Hilbertian Galois extension N of a Hilbertian field K is not the compositum of two Galois extensions of K which are properly contained in N. For example, K_s is not the compositum of two proper subfields which are Galois over K.

LEMMA 13.8.1: Let $K \subseteq L_0 \subseteq L$ be fields with L/K finite and Galois. Let $c_1, \ldots, c_n$ be a basis of L_0/K and $t_1, \ldots, t_n$ algebraically independent elements over K. Let $f \in L_0[U, X]$ be an absolutely irreducible polynomial which is monic in X and Galois over $L(U)$ with $\deg_X f \geq 2$. Put $G = \mathrm{Gal}(L/K) = \mathrm{Gal}(L(\mathbf{t})/K(\mathbf{t}))$, $G_0 = \mathrm{Gal}(L/L_0) = \mathrm{Gal}(L(\mathbf{t})/L_0(\mathbf{t}))$, and $A = \mathrm{Gal}(f(U,X), L(U))$. Then G_0 acts on A and there exist fields F and $\hat{F}$ with the following properties:
(a) F is a regular extension of L. Moreover, $F = L(\mathbf{t}, z)$ with $\mathrm{irr}(z, L(\mathbf{t})) = f(\sum_{i=1}^n c_i t_i, Z) \in L_0[\mathbf{t}, Z]$.
(b) $K(\mathbf{t}), L_0(\mathbf{t}), L(\mathbf{t}), F, \hat{F}$ realize $A\operatorname{wr}_{G_0} G$.

Proof: Fix $x \in L(U)_s$ with $f(U, x) = 0$. Since f is absolutely irreducible, $[L(U, x) : L(U)] = [L_0(U, x) : L_0(U)]$, so $L(U)$ is linearly disjoint from $L_0(U, x)$ over $L_0(U)$ and we may identify both $\mathrm{Gal}(L(U)/L_0(U))$ and $\mathrm{Gal}(L(U, x), L_0(U, x))$ with G_0 via restriction. Since $f(U, X)$ is Galois over $L(U)$ with respect to X, the extension $L(U, x)/L(U)$ is Galois with Galois group A. In addition, $A \neq 1$ because $\deg_X f \geq 2$. The fixed field in $L(U, x)$ of the subgroup of $\mathrm{Aut}(L(U, x))$ generated by A and G_0 is $L_0(U)$. By Artin [Lang7, p. 264, Thm. 1.8], $L(U, x)/L_0(U)$ is Galois and

$$\mathrm{Gal}(L(U, x)/L_0(U)) \cong \mathrm{Gal}(L(U, x)/L_0(U, x)) \ltimes \mathrm{Gal}(L(U, x)/L(U)) \cong G_0 \ltimes A.$$

This defines an action of G_0 on A.

As in Section 13.7, choose a system of representatives Σ for the right cosets $G_0\sigma$ of G_0 in G. Let $\mathcal{S} = \{G_0\sigma \mid \sigma \in \Sigma\}$. Choose algebraically independent elements u^S, $S \in \mathcal{S}$, over K. Write $Q = L(u^S \mid S \in \mathcal{S})$. For

13.8 The Diamond Theorem

each $S \in \mathcal{S}$ write $c_i^S = c_i^\sigma$ and $f^S = f^\sigma$ for some $\sigma \in S$. Since $c_i \in L_0$ and $f \in L_0(U, X)$, both c_i^S and f^S are independent of σ. Then choose a root z^S of $f^S(u^S, Z)$ in Q_s and write $F^S = Q(z^S)$. Since f^S is absolutely irreducible, $L_0^S(u^S, z^S)$ is a regular extension of L_0^S (Example 2.6.11). Hence, $L(u^S, z^S)$ is a regular extension of L. Since $L(u^S, z^S)$, $S \in \mathcal{S}$, are algebraically independent over L, they are linearly disjoint over L (Lemma 2.6.7). Moreover, their compositum $\hat{F} = L(u^S, z^S \mid S \in \mathcal{S}) = Q(z^S \mid S \in \mathcal{S})$ is a regular extension of L. By Lemma 2.5.11, in each rectangle of the following diagram, the fields lying in the left upper corner and the right lower corner are linearly disjoint over the field in the left lower corner and their compositum is the field in the right upper corner:

$$
\begin{array}{ccccc}
L(u^{S'}, z^{S'} \mid S' \neq S) & \longrightarrow & L(\mathbf{u})(z^{S'} \mid S' \neq S) & \xrightarrow{A} & L(\mathbf{u}^{S'}, z^{S'} \mid S' \in \mathcal{S}) = \hat{F} \\
\big| & & \big| & & \big| \\
L(u^{S'} \mid S' \neq S) & \longrightarrow & L(\mathbf{t}) = L(\mathbf{u}) = Q & \xrightarrow{A} & L(\mathbf{u}, z^S) = F^S \\
\big| & & \big| & & \big| \\
L & \longrightarrow & L(u^S) & \xrightarrow{A} & L(u^S, z^S) \\
\big| & & \big| & & \big| \\
L_0^S & \longrightarrow & L_0^S(u^S) & \xrightarrow{A} & L_0^S(u^S, z^S)
\end{array}
$$

It follows that, F^S/Q is a Galois extension with Galois group isomorphic to A, the F^S, $S \in \mathcal{S}$, are linearly disjoint over Q, and $\hat{F}$ is their compositum. Thus, $\hat{F}/Q$ is Galois.

The matrix $(c_i^S) \in M_n(L)$ is invertible [Lang7, p. 286, Cor. 5.4]. Hence, the system of equations

(1) $$\sum_{i=1}^{n} c_i^S T_i = u^S, \qquad S \in \mathcal{S}$$

has a unique solution $t'_1, \ldots, t'_n$. It satisfies $L(t'_1, \ldots, t'_n) = L(u^S \mid S \in \mathcal{S}) = Q$. Since trans.deg$(Q/K) = n$, the elements $t'_1, \ldots, t'_n$ are algebraically independent over K. We may therefore assume that $t'_i = t_i$, $i = 1, \ldots, n$, so $Q = L(\mathbf{t})$.

Use the linear disjointness of the F^S's over Q to extend the action of G on L to an action on $\hat{F}$: $(u^S)^\tau = u^{S\tau}$ and $(z^S)^\tau = z^{S\tau}$. In particular, τ permutes the equations of (1). Hence, $(t_1^\tau, \ldots, t_n^\tau)$ is a solution of (1), so it coincides with $(t_1, \ldots, t_n)$. In other words, the action of G on Q is the unique extension of the given action on L that fixes $t_1, \ldots, t_n$. In particular, $K(\mathbf{t})$ is the fixed field of G in Q.

Now write $u = u^{G_0 \cdot 1}$, $z = z^{G_0 \cdot 1}$, and $F = F^{G_0 \cdot 1}$. Then $F^S = F^\sigma$ for all $S \in \mathcal{S}$ and each $\sigma \in S$. Hence, $\hat{F} = \prod_{\sigma \in \Sigma} F^\sigma$. Moreover, by (1), $u = \sum_{i=1}^{n} c_i t_i \in L_0(\mathbf{t})$, u is transcendental over K, $f(u, z) = 0$, and $F = Q(z)$.

Hence, $f(u, Z) \in L_0(u)[Z]$, and $f(u, Z)$ is Galois over $L(u)$. Since $Q = L(\mathbf{u})$ is a purely transcendental extension of $L(u)$, $f(u, Z)$ is irreducible and Galois over Q. Therefore, $f(u, Z) = \mathrm{irr}(z, Q)$ (this settles (a)), and F is a Galois extension of $L_0(\mathbf{t})$.

Finally, since $\hat{F}/L(\mathbf{t})$ and $L(\mathbf{t})/K(\mathbf{t})$ are Galois and each $\tau \in G = \mathrm{Gal}(L(\mathbf{t})/K(\mathbf{t}))$ extends to an automorphism of $\hat{F}$, the extension $\hat{F}/K(\mathbf{t})$ is Galois. Let $\hat{K}$ be the fixed field of G in $\hat{F}$. Then $\hat{K} \cap L(\mathbf{t}) = K(\mathbf{t})$ is the fixed field of G in $L(\mathbf{t})$ and $L(\mathbf{t})\hat{K} = \hat{F}$. By Remark 13.7.6(b), $K(\mathbf{t}), \hat{K}, L_0(\mathbf{t}), L(\mathbf{t}), F, \hat{F}$ realize $A\,\mathrm{wr}_{G_0} G$. □

PROPOSITION 13.8.2 (Realization of Twisted Wreath Products): *Let $K \subseteq L_0 \subseteq L$ be a tower of fields with K Hilbertian and L/K finite Galois. Consider an absolutely irreducible polynomial $f(T, X) \in L_0[T, X]$ which is Galois over $L(T)$. Let $G = \mathrm{Gal}(L/K)$, $G_0 = \mathrm{Gal}(L/L_0)$, and $A = \mathrm{Gal}(f(T, X), L(T))$. Then G_0 acts on A and there are fields M, N, such that K, L_0, L, M, N realize $A\,\mathrm{wr}_{G_0} G$.*

Proof: Let $t_1, \ldots, t_n$ be algebraically independent elements over K with $n = [L_0 : K]$. Lemma 13.8.1 gives fields $\hat{K}$, F, and $\hat{F}$ such that $K(\mathbf{t}), \hat{K}, L_0(\mathbf{t}), L(\mathbf{t}), F, \hat{F}$ realize $A\,\mathrm{wr}_{G_0} G$. Thus, Conditions (5a)-(5d) of Section 13.7 hold for $K(\mathbf{t}), L_0(\mathbf{t}), L(\mathbf{t})$ instead of for K, L_0, L. Since K is Hilbertian, Lemma 13.1.1 gives $\mathbf{a} \in K^r$ such that those conditions hold for the residue fields under each L-place φ of $\hat{F}$ with $\varphi(K(\mathbf{t})) = K \cup \{\infty\}$. Lemma 2.6.6 gives such a place. Moreover, the residue fields of $L_0(\mathbf{t})$ and $L(\mathbf{t})$ under φ are L_0 and L, respectively. Let $\hat{K}'$, F', and $\hat{F}'$ be the residue fields of $\hat{K}$, F, and $\hat{F}$, respectively, under φ. By Remark 13.7.6(b), they realize $A\,\mathrm{wr}_{G_0} G$. □

THEOREM 13.8.3 (Diamond Theorem [Haran4, Thm. 4.1]): *Let K be a Hilbertian field, M_1 and M_2 Galois extensions of K, and M an intermediate field of $M_1 M_2/K$. Suppose that $M \not\subseteq M_1$ and $M \not\subseteq M_2$. Then M is Hilbertian.*

Proof: Corollary 12.2.3 allows us to assume that $[M : K] = \infty$. Part A of the proof strengthens this assumption:

PART A: *We may assume $[M : (M_1 \cap M)] = \infty$.* Otherwise, $[M : (M_1 \cap M)] < \infty$. Then K has a finite Galois extension M_2' such that $M \subseteq (M_1 \cap M)M_2'$. Then $M \subseteq M_1 M_2'$ and $[M : M \cap M_2'] = \infty$. Replace M_1 by M_2' and M_2 by M_1, if necessary, to restore our assumption.

PART B: *Reduction to an absolutely irreducible Galois polynomial.* By Lemma 13.1.4, each separable Hilbert subset of M contains a subset of the form $M \cap H_{M'}(f)$, where M' is a finite Galois extension of M and $f \in M[T, X]$ is an absolutely irreducible polynomial which is monic and separable in X and $f(T, X)$ is Galois over $M'(T)$. It suffices to find $a \in M$ such that $f(a, X)$ is irreducible over M'.

We may assume $M' \subseteq M_1 M_2$. Indeed, K has a finite Galois extension K' such that $M' \subseteq MK'$. If $M' \not\subseteq M_2K'$, replace M_2 by M_2K'. If $M' \subseteq M_2K'$,

13.8 The Diamond Theorem

replace M_1 by K'. In the latter case we still have $[M : (K' \cap M)] = \infty$, because $[M : K] = \infty$ and $[K' : K] < \infty$.

Thus, we may assume that $f(T, X)$ is Galois over $M_1 M_2$. It suffices to produce $a \in M$ such that $f(a, X)$ is irreducible over $M_1 M_2$.

PART C: *Finite Galois extensions.* Write $M_0 = M$ and $N = M_1 M_2$. Then N/K is Galois. For each finite Galois extension L of K in N let $L_i = M_i \cap L$, $i = 0, 1, 2$. Then L_i/K is Galois, $i = 1, 2$. Use the assumptions $M_0 \not\subseteq M_i$, $i = 1, 2$ and $[M_0 : M_1 \cap M_0] = \infty$ (Part A) to choose L large with $L_0 \not\subseteq L_i$ for $i = 1, 2$, $[L_0 : L_1 \cap L_0] > 2$, and
(2) $f \in L_0[T, X]$ and $f(T, X)$ is Galois over $L(T)$.

The conditions on the fields L_i translate into conditions on the groups $G_i = \mathrm{Gal}(L/L_i)$, $i = 0, 1, 2$:
(3a) $G_1, G_2 \not\leq G_0$.
(3b) $(G_0 G_1 : G_0) > 2$.

PART D: *Twisted wreath products.* Let

$$A' = \mathrm{Gal}(f, L(T)) = \mathrm{Gal}(f, K_s(T)).$$

Choose a basis $b_1, \ldots, b_n$ for L_0/K and algebraically independent elements $t_1, \ldots, t_n$ over K. By (2) and Lemma 13.8.1, the group G_0 acts on A'. Moreover, there are fields $F, \hat{F}$ such that
(4a) $K(\mathbf{t}), L_0(\mathbf{t}), L(\mathbf{t}), F, \hat{F}$ realize $A' \mathrm{wr}_{G_0} G$ and
(4b) $F = L(\mathbf{t}, z)$ with $\mathrm{irr}(z, L(\mathbf{t})) = f(\sum_{i=1}^n b_i t_i, X)$.

Since K is Hilbertian, Lemma 13.1.1 gives $c_1, \ldots, c_n \in K$ such that the specialization $\mathbf{t} \to \mathbf{c}$ extends to an L-place of $\hat{F}$ onto a Galois extension $\hat{F}'$ of K with Galois group isomorphic to $\mathrm{Gal}(\hat{F}/K(\mathbf{t}))$. Thus, $\hat{F}'$ has a subfield F' with these properties:
(5a) $K, L_0, L, F', \hat{F}'$ realize $A' \mathrm{wr}_{G_0} G$ and
(5b) $F' = L(z')$ with $\mathrm{irr}(z', L) = f(\sum_{i=1}^n b_i c_i, X)$.

Let $a = \sum_{i=1}^n b_i c_i$. Then a is in L_0, hence in M. If we prove $N \cap F' = L$, it will follow from (5b) that $[N(z') : N] = [F' : L] = \deg(f(a, X))$. Thus, $f(a, X)$ will be irreducible over N, as desired.

PART E: *Conclusion of the proof.* Let $E = N \cap F'$ and $A = \mathrm{Gal}(E/L)$. By Remark 13.7.6(c), G_0 acts on A and there is a Galois extension $\hat{E}$ of K such that
(6) $K, L_0, L, E, \hat{E}$ realize $H = A \mathrm{wr}_{G_0} G$.
In particular, all conjugates of E over K are in N. Hence, $\hat{E} \subset N$.

Identify H with $\mathrm{Gal}(\hat{E}/K)$ such that $\mathrm{res}_{\hat{E}/L} \colon \mathrm{Gal}(\hat{E}/K) \to \mathrm{Gal}(L/K)$ coincides with the projection $\pi \colon H \to G$. Then $\pi \circ \mathrm{res}_{N/\hat{E}} = \mathrm{res}_{N/L}$.

For $i = 1, 2$ let $H_i = \mathrm{res}_{N/\hat{E}}(\mathrm{Gal}(N/M_i))$. Then $H_i \triangleleft H$ and $\pi(H_i) = \mathrm{res}_{N/L}(\mathrm{Gal}(N/M_i)) = G_i$. Since $\mathrm{Gal}(N/M_1)$ and $\mathrm{Gal}(N/M_2)$ are normal subgroups of $\mathrm{Gal}(N/K)$ with a trivial intersection, they commute. Hence, H_1 and H_2 commute.

By (3a) there exists $h_2 \in H_2$ with $\pi(h_2) \notin G_0$. If A were not trivial, then by (3b) and Lemma 13.7.4 there would be $h_1 \in H_1$ which does not commute with h_2. We conclude from this contradiction that $A = 1$ and therefore $N \cap F' = L$, as desired. □

We conclude this section with an application of the diamond theorem that solves Problems 12.18 and 12.19 of [Fried-Jarden3]:

COROLLARY 13.8.4 ([Haran-Jarden5]): *Let K be a Hilbertian field and let N be a Galois extension of K which is not Hilbertian. Then N is not the compositum of two Galois extensions of K neither of which is contained in the other. In particular, this conclusion holds for K_s.*

13.9 Weissauer's Theorem

The most useful application of the diamond theorem is part (b) of the following result:

THEOREM 13.9.1 (Weissauer): *Let K be a Hilbertian field.*
(a) *Let M be a separable algebraic extension of K and M' a finite proper separable extension of M. Suppose M' is not contained in the Galois closure of M/K. Then M' is Hilbertian.*
(b) *Let N be a Galois extension of K and N' a finite proper separable extension of N. Then N' is Hilbertian.*
(c) *Let N be a Galois extension of a Hilbertian field K and L a finite proper separable extension of K. Suppose that $N \cap L = K$. Then NL is Hilbertian.*

Proof: Statement (c) is a special case of Statement (b). Statement (b) is a special case of Statement (a). Statement (a) follows from Corollary 12.2.3 if $[M : K] < \infty$. Suppose therefore that $[M : K] = \infty$. Then K has a finite Galois extension L with $M' \subseteq ML$. Let N be the Galois closure of M/K. Then M' is contained neither in L nor in N. By Theorem 13.8.3, M' is Hilbertian. □

Proposition 13.9.3 below is a stronger version of Theorem 13.9.1(c) which gives information about the Hilbert sets of NL. That version implies Theorem 13.9.1(a), hence also Theorem 13.9.1(b). The proof of Proposition 13.9.3 depends on the following lemma rather that on the diamond theorem:

LEMMA 13.9.2: *Let $L = K(\alpha)$ be a finite proper separable extension of a field K and let $\hat{L}$ be the Galois hull of L/K. Consider an absolutely irreducible polynomial $h \in L[T, X]$ with $\deg_X(h) > 1$ which is separable with respect to X. For u, v algebraically independent elements over K, let E be a Galois extension of $K(u, v)$ such that $E \cap L(u, v) = K(u, v)$. Then the polynomial $h(u + \alpha v, X)$ has no root in the field $E\hat{L}$.*

Proof: Put $t = u + \alpha v$ and $F = E\hat{L}$. Then choose $\sigma \in \text{Gal}(F/E)$ with $\alpha' = \sigma \alpha \neq \alpha$ and put $t' = \sigma t = u + \alpha' v$ and $h' = \sigma h$. Assume that there exists $x \in F$ with $h(t,x) = 0$. Then $x' = \sigma x \in F$ and $h'(t', x') = 0$. Since $\alpha \neq \alpha'$, we have $\hat{L}(t, t') = \hat{L}(u, v)$. Hence, t, t' are algebraically independent over K.

$$\begin{array}{ccccc}
\hat{L}(t,x) & \!\!\!\!\!\!\! \text{---} \!\!\!\!\!\!\! & \hat{L}(t,t',x) & \!\!\!\!\!\!\! \text{---} \!\!\!\!\!\!\! & F \\
| & & | & & | \\
\hat{L}(t) & \!\!\!\!\!\!\! \text{---} \!\!\!\!\!\!\! & \hat{L}(t,t') & \!\!\!\!\!\!\! \text{---} \!\!\!\!\!\!\! & \hat{L}(t,t',x') \\
| & & | & & | \\
\hat{L} & \!\!\!\!\!\!\! \text{---} \!\!\!\!\!\!\! & \hat{L}(t') & \!\!\!\!\!\!\! \text{---} \!\!\!\!\!\!\! & \hat{L}(t',x')
\end{array}$$

Therefore, $\hat{L}(t,x)$ and $\hat{L}(t',x')$ are algebraically independent over $\hat{L}$. Since they are regular extensions of $\hat{L}$ they are linearly disjoint (Lemma 2.6.7). By Lemma 2.5.11, $\hat{L}(t,t',x)$ and $\hat{L}(t,t',x')$ are linearly disjoint over $\hat{L}(t,t')$. In particular,

(1) $$\hat{L}(t,t',x) \cap \hat{L}(t,t',x') = L(t,t').$$

On the other hand, with $E_0 = E \cap \hat{L}(t,t')$, we have $\text{Gal}(F/E_0) = \text{Gal}(F/E) \times \text{Gal}(F/\hat{L}(t,t'))$. Hence, $\sigma\tau = \tau\sigma$ for all $\tau \in \text{Gal}(F/\hat{L}(t,t',x))$. In particular, $\tau x' = \tau\sigma x = \sigma\tau x = \sigma x = x'$. Therefore, $x' \in \hat{L}(t,t',x)$. By (1), $\hat{L}(t,t',x') = \hat{L}(t,t')$, a contradiction to $\deg_X(h') > 1$. $\square$

PROPOSITION 13.9.3: *Let L be a finite proper separable extension of a Hilbertian field K and let M be a Galois extension of K such that $M \cap L = K$. Then the field $N = ML$ is Hilbertian. Moreover, every separable Hilbert subset of N contains elements of L.*

Proof: By Lemma 13.1.2, it suffices to consider absolutely irreducible polynomials $h_1, \ldots, h_m \in N[T, X]$, separable monic and of degree at least 2 in X and to find $c \in L$ such that $h_i(c, X)$ has no root in N, $i = 1, \ldots, m$.

Let u and v be algebraically independent over K. Choose a primitive element α for L/K and let L' be a finite extension of L which is contained in N and contains all coefficients of $h_1, \ldots, h_m$. Put $K' = M \cap L'$ and let F be a finite Galois extension of $K'(u,v)$ that contains $L'(u,v)$ and over which all polynomials $h_1(u + \alpha v, X), \ldots, h_m(u + \alpha v, X)$ decompose into linear factors. Let $g \in K[u, v]$ be the product of the discriminants of $h_1(u + \alpha v, X), \ldots, h_m(u + \alpha v, X)$ with respect to X. Let B' be the set of all $(a,b) \in (K')^2$ with $g(a,b) \neq 0$ satisfying the following condition:

(2) The L'-specialization $(u,v) \to (a,b)$ extends to a place φ of F which induces an isomorphism φ': $\text{Gal}(F/K'(u,v)) \to \text{Gal}(F'/K')$ (with F' being the residue field of F) such that $(\varphi'\sigma)(\varphi x) = \varphi(\sigma x)$ for all $\sigma \in \text{Gal}(F/K'(u,v))$ and $x \in F$ with $\varphi x \neq \infty$. In particular φ maps the set of all zeros of $h_i(u + \alpha v, X)$ bijectively onto the set of zeros of $h_i(a + \alpha b, X)$, $i = 1 \ldots, m$.

By Lemma 13.1.1 and Example 2.6.10, B' contains a separable Hilbert subset A' of $(K')^2$. By Corollary 12.2.3, A' contains a separable Hilbert subset of K^2.

Thus, there are $a, b \in K$ satisfying (2). Let $c = a + \alpha b$. Assume there is an i between 1 and m such that the polynomial $h_i(c, X)$ has a root $\bar{x}$ in N. Then $\bar{x} \in F' \cap N$. Since res: $\mathrm{Gal}(N/L') \to \mathrm{Gal}(M/K')$ is an isomorphism, there exists a Galois extension E' of K' contained in M such that $E'L' = F' \cap N$. Since $E' \cap L' = K' = M \cap L'$,

(3)
$$\mathrm{Gal}(F'/E') \cdot \mathrm{Gal}(F'/L') = \mathrm{Gal}(F'/K')$$
$$\mathrm{Gal}(F'/E') \cap \mathrm{Gal}(F'/L') = \mathrm{Gal}(F'/F' \cap N).$$

Therefore, $K'(u, v)$ has a Galois extension E in F such that $\varphi'(\mathrm{Gal}(F/E)) = \mathrm{Gal}(F'/E')$. From (3), $E \cap L'(u, v) = K'(u, v)$ and $\varphi'(\mathrm{Gal}(F/EL')) = \mathrm{Gal}(F'/F' \cap N)$. Moreover, the polynomial $h_i(u + \alpha v, X)$ has a root x such that $\varphi(x) = \bar{x}$. For each $\sigma \in \mathrm{Gal}(F/EL')$ we have $\varphi(\sigma x) = \varphi'(\sigma)(\bar{x}) = \bar{x} = \varphi(x)$. Hence, $\sigma x = x$. In particular, $x \in EL'$, a contradiction to Lemma 13.9.2. □

We deduce Theorem 13.9.1(a) from Proposition 13.9.3:

PROPOSITION 13.9.4: *Let M be a separable algebraic extension of a Hilbertian field K and M' a proper finite separable extension of M. Suppose M' is not contained in the Galois closure of M/K. Then M' is Hilbertian.*

Proof: Denote the Galois closure of M/K by $\hat{M}$. With no loss replace M by $\hat{M} \cap M'$. Choose a primitive element α for M'/M. Put $L' = K(\alpha)$, and $K' = \hat{M} \cap L'$. Then the conditions of Proposition 13.9.3 are satisfied for $K', L', \hat{M}$ replacing K, L, M.

Let $f \in M'[T, X]$ be an irreducible polynomial, monic and separable in X with $\deg_X(f) > 1$. By Lemma 13.1.2, there exist $h_1, \ldots, h_m \in M'[T, X]$ absolutely irreducible and separable in X such that $H'_{M'}(h_1, \ldots, h_m) \subseteq H_{M'}(f)$. Since $h_1, \ldots, h_m$ are irreducible in $\hat{M}'[T, X]$, Proposition 13.9.3, gives $a \in L'$ with $h_1(a, X), \ldots, h_m(a, X)$ irreducible in $\hat{M}(\alpha)[X]$ and therefore in $M'[X]$. By Proposition 13.2.2, M' is Hilbertian. □

Example 13.9.5: *Non-Hilbertian subfield of $\tilde{\mathbb{Q}}$ whose finite proper extensions are Hilbertian.* Denote the compositum of all finite solvable extensions of $\mathbb{Q}$ by $\mathbb{Q}_{\mathrm{solv}}$. It is not Hilbertian; there exists no $a \in \mathbb{Q}_{\mathrm{solv}}$ such that $X^2 - a$ is irreducible over $\mathbb{Q}_{\mathrm{solv}}$. However, $\mathbb{Q}_{\mathrm{solv}}$ is a Galois extension of $\mathbb{Q}$. Hence, by Corollary 13.9.1(b), every finite proper extension of $\mathbb{Q}_{\mathrm{solv}}$ is Hilbertian. By Corollary 13.8.4, $\mathbb{Q}_{\mathrm{solv}}$ is not the compositum of two Galois extensions of $\mathbb{Q}$ neither of which is contained in the other. □

Exercises

1. [Fried10, §2] This exercise shows that it is not always possible to take $m = 1$ in Lemma 13.1.2. Consider an irreducible polynomial $f \in \mathbb{Q}[T, X]$, monic in X, and $g_1, g_2 \in \mathbb{Q}[Y]$, polynomials of positive degree for which $f(g_i(Y), X)$

Exercises

is reducible, $i = 1, 2$. In addition, suppose there exists an irreducible $h \in \mathbb{Q}[T, X]$ with $\deg_X(h) > 1$ such that

$$H'_\mathbb{Q}(h) = \{a \in \mathbb{Q} \mid h(a, X) \text{ has no zero in } \mathbb{Q}\} \subseteq H_\mathbb{Q}(f).$$

(a) Use Exercise 1 of Chapter 12 to conclude that $h(g_i(Y), X)$ has a factor of degree 1 in X, say $X - m_i(Y)$, where $m_i \in \mathbb{Q}(Y)$, $i = 1, 2$.

(b) Use field theory to interpret (a): Let t be an indeterminate, y_i a zero of $g_i(Y) - t$, and $x_i = m_i(y_i)$, $i = 1, 2$. Note that x_1 and x_2 are both zeros of $h(t, X)$, so that they are conjugate over $\mathbb{Q}(t)$. Conclude that $[\mathbb{Q}(y_1) \cap \mathbb{Q}(y'_2) : \mathbb{Q}(t)] > 1$ for some conjugate y'_2 of y_2 over $\mathbb{Q}(t)$.

(c) Consider the case $f(T, X) = X^4 + 2X^2 - T$ and $g_1(Y) = Y^4 + 2Y^2$, $g_2(Y) = -4Y^4 - 4Y^2 - 1$. Prove that the splitting field of $g_i(Y) - t$ over $\mathbb{Q}(t)$ has the dihedral group of order 8 as its Galois group over $\mathbb{Q}(t)$. Observe that since $\mathbb{Q}(y_1)/\mathbb{Q}(t)$ is a nonnormal extension of degree 4, $\mathbb{Q}(y_1^2)/\mathbb{Q}(t)$ is its only quadratic subextension. Prove that the prime of $\mathbb{Q}(t)$ corresponding to the specialization $t \to 0$ is unramified in $\mathbb{Q}(y_1^2)$ but ramified in $\mathbb{Q}(y_2^2)$. Conclude that there is no irreducible $h \in \mathbb{Q}[T, X]$ with $\deg_X(h) > 1$ and $H'_\mathbb{Q}(h) \subseteq H_\mathbb{Q}(f)$.

2. Let $f_1(T, X), \ldots, f_m(T, X)$, with $\deg_X(f_i) > 1$, $i = 1, \ldots, m$, be absolutely irreducible polynomials, separable in X, with coefficients in a global field K. Let t be transcendental over K and denote the splitting fields of $f_1(t, X), \ldots, f_m(t, X)$, respectively, over $E = K(t)$ by $F_1, \ldots, F_m$. Assume $F_1, \ldots, F_m$ are linearly disjoint over E and let $F = F_1 \ldots F_m$. Observe that there is a $\sigma \in \mathrm{Gal}(F/E)$ that fixes none of the roots of $f_1(t, X) \ldots f_m(t, X)$ and improve Lemma 13.3.4. Show there exists a prime ideal $\mathfrak{p}$ of O_K and an element $a_\mathfrak{p} \in O_K$ such that for $a \in O_K$, if $a \equiv a_\mathfrak{p} \bmod \mathfrak{p}$, then $f_i(a, X)$ has no zeros in K, $i = 1, \ldots, m$.

3. Consider the three absolutely irreducible polynomials

$$f_1(T, X) = X^2 - T, f_2(T, X) = X^2 - (T + 1), f_3(T, X) = X^2 - T(T + 1)$$

and let $H = H_\mathbb{Q}(f_1, f_2, f_3)$. Choose a prime number p and an integer a such that both a and $a + 1$ are quadratic nonresidues modulo p. Prove that $a + p\mathbb{Z} \subseteq H$, even though F_1, F_2, and F_3 (in the notation of Exercise 2) are not linearly disjoint over E.

4. Let K be a Hilbertian field with valuation v. Prove that every Hilbert subset H of K^r is v-dense in K^r.

Hint: Let $H = H_K(f_1, \ldots, f_m)$, where $f_1, \ldots, f_m$ are irreducible polynomials in $K(T_1, \ldots, T_r)[X_1, \ldots, X_n]$. For $(a_1, \ldots, a_r) \in K^r$, and $\gamma = v(c)$ an element of the value group, each polynomial in the set

$$\{f_i(a_1 + cT_1^{\varepsilon_1}, \ldots, a_r + cT_r^{\varepsilon_r}, \mathbf{X}) \mid 1 \leq i \leq m \text{ and } \varepsilon_1, \ldots, \varepsilon_r \in \{\pm 1\}\},$$

is irreducible in $K(\mathbf{T})[\mathbf{X}]$. Substitute elements $t_1, \ldots, t_r$ for $T_1, \ldots, T_r$ so that these polynomials remain irreducible in $K(\mathbf{X})$. Thus, find $(b_1, \ldots, b_r) \in H$ such that $v(b_i - a_i) \geq \gamma$, $i = 1, \ldots, r$.

5. Suppose K is the quotient field of a Hilbertian ring R. Let M_1, M_2 be Galois extensions of K and M an extension of K in $M_1 M_2$ which is contained in neither M_1 nor in M_2. Prove the integral closure of R in M is Hilbertian.

Hint: In the proof of Theorem 13.8.3 choose $b_1, \ldots, b_n$ integral over R. Then choose $c_1, \ldots, c_n$ in R.

Notes

Hilbert [Hilbert, p. 280] proves Lemma 13.1.1(b) for number fields. Our proof of Proposition 13.2.1 is a version of Inaba's proof [Inaba]. Theorem 13.3.5(a) for number fields appears in [Eichler]. We follow [Fried5]. Lang [Lang3, p. 152] reproduces Franz's power series expansion proof [Franz].

Theorem 13.6.2 for $g = 0$ and K a number field appears in [Corvaja-Zannier]. Generalization to arbitrary g and arbitrary countable Hilbertian field K of characteristic 0 appears in [Fried-Jarden4]. The proof uses a deep group theoretic result due to Guralnick, Thompson, et al. and the Riemann existence theorem. Thus, one cannot generalize the proof to positive characteristic. An elementary proof of the theorem for arbitrary g but still for K a number field can be found in [Zannier]. Our proof generalizes that of Zannier. We replace the polynomial $Y^2 - X^{2q} + 2$ which Zannier uses by the polynomial $Y^l - X^{lq} + u$ in order to make the proof works for each global field K.

Weissauer proves that every finite proper separable extension of a Galois extension of a Hilbertian field is Hilbertian [Weissauer, Satz 9.7]. Section 13.9 replaces Weissauer's "nonstandard" proof by a simpler "standard" one still using Weissauer's auxiliary variable trick as in Lemma 13.9.2, [Fried10, Thm. 1.3]. Haran's diamond theorem 13.8.3 generalizes Weissauer's theorem (Theorem 13.9.1) and Corollary 13.8.5 [Haran-Jarden5]. Its proof is an offspring of the proofs of the two earlier results.

Proposition 13.4.1 says the ring of integers O_K of a number field K is Hilbertian. Thus, $O_K \cap H_K(f)$ is an infinite set for each irreducible polynomial $f \in K[T, X]$. One may further ask when $O_K \smallsetminus H_K(f)$ is finite. This is not always the case. For example, $O_K \smallsetminus H_K(X^2 - T)$ is the infinite set of all squares in O_K. But, for $K = \mathbb{Q}$, Müller proves the set $\mathbb{Z} \smallsetminus H_\mathbb{Q}(f)$ is finite in each of the following cases: $\deg_X(f)$ is a prime number and the curve $f(T, X) = 0$ has positive genus or $\mathrm{Gal}(f(T, X), \mathbb{Q}(T)) \cong S_n$ for some positive integer $n \neq 5$ [Müller, Thm. 1.2]. The proof uses Siegel's theorem about integral points on algebraic curves and classical results about finite groups.

Chapter 14. Nonstandard Structures

A. Robinson invented "nonstandard" methods in order to supplement the Weierstrass ε, δ formalism of the calculus by a rigorous version of the classical calculus of infinitesimals in the spirit of Leibniz and other formalists. We will apply the nonstandard approach to algebra in Chapter 15 in order to find new Hilbertian fields. Its main virtue, from an algebraic point of view, is that it creates additional algebraic structures to which well known theorems can be applied.

Here we present the basics of the nonstandard method: the higher order structure on a set M (Section 14.1); the concept of an enlargement M^* of M (Sections 14.2 and 14.3); and the existence of M^* (Section 14.4) via ultraproducts.

14.1 Higher Order Predicate Calculus

Sentences that speak of arbitrary subsets, functions, relations, etc., are common in mathematics, even though they are usually outside the scope of first order languages. Here we introduce a language which includes such sentences.

First the notion of a **type** (of a higher order object) is inductively defined by the following rules:
(1a) The number 0 is a type.
(1b) If n is a positive integer and $\tau(1), \ldots, \tau(n)$ are types, then the sequence $(\tau(1), \ldots, \tau(n))$ is a type.

Denote the set of all types by T. Given a set M, attach a set M_τ to each type τ as follows:
(2a) $M_0 = M$.
(2b) If $\tau(1), \ldots, \tau(n)$ are types and $\tau = (\tau(1), \ldots, \tau(n))$, then

$$M_\tau = \mathcal{P}(M_{\tau(1)} \times \cdots \times M_{\tau(n)}) = \{\text{all subsets of } M_{\tau(1)} \times \cdots \times M_{\tau(n)}\}.$$

Elements of M_τ are **objects of type** τ over M. We call them **sets** (or **relations**) **of type** τ if $\tau \neq 0$. A **higher order set** is a set of type τ for some $\tau \neq 0$. Call the system $\mathcal{M} = \langle M_\tau \mid \tau \in T \rangle$ the **higher order structure over** M.

For each higher order set A of M introduce a sequence of variables $X_{A1}, X_{A2}, \ldots$. Inductively define formulas of the higher order language $\mathcal{L}_\infty(M)$ as follows:
(3a) $X_{Ai} = X_{Aj}$, $X_{Ai} = a$, and $a = b$ are formulas for each higher order set A, all $a, b \in A$, and $i, j \in \mathbb{N}$.
(3b) If $A(1), \ldots, A(n)$ are higher order sets, $A(0)$ is a subset of $\mathcal{P}(A(1) \times \cdots \times A(n))$, and for ν between 0 and n either t_ν is a variable $X_{A(\nu), i(\nu)}$ or an element of $A(\nu)$, then $(t_1, \ldots, t_n) \in t_0$ is a formula.

(3c) Negations and disjunctions (hence also conjunctions and implications) of formulas are formulas.

(3d) If φ is a formula, then $(\exists X_{Ai} \in A)[\varphi]$ (hence also $(\forall X_{Ai} \in A)[\varphi]$) is a formula for all higher order set A and $i \in \mathbb{N}$.

Define **free variables** of a formula as usual (Section 7.1). In particular, a sentence is a formula without free variables. Likewise, a **substitution** is a function f that, for each higher order set A of M, replaces each X_{Ai} by an element x_{Ai} of A and fixes the elements of A. Interpret the **truth** of a formula φ **under** f by reading "$(t_1,\ldots,t_n) \in t_0$" as "$(f(t_1),\cdots,f(t_n))$ belongs to $f(t_0)$."

Note that each object t of type $\tau = (\tau(1),\ldots,\tau(n)) \neq 0$ has two roles in the language $\mathcal{L}_\infty(M)$. First, it is a constant: t is equal or not equal to another object of the same type. Second, it is a relation between objects of types $\tau(1),\ldots,\tau(n)$.

If the set M has a first order structure, embed this in the higher order structure of M in a natural way. For example, if M is a field, then addition corresponds to a subset A of $M \times M \times M$, and $X_1 + b = c$ in the first order language becomes $(X_{M1}, b, c) \in A$.

Remark: Let $\mathbb{N}^*$ be a proper elementary extension of $\mathbb{N}$. Then $\mathbb{N}_{(0)}$ (resp. $(\mathbb{N}^*)_{(0)}$) is the collection of subsets of $\mathbb{N}$ (resp. $\mathbb{N}^*$). The second order structure $\mathbb{N}^* \cup (\mathbb{N}^*)_{(0)}$ is not an elementary extension of the second order structure $\mathbb{N} \cup \mathbb{N}_{(0)}$. For example the induction axiom

$$(\forall X \in \mathbb{N}_{(0)})[[1 \in X \wedge (\forall x \in \mathbb{N})[x \in X \to x+1 \in X]] \to (\forall x \in \mathbb{N})[x \in X]]$$

holds in $\mathbb{N} \cup \mathbb{N}_{(0)}$ but it fails in $\mathbb{N}^* \cup (\mathbb{N}^*)_{(0)}$ if we replace $\mathbb{N}$ and $\mathbb{N}_{(0)}$ by $\mathbb{N}^*$ and $(\mathbb{N}^*)_{(0)}$. Indeed, the induction axiom does not hold for $X = \mathbb{N}$. In order to restore the elementary extension property for higher order structures of the elementary extension $\mathbb{N}^*$ of $\mathbb{N}$, we must restrict quantification of subsets of $\mathbb{N}^*$ to a proper subcollection of $(\mathbb{N}^*)_{(0)}$, to be denoted $\mathbb{N}^*_{(0)}$, the "inner" subset of $\mathbb{N}^*$. We define these in the next section. □

14.2 Enlargements

We consider a set M together with its higher order structure and define an **enlargement** of M as a special model of the higher order theory of M satisfying Conditions I and II below and III of Section 14.3. The enlargement will be saturated with respect to all higher order relations.

The underlying set M^* of this model has the property that to each object a of type τ of M there corresponds an object a^* of M^* of the same type (i.e. a^* is an element of $(M^*)_\tau$). Call such an a^* **standard**. In particular, M_τ itself, an element of $M_{(\tau)}$, corresponds to an element $(M_\tau)^*$ of $(M^*)_{(\tau)}$. This means $(M_\tau)^*$ is a subset of $(M^*)_\tau$. Call the elements of $(M_\tau)^*$ the **internal** objects of type τ of M^*. All other elements of $(M^*)_\tau$ are **external**.

14.2 Enlargements

Note: For $\tau = 0$ we have $(M_0)^* = (M^*)_0 = M^*$. Thus, every element of M^* is internal, although $(M^*)_{(0)}$ itself may have external subsets.

To simplify notation write M_τ^* for $(M_\tau)^*$. If $\tau(1), \ldots, \tau(n)$ are types and if $R \subseteq M_{\tau(1)} \times \cdots \times M_{\tau(n)}$, then $R \in M_\tau$ for $\tau = (\tau(1), \ldots, \tau(n))$. Hence, $R^* \in M_\tau^*$, that is, $R^* \subseteq (M^*)_{\tau(1)} \times \cdots \times (M^*)_{\tau(n)}$.

We demand, however, that an enlargement satisfy a stronger condition:

I. *Internal condition on n-ary relations.* If $\tau(1), \ldots, \tau(n)$ are types and if $R \subseteq M_{\tau(1)} \times \cdots \times M_{\tau(n)}$, then $R^* \subseteq M_{\tau(1)}^* \times \cdots \times M_{\tau(n)}^*$.

That is, the elements of a standard n-ary relation are n-tuples with internal coordinates. Call them **internal n-tuples**.

Suppose τ is a type and A is a subset of M_τ. By Condition I, $A^* \subseteq M_\tau^*$. So, each element of A^* is an element of M_τ^*, hence internal. Consequently, each element of a standard set is internal.

On the other hand, each subset of A^* is an element of $(M^*)_{(\tau)}$. It is internal exactly when it belongs to $M_{(\tau)}^*$.

To interpret a formula of $\mathcal{L}_\infty(M)$ in M^* consider only **internal substitutions** f. These satisfy the condition $f(X_{Ai}) \in A^*$ for each higher order set A. Define the truth value of the formula φ under f first by placing an asterisk to the right of each constant or relation symbol that appears in φ to obtain a formula φ^*. Then interpret the formula φ^* (under f) as usual. Note: If $(\exists X \in A)$ is part of φ, then $(\exists X \in A^*)$ is a part of φ^*. Since each element of A^* is internal, this means one quantifies only on internal objects.

For example, consider a formula $\varphi(X_1, \ldots, X_n)$ where $X_\nu = X_{M_{\tau(\nu)}, i(\nu)}$, $\nu = 1, \ldots, n$. Let

$$R = \{(a_1, \ldots, a_n) \in M_{\tau(1)} \times \cdots \times M_{\tau(n)} \mid M \models \varphi(a_1, \ldots, a_n)\}.$$

Then the sentence θ

$$(\forall X_1 \in M_{\tau(1)}) \cdots (\forall X_n \in M_{\tau(n)})[(X_1, \ldots, X_n) \in R \leftrightarrow \varphi(X_1, \ldots, X_n)]$$

is true in M. The sentence θ^* takes the form

$$(\forall X_1 \in M_{\tau(1)}^*) \ldots (\forall X_n \in M_{\tau(n)}^*)[(X_1, \ldots, X_n) \in R^* \leftrightarrow \varphi^*(X_1, \ldots, X_n)].$$

It is reasonable to ask that θ^* will be true in M^*. Since Condition I implies that R^* contains only elements with internal coordinates, we may rephrase θ^* as

$$R^* = \{(a_1, \ldots, a_n) \in M_{\tau(1)}^* \times \cdots \times M_{\tau(n)}^* \mid M^* \models \varphi^*(a_1, \ldots, a_n)\}.$$

Here is the condition guaranteeing this indeed holds:

II. *Elementary extension condition.* Let $A_1, \ldots, A_n$ be higher order sets, $\varphi(X_{A_1,i(1)}, \ldots, X_{A_n,i(n)})$ a formula of $\mathcal{L}_\infty(M)$, and $a_\nu \in A_\nu$, $\nu = 1, \ldots, n$. Then $\varphi(a_1, \ldots, a_n)$ is true in M if and only if $\varphi^*(a_1^*, \ldots, a_n^*)$ is true in M^*.

For example, if $a \in M_\tau$, then $a^* \in M_\tau^*$. Thus, standard objects are internal. Also, if $a, b \in M_\tau$ and $a \neq b$, then $a^* \neq b^*$: the canonical map $a \mapsto a^*$ of M_τ into M_τ^* is injective. Therefore, we occasionally regard M_τ as a subset of M_τ^*. In addition, for $R \subseteq A_1 \times \cdots \times A_n$, if $\mathbf{a} \in A_1 \times \cdots \times A_n$ belongs to R^*, then $\mathbf{a} \in R$. Thus, $R^* \cap (A_1 \times \cdots \times A_n) = R$ and the relation R^* is an extension of R. In particular, if M has a first order structure, then it is an elementary substructure of a natural extension to M^*.

Now consider the notion of an **internal function**. Let A and B be two higher order sets. We view a function $f \colon A \to B$ as a subset of $A \times B$ that satisfies these two conditions:

(1a) $\qquad\qquad\qquad (\forall a \in A)(\exists b \in B)[(a,b) \in f]$;

(1b) $\qquad (\forall a \in A)(\forall b_1, b_2 \in B)[(a, b_1) \in f \wedge (a, b_2) \in f \to b_1 = b_2]$.

The subset f^* of $A^* \times B^*$ satisfies the corresponding conditions: it is a function from A^* to B^*. Call it a **standard function**. For F, the set of all functions from A to B, the elements of F^* are the **internal functions** from A^* to B^*.

LEMMA 14.2.1: *The image of each internal subset of A^* under an internal function $f \colon A^* \to B^*$ is an internal subset of B^*.*

Proof: For each $f \in F$ the following statement holds in M:

(3) $\quad (\forall f \in F)(\forall X \subseteq A)(\exists Y \subseteq B)\big[[(\forall a \in X)\, f(a) \in Y]$
$\qquad\qquad\qquad\qquad\qquad\qquad \wedge\, [(\forall b \in Y)(\exists a \in X)\, f(a) = b]\big]$

The close "$\forall X \subseteq A$" is not part of the language $\mathcal{L}_\infty(M)$. However, it can be reinterpreted as "$(\forall X \in M_{(\tau)})(\forall x \in M_\tau)[x \in X \to x \in A]$", assuming $A \subseteq M_\tau$ and with $X = X_{M_{(\tau)},1}$ and $x = X_{M_\tau,1}$. The interpretation of this in M^* is "$(\forall X \in M_{(\tau)}^*)(\forall x \in M_\tau^*)[x \in X \to X \in A^*]$". This means "for all internal subsets X of A^*". Similarly, "$\exists Y \subseteq B$" interprets in M^* as "there is an internal subset of B^*". Thus, our claim is the interpretation of (3) in M^*. □

14.3 Concurrent Relations

Suppose a higher order relation R of M is finite with elements $a_1, \ldots, a_n$. Then $M \models (\forall X \in R)[X = a_1 \vee \cdots \vee X = a_n]$, so that $M^* \models (\forall X \in R^*)[X = a_1^* \vee \cdots \vee X = a_n^*]$. That is, each element of R^* is standard. We impose a final condition on M^* guaranteeing R^* will be a true "enlargement" of R if R is an infinite relation.

14.3 Concurrent Relations

DEFINITION 14.3.1: *Let A and B be two higher order sets of M. Call a binary relation $R \subseteq A \times B$* **concurrent** *if for all $a_1, \ldots, a_n \in A$ there exists $b \in B$ with $(a_i, b) \in R$, $i = 1, \ldots, n$.*

III. *Compactness (or saturation) condition.* If A, B are two higher order sets of M and if $R \subseteq A \times B$ is a concurrent relation, then there exists $b \in B^*$ such that $(a, b) \in R^*$ for each $a \in A$.

COROLLARY 14.3.2: *Let A be a higher order set of M.*
(a) *If $A = \{a_1, \ldots, a_n\}$ is a finite set, then $A^* = \{a_1^*, \ldots, a_n^*\}$ and A is identified with A^* as higher order sets.*
(b) *If A is infinite, then A is properly contained in A^*.*

Proof of (a): The sentence $(\forall X_{A,1} \in A)[\bigvee_{i=1}^{n} X_{A,1} = a_1]$ holds in M. Hence, by Condition II of Section 14.2, $(\forall X_{A,1} \in A^*)[\bigvee_{i=1}^{n} X_{A,1} = a_1^*]$ is true in M^*. Thus, $A^* = \{a_1^*, \ldots, a_n^*\}$.

Proof of (b): For every $a_1, \ldots, a_n$ in A there exists $b \in A$ with $a_i \neq b$, $i = 1, \ldots, n$. That is, the inequality relation on A is concurrent. Hence, there exists $b \in A^*$ such that $a \neq b$ for each $a \in A$ (although not for each $a \in A^*$). □

Remark 14.3.3: Warning. Let A be a higher order set of M. Suppose A is an element of another higher order set B. In Section 14.2, we have identified B with a subset of B^*. Under this identification we have identified the element A of B with the element A^* of B^*. Corollary 14.3.2 shows that the latter identification identifies A with A^* as higher order sets if and only if A is finite. □

The case $\mathbb{N} = M$, the natural numbers, provides the first example of an external object. Replacing M with $\mathbb{N} \cup M$, if necessary, we tacitly assume from now on that $\mathbb{N} \subseteq M$.

LEMMA 14.3.4: *The set $\mathbb{N}$ is an external subset of $\mathbb{N}^*$.*

Proof: By Corollary 14.3.2, $\mathbb{N}^*$ contains a nonstandard element c. Since there are no elements of $\mathbb{N}$ between n and $n + 1$, the same is true for $\mathbb{N}^*$. In the extension $<^*$ of the order relation to $\mathbb{N}^*$, the element c is therefore greater than every element of $\mathbb{N}$. That is, $\mathbb{N}$ is a bounded subset of $\mathbb{N}^*$.

On the other hand, each bounded subset of $\mathbb{N}$ has a maximal element. Formulate this statement in the language $\mathcal{L}_\infty(M)$,

$$(\forall A \in \mathbb{N}_{(0)})[(\exists b \in \mathbb{N})(\forall a \in A)[a < b] \to (\exists c \in A)(\forall a \in A)[a \leq c]].$$

Interpret this in $\mathbb{N}^*$ to conclude that every internal bounded subset of $\mathbb{N}^*$ has a maximal element. Since $\mathbb{N}$ has no maximal element, $\mathbb{N}$ is not an internal subset of $\mathbb{N}^*$. □

As a consequent, we generalize Lemma 14.3.4 and refine Corollary 14.3.2:

LEMMA 14.3.5: *Let A be a higher order set.*
(a) *If A is infinite, then A is an external subset of A^*.*
(b) *If A is an internal subset of A^* and all elements of A are standard, then A is finite.*

Proof of (a): Since A is infinite, there exists a surjective map $f \colon A \to \mathbb{N}$. Hence, f extends to a standard map $f^* \colon A^* \to \mathbb{N}^*$. By Lemma 14.2.1, f^* maps internal subsets of A^* onto internal subsets of $\mathbb{N}^*$. Since $f(A) = \mathbb{N}$ and $\mathbb{N}$ is external (Lemma 14.3.4), A must be external.

Proof of (b): Since all elements of A are standard, A is standard. Hence, by (a), A is finite. □

Here is another consequence of the compactness condition:

PROPOSITION 14.3.6: *For each i in a set I let $\varphi_i(X_1,\ldots,X_n)$ be a formula of $\mathcal{L}_\infty(M)$ with $X_\nu = X_{M_{\tau(\nu)},j(\nu)}$, $\nu = 1,\ldots,n$. Suppose for each finite subset I_0 of I there exists $(x_1,\ldots,x_n) \in M_{\tau(1)} \times \cdots \times M_{\tau(n)}$ with $M \models \bigwedge_{i \in I_0} \varphi_i(x_1,\ldots,x_n)$. Then there exists $(x_1,\ldots,x_n) \in M^*_{\tau(1)} \times \cdots \times M^*_{\tau(n)}$ with $M^* \models \varphi_i^*(x_1,\ldots,x_n)$ for each $i \in I$.*

Proof: Let $N = I \cup M$ and $B = M_{\tau(1)} \times \cdots \times M_{\tau(n)}$. Define a binary higher order relation $R \subseteq I \times B$ of N:

$$R = \{(i, \mathbf{x}) \in I \times B \mid M \models \varphi_i(\mathbf{x})\}.$$

By assumption, R is concurrent. Hence, by Condition III applied to N, there exists $\mathbf{x} \in B^*$ with $(i, \mathbf{x}) \in R^*$, for each $i \in I$. □

14.4 The Existence of Enlargements

The compactness Theorem (Proposition 7.7.6) for ultraproducts allows us to construct enlargements of the higher order structure of a given set M. We first make some preliminary observations.

Let $\mathcal{D}$ be an ultrafilter of a set I. Consider a set B along with its power set, $\mathcal{P}(B)$. Identify the ultrapower $\mathcal{P}(B)^I/\mathcal{D}$ with a subset of the power set $\mathcal{P}(B^I/\mathcal{D})$ in the following way. For A an element of $\mathcal{P}(B)^I/\mathcal{D}$ let A_i be the ith coordinate of a representative of A. Identify A with the set A' of all $a \in B^I/\mathcal{D}$ for which $\{i \in I \mid a_i \in A_i\} \in \mathcal{D}$. In the construction of the enlargement below, the standard subsets of $\mathcal{P}(B^I/\mathcal{D})$ are those of the form $A^I/\mathcal{D}$ for $A \in \mathcal{P}(B)$, the internal subsets are those that belong to $\mathcal{P}(B)^I/\mathcal{D}$, and the rest are external subsets of $\mathcal{P}(B^I/\mathcal{D})$. Thus, if B is an infinite set, its "diagonal" identification with a subset of $B^I/\mathcal{D}$ is an external subset (Lemma 14.3.5(a)).

The family $\mathcal{P}(B)^I/\mathcal{D}$ of internal sets inherits the structure of a Boolean algebra from $\mathcal{P}(B)$, and the identification of $\mathcal{P}(B)^I/\mathcal{D}$ with a subset of $\mathcal{P}(B^I/\mathcal{D})$ preserves this structure.

14.4 The Existence of Enlargements

Similarly, the ultrapower construction commutes with finite direct products: If $B_1, \ldots, B_n$ are sets, then canonically identify $(B_1 \times \cdots \times B_n)^I/\mathcal{D}$ with $B_1^I/\mathcal{D} \times \cdots \times B_n^I/\mathcal{D}$. In particular, if $R \subseteq B_1 \times \cdots \times B_n$, then $(B_1 \times \cdots \times B_n) \cap R^I/\mathcal{D} = R$. Note: An injective map $f: A \to B$ of sets canonically defines an injective map $f: \mathcal{P}(A) \to \mathcal{P}(B)$ of their respective power sets.

We are now ready to construct the enlargement of M. Define W to be the set of all quadruples (A, B, R, a), where A and B are higher order sets, $R \subseteq A \times B$ is a concurrent binary relation and $a \in A$. Consider the collection I, of all finite subsets of W. For each $i \in I$ define the **cone** over i as the set $D_i = \{j \in I \mid i \subseteq j\}$. Since $D_i \cap D_{i'} = D_{i \cup i'}$, for all $i, i' \in I$, there exists an ultrafilter $\mathcal{D}$ of I containing all cones D_i (Corollary 7.5.3).

Now define $M^* = M_0^*$ to be the ultrapower $M^I/\mathcal{D}$. Do an induction over types $\tau \neq 0$ in order to identify the ultrapower $M_\tau^I/\mathcal{D}$ with a subset M_τ^* of $(M^*)_\tau$. Indeed, suppose that $\tau = (\tau(1), \ldots, \tau(n))$ and that $M_{\tau(j)}^I/\mathcal{D}$ has been identified with $M_{\tau(j)}^*$, $j = 1, \ldots, n$. Then, start with the equality $M_\tau = \mathcal{P}(M_{\tau(1)} \times \cdots \times M_{\tau(n)})$ and let

$$M_\tau^* = M_\tau^I/\mathcal{D} \subseteq \mathcal{P}(M_{\tau(1)}^I/\mathcal{D} \times \cdots \times M_{\tau(n)}^I/\mathcal{D}) = \mathcal{P}(M_{\tau(1)}^* \times \cdots \times M_{\tau(n)}^*)$$
$$\subseteq \mathcal{P}((M^*)_{\tau(1)} \times \cdots \times (M^*)_{\tau(n)}) = (M^*)_\tau.$$

If $R \in M_\tau$ (i.e. $R \subseteq M_{\tau(1)} \times \cdots \times M_{\tau(n)}$), define R^* to be the ultrapower $R^I/\mathcal{D}$, and regard it as a subset of $M_{\tau(1)}^I/\mathcal{D} \times \cdots \times M_{\tau(n)}^I/\mathcal{D} = M_{\tau(1)}^* \times \cdots \times M_{\tau(n)}^*$. That is, R^* is an element of $(M^*)_\tau$ and Condition I of Section 14.2 holds. In addition, the canonical embedding $M_\tau \to M_\tau^I/\mathcal{D}$, identifies R with an element of $M_\tau^I/\mathcal{D} = M_\tau^*$. We have shown this element to be R^*. This works even in the case $\tau = 0$. Thus, if $a \in M$, then a^* is the corresponding element of $M^* = M^I/\mathcal{D}$.

The proof of Condition II requires an analog of Proposition 7.7.1:

LEMMA 14.4.1: Let $A_1, \ldots, A_n$ be higher order sets, $a_\nu \in A_\nu^*$, $\nu = 1, \ldots, n$, and $\varphi(X_{A_1,j(1)}, \ldots, X_{A_n,j(n)})$ a formula of $\mathcal{L}_\infty(M)$. Then

(1) $\qquad M^* \models \varphi^*(a_1, \ldots, a_n) \iff \{i \in I \mid M \models \varphi(a_{1i}, \ldots, a_{ni})\} \in \mathcal{D}.$

Here $a_{\nu i}$ is the ith coordinate of a representative of $a_\nu \in A_\nu^I/\mathcal{D} = A_\nu^*$. In particular, if $a_\nu \in A_\nu$, $\nu = 1, \ldots, n$, then

(2) $\qquad M^* \models \varphi^*(a_1^*, \ldots, a_n^*) \iff M \models \varphi(a_1, \ldots, a_n).$

Proof: We prove (1) by induction on the structure of φ according to definition (3) of Section 14.1. When φ is defined by either (3a) or (3b) of Section 14.1, (1) follows from the basic properties of ultraproducts. The induction step of (3c) of Section 14.1 also presents no difficulty. So we assume (1) is true for the formula $\varphi(X_{A_1,j(1)}, \ldots, X_{A_n,j(n)})$ and for all $a_\nu \in A_\nu^*$, $\nu = 1, \ldots, n$. Let $a_\nu \in A_\nu^*$, $\nu = 2, \ldots, n$ and suppose that

(3) $\qquad \{i \in I \mid M \models (\exists X_{A_1,j(1)} \in A_1)[\varphi(X_{A_1,j(1)}, a_{2i}, \ldots, a_{ni})]\} \in \mathcal{D}.$

For each i in the latter set choose $a_{1i} \in A_1$ such that $M \models \varphi(a_{1i}, a_{2i}, \ldots, a_{ni})$. The a_{1i}'s define an element $a_1 \in A_1^*$ such that $M^* \models \varphi^*(a_1, a_2, \ldots, a_n)$. Thus, $M^* \models (\exists X_{A_1, j(1)} \in A_1^*)[\varphi(X_{A_1, j(1)}, a_2, \ldots, a_n)]$. Prove "$\Leftarrow$" of (1) similarly for $(\exists X_{A_1, j(1)} \in A_1)[\varphi(X_{A_1, j(1)}, \ldots, X_{A_n, j(n)})]$.

Finally, consider a concurrent relation $R \subseteq A \times B$. For each $i \in I$ choose an element $b_i \in B$ such that if a quadruple (A, B, R, a) belongs to i, then $(a, b_i) \in R$. The element b_i is the ith coordinate of a representative of an element $b \in B^* = B^I/\mathcal{D}$. If $a_0 \in A$, then $i_0 = \{(A, B, R, a_0)\}$ is an element of I. This element belongs to each i in the cone D_{i_0} over i_0, so $(a_0, b_i) \in R$. By the definition of $\mathcal{D}$ we have $\{i \in I \mid (a_0, b_i) \in R\} \in \mathcal{D}$, so $(a_0, b) \in R^*$. Therefore, Condition III holds.

Consequently, M^* is an enlargement of the higher order structure of M.

14.5 Examples

We give examples of enlargements of three well known structures.

Example 14.5.1: Starfinite summation. Let P be a set and K a field. Denote the set of all functions from subsets I belonging to $\mathcal{P}_{\text{fin}}(P)$ to K by A. For each $a \in A$ write the image of a at $i \in I$ by a_i. Then consider the sum $\sum_{i \in I} a_i$. Regard the summation symbol as a function $\sum \colon A \to K$. It satisfies well known rules like $\sum_{i \in I} a_i + \sum_{i \in I} b_i = \sum_{i \in I} (a_i + b_i)$.

Consider an enlargement of a higher order structure that includes both P and K. Call the elements of $\mathcal{P}_{\text{fin}}^*(P)$ the **starfinite subsets** of P^*. They are, in particular, internal subsets of P^* and they share many properties of finite sets. For example, an internal function f of a starfinite set I onto itself is necessarily injective. The family $\mathcal{P}_{\text{fin}}^*(P)$ is closed under Boolean operations and it contains all finite subsets of P^*. The elements of A^* are the internal functions from starfinite sets into K^*. The summation function extends to a function $\sum$ from A^* into K^* that has much the same properties the original summation has. For example, if I, J are disjoint starfinite sets and $a \colon I \cup J \to K^*$ is an internal function, then $\sum_{i \in I} a_i + \sum_{j \in J} a_j = \sum_{i \in I \cup J} a_i$. We may define starfinite products similarly. $\square$

Example 14.5.2: Infinitesimal elements. Consider an enlargement $\mathbb{R}^*$ of the field $\mathbb{R}$ of real numbers. It is an ordered field and the subring

$$\mathbb{R}_{\text{fin}} = \{x \in \mathbb{R}^* \mid |x| < r \text{ for some } r \in \mathbb{R}\}$$

is **convex**: If $x, y \in \mathbb{R}_{\text{fin}}$ and $z \in \mathbb{R}^*$ satisfy $x < z < y$, then $z \in \mathbb{R}_{\text{fin}}$. The subset

$$\mathbb{R}_0 = \{x \in \mathbb{R}^* \mid |x| < r \text{ for all positive } r \in \mathbb{R}\}$$

is an ideal of $\mathbb{R}_{\text{fin}}$, consisting of the **infinitesimal** elements. Each $\alpha \in \mathbb{R}_{\text{fin}}$ corresponds to the real number $\bar{\alpha} = \sup\{x \in \mathbb{R} \mid x \leq \alpha\}$. Then the map $\alpha \mapsto \bar{\alpha}$ induces an isomorphism of the residue field $\mathbb{R}_{\text{fin}}/\mathbb{R}_0$ onto $\mathbb{R}$.

Consider the (additive) quotient group $\dot{\mathbb{R}} = \mathbb{R}^*/\mathbb{R}_{\text{fin}}$. Denote the image of $\alpha \in \mathbb{R}^*$ under the quotient map $\mathbb{R}^* \to \mathbb{R}^*/\mathbb{R}^{\text{fin}}$ by $\dot{\alpha}$. For $\alpha, \beta \in \mathbb{R}^*$ consider the relation
$$\dot{\alpha} \leq \dot{\beta} \iff (\exists r \in \mathbb{R})[\alpha \leq \beta + r].$$
Then $\dot{\alpha} < \dot{\beta}$ means that $\alpha < \beta + r$ for each $r \in \mathbb{R}$.

Occasionally we speak about the **order of magnitude** of the elements of $\mathbb{R}^*$ and we write $\alpha \dot{=} \beta$, $\alpha \dot{\leq} \beta$ and $\alpha \dot{<} \beta$ instead of $\dot{\alpha} = \dot{\beta}$, $\dot{\alpha} \leq \dot{\beta}$ and $\dot{\alpha} < \dot{\beta}$, respectively. $\square$

Example 14.5.3: *Nonstandard field theory.* Let K^* be a nonstandard extension of a field K. Then K^* is an elementary extension of K. By Example 7.3.3, K^* is a regular extension of K.

Suppose K is PAC and V is a variety over K. Then the set $V(K)$ contains points in the complement of U, for each proper K-closed subset U of V (Proposition 11.1.1). By the compactness condition, there exists $\mathbf{x} \in V(K^*)$ that belongs to no proper K-closed subset of V. That is, $\mathbf{x}$ is a generic point of V over K. Consequently, every finitely generated regular extension of K can be embedded in K^*. $\square$

Exercises

1. Replace M in Section 14.3 by $\mathbb{N} \cup M$. Show that the statement "there exists $n \in \mathbb{N}$ and there exist elements $a_1, \ldots, a_n \in A$" is equivalent to a formula in the language $\mathcal{L}_\infty(\mathbb{N} \cup M)$. For G a group express the statement "G is solvable" as a sentence of $\mathcal{L}_\infty(\mathbb{N} \cup G)$.

2. Use the discussion preceding Condition II (Section 14.2) to prove that if $R, S \subseteq M_{\tau(1)} \times \cdots \times M_{\tau(n)}$, then $(R \cup S)^* = R^* \cup S^*$, $(R \cap S)^* = R^* \cap S^*$, $(R \smallsetminus S)^* = R^* \smallsetminus S^*$, and $(R \times S)^* = R^* \times S^*$. Show that $R \subseteq S$ implies $R^* \subseteq S^*$ and that if R is empty, then R^* is also empty.

3. Extension rule for internal sets: Suppose $\boldsymbol{\tau} = (\tau(1), \ldots, \tau(n))$ and $R, S \in M^*_{\boldsymbol{\tau}}$. Prove that $R = S$ if and only if for each $a_i \in M^*_{\tau(i)}$, $i = 1, \ldots, n$, the n-tuple $(a_1, \ldots, a_n)$ is in R if and only if it is in S. That is, internal sets are determined by their internal elements.

4. For $\boldsymbol{\tau} = (\tau(1), \ldots, \tau(n))$ and $R, S \in M^*_{\boldsymbol{\tau}}$, show that $R \cup S$, $R \cap S$ and $R \smallsetminus S$ are in $M^*_{\boldsymbol{\tau}}$ and $R \times S \in M^*_{(\tau(1),\ldots,\tau(n),\tau(1),\ldots,\tau(n))}$. That is, Boolean combinations of internal sets are internal.

5. Let $\tau(1), \ldots, \tau(m), \tau(m+1), \ldots, \tau(n)$ be types. Let $\varphi(X_1, \ldots, X_n)$ be a formula of $\mathcal{L}_\infty(M)$ with $X_\nu = X_{M_{\tau(\nu)}, i(\nu)}$, $\nu = 1, \ldots, n$, and $a_j \in M^*_{\tau(j)}$, $j = m+1, \ldots, n$. Prove that the set
$$\{(a_1, \ldots, a_m) \in M^*_{\tau(1)} \times \cdots \times M^*_{\tau(m)} \mid M^* \models \varphi^*(a_1, \ldots, a_m, \ldots, a_n)\}$$
is internal (i.e. it is in $M^*_{(\tau(1),\ldots,\tau(n))}$).

6. Show, in Example 14.5.1, that $\mathcal{P}^*_{\text{fin}}(P)$ contains all finite subsets of P^*.

7. Show for an internal subset A of $\mathbb{N}^*$ that A is starfinite (Example 14.5.1) if and only if A is bounded.

8. Prove (in Example 14.5.2) that $\mathbb{R}_{\text{fin}}$, $\mathbb{R}_0$, and $\dot{\mathbb{R}}$ are external objects of $\mathbb{R}^*$. Hint: Regard $\dot{\mathbb{R}}$ as a collection of subsets of $\mathbb{R}^*$.

9. As in example 14.5.3, prove for any field K and any irreducible $f(X, Y) \in K[X, Y]$ that the function field of the curve $f(X, Y) = 0$ embeds in K^* if and only if the curve has infinitely many K-rational points.

10. Let $\{a_n \mid n \in \mathbb{N}^*\}$ be an internal subset of $\mathbb{R}^*$ with a_n infinitesimal for each $n \in \mathbb{N}$. Prove that there exists $t \in \mathbb{N}^* \smallsetminus \mathbb{N}$ such that a_n is infinitesimal for each $n < t$. Hint: Consider the set $\{t \in \mathbb{N}^* \mid (\forall n \in \mathbb{N}^*)[n < t \Longrightarrow |ta_n| < 1]\}$.

11. (a) Prove that every internal ideal of $\mathbb{Z}^*$ is principal.

    (b) Prove that every finitely generated ideal of $\mathbb{Z}^*$ is principal (that is, $\mathbb{Z}^*$ is a **Bezout ring**).

12. Follow these steps to show that $\mathbb{Z}^*$ is not a principal ideal domain.

    (a) Choose $t \in \mathbb{N}^* \smallsetminus \mathbb{N}$. For each $n \in \mathbb{N}$ define $x_n \in \mathbb{Z}^*$ by a starfinite product (Example 14.5.1), $x_n = \prod_{n \leq x < t} x$.

    (b) Prove that $x_2 \mathbb{Z}^* < x_3 \mathbb{Z}^* < x_4 \mathbb{Z}^* < \cdots$ is a strictly ascending sequence of ideals of $\mathbb{Z}^*$ (thus $\mathbb{Z}^*$ is not a Noetherian domain) and that the ideal generated by all x_i's is not principal.

13. Let P be the set of rational primes. Each $p \in P^* \smallsetminus P$ is an **infinite prime** of $\mathbb{Z}^*$. Show that for each infinite prime p the quotient ring $\mathbb{Z}^*/p\mathbb{Z}^*$ is a PAC field. Hint: For each $d \in \mathbb{N}$ the set of all $p \in P$ such that $f(X, Y) = 0$ has no solution in $\mathbb{Z}/p\mathbb{Z}$ for some absolutely irreducible polynomial f of degree d with coefficients in $\mathbb{Z}/p\mathbb{Z}$ is finite.

14. Follow these steps to construct an infinite descending sequence of prime ideals of $\mathbb{Z}^*$.

    (a) For a standard prime p extend the p-adic valuation to a valuation $v_p \colon \mathbb{Z}^* \to \mathbb{N}^* \cup \{0, \infty\}$. Find an element $z \in \mathbb{Z}^*$ such that $\zeta = v_p(z) > n$ for all $n \in \mathbb{N}$.

    (b) Prove for each $0 \leq k \in \mathbb{Z}$ that $I_k = \{x \in \mathbb{Z}^* \mid v_p(x) > n\zeta^k \text{ for all } n \in \mathbb{N}\}$ is a prime ideal of $\mathbb{Z}^*$.

    (c) Finally, observe that $p\mathbb{Z}^* > I_0 > I_1 > I_2 > \cdots$ is a strictly descending sequence.

15. (a) Prove that $\mathbb{N}^* \smallsetminus \mathbb{N}$ is not an internal subset of $\mathbb{N}^*$.

    (b) Prove that for each $a \in \mathbb{N}$ the set $\{x \in \mathbb{N}^* \mid x > a\}$ is internal.

Notes

This chapter is in the spirit of Chapter 2 of the [Robinson-Roquette], which contains a nonstandard proof of the Siegel-Mahler theorem.

Chapter 15.
Nonstandard Approach to Hilbert's Irreducibility Theorem

We use the nonstandard methods of Chapter 14 to give a new criterion for a field K to be Hilbertian: There exists a nonstandard element t of an enlargement K^* of K such that t has only finitely many poles in $K(t)_s \cap K^*$. From this there results a second and uniform proof (Theorem 15.3.4) that the classical Hilbertian fields are indeed Hilbertian. In addition, a formal power series field, $K_0((X_1, \ldots, X_n))$ of $n \geq 2$ variables over an arbitrary field K_0, is also Hilbertian (Example 15.5.2).

15.1 Criteria for Hilbertianity

We give two criteria for a field K to be Hilbertian in terms of an enlargement K^* of K. The first is a straightforward application of the compactness property of K.

PROPOSITION 15.1.1 (Gilmore-Robinson): *The field K is Hilbertian if and only if there exists $t \in K^* \smallsetminus K$ such that $K(t)_s \cap K^* = K(t)$.*

Proof: Suppose K is Hilbertian. If $f_1, \ldots, f_m \in K[T,X]$ irreducible polynomials which are separable in X and $g_1, \ldots, g_m \in K[T]$ are nonzero polynomials, then there exists $a \in K$ with $f_i(a, X)$ irreducible $K[X]$ and $g_i(a) \neq 0$, $i = 1, \ldots, m$. The compactness property (Condition III of Section 14.3) gives $t \in K^*$ such that for each irreducible $f \in K[T, X]$ which is separable in X and each $0 \neq g \in K[T]$, the polynomial $f(t, X)$ is irreducible in $K^*[X]$ and $g(t) \neq 0$. The second condition implies $t \notin K$. Consider $x \in K(t)_s \cap K^*$. Let $f \in K[T, X]$ be an irreducible polynomial which is separable in X with $f(t, x) = 0$. Since $f(t, X)$ is irreducible over K^*, it is linear. Hence, $x \in K(t)$.

Conversely, suppose t is an element of $K^* \smallsetminus K$ with $K(t)_s \cap K^* = K$. Let $f_1, \ldots, f_m \in K[T, X]$ be irreducible polynomials which are separable, monic, and of degree at least 2 in X and let $0 \neq g \in K[T]$. If for some i between 1 and m, $f_i(t, X)$ is reducible over K^*, the coefficients of its factors lie in $K(t)_s \cap K^* = K(t)$. Thus, $f_i(t, X)$ is reducible over $K(t)$. But since t is transcendental over K, this means $f_i(T, X)$ is reducible over K, contrary to our assumption. Finally, since K^* is an elementary extension of K, there exists $a \in K$ such that $g(a) \neq 0$ and $f_i(a, X)$ is irreducible over K, $i = 1, \ldots, m$. By Lemma 12.1.4, K is Hilbertian. $\square$

One may reapproach the classical Hilbertian fields through Proposition 15.1.1 [Roquette2], but it is easier to apply the following weaker condition of Weissauer.

Consider $t \in K^* \smallsetminus K$ and let $\Omega_t = K(t)_s \cap K^*$. Then Ω_t is a separable algebraic (possibly infinite) extension of $K(t)$. Since Ω_t is regular over K

(Example 14.5.3) consider it as a generalized function field of one variable over K. It is a union of function fields of one variable over K. Call an equivalence class of valuations of Ω_t which are trivial on K a **prime divisor** of Ω_t/K. Refer to a prime divisor as a **pole** of t if t has a negative value. Finally, call t **polefinite** if t has only finitely many poles in Ω_t. In other words, there is an integer m such that the number of poles of t in any function field F with $K(t) \subseteq F \subseteq \Omega_t$ is at most m.

PROPOSITION 15.1.2 ([Weissauer, Folgerung 3.2]): *If K^* contains a polefinite element, then K is Hilbertian.*

Proof: The existence of a polefinite element implies $K \neq K^*$. Hence, K is infinite (discussion preceding Corollary 14.3.2). Assume K is not Hilbertian. Then Proposition 13.2.2 gives an irreducible polynomial $f \in K[T, X]$, separable, monic, and of degree at least 2 in X, such that $f(a, X)$ is reducible over K for each $a \in K$. The same statement holds in K^*. In particular, for each $t \in K^* \smallsetminus K$,

(1) $f(t, X)$ is irreducible over $K(t)$, but reducible over K^*.

Consider now an element t of $K^* \smallsetminus K$. The remaining parts of the proof consider properties of the prime divisors of $K(t)$.

PART A: *Removing ramification over ∞.* Let F be the splitting field of $f(t, X)$ over $K(t)$. Suppose $\deg_t(f) = d$. For each $a \in \tilde{K} \cup \{\infty\}$ let $\mathfrak{p}_a$ be the prime divisor of $K(t)/K$ defined by $t \mapsto a$. In particular, $\mathfrak{p}_\infty$ is the unique pole of t in $K(t)/K$. Since K is infinite, we may choose $a \in K$ with $\mathfrak{p}_a$ unramified in F and with residue field $\bar{F}$ Galois over K. Apply the K-automorphism of $K(t)$ defined by $t \mapsto \frac{1}{t-a}$ to replace $f(T, X)$ by $T^d g(T, X) = f(\frac{1}{T-a}, X)$, the field F by the splitting field of $g(t, X)$, and $\mathfrak{p}_a$ by $\mathfrak{p}_\infty$ (Section 2.2). Thus assume, along with (1), that

(2) $\mathfrak{p}_\infty$ is unramified in F.

Each $a \in K$ defines a K-automorphism σ_a of $K(t)$ by $t \mapsto t + a$. It fixes $\mathfrak{p}_\infty$. Extend σ_a to an automorphism of $K(t)_s$. Denote $\sigma_a(F)$ by F_a. Thus, $\mathfrak{p}_\infty$ is unramified in F_a and $\bar{F}_a$ is Galois over K. By Corollary 2.3.7(c),

(3) $\mathfrak{p}_\infty$ is unramified in the compositum, F', of F_a for all $a \in K$.

PART B: *The finiteness of F' over ∞.* Let $\mathfrak{P}$ be a prime divisor of F' over $\mathfrak{p}_\infty$. Denote reduction modulo $\mathfrak{P}$ by a bar. The residue fields $\bar{F}$ and $\bar{F}_a$ are conjugate over K (Section 2.3). Since both are Galois extensions of K (Lemma 6.1.1), they coincide. The compositum of the residue class fields of unramified extensions of $K(t)$ is the residue class field of their compositum (Lemma 2.4.8). Thus, $\bar{F}' = \bar{F}$ is a finite extension of K which is independent of $\mathfrak{P}$.

PART C: *The infinitude of $[F' \cap K^* : K(t)]$.* By (1), F_a is not linearly disjoint from K^* over $K(t)$. Hence, $E_a = F_a \cap K^*$, a regular extension of K contained in Ω_t, is a proper extension of $K(t)$. Let $\mathcal{R}(E_a)$ (resp. $\mathcal{R}(F_a)$)

be the set of prime divisors of $K(t)/K$ that ramify in E_a (resp. in F_a). By Remark 3.6.2(b), $\mathcal{R}(E_a)$ is finite and nonempty.

Note: $\mathfrak{p}_c = \mathfrak{p}_{c'}$ if and only if c' is conjugate to c over K. If $\mathcal{R}(F) = \{\mathfrak{p}_{c_1}, \ldots, \mathfrak{p}_{c_n}\}$, then $\mathcal{R}(F_a) = \{\mathfrak{p}_{c_1-a}, \ldots, \mathfrak{p}_{c_n-a}\}$. For L a finite separable extension of $K(t)$ let $\mathcal{R}(L)$ be $\{\mathfrak{p}_{d_1}, \ldots, \mathfrak{p}_{d_r}\}$. Since K is an infinite field, we may choose $a \in K$ with $c_i - a \neq \tau(d_j)$ for all i and j and for every $K(t)$-isomorphism τ of L. Hence $\mathcal{R}(E_a) \cap \mathcal{R}(L) \subseteq \mathcal{R}(F_a) \cap \mathcal{R}(L) = \emptyset$. Since $\mathcal{R}(E_a) \neq \emptyset$, the field E_a is not contained in L. Therefore, the compositum E' of E_a for all $a \in K$ is an infinite extension of $K(t)$ contained in Ω_t.

PART D: *Conclusion of the proof.* Assume t has only m poles in Ω_t. Use Part C to choose a finite extension N of $K(t)$ in E' with $[N : K(t)] > m[\bar{F} : K]$. Let $\mathfrak{q}_1, \ldots, \mathfrak{q}_k$ be all extensions of $\mathfrak{p}_\infty$ to N. Each of them extends to a pole of t in Ω_t, so $k \leq m$. By Part B, $\bar{N}_{\mathfrak{q}_i} \subseteq \bar{F}$. Hence, by Proposition 2.3.2,

$$[N : K(t)] = \sum_{i=1}^{k}[\bar{N}_{\mathfrak{q}_i} : K] \leq m[\bar{F} : K].$$

This contradiction to the choice of N prove that t is not a polefinite element. □

15.2 Arithmetical Primes Versus Functional Primes

To apply Proposition 15.1.2 we start with a field K that carries arithmetic structure and extend this structure to an enlargement K^* of K. For each $t \in K^* \smallsetminus K$ we consider finite extensions F of $K(t)$ in K^*. Then we compare the function field structure of F/K with the arithmetic structure induced on F from K^*. The goal is to find conditions on t to be a polefinite element.

Let S is a set of primes of K. Thus, S is a set of equivalent classes of absolute values of K (Section 13.3). For each $\mathfrak{p} \in S$ choose a representative $|\ |_\mathfrak{p}$. Define a map $v_\mathfrak{p}: K \to \mathbb{R} \cup \{\infty\}$ by $v_\mathfrak{p}(a) = -\log(|a|_\mathfrak{p})$. Conditions (3a)–(3d) of Section 13.3 on the valuation translate into the following properties of $v_\mathfrak{p}$:
(1a) $v_\mathfrak{p}(a) = \infty$ if and only if $a = 0$.
(1b) $v_\mathfrak{p}(ab) = v_\mathfrak{p}(a) + v_\mathfrak{p}(b)$.
(1c) $v_\mathfrak{p}(a+b) \geq \min(v_\mathfrak{p}(a), v_\mathfrak{p}(b)) - \log 2$.
(1d) There is an $a \in K^\times$ with $v_\mathfrak{p}(a) \neq 0$.

We call $v_\mathfrak{p}$ an **additive absolute value** of K. If $\mathfrak{p}$ is ultrametric, then $|a+b|_\mathfrak{p} \leq \max(|a|_\mathfrak{p}, |b|_\mathfrak{p})$, hence $v_\mathfrak{p}(a+b) \geq \min(v_\mathfrak{p}(a), v_\mathfrak{p}(b))$ for all $a, b \in K$. Thus, $v_\mathfrak{p}$ is a valuation of K. The following lemma gives a simple criterion for $\mathfrak{p}$ to be metric:

LEMMA 15.2.1: *A prime $\mathfrak{p}$ of a field K is metric if and only if $v_\mathfrak{p}(2) < 0$.*

Proof: Suppose first that $\mathfrak{p}$ is metric. By Section 13.3, there is a positive integer n with $|n|_\mathfrak{p} > 1$. Hence, $\text{char}(K) = 0$ and the restriction of $|\ |_\mathfrak{p}$ to $\mathbb{Q}$ is a metric absolute value. By a theorem of Ostrowski, $|\ |_\mathfrak{p}$ is equivalent to the ordinary absolute value of $\mathbb{Q}$. Thus, there is a positive real number c such that $|x|_\mathfrak{p} = |x|^c$ for each $x \in \mathbb{Q}$. In particular, $|2|_\mathfrak{p} = |2|^c = |n^{\log 2/\log n}|^c > 1$. Consequently, $v_\mathfrak{p}(2) < 0$.

Conversely, if $v_\mathfrak{p}(2) < 0$, then $|2|_\mathfrak{p} > 1$, so $\mathfrak{p}$ is metric. $\square$

We assume now that S satisfies the following finiteness condition:
(2) If $a \in K^\times$, then $\{\mathfrak{p} \in S \mid v_\mathfrak{p}(a) \neq 0\}$ is a finite set.
By Lemma 15.2.1, there are only finitely many archimedean primes in S.

Consider again an enlargement of K^* of K. Then S extends to a set S^* of **arithmetical primes** of K^*. To each $\mathfrak{p} \in S^*$ there corresponds a **star-additive absolute value** $v_\mathfrak{p}: (K^*)^\times \to \mathbb{R}^*$.

In particular, each prime in S extends to an element $\mathfrak{p}$ of S^*, a **standard prime**. The elements of $S^* \smallsetminus S$ are the nonstandard primes. Condition (2) becomes:
(2)* If $a \neq a \in K^*$, then $\{\mathfrak{p} \in S^* \mid v_\mathfrak{p}(a) \neq 0\}$ is starfinite (Example 14.5.1).
In particular, if $0 \neq a \in K$, then the finite set $\{\mathfrak{p} \in S \mid v_\mathfrak{p}(a) \neq 0\}$ does not grow in the enlargement Thus, $v_\mathfrak{p}(a) = 0$ for all $\mathfrak{p} \in S^* \smallsetminus S$ and $a \in K^\times$. Therefore, by (2), there are only finitely many archimedean primes in S^*; they are all standard.

For an arbitrary prime $\mathfrak{p} \in S^*$ consider the **convex hull** of $v_\mathfrak{p}(K^\times)$:

$$\Gamma_\mathfrak{p} = \{r \in \mathbb{R}^* \mid \exists x \in K^\times \text{ with } |r| \leq v_\mathfrak{p}(x)\}.$$

In the notation of Example 14.5.2,
(3) $\Gamma_\mathfrak{p} = \mathbb{R}_\mathrm{fin}$ if $\mathfrak{p} \in S$ and $\Gamma_\mathfrak{p} = 0$ if $\mathfrak{p} \in S^* \smallsetminus S$.

Unlike nonstandard primes, standard primes form valuations that are nontrivial on K. We modify them so that they will "behave" like the nonstandard primes. However, these "modified" primes will not be internal objects. Combining internal and external objects is the key to the nonstandard machinery.

For $\mathfrak{p} \in S^*$, consider the ordered group $\dot{\mathbb{R}}_\mathfrak{p} = \mathbb{R}^*/\Gamma_\mathfrak{p}$. By (3), $\dot{\mathbb{R}}_\mathfrak{p} = \dot{\mathbb{R}}$ (Example 14.5.2) if $\mathfrak{p}$ is standard and $\dot{\mathbb{R}}_\mathfrak{p} = \mathbb{R}^*$ otherwise. Define the **modified valuation** $\dot{v}_\mathfrak{p}: (K^*)^\times \to \dot{\mathbb{R}}_\mathfrak{p}$ by $\dot{v}_\mathfrak{p}(x) = v_\mathfrak{p}(x) + \Gamma_\mathfrak{p}$.

Suppose $\mathfrak{p}$ is standard. Then $\mathfrak{p}$ may be metric or ultrametric. In the second case $v_\mathfrak{p}$ is a valuation, whereas in the first case it is not. In both cases, however, $v_\mathfrak{p}$ satisfies (3). Since $\log 2 \in \mathbb{R} \subseteq \Gamma_\mathfrak{p}$, $\dot{v}_\mathfrak{p}(a+b) \geq \min\{\dot{v}_\mathfrak{p}(a), \dot{v}_\mathfrak{p}(b)\}$. Hence, $\dot{v}_\mathfrak{p}$ is a valuation. In addition, if $0 \neq x \in K$, then $v_\mathfrak{p}(x) \in \mathbb{R}$. Hence, $\dot{v}_\mathfrak{p}(x) = 0$. In summary:

LEMMA 15.2.2: *For each $\mathfrak{p} \in S^*$ the function $\dot{v}_\mathfrak{p}: (K^*)^\times \to \dot{\mathbb{R}}_\mathfrak{p}$ defined by $\dot{v}_\mathfrak{p}(x) = v_\mathfrak{p}(x) + \Gamma_\mathfrak{p}$ is a valuation which is trivial on K. Moreover, if $\mathfrak{p} \in S^* \smallsetminus S$, then $\dot{v}_\mathfrak{p} = v_\mathfrak{p}$.*

Consider an element $t \in K^* \smallsetminus K$. Let F be a finite extension of $K(t)$ in K^*. By Example 14.5.3, F is a regular extension of K, hence a function field over K. Consider $\mathfrak{p} \in S^*$ with $\dot{v}_\mathfrak{p}$ not vanishing on $F^\times$. Then, by Lemma 15.2.2, the restriction of $\dot{v}_\mathfrak{p}$ to F is a valuation which is trivial valuation on K. Thus, it defines a prime P of F/K. We say P is **induced** from $\mathfrak{p}$.

Conversely, we may ask if a prime divisor P of F/K (i.e. a **functional prime**) if it is induced by an arithmetical prime. This question is inspired by the following observation.

PROPOSITION 15.2.3: *Let S be a set of primes of a field K with $\{\mathfrak{p} \in S \mid v_\mathfrak{p}(a) \neq 0\}$ being a finite set for each $a \in K^\times$. Let t be a nonstandard element of an enlargement K^* of K. Suppose $S(t) = \{\mathfrak{p} \in S^* \mid \dot{v}_\mathfrak{p}(t) < 0\}$ is a finite set and for each finite separable extension F of $K(t)$ in K^* all poles of t in F are induced by arithmetical primes (i.e. elements of S^*). Then t is a polefinite element.*

Proof: Let P be a pole of t in Ω_t. For each finite extension F of K in Ω_t the restriction of P to F is also a pole of t. Hence, there is a $\mathfrak{p}_F \in S^*$ which induces $P|_F$. In particular, $\dot{v}_{\mathfrak{p}_F}(t) < 0$, so $\mathfrak{p}_F \in S(t)$. Since $S(t)$ is finite, there is a $\mathfrak{p} \in S(t)$ which induces $P|_F$ for each F as above (Use compactness of $S(t)$.) Therefore, $\mathfrak{p}$ determines P.

Thus, the number of poles of t in Ω_t is at most $|S(t)|$. Consequently, t is a polefinite element. $\square$

15.3 Fields with the Product Formula

As in Section 15.2, suppose S is a nonempty set of primes of K. For each $\mathfrak{p} \in S$ choose an absolute value $|\ |_\mathfrak{p}$ representing $\mathfrak{p}$ and let $v_\mathfrak{p}$ be the corresponding additive absolute value. We say S satisfies a **product formula** if the following statement holds: for each $\mathfrak{p} \in S$ there exists a positive real number $\lambda_\mathfrak{p}$ with the following property:

(1) For each $a \in K^\times$ the set $\{\mathfrak{p} \in S \mid |a|_\mathfrak{p} \neq 1\}$ is finite and $\prod_{\mathfrak{p} \in S} |a|_\mathfrak{p}^{\lambda_\mathfrak{p}} = 1$.

In this case call K a **field with a product formula**.

Example 15.3.1: *Basic examples of product formulas.* For $K = \mathbb{Q}$ and p a prime number, define $|\ |_p$ by $|a|_p = p^{-r}$ for $a = \frac{x}{y} \cdot p^r$ with $r \in \mathbb{Z}$ and $x, y \in \mathbb{Z}$ relatively prime to p. The **infinite absolute value** $|\ |_\infty$ is the usual absolute value. This set S of primes satisfies the product formula with $\lambda_\mathfrak{p} = 1$ for each $\mathfrak{p} \in S$.

For $K_0(t)$, a rational function field over an arbitrary field K_0, choose a real number $0 < c < 1$. For each irreducible polynomial $p \in K_0[t]$ define the absolute value $|\ |_p$ by $|u|_p = c^{-r}$ for $u = \frac{f}{g} \cdot p^r$ with f, g polynomials relatively prime to p. Let λ_p be $\deg(p)$. For a quotient f/g of polynomials, define the infinite absolute value as $|f/g|_\infty = c^{\deg(f) - \deg(g)}$, and let $\lambda_\infty = 1$. Check that the corresponding set of primes of $K_0(t)$ satisfies the product formula. $\square$

If a set S of primes of a field K is a field with a the product formula and if S' is the set of primes of a finite extension K'/K lying over S, then

S' satisfies the product formula [Lang3, p. 20]. We conclude from Example 15.3.1 that every global field is a field with a product formula.

Again let K be any field with the product formula with respect to a nonempty set of primes S. Rewrite condition (1) additively:

(2) For each $a \in K^\times$, $\{\mathfrak{p} \in S \mid v_\mathfrak{p}(a) \neq 0\}$ is finite and $\sum_{\mathfrak{p} \in S} \lambda_\mathfrak{p} v_\mathfrak{p}(a) = 0$.

For an enlargement K^* of K, Condition (2) has a similar form:

(2)* For each $a \in (K^*)^\times$, the set $\{\mathfrak{p} \in S^* \mid v_\mathfrak{p}(a) \neq 0\}$ is starfinite and $\sum_{\mathfrak{p} \in S^*} \lambda_\mathfrak{p} v_\mathfrak{p}(a) \doteq 0$, where $\lambda_\mathfrak{p} \in \mathbb{R}^*$ and $\lambda_\mathfrak{p} > 0$ are independent of $\mathfrak{p}$ and $\sum$ here is the starfinite summation (Example 14.5.1).

PROPOSITION 15.3.2: *Let t be a nonstandard element of K^* and F a finite separable extension of $K(t)$ in K^*. If there exists $\mathfrak{q} \in S^*$ with $\dot{v}_\mathfrak{q}(t) < 0$, then each functional prime of F is induced by an arithmetical prime of K^*.*

Proof: Denote the set of all functional primes of F which are induced by elements of S^* by $\mathcal{P}$. Let

$$D = \{x \in F \mid v_P(x) \geq 0 \text{ for all } P \in \mathcal{P}\} = \{x \in F \mid \dot{v}_\mathfrak{p}(x) \geq 0 \text{ for all } \mathfrak{p} \in S^*\}$$

be the holomorphy ring of $\mathcal{P}$. The assumption $\dot{v}_\mathfrak{q}(t) < 0$ implies that $t \notin D$ and $\mathcal{P} \neq \emptyset$, so $D \neq F$. Assume $\mathcal{P}$ does not contain all the functional primes of F. Then $F = \mathrm{Quot}(D)$ (Proposition 3.3.2(a)). We now show that D is a field. This will give a contradiction to $D \neq F$ and establish the theorem.

We need to prove that if $x \in D^\times$ and $x \neq 0$, then $x^{-1} \in D$. Indeed, by (2)* the set $S(x) = \{\mathfrak{p} \in S^* \mid v_\mathfrak{p}(x) < 0\}$ is starfinite. For $\mathfrak{p} \in S^* \smallsetminus S$, Lemma 15.2.2 shows that $v_\mathfrak{p}(x) = \dot{v}_\mathfrak{p}(x) \geq 0$. This means $S(x)$ contains only standard primes. Hence, by Corollary 14.3.4(a), $S(x)$ is a finite set. For $\mathfrak{p} \in S(x)$ we have $v_\mathfrak{p}(x) < 0$ and $\dot{v}_\mathfrak{p}(x) \geq 0$, so $\dot{v}_\mathfrak{p}(x) = 0$. Now use the summation formula (2)* and the additivity of starfinite summation (Example 14.5.1):

(3) $$\sum_{\mathfrak{p} \in S^* \smallsetminus S(x)} \lambda_\mathfrak{p} v_\mathfrak{p}(x) \doteq \sum_{\mathfrak{p} \in S^*} \lambda_\mathfrak{p} v_\mathfrak{p}(x) \doteq 0.$$

If $\mathfrak{p}' \in S^* \smallsetminus S(x)$ and $\dot{v}_{\mathfrak{p}'}(x) \neq 0$, then $\dot{v}_{\mathfrak{p}'}(x) > 0$. This means

$$\sum_{\mathfrak{p} \in S^* - S(x)} \lambda_\mathfrak{p} v_\mathfrak{p}(x) \geq \lambda_{\mathfrak{p}'} v_{\mathfrak{p}'}(x) \dot{>} 0,$$

a contradiction to (3). Thus, $\dot{v}_\mathfrak{p}(x) = 0$ for each $\mathfrak{p} \in S^* \smallsetminus S(x)$ and therefore for each $\mathfrak{p} \in S^*$. It follows that $\dot{v}_\mathfrak{p}(x^{-1}) = 0$ for each $\mathfrak{p} \in S^*$. Consequently, $x^{-1} \in D$, as desired. □

THEOREM 15.3.3 ([Weissauer, Satz 6.2]): *Every field K with the product formula is Hilbertian.*

Proof: Let S be a nonempty set of primes of a field K satisfying the product formula. Choose $\mathfrak{q} \in S$ and $a \in K$ with $v_\mathfrak{q}(a) < 0$. In an enlargement

15.4 Generalized Krull Domains

K^* of K, pick a nonstandard positive integer $\omega \in \mathbb{N}^* \smallsetminus \mathbb{N}$. We prove the nonstandard element $t = a^\omega$ of K^* is polefinite. Indeed,

$$S(t) = \{\mathfrak{p} \in S^* \mid v_\mathfrak{p}(t) < 0\} = \{\mathfrak{p} \in S^* \mid v_\mathfrak{p}(a) < 0\} = \{\mathfrak{p} \in S^* \mid \hat{v}_\mathfrak{p}(t) < 0\}.$$

Therefore, $S(t)$ is a starfinite set that contains $\mathfrak{q}$. In particular, $\hat{v}_\mathfrak{q}(t) < 0$. By Lemma 15.1.2, if $\mathfrak{p} \in S^* \smallsetminus S$, then $\hat{v}_\mathfrak{p}(a) = 0$. Thus, $S(t)$ consists only of standard primes, so, by Corollary 14.3.4(b), $S(t)$ is a finite set.

Suppose F is a finite separable extension of $K(t)$ in K^*. By Propositions 15.3.2, each functional prime P of $F/K(t)$ is induced by some $\mathfrak{p} \in S^*$. If P is a pole of t, then $\mathfrak{p} \in S(t)$. By Proposition 15.2.3, t is polefinite. Consequently, by Proposition 15.1.2, K is Hilbertian. □

COROLLARY 15.3.4: *Every number field and every function field of several variables over an arbitrary field is Hilbertian.*

15.4 Generalized Krull Domains

Non-standard methods produce new Hilbertian fields: The quotient field of each 'generalized Krull domain of dimension at least 2' is Hilbertian (Theorem 15.4.6). In particular, fields of formal power series of at least two variables over arbitrary fields are Hilbertian (Theorem 15.4.6).

Let R be an integral domain with quotient field K. We call R a **generalized Krull domain** if K has a nonempty set S of primes satisfying the following conditions:
(1a) For each $\mathfrak{p} \in S$, $v_\mathfrak{p}$ is a real valuation.
(1b) The valuation ring $O_\mathfrak{p}$ of $v_\mathfrak{p}$ is the local ring of R relative to $\mathfrak{m}_\mathfrak{p} = \{a \in R \mid v_\mathfrak{p}(a) > 0\}$.
(1c) $R = \bigcap_{\mathfrak{p} \in S} O_\mathfrak{p}$.
(1d) For each $a \in K^\times$ the set $\{\mathfrak{p} \in S \mid v_\mathfrak{p}(a) \neq 0\}$ is finite.

The **dimension** of a ring R is the maximal integer n for which there is a sequence $\mathfrak{p}_0 \subset \mathfrak{p}_1 \subset \cdots \subset \mathfrak{p}_n$ of $n+1$ distinct prime ideals. Thus, $\dim(R) \geq 2$ if and only if
(1e) R has a maximal ideal M which properly contains a nonzero prime ideal.

Thus, R is a generalized Krull ring of dimension exceeding 1 if and only if it satisfies Condition (1).

LEMMA 15.4.1: *Let R be an integral domain satisfying Condition (1). Then:*
(a) *For each nonunit b of R, $b \neq 0$, there exists $\mathfrak{p} \in S$ with $v_\mathfrak{p}(b) > 0$.*
(b) *For each $\mathfrak{p} \in S$, $\mathfrak{m}_\mathfrak{p}$ is minimal among nonzero prime ideals of R.*
(c) *If $\mathfrak{p}, \mathfrak{q}$ are distinct primes in S, then $\mathfrak{m}_\mathfrak{p} \not\subseteq \mathfrak{m}_\mathfrak{q}$ and $\mathfrak{m}_\mathfrak{p} \neq M$.*

Proof of (a): Otherwise $b^{-1} \in \bigcap_{\mathfrak{p} \in S} O_\mathfrak{p} = R$.

Proof of (2b): Let $\mathfrak{n} \subseteq \mathfrak{m}_\mathfrak{p}$ be a nonzero prime ideal. Choose $0 \neq a \in \mathfrak{n}$. Since $v_\mathfrak{p}$ is real, for each $b \in \mathfrak{m}_\mathfrak{p}$ there is a positive integer n with $v_\mathfrak{p}(a) \leq$

$nv_\mathfrak{p}(b) = v_\mathfrak{p}(b^n)$, so $\frac{b^n}{a} \in O_\mathfrak{p}$. By (1b), there are $c,d \in R$ with $\frac{b^n}{a} = \frac{c}{d}$ and $d \notin \mathfrak{m}_\mathfrak{p}$, so $db^n = ac \in \mathfrak{n}$. Since $d \notin \mathfrak{n}$, we have $b \in \mathfrak{n}$. Consequently, $\mathfrak{n} = \mathfrak{m}_\mathfrak{p}$.

Proof of (c): By (1b), $\mathfrak{m}_\mathfrak{p} \neq \mathfrak{m}_\mathfrak{q}$. By (b), $\mathfrak{m}_\mathfrak{p} \not\subset \mathfrak{m}_\mathfrak{q}$ Hence, $\mathfrak{m}_\mathfrak{p} \not\subseteq \mathfrak{m}_\mathfrak{q}$. In addition, by (b) and (1e), $\mathfrak{m} \neq M$. □

LEMMA 15.4.2: *The local ring R_M of R at M is a generalized Krull domain of dimension exceeding 1 with respect to $S' = \{\mathfrak{p} \in S \mid \mathfrak{m}_\mathfrak{p} \subset M\}$.*

Proof: Conditions (1a), (1b), (1d), and (1e) follow from the basic definitions of the local ring R_M. It remains to prove that $R_M = \bigcap_{\mathfrak{p} \in S'} O_\mathfrak{p}$. This follows if we show, for $x \in K$ with $v_\mathfrak{p}(x) \geq 0$ for each $\mathfrak{p} \in S'$, that $x \in R_M$.

By (1d), the set $T = \{\mathfrak{q} \in S \mid v_\mathfrak{q}(x) < 0\}$ is finite. If $\mathfrak{q} \in T$, then $\mathfrak{q} \notin S'$, so $\mathfrak{m}_\mathfrak{q} \not\subseteq M$. Choose $a_\mathfrak{q} \in \mathfrak{m}_\mathfrak{q} \smallsetminus M$. By (1a) there is a positive integer $n(\mathfrak{q})$ with $v_\mathfrak{q}(a_\mathfrak{q}^{n(\mathfrak{q})}) > -v_\mathfrak{q}(x)$. Let $a = \prod_{\mathfrak{q} \in T} a_\mathfrak{q}^{n(\mathfrak{q})}$ and $y = ax$. Then $a \in R \smallsetminus M$ and $v_\mathfrak{p}(y) \geq 0$ for each $\mathfrak{p} \in S$. Therefore, by (1c), $y \in R$. Consequently, $x \in R_M$. □

For the goal of this section - a proof that K is Hilbertian - we may replace R by R_M to assume that

(2) *R is a local ring and M is its maximal ideal.*

Consider an enlargement K^* of K. Then R^* is a local ring with maximal ideal M^*. Also, S^* has these properties:

(1a)* For $\mathfrak{p} \in S^*$, $v_\mathfrak{p}$ is a valuation of K^* with values in $\mathbb{R}^*$.
(1b)* The valuation ring $O_\mathfrak{p}^*$ of $v_\mathfrak{p}$ is the local ring of R^* relative to $\mathfrak{m}_\mathfrak{p}^* = \{a \in R^* \mid v_\mathfrak{p}(a) > 0\}$.
(1c)* $R^* = \bigcap_{\mathfrak{p} \in S^*} O_\mathfrak{p}^*$.
(1d)* For each $a \in K^* \smallsetminus \{0\}$ the set $\{\mathfrak{p} \in S^* \mid v_\mathfrak{p}(a) \neq 0\}$ is starfinite.
(1e)* M^* properly contains the prime ideals $\mathfrak{m}_\mathfrak{p}^*$.

As with fields with a product formula, consider the modified valuations $\dot{v}_\mathfrak{p}$ and their valuation rings $\dot{O}_\mathfrak{p} = \{x \in K^* \mid \dot{v}_\mathfrak{p}(x) \geq 0\}$. The holomorphy ring of these valuations, $\dot{R} = \bigcap_{\mathfrak{p} \in S^*} \dot{O}_\mathfrak{p}$, contains both K (Lemma 15.2.2) and R^*. Moreover:

LEMMA 15.4.3: *The ring $\dot{R}$ equals $K \cdot R^*$.*

Proof: We have to show only that each $x \in \dot{R}$ lies in the composite $K \cdot R^*$. Indeed, $\dot{v}_\mathfrak{p}(x) \geq 0$ for each $\mathfrak{p} \in S^*$. From (1d)* and lemma 15.2.2 the starfinite set $S(x) = \{\mathfrak{p} \in S^* \mid v_\mathfrak{p}(x) < 0\}$ contains only standard primes. Hence, (Corollary 14.3.5(b)), it is actually finite. For each $\mathfrak{p} \in S(x)$ there exists $r_\mathfrak{p} \in \mathbb{R}$ with $v_\mathfrak{p}(x) \geq r_\mathfrak{p}$ (because $\dot{v}_\mathfrak{p}(x) \geq 0$). By (1b) we may find $0 \neq a_\mathfrak{p} \in R$ with $v_\mathfrak{p}(a_\mathfrak{p}) \geq -v_\mathfrak{p}(x)$. Let $a = \prod_{\mathfrak{p} \in S(x)} a_\mathfrak{p}$ and $y = ax$. Then $v_\mathfrak{p}(y) \geq 0$ for each $\mathfrak{p} \in S^*$. By (1c)*, $y \in R^*$. Consequently, $x = a^{-1}y \in K \cdot R^*$. □

LEMMA 15.4.4: *Let x and y be nonunits of $\dot{R}$. Suppose for each $\mathfrak{p} \in S$ we have $\dot{v}_\mathfrak{p}(x) > 0$ or $\dot{v}_\mathfrak{p}(y) > 0$. Then $\alpha x + \beta y \neq 1$ for all $\alpha, \beta \in \dot{R}$.*

15.4 Generalized Krull Domains

Proof: Assume there exist $\alpha, \beta \in \dot{R}$ with $\alpha x + \beta y = 1$. Apply Lemma 15.4.3 to write

(3) $$\frac{ax}{c} + \frac{by}{c} = 1 \text{ with } a, b \in R^* \text{ and } c \in K^\times.$$

We show that both summands on the left hand side of (3) belong to M^*. This contradiction to (3) will conclude the proof of the lemma.

Indeed, let $\mathfrak{p} \in S^*$ with $v_\mathfrak{p}(c) > 0$. Then $\dot{v}_\mathfrak{p}(c) = 0$ and $\mathfrak{p}$ is standard (Lemma 15.2.2). Hence, $\dot{v}_\mathfrak{p}(x) > 0$ or $\dot{v}_\mathfrak{p}(y) > 0$. If $\dot{v}_\mathfrak{p}(x) > 0$, then $\dot{v}_\mathfrak{p}(\frac{ax}{c}) > 0$. Hence, $v_\mathfrak{p}(\frac{ax}{c}) \dot{>} 0$. So, by (3), $v_\mathfrak{p}(\frac{by}{c}) = 0$. Similarly, if $\dot{v}_\mathfrak{p}(y) > 0$, then $v_\mathfrak{p}(\frac{by}{c}) \dot{>} 0$ and $v_\mathfrak{p}(\frac{ax}{c}) = 0$.

Next consider $\mathfrak{p} \in S^*$ with $v_\mathfrak{p}(c) = 0$. Then $v_\mathfrak{p}(\frac{ax}{c}) \geq 0$ and $v_\mathfrak{p}(\frac{by}{c}) \geq 0$. It follows from (1c)* that both $\frac{ax}{c}$ and $\frac{by}{c}$ belong to R^*.

Since x is a nonunit in $\dot{R}$, there exists $\mathfrak{p} \in S^*$ with $\dot{v}_\mathfrak{p}(x) > 0$. Hence $\dot{v}_\mathfrak{p}(\frac{ax}{c}) > 0$ and therefore $\frac{ax}{c} \in M^*$. Similarly $\frac{by}{c} \in M^*$. □

Our next lemma is "standard":

LEMMA 15.4.5: *Let $\{\mathfrak{p}_1, \ldots, \mathfrak{p}_m\}$ and $\{\mathfrak{q}_1, \ldots, \mathfrak{q}_n\}$ be two disjoint finite subsets of S. Then there exists an element $a \in R$ such that*
(5) $v_{\mathfrak{p}_i}(a) = 0$, $i = 1, \ldots, m$, and $v_{\mathfrak{q}_j}(a) > 0$, $j = 1, \ldots, n$.

Proof: Proceed by induction on m. Suppose $m = 1$. By Lemma 15.4.1(c), $\mathfrak{m}_{\mathfrak{q}_j} \not\subseteq \mathfrak{m}_{\mathfrak{p}_1}$, $1 \leq j \leq n$. Let $a_j \in \mathfrak{m}_{\mathfrak{q}_j} \smallsetminus \mathfrak{m}_{\mathfrak{p}_1}$. Then $a = a_1 \cdots a_n$ satisfies (5).

Suppose $m > 1$. The induction hypothesis gives $a_1 \in R$ with
$$v_{\mathfrak{p}_1}(a_1) = \cdots = v_{\mathfrak{p}_{m-1}}(a_1) = 0 \text{ and } v_{\mathfrak{p}_m}(a_1), v_{\mathfrak{q}_1}(a_1), \ldots, v_{\mathfrak{q}_n}(a_1) > 0.$$

By the case $m = 1$, there exists $a_2 \in R$ with
$$v_{\mathfrak{p}_m}(a_2) = 0 \text{ and } v_{\mathfrak{p}_1}(a_2), \ldots, v_{\mathfrak{p}_{m-1}}(a_2), v_{\mathfrak{q}_1}(a_2), \ldots, v_{\mathfrak{q}_n}(a_2) > 0.$$

The element $a = a_1 + a_2$ satisfies (5). □

We now prove the main theorem of this section.

THEOREM 15.4.6 (Weissauer): *The quotient field of a generalized Krull domain of dimension exceeding 1 is Hilbertian.*

Proof: As previously, let K, R, S, M satisfy (1)-(2). Let $\mathfrak{q} \in S$ and choose $b \in M \smallsetminus \mathfrak{m}_\mathfrak{q}$ (by (Lemma 15.4.1(c))). By (1d) and Lemma 15.4.1(a), the set $T(b) = \{\mathfrak{p} \in S | v_\mathfrak{p}(b) > 0\}$ is finite and nonempty. For each finite set T with $T(b) \subseteq T \subseteq S$ there exists $a_T \in R$ such that $v_\mathfrak{p}(a_T) = 0$ for $\mathfrak{p} \in T(b)$ and $v_\mathfrak{p}(a_T) > 0$ for each $\mathfrak{p} \in T \smallsetminus T(b)$ (Lemma 15.4.5). Proposition 14.3.6 gives $a \in R^*$ such that $v_\mathfrak{p}(a) = 0$ for each $\mathfrak{p} \in T(b)$ and $v_\mathfrak{p}(a) > 0$ for each $\mathfrak{p} \in S \smallsetminus T(b)$. By Lemma 15.2.2, $v_\mathfrak{p}(b) = 0$ for each $\mathfrak{p} \in S^* \smallsetminus S$. Thus,
(5) $T(a) = \{\mathfrak{p} \in S^* | v_\mathfrak{p}(a) > 0\}$ is disjoint from $T(b) = \{\mathfrak{p} \in S^* | v_\mathfrak{p}(b) > 0\}$.

Choose $\omega \in \mathbb{N}^* \smallsetminus \mathbb{N}$. Put $x = a^\omega$ and $y = b^\omega$. We conclude the proof in parts, from Lemma 15.1.2, by showing that $t = \frac{x}{y}$ is polefinite.

PART A: $S(t) = \{\mathfrak{p} \in S^* \mid v_\mathfrak{p}(t) < 0\}$ *is a finite set.* Indeed, as $\mathfrak{q} \in S \smallsetminus T(b)$, we have $v_\mathfrak{q}(a) > 0$. Hence, $\dot{v}_\mathfrak{q}(x) > 0$. Next choose $\mathfrak{q}' \in T(b)$. Then $v_{\mathfrak{q}'}(b) > 0$ and therefore $\dot{v}_{\mathfrak{q}'}(y) > 0$. Thus, x and y are nonunits of $\dot{R}$. Moreover, if $\mathfrak{p} \in T(b)$, then $\dot{v}_\mathfrak{p}(y) > 0$ and if $\mathfrak{p} \in S \smallsetminus T(b)$, then $\dot{v}_\mathfrak{p}(x) > 0$. By Lemma 15.4.4,

(6) $$\alpha x + \beta y \neq 1 \text{ for all } \alpha, \beta \in \dot{R}.$$

In particular, $\alpha t + \beta \neq 1$ for all $\alpha, \beta \in K$, so $t \notin K$.

For each $\mathfrak{p} \in S^*$, (5) gives

(7a) $\quad v_\mathfrak{p}(x) > 0 \Longrightarrow v_\mathfrak{p}(y) = 0$ and $v_\mathfrak{p}(t) = v_\mathfrak{p}(x)$; and
(7b) $\quad v_\mathfrak{p}(y) > 0 \Longrightarrow v_\mathfrak{p}(x) = 0$ and $v_\mathfrak{p}(t) = -v_\mathfrak{p}(y)$.

Therefore, $S(t) = T(b)$ is a finite set.

PART B: *Application of Proposition 15.2.3.* To complete the proof it suffices to show that if $F/K(t)$ is a finite separable extension in K^*, then the functional primes of F are induced by arithmetical primes.

Let $\mathcal{P}$ be the set of functional primes of F which are induced by arithmetical primes. Let $D = \bigcap_{P \in \mathcal{P}} O_P$ be the corresponding holomorphy ring in F. Since $v_\mathfrak{q}(t) = \dot{v}_\mathfrak{q}(x) > 0$, the set $\mathcal{P}$ is nonempty. Assume $\mathcal{P}$ excludes a functional prime of F. Then the holomorphy ring theorem (Proposition 3.3.2) says that D is a Dedekind domain. By (7),

$$A = \{z \in D \mid \dot{v}_\mathfrak{p}(z) \geq \dot{v}_\mathfrak{p}(x) \text{ for all } \mathfrak{p} \in S^*\}$$
$$= \{z \in D \mid v_P(z) \geq \max(v_P(t), 0) \text{ for all } P \in \mathcal{P}\}; \text{ and}$$
$$B = \{z \in D \mid \dot{v}_\mathfrak{p}(z) \geq \dot{v}_\mathfrak{p}(y) \text{ for all } \mathfrak{p} \in S^*\}$$
$$= \{z \in D \mid v_P(z) \geq \max(-v_P(t), 0) \text{ for all } P \in \mathcal{P}\}.$$

Every maximal ideal of D is the center of a prime $P \in \mathcal{P}$ (Proposition 3.3.2(d)). Thus, A is the ideal of zeros of t and B is the ideal of poles of t. That is, $A = P_1^{k_1} \cdots P_r^{k_r}$ and $B = Q_1^{l_1} \cdots Q_s^{l_s}$, where $P_1, \ldots, P_r, Q_1, \ldots, Q_s$ are distinct maximal ideals of D and $k_1, \ldots, k_r, l_1, \ldots, l_s$ are positive integers with $v_{P_i}(t) = k_i$, $i = 1, \ldots, r$ and $v_{Q_j}(t) = -l_j$, $j = 1, \ldots, s$. In particular, A and B are relatively prime ideals of D. Hence, $A + B = D$. Thus, there exist $\lambda \in A$ and $\mu \in B$ such that $\lambda + \mu = 1$. By definition, $A \subseteq x\dot{R}$ and $B \subseteq y\dot{R}$. Thus, $\lambda = \alpha x$ and $\mu = \beta y$ with $\alpha, \beta \in \dot{R}$, a contradiction to (6). $\square$

15.5 Examples

Let R be an integral domain with quotient field K. Suppose S is a nonempty set of primes of K which satisfies Conditions (1b)–(1d) of Section 15.4. Suppose in addition $v_\mathfrak{p}$ is a discrete valuation, $\mathfrak{p} \in S$. Then R is a **Krull domain**, hence a generalized Krull domain.

15.5 Examples

Every Dedekind ring R is a Krull domain with S being the set of primes of K associated with the maximal ideals of R. Since each nonzero prime ideal of R is maximal, $\dim(R) = 1$. Thus, R does not satisfy Condition (1e) of Section 15.4.

Example 15.5.1: Polynomial rings over fields. Every unique factorization domain R is a Krull domain. Here S corresponds to the set of nonzero prime ideals of R. For example, the polynomial ring $R = K_0[X_1, \ldots, X_n]$ over an arbitrary field K_0 is a unique factorization domain [Zariski-Samuel1, p. 38, Thm. 13]. When $n \geq 2$, $R/(RX_1 + RX_2) \cong K_0[X_3, \ldots, X_n]$ is an integral domain. Hence, $RX_1 + RX_2$ is a prime ideal of R which properly contains each of the prime ideals RX_1 and RX_2. Thus, $\dim(R) \geq 2$. (Indeed, $\dim(R) = n$ [Matsumura, p. 117, Thm. 15.4].) By Theorem 15.4.6, $K_0(X_1, \ldots, X_n)$ is Hilbertian. This is a weaker result than Theorem 13.4.2 which says that $K_0(X_1, \ldots, X_n)$ is Hilbertian for each $n \geq 1$. □

Example 15.5.2 [Weissauer p. 203]: Formal power series over a field. Let K_0 be a field and $n \geq 2$. The ring of formal power series $R = K_0[[X_1, \ldots, X_n]]$ is a local integral domain with the maximal ideal

$$M = \{\sum_{i=1}^{\infty} f_i \mid f_i \in K_0[X_1, \ldots, X_n] \text{ is a form of degree } i\}.$$

Denote its quotient field by $K_0((X_1, \ldots, X_n))$: the **field of formal power series over** K_0 **in** $X_1, \ldots, X_n$.

The Weierstrass preparation theorem implies that R is a unique factorization domain [Zariski-Samuel2, p. 148]. For each prime element p of R one of the elements X_1 or X_2 does not belong to Rp. Thus, Rp is properly contained in M, so $\dim(R) \geq 2$. (Again, $\dim(R) = n$ [Zariski-Samuel2, p. 218 or Matsumura, p. 117, Thm. 15.4].) By Theorem 15.4.6, K is Hilbertian.

This settles a problem of [Lang3, p. 142, end of third paragraph]. □

Example 15.5.3: Formal power series over a ring. Let A be a Krull domain and $n \geq 1$. Then the ring of polynomials $A[X_1, \ldots, X_n]$ and the ring of formal power series $A[[X_1, \ldots, X_n]]$ in the variables $X_1, \ldots, X_n$ over A are Krull domains (see [Matsumura, p. 89] for the case $n = 1$; the general case follows by induction).

Suppose in addition that A is not a field. Let $\mathfrak{p}_0$ be a nonzero prime ideal of A. Then

$$P = \{\sum_{i=1}^{\infty} f_i \in A[X_1, \ldots, X_n] \text{ is a form of degree } i\}$$

$$P_0 = \{\sum_{i=0}^{\infty} f_i \in A[X_1, \ldots, X_n] \text{ is a form of degree } i \text{ and } f_0 \in \mathfrak{p}_0\}$$

are nonzero prime ideals of $R = A[[X_1,\ldots,X_n]]$ and $P_0 \subset P$. Indeed, $A/P \cong A$ and $A/P \cong A/\mathfrak{p}_0$ are integral domains. Thus, $\dim(A[[X_1,\ldots,X_n]]) \geq 2$ (Actually, $\dim(A) = \dim(A) + n$ [Matsumura, p. 117]). Consequently, by Theorem 15.4.6, the quotient field of $A[[X_1,\ldots,X_n]]$ is a Hilbertian field.

For example, the quotient fields of $\mathbb{Z}[[X_1,\ldots,X_n]]$ and $O[[X_1,\ldots,X_n]]$, where O is a discrete valuation rings, are Hilbertian. □

LEMMA 15.5.4: *No Henselian field is Hilbertian.*

Proof: Let K be a Henselian field with valuation ring R and maximal ideal $\mathfrak{m}$. Choose $m \in \mathfrak{m}$, $m \neq 0$ and a prime number $p \neq \mathrm{char}(K)$. Consider the irreducible polynomials $f(T,X) = X^p + mT - 1$ and $g(T,X) = X^p + T^{-1} - 1$ of $K(T)[X]$. Assume K is Hilbertian. Then there exists $a \in K^\times$ with both $f(a,X)$ and $g(a,X)$ irreducible in $K[X]$. In particular, none of them has a zero in K.

But either $a \in R$ or $a^{-1} \in \mathfrak{m}$. Suppose first $a \in R$. Then $f(a,1) \equiv 0 \bmod \mathfrak{m}$ and $\frac{\partial f}{\partial X}(a,1) \not\equiv 0 \bmod \mathfrak{m}$. Since K is Henselian, $f(a,X)$ has a zero in K. Similarly, if $a^{-1} \in \mathfrak{m}$, then $g(a,X)$ has a zero in K. This contradiction to the preceding paragraph proves that K is not Hilbertian. □

Example 15.5.5: $\mathbb{Q}_p$ and the formal power series $K_0((X))$ are complete discrete valuation fields. Hence, they are Henselian (Proposition 3.5.2). By Lemma 15.5.4, they are not Hilbertian. Thus, the assumption "$n \geq 2$" in Example 15.5.2 is necessary. □

Example 15.5.6 (Geyer): In the notation of Example 15.5.2, the ring $R = K_0[[X_1,\ldots,X_n]]$ of formal power series is not Hilbertian. Indeed, let $K = \mathrm{Quot}(R)$ and check for $\mathrm{char}(K_0) \neq 2$ (resp. $\mathrm{char}(K_0) \neq 3$) that every power series $f(X_1,X_2) = 1 + \sum_{i+j>0} a_{ij} X_1^i X_2^j$ is a square (resp. a cube) in R. Thus, the polynomial $Z^2 - (1+X_1T)$ (resp. $Z^3 - (1+X_1T)$) is irreducible in $K[T,Z]$ but $Z^2 - (1+X_1t)$ (resp. $Z^3 - (1+X_1t)$) is reducible for each $t \in R$. □

LEMMA 15.5.7: *Let R_0 be a unique factorization domain with quotient field K_0. Consider a set P of unequivalent prime elements of R_0. For each $p \in P$ let v_p be the corresponding discrete valuation of K_0. Let K be an algebraic extension of K_0. Suppose each v_p with $p \in P$ is unramified in K. For each $p \in P$ choose an extension w_p of v_p to K. Then the holomorphy ring $R = \{x \in K \mid w_p(x) \geq 0 \text{ for each } p \in P\}$ is a unique factorization domain with quotient field K.*

Proof: The assumptions imply $w_p(p) = 1$ and $w_{p'}(p) = 0$ for all distinct $p,p' \in P$. Consider $x \in R$. Then $w_p(x) \geq 0$ for each $p \in P$. Moreover, there are only finitely many $p \in P$ with $w_p(x) > 0$. Indeed, if $f = \mathrm{irr}(x,K_0)$ and $v_p(f(0)) = 0$, then $w_p(x) = 0$. Thus, $u = x \prod_{p \in P} p^{-w_p(x)}$ is an element of K and $w_p(u) = 0$ for each $p \in P$. Hence, u is a unit of R and $x = u \prod_{p \in P} p^{w_p(x)}$ is the desired decomposition of x. Finally, observe that R contains the integral closure of R_0 in K. Therefore, $K = \mathrm{Quot}(R)$. □

The following example generalizes [Corvaja-Zannier, Theorem 1 (i) and (ii)]:

Example 15.5.8: *Unique factorization domain with a non-Hilbertian quotient field.* Let R_0 be either $\mathbb{Z}$ or $F[t]$ for some finite field F. It is a unique factorization domain. Put $K_0 = \text{Quot}(R_0)$. The proof of Theorem 13.6.2 (with K_0 replacing K) gives prime numbers l and q, an element $u \in K^\times$, an infinite set P of nonequivalent prime elements of R_0, and a field extension K (in the notation of Theorem 13.6.2 equals to $M(K_0)$) with these properties:
(1a) For every $x \in K$ there is a $y \in K$ with $y^l - x^{lq} + u = 0$.
(1b) For each $p \in P$, the discrete valuation v_p is unramified in K.

Since $Y^l - X^{lq} + u$ is absolutely irreducible, Condition (1a) implies K is not Hilbertian. Let w_p, $p \in P$, and R be as in Lemma 15.5.7. Then $K = \text{Quot}(R)$, $\{w_p \mid p \in P\}$ is an infinite set of discrete valuations, and R is a unique factorization domain.

This answers negatively the questions posed in Problems 14.20 and 14.21 of [Fried-Jarden3]. □

The following problem asks for a generalization of Example 15.5.3:

Problem 15.5.9: Let A be a generalized Krull domain which is not a field.
(a) Is $A[[X]]$ a generalized Krull domain?
(b) Is $\text{Quot}(A[[X]])$ Hilbertian?

By Theorem 15.4.6, an affirmative answer to Problem 15.5.9(a) will give an affirmative answer to Problem 15.5.9(b).

Exercises

1. Let K be a countable Hilbertian field. List the irreducible polynomials in $K[T, X]$ as $f_1(T, X), f_2(T, X), \ldots$. For each i choose $t_i \in K$ for which $f_1(t_i, Y), \ldots, f_i(t_i, Y)$ are irreducible in $K[Y]$.
   (a) Observe that the infinite set $H = \{t_1, t_2, \ldots\}$ has the **universal Hilbert subset property**: $H \smallsetminus H_K(g_1, \ldots, g_m)$ is finite for every collection $\{g_1, \ldots, g_m\}$ of irreducible polynomials in $K[T, X]$.
   (b) Let K^* be an enlargement of K. Prove that if $t \in H^* \smallsetminus H$, then $K(t)$ is algebraically closed in K^*.

2. Let $A = \{1, 2^{2!}, 3^{3!}, \ldots\}$ and consider an enlargement $\mathbb{Q}^*$ of $\mathbb{Q}$. Prove for each nonstandard element $t \in A^*$ that $\bigcup_{n=1}^\infty \mathbb{Q}(t^{1/n}) \subseteq \mathbb{Q}^*$.

3. Let $f_1, f_2, f_3, \ldots$ be a sequence in $\mathbb{Q}(Y)$. Put
$$g_n(Y) = f_1(\ldots(f_{n-1}(f_n(Y)))).$$
Suppose that if y is transcendental over $\mathbb{Q}$ and $x = g_n(y)$, then x has at least n distinct poles in $\mathbb{Q}(y)$. Let $A = \{g_1(1), g_2(2), g_3(3), \ldots\}$. Consider a nonstandard element t of A^*. Prove that the equation $t = g_n(Y)$ is solvable in $\mathbb{Q}^*$ for each n. Conclude that t is not polefinite (Section 15.1).

4. [Weissauer, Satz 2.3] Generalize the Gilmore-Robinson criterion as follows. Let K^* be an enlargement of a field K. Then K is Hilbertian if and only if there is a Hilbertian field F with $K \subseteq F \subseteq K^*$ and F is separably closed in K^*.

5. Consider the set $A = \{2, 3!, 4!, 5!, \ldots\}$ and let t be a nonstandard element of A^*. Show that the metric absolute values of $\mathbb{Q}$ induces via $\mathbb{Q}^*$ (Section 15.2) the infinite prime, $t \to \infty$ of $\mathbb{Q}(t)$, while every ultrametric absolute value of $\mathbb{Q}$ induces the prime, $t \to 0$, of $\mathbb{Q}(t)$. Conclude from Proposition 15.3.2, that all other primes of $\mathbb{Q}(t)$ are induced by nonstandard arithmetical primes of $\mathbb{Q}^*$.

6. Consider the field $K = K_0((X_1, \ldots, X_n))$ of formal power series in n variables over a basic field K_0. Construct a K_0-place $\varphi \colon K \to K_0 \cup \{\infty\}$ with $\varphi(X_i) = 0$, $i = 1, \ldots, n$. Conclude that K/K_0 is a regular extension. Hint: The case $n = 1$ is easy. For arbitrary n embed K in the field of iterated power series $L = K_0((X_1))((X_2)) \cdots ((X_n))$.

Notes

The nonstandard characterization of Hilbertian fields appears in [Gilmore-Robinson]. [Roquette2] has exploited the Gilmore-Robinson criterion through a nonstandard interpretation of the Siegel-Mahler theorem (compare with the remarks at the beginning of Section 13.3).

Most of this Chapter is from Weissauer's thesis [Weissauer]. The proof of 15.3.2 for number field that appear in [Robinson] based on nonstandard interpretation of the "distributions" that appear in Weil's thesis [Weil1]. The influence of [Robinson-Roquette] in Section 15.2 and parts of Section 15.3 should be obvious.

The polefinite property of Section 5.1 and its relation to Hilbert's irreducibility theorem appears in a standard form based on p-adic analysis in the case $K = \mathbb{Q}$ in [Sprindžuk1]. Standard simplified proofs of both Sprindžuk's result and Weissauer's approach to Hilbertianity of fields with a product formula (Section 15.3), featuring their common concepts, appear in [Fried10] which also gives the concept of a universal Hilbert subset (e.g. in Exercise 1) whose existence is a consequence of the Gilmore-Robinson observation (Proposition 15.1). As far as we know [Sprindžuk2] is the first to give an explicit universal Hilbert subset (over $\mathbb{Q}$):

$$H = \{[\exp(\sqrt{\log(\log(m))})] + m!2^{m^2}\}_{m=1}^{\infty}.$$

It has the property that $H \smallsetminus H_{\mathbb{Q}}(g_1, \ldots, g_t)$ is finite for every collection $\{g_1, \ldots, g_t\}$ of irreducible polynomials in $\mathbb{Q}[T, X]$ [Fried10, Thm. 4.9].

Finally, [Klein] gives "standard" proofs of Weissauer's results, Theorems 15.3.3 and 15.4.6.

Chapter 16.
Galois Groups over Hilbertian Fields

Given a field K, one may ask which finite groups occur as Galois groups over K. If K is Hilbertian, then every finite group that occurs over $K(t)$, with t transcendental over K, also occurs over K. Moreover, suppose $F/K(t)$ is Galois with Galois group G and F/K is regular. Then K has a linearly disjoint sequence of Galois extensions, $L_1, L_2, L_3, \ldots$, with $\mathrm{Gal}(L_i/K) \cong G$, $i = 1, 2, 3, \ldots$ (Lemma 16.2.6).

We prove that this set up occurs for symmetric groups (Corollary 16.2.7), Abelian groups (Proposition 16.3.5), and when $\mathrm{char}(K) = 0$ for alternating groups (Proposition 16.7.6). If K is PAC (but not necessarily Hilbertian), every finite group is regular over K (Proposition 16.12.2).

If K is Hilbertian, then $\mathbb{Z}_p$ occurs over K (Corollary 16.6.7) but is not necessarily regular over K. For example, $\mathbb{Z}_p$ is not regular over $\mathbb{Q}$ (Corollary 16.6.11).

Realization of a finite nonsimple group over K is usually done in steps. First one realizes a quotient of the group and then embeds the solution field in a larger Galois extension with the given Galois group. The latter step is always possible when K is Hilbertian and the kernel is a product of non-Abelian simple groups each of which has a GAR realization over K (Sections 16.8 and 16.9). For example, A_n with $n = 5$ or $n \geq 7$ is GAR over K when $\mathrm{char}(K) \nmid (n-1)n$ (Corollary 16.9.2).

A $\mathbb{Z}_p$-extension N of K is an example of a Galois extension with finitely generated Galois groups. If K is Hilbertian, then so is N (Proposition 16.11.1). This is one ingredient of the proof that each Abelian extension of K is Hilbertian (Theorem 16.11.3).

Finally, the regularity of $\mathbb{Z}/p\mathbb{Z}$ over a Hilbertian field K has some implications to the structure of $\mathrm{Gal}(K)$. For example, $\mathrm{Gal}(K)$ has no closed prosolvable normal subgroup. In particular, the center of $\mathrm{Gal}(K)$ is trivial. Chapter 18 will exploit the result about the regularity of S_n.

16.1 Galois Groups of Polynomials

We prove two theorems about preservation of Galois groups of polynomials under specializations of parameters. One of them (Proposition 16.1.5) is a polynomial analog of Lemma 13.1.1(b). It assumes the ground field to be Hilbertian. The other one (Proposition 16.1.4) is an application of Bertini-Noether. Here the ground field is arbitrary but the polynomial in question is absolutely irreducible and Galois.

We start with an analog of Lemma 13.1.1(a) for Galois groups of polynomials:

Let $f \in K[X]$ be a separable polynomial of degree n. By definition, f

has n distinct roots $x_1,\ldots,x_n$ in K_s. Thus, $L = K(x_1,\ldots,x_n)$ is a finite Galois extension. Restriction of elements of $\mathrm{Gal}(L/K)$ to $\{x_1,\ldots,x_n\}$ gives an embedding of $\mathrm{Gal}(L/K)$ into the group of all permutations of $\{x_1,\ldots,x_n\}$, called a **permutation representation**. The image of $\mathrm{Gal}(L/K)$ under this representation is $\mathrm{Gal}(f,K)$.

Next consider another field $\bar{K}$ and a separable polynomial $\bar{f} \in \bar{K}[X]$ of degree n. We write $\mathrm{Gal}(f,K) \cong \mathrm{Gal}(\bar{f},\bar{K})$ to indicate that the two groups in question are isomorphic as permutation groups. Thus, there exists an isomorphism $\sigma \mapsto \bar{\sigma}$ of $\mathrm{Gal}(f,K)$ and $\mathrm{Gal}(\bar{f},\bar{K})$ as abstract groups and there is a listing $\bar{x}_1,\ldots,\bar{x}_n$ of the roots of $\bar{f}$ with $\overline{\sigma x_i} = \bar{\sigma}\bar{x}_i$, $\sigma \in \mathrm{Gal}(f,K)$, $i = 1,\ldots,n$. Similarly we speak about an embedding of $\mathrm{Gal}(\bar{f},\bar{K})$ into $\mathrm{Gal}(f,K)$ as permutation groups. Finally, we write $\mathrm{Gal}(f,K) \cong G$ for a finite group G when $\mathrm{Gal}(f,K)$ and G are isomorphic as abstract groups.

LEMMA 16.1.1: *Let L/K be a finite Galois extension, $f \in K[X]$ a separable polynomial, and φ a place of L. Denote the residue field of K (resp. L) under φ by $\bar{K}$ (resp. $\bar{L}$). Suppose L is the splitting field of f over K and $\bar{f} = \varphi(f)$ is a separable polynomial in $\bar{K}[X]$ with $\deg(\bar{f}) = \deg(f)$.*

(a) *Then there is an embedding $\varphi^*\colon \mathrm{Gal}(\bar{f},\bar{K}) \to \mathrm{Gal}(f,K)$.*

(b) *Suppose in addition, f is irreducible and Galois, and $\bar{f}$ is irreducible. Then $\bar{f}$ is Galois, φ^* is an isomorphism, and $\bar{L}$ is the splitting field of $\bar{f}$ over $\bar{K}$.*

(c) *Alternatively, suppose $\bar{L}/\bar{K}$ is separable. Then $\bar{L}$ is the splitting field of $\bar{f}$ over $\bar{K}$.*

Proof of (a): Let a be the leading coefficient of f. Then $f(X) = a\prod_{i=1}^n(X - x_i)$ with distinct $x_1,\ldots,x_n \in K_s$ and $L = K(x_1,\ldots,x_n)$ is the splitting field of f over K. Extend φ to a place of $L(X)$ with the same notation such that $\varphi(X) = X$. Since $\deg(\bar{f}) = \deg(f)$, we have $\bar{a} = \varphi(a) \in \bar{K}^\times$. Hence, $\bar{x}_i = \varphi(x_i)$, $i = 1,\ldots,n$, are in $\bar{K}_s^\times$ and $\bar{f}(X) = \bar{a}\prod_{i=1}^n(X - \bar{x}_i)$. By assumption, $\bar{x}_1,\ldots,\bar{x}_n$ are distinct. Therefore, the map $x_i \mapsto \bar{x}_i$, $i = 1,\ldots,n$, is bijective.

Lemma 6.1.1 gives an epimorphism $\sigma \mapsto \bar{\sigma}$ of the decomposition group D_φ onto $\mathrm{Aut}(\bar{L}/\bar{K})$. It satisfies $\bar{\sigma}\bar{y} = \overline{\sigma y}$ for each $y \in K$ with $\bar{y} = \varphi(y) \in \bar{L}$. Suppose $\bar{\sigma} = 1$. Then $\overline{\sigma x_i} = \bar{\sigma}\bar{x}_i = \bar{x}_i$. Since σ permutes $x_1,\ldots,x_n$, we have $\sigma x_i = x_i$, $i = 1,\ldots,n$. Hence, $\sigma = 1$.

Moreover, let $\tau \in \mathrm{Aut}(\bar{L}/\bar{K}(\bar{x}_1,\ldots,\bar{x}_n))$. Then there is a $\sigma \in D_\varphi$ with $\bar{\sigma} = \tau$. By the preceding paragraph, $\bar{\sigma} = 1$. It follows that res: $\mathrm{Aut}(\bar{L}/\bar{K}) \to \mathrm{Gal}(\bar{K}(\bar{x}_1,\ldots,\bar{x}_n)/\bar{K})$ is an isomorphism.

Finally, consider each $\sigma \in \mathrm{Gal}(L/K)$ as a permutation of $\{x_1,\ldots,x_n\}$. Likewise, consider $\bar{\sigma}$ as a permutation of $\{\bar{x}_1,\ldots,\bar{x}_n\}$. Then $\bar{\sigma} \mapsto \sigma$ is an embedding $\varphi^*\colon \mathrm{Gal}(\bar{f},\bar{K}) \to \mathrm{Gal}(f,K)$.

Proof of (b): Since f is irreducible and Galois, $\deg(f) = |\mathrm{Gal}(f,K)|$. Since

16.1 Galois Groups of Polynomials

$\bar{f}$ is irreducible, $\deg(\bar{f}) \leq |\text{Gal}(\bar{f}, \bar{K})|$. By (a),

$$\deg(f) = \deg(\bar{f}) \leq |\text{Gal}(\bar{f}, \bar{K})| = [\bar{K}(\bar{x}_1, \ldots, \bar{x}_n) : \bar{K}]$$
$$\leq [\bar{L} : \bar{K}] \leq [L : K] = |\text{Gal}(f, K)| = \deg(f).$$

Hence, $|\text{Gal}(\bar{f}, \bar{K})| = |\text{Gal}(f, K)|$ and $\deg(\bar{f}) = [\bar{K}(\bar{x}_1, \ldots, \bar{x}_n) : \bar{K}]$. Thus, $\bar{f}$ is Galois and φ^* is an isomorphism. In addition, $\bar{L} = \bar{K}(\bar{x}_1, \ldots, \bar{x}_n)$ is Galois over $\bar{K}$.

Proof of (c): Since $\bar{L}/\bar{K}$ is normal and separable, it is Galois. By the proof of (a), $\text{Gal}(\bar{L}/\bar{K}(\bar{x}_1, \ldots, \bar{x}_n)) = 1$. Therefore, $\bar{L} = \bar{K}(\bar{x}_1, \ldots, \bar{x}_n)$. □

An essential assumption in Lemma 16.1.1 is the irreducibility of $\bar{f}$. Lemma 16.1.4 below uses Bertini-Noether to satisfy this assumption. Proposition 16.1.5 applies Hilbert irreducibility theorem to achieve the same goal.

LEMMA 16.1.2: *Let V be a variety in $\mathbb{A}^n$ over a field K and $\mathbf{x}$ a generic point of V over K. Then V has a nonempty Zariski K-open subset V_0 with the following property: For each $\mathbf{a} \in V_0(\tilde{K})$ there is a K-place of $K(\mathbf{x})$ with $\varphi(\mathbf{x}) = \mathbf{a}$ and with residue field $K(\mathbf{a})$.*

Proof: By Corollary 10.2.2(a), $K(\mathbf{x})/K$ is a regular extension. Assume without loss that $x_1, \ldots, x_r$ form a separating transcendence base for $K(\mathbf{x})/K$. Example 2.6.10 produces a K-place φ_0 of $K(x_1, \ldots, x_r)$ with $\varphi(x_i) = a_i$, $i = 1, \ldots, r$, and with residue field $K(a_1, \ldots, a_r)$.

Use Remark 6.1.5 to find a nonzero polynomial $g \in K[X_1, \ldots, X_r]$ such that $K[x_1, \ldots, x_{i+1}, g(x_1, \ldots, x_r)^{-1}]/K[x_1, \ldots, x_i, g(x_1, \ldots, x_r)^{-1}]$ is a ring cover, $i = r, \ldots, n - 1$. Let $V_0 = \{\mathbf{a} \in V \mid g(a_1, \ldots, a_r) \neq 0\}$.

Suppose $\mathbf{a} \in V_0(\tilde{K})$. Let φ a place of $K(\mathbf{x})$ which extends φ_0 with $\varphi(\mathbf{x}) = \mathbf{a}$. By Remark 6.1.7, the residue field of $K(\mathbf{x})$ at φ is $K(\mathbf{a})$. □

Remark 16.1.3: *Simple points.* The conclusion of Lemma 16.1.2 is actually true for each simple point $\mathbf{a}$ of V [Jarden-Roquette, Cor. A2]. □

PROPOSITION 16.1.4: *Let V be a variety over a field K_0 in $\mathbb{A}^m$, $\mathbf{u}$ a generic point of V over K_0, $K = K_0(\mathbf{u})$, and*

$$h_1, \ldots, h_k \in K_0[U_1, \ldots, U_m, T_1, \ldots, T_r, X]$$

polynomials. Suppose $h_j(\mathbf{u}, \mathbf{T}, X)$ are absolutely irreducible as polynomials in $(\mathbf{T}, X)$ over K and Galois as polynomial in X over $K(\mathbf{T})$. Then there exists a nonempty Zariski K_0-open subset V_0 of V with the following property: For each $\mathbf{a} \in V_0(K_0)$, $h_j(\mathbf{a}, \mathbf{T}, X)$ is absolutely irreducible, Galois over $K_0(\mathbf{T})$, and

$$\text{Gal}(h_j(\mathbf{a}, \mathbf{T}, X), K_0(\mathbf{T})) \cong \text{Gal}(h_j(\mathbf{u}, \mathbf{T}, X), K(\mathbf{T})),$$

$j = 1, \ldots, k$.

Proof: Bertini-Noether (Proposition 9.4.3) gives a nonempty Zariski K_0-open subset V_1 of V such that $h_j(\mathbf{a}, \mathbf{T}, X)$ are absolutely irreducible for each

$\mathbf{a} \in V_1(\tilde{K}_0)$, $j = 1, \ldots, k$. Lemma 16.1.2, applied to V over $K_0(\mathbf{T})$, gives a nonempty Zariski $K_0(\mathbf{T})$-open subset V_2' of V satisfying this: For each $\mathbf{a} \in V_2'(\widetilde{K_0(\mathbf{T})})$ there is a $K_0(\mathbf{T})$-place φ of $K_0(\mathbf{u}, \mathbf{T})$ with residue field $K_0(\mathbf{a}, T)$. Choose a nonempty Zariski K_0-open subset V_2 of V such that $V_2(\tilde{K}_0) \subseteq V_2'(\tilde{K}_0)$. Finally, there is a nonempty Zariski K_0-open subset V_3 of V with $h_j(\mathbf{a}, \mathbf{T}, X)$ separable and $\deg_X(h_j(\mathbf{a}, \mathbf{T}, X)) = \deg_X(h_j(\mathbf{u}, \mathbf{T}, X))$ for each $\mathbf{a} \in V_3(\tilde{K}_0)$, $j = 1, \ldots, k$.

$V_0 = V_1 \cap V_2 \cap V_3$ is a nonempty Zariski K_0-open subset of V. For each $\mathbf{a} \in V_0(K)$ let φ be a K_0-place of $K(\mathbf{T})$ with residue field $K_0(\mathbf{T})$. By Lemma 16.1.1, $h_j(\mathbf{a}, \mathbf{T}, X)$ is Galois over $K_0(\mathbf{T})$ and (1) holds. □

PROPOSITION 16.1.5: *Let K be a Hilbertian field and $h_j \in K[T_1, \ldots, T_r, X]$ a separable polynomial in X, $j = 1, \ldots, k$. Then K^r has a separable Hilbert subset H with $\mathrm{Gal}(h_j(\mathbf{a}, X), K) \cong \mathrm{Gal}(h_j(\mathbf{T}, K), K(\mathbf{T}))$ for each $\mathbf{a} \in H$, $j = 1, \ldots, k$.*

Proof: Denote the splitting field of $h_j(\mathbf{T}, X)$ over $K(\mathbf{T})$ by F_j. Proposition 13.1.1(a) gives a nonempty Zariski K-open subset U_1 of $\mathbb{A}^r$ satisfying the following assertion: For each $\mathbf{a} \in U_1(K)$ there is a K-place φ_j of F_j extending $\mathbf{T} \mapsto \mathbf{a}$ such that the residue field $\bar{F}_j$ of F_j under φ_j is Galois over K, $j = 1, \ldots, k$.

Proposition 13.1.1(b) gives a separable Hilbert subset H' of K^r such that for each $\mathbf{a} \in H' \cap U_1(K)$ and every K-place φ of F_j with residue field $\bar{F}_j$ we have $\mathrm{Gal}(\bar{F}_j/K) \cong \mathrm{Gal}(F_j/K(\mathbf{T}))$.

Finally, there is a nonempty Zariski open subset U_2 of $\mathbb{A}^r$ satisfying this: $h_j(\mathbf{a}, X)$ is separable and $\deg_X(h_j(\mathbf{a}, X)) = \deg_X(h_j(\mathbf{T}, X))$ for each $\mathbf{a} \in U_2(K)$, $j = 1, \ldots, k$.

Put $H = H' \cap U_1(K) \cap U_2(K)$. Consider $\mathbf{a} \in H$ and let φ_j be as above. Then Lemma 16.1.1 gives an isomorphism

$$\varphi_j^* \colon \mathrm{Gal}(h_j(\mathbf{a}, X), K) \to \mathrm{Gal}(h_j(\mathbf{T}, X), K(\mathbf{T})),$$

$j = 1, \ldots, k$, as claimed. □

16.2 Stable Polynomials

Let K be a field and G a profinite group. Suppose K has a Galois extension L with $\mathrm{Gal}(L/K) \cong G$. Then, G **occurs** (or is **realizable**) over K and L is a G-**extension** of K. If G belongs to a family $\mathcal{G}$ of profinite groups, we also say that L is a $\mathcal{G}$-**extension** of K. For example, when G is Abelian, we say L is an **Abelian extension** of K. If G is an inverse limit of finite solvable groups, we say L is a **prosolvable extension** of K.

The main problem of Galois theory is whether every finite group occurs over $\mathbb{Q}$. Even if this holds, it is not clear whether the same holds for every Hilbertian field.

16.2 Stable Polynomials

To approach the latter problem, consider algebraically independent elements $t_1, \ldots, t_r$ over K. If K is Hilbertian, G is finite, and G occurs over $K(\mathbf{t})$, then G occurs over K (Lemma 13.1.1(b)). If we want G to occur more than once over K, we have to assume more, as we now explain.

We say G is **regular** over K if there exist algebraically independent elements $t_1, \ldots, t_r$ over K such that $K(\mathbf{t})$ has a Galois extension F which is regular over K with $\mathrm{Gal}(F/K(\mathbf{t})) \cong G$. This stronger property is inherited by all extensions of K:

LEMMA 16.2.1: *Let K be a field and G a profinite group. Suppose G is regular over K. Then G is regular over every extension L of K.*

Proof: Let $\mathbf{t} = (t_1, \ldots, t_r)$ and F be as above. Consider a field extension L of K. Assume without loss that $t_1, \ldots, t_r$ are algebraically independent over L. Then F is linearly disjoint from L over K (Lemma 2.6.7). Hence, FL is a regular extension of L (Corollary 2.6.8), F is linearly disjoint from $L(\mathbf{t})$ over $K(\mathbf{t})$ (Lemma 2.5.3), and $\mathrm{Gal}(FL/L(\mathbf{t})) \cong G$. Consequently, G is regular over L. □

The **regular inverse Galois problem** asks whether every finite group is regular over every field. By Lemma 16.2.1, it suffices to check the problem over $\mathbb{Q}$ and over each of the fields $\mathbb{F}_p$. In this generality the problem is still wide open. Nevertheless, there are many special cases where groups G are proved to be regular over specific fields. This chapter discusses cases when this happens.

Remark 16.2.2: Reinterpretation of 'regularity' by polynomials. Consider an irreducible polynomial $f \in K[T_1, \ldots, T_r, X]$, separable with respect to X. Call f X**-stable** over K if

(1) $\qquad \mathrm{Gal}(f(\mathbf{T}, X), K(\mathbf{T})) \cong \mathrm{Gal}(f(\mathbf{T}, X), L(\mathbf{T}))$

for every extension L of K. In this case denote the splitting field of $f(\mathbf{T}, X)$ over $K(\mathbf{T})$ by F. Then $\mathrm{Gal}(F/K(\mathbf{T})) \cong \mathrm{Gal}(FL/L(\mathbf{T}))$. Hence, F is linearly disjoint from L over K (Lemma 2.5.3). Thus, F is a regular extension of K and G is regular over K. In addition, f is absolutely irreducible (Corollary 10.2.2).

Conversely, suppose $K(T_1, \ldots, T_r)$ has a Galois extension F which is regular over K and with Galois group G. Choose a primitive element x for $F/K(\mathbf{T})$ and let $f \in K[\mathbf{T}, X]$ be an irreducible polynomial with $f(\mathbf{T}, x) = 0$. Then $f(\mathbf{T}, X)$ is X-stable over K, Galois with respect to X over $K(\mathbf{T})$, and $\mathrm{Gal}(f(\mathbf{T}, X), K(\mathbf{T})) \cong G$. □

LEMMA 16.2.3: *Let K be a field and f a polynomial in $K[T_1, \ldots, T_r, X]$, separable in X. Suppose $\mathrm{Gal}(f(\mathbf{T}, X), K(\mathbf{T})) \cong \mathrm{Gal}(f(\mathbf{T}, X), K_s(\mathbf{T}))$. Then f is X-stable over K.*

Proof: Condition (1) holds when L is a purely inseparable or a regular extension of K. Hence, we have to consider only the case where L is a separable algebraic extension of K. In this case both maps

$$\text{res}: \text{Gal}(f(\mathbf{T}, X), K_s(\mathbf{T})) \to \text{Gal}(f(\mathbf{T}, X), L(\mathbf{T}))$$
$$\text{res}: \text{Gal}(f(\mathbf{T}, X), L(\mathbf{T})) \to \text{Gal}(f(\mathbf{T}, X), K(\mathbf{T}))$$

are injective. By assumption, their compositum is bijective. Hence, each of them is bijective. ☐

LEMMA 16.2.4: *Let K be a field and N a Galois extension of K.*
(a) *Suppose f is a polynomial in $N[T_1, \ldots, T_r, X]$ separable in X. Then K has a finite Galois extension L in N with $f \in L[\mathbf{T}, X]$ and*

$$\text{Gal}(f(\mathbf{T}, X), L(\mathbf{T})) \cong \text{Gal}(f(\mathbf{T}, X), N(\mathbf{T})).$$

(b) *If f is Galois over $N(\mathbf{T})$, then f is Galois over $L(\mathbf{T})$.*
(c) *If f is X-stable over N, then f is X-stable over L.*

Proof of (a): Choose a finite extension L_0 of K in N which contains the coefficients of f. Denote the splitting field of $f(\mathbf{T}, X)$ over $L_0(\mathbf{T})$ by F. Then $L_1 = F \cap N$ is a finitely generated extension of K (Lemma 10.5.1). In addition, L_1/K is a separable algebraic extension. Hence, L_1/K is finite. Also, $\text{Gal}(F/L_1(\mathbf{T})) \cong \text{Gal}(FN/N(\mathbf{T}))$. Finally, let L be the Galois closure of L_1/K. Then L satisfies the requirements of (a).

Proof of (b): Suppose f and L satisfy the conclusion of (a) and f is Galois over $N(\mathbf{T})$. Let $x_1, \ldots, x_n$ be the roots of $f(\mathbf{T}, X)$ in $K(\mathbf{T})_s$. Consider $\sigma \in \text{Gal}(f(\mathbf{T}, X), L(\mathbf{T}))$. Suppose $x_1^\sigma = x_1$. By (a), σ extends to an element $\bar{\sigma}$ in $\text{Gal}(f(\mathbf{T}, X), N(\mathbf{T}))$. Hence, $x_i^\sigma = x_i$, $i = 1, \ldots, n$. Therefore, f is Galois over $L(\mathbf{T})$.

Proof of (c): By assumption, $\text{Gal}(f(\mathbf{T}, X), N(\mathbf{T})) \cong \text{Gal}(f(\mathbf{T}, X), K_s(\mathbf{T}))$. Hence, by (a), $\text{Gal}(f(\mathbf{T}, X)), L(\mathbf{T})) \cong \text{Gal}(f(\mathbf{T}, X), K_s(\mathbf{T}))$. By Lemma 16.2.3, f is X-stable over L. ☐

Example 16.2.5: *Stable polynomials.*
  (a) The **general polynomial of degree** n is

$$f(T_1, \ldots, T_n, X) = X^n + T_1 X^{n-1} + \cdots + T_n.$$

It satisfies $\text{Gal}(f(\mathbf{T}, X), K(\mathbf{T})) \cong S_n$ for every positive integer n and for every field K [Lang7, p. 272, Example 4]. Thus, f is X-stable over every field.
  (b) The polynomial $X^n - T$ satisfies $\text{Gal}(X^n - T, K(T)) \cong \mathbb{Z}/n\mathbb{Z}$ for every field K with $\text{char}(K) \nmid n$ that contains ζ_n. Thus, $X^n - T$ is X-stable over K. If however $\zeta_n \notin K$, then $\text{Gal}(X^n - T, K(T)) \cong \text{Gal}(K(\zeta_n)/K) \ltimes \mathbb{Z}/n\mathbb{Z}$. Therefore, $X^n - T$ is not X-stable over K.
  (c) The polynomial $X^p - X - T$ satisfies $\text{Gal}(X^p - X - T, K(T)) \cong \mathbb{Z}/p\mathbb{Z}$ for every field K of characteristic p [Lang7, p. 290, Thm. 6.4]. Therefore, f is X-stable over K.

16.2 Stable Polynomials

(d) Suppose $\text{Gal}(f(\mathbf{T}, X), K(\mathbf{T}))$ is simple. Let F be the splitting field of $f(\mathbf{T}, X)$ over $K(\mathbf{T})$. Suppose $F \not\subseteq K_s(\mathbf{T})$. Then $F \cap K_s(\mathbf{T}) = K(\mathbf{T})$. Therefore, f is X-stable over K.

(e) Suppose $f \in K[\mathbf{T}, X]$ is absolutely irreducible and Galois over $K(\mathbf{T})$. Then f is X-stable over K. Indeed, let L be an extension of K. Then $\text{Gal}(f(\mathbf{T}, X), L(\mathbf{T}))$ is a subgroup of $\text{Gal}(f(\mathbf{T}, X), K(\mathbf{T}))$. On the other hand, $f(\mathbf{T}, X)$ is also Galois over $L(\mathbf{T})$. Hence,

$$|\text{Gal}(f(\mathbf{T}, X), L(\mathbf{T}))| = \deg_X(f(\mathbf{T}, X)) = |\text{Gal}(f(\mathbf{T}, X), K(\mathbf{T}))|.$$

Thus, $\text{Gal}(f(\mathbf{T}, X), L(\mathbf{T})) \cong \text{Gal}(f(\mathbf{T}, X), K(\mathbf{T}))$, as claimed. □

We use stable polynomial to construct linearly disjoint sequences of Galois extensions over Hilbertian fields with a given Galois groups:

LEMMA 16.2.6: *Let $f_i(T_1, \ldots, T_r, X)$ be an X-stable polynomial over a Hilbertian field K and $G_i = \text{Gal}(f_i(\mathbf{T}, X), K(\mathbf{T}))$, $i = 1, 2, 3, \ldots$. Given a finite separable extension L_0 of K, there is a sequence, $L_1, L_2, L_3, \ldots$ of Galois extensions of K with the following properties:*

(a) $\text{Gal}(L_i/K) \cong G_i$, *and f_i has an L_i-rational zero $\mathbf{c}_i$, $i = 1, 2, 3, \ldots$.*
(b) $\mathbf{c}_1, \mathbf{c}_2, \mathbf{c}_3, \ldots$ *are distinct.*
(c) *The sequence $L_0, L_1, L_2, \ldots$ is linearly disjoint over K.*

Proof: Suppose by induction, there are Galois extensions $L_1, \ldots, L_n$ of K such that $\text{Gal}(L_i/K) \cong G_i$, $i = 1, \ldots, n$, and $L_0, L_1, \ldots, L_n$ are linearly disjoint over K. In addition suppose there are $\mathbf{a}_i \in K^r$ and $b_i \in L_i$ such that $f_i(\mathbf{a}_i, b_i) = 0$ and $\mathbf{a}_1, \ldots, \mathbf{a}_n$ are distinct. Then $L = L_0 L_1 \cdots L_n$ is a finite separable extension of K. By assumption, $\text{Gal}(f_{n+1}(\mathbf{T}, X), L(\mathbf{T})) \cong \text{Gal}(f_{n+1}(\mathbf{T}, X), K(\mathbf{T})) = G_{n+1}$. Apply Lemma 16.1.5 and Corollary 12.2.3 to find $\mathbf{a}_{n+1} \in K^r$ with

$$\text{Gal}(f_{n+1}(\mathbf{a}_{n+1}, X), L) \cong \text{Gal}(f_{n+1}(\mathbf{a}_{n+1}, X), K) \cong G_{n+1}$$

and $\mathbf{a}_{n+1} \neq \mathbf{a}_1, \ldots, \mathbf{a}_n$. Let L_{n+1} be the splitting field of $f(\mathbf{a}, X)$ over K. Then $\text{Gal}(L_{n+1}/K) \cong G_{n+1}$ and L_{n+1} is linearly disjoint from L over K. In particular, $L_0, \ldots, L_{n+1}$ are linearly disjoint over K. Finally, choose a zero b_{n+1} of $f_{n+1}(\mathbf{a}_{n+1}, X)$. Then $b_{n+1} \in L_{n+1}$. □

Our next result is an immediate application of this lemma to the polynomials of Example 16.2.5(a),(b),(c):

COROLLARY 16.2.7: *Let K be a Hilbertian field and G a finite group. Then K has a linearly disjoint sequence of Galois extensions with G as Galois group in each of the following cases:*
(a) $G = S_n$, $n \in \mathbb{N}$.
(b) $\zeta_n \in K$ *and* $G = \mathbb{Z}/n\mathbb{Z}$, $n \in \mathbb{N}$, $\text{char}(K) \nmid n$.

(c) $\mathrm{char}(K) = p > 0$ and $G = \mathbb{Z}/p\mathbb{Z}$.

We prove in Section 16.4 that every finite Abelian group A satisfies the conclusion of Corollary 16.2.7. But this simple corollary already has interesting implications to closed normal subgroups of $\mathrm{Gal}(K)$ (Section 16.12).

The following result shows that in many cases it suffices to take $r = 1$ in the definition of "G is regular over K":

PROPOSITION 16.2.8: *Let K be an infinite field and G a finite group. Suppose G is K-regular. Then:*
(a) *There is an X-stable polynomial $h \in K[T, X]$ which is Galois with respect to X and with $\mathrm{Gal}(h(T, X), K(T)) \cong G$.*
(b) *If in addition K is Hilbertian, then K has a linearly disjoint sequence $L_1, L_2, L_3, \ldots$ of Galois extensions such that $\mathrm{Gal}(L_i/K) \cong G$ for all i.*

Proof: Statement (b) follows from (a) by Lemma 16.2.6. We prove (a):

Remark 16.2.2 gives a polynomial $f \in K[T_1, \ldots, T_r, X]$ which is X-stable over K with $r \geq 2$ and Galois with respect to X over $K(\mathbf{T})$ such that $\mathrm{Gal}(f(\mathbf{T}, X), K(\mathbf{T})) \cong G$. Assume without loss f is monic with respect to X.

Let u_1, u_2 be algebraically independent elements over K. Put $L = K(u_1, u_2)$, $\mathbf{T}' = (T_1, \ldots, T_{r-1})$, and

$$g(\mathbf{T}', X) = f(T_1, \ldots, T_{r-1}, u_1 + u_2 T_{r-1}, X).$$

By Proposition 10.5.4, $g(\mathbf{T}', X)$ is absolutely irreducible. Extend the map $\mathbf{T} \to (\mathbf{T}', u_1 + u_2 T_{r-1})$ to an $L(\mathbf{T}')$-place of $L(\mathbf{T})$ with residue field $L(\mathbf{T}')$ (Lemma 2.2.7). By Lemma 16.1.1,

$$G \cong \mathrm{Gal}(f(\mathbf{T}, X), L(\mathbf{T})) \cong \mathrm{Gal}(g(\mathbf{T}', X), L(\mathbf{T}')).$$

By assumption, K is infinite. Hence, Proposition 16.1.4 gives $b_1, b_2 \in K$ with $\bar{g}(\mathbf{T}', X) = f(T_1, \ldots, T_{r-1}, b_1 + b_2 T_{r-1}, X)$ absolutely irreducible, Galois over $K(\mathbf{T}')$, and $\mathrm{Gal}(\bar{g}(\mathbf{T}', X), K(\mathbf{T}')) \cong \mathrm{Gal}(g(\mathbf{T}', X), L(\mathbf{T}')) \cong G$.

Finally, use induction on r to find an absolutely irreducible polynomial $h \in K[T_1, X]$ with $\mathrm{Gal}(h(T_1, X), K(T_1)) \cong G$. □

Problem 16.2.9: Does Proposition 16.2.8 hold if K is finite?

16.3 Regular Realization of Finite Abelian Groups

The inverse problem of Galois theory has an affirmative solutions for every finite Abelian group A and every Hilbertian field K (Corollary 16.3.6). Moreover, every finite Abelian group is even regular over K (Proposition 16.3.5). The proof of the latter result moves from the case where A is cyclic to the general case. In the cyclic case we have to distinguish between the case where $\mathrm{char}(K)$ does not divide the order of the group and where the order of the group is a power of $\mathrm{char}(K)$. In the general case, we distinguish between the cases where K is finite and K is infinite.

16.3 Regular Realization of Finite Abelian Groups

LEMMA 16.3.1: *Let K be a field, n a positive integer with $\mathrm{char}(K) \nmid n$, and t an indeterminate. Then $K(t)$ has a cyclic extension F of degree n which is contained in $K((t))$.*

Proof: Choose a root of unity ζ_n of order n in K_s. Let $L = K(\zeta_n)$ and $G = \mathrm{Gal}(L/K)$. Then there is a map $\chi \colon G \to \{1, \ldots, n-1\}$ such that $\sigma(\zeta_n) = \zeta_n^{\chi(\sigma)}$. Then $\gcd(\chi(\sigma), n) = 1$ and

(1) $$\chi(\sigma\tau) \equiv \chi(\sigma)\chi(\tau) \bmod n$$

for all $\sigma, \tau \in G$. By Example 3.5.1, $K((t))$ is a regular extension of K and $L((t)) = K((t))(\zeta_n)$. Thus, we may identify G with $\mathrm{Gal}(L((t))/K((t)))$.

Choose a primitive element c of L/K. Consider the element

$$g(t) = \prod_{\sigma \in G} (1 + \sigma(c)t)^{\chi(\sigma^{-1})}$$

of $L[t]$. Since $\mathrm{char}(K) \nmid n$, Hensel's lemma (Proposition 3.5.2) gives an $x \in L[[t]]$ with $x^n = 1 + ct$. Then $y = \prod_{\sigma \in G} \sigma(x)^{\chi(\sigma^{-1})} \in L[[t]]$ and $y^n = \prod_{\sigma \in G} \sigma(x^n)^{\chi(\sigma^{-1})} = \prod_{\sigma \in G}(1 + \sigma(c)t)^{\chi(\sigma^{-1})} = g(t)$. Since $\zeta_n \in L$, $F = L(t, y)$ is a cyclic extension of degree d of $L(t)$, where $d | n$ and $y^d \in L(t)$ [Lang7, p. 289, Thm. 6.2(ii)]. Since $\chi(\sigma^{-1})$ is relatively prime to n, we must have $d = n$. The Galois group $\mathrm{Gal}(F/L(t))$ is generated by an element ω satisfying $\omega(y) = \zeta_n y$.

By (1) there exist for each $\tau, \rho \in G$ a positive integer $k(\tau, \rho)$ and a polynomial $f_\tau(t) \in L[t]$ such that $\tau(y) = \prod_{\sigma \in G} \tau\sigma(x)^{\chi(\sigma^{-1})} = \prod_{\rho \in G} \rho(x)^{\chi(\rho^{-1}\tau)} = \prod_{\rho \in G} \rho(x)^{\chi(\rho^{-1})\chi(\tau)+k(\tau,\rho)n} = y^{\chi(\tau)} \prod_{\rho \in G}(1 + \rho(c)t)^{k(\tau,\rho)} = y^{\chi(\tau)} f_\tau(t)$. It follows that G leaves F invariant. Let E be the fixed field of G in F.

$$\begin{array}{ccc} K((t)) & \!\!\!\text{---}\!\!\! & L((t)) \\ | & & | \\ E & \!\!\!\text{---}\!\!\! & F = L(t,y) \\ | & & | \\ K(t) & \!\!\!\text{---}\!\!\! & L(t) \\ | & & | \\ K & \!\!\!\text{---}\!\!\! & L = K(\zeta_n) \end{array}$$

Denote the subgroup of $\mathrm{Aut}(F/K(t))$ generated by G and $\mathrm{Gal}(F/L(t))$ by H. Then the fixed field of H is $K(t)$, so $F/K(t)$ is a Galois extension with $\mathrm{Gal}(F/K(t)) = G \cdot \mathrm{Gal}(F/L(t))$. Moreover, given $\tau \in G$, put $m = \chi(\tau)$. Then $\tau\omega(y) = \tau(\zeta_n y) = \zeta_n^m y^m f_\tau(t) = \omega(y)^m f_\tau(t) = \omega(y^m f_\tau(t)) = \omega\tau(y)$. Thus, $\tau\omega = \omega\tau$, so G commutes with $\mathrm{Gal}(F/L(t))$. Therefore, $E/K(t)$ is a Galois extension with $\mathrm{Gal}(E/K(t)) \cong \mathrm{Gal}(F/L(t)) \cong \mathbb{Z}/n\mathbb{Z}$. $\square$

LEMMA 16.3.2: *Suppose $p = \mathrm{char}(K)$. Let L be a cyclic extension of degree p^n, $n \geq 1$, of K. Then K has a $\mathbb{Z}/p^{n+1}\mathbb{Z}$-extension L' which contains L.*

Proof: Define L' to be $L(x)$ where x is a zero of $X^p - X - a$ with $a \in L$. The three parts of the proof produce a, and then show L' has the desired properties.

PART A: *Construction of a.* Since L/K is separable, there is a $b_1 \in L$ with $c = \mathrm{trace}_{L/K}(b_1) \neq 0$ [Lang7, p. 286, Thm. 5.2]. Put $b = \frac{b_1}{c}$. Then $\mathrm{trace}_{L/K}(b) = 1$ and $\mathrm{trace}_{L/K}(b^p - b) = (\mathrm{trace}_{L/K}(b))^p - \mathrm{trace}_{L/K}(b) = 0$. With σ a generator of $\mathrm{Gal}(L/K)$, the additive form of Hilbert's Theorem 90 [Lang7, p. 290, Thm. 6.3] gives $a \in L$ with

(2) $$\sigma a - a = b^p - b.$$

PART B: *Irreducibility of $X^p - X - a$.* Assume $X^p - X - a$ is reducible over L. Then $x \in L$ [Lang7, p. 290, Thm. 6.4(b)]. Thus

(3) $(\sigma x - x)^p - (\sigma x - x) - (b^p - b) = (\sigma x - x)^p - (\sigma x - x) - (\sigma a - a)$
$$= (\sigma x^p - \sigma x - \sigma a) - (x^p - x - a) = 0$$

Since b is a root of $X^p - X - (b^p - b)$, there is an i with $\sigma x - x = b + i$ [Lang7, p. 290, Thm. 6.4(b)]. Apply $\mathrm{trace}_{L/K}$ to both sides to get 0 on the left and 1 on the right. This contradiction proves $X^p - X - a$ is irreducible.

PART C: *Extension of σ to σ' that maps x to $x+b$.* Equality (2) implies $x+b$ is a zero of $X^p - X - \sigma a$. Thus, by Part B, σ extends to an automorphism σ' of L' with $\sigma'(x) = x + b$. We need only prove that σ' has order p^{n+1}. Induction shows $(\sigma')^j(x) = x + b + \sigma b + \cdots + \sigma^{j-1} b$. In particular,

(4) $$(\sigma')^{p^n}(x) = x + \mathrm{trace}_{L/K}(b) = x + 1.$$

Hence, $(\sigma')^{ip^n}(x) = x + i$, $i = 1, \ldots, p$. Therefore, the order of σ' is p^{n+1}, as contended. □

Remark 16.3.3: *Lemma 16.3.2 is a special case of a theorem of Witt.* Suppose $\mathrm{char}(K) = p$. Then $\mathrm{AS}(K) = \{x^p - x \mid x \in K\}$ is a subgroup of the additive group of K and $K/\mathrm{AS}(K)$ is a vector space over $\mathbb{F}_p$ of dimension, say, r. Consider an embedding problem $G \to \mathrm{Gal}(L/K)$ over K with G a finite p-group which is generated by r elements. Witt's theorem says this problem is solvable (If $r = \infty$, there is no restriction on the number of generators of G.) The technique of Galois cohomology [Ribes, p. 257, Cor. 3.4] simplifies Witt's original proof [Witt]. □

LEMMA 16.3.4: *Let K be a field, t an indeterminate, and A a finite cyclic group. Then $K(t)$ has a Galois extension F such that $\mathrm{Gal}(F/K(t)) \cong A$ and F/K is regular.*

Proof: We put $p = \mathrm{char}(K)$ and divide the proof into three parts:

16.3 Regular Realization of Finite Abelian Groups

PART A: $A \cong \mathbb{Z}/m\mathbb{Z}$ and $p \nmid m$. By Lemma 16.3.1, $K(t)$ has a cyclic extension E_m of degree m which is contained in $K((t))$. By Example 3.5.1, $K((t))$ is a regular extension of K. Hence, so is E_m (Corollary 2.6.5(b)).

PART B: $A \cong \mathbb{Z}/p^k\mathbb{Z}$. Assume without loss that $k \geq 1$. By Eisenstein's criterion and Gauss' lemma, the polynomial $X^p - X - t$ is irreducible over $\tilde{K}(t)$. Let x be a root of $X^p - X - t$ in $K(t)_s$. Then, by Artin-Schreier, [Lang7, p. 290, Thm. 6.4(b)], $K(x)$ is a cyclic extension of degree p of $K(t)$. Lemma 16.3.2 gives a cyclic extension E_{p^k} of $K(t)$ of degree p^k which contains $K(x)$.

By the preceding paragraph, $K(x) \cap \tilde{K}(t) = K(t)$. Since $\mathrm{Gal}(E_{p^k}/K(t))$ is a cyclic group of order p^k, each subextension of E_{p^k} which properly contains $K(t)$ must contain $K(x)$. Hence, $E_{p^k} \cap \tilde{K}(t) = K(t)$. Thus, E_{p^k} is linearly disjoint from $\tilde{K}(t)$ over $K(t)$. By the tower property (Lemma 2.5.3), E_{p^k} is linearly disjoint from $\tilde{K}$ over K; that is, E_{p^k}/K is regular.

PART C: $A \cong \mathbb{Z}/n\mathbb{Z}$, $n = mp^k$, $p \nmid m$. The compositum $E_n = E_m E_{p^k}$ is a cyclic extension of $K(t)$ of degree n. Moreover, $E_n \cap \tilde{K}(t)$ decomposes into a cyclic extension of $K(t)$ of degree which divides m and a cyclic extension of $K(t)$ degree dividing p^k. By Parts A and B, both subextensions must be $K(t)$. It follows that E_n is a regular extension of K. □

We generalize Lemma 16.3.4 from cyclic groups to arbitrary Abelian groups:

PROPOSITION 16.3.5: *Let K be a field, t an indeterminate, and A a finite Abelian group. Then $K(t)$ has a Galois extension F such that $\mathrm{Gal}(F/K(t)) \cong A$ and F/K is regular.*

Proof: The first two parts of the proof prove the proposition in the case where $A = (\mathbb{Z}/q\mathbb{Z})^n$, $q = p^r$, p is a prime number, and n, r are positive integers.

PART A: $A = (\mathbb{Z}/q\mathbb{Z})^n$ *is as above and K is infinite*. Choose algebraically independent elements $t_1, \ldots, t_n$ over K. For each i between 1 and n, Lemma 16.3.4 gives a finite cyclic extension E_i of $K(t_i)$ of degree q which is regular over K. Then $E_1, \ldots, E_n$ are algebraically independent over K. Hence, by Corollary 2.6.8, $E = E_1 \cdots E_n$ is a regular extension of K. In addition, by Lemma 2.6.7, $E_1, \ldots, E_n$ are linearly disjoint over K. Hence, by Lemma 2.5.11, $E_1(\mathbf{t}), \ldots, E_n(\mathbf{t})$ are linearly disjoint over $K(\mathbf{t})$ and each $E_i(\mathbf{t})$ is a cyclic extension of $K(\mathbf{t})$ of degree q. It follows that $\mathrm{Gal}(E/K(\mathbf{t})) \cong (\mathbb{Z}/q\mathbb{Z})^n$. Consequently, by Proposition 16.2.8, $K(t)$ has a cyclic extension F of degree q which is regular over K.

PART B: $A = (\mathbb{Z}/q\mathbb{Z})^n$ *is as above and K is finite*. Choose a transcendental element u over $K(t)$. Lemma 16.3.4 gives a cyclic extension E of $K(t, u)$ of degree q which is regular over $K(t)$. Choose a polynomial $h \in K(t)[u, X]$ which is X-stable with $\mathrm{Gal}(h(u, X), K(t, u)) \cong \mathbb{Z}/q\mathbb{Z}$ (e.g. use Proposition

16.2.8). Denote the unique extension of K of degree q by K_q. By Theorem 13.4.2, $K(t)$ is Hilbertian. Hence, by Lemma 16.2.6, $K(t)$ has sequence $F_1, F_2, F_3, \ldots$ of cyclic extensions of degree q such that $K_q(t), F_1, F_2, \ldots, F_n$ are linearly disjoint over $K(t)$. Then $F = F_1 F_2 \cdots F_n$ is a Galois extension of $K(t)$ with $\mathrm{Gal}(F/K(t)) \cong (\mathbb{Z}/q\mathbb{Z})^n$. Moreover, $F \cap K_q(t) = K(t)$.

The group $A = \mathrm{Gal}(F \cap \tilde{K}(t)/K)$ is cyclic quotient of $\mathrm{Gal}(F/K(t))$, so A is of exponent q. Hence, $[F \cap \tilde{K}(t) : K(t)] = p^j$ with $j \leq r$. Therefore, $F \cap \tilde{K}(t) \subseteq K_q(t)$. It follows from the preceding paragraph that $F \cap \tilde{K}(t) = K(t)$. Consequently, F/K is regular.

PART C: $A \cong \prod_{i=1}^{m} (\mathbb{Z}/q_i\mathbb{Z})^{n_i}$, where $q_i = p_i^{r_i}$, m, n_i, r_i are positive integers, and $p_1, \ldots, p_r$ are distinct prime numbers. Parts A and B give for each i a Galois extension F_i of $K(t)$ which is regular over K and with $\mathrm{Gal}(F_i/K(t)) \cong (\mathbb{Z}/q_i\mathbb{Z})^{n_i}$. Since $q_1, \ldots, q_m$ are pairwise relatively prime, $F_1, \ldots, F_m$ are linearly disjoint over $K(t)$. Hence, $F = F_1 \cdots F_m$ is a Galois extension with $\mathrm{Gal}(F/K(t)) \cong A$.

Let $E = F \cap \tilde{K}(t)$. By the preceding paragraph, $F_i \cap E = K(t)$, so $q_i^{n_i} | [F : E]$ for $i = 1, \ldots, m$. Since $q_1, \ldots, q_m$ are pairwise relatively prime, $[F : K(t)] = \prod_{i=1}^{m} q_i^{n_i}$ divides $[F : E]$. Hence, $E = K(t)$. Consequently, F is a regular extension of K.

PART D: A is an arbitrary finite Abelian group. Then

$$A = \prod_{i=1}^{m} \prod_{j=1}^{m_i} (\mathbb{Z}/p_i^{r_{ij}}\mathbb{Z})^{n_{ij}}$$

where $p_1, \ldots, p_m$ are distinct prime numbers, $m \geq 0$, and m_i, n_{ij} are positive integers. For each i let $r_i = \max(r_{i1}, \ldots, r_{i,m_i})$ and $n_i = \max(n_{i1}, \ldots, n_{i,m_i})$. Part C gives a Galois extension $\hat{F}$ of $K(t)$ which is regular over K and with $\mathrm{Gal}(\hat{F}/K(t)) \cong \prod_{i=1}^{m} (\mathbb{Z}/p_i^{r_i}\mathbb{Z})^{n_i}$. By construction, A is a quotient of $\mathrm{Gal}(\hat{F}/K(t))$. Hence, $K(t)$ has a Galois extension F in $\hat{F}$ with $\mathrm{Gal}(F/K(t)) \cong A$. Since $\hat{F}/K$ is regular, so is F/K. □

COROLLARY 16.3.6: *Let K be a Hilbertian field and A a finite Abelian group. Then A is realizable over K. Moreover, K has a linearly disjoint sequence $L_1, L_2, L_3, \ldots$ of Galois extensions with $\mathrm{Gal}(L_i/K) \cong A$ for each i.*

Proof: By Proposition 16.3.5, A is regular over K. Hence, by Proposition 16.2.8, K has a linearly disjoint sequence $L_1, L_2, L_3, \ldots$ of Galois extensions with $\mathrm{Gal}(L_i/K) \cong A$ for each i. □

16.4 Split Embedding Problems with Abelian Kernels

Attempts to realize a finite Abelian group A over a Hilbertian field K usually lead to an extension of K with roots of unity. This gives a "split embedding problem with Abelian kernel". The main result of this section is that each

16.4 Split Embedding Problems with Abelian Kernels

such problem is solvable. Consequently, A occurs over K. Indeed, for A to be regular over K it is not necessary that K is Hilbertian. We prove that it is true for arbitrary field.

Definition 16.4.1: Embedding problems. Let L/K be a Galois extension, G a profinite group, and $\alpha\colon G \to \mathrm{Gal}(L/K)$ an epimorphism. The **embedding problem** associated with α consists of embedding L in a Galois extension N of K with an isomorphism $\beta\colon \mathrm{Gal}(N/K) \to G$ satisfying $\alpha \circ \beta = \mathrm{res}_{N/L}$. Refer to α as an **embedding problem over** K, β its **solution**, and N the **solution field**.

Call the problem **finite** if G is finite. The problem **splits** if α has a section, that is, an embedding $\alpha'\colon \mathrm{Gal}(L/K) \to G$ with $\alpha \circ \alpha' = \mathrm{id}_{\mathrm{Gal}(L/K)}$. The latter case occurs when $G = \mathrm{Gal}(L/K) \ltimes \mathrm{Ker}(\alpha)$ and α is the projection of G onto $\mathrm{Gal}(L/K)$.

Let $t_1,\ldots,t_n$ be algebraically independent elements over K. Then $\mathrm{res}\colon \mathrm{Gal}(L(\mathbf{t})/K(\mathbf{t})) \to \mathrm{Gal}(L/K)$ is an isomorphism. Hence, $\alpha\colon G \to \mathrm{Gal}(L/K)$ gives rise to an embedding problem $\alpha_\mathbf{t}\colon G \to \mathrm{Gal}(L(\mathbf{t})/K(\mathbf{t}))$ over $K(\mathbf{t})$ with $\mathrm{res}_{L(\mathbf{t})/L} \circ \alpha_\mathbf{t} = \alpha$. Refer to a solution of $\alpha_\mathbf{t}$ as a **solution of α over $K(\mathbf{t})$**. Refer to a solution field F of $\alpha_\mathbf{t}$ as a **regular solution** of α if F/L is regular. We say α is **regularly solvable** if there are $t_1,\ldots,t_n$ as above and $\alpha_\mathbf{t}$ has a solution field F which is regular over L. □

LEMMA 16.4.2: Let K be a Hilbertian field, $\alpha\colon H \to \mathrm{Gal}(L/K)$ a finite embedding problem, $t_1,\ldots,t_n$ algebraically independent elements over K, and M a finite separable extension of L. If α is solvable over $K(\mathbf{t})$, then α is also solvable over K. If α is regularly solvable, then α has a solution field N over K which is linearly disjoint from M over L.

Proof: Let F be a solution field of α over $K(\mathbf{t})$. Thus, F is a Galois extension of $K(\mathbf{t})$ which contains L and there is an isomorphism $\theta\colon \mathrm{Gal}(F/K(\mathbf{t})) \to H$ with $\alpha \circ \theta = \mathrm{res}_{F/L}$. Lemma 13.1.1 gives a separable Hilbert subset A of K^r having the following property: For each $\mathbf{a} \in A$ there is a K-place $\varphi\colon F \to \tilde{K} \cup \{\infty\}$ satisfying these conditions:

(3a) $\varphi(\mathbf{t}) = \mathbf{a}$, $\overline{K(\mathbf{t})}_\varphi = K$, and $\overline{L(\mathbf{t})}_\varphi = L$.

(3b) There is an isomorphism $\varphi^*\colon \mathrm{Gal}(\bar{F}/K) \to \mathrm{Gal}(F/K(\mathbf{t}))$ with $\mathrm{res}_{F/L} \circ \varphi^* = \mathrm{res}_{\bar{F}/L}$. Here, $\bar{F} = \bar{F}_\varphi$.

The map $\beta = \theta \circ \varphi^*$ solves embedding problem α.

Suppose now F/L is regular. Then $\mathrm{Gal}(FM/M(\mathbf{t})) \cong \mathrm{Gal}(F/L(\mathbf{t}))$. Lemma 13.1.1 gives a separable Hilbert subset A' of M^r satisfying this: For each $\mathbf{a} \in A \cap A'$ and each M-place φ of FM satisfying (3), $\mathrm{Gal}(FM/M(\mathbf{t})) \cong \mathrm{Gal}(\bar{F}M/M)$. By Corollary 12.2.3, there is an $\mathbf{a} \in A \cap A'$. The preceding paragraph gives φ satisfying (3). The corresponding field $\bar{F}$ is linearly disjoint from M over L. □

Let A and G be finite groups. Recall that A^G is the group of all functions $f\colon G \to A$ and G acts on A^G by the rule $f^\tau(\sigma) = f(\tau\sigma)$. The semidirect product $G \ltimes A^G$ is the wreath product $A \,\mathrm{wr}\, G$ (Remark 13.7.7).

LEMMA 16.4.3: *Let G and A be finite groups with A Abelian and G acting on A. Then there is an epimorphism $A \operatorname{wr} G \to G \ltimes A$ with kernel in A^G.*

Proof: Define a map $\alpha \colon A^G \to A$ by $\alpha(f) = \prod_{\sigma \in G} f(\sigma)^{\sigma^{-1}}$. Since A is Abelian, the right hand side is a well defined homomorphism. For each $a \in A$ define a function $f_a \colon G \to A$ by $f_a(1) = a$ and $f_a(\sigma) = 1$ for $\sigma \in G$, $\sigma \neq 1$. Then $\alpha(f_a) = a$, so α is surjective.

Next consider $\tau \in G$. Then

$$\alpha(f^\tau) = \prod_{\sigma \in G} f^\tau(\sigma)^{\sigma^{-1}} = \prod_{\sigma \in G} f(\tau\sigma)^{\sigma^{-1}} = \prod_{\rho \in G} f(\rho)^{\rho^{-1}\tau} = \alpha(f)^\tau.$$

Thus, α respects the action of G. Therefore, α extends to an epimorphism $\alpha \colon G \ltimes A^G \to G \ltimes A$ satisfying $\alpha(\sigma f) = \sigma \alpha(f)$ for all $\sigma \in G$ and $f \in A^G$. This gives rise to the following commutative diagram:

$$\begin{array}{ccccccccc} 1 & \longrightarrow & A^G & \longrightarrow & G \ltimes A^G & \longrightarrow & G & \longrightarrow & 1 \\ & & \alpha \downarrow & & \alpha \downarrow & & \| & & \\ 1 & \longrightarrow & A & \longrightarrow & G \ltimes A & \longrightarrow & G & \longrightarrow & 1 \end{array}$$

By definition, $\operatorname{Ker}(\alpha) \leq A^G$. □

PROPOSITION 16.4.4: *Let L/K be a finite Galois extension of degree n with Galois group G. Suppose G acts on a finite Abelian group A. Let $\pi \colon G \ltimes A \to G$ be the projection map. Then π is regularly solvable.*

Proof: By Proposition 16.3.5, $K(T)$ has a Galois extension N such that $\operatorname{Gal}(N/K(T)) \cong A$ and N/K is regular. By Proposition 16.2.8, there exists an X-stable polynomial $h \in K[T, X]$ with $\operatorname{Gal}(h(T, X), K(T)) \cong A$. In particular, $h(T, X)$ is absolutely irreducible and $\operatorname{Gal}(h(T, X), L(T)) \cong A$. Let $\hat{\pi} \colon G \ltimes A^G \to G$ be the projection on G. Let $t_1, \ldots, t_n$ be algebraically independent elements over K with $n = [L : K]$. Lemma 13.8.1, with G_0 trivial, gives a Galois extension $\hat{F}$ of $K(\mathbf{t})$ which contains L and an isomorphism $\gamma \colon \operatorname{Gal}(\hat{F}/K(\mathbf{t})) \to G \ltimes A^G$ with $\hat{\pi} \circ \gamma = \operatorname{res}$. Moreover, $\hat{F}$ is a regular extension of L.

Lemma 16.4.3 gives an epimorphism $\alpha \colon G \ltimes A^G \to G \ltimes A$ which is the identity map on G. Thus, $\hat{\pi} = \pi \circ \alpha$. Let F be the fixed field in $\hat{F}$ of $\operatorname{Ker}(\alpha \circ \gamma)$. Then there is an epimorphism $\beta \colon \operatorname{Gal}(F/K(\mathbf{t})) \to G \ltimes A$ with $\alpha \circ \gamma = \beta \circ \operatorname{res}_{\hat{F}/F}$ and $\pi \circ \beta = \operatorname{res}_{F/L}$. Thus, β is a solution of π over $K(\mathbf{t})$ with F being the solution field. By construction, $L \subseteq F \subseteq \hat{F}$. Since $\hat{F}/L$ is regular, so is F/L (Corollary 2.6.5(b)). □

PROPOSITION 16.4.5 ([Ikeda, p. 126]): *Let K be a Hilbertian field. Then every finite split embedding problem over K with Abelian kernel is solvable*

Proof: Combine Propositions 16.4.2 and 16.4.4. □

16.4 Split Embedding Problems with Abelian Kernels

Remark 16.4.6: Proposition 16.4.5 does not hold for an arbitrary profinite Abelian group. For example, $\mathbb{Z}_p$ is regular over $\mathbb{Q}$ for no p (Proposition 16.6.10), but it is not known if every finite p-group is regular over $\mathbb{Q}$. Nevertheless, every finite p-group occurs over $\mathbb{Q}$. More generally, every finite solvable group occurs over every global field. This is a theorem of Shafarevich. Its proof does not use Hilbert irreducibility theorem but rather class field theory and complicated combinatorial arguments. See [Neukirch-Schmidt-Wingberg, Section 9.5] for a proof that uses cohomological arguments and for a reference to the original articles.

Class field theory is not available over an arbitrary Hilbertian field K. Thus, Shafarevich's proof does not apply to K. However, when $p = \mathrm{char}(K) > 0$, every finite p-group occurs over K. This follows from a theorem of Witt (Remark 16.3.3), but it is not known if each finite p-group occurs over K when $\mathrm{char}(K) = 0$. □

Amazingly enough, both realization problems raised in Remark 16.4.6 are easy consequences of Shafarevich's theorem when $\mathrm{char}(K) > 0$. This is the content of the following result:

THEOREM 16.4.7: *Let G be a finite p-group and K a field of positive characteristic. Then G is regular over K. If in addition K is Hilbertian, then G is realizable over K.*

Proof: The second statement of the theorem follows from the first one by Hilbert (Lemma 13.1.1). In order to prove that G is regular over every field of positive characteristic, it suffices to prove that G is regular over every finite field K (Lemma 16.2.1).

Assume without loss G is nontrivial. Let t be an indeterminate. By Shafarevich, $G \times G$ occurs over $K(t)$. Thus, $K(t)$ has linearly disjoint Galois extensions F_1 and F_2 with $\mathrm{Gal}(F_i/K(t)) \cong G$, $i = 1, 2$. Assume none of them is regular over K. Then, for $i = 1, 2$, the field $\tilde{K} \cap F_i$ is a cyclic extension of K of degree p^{k_i} with $k_i \geq 1$. Hence, F_i contains the unique extension K_p of K of degree p. Therefore, both F_1 and F_2 contain $K_p(t)$ which is a proper extension of $K(t)$. This contradiction to the linear disjointness of F_1 and F_2 over $K(t)$ proves that one of them is regular over K. □

COROLLARY 16.4.8: *Let K be a field, G a finite group, and A a finite Abelian group. Suppose G is regular over K and G acts on A. Then $G \ltimes A$ is regular over K.*

Proof: There exists $t_1, \ldots, t_n$ and a finite Galois extension E of $K(\mathbf{t})$ such that $\mathrm{Gal}(E/K(\mathbf{t})) \cong G$ and E/K is regular. Choose an indeterminate u. Proposition 16.4.4 gives a Galois extension F of $K(\mathbf{t}, u)$ such that $\mathrm{Gal}(F/K(\mathbf{t}, u)) \cong G \ltimes A$ and F is a regular extension of E. By Corollary 6.2.5, F is a regular extension of K. Thus, $G \ltimes A$ is regular over K. □

Remark 16.4.9: Realization of p-groups of low order. The results of this section imply that each p-group of order at most p^4 is regular over every field K.

To this end let $\mathcal{A}$ be the smallest family of all finite groups satisfying this:
(4a) Every Abelian group belongs to $\mathcal{A}$.
(4b) Suppose $H = G \cdot A$ with $G \in \mathcal{A}$ and A Abelian and normal. Then $H \in \mathcal{A}$.

Note that the group H in (4b) is a quotient of $G \ltimes A$. Hence, if G is regular over K, so is H (Proposition 16.4.8). It follows by induction on the order of the group that each $G \in \mathcal{A}$ is regular over K. Thus, if K is Hilbertian, every $G \in \mathcal{A}$ occurs over K (Lemma 13.1.1(b)).

The family $\mathcal{A}$ contains each finite group G which satisfies one of the following conditions:
(5a) G has nilpotence class at most 2, i.e. $[G,G] \leq Z(G)$ (Thompson [Malle-Matzat, p. 277, Prop. 2.9(a)]).
(5b) G is solvable and every Sylow subgroup of G is Abelian (Thompson [Malle-Matzat, p. 277, Prop. 2.9(b)]).
(5c) G is a p-group of order at most p^4 (Dentzer [Malle-Matzat, p. 278, Cor. 2.10].
(5d) G is a 2-group of order 2^5 (Dentzer [Malle-Matzat, p. 278, Cor. 2.10]).

However, there are groups of order p^5 (for $p \neq 2$) and 2^6 which do not belong to G [Malle-Matzat, p. 278]. $\square$

16.5 Embedding Quadratic Extensions in $\mathbb{Z}/2^n\mathbb{Z}$-Extensions

Nonsplit embedding problems with Abelian kernel over Hilbertian fields need not be solvable. For example $\mathbb{Z}/4\mathbb{Z} \to \mathrm{Gal}(\mathbb{Q}(\sqrt{-1})/\mathbb{Q})$ is not solvable. Otherwise, $\mathbb{Q}$ has a Galois extension N containing $\mathbb{Q}(\sqrt{-1})$ with $\mathrm{Gal}(N/\mathbb{Q}) \cong \mathbb{Z}/4\mathbb{Z}$. The only subfields of N are $\mathbb{Q}$, $\mathbb{Q}(\sqrt{-1})$, and N. Hence, $N \cap \mathbb{R} = \mathbb{Q}$. Hence, $[N\mathbb{R} : \mathbb{R}] = 4$, which is a contradiction.

Here is a general criterion for a quadratic extension to be embeddable in a $\mathbb{Z}/4\mathbb{Z}$-extension.

PROPOSITION 16.5.1: *Let K be a field with $\mathrm{char}(K) \neq 2$ and a a nonsquare in K. Then the embedding problem $\mathbb{Z}/4\mathbb{Z} \to \mathrm{Gal}(K(\sqrt{a})/K)$ is solvable if and only if there are $x, y \in K$ with $a = x^2 + y^2$.*

Proof: By assumption, $L = K(\sqrt{a})$ is a quadratic extension of K. Suppose K has a Galois extension N containing L with $\mathrm{Gal}(N/K) \cong \mathbb{Z}/4\mathbb{Z}$. Then there is a $u \in L$ with $N = L(\sqrt{u})$. Let σ be a generator of $\mathrm{Gal}(N/K)$. Then $L(\sqrt{\sigma u}) = L(\sqrt{u})$. By Kummer theory, $\sigma\sqrt{u} = v\sqrt{u}$ with $v \in L$. Hence, $\sigma^2 \sqrt{u} = v \cdot \sigma v \sqrt{u} = b\sqrt{u}$ with $b = v \cdot \sigma v$. Thus, b is an element of K which is a norm from L. On the other had, σ^2 generates $\mathrm{Gal}(L(\sqrt{u})/L)$, so $\sigma^2 \sqrt{u} = -\sqrt{u}$. Therefore, $b = -1$, $v \cdot \sigma v \sqrt{u} = -\sqrt{u}$, and $v \cdot \sigma v = -1$.

Finally, write $v = c + d\sqrt{a}$ with $c, d \in K$. First suppose $d = 0$. Then $v \in K$ and $v^2 \sqrt{u} = -\sqrt{u}$. Hence, $v^2 = -1$ and $\sqrt{-1} \in K$. Therefore, the

16.5 Embedding Quadratic Extensions in $\mathbb{Z}/2^n\mathbb{Z}$-Extensions

identity $a = \left(\sqrt{-1}(1 - \frac{a}{4})\right)^2 + \left(1 + \frac{a}{4}\right)^2$ yields a representation of a as a sum of two squares in K. Now suppose $d \neq 0$. Then $-1 = v \cdot \sigma v = c^2 - ad^2$. Hence, $a = \left(\frac{c}{d}\right)^2 + \left(\frac{1}{d}\right)^2$, as desired.

Conversely, suppose $a = x^2 + y^2$ with $x, y \in K$. Then $x, y \neq 0$ and $-1 = c^2 - ad^2$, where $c = \frac{x}{y}$ and $d = \frac{1}{y}$. Put $v = c + d\sqrt{a}$. Then $\mathrm{norm}_{L/K} v = c^2 - ad^2 = -1$. Hence, $\mathrm{norm}_{L/K} v^2 = 1$. Let σ be an element of $\mathrm{Gal}(K)$ whose restriction to L generates $\mathrm{Gal}(L/K)$. Hilbert's Theorem 90 gives $u \in L$ with $v^2 = \frac{\sigma u}{u}$ [Lang7, p. 288, Thm. 6.1]. Thus, $v^2 = \left(\frac{\sigma\sqrt{u}}{\sqrt{u}}\right)^2$ and $\frac{\sigma\sqrt{u}}{\sqrt{u}} = \pm v$. Replacing v by $-v$, if necessary, we may assume $\sigma\sqrt{u} = v\sqrt{u}$, $\sigma^2\sqrt{u} = v \cdot \sigma v \sqrt{u} = -\sqrt{u}$, $\sigma^3\sqrt{u} = -v\sqrt{u}$, and $\sigma^4\sqrt{u} = \sqrt{u}$. Thus, $\sqrt{u}$, $\sigma\sqrt{u}$, $\sigma^2\sqrt{u}$, $\sigma^3\sqrt{u}$ are distinct conjugates of $\sqrt{u}$ over K. All of them belong to $N = L(\sqrt{u})$. Therefore, N is a cyclic extension of degree 4 of K which contains L. Consequently, $\mathbb{Z}/4\mathbb{Z} \to \mathrm{Gal}(K(\sqrt{a})/K)$ is solvable. □

Remark 16.5.2: Embedding in cyclic extensions of higher order. It is possible to slightly generalize Proposition 16.5.1: Suppose K is a field and a is a nonsquare in K. Then $K(\sqrt{a})$ can be embedded in a $\mathbb{Z}/8\mathbb{Z}$-extension of K if and only if a is a sum of two squares in K and there are x_1, x_2, x_3, x_4 in K, not all zero, with $x_1^2 + 2x_2^2 + ax_3^2 - 2ax_4^2 = 0$. This result is proved in [Kiming, Thm. 3] in a slightly different form. The formulation we give appears in [Geyer-Jensen2, 20°].

Let K be a number field and $a \in K$. Suppose -1 is a nonsquare in K and $K(\sqrt{-1})$ is embeddable in a $\mathbb{Z}/16\mathbb{Z}$-extension of K. Then $K(\sqrt{-1})$ is embeddable in a $\mathbb{Z}/2^n\mathbb{Z}$-extension of K for each positive integer n [Geyer-Jensen2, Thm. 4]. The proof requires class field theory.

As an example, let $K = \mathbb{Q}(\sqrt{-14})$. Then $K(\sqrt{-1})$ is embeddable in a $\mathbb{Z}/8\mathbb{Z}$-extension of K [Arason-Fein-Schacher-Sonn, p. 846, Cor. 4 or Geyer-Jensen2, Remark 19°]. But $K(\sqrt{-1})$ is not embeddable in a $\mathbb{Z}/16\mathbb{Z}$-extension of K [Geyer-Jensen2, Example to Proposition 3].

Let O_K be the ring of integers of K. Consider a nonzero prime ideal $\mathfrak{p}$ of O_K such that

(1) -1 a nonsquare but -1 is the sum of two squares in the completion $\hat{K}_\mathfrak{p}$.

Since $X^2 + Y^2 + 1$ is absolutely irreducible. Hence, -1 is a sum of two squares in $\mathbb{F}_p$ for all large p. By Hensel's lemma, -1 is a sum of two squares in $\hat{K}_\mathfrak{p}$ for all but finitely many $\mathfrak{p}$. Hence, by Chebotarev, Condition (1) holds with density $\frac{1}{2}$. Then $\hat{K}_\mathfrak{p}(\sqrt{-1})$ is embeddable in a $\mathbb{Z}/2^n\mathbb{Z}$-extension of $\hat{K}_\mathfrak{p}$ for all n [Geyer-Jensen2, Prop. 3]. Thus, there could be no criterion for embedding a quadratic extension of a field in a $\mathbb{Z}/16\mathbb{Z}$-extension similar to the one we gave above for embedding in a $\mathbb{Z}/8\mathbb{Z}$-extension.

Consider now the field $K = \mathbb{Q}(\sqrt{-17})$. Then $K(\sqrt{-1})$ is embeddable in a $\mathbb{Z}/2^n\mathbb{Z}$-extension for each positive integer n but is not embeddable in a $\mathbb{Z}_2$-extension [Geyer-Jensen1, p. 371]. □

16.6 $\mathbb{Z}_p$-Extensions of Hilbertian Fields

Lemma 16.3.2 implies that if K is a field of characteristic p, then every $\mathbb{Z}/p\mathbb{Z}$-extension of K can be embedding in a $\mathbb{Z}_p$-extension. This result does not generalize to the general case. Remark 16.5.2 gives an example of a quadratic extension L/K (with char$(K) \neq 2$) which can not be embedded in a $\mathbb{Z}_2$-extension. So, we have to settle for less. We fix a field K, a prime number p, and let $q = p$ if $p = 2$ and $q = 4$ if $p = 2$. We ask when $\mathbb{Z}_p$ occurs over K. We prove this is the case when $\mathbb{Z}/q\mathbb{Z}$ occurs over K. This condition is satisfied when K is Hilbertian.

LEMMA 16.6.1: Suppose $p \neq $ char(K). Let L be a cyclic extension of K of degree p^n with $n \geq 1$.
(a) Suppose $\zeta_{p^{n+1}} \in K$. Then K has a $\mathbb{Z}/p^{n+1}\mathbb{Z}$-extension which contains L.
(b) Suppose $\zeta_{p^k} \in K$ for all k. Then K has a $\mathbb{Z}_p$-extension that contains L.

Proof: Statement (b) follows from (a) by induction and taking inverse limit. It remains to prove Statement (a).

The theory of cyclic extensions gives $a \in K$ with $L = K(a^{1/p^n})$ [Lang7, p. 289, Thm. 6.2(i)]. In particular, $a^{1/p} \notin K$. The field $L' = K(a^{1/p^{n+1}})$ contains L. By [Lang7, p. 289, Thm. 6.2(ii)], L' is a cyclic extension of degree p^k for some $k \leq n+1$ and $a^{1/p^{n+1-k}} \in K$. If $k \leq n$, then $a^{1/p} = (a^{1/p^{n+1-k}})^{p^{n-k}} \in K$. We conclude from this contradiction that $k = n+1$, as desired. □

Remark 16.6.2: Abelian pro-p groups as $\mathbb{Z}_p$-modules. Let A be a finite Abelian additive p-group of exponent p^m. Then $\mathbb{Z}/p^m\mathbb{Z}$ acts on A through the rule $(k + p^m\mathbb{Z})a = ka$, $a \in A$. Since $\mathbb{Z}/p^m\mathbb{Z}$ is a quotient of $\mathbb{Z}_p$, this defines a continuous action of $\mathbb{Z}_p$ on A with the discrete topology. This action commutes with homomorphisms of finite Abelian p-groups. Hence, it defines a continuous action of $\mathbb{Z}_p$ on projective limits of Abelian p-groups (also called **Abelian pro-p groups** which commutes with homomorphisms.

Consider elements $a_1, \ldots, a_r$ of A. Then the map $\mathbb{Z}_p^r \to A$ given by $(z_1, \ldots, z_r) \mapsto \sum_{i=1}^r z_i a_i$ is continuous. Its image $\sum_{i=1}^r \mathbb{Z}_p a_i$ is compact, hence closed. On the other hand, one can approximate each sum $\sum_{i=1}^r z_i a_i$ with $z_i \in \mathbb{Z}_p$ by sums $\sum_{i=1}^n k_i a_i$ with $k_i \in \mathbb{Z}$. Therefore, the closed subgroup of A generated by $a_1, \ldots, a_r$ coincides with the $\mathbb{Z}_p$-submodule of A generated by $a_1, \ldots, a_r$. □

LEMMA 16.6.3: Let $0 \to B \to A \to \mathbb{Z}/p\mathbb{Z} \to 0$ be an exact sequence of Abelian pro-p groups with $B \cong \mathbb{Z}_p$. Then either $A \cong \mathbb{Z}_p$ or $A = B \oplus A_0$ with $A_0 \cong \mathbb{Z}/p\mathbb{Z}$.

Proof: Consider A as a $\mathbb{Z}_p$-module (Remark 16.6.2). The exact sequence yields $pA \leq \mathbb{Z}_p$. By Lemma 1.4.2(e), pA is generated by one element.

On the other hand, A is generated by two elements, one which generates B and the other with image in $\mathbb{Z}/p\mathbb{Z}$ generating that module. Assume $A \not\cong \mathbb{Z}_p$ and $A \not\cong \mathbb{Z}_p \oplus \mathbb{Z}/p\mathbb{Z}$. Then, by Proposition 2.2.3, $A \cong \mathbb{Z}_p \oplus \mathbb{Z}/p^k\mathbb{Z}$ with $k \geq 2$

16.6 $\mathbb{Z}_p$-Extensions of Hilbertian Fields

or $A \cong \mathbb{Z}_p \oplus \mathbb{Z}_p$. In each cases $pA \cong \mathbb{Z}/p\mathbb{Z} \oplus \mathbb{Z}/p\mathbb{Z}$, so pA is not generated by one element, contradicting the preceding paragraph. Therefore, $A \cong \mathbb{Z}_p$ or $A \cong \mathbb{Z}_p \oplus \mathbb{Z}/p\mathbb{Z}$. In the second case A has a subgroup A_0 isomorphic to $\mathbb{Z}/p\mathbb{Z}$. Since $\mathbb{Z}_p$ contain no nontrivial closed subgroups of finite order (Lemma 1.4.2(d)), A_0 is not contained in B. Consequently, $A = B \oplus A_0$. □

LEMMA 16.6.4: Suppose $p \neq \text{char}(K)$. Let $L = K(\zeta_{p^n} \mid n = 1, 2, 3, \ldots)$. Then L/K is an Abelian extension and there is a field K_∞ satisfying the following conditions:
(a) $K_\infty(\zeta_q) = L$ and $K_\infty \cap K(\zeta_q) = L$.
(b) $\text{Gal}(L/K) = \text{Gal}(L/K(\zeta_q)) \times \text{Gal}(L/K_\infty)$.
(c) $\text{Gal}(L/K_\infty) \cong \text{Gal}(K(\zeta_q)/K)$ and $\text{Gal}(K(\zeta_q)/K)$ is isomorphic to a subgroup of $\mathbb{Z}/(p-1)\mathbb{Z}$ (resp. $\mathbb{Z}/2\mathbb{Z}$) if $p \neq 2$ (resp. $p = 2$).
(d) If $L \neq K(\zeta_q)$, then $\text{Gal}(K_\infty/K) \cong \text{Gal}(L/K(\zeta_q)) \cong \mathbb{Z}_p$.

Proof: Let n be a positive integer. Then $K(\zeta_{p^n})$ is the splitting field of the separable polynomial $X^{p^n} - 1$ over K. Hence, $K(\zeta_{p^n})/K$ is Galois. Embed $\text{Gal}(K(\zeta_{p^n})/K)$ into $(\mathbb{Z}/p^n\mathbb{Z})^\times$ by mapping each $\sigma \in \text{Gal}(K(\zeta_{p^n})/K)$ onto the element $s(\sigma)$ of $(\mathbb{Z}/p^n\mathbb{Z})^\times$ satisfying $\zeta_{p^n}^\sigma = \zeta_{p^n}^{s(\sigma)}$. Thus, $K(\zeta_{p^n})/K$ is Abelian. Hence, L/K is also Abelian.

In particular, $\text{Gal}(K(\zeta_q)/K)$ is isomorphic to a subgroup of $\mathbb{Z}/(p-1)\mathbb{Z}$ if $p \neq 2$ and of $\mathbb{Z}/2\mathbb{Z}$ if $p = 2$. If $K(\zeta_q) = L$, then $K_\infty = K$ satisfies Conditions (a)-(c).

Assume from now on $K(\zeta_q) \neq L$. Let m be the maximal positive integer with $\zeta_{p^m} \in K(\zeta_q)$. Then $\zeta_{p^m} \neq x^p$ for all $x \in K(\zeta_q)$. If $p = 2$, then $m \geq 2$ and $-1 = \zeta_4^2$ is a square in $K(\zeta_q)$, so $\zeta_{p^m} \neq -4y^4$ for all $y \in K(\zeta_4)$. Let $n \geq m$. Then $K(\zeta_{p^n}) = K(\zeta_q)(\zeta_{p^m}^{1/p^{n-m}})$ and $X^{p^{n-m}} - \zeta_{p^m}$ is irreducible over $K(\zeta_q)$ [Lang7, p. 297, Thm. 9.1]. So,

(1) $$[K(\zeta_{p^n}) : K(\zeta_q)] = p^{n-m}.$$

Divide the rest of the proof into two parts.

PART A: Suppose $p \neq 2$ and $q = p$. Then $(\mathbb{Z}/p^n\mathbb{Z})^\times \cong \mathbb{Z}/(p-1)\mathbb{Z} \times \mathbb{Z}/p^{n-1}\mathbb{Z}$. Let $\sigma \in \text{Gal}(K(\zeta_{p^n})/K(\zeta_p))$. Then $\zeta_p^{s(\sigma)} = \zeta_p^\sigma = \zeta_p$. Hence, $s(\sigma) \equiv 1 \bmod p$. Therefore, $s(\sigma)^{p^{n-1}} \equiv 1 \bmod p^n$ [LeVeque, p. 50, Thm. 4-5]. Thus, $\text{Gal}(K(\zeta_{p^n})/K(\zeta_p))$ is isomorphic to a subgroup of $\mathbb{Z}/p^{n-1}\mathbb{Z}$. It follows from (1) that $\text{Gal}(K(\zeta_{p^n})/K(\zeta_p)) \cong \mathbb{Z}/p^{n-m}\mathbb{Z}$.

By the beginning of the preceding paragraph, $\text{Gal}(K(\zeta_{p^n})/K)$ is cyclic of order dp^{n-m} with $d = [K(\zeta_p) : K]$. Let τ be a generator of this group. Then $\text{ord}(\tau^{p^{n-m}}) = d$. Denote the fixed field of $\tau^{p^{n-m}}$ in $K(\zeta_{p^n})$ by K_n. Then $K_n \cap K(\zeta_p) = K$ and $K_n(\zeta_p) = K(\zeta_{p^n})$. Moreover, K_n is the unique field with these properties. Hence, $K_n \subseteq K_{n'}$ for $n' \geq m$. Now let $K_\infty = \bigcup_{n=m}^\infty K_n$. Then $K_\infty \cap K(\zeta_p) = K$ and $K_\infty(\zeta_p) = L$. It follows from the preceding paragraph that $\text{Gal}(L/K_\infty) \cong \text{Gal}(K(\zeta_p)/K)$ and $\text{Gal}(K_\infty/K) \cong \text{Gal}(L/K(\zeta_p)) \cong \mathbb{Z}_p$.

PART B: *Suppose $p = 2$ and $q = 4$*. Let $n \geq \max(3, m)$. Then $5^{2^{n-3}} \equiv 1 + 2^{n-1} \mod 2^n$ and $\text{ord}_{2^n} 5 = 2^{n-2}$ [LeVeque, p. 54]. In addition, $-1 + 2^n \mathbb{Z}$ does not belong to the subgroup of $(\mathbb{Z}/2^n\mathbb{Z})^\times$ generated by $5 + 2^n \mathbb{Z}$. Otherwise, $5^k \equiv -1 \mod 2^n$ for some $1 \leq k < 2^{n-2}$. Raising both sides to an odd power, we may assume $k = 2^l$ with $1 \leq l \leq n-3$. Then $5^{2^{l+1}} \equiv 1 \mod 2^n$. Hence, $n - 2 \leq l + 1$, so $l = n - 3$ and $5^{2^{n-3}} \equiv -1 \mod 2^n$. It follows that $-1 \equiv 1 + 2^{n-1} \mod 2^n$, which is a contradiction. Consequently, $(\mathbb{Z}/2^n\mathbb{Z})^\times \cong \mathbb{Z}/2\mathbb{Z} \times \mathbb{Z}/2^{n-2}\mathbb{Z}$ with -1 generating the first factor and 5 generating the second factor.

Consider $\sigma \in \text{Gal}(K(\zeta_{2^n})/K(\zeta_4))$. Then $\zeta_4^{s(\sigma)} = \zeta_4^\sigma = \zeta_4$. Hence, $s(\sigma) \equiv 1 \mod 4$. By the preceding paragraph, $s(\sigma) \equiv (-1)^i 5^j \mod 2^n$ with $0 \leq i \leq 1$ and $0 \leq j \leq n-2$. Then $(-1)^i \equiv 1 \mod 4$, so $i = 0$ and $s(\sigma) \equiv 5^j \mod 2^n$.

Thus, $\text{Gal}(K(\zeta_{2^n})/K(\zeta_4))$ is isomorphic to a subgroup of $\mathbb{Z}/2^{n-2}\mathbb{Z}$. By (1), $\text{Gal}(K(\zeta_{2^n}/K(\zeta_4))) \cong \mathbb{Z}/2^{n-m}\mathbb{Z}$.

Taking inverse limit, we get $\text{Gal}(L/K(\zeta_4)) \cong \mathbb{Z}_2$. This gives a short exact sequence of Abelian pro-2 groups $0 \to \mathbb{Z}_2 \to \text{Gal}(L/K) \to \text{Gal}(K(\zeta_4)/K) \to 1$ with $\text{Gal}(K(\zeta_4)/K)$ trivial or isomorphic to $\mathbb{Z}/2\mathbb{Z}$. By Lemma 16.6.3, $\text{Gal}(L/K) = \text{Gal}(L/K(\zeta_4)) \times A_0$ with A_0 trivial or of order 2. Denote the fixed field of A_0 in L by K_∞. It satisfies Conditions (a)–(d). □

The case where $p \neq \text{char}(K)$, and $1 < [L : K] < \infty$ is the most complicated. We need some concepts and facts from group theory.

Let G be a profinite group and $x, y, z \in G$. Define the **commutator** of x, y by $[x, y] = x^{-1}y^{-1}xy$. It satisfies the following identities:

(2) $$[x,y]^{-1} = [y,x], \quad [x,y]^z = [x^z, y^z],$$
$$[x, yz] = [x,z][x,y]^z, \quad [xy, z] = [x,z]^y [y,z]$$

The **commutator subgroup** of G is $[G, G] = \langle [x, y] \mid x, y \in G \rangle$. Suppose N is a closed subgroup of G which contains $[G, G]$. For each $n \in N$ and $g \in G$ we have $n^g = n[n, g]$. Hence, $N \triangleleft G$ and G/N is Abelian. Conversely, if N is a closed normal subgroup of G and G/N is Abelian, then $[G, G] \leq N$.

Consider now a profinite group C and an epimorphism $g: G \to C$. Suppose $A = \text{Ker}(g)$ is Abelian. Define an action of C on A in the following way: For each $\gamma \in C$ choose $\tilde{\gamma} \in G$ with $g(\tilde{\gamma}) = \gamma$. Then let $a^\gamma = \tilde{\gamma}^{-1} a \tilde{\gamma}$ for $a \in A$. Since A is Abelian, this action is independent of $\tilde{\gamma}$.

Suppose C is finite. The **group ring** $\mathbb{Z}[C]$ consists of all formal sums $\sum_{\gamma \in C} k_\gamma \gamma$ with $k_\gamma \in \mathbb{Z}$. Addition is defined componentwise. Multiplication in $\mathbb{Z}[C]$ is a linear extension of multiplication in C. Thus,

$$\sum_{\gamma \in C} k_\gamma \gamma \cdot \sum_{\delta \in C} l_\delta \delta = \sum_{\varepsilon \in C} \Big(\sum_{\gamma \delta = \varepsilon} k_\gamma l_\delta \Big) \varepsilon.$$

The action of C on A naturally extends to an action of $\mathbb{Z}[C]$ on A:

$$a^{\sum_{\gamma \in C} k_\gamma \gamma} = \prod_{\gamma \in C} (a^{k_\gamma})^\gamma.$$

16.6 $\mathbb{Z}_p$-Extensions of Hilbertian Fields

Let γ be an element of C with $\gamma^m = 1$. Put $c = \sum_{i=0}^{m-1} \gamma^i$. Then $(1-\gamma)c = 1 - \gamma^m = 0$. Hence,

$$(3) \qquad (1-\gamma^k)c = \sum_{i=0}^{k-1} \gamma^i(1-\gamma)c = 0, \quad k = 0, \ldots, m-1$$

LEMMA 16.6.5: *Let p be a prime number, C a finite cyclic group, G a profinite group, and $g\colon G \to C$ an epimorphism. Suppose $A = \mathrm{Ker}(g)$ is an Abelian pro-p group and $f_0\colon A \to \mathbb{Z}_p$ is an epimorphism. Let q' be a power of p and let $\pi\colon \mathbb{Z}_p \to \mathbb{Z}/q'\mathbb{Z}$ an epimorphism. Put $\alpha = \pi \circ f_0$. Suppose one of the following conditions holds:*
(a) $q' \neq 1$ and $p \nmid |C|$.
(b) $p = |C| = 2$ and $q' \geq 4$.
(c) $p = q' = |C| = 2$ and α extends to an epimorphism $\beta\colon G \to \mathbb{Z}/4\mathbb{Z}$.

Then there exists an epimorphism $f\colon A \to \mathbb{Z}_p$ with $\mathrm{Ker}(f) \triangleleft G$ such that $G/\mathrm{Ker}(f)$ is Abelian.

Proof: The first two parts of the proof are common to all cases. The rest of the proof handles each case separately.

PART A: *The commutator of G.* As A is Abelian, C acts on A by lifting and conjugating. Extend this to an action of $\mathbb{Z}[C]$ on A. Since C is Abelian, $[G, G] \leq A$. Moreover,

$$(4) \qquad [G, G] = \langle a^{1-\gamma} \mid a \in A,\ \gamma \in C \rangle.$$

Indeed, choose a generator γ_0 of C and an element $\tilde{\gamma}_0$ in G with $g(\tilde{\gamma}_0) = \gamma_0$. For each i let $\widetilde{\gamma_0^i} = \tilde{\gamma}_0^i$. Then $[\tilde{\gamma}, \tilde{\delta}] = 1$ for all $\gamma, \delta \in C$ and

$$(5) \qquad a^{1-\gamma} = [\tilde{\gamma}, a] \quad \text{for each } a \in A.$$

Thus, the right hand side of (4) (which we denote by G_1) is contained in $[G, G]$.

To prove the other inclusion, it suffices to prove $[u, v] \in G_1$ for all $u, v \in G$. To this end, write $u = a\tilde{\gamma}$ and $v = b\tilde{\delta}$ with $a, b \in A$ and $\gamma, \delta \in C$. Now use (2), (5) and the hypothesis that A is Abelian:

$$[u, v] = [a\tilde{\gamma}, b\tilde{\delta}] = [a\tilde{\gamma}, \tilde{\delta}][a\tilde{\gamma}, b]^{\tilde{\delta}}$$
$$= [a, \tilde{\delta}]^{\tilde{\gamma}}[\tilde{\gamma}, \tilde{\delta}][a, b]^{\tilde{\gamma}\tilde{\delta}}[\tilde{\gamma}, b]^{\tilde{\delta}} = [a^{\gamma}, \tilde{\delta}][\tilde{\gamma}, b^{\delta}] \in G_1$$

PART B: *Twist of A.* Let $m = |C|$ and put $c = \sum_{i=0}^{m-1} \gamma_0^i$. Let $\mu\colon A \to A$ be the homomorphism given by $\mu(a) = a^c$. Since $\gamma_0^m = 1$, (3) implies $(1-\gamma)c = 0$ for each $\gamma \in C$. Hence, $(a^{1-\gamma})^c = a^{(1-\gamma)c} = 1$ for each $a \in A$. Therefore, by (4), $\mu(w) = w^c = 1$ for each $w \in [G, G]$.

By (5), $\alpha(a^{1-\gamma}) = \alpha([\tilde{\gamma}, a]) = 0$. Hence, $\alpha(a^{\gamma}) = \alpha(a)$ for all $a \in A$ and $\gamma \in C$. Consequently, $\alpha(\mu(a)) = \sum_{i=0}^{m-1} \alpha(a^{\gamma_0^i}) = m\alpha(a)$.

312   Chapter 16. Galois Groups over Hilbertian Fields

PART C: *Suppose $q' > 1$ and $p \nmid |C|$.* Then m is an invertible element of $\mathbb{Z}_p$ and $\mathbb{Z}/q'\mathbb{Z}$. Let $m^{-1}: \mathbb{Z}_p \to \mathbb{Z}_p$ and $m^{-1}: \mathbb{Z}/q'\mathbb{Z} \to \mathbb{Z}/q'\mathbb{Z}$ be multiplications by m^{-1}. Define a homomorphism $f: A \to \mathbb{Z}_p$ by $f = m^{-1} \circ f_0 \circ \mu$. By Part B, $[G, G] \leq \mathrm{Ker}(\mu) \leq \mathrm{Ker}(f)$. Thus, $\mathrm{Ker}(f) \triangleleft G$ and $G/\mathrm{Ker}(f)$ is Abelian.

By Part B, this establishes the following commutative diagram:

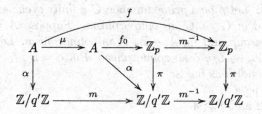

In particular, $\alpha = \pi \circ f$. Thus, $\pi(f(A)) = \alpha(A) = \mathbb{Z}/q'\mathbb{Z}$. It follows from Lemma 1.4.3 that f is surjective.

PART D: *Suppose $p = |C| = 2$ and $q' \geq 4$.* In this case Part B gives the following commutative diagram:

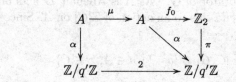

Let $H = f_0(\mu(A))$. By Remark 1.2.1(e), H is a closed subgroup of $\mathbb{Z}_2$. It satisfies, $\pi(H) = 2(\mathbb{Z}/q'\mathbb{Z}) = \pi(2\mathbb{Z}_2)$. Hence, $H + q'\mathbb{Z}_2 = 2\mathbb{Z}_2 + q'\mathbb{Z}_2 = 2\mathbb{Z}_2$. By Lemma 1.4.2, H is trivial or $H = 2^n\mathbb{Z}_2$ for some positive integer n. It follows from $q' \geq 4$ that $H = 2\mathbb{Z}_2$.

PART E: *Suppose $p = q' = 2 = |C|$ and α extends to an epimorphism $\beta: G \to \mathbb{Z}/4\mathbb{Z}$.* Then $(G : A) = |C| = 2$. Hence, in the notation of Part B, $a = \tilde{\gamma}_0^2 \in A$. So, $a^{\gamma_0} = a$. Hence, $\mu(a) = aa^{\gamma_0} = a^2$.

The existence of β gives a commutative diagram

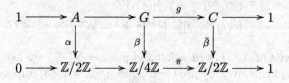

with both rows exact. In particular, $\bar{\pi}(\beta(\tilde{\gamma}_0)) = \bar{\beta}(g(\tilde{\gamma}_0)) = 1 + 2\mathbb{Z}$. Hence, $\beta(\tilde{\gamma}_0) = \pm 1 + 4\mathbb{Z}$. So, $\pi(f_0(a)) = \alpha(a) = \beta(\tilde{\gamma}_0^2) = 1 + 2\mathbb{Z}$. Therefore, by Lemma 1.4.3, $\langle f_0(a) \rangle = \mathbb{Z}_2$. Hence, $\langle f_0(\mu(a)) \rangle = \langle f_0(a^2) \rangle = 2\mathbb{Z}_2$. It follows that $f_0(\mu(A)) = \mathbb{Z}_2$ or $f_0(\mu(A)) = 2\mathbb{Z}_2$.

16.6 $\mathbb{Z}_p$-Extensions of Hilbertian Fields

Suppose first $f_0(\mu(A)) = \mathbb{Z}_2$. Then $f = f_0 \circ \mu$ maps A onto $\mathbb{Z}_2$. For each $w \in [G,G]$, Part B implies $f(w) = f_0(\mu(w)) = 1$. Therefore, $[G,G] \leq \text{Ker}(f)$, $\text{Ker}(f) \triangleleft G$, and $G/\text{Ker}(f)$ is Abelian.

The case $f_0(\mu(A)) = 2\mathbb{Z}_2$ (which we henceforth assume), will be handled in Part F.

PART F: *Conclusion of the proof in cases (b) and (c).* Multiplication by 2 gives an isomorphism of $\mathbb{Z}_2$ onto $2\mathbb{Z}_2$. Hence, by Parts D and E, there is an epimorphism $f: A \to \mathbb{Z}_2$ with $2f = f_0 \circ \mu$. For each $w \in [G,G]$, Part B implies $2f(w) = f_0(\mu(w)) = 0$. Hence, $f(w) = 0$. Therefore, $[G,G] \leq \text{Ker}(f)$, $\text{Ker}(f) \triangleleft G$, and $G/\text{Ker}(f)$ is Abelian. ☐

THEOREM 16.6.6 ([Whaples, Thm. 2]): *Let K be a field and p a prime number. Put $q = p$ if $p \neq 2$ and $q = 4$ if $p = 2$. Suppose $\mathbb{Z}/q\mathbb{Z}$ occurs over K. Then $\mathbb{Z}_p$ occurs over K.*

Proof: Let K' be a $\mathbb{Z}/q\mathbb{Z}$-extension of K. Suppose first $p = \text{char}(K)$. Lemma 16.3.2 embeds K' in a $\mathbb{Z}_p$-extension of K. Assume from now on, $p \neq \text{char}(K)$.

Let $L = K(\zeta_{p^n} \mid n = 1, 2, 3, \ldots)$. When $L = K$, Lemma 16.6.1 gives a $\mathbb{Z}_p$-extension of K. When $L \neq K(\zeta_q)$, Lemma 16.6.4 gives a $\mathbb{Z}_p$-extension of K. Assume from now on, $L = K(\zeta_q)$ and $L \neq K$. Let $C = \text{Gal}(L/K)$. Then C is a cyclic group of order which divides $p - 1$ if $p \neq 2$ and of order 2 of $p = 2$.

Put $L' = LK'$. Then L' is a finite Abelian extension of K. Moreover, L'/L is a cyclic extension of degree q' which is a power of p. We distinguish between three cases:

CASE A: $p \neq 2$. Then K'/K is a cyclic extension of degree p and $[L:K] \mid p - 1$. Hence, $q' = [L':L] = p$ and $p \nmid |C|$.

CASE B: $p = 2$ and $L \cap K' = K$. Then K'/K is a cyclic extension of degree 4, $|C| = 2$, and $q' = [L':L] = 4$.

CASE C: $p = 2$ and $L \cap K' \not\subseteq K$. Then, $L \subseteq K'$, $K' = L'$, and $q' = |C| = 2$.

In each case Lemma 16.6.1 gives a $\mathbb{Z}_p$-extension F of L which contains L'. Denote the compositum of all finite Abelian extensions of L of a p-power order by N. In particular, $F \subseteq N$. Since L/K is Galois, so is N/K. Put $G = \text{Gal}(N/K)$ and $A = \text{Gal}(N/L)$. Let f_0 be $\text{res}_{N/F}: \text{Gal}(N/L) \to \text{Gal}(F/L)$ and let π be $\text{res}_{F/L'}: \text{Gal}(F/L) \to \text{Gal}(L'/L)$. Then $\alpha = \pi \circ f_0 = \text{res}_{N/L'}: \text{Gal}(N/L) \to \text{Gal}(L'/L)$. In Case C, $\text{Gal}(K'/K) \cong \mathbb{Z}/4\mathbb{Z}$ and $\beta = \text{res}_{N/L'}: \text{Gal}(N/K) \to \text{Gal}(L'/K)$ extends α.

In each case, Lemma 16.6.5 gives an extension F' of L in N with F'/K Abelian and $\text{Gal}(F'/L) \cong \mathbb{Z}_p$. Suppose first $p \neq 2$. Then $[L:K] \mid p-1$. So, $\text{Gal}(F'/K)$ has a unique subgroup of order $p - 1$. This is a special case of a profinite version of the Schur-Zassenhaus theorem (Lemma 22.10.1). Its fixed field E in F' satisfies $\text{Gal}(E/K) \cong \mathbb{Z}_p$. Now suppose $p = 2$. Lemma

16.6.3 gives a $\mathbb{Z}_2$-extension E of K in F'. Consequently, in each case K has a $\mathbb{Z}_p$-extension. □

COROLLARY 16.6.7: *Let K be a Hilbertian field and p a prime number. Then $\mathbb{Z}_p$-occurs over K.*

Proof: By Corollary 16.3.6, $\mathbb{Z}/p\mathbb{Z}$ and $\mathbb{Z}/p^2\mathbb{Z}$ occur over K. Hence, by Theorem 16.6.6, $\mathbb{Z}_p$ occurs over K. □

Remark 16.6.8: The assumption in Theorem 16.6.6 that $\mathbb{Z}/4\mathbb{Z}$ rather than $\mathbb{Z}/2\mathbb{Z}$ occurs over K is necessary for the theorem to hold. Indeed, $\text{Gal}(\mathbb{C}/\mathbb{R}) \cong \mathbb{Z}/2\mathbb{Z}$ but $\mathbb{Z}_2$ does not occur over $\mathbb{R}$. □

Proposition 16.6.10 below shows it is impossible to conclude that $\mathbb{Z}_p$ is regular over K in Corollary 16.6.7.

LEMMA 16.6.9: *Let E be a field, p a prime number, v a discrete valuation of E, and F a $\mathbb{Z}_p$-extension of E. Suppose $p \nmid \text{char}(\bar{E}_v)$ and $\bar{E}_v(\zeta_p, \zeta_{p^2}, \zeta_{p^3}, \ldots)$ is an infinite extension of $\bar{E}_v$. Then v is unramified in F.*

Proof: Let w be a valuation of F extending v. Denote reduction at w by a bar. Assume w is ramified over E. Then its inertia group $I_{w/v}$ is nontrivial. By Lemma 1.4.2, $I_{w/v}$ is an open subgroup of $\text{Gal}(F/E)$. Replace E by the fixed field of $I_{w/v}$ in F, if necessary, to assume $I_{w/v} = \text{Gal}(F/E)$.

Let n be a positive integer. Denote the unique extension of E in F of degree p^n by E_n. Let v_n be a normalized valuation of E_n over v. It is totally ramified over E. Consider the completion $\hat{E}$ of E under v. Then $\hat{E}_n = E_n\hat{E}$ is the completion of E_n under v_n (Proposition 3.5.3) and $\text{Gal}(\hat{E}_n/\hat{E}) \cong \text{Gal}(E_n/E) \cong \mathbb{Z}/p^n\mathbb{Z}$ [Cassels-Fröhlich, p. 41, Prop. 3].

Choose $y \in \hat{E}_n$ and $x \in \hat{E}$ with $v_n(y) = 1$ and $v(x) = 1$. Then $v_n(x) = p^n = v_n(y^{p^n})$. Hence, $y^{p^n} = ux$ with $u \in E_n$ and $v_n(u) = 0$. Since $\bar{E}_n = \bar{E}$, there is an $a \in E$ with $\bar{a} = \bar{u}$. Let $u' = ua^{-1}$ and $x_1 = ax$. Then $v_n(u'-1) > 0$, and $x_1 \in E$. Since $p \nmid \text{char}(\bar{E}_v)$, Hensel's lemma (Proposition 3.5.2(a)) gives $u_1 \in E_n$ with $u_1^{p^n} = u'$. Put $y_1 = u_1^{-1}y$. Then $y_1^{p^n} = x_1$. Since E_n/E is cyclic of degree n, this implies $\zeta_{p^n} \in E_n$. Taking residues, we find $\zeta_{p^n} \in \bar{E}_v$. Thus, $\bar{E}_v(\zeta_p, \zeta_{p^2}, \zeta_{p^3}, \ldots) = \bar{E}_v$, contrary to the assumption of the lemma. Consequently, v is unramified in F. □

PROPOSITION 16.6.10: *Let K be a field and p a prime number. Suppose $p \neq \text{char}(K)$ and $K(\zeta_p, \zeta_{p^2}, \zeta_{p^3}, \ldots)$ is an infinite extension of K. Then $\mathbb{Z}_p$ is not regular over K.*

Proof: Assume $\mathbb{Z}_p$ is regular over K. Then there are algebraically independent elements $t_1, \ldots, t_r$ over K and there is a $\mathbb{Z}_p$-extension F of $E = K(t_1, \ldots, t_r)$. Induction on r proves it suffices to consider the case where $r = 1$. Put $t = t_1$.

Let E_1 be the unique extension of E in F of degree p. Then E_1/K is regular. Remark 3.6.2(b) gives a prime divisor $\mathfrak{p}$ of E/K which is ramified in E_1. The valuation $v_\mathfrak{p}$ of E associated with $\mathfrak{p}$ is discrete (Section 3.1). Its

16.7 Symmetric and Alternating Groups over Hilbertian Fields

residue field $\bar{E}_{\mathfrak{p}}$ is a finite extension of K. Hence, $\bar{E}_{\mathfrak{p}}(\zeta_p, \zeta_{p^2}, \zeta_{p^3}, \ldots)$ is an infinite extension of K. In addition, $\mathrm{char}(\bar{E}_v) = \mathrm{char}(K) \nmid p$. Hence, by Lemma 16.6.9, $\mathfrak{p}$ is unramified in F. In particular, $\mathfrak{p}$ is unramified in E_1. It follows from this contradiction that $\mathbb{Z}_p$ is not regular over K. □

COROLLARY 16.6.11: *Let K be a field and p a prime number. Suppose $p \neq \mathrm{char}(K)$ and K is a finitely generated over its prime field. Then $\mathbb{Z}_p$ is not regular over K.*

16.7 Symmetric and Alternating Groups over Hilbertian Fields

The Galois group of the general polynomial of degree n is S_n (Example 16.2.5(a)). There is a standard strategy to construct special polynomials with Galois groups S_n.

LEMMA 16.7.1: *Let K be a field, $t_1, \ldots, t_r$ algebraically independent elements over K, and F a separable extension of $K(\mathbf{t})$ of degree n. Denote the Galois closure of $F/K(\mathbf{t})$ by $\hat{F}$. Suppose $\mathrm{Gal}(\hat{F}\tilde{K}/\tilde{K}(\mathbf{t})) \cong S_n$. Then, $\mathrm{Gal}(\hat{F}/K(\mathbf{t})) \cong S_n$.*

Proof: Since $[F : K(\mathbf{t})] = n$, the group $G = \mathrm{Gal}(\hat{F}/K(\mathbf{t}))$ is isomorphic to a subgroup of S_n, On the other hand, $|S_n| = [\hat{F}\tilde{K} : \tilde{K}(\mathbf{t})] \leq [\hat{F} : K(\mathbf{t})] = |G|$. Consequently, $G \cong S_n$. □

Consider for example the case when $r = 1$. Put $t = t_1$. Let x be a primitive element of $F/K(t)$, and $f \in K[T, X]$ an absolutely irreducible polynomial which separable and of degree n in X with $f(t, x) = 0$. Then $\tilde{G} = \mathrm{Gal}(f(t, X), \tilde{K}(t))$ acts transitively on the n distinct roots of $f(t, X)$ in $\widetilde{K(t)}$. Inspecting inertia groups of prime divisors of $F\tilde{K}/\tilde{K}(t)$ in $\hat{F}\tilde{K}$, one may prove that $\tilde{G}$ contains cycles which generate S_n. Then, by Lemma 16.7.1, $\mathrm{Gal}(\hat{F}/K(t)) \cong S_n$.

When the Galois group is S_n, one may consider the fixed field E of A_n in $\hat{F}$. It is a quadratic extension of $K(t)$. If we prove that E is rational over K, then A_n becomes regular over K. We are able to do it, for example, if $\mathrm{char}(K) \neq 2$ and $K(t)/K$ has at most three prime divisors which ramify in F, each of degree 1 (Lemma 16.7.5).

We start with a valuation theoretic condition that yields e-cycles for the Galois group of a polynomial. By an **e-cycle** of S_n we mean a cycle of length e.

LEMMA 16.7.2: *Let (E, v) be a discrete valued field, $f \in O_v[X]$ a monic separable irreducible polynomial over E, and x a root of f in E_s. Denote reduction at v by a bar. Suppose $\bar{E}$ is separably closed, $\bar{f}(X) = (X - \bar{a})^e \eta(X)$ with $a \in O_v$, $e \geq 1$, $\mathrm{char}(\bar{E}) \nmid e$, and $\eta \in \bar{E}[X]$ monic separable polynomial with $\eta(\bar{a}) \neq 0$. Suppose v extends to a valuation w of $E(x)$ with $w(x - a) > 0$ and $e_{w/v} = e$. Then $\mathrm{Gal}(f, E)$ contains an e-cycle.*

Proof: Denote the completion of (E, v) by $(\hat{E}, \hat{v})$. Embed E_s into $\hat{E}_s$ and extend $\hat{v}$ to a valuation $\hat{v}$ of $\hat{E}_s$. Let $\hat{O}$ be the valuation ring of $\hat{E}$. By assumption, $\eta(X)$ is relatively prime to $(X - \bar{a})^e$. Hence, Hensel's lemma (Proposition 3.5.2(b)) gives a factorization $f(X) = g(X)h(X)$ with $g, h \in \hat{O}[X]$ monic, $\bar{g}(X) = (X - \bar{a})^e$ and $\bar{h}(X) = \eta(X)$.

Let $g(X) = g_1(X) \cdots g_k(X)$ with $g_1, \ldots, g_k$ distinct monic irreducible polynomials in $\hat{E}[X]$. For each i choose a root y_i of g_i in E_s. Then there is a $\sigma_i \in \text{Gal}(E)$ with $\sigma_i x = y_i$. The formula $w_i(z) = \hat{v}(\sigma_i z)$ for $z \in E(x)$ defines a valuation w_i of $E(x)$ which extends v and with $\hat{E}(y_i)$ as completion (Proposition 3.5.3(a),(b)).

Similarly, for each root y of $h(X)$ there is a $\sigma \in \text{Gal}(E)$ with $\sigma x = y$ and $v'(z) = \hat{v}(\sigma z)$ for $z \in E(x)$ defines an extension of v. By assumption, $\bar{h}(\bar{y}) = 0$ and $\bar{h}(\bar{a}) \neq 0$. Hence, $v'(x - a) = \hat{v}(y - a) = 0$. Since $w(x - a) > 0$, this implies, $w \neq v'$. Thus, w is one of the w_i's, say $w = w_1$ (Proposition 3.5.3(c)). Therefore,

$$e = e_{w/v} \leq \sum_{i=1}^{k} e_{w_i/v} \leq \sum_{i=1}^{k} [\hat{E}(y_i) : \hat{E}] = \sum_{i=1}^{k} \deg(g_i) = \deg(g) = e.$$

It follows that $k = 1$ and g is irreducible over $\hat{E}$. List the roots of g in E_s as $x_1, \ldots, x_e$.

By the preceding paragraph, $[\hat{E}(x) : \hat{E}] = \deg(g) = e$. Since $\text{char}(\bar{E}) \nmid e$, $\hat{E}(x)/\hat{E}$ is tamely ramified. Hence, $\hat{E}(x) = \hat{E}(\pi^{1/e})$ for some $\pi \in \hat{E}$ [Lang5, p. 52]. Since $\bar{E}$ is separably closed, it contains ζ_e. By Hensel's lemma, $\zeta_e \in \hat{E}$. Hence, $\hat{E}(x)/\hat{E}$ is a cyclic group of order e. In other words, $\text{Gal}(g, \hat{E})$ is a cyclic group of order e. It acts transitively on $x_1, \ldots, x_e$. Therefore, it is generated by an e-cycle, say $(x_1 x_2 \ldots x_e)$.

Finally, by assumption, $\bar{h}(X)$ is separable. Hence, it is a product of distinct monic linear factors in $\bar{E}[X]$. By Hensel's lemma, $h(X)$ decomposes into distinct monic linear factors in $\hat{E}[X]$, say, $h(X) = \prod_{i=e+1}^{n}(X - x_i)$. Therefore, $\text{Gal}(\hat{E}(x)/\hat{E})$ acts trivially on $x_{e+1}, \ldots, x_n$. Thus, $\text{Gal}(f, \hat{E})$ is generated by the cycle $(x_1 x_2 \ldots x_e)$. Since $\text{Gal}(f, \hat{E}) \leq \text{Gal}(f, E)$, the group $\text{Gal}(f, E)$ contains an e-cycle $\square$

LEMMA 16.7.3: Let K be a field, $q \in K[X]$ with $\deg(q) = n$, and x a transcendental element over K. Suppose $\text{char}(K) \nmid n$. Put $t = q(x)$ and $f(T, X) = q(X) - T$. Then:
(a) $f(t, X)$ is irreducible over $\tilde{K}(t)$, separable in X, and $[K(x) : K(t)] = n$.
(b) Denote the derivative of q by q'. Let φ be a K-place of $K(t)$. Denote the corresponding prime divisor of $K(t)/K$ by $\mathfrak{p}$. Suppose $\varphi(t) \neq \infty$ and $\varphi(t) \neq q(b)$ for each root b of $q'(X)$. Then $\mathfrak{p}$ is unramified in $K(x)$.
(c) $\text{Gal}(f(t, X), \tilde{K}(t))$ contains an e-cycle.
(d) Suppose there is a $b \in \tilde{K}$ such that $q(X) - q(b) = (X - b)^e \eta(X)$ with $e \geq 2$, $\text{char}(K) \nmid e$, $\eta(X) \in \tilde{K}[X]$, and $\eta(b) \neq 0$. Then $\text{Gal}(f(t, X), \tilde{K})$ contains an e-cycle.

16.7 Symmetric and Alternating Groups over Hilbertian Fields

(e) *Suppose $q'(X) = (X - b)\lambda(X)$ with $b \in \tilde{K}$, $\lambda \in \tilde{K}[X]$, and $\lambda(b) \neq 0$. Then $\mathrm{Gal}(f(t,X), \tilde{K}(t))$ contains a 2-cycle.*

Proof: Let c be the coefficient of X^n in q. Replace x by cx and t by $c^{n-1}t$ to assume q is monic.

Proof of (a): Since $f(T, X)$ is linear in T, it is absolutely irreducible. Hence, $f(t, X)$ is irreducible over $K(t)$ and over $\tilde{K}(t)$. Therefore, $[K(x) : K(t)] = n$. The separability of $f(t, X)$ follows from the assumption $\mathrm{char}(K) \nmid n$.

Proof of (b): Extend φ to a place ψ of $K(x)$. Then $\varphi(t) = \psi(q(x)) = q(\psi(x))$. Hence, by assumption, $\psi(q'(x)) = q'(\psi(x)) \neq 0$. Therefore, the value of φ at the product of all $K(t)$-conjugates of $q'(x)$ is not 0. In other words, $\varphi(N_{K(x)/K(t)}q'(x)) \neq 0$. By (a), $f(t, X) = \mathrm{irr}(x, K(t))$. In addition, $q'(x) = \frac{\partial f}{\partial X}(t, x)$. It follows from Lemma 6.1.8 that $\mathfrak{p}$ is unramified in $K(x)$.

Proof of (c): Let $q(X) = X^n + c_{n-1}X^{n-1} + \cdots + c_0$. Then $x^n + c_{n-1}x^{n-1} + \cdots + c_0 = t$. Let v_∞ be the valuation of $\tilde{K}(t)/\tilde{K}$ with $v_\infty(t) = -1$ and let w be an extension of v_∞ to $\tilde{K}(x)$. Put $e = e_{w/v_\infty}$. By Example 2.3.11, $w(x) = -1$ and $e = n$.

Now make a change of variables: $u = t^{-1}$ and $y = t^{-1}x$. Put $g(U, Y) = Y^n + c_{n-1}UY^{n-1} + \cdots + c_0U^n - U^{n-1}$. Then $\tilde{K}(u) = \tilde{K}(t)$, $\tilde{K}(y) = \tilde{K}(x)$, $w(u) = n$, $w(y) = n - 1$, $g(u, Y)$ is irreducible over $\tilde{K}(u)$, separable in Y, $g(u, y) = 0$, and $g(0, Y) = Y^n$. By Lemma 16.7.2, $\mathrm{Gal}(g(u, Y), \tilde{K}(u))$ contains an n-cycle. Therefore, $\mathrm{Gal}(f(t, X), \tilde{K}(t))$ which is isomorphic to $\mathrm{Gal}(g(u, X), \tilde{K}(u))$ has an n-cycle.

Proof of (d): Put $q(b) = a$. Let v_a be the valuation of $\tilde{K}(t)/\tilde{K}$ with $v_a(t - a) = 1$. By assumption, $f(a, X) = (X - b)^e \eta(X)$ with $\eta \in \tilde{K}[X]$, $\eta(b) \neq 0$. Denote the completion of $(\tilde{K}(t), v_a)$ by $\hat{E}$ (It is $\tilde{K}((t-a))$.) Let $\hat{O}$ be the valuation ring of $\hat{E}$. By Hensel's lemma, $f(t, X) = g(X)h(X)$ with g, h monic polynomials in $\hat{O}[X]$ whose residue at v_a are $(X - b)^e, \eta(X)$, respectively. Embed $K(x)$ in $\hat{E}_s$ such that $\bar{x} = b$. Let w be the corresponding extension of v_a to $K(x)$. Then $e_{w/v_a} \leq [\hat{E}(x) : \hat{E}] \leq \deg(g) = e$.

By the preceding paragraph, $(x - b)^e \eta(x) = q(x) - a = t - a$. Since $\eta(b) \neq 0$, we have $w(\eta(x)) = 0$. Hence, $ew(x - b) = e_{w/v_a} \leq e$. Thus, $w(x - b) = 1$ and $e_{w/v_a} = e$.

We conclude from Lemma 16.7.2 that $\mathrm{Gal}(f(t, X), \tilde{K}(t))$ has an e-cycle.

Proof of (e): Write $q(X) - q(b) = (X - b)^e \eta(X)$ with $e \geq 1$, $\eta \in \tilde{K}[X]$, and $\eta(b) \neq 0$. Then

(1) $$q'(X) = (X - b)^{e-1}\bigl(e\eta(X) + (X - b)\eta'(X)\bigr)$$

If $e = \mathrm{char}(K)$, then $q'(X) = (X - b)^e \eta'(X)$. Hence, $e = 1$, which is impossible. Therefore, $e \neq \mathrm{char}(K)$, so the second factor on the right hand side of (1) does not vanish in b. On the other hand, $q'(X) = (X - b)\lambda(X)$ with $\lambda(b) \neq 0$. Hence, $e = 2$. Consequently, by (d), $\mathrm{Gal}(f(t, X), \tilde{K}(t))$ has a 2-cycle. □

LEMMA 16.7.4: *Let K be a field, $n \geq 2$ an integer, x an indeterminate, and $t = x^n - x^{n-1}$. Suppose $\mathrm{char}(K) \nmid (n-1)n$. Then $X^n - X^{n-1} - t$ is separable, irreducible over both $K(t)$ and $\tilde{K}(t)$, and*

$$\mathrm{Gal}(X^n - X^{n-1} - t, K(t)) = \mathrm{Gal}(X^n - X^{n-1} - t, \tilde{K}(t)) = S_n.$$

Moreover, $K(t)$ has at most three prime divisors which ramify in $K(x)$. Each of them has degree 1.

Proof: The polynomial $q(X) = X^n - X^{n-1}$ factors as $q(X) = X^{n-1}(X-1)$. By Lemma 16.7.3(d) (with $b = 0$), $\mathrm{Gal}(X^n - X^{n-1} - t, \tilde{K}(t))$ contains an $(n-1)$-cycle. The derivative of $q(X)$ is $q'(X) = nX^{n-2}(X - \frac{n-1}{n})$. By Lemma 16.7.3(e) (with $b = \frac{n-1}{n}$), $\mathrm{Gal}(X^n - X^{n-1} - t, \tilde{K}(t))$ contains a 2-cycle. Hence, by [Waerden1, p. 199], $\mathrm{Gal}(X^n - X^{n-1} - t, \tilde{K}(t)) = S_n$. It follows from Lemma 16.7.1 that $\mathrm{Gal}(X^n - X^{n-1} - t, \tilde{K}(t)) = S_n$.

The derivative $q'(X)$ has exactly two roots, 0 and $\frac{n-1}{n}$. Put $q(\frac{n-1}{n}) = a$ and observe that $q(0) = 0$. Let φ be a K-place of $K(t)$ and $\mathfrak{p}$ the prime divisor of $K(t)/K$ corresponding to φ. Suppose first that $\varphi(t) \neq a, 0, \infty$. By Lemma 16.7.3(b), $\mathfrak{p}$ is unramified in $K(x)$. Thus, $\mathfrak{p}$ ramifies in $K(x)$ in at most three cases, when $\varphi(t)$ is a, or 0, or ∞. In each of these cases $\deg(\mathfrak{p}) = 1$. □

LEMMA 16.7.5: *Let K be a field, t an indeterminate, and E a quadratic extension of $K(t)$. Suppose $\mathrm{char}(K) \neq 2$, E/K is regular, and $K(t)/K$ has at most three prime divisors which ramify in E, each of degree 1. Then $E = K(u)$ with u transcendental over K.*

Proof: Denote the genus of E by g. Let $\mathfrak{p}_1, \ldots, \mathfrak{p}_k$ be the number of prime divisors of $K(t)/K$ which ramify in E. By assumption, $k \leq 3$ and $\deg(\mathfrak{p}_i) = 1$, $i = 1, \ldots, k$. Since $[E : K(t)] = 2$, each $\mathfrak{p}_i$ extends uniquely to a prime divisor $\mathfrak{q}_i$ of E/K of degree 1. Since $\mathrm{char}(K) \neq 2$, the ramification of $\mathfrak{p}_i$ in E is tame. By Riemann-Hurwitz (Remark 3.6.2(c)), $2g - 2 = -4 + k \leq -1$. Hence, $g \leq \frac{1}{2}$. This implies, $g = 0$ and $k = 2$. It follows from Example 3.2.4 that $E = K(u)$ with u transcendental over K. □

PROPOSITION 16.7.6: *Let K be a field and an integer $n \geq 2$ with $\mathrm{char}(K) \nmid (n-1)n$. Then A_n is regular over K.*

Specifically, there is a tower of fields $K(t) \subseteq K(u) \subseteq F$ satisfying: F/K is regular, $F/K(t)$ is Galois, $\mathrm{Gal}(F/K(t)) \cong S_n$, $\mathrm{Gal}(F/K(u)) \cong A_n$, and $K(t)/K$ has at most three prime divisors which ramify in F, each of degree 1.

Proof: Let t and x be transcendental elements over K with $x^n - x^{n-1} = t$. By Proposition 16.7.4, $K(x)/K(t)$ is separable. Denote the Galois closure of $K(x)/K(t)$ by F. By Proposition 16.7.4, $\mathrm{Gal}(F/K(t)) \cong \mathrm{Gal}(F\tilde{K}/\tilde{K}(t)) \cong S_n$. In particular, F/K is regular. In addition, $K(t)/K$ has at most three prime divisors $\mathfrak{p}_1, \mathfrak{p}_2, \mathfrak{p}_3$ which may ramify in $K(x)$, each of degree 1. Denote the fixed field of A_n in F by E. By Corollary 2.3.7(c), $\mathfrak{p}_1, \mathfrak{p}_2, \mathfrak{p}_3$ are the

16.7 Symmetric and Alternating Groups over Hilbertian Fields

only prime divisors of $K(t)/K$ which may ramify in F. Hence, they are the only prime divisors of $K(t)/K$ which may ramify in E. By Lemma 16.7.5, $E = K(u)$ with u transcendental over K and $\mathrm{Gal}(F/K(u)) \cong A_n$, as desired. $\square$

Next we use the polynomial $X^n - X^{n-1} - t$ to solve embedding problems of the form $S_n \to \mathrm{Gal}(K(\sqrt{a})/K)$ when K is Hilbertian.

Remark 16.7.7: On the discriminant of a polynomial. Let K be a field and $f \in K[X]$ a monic separable polynomial of degree n. Write $f(X) = \prod_{i=1}^{n}(X - x_i)$ with $x_i \in K_s$. Put $N = K(x_1, \ldots, x_n)$ and embed $\mathrm{Gal}(N/K)$ into S_n by $\sigma(i) = j$ if $\sigma(x_i) = x_j$. Formula (1) of Section 6.1 for the discriminant $\mathrm{disc}(f)$ of f is

$$(2) \qquad \mathrm{disc}(f) = (-1)^{\frac{n(n-1)}{2}} \prod_{i \neq j}(x_i - x_j) = (-1)^{\frac{n(n-1)}{2}} \prod_{j=1}^{n} f'(x_j),$$

with f' the derivative of f. It can be rewritten as $\mathrm{disc}(f) = \prod_{i<j}(x_i - x_j)^2$. Thus,

$$\sqrt{\mathrm{disc}(f)} = \prod_{i<j}(x_i - x_j).$$

For each $\sigma \in \mathrm{Gal}(N/K)$ we have

$$\sigma(\sqrt{\mathrm{disc}(f)}) = \prod_{i<j}(x_{\sigma(i)} - x_{\sigma(j)}) = (-1)^{\mathrm{sgn}(\sigma)}\sqrt{\mathrm{disc}(f)}.$$

where $\mathrm{sgn}(\sigma)$ is the number of pairs (i,j) with $i < j$ and $\sigma(i) > \sigma(j)$. If $\mathrm{char}(K) \neq 2$, then $(-1)^{\mathrm{sgn}(\sigma)} \neq 1$ for odd σ and $(-1)^{\mathrm{sgn}(\sigma)} = 1$ for even σ. In this case,

$$\mathrm{Gal}(N/K(\sqrt{\mathrm{disc}(f)})) = A_n \cap \mathrm{Gal}(N/K).$$

In particular, if $\mathrm{Gal}(N/K) = S_n$, then $K(\sqrt{\mathrm{disc}(f)})$ is the unique quadratic extension of K in N and $\mathrm{Gal}(N/K(\sqrt{\mathrm{disc}(f)})) = A_n$.

Suppose now $\mathrm{char}(K) \nmid n$. Then $f'(X) = n\prod_{j=1}^{n-1}(X - y_j)$ with $y_j \in \tilde{K}$. Substituting in (2) and changing the order of multiplication gives an alternative formula for the discriminant:

$$(3) \qquad \mathrm{disc}(f) = (-1)^{\frac{n(n-1)}{2}} n^n \prod_{j=1}^{n-1} f(y_j). \qquad \square$$

The proof of part (a) of the following result gives an explicit polynomial over $K(t)$ with Galois group A_n if $\mathrm{char}(K) \nmid (n-1)n$.

PROPOSITION 16.7.8 (Brink): *Let K be a Hilbertian field, a a nonsquare in K, and $n \geq 2$ an integer with $\mathrm{char}(K) \nmid (n-1)n$. Then*
(a) *A_n is regular over K and*
(b) *the embedding problem $S_n \to \mathrm{Gal}(K(\sqrt{a})/K)$ is solvable.*

Proof: Let $f(t, X) = X^n - X^{n-1} - t$ and F be the splitting field of $f(t, X)$ over $K(t)$. By Proposition 16.7.4, $\mathrm{Gal}(F/K(t)) \cong S_n$. Since $\mathrm{char}(K) \nmid n$, we have $f'(t, X) = n\left(X - \frac{n-1}{n}\right)X^{n-2}$. Hence, by (3),

$$\mathrm{disc}(f) = (-1)^{\frac{n(n-1)}{2}} n^n f\left(\frac{n-1}{n}\right) f(0)^{n-2}$$

$$= (-1)^{\frac{n(n-1)}{2}} n^n (-t)^{n-2} \left[\left(\frac{n-1}{n}\right)^n - \left(\frac{n-1}{n}\right)^{n-1} - t\right]$$

$$= (-1)^{\frac{(n-1)(n+2)}{2}} [(n-1)^{n-1} t^{n-2} + n^n t^{n-1}].$$

By Remark 16.7.7, $K(\sqrt{\mathrm{disc}(f)})$ is a quadratic extension of $K(t)$ in F and $\mathrm{Gal}(F/K(\sqrt{\mathrm{disc}(f)})) \cong A_n$. Let

$$u = \begin{cases} \mathrm{disc}(f) t^{1-n} = (-1)^{\frac{(n-1)(n+2)}{2}} [(n-1)^{n-1} t^{-1} + n^n] & \text{if } n \text{ is odd} \\ \mathrm{disc}(f) t^{2-n} = (-1)^{\frac{(n-1)(n+2)}{2}} [(n-1)^{n-1} + n^n t] & \text{if } n \text{ is even} \end{cases}$$

and $u' = \sqrt{u}$. Then $K(u) = K(t)$, so both u and u' are transcendental over K. Moreover, u is a product of $\mathrm{disc}(f)$ with a square of $K(t)$, so $K(u') = K(\sqrt{\mathrm{disc}(f)})$. Thus, $\mathrm{Gal}(F/K(u')) \cong A_n$. This proves (a).

Next, express t in terms of u and substitute in $f(t, X)$ to get the irreducible polynomial for the roots of $f(t, X)$ in terms of u:

$$g(u, X) = \begin{cases} X^n - X^{n-1} - \frac{(n-1)^{n-1}}{(-1)^{(n-1)(n-2)/2} u - n^n} & \text{if } n \text{ is odd} \\ X^n - X^{n-1} - \frac{(-1)^{(n-1)(n-2)/2} u - (n-1)^{n-1}}{n^n} & \text{if } n \text{ is even.} \end{cases}$$

Then $\mathrm{Gal}(g(u, X), K(u)) = S_n$ and $\mathrm{Gal}(g((u')^2, X), K(u')) = A_n$.

Let $v = ua$ and $v' = u'\sqrt{a}$. The map $u \mapsto v$ extends to an automorphism of $K(u)$ over K. This automorphism further extends to an isomorphism of $K(u')$ onto $K(v')$ and further to an isomorphism of F onto a field F'. Thus, $\mathrm{Gal}(F'/K(u)) \cong S_n$ and $\mathrm{Gal}(F'/K(v')) \cong A_n$. Since a is not a square in K, there is no $c \in K$ with $v = c^2 u$. Hence, $K(u')$ and $K(v')$ are distinct quadratic extensions of $K(u)$. Therefore, $K(u') \not\subseteq F'$. Put $F'' = K(u')F'$. Then $\mathrm{Gal}(F''/K(u')) \cong S_n$ and $\mathrm{Gal}(F''/K(u', v')) \cong A_n$. In addition, $K(u', v') = K(u', \sqrt{a})$. By Lemma 16.4.2, the embedding problem $S_n \to \mathrm{Gal}(K(\sqrt{a})/K)$ is solvable. □

16.8 GAR-Realizations

Let K be a field, L a finite Galois extension of K, and

(1) $$1 \longrightarrow C \longrightarrow G \xrightarrow{\alpha} \text{Gal}(L/K) \longrightarrow 1$$

an embedding problem over K with kernel C having a trivial center. We give a sufficient condition for the embedding problem to have a regular solution. As usual, $\text{Aut}(C)$ denotes the group of all automorphisms of C. For each $c \in C$ let $\iota(c)$ be the inner automorphism of C induced by conjugation with c. The map $c \mapsto \iota(c)$ identifies C with the group $\text{Inn}(C)$ of all inner automorphisms of C.

Let F/K be an extension of fields. We say F is **rational** over K (or F is K-**rational**) if $F = K(T)$ with T being a set of algebraically independent elements over K.

Definition 16.8.1: GAR-Realizations. Let C be a finite group with a trivial center. We say C is **GA** over K if there are algebraically independent elements $t_1, \ldots, t_r$ over K satisfying:
(2) $K(\mathbf{t})$ has a finite extension F which is regular over K such that $\text{Aut}(C)$ acts on F with $K(\mathbf{t})$ being the fixed field of C.

Denote the fixed field of $\text{Aut}(C)$ in F by E. We say C is **GAR** over K if in addition to (2) this holds:
(3) Every extension E' of E satisfying $E'K_s = K_s(\mathbf{t})$ is a purely transcendental extension of $E' \cap K_s$. □

Remark 16.8.2:
(a) The "G" in GAR abbreviates "Galois", "A" abbreviates "Automorphisms", and "R" abbreviates "Rational".

(b) The GAR-condition implies that the fixed field E of $\text{Aut}(C)$ is regular over K. But, it does not require E to be rational over K. This is however the case when $r = 1$ (Lüroth's theorem, Remark 3.6.2(a)). □

LEMMA 16.8.3: *Let $C_1, \ldots, C_r$ be finite non-Abelian simple groups. Put $C = \prod_{i=1}^r C_i$. Then:*
(a) *Each normal simple subgroup N of C coincides with C_i for some i between 1 and r.*
(b) *Suppose $\alpha_i : C_1 \to C_i$ is an isomorphism, $i = 1, \ldots, r$. Then $\text{Aut}(C) \cong S_r \ltimes \prod_{i=1}^r \text{Aut}(C_i)$.*

Proof of (a): Choose $n \in N$, $n \neq 1$. Write $n = c_1 \cdots c_r$ with $c_i \in C_i$. Assume, without loss, $c_1 \neq 1$. Since C_1 is simple, its center is trivial. Hence, there is a $c_1' \in C_1$ with $c_1 c_1' \neq c_1' c_1$. For each $i \geq 2$ we have $c_1' c_i = c_i c_1'$. Therefore, $nc_1' \neq c_1' n$ and $(c_1')^{-1} n^{-1} c_1' n_1 \neq 1$. It follows that, $N \cap C_1 \neq 1$. Consequently, $N = C_1$.

Proof of (b): To simplify notation identify C_i with C_1 via α_i. Thus, each element of C is an r-tuple $\mathbf{c} = (c_1, \ldots, c_r)$ with $c_i \in C_1$. Each C_i is then the group of all c with $c_j = 1$ for $j \neq i$. We embed $\mathrm{Aut}(C_1)^r$ and S_r in $\mathrm{Aut}(C)$ by the rules:

$$\mathbf{c}^\gamma = (c_1^{\gamma_1}, \ldots, c_r^{\gamma_r}) \quad \text{and} \quad (c_1, \ldots, c_r)^\sigma = (c_{1\sigma^{-1}}, \ldots, c_{r\sigma^{-1}})$$

This gives:

(4) $$\sigma^{-1}\gamma\sigma = (\gamma_{1\sigma^{-1}}, \ldots, \gamma_{r\sigma^{-1}}).$$

Thus, S_r normalizes $\mathrm{Aut}(C_1)^r$. In addition observe that $\mathrm{Aut}(C_1)^r \cap S_r = 1$.

Finally, consider $\gamma \in \mathrm{Aut}(C)$. By (a), there is a unique $\sigma \in S_r$ with $C_i^\gamma = C_{i\sigma}$, $i = 1, \ldots, r$. So, there are $\gamma_1, \ldots, \gamma_r$ in $\mathrm{Aut}(C_1)$ with $\gamma = (\gamma_1, \ldots, \gamma_r)\sigma^{-1}$. Thus, $\mathrm{Aut}(C)$ is the semidirect product of $\mathrm{Aut}(C_1)^r$ with S_r, where the action of S_r on $\mathrm{Aut}(C_1)^r$ given by (4). □

Remark 16.8.4: Minimal normal subgroups. Let G be a finite group and C a normal subgroup. Suppose C is a **minimal normal subgroup** of G. Thus, $C \neq 1$, $C \triangleleft G$, and G has no normal subgroup in C other than 1 and C. Let C_1 be a simple normal subgroup of C. The conjugates $C_1, C_2, \ldots, C_r$ of C_1 in G generate a nontrivial normal subgroup of G which is contained in C. Hence, they generate C.

Suppose in addition, C_1 is non-Abelian. Let k be an integer between 0 and $r-1$. Suppose by induction $\langle C_1, \ldots, C_k \rangle = \prod_{i=1}^k C_i$. By Lemma 16.8.3(a), $C_{k+1} \not\leq \prod_{i=1}^k C_i$. Hence, $\langle C_1, \ldots, C_{k+1} \rangle = \prod_{i=1}^{k+1} C_i$. In particular, $C = \prod_{i=1}^r C_i$.

By construction, G acts transitively on the set $\{C_1, \ldots, C_r\}$ by conjugation. □

LEMMA 16.8.5 (Semilinear Rationality Criterion): *Let K, L, E, F be fields with L/K Galois, $K \subseteq E$, $E \cap L = K$, and $EL = F$. Let V be a L-subspace of F. Consider the K-subspace $U = \{v \in V \mid v^\sigma = v \text{ for all } \sigma \in \mathrm{Gal}(F/E)\}$ of E.*
(a) *Suppose V is $\mathrm{Gal}(F/E)$-invariant. Then each K-basis of U is an L-basis of V. Thus, $V \cong U \otimes_K L$.*
(b) *Suppose, in addition to the assumption of (a), $F = L(V)$. Then $E = K(U)$.*
(c) *Suppose, in addition to the assumptions of (a) and (b), $\dim(V)$ is finite and equal $\mathrm{trans.deg}(F/L)$. Then F is L-rational and E is K-rational.*

Proof of (a): By assumption, E is linearly disjoint from L over K. Thus, each K-basis of U is linearly independent over L. Hence, it suffices to prove that U spans V over L.

To this end consider $v \in V$. Choose a finite Galois extension F_0 of E in F containing v. Let $L_0 = F_0 \cap L$. By assumption, $F_0 L = F$ and res: $\mathrm{Gal}(F/E) \to \mathrm{Gal}(L/K)$ is an isomorphism. Hence, $\mathrm{res}(\mathrm{Gal}(F/F_0)) = \mathrm{Gal}(L/L_0)$ and res: $\mathrm{Gal}(F_0/E) \to \mathrm{Gal}(L_0/K)$ is an isomorphism. Put $m = [F_0 : E] = [L_0 : K]$ and $G = \mathrm{Gal}(F_0/E)$.

16.8 GAR-Realizations

Choose a basis $c_1, \ldots, c_m$ for L_0/K. For each i let $u_i = \sum_{\sigma \in G} c_i^\sigma v^\sigma$. Then $u_i \in U$. Since $\det(c_i^\sigma) \neq 0$ [Lang7, p. 286, Cor. 5.4] and $|G| = m$, each v^σ is a linear combination of the u_i with coefficients which are rational functions in c_i^τ, $\tau \in G$, $i = 1, \ldots, m$. In particular, v is in the L-vector space spanned by U, as desired.

Proof of (b): By the tower property (Lemma 2.5.3), E is linearly disjoint from $L(U)$ over $K(U)$. By (a), $F = L(V) = L(U)$. Therefore, $E = K(U)$.

Proof of (c): Let $u_1, \ldots, u_n$ be a K-basis of U. By (a), $u_1, \ldots, u_n$ is an L-basis of V. By (b), $F = L(u_1, \ldots, u_n)$. Since $n = \text{trans.deg}(F/L)$, the elements $u_1, \ldots, u_n$ are algebraically independent over L. Therefore, F is L-rational.

By (b), $E = K(u_1, \ldots, u_n)$. Since F/E and L/K are algebraic,

$$\text{trans.deg}(E/K) = \text{trans.deg}(F/L) = n.$$

Consequently, E is K-rational. □

PROPOSITION 16.8.6 (Matzat): *Consider an embedding problem (1), where C a minimal normal subgroup of G and $C = \prod_{i=1}^r C_i$ with C_i simple non-Abelian and conjugate to C_1 in G, $i = 1, \ldots, r$. Suppose C_1 is GAR over K. Then embedding problem (1) is regularly solvable over K.*

Proof: We break the proof into several parts.

PART A: *Group theory.* The map $g \mapsto (\iota(g), \alpha(g))$ embeds G into $\text{Aut}(C) \times \text{Gal}(L/K)$ with

(5) $\qquad G \cap \text{Aut}(C) = C, \quad \alpha = \text{pr}_2, \quad \text{pr}_2(G) = \text{Gal}(L/K).$

For each i let $D_i = N_G(C_i)$. Put $D = \bigcap_{i=1}^r D_i$. Then $C \leq D \leq D_i \leq G$. By (5),

(6) $\qquad D_i \cap \text{Aut}(C_i) = C_i.$

For each $\sigma \in G$, Remark 16.8.4 gives a unique $\pi \in S_r$ with $C_i^\sigma = C_{i^\pi}$, $i = 1, \ldots, r$. Thus, $D_i^\sigma = D_{i^\pi}$, $i = 1, \ldots, r$. Therefore, D is normal in G.

Finally, as in Lemma 16.8.3(b), we embed $\prod_{i=1}^r \text{Aut}(C_i)$ in $\text{Aut}(C)$ by the rule:

$$(c_1, \ldots, c_r)^{(\gamma_1, \ldots, \gamma_r)} = (c_1^{\gamma_1}, \ldots, c_r^{\gamma_r}).$$

PART B: *GA-realization of C.* The assumption on C_1 gives algebraically independent elements $t_1, \ldots, t_m$ over K and fields E_1, F_1 satisfying this:
(7a) $K \subseteq E_1 \subseteq K(t_1, \ldots, t_m) \subseteq F_1$, F_1/K regular, F_1/E_1 Galois,
(7b) $\text{Gal}(F_1/E_1) = \text{Aut}(C_1)$, and $\text{Gal}(F_1/K(t_1, \ldots, t_m)) = C_1$.
(7c) Suppose M is an extension of E_1 with $MK_s = K_s(t_1, \ldots, t_m)$. Then M is a field of rational functions over $M \cap K_s$.

Choose algebraically independent elements t_{ij}, $i = 1,\ldots,r$, $j = 1,\ldots,m$ over K with $t_{1j} = t_j$, $j = 1,\ldots,m$. For each i let $\mathbf{t}_i = (t_{i1},\ldots,t_{im})$. Also, let $\alpha_i\colon K(\mathbf{t}_1) \to K(\mathbf{t}_i)$ be the K-isomorphism with $t_j^{\alpha_i} = t_{ij}$, $j = 1,\ldots,m$. Put $E_i = E_1^{\alpha_i}$ and extend α_i to an isomorphism of F_1 on a field F_i. The latter induces an isomorphism of $\mathrm{Gal}(F_1/K(\mathbf{t}_1))$ and $\mathrm{Gal}(F_1/E_1)$ onto $\mathrm{Gal}(F_i/K(\mathbf{t}_i))$ and $\mathrm{Gal}(F_i/E_i)$, respectively. They are given by $\gamma \mapsto \alpha_i^{-1}\gamma\alpha_i$. Identify $\mathrm{Gal}(F_i/K(\mathbf{t}_i))$ and $\mathrm{Gal}(F_i/E_i)$ through this isomorphism with C_i and $\mathrm{Aut}(C_i)$, respectively. Hence, $E_i, K(\mathbf{t}_i), F_i$ satisfy (7) with i replacing 1. Also, $F_1,\ldots,F_r$ are algebraically independent over K and regular over K (by (7a)). Therefore, they are linearly disjoint over K (Lemma 2.6.7).

Next let $E = E_1 \cdots E_r$, $\mathbf{t} = (\mathbf{t}_1,\ldots,\mathbf{t}_r)$, and $F = F_1 \cdots F_r$. Then
(8a) $K \subseteq E \subseteq K(\mathbf{t}) \subseteq F$, F/K is regular, F/E is Galois,
(8b) $\mathrm{Gal}(F/E) = \prod_{i=1}^r \mathrm{Aut}(C_i)$, and $\mathrm{Gal}(F/K(\mathbf{t})) = \prod_{i=1}^r C_i$.

The group S_r acts on F by permuting the triples $(E_i, \mathbf{t}_i, F_i)$, $i = 1,\ldots,r$. Namely, $x^\pi = x^{\alpha_i^{-1}\alpha_j}$ for $\pi \in S_r$ with $i^\pi = j$ and for $x \in F_i$. This induces an action on $\mathrm{Gal}(F/E)$ which coincides with the action given by (4). (Recall: To obtain (4) we identify C_i with C_1 via α_i and then identify α_i with the identity map.) So, by Lemma 16.8.3(b), the action of $\prod_{i=1}^r \mathrm{Aut}(C_i)$ on F extends to an action of $\mathrm{Aut}(C)$ on F over K. Denote the fixed field of $\mathrm{Aut}(C)$ in F by E_0.

PART C: *The field crossing argument.* By 7(a), F is linearly disjoint from L over K. Put $Q = FL$. Then $\mathrm{Gal}(Q/E_0) \cong \mathrm{Aut}(C) \times \mathrm{Gal}(L/K)$ with $\mathrm{pr}_2 = \mathrm{res}_{Q/L}$. Also, Q/L is regular. Identify $\mathrm{Gal}(Q/E_0L)$ with $\mathrm{Aut}(C)$ and $\mathrm{Gal}(Q/L(\mathbf{t}))$ with C. Part A identifies G with a subgroup of $\mathrm{Gal}(Q/E_0)$. Denote the fixed field of G in Q by P. By (5), $\mathrm{Gal}(Q/P) = G$ and restriction of G into $\mathrm{Gal}(L/K)$ is surjective.

(9) $$PL = L(\mathbf{t}) \quad \text{and} \quad P \cap L = K.$$

Thus, all that remains to be proved is the rationality of P over K.

PART D: *New transcendence basis for $L(\mathbf{t})/L$.* The group D_1 acts on L because L/K is Galois. Denote the fixed field by M_0. Each $\sigma \in D_1$ satisfies $C_1^\sigma = C_1$. Hence, the permutation of $\{1,\ldots,r\}$ corresponding to σ fixes 1, so $F_1^\sigma = F_1$. Thus, D_1 acts on F_1L. Denote the fixed field by M. By (6), $D_1 \cap \mathrm{Aut}(C_1) = C_1$. The fixed field of C_1 in F_1L is $L(\mathbf{t}_1)$ and that of $\mathrm{Aut}(C_1)$ in F_1L is E_1L. Therefore, $ME_1L = L(\mathbf{t}_1)$. Next observe that the restriction of D_1 to F_1 maps D_1 into $\mathrm{Aut}(C_1)$, hence into $\mathrm{Gal}(F_1/E_1)$. Hence, $E_1 \subseteq M$. Thus, $ML = L(\mathbf{t}_1)$, so $MK_s = K_s(\mathbf{t}_1)$. By construction, $M \cap L = M_0$. It follows from (7c) that there are algebraically independent elements $v_1,\ldots,v_m$ over K with $M = M_0(v_1,\ldots,v_m)$. They satisfy

(10) $$L(v_1,\ldots,v_m) = L(\mathbf{t}_1).$$

16.9 Embedding Problems over Hilbertian Fields

Here is a partial diagram of the fields involved:

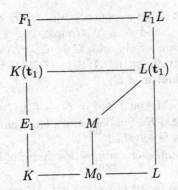

For each i between 1 and r choose $\sigma_i \in G$ with $C_1^{\sigma_i} = C_i$ (Remark 16.8.4). Then $D_1^{\sigma_i} = D_i$. Put $v_{ij} = v_j^{\sigma_i}$, $j = 1,\ldots,m$, and $\mathbf{v}_i = (v_{i1},\ldots,v_{im})$. Then let $\mathbf{v} = (\mathbf{v}_1,\ldots,\mathbf{v}_r)$. By (10), $L(\mathbf{v}_i) = L(\mathbf{t}_i)$, $i = 1,\ldots,r$, so $L(\mathbf{v}) = L(\mathbf{t})$. Hence, the v_{ij} are algebraically independent over K.

Since $v_j \in M$, we have $v_j^\delta = v_j$ for all $\delta \in D_1$ and $j = 1,\ldots,m$. If $1 \le i \le r$ and $\delta \in D_i$, then $\sigma_i \delta \sigma_i^{-1} \in D_1$, hence $v_{ij}^\delta = v_j^{(\sigma_i \delta \sigma_i^{-1})\sigma_i} = v_j^{\sigma_j} = v_{ij}$. It follows for $\delta = D = \bigcap_{i=1}^r D_i$ that $v_{ij}^\delta = v_{ij}$ for all i and j.

PART E: *Rationality of P.* Denote the fixed field of D in Q by N. By Part A, N/P is Galois. Put $N_0 = N \cap L$. By Part D, $v_{ij}^\delta = v_{ij}$ for each $\delta \in D$ and all i,j. Thus, $N_0(\mathbf{v}) \subseteq N$. As $[L(\mathbf{v}) : N_0(\mathbf{v})] = [L : N_0] = [L(\mathbf{v}) : N]$, we have $N = N_0(\mathbf{v})$. By (9), $PN_0 = N$.

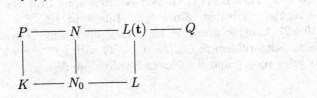

Let V be the vector space spanned by the v_{ij}'s over N_0. For each $\sigma \in G$ and each i there exists i' with $C_i^\sigma = C_{i'} = C_i^{\sigma_{i'}^{-1}\sigma_{i'}}$. Hence, $\sigma\sigma_{i'}^{-1}\sigma_i \in D_i$, so $v_{ij}^\sigma = v_{ij}^{(\sigma\sigma_{i'}^{-1}\sigma_i)\sigma_i^{-1}\sigma_{i'}} = v_{ij}^{\sigma_i^{-1}\sigma_{i'}} = v_{i',j}$. Thus, $\mathrm{Gal}(N/P)$ leaves V invariant. By Part D, $\dim(V) = rm = \mathrm{trans.deg}(P/K)$. It follows from Lemma 16.8.5 that P is K-rational. $\square$

16.9 Embedding Problems over Hilbertian Fields

An affirmative solution to the inverse problem of Galois theory, the realization of all finite groups over $\mathbb{Q}$, seems at present to be out of reach. Matzat's method of GAR-realization gives an effective tool for a partial solution of the problem, namely for the of realization of finite groups with non-Abelian composition factors.

Every finite group G has a sequence $N_0, \ldots, N_n$ of subgroups with $N_0 = G$, $N_n = 1$, $N_i \triangleleft N_{i-1}$, and N_{i-1}/N_i simple, $i = 1, \ldots, n$. For each finite simple group C, the number of i with $C \cong N_{i-1}/N_i$ depends only on G and not on the sequence [Huppert, p. 64, Satz 11.7]. If this number is positive, C is a **composition factor** of G.

PROPOSITION 16.9.1 (Matzat): *Let $\alpha: G \to \mathrm{Gal}(L/K)$ be a finite embedding problem over a field K. Suppose every finite embedding problem for G with an Abelian kernel is solvable.*
(a) *If K is Hilbertian and each composition factor of $\mathrm{Ker}(\alpha)$ is GAR over K, then α is solvable.*
(b) *If each composition factor of $\mathrm{Ker}(\alpha)$ is GAR over every extension of K, then α is regularly solvable.*

Proof: Assume without loss that $\mathrm{Ker}(\alpha)$ is nontrivial. Choose a minimal normal subgroup C of G in $\mathrm{Ker}(\alpha)$. Then α induces an isomorphism $\alpha': G/C \to \mathrm{Gal}(L/K)$. The rest of the proof splits into two parts.

PART A: *Proof of (a).* An induction hypothesis gives a Galois extension L' of K containing L and an isomorphism $\beta': \mathrm{Gal}(L'/K) \to G/C$ with $\alpha' \circ \beta' = \mathrm{res}_{L'/L}$.

Let $\pi: G \to G/C$ be the quotient map. The kernel of $(\beta')^{-1} \circ \pi: G \to \mathrm{Gal}(L'/K)$ is C. By Remark 16.8.4, $C \cong \prod_{i=1}^{r} C_i$ with C_i simple non-Abelian and C_i conjugate to C_1 in G, $i = 1, \ldots, r$. In particular, C_1 is a composition factor of G. By assumption, α is solvable if C_1 is Abelian. Suppose C_1 is non-Abelian. Then C_1 is GAR over K. Hence, by Proposition 16.8.6, $(\beta')^{-1} \circ \pi$ is regularly solvable. Since K is Hilbertian, $(\beta')^{-1} \circ \pi$ is solvable (Lemma 16.4.2). In other words, K has a Galois extension N containing L' and there is an isomorphism $\beta: \mathrm{Gal}(N/K) \to G$ with $(\beta')^{-1} \circ \pi \circ \beta = \mathrm{res}_{N/L'}$. Then, $\alpha \circ \beta = \mathrm{res}_{N/L}$ and β solves embedding problem α.

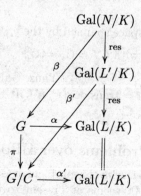

PART B: *Proof of (b).* An induction hypothesis gives algebraically independent elements $t_1,\ldots,t_m$ over K, a Galois extension E of $K(\mathbf{t})$, and an epimorphism $\beta'\colon \mathrm{Gal}(E/K(\mathbf{t})) \to G/C$ satisfying: E is a regular extension of L and $\alpha' \circ \beta' = \mathrm{res}_{E/L}$. As above, let $\pi\colon G \to G/C$ be the quotient map.

By assumption, each composition factor of C is regular over $K(\mathbf{t})$. Proposition 16.8.6 gives algebraically independent elements $u_1,\ldots,u_n$ over $K(\mathbf{t})$, a Galois extension F of $K(\mathbf{t}, \mathbf{u})$, and an isomorphism $\beta\colon \mathrm{Gal}(F/K(\mathbf{t}, \mathbf{u})) \to G$ satisfying: F is a regular extension of E and $(\beta')^{-1} \circ \pi \circ \beta = \mathrm{res}_{F/E}$. Then F is a regular extension of L and β is a regular solution of embedding problem α. □

PROPOSITION 16.9.2: *Let K be a field and n a positive integer with $\mathrm{char}(K) \nmid (n-1)n$ and $n \neq 2,3,6$. Then A_n is GAR over K.*

Proof: Use the notation of Proposition 16.7.6. The assumption $n \neq 2$ implies that the center of S_n is trivial. The assumption $n \neq 2, 3, 6$ implies $S_n = \mathrm{Aut}(A_n)$ [Suzuki, p. 299, Statement 2.17]. Hence, $F/K(u)$ is a GA-realization of A_n. We must still prove Condition (3) of Section 16.8.

Consider an extension E' of $K(t)$ with $E'K_s = K_s(u)$. Let $L = E' \cap K_s$. We have to prove E' is L-rational.

To begin, note that $E' \cap K_s(t) = L(t)$ and $[E' : L(t)] = [K_s(u) : K_s(t)] = [K(u) : K(t)] = 2$. In particular, E' is a function field of one variable over L. Since the genus of $K_s(u)$ is 0, so is the genus of E' (Proposition 3.4.2(b)). Suppose $\mathfrak{p}'$ is a prime divisor of $L(t)/L$ which ramifies in E'. Then $\deg(\mathfrak{p}') = 1$. Hence, there are exactly one prime divisor $\mathfrak{p}_s$ of $K_s(t)/K_s$ over $\mathfrak{p}'$ (Proposition 3.4.2(e)) and $\mathfrak{p}_s$ ramifies in $K_s(u)$ (Proposition 3.4.2(c)). Let $\mathfrak{p}$ be the common restriction of $\mathfrak{p}_s$ and of $\mathfrak{p}'$ to $K(t)$. Then $\mathfrak{p}$ ramifies in $K(u)$ (Proposition 3.4.2(c)). But there are at most three such $\mathfrak{p}$ and each of them is of degree 1. Hence, there are at most three possibilities for $\mathfrak{p}'$ (Proposition 3.4.2(e)). It follows from Lemma 16.7.5 that E' is L-rational. □

The combination of Propositions 16.9.1 and 16.9.2 gives concrete solvable embedding problems with non-Abelian kernels.

PROPOSITION 16.9.3: *Let K be a field and $\alpha\colon G \to \mathrm{Gal}(L/K)$ a finite embedding problem over K. Suppose each composition factor of $\mathrm{Ker}(\alpha)$ is A_n with $\mathrm{char}(K) \nmid (n-1)n$ and $n \neq 6$. Then α is regularly solvable over K. If, in addition, K is Hilbertian, then α is solvable over K.*

Take $L = K$ in Proposition 16.9.3:

COROLLARY 16.9.4: *Let K be a field and G a finite group. Suppose each composition factor of G is A_n with $\mathrm{char}(K) \nmid (n-1)n$ and $n \neq 6$. Then G is regular over K. If, in addition, K is Hilbertian, then G is realizable over K.*

Remark 16.9.5: More GAR-realizations. There is a long list of finite non-Abelian simple groups which are known to be GAR over $\mathbb{Q}$. Beside A_n (with $n \neq 6$), this list includes $\mathrm{PSL}_2(\mathbb{F}_p)$ with p odd and $p \not\equiv \pm 1 \bmod 24$ and all sporadic groups with the possible exception of M_{23}. See [Malle-Matzat, Thm. IV.4.3].

The list becomes longer over the maximal Abelian extension $\mathbb{Q}_{ab}$ of $\mathbb{Q}$. In addition to the groups that are GAR over $\mathbb{Q}$ it contains A_6, $\mathrm{PSL}_2(\mathbb{F}_p)$ with p odd, and M_{23}. See [Malle-Matzat, Thm. IV.4.6].

It is still unknown whether every finite non-Abelian simple group is GAR over $\mathbb{Q}_{ab}$. If this is the case, each finite embedding problem over $\mathbb{Q}_{ab}$ would be solvable, as we will see in Example 24.8.5.

It is even unknown if every finite non-Abelian simple group is GAR over each field K containing $\tilde{\mathbb{Q}}$. An affirmative answer to this question would enable us to solve embedding problems over the field $\tilde{\mathbb{Q}}((t_1, t_2))$ which is Hilbertian by Example 15.5.2.

The lists over $\mathbb{Q}$ and $\mathbb{Q}_{ab}$ have been established in large part by using the Riemann Existence Theorem. This partially explains the lack of knowledge of GAR realizations in characteristic p.

An exception to our lack of knowledge of GAR realization is the family A_n of alternative groups. Theorem 15 of [Brink] improves Proposition 16.9.2 and proves that A_n is GAR over every field K if $n \neq 2, 6$ and $\mathrm{char}(K) \neq 2$. The case $\mathrm{char}(K) = 2$ is left open. □

16.10 Finitely Generated Profinite Groups

Let S be a subset of a profinite group G. Denote the closed subgroup generated by S by $\langle S \rangle$. We say S **generates** G if $\langle S \rangle = G$. In this case each map φ_0 of S into a profinite group H has at most one extension to a (continuous) homomorphism $\varphi \colon G \to H$.

A profinite group G is **finitely generated** if it has a finite set of generators. In this case, the minimal number of generators of G is the **rank** of G. Note that the rank of a quotient of G does not exceed the rank of G.

Example 16.10.1: Consider the group $(\mathbb{Z}/p\mathbb{Z})^n$ as a vector space over $\mathbb{F}_p$ of dimension n. Each group theoretic set of generators of $(\mathbb{Z}/p\mathbb{Z})^n$ generates $(\mathbb{Z}/p\mathbb{Z})^n$ as a vector space over $\mathbb{F}_p$. Hence, $\mathrm{rank}(\mathbb{Z}/p\mathbb{Z})^n = n$. Since $(\mathbb{Z}/p\mathbb{Z})^n$ is a quotient of $\mathbb{Z}_p^n$, the rank of the latter group is at least n.

For each i between 1 and n consider the element $e_i = (0, \ldots, 1, \ldots, 0)$ with 1 in the ith coordinate and 0 elsewhere. Then $e_1, \ldots, e_n$ generates $\mathbb{Z}_p^n$. It follows that $\mathrm{rank}(\mathbb{Z}_p^n) = n$. □

LEMMA 16.10.2: *A finitely generated profinite group G has, for each positive integer n, only finitely many open subgroups of index at most n.*

Proof: Each open subgroup M of G of index $\leq n$ contains an open normal subgroup N with G/N isomorphic to a subgroup of S_n.

Indeed, suppose $m = (G : M) \leq n$. Then, $B = \{gM \mid g \in G\}$ is a set of order m. Multiplication from the left with an element x of G induces a permutation $\pi(x)$ of B. Specifically, $\pi(x)(gM) = xgM$. Thus, π is a

16.10 Finitely Generated Profinite Groups

homomorphism of G into S_n with $\mathrm{Ker}(\pi) = \bigcap_{x \in G} M^x$. Therefore, $\mathrm{Ker}(\pi)$ is an open normal subgroup of G which is contained in M. Moreover, $G/\mathrm{Ker}(\pi)$ is isomorphic to a subgroup of S_m, hence to a subgroup of S_n.

Thus, it suffices to prove G has only finitely many open normal subgroups N with G/N isomorphic to a subgroup of S_n.

The map $\alpha \mapsto \mathrm{Ker}(\alpha)$ maps the set of all homomorphisms $\alpha \colon G \to S_n$ onto the set of all open normal subgroups N of G such that G/N is isomorphic to a subgroup of S_n. Hence, the number ν of those N's does not exceed the number of the α's. Let S be a finite set of generators of G. Then every homomorphism $\alpha \colon G \to S_n$ is determined by its values on S. Therefore, $\nu \leq (n!)^{|S|}$. $\square$

We call a profinite group G **small** if for each positive integer n the group G has only finitely many open subgroups of index n. By Lemma 16.10.2, every finitely generated profinite group is small. Thus, each of the results we prove in this section for small profinite groups holds for finitely generated profinite groups.

Remark 16.10.3: Small profinite groups and open subgroups. Let G be a profinite group.

(a) Denote the intersection of all open subgroups of G of index at most n by G_n. Then G_n is a closed normal subgroup of G. Moreover, G is small if and only if G_n is open in G for all n.

(b) Let H be an open subgroup of index m of G. Consider open subgroups G' and H' of G and H, respectively. Then $(G : H') = (G : H)(H : H')$ and $(H : H \cap G') \leq (G : G')$. Thus, $G_{mn} \leq H_n \leq G_n$. By (a), G is small if and only if H is small.

(c) Let G and H be as in (b). Suppose H is finitely generated, say by $h_1, \ldots, h_d$. Let $g_1, \ldots, g_n$ be representatives of G/H. Then $g_1, \ldots, g_n, h_1, \ldots, h_d$ generate G.

Conversely, if G is finitely generated, then H is finitely generated (Exercise 7). We prove a qualitative version of this result in Section 17.6.

(d) Let $\alpha \colon G \to H$ be an epimorphism of profinite groups. If $g_1, \ldots, g_n$ are generators of G, then $\alpha(g_1), \ldots, \alpha(g_n)$ are generators of H. If G is small, so is H. Indeed, let n be a positive integer and H_0 an open subgroup of H of index n. Then $G_0 = \alpha^{-1}(H_0)$ is an open subgroup of G of index n. The map $H_0 \mapsto \alpha^{-1}(H_0)$ is injective. Since there are only finitely many G_0's, there are only finitely many H_0's. $\square$

Example 16.10.4: A small profinite group which is not finitely generated. Let $A = \prod \mathbb{Z}_p^p$ with p ranging over all prime numbers. Consider an open subgroup N of A of index n with $n = \prod_{p \leq m} p^{k_p}$. Then N contains the open subgroup $\prod_{p \leq m} p^{k_p} \mathbb{Z}_p^p \times \prod_{p > m} \mathbb{Z}_p^p$. Hence, there are only finitely many possibilities for N. Thus, A is small.

On the other hand, $\mathrm{rank}(A) \geq \mathrm{rank}(\mathbb{Z}_p^p) = p$ for each p (Example 16.10.1). Therefore, A is not finitely generated. $\square$

Remark 16.10.5: Characteristic subgroups. A closed subgroup N of a profinite group G is **characteristic** if it is invariant under every automorphism of G. In particular, N is normal in G.

Suppose G is small. Then, G_n is a characteristic open subgroup. The decreasing sequence $G \geq G_2 \geq G_3 \geq \cdots$ intersects in 1. Thus, it consists of a basis of open neighborhoods of 1 in G.

Let N be an open characteristic subgroup of G. Then $\text{Aut}(G/N)$ is a finite group. For each $\alpha \in \text{Aut}(G)$ define $\alpha_N \in \text{Aut}(G/N)$ by $\alpha_N(gN) = \alpha(g)N$. The map $\alpha \mapsto \alpha_N$ gives a homomorphism $\Phi_N \colon \text{Aut}(G) \to \text{Aut}(G/N)$. By the preceding paragraph, the intersection of all these N is the trivial group. Hence, the Φ_N combine to an embedding $\text{Aut}(G) \to \varprojlim \text{Aut}(G/N)$ which is actually surjective. Thus, $\text{Aut}(G)$ is a profinite group. $\square$

The most useful properties of small profinite groups are embodied in the following result:

PROPOSITION 16.10.6: *Let G be a small profinite group. Then:*
(a) *Every epimorphism of G onto itself is an automorphism.*
(b) *Let $\alpha \colon G \to H$ and $\beta \colon H \to G$ be epimorphisms. Then both α and β are isomorphisms.*

Proof of (a): Let $\theta \colon G \to G$ be an epimorphism. Let $\mathcal{G}_n$ be the finite set of all open subgroups of G of index at most n. The map $H \mapsto \theta^{-1}(H)$ maps $\mathcal{G}_n$ injectively into itself. Hence, it maps $\mathcal{G}_n$ onto itself. Therefore, in the notation of Remark 16.10.3

$$G_n = \bigcap_{H \in \mathcal{G}_n} H = \bigcap_{H \in \mathcal{G}_n} \theta^{-1}(H) = \theta^{-1}(\bigcap_{H \in \mathcal{G}_n} H) = \theta^{-1}(G_n).$$

If $\theta(g) = 1$, then $g \in \bigcap_{n=1}^\infty \theta^{-1}(G_n) = \bigcap_{n=1}^\infty G_n = 1$. Therefore, θ is injective.

Proof of (b): By (a), $\beta \circ \alpha$ is an isomorphism. Hence, both α and β are injective. $\square$

For a profinite group G, we denote the set of all finite quotients (up to an isomorphism) of G by $\text{Im}(G)$.

PROPOSITION 16.10.7: *Let G and H be profinite groups with G small.*
(a) *If $\text{Im}(H) \subseteq \text{Im}(G)$, then H is a quotient of G.*
(b) *If $\text{Im}(H) = \text{Im}(G)$, then H is isomorphic to G.*

Proof of (a): First we prove H is small. Indeed, let n be a positive number and $B_1, \ldots, B_r$ distinct open subgroups of H of index at most n. Choose an open normal subgroup N of H which is contained in $\bigcap_{i=1}^r B_i$. Then $H/N \in \text{Im}(H)$. By assumption, G has an open normal subgroup M with $G/M \cong H/N$. Hence, G has r open subgroups of index at most n. Consequently, r is bounded.

In the notation of Remark 16.10.3(a), the finite group H/H_n belongs to $\text{Im}(H)$, and therefore to $\text{Im}(G)$. Thus, G has an open normal subgroup K

16.10 Finitely Generated Profinite Groups

with $G/K \cong H/H_n$. In particular, K is an intersection of open subgroups of index $\leq n$. Hence, $G_n \leq K$. Therefore, there is an epimorphism from G/G_n to H/H_n. Denote the finite nonempty set of all epimorphisms of $G/G_n \to H/H_n$ by Φ_n.

Let $\varphi\colon G/G_{n+1} \to H/H_{n+1}$ be an epimorphism. It maps the set of all subgroups of G/G_n of index at most n onto the set of all subgroups of H/H_n of index at most n. Hence, $\varphi(G_n/G_{n+1}) \leq H_n/H_{n+1}$. Therefore, φ induces an epimorphism $\bar{\varphi}\colon G/G_n \to H/H_n$. This defines a map $\Phi_{n+1} \to \Phi_n$.

By Corollary 1.1.4, $\varprojlim \Phi_n$ is nonempty. Each element in $\varprojlim \Phi_n$ gives a compatible system of epimorphisms $\beta_n\colon G/G_n \to H/H_n$. It defines an epimorphism $\beta\colon G \to H$, as desired.

Proof of (b): Statement (a) gives epimorphisms $\varphi\colon G \to H$ and $\psi\colon H \to G$. Thus, $\psi \circ \varphi\colon G \to G$ and $\varphi \circ \psi\colon H \to H$ are epimorphisms. By (a), H is small. Hence, by Proposition 16.10.6, both $\psi\circ\varphi$ and $\varphi\circ\psi$ are automorphisms. Consequently, both φ and ψ are isomorphisms. □

COROLLARY 16.10.8: *Let $\alpha\colon G \to H$ be an epimorphism of profinite groups. Suppose G is small and $\mathrm{Im}(G) \subseteq \mathrm{Im}(H)$. Then α is an isomorphism.*

Proof: Since G is small, so is H (remark 16.10.3(d)). By Proposition 16.10.7(a), G is a quotient of H. Therefore, by Proposition 16.10.6(b), α is an isomorphism. □

Example 16.10.9: Small Galois groups.

(a) For each finite field K, $\mathrm{Gal}(K) \cong \tilde{\mathbb{Z}}$ (Section 1.5). Thus, $\mathrm{Gal}(K)$ is generated by one element.

(b) Let p be a prime number. The local compactness of $\mathbb{Q}_p$ and Krasner's lemma imply that $\mathrm{Gal}(\mathbb{Q}_p)$ is small [Lang5, p. 54, Prop. 14]. Deeper arguments show $\mathrm{Gal}(\mathbb{Q}_p)$ is generated by 4 elements [Jannsen, Satz 3.6].

(c) Let K be a number field, O_K its ring of integers, and S a finite number of prime ideals of O_K. Denote the maximal algebraic extension of K unramified outside S by K_S. It is a Galois extension of K (Corollary 2.3.7(c)). We prove $\mathrm{Gal}(K_S/K)$ is a small profinite group.

Let T be the set of all prime numbers which ramify in K or lie under a prime ideal belonging to S. Then T is finite and $K_S \subseteq \mathbb{Q}_T$. Suppose we already know that $\mathrm{Gal}(\mathbb{Q}_T/\mathbb{Q})$ is small. Then, by Remark 16.10.3(b),(d), $\mathrm{Gal}(\mathbb{Q}_T/K)$ is small. Hence, $\mathrm{Gal}(K_S/K)$ is also small. We may therefore assume $K = \mathbb{Q}$ and S is a finite set of prime numbers.

Suppose L is a finite extension of $\mathbb{Q}$ in $\mathbb{Q}_S$ of degree at most n. By [Serre5, p. 130, Proposition 6],

$$\log \mathrm{discriminant}(L/\mathbb{Q}) \leq (n-1) \sum_{p \in S} \log p + n|S| \log n.$$

Thus, $\mathrm{discriminant}(L/\mathbb{Q})$ is bounded. By Hermite-Minkowski [Lang5, p. 121, Thm. 5], there are only finitely many extensions of $\mathbb{Q}$ a given discriminant. Consequently, there are only finitely many possibilities for L.

Alternatively, one may follow [Serre9, p. 107] and first observe that $d = $ discriminant$(L/\mathbb{Q})$ is divisible only by $p \in S$. For each such p, $[L\mathbb{Q}_p : \mathbb{Q}_p] \leq n$. By (b), there are only finitely many possibilities for $L\mathbb{Q}_p$, hence for the pth part of the different of $L/\mathbb{Q}$. Therefore, there are only finitely many possibilities for d.

It is, however, not clear whether $\mathrm{Gal}(K_S/K)$ is finitely generated [Shafarevich1, §3].

(d) There are small absolute Galois groups which are not finitely generated. The group $A = \prod \mathbb{Z}_p^p$ of Example 16.10.4 is one example.

To construct a field with absolute Galois group A, we start from a field K of characteristic 0 that contains all roots of unity. Then $\mathrm{Gal}(K((t))) \cong \mathrm{Gal}(K) \times \hat{\mathbb{Z}}$ [Geyer-Jarden2, Cor. 4.2]. Thus, for each p there is an algebraic extension of $K((t))$ with absolute Galois group $\mathrm{Gal}(K) \times \mathbb{Z}_p$. In particular, taking K to be algebraically closed, we find a field K' with $\mathrm{Gal}(K') \cong \mathbb{Z}_p$.

Induction gives fields $K_{p,i}$ with p ranging over all prime numbers and $i = 1, \ldots, p$ satisfying these conditions: $\mathrm{Gal}(K_{p,i}) = \prod_{l<p} \mathbb{Z}_l^l \times \mathbb{Z}_p^i$ and $K_{l,i} \subseteq K_{p,j}$ if $l < p$ or $l = p$ and $i \leq j$. Let $L = \bigcup_p \bigcup_{i=1}^p K_{p,i}$. Then $\mathrm{Gal}(L) = \varprojlim \mathrm{Gal}(K_{p,i}) \cong A$. □

16.11 Abelian Extensions of Hilbertian Fields

We give here some results about Galois extensions of Hilbertian fields which involve small groups:

PROPOSITION 16.11.1: *Let N be a Galois extension of a Hilbertian field K. Suppose $\mathrm{Gal}(N/K)$ is small. Then each separable Hilbert subset H of N^r contains a separable Hilbert subset of K^r. In particular, N is Hilbertian.*

Proof: By definition $H = H_K(f_1, \ldots, f_k; g)$, where $f_i \in N(T_1, \ldots, T_r)[X]$ is irreducible and separable, $i = 1, \ldots, k$, and $g \in N[\mathbf{T}]$, $g \neq 0$. Let $n = \max(\deg_X(f_1), \ldots, \deg_X(f_k))$. Choose a finite extension L of K in N that contains all coefficients of $f_1, \ldots, f_k, g$. Put $m = [L:K]$. Denote the composite of all extensions of K in N of degree at most mn by M. By assumption, $[M:K] < \infty$. By Corollary 12.2.3, $H_M(f_1, \ldots, f_k; g)$ contains a separable Hilbert subset H_K of K^r.

Let $\mathbf{a} \in H_K$. Consider i between 1 and k. Then $g(\mathbf{a}) \neq 0$ and $f_i(\mathbf{a}, X)$ is irreducible over M. Let b be a zero in $f_i(\mathbf{a}, X)$. Then $L(b)$ is linearly disjoint from M over L. But, $[N \cap L(b) : K] \leq mn$. Hence, $N \cap L(b) \subseteq M \cap L(b) = L$. Thus, $f_i(\mathbf{a}, X)$ is irreducible over N. Consequently, $\mathbf{a} \in H$. □

THEOREM 16.11.2: *Let K be a Hilbertian field and A a finite Abelian group. Then $\hat{\mathbb{Z}} \times A$ occurs over K.*

Proof: For each prime number p Corollary 16.6.7 gives a $\mathbb{Z}_p$-extension L_p of K. The sequence of all these extensions is linearly disjoint over K. Hence, the compositum L of all L_p is a $\hat{\mathbb{Z}}$-extension of K (Lemma 1.4.5).

16.11 Abelian Extensions of Hilbertian Fields

By Proposition 16.3.5 and Remark 16.2.2, there is an X-stable polynomial $f \in K[T, X]$ which is Galois in X with $\mathrm{Gal}(f(T, X), K(T)) \cong A$. In particular, $\mathrm{Gal}(f(T, X), L(T)) \cong A$. Let H_1 (resp. H_2) be the set of all $a \in K$ with $\mathrm{Gal}(f(a, X), K) \cong A$ (resp. $\mathrm{Gal}(f(a, X), L) \cong A$). Since $\hat{\mathbb{Z}}$ is small, L is Hilbertian (Proposition 16.11.1). Moreover, each Hilbert subset of L contains a Hilbert subset of K. By Proposition 13.1.1, H_1 (resp. H_2) contains a Hilbert subset of K (resp. L). Hence, $H_1 \cap H_2$ contains a Hilbert subset H of K.

Choose $a \in H$. Let M be the splitting field of $f(a, X)$ over K. Put $N = LM$. Then $\mathrm{Gal}(M/K) \cong \mathrm{Gal}(N/L) \cong A$. Therefore, $\mathrm{Gal}(N/K) \cong \hat{\mathbb{Z}} \times A$. □

THEOREM 16.11.3 ([Kuyk2, p. 113]): *Every Abelian extension N of a Hilbertian field K is Hilbertian.*

Proof ([Weissauer, Satz 9.8]): We may assume that $N \neq K$ and choose $\sigma \in \mathrm{Gal}(N/K)$, $\sigma \neq 1$. The fixed field L of σ in N is a proper subfield of N. Hence, it has a finite proper extension M in N. Let $[M : L] = m$. Since $\mathrm{Gal}(N/L)$ is a closed subgroup of an Abelian group, L/K is Galois. By Theorem 13.9.1(b), M is a Hilbertian field. Now observe that σ^m generates $\mathrm{Gal}(N/M)$. Therefore, by Proposition 16.11.1, N is Hilbertian. □

Remark 16.11.4: *We cannot draw the conclusion, in Theorem 16.11.3, that every separable Hilbert subset of N contains a separable Hilbert subset of K.* Indeed, consider $N = K_{\mathrm{ab}}$, the maximal Abelian extension of K. Suppose $\mathrm{char}(K) \neq 2$. Then there exists no $a \in K$ with $X^2 - a$ irreducible over K_{ab}, even though $X^2 - T$ is absolutely irreducible. When $\mathrm{char}(K) = 2$, there is no $a \in K$ with $X^2 - X - a$ irreducible over K_{ab} although $X^2 - X - T$ is absolutely irreducible and separable in X. □

LEMMA 16.11.5: *Let K be a Hilbertian field. Then $\mathrm{Gal}(K)$ is neither prosolvable nor small.*

Proof: By Corollary 16.2.7(a), K has a Galois extension L with Galois group S_5. Hence, $\mathrm{Gal}(K)$ is not prosolvable.

Again, by Corollary 16.2.7(b),(c), K has infinitely many quadratic extensions. Therefore, $\mathrm{Gal}(K)$ is not small. □

PROPOSITION 16.11.6: *Let N be a Galois extension of a Hilbertian field K. Suppose $N \neq K_s$. Then $\mathrm{Gal}(N)$ is neither prosolvable ([F.K.Schmidt], [Kuyk2, Thm. 2]) nor it is contained in a closed small subgroup of $\mathrm{Gal}(K)$.*

Proof: By assumption, N has a proper finite separable extension N'. By Theorem 13.9.1(b), N' is Hilbertian. Hence, by Lemma 16.11.5, $\mathrm{Gal}(N')$ is not prosolvable. Consequently, $\mathrm{Gal}(N)$ is not prosolvable.

Now consider an extension M of K in N. Let M' be a finite extension of M with $NM' = N'$. In particular, $M' \not\subseteq N$. Hence, by Theorem 13.9.1(a), M' is Hilbertian. So, by Lemma 16.11.5, $\mathrm{Gal}(M')$ is not small. It follows from Remark 16.10.3(b) that $\mathrm{Gal}(M)$ is not small. □

Example 16.11.7: A Hilbertian field which is not a proper finite extension of any field. For each p, $\mathbb{Q}$ has a Galois extension N with $\text{Gal}(N/\mathbb{Q}) \cong \mathbb{Z}_p$ (Lemma 16.6.4 or Corollary 16.6.7). By Proposition 16.11.1, N is Hilbertian. Since $\mathbb{Z}_p$ has no nontrivial finite subgroups (Lemma 1.4.2(c)), N is not a proper finite extension of any field. □

16.12 Regularity of Finite Groups over Complete Discrete Valued Fields

The results of this section depends on a theorem of Harbater whose proof is unfortunately outside the scope of this book.

PROPOSITION 16.12.1: *Suppose K is a complete field under a discrete valuation and G is a finite group. Then G is regular over K.*

On the proof: Harbater's proof of the proposition uses the language of formal schemes [Harbater1, Thm. 2.3]. Liu [Liu1] and Serre [Serre8, Thm. 8.4.6] translate the proof into the language of rigid analytic geometry. Both approaches rely on general GAGA theorems relating formal (resp. rigid analytic) geometry to algebraic geometry. A short cut proof which is more algebraic than the former ones appears in [Haran-Völklein] and in [Völklein, p. 239]. Each of these proofs actually provides a Galois extension F of $K(t)$ (with t transcendental over K), regular over K, with $\text{Gal}(F/K(t)) \cong G$. □

PROPOSITION 16.12.2: *Let K be a PAC field and G a finite group. Then G is regular over K.*

Proof: The field $\hat{K} = K((z))$ of formal power series in z is complete under a discrete valuation (Example 3.5.1). By Proposition 16.12.1, G is regular over $\hat{K}$. By Proposition 16.2.8, there exists an absolutely irreducible polynomial $f \in \hat{K}[T, X]$, monic and Galois in X, with $\text{Gal}(f(T,X), \hat{K}(T)) \cong G$. Choose $u_1, \ldots, u_n \in \hat{K}$ such that $K[\mathbf{u}]$ contains the coefficients of f and f is Galois over $K(\mathbf{u}, T)$ with $\text{Gal}(f(\mathbf{u}, T, X), K(\mathbf{u})) \cong G$. Write $f(T,X) = g(\mathbf{u}, T, X)$ with g a polynomial with coefficients in K.

Denote the K-variety that $\mathbf{u}$ generates in $\mathbb{A}^n$ by V. Since $\hat{K}/K$ is regular (Example 3.5.1), so is $K(\mathbf{u})/K$ (Corollary 2.6.5(b)). Hence, V is absolutely irreducible (Corollary 10.2.2(a)). Proposition 16.1.4 gives a nonempty Zariski open subset V_0 of V satisfying: For each $\mathbf{a} \in V_0(K)$, the polynomial $g(\mathbf{a}, T, X)$ is absolutely irreducible and Galois over $K(T)$. Moreover,

$$\text{Gal}(g(\mathbf{a}, T, X), K(T)) \cong \text{Gal}(g(\mathbf{u}, T, X), K(\mathbf{u}, T)) \cong G.$$

Since K is PAC, there exists an $\mathbf{a} \in V_0(K)$. By the preceding paragraph, $g(\mathbf{a}, T, X)$ is absolutely irreducible, monic and Galois in X, and

$$\text{Gal}(g(\mathbf{a}, T, X), K(T)) \cong G.$$

Consequently, G is regular over K. □

Remark 16.12.3: Ample fields. The assumption "K is PAC" in the proof of Proposition 16.12.2 is used only to find a K-rational point of V_0. In many instances the condition "K is PAC" can be replace by a weaker one: We say that a field K is **ample** if K is existentially closed in $K((t))$ (cf. Proposition 11.3.5). For example, PAC fields and Henselian fields are ample. For more details see [Haran-Jarden6, §6]. □

PROPOSITION 16.12.4: *Let K be a field and G a finite group.*
(a) *Then G is regular over a finite Galois extension L of K.*
(b) *Moreover, suppose K is Hilbertian. Then L has a finite Galois extension with $\mathrm{Gal}(N/L) \cong G$.*

Proof of (a): K_s is PAC (Section 11.1). By Proposition 16.12.2, G is regular over K_s. Hence, there is an irreducible Galois polynomial $f \in K_s[\mathbf{T}, X]$ in X with $\mathbf{T} = (T_1, \ldots, T_r)$ and $\mathrm{Gal}(f(\mathbf{T}, X), K_s(\mathbf{T})) \cong G$. By Lemma 16.2.4(a),(b), there is a finite Galois extension L of K such that $f(\mathbf{T}, X)$ is Galois over $L(\mathbf{T})$ and $\mathrm{Gal}(f(\mathbf{T}, X), L(\mathbf{T})) \cong G$. Consequently, by Lemma 16.2.4(c), G is regular over L.

Proof of (b): By Corollary 12.2.3, L is Hilbertian. Proposition 16.1.5 gives $\mathbf{a} \in L^r$ with $f(\mathbf{a}, X)$ Galois over L and $\mathrm{Gal}(f(\mathbf{a}, X), L) \cong G$. Let N be the splitting field of $f(\mathbf{a}, X)$ over L. Then $\mathrm{Gal}(N/L) \cong G$. □

Remark 16.12.5: An evaluation of Proposition 16.12.4(b). The proof of Proposition 16.12.4(b) relies on the deep result, Proposition 16.12.1. If we ask L/K only to be finite and separable, the proof becomes quite elementary. To this end embed G in S_n, say, for $n = |G|$. Then use the general polynomial of degree n to find a Galois extension N of K with $\mathrm{Gal}(N/K) \cong S_n$. Denote the fixed field of G in N by L. Then $\mathrm{Gal}(N/L) \cong G$. However, unless G is trivial, $G \cong \mathbb{Z}/2\mathbb{Z}$, or $G = S_n$ itself, L will not be Galois over K. □

Exercises

1. Let K be a $\mathbb{Z}_p$ extension of $\mathbb{Q}$ and $l \neq p$ a prime number. Use Lemma 16.6.9 to prove that l is unramified in K.

2. Prove that $\mathbb{Z}_p \times \mathbb{Z}_p$ does not occur over $\mathbb{Q}$. Hint: Use Exercise 1 and Kronecker-Weber: Every finite Abelian extension of $\mathbb{Q}$ is contained in $\mathbb{Q}(\zeta_n)$ for some positive integer n. Alternatively, use Lemma 16.6.4 to prove that $\mathrm{Gal}(\mathbb{Q}_{\mathrm{ab}}/\mathbb{Q}) \cong G \times \hat{\mathbb{Z}}$ where G is a product of finite groups. Then use Kronecker-Weber.

3. Suppose each of the direct powers G^n of a finite group G can be realized over the field K. Prove there exists a linearly disjoint sequence $L_1, L_2, L_3, \ldots$ of Galois extensions of K with $\mathrm{Gal}(L_i/K) \cong G$, $i = 1, 2, 3, \ldots$.

4. Use Corollary 16.2.7, to prove that if K is a Hilbertian field, then the group $\mathrm{Gal}(K_{\mathrm{ab}}/K)$ is not finitely generated.

5. Strengthen Theorem 16.11.6 to prove that if L is a separable algebraic extension of a Hilbertian field K and $\mathrm{Gal}(L)$ is a prosolvable group, then the Galois closure of L/K is K_s.

6. Let G be a profinite group, N be a closed characteristic subgroup of G, and α an automorphism of G. Suppose the centralizer of N in G is trivial. Prove that the map $\alpha \mapsto \alpha|_N$ is an embedding of $\mathrm{Aut}(G)$ into $\mathrm{Aut}(N)$. Hint: Use the identity $\alpha(n^{-1}gn) = \alpha(n)^{-1}\alpha(g)\alpha(n)$ for $g \in G$ and $n \in N$.

7. Let G be a finitely generated group and H an open subgroup. Prove that H is finitely generated. Hint: Let $x_1, \ldots, x_e$ be generators of G and write $G = \bigcup_{i=1}^n Hg_i$ with $g_1 = 1$. Prove that H is generated by all elements $g_i x_j^{\pm 1} g_k^{-1}$ which belong to H.

Notes

Let $\alpha\colon G \to \mathrm{Gal}(L/K)$ be an embedding problem over a field K. Put $\tilde{G} = \mathrm{Gal}(K)$, $A = \mathrm{Gal}(L/K)$, and $\varphi = \mathrm{res}_{\tilde{K}/L}$. We call a homomorphism $\gamma\colon \tilde{G} \to G$ satisfying $\alpha \circ \gamma = \varphi$ a solution of the embedding problem if γ is surjective, because we are exclusively looking for solutions of embedding problems over fields. However, in the framework of profinite groups, it is of great interest to consider also homomorphisms $\gamma\colon \tilde{G} \to A$ satisfying $\alpha \circ \gamma = \varphi$ which are not necessarily surjective. In Chapter 22 we call such γ's weak solutions of the embedding problem (φ, α) and define $\tilde{G}$ to be projective if every finite embedding problem for $\tilde{G}$ has a weak solution. Other authors (e.g. [Malle-Matzat, p. 296]) prefer the terminology "solution" and "proper solution" rather than "weak solution" and "solution". Likewise, what we call a "regular solution" of the embedding problem $\alpha\colon G \to \mathrm{Gal}(L(\mathbf{t})/K(\mathbf{t}))$ in Definition 16.4.1, is called a "proper parametric solution" in [Malle-Matzat, p. 296].

The use of the group epimorphism $A \mathrm{\,wr\,} G \to G \ltimes A$ (Lemma 16.4.3) replaces Uchida's proof of Proposition 16.4.4 as presented in [Fried-Jarden3, Lemma 24.46]. We have borrowed this approach from [Malle-Matzat, Section IV.2.2]

[Serre8, Thm. 1.2.1] gives a cohomological proof of Proposition 16.5.1.

Our proof of Whaple's theorem (Theorem 16.6.6) is an elaboration of [Kuyk-Lenstra].

The special case of Proposition 16.6.10 when $K = \mathbb{Q}$ appears in [Serre8, p. 36, Exer. 1]. The article [Geyer-Jensen1] studies the number of linearly disjoint $\mathbb{Z}_p$-extensions of a given field K. Theorem 2.7 of that article says this number does not grow in a finitely generated regular extension of a finitely generated extension of $\mathbb{Q}$. In characteristic 0 this is essentially Proposition 16.6.10.

The realization of symmetric and alternating groups over $\mathbb{Q}$ using Hilbert irreducibility theorem goes back to [Hilbert]. Our treatment of this subject in Section 16.7 is a workout of [Serre10, pp. 42-43].

Notes

The concept of GAR-realization and its application to solve embedding problems with kernels having trivial centers go back to [Matzat]. See also [Malle-Matzat, Chap. IV.3].

The concept of small profinite groups appears in [Klingen1].

Theorem 22.9.7 partially generalizes Theorem 16.11.3 to pronilpotent extensions of K.

Geyer (private communication) uses wreath products to prove that if K is a Hilbertian field, then $\mathrm{Gal}(K)$ has no nontrivial finitely generated normal closed subgroups. Proposition 16.11.6 generalizes this result.

Example 16.11.7 settles a question of Sonn.

The proof of Proposition 16.12.2 essentially depends on the following result:

(1) For a field K and a finite group G there is an absolutely irreducible variety V over K satisfying this: If $V(K) \neq \emptyset$, then G is regular over K.

[Fried-Völklein1, Thm. 2] proves (1) in characteristic 0 as an application of the parametrization of Galois extensions of $\mathbb{C}(t)$ by moduli spaces. The proof of Proposition 16.12.2 assumes G is regular over a regular extension of K (e.g. over $K((t))$) and construct V satisfying the conclusion of (1).

Pop observed that the condition "K is existentially closed in $K((t))$" is common to PAC fields and to Henselian fields as well as to other families of fields. He then observed that "G is regular over K" is an elementary property of K and deduced Proposition 16.12.2 for ample fields. This is included in a letter from P. Roquette to W.-D. Geyer in February, 1991. Pop's observation replaces the use of Bertini-Noether in the proof of Proposition 16.12.2.

Pop actually uses the terminology "large fields" instead of "ample fields" [Pop, p. 4]. We prefer the latter terminology because the adjective "large" in the naive sense has been attached to algebraic extensions of Hilbertian fields in several works (e.g. in [Frey-Jarden, Geyer-Jarden1, Jarden2, Jarden4, Jarden7, Jarden12, Jacobson-Jarden, Lubotzky-v.d.Dries]).

Chapter 17. Free Profinite Groups

We continue the discussion on profinite groups of Chapters 1 and 16. Central to this chapter is a discussion on free profinite groups. In particular, we prove that an open subgroup of a free profinite group is free (Proposition 17.6.2).

17.1 The Rank of a Profinite Group

The rank of a finitely generated profinite group is defined to be the minimal number of generators of the sets (Section 16.10). The definition of the rank of arbitrary profinite group G puts a topological condition on the minimal set of generators of G. Once this condition is satisfied, the cardinality m of that set depends only on the group. Indeed, if G is not finitely generated, then m is the cardinality of the set of all open subgroups of G (Proposition 17.1.2) and rank(G) is defined to be m.

The above mentioned topological condition generalizes convergence of a sequence. A subset X of a profinite group G is said to **converge to** 1, if $X \smallsetminus N$ is a finite set for every open normal subgroup N of G.

PROPOSITION 17.1.1 ([Douady]): *Every profinite group G has a system of generators that converges to 1.*

Proof: Without loss assume G is infinite. Well order the set $\mathcal{N}$ of open normal subgroups of G: $\mathcal{N} = \{N_\alpha \mid \alpha < m\}$, where m is an infinite cardinal number. For $\beta \leq m$ let $M_\beta = \bigcap_{\alpha < \beta} N_\alpha$. Choose a set I of cardinality greater than m.

We break the rest of the proof into parts to do a transfinite induction on $\beta \leq m$ and define $X_\beta = \{x_{\beta i} \mid i \in I\} \subseteq G$ with these properties:
(1a) X_β generates G modulo M_β.
(1b) For each open normal subgroup N of G that contains M_β the set $\{i \in I \mid x_{\beta i} \notin N\}$ is finite.
(1c) $x_{\alpha i} \equiv x_{\beta i} \bmod M_\alpha$, for all $\alpha < \beta$ and $i \in I$.
In particular, X_m is a system of generators of G that converges to 1.

PART A: *Induction for a successor ordinal.* Let $\gamma = \beta + 1 < m$ and suppose we have defined X_α for each $\alpha \leq \beta$. The set $I_\beta = \{i \in I \mid x_{\beta i} \notin M_\beta\} = \bigcup_{\alpha < \beta} \{i \in I \mid x_{\beta i} \notin N_\alpha\}$ is a union of no more than $|\beta|$ finite sets. Hence, $|I_\beta| \leq \aleph_0 \cdot |\beta| \leq m < |I|$. Let $y_1, \ldots, y_k$ be generators of M_β modulo M_γ. Choose k distinct elements $i_1, \ldots, i_k$ in $I \smallsetminus I_\beta$. Define $x_{\gamma i}$ as follows: For the finite number of i's with $x_{\beta i} \in G \smallsetminus M_\beta N_\beta$, let $x_{\gamma i} = x_{\beta i}$. For $x_{\beta i} \in M_\beta N_\beta \smallsetminus M_\beta$, choose $x_{\gamma i} \in N_\beta$ such that $x_{\gamma i} \equiv x_{\beta i} \bmod M_\beta$. For $i = i_j$ and $j \in \{1, 2, \ldots, k\}$ choose $x_{\gamma i} = y_j$. Finally, for all other i, define $x_{\gamma i} = 1$. Then X_γ satisfies (1) with γ replacing β. For example, in order to prove (1b), consider an open normal subgroup N of G that contains M_γ. By definition, $M_\beta \cap N_\beta = M_\gamma$. Hence, by Lemma 1.2.5, $M_\beta N_\beta$ has an open

17.1 The Rank of a Profinite Group

normal subgroup U containing M_β such that $U \cap N_\beta \leq N$. For all but finitely many $i \in I$ we have $x_{\beta i} \in U \smallsetminus M_\beta$ or $x_{\gamma i} = 1$. In the former case $x_{\gamma i} \in N_\beta$ and $x_{\gamma i} \equiv x_{\beta i} \bmod M_\beta$, so $x_{\gamma i} \in U \cap N_\beta \leq N$. Thus, all but finitely many $i \in I$ belong to N.

PART B: *Induction for a limit ordinal.* Assume γ is a limit ordinal $\leq m$ and X_α has been defined for each $\alpha < \gamma$. Let $i \in I$. By (1c), the collection $\{x_{\alpha i} M_\alpha \mid \alpha < \gamma\}$ of closed sets satisfies the finite intersection property. Therefore, there is an $x_{\gamma i} \in G$ with $x_{\gamma i} \equiv x_{\alpha i} \bmod M_\alpha$ for each $\alpha < \gamma$.

Suppose N is an open normal subgroup of G that contains M_γ. Then there is an $\alpha < \gamma$ with $M_\alpha \subseteq N$ (Lemma 1.2.2(a)). Hence, $x_{\gamma i} \in N$ for all but finitely many $i \in I$. Finally, for each $\alpha < \gamma$, the set X_γ generates G modulo M_α. Put $H = \langle x_{\gamma i} \mid i \in I \rangle$. By Lemma 1.2.2(b) and by induction, $HM_\gamma = H\bigcap_{\beta<\gamma} M_\beta = \bigcap_{\beta<\gamma} HM_\beta = \bigcap_{\beta<\gamma} G = G$. This concludes the transfinite induction. $\square$

We define the **rank** of a nonfinitely generated profinite group G as the cardinality of a system of generators of G that converges to 1. The following result proves this definition to be independent of the particular system of generators:

PROPOSITION 17.1.2: *Let G be a nonfinitely generated profinite group. Denote the family of all open (resp. open normal) subgroups of G by $\mathcal{M}$ (resp. $\mathcal{N}$). Suppose X is a system of generators of G that converges to 1. Then $|X| = |\mathcal{M}| = |\mathcal{N}|$.*

Proof: First note that G has only finitely many open subgroups containing each $N \in \mathcal{N}$. Hence, $|\mathcal{M}| = |\mathcal{N}|$. It remains to prove that $|X| = |\mathcal{N}|$.

By definition, $X = \bigcup_{N \in \mathcal{N}} (X \smallsetminus N)$ is a union of $|\mathcal{N}|$ finite sets. Hence, $|X| \leq |\mathcal{N}|$. For each finite subset A of X, let $\mathcal{N}(A) = \{N \in \mathcal{N} \mid X \smallsetminus A \subseteq N\}$. If we show that $\mathcal{N}(A)$ is countable, then $\mathcal{N} = \bigcup_A \mathcal{N}(A)$ would be a union of no more than $|X|$ countable sets. Thus, $|\mathcal{N}| \leq |X|$, and we are done.

Let $K(A)$ be the minimal closed normal subgroup of G that contains $X \smallsetminus A$. Then $G/K(A)$ is finitely generated and there exists a bijective correspondence between $\mathcal{N}(A)$ and the family of open normal subgroups of $G/K(A)$. By Lemma 16.10.2, both are therefore countable. $\square$

Remark 17.1.3: On the definition of the rank. If G is an infinite finitely generated profinite group, then G has infinitely many open normal subgroups. Thus, Proposition 17.1.2, does not hold in this case. Moreover, $\{2,3\}$ is a minimal set of generators of $\hat{\mathbb{Z}}$, although $\text{rank}(\hat{\mathbb{Z}}) = 1$. Thus, the definitions we gave to the rank of finitely generated and nonfinitely generated groups differ from each other. A unified definition of $\text{rank}(G)$ in both cases could be taken as the minimal cardinality of a system of generators that converges to 1. $\square$

COROLLARY 17.1.4: *Let $\varphi: G \to A$ be an epimorphism of profinite groups. Then $\text{rank}(A) \leq \text{rank}(G)$.*

Proof: First suppose $\mathrm{rank}(G) < \infty$. Then φ maps each finite set of generators of G onto a set of generators of A. Hence, $\mathrm{rank}(A) \leq \mathrm{rank}(G)$.

Now suppose both ranks are infinite. Then the map $N \mapsto \varphi^{-1}(N)$ maps the set of all open normal subgroups of A injectively into the set of all open normal subgroups of G. Therefore, by Proposition 17.1.2, $\mathrm{rank}(A) \leq \mathrm{rank}(G)$. □

COROLLARY 17.1.5: *Let G be a profinite group of infinite rank and H a closed subgroup. Then $\mathrm{rank}(H) \leq \mathrm{rank}(G)$. If H is open, then $\mathrm{rank}(H) = \mathrm{rank}(G)$.*

Proof: The map $G_0 \mapsto G_0 \cap H$ surjects the set of all open subgroups of G onto the set of all open subgroups of H (Lemma 1.2.5). Hence, by Proposition 17.1.2, $\mathrm{rank}(H) \leq \mathrm{rank}(G)$. If H is open, the fibers of that map are finite. Hence, by Proposition 17.1.2, $\mathrm{rank}(H) \geq \mathrm{rank}(G)$. Consequently, in this case, $\mathrm{rank}(H) = \mathrm{rank}(G)$. □

In Section 17.6 we prove $\mathrm{rank}(H)$ to be finite if $\mathrm{rank}(G)$ is finite. Moreover, we estimate $\mathrm{rank}(H)$ in terms of $\mathrm{rank}(G)$ and $(G : H)$.

COROLLARY 17.1.6: *Let G be a profinite group, m an infinite cardinal number, and I a set of cardinality at most m.*
(a) *For each $i \in I$ let N_i be a closed normal subgroup of G. Suppose $\bigcap_{i \in I} N_i = 1$ and $\mathrm{rank}(G/N_i) \leq m$ for each $i \in I$. Then $\mathrm{rank}(G) \leq m$.*
(b) *Suppose $G = \prod_{i \in I} G_i$ with $\mathrm{rank}(G_i) \leq m$. Then $\mathrm{rank}(G) \leq m$. Moreover, $\mathrm{rank}(G) = m$ if $\mathrm{rank}(G_j) = m$ for at least one $j \in I$ or each G_i is nontrivial and $|I| = m$.*

Proof of (a): Denote the set of all open normal subgroups of G which contains N_i by $\mathcal{N}_i$. Each open normal subgroup of G is in $\mathcal{N}_i$ for some $i \in I$ (Lemma 1.2.2(a)). Hence, by Proposition 17.1.2, $\mathrm{rank}(G) \leq |\bigcup_{i \in I} \mathcal{N}_i| \leq m$.

Proof of (b): For each $i \in I$, let N_i be the kernel of the projection $G \to G_i$. By (a), $\mathrm{rank}(G) \leq m$.

If $\mathrm{rank}(G_j) = m$ for at least one $j \in I$, then $\mathrm{rank}(G) \geq m$ (Corollary 17.1.5). The same conclusion holds if $|I| = m$. Thus, $\mathrm{rank}(G) = m$. □

Example 17.1.7: Countable rank groups.
(a) The rank of a profinite group G is $\leq \aleph_0$ if and only if G has a descending sequence $G = G_0 \geq G_1 \geq G_2 \geq \ldots$ of open normal subgroups with a trivial intersection.

(b) If K is a countable field, then it has at most countably many finite Galois extensions. Hence, $\mathrm{rank}(\mathrm{Gal}(K)) \leq \aleph_0$. □

17.2 Profinite Completions of Groups

We have already seen that profinite groups appear as Galois groups of Galois extensions (Corollary 1.3.4). In this section we study profinite groups that arise as completions of abstract groups.

17.2 Profinite Completions of Groups

Consider a **directed** family $\mathcal{N}$ of normal subgroups of finite index in a group G. Thus,

(1) for all $N_1, \ldots, N_r \in \mathcal{N}$ there is an $N \in \mathcal{N}$ with $N \leq \bigcap_{i=1}^{r} N_i$.

To each pair $N_1 \leq N_2$ of groups in $\mathcal{N}$ associate the quotient map $G/N_1 \to G/N_2$ to obtain a compatible system of maps that defines a profinite group $\hat{G} = \varprojlim G/N$. We call $\hat{G}$ the **profinite completion of G with respect to $\mathcal{N}$**. If $\mathcal{N}$ is the family of all normal subgroups of G of finite index, then $\hat{G}$ is the **profinite completion** of G.

The map $g \mapsto (gN)_{N \in \mathcal{N}}$ is a canonical homomorphism $\theta \colon G \to \hat{G}$ whose kernel N_∞ is the intersection of all $N \in \mathcal{N}$. For each $N \in \mathcal{N}$ denote the projection of $\hat{G}$ onto G/N by π_N. Thus, $\pi_N((x_{N'}N')_{N' \in \mathcal{N}}) = x_N N$ and $\pi_N(\theta(g)) = gN$ for each $g \in G$. Note that π_N is surjective. Finally, let $\mathcal{M}$ be the family of all subgroups M of G which contain some $N \in \mathcal{N}$. For each $M \in \mathcal{M}$ let $\hat{M}$ be the closure of $\theta(M)$ in $\hat{G}$. The ambiguity in the definition of $\hat{G}$ is settled in Part (a4) of the following lemma:

LEMMA 17.2.1:
(a) *The following statements hold for each $N \in \mathcal{N}$:*
   (a1) $\hat{N} = \mathrm{Ker}(\pi_N)$.
   (a2) π_N *induces an isomorphism* $\bar{\pi}_N \colon \hat{G}/\hat{N} \to G/N$.
   (a3) *The set* $\{\hat{N} \mid N \in \mathcal{N}\}$ *is a basis for the open neighborhoods of 1 in $\hat{G}$.*
   (a4) $\theta(G)$ *is dense in* $\hat{G}$.
(b) *The map $M \mapsto \hat{M}$ maps $\mathcal{M}$ bijectively onto the family of open subgroups of $\hat{G}$. For each $M \in \mathcal{M}$ we have* $\theta^{-1}(\hat{M}) = M$.
(c) *The following statements hold for all $M_1, M_2 \in \mathcal{M}$:*
   (c1) $M_1 \leq M_2$ *if and only if* $\hat{M}_1 \leq \hat{M}_2$.
   (c2) *If* $M_1 \leq M_2$, *then* $(M_2 : M_1) = (\hat{M}_2 : \hat{M}_1)$.
   (c3) $M_1 \triangleleft M_2$ *if and only if* $\hat{M}_1 \triangleleft \hat{M}_2$.
   (c4) *If* $M_1 \triangleleft M_2$, *then* $M_2/M_1 \cong \hat{M}_2/\hat{M}_1$.
(d) *Let $M \in \mathcal{M}$. Put $\mathcal{N}_M = \{N \in \mathcal{N} \mid N \leq M\}$. Then $\hat{M} \cong \varprojlim M/N$, where N ranges over $\mathcal{N}_M$.*

Proof: Consider $N \in \mathcal{N}$.

PART A: We prove: $\hat{N} = \mathrm{Ker}(\pi_N)$. Since $\theta(N)$ is contained in $\mathrm{Ker}(\pi_N)$ and $\mathrm{Ker}(\pi_N)$ is closed, $\hat{N} \leq \mathrm{Ker}(\pi_N)$. Conversely, consider $x \in \mathrm{Ker}(\pi_N)$. Every open neighborhood of x contains a coset of the form $x \cdot \mathrm{Ker}(\pi_{N'})$ with $N' \in \mathcal{N}$. Choose $N'' \in \mathcal{N}$ with $N'' \leq N \cap N'$ and let $\pi_{N''}(x) = yN''$ with $y \in G$. Then $yN = \pi_N(x) = N$ and $yN' = \pi_{N'}(x)$. Hence, $\theta(y) \in \theta(N) \cap x \cdot \mathrm{Ker}(\pi_{N'})$. Thus, every open neighborhood of x in $\hat{G}$ meets $\theta(N)$. It follows that x belongs to the closure $\hat{N}$ of $\theta(N)$ in $\hat{G}$.

Applying the arguments of the latter paragraph to G, we conclude that G is dense in $\hat{G}$.

Statement (a2) follows from statement (a1) and (a3) holds because $\{\mathrm{Ker}(\pi_N) \mid N \in \mathcal{N}\}$ is a basis for the open neighborhoods of 1 in $\hat{G}$.

PART B: *The map* $\theta_N\colon G/N \to \hat{G}/\hat{N}$. By definition, $\mathrm{Ker}(\theta) \leq N$. Hence, the map $gN \mapsto \theta(g)\theta(N)$ is an isomorphism of G/N onto $\theta(G)/\theta(N)$.

Let $g \in G$ with $\theta(g) \in \hat{N}$. Then, by Part A, $gN = \pi_N(\theta(g)) = N$. Hence, $g \in N$. Therefore, $\theta(G) \cap \hat{N} = \theta(N)$.

For each $x \in \hat{G}$ there is a $g \in G$ with $\pi_N(x) = gN = \pi_N(\theta(g))$. Hence, $x \in \theta(g)\hat{N}$. Therefore, $\hat{G} = \theta(G)\hat{N}$.

Thus, the map $\theta(g)\theta(N) \mapsto \theta(g)\hat{N}$ is an isomorphism of $\theta(G)/\theta(N)$ onto $\hat{G}/\hat{N}$. It follows that the map $\theta_N\colon G/N \to \hat{G}/\hat{N}$ defined by $\theta_N(gN) = \theta(g)\hat{N}$ is an isomorphism. By definition, θ_N is the inverse of $\bar{\pi}_N$.

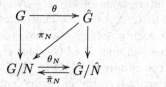

PART C: *A bijection of families of subgroups.* Let M be a subgroup of G containing some $N \in \mathcal{N}$. Then $\theta(M)\hat{N}$ is an open subgroup of $\hat{G}$, hence closed, which contains $\theta(M)$. Hence, $\hat{M} \leq \theta(M)\hat{N}$. On the other hand, from $N \leq M$ follows $\theta(M)\hat{N} \leq \hat{M}$, so $\hat{M} = \theta(M)\hat{N}$.

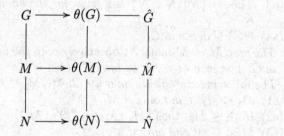

Finally, $\mathrm{Ker}(\theta) \leq N \leq M$. Hence, $\theta^{-1}(\hat{M}) = \theta^{-1}(\theta(M)) = M$. It follows from Part B that the map $M \mapsto \hat{M}$ is a bijection of all subgroups of G which contain N onto all subgroups of $\hat{G}$ which contain $\hat{N}$. In particular, (c1)-(c4) hold when $N \leq M_1, M_2$.

PART D: *Conclusion of the proof.* Consider open subgroups M_1' and M_2' of $\hat{G}$. By (a3), there exist $N \in \mathcal{N}$ with $\hat{N} \leq M_1' \cap M_2'$. By Part C, there are open subgroups M_1, M_2 of G which contain N such that $M_i' = \hat{M}_i$, $i = 1, 2$. This concludes the proof of (b) and (c).

Finally, consider $M \in \mathcal{M}$. By (a3), $\{\hat{N} \mid N \in \mathcal{N}_M\}$ is a directed family of open normal subgroups of $\hat{M}$ and $\bigcap_{N \in \mathcal{N}_M} \hat{N} = 1$. Hence, by Lemma 1.2.4, $\hat{M} \cong \varprojlim \hat{M}/\hat{N}$, where N ranges over $\mathcal{N}_M$. By (c4), $\hat{M} \cong \varprojlim M/N$, where N ranges over $\mathcal{N}_M$. □

The completion of a group is universal in the sense phrased by the following result:

17.2 Profinite Completions of Groups

LEMMA 17.2.2: *Let G be a group, $\mathcal{N}$ a directed family of normal subgroups of G of finite index, H a profinite group, and $\alpha\colon G \to H$ a homomorphism. Suppose $\alpha^{-1}(H_0)$ contains some $N \in \mathcal{N}$ for each open normal subgroup H_0 of H. Let $\hat{G} = \varprojlim G/N$ be the profinite completion of G with respect to $\mathcal{N}$ and $\theta\colon G \to \hat{G}$ the canonical map of G into $\hat{G}$. Then, there is a unique homomorphism $\hat{\alpha}\colon \hat{G} \to H$ satisfying $\hat{\alpha} \circ \theta = \alpha$.*

Proof: The uniqueness of $\hat{\alpha}$ follows from the density of $\theta(G)$ in $\hat{G}$. We prove the existence of $\hat{\alpha}$.

Consider an open normal subgroup H_0 of H. Choose $N \in \mathcal{N}$ with $N \leq \alpha^{-1}(H_0)$. Define a homomorphism $\alpha_{H_0}\colon \hat{G} \to H/H_0$ by $\alpha_{H_0}(\hat{g}) = \alpha(\pi_N(\hat{g}))H_0$, $\hat{g} \in \hat{G}$. In particular, $\alpha_{H_0}(\theta(g)) = \alpha(g)H_0$ for each $g \in G$.

The definition of α_{H_0} is independent of N. Indeed, if $N' \in \mathcal{N}$ and $N' \leq N$, then $\pi_{N'}(\hat{g})N = \pi_N(\hat{g})N$, hence $\pi_{N'}(\hat{g})\alpha^{-1}(H_0) = \pi_N(\hat{g})\alpha^{-1}(H_0)$, so $\alpha(\pi_{N'}(\hat{g}))H_0 = \alpha(\pi_N(\hat{g}))H_0$.

Consider now an additional open normal subgroup H_1 of H which is contained in H_0. Let $\pi_{H_1,H_0}\colon H/H_1 \to H/H_0$ be the quotient map. Choose $N \in \mathcal{N}$ with $N \leq \alpha^{-1}(H_1)$ and use N to define α_{H_1}. Then $\pi_{H_1,H_0} \circ \alpha_{H_1} = \alpha_{H_0}$.

The latter conclusion yields a homomorphism $\hat{\alpha}\colon \hat{G} \to H$ with $\alpha_{H_0}(\hat{g}) = \hat{\alpha}(\hat{g})H_0$ for each open normal subgroup H_0 of H and every $\hat{g} \in \hat{G}$. Consequently, $\hat{\alpha}(\theta(g)) = \alpha(g)$ for each $g \in G$. $\square$

As a corollary of Lemma 17.2.2 we prove that the completion of a group is unique:

LEMMA 17.2.3: *Let $\mathcal{N}$ and $\mathcal{N}'$ be directed families of normal subgroups of finite index of a group G. Suppose $\mathcal{N}$ and $\mathcal{N}'$ are **cofinite in each other**; that is, each $N \in \mathcal{N}$ contains some $N' \in \mathcal{N}'$ and each $N' \in \mathcal{N}'$ contains some $N \in \mathcal{N}$. Let $\hat{G} = \varprojlim G/N$ (resp. $\hat{G}' = \varprojlim G/N'$) be the completion of G with respect to $\mathcal{N}$ (resp. $\mathcal{N}'$) and let $\theta\colon G \to \hat{G}$ (resp. $\theta'\colon G \to \hat{G}'$) be the corresponding canonical homomorphism. Then there exists a unique isomorphism $\alpha\colon \hat{G} \to \hat{G}'$ such that $\alpha \circ \theta = \theta'$.*

Proof: By Lemma 17.2.1, each open normal subgroup of $\hat{G}'$ is the closure $\hat{N}'$ in $\hat{G}'$ of $\theta'(N')$ for some $N' \in \mathcal{N}'$. Moreover, $N' = (\theta')^{-1}(\hat{N}')$. Lemma 17.2.2 gives a unique homomorphism $\alpha\colon \hat{G} \to \hat{G}'$ with $\alpha \circ \theta = \theta'$. By symmetry, there exists a unique homomorphism $\alpha'\colon \hat{G}' \to \hat{G}$ such that $\alpha' \circ \theta' = \theta$. Thus, $\alpha' \circ \alpha \circ \theta = \alpha' \circ \theta' = \theta = \mathrm{id}_{\hat{G}} \circ \theta$. The uniqueness part of Lemma 17.2.2 applied to the map $\theta\colon G \to \hat{G}$ implies that $\alpha' \circ \alpha = \mathrm{id}_{\hat{G}}$. By symmetry, $\alpha \circ \alpha' = \mathrm{id}_{\hat{G}'}$. Consequently, α is an isomorphism. $\square$

Remark 17.2.4: The group $\mathbb{Z}_p$ is the completion of $\mathbb{Z}$ with respect to the family $\{p^i\mathbb{Z} \mid i \in \mathbb{N}\}$, while the group $\hat{\mathbb{Z}}$ is the completion of $\mathbb{Z}$ with respect to the family $\{n\mathbb{Z} \mid n \in \mathbb{N}\}$. $\square$

17.3 Formations of Finite Groups

We classify profinite groups according to their ranks and their finite quotients. To this end consider a family $\mathcal{C}$ of finite groups containing the trivial group. Each group in $\mathcal{C}$ is called a $\mathcal{C}$-group. We call $\mathcal{C}$ a **formation** if it is closed under taking quotients and fiber products (Section 22.2). In other words, $\mathcal{C}$ satisfies the following conditions:

(1a) If $G \in \mathcal{C}$ and $\bar{G}$ is a homomorphic image of G, then $\bar{G} \in \mathcal{C}$.

(1b) Let G be a finite group and $N_1, N_2 \triangleleft G$. Suppose $G/N_1, G/N_2 \in \mathcal{C}$ and $N_1 \cap N_2 = 1$. Then $G \in \mathcal{C}$.

We call $\mathcal{C}$ a **Melnikov formation** if it is closed under taking quotients, normal subgroups and extensions. This means that $\mathcal{C}$ satisfies this condition:

(1c) Let $1 \to N \to G \to \bar{G} \to 1$ be a short exact sequence of finite groups. Then $G \in \mathcal{C}$ if and only if $N, \bar{G} \in \mathcal{C}$.

In particular, $\mathcal{C}$ satisfies (1a) and (1b). Indeed, under the assumptions of (1b), N_2 is isomorphic to $N_1 N_2 / N_1$ and $N_1 N_2 / N_1$ is a normal subgroup of G/N_1, so $N_2 \in \mathcal{C}$. Since also $G/N_2 \in \mathcal{C}$, we have $G \in \mathcal{C}$. It follows that every Melnikov formation is a formation.

We say $\mathcal{C}$ is a **full formation** if it is closed under taking quotients, subgroups, and extensions. In other words, $\mathcal{C}$ is a Melnikov formation which satisfies the following condition:

(1d) If $G \in \mathcal{C}$ and $H \leq G$, then $H \in \mathcal{C}$.

A **pro-$\mathcal{C}$ group** is an inverse limit $G = \varprojlim G_i$ of $\mathcal{C}$-groups for which the connecting homomorphisms $G_j \to G_i$ are epimorphisms. Thus, G has a collection $\mathcal{N}$ of open normal subgroups with $G/N \in \mathcal{C}$ for each $N \in \mathcal{N}$ and $\bigcap_{N \in \mathcal{N}} N = 1$.

LEMMA 17.3.1: *Let $\mathcal{C}$ be a family of finite groups satisfying (1a). Consider a profinite group G. Then this holds:*

(a) *If G is pro-$\mathcal{C}$, then each homomorphic image of G is pro-$\mathcal{C}$.*

(b) *Suppose $\mathcal{C}$ is closed under taking normal subgroups and G is pro-$\mathcal{C}$. Then every closed normal subgroup N of G is pro-$\mathcal{C}$.*

(c) *Suppose $\mathcal{C}$ is closed under taking subgroups and G is pro-$\mathcal{C}$. Then every closed subgroup H of G is pro-$\mathcal{C}$.*

(d) *Suppose (1b) holds. Let N_1, N_2 be closed normal subgroups of G. Suppose $G/N_1, G/N_2$ are pro-$\mathcal{C}$. Then $G/N_1 \cap N_2$ is pro-$\mathcal{C}$.*

(e) *Suppose (1c) holds. Let N be a closed normal subgroup of G. Suppose both N and G/N are pro-$\mathcal{C}$. Then G is pro-$\mathcal{C}$.*

(f) *Suppose $\mathcal{C}$ is closed under taking normal subgroups. Then inverse limits of pro-$\mathcal{C}$ groups are pro-$\mathcal{C}$ groups.*

Proof of (a): Let N be a closed normal subgroup of G. First suppose N is open. Then G has an open normal subgroup M in N with $G/M \in \mathcal{C}$. Since G/N is a homomorphic image of G/M, it belongs to $\mathcal{C}$.

17.3 Formations of Finite Groups

In the general case $N = \bigcap_{j \in J} N_j$ with N_j normal and open in G. By the preceding paragraph, $G/N_j \in \mathcal{C}$ for each $j \in J$. Thus, $G/N = \varprojlim G/N_j$ is pro-$\mathcal{C}$.

Proof of (b): Let N be a closed normal subgroup of G. Then $N \cong \varprojlim N/H \cap N$ with H ranging over all open normal subgroups H of G. Consider an open normal subgroup H of G. By (a), $G/H \in \mathcal{C}$. Also, $HN/H \triangleleft G/H$. Hence, $N/H \cap N \cong HN/H \in \mathcal{C}$. Consequently, N is pro-$\mathcal{C}$.

Proof of (c): As the proof of (b).

Proof of (d): Let H be an open normal subgroup of G. By (a), $G/N_i H \in \mathcal{C}$, $i = 1, 2$. Hence, by (1b), $G/N_1 H \cap N_2 H \in \mathcal{C}$. By Lemma 1.2.2(b), $\bigcap_H N_1 H \cap N_2 H = N_1 \cap N_2$. Hence, $G/N_1 \cap N_2$ is a pro-$\mathcal{C}$ group.

Proof of (e): Let H be an open normal subgroup of G. Then G/HN is a quotient of G/N. Hence, $G/HN \in \mathcal{C}$. Also, $HN/H \cong N/H \cap N \in \mathcal{C}$. Hence, by (1c), $G/H \in \mathcal{C}$. Consequently, G is a pro-$\mathcal{C}$ group.

Proof of (f): It suffices to prove that if G is a profinite group and $\mathcal{N}$ is a directed family of closed normal subgroups such that G/N is pro-$\mathcal{C}$ for each $N \in \mathcal{N}$ and the intersection of all $N \in \mathcal{N}$ is trivial, then G is pro-$\mathcal{C}$. Indeed, $G \cong \varprojlim G/H$, where H ranges over all open normal subgroups of G which contain N for some $N \in \mathcal{N}$. By (a), $G/H \in \mathcal{C}$ for each such H. Hence, G is pro-$\mathcal{C}$. □

In particular, the consequences of Lemma 17.3.1 except (c) hold if $\mathcal{C}$ is a formation and all consequences of Lemma 17.3.1 hold if $\mathcal{C}$ is full.

Definition 17.3.2: *Maximal pro-$\mathcal{C}$ quotients.* Let $\mathcal{C}$ be a formation of finite groups and G a profinite group. Denote the family of all closed normal subgroups N of G with $G/N \in \mathcal{C}$ by $\mathcal{N}$. By Lemma 17.3.1(d), $\mathcal{N}$ is closed under intersections. Put $N_\mathcal{C} = \bigcap_{N \in \mathcal{N}} N$. Then $G/N_\mathcal{C}$ is pro-$\mathcal{C}$. Moreover, suppose $\bar{G}$ is a quotient of G and $\bar{G} \in \mathcal{C}$. Then $\bar{G}$ is a quotient of $G/N_\mathcal{C}$. We call $G/N_\mathcal{C}$ the **maximal pro-$\mathcal{C}$ quotient** of G. □

Example 17.3.3: Standard families.

(a) $\mathcal{C}$ contains all finite groups: pro-$\mathcal{C}$ groups are just profinite groups.

(b) $\mathcal{C}$ consists of all finite p-groups, p a prime: pro-$\mathcal{C}$ groups are then called **pro-p groups**.

(c) $\mathcal{C}$ consist of all solvable groups: pro-$\mathcal{C}$ groups are **prosolvable groups**.

Each of the families (a), (b), and (c) is full.

(d) $\mathcal{C}$ consists of all finite Abelian (resp. nilpotent) groups: pro-$\mathcal{C}$ groups are **Abelian profinite** (resp. pronilpotent). In this case (1a) and (1b) hold. Thus, $\mathcal{C}$ is a formation. But (1c) is false. For example, in the short exact sequence $1 \to A_3 \to S_3 \to C_2 \to 1$ both A_3 and C_2 are Abelian but S_3 is not nilpotent. Therefore, $\mathcal{C}$ is not Melnikov. □

Remark 17.3.4: Melnikov formations. Let $\mathcal{D}$ be a nonempty family of finite simple groups. Denote the family of all finite groups whose composition factors belong to $\mathcal{D}$ by $\mathcal{C}$. If $1 \to N \to G \to \bar{G} \to 1$ is a short exact sequence of finite groups and the composition factors of both N and $\bar{G}$ belong to $\mathcal{D}$, then each composition factor of G belongs to $\mathcal{D}$. Thus, $\mathcal{C}$ is a Melnikov formation. We call $\mathcal{C}$ the **formation generated** by $\mathcal{D}$.

Conversely, let $\mathcal{C}$ be a Melnikov formation. Denote the family of all composition factors of groups belonging to $\mathcal{C}$ by $\mathcal{D}$. Then $\mathcal{D}$ generates $\mathcal{C}$. We call $\mathcal{D}$ the **underlying set of simple groups** of $\mathcal{C}$. By the preceding paragraph, the consequences of Lemma 17.3.1(a),(b),(d),(e),(f) hold for pro-$\mathcal{C}$ groups.

If $\mathcal{D}$ consists of a single simple group S, we call a pro-$\mathcal{C}$ group also a **pro-S group**. In particular, if $S = \mathbb{Z}/p\mathbb{Z}$, pro-$\mathcal{C}$ groups are just pro-p groups. If S is a simple non-Abelian group, no nontrivial subgroup is an S-group. Hence, $\mathcal{C}$ is not full. When $\mathcal{D}$ is the set of all finite groups, pro-$\mathcal{C}$ groups are just profinite groups. □

17.4 Free pro-$\mathcal{C}$ Groups

Free pro-$\mathcal{C}$ groups arise as completions of free abstract groups. Like the latter, they are defined by a universal property in the category of pro-$\mathcal{C}$ groups:

Definition 17.4.1: Let X be a set and G a profinite group. A map $\varphi \colon X \to G$ is **convergent to** 1 if $X \smallsetminus \varphi^{-1}(H)$ is a finite set for each open normal subgroup H of G.

Let $\mathcal{C}$ be a formation of finite groups and X a set. A **free pro-$\mathcal{C}$ group on X** is a pro-$\mathcal{C}$ group $\hat{F}$ with a map $\iota \colon X \to \hat{F}$ satisfying:
(1a) ι converges to 1 and $\iota(X)$ generates $\hat{F}$.
(1b) For each map φ of X into a pro-$\mathcal{C}$ group G which converges to 1 and satisfies $G = \langle \varphi(X) \rangle$ there exists a unique epimorphism $\hat{\varphi} \colon \hat{F} \to G$ with $\hat{\varphi} \circ \iota = \varphi$. □

When $\mathcal{C}$ is full, each closed subgroup of G is pro-$\mathcal{C}$. Hence, the condition on $\varphi(X)$ in (1b) to generate G is redundant.

We construct free pro-$\mathcal{C}$ groups from free abstract groups. Recall that an abstract group F is **free** on a subset X if X generates F and every map φ_0 of X into an abstract group G uniquely extends to a homomorphism $\varphi \colon F \to G$. We call X a **basis** of F or also a **free set of generators** of F.

If X is infinite, then $|X| = |F|$. If X is finite, say of cardinality n, then there is an epimorphism of F onto $(\mathbb{Z}/2\mathbb{Z})^n$. Since the latter group is not generated by less than n elements, n is the minimal number of generators of F. We call $|X|$ the **rank** of F.

PROPOSITION 17.4.2: *Let $\mathcal{C}$ be a formation of finite groups and X a set. Then there exists a free pro-$\mathcal{C}$ group $\hat{F}$ on X.*

More precisely: Let F be the free abstract group on X. Denote the family of normal subgroups N of F such that $F/N \in \mathcal{C}$ and $X \smallsetminus N$ is finite

17.4 Free pro-$\mathcal{C}$ Groups

by $\mathcal{N}$. Let $\hat{F} = \varprojlim F/N$ be the profinite completion of F with respect to $\mathcal{N}$, $\theta \colon F \to \hat{F}$ the canonical homomorphism, and $\iota \colon X \to \hat{F}$ the restriction of θ to X. Then $\hat{F}$ is a free pro-$\mathcal{C}$ group on X with ι the associated map.

Proof: By (1) of 17.3, $\mathcal{N}$ is a directed family (Section 17.2). Hence, $\hat{F}$ is a pro-$\mathcal{C}$ group.

By Lemma 17.2.1, $\hat{F} = \langle \iota(X) \rangle$. If $\hat{H}$ is an open normal subgroup of $\hat{F}$, then $\theta^{-1}(\hat{H}) \in \mathcal{N}$, so $X \smallsetminus \iota^{-1}(\hat{H})$ is finite. Therefore, ι converges to 1.

Next consider a map φ of X into a pro-$\mathcal{C}$ group G which converges to 1 such that $\varphi(X)$ generates G. Extend φ in the unique possible way to a homomorphism $\varphi \colon F \to G$ of abstract groups. Let H be an open normal subgroup of G. Then $\varphi^{-1}(H)$ is a normal subgroup of F and $\varphi(F)H = G$, because G is the closure of $\varphi(F)$. Hence, $F/\varphi^{-1}(H) \cong G/H \in \mathcal{C}$. Also, $X \smallsetminus \varphi^{-1}(H)$ is finite. Therefore, $\varphi^{-1}(H) \in \mathcal{N}$. Lemma 17.2.2 gives a homomorphism $\hat{\varphi} \colon \hat{F} \to G$ satisfying $\hat{\varphi} \circ \theta = \varphi$, so $\hat{\varphi} \circ \iota = \varphi$.

Consequently, $\hat{F}$ is a free pro-$\mathcal{C}$ group on X. □

The universal property of a free pro-$\mathcal{C}$ group on a set X implies its uniqueness:

LEMMA 17.4.3: *Let $\mathcal{C}$ be a formation of finite groups. Consider a free pro-$\mathcal{C}$ group $\hat{F}$ on a set X and let $\iota \colon X \to \hat{F}$ be the associated map. Let $\kappa \colon X \to X'$ be a bijection of sets.*
(a) *Then $\hat{F}$ is a free pro-$\mathcal{C}$ group on X' with the associated map $\iota \circ \kappa^{-1} \colon X' \to \hat{F}$.*
(b) *Let F' be a free pro-$\mathcal{C}$ group on X' with an associated map $\iota' \colon X' \to F'$. Then there exists a unique isomorphism $\hat{\kappa} \colon \hat{F} \to F'$ with $\iota' \circ \kappa = \hat{\kappa} \circ \iota$.*

Proof: We prove only (b), leaving (a) to the reader. By definition, $\iota'(\kappa(X)) = \iota'(X')$ generates F'. Also, the map $\iota' \circ \kappa \colon X \to F'$ converges to 1. Indeed, if H' is an open subgroup of F', then $X' \smallsetminus (\iota')^{-1}(H')$ is finite. Since κ is bijective, $X \smallsetminus (\iota' \circ \kappa)^{-1}(H') = \kappa^{-1}(X' \smallsetminus (\iota')^{-1}(H'))$ is finite. It follows that there exists a unique epimorphism $\hat{\kappa} \colon \hat{F} \to F'$ with $\hat{\kappa} \circ \iota = \iota' \circ \kappa$. Similarly, there exists an epimorphism $\kappa' \colon F' \to \hat{F}$ with $\kappa' \circ \iota' = \iota \circ \kappa^{-1}$. The uniqueness part of (1b) implies that both $\kappa' \circ \hat{\kappa}$ and $\hat{\kappa} \circ \kappa'$ are the identity maps. Consequently, $\hat{\kappa}$ is an isomorphism. □

Following Proposition 17.4.2 and Lemma 17.4.3, we denote the unique (up to isomorphism) free pro-$\mathcal{C}$ group on a set X by $\hat{F}_X(\mathcal{C})$. If X is empty, then $\hat{F}_X(\mathcal{C}) = 1$.

The next result shows that in general the map ι of (1) is injective:

LEMMA 17.4.4: *Let $\mathcal{C}$ be a formation of finite groups, X a set, and $\iota \colon X \to \hat{F}_X(\mathcal{C})$ be the associated map into the free pro-$\mathcal{C}$ group on X. Suppose there is a nontrivial group $C \in \mathcal{C}$ with $\mathrm{rank}(C) \le |X|$. Then ι is injective. Moreover, $\iota(x) \ne 1$ for each $x \in X$.*

Proof: Consider distinct elements x and y of X. Put $r = \mathrm{rank}(C)$. Let $c_1, \ldots, c_r$ be generators of C. Then $c_1, \ldots, c_r$ are distinct and $c_i \ne 1$ for each i. When $r = 1$, define $\varphi(x) = c_1$ and $\varphi(z) = 1$ for all $z \in X \smallsetminus \{x\}$. When $r \geq 2$, choose $x_3, \ldots, x_r \in X \smallsetminus \{x, y\}$. Then let $\varphi(x) = c_1$, $\varphi(y) = c_2$, $\varphi(x_i) = c_i$ for $i = 3, \ldots, r$, and $\varphi(z) = 1$ for $z \in X \smallsetminus \{x, y, x_3, \ldots, x_r\}$. In each case $\langle \varphi(X) \rangle = C$ and $\varphi \colon X \to C$ converges to 1. Hence, there is an epimorphism $\hat{\varphi} \colon \hat{F} \to C$ with $\hat{\varphi} \circ \iota = \varphi$. By definition, $\varphi(x) \neq \varphi(y)$. Hence, $\iota(x) \neq \iota(y)$. Moreover, $\varphi(x) \neq 1$, hence $\iota(x) \neq 1$. $\square$

Assuming $\mathcal{C}$, X, ι, and C are as in Lemma 17.4.4, we may identify X with a subset of $\hat{F}_X(\mathcal{C})$ which does not contain 1 and $\iota \colon X \to \hat{F}_X(\mathcal{C})$ with the inclusion map. Then $\hat{F}_X(\mathcal{C})$ becomes a free pro-$\mathcal{C}$ group with basis X in the following sense:

Definition 17.4.5: Let $\mathcal{C}$ be a formation of finite groups, $\hat{F}$ a pro-$\mathcal{C}$ group, and X a subset of $\hat{F}$ which does not contain 1. We say that $\hat{F}$ is a **free pro-$\mathcal{C}$ group with basis X** if the following hold:
(2a) X generates $\hat{F}$ and converges to 1.
(2b) Each map of X into a pro-$\mathcal{C}$ group G which converges to 1 and satisfies $G = \langle \varphi(X) \rangle$ extends uniquely to an epimorphism $\hat{\varphi} \colon \hat{F} \to G$. $\square$

LEMMA 17.4.6: *Let $\mathcal{C}$ be a formation of finite groups and $\hat{F}$ a free pro-$\mathcal{C}$ group with a basis X. Suppose $\mathcal{C}$ contains a nontrivial group of rank at most $|X|$. Then:*
(a) $\mathrm{rank}(\hat{F}) = |X|$.
(b) *Suppose $e = |X| < \infty$. Then every set of generators of $\hat{F}$ of e elements is a basis of $\hat{F}$.*
(c) *Let F be the free abstract group on X and $\mathcal{N}(X)$ the set of all normal subgroups N of F with $F/N \in \mathcal{C}$ and $X \smallsetminus N$ finite. Then $\hat{F}$ is the profinite completion of F with respect to $\mathcal{N}(X)$ and the canonical map $\theta \colon F \to \hat{F}$ maps each $x \in X$ onto itself.*

Proof of (a) when X is finite: Suppose $X = \{x_1, \ldots, x_e\}$ with $x_1, \ldots, x_e$ distinct. Let $\{s_1, \ldots, s_d\}$ be a system of generators of $\hat{F}$, with $d \leq e$. Put $s_{d+1} = \cdots = s_e = 1$. Then the map $x_i \mapsto s_i$, $i = 1, \ldots, e$ extends to an epimorphism φ of $\hat{F}$ onto itself. By Proposition 16.10.6, φ is an automorphism. Since $x_i \neq 1$ for all i, we have $d = e$, so $\{s_1, \ldots, s_e\}$ is a basis of $\hat{F}$. This proves (a) for finite X and proves (b).

Proof of (a) when X is infinite: Suppose X is infinite. By definition, X converges to 1. Hence, $|X| = \mathrm{rank}(\hat{F})$ (Section 17.1). This concludes the proof of (a).

Proof of (c): By Propositions 17.4.2 and 17.4.4, the free pro-$\mathcal{C}$ group F' with basis X is the completion of F with respect to $\mathcal{N}(X)$. Hence, by Lemma 17.4.3, the identity map of X extends to an isomorphism $F' \to \hat{F}$. This proves the claim. $\square$

Remark 17.4.7: Let $\mathcal{C}$ be a formation of finite groups and $m \geq 1$ a cardinal number. Suppose $\mathcal{C}$ contains a group of rank at most m (e.g. when $\mathcal{C}$ is full). Then, Propositions 17.4.2 and 17.4.4 give a free pro-$\mathcal{C}$ group $\hat{F}_X(\mathcal{C})$ with a

17.4 Free pro-$\mathcal{C}$ Groups

basis X of cardinality m. By Lemma 17.4.3, $\hat{F}_X(\mathcal{C})$ is uniquely determined up to an isomorphism by m and $\mathcal{C}$. We denote $\hat{F}_X(\mathcal{C})$ by $\hat{F}_m(\mathcal{C})$ and call it the **free pro-$\mathcal{C}$ group of rank** m. In particular, we write $\hat{F}_e(\mathcal{C})$ when m is a positive integer e and $\hat{F}_\omega(\mathcal{C})$ for the free pro-$\mathcal{C}$ group of rank $\aleph_0$. For $\mathcal{C}$ the family of all finite groups, we simplify $\hat{F}_m(\mathcal{C})$ to $\hat{F}_m$ (e.g. $\hat{F}_1 = \hat{\mathbb{Z}}$). When $\mathcal{C}$ is the family of all p-groups (resp. solvable groups), we use $\hat{F}_m(p)$ (resp. $\hat{F}_m(\text{solv})$) for $\hat{F}_m(\mathcal{C})$ (e.g. $\hat{F}_1(p) = \mathbb{Z}_p$).

Now assume $\text{rank}(G) > m$ for all nontrivial $G \in \mathcal{C}$ (so m is finite). Let F be a free abstract group on a set X of m elements. Then F has no proper normal subgroup N with $F/N \in \mathcal{C}$. Thus, $\hat{F}_X(\mathcal{C})$ is trivial and $\hat{F}_m(\mathcal{C})$ is undefined. To avoid this situation, whenever we say "let $\hat{F}_m(\mathcal{C})$ be a free pro-$\mathcal{C}$ group of rank m", we tacitly assume $\mathcal{C}$ has nontrivial groups of rank at most m.

For example, consider the case where $\mathcal{C}$ is a Melnikov formation which contains only non-Abelian groups. In this case, $\hat{F}_1(\mathcal{C})$ is undefined. By the classification of finite simple groups, each finite simple group is generated by two elements [Aschbacher-Guralnick, Thm. B]. Hence, for each cardinal number $m \geq 2$ there is a free pro-$\mathcal{C}$ group on a set of cardinality m. Nevertheless, we will not use this application of the Classification in this book. □

PROPOSITION 17.4.8: *Let $\mathcal{C}$ be a formation of finite groups. Then every pro-$\mathcal{C}$ group G is a quotient of a free pro-$\mathcal{C}$ group; if $\text{rank}(G) = m$, then G is a quotient of $\hat{F}_m(\mathcal{C})$.*

Proof: Use Douady (Proposition 17.1.1). □

Next we show that every free pro-$\mathcal{C}$ group of rank $\leq m$ appears both as a closed subgroup and as a quotient of $\hat{F}_m(\mathcal{C})$:

LEMMA 17.4.9: *Let $\mathcal{C}$ be a formation of finite groups, F a free pro-$\mathcal{C}$ group with a basis X, C a nontrivial group in $\mathcal{C}$.*

(a) *Let A be a subset of X of cardinality at least $\text{rank}(C)$. Then $\langle A \rangle$ is a free pro-$\mathcal{C}$ group with basis A. Moreover, denote the minimal closed normal subgroup of F containing $X \smallsetminus A$ by N. Then $F = \langle A \rangle \ltimes N$. Thus, F/N is isomorphic to the free pro-$\mathcal{C}$ group on A.*

(b) *F is the inverse limit of free pro-$\mathcal{C}$ groups on A where A ranges over all finite subsets of X of cardinality at least $\text{rank}(C)$.*

Proof of (a): Since $X \smallsetminus A \subseteq N$, $N \cdot \langle A \rangle = F$. Define a map $\varphi_0 \colon X \to \hat{F}_A(\mathcal{C})$ by $\varphi_0(x) = x$ if $x \in A$ and $\varphi_0(x) = 1$ if $x \in X \smallsetminus A$. By (1), φ_0 extends to an epimorphism $\varphi_A \colon F \to \hat{F}_A(\mathcal{C})$ which is the identity on $\langle A \rangle$. Since $X \smallsetminus A \leq \text{Ker}(\varphi_A)$, $N \leq \text{Ker}(\varphi_A)$. Hence, $N \cap \langle A \rangle = 1$.

Conversely, let $k \in \text{Ker}(\varphi_A)$. Then, $k = zn$, with $n \in N$ and $z \in \langle A \rangle$. Apply φ_A to obtain $1 = z$. Thus, $k \in N$. It follows that $N = \text{Ker}(\varphi_A)$. Consequently, $F = \langle A \rangle \ltimes N$ and $\langle A \rangle \cong F/N \cong \hat{F}_A(\mathcal{C})$.

Proof of (b): Suppose $A \subseteq B$ are finite subsets of X with $\text{rank}(C) \leq |A|$. By (a), there is an epimorphism $\varphi_{B,A} \colon \hat{F}_B(\mathcal{C}) \to \hat{F}_A(\mathcal{C})$ with $\varphi_{B,A}(x) = x$ for $x \in A$ and $\varphi_{B,A}(x) = 1$ for $x \in B \smallsetminus A$. Hence, $\varphi_{B,A} \circ \varphi_B = \varphi_A$.

Put $F' = \varprojlim \hat{F}_A(\mathcal{C})$ and let $\pi_A\colon F' \to \hat{F}_A(\mathcal{C})$ be the canonical projection. The maps $\varphi_A\colon F \to \hat{F}_A(\mathcal{C})$ of (a) define an epimorphism $\varphi\colon F \to F'$. For each A we have $\varphi_A = \pi_A \circ \varphi$. Hence, φ is injective on $\langle A \rangle$. But the groups $\langle A \rangle$ generate F. Consequently, φ is an isomorphism. $\square$

LEMMA 17.4.10: *Let $\mathcal{B} \subseteq \mathcal{C}$ be formations of finite groups and m a cardinal number. Suppose there is a $B \in \mathcal{B}$ with $\mathrm{rank}(B) \le m$. Then $\hat{F}_m(\mathcal{B})$ is the maximal pro-$\mathcal{B}$ quotient of $\hat{F}_m(\mathcal{C})$.*

In other words, let N be the intersection of all open normal subgroups H of $\hat{F}_m(\mathcal{C})$ with $\hat{F}_m(\mathcal{C})/H \in \mathcal{B}$. Then $\hat{F}_m(\mathcal{C})/N \cong \hat{F}_m(\mathcal{B})$. If X is a basis of $\hat{F}_m(\mathcal{C})$, then X/N is a basis of $\hat{F}_m(\mathcal{B})$.

Suppose in addition, $\mathcal{B}$ is Melnikov. Then N has no proper open normal subgroup M with $N/M \in \mathcal{B}$.

Proof: Choose bases X and Y of cardinality m for $\hat{F}_m(\mathcal{B})$ and $\hat{F}_m(\mathcal{C})$, respectively. Let $\varphi_0\colon Y \to X$ be a bijective map. Since each pro-$\mathcal{B}$ group is a pro-$\mathcal{C}$ group, φ_0 extends to an epimorphism $\varphi\colon \hat{F}_m(\mathcal{C}) \to \hat{F}_m(\mathcal{B})$.

Conversely, let G be a pro-$\mathcal{B}$ group and $\gamma\colon \hat{F}_m(\mathcal{C}) \to G$ an epimorphism. Then $\gamma \circ \varphi_0^{-1}(X)$ converges to 1 and generates G. Hence, $\gamma \circ \varphi_0^{-1}$ extends to an epimorphism $\psi\colon \hat{F}_m(\mathcal{B}) \to G$ such that $\psi \circ \varphi = \gamma$. According to Definition 17.3.2, $\hat{F}_m(\mathcal{B})$ is the maximal pro-$\mathcal{B}$ quotient of $\hat{F}_m(\mathcal{C})$. Hence, $N = \mathrm{Ker}(\varphi)$.

Finally, suppose $\mathcal{B}$ is Melnikov and M is a proper open normal subgroup of N with $N/M \in \mathcal{B}$. Then $M = H \cap N$ for some open subgroup H of $\hat{F}_m(\mathcal{C})$ (Lemma 1.2.5) and $M \triangleleft H$. Write $\hat{F}_m(\mathcal{C}) = \bigcup_{i=1}^n Hz_i$. Then $M_0 = \bigcap_{i=1}^n M^{z_i}$ is an open subgroup of N which is normal in $\hat{F}_m(\mathcal{C})$. Moreover, $N/M_0 \in \mathcal{B}$. Hence, by Lemma 17.3.1(e), $\hat{F}_m(\mathcal{C})/M_0$ is a pro-$\mathcal{B}$ group. This contradicts the minimality of N. $\square$

LEMMA 17.4.11: *Let $\mathcal{C}$ be a formation of finite groups, e a positive integer, and G a pro-$\mathcal{C}$ group of rank at most e. Then each epimorphism $\beta\colon G \to \hat{F}_e(\mathcal{C})$ is an isomorphism.*

Proof: By assumption, there is an epimorphism $\alpha\colon \hat{F}_e(\mathcal{C}) \to G$. By Proposition 16.10.6(b), β is an isomorphism. $\square$

17.5 Subgroups of Free Discrete Groups

Free profinite groups are introduced in Proposition 17.4.2 as inverse limits of finite quotients of free abstract groups. It is therefore reassuring that the study of closed subgroups of free profinite groups depends on the classical result that subgroups of free abstract groups are free.

Consider a free abstract group F on a set X. With each $x \in X$ associate a new letter x^{-1} and put $X^{-1} = \{x^{-1} \mid x \in X\}$. Each element $f \in F$ can be presented as a **word** in the letters of $X \cup X^{-1}$:

(1) $$f = s_1 s_2 \cdots s_n, \text{ with } s_1, \ldots, s_n \in X \cup X^{-1}$$

17.5 Subgroups of Free Discrete Groups

Among all possible words that express f there is a unique one in which n is minimal. In this case $s_i \neq s_{i+1}^{-1}$ for each i, $1 \leq i \leq n-1$, and the word $s_1 s_2 \cdots s_n$ is said to be **reduced**. We say that f **starts with** s_1 and **ends with** s_n. Define $\text{length}_X(f)$ to be n.

Now let H be a subgroup of F and define the length of the right coset Hf for $f \in F$:

$$\text{length}_X(Hf) = \min\{\text{length}_X(hf) \mid h \in H\}.$$

Definition 17.5.1: Representation of right cosets. Let R be a system of representatives of the right cosets of F modulo H. Thus, $F = \bigcup_{r \in R} Hr$. For each $f \in F$ let $r = \rho_R(f)$ be the unique element of R satisfying $Hf = Hr$. The function $\rho = \rho_R \colon F \to R$ satisfies the following conditions:
(2a) $\rho(f) \in Hf$.
(2b) $\rho(hf) = \rho(f)$ for all $h \in H$ and $f \in F$.
(2c) $R = \rho(F)$.
These conditions imply that
(3a) $\rho(\rho(f)g) = \rho(fg)$, for all $f, g \in F$; and
(3b) $\rho(r) = r$ for each $r \in R$.

Conversely, if a function $\rho \colon F \to R$ satisfies (2), then R is a system of representatives of the right cosets of F modulo H and $\rho_R = \rho$. □

LEMMA 17.5.2: *Let t be an element of $X \cup X^{-1}$ not in H. Then there exists a system of representatives R for the right cosets of F modulo H with the following properties:*
(4a) $\text{length}_X(\rho_R(f)) = \text{length}_X(Hf)$ *for each $f \in F$.*
(4b) *If $\rho_R(f) = s_1 s_2 \cdots s_n$ is a reduced presentation of $\rho_R(f)$, then $s_1 s_2 \cdots s_i \in R$ for each i between 1 and n.*
(4c) $1, t \in R$.

Proof: It suffices to define a function $\rho \colon F \to R$ satisfying (2) and (4).

We define $\rho(1) = 1$ and $\rho(t) = t$. Then,

$$\text{length}_X(\rho(f)) = 1 = \text{length}_X(Hf)$$

for $f = 1$ and $f = t$.

Assume by induction, that in addition to the right cosets H and Ht, we have chosen representatives for all right cosets of length smaller than a positive integer k. Let Hf be a right coset of length k such that $Hf \neq Ht$. Without loss suppose that $\text{length}_X(f) = k$ and let $f = s_1 s_2 \cdots s_k$ be a reduced presentation of f. Then for $f' = s_1 s_2 \cdots s_{k-1}$, $\text{length}_X(Hf') \leq k - 1$. Thus, $\rho(f')$ is already defined and it has a reduced presentation $\rho(f') = s'_1 s'_2 \cdots s'_l$, with $l \leq k - 1$, such that $s'_1 s'_2 \cdots s'_i \in R$ for each i, $1 \leq i \leq l$. In particular, there exists $h \in H$ such that $f' = h\rho(f')$. Hence, $f = f' s_k = h s'_1 \cdots s'_l s_k$. The assumption $\text{length}_X(f) = \text{length}_X(Hf)$ implies that $l = k - 1$ and that $s'_1 \cdots s'_l s_k$ is a reduced word. Define $\rho(f) = s'_1 \cdots s'_l s_k$ and observe that conditions (4a) and (4b) are satisfied. This completes the induction. □

Notation 17.5.3: The Schreier construction [Kurosh, §36]. Let R be a system of representatives of the right cosets of F modulo H and t an element of $X \cup X^{-1} \smallsetminus H$. We call R a **Schreier system** (with respect to X, H, t) if it satisfies Condition (4) appearing in Lemma 17.5.2. In this case we put $\rho = \rho_R$ and define a map $\varphi_R \colon R \times X \to F$ by

$$\varphi_R(r, x) = rx\rho(rx)^{-1}, \qquad r \in R, \; x \in X.$$

By (3b), $\varphi_R(r, x) = 1$ if and only if $rx \in R$. Put

$$Y = Y_R = \{\varphi_R(r, x) \mid r \in R, \; x \in X\} \smallsetminus \{1\}.$$

Since $Hrx = H\rho(rx)$, Y is a subset of H. We show below that H is free and Y is a basis of H. We call Y a **Schreier basis** of H (with respect to X, t). □

Use the convention $\prod_{i=1}^{n} f_i = f_1 f_2 \cdots f_n$ for products of elements of a group.

LEMMA 17.5.4: *The following hold for the objects F, X, H, t, R, Y constructed above:*
(a) *The set Y generates H.*
(b) $\text{length}_Y(h) \leq \text{length}_X(h)$, *for each $h \in H$.*
(c) *If the reduced presentation of an element $h \in H$ starts with t, then $\text{length}_Y(h) < \text{length}_X(h)$.*
(d) $X \cap H \subseteq Y$.
(e) *Let N be a subgroup of H and $x \in X \cap N$. Suppose N is normal in F. Then $\varphi_R(r, x) = rxr^{-1}$ for each $r \in R$.*
(f) *Suppose in addition to (e) that $(F : H) < \infty$ and $X \smallsetminus N$ is finite. Then $Y \smallsetminus N$ is finite.*

Proof of (a): Put $\rho = \rho_R$ and $\varphi = \varphi_R$. Let

$$(5) \qquad h = s_1 s_2 \cdots s_n$$

be a presentation of an element $h \in H$ as a (not necessarily reduced) word. For $0 \leq i \leq n$ put $g_i = s_1 s_2 \cdots s_i$. Then $g_0 = 1$ and $g_n = h$. In particular, $\rho(g_0) = \rho(g_n) = 1$. Therefore

$$(6) \qquad h = \prod_{i=1}^{n} \rho(g_{i-1}) s_i \rho(g_i)^{-1}.$$

For each i, $1 \leq i \leq n$, either $s_i \in X$ or $s_i^{-1} \in X$. In the first case, by (3), the ith factor in the right hand side of (6) is

$$\rho(g_{i-1}) s_i \rho(g_i)^{-1} = \rho(g_{i-1}) s_i \rho(g_{i-1} s_i)^{-1}$$
$$= \rho(g_{i-1}) s_i \rho(\rho(g_{i-1}) s_i)^{-1} = \varphi(\rho(g_{i-1}), s_i).$$

17.5 Subgroups of Free Discrete Groups

In the second case,

$$\rho(g_{i-1})s_i\rho(g_i)^{-1} = \rho(g_is_i^{-1})s_i\rho(g_i)^{-1} = \rho(\rho(g_i)s_i^{-1})s_i\rho(g_i)^{-1}$$
$$= \left(\rho(g_i)s_i^{-1}\rho(\rho(g_i)s_i^{-1})^{-1}\right)^{-1} = \varphi(\rho(g_i), s_i^{-1})^{-1}.$$

Thus, we may rewrite (6) as

$$(7) \qquad h = \prod_{i=1}^{n} \varphi(r_i, x_i)^{\varepsilon_i},$$

where

$$(8) \qquad (r_i, x_i, \varepsilon_i) = \begin{cases} (\rho(g_{i-1}), s_i, 1) & \text{if } s_i \in X \\ (\rho(g_i), s_i^{-1}, -1) & \text{if } s_i \in X^{-1}. \end{cases}$$

This is a presentation of h as a product of elements of $Y \cup Y^{-1}$.

Proof of (b): Suppose now that the presentation (5) of h is reduced. Then (7) implies that $\text{length}_Y(h) \leq n = \text{length}_X(h)$.

Proof of (c): If $s_1 = t$ in (5), then (4c) gives $\rho(g_0)s_1\rho(g_1)^{-1} = t \cdot t^{-1} = 1$. Hence, $\text{length}_Y(h) \leq n - 1$, as asserted.

Proof of (d): If $x \in X \cap H$, then $\rho(x) = 1$, and $x = \varphi(1, x) \in Y$.

Proof of (e): If $x \in X \cap N$, then $H\rho(rx) = Hrx = H(rxr^{-1})r = Hr$. Hence, $\rho(rx) = r$ and $\varphi(r, x) = rxr^{-1}$.

Proof of (f): Here $|R| = (F : H) < \infty$. By (e), $\varphi(r, x) = rxr^{-1} \in N$ for all $r \in R$ and $x \in X \cap N$. Therefore, $Y \smallsetminus N$ is contained in the finite set $\{\varphi(r, x) \mid r \in R, x \in X \smallsetminus N\}$. □

LEMMA 17.5.5: *H is a free group on Y.*

Proof: Put $\rho = \rho_R$ and $\varphi = \varphi_R$. To each pair $(r, x) \in R \times X$ with $\varphi(r, x) \neq 1$ attach a letter $\psi(r, x)$. For $(r, x) \in R \times X$ with $\varphi(r, x) = 1$ put $\psi(r, x) = 1$. Then let Y^* be the set of all $\psi(r, x)$ with $(r, x) \in R \times X$ and $\varphi(r, x) \neq 1$. Denote the free abstract group on Y^* by H^*. Map H into H^* by mapping $h \in H$, presented by (5), onto

$$(9) \qquad h^* = \prod_{i=1}^{n} \psi(r_i, x_i)^{\varepsilon_i}, \text{ with } r_i, x_i, \varepsilon_i \text{ defined by (8)}, i = 1, \ldots, n.$$

We prove that the map $h \mapsto h^*$ is a homomorphism of H onto H^* mapping Y bijectively onto Y^*. It will follow that this map is an isomorphism.

PART A: *The map $h \mapsto h^*$ is well defined.* If $s_{i+1} = s_i^{-1}$, then $g_{i+1} = g_{i-1}s_is_{i+1} = g_{i-1}$. Therefore, if $s_i \in X$, then $(r_i, x_i, \varepsilon_i) = (\rho(g_{i-1}), s_i, 1)$ and $(r_{i+1}, x_{i+1}, \varepsilon_{i+1}) = (\rho(g_{i+1}), s_{i+1}^{-1}, -1)$. Thus, $(r_i, x_i) = (r_{i+1}, x_{i+1})$, $\varepsilon_{i+1} = -\varepsilon_i$, and $\psi(r_{i+1}, x_{i+1}) = \psi(r_i, x_i)$. Similarly, the latter two equalities hold if $s_i \in X^{-1}$.

Two presentations of h as words in $X \cup X^{-1}$ are obtained from each other by inserting or canceling finitely many factors of the form ss^{-1}, with $s \in X \cup X^{-1}$. Thus, the preceding paragraph shows that the definition of h^* is independent of the presentation (5) of h.

PART B: *The map $h \mapsto h^*$ is a homomorphism.* Let $h' = s'_1 s'_2 \cdots s'_{n'}$ be another element of H with $s'_1, s'_2, \ldots, s'_{n'} \in X \cup X^{-1}$. By (2b),

$$\rho(s_1 \cdots s_n s'_1 \cdots s'_j) = \rho(hs'_1 \cdots s'_j) = \rho(s'_1 \cdots s'_j).$$

Therefore, the presentation (6) for hh' is the product of the presentation (6) for h followed by the presentation (6) for h'. This implies $(hh')^* = h^*(h')^*$. In other words, the map $h \mapsto h^*$ is a homomorphism of H into H^*.

PART C: We prove that $\psi(r, x) = \varphi(r, x)^*$ for each $(r, x) \in R \times X$ with $\varphi(r, x) \neq 1$. Let $r = u_1 u_2 \cdots u_k$ and $\rho(rx) = v_1 v_2 \cdots v_l$ be the reduced presentations of r and $\rho(rx)$ as words in the letters of $X \cup X^{-1}$. Apply (5) and (6) to $h = \varphi(r, x)$:

$$\varphi(r, x) = rx\rho(rx)^{-1} = \prod_{i=1}^{k} u_i \cdot x \cdot \prod_{j=0}^{l-1} v_{l-j}^{-1}$$

$$= \prod_{i=1}^{k} \rho(u_1 \cdots u_{i-1}) u_i \rho(u_1 \cdots u_{i-1} u_i)^{-1}$$

$$\cdot u_1 \cdots u_k x \rho(rx)^{-1}$$

$$\cdot \prod_{j=0}^{l-1} \rho(rxv_l^{-1} \cdots v_{l-j+1}^{-1}) v_{l-j}^{-1} \rho(rxv_l^{-1} \cdots v_{l-j+1}^{-1} v_{l-j}^{-1})^{-1}.$$

Here, for $j = 0$, regard $v_l^{-1} \cdots v_{l-j+1}^{-1}$ as an empty product with the value 1. Also, for $j = l-1$ we have $\rho(rxv_l^{-1} \cdots v_1^{-1}) = \rho(h) = 1$. Use (3) to define appropriate r_i, x_i, ε_i and $r'_j, x'_j, \varepsilon'_j$ as in (8) such that

$$\varphi(r, x) = \prod_{i=1}^{k} \varphi(r_i, x_i)^{\varepsilon_i} \cdot \varphi(r, x) \cdot \prod_{j=0}^{l-1} \varphi(r'_j, x'_j)^{\varepsilon'_j}.$$

From definition (9)

$$\varphi(r, x)^* = \prod_{i=1}^{k} \psi(r_i, x_i)^{\varepsilon_i} \cdot \psi(r, x) \cdot \prod_{j=0}^{l-1} \psi(r'_j, x'_j)^{\varepsilon'_j}.$$

17.5 Subgroups of Free Discrete Groups

For each i, $1 \leq i \leq k$, compute from (3a), (3b), and (4b) that

$$\varphi(r_i, x_i)^{\varepsilon_i} = \rho(u_1 \cdots u_{i-1}) u_i \rho(u_1 \cdots u_{i-1} u_i)^{-1}$$
$$= (u_1 \cdots u_{i-1}) u_i (u_1 \cdots u_{i-1} u_i)^{-1} = 1,$$

so $\psi(r_i, x_i) = 1$. Similarly, for each j, $0 \leq j \leq l-1$,

$$\rho(rxv_l^{-1} \cdots v_{l-j+1}^{-1}) = \rho(\rho(rx) v_l^{-1} \cdots v_{l-j+1}^{-1})$$
$$= \rho(v_1 v_2 \cdots v_{l-j}) = v_1 v_2 \cdots v_{l-j}.$$

Hence

$$\varphi(r'_j, x'_j)^{\varepsilon'_j} = \rho(rxv_l^{-1} \cdots v_{l-j+1}^{-1}) v_{l-j}^{-1} \rho(rxv_l^{-1} \cdots v_{l-j+1}^{-1} v_{l-j}^{-1})^{-1}$$
$$= (v_1 \cdots v_{l-j}) v_{l-j}^{-1} (v_1 \cdots v_{l-j-1})^{-1} = 1.$$

Thus, $\psi(r'_j, x'_j) = 1$. Consequently, $\varphi(r,x)^* = \psi(r,x)$. This concludes the proof of the lemma. □

We summarize the information achieved in this section up to this point:

PROPOSITION 17.5.6 (Schreier): *Let F be a free abstract group on a set X and H a subgroup of F. Then H is free. Moreover, for each $t \in X \cup X^{-1} \smallsetminus H$ there exist a Schreier system of representatives R for the right cosets of F modulo H and a Schreier basis $Y = Y_R$ for X, H, t. By definition, R and Y satisfy all assertions of Lemma 17.5.4:*
(a) $\text{length}_Y(h) \leq \text{length}_X(h)$, *for each $h \in H$.*
(b) *If the reduced presentation of an element $h \in H$ starts with t, then $\text{length}_Y(h) < \text{length}_X(h)$.*
(c) $X \cap H \subseteq Y$.
(d) *Let N be a subgroup of H and $x \in X \cap N$. Suppose N is normal in F. Then $\varphi_R(r,x) = rxr^{-1}$ for each $r \in R$.*
(e) *Moreover, suppose $(F:H) < \infty$ and $X \smallsetminus N$ is finite. Then $Y \smallsetminus N$ is finite.*

PROPOSITION 17.5.7 (Nielsen-Schreier Formula): *Let F be a free abstract group of rank e and let H be a subgroup of index n. Then H is free of rank $1 + n(e-1)$.*

Proof: Continue to use the notation of the previous lemmas. In this case the basis X of F contains e elements and the system of representatives R of F modulo H has n elements. Therefore, the number of elements in the basis Y of H (Lemma 17.5.6) equals ne minus the number of elements of the set

$$\Gamma = \{(r,x) \in R \times X \mid rx = \rho(rx)\}.$$

If we prove that $|\Gamma| = n - 1$, then

$$\text{rank}(H) = ne - (n-1) = 1 + n(e-1),$$

and we are done.

To this end, define two subsets of R for each positive integer k:

$$R_k = \{r \in R \mid \text{length}_X(r) = k \text{ and } r \text{ ends with some } x \in X\}; \text{ and}$$
$$R'_k = \{r \in R \mid \text{length}_X(r) = k \text{ and } r \text{ ends with some } x^{-1} \in X^{-1}\}.$$

In addition, let $R'_0 = \emptyset$ and $R' = \bigcup_{k \geq 0} R'_k \cup R_{k+1}$. Then $R = \{1\} \cup R'$. Thus, $|R'| = n - 1$.

On the other hand, for each $k \geq 0$ define two subsets of Γ:

$$\Gamma_k = \{(r, x) \in \Gamma \mid \text{length}_X(r) = k \text{ and } \text{length}_X(rx) = k + 1\}; \text{ and}$$
$$\Gamma'_k = \{(r, x) \in \Gamma \mid \text{length}_X(r) = k \text{ and } \text{length}_X(rx) = k - 1\}.$$

Also let $\Gamma'_0 = \emptyset$. Then $\Gamma = \bigcup_{k \geq 0} \Gamma'_k \cup \Gamma_k$.

Now define a bijective map $\tau \colon \Gamma \to R'$. Suppose $k > 0$ and $(r, x) \in \Gamma'_k$. Then r ends with x^{-1}. Define $\tau(r, x)$ to be r. Conversely, if $r \in R'_k$, then r ends with some $x^{-1} \in X^{-1}$. Hence, $\text{length}_X(rx) = k - 1$. Therefore, by (4b), $rx \in R$, so $\rho(rx) = rx$ and $(r, x) \in \Gamma'_k$. Consequently, τ maps Γ'_k bijectively onto R'_k.

Finally, suppose $k \geq 0$ and $(r, x) \in \Gamma_k$. Define $\tau(r, x) = rx$ and note that $rx \in R_{k+1}$. Conversely, if $r' \in R_{k+1}$, then, by (4b), $r' = rx$, where $x \in X$ and $r \in R$ is of length k. By (3b), $rx = \rho(rx)$. Hence, $(r, x) \in \Gamma_k$. Consequently, τ maps Γ_k bijectively onto R_{k+1}.

The definition of τ is now complete. We conclude that $|\Gamma| = |R'| = n - 1$, as desired. □

We define the **rank** of an abstract group G as the minimal cardinality of a set of generators of G. If G is finite, than the rank of G as an abstract group is equal to the rank of G as a profinite group. However, if G is infinite, then the rank of G as a profinite group may be less than the rank of G as an abstract group. For example, the rank of $\hat{\mathbb{Z}}$ as a profinite group is 1 while its rank as an abstract group is $2^{\aleph_0}$. Nevertheless, it should be always clear from the context in which sense we use $\text{rank}(G)$.

COROLLARY 17.5.8: *Let G_0 be a subgroup of index n of an abstract group G of rank e. Then $\text{rank}(G_0) \leq 1 + n(e - 1)$.*

Proof: There is an epimorphism $\alpha \colon F \to G$, where F is the free abstract group on e generators. Put $F_0 = \alpha^{-1}(G_0)$. Then $(F : F_0) = n$. By Proposition 17.5.7, $\text{rank}(F_0) = 1 + n(e - 1)$. Therefore, $\text{rank}(G_0) \leq 1 + n(e - 1)$. □

Remark 17.5.9: Residually finite groups. Let G be an abstract group and $\mathcal{C}$ a family of finite groups. Then G is **residually-$\mathcal{C}$** if the intersection of all normal subgroups N of G with $G/N \in \mathcal{C}$ is 1. When $\mathcal{C}$ consists of all finite groups, we use **residually finite** for residually-$\mathcal{C}$.

17.5 Subgroups of Free Discrete Groups

Suppose $\mathcal{C}$ is a formation of finite groups. Denote the family of all $N \triangleleft G$ with $G/N \in \mathcal{C}$ by $\mathcal{N}$. Then $\mathcal{N}$ is directed. Let θ be the canonical map of G into its completion $\hat{G}$ with respect to $\mathcal{N}$. Then G is residually $\mathcal{C}$ if and only if θ is injective. $\square$

Let $x_1, \ldots, x_e$ be free generators of a free abstract group F and w a nonempty reduced word in $\{x_1, \ldots, x_e\}$ of length n. Then there exists a homomorphism $\varphi \colon F \to S_{n+1}$ with $\varphi(w) \neq 1$ [Kurosh, p. 42]. Thus, F is residually finite. Proposition 17.5.11 below generalizes this conclusion considerably.

LEMMA 17.5.10 (Levi): *Let F be a free abstract group and $F > F_1 > F_2 > \cdots$ a decreasing sequence of subgroups. Suppose each F_{i+1} is a characteristic subgroup of F_i. Then $\bigcap_{i=1}^{\infty} F_i = 1$.*

Proof (Lubotzky): Assume $\bigcap_{i=1}^{\infty} F_i$ contains a nontrivial element w. By Proposition 17.5.6, each F_i is free. Let $\text{length}_i(w) = \min(\text{length}_X(w))$ where X ranges on all bases of F_i. If we prove that $\text{length}_{i+1}(w) < \text{length}_i(w)$, we will draw a contradiction.

Indeed, let X be a basis of F_i with $\text{length}_X(w) = \text{length}_i(w)$. Let $w = s_1 s_2 \cdots s_n$ be a reduced word in $X \cup X^{-1}$. Assume $s_1 \in F_{i+1}$. Replacing s_1 with its inverse, if necessary, we may assume $s_1 \in X$. Let $x \in X$. The bijective map $\varphi_0 \colon X \to X$ which permutes x and s_1 and fixes all other elements of X extends to an automorphism $\varphi \colon F_i \to F_i$. Since F_{i+1} is a characteristic subgroup of F_i, φ maps F_{i+1} onto itself. In particular, $x \in F_{i+1}$. It follows that $F_{i+1} = F_i$. This contradiction proves that $s_1 \notin F_{i+1}$.

Proposition 17.5.6 gives a basis Y of F_{i+1} such that $\text{length}_Y(w) < \text{length}_X(w)$. Hence, $\text{length}_{i+1}(w) < \text{length}_i(w)$. We conclude that $\bigcap_{i=1}^{\infty} F_i = 1$. $\square$

PROPOSITION 17.5.11 ([Ribes-Zalesskii, Prop. 3.3.15]): *Let $\mathcal{C}$ be a Melnikov formation of finite groups, m a cardinal number, F the free abstract group of rank m, and $\hat{F} = \hat{F}_m(\mathcal{C})$ the free pro-$\mathcal{C}$ group of rank m. Then F is residually-$\mathcal{C}$ and the canonical map $\theta \colon F \to \hat{F}$ is injective.*

Proof: The assumption on the existence of $\hat{F}$ yields $G \in \mathcal{C}$ with $\text{rank}(G) \leq m$. Let F_1 be the intersection of all normal subgroups N of F with $F/N \cong G$. Then F_1 is a characteristic subgroup of F. By Proposition 17.5.6, F_1 is free.

To estimate the rank of F_1 we choose a prime number p such that $\mathbb{Z}/p\mathbb{Z}$ is not a composition factor of G. Then F has a normal subgroup N with $F/N \cong (\mathbb{Z}/p\mathbb{Z})^m$ and $F_1/N \cap F_1 \cong F/N$. Hence, $\text{rank}(F_1) \geq m$.

We may therefore define F_2 to be the intersection of all normal subgroups N of F_1 with $F_1/N \cong G$. Then F_2 is characteristic in F_1, hence in F, and $\text{rank}(F_2) \geq m$. Continue by induction to construct an descending sequence $F > F_1 > F_2 > \cdots$ of characteristic subgroups of F with $\text{rank}(F_i) \geq m$. By Levi's result (Lemma 17.5.10), $\bigcap_{i=1}^{\infty} F_i = 1$. Therefore, F is residually-$\mathcal{C}$. $\square$

Remark 17.5.12: Let F be the free abstract group of rank $m \geq 2$ and let $\mathcal{C}$ be an infinite family of finite non-Abelian simple groups. By [Weigel1,2,3] or [Dixon-Pyber-Seress-Shalev], F is residually-$\mathcal{C}$. Note that the assumption of the groups in $\mathcal{C}$ to be non-Abelian is essential. If $\mathcal{C}$ consists of all groups $\mathbb{Z}/p\mathbb{Z}$, then the intersection of all normal subgroups N of F with $F/N \in \mathcal{C}$ contains the commutator subgroup $[F, F]$ of F which is not trivial. $\square$

17.6 Open Subgroups of Free Profinite Groups

We prove in this section that an open subgroup of a free profinite group is free and satisfies the Nielsen-Schreier formula for the ranks:

Remark 17.6.1: *Identifying F as a subgroup of $\hat{F}$.* Let $\mathcal{C}$ be a Melnikov formation of finite group and F the free abstract group on a set X. Suppose there is a nontrivial group $C \in \mathcal{C}$ with $\text{rank}(C) \leq |X|$. Let $\hat{F}_X(\mathcal{C})$ be the corresponding profinite completion. We use Proposition 17.5.11 to identify F with a dense abstract subgroup of $\hat{F}_X(\mathcal{C})$ and to identify the canonical map $\theta \colon F \to \hat{F}_X(\mathcal{C})$ with the inclusion map.

Conversely, let $\hat{F}$ be a free pro-$\mathcal{C}$ group with a basis X. Lemma 17.4.3 extends the identity map $X \to X$ to an isomorphism $\hat{F}_X(\mathcal{C}) \to \hat{F}$. The latter maps F onto the abstract subgroup F_0 of $\hat{F}$ generated by X. It follows that F_0 is free on X. $\square$

PROPOSITION 17.6.2: Let $\mathcal{C}$ be a Melnikov formation of finite groups, $\hat{F}$ a free pro-$\mathcal{C}$-group, and $\hat{H}$ an open subgroup of $\hat{F}$. Suppose $\hat{H}$ is pro-$\mathcal{C}$ (e.g. $\mathcal{C}$ is full or $\hat{H} \triangleleft \hat{F}$). Then $\hat{H}$ is a free pro-$\mathcal{C}$ group. Moreover, if $e = \text{rank}(\hat{F})$ is finite, then $\text{rank}(\hat{H}) = 1 + (\hat{F} : \hat{H})(e-1)$. If $m = \text{rank}(\hat{F})$ is infinite, then $\text{rank}(\hat{H}) = m$.

Proof: Let X be a basis for $\hat{F}$ and F the free abstract group on X. Consider the family $\mathcal{N}(X)$ of all normal subgroups N of F with $X \smallsetminus N$ finite and $F/N \in \mathcal{C}$. By Lemma 17.4.6(c), $\hat{F}$ is the profinite completion of F with respect to $\mathcal{N}(X)$. By Proposition 17.5.11, we may assume that F is contained in $\hat{F}$ and $\theta \colon F \to \hat{F}$ is the inclusion map. Lemma 17.2.1 gives a subgroup H of F and a $M \in \mathcal{N}(X)$ such that $M \leq H$ and $\hat{H}$ is the closure of H in $\hat{F}$. Moreover, $(F : H) = (\hat{F} : \hat{H})$ is finite and $\hat{H} = \varprojlim H/N$, where N ranges over $\mathcal{N}_0(X) = \{N \in \mathcal{N}(X) \mid N \leq H\}$.

Let Y, R, and φ be as in Proposition 17.5.6. Then R is finite, H is free, Y is a basis of H, and $Y \smallsetminus N$ is finite for each $N \in \mathcal{N}_0(X)$. By Proposition 17.4.2, the free pro-$\mathcal{C}$-group on Y is isomorphic to the inverse limit $\varprojlim H/N$, where N ranges over the family $\mathcal{N}(Y)$ of all open normal subgroups N of H such that $Y \smallsetminus N$ is finite and $H/N \in \mathcal{C}$. We show that $\mathcal{N}_0(X)$ and $\mathcal{N}(Y)$ are cofinite in each other. By Lemma 17.2.3, this will prove that $\hat{H}$ is the free pro-$\mathcal{C}$-group on Y.

Indeed, let $N \in \mathcal{N}_0(X)$. Denote the closure of N in $\hat{F}$ by $\hat{N}$. By Lemma 17.2.1, $\hat{N} \triangleleft \hat{H}$ and $H/N \cong \hat{H}/\hat{N}$. Since $\hat{H}$ is pro-$\mathcal{C}$, we have $H/N \in \mathcal{C}$, so

$N \in \mathcal{N}(Y)$.

Conversely, let $N \in \mathcal{N}(Y)$. Then $H/N \in \mathcal{C}$ and $Y \smallsetminus N$ is finite. Let $N_0 = \bigcap_{r \in R} N^r$ and $H_0 = \bigcap_{r \in R} H^r$. Then N_0 and H_0 are normal subgroups of F, $M \leq H_0$, and $N_0 \leq H_0$. Since NH_0/N is normal in H/N, it belong to $\mathcal{C}$. Since $H_0/(N \cap H_0)$ is isomorphic to NH_0/N, it belongs to $\mathcal{C}$. Hence, $H_0/(N^r \cap H_0) \in \mathcal{C}$ for each $r \in R$. Since R is finite, $H_0/N_0 = H_0/\bigcap_{r \in R}(N^r \cap H_0) \in \mathcal{C}$ (Condition (1b) of Section 17.3)). In addition, $F/H_0 \in \mathcal{C}$ because $F/M \in \mathcal{C}$ and F/H_0 is a quotient of F/M. Finally, since $\mathcal{C}$ is Melnikov, $F/N_0 \in \mathcal{C}$.

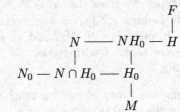

If we prove that $X \smallsetminus N_0$ is finite, it will follow that $N_0 \in \mathcal{N}_0(X)$. This will complete the proof of the cofiniteness of $\mathcal{N}(Y)$ and $\mathcal{N}_0(X)$ in each other.

Indeed, $X \smallsetminus M$, $Y \smallsetminus N$, and R are finite sets. Thus, almost all $x \in X$ belong to M. For each $x \in X \cap M$ and $r \in R$ we have $\varphi(r, x) = rxr^{-1}$ (Proposition 17.5.6(d)). For each $r \in R$ the map $x \mapsto rxr^{-1}$ of $X \cap M$ into Y is injective. Hence, only finitely many $x \in X \cap M$ satisfy $rxr^{-1} \in Y \smallsetminus N$. Therefore, $X_0 = \{x \in X \cap M \mid \varphi(r, x) = rxr^{-1}$ and $rxr^{-1} \in N$ for all $r \in R\}$ is cofinite in X. It follows that almost every $y \in Y$ can be written as $y = vxv^{-1}$ with $x \in X_0$ and $v \in R$. For each $u \in R$ there exist $r \in R$ and $h \in H$ with $uv = hr$. Therefore, $uyu^{-1} = uvxv^{-1}u^{-1} = hrxr^{-1}h^{-1} \in hNh^{-1} = N$, so, $y \in N^u$. It follows that almost every $y \in Y$ belongs to N_0. By Lemma 17.5.4(d), $X \smallsetminus N_0 = (X \smallsetminus H) \cup (X \cap H \smallsetminus N_0) \subseteq (X \smallsetminus H) \cup (Y \smallsetminus N_0)$ is finite, as claimed.

The rank formula for the finitely generated case now follows from Proposition 17.5.7. When m is infinite, use Corollary 17.1.5. □

COROLLARY 17.6.3: *Let G be a profinite group of rank at most e and H an open subgroup of G of index n. Then* $\mathrm{rank}(H) \leq 1 + n(e - 1)$.

Proof: There exists an epimorphism $\theta \colon \hat{F}_e \to G$ and $(\hat{F}_e : \theta^{-1}(H)) = n$. By Proposition 17.6.2, $\mathrm{rank}(\theta^{-1}(H)) = 1 + n(e - 1)$. Hence, by Corollary 17.1.4, $\mathrm{rank}(H) \leq 1 + n(e - 1)$. □

The following result is used in Section 25.7 to prove that a closed normal subgroup $\hat{N}$ of a free pro-$\mathcal{C}$ group is free pro-$\mathcal{C}$ if $\hat{N}$ contains nontrivial elements of the underlying abstract free group.

PROPOSITION 17.6.4: *Let $\mathcal{C}$ be a Melnikov formation, $\hat{F}$ a free pro-$\mathcal{C}$ group of rank $m \geq 2$, X a basis of $\hat{F}$, F the abstract group of $\hat{F}$ generated by X, and $\hat{N}$ a closed normal subgroup of $\hat{F}$. Suppose $N = F \cap \hat{N} \neq 1$. Then:*

(a) $\hat{F}$ has an open normal subgroup $\hat{E}$ containing $\hat{N}$ with a Schreier basis Y with respect to X such that $Y \cap N \neq \emptyset$.

(b) If $m < \infty$ than $\hat{F}$ has an open normal subgroup $\hat{E}$ containing $\hat{N}$ with $\mathrm{rank}(\hat{E}/\hat{N}) < 1 + (\hat{F}:\hat{E})(m-1)$.

Proof of (a): Let $t \in F \cap \hat{N}$, $t \neq 1$. Denote the set of all open normal subgroups of $\hat{F}$ which contain $\hat{N}$ by $\mathcal{H}$. For each $\hat{E} \in \mathcal{H}$ put $E = F \cap \hat{E}$. Find $\hat{E} \in \mathcal{H}$, a Schreier basis Y of E with respect to X (Notation 17.5.3), and a presentation $t = s_1 s_2 \cdots s_n$ as a word in Y with a minimal length among all presentations of this type. We show that $s_1 \in N$.

Assume $s_1 \notin N$. By Lemma 1.2.3, $\hat{N}$ is the intersection of all $\hat{H} \in \mathcal{H}$. Hence, there is an $\hat{H} \in \mathcal{H}$ such that $\hat{H} \leq \hat{E}$ and $s_1 \notin \hat{H}$. By Proposition 17.5.6(b), H has a Schreier basis Z with respect to Y with $\mathrm{length}_Z(t) < \mathrm{length}_Y(t)$, a contradiction.

Proof of (b): Continuing the proof of (a), we conclude from $s_1 \in Y \cap \hat{N}$ and from Nielsen-Schreier that $\mathrm{rank}(\hat{E}/\hat{N}) < \mathrm{rank}(\hat{E}) = 1 + (\hat{F}:\hat{E})(m-1)$. $\square$

COROLLARY 17.6.5: *Let $\mathcal{C}$ be a Melnikov formation and $\hat{F}$ a free pro-$\mathcal{C}$-group of positive rank. Then $\hat{F}$ is infinite.*

Proof: Assume $\hat{F}$ is finite. Then the trivial group is a free pro-$\mathcal{C}$-group of positive rank (Proposition 17.6.2). This is a contradiction. $\square$

Example 26.1.10 shows that Corollary 17.6.5 does not hold for arbitrary formations.

17.7 An Embedding Property

Here is a characterization of free pro-$\mathcal{C}$ groups of finite rank in terms of their finite quotients:

LEMMA 17.7.1: *Let $\mathcal{C}$ be a formation of finite groups, e a positive integer, and F a pro-$\mathcal{C}$ group. Then $F \cong \hat{F}_e(\mathcal{C})$ if and only if F satisfies the following conditions:*

(a) *Every finite homomorphic image of F is generated by e elements.*

(b) *Every $\mathcal{C}$-group of rank at most e is a homomorphic image of F.*

Proof: A compactness argument shows that (a) holds if and only if $\mathrm{rank}(F) \leq e$. Now apply Proposition 16.10.7(b). $\square$

We conclude this section by establishing an "embedding property" of $\hat{F}_e(\mathcal{C})$. It depends on a surprising lemma that allows us to "lift generators" of a homomorphic image:

LEMMA 17.7.2 (Gaschütz): *Let $\pi\colon G \to H$ be an epimorphism of profinite groups with $\mathrm{rank}(G) \leq e$. Let $h_1, \ldots, h_e$ be a system of generators of H.*

Then there exists a system of generators $g_1, \ldots, g_e$ of G such that $\pi(g_i) = h_i$, $i = 1, \ldots, e$.

Proof (Roquette): We start with the crucial case: G is a finite group. For each subgroup C of G satisfying $\pi(C) = H$ and all systems of generators $a_1, \ldots, a_e$ of H denote the number of e-tuples $\mathbf{c} \in C^e$ that generate C and satisfy $\pi(\mathbf{c}) = \mathbf{a}$ by $\varphi_C(\mathbf{a})$. We prove by induction on $|C|$ that $\varphi_C(\mathbf{a})$ is independent of $\mathbf{a}$.

Assume $\varphi_B(\mathbf{a})$ is independent of $\mathbf{a}$ for every proper subgroup B of C satisfying $\pi(B) = H$. Let $m = \frac{|C|}{|H|}$. Then there are exactly m^e elements $\mathbf{b} \in C^e$ with $\pi(\mathbf{b}) = \mathbf{a}$. Each such $\mathbf{b}$ generates a subgroup B of C satisfying $\pi(B) = H$. Hence,

$$(1) \qquad m^e = \varphi_C(\mathbf{a}) + {\sum}'_{B < C} \varphi_B(\mathbf{a}),$$

where $\sum'$ indicates a sum over groups with $\pi(B) = H$. By assumption, the $\sum'$ is independent of $\mathbf{a}$. Therefore, so is $\varphi_C(\mathbf{a})$.

Now choose a system of generators $g'_1, \ldots, g'_e$ for G. Then $\pi(\mathbf{g}') = \mathbf{h}'$ generates H. By the preceding paragraph, $\varphi_G(\mathbf{h}) = \varphi_G(\mathbf{h}') \geq 1$. Consequently, G has a system of generators $g_1, \ldots, g_e$ such that $\pi(\mathbf{g}) = \mathbf{h}$.

For G general present $\pi: G \to H$ as an inverse limit of epimorphisms of finite groups $\pi_i: G_i \to H_i$, $i \in I$. Specifically, if $j \geq i$, then there are epimorphisms $\xi_{ji}: G_j \to G_i$ and $\eta_{ji}: H_j \to H_i$ such that $\pi_i \circ \xi_{ji} = \eta_{ji} \circ \pi_j$. In addition, there are epimorphisms $\xi_i: G \to G_i$ and $\eta_i: H \to H_i$ such that $\pi_i \circ \xi_i = \eta_i \circ \pi$. For each $i \in I$ denote the set of e-tuples $\mathbf{x} \in G_i^e$ that generate G_i and satisfy $\pi_i(\mathbf{x}) = \eta_i(\mathbf{h})$ by A_i. By the case where G is finite, A_i is nonempty. In addition, A_i is finite. If $j \geq i$ and $\mathbf{y} \in A_j$, then $\xi_{ji}(\mathbf{y}) \in A_i$. By Corollary 1.1.4, the inverse limit A of the inverse system $\langle A_i, \xi_{ji} \rangle_{i,j \in I}$ is nonempty. Each element in A defines a system of generators $g_1, \ldots, g_e$ of G with $\pi(\mathbf{g}) = \mathbf{h}$. $\square$

PROPOSITION 17.7.3: *Let $\mathcal{C}$ be a formation of finite groups, e a positive integer, and $(\varphi: \hat{F}_e(\mathcal{C}) \to H, \alpha: G \to H)$ a pair of epimorphisms, where G is a pro-$\mathcal{C}$ group and $\mathrm{rank}(G) \leq e$. Then there exists an epimorphism $\gamma: \hat{F}_e(\mathcal{C}) \to G$ such that $\alpha \circ \gamma = \varphi$.*

Proof: Let $x_1, \ldots, x_e$ be a basis of $\hat{F}_e(\mathcal{C})$. Then $\varphi(x_1), \ldots, \varphi(x_e)$ generate H. By Lemma 17.7.2, there are $g_1, \ldots, g_e$ which generate G such that $\alpha(g_i) = \varphi(x_i)$, $i = 1, \ldots, e$. The map $x_i \mapsto g_i$, $i = 1, \ldots, e$, extends to an epimorphism $\varphi: \hat{F}_e(\mathcal{C}) \to G$ such that $\alpha \circ \gamma = \varphi$. $\square$

Exercises

1. Let $\theta: G \to H$ be an epimorphism of profinite groups.
    (a) Prove that θ maps subsets of G that converge to 1 onto subsets of H that converge to 1.

(b) Let $\{h_i \mid i \in I\}$ be a subset of H that converges to 1. Imitate the proof of Proposition 17.1.1 to prove that G contains a subset $\{g_i \mid i \in I\}$ which converges to 1 with $\theta(g_i) = h_i$ for each $i \in I$.

2. Let G be a profinite group, H an open subgroup, and G_0 a dense subset of G. Prove that $G_0 \cap H$ is dense in H.

3. Let $\mathcal{D}$ be a family of finite simple groups and $\mathcal{C}$ the Melnikov formation generated by $\mathcal{C}$ (Remark 17.3.4). Suppose every subgroup of each $S \in \mathcal{D}$ belongs to $\mathcal{C}$. Prove that $\mathcal{C}$ is full. Hint: Let $G \in \mathcal{C}$. Choose a composition series of G and apply induction on its length.

4. Use Lemma 17.4.10 to prove that the quotient of $\hat{F}_e$ modulo its commutator subgroup $(\hat{F}_e)'$ is isomorphic to $\hat{\mathbb{Z}}^e$.

5. Let $\hat{G}$ be a profinite completion of a finitely generated group G. Prove that $\mathrm{rank}(\hat{G}) \leq \mathrm{rank}(G)$.

6. For $f \geq e > 1$ use Nielsen-Schreier to prove for each $x_1, \ldots, x_f \in \hat{F}_e$ that the inequality $(\hat{F}_e : \langle \mathbf{x} \rangle) < \infty$ implies $(\hat{F}_e : \langle \mathbf{x} \rangle) \leq \frac{f-1}{e-1}$. Conclude for $x_1, \ldots, x_e \in \hat{F}_e$ that either $\langle \mathbf{x} \rangle = \hat{F}_e$ or $(\hat{F}_e : \langle \mathbf{x} \rangle) = \infty$.

Notes

Ershov [Ershov2, Lemma 1, and Exercise 9 of Chapter 21] suggests an alternative proof of Douady's theorem (Proposition 17.1.1).

The open subgroup theorem (Proposition 17.6.2) for a free profinite group is a special case of the subgroup theorem for free products of profinite groups appearing in [Binz-Neukirch-Wenzel]. The latter, is based on the deeper Kurosh subgroup theorem for free products of abstract groups [Specht, p. 189, Satz 8] while our proof of Theorem 17.6.2 is based on the simpler Schreier construction of Section 17.5.

Gaschütz's proof of Lemma 17.7.2 [Gaschütz] provides a (complicated) formula for $\varphi_G(\mathbf{a})$.

The proof of Levi's Lemma 17.5.10 depends on Schreier's construction of a basis of a subgroup of a free abstract group rather than on Nielsen's construction as in [Lyndon-Schupp, p. 14, Prop. 3.3]. The latter proof considers a nontrivial element w in F_i (notation as in Lemma 17.5.10) and a basis X of F and proves that $\mathrm{length}_X(w) \geq i$. Our proof constructs a basis Y of F_i such that $\mathrm{length}_Y(w) \leq \max(0, \mathrm{length}_X(w) - i)$.

Chapter 18. The Haar Measure

It is well known that every locally compact group admits a (one sided) translation invariant Haar measure. Applications of the Haar measure in algebraic number theory to local fields and adelic groups appear in [Cassels-Fröhlich, Chap. II] and [Weil6]. Here we use it to investigate absolute Galois groups of fields. Since these groups are compact, the Haar measure is a two sided invariant. We provide a simple direct proof of the existence and uniqueness of the Haar measure of profinite groups (Sections 18.1 and 18.2).

We normalize the Haar measure μ of a compact group G so that $\mu(G) = 1$. Thus, μ is a probability measure. When $G = \text{Gal}(K)$ is the absolute Galois group of a field K, μ-independence of sets is related to linear disjointness of fields (Section 18.5). For K a Hilbertian field, we prove that $\langle \sigma_1, \ldots, \sigma_e \rangle \cong \hat{F}_e$ for almost all $\boldsymbol{\sigma} \in \text{Gal}(K)^e$. In addition, if K is countable, then the fixed fields $K_s(\sigma_1, \ldots, \sigma_e)$ are PAC for almost all $\boldsymbol{\sigma} \in \text{Gal}(K)^e$ (Section 18.6). In the uncountable case we provide examples where the set of $\boldsymbol{\sigma} \in \text{Gal}(K)^e$ such that $K_s(\sigma_1, \ldots, \sigma_e)$ is PAC, is nonmeasurable.

The complete proof that every field is stable (Definition 18.9.1) lies unfortunately outside the scope of this book. Section 18.9 outlines the main ingredients and steps of the proof. Once this is done, we are able to construct for each countable Hilbertian field an abundance of Galois extensions which are PAC (Theorem 18.10.2).

18.1 The Haar Measure of a Profinite Group

Let G be a profinite group. We define the completed Haar measure of G and prove its uniqueness. In the next section we prove the existence of the Haar measure.

Consider a collection $\mathcal{A}$ of subsets of G. The **σ-algebra generated** by $\mathcal{A}$ is the smallest collection $\mathcal{A}'$ of subsets of G which contains $\mathcal{A}$ and is closed under taking complements and countable unions. When $\mathcal{A}$ is a Boolean algebra, $\mathcal{A}'$ is also the smallest collection of subsets of G which contains $\mathcal{A}$ and is **monotone**. That is, $\mathcal{A}'$ is closed under countable increasing unions and countable decreasing intersections [Halmos, p. 27, Thm. B]

The **Borel field** of G is the σ-algebra generated by all closed (= compact) subsets of G. We denote it by $\mathcal{B}$ or also by $\mathcal{B}(G)$ if reference to G is needed.

Consider a function $\mu \colon \mathcal{B} \to \mathbb{R}$ satisfying:
(1a) $0 \leq \mu(B) \leq 1$ for each $B \in \mathcal{B}$.
(1b) $\mu(\emptyset) = 0$ and $\mu(G) = 1$.
(1c) Let $B_1, B_2, B_3, \ldots$ be pairwise disjoint Borel sets. Then $\mu(\bigcup_{i=1}^\infty B_i) = \sum_{i=1}^\infty \mu(B_i)$ (σ- **additivity**).

(1d) If $B \in \mathcal{B}$ and $g \in G$, then $\mu(gB) = \mu(Bg) = \mu(B)$ (**translation invariance**).

(1e) For all $B \in \mathcal{B}$ and each $\varepsilon > 0$ there are an open set U and a closed set C satisfying $C \subseteq B \subseteq U$ and $\mu(U \smallsetminus C) < \varepsilon$ (**regularity**).

Condition (1) has the following immediate consequences:

(2a) Let $A \subseteq B$ be Borel sets. Then $\mu(B \smallsetminus A) = \mu(B) - \mu(A)$. In particular, $\mu(G \smallsetminus A) = 1 - \mu(A)$.

(2b) Let $B_1, B_2, B_3, \ldots$ be Borel sets. Then $\mu(\bigcup_{i=1}^{\infty} B_i) \leq \sum_{i=1}^{\infty} \mu(B_i)$. (Write $B'_n = B_n \smallsetminus \bigcup_{i=1}^{n-1} B_i$ and apply (1c).)

(2c) Let $B_1, B_2, B_3, \ldots$ be Borel sets satisfying $\mu(B_i) = 0$, $i = 1, 2, 3, \ldots$. Then $\mu(\bigcup_{i=1}^{\infty} B_i) = 0$ (use 2b).

(2d) Let $B_1, B_2, B_3, \ldots$ be Borel sets satisfying $\mu(B_i) = 1$, $i = 1, 2, 3, \ldots$. Then $\mu(\bigcap_{i=1}^{\infty} B_i) = 1$ (use (2a) and (2c)).

(2e) Suppose $A_1 \subseteq A_2 \subseteq A_3 \subseteq \cdots$ is an increasing sequence of Borel sets. Then $\mu(\bigcup_{i=1}^{\infty} A_i) = \lim_{i \to \infty} \mu(A_i)$ (use (1c)).

(2f) Suppose $A_1 \supseteq A_2 \supseteq A_3 \supseteq \cdots$ is a decreasing sequence of Borel sets. Then $\mu(\bigcap_{i=1}^{\infty} A_i) = \lim_{i \to \infty} \mu(A_i)$ (take complements and use (2a) and (2e)).

(2g) Suppose H is an open subgroup of G of index n. Then $\mu(H) = \frac{1}{n}$ (write $G = \bigcup_{i=1}^{n} g_i H$ and apply (1c) and (1d)). □

LEMMA 18.1.1:
(a) If H is a closed subgroup of G, then $\mu(H) = \frac{1}{(G:H)}$.
(b) Let N be an open normal subgroup of G, $\bar{A}$ a subset of G/N, and $A = \{g \in G \mid gN \in \bar{A}\}$. Then $\mu(A) = \frac{|\bar{A}|}{(G:N)}$.
(c) Finally, if U is a nonempty open subset of G, then $\mu(U) > 0$.

Proof of (a): First suppose $(G : H) = n$ is finite and write $G = \bigcup_{i=1}^{n} g_i H$. By (1), $1 = \mu(G) = \sum_{i=1}^{n} \mu(g_i H) = n\mu(H)$. Hence, $\mu(H) = \frac{1}{n}$, as claimed.

Now assume H has infinite index. Then H is contained in the intersection of a decreasing sequence of open subgroups, $H_1 > H_2 > \ldots$. Thus, $\mu(H) \leq \lim_{i \to \infty} \frac{1}{(G:H_i)} = 0$.

Proof of (b): Use (a) and (1d).

Proof of (c): By definition, U contains a coset gH, where H is an open subgroup of G. By (a) and (1d), $\mu(U) \geq \mu(gH) = \mu(H) = \frac{1}{(G:H)} > 0$. □

Definition 18.1.2: Zero sets. Call a subset A of G a **zero set** (with respect to $(\mathcal{B}, \mu)$) if $A \subseteq B \in \mathcal{B}$ with $\mu(B) = 0$. □

Let $\hat{\mathcal{B}}$ be the σ-algebra of G generated by $\mathcal{B}$ and the zero sets. Write $\hat{\mathcal{B}}_\mu$, $\hat{\mathcal{B}}(G)$, or $\hat{\mathcal{B}}_\mu(G)$ if a reference to μ or G is needed. Each set $\hat{B}$ of $\hat{\mathcal{B}}$ is the union of a set $B \in \mathcal{B}$ and a zero set. Extend μ to a function $\hat{\mu}: \hat{\mathcal{B}} \to \mathbb{R}$ by defining $\hat{\mu}(\hat{B})$ to be $\mu(B)$. The extended measure $\hat{\mu}$ has properties (1) and (2) with $\mathcal{B}$ replaced by $\hat{\mathcal{B}}$ and also the following:

18.1 The Haar Measure of a Profinite Group

(3a) $\hat{\mathcal{B}}$ contains all zero sets with respect to $(\hat{\mathcal{B}}, \hat{\mu})$ ($\hat{\mu}$ is **complete**).

(3b) For each $B \in \hat{\mathcal{B}}$ there are $B_1, B_2 \in \mathcal{B}$ with $B_1 \subseteq B \subseteq B_2$ and $\mu(B_2 \smallsetminus B_1) = 0$ (Take $B_1 = \bigcup_{n=1}^\infty B_{1,n}$ and $B_2 = \bigcap_{n=1}^\infty B_{2,n}$ with $B_{1,n}, B_{2,n} \in \mathcal{B}$ satisfying $B_{1,n} \subseteq B_{1,n+1} \subseteq B \subseteq B_{2,n+1} \subseteq B_{2,n}$ and $\mu(B_{2,n} \smallsetminus B_{1,n}) < \frac{1}{n}$.)

The following result shows that $\hat{\mu}$ is determined by its values on open-closed sets. A compactness argument shows that each open-closed set is a union of finitely many left cosets of open normal subgroups of G. The latter union can be made disjoint. Hence, $\hat{\mu}$ is unique.

PROPOSITION 18.1.3: *Let $\mathcal{B}_0$ be the Boolean algebra of open-closed sets in G and $\mathcal{B}_1$ the σ-algebra generated by $\mathcal{B}_0$. Then:*
(a) *For every $U \in \hat{\mathcal{B}}_\mu$ there exist $A, B \in \mathcal{B}_1$ with $A \subseteq U \subseteq B$ and $\mu(B \smallsetminus A) = 0$.*
(b) *Suppose μ and ν are functions from $\mathcal{B}$ to $\mathbb{R}$ satisfying (1). Then $(\hat{\mathcal{B}}_\mu, \hat{\mu}) = (\hat{\mathcal{B}}_\nu, \hat{\nu})$.*

Proof of (a): It suffices to prove (a) for U open. We write $U = \bigcup_{i \in I} x_i M_i$ with open normal subgroups M_i of G and $x_i \in G$ and let α be the supremum of all $\mu(\bigcup_{i \in I'} x_i M_i)$ where I' is a countable subset of I. For each n we choose a countable subset J_n of I such that $\alpha - \mu(\bigcup_{i \in J_n} x_i M_i) < \frac{1}{n}$. Let $J = \bigcup_{n=1}^\infty J_n$. Then $A = \bigcup_{j \in J} x_j M_j \subseteq U$, $A \in \mathcal{B}_1$, and $\mu(A) = \alpha$.

Now we consider the closed normal subgroup $N = \bigcap_{j \in J} M_j$ of G. Then G/N has a countable basis for its topology. Let $\pi \colon G \to G/N$ be the quotient map. The sets $\pi(x_i M) = x_i M_i N/N$ are open in G/N and their union is $\pi(U)$. By a lemma of Lindelöf [Hocking-Young, p. 66], I has a countable subset K such that $\pi(U) = \bigcup_{k \in K} \pi(x_k M_k)$. In addition, $\pi^{-1}(\pi(A)) = A$ and $\pi(U \smallsetminus A) = \pi(U) \smallsetminus \pi(A)$. Hence,

$$U \smallsetminus A \subseteq \pi^{-1}(\pi(U \smallsetminus A)) = \bigcup_{k \in K} (x_k M_k N \smallsetminus A).$$

The right hand side of this equality, which we denote by B_0, belongs to $\mathcal{B}_1$. If we prove that $\mu(B_0) = 0$, then $B = A \cup B_0 \in \mathcal{B}_1$ and $\mu(B \smallsetminus A) = 0$, as needed. It suffices to prove that $\mu(x_k M_k N \smallsetminus A) = 0$ for each $k \in K$.

Assume there exists $k \in K$ with $\mu(x_k M_k N \smallsetminus A) > 0$. Write $N = \bigcup_{\rho=1}^r (N \cap M_k) n_\rho$. Then $x_k M_k N \smallsetminus A = \bigcup_{\rho=1}^r (x_k M_k n_\rho \smallsetminus A)$, so there exists ρ with $\mu(x_k M_k n_\rho \smallsetminus A) > 0$. Note that $An_\rho = A$, because $n_\rho \in M_j$ for each $j \in J$. Hence, $\mu(x_k M_k \smallsetminus A) = \mu((x_k M_k \smallsetminus A)n_\rho) = \mu(x_k M_k n_\rho \smallsetminus A) > 0$. It follows that $\mu\left(\bigcup_{j \in J} x_j M_j \cup x_k M_k\right) = \mu(A) + \mu(x_k M_k \smallsetminus A) > \alpha$. This contradiction to the definition of α concludes the proof of (a).

Proof of (b): Denote the collection of all Borel sets on which μ and ν coincide by $\mathcal{B}_2$. Lemma 18.1.1 implies each open subgroup of G belongs to $\mathcal{B}_2$. Hence, μ and ν coincide on all sets $\bigcup_{i=1}^r g_i N$ with N open normal in G. In other words, μ and ν coincide on $\mathcal{B}_0$. By (2e) and (2f), $\mathcal{B}_2$ is monotone. It follows that μ and ν coincide on $\mathcal{B}_1$.

Consider now a closed subset C of G. The proof of (a) gives a set $D \in \mathcal{B}_1$ with $C \subseteq D$ and $\mu(C) = \mu(D)$. By the preceding paragraph, $\nu(D) = \mu(D)$. Thus, $\nu(C) \leq \nu(D) = \mu(C)$. By symmetry, $\mu(C) \leq \nu(C)$. Hence, $\mu(C) = \nu(C)$, so $C \in \mathcal{B}_2$. It follows that $\mathcal{B} = \mathcal{B}_2$. In particular, $\mu(B) = 0$ if and only if $\nu(B) = 0$ for each $B \in \mathcal{B}$. Hence, a subset A of G is a zero set with respect to $(\mathcal{B}, \mu)$ if and only if A is a zero set with respect to $(\mathcal{B}, \nu)$. Therefore, $\hat{\mathcal{B}}_\mu = \hat{\mathcal{B}}_\nu$ and $\hat{\mu} = \hat{\nu}$. □

Call each set in $\hat{\mathcal{B}}$ **measurable**.

Example 18.1.4: Haar measure on finite groups. Let G be a finite group equipped with the discrete topology. Then each subset B of G is open and $\mu(B) = \frac{|B|}{|G|}$ is the unique Haar measure of G. □

Example 18.1.5: Infinite profinite groups have nonmeasurable subsets. If G is an infinite profinite group, it has a countable abstract subgroup H. Let R be a set of representatives for the left cosets of G modulo H. Then $G = \bigcup_{h \in H} hR$. Assume R is measurable. Then $1 = \sum_{h \in H} \mu(hR)$. Since $\mu(hR) = \mu(R)$ for each $h \in H$, this is impossible. Therefore, R is nonmeasurable. □

18.2 Existence of the Haar Measure

Proposition 18.1.3 gives the uniqueness of the Haar measure on a profinite group. The definition of μ on the Boolean algebra $\mathcal{B}_0$ of open-closed sets in G is a straightforward application of the invariance property of μ. A theorem of Caratheodory [Halmos, p. 42, Thm. A] then extends μ to the σ-algebra $\hat{\mathcal{B}}_0$ generated by $\mathcal{B}_0$. If $\mathrm{rank}(G) \leq \aleph_0$, $\hat{\mathcal{B}}_0$ contains every open set and therefore every Borel set of G. In the general case one has to work harder. The proof that we give here for arbitrary G avoids Caratheodory's theorem.

PROPOSITION 18.2.1: *Every profinite group G has a unique Haar measure.*

Proof: Proposition 18.1.3 asserts the uniqueness of μ. The proof of the existence of μ divides into five parts. It uses the notation of Proposition 18.1.3.

PART A: *A measure on $\mathcal{B}_0$.* Represent $B \in \mathcal{B}_0$ as $B = \bigcup_{i=1}^m g_i N$ with N an open normal subgroup. Define $\mu_0(B)$ to be $\frac{m}{(G:N)}$. We have to show that this is independent of the choice of N. Suppose N' is an open normal subgroup of G contained in N. Put $n = (N:N')$. Then $N = \bigcup_{j=1}^n h_j N'$ and $B = \bigcup_{i=1}^m \bigcup_{j=1}^n g_i h_j N'$. The computation of $\mu_0(B)$ with respect to N' gives $\frac{mn}{(G:N')} = \frac{m}{(G:N)}$. Thus, $\mu_0(B)$ is well defined. By definition, $\mu_0(G) = 1$. Finite additivity and translation invariance of μ_0 on $\mathcal{B}_0$ are easy exercises.

18.2 Existence of the Haar Measure

PART B: *Extension of μ_0 to open sets.* For each open subset U of G let

$$\mu_1(U) = \sup(\mu_0(A) \mid A \in \mathcal{B}_0 \text{ and } A \subseteq U).$$

The function μ_1 extends μ_0 and has these properties:
(1a) U_1 and U_2 open $\implies \mu_1(U_1 \cup U_2) \leq \mu_1(U_1) + \mu_1(U_2)$.
(1b) U_1 and U_2 open disjoint $\implies \mu_1(U_1 \cup U_2) = \mu_1(U_1) + \mu_1(U_2)$.
(1c) $U_1 \subseteq U_2$ open $\implies \mu_1(U_1) \leq \mu_1(U_2)$.
(1d) U_i is open, $i = 1, 2, 3, \ldots \implies \mu_1(\bigcup_{i=1}^{\infty} U_i) \leq \sum_{i=1}^{\infty} \mu_1(U_i)$.
(1e) U open and $x \in G \implies \mu_1(xU) = \mu_1(Ux) = \mu_1(U)$.

To prove (1a), denote $U_1 \cup U_2$ by U. For $\varepsilon > 0$ let $A \in \mathcal{B}_0$ with $A \subseteq U$ and $\mu_1(U) < \mu_0(A) + \varepsilon$. The open set U_i is a union of sets in $\mathcal{B}_0$, $i = 1, 2$. Since A is compact, A is contained in a union $B_1 \cup \cdots \cup B_n$ such that for each i, $B_i \in \mathcal{B}_0$ and $B_i \subseteq U_1$ or $B_i \in \mathcal{B}_0$ and $B_i \subseteq U_2$. Let C_1 be the union of all B_i which are contained in U_1. Let C_2 be the union of all B_i which are contained in U_2. Then $C_1 \cup C_2 = B_1 \cup \cdots \cup B_n$ and

$$\mu_1(U) - \varepsilon < \mu_0(A) \leq \mu_0(C_1 \cup C_2) \leq \mu_0(C_1) + \mu_0(C_2) \leq \mu_1(U_1) + \mu_1(U_2).$$

Since this inequality holds for each $\varepsilon > 0$, $\mu_1(U) \leq \mu_1(U_1) + \mu_1(U_2)$.

To prove (1b), let $\varepsilon > 0$. Choose $A_i \in \mathcal{B}_0$, $A_i \subseteq U_i$ such that $\mu_1(U_i) < \mu_0(A_i) + \frac{\varepsilon}{2}$, $i = 1, 2$. Since μ_0 is additive,

$$\mu_1(U_1) + \mu_1(U_2) < \mu_0(A_1) + \mu_0(A_2) + \varepsilon = \mu_0(A_1 \cup A_2) + \varepsilon \leq \mu_1(U_1 \cup U_2) + \varepsilon.$$

Hence, $\mu_1(U_1) + \mu_1(U_2) \leq \mu_1(U_1 \cup U_2)$. Combining this with (1a) gives (1b).

For (1c), let $\varepsilon > 0$ and choose $A \in \mathcal{B}_0$ such that $A \subseteq U_1$ and $\mu_1(U_1) < \mu_0(A) + \varepsilon$. Since $A \subseteq U_2$, $\mu_1(U_1) < \mu_1(U_2) + \varepsilon$. Hence, $\mu_1(U_1) \leq \mu_1(U_2)$.

To prove (1d), let $U = \bigcup_{i=1}^{\infty} U_i$ and let $\varepsilon > 0$. Choose $A \in \mathcal{B}_0$ with $A \subseteq U$ and $\mu_1(U) < \mu_0(A) + \varepsilon$. Since A is compact, there exists n with $A \subseteq \bigcup_{i=1}^{n} U_i$. Hence, by (1c) and (1a),

$$\mu_1(U) \leq \mu_1(\bigcup_{i=1}^{n} U_i) + \varepsilon \leq \sum_{i=1}^{n} \mu_1(U_i) + \varepsilon \leq \sum_{i=1}^{\infty} \mu_1(U_i) + \varepsilon,$$

and (1d) follows.

Finally, multiplication from the left (resp. right) with an element $x \in G$ is a homeomorphism of G onto itself. Therefore, (1e) follows from the translation invariance of μ_0.

PART C: *An outer measure.* For each $E \subseteq G$ let

$$\mu_2(E) = \inf(\mu_1(U) \mid U \text{ open and } E \subseteq U).$$

The function μ_2 extends μ_1 and has these properties:
(2a) $E_1 \subseteq E_2 \implies \mu_2(E_1) \leq \mu_2(E_2)$.

(2b) $\mu_2(\bigcup_{i=1}^{\infty} E_i) \leq \sum_{i=1}^{\infty} \mu_2(E_i)$.
(2c) $\mu_2(xE) = \mu_2(Ex) = \mu_2(E)$, $x \in G$.

Both (2a) and (2c) are easy exercises. For (2b), let $\varepsilon > 0$. Choose U_i open such that $E_i \subseteq U_i$ and $\mu_1(U_i) \leq \mu_2(E_i) + \frac{\varepsilon}{2^i}$. By (1d) and (2a)

$$\mu_2(\bigcup_{i=1}^{\infty} E_i) \leq \mu_1(\bigcup_{i=1}^{\infty} U_i) \leq \sum_{i=1}^{\infty} \mu_1(U_i) \leq \sum_{i=1}^{\infty} \mu_2(E_i) + \varepsilon,$$

and (2b) follows.

PART D: *Measurable sets.* Call a subset E of G **measurable** if for each $A \subseteq G$ we have $\mu_2(A) = \mu_2(A \cap E) + \mu_2(A \smallsetminus E)$.

The following rules hold:
(3a) E is measurable $\Longrightarrow$ $G \smallsetminus E$ is measurable.
(3b) E_1 and E_2 are measurable $\Longrightarrow$ $E_1 \cup E_2$ is measurable.
(3c) $\mu_2(E) = 0 \Longrightarrow E$ is measurable.
(3d) $E_1, \ldots, E_n$ are measurable and mutually disjoint $\Longrightarrow \mu_2(A \cap \bigcup_{i=1}^{n} E_i) = \sum_{i=1}^{n} \mu_2(A \cap E_i)$ for each $A \subseteq G$.
(3e) $E_1, E_2, E_3, \ldots$ are measurable and mutually disjoint $\Longrightarrow E = \bigcup_{i=1}^{\infty} E_i$ is measurable and $\mu_2(E) = \sum_{i=1}^{\infty} \mu_2(E_i)$.
(3f) Each $B \in \mathcal{B}_0$ is measurable.
(3g) Each open set is measurable.
(3h) E is measurable and $x \in G \Longrightarrow xE$ and Ex are measurable.

Rule (3a) follows from the definition. To prove (3b), let $A \subseteq G$. By (2b)

$$\mu_2(A) \leq \mu_2(A \cap (E_1 \cup E_2)) + \mu_2(A \smallsetminus (E_1 \cup E_2))$$
$$= \mu_2((A \cap E_1) \cup ((A \smallsetminus E_1) \cap E_2)) + \mu_2((A \smallsetminus E_1) \smallsetminus E_2)$$
$$\leq \mu_2(A \cap E_1) + \mu_2((A \smallsetminus E_1) \cap E_2) + \mu_2((A \smallsetminus E_1) \smallsetminus E_2)$$
$$= \mu_2(A \cap E_1) + \mu_2(A \smallsetminus E_1) = \mu_2(A).$$

Thus, these inequalities are in fact equalities. Consequently, $E_1 \cup E_2$ is measurable.

To prove (3c), let $A \subseteq G$. By (2a) and (2b), $\mu_2(A) \leq \mu_2(A \cap E) + \mu_2(A \smallsetminus E) \leq \mu_2(E) + \mu_2(A) = \mu_2(A)$. Hence, $\mu_2(A) = \mu_2(A \cap E) + \mu_2(A \smallsetminus E)$. Thus, E is measurable.

The proof of (3d) is done by induction. It is trivial for $n = 1$. Assume it holds for $n - 1$. Then

$$\mu_2\Big(A \cap \bigcup_{i=1}^{n} E_i\Big) = \mu_2\Big((A \cap \bigcup_{i=1}^{n} E_i) \cap E_n\Big) + \mu_2\Big((A \cap \bigcup_{i=1}^{n} E_i) \smallsetminus E_n\Big)$$
$$= \mu_2(A \cap E_n) + \mu_2\Big(A \cap \bigcup_{i=1}^{n-1} E_i\Big)$$
$$= \mu_2(A \cap E_n) + \sum_{i=1}^{n-1} \mu_2(A \cap E_i).$$

18.2 Existence of the Haar Measure

For (3e), (3b) implies that $F_n = \bigcup_{i=1}^n E_i$ is measurable, $n = 1, 2, 3, \ldots$. Hence, for each $A \subseteq G$, (3d) implies

$$\mu_2(A) = \mu_2(A \cap F_n) + \mu_2(A \smallsetminus F_n) \geq \mu_2(A \cap F_n) + \mu_2(A \smallsetminus E)$$
$$= \sum_{i=1}^n \mu_2(A \cap E_i) + \mu_2(A \smallsetminus E).$$

Therefore, by (2b),

$$\mu_2(A) \geq \sum_{i=1}^\infty \mu_2(A \cap E_i) + \mu_2(A \smallsetminus E) \geq \mu_2(A \cap E) + \mu_2(A \smallsetminus E) \geq \mu_2(A).$$

It follows that E is measurable.

By (3d), $\mu_2(E) \geq \mu_2(\bigcup_{i=1}^n E_i) = \sum_{i=1}^n \mu_2(E_i)$. Hence, by (2b), $\mu_2(E) = \sum_{i=1}^\infty \mu_2(E_i)$.

To prove (3f), let $A \subseteq G$ and U be an open set that contains A. Then $U \cap B$ and $U \smallsetminus B$ are open disjoint sets. By (2a) and (1b)

$$\mu_2(A \cap B) + \mu_2(A \smallsetminus B) \leq \mu_2(U \cap B) + \mu_2(U \smallsetminus B) = \mu_1(U).$$

Take the infimum on all U to conclude that $\mu_2(A) \leq \mu_2(A \cap B) + \mu_2(A \smallsetminus B) \leq \mu_2(A)$. Hence, B is measurable.

Now we prove (3g): Let U be an open set. Consider $A \subseteq G$ and $\varepsilon > 0$. Since $\mu_2(U) = \mu_1(U)$ (Part C), there is a $B \in \mathcal{B}_0$ with $B \subseteq U$ and $\mu_2(U) - \mu_2(B) < \varepsilon$. Since B and $U \smallsetminus B$ are disjoint open sets, (1b) implies $\mu_2(U) = \mu_2(B) + \mu_2(U \smallsetminus B)$. Hence, $\mu_2(U \smallsetminus B) < \varepsilon$. By (3f), $\mu_2(A) = \mu_2(A \cap B) + \mu_2(A \smallsetminus B)$. Therefore,

$$\mu_2(A) \leq \mu_2(A \cap U) + \mu_2(A \smallsetminus U)$$
$$\leq \mu_2(A \cap (U \smallsetminus B)) + \mu_2(A \cap B) + \mu_2(A \smallsetminus B)$$
$$\leq \mu_2(U \smallsetminus B) + \mu_2(A) < \varepsilon + \mu_2(A)$$

It follows that $\mu_2(A) = \mu_2(A \cap U) + \mu_2(A \smallsetminus U)$. Thus, U is measurable.

PART E: *The Baire field $\hat{\mathcal{B}}$.* Let $\hat{\mathcal{B}}$ be the collection of all measurable sets. By (3a), (3b) and (3e), $\hat{\mathcal{B}}$ is a σ-algebra. Since $\hat{\mathcal{B}}$ contains each open set, it contains each Borel set. Denote the restriction of μ_2 to $\hat{\mathcal{B}}$ by μ. By (3e), (3h), and (2c), μ is a σ-additive invariant measure. For each $E \in \hat{\mathcal{B}}$ and every $\varepsilon > 0$ there is an open set U with $E \subseteq U$ and $\mu_2(U \smallsetminus E) < \varepsilon$. Apply this to $G \smallsetminus E$ to conclude the existence of a closed set $C \subseteq E$ such that $\mu(E \smallsetminus C) < \varepsilon$. Thus, μ is regular. In particular, each $E \in \hat{\mathcal{B}}$ is contained in a Borel set $F = \bigcap_{i=1}^\infty U_i$ with U_i open, $i = 1, 2, 3, \ldots$ such that $\mu(F \smallsetminus E) = 0$. Combining this with (3c), we conclude that μ is complete on $\hat{\mathcal{B}}$ and $(\hat{\mathcal{B}}, \mu)$ is the completion of the Borel field $\mathcal{B}$ with respect to the restriction of μ to $\mathcal{B}$. Thus, μ is the desired Haar measure of G. $\square$

PROPOSITION 18.2.2: *Let $\pi: G \to H$ be an epimorphism of profinite groups and μ_G, μ_H the corresponding Haar measures. Then $\mu_H(B) = \mu_G(\pi^{-1}(B))$ for each measurable subset B of H.*

Proof: The map $B \mapsto \pi^{-1}(B)$ maps closed subgroups of H onto closed subgroups of G. In addition, it commutes with complements and unions. Hence, it maps $\mathcal{B}(H)$ into $\mathcal{B}(G)$.

The function $\mu_G \circ \pi^{-1}$ satisfies Condition (1) of Section 18.1 for $B \in \mathcal{B}(H)$. This is clear for (1a)-(1d) of Section 18.1. To prove (1e) of Section 18.1, consider $B \in \mathcal{B}(H)$ and $\varepsilon > 0$. Then G has a closed subset C with $C \subseteq \pi^{-1}(B)$ and $\mu_G(\pi^{-1}(B) \smallsetminus C) < \varepsilon$. Then $\pi(C)$ is closed in H and $\pi(C) \subseteq B$. Thus, $\pi^{-1}(\pi(C)) \subseteq \pi^{-1}(B)$ and $\mu_G(\pi^{-1}(B \smallsetminus \pi(C))) \leq \mu_G(\pi^{-1}(B) \smallsetminus C) < \varepsilon$. Applying this result to $H \smallsetminus B$ and taking complements, we get an open subset U of H with $B \subseteq U$ and $\mu_G \circ \pi^{-1}(U \smallsetminus B) < \varepsilon$.

By Proposition 18.1.3, the completion of $(\mathcal{B}(H), \mu_G \circ \pi^{-1})$ coincides with $(\hat{\mathcal{B}}(H), \mu_H)$.

It remains to prove that $\mu_H(B) = \mu_G(\pi^{-1}(B))$ for each $B \in \hat{\mathcal{B}}(H)$. If A is a μ_H-zero set, then $A \subseteq A_1$ for some $A_1 \in \mathcal{B}(H)$ with $\mu_H(A_1) = 0$. Then $\mu_G(\pi^{-1}(A)) \leq \mu_G(\pi^{-1}(A_1)) = \mu_H(A_1) = 0$, so $\mu_G(\pi^{-1}(A)) = 0$. An arbitrary $B \in \hat{\mathcal{B}}(H)$ differs from a set in $\mathcal{B}(H)$ by a μ_H-zero set. Hence, $\mu_H(B) = \mu_G(\pi^{-1}(B))$ holds for B. □

Example 18.2.3: Finite quotients. Let π be an epimorphism of a profinite group G onto a finite group H. Then, by Example 18.1.4 and Proposition 18.2.2, $\mu_G(\pi^{-1}(B)) = \frac{|B|}{|H|}$ for each subset B of H. □

PROPOSITION 18.2.4: *Let H be an open subgroup of a profinite group G. Then $\mu_H(B) = (G : H)\mu_G(B)$ for each $B \in \hat{\mathcal{B}}(H)$.*

Proof: The restriction of $(G : H)\mu_G$ to $\hat{\mathcal{B}}(H)$ is a Haar measure of H, so it coincides with μ_H. □

18.3 Independence

Now that we have a unique normalized Haar measure μ on G, we may regard G as a probability space. Recall that a family $\{A_i \mid i \in I\}$ of measurable subsets of G is μ-**independent** if $\mu(\bigcap_{i \in J} A_i) = \prod_{i \in J} \mu(A_i)$ for each finite subset J of I.

LEMMA 18.3.1: *The following hold for measurable subsets $A_1, \ldots, A_n$ of G:*
(a) $\mu(\bigcup_{i=1}^n A_i) = \sum_{k=1}^n (-1)^{k-1} \sum_{1 \leq i_1 < \cdots < i_k \leq n} \mu(A_{i_1} \cap \cdots \cap A_{i_k})$
(inclusion-exclusion principle).
(b) *Suppose $A_1, \ldots, A_n$ are μ-independent. Then*

$$(1) \quad \mu\Big(\bigcup_{i=1}^n A_i\Big) = \sum_{k=1}^n (-1)^{k-1} \sum_{1 \leq i_1 < \cdots < i_k \leq n} \mu(A_{i_1}) \cdots \mu(A_{i_k})$$

18.3 Independence

and $G \smallsetminus A_1, \ldots, G \smallsetminus A_n$ are μ-independent.

Proof of (a): The principle is trivial for $n = 1$. For $n = 2$ it follows from the identity $A_1 \cup A_2 = (A_1 \smallsetminus A_1 \cap A_2) \cup A_2$. Suppose it holds for $n - 1$. Then

$$\mu(\bigcup_{i=1}^{n} A_i) = \mu(\bigcup_{i=1}^{n-1} A_i \cup A_n)$$

$$= \mu(\bigcup_{i=1}^{n-1} A_i) + \mu(A_n) - \mu(\bigcup_{i=1}^{n-1} A_i \cap A_n)$$

$$= \sum_{k=1}^{n-1} (-1)^{k-1} \sum_{1 \leq i_1 < \cdots < i_k \leq n-1} \mu(A_{i_1} \cap \cdots \cap A_{i_k}) + \mu(A_n)$$

$$- \sum_{k=1}^{n-1} (-1)^{k-1} \sum_{1 \leq i_1 < \cdots < i_k \leq n-1} \mu(A_{i_1} \cap \cdots \cap A_{i_k} \cap A_n)$$

$$= \sum_{k=1}^{n} (-1)^{k-1} \sum_{1 \leq i_1 < \cdots < i_k \leq n} \mu(A_{i_1} \cap \cdots \cap A_{i_k})$$

Proof of (b): Since $\mu(A_{i_1} \cap \cdots \cap A_{i_k}) = \mu(A_{i_1}) \cdots \mu(A_{i_k})$, (1) follows from (a).

It remains to prove that $G \smallsetminus A_1, \ldots, G \smallsetminus A_n$ are independent. By symmetry, it suffices to prove the independence condition for $G \smallsetminus A_1, \ldots, G \smallsetminus A_m$, for all $m \leq n$. Indeed,

$$\mu(\bigcap_{i=1}^{m} G \smallsetminus A_i) = \mu(G \smallsetminus \bigcup_{i=1}^{m} A_i) = 1 - \mu(\bigcup_{i=1}^{n} A_i)$$

$$= \sum_{k=0}^{m} (-1)^k \sum_{1 \leq i_1 < \cdots < i_k \leq m} \mu(A_{i_1}) \cdots \mu(A_{i_k})$$

$$= \prod_{i=1}^{m} (1 - \mu(A_i)) = \prod_{i=1}^{m} \mu(G \smallsetminus A_i).$$

Consequently, $G \smallsetminus A_1, \ldots, G \smallsetminus A_m$ are μ-independent. $\square$

Let $a_1, a_2, a_3, \ldots$ be a sequence of nonzero real numbers. Define $\prod_{i=1}^{\infty} a_i$ to be $\lim_{n \to \infty} \prod_{i=1}^{n} a_i$, if the limit exists.

LEMMA 18.3.2: Suppose $0 < a_i < 1$, $i = 1, 2, 3, \ldots$. Then $\prod_{i=1}^{\infty} (1 - a_i) = 0$ if and only if $\sum_{i=1}^{\infty} a_i = \infty$.

Proof: Consider $0 < x < 1$. Then $\log(1 - x) = \sum_{i=1}^{\infty} -\frac{x^i}{i} < -x$. On the other hand, $\log(1 - x) \geq -\sum_{i=1}^{\infty} x^i = -\frac{x}{1-x}$.

Now suppose that $\sum_{i=1}^{\infty} a_i = \infty$. Then

$$\log \prod_{i=1}^{n}(1-a_i) = \sum_{i=1}^{n}\log(1-a_i) < -\sum_{i=1}^{n} a_i.$$

Hence, $\log\lim_{n\to\infty}\prod_{i=1}^{n}(1-a_i) = \lim_{n\to\infty}\log\prod_{i=1}^{n}(1-a_i) = -\infty$. Therefore, $\prod_{i=1}^{\infty}(1-a_i) = 0$.

Conversely, suppose $\sum_{i=1}^{\infty} a_i < \infty$. Then $\lim_{i\to\infty} a_i = 0$. Hence, there is a $c > 0$ with $1 - a_i > c$, $i = 1, 2, \ldots$. Thus, $\log(1-a_i) \geq -\frac{a_i}{1-a_i} \geq -\frac{a_i}{c}$, $i = 1, 2 \ldots$. Therefore, $\log\prod_{i=1}^{\infty}(1-a_i) \geq -\frac{1}{c}\sum_{i=1}^{\infty} a_i > -\infty$. Consequently, $\prod_{i=1}^{\infty}(1-a_i) > 0$. □

Example 18.3.3: Zeta function. The most prominent example of infinite products is the Euler product for the Riemann zeta function:

$$\zeta(s) = \sum_{n=1}^{\infty}\frac{1}{n^s} = \prod_{p}\left(1 - \frac{1}{p^s}\right)^{-1}, \quad s > 1$$

with p ranging over all prime numbers. Here $\sum_{p}\frac{1}{p^s} < \infty$ for $s > 1$. Hence, by Lemma 18.3.2, $\prod_{p}(1 - \frac{1}{p^s}) > 0$. For $s = 1$ we have $\sum \frac{1}{p} = \infty$ [LeVeque, Thm. 6-13] and $\prod(1 - \frac{1}{p}) = 0$. □

LEMMA 18.3.4: *Let $A_1, A_2, A_3, \ldots$ be a μ-independent sequence of measurable subsets of a profinite group G. If $\sum_{i=1}^{\infty}\mu(A_i) = \infty$, then $\mu(\bigcup_{i=1}^{\infty} A_i) = 1$.*

Proof: By Lemma 18.3.2, $\prod_{i=1}^{\infty}(1 - \mu(A_i)) = 0$. By Lemma 18.3.1, $G \smallsetminus A_1, G \smallsetminus A_2, \ldots$ are μ-independent. Hence,

$$\mu(G \smallsetminus \bigcup_{i=1}^{\infty} A_i) = \mu(\bigcap_{i=1}^{\infty}(G \smallsetminus A_i)) = \lim_{n\to\infty}\mu(\bigcap_{i=1}^{n} G \smallsetminus A_i)$$

$$= \lim_{n\to\infty}\prod_{i=1}^{n}(1 - \mu(A_i)) = \prod_{i=1}^{\infty}(1 - \mu(A_i)) = 0.$$

Thus, $\mu(\bigcup_{i=1}^{\infty} A_i) = 1$. □

Of fundamental importance is the following strengthening of Lemma 18.3.4:

LEMMA 18.3.5 (Borel-Cantelli): *Let $A_1, A_2, A_3, \ldots$ be a sequence of measurable subsets of a profinite group G. Put*

$$A = \bigcap_{n=1}^{\infty}\bigcup_{i=n}^{\infty} A_i = \{x \in G \mid x \text{ belongs to infinitely many } A_i\text{'s}\}$$

$$A' = \bigcup_{n=1}^{\infty}\bigcap_{i=n}^{\infty} A_i = \{x \in G \mid x \text{ belongs to all but finitely many } A_i\text{'s}\}.$$

18.3 Independence

Then the following hold:
(a) $\sum_{i=1}^{\infty} \mu(A_i) < \infty \implies \mu(A) = 0$.
(b) *Suppose $A_1, A_2, A_3, \ldots$ are μ-independent and $\sum_{i=1}^{\infty} \mu(A_i) = \infty$. Then $\mu(A) = 1$.*
(c) $\prod_{i=1}^{\infty} \mu(A_i) > 0 \implies \mu(A') = 1$.
(d) *Suppose $A_1, A_2, A_3, \ldots$ are μ-independent, $\mu(A_i) > 0$, and $\prod_{i=1}^{\infty} \mu(A_i) = 0$. Then $\mu(A') = 0$.*

Proof: For each positive n, $\mu(A) \leq \mu(\bigcup_{i=n}^{\infty} A_i) \leq \sum_{i=n}^{\infty} \mu(A_i)$. If $\sum_{i=1}^{\infty} \mu(A_i) < \infty$, the right hand side approaches 0 as n tends to infinity. Thus, $\mu(A) = 0$ and (a) follows. By Lemma 18.3.4, the hypotheses of (b) implies $\mu(\bigcup_{i=n}^{\infty} A_i) = 1$ for every positive n. Therefore, $\mu(A) = 1$.

Now suppose $\prod_{i=1}^{\infty} \mu(A_i) > 0$. Then $\prod_{i=1}^{\infty} \left(1 - \mu(G \smallsetminus A_i)\right) > 0$. By Lemma 18.3.2, $\sum_{i=1}^{\infty} \mu(G \smallsetminus A_i) < \infty$. Hence, by (a), almost no $x \in G$ belongs to infinitely many $(G \smallsetminus A_i)$'s. Therefore, almost all $x \in G$ belong to all but finitely many A_i's. This means, $\mu(A') = 1$.

Finally, suppose $A_1, A_2, A_3, \ldots$ are μ-independent, $\mu(A_i) > 0$ for each i, and $\prod_{i=1}^{\infty} \mu(A_i) = 0$. Then $\mu(\bigcap_{i=n}^{\infty} A_i) = \prod_{i=n}^{\infty} \mu(A_i) = 0$ for each n. So, $\mu(A') = 0$. □

Remark 18.3.6:
(a) Chebyshev's inequality [Rényi, p. 391] sharpens (b) of Lemma 18.3.5 by allowing one to assume only that the sets $A_1, A_2, \ldots$ are pairwise μ-independent.

(b) The most frequently used cases of Lemma 18.3.5 are when
(b1) $\mu(A_i)$ is a positive constant, where $\sum_{i=1}^{\infty} \mu(A_i)$ obviously diverges, or
(b2) $\mu(A_p) = \frac{1}{p^e}$ where p ranges over all prime numbers; then $\sum \frac{1}{p^e}$ diverges for $e = 1$ (Example 18.3.3) and converges for $e \geq 2$. □

Sequences $A_1, A_2, \ldots$ with $\mu(A_i) = 1$ (resp. $\mu(A_i) = 0$), $i = 1, 2, \ldots$ are clearly μ-independent. The next lemma ties independence to the group structure on G:

LEMMA 18.3.7: *Open subgroups $H_1, \ldots, H_n$ of a profinite group G are μ-independent if and only if*

(2) $$(G : \bigcap_{i=1}^{n} H_i) = \prod_{i=1}^{n} (G : H_i).$$

Suppose $H_1, \ldots, H_n$ are μ-independent open subgroups of G. For each i let $\bar{A}_i$ be a set of left cosets of H_i in G. Put $A_i = \{g \in G \mid gH_i \in \bar{A}_i\}$. Then $A_1, \ldots, A_n$ are μ-independent and

(3) $$\mu(\bigcap_{i=1}^{n} A_i) = \prod_{i=1}^{n} \frac{|\bar{A}_i|}{(G : H_i)}.$$

374   Chapter 18. The Haar Measure

Proof: Suppose (2) holds and let I be a subset of $\{1, \ldots, n\}$. With no loss assume $I = \{1, \ldots, m\}$ with $m \leq n$. Let $K = \bigcap_{i=1}^m H_i$ and $H = \bigcap_{i=1}^n H_i$. The canonical map $gK \mapsto (gH_1, \ldots, gH_m)$ embeds the coset space G/K into $G/H_1 \times \cdots \times G/H_m$. Hence, $(G : K) \leq \prod_{i=1}^m (G : H_i)$. Similarly,

$$(K : H) \leq \prod_{i=m+1}^n (K : K \cap H_i) \leq \prod_{i=m+1}^n (G : H_i).$$

Combining these with $(G : H) = (G : K)(K : H)$ and (2), we conclude the equality $(G : K) = \prod_{i=1}^m (G : H_i)$. Thus, (2) gives $\mu(\bigcap_{i=1}^m H_i) = \prod_{i=1}^m \mu(H_i)$, as desired. Necessity of (2) for μ-independence is clear.

Now suppose $H_1, \ldots, H_n$ are μ-independent. Put $A = A_1 \cap \cdots \cap A_n$ and $\bar{A} = \{aH \mid a \in A\}$. If $g \in G$ and $gH \in \bar{A}$, then $gH = aH$ for some $a \in A$. Hence, $gH_i \in \bar{A}_i$, so $g \in A_i$, $i = 1, \ldots, n$ and $g \in A$. It follows that $A = \{gH \mid gH \in \bar{A}\}$. List the elements of $\bar{A}$ as $a_1 H, \ldots, a_m H$. Then $A = \bigcup_{i=1}^m a_i H$ and $\mu(A) = |\bar{A}|\mu(H)$.

Next define a map $\alpha \colon G/H \to G/H_1 \times \cdots \times G/H_n$ by

$$\alpha(gH) = (gH_1, \ldots, gH_n).$$

By (2), α is bijective. In particular, α maps $\bar{A}$ bijectively onto $\bar{A}_1 \times \cdots \times \bar{A}_n$, so $|\bar{A}| = |\bar{A}_1| \cdots |\bar{A}_n|$. In addition, $\mu(H) = \mu(H_1) \cdots \mu(H_n)$. Hence, $\mu(A) = |\bar{A}|\mu(H) = \prod_{i=1}^n |\bar{A}_i|\mu(H_i) = \prod_{i=1}^n \mu(A_i)$. Thus, $A_1, \ldots, A_n$ are μ-independent and (3) holds. □

Example 18.3.8: *Relatively prime indices.* Let $H_1, \ldots, H_n$ be open subgroups of a profinite group G. Suppose $(G : H_1), \ldots, (G : H_n)$ are relatively primes in pairs. Then (2) holds. Hence, by Lemma 18.3.7, $H_1, \ldots, H_n$ are μ-independent. □

Part (b) of the following lemma generalizes Lemma 16.8.3(a).

LEMMA 18.3.9: *Let $G = \prod_{i \in I} S_i$ be a direct product of finite non-Abelian simple groups S_i and N a closed normal subgroup of G. For each $i \in I$ let $\pi_i \colon G \to S_i$ be the projection on the ith factor. Then the following hold:*
(a) *$\pi_j(N) = S_j$ if and only if $S_j \leq N$.*
(b) *$N = \prod_{i \in I_0} S_i$, where $I_0 = \{i \in I \mid S_i \leq N\}$. Moreover, $G = N \times N'$ with $N' = \prod_{i \in I \smallsetminus I_0} S_i$.*
(c) *Let α be an automorphism of G with $\alpha(N) = N$. Then $\alpha(N') = N'$.*
(d) *Suppose G is a normal subgroup of a profinite group F and $N \triangleleft F$. Then $N' \triangleleft F$.*

Proof of (a): First suppose $S_j \leq N$. Since π_j is the identity map on S_j, we have $\pi_j(N) = S_j$.

Now suppose $S_j \not\leq N$. Since S_j is simple, $S_j \cap N = 1$. Hence, $[n, s] = 1$ for all $n \in N$ and $s \in S_j$. Therefore, $[\pi_j(n), s] = 1$. Since S_j is non-Abelian, its center is trivial. Consequently, $\pi_j(n) = 1$.

18.3 Independence

Proof of (b): By definition, $\prod_{i\in I_0} S_i \leq N$. Conversely, let $n \in N$. If $j \in I \smallsetminus I_0$, then $\pi_j(n) = 1$, by (a). Hence, $n \in \prod_{i \in I_0} S_i$.

Proof of (c): $S_i \leq N$ if and only if $\alpha(S_i) \leq N$, so $\alpha(N') = \prod_{S_i \not\leq N} \alpha(S_i) = \prod_{S_i \not\leq N} S_i = N'$.

Proof of (d): Apply (c) to conjugation of G by elements of F. □

LEMMA 18.3.10: *Let $G = \prod_{i=1}^r S_i$ be a direct product of r finite simple non-Abelian groups S_i. Then G has exactly r normal subgroups N with G/N simple.*

Proof: Let N be a normal subgroup of G such that G/N is simple. By Lemma 18.3.9, $N = \prod_{i\in I_0} S_i$, where I_0 is a subset of $\{1, 2, \ldots, r\}$. Then $G/N = \prod_{i \in I \smallsetminus I_0} S_i$. Since G/N is simple, $I \smallsetminus I_0 = \{j\}$ for some j between 1 and r. Thus, $N = \prod_{i \neq j} S_i$. Consequently, there are exactly r possibilities for N. □

LEMMA 18.3.11: *Let G be a profinite group and λ an ordinal number. For each $\alpha < \lambda$ let N_α be an open normal subgroup of G. Suppose G/N_α is a simple non-Abelian group and $N_\alpha \neq N_{\alpha'}$ if $\alpha \neq \alpha'$. Then $G/\bigcap_{\alpha<\lambda} N_\alpha \cong \prod_{\alpha<\lambda} G/N_\alpha$ and the N_α's are μ-independent. Moreover, if N is an open subgroup of G containing $\bigcap_{\alpha<\lambda} N_\alpha$ and G/N is simple, then $N = N_\alpha$ for some $\alpha < \lambda$.*

Proof: For each $\gamma \leq \lambda$ let $M_\gamma = \bigcap_{\alpha<\gamma} N_\alpha$. Suppose by transfinite induction for each $\beta < \gamma$ there exists an isomorphism $\varphi_\beta \colon G/M_\beta \to \prod_{\alpha<\beta} G/N_\alpha$ such that the following diagram commutes for all $\beta < \beta' < \gamma$:

$$
(4) \qquad \begin{array}{ccc} G/M_{\beta'} & \xrightarrow{\varphi_{\beta'}} & \prod_{\alpha<\beta'} G/M_\alpha \\ \downarrow & & \downarrow \\ G/M_\beta & \xrightarrow{\varphi_\beta} & \prod_{\alpha<\beta} G/M_\alpha \end{array}
$$

Here the left vertical map is the quotient map and the right vertical map is the projection on the first β coordinates. If γ is a limit ordinal, then $M_\gamma = \bigcap_{\beta<\gamma} M_\beta$. Hence, $G/M_\gamma \cong \varprojlim G/M_\beta \cong \varprojlim \prod_{\alpha<\beta} G/M_\alpha \cong \prod_{\alpha<\gamma} G/M_\alpha$. Moreover, Diagram (4) is commutative with γ replacing β'.

Now suppose $\gamma = \beta + 1$. Let N be an open normal subgroup of G containing M_β such that G/N is simple. By Lemma 1.2.2(a), there are $\alpha_1, \ldots, \alpha_r \leq \beta$ such that $\bigcap_{i=1}^r N_{\alpha_i} \leq N$. By the induction hypothesis, $G/\bigcap_{i=1}^r N_{\alpha_i} \cong \prod_{i=1}^r G/N_{\alpha_i}$. Hence, by Lemma 18.3.10, $N_{\alpha_1}, \ldots, N_{\alpha_r}$ are all open normal subgroups of G containing $\bigcap_{i=1}^r N_{\alpha_i}$ with a simple coquotient. Thus, $N = N_{\alpha_i}$ for some i between 1 and r. It follows that $M_\beta \not\leq N_\beta$. Therefore, $M_\beta N_\beta = G$ (because G/N_β is simple) and $G/M_\gamma \cong G/M_\beta \times G/N_\beta \cong \prod_{\alpha<\gamma} G/M_\alpha$, as claimed.

The μ-independence of the N_α's follows now from Lemma 18.3.7. □

18.4 Cartesian Product of Haar Measures

The direct product $G \times H$ of profinite groups G and H is a profinite group. By Propositions 18.2.1, $G \times H$ has a unique Haar measure $\mu_{G \times H}$. Consider also the **product measure** $\mu_G \times \mu_H$ of $G \times H$ [Halmos, Chapter VII]. It is first defined on sets of the form $A \times B$ with A a μ_G-measurable set and B a μ_H-measurable set (i.e. **measurable rectangles**) by the rule $(\mu_G \times \mu_H)(A \times B) = \mu_G(A)\mu_H(B)$. Then $\mu_G \times \mu_H$ extends to a σ-additive measure on the σ-algebra generated by these rectangles [Halmos, p. 144, Thm. B]. Finally, one completes $\mu_G \times \mu_H$ by adding all zero sets. We prove that $\mu_{G \times H}$ coincides with $\mu_G \times \mu_H$.

Given families $\mathcal{A}$ and $\mathcal{B}$ of subsets of G and H, we denote the set of all pairs (A, B) with $A \in \mathcal{A}$ and $B \in \mathcal{B}$, as usual, by $\mathcal{A} \times \mathcal{B}$. Following [Neveu, p. 71], we denote the σ-algebra of $G \times H$ generated by all of the sets $A \times B$ with $(A, B) \in \mathcal{A} \times \mathcal{B}$ by $\mathcal{A} \otimes \mathcal{B}$. The latter notation is partially justified by the following result.

LEMMA 18.4.1: *Let G and H be sets, $\mathcal{A}$ a collection of subsets of G, and $\mathcal{B}$ a collection of subsets of H. Let $\mathcal{A}'$ be the σ-algebra of G generated by $\mathcal{A}$, $\mathcal{B}'$ the σ-algebra of H generated of $\mathcal{B}$, and $\mathcal{C}$ a σ-algebra of $G \times H$. Suppose $A \times H$ and $G \times B$ are in $\mathcal{C}$ for all $(A, B) \in \mathcal{A} \times \mathcal{B}$. Then $\mathcal{A}' \otimes \mathcal{B}' \subseteq \mathcal{C}$.*

Proof: Let $\mathcal{A}_1 = \{A \subseteq G \mid A \times H \in \mathcal{C}\}$. Then $\mathcal{A}_1$ contains $\mathcal{A}$ and is closed under countable unions and taking complements. Hence, $\mathcal{A}' \subseteq \mathcal{A}_1$. In other words, $A \times H \in \mathcal{C}$ for each $A \in \mathcal{A}'$. Similarly, $G \times B \in \mathcal{C}$ for each $B \in \mathcal{B}'$. It follows that $A \times B = (A \times H) \cap (G \times B) \in \mathcal{C}$ for all $(A, B) \in \mathcal{A}' \times \mathcal{B}'$. Consequently, $\mathcal{A}' \otimes \mathcal{B}' \subseteq \mathcal{C}$. □

PROPOSITION 18.4.2: *The Haar measure on $G \times H$ coincides with $\mu_G \times \mu_H$.*

Proof: Let $\mathcal{B}_0(G)$ be the Boolean algebra of all open-closed subsets of G, $\mathcal{B}_1(G)$ the σ-algebra generated by $\mathcal{B}_0(G)$, $\mathcal{B}(G)$ the Borel field of G, and $\hat{\mathcal{B}}(G)$ the family of all measurable subsets of G. Use the corresponding notation for H and for $G \times H$. In addition let $\mathcal{R}_0$ be the Boolean algebra of $G \times H$ generated by all of the sets $A \times B$ with $(A, B) \in \mathcal{B}_0(G) \times \mathcal{B}_0(H)$, $\mathcal{R}_1 = \mathcal{B}_0(G) \otimes \mathcal{B}_0(H)$, $\mathcal{R} = \mathcal{B}(G) \otimes \mathcal{B}(H)$, and $\hat{\mathcal{R}}$ be the completion of $\mathcal{R}$ with respect to $\mu_G \times \mu_H$. In four parts we prove that $\hat{\mathcal{R}} = \hat{\mathcal{B}}(G \times H)$ and $(\mu_G \times \mu_H)(C) = \mu_{G \times H}(C)$ for each $C \in \hat{\mathcal{R}}$.

PART A: $\mathcal{R}_0 = \mathcal{B}_0(G \times H)$, $\mathcal{R}_1 = \mathcal{B}_1(G \times H)$, *and* $\mathcal{B}_1(G) \otimes \mathcal{B}_1(H) \subseteq \mathcal{R}_1$. If G_0 is an open subgroup of G and H_0 is an open subgroup of H, then $G_0 \times H_0$ is an open subgroup of $G \times H$. Therefore, $\mathcal{R}_0 \subseteq \mathcal{B}_0(G \times H)$.

Conversely, let M be an open subgroup of $G \times H$. Then $M \cap G$ is an open subgroup of G, $M \cap H$ is an open subgroup of H, and $L = (M \cap G) \times (M \cap H) \leq M$. Every union of cosets of M is a union of cosets of L. Thus, $\mathcal{B}_0(G \times H) \subseteq \mathcal{R}_0$.

The combination of the preceding two paragraphs proves that $\mathcal{R}_0 = \mathcal{B}_0(G \times H)$. Closing under countable unions and taking complements, we find that $\mathcal{R}_1 = \mathcal{B}_1(G \times H)$.

18.4 Cartesian Product of Haar Measures

Finally, by Lemma 18.4.1, $\mathcal{B}_1(G) \otimes \mathcal{B}_1(H) \subseteq \mathcal{R}_1$.

PART B: $\mu_G \times \mu_H$ and $\mu_{G \times H}$ coincide on $\mathcal{R}_1$. Let $A \in \mathcal{B}_0(G)$ and $B \in \mathcal{B}_0(H)$. Then, there exist open normal subgroups M and N of G and H, respectively, such that $A = \bigcup_{i=1}^{r} g_i M$ and $B = \bigcup_{j=1}^{s} h_j N$. Hence,

$$\mu_{G \times H}(A \times B) = \mu_{G \times H}(\bigcup_{i,j}(g_i, h_j)M \times N) = \frac{rs}{(G:M)(H:N)}$$

$$= \mu_G(\bigcup_{i=1}^{r} g_i M) \cdot \mu_H(\bigcup_{j=1}^{s} h_j N) = (\mu_G \times \mu_H)(A \times B).$$

It follows that $A \times B$ belongs to the set

$$\mathcal{S} = \{C \in \hat{\mathcal{R}} \cap \hat{\mathcal{B}}(G \times H) \mid (\mu_G \times \mu_H)(C) = \mu_{G \times H}(C)\}.$$

Observe that $\mathcal{S}$ is closed under countable unions and taking complements. Therefore, $\mathcal{R}_1 \subseteq \mathcal{S}$. In addition, if $E_0 \subseteq E_1$, $E_1 \in \mathcal{S}$, and $(\mu_G \times \mu_H)(E_1) = 0$, then $E_0 \in \mathcal{S}$.

PART C: $\mathcal{S}$ contains $\mathcal{R}$. Let A be a closed subset of G and B a closed subsets of H. By Proposition 18.1.3, there exists an $A_1 \in \mathcal{B}_1(G)$ with $A \subseteq A_1$ and $\mu_G(A) = \mu_G(A_1)$. Similarly, there exists a $B_1 \in \mathcal{B}_1(H)$ with $B \subseteq B_1$ and $\mu_H(B) = \mu_H(B_1)$. By Part A, $A_1 \times B_1 \in \mathcal{R}_1$. By Part B, $A_1 \times B_1 \in \mathcal{S}$. Hence, $\mu_{G \times H}(A \times B) \leq \mu_{G \times H}(A_1 \times B_1) = (\mu_G \times \mu_H)(A_1 \times B_1) = (\mu_G \times \mu_H)(A \times B)$.

Conversely, since $A \times B$ is closed in $G \times H$, Proposition 18.1.3 gives $C_1 \in \mathcal{B}_1(G \times H)$ with $A \times B \subseteq C_1$ and $\mu_{G \times H}(A \times B) = \mu_{G \times H}(C_1)$. Hence, by Parts A and B, $(\mu_G \times \mu_H)(A \times B) \leq (\mu_G \times \mu_H)(C_1) = \mu_{G \times H}(C_1) = \mu_{G \times H}(A \times B)$. It follows that $(\mu_G \times \mu_H)(A \times B) = \mu_{G \times H}(A \times B)$. In other words, $A \times B \in \mathcal{S}$.

By Lemma 18.4.1, $\mathcal{R} = \mathcal{B}(G) \otimes \mathcal{B}(H) \subseteq \mathcal{S}$.

PART D: Conclusion of the proof. Each $C \in \hat{\mathcal{R}}$ can be written as $C = C_0 \cup C'$ with $C_0 \subseteq C_1$, $C_1 \in \mathcal{R}$, $(\mu_G \times \mu_H)(C_1) = 0$, and $C' \in \mathcal{R}$. By Part C, both C_1 and C' are in $\mathcal{S}$. It follows that $C \in \mathcal{S}$. Therefore, $\hat{\mathcal{R}} \subseteq \mathcal{S}$.

By Proposition 18.1.3, each $D \in \hat{\mathcal{B}}(G \times H)$ can be written as $D = D_0 \cup D'$ with $D_0 \subseteq D_1$, $D_1 \in \mathcal{B}_1(G \times H)$, $\mu_{G \times H}(D_1) = 0$, and $D' \in \mathcal{B}_1(G \times H)$. By Part A, $\mathcal{B}_1(G \times H) = \mathcal{R}_1$, so, by Part B, $\mathcal{B}_1(G \times H) \subseteq \mathcal{S}$. Therefore, $D \in \mathcal{S}$. Thus, $\hat{\mathcal{B}}(G \times H) \subseteq \mathcal{S}$.

It follows that $\hat{\mathcal{R}} = \mathcal{S} = \hat{\mathcal{B}}(G \times H)$, as claimed. □

We apply Proposition 18.4.2 to the product G^e of e copies of G. For simplicity, we also denote the Haar measure on G^e by μ. It has this property: for $A_1, \ldots, A_e$ measurable subsets of G,

(1) $$\mu(A_1 \times \cdots \times A_e) = \mu(A_1) \cdots \mu(A_e),$$

where μ on the right (resp. left) denotes Haar measure on G (resp. G^e). As an immediate consequence of (1) we find:

(2) Suppose $A_{1i}, A_{2i}, A_{3i}, \ldots$ is a sequence of μ-independent subsets of G, $i = 1, \ldots, e$. Then $\prod_{i=1}^{e} A_{1i}, \prod_{i=1}^{e} A_{2i}, \prod_{i=1}^{e} A_{3i}, \ldots$ is a μ-independent sequence of subsets of G^e.

18.5 The Haar Measure of the Absolute Galois Group

We consider the Haar measure μ on the absolute Galois group, $\text{Gal}(K)$, of a fixed field K. If L is a finite separable extension of K, then $(\text{Gal}(K) : \text{Gal}(L)) = [L : K]$. This enables us to rephrase the first part of Lemma 18.3.7 as follows:

LEMMA 18.5.1: Let $L_1, \ldots, L_n$ be finite separable extensions of K. Then the following conditions are equivalent:
(a) $\text{Gal}(L_1), \ldots, \text{Gal}(L_n)$ are μ-independent.
(b) $[L_1 \cdots L_n : K] = \prod_{i=1}^n [L_i : K]$.
(c) $L_1, \ldots, L_n$ are linearly disjoint over K.

Now use Lemma 18.5.1 to rephrase Lemma 18.3.5 and the second part of Lemma 18.3.7 in terms of Galois theory:

LEMMA 18.5.2: Let $L_1, L_2, L_3, \ldots$ be a linearly disjoint sequence of finite separable extensions of K. For each $i \geq 1$ let $\bar{A}_i$ be a set of left cosets of $\text{Gal}(L_i)^e$ in $\text{Gal}(K)^e$ and $A_i = \{\sigma \in \text{Gal}(K)^e \mid \sigma \text{Gal}(L_i)^e \in \bar{A}_i\}$. Then the sequence $A_1, A_2, A_3, \ldots$ is μ-independent. If $\sum_{i=1}^\infty \frac{|\bar{A}_i|}{[L_i:K]^e} = \infty$, then $\mu(\bigcup_{i=1}^\infty A_i) = 1$.

Proof: By Lemma 18.5.1, $\text{Gal}(L_1), \text{Gal}(L_2), \text{Gal}(L_3), \ldots$ are μ-independent. Hence, by (2) of Section 18.4, so are $\text{Gal}(L_1)^e, \text{Gal}(L_2)^e, \text{Gal}(L_3)^e, \ldots$. By Lemma 18.3.7, applied to $G = \text{Gal}(K)^e$, $A_1, A_2, A_3, \ldots$ are μ-independent.

Now suppose $\sum_{i=1}^\infty \frac{|\bar{A}_i|}{[L_i:K]^e} = \infty$. Then, in view of $\mu(A_i) = \frac{|\bar{A}_i|}{[L_i:K]^e}$, Lemma 18.3.4 implies $\mu(\bigcup_{i=1}^\infty A_i) = 1$. □

Lemma 18.3.5 has a similar interpretation:

LEMMA 18.5.3: Let $L_1, L_2, L_3, \ldots$ be finite separable extensions of K. For each $i \geq 1$ let $\bar{A}_i$ be a set of left cosets of $\text{Gal}(L_i)^e$ in $\text{Gal}(K)^e$ and $A_i = \{\sigma \in \text{Gal}(K)^e \mid \sigma \text{Gal}(L_i)^e \in \bar{A}_i\}$. Let A be the set of all $\sigma \in \text{Gal}(K)^e$ which belong to infinitely many A_i's.
(a) If $\sum_{i=1}^\infty \frac{|\bar{A}_i|}{[L_i:K]^e} < \infty$, then $\mu(A) = 0$.
(b) Suppose $L_1, L_2, L_3, \ldots$ are linearly disjoint over K and $\sum_{i=1}^\infty \frac{|\bar{A}_i|}{[L_i:K]^e} = \infty$, then $\mu(A) = 1$.

Remark 18.5.4: If L_i is a Galois extension of K, replace $\text{Gal}(K)^e/\text{Gal}(L_i)^e$ in the foregoing lemmas by $\text{Gal}(L_i/K)^e$ and A_i by $\{\sigma \in \text{Gal}(K)^e \mid \text{res}_{L_i}\sigma \in \bar{A}_i\}$. □

If K is a Hilbertian field, we can construct linearly disjoint sequences $L_1, L_2, \ldots$ of finite Galois extensions of K with $\text{Gal}(L_i/K) \cong S_n$, $i = 1, 2, \ldots$ (Corollary 16.2.7). We interpret this in terms of μ-independence, abbreviating "for all but a set of measure 0" by "almost all":

18.5 The Haar Measure of the Absolute Galois Group

LEMMA 18.5.5: *Suppose K is Hilbertian. Let $\pi_1, \ldots, \pi_e$ be e elements of the symmetric group S_n. Then for almost all $\sigma \in \mathrm{Gal}(K)^e$ there exists a continuous homomorphism $\rho \colon \mathrm{Gal}(K) \to S_n$ with $\rho(\sigma_j) = \pi_j$, $j = 1, \ldots, e$.*

Proof: Corollary 16.2.7 gives a linearly disjoint sequence, $L_1, L_2, L_3, \ldots$ of Galois extensions of K with $\mathrm{Gal}(L_i/K) \cong S_n$, $i = 1, 2, 3, \ldots$. For each i choose an isomorphism $\rho_i \colon \mathrm{Gal}(L_i/K) \to S_n$ and $\sigma_{i1}, \ldots, \sigma_{ie} \in \mathrm{Gal}(L_i/K)$ with $\rho_i(\sigma_{ij}) = \pi_j$, $j = 1, \ldots, e$. Put $\bar\sigma_i = (\sigma_{i1}, \ldots, \sigma_{ie})$, $\bar{A}_i = \{\bar\sigma_i\}$, and $A_i = \{\sigma \in \mathrm{Gal}(K)^e \mid \mathrm{res}_{L_i}(\sigma) = \bar\sigma_i\}$. For each i and each $\sigma \in A_i$, $\rho_i \circ \mathrm{res}_{L_i}$ is a homomorphism from $\mathrm{Gal}(K)$ into S_n that maps σ_{ij} onto π_j, $j = 1, \ldots, e$. The assumption of Lemma 18.5.2 holds here trivially: $\sum_{i=1}^\infty \frac{|\bar{A}_i|}{[L_i:K]^e} = \sum_{i=1}^\infty \frac{1}{(n!)^e} = \infty$. Consequently, $\mu(\bigcup_{i=1}^\infty A_i) = 1$. $\square$

THEOREM 18.5.6 (The Free Generators Theorem [Jarden3, Thm. 5.1]): *Let K be a Hilbertian field and e a positive integer. Then $\langle \sigma_1, \ldots, \sigma_e \rangle \cong \hat{F}_e$ for almost all $(\sigma_1, \ldots, \sigma_e) \in \mathrm{Gal}(K)^e$.*

Proof: Let G be a finite group generated by e elements. Embed G in S_n for some positive integer n. Choose $\pi_1, \ldots, \pi_e \in S_n$ which generate the image of G in S_n. For almost all $\sigma \in \mathrm{Gal}(K)^e$ Lemma 18.5.5 gives a homomorphism $\rho \colon \mathrm{Gal}(K) \to S_n$ with $\rho(\sigma_j) = \pi_j$, $j = 1, \ldots, e$. Thus, $S(G) = \{\sigma \in \mathrm{Gal}(K)^e \mid G \text{ is a homomorphic image of } \langle \sigma \rangle\}$ has measure 1. Since there are only countably many finite groups, the intersection of all $S(G)$'s has measure 1. In other words, for almost all $\sigma \in \mathrm{Gal}(K)^e$, each group with e generators is a quotient of $\langle \sigma \rangle$. By Lemma 17.7.1, $\langle \sigma \rangle \cong \hat{F}_e$. $\square$

The group generated by $\sigma_1 = \cdots = \sigma_e = 1$, is certainly not isomorphic to $\hat{F}_e$. The same holds when $\sigma_1, \ldots, \sigma_e$ equal to the same involution of $\mathrm{Gal}(K)$. Thus, we cannot remove the phrase "almost all" from Theorem 18.5.6. We continue now with more results that indicate the behavior of e-tuples σ excluding a subset of $\mathrm{Gal}(K)^e$ of measure zero.

Definition 18.5.7: *A field M is called e-free if $\mathrm{Gal}(M) \cong \hat{F}_e$.* $\square$

For a field K and for a $\sigma \in \mathrm{Gal}(K)^e$ denote the fixed field in K_s of the entries of σ by $K_s(\sigma)$. From Theorem 18.5.6, if K is Hilbertian, then $K_s(\sigma)$ is e-free for almost all $\sigma \in \mathrm{Gal}(K)^e$.

Let K be a finite field. Although K is not Hilbertian, Theorem 18.5.6 holds for $e = 1$. For $e \geq 2$ we obtain an entirely different result:

LEMMA 18.5.8:
(a) *For almost all $z \in \hat{\mathbb{Z}}$ we have $\langle z \rangle \cong \hat{\mathbb{Z}}$ and $(\hat{\mathbb{Z}} : \langle z \rangle) = \infty$.*
(b) *Suppose $e \geq 2$. Then, for almost all $\mathbf{z} \in \hat{\mathbb{Z}}^e$, we have $\langle \mathbf{z} \rangle \cong \hat{\mathbb{Z}}$ and $(\hat{\mathbb{Z}} : \langle \mathbf{z} \rangle) < \infty$.*

Proof of (a): For each prime number l let $\mathbb{Z}'_l = \bigcap_{i=1}^\infty l^i \hat{\mathbb{Z}}$. Then $\mathbb{Z}'_l \cong \prod \mathbb{Z}_p$ where p ranges over all prime numbers excluding l, and $\hat{\mathbb{Z}}/\mathbb{Z}'_l \cong \mathbb{Z}_l$. Hence, $\mu(\mathbb{Z}'_l) = 0$. Therefore, $\mu(\bigcap_l (\hat{\mathbb{Z}} \smallsetminus \mathbb{Z}'_l)) = 1$.

Suppose $z \in \bigcap_l (\hat{\mathbb{Z}} \smallsetminus \mathbb{Z}'_l)$. Then for each prime l, $\mathbb{Z}'_l$ is a proper closed subgroup of $\langle z, \mathbb{Z}'_l \rangle$. By Lemma 1.4.2(e), $\langle z \rangle / \langle z \rangle \cap \mathbb{Z}'_l \cong \langle z, \mathbb{Z}'_l \rangle / \mathbb{Z}'_l \cong \mathbb{Z}_l$. Hence, $\mathbb{Z}/l^i \mathbb{Z}$ is a homomorphic image of $\langle z \rangle$ for each positive integer i. By the Chinese remainder theorem the same holds for $\mathbb{Z}/n\mathbb{Z}$ for each positive integer n. We conclude from Lemma 17.7.1 that $\langle z \rangle \cong \hat{\mathbb{Z}}$

To prove $(\hat{\mathbb{Z}} : \langle z \rangle) = \infty$ for almost all $z \in \hat{\mathbb{Z}}$, recall that the sum $\sum_l \mu(l\hat{\mathbb{Z}}) = \sum \frac{1}{l}$ over all primes l diverges [LeVeque, p. 100]. By Example 18.3.8, the groups $l\hat{\mathbb{Z}}$ are μ-independent. By Borel-Cantelli (Lemma 18.3.5(b) with $A_i = l_i \hat{\mathbb{Z}}$ where l_i is the ith prime) the set of z that belongs to infinitely many of the groups $l\hat{\mathbb{Z}}$ has measure 1. Since the intersection of infinitely many groups $l\hat{\mathbb{Z}}$ has infinite index, $(\hat{\mathbb{Z}} : \langle z \rangle) = \infty$ for each z in that set.

Proof of (b): Under the assumption $e \geq 2$ we have, $\sum_{n=1}^{\infty} \mu((n\hat{\mathbb{Z}})^e) = \sum_{n=1}^{\infty} n^{-e} < \infty$. By Lemma 18.3.5(a), almost all e-tuples $\mathbf{z} \in \hat{\mathbb{Z}}^e$ belong to $(n\hat{\mathbb{Z}})^e$ for only finitely many n. For each such z, $(\hat{\mathbb{Z}} : \langle \mathbf{z} \rangle) < \infty$. By Lemma 1.4.4, every open subgroup of $\hat{\mathbb{Z}}$ is isomorphic to $\hat{\mathbb{Z}}$. Thus, $\langle \mathbf{z} \rangle \cong \hat{\mathbb{Z}}$ for almost all $\mathbf{z} \in \hat{\mathbb{Z}}^e$. □

For K a finite field, $\mathrm{Gal}(K) \cong \hat{\mathbb{Z}}$. We reformulate Lemma 18.5.8 for $\mathrm{Gal}(K)$:

COROLLARY 18.5.9 ([Jarden4, p. 122]): *Let K be a finite field. For almost all $\sigma \in \mathrm{Gal}(K)$ the field $\tilde{K}(\sigma)$ is 1-free and infinite. If $e \geq 2$, then for almost all $\boldsymbol{\sigma} \in \mathrm{Gal}(K)^e$, the field $\tilde{K}(\boldsymbol{\sigma})$ is finite (and therefore 1-free).*

Remark 18.5.10: Chapter 26 generalizes Lemma 18.5.8 to the groups $\hat{\mathbb{Z}}^n$ and $\hat{F}_n$. □

18.6 The PAC Nullstellensatz

In this section we use measure theory to produce an abundance of algebraic PAC fields. They arise as the fixed fields, $K_s(\boldsymbol{\sigma})$, of $\boldsymbol{\sigma} \in \mathrm{Gal}(K)^e$ for K a countable Hilbertian field:

THEOREM 18.6.1 (PAC Nullstellensatz [Jarden2, p. 76]): *Let K be a countable Hilbertian field and e a positive integer. Then $K_s(\boldsymbol{\sigma})$ is a PAC field for almost all $\boldsymbol{\sigma} \in \mathrm{Gal}(K)^e$.*

Proof: The proof has two parts. We use countability of K only in the second.

PART A: *Zeros of one polynomial.* Let $f \in K[T, X]$ be a nonconstant absolutely irreducible polynomial. Denote the set of all $\boldsymbol{\sigma} \in \mathrm{Gal}(K)^e$ such that f has a zero in $K_s(\boldsymbol{\sigma})$ by $S(f)$. Our goal is to prove that $\mu(S(f)) = 1$. Without loss assume f is separable in X and let $d = \deg_X(f)$. We use induction to construct a linearly disjoint sequence $L_1, L_2, L_3, \ldots$ of separable extensions of K of degree d such that f has a zero in each of the L_i's.

18.6 The PAC Nullstellensatz

Assume $L_1, \ldots, L_n$ are already defined. Then $L = L_1 \cdots L_n$ is a finite separable extension of K and f is irreducible over L. By Corollary 12.2.3, there exists $a \in K$ such that $f(a, X)$ is an irreducible separable polynomial of degree d over L. For b a zero of $f(a, X)$, the field $L_{n+1} = K(b)$ is a separable extension of K of degree d. It contains a zero of f and it is linearly disjoint from L over K. This completes our induction. By construction, $S(f)$ contains $\bigcup_{i=1}^\infty \mathrm{Gal}(L_i)^e$. Lemma 18.5.2, with $A_i = \mathrm{Gal}(L_i)^e$, gives $\mu(\bigcup_{i=1}^\infty \mathrm{Gal}(L_i)) = 1$. Hence, $\mu(S(f)) = 1$.

PART B: *Countability.* There are countably many nonconstant absolutely irreducible polynomials $f \in K[T, X]$. Thus, the intersection S of all the sets $S(f)$ has measure 1. By Theorem 11.2.3, the fixed field $K_s(\boldsymbol{\sigma})$ of each e-tuple $\boldsymbol{\sigma} \in S$ is a PAC field. Therefore, $K_s(\boldsymbol{\sigma})$ is a PAC field for almost all $\boldsymbol{\sigma} \in \mathrm{Gal}(K)^e$. $\square$

Remark 18.6.2: We cannot drop the phrase "almost all" from Theorem 18.6.1. For example, if σ is the automorphism of $\tilde{\mathbb{Q}}$ induced by complex conjugation, $\tilde{\mathbb{Q}}(\sigma)$ is a real closed field and not PAC (Theorem 11.5.1). The next example [Ax2, p. 269] arises from the valuation of $\tilde{\mathbb{Q}}$ given by the prime p.

Let $\mathbb{Q}_{p,\mathrm{ur}}$ be the maximal unramified extension of $\mathbb{Q}_p$. It is well known that $\mathrm{Gal}(\mathbb{Q}_{p,\mathrm{ur}}/\mathbb{Q}_p) \cong \hat{\mathbb{Z}}$ [Cassels-Fröhlich, p. 28]. Choose a generator $\bar{\sigma}$ of $\mathrm{Gal}(\mathbb{Q}_{p,\mathrm{ur}}/\mathbb{Q}_p)$ and extend $\bar{\sigma}$ to an automorphism σ of $\tilde{\mathbb{Q}}_p$. Then the field $E = \tilde{\mathbb{Q}}_p(\sigma)$ is a totally ramified extension of $\mathbb{Q}_p$. It therefore has $\mathbb{F}_p$ as its residue field. Also, $\mathrm{Gal}(E) = \langle\sigma\rangle \cong \hat{\mathbb{Z}}$ (Lemma 17.4.11). It is also well known that $\tilde{\mathbb{Q}} \cdot \mathbb{Q}_p = \tilde{\mathbb{Q}}_p$ (a corollary of Krasner's Lemma [Ribenboim, p. 190]). Thus, with $\sigma_0 = \mathrm{res}_{\tilde{\mathbb{Q}}}(\sigma)$, we have that $\langle\sigma_0\rangle \cong \langle\sigma\rangle \cong \hat{\mathbb{Z}}$. Hence, $\tilde{\mathbb{Q}}(\sigma_0)$ is a 1-free field that admits a valuation with a finite residue field. By Corollary 11.5.5 (or Exercise 7 of Chapter 11), $\tilde{\mathbb{Q}}(\sigma_0)$ is not PAC.

More generally, if L is an algebraic extension of $\tilde{\mathbb{Q}}(\sigma_0)$ and $L \neq \tilde{\mathbb{Q}}$, then L is Henselian with respect to a rank-1 valuation. By Theorem 11.5.5, L is not PAC. By [Jarden14, Thm. 21.3], $M = \tilde{\mathbb{Q}}(\sigma_0^{\tau_1}, \ldots, \sigma_0^{\tau_e})$ is an e-free field for almost all $(\tau_1, \ldots, \tau_e) \in \mathrm{Gal}(\mathbb{Q})^e$. Nevertheless M is not a PAC field, since the residue field of M with respect to the p-adic valuation is $\mathbb{F}_p$. $\square$

PROBLEM 18.6.3: Let $\sigma \in \mathrm{Gal}(\mathbb{Q})$ for which $\tilde{\mathbb{Q}}(\sigma)$ is neither PAC nor formally real. Does $\tilde{\mathbb{Q}}(\sigma)$ admit a valuation with a non-algebraically closed completion?

Finite fields are not Hilbertian. Still, Theorem 18.6.1 holds for them with $e = 1$:

PROPOSITION 18.6.4: *Let K be a finite field. Then $\tilde{K}(\sigma)$ is a PAC field for almost all $\sigma \in \mathrm{Gal}(K)$.*

Proof: For almost all $\sigma \in \mathrm{Gal}(K)$ the field $\tilde{K}(\sigma)$ is an infinite extension of K (Corollary 18.5.9). Hence, by Corollary 11.2.4, $\tilde{K}(\sigma)$ is a PAC field. $\square$

REMARK 18.6.5: *K finite and $e \geq 2$.* The analog of Proposition 18.6.4 for $e \geq 2$ is false. Indeed, by Corollary 18.5.9, for almost all $\sigma \in \text{Gal}(K)^e$, the field $\tilde{K}(\sigma)$ is finite. Hence, by Proposition 11.1.1, $\tilde{K}(\sigma)$ is not PAC. ☐

REMARK 18.6.6: *Let K be a countable infinite field such that $K_s(\sigma)$ is a PAC e-free field for all positive integers e and almost all $\sigma \in \text{Gal}(K)^e$. One may ask whether K is necessarily Hilbertian? Example 26.1.11 gives an example where this is not the case.* ☐

18.7 The Bottom Theorem

We supplement the free generators theorem for a Hilbertian field K. Not only is it true that a σ selected at random in $\text{Gal}(K)^e$ generates a group isomorphic to $\hat{F}_e$, but also, the field $K_s(\sigma)$ has no proper subfields of finite codegree (Theorem 18.7.7). Our first concern is to eliminate the possibility that $K_s(\sigma)$ has a proper subfield of finite codegree which is the fixed field of some involution of $\text{Gal}(K)$ acting on $K_s(\sigma)$ (Proposition 18.7.5):

LEMMA 18.7.1: *Let K be a field of characteristic 0 and*

$$f \in K(T_1, \ldots, T_r)[X_1, \ldots, X_n]$$

an irreducible polynomial. Then the polynomial

(1) $$f(\sum_{j=1}^{3} Y_{1j}^2, \ldots, \sum_{j=1}^{3} Y_{rj}^2, X_1, \ldots, X_n)$$

is defined and irreducible in $K(\mathbf{Y})[\mathbf{X}]$.

Proof: Assume without loss that $T_1, \ldots, T_r$ occur in F. Then $\frac{\partial f}{\partial T_i} \neq 0$ for $i = 1, \ldots, r$. Take $g_i(\mathbf{Y}_i) = Y_{i1}^2 + Y_{i2}^2 + Y_{i3}^2$ in Corollary 10.3.2. Then $g_i(\mathbf{Y}_i) + c$ is absolutely irreducible for all $c \in \tilde{K}$. Therefore, the polynomial in (1) is irreducible in $K(\mathbf{Y})[\mathbf{X}]$. ☐

LEMMA 18.7.2: *Let K be a formally real Hilbertian field, H a Hilbert subset of K^r and $a_1 < b_1, \ldots, a_r < b_r$ rational numbers. Then there is a $(z_1, \ldots, z_r) \in H$ with $a_i < z_i < b_i$ in each ordering of K, $i = 1, \ldots, r$.*

Proof: We prove the lemma only for the case $r = 1$. The general case is analogous.

Write H as $H_K(f_1, \ldots, f_l)$ with $f_j \in K(T)[X_1, \ldots, X_n]$, irreducible, $j = 1, \ldots, l$. Let $a < b$ be rational numbers. Put $c = \frac{1}{b-a}$. Then $f_j(a + \frac{1}{c+T}, \mathbf{X})$ is irreducible. Hence, so is $f_j(a + \frac{1}{c+Y_1^2+Y_2^2+Y_3^2}, \mathbf{X})$ (Lemma 18.7.1). Since K is Hilbertian there exist $y_1, y_2, y_3 \in K$ with $y_1 \neq 0$ such that $f_j(a + \frac{1}{c+y_1^2+y_2^2+y_3^2}, \mathbf{X})$ is defined and irreducible in $K[\mathbf{X}]$, $j = 1, \ldots, l$. Let $z = a + \frac{1}{c+y_1^2+y_2^2+y_3^2}$. Then $a < z < b$ in every ordering of K and $f_j(z, \mathbf{X})$ is defined and irreducible in $K[\mathbf{X}]$, $j = 1, \ldots, l$. ☐

18.7 The Bottom Theorem

The following result strengthens Corollary 16.2.7 in the case where $\text{char}(K) = 0$:

LEMMA 18.7.3: Let K be a formally real Hilbertian field and $m \geq 2$ a positive integer. Then there exists a linearly disjoint sequence $\{K_i/K\}_{i=1}^{\infty}$ of Galois extensions such that $\text{Gal}(K_i/K) \cong S_m$ and K_i/K has a totally imaginary quadratic subextension E_i/K (i.e. E_i is contained in no real closure of K), $i = 1, 2, 3, \ldots$.

Proof: Suppose by induction $K_1, \ldots, K_n$ and $E_1, \ldots, E_n$ have already been constructed. Put $L = K_1 \cdots K_n$. Consider the general polynomial of degree m,

$$f(\mathbf{T}, X) = X^m + T_1 X^{m-1} + \cdots + T_m = \prod_{i=1}^{m}(X - x_i).$$

Put $\Delta(\mathbf{T}) = \prod_{i<j}(x_i - x_j)^2$. Now let $c_1, \ldots, c_m \in \mathbb{R}$ with

$$(X^2 + 1) \prod_{i=1}^{m-2}(X - i) = X^m + c_1 X^{m-1} + \cdots + c_m = f(\mathbf{c}, X).$$

Then c_i is an integer, $i = 1, \ldots, m$, and $\Delta(\mathbf{c}) < 0$. Since $\Delta(\mathbf{T})$ is a polynomial with integral coefficients, there exist rational numbers a_i, b_i with $a_i < c_i < b_i$, $i = 1, \ldots, m$, such that for all $t_1, \ldots, t_m \in \mathbb{R}$

(2) $\qquad a_i < t_i < b_i, \; i = 1, \ldots, m, \quad \text{implies } \Delta(\mathbf{t}) < 0.$

Since each real closure of K is an elementary equivalent to $\mathbb{R}$ [Prestel2, p. 51], (2) holds for all $t_1, \ldots, t_m \in K$ and for every ordering $<$ of K. By Lemma 13.1.1 and Corollary 12.2.3 there exists a Hilbert subset H of K^m such that for each $\mathbf{t} \in H$, $\text{Gal}(f(\mathbf{t}, X), K) \cong \text{Gal}(f(\mathbf{t}, X), L) \cong S_m$. Choose $\mathbf{t} \in H$ such that $a_i < t_i < b_i$, $i = 1, \ldots, m$, in each ordering $<$ of K (Lemma 18.7.2). Let K_{n+1} be the splitting field of $f(\mathbf{t}, X)$ over K. Then K_{n+1} is linearly disjoint from L over K, $\text{Gal}(K_{n+1}/K) \cong S_m$ and $E_{n+1} = K(\sqrt{\Delta(\mathbf{t})})$ is a totally imaginary quadratic extension of K contained in K_{n+1}. This completes the induction. □

LEMMA 18.7.4: Let p be an odd prime, γ be the cycle $(12 \ldots p)$ in S_p, N the normalizer of $\langle \gamma \rangle$ in S_p, and π an element of N of order 2. Then $\pi \in A_p$ if and only if $p \equiv 1 \mod 4$.

Proof (Cherlin): The element $\tau = (1\,p)(2\,p-1) \cdots (\frac{p-1}{2}\,\frac{p+3}{2})$ of S_p is a product of $\frac{p-1}{2}$ transpositions. Hence, $\tau \in A_p$ if and only if $p \equiv 1 \mod 4$. It satisfies $\tau^{-1}\gamma\tau = \gamma^{-1}$. Hence, $\tau \in N$. Since p is odd, γ is in A_p. It suffices therefore to check that each element $\pi \in N$ of order 2 lies in the coset $\langle \gamma \rangle \tau$.

Observe that $\langle \gamma \rangle$ is its own centralizer in G. Indeed, suppose σ commutes with γ. Find i with $\gamma^i(1) = \sigma^{-1}(1)$. Then prove inductively that $\gamma^i(k) = \sigma^{-1}(k)$ for all k. Therefore, $\sigma = \gamma^{-i}$.

Consider $\pi \in N$ of order 2. Then $\pi^{-1}\gamma\pi = \gamma^j$ for some $j \in \mathbb{N}$ with $j^2 \equiv 1 \mod p$. Hence, $j \equiv \pm 1 \mod p$. If $j \equiv 1 \mod p$, then π commutes with γ. By the preceding paragraph $\pi \in \langle\gamma\rangle$. Since p is odd, this is a contradiction. Thus, $\pi^{-1}\gamma\pi = \gamma^{-1}$. Hence, $(\pi\tau^{-1})^{-1}\gamma(\pi\tau^{-1}) = \gamma$. Consequently, $\pi\tau^{-1} \in \langle\gamma\rangle$, so $\pi \in \langle\gamma\rangle\tau$, as required. $\square$

PROPOSITION 18.7.5 ([Jarden3, p. 299]): *Let K be a Hilbertian field. Then for almost all $\boldsymbol{\sigma} \in \mathrm{Gal}(K)^e$ there exists no $\tau \in \mathrm{Gal}(K)$, $\tau \neq 1$, of finite order such that $[K_s(\boldsymbol{\sigma}) : K_s(\boldsymbol{\sigma},\tau)] < \infty$.*

Proof: By Artin-Schreier we only have to prove the proposition for K a formally real field, $\tau^2 = 1$, and $\tau \neq 1$ [Lang7, p. 299, Cor. 9.3]. Moreover, it suffices to prove for each $n \in \mathbb{N}$, that for almost all $\boldsymbol{\sigma} \in \mathrm{Gal}(K)^e$ there exists no $\tau \in \mathrm{Gal}(K)$ with $\tau^2 = 1$, $\tau \neq 1$, and $[\tilde{K}(\boldsymbol{\sigma}) : \tilde{K}(\boldsymbol{\sigma},\tau)] = n$.

Choose a prime $p \equiv 1 \mod 4$ with $p \geq n$. Consider the sequence $\{K_i/K\}_{i=1}^\infty$ of fields constructed in Lemma 18.7.3 for $m = p$. In particular, there is an isomorphism $\theta_i \colon S_p \to \mathrm{Gal}(K_i/K)$, $i = 1, 2, 3, \ldots$. Let $\gamma_i = \theta_i((1\, 2\, \ldots\, p))$. Denote the set of all $\boldsymbol{\sigma} \in \mathrm{Gal}(K)^e$ for which there exists i with $\mathrm{res}_{K_i}(\sigma_j) = \gamma_i$, $j = 1,\ldots,e$ by S. By Lemma 18.5.2, $\mu(S) = 1$. We prove that each $\boldsymbol{\sigma} \in S$ has the desired property.

Assume there is a $\tau \in \mathrm{Gal}(K)$ with $\tau^2 = 1$, $\tau \neq 1$, and $[\tilde{K}(\boldsymbol{\sigma}) : \tilde{K}(\boldsymbol{\sigma},\tau)] = n$. Let L be the smallest normal extension of $\tilde{K}(\boldsymbol{\sigma},\tau)$ that contains $\tilde{K}(\boldsymbol{\sigma})$. Then $[L : \tilde{K}(\boldsymbol{\sigma},\tau)]$ divides $n!$, so $[L : \tilde{K}(\boldsymbol{\sigma})]$ divides $(n-1)!$. Hence, p does not divide $[L : \tilde{K}(\boldsymbol{\sigma})]$. In addition, there is an i with $\mathrm{res}_{K_i}(\sigma_j) = \gamma_i$, $j = 1,\ldots,e$. Then $K_i \cap \tilde{K}(\boldsymbol{\sigma}) = K_i(\gamma_i)$ and $[K_i \cap L : K_i(\gamma_i)] = [(K_i \cap L)\tilde{K}(\boldsymbol{\sigma}) : \tilde{K}(\boldsymbol{\sigma})]$. Thus, $[K_i \cap L : K_i(\gamma_i)]$ divides the relatively prime numbers $[K_i : K_i(\gamma_i)] = p$ and $[L : \tilde{K}(\boldsymbol{\sigma})]$. Hence, $K_i \cap L = K_i(\gamma_i)$. It follows for $\bar{\tau} = \mathrm{res}_{K_i}(\tau)$ that $K_i(\gamma_i)/K_i(\gamma_i,\bar{\tau})$ is a Galois extension. That is, $\bar{\tau}$ belongs to the normalizer of $\langle\gamma_i\rangle$ in $\mathrm{Gal}(K_i/K)$. Since $\bar{\tau}^2 = 1$ and $p \equiv 1 \mod 4$, Lemma 18.7.4 implies that $\bar{\tau}$ belongs to $\theta_i(A_p)$, a subgroup of $\mathrm{Gal}(K_i/K)$ of index 2. Since S_p has only one subgroup of index 2, $\theta_i(A_p) = \mathrm{Gal}(K_i/E_i)$. Thus, E_i is contained in $\tilde{K}(\tau)$, a real closed field [Lang7, p. 299, Cor. 9.3 and p. 452, Prop. 2.4]. This contradicts the choice made for K_i/K from Lemma 18.7.3 that E_i/K is a totally imaginary extension. Consequently, a τ as above does not exist. $\square$

LEMMA 18.7.6: *Let K be a Hilbertian field. Write $\hat{F}_e$ as the inverse limit $\cdots \xrightarrow{\pi_3} H_3 \xrightarrow{\pi_2} H_2 \xrightarrow{\pi_1} H_1$ of epimorphisms of finite groups. Then there exists a sequence $\cdots \xrightarrow{\pi_3} C_3 \xrightarrow{\pi_2} C_2 \xrightarrow{\pi_1} C_1$ of finite groups such that for each $k \in \mathbb{N}$ the following hold:*

(3a) *H_k is a subgroup of C_k, and $\pi_k \colon C_{k+1} \to C_k$ extends $\pi_k \colon H_{k+1} \to H_k$.*

(3b) *There exists a subgroup B_k of C_k such that $C_k = B_k H_k = \{bh \mid b \in B_k, h \in H_k\}$ and $B_k \cap H_k = 1$.*

(3c) *For almost all $\boldsymbol{\sigma} \in \mathrm{Gal}(K)^e$ there exists a homomorphism $\rho \colon \mathrm{Gal}(K) \to C_k$ such that $\rho(\langle\boldsymbol{\sigma}\rangle) = H_k$.*

18.7 The Bottom Theorem

Proof: For each $k \in \mathbb{N}$ let $n_k = |H_k|$. Multiplication from the right gives an embedding α_k of H_k into the symmetric group S_{n_k} satisfying:
(4) For all $1 \leq i, j \leq n_k$ there is a unique $h \in H_k$ such that $\alpha_k(h)(i) = j$.

Let $C_k = S_{n_1} \times \cdots \times S_{n_k}$ and define $\lambda_k \colon H_k \to C_k$ by

$$\lambda_k(h) = ((\alpha_1 \circ \pi_1 \circ \cdots \circ \pi_{k-1})(h), (\alpha_2 \circ \pi_2 \circ \cdots \circ \pi_{k-1})(h), \ldots, \alpha_k(h)).$$

This is an embedding, so identify H_k with its image in C_k. Let $\pi_k \colon C_{k+1} \to C_k$ be projection on the first k coordinates. Then (3a) holds. Assertion (3b) follows from (4) for $B_k = \{(\tau_1, \ldots, \tau_k) \in C_k \mid \tau_k(1) = 1\}$.

To prove (3c), choose $\bar{\sigma}_1, \ldots, \bar{\sigma}_e \in H_k$ with $H_k = \langle \bar{\sigma}_1, \ldots, \bar{\sigma}_e \rangle$. For each i, $1 \leq i \leq e$, denote the jth component of $\bar{\sigma}_i$, viewed as an element of $C_k = S_{n_1} \times \cdots \times S_{n_k}$, by $\bar{\sigma}_{ij}$. For almost all $\boldsymbol{\sigma} \in \mathrm{Gal}(K)^e$ and each j, $1 \leq j \leq k$, there is a continuous homomorphism $\rho_j \colon \mathrm{Gal}(K) \to S_{n_j}$ with $\rho_j(\sigma_1, \ldots, \sigma_e) = (\bar{\sigma}_{1j}, \ldots, \bar{\sigma}_{ej})$ (Lemma 18.5.5). Combine $\rho_1, \ldots, \rho_k$ to a homomorphism $\rho \colon \mathrm{Gal}(K) \to C_k$ with $\rho(\sigma_i) = (\bar{\sigma}_{i1}, \ldots, \bar{\sigma}_{ik}) = \bar{\sigma}_i$ for $i = 1, \ldots, e$. Thus, $\rho(\langle \boldsymbol{\sigma} \rangle) = H_k$. $\square$

THEOREM 18.7.7 (The Bottom Theorem [Haran2]): *Let K be a Hilbertian field and $e \in \mathbb{N}$. Then for almost all $\boldsymbol{\sigma} \in \mathrm{Gal}(K)^e$, $K_s(\boldsymbol{\sigma})$ is a finite extension of no proper subfield that contains K.*

Proof: Let S be the set of all $\boldsymbol{\sigma} \in \mathrm{Gal}(K)^e$ satisfying this:
(5a) $\langle \boldsymbol{\sigma} \rangle \cong \hat{F}_e$.
(5b) There is no $\tau \in \mathrm{Gal}(K)$, $\tau \neq 1$, of finite order with $(\langle \boldsymbol{\sigma}, \tau \rangle : \langle \boldsymbol{\sigma} \rangle) < \infty$.
(5c) For each positive integer k there is a homomorphism $\rho \colon \mathrm{Gal}(K) \to C_k$ with $\rho(\langle \boldsymbol{\sigma} \rangle) = H_k$. Here we use the notation of Lemma 18.7.6.

Theorem 18.5.6, Proposition 18.7.5, and Lemma 18.7.6 imply that S is the intersection of three sets of measure 1. Thus, $\mu(S) = 1$. We show that each $\boldsymbol{\sigma} \in S$ satisfies the conclusion of the theorem.

Put $H = \langle \boldsymbol{\sigma} \rangle$. Assume H is an open subgroup of index n of a closed subgroup G of $\mathrm{Gal}(K)$. By (5b), G is torsion free. For each positive integer k denote the set of all homomorphisms $\rho \colon G \to C_k$ with $\rho(H) = H_k$ by R_k. By (5c), R_k is nonempty. Since $(G : H) < \infty$, G is finitely generated. Hence, R_k is finite.

The homomorphisms π_k of (3a) induce an inverse system

$$\cdots \xrightarrow{\pi_3} R_3 \xrightarrow{\pi_2} R_2 \xrightarrow{\pi_1} R_1$$

whose inverse limit is nonempty. By Corollary 1.1.4, there are $\rho_k \in R_k$ with $\pi_k \circ \rho_{k+1} = \rho_k$, $k = 1, 2, 3, \ldots$. Let $G_k = \rho_k(G)$, $k = 1, 2, 3, \ldots$. The ρ_k's define an epimorphism $\rho \colon G \to \varprojlim G_k$ with $\rho(H) = \varprojlim H_k$. From the conditions of Lemma 18.7.6 and Condition (5a), $\rho(H) = \hat{F}_e \cong H$. By Lemma 17.4.11, $H \cap \mathrm{Ker}(\rho) = 1$. But then $|\mathrm{Ker}(\rho)| = (H \cdot \mathrm{Ker}(\rho) : H) \leq (G : H) < \infty$. Since G is torsion free, $\mathrm{Ker}(\rho) = 1$. Hence, ρ maps G (resp. H) isomorphically onto $\varprojlim G_k$ (resp. $\varprojlim H_k$). This gives k_0 with

(6) $(G_k : H_k) = (G : H) = n$ for all $k \geq k_0$.

Assume without loss that $k_0 = 1$.

For each $k \in \mathbb{N}$ let $\mathcal{B}_k$ be the set of all subgroups B'_k of G_k satisfying

(7) $B'_k H_k = G_k$ and $B'_k \cap H_k = 1$.

The condition $B_k H_k = C_k$ (Lemma 18.7.6(3b)) implies $(B_k \cap C_k) H_k = G_k$. Thus, $\mathcal{B}_k$ is nonempty and finite. By (6) and (7), the order of each $B'_k \in \mathcal{B}_k$ is n. Moreover,

$$\pi_{k-1}(B'_k) H_{k-1} = \pi_{k-1}(B'_k) \pi_{k-1}(H_k) = \pi_{k-1}(G_k) = G_{k-1}.$$

Hence,

$$n = (G_{k-1} : H_{k-1}) = (\pi_{k-1}(B'_k) H_{k-1} : H_{k-1}) = \frac{|\pi_{k-1}(B'_k)|}{|\pi_{k-1}(B'_k) \cap H_{k-1}|}.$$

Since $|\pi_{k-1}(B'_k)| \leq |B'_k| = n$, this implies $|\pi_{k-1}(B'_k) \cap H_{k-1}| = 1$. Therefore, $\pi_{k-1}(B'_k) \in \mathcal{B}_{k-1}$. Thus, $\cdots \xrightarrow{\pi_3} \mathcal{B}_3 \xrightarrow{\pi_2} \mathcal{B}_2 \xrightarrow{\pi_1} \mathcal{B}_1$, is an inverse system with a nonempty inverse limit. In other words, there is a $B'_k \leq G_k$ of order n with $\pi_{k-1}(B'_k) = B'_{k-1}$, $k = 2, 3, \ldots$. The inverse limit $B' = \varprojlim B'_k$ is a subgroup of G of order n. But G is torsion free. Consequently, $n = 1$. □

PROBLEM 18.7.8: *Is the following generalization of the bottom theorem true: Let K be a Hilbertian field and e a positive integer. Then for almost all $\boldsymbol{\sigma} \in \mathrm{Gal}(K)^e$, $K_s(\boldsymbol{\sigma})$ is a finite extension of no proper subfield.*

18.8 PAC Fields over Uncountable Hilbertian Fields

In contrast to Theorem 18.6.1, if K is an uncountable Hilbertian field, it is possible that the set

(1) $\qquad S_e(K) = \{\boldsymbol{\sigma} \in \mathrm{Gal}(K)^e \mid K_s(\boldsymbol{\sigma}) \text{ is a PAC field}\}$

is not even measurable. The main theorem of this section says this happens if K is a rational function field in uncountably many variables over a field E_0 or if $K = \mathbb{C}(t)$.

LEMMA 18.8.1: *Let G be a profinite group and S a subset of G. Denote the Boolean algebra of all open-closed subsets of G by $\mathcal{B}_0$ and the σ-algebra generated by $\mathcal{B}_0$ by $\mathcal{B}_1$. Suppose for each $B \in \mathcal{B}_1$ there exists an epimorphism r of G onto a profinite group H such that*

(2) $\qquad B = r^{-1}(r(B))$ and $\mu_H(r(S)) = 1$.

Then $G \smallsetminus S$ contains no subset of positive measure.

Proof: Let B' be a measurable subset of $G \smallsetminus S$. Proposition 18.1.3 gives a set $B \in \mathcal{B}_1$ with $B \subseteq B'$ and $\mu(B' \smallsetminus B) = 0$. Let $r \colon G \to H$ be an epimorphism satisfying (2). Since $B \subseteq G \smallsetminus S$, we have $S \subseteq G \smallsetminus B$ and $\mu_H(r(G \smallsetminus B)) = 1$. In addition, $G \smallsetminus B = r^{-1}(r(G \smallsetminus B))$. By Proposition 18.2.2, $\mu(G \smallsetminus B) = 1$. Hence, $\mu(B') = 0$. □

18.8 PAC Fields over Uncountable Hilbertian Fields

LEMMA 18.8.2: *Let E be a field, A an infinite subset of E, and $\mathcal{F} \subseteq E[X_1, \ldots, X_n]$ a set of nonzero polynomials with $|\mathcal{F}| < |A|$. Then the cardinality of the set $C = \{\mathbf{a} \in A^n | f(\mathbf{a}) \neq 0 \text{ for each } f \in \mathcal{F}\}$ is equal to $|A|$. In particular, if $\{U_i | i \in I\}$ is a family of nonempty Zariski E-open sets in $\mathbb{A}^n$ and $|I| < |A|$, then $|A^n \cap (\bigcap_{i \in I} U_i)| = |A|$.*

Proof: Our assertion is true for $n=1$, because every polynomial $f \in \mathcal{F}$ has only finitely many zeros. Assume, by induction, the assertion is true for $n-1$, where $n \geq 2$. Each $f \in \mathcal{F}$ has a nonzero coefficient $g \in E[X_1, \ldots, X_{n-1}]$. Hence, by the induction hypothesis, applied to the set of those g's, the cardinality of the set

$$B = \{(a_1, \ldots, a_{n-1}) \in A^{n-1} | f(a_1, \ldots, a_{n-1}, X_n) \neq 0 \text{ for every } f \in \mathcal{F}\}$$

is equal to that of A. For each $(a_1, \ldots, a_{n-1}) \in B$ the case $n=1$ gives an element $a_n \in A$ such that $f(a_1, \ldots, a_n) \neq 0$ for each $f \in \mathcal{F}$. This gives an embedding of B into C. Consequently, $|C| = |A|$.

Since each Zariski open set is defined by finitely many polynomial inequalities, the last assertion follows from the above. □

A finite separable extension has only finitely many subextensions. Define the **rank** of an infinite separable algebraic extension of fields as the cardinality of the family of all finite subextensions. Thus, if L is the compositum of m finite separable extensions of a field K and m is an infinite cardinal number, then $\text{rank}(L/K) = m$. If L/K is a Galois extension and $\text{Gal}(L/K)$ is not finitely generated, then $\text{rank}(L/K) = \text{rank}(\text{Gal}(L/K))$ (Proposition 17.1.2).

LEMMA 18.8.3: *Let $K = E(t)$ be a rational function field in one variable over an infinite field E. Consider a separable extension L of K with $\text{rank}(L/K) < |E|$. Let $f \in K[X, Y]$ be an irreducible polynomial in $L[X, Y]$, separable in Y. Then there exists an $x \in K$ such that $f(x, Y)$ is irreducible and separable in $L[Y]$.*

Proof: Let $\{K_i | i \in I\}$ be the family of all finite extensions of K contained in L. By assumption, $|I| < |E|$. By Corollary 12.2.3, every separable Hilbert subset of K_i contains a separable Hilbert subset of K. Therefore, by Proposition 13.2.1, there exists a nonempty Zariski E-open set $U_i \subseteq \mathbb{A}^2$ with $f(a + bt, Y)$ irreducible and separable in $K_i[Y]$ for each $(a, b) \in U_i(E)$. By Lemma 18.8.2, $\bigcap_{i \in I} U_i(E)$ is nonempty. If (a, b) lies in this intersection and $x = a + bt$, then $f(x, Y)$ is irreducible and separable over every K_i, hence also over L. □

LEMMA 18.8.4: *Let K be as in Lemma 18.8.3 and N a Galois extension of K with $\text{rank}(N/K) < |K|$. Then each $\boldsymbol{\sigma} = (\sigma_1, \ldots, \sigma_e) \in \text{Gal}(N/K)^e$ extends to $\boldsymbol{\tau} = (\tau_1, \ldots, \tau_e) \in \text{Gal}(K)^e$ such that $K_s(\boldsymbol{\tau})$ is a PAC field.*

Proof: Wellorder the absolutely irreducible polynomials of $K[X, Y]$ which are separable in Y in a transfinite sequence $\{f_\alpha | \alpha < m\}$, where $m = |K|$.

For each $\alpha < m$ we produce a finite separable extension K_α of K containing a zero of f_α and satisfying this: For each $\beta < m$ the set of fields $\{N\}\cup\{K_\alpha \mid \alpha < \beta\}$ is linearly disjoint over K.

Indeed, let $\beta < m$. Assume that K_α has been defined for each $\alpha < \beta$. Let L be the compositum of N and of all fields K_α with $\alpha < \beta$. Then L is a separable extension of K with $\mathrm{rank}(L/K) < m$. Lemma 18.8.3 gives an $x \in K$ with $f_\beta(x,Y)$ irreducible and separable in $L[Y]$. Choose $y \in K_s$ with $f(x,y) = 0$. Then $K_\beta = K(y)$ is linearly disjoint from L over K.

The compositum M of all K_α is a separable algebraic extension of K which is linearly disjoint from N. Hence, res: $\mathrm{Gal}(M) \to \mathrm{Gal}(N/K)$ is an epimorphism. By Theorem 11.2.3, M is PAC. Extend $\sigma_1,\ldots,\sigma_e$ to automorphisms $\tau_1,\ldots,\tau_e \in \mathrm{Gal}(M)$. Their fixed field $K_s(\tau)$ is an algebraic extension of M. From Corollary 11.2.5, $K_s(\tau)$ is a PAC field. □

The next result is a weak form of Theorem 18.6.1 for $K = E(t)$.

PROPOSITION 18.8.5 ([Jarden-Shelah]): *Let $K = E(t)$ be a rational function field in one variable over an uncountable field E. For each $e \geq 1$, the complement of $S_e(K)$ in $\mathrm{Gal}(K)^e$ contains no subset of positive measure.*

Proof: Use Lemma 18.8.1 with $G = \mathrm{Gal}(K)^e$. Note first that every open-closed subset U of $\mathrm{Gal}(K)^e$ is a union of finitely many products of the form $U_1 \times \cdots \times U_e$ where each U_i is the union of translates of $\mathrm{Gal}(L)$ for L (independent of i) a finite Galois extension of K. For $B \in \mathcal{B}_1$, let N be the compositum of all fields L associated with those products $U_1 \times \cdots \times U_e$ that are required to express B as an element of $\mathcal{B}_1$. Since there are at most countable such L's, $\mathrm{rank}(N/K) \leq \aleph_0$. Moreover, $r^{-1}(r(B)) = B$, where $r: \mathrm{Gal}(K)^e \to \mathrm{Gal}(N/K)^e$ is the restriction map.

By Lemma 18.8.4, $r(S_e(K)) = \mathrm{Gal}(N/K)^e$. Thus, Lemma 18.8.1 implies that $\mathrm{Gal}(K)^e \smallsetminus S_e(K)$ contains no set of positive measure. □

When K is a rational function field in uncountably many variables over a field E_0, we are nearly ready to conclude that $S_e(K)$ is nonmeasurable.

LEMMA 18.8.6:
(a) *Let L/K_0 be a field extension with $L \neq L_s$ and v a Henselian valuation of L which is trivial on K_0 such that $\bar{L}_v = K_0$. Then each $\sigma \in \mathrm{Gal}(K_0)^e \smallsetminus \{1\}$ extends to $\rho \in \mathrm{Gal}(L)^e$ with $L_s(\rho)$ non-PAC.*
(b) *Let $K = E(t)$ be a rational function field in one variable over a field E. Consider a subfield K_0 of K and $\sigma \in \mathrm{Gal}(K_0)^e$. Then σ extends to $\rho \in \mathrm{Gal}(K)^e$ with $K_s(\rho)$ non-PAC in each of the following cases:*
   (b1) $K_0 = E$.
   (b2) $K_0 = E_0(t)$ with E_0 algebraically closed and $E = \widetilde{E_0(u)}$ with u transcendental over E_0.

Proof of (a): By Lemma 2.6.9, L is a regular extension of K_0. Hence, each $\sigma \in \mathrm{Gal}(K_0)^e \smallsetminus \{1\}$ extends to $\rho \in \mathrm{Gal}(L)^e$ with $L_s(\rho) \neq L_s$. Then $L_s(\rho)$ is

Henselian (Section 11.5) but not separably closed. It follows from Corollary 11.5.6 that $L_s(\rho)$ is not PAC.

Proof of (b): By (a), it suffices to construct an algebraic extension L of K with $L \neq L_s$ and a Henselian valuation v of L which is trivial on K_0 such that $\bar{L}_v = K_0$.

First suppose $K_0 = E$. Let v be the valuation of K/E with $v(t) = 1$. Choose a Henselian closure (L, v) of (K, v). Then $\bar{L}_v = E$ and $L \neq L_s$, as desired.

Now suppose K_0 and E are as in (b2). Let v_0 be the valuation of $E_0(u)$ over E_0 with $v_0(u) = 1$. Extend v_0 to a valuation $\tilde{v}_0$ of E. As E_0 is algebraically closed, $E_0 = \bar{E}_{\tilde{v}_0}$. Next extend $\tilde{v}_0$ to a valuation v of K/K_0 with $\bar{K}_v = K_0$ (Example 2.3.3). Then let (L, v) be the Henselian closure of (K, v). In particular, $\bar{L}_v = K_0$ is not separably closed. Hence, $L \neq L_s$, as desired. □

LEMMA 18.8.7: *Let F be a field, e a positive integer, and $\mathcal{B}_1$ the σ-algebra generated by all open-closed subsets of $\mathrm{Gal}(F)^e$. Let $\mathcal{F} = \{F_i \mid i \in I\}$ be a family of subfields of F. For each $i \in I$, let $r_i \colon \mathrm{Gal}(F)^e \to \mathrm{Gal}(F_i)^e$ be the restriction map. Suppose F_i is algebraically closed in F. Suppose also, $F = \bigcup_{i \in I} F_i$ and for each subset I_0 of I with $|I_0| \leq \aleph_0$ there exists $j \in I$ such that $\bigcup_{i \in I_0} F_i \subseteq F_j$. Then, for each $B \in \mathcal{B}_1$ there is an $i \in I$ with $B = r_i^{-1}(r_i(B))$.*

Proof: First suppose B is open-closed. Then there is a finite Galois extension F' of F such that B is the lifting to $\mathrm{Gal}(F)^e$ of some subset $\bar{B}$ of $\mathrm{Gal}(F'/F)^e$. Then there is an $i \in I$ and a Galois extension F'_i of F_i with $F'_i F = F'$. By assumption, $F'_i \cap F = F_i$. Hence, res: $\mathrm{Gal}(F'/F)^e \to \mathrm{Gal}(F'_i/F_i)$ is an isomorphism. Therefore, $B = r_i^{-1}(r_i(B))$.

Next let $\mathcal{B}'_1$ be the union of all σ-algebras generated by countable collections of open-closed subsets of $\mathrm{Gal}(K)^e$. Then $\mathcal{B}'_1$ is a sub-σ-algebra of $\mathcal{B}_1$ which contains all open-closed subsets. Therefore, $\mathcal{B}'_1 = \mathcal{B}_1$.

Now consider $B \in \mathcal{B}_1$. The preceding paragraph gives open-closed sets $B_1, B_2, B_3, \ldots$ which generate a σ-algebra containing B. For each k the first paragraph gives $i_k \in I$ such that $B_k = r_{i_k}^{-1}(r_{i_k}(B_k))$. By assumption, there is an $i \in I$ with $\bigcup_{k=1}^{\infty} F_{i_k} = F_i$. For each k we have $r_i^{-1}(r_i(B_k)) = B_k$. The family of all subsets A of $\mathrm{Gal}(F)^e$ satisfying $r_i^{-1}(r_i(A)) = A$ is closed under countable unions and taking complements. Hence, it contains B, as contended. □

THEOREM 18.8.8: *Let K be one of the following types of fields:*
(a) *$K = E_0(T)$, where E_0 is a field and T is an uncountable set of algebraically independent elements over E_0.*
(b) *$K = E(t)$, where E is an uncountable algebraically closed field and t is a transcendental element over E.*

Then, for each positive integer e neither $S_e(K)$ nor $\mathrm{Gal}(K)^e \smallsetminus S_e(K)$ contains a set of positive measure.

Proof: In both cases $K = E(t)$ with t transcendental over an uncountable field E. Thus, from Proposition 18.8.5 we need only to prove that $S_e(K)$ contains no set of positive measure. This will be achieved by applying Lemma 18.8.1 to $G = \mathrm{Gal}(K)^e$ and $S = \mathrm{Gal}(K)^e \smallsetminus S_e(K)$. So, consider $B \in \mathcal{B}_1$.

First suppose that (a) holds. Then K is the union of all $E_0(T_0)$ where T_0 ranges over all countable subsets of T. By Lemma 18.8.7, there is a countable subset T_1 of T such that $B = r^{-1}(r(B))$ where $r\colon \mathrm{Gal}(K)^e \to \mathrm{Gal}(E_0(T_1))^e$ is the restriction map. Since T is uncountable there is a $t \in T \smallsetminus T_1$. Let $E = E_0(T \smallsetminus \{t\})$ and let $r_1\colon \mathrm{Gal}(K)^e \to \mathrm{Gal}(E)^e$ be the restriction map. By Lemma 18.8.6, $r_1(\mathrm{Gal}(K)^e \smallsetminus S_e(K)) \supseteq \mathrm{Gal}(E)^e \smallsetminus \{1\}$. Hence, $r(\mathrm{Gal}(K)^e \smallsetminus S_e(K)) \supseteq \mathrm{Gal}(E_0(T_1))^e \smallsetminus \{1\}$.

Now suppose (b) holds. Then K is the union of all fields $E_0(t)$ with E_0 a countable algebraically closed subfields of E. Lemma 18.8.7 gives a countable algebraically closed subfield E_1 of E such that $B = r_1^{-1}(r_1(B))$, where $r_1\colon \mathrm{Gal}(K)^e \to \mathrm{Gal}(E_1(t))^e$ is the restriction map. Since E is uncountable, it has a nonempty transcendence basis U over E_1. Choose $u \in U$ and let E_2 be the algebraic closure of $E_1(U \smallsetminus \{u\})$. Then E is the algebraic closure of $E_2(u)$. Moreover, with $r\colon \mathrm{Gal}(K)^e \to \mathrm{Gal}(E_2(t))^e$ the restriction map, $r^{-1}(r(B)) = B$. By Lemma 18.8.6(b2), $r(\mathrm{Gal}(K)^e \smallsetminus S_e(K)) \supseteq \mathrm{Gal}(E_2(t))^e \smallsetminus \{1\}$.

In both cases we conclude from Lemma 18.8.1 that $S_e(K)$ contains no set of positive measure. ☐

COROLLARY 18.8.9: *Let K be a field of type (a) or (b), as in Theorem 18.8.8. Then, for each finite separable extension K' of K, the set $S_e(K')$ is nonmeasurable.*

Proof: Note that $S_e(K') = S_e(K) \cap \mathrm{Gal}(K')^e$. Assume $S_e(K')$ is measurable. If $\mu(S_e(K')) > 0$, then $\mu(S_e(K)) > 0$. Otherwise, $\mu(S_e(K')) = 0$. Hence, $\mu(\mathrm{Gal}(K)^e \smallsetminus S_e(K)) \geq \mu(\mathrm{Gal}(K')^e \smallsetminus S_e(K')) = \frac{1}{[K':K]^e} > 0$. In both cases we get a contradiction to Theorem 18.8.8. ☐

Proposition 13.4.6 produces an uncountable Hilbertian PAC field K. Since every algebraic extension of K is again a PAC field, $S_e(K) = \mathrm{Gal}(K)^e$ for each positive integer e. Thus, Theorem 18.8.8 is false for arbitrary uncountable Hilbertian fields. It is still open if case (b) of Theorem 18.8.8 holds without the assumption that E is algebraically closed.

PROBLEM 18.8.10: *Let $E(t)$ be a rational function field in one variable over an uncountable field E. Is $S_e(E(t))$ nonmeasurable?*

18.9 On the Stability of Fields

We defined stable polynomials in Section 16.2 in order to realize certain groups over Hilbertian fields. Here we interpret stability in terms of function field properties in order to construct special PAC fields:

18.9 On the Stability of Fields

Definition 18.9.1: Let F/K be a field extension. Call F/K a **function field of several variables** if F/K is finitely generated and regular. We say F/K is **stable** if it has a separating transcendence base $t_1, \ldots, t_r$ such that the Galois closure, $\hat{F}$, of the (separable) extension $F/K(\mathbf{t})$ is regular over K. This is equivalent to $\mathrm{Gal}(\hat{F}/K(\mathbf{t})) \cong \mathrm{Gal}(\hat{F}\tilde{K}/\tilde{K}(\mathbf{t}))$. Refer to $t_1, \ldots, t_r$ as a **stabilizing base** for F/K. Call K **stable** if every finitely generated regular extension F/K is stable. □

The following sufficient condition for stability is an analog of Remark 16.2.2:

LEMMA 18.9.2: *Let $t_1, \ldots, t_r$ be a separating transcendence base of a finitely generated regular extension F/K. Let $n = [F : K(\mathbf{t})]$. Denote the Galois closure of $F/K(\mathbf{t})$ by $\hat{F}$. If $\mathrm{Gal}(\hat{F}\tilde{K}/\tilde{K}(\mathbf{t})) \cong S_n$, then $t_1, \ldots, t_r$ stabilizes F/K. Thus, F/K is stable.*

Proof: By Galois theory $\mathrm{Gal}(\hat{F}\tilde{K}/\tilde{K}(\mathbf{t})) \cong \mathrm{Gal}(\hat{F}/\hat{F} \cap \tilde{K}(\mathbf{t}))$. Since $[F : K(\mathbf{t})] = n$, $n! = [\hat{F} : \hat{F} \cap \tilde{K}(\mathbf{t})] \leq [\hat{F} : K(\mathbf{t})] \leq n!$. Hence, $\hat{F} \cap \tilde{K}(\mathbf{t}) = K(\mathbf{t})$ and $\mathrm{Gal}(\hat{F}\tilde{K}/\tilde{K}(\mathbf{t})) \cong \mathrm{Gal}(\hat{F}/K(\mathbf{t}))$. Thus, $t_1, \ldots, t_r$ is a stabilizing base for F/K. □

Refer to the transcendence base $t_1, \ldots, t_r$ of Lemma 18.9.2 as a **symmetrically stabilizing base** and to n as its **degree**.

THEOREM 18.9.3 ([Fried-Jarden1,2, Geyer-Jarden3, Madan-Madden, Geyer2, Neumann]): *Every field is stable. Moreover, every function field of several variables has a symmetrically stabilizing base of arbitrarily high degree.*

Outline of proof: As the list of creditors indicates, the proof was developed over a long period of time, each new step covering more fields and having an increased level of difficulty.

Each of the proofs considers a function field F of one variable over K and produces a symmetrically stabilizing element t. Then it applies Lemma 18.9.2 to conclude F/K is stable. The production of a symmetrically stabilizing base for arbitrary function fields is reduced to the case $r = 1$.

We survey the various proofs and indicate the new ingredients that enter in each of them:

PART A: *Reduction to function fields of one variable.* Suppose F/K is a function field of transcendence degree $r > 1$. If K is infinite, Matsusaka-Zariski (Proposition 10.5.2) gives $x, y \in F$ and $c \in K$ with $F/K(x + cy)$ regular of transcendence degree $r - 1$. In the general case there are $x, y \in F$ and a positive integer n with $F/K(x + y^n)$ regular [Neumann, Lemma 2.1, due to Geyer]. Let t_1 be $x + cy$ in the former case or $x + y^n$ in the general case. Induction gives a symmetrically stabilizing base $t_2, \ldots, t_r$ for $F/K(t_1)$ of large degree n. Then $t_1, t_2, \ldots, t_r$ form a symmetrically stabilizing base of degree n for F/K.

Assume from now on F/K is a function field of one variable.

PART B: *K is PAC [Fried-Jarden1]*. Since K is PAC, F has a prime divisor $\mathfrak{p}$ of degree 1. Let g be the genus of F/K and choose a prime number $l > 2g - 1$ with $\text{char}(K) \nmid l(l-2)$. Lemma 3.2.3 gives an element $x \in F$ with $\text{div}_\infty(x) = (l-2)\mathfrak{p}$. Now use that K is infinite to find a K-rational prime divisor $\mathfrak{q}$ of $K(x)/K$ unramified in F. Let $\mathfrak{q} = \sum_{i=1}^m \mathfrak{q}_i$ be its decomposition in F. Then $\sum_{i=1}^m \deg(\mathfrak{q}_i) = [F:K(x)] = \deg(\text{div}_\infty(x)) = l - 2$. The divisor $\mathfrak{d} = \mathfrak{q}_1 + \cdots + \mathfrak{q}_m + 2\mathfrak{p}$ of F/K satisfies $\deg(\mathfrak{d}) = l > 2g - 1$. By Exercise 3 of Chapter 3 there exists $t \in F$ with $\text{div}_\infty(t) = \mathfrak{d}$. Hence, $[F:K(t)] = l$.

Now extend the field of constants from K to $\tilde{K}$. The pole divisor of t in $\tilde{K}(t)/\tilde{K}$ decomposes in $F\tilde{K}$ as $\mathfrak{p}_1 + \ldots + \mathfrak{p}_{l-2} + 2\mathfrak{p}'$ where $\mathfrak{p}_1, \ldots, \mathfrak{p}_{l-2}, \mathfrak{p}'$ are distinct and $\mathfrak{p}'$ is the unique prime divisor of $\tilde{K}F/\tilde{K}$ that lies over $\mathfrak{p}$. Consider the Galois closure $\hat{F}\tilde{K}$ of $F\tilde{K}/\tilde{K}(t)$. Let $\hat{\mathfrak{p}}$ be a prime divisor of $\tilde{K}\hat{F}$ that extends $\mathfrak{p}'$. The Galois group $\text{Gal}(\tilde{K}\hat{F}/\tilde{K}(t))$ acts faithfully on the cosets of $\text{Gal}(\tilde{K}\hat{F}/\tilde{K}F)$ and the inertia group of $\hat{\mathfrak{p}}$ contains a transposition [Fried-Jarden1, p. 777]. Since $[\tilde{K}F : \tilde{K}(t)] = l$ is a prime, it is an elementary lemma of group theory that $\text{Gal}(\tilde{K}\hat{F}/\tilde{K}(t)) \cong S_l$ [Waerden1, p. 201], as desired.

PART C: $\text{char}(K) = 0$ *[Fried-Jarden2]*. In this case F has a projective plane curve model C defined over K having only **nodes** (i.e. singular points of multiplicity 2) as singularities [Hartshorne, p. 313]. Like any other irreducible plane curve in characteristic 0, C has these properties (over $\tilde{K}$):

(1a) C has only finitely many **inflection points** (i.e. points at which the tangent cuts C with multiplicity at least 3).

(1b) C has only finitely many **double tangents** (i.e. tangents in at least two distinct points).

(1c) Only finitely many tangents to C pass through each point of $\mathbb{P}^2(\tilde{K})$ (i.e. C has no strange points).

Let $n = \deg(C)$. Since K is infinite, there is an $\mathbf{o} \in \mathbb{P}^2(K) \smallsetminus C(K)$ outside the finitely many lines involved in (1). Each such $\mathbf{o}$ satisfies this:

(2) Every line that passes through $\mathbf{o}$ cuts C in at least $n-1$ points, and all but finitely many lines that pass through $\mathbf{o}$ cut C in n points.

Use $\mathbf{o}$ to project C onto $\mathbb{P}^1$. In field theoretic terms this means there exists $t \in F$ with $[F:K(t)] = n$ such that all inertial groups of places of $\hat{F}\tilde{K}/\tilde{K}(t)$ are generated by 2-cycles. As in Part B, consider $\text{Gal}(\hat{F}\tilde{K}/\tilde{K}(t))$ as a transitive permutation group on the set of cosets of $\text{Gal}(\hat{F}\tilde{K}/\tilde{K}F)$. The fixed field in $\hat{F}\tilde{K}$ of the group generated by the inertial groups of the extension $\hat{F}\tilde{K}/\tilde{K}(t)$ is an unramified extension of $\tilde{K}(t)$. Thus, by Remark 3.6.2(b), this fixed field is $\tilde{K}(t)$. Hence, $\text{Gal}(\hat{F}\tilde{K}/\tilde{K}(t))$ is generated by its inertial groups. Another elementary lemma of group theory [Fried-Jarden2, Lemma 1.4] implies that this group is S_n, as desired.

PART D: *K is infinite and perfect [Geyer-Jarden1]*. Not all plane curves in positive characteristic satisfy Condition (1). For example, if $\text{char}(K) = p \neq 0$ and C is defined by $X_0^{p-1}X_2 + X_1^p = 0$, then all tangents to C at finite points pass through $(0:1:0)$. Thus, $(0:1:0)$ is a strange point of C. The construction of a projective plane node model for F/K satisfying (1) must

18.9 On the Stability of Fields

therefore be done with care. It is carried out in few steps. For simplicity assume genus(F/K) > 0 and char(K) ≠ 2.

The reduction step from arbitrary finitely generated regular extensions F/K to function fields of one variable forces us to prove the latter case not only for K perfect but for an arbitrary infinite field K and for F/K **conservative**. That is, genus(F/K) = genus($F\tilde{K}/\tilde{K}$).

(3a) Choose $x, y \in F$ with $F = K(x, y)$ separable over $K(x)$. Then choose $u \in K$ such that the affine curve Γ generated over K by $(x, y + ux^2)$ has only finitely many inflection points. Then Γ has only finitely many double tangents and no strange points.

(3b) Let $(x_0{:}x_1{:}x_2)$ be a generic homogeneous point of Γ. For $i = 0, 1, 2$ let $y_i = x_i$. Then use the condition "F/K is conservative" to choose $y_3, \ldots, y_n \in F$ with $n \geq 4$ large such that $\mathbf{y} = (y_0{:}y_1{:}\cdots{:}y_n)$ generates a smooth curve Δ in $\mathbb{P}^n$ over K. The projection on the first three coordinates is a rational morphism which fails to be regular only at finitely many points. Use this projection to prove that Δ has only finitely many inflection points, finitely many double tangents and no strange points.

(3c) Choose a noninflection point $\mathbf{a} \in \Delta(\tilde{K})$ such that the tangent through $\mathbf{a}$ is not double and it intersects only finitely many tangents to Δ.

(3d) Choose a point $\mathbf{o} \in \mathbb{P}^n(K)$ not on Δ such that the projection π from $\mathbf{o}$ into $\mathbb{P}^{n-1}$ maps Δ onto a model Λ of F/K in $\mathbb{P}^{n-1}$ such that $\pi(\mathbf{a})$ is a simple noninflection point of Λ with a nondouble tangent.

(3e) If $n \geq 4$, choose $\mathbf{o}$ such that, in addition, Λ is smooth and the tangent to Λ at $\pi(\mathbf{a})$ intersects only finitely many tangents of Λ. Then repeat step (3d).

(3f) If $n = 3$, choose $\mathbf{o}$ such that, in addition, the only singular points of Λ are nodes and the tangent at $\pi(\mathbf{a})$ goes through no singular point. Consequently, $C = \Lambda$ satisfies (1).

The condition "F/K is conservative" is equivalent to "F/K has a **normal** projective model Γ". That is, the local ring of each point of $\Gamma(\tilde{K})$ in $F\tilde{K}$ is integrally closed [Rosenlicht, Thm. 12]. Suppose now F/K is a function field of several variables. Define F/K to be **conservative** if it satisfies the above condition. This is always the case when K is perfect. However, in the reduction process of Part A, $K(t_1)$ is no longer perfect, so one constructs t_1 such that $F/K(t_1)$ is conservative. Then proceed by induction.

PART E: *K is finite [Madan-Madden, Geyer2].* Following [Geyer2] choose large prime numbers k, l with $l - k \geq 3$. Write $l = k + 2m$. The Riemann Hypothesis for function fields (Theorem 4.5.2) gives $t \in F \smallsetminus K$ with $F/K(t)$ separable whose pole divisor (over $\tilde{K}$) factors as $\sum_{i=1}^{m} 2\mathfrak{p}_i + \sum_{i=1}^{k} \mathfrak{q}_i$ with $\mathfrak{p}_1, \ldots, \mathfrak{p}_m, \mathfrak{q}_1, \ldots, \mathfrak{q}_k$ distinct prime divisors of $F\tilde{K}/\tilde{K}$. Let $\hat{F}$ be the Galois closure of F/K and $G = \text{Gal}(\hat{F}\tilde{K}/\tilde{K})$. Then $[F : K(t)] = [F\tilde{K} : \tilde{K}(t)] = l$. Hence, G may be considered as a primitive subgroup of S_l. As such G contains a cycle of length k. By a theorem of Jordan [Wielandt, Thm. 13.9], $A_l \leq G$. Also, G contains an odd permutation. Therefore, $G = S_l$.

PART F: *K is infinite [Neumann].* In this case, F/K need not be conservative. Hence, F/K need not have a projective plane node model (as in Part D) [Rosenlicht, Thm. 12]. Instead, [Neumann, Prop. 2.11] considers a large prime number l and constructs a plane model C for F/K with $n = \deg(C)$ large satisfying, in addition to (1), the following conditions:

(4a) All singular points of C are either nodes or cusps of multiplicity at most l.

(4b) C has at least one cusp of multiplicity l.

Here a **cusp** is a singular point of C whose local ring in $F\tilde{K}$ is contained in a unique valuation ring.

As in Part C, use a point $\mathbf{o} \in \mathbb{P}^2(K)$ outside C and the finitely many tangents involved in (1). Projection from $\mathbf{o}$ onto $\mathbb{P}^1$ produces an element $t \in F \smallsetminus K$ satisfying this:

(5) Let $\mathfrak{p}$ be a prime divisor of $\tilde{K}(t)/\tilde{K}$. Then $\mathfrak{p} = \sum_{i=1}^{n-k} \mathfrak{p}_i + k\mathfrak{q}$, where $\mathfrak{p}_1, \ldots, \mathfrak{p}_{n-k}, \mathfrak{q}$ are distinct prime divisors of F/K. The coefficient k takes the value 0 for all but finitely many $\mathfrak{p}$, it is at most l, and it takes each of the values 2 and l at least once.

This guarantees that $\mathrm{Gal}(\hat{F}\tilde{K}/\tilde{K}) \cong S_n$ [Neumann, Prop. 2.7]. The construction of C has the same pattern as the construction of Part D, although it is more refined due to the possible lack of smooth model of F/K:

(6a) Construct a projective plane model Γ_1 for F/K satisfying (1) with a cusp $\mathbf{p}_1$ of multiplicity l [Neumann, Lemma 2.7].

(6b) Construct a projective model Γ_2 over K in $\mathbb{P}^m$ for some m with a point $\mathbf{p}_2$ whose local ring (over K) coincides with that of $\mathbf{p}_1$. The local rings (over K) of all other points are integrally closed [Rosenlicht, p. 174, Thm. 5].

(6c) Let $\varphi \colon \Gamma_1 \to \Gamma_2$ be a birational map over K which maps $\mathbf{p}_1$ biregularly onto $\mathbf{p}_2$. Denote the graph of φ by Γ. It is a projective curve in some $\mathbb{P}^m$ which satisfies (1) and has a cusp $\mathbf{p}$ of multiplicity l. Moreover, all singular points of Γ over $\tilde{K}$ are cusps of multiplicity at most l. This construction replaces the construction in (3b) which assumes that F/K is conservative.

(6d) Project Γ from a suitable point of $\mathbb{P}^m(K)$ into $\mathbb{P}^{m-1}$, being careful to preserve all properties of Γ except the following: When $m = 3$, the projected curve may have also nodes as singularities. Indeed, it must have at least one node. Infinitely many such projections produce the desired curve C. $\square$

18.10 PAC Galois Extensions of Hilbertian Fields

Let K be a countable Hilbertian field. We know $K_s(\boldsymbol{\sigma})$ is PAC for almost all $\boldsymbol{\sigma} \in \mathrm{Gal}(K)^e$ (Theorem 18.6.1). Hence, each algebraic extension of those

18.10 PAC Galois Extensions of Hilbertian Fields

$K_s(\sigma)$'s is PAC (Corollary 11.2.5). However, since $\mathrm{Gal}(K_s(\sigma)) = \langle\sigma\rangle$ is finitely generated, none of these extensions except K_s is Galois over K (Proposition 16.11.6). In this section we use the stability of fields to construct an abundance of Galois extensions of K which are PAC

LEMMA 18.10.1: *Let K be a field and M an algebraic extension of K. Suppose each X-stable polynomial $f \in K[T, X]$ with $\mathrm{Gal}(f(T, X), K(T)) \cong S_n$ for some positive integer n has infinitely many zeros in M. Then M is a PAC field.*

Proof: Apply Theorem 11.2.3. We need to prove that every nonconstant absolutely irreducible polynomial $g \in K[X, Y]$ has an M-zero. Let (x, y) be a generic point of $V(g)$. Then $F = K(x, y)$ is a regular extension of K. Theorem 18.9.3 gives a symmetrically stabilizing element t for F/K. That is, with $\hat{F}$ being the Galois closure of $F/K(t)$ we have $\mathrm{Gal}(\hat{F}/K(t)) \cong S_n$ for some positive integer n. Let z be a primitive element for $\hat{F}/K(t)$. Choose z to be integral over $K[t]$. Put $f(t, Z) = \mathrm{irr}(z, K(t))$. Then f is Z-stable over K, $x = \frac{h_1(t,z)}{h_0(t)}$, and $y = \frac{h_2(t,z)}{h_0(t)}$ for suitable polynomials h_0, h_1, h_2 with coefficients in K such that $h_0(t) \neq 0$. By assumption, f has a zero (a, b) in M with $h_0(a) \neq 0$. Then $\left(\frac{h_1(a,b)}{h_0(a)}, \frac{h_2(a,b)}{h_0(a)}\right)$ is an M-zero of g. □

Let K be a field and $\sigma_1, \ldots, \sigma_e \in \mathrm{Gal}(K)$. Denote the maximal Galois extension of K in $K_s(\sigma)$ by $K_s[\sigma]$. Note that $\mathrm{Gal}(K_s[\sigma]) = \langle\sigma^\tau \mid \tau \in \mathrm{Gal}(K)\rangle$. Thus, $\mathrm{Gal}(K_s[\sigma])$ is the smallest closed normal subgroup of $\mathrm{Gal}(K)$ containing $\langle\sigma\rangle$. We denote that group by $[\sigma]$.

THEOREM 18.10.2: *Let K be a countable Hilbertian field and e a positive integer. Then $K_s[\sigma]$ is PAC for almost all $\sigma \in \mathrm{Gal}(K)^e$.*

Proof: Denote the set of all polynomials $f \in K[T, X]$ which are X-stable by $\mathcal{F}$. For each $f \in \mathcal{F}$, Lemma 16.2.6 gives a linearly disjoint sequence $L_1, L_2, L_3, \ldots$ of Galois extensions of K and a sequence $\mathbf{a}_1, \mathbf{a}_2, \mathbf{a}_3, \ldots$ of distinct zeros of f such that for each i, $\mathbf{a}_i$ is L_i-rational and $\mathrm{Gal}(L_i/K) \cong \mathrm{Gal}(f(T, X), K(T))$.

Denote the set of $\sigma \in \mathrm{Gal}(K)^e$ which are contained in infinitely many of the sets $\mathrm{Gal}(L_i)^e$ by $S(f)$. By Borel-Cantelli (Lemma 18.5.3(b)), $\mu(S(f)) = 1$. Hence, $S = \bigcap_{f \in \mathcal{F}} S(f)$ also has measure 1. If $\sigma \in S$, then there is an infinite set I of positive integers such that $L_i \subseteq K_s(\sigma)$. By definition, $L_i \subseteq K_s[\sigma]$. Thus, $\mathbf{a}_i$ is a $K_s[\sigma]$-rational point of f. It follows from Lemma 18.10.1 that $K_s[\sigma]$ is PAC. □

In Section 25.10 we prove that almost all $\tilde{K}[\sigma]$ mentioned in Theorem 18.10.2 are Hilbertian. Here we give an example of a PAC Galois extension of K which is Hilbertian and has a special Galois group.

THEOREM 18.10.3 ([Fried-Jarden2, Thm. 4.4 and Fried-Völklein2, p. 475, Cor. 1]): *Let K be a countable Hilbertian field K. Then K has a Galois extension N which is Hilbertian and PAC with $\mathrm{Gal}(N/K) \cong \prod_{k=1}^{\infty} S_k$.*

Proof: List all plane curves over K in a sequence $C_1, C_2, C_3, \ldots$. By induction construct a sequence $k_1 < k_2 < k_3 < \cdots$ of positive integers, a sequence $\mathbf{p}_1, \mathbf{p}_2, \mathbf{p}_3, \ldots$ of points in $\mathbb{A}^2(K_s)$, and a linearly disjoint sequence $L_1, L_2, L_3, \ldots$ of Galois extensions of K satisfying this:
(1a) $\mathrm{Gal}(L_k/K) \cong S_k$, $k = 1, 2, 3, \ldots$.
(1b) $\mathbf{p}_i \in C_i(L_{k_i})$.
(1c) The points $\mathbf{p}_1, \mathbf{p}_2, \mathbf{p}_3, \ldots$ are distinct.

Suppose $k_1, \ldots, k_{n-1}$, $\mathbf{p}_1, \ldots, \mathbf{p}_{n-1}$, $L_1, \ldots, L_{n-1}$, and $L_{k_1}, \ldots, L_{k_{n-1}}$ have been constructed with the above mentioned properties. Choose a generic point (x, y) for C_n over K. Then $F = K(x, y)$ is a function field of one variable over K. Theorem 18.9.3 gives a symmetrically stabilizing element t for F/K of degree k_n greater than k_{n-1}. Denote the Galois closure of $F/K(t)$ by $\hat{F}$. Then $\hat{F}/K$ is regular and $\mathrm{Gal}(\hat{F}/K(t)) \cong S_{k_n}$.

Now use Lemma 13.1.1: Specialize t to an element $a \in K$ and extend the specialization to a K-place φ of $\hat{F}$ which is finite at x, y such that the residue field L_{k_n} of F is Galois over K, linearly disjoint from $L_1 \cdots L_{n-1} L_{k_1} \cdots L_{k_{n-1}}$, and $\mathrm{Gal}(L_{k_n}/K) \cong S_{k_n}$. Then $\mathbf{p}_n = (\varphi(x), \varphi(y))$ is an L_{k_n}-rational point of C_n. A careful choice of a will place $\mathbf{p}_n$ outside the set $\{\mathbf{p}_1, \ldots, \mathbf{p}_{n-1}\}$.

If n belongs to $\{k_1, \ldots, k_n\}$, then L_n is already defined. Otherwise, use Corollary 16.2.7 to construct a Galois extension L_n of K which is linearly disjoint from $L_1 \cdots L_{n-1} L_{k_1} \cdots L_{k_n}$ over K and with $\mathrm{Gal}(L_n/K) \cong S_n$. This completes the induction.

Put $N = \prod_{k=1}^{\infty} L_k$. Then N is a Galois extension and $\mathrm{Gal}(N/K) \cong \prod_{k=1}^{\infty} S_k$. Moreover, N is a finite proper extension of the Galois extension $\prod_{k=2}^{\infty} L_k$ of K. By Weissauer (Theorem 13.9.1(b)), N is Hilbertian. Finally, each plane curve over K has an N-rational zero. It follows from Lemma 11.2.3 that N is a PAC field. $\square$

We generalize that part of Theorem 18.10.3 which states that N is Hilbertian.

For each field K we denote the compositum of all Galois extensions L of K with $\mathrm{Gal}(L/K) \cong S_n$ for some n by K_{symm}. Similarly K_{alt} denotes the compositum of all Galois extensions L of K with $\mathrm{Gal}(L/K) \cong A_n$ for some n.

THEOREM 18.10.4: *Let K be a Hilbertian field. Then K_{symm} is PAC and Hilbertian.*

Proof: In order to prove that K_{symm} is PAC, it suffices to prove that every absolutely plane curve C defined over K has a K_{symm}-rational point (Theorem 11.2.3).

Indeed, let (x, y) be a generic point over K and let $F = K(x, y)$ be the corresponding function field of C over K. By Theorem 18.9.3, F/K has a symmetrically stabilizing element t. Denote the Galois closure of $F/K(t)$ by $\hat{F}$. Then $\mathrm{Gal}(\hat{F}/K) \cong S_n$ with $n = [F : K(t)]$. By Lemma 13.1.1, $\hat{F}$ has a K-place φ such that $a = \varphi(t) \in K$, $b = \varphi(x) \neq \infty$, $c = \varphi(y) \neq \infty$, and the residue field of $\hat{F}$ at φ is Galois over K with Galois group S_n. It follows that $(b,c) \in C(K_{\text{symm}})$, as desired.

Next let K' be the compositum of all finite Galois extensions L of K with $\text{Gal}(L/K) \cong S_{2n+1}$ for some positive integer n. Let K'' be the compositum of all finite Galois extensions L of K with $\text{Gal}(L/K) \cong S_{2n}$ for some positive integer n. Then both K' and K'' are Galois extensions of K and $K_{\text{symm}} = K'K''$. By Corollary 16.2.7(a), K has a finite Galois extension L in K' with $\text{Gal}(L/K) \cong S_5$. If $K' \subseteq K''$, then the simple group A_5 is a composition factor of $\text{Gal}(K''/K)$, hence also of S_{2n} for some positive integer n. This means that $A_5 = A_{2n}$, so $5 = 2n$. This contradiction implies that $K' \not\subseteq K''$. Exchanging the roles of K' and K'' and replacing 5 by 6 in the above argument proves that $K'' \not\subseteq K'$. It follows from Corollary 13.8.4 that K_{symm} is Hilbertian. □

A similar argument as in the last paragraph of the proof of Theorem 18.10.4 that applies Proposition 16.7.8 to the groups A_n rather than to the groups S_n proves that K_{symm} is Hilbertian if K is. However, in contrast to K_{symm}, we do not know if K_{alt} is PAC.

PROBLEM 18.10.6 ([Fried-Völklein2, p. 476]): *Let K be a Hilbertian field. Is K_{alt} PAC?*

Definition 18.10.6: Let M be a field and let w be a valuation of $\tilde{M}$. Then M is said to have the **density property** with respect to w if $V(M)$ is w-dense in $V(\tilde{M})$ for every variety V defined over M. □

As an application of Lemma 18.9.3 (using only the case where the ground field is PAC) [Fried-Jarden1, p. 785] proves the following strong density property for almost all fields $\tilde{K}(\boldsymbol{\sigma})$:

THEOREM 18.10.7: *Let K be a countable Hilbertian field equipped with a valuation v. Denote the set of all extensions of v to $\tilde{K}$ by W and let e be a positive integer. Then for almost all $\boldsymbol{\sigma} \in \text{Gal}(K)^e$, the field $K_s(\boldsymbol{\sigma})$ has the density property with respect to each $w \in W$.*

If a field M has the density property with respect to a valuation w of $\tilde{M}$, then M is PAC, so Theorem 18.10.4 strengthens Theorem 18.6.1. Problem 11.5.4 asks whether every PAC field has the density property with respect to every valuation w of $\tilde{M}$.

18.11 Algebraic Groups

In this section, K denotes a field finitely generated over its prime field. In particular, if K is infinite, then K is Hilbertian (Theorem 13.4.2). We list some results about algebraic groups over the fields $K_s(\boldsymbol{\sigma})$ partially extending the PAC Nullstellensatz. In this survey we assume familiarity with the concepts of an algebraic group and of an Abelian variety. For an algebraic group G defined over K, we let $G_{\text{tor}}(K)$ be the set of all $\mathbf{p} \in G(K)$ of finite order, $G_n(K) = \{\mathbf{p} \in \text{Gal}(K) \mid \mathbf{p}^n = 1\}$, and for each prime number l, $G_{l^\infty}(K) = \bigcup_{i=1}^{\infty} G_{l^i}(K)$. In particular, if A is an Abelian variety defined over an extension L of K, then $A(L)$ is an Abelian group and rank$(A(L))$ is its rank as a $\mathbb{Z}$-module.

THEOREM 18.11.1 ([Frey-Jarden]): *Suppose K is infinite. Let e be a positive integer. Then almost all $\sigma \in \text{Gal}(K)^e$ have this property: For each Abelian variety A of positive dimension defined over $K_s(\sigma)$, rank$(A(K_s(\sigma))) = \infty$. Thus, $A(K_s(\sigma))$ contains points $\mathbf{a}_1, \mathbf{a}_2, \mathbf{a}_3, \ldots$ which are linearly independent over $\mathbb{Z}$.*

The celebrated Mordell-Weil Theorem asserts that $A(K)$ is a finitely generated Abelian group [Lang3, p. 71, Thm. 1]. In particular, rank$(A(K))$ is finite and $A_{\text{tor}}(K)$ is a finite group. By Theorem 18.11.1, $K_s(\sigma)$ is a "large algebraic extension of K" as far as the ranks of Abelian groups are concerned. In contrast, the following conjecture draws a line between 1 and the integers greater than 1. Whereas, $\tilde{K}(\sigma)$ are "large" with respect to torsion if $e = 1$, they seem to be "small" if $e \geq 2$.

In what follows we let l vary on the set of prime numbers.

CONJECTURE 18.11.2 ([Geyer-Jarden1, p. 620]): *For almost all $\sigma \in \text{Gal}(K)^e$ each Abelian variety A with $\dim(A) \geq 1$ defined over $\tilde{K}(\sigma)$ has these properties:*
(a) *If $e = 1$, there exist infinitely many l's with $A_l(\tilde{K}(\sigma)) \neq 0$.*
(b) *If $e \geq 2$, then there exist only finitely l's with $A_l(\tilde{K}(\sigma)) \neq 0$.*
(c) *If $e = 1$, then $A_{l^\infty}(\tilde{K}(\sigma))$ is finite for every l.*

Remark 18.11.3: Partial results. [Geyer-Jarden1] proves Conjecture 18.11.2 for elliptic curves, that is, when $\dim(A) = 1$. In the general case the conjecture is true if K is finite [Jacobson-Jarden1, p. 114]. The article [Jacobson-Jarden2] proves part (c) of the conjecture in each case and part (b) for char$(K) = 0$. [Geyer-Jarden6] proves a weak version of (a) when K is a number field and A is defined over K: There exists a finite Galois extension L over K such that almost all $\sigma \in \text{Gal}(L)^e$ satisfy (a), (b), and (c) of Conjecture 18.11.2. Moreover, if End$(A) = \mathbb{Z}$ and $\dim(A)$ is either odd or 2, or 6, L may be taken to be K. All other cases of the conjecture are still open. $\square$

Analogous results hold for linear algebraic groups:

THEOREM 18.11.4 ([Jarden7, Thms. B and E]): *For almost all $\sigma \in \text{Gal}(K)^e$ each linear algebraic group G defined over $\tilde{K}(\sigma)$ has the following properties:*
(a) *If $e = 1$ and the order of the torsion of $G(\tilde{K})$ is not bounded, then there are infinitely many l's with $G_l(\tilde{K}(\sigma)) \neq 1$.*
(b) *If $e \geq 2$, then the order of the torsion of $G(\tilde{K}(\sigma))$ is bounded.*

For a later use we prove Theorem 18.11.4 in the case where G is the multiplicative group $\mathbb{G}_m$ of a field. In this case the torsion part is the group

18.11 Algebraic Groups

U of roots of unity.

LEMMA 18.11.5: *Let K be a finitely generated extension of $\mathbb{Q}$ and e a positive integer. Then the following statements hold for almost all $\sigma \in \mathrm{Gal}(K)^e$:*
(a) *If $e = 1$, then there are infinitely many l's with $\zeta_l \in \tilde{K}(\sigma)$.*
(b) *If $e \geq 2$, there are only finitely many l's with $\zeta_l \in \tilde{K}(\sigma)$.*
(c) *If $e \geq 1$, then for each l there are only finitely many positive integers i with $\zeta_{l^i} \in \tilde{K}(\sigma)$.*

Proof of (a) and (b): Let $S_e(K)$ be the set of all $\sigma \in \mathrm{Gal}(K)^e$ for which there are infinitely many l with $\zeta_l \in \tilde{K}(\sigma)$. For $K = \mathbb{Q}$ the series

$$\sum_{l \geq 3} \frac{1}{[\mathbb{Q}(\zeta_l) : \mathbb{Q}]^e} = \sum_{l \geq 3} \frac{1}{(l-1)^e}$$

diverges if $e = 1$ and converges if $e \geq 2$. In addition, the fields $\mathbb{Q}(\zeta_l)$, with l ranging over all prime numbers, are linearly disjoint over $\mathbb{Q}$ (Example 2.5.9). Hence, by Lemma 18.5.3, almost all $\sigma \in \mathrm{Gal}(K)$ belong to infinitely many $\mathrm{Gal}(\mathbb{Q}(\zeta_l))$ and almost no $\sigma \in \mathrm{Gal}(K)^e$ belong to infinitely many $\mathrm{Gal}(\mathbb{Q}(\zeta_l))^e$ if $e \geq 2$. Thus, $\mu_{\mathbb{Q}}(S_1(\mathbb{Q})) = 1$ and $\mu_{\mathbb{Q}}(S_e(\mathbb{Q})) = 0$ if $e \geq 2$.

If K is a finite extension of $\mathbb{Q}$, then $S_e(K) = S_e(\mathbb{Q}) \cap \mathrm{Gal}(K)^e$. Hence, $\mu_K(S_1(K)) = 1$ and $\mu_K(S_e(K)) = 0$ if $e \geq 2$. Finally, in the general case, $K_0 = \tilde{\mathbb{Q}} \cap K$ is a finite extension of $\mathbb{Q}$. Let $\rho \colon \mathrm{Gal}(K)^e \to \mathrm{Gal}(K_0)^e$ be the restriction map. If $\sigma \in \mathrm{Gal}(K)^e$ and $\sigma_0 = \rho(\sigma)$, then $\zeta_l \in \tilde{K}(\sigma)$ if and only if $\zeta_l \in \tilde{K}_0(\sigma_0)$. Hence, $S_e(K) = \rho^{-1}(S_e(K_0))$. By Proposition 18.2.2 and the preceding paragraph, $\mu_K(S_1(K)) = 1$ and $\mu_K(S_e(K)) = 0$ if $e \geq 2$.

Proof of (c): Since the intersection of countably many sets of measure 1 is a set of measure 1, it suffices to fix l and to prove that for almost all $\sigma \in \mathrm{Gal}(K)^e$ there are only finitely many i's with $\zeta_{l^i} \in \tilde{K}(\sigma)$.

Put $K(\zeta_{l^\infty}) = \bigcup_{i=1}^{\infty} K(\zeta_{l^i})$. Since $[\mathbb{Q}(\zeta_{l^i}) : \mathbb{Q}]$ is $(l-1)l^{i-1}$ if $l \neq 2$ and 2^{i-2} if $l = 2$ and $i \geq 2$, the field $\mathbb{Q}(\zeta_{l^\infty})$ is an infinite extension of $\mathbb{Q}$. Since $\tilde{\mathbb{Q}} \cap K$ is a finite extension of $\mathbb{Q}$, the field $K(\zeta_{l^\infty})$ is an infinite extension of K. Hence, $\mu_K(\mathrm{Gal}(K(\zeta_{l^\infty}))^e) = 0$ (Lemma 18.1.1). If $\sigma \in \mathrm{Gal}(K)^e$ and there are infinitely many i's with $\zeta_{l^i} \in \tilde{K}(\sigma)$, then $\zeta_{l^i} \in \tilde{K}(\sigma)$ for all i, so $\sigma \in \mathrm{Gal}(K(\zeta_{l^\infty}))^e$. Consequently, for almost all $\sigma \in \mathrm{Gal}(K)^e$ there are only finitely many i's with $\zeta_{l^i} \in \tilde{K}(\sigma)$. $\square$

We denote the group of roots of unity of a field F by $U(F)$.

LEMMA 18.11.6: *Let F be a field. Suppose $\zeta_l \in F$ for only finitely many l's. For each l suppose in addition that F contains ζ_{l^i} for only finitely many i's. Then $U(F)$ is finite.*

Proof: Denote the set of all positive integers n with $\zeta_n \in F$ by N. If $n \in N$ and $l^i | n$, then $\zeta_{l^i} \in F$. Hence, by assumption, only finitely many prime numbers l divide integers n belonging to N. Moreover, the powers of those l's are bounded. Therefore, N is finite. $\square$

THEOREM 18.11.7: *For almost all* $\sigma \in \operatorname{Gal}(K)^e$, $U(\tilde{K}(\sigma))$ *is infinite if* $e = 1$ *and finite if* $e \geq 2$.

Proof: If K is a finitely generated extension of $\mathbb{Q}$, the theorem is a consequence of Lemmas 18.11.5 and 18.11.6. Suppose K is a finitely generated extension of $\mathbb{F}_p$. Then $K_0 = \tilde{\mathbb{F}}_p \cap K$ is a finite field. For almost all $\sigma \in \operatorname{Gal}(K_0)^e$ the field $\tilde{K}_0(\sigma)$ is infinite if $e = 1$ and finite if $e \geq 2$ (Corollary 18.5.9). Also, $U(\tilde{K}_0(\sigma)) = \tilde{K}_0(\sigma)^\times$. Applying Proposition 18.2.2 to the map res: $\operatorname{Gal}(K)^e \to \operatorname{Gal}(K_0)^e$, we conclude that for almost all $\sigma \in \operatorname{Gal}(K)^e$, $U(\tilde{K}(\sigma))$ is infinite if $e = 1$ and finite if $e \geq 2$. □

Exercises

1. Let $B_1, B_2, B_3, \ldots$ be measurable subsets of a profinite group G. Suppose $\mu(B_i \cap B_j) = 0$ for $i \neq j$. Prove: $\mu(\bigcup_{i=1}^\infty B_i) = \sum_{i=1}^\infty \mu(B_i)$.

2. Let μ be the Haar measure on a profinite group G. Prove that $\mu(A^{-1}) = \mu(A)$ for every measurable subset A of G. Hint: Use the uniqueness of μ.

3. Let L/K be a finite Galois extension and $\sigma_0 \in \operatorname{Gal}(L/K)^e$. Consider a linearly disjoint sequence of finite extensions $M_1, M_2, M_3, \ldots$ of L satisfying this: M_i/K is Galois, $i = 1, 2, 3, \ldots$ and $\sum_{i=1}^\infty [M_i : K]^{-e} = \infty$. Let $\sigma_i \in \operatorname{Gal}(M_i/K)^e$ with $\operatorname{res}_L(\sigma_i) = \sigma_0$. Put $A_i = \{\sigma \in \operatorname{Gal}(K)^e \mid \operatorname{res}_{M_i}(\sigma) = \sigma_i\}$. Prove: $\mu(\bigcup_{i=1}^\infty A_i) = [L : K]^{-e}$. Hint: Extend σ_0 to $\rho \in \operatorname{Gal}(K)^e$ and consider the sets $\rho^{-1} A_i$, $i = 1, 2, 3, \ldots$.

4. Follow the following outline to give an alternative proof of Theorem 18.6.1 based on Exercise 3 of Chapter 11 rather than on the stronger Theorem 11.2.3.

    (a) Consider a finite separable extension L of K, an absolutely irreducible polynomial $f \in L[T_1, \ldots, T_r, X]$, separable in X, and a nonzero $g \in L[T_1, \ldots, T_r]$. Denote the set of $\sigma \in \operatorname{Gal}(K)^e$ for which there are $a_1, \ldots, a_r, b$ in $K_s(\sigma)$ with $f(\mathbf{a}, b) = 0$ and $g(\mathbf{a}) \neq 0$ by $S(L, f, g)$. Imitate the construction of Part A of the proof of Theorem 18.6.1 to prove that $S(L, f, g)$ has measure 1 in $\operatorname{Gal}(L)^e$. Deduce from Proposition 18.2.4 that $\operatorname{Gal}(K)^e \smallsetminus S(L, f, g)$ is a measure zero subset of $\operatorname{Gal}(K)^e$.

    (b) Conclude the proof of Theorem 18.6.1 by considering all possible L, f and g.

5. For K a countable Hilbertian field, consider a finite proper separable extension L_0 of K. Order the nonconstant absolutely irreducible polynomials $f \in K[T, X]$ in a sequence. Use it to construct a separable algebraic extension L of K which is PAC and linearly disjoint from L_0. In particular, $L \neq K_s$.

6. Let K be a countable Hilbertian PAC field with $\operatorname{char}(K) = 0$. Let $\mathcal{F}$ be as in the proof of Theorem 18.10.2. Order the elements of $\mathcal{F}$ in two sequences such that each polynomial $f \in \mathcal{F}$ appears infinitely many times in both of them. Use these sequences to construct two linearly disjoint Galois extensions N_1 and N_2 of K such that each of them is Hilbertian and PAC [Dries-Smith].

7. Let K and $\mathcal{F}$ be as in Exercise 6. List all finite extensions of K as $K_1, K_2, K_3, \ldots$. Use Lemma 2.5.6 to select suitable subsequences of $\mathcal{F}$ and to construct by induction a descending sequence $N_1 \supseteq N_2 \supseteq N_3 \supseteq \cdots$ of Galois extensions of K which are Hilbertian and PAC and such that $K_i \cap N_i = K$ for $i = 1, 2, 3, \ldots$. Observe that $\bigcap_{i=1}^{\infty} N_i = K$ [Dries-Smith].

8. Prove: The set of all $\sigma \in \text{Gal}(\mathbb{Q})$ with $\tilde{\mathbb{Q}}(\sigma)$ non-PAC is dense in $\text{Gal}(\mathbb{Q})$. Hint: Use the arguments of Remark 18.6.2

9. Let B be a subset of positive Haar measure of a profinite group G. Prove: There is an epimorphism π of G onto a profinite group $\bar{G}$ with $\text{rank}(\bar{G}) \leq \aleph_0$ and $\mu_{\bar{G}}(\pi(B)) = \mu_G(B)$.

10. Let K be a Hilbertian field. Use Corollary 16.3.6 and the proof of the free generators theorem (Theorem 18.5.6) to prove for each positive integer e that $\langle \sigma \rangle \cong \hat{\mathbb{Z}}^e$ for almost all $\sigma \in \text{Gal}(K_{ab}/K)^e$.

Notes

The proof of the existence of the Haar measure for a profinite group (Proposition 18.2.1) is easier than the proof of the existence of the Haar measure for arbitrary locally compact groups [Halmos, Chap. XI]. It was worked out by Comay.

Let G and H be profinite groups and $\mathcal{A}$ and $\mathcal{B}$ famailies of subsets of G and H, respectively. Following [Halmos1, p. 140] but in contradiction to the usual convention of set theory, [Fried-Jarden5, Sec. 18.4] uses $\mathcal{A} \times \mathcal{B}$ for the σ-algebra generated by all of the sets $A \times B$ with $A \in \mathcal{A}$ and $B \in \mathcal{B}$. Section 18.4 uses now the more appropriate notation $\mathcal{A} \otimes \mathcal{B}$.

[Ax1, p. 177] proves the free generators theorem (Theorem 18.5.6) in the special case $K = \mathbb{Q}$ and $e = 1$. His method of proof is restricted to this case and is more in the nature of the proof of Lemma 18.5.8.

Theorem 18.6.1 (the PAC Nullstellensatz) was originally proved in [Jarden1] in respond to a problem of [Ax2, p. 269] to produce a proper subfield of $\tilde{\mathbb{Q}}$ which is PAC.

The bottom theorem (Theorem 18.7.7) was conjectured in [Jarden3, p. 300] and was proved there for $e = 1$. A proof for $e \leq 5$ is given in [Jarden6]. [Chatzidakis2] proves the bottom theorem for countable Hilbertian fields. A pro-p theoretic analog to the bottom theorem is Serre's theorem [Serre1]:

If a torsion free finitely generated pro-p group G contains an open free pro-p group F, then G is pro-p free.

[Stallings] proves an analogous theorem for abstract groups. It is shown in [Jarden3, p. 300] that the profinite analog of the Serre-Stallings theorem would imply the bottom theorem. In [Jarden11] it is proved that if G is a torsion-free profinite group which contains a free profinite subgroup F of rank e and $2 \leq e < \infty$, then $(G : F)$ divides $e - 1$. In particular, for $e = 2$, $G = F$. So the profinite analog of the Serre-Stalling theorem holds for $e = 2$.

For $e \neq 2$ the analog is false. Brandis [Jarden11, p. 32] gives a counter example for $e = 1$. [Haran1] and Melnikov (private communication) give counter examples for $e \geq 3$.

Theorem 18.8.8(a) was originally proved in [Jarden-Shelah]. Theorem 18.8.8(b) was left in [Jarden-Shelah] as an open problem. It was settled in [Fried-Jarden3] and later also in [Ershov6].

We consider the stability of fields (Theorem 18.9.3) as one of the cornerstones of Field Arithmetic. The proof of this theorem which was started by [Fried-Jarden1] in 1976, and continued in [Fried-Jarden2], [Geyer-Jarden3], [Madan-Madden], and [Geyer2] was brought to a successful end by [Neumann] in 1998.

Chapter 19. Effective Field Theory and Algebraic Geometry

Present fashion in field theory and algebraic geometry is to replace classical constructive proofs by elegant existence proofs. For example, it is rare for students to see an actual procedure for factoring polynomials in $\mathbb{Q}[X_1,\ldots,X_n]$ in the course of finding out that it is a unique factorization domain. But constructive factorization is the essential backbone of constructive demonstrations that every K-closed set is the union of finitely many K-varieties, (i.e. Hilbert's basis theorem) and that every K-variety can be normalized.

Since one of the goals of this book is to present decision procedures for theories of PAC fields, we require a preliminary concept, "a presented field with elimination theory", in order to display the essential (for our purposes), basic explicit operations of field theory and algebraic geometry. In particular, all fields finitely generated over their prime fields, and their algebraic closures, are fields with elimination theory (Corollary 19.2.10).

19.1 Presented Rings and Fields

The basic structures are the prime fields $\mathbb{Q}$, $\mathbb{F}_p$, finite algebraic extensions of these fields, the ring $\mathbb{Z}$ and rings of polynomials over these rings. We consider elements of these rings as **recognizable** and we may perform computations with them effectively. To make this precise we now employ the concept of primitive recursive functions (Sections 8.3 and 8.4).

Consider a sequence $(\xi_1, \xi_2, \xi_3, \ldots)$ of symbols. Define **polynomial words** inductively: Each element of $\mathbb{Z}$ and each ξ_i is a polynomial word. If t_1 and t_2 are polynomial words and $n \in \mathbb{Z}$, then $n \cdot t_1$, $(t_1 + t_2)$, and $(t_1 \cdot t_2)$ are polynomial words. Denote the set of all formal quotients of polynomial words by Λ. For example, $((\xi_1 + \xi_2) + \xi_3)$ and $(\xi_1 + (\xi_2 + \xi_3))$ are two distinct polynomial words and $((5 \cdot \xi_1) + (\xi_2 \cdot \xi_2))/(\xi_2 + (-1 \cdot \xi_2))$ is an element of Λ.

Writing each $n \in \mathbb{N}$ in its decimal form and ξ_i as $\xi[i]$, we view Λ as a subset of the set Λ' of all finite strings in the finite alphabet

$$(\alpha_1, \ldots, \alpha_{19}) = (0, 1, \ldots, 9, \xi, /, +, \cdot, -, (,), [,]).$$

A **Gödel numbering** (or a **primitive recursive indexing** [Rabin]) on Λ' (resp. Λ) is given by the injective map $i' \colon \Lambda' \to \mathbb{N}$ (resp. its restriction i to Λ) defined by

(1) $$i'(\alpha_{j(1)} \alpha_{j(2)} \cdots \alpha_{j(n)}) = p_1^{j(1)} p_2^{j(2)} \cdots p_n^{j(n)},$$

where $2 = p_1 < p_2 < p_3 < \cdots$ is the sequence of prime numbers. Then $i(\Lambda)$ is a primitive recursive subset of $\mathbb{N}$. For each $n \in \mathbb{N}$, let $i^{(n)} \colon \Lambda^n \to \mathbb{N}^n$ be the nth power if i.

To each function $f\colon \Lambda^n \to \Lambda$ there corresponds a unique function $f'\colon \mathbb{N}^n \to \mathbb{N}$ such that $f' \circ i^{(n)} = i \circ f$ and f' is identically 1 on $\mathbb{N}^n \smallsetminus i^{(n)}(\Lambda^{(n)})$.

Call f **primitive recursive** if f' is primitive recursive. Similarly, a subset Δ of Λ^n is primitive recursive if $i^{(n)}(\Delta)$ is primitive recursive. The following sets and functions over Λ are primitive recursive:

(2a) $\mathbb{N} = \{1, 2, \ldots\}$ and its primitive recursive functions and sets.
(2b) $\mathbb{Z} = \{0, \pm 1, \pm 2, \ldots\}$.
(2c) All sets $\{\xi_j \mid j \in S\}$ with S a primitive recursive subset of $\mathbb{N}$.
(2d) The set Π of all polynomial words.
(2e) For each $n \in \mathbb{N}$, the set Π_n of all polynomial words in $\xi_1, \ldots, \xi_n$.
(2f) The function from Π to $\mathbb{N}$ that maps $\pi \in \Pi$ to the smallest n with $\pi \in \Pi_n$.
(2g) The subset Π'_n of Π_n of all polynomial words $\sum a_i \xi_1^{i_1} \xi_2^{i_2} \cdots \xi_n^{i_n}$ having a canonical form, where $a_i \in \mathbb{Z}$, the $i_1, i_2, \ldots, i_n$ are nonnegative integers, and the monomials are ordered, say, lexicographically.
(2h) The function that maps $\pi \in \Pi_n$ onto its canonical form in Π'_n.
(2i) The various degree functions on Π'_n.
(2j) The various coefficient functions on Π_n.
(2k) Addition and multiplication on the union of the sets Π'_n, $n = 1, 2, \ldots$.

In short, the usual information on polynomials is available by means of primitive recursive functions. Refer to all above mentioned functions and sets, as well as the constant functions and projections from Λ^n onto Λ, as **basic functions** and **sets**. These give the minimal framework for field theoretic operations.

Definition 19.1.1: A ring R is said to be **presented** if there exists an injective map $j\colon R \to \Lambda$ such that $j(R)$ is a primitive recursive subset of Λ and addition and multiplication are primitive recursive functions over R (via j). In addition the primitive recursive construction of this data from basic functions should be "given explicitly" (i.e. in a naive sense suitable for practical purposes). □

A field F is **presented** if, in addition, the characteristic of F is "given explicitly", and the inverse function on $F^\times$ is a primitive recursive function, "explicitly given" in terms of basic functions.

An integral domain R is **presented in its quotient field** F if both R and F are presented with R a primitive recursive subset of F. Note that a presented ring is countable and that $\mathbb{Z}$, $\mathbb{Q}$, and $\mathbb{F}_p$ can be presented.

In order to work with polynomials over presented rings, introduce the set Σ of all polynomial words in $X_1, X_2, X_3, \ldots$ with coefficients in Λ. Define primitive recursive functions as above. In particular, Λ should be a primitive recursive subset of Σ. For R a presented ring with a presentation $j\colon R \to \Lambda$, extend j to an embedding of $S = R[X_1, X_2, X_3, \ldots]$ into Σ by mapping each polynomial in S to its canonical form in Σ. Thus, we may refer to primitive recursive functions on S.

19.1 Presented Rings and Fields

An **effective algorithm** over a presented ring R is an explicitly given (in terms of basic functions) primitive recursive map $\lambda\colon A \to B$, where A and B are explicitly given primitive recursive subsets of S^n and S^m, respectively.

Definition 19.1.2: A presented field K is said to have the **splitting algorithm** if K has an effective algorithm for factoring each elements of $K[X]$ into a product of irreducible factors. Lemma 19.1.3 shows that $\mathbb{Q}$ and each of the fields $\mathbb{F}_p$ have a splitting algorithm. □

Of course, sole reliance on this definition would result in tedious descriptions of algorithms. In practice we work in terms of previously defined algorithms by using basic operations such as composition, induction, case distinction, etc. but not the unbounded μ operator (Definition 8.5.1). We accept without proof the effectiveness of the following well known algorithms:

(3a) Finding the remainder and quotient of the division of two elements of $\mathbb{N}$.
(3b) Writing an element in $\mathbb{N}$ as a product of prime powers.
(3c) The Euclidean algorithm to find the greatest common divisor of polynomials over a presented field.
(3d) The Gauss elimination procedure to solve systems of linear equations over a presented field.
(3e) For K a presented infinite field and $f \in K[X_1,\ldots,X_n]$ an explicitly given nonzero polynomial, finding $a_1,\ldots,a_n$ with $f(a_1,\ldots,a_n) \neq 0$.

The following lemma of Kronecker gives the foundational effectiveness results in algebra:

LEMMA 19.1.3: *The following algorithms are effective:*
(a) *Factoring an element of $\mathbb{Z}[X]$ into a product of irreducible factors.*
(b) *Factoring an element of $\mathbb{Q}[X]$ into a product of irreducible factors.*
(c) *Factoring an element of $K[X_1,\ldots,X_n]$ into a product of irreducible factors, with K a presented field with a splitting algorithm.*

Proof of (a): Given a polynomial $f \in \mathbb{Z}[X]$ of degree n, factor out the greatest common divisor of its coefficients to assume f is primitive. It suffices to decide for each $0 < m < n$ if f has a factor $g \in \mathbb{Z}[X]$ of degree m and in the affirmative case to compute g.

To this end, compute all factors of $f(0), f(1),\ldots,f(m)$ (include ± 1). Each $(m+1)$-tuple of possibilities for $g(0), g(1),\ldots,g(m)$ arises by choosing a factor of $f(i)$, $i = 0,\ldots,m$, for $g(i)$. Each such possibility yields a unique polynomial $b_0 + b_1 X + \cdots + b_m X^m = g_1(X)$ for which $g_1(i) = g(i)$, $i = 0,\ldots,m$. Indeed, this gives a system of linear equations for $b_0,\ldots,b_m$

$$b_0 + b_1 i + \cdots + b_m i^m = g(i), \quad i = 0,\ldots,m,$$

whose coefficients form a Vandermonde matrix. If $b_0, b_1,\ldots,b_m$ happen to be in $\mathbb{Z}$, check if $g_1(X)$ divides $f(X)$.

If none of the above cases leads to a factor of $f(X)$, then $f(X)$ has no factor of degree m.

Proof of (b): Write each polynomial in $\mathbb{Q}[X]$ as $r \cdot f(X)$, where $r \in \mathbb{Q}$ and $f(X)$ is a primitive element of $\mathbb{Z}[X]$. By Gauss' Lemma [Lang7, p. 181, Thm. 2.1] each decomposition of $f(X)$ into irreducible factors in $\mathbb{Q}[X]$ corresponds to a decomposition of $f(X)$ in $\mathbb{Z}[X]$. Now apply (a).

Proof of (c): The Kronecker substitution S_d (Section 11.3) applied to $f \in K[X_1, \ldots, X_n]$ with $d > \deg_{X_i}(f)$, $i = 1, \ldots, n$, gives $S_d(f) \in K[Y]$. If $S_d(f)$ has a nontrivial monic irreducible factor $p \in K[Y]$, then there is a unique polynomial $g \in K[X_1, \ldots, X_n]$ with $S_d(g) = p$. Check if g divides f.

If g fails to divide f for all possible monic p, then f is irreducible. □

19.2 Extensions of Presented Fields

We generalize Lemma 19.1.3(b) to fields finitely generated over their prime fields. Let F be a field extension of a presented field K. Call an element x of F **presented over** K if either x is algebraic over K and $\mathrm{irr}(x, K)$ is explicitly given or it is known that x is transcendental over K.

LEMMA 19.2.1: *If an element x is presented over a presented field K, then $K(x)$ is also a presented field.*

Proof: Let $j \colon K \to \Lambda$ be the presentation of K. Apply the transformation $\xi_i \mapsto \xi_{i+1}$, $i = 1, 2, 3, \ldots$, to assume that $\xi_1 \notin j(K)$.

First suppose x is algebraic over K of degree n. Use the Euclidean algorithm to assign to each element y of $K[x]$ a polynomial g_y in ξ_1 of degree at most $n - 1$ with coefficients in $j(K)$ for which $j^{-1}(g_y)(x) = y$. This presents $K(x)$ as a ring.

If $g \in K[X]$ is a nonzero polynomial of degree at most $n - 1$, then we can write the inverse of $g(x)$ as $c_{n-1}x^{n-1} + \cdots + c_0$, with $c_0, \ldots, c_{n-1} \in K$, as follows: Write $g(x)(c_{n-1}x^{n-1} + \cdots + c_0)$ as a polynomial $h(x)$ of degree at most $n - 1$. Then, with $h(x)$ set equal to 1, solve the system of n linear equations in the unknowns $c_0, \ldots, c_{n-1}$ that results from equating the coefficients of the powers of x on both sides. Therefore, $K(x)$ is a presented field.

If x is transcendental over K, then assign to each element in $K(x)$ the corresponding function $p(\xi_1)/q(\xi_1)$, where $\gcd(p, q) = 1$ and q is monic. □

The construction of the presentation of $K(x)$ in Lemma 19.2.1 requires a change in the presentation of K. The essential point, however, is that K is a primitive recursive subset of $K(x)$. We say $K(x)$ is **presented over** K.

Let x be a presented algebraic element over K with $[K(x) : K] = n$. Each $y \in K(x)$ can be **presented** (with respect to x) as $y = a_0 + a_1 x + \cdots + a_{n-1}x^{n-1}$, with $a_0, a_1, \ldots, a_{n-1}$ presented elements of K. Then we can effectively find the norm, $N_{K(x)/K}(y)$, as the determinant of the matrix, relative to the basis $1, x, \ldots, x^{n-1}$, of the linear operator given by multiplication

19.2 Extensions of Presented Fields

by y. Thus, the norm function $\mathrm{norm}_{K(x)/K}$ is primitive recursive. To find $\mathrm{irr}(y, K)$, check which of $1, y, y^2, \ldots, y^{n-1}$ is the first that linearly depends (over K) on the previous elements.

An n-tuple $(x_1, x_2, \ldots, x_n)$ is said to be **presented over** K if x_i is presented over $K(x_1, \ldots, x_{i-1})$, $i = 1, 2, \ldots, n$. A sequence $x_1, x_2, x_3, \ldots$ of elements of an extension F of K is said to be **presented over** K if the function $n \mapsto \mathrm{irr}(x_n, K(x_1, \ldots, x_{n-1}))$ is primitive recursive. In both cases $K(x_1, x_2, \ldots)$ is said to be **presented** over K. In this context, for x transcendental over K, define $\mathrm{irr}(x, K)$ to be the zero polynomial. In all cases $K(x_1, x_2, \ldots)$ is presented over $K(x_1, \ldots, x_m)$ for $m = 0, 1, 2, \ldots$. For the case of the infinite sequence, change, if necessary, the presentation $j\colon K \mapsto \Lambda$ so that infinitely many of ξ_i do not appear in $j(K)$, say, by using the transformation $\xi_i \mapsto \xi_{2i}$.

LEMMA 19.2.2: *If K is a field with a splitting algorithm and x is separable (this includes the transcendental case) and presented over K, then $K(x)$ also has a splitting algorithm.*

Proof: If x is transcendental over K, then Lemma 19.1.3(c) provides an algorithm for factorization in the ring $K[x, X]$. By Gauss' Lemma [Lang7, p. 181, Thm. 2.1] this algorithm extends to $K(x)[X]$. For the remaining three parts of the proof assume x is separably algebraic over K.

PART A: *Adjoining two more variables.* Let $f \in K(x)[Z]$ be a monic polynomial of degree m. Consider the following polynomial of $K(x)[Y, Z]$:

$$(1) \qquad g(Y, Z) = f(Y - xZ) = (Y - xZ)^m + g^*(Y, Z)$$

where $g^* \in K(x)[Y, Z]$ is of total degree $< m$ and g is monic of degree m as a polynomial in Y over $K(x, Z)$. The norm, $\mathrm{norm}_{K(x)/K}(g(Y, Z)) = h(Y, Z)$, of g is a monic polynomial in Y over $K(Z)$ of degree $m \cdot [K(x) : K]$ of the form

$$h(Y, Z) = \mathrm{norm}_{K(x)/K}(Y - xZ)^m + h^*(Y, Z),$$

where $h^* \in K[Y, Z]$ is of total degree less than $m \cdot [K(x) : K] = \deg(h)$.

Decompose h (Lemma 19.1.3(c)) into irreducible factors in $K[Y, Z]$: $h(Y, Z) = h_1(Y, Z) \cdots h_r(Y, Z)$. By (1), the highest homogeneous part in Y, Z of each of the factors h_i, is a factor of $\mathrm{norm}_{K(x)/K}(Y - xZ)^m$. Since each such monomial is defined over K, it must be a power of $\mathrm{norm}_{K(x)/K}(Y - xZ)$. Thus

$$(2) \qquad h_i(Y, Z) = \mathrm{norm}_{K(x)/K}(Y - xZ)^{m_i} + h_i^*(Y, Z)$$

where $1 \leq m_i \leq m$ and $h_i^* \in K[Y, Z]$ has degree smaller than $m_i \cdot [K(x) : K] = \deg(h_i)$, $i = 1, \ldots, r$.

For each i apply the Euclidean algorithm to find the greatest common divisor of g and h_i in $K(x, Z)[Y]$. Since both g and h_i are monic in Y, all polynomials involved in the procedure belong to $K(x)[Y, Z]$. Since g divides the product h of the h_i's, at least one of the resulting polynomials has positive degree. Thus, for some i, either g divides h_i or $0 < \deg(\gcd(g, h_i)) < \deg(g)$.

PART B: *The case where $g(Y, Z)q(Y, Z) = h_i(Y, Z)$ with $q \in K(x)[Y, Z]$.* Compare the highest homogeneous parts in Y and Z on each side of this equation. By (1) and (2):

$$(3) \qquad (Y - xZ)^m q^*(Y, Z) = \text{norm}_{K(x)/K}(Y - xZ)^{m_i}$$

with $q^* \in K(x)[Y, Z]$ a homogeneous polynomial. From the separability of x over K conclude that the factor $Y - xZ$ appears on the right hand side of (3) with exact multiplicity m_i. Hence, $m \leq m_i$. Therefore, $m = m_i$ and $h = h_i$ is an irreducible polynomial in $K[Y, Z]$. It follows that g is an irreducible element of $K(x)[Y, Z]$. Consequently, f is an irreducible element of $K(x)[X]$.

PART C: *There is an i with $0 < \deg(\gcd(g, h_i)) < \deg(g)$.*
Then $g(Y, Z) = g_1(Y, Z)g_2(Y, Z)$ with $g_1, g_2 \in K(x)[Z, Y]$ monic in Y, and $\deg_Y(g_1), \deg_Y(g_2) > 1$.

Substitute $Z = 0$ and $Y = X$ to obtain a nontrivial factorization $f(X) = g_1(X, 0)g_2(X, 0)$ for f in $K(x)[X]$. Continue by induction on the degree of f to complete the factorization of f into irreducible factors in $K(x)[X]$. □

From the next lemma we derive a simple conclusion about a presented n-tuple $(x_1, \ldots, x_n)$ over a field K: If each x_i is separable over $K(x_1, \ldots, x_{i-1})$, then the order of the entries of $(x_1, \ldots, x_n)$ is unimportant.

LEMMA 19.2.3: *Let (x, y) be a presented pair over a field K with a splitting algorithm. Then we may effectively present y over K. If $K(y)$ has a splitting algorithm, then we may effectively present x over $K(y)$.*

Proof: First assume x is transcendental over K. If $\text{irr}(y, K(x)) \in K[X]$, then $\text{irr}(y, K) = \text{irr}(y, K(x))$ and x is transcendental over $K(y)$. Otherwise, y is transcendental over K and $\text{irr}(y, K(x))$ involves x. Write $\text{irr}(y, K(x))$ as a monic polynomial in Y with coefficients in $K(x)$. Multiply $\text{irr}(y, K(x))$ with the least common multiple of the denominators of the coefficients to obtain an irreducible polynomial $f(x, Y)$ in $K[x, Y]$. Now replace x by X and Y by y and rewrite $f(X, y)$ as a polynomial in X with coefficients in $K[y]$. Finally, divide $f(X, y)$ by the leading coefficient to obtain $\text{irr}(x, K(y))$.

Now assume x is algebraic over K. If y is transcendental over $K(x)$, then y is transcendental over K, so assume y is algebraic over $K(x)$. Then $\text{irr}(y, K) | N_{K(x)/K}(\text{irr}(y, K(x)))$. Decompose $N_{K(x)/K}(\text{irr}(y, K(x)))$ into irreducible factors over K and check which of the factors is zero when evaluated at y. This factor will be $\text{irr}(y, K)$. Also, $\text{irr}(x, K(y))$ divides $\text{irr}(x, K)$. If $K(y)$ has a splitting algorithm, we can factor $\text{irr}(x, K)$ over $K(y)$ and check which of the factors is annihilated by x. This factor will be $\text{irr}(x, K(y))$.
□

19.2 Extensions of Presented Fields

It may, however, happen that for a presented n-tuple $(x_1, \ldots, x_n)$ over a field K the field extension $K(x_1, \ldots, x_n)/K$ is separable but there exists i with x_i inseparable over $K(x_1, \ldots, x_{i-1})$. The next lemma handles this inconvenient situation:

LEMMA 19.2.4: Let $\mathbf{x} = (x_1, \ldots, x_n)$ be a presented n-tuple over a field K with a splitting algorithm such that $K(\mathbf{x})/K$ is separable. Then:
(a) For each explicitly given permutation π of $\{1, 2, \ldots, n\}$ we can present $(x_{\pi(1)}, \ldots, x_{\pi(n)})$ over K.
(b) We can effectively find π and $r \leq n$ such that $\{x_{\pi(1)}, \ldots, x_{\pi(r)}\}$ is a separating transcendence base for $K(\mathbf{x})/K$.
(c) $K(\mathbf{x})$ has a splitting algorithm.

Proof: Statement (c) follows from (a) and (b) by Lemma 19.2.2. To prove (a) and (b), we do an induction on n and let π be a permutation of $\{1, 2, \ldots, n\}$. Since π is a product of permutations of two successive numbers we may assume π permutes k and $k+1$, and leaves all other i unchanged. By assumption, (x_k, x_{k+1}) is presented over $K(x_1, \ldots, x_{k-1})$. By induction, $K(x_1, \ldots, x_{k-1})$ has a splitting algorithm. Hence, by Lemma 19.2.3, we can effectively present x_{k+1} over $K(x_1, \ldots, x_{k-1})$. Again, by induction, $K(x_1, \ldots, x_{k-1}, x_{k+1})$ has a splitting algorithm. Therefore, by Lemma 19.2.3, we can effectively present x_k over $K(x_1, \ldots, x_{k-1}, x_{k+1})$. Consequently, $(x_{\pi(1)}, \ldots, x_{\pi(n)})$ can be presented over K.

To construct a separating transcendence base for $K(\mathbf{x})/K$ find a maximal sequence of integers $\tau_1 < \cdots < \tau_r$ between 1 and n such that x_{τ_i} is transcendental over $K(x_1, \ldots, x_{\tau_i - 1})$, $i = 1, \ldots, r$. Then $\{x_{\tau_1}, \ldots, x_{\tau_r}\}$ is a transcendence base for $K(\mathbf{x})/K$. By the preceding paragraph, we may reorder the x_i's to assume $x_1, \ldots, x_r$ is a transcendence base for $K(\mathbf{x})/K$.

If $r = n$ or if $\mathrm{char}(K) = 0$, we are done, so assume $\mathrm{char}(K) = p$ and $r < n$. Find $f \in K[X_1, \ldots, X_n]$ of minimal degree with $f(\mathbf{x}) = 0$. If f is a polynomial in $X_1^p, \ldots, X_n^p$, then $f(\mathbf{x}) = 0$ is a nontrivial relation $\sum_{i=1}^m c_i M_i^p = 0$, with $M_1, \ldots, M_m$ monomials in $x_1, \ldots, x_n$ and $c_1, \ldots, c_m \in K$. Since $K(\mathbf{x})$ and $K^{1/p}$ are linearly disjoint over K, the linear dependence of $M_1, \ldots, M_m$ over $K^{1/p}$ gives a linear dependence of $M_1, \ldots, M_m$ over K. This gives an equation $g(\mathbf{x}) = 0$ with $g \in K[X_1, \ldots, X_n]$ of lower degree than f. This is a contradiction.

It follows that there is some X_i, say X_1, that appears in f to a power not divisible by p. The element x_1 is separably algebraic over $K(x_2, \ldots, x_n)$, and we may find $\mathrm{irr}(x_1, K(x_2, \ldots, x_n))$.

Now, use the induction hypothesis to find a permutation π of $\{2, \ldots, n\}$ such that $(x_{\pi(2)}, \ldots, x_{\pi(r+1)})$ is a separating transcendence base for $K(x_2, \ldots, x_n)/K$. $\square$

LEMMA 19.2.5: Let E and F be algebraically independent extensions of a field K. Suppose E/K is separable. Then:
(a) EF/F is separable.

(b) *If in addition F/K is separable, then EF/K is separable.*

Proof: Statement (b) follows from (a) by Corollary 2.6.2(a). In order to prove (a), we may assume E/K is finitely generated. Let $t_1, \ldots, t_r$ be a separating transcendence base for E/K. Then $t_1, \ldots, t_r$ are algebraically independent over F and $EF/F(\mathbf{t})$ is separable algebraic. Therefore, EF/F is separable. ☐

LEMMA 19.2.6: *Let $\mathbf{x} = (x_1, \ldots, x_m)$ and $\mathbf{y} = (y_1, \ldots, y_n)$ be presented tuples over a field K with a splitting algorithm. Suppose $K(\mathbf{x})$ and $K(\mathbf{y})$ are separable over K and algebraically independent. Then $(\mathbf{x}, \mathbf{y})$ can be presented over K and $K(\mathbf{x}, \mathbf{y})$ has a splitting algorithm.*

Proof: Suppose we have already presented $(\mathbf{x}, \mathbf{y})$ over K. By Lemma 19.2.5, $K(\mathbf{x}, \mathbf{y})$ is separable over K. Hence, by Lemma 19.2.4, $K(\mathbf{x}, \mathbf{y})$ has a splitting algorithm.

Use induction on n to present $(\mathbf{x}, \mathbf{y})$ over K. First suppose, $n = 1$. If y_1 is transcendental over K, then y_1 is also transcendental over $K(\mathbf{x})$. Suppose y_1 is algebraic over K. By Lemma 19.2.4, $K(\mathbf{x})$ has a splitting algorithm. Decompose $\mathrm{irr}(y_1, K(\mathbf{x}))$ into irreducible factors over $K(\mathbf{x})$ and check which has y_1 as a root. This will be $\mathrm{irr}(y_1, K(\mathbf{x}))$.

Now suppose $n > 1$ and there is a presentation of $(\mathbf{x}, y_1, \ldots, y_{n-1})$ over K. Lemma 19.2.4 presents $(\mathbf{x}, y_1, \ldots, y_{n-1})$ over $K(y_1, \ldots, y_{n-1})$. Since $K(\mathbf{y})$ is algebraically independent from $K(\mathbf{x})$ over K, it is algebraically independent from $K(\mathbf{x}, y_1, \ldots, y_{n-1})$ over $K(y_1, \ldots, y_{n-1})$ (Section 2.6, discussion preceding Lemma 2.6.5). The case $n = 1$ gives a presentation of $(\mathbf{x}, \mathbf{y})$ over $K(\mathbf{y})$, hence also over K. ☐

Every field extension of a perfect field is separable. This yields another application of Lemma 19.2.4:

LEMMA 19.2.7: *Let K be a perfect field with a splitting algorithm. If an n-tuple $(x_1, \ldots, x_n)$ is presented over K, then $L = K(x_1, \ldots, x_n)$ has a splitting algorithm.*

Thus, every finitely generated presented extension of K has a splitting algorithm. This applies, in particular, to the case where K is a finite field.

Definition 19.2.8: A presented field K is said to have an **elimination theory** if every finitely generated presented extension F of K has a splitting algorithm. Consequently, F itself has an elimination theory. ☐

LEMMA 19.2.9: *Let $(x_1, \ldots, x_n)$ be a presented n-tuple over a field K with an elimination theory and π a permutation of $(1, 2, \ldots, n)$. Then we can present $(x_{\pi(1)}, \ldots, x_{\pi(n)})$ over K. Moreover, we can find π such that $x_{\pi(1)}, \ldots, x_{\pi(r)}$ is a transcendence base of $K(\mathbf{x})/K$.*

Proof: As in the proof of Lemma 19.2.4, we may assume that π permutes k and $k+1$ and fixes all other i. By assumption, $(x_1, \ldots, x_{k-1})$ is presented

19.3 Galois Extensions of Presented Fields

over K. Since K has elimination theory, $K(x_1, \ldots, x_{k-1})$ has a splitting algorithm. By assumption, (x_k, x_{k+1}) is presented over $K(x_1, \ldots, x_{k-1})$. Hence, by Lemma 19.2.3, x_k can be presented over $K(x_1, \ldots, x_{k-1}, x_{k+1})$. Again, by assumption, $(x_{k+2}, \ldots, x_n)$ is presented over $K(x_1, \ldots, x_{n+1})$. Hence, $(x_1, \ldots, x_{k-1}, x_{k+1}, x_k, x_{k+2}, \ldots, x_n)$ is presented over K. □

COROLLARY 19.2.10: *Every presented perfect field K with a splitting algorithm has an elimination theory. Hence, every presented field K, finitely generated over its prime field, has an elimination theory.*

Remark 19.2.11: The phrase "elimination theory" derives from the observation that there are effective versions of the classical elimination theory algorithms of algebraic geometry if and only if K satisfies the condition of Definition 19.2.8.

Note, that in contrast to Lemma 19.2.7, there are fields with a splitting algorithm which do not have an elimination theory [Fröhlich-Shepherdson, Thm. 7.27]. Such fields have finite purely inseparable extensions that do not have a splitting algorithm. In addition, a presented infinite algebraic extension of $\mathbb{Q}$ need not have a splitting algorithm [Fröhlich-Shepherdson, Thm. 7.12]. □

19.3 Galois Extensions of Presented Fields

Among the infinite extensions of a presented field which can be presented are the algebraic closure and the separable closure. First we give a constructive version of the primitive element theorem:

LEMMA 19.3.1 (Kronecker): *Let $L = K(x_1, \ldots, x_n)$ be a separable algebraic presented extension of a field K with a splitting algorithm. Then one can effectively find $z \in L$ such that $K(z) = L$.*

Proof: It suffices to consider the case where K is infinite: Let $T_1, \ldots, T_n$ be indeterminates and consider the element $x = T_1 x_1 + \cdots + T_n x_n$ from the rational function field $L(\mathbf{T})$. Compute a nonzero polynomial $f \in K[\mathbf{T}, X]$ of lowest degree with $f(\mathbf{T}, x) = 0$. Since x is separable over $K(\mathbf{T})$, $\frac{\partial f}{\partial X}(\mathbf{T}, x) \neq 0$. Since K is infinite, we can find $a_1, \ldots, a_n \in K$ with $\frac{\partial f}{\partial X}(\mathbf{a}, a_1 x_1 + \cdots + a_n x_n) \neq 0$. With $z = a_1 x_1 + \cdots + a_n x_n$, express $x_1, \ldots, x_n$ as elements of $K(z)$. Indeed, the partial derivative with respect to T_i of the identity $f(\mathbf{T}, T_1 x_1 + \cdots + T_n x_n) = 0$ gives

$$\frac{\partial f}{\partial T_i}(\mathbf{T}, T_1 x_1 + \cdots + T_n x_n) + x_i \cdot \frac{\partial f}{\partial X}(\mathbf{T}, T_1 x_1 + \cdots + T_n x_n) = 0,$$

$i = 1, \ldots, n$. Substituting $a_1, \ldots, a_n$ for $T_1, \ldots, T_n$, respectively, we conclude that $x_i = -\frac{\partial f}{\partial T_i}(\mathbf{a}, z) / \frac{\partial f}{\partial X}(\mathbf{a}, z)$, $i = 1, \ldots, n$. Thus, $K(z) = L$. □

LEMMA 19.3.2: *Given a separable polynomial $f(X)$ of degree n over a field K with a splitting algorithm, we may present the n-tuple $(x_1, \ldots, x_n)$ of all roots of $f(X)$, and therefore the splitting field L of $f(X)$ over K. Moreover, we can effectively compute the Galois group $G = \mathrm{Gal}(f, K)$, through its action on L, as a group of permutations of $\{x_1, \ldots, x_n\}$. If H is a presented subgroup of G, then we can compute the fixed field $L(H)$ of H acting on L.*

Note: An **effective computation** of $\mathrm{Gal}(f, K)$ is a primitive recursive map $f \mapsto \mathrm{Gal}(f, K)$ from the above polynomials in $K[X]$ of degree n into subgroups of S_n.

Proof: Let u be transcendental over K. Then $\mathrm{Gal}(f, K) = \mathrm{Gal}(f, K(u))$. Replacing K by $K(u)$ if necessary, we may assume K is infinite.

Let $g \in K[X]$ be an irreducible factor of $f(X)$ having x_1 as a zero. Then (Lemma 19.2.2) $K(x_1)$ has a splitting algorithm. Proceeding by induction we may present the splitting field of $f(X)/(X - x_1)$ over $K(x_1)$, hence the splitting field of f over K.

To compute $\mathrm{Gal}(f, K)$ we may assume f has no multiple roots. Consider the element $x = x_1 T_1 + \ldots + x_n T_n$ which appears in the proof of Lemma 19.3.1. Let $E = K(\mathbf{T})$, $F = L(\mathbf{T})$. Consider each element σ of $\mathrm{Gal}(L/K)$ as an element of S_n through the formula $\sigma(x_i) = x_{\sigma(i)}$, $i = 1, \ldots, n$. On the other hand, σ uniquely extends to an element of $\mathrm{Gal}(F/E)$.

Embed S_n into $\mathrm{Aut}(F/L)$ through the formula $(T_i)^\pi = T_{\pi^{-1}(i)}$ for $\pi \in S_n$ and $i = 1, \ldots, n$. Thus, for $\sigma \in \mathrm{Gal}(L/K)$,

$$\sigma(x) = x_{\sigma(1)} T_1 + \cdots + x_{\sigma(n)} T_n = x_1 T_{\sigma^{-1}(1)} + \cdots + x_n T_{\sigma^{-1}(n)} = x^\sigma.$$

Compute the polynomial $h(X) = \mathrm{irr}(x, E)$. Act on $h(X)$ by S_n from the right through the action on the indeterminates. We show that $\mathrm{Gal}(f, K) = \{\pi \in S_n \mid h^\pi = h\}$.

Indeed, if $h^\pi = h$ for some $\pi \in S_n$, then x and x^π are conjugate over E. Hence, there exists $\sigma \in \mathrm{Gal}(f, K)$ with $x^\sigma = \sigma(x) = x^\pi$. Thus, $T_{\sigma^{-1}(i)} = T_{\pi^{-1}(i)}$, $i = 1, \ldots, n$. Therefore, $\sigma = \pi$. Conversely, if $\sigma \in \mathrm{Gal}(f, K)$, then x and $\sigma(x)$ are conjugate over E. Hence, $h^\sigma = \mathrm{irr}(x, E) = h$.

Finally, let H be a subgroup of G. To find $L(H)$, we may assume f is irreducible in $K[X]$ and $L = K(x_1)$ (Lemma 19.3.1). Let $\{x_1, \ldots, x_m\}$ be an H-orbit in the action of H on $\{x_1, \ldots, x_n\}$. Compute $p(X) = \prod_{i=1}^{m}(X - x_i)$ and adjoin its coefficients to K. The field K' so obtained is in $L(H)$. In addition, $L = K'(x_1)$. Hence, $[L : K'] \leq m = |H|$. Consequently, $K' = L(H)$. $\square$

19.4 The Algebraic and Separable Closures of Presented Fields

We show how to present the algebraic and separable closures of fields with elimination theory:

LEMMA 19.4.1: *If K is a field with a splitting algorithm, then we can effectively find a sequence $x_1, x_2, x_3, \ldots$ of elements of K_s that presents K_s over K as a field with a splitting algorithm. If K has an elimination theory, then $\tilde{K}$ has one too.*

Proof: Order the set of all nonconstant separable polynomials of $K[X]$ in a primitive recursive sequence $p_1, p_2, p_3, \ldots$. Inductively construct a sequence $x_1, x_2, x_3, \ldots$ of separable algebraic elements over K, a sequence $n_1 \leq n_2 \leq n_3 \leq \cdots$ of positive integers, and a sequence $K \subseteq K_1 \subseteq K_2 \subseteq \cdots$ of field extensions such that $K_k = K(x_1, x_2, \ldots, x_{n_k})$ is the splitting field of $p_1 \ldots p_k$. Then $K(x_1, x_2, x_3, \ldots)$ is the separable closure of K and it is presented by the sequence $x_1, x_2, x_3, \ldots$.

We must still show how to factor polynomials over K_s into irreducible factors. Let $f \in K_s[X]$ be a nonconstant polynomial. Then f has coefficients in $L = K(x_1, \ldots, x_j)$, for some j. Compute $g(X) = \text{norm}_{L/K}(f(X))$. Factor g over K, $g(X) = \prod_{i=1}^r g_i(X^{q_i})$, with $g_i \in K[X]$ separable, irreducible, and q_i a power of $\text{char}(K)$ (or $q_i = 1$ if $\text{char}(K) = 0$), $i = 1, \ldots, r$. Then find $k \geq j$ with $g_1, \ldots, g_r \in \{p_1, \ldots, p_k\}$. The factorization of $f(X)$ into irreducible polynomials over K_k is the desired factorization of f over K_s.

If K is a field with elimination theory, an analogous construction, with $p_1, p_2, p_3, \ldots$ now all nonconstant irreducible polynomials of $K[X]$, proves that $\tilde{K}$ has a splitting algorithm. Since $\tilde{K}$ is perfect, $\tilde{K}$ is a field with elimination theory. □

Relative algebraic closures also occur in several constructions:

LEMMA 19.4.2: *Let K be a field with a splitting algorithm and let $F = K(x_1, \ldots, x_n)$ be a separable presented extension of K. Then one can effectively find the separable algebraic closure L of K in F.*

Proof: Use Lemma 19.2.4 to rearrange $(x_1, \ldots, x_n)$ so that $(x_1, \ldots, x_r)$ is a separating transcendence base for the extension F/K. By Lemma 19.3.1 we may assume $n = r + 1$. Then

$$f(X) = \text{irr}(x_{r+1}, L(x_1, \ldots, x_r)) \text{ is also } \text{irr}(x_{r+1}, K_s(x_1, \ldots, x_r)).$$

Apply Lemma 19.4.1 to compute $f(X)$ effectively. Write its coefficients as rational functions $\frac{p_i(x_1, \ldots, x_r)}{q_i(x_1, \ldots, x_r)}$, where p_i, q_i are relatively prime polynomials in $K_s[X_1, \ldots, X_r]$ and one of the coefficients of q_i is 1, $i = 0, \ldots, k-1$. Then the coefficients $b_1, \ldots, b_m$ of $p_0, q_0, \ldots, p_{k-1}, q_{k-1}$, belong to L. Since F is linearly disjoint from K_s over $K(b_1, \ldots, b_m)$, we have $L = K(b_1, \ldots, b_m)$. □

19.5 Constructive Algebraic Geometry

Throughout this section we fix a field K with elimination theory. We show how to carry out the basic operations of algebraic geometry over K:

414   Chapter 19. Effective Field Theory and Algebraic Geometry

LEMMA 19.5.1 (Noether Normalization Theorem): Let $(x_1,\ldots,x_n)$ be a presented n-tuple over K and assume that K is infinite. Then we can effectively compute the transcendence degree r of $K(\mathbf{x})$ over K. Moreover, we can effectively find linear forms

(1) $$t_i = \sum_{j=1}^{n} a_{ij}x_j, \quad i = 1,\ldots,n$$

with $a_{ij} \in K$ and $\det(a_{ij}) \neq 0$ such that $t_1,\ldots,t_r$ is a transcendence base for $K(\mathbf{x})/K$, and $t_1,\ldots,t_n$ (as well as $x_1,\ldots,x_n$) are integral over $K[t_1,\ldots,t_r]$ and presented over $K(t_1,\ldots,t_r)$.

Proof: If each x_i is transcendental over $K(x_1,\ldots,x_{i-1})$, then take $r = n$ and $t_i = x_i$, $i = 1,\ldots,n$.

Otherwise, assume x_n is algebraic over $K(x_1,\ldots,x_{n-1})$ (Lemma 19.2.9). Find an irreducible polynomial $f \in K[\mathbf{X}]$ such that $f(\mathbf{x}) = 0$ in which X_n occurs. Write f as the sum of its homogeneous parts, $f(\mathbf{X}) = \sum_{j=0}^{d} f_j(\mathbf{X})$, where $f_d(\mathbf{X}) \neq 0$ (so that $f_d(X_1,\ldots,X_{n-1},1) \neq 0$). Find $c_1,\ldots,c_{n-1} \in K$ with $f_d(c_1,\ldots,c_{n-1},1) \neq 0$ (K is infinite). Then put $t_i = x_i - c_ix_n$, $i = 1,\ldots,n-1$, and $t_n = x_n$. We claim t_n is integral over $K[t_1,\ldots,t_{n-1}]$ and presented over $K(t_1,\ldots,t_{n-1})$. Hence, so are $x_1,\ldots,x_n$.

Indeed, write $f_d(\mathbf{X}) = \sum_{\mathbf{k}} b_{\mathbf{k}} X_1^{k_1} \cdots X_n^{k_n}$, where $k_1 + \cdots + k_n = d$ for each $\mathbf{k} = (k_1,\ldots,k_n)$. Then the coefficient of X_n^d in the transformed polynomial $f(T_1 + c_1X_n,\ldots,T_{n-1} + c_{n-1}X_n, X_n) = g(T_1,\ldots,T_{n-1},X_n)$ is $\sum b_{\mathbf{k}} c_1^{k_1} \cdots c_{n-1}^{k_{n-1}} = f_d(c_1,\ldots,c_{n-1},1) \neq 0$ and $\deg_{X_n}(g) = d$. Since $(t_1,\ldots,t_{n-1},x_n)$ is a zero of $g(T_1,\ldots,T_{n-1},X_n)$, we may compute $\mathrm{irr}(x_n, K(t_1,\ldots,t_{n-1}))$. Thus, $x_n = t_n$ and $x_i = t_i + c_ix_n$, $i = 1,\ldots,n-1$, are integral over $K[t_1,\ldots,t_{n-1}]$. Now apply the induction hypothesis to $(t_1,\ldots,t_{n-1})$ and use transitivity of integral dependence to conclude the lemma. $\square$

LEMMA 19.5.2: Let $(x_1,\ldots,x_n)$ be a presented n-tuple over K. Then we can effectively find a finite set of polynomials $g_i \in K[X_1,\ldots,X_n]$, $i \in I$, such that $V(g_i \mid i \in I)$ is the K-variety generated by $\mathbf{x}$.

Proof: The first two parts of the proof handle the case where K is infinite. The third part reduces the finite case to the infinite case.

PART A: Finding the g_i when K is infinite. Apply Lemma 19.5.1 to compute the transcendence degree r of $K(\mathbf{x})/K$. Assume, after applying an invertible linear transformation, that $x_1,\ldots,x_r$ form a transcendence base for $K(\mathbf{x})/K$ and that $x_1,\ldots,x_n$ are integral over $K[x_1,\ldots,x_r]$. Let $t_1,\ldots,t_n$ be algebraically independent over $K(\mathbf{x})$. Then $z = \sum_{i=1}^{n} x_it_i$ is integral over $K[\mathbf{t}, x_1,\ldots,x_r]$. Thus, we can find a polynomial $f(\mathbf{T}, X_1,\ldots,X_r, Z)$, irreducible in $K[\mathbf{T}, X_1,\ldots,X_r, Z]$ and monic in Z, such that $f(\mathbf{t}, x_1,\ldots,x_r,z) = 0$. Write $f(\mathbf{T}, X_1,\ldots,X_r, \sum_{i=1}^{n} X_iT_i) = \sum_{i \in I} g_i(\mathbf{X})T_1^{i_1} \cdots T_n^{i_n}$, with $g_i \in K[\mathbf{X}]$, $i \in I$.

19.5 Constructive Algebraic Geometry

PART B: $V(g_i \mid i \in I)$ is the K-variety V generated by $\mathbf{x}$. By definition, $f(\mathbf{t}, x_1, \ldots, x_r, z) = \sum_{i \in I} g_i(\mathbf{x}) t_1^{i_1} \cdots t_n^{i_n} = 0$. As $t_1, \ldots, t_n$ are algebraically independent over $K(\mathbf{x})$, $g_i(\mathbf{x}) = 0$ for each $i \in I$. Hence, $V \subseteq V(g_i \mid i \in I)$.

Conversely, let $\mathbf{x}' \in \tilde{K}^n$ be a common zero of the g_i. We show $\mathbf{x} \to \mathbf{x}'$ is a K-specialization. First, extend the K-specialization $(\mathbf{t}, x_1, \ldots, x_r) \to (\mathbf{t}, x_1', \ldots, x_r')$ to a K-specialization

$$(2) \qquad (\mathbf{t}, x_1, \ldots, x_r, z) \to (\mathbf{t}, x_1', \ldots, x_r', z')$$

where $z' = \sum_{i=1}^n x_i' t_i$. Indeed, f is irreducible and monic. Hence, if $h \in K[\mathbf{T}, X_1, \ldots, X_r, Z]$ is a polynomial with $h(\mathbf{t}, x_1, \ldots, x_r, z) = 0$, then h is a multiple of f in $K[\mathbf{T}, X_1, \ldots, X_r, Z]$ (by Gauss' Lemma). In addition, $f(\mathbf{t}, x_1', \ldots, x_r', z') = \sum_{i \in I} g(\mathbf{x}') t_1^{i_1} \cdots t_n^{i_n} = 0$. Hence, $h(\mathbf{t}, x_n', \ldots, x_r', z') = 0$. Therefore, (2) holds.

Now use that $x_{r+1}, \ldots, x_n$ are integral over $K[x_1, \ldots, x_r]$ to extend (2) to a specialization

$$(3) \qquad (\mathbf{t}, x_1, \ldots, x_r, x_{r+1}, \ldots, x_n, z) \to (\mathbf{t}, x_1', \ldots, x_r', x_{r+1}'', \ldots, x_n'', z'),$$

with $x_{r+1}'', \ldots, x_n'' \in \tilde{K}$. In particular,

$$z' = x_1' t_1 + \cdots + x_r' t_r + x_{r+1}'' t_{r+1} + \cdots + x_n'' t_n.$$

Combining this with $z' = \sum_{i=1}^n x_i' t_i$ we conclude that $x_i'' = x_i'$, $i = r+1, \ldots, n$. This gives the K-specialization $\mathbf{x} \to \mathbf{x}'$.

PART C: *K is finite.* Extend K by an element u, transcendental over $K(\mathbf{x})$. By Parts A and B, we can effectively find $g_i \in K(u)[\mathbf{X}]$, for $i \in I$, such that $V(g_i \mid i \in I)$ is the $K(u)$-variety generated by $\mathbf{x}$. Without loss write the g_i as $g_i(\mathbf{X}) = \sum_{j \in J_i} g_{ij}(\mathbf{X}) u^j$, with $g_{ij} \in K[\mathbf{X}]$. Then $V(g_{ij} \mid i \in I, j \in J_i)$ is the K-variety generated by $\mathbf{x}$. □

We now present the main theorem of classical elimination theory:

LEMMA 19.5.3: *Let $f_1(\mathbf{X}, Y), \ldots, f_s(\mathbf{X}, Y)$ be explicitly given nonzero polynomials in $K[X_1, \ldots, X_n, Y]$. Then we can effectively compute a collection $g_1(\mathbf{X}), \ldots, g_r(\mathbf{X})$ of nonzero polynomials in $K[\mathbf{X}]$, called a **resultant system** of $f_1, \ldots, f_s$ with respect to Y satisfying: For each $\mathbf{x} \in \mathbb{A}^n$, $g_1(\mathbf{x}) = \cdots = g_r(\mathbf{x}) = 0$ if and only if*

(4a) *$f_1(\mathbf{x}, Y), \ldots, f_s(\mathbf{x}, Y)$ have a common zero in the algebraic closure of $K(\mathbf{x})$; or*

(4b) *the leading coefficient of f_i (as a polynomial in Y) vanishes at $\mathbf{x}$, $i = 1, \ldots, s$.*

Proof: Let $f(X) = a_0 X^k + \cdots + a_k$ and $g(X) = b_0 X^l + \cdots + b_l$ be polynomials with coefficients in an integral domain R. Then Resultant(f, g) is the determinant of an explicit $(k+l) \times (k+l)$ matrix whose entries are the

a_i's, b_j's and zeros [Lang7, p. 200]. Moreover, Resultant$(f,g) = 0$ if and only if f and g have a common root (in the algebraic closure of Quot(R)) or $a_0 = b_0 = 0$.

Returning to the original f_i's, let $d_i = \deg_Y(f_i)$ and let $a_i(\mathbf{X})$ be the leading coefficient of $f_i(\mathbf{X}, Y)$ as a polynomial in Y, $i = 1, \ldots, s$. Choose $d \geq d_i$, $i = 1, \ldots, s$ and let $u_1, \ldots, u_s, v_1, \ldots, v_s$ be algebraically independent elements over the universal domain containing K. Compute the resultant of the polynomials

(5)
$$p(Y) = u_1 Y^{d-d_1} f_1(\mathbf{X}, Y) + \cdots + u_s Y^{d-d_s} f_s(\mathbf{X}, Y)$$
$$q(Y) = v_1(Y-1)^{d-d_1} f_1(\mathbf{X}, Y) + \cdots + v_s(Y-1)^{d-d_s} f_s(\mathbf{X}, Y)$$

with respect to Y. It is a polynomial $g \in K[\mathbf{u}, \mathbf{v}, \mathbf{X}]$. Write it in the form $g(\mathbf{u}, \mathbf{v}, \mathbf{X}) = \sum_{j=1}^{r} g_j(\mathbf{X}) m_j(\mathbf{u}, \mathbf{v})$ where the $m_j(\mathbf{u}, \mathbf{v})$ are distinct monomials in $\mathbf{u}, \mathbf{v}$. Note that $\sum_{i=1}^{s} u_i a_i(\mathbf{X})$ is the leading coefficient of $p(Y)$ while $\sum_{i=1}^{s} v_i a_i(\mathbf{X})$ is the leading coefficient of $q(Y)$.

CLAIM: $g_1(\mathbf{X}), \ldots, g_r(\mathbf{X})$ is a resultant system of $f_1(\mathbf{X}, Y), \ldots, f_s(\mathbf{X}, Y)$ with respect to Y. Indeed, let $\mathbf{x} \in \mathbb{A}^n$. Suppose $g_1(\mathbf{x}) = \cdots = g_r(\mathbf{x}) = 0$. Then $g(\mathbf{u}, \mathbf{v}, \mathbf{x}) = \sum_{j=1}^{r} g_j(\mathbf{x}) m_j(\mathbf{u}, \mathbf{v}) = 0$. Hence, there is a y with

(6) $\displaystyle\sum_{i=1}^{s} u_i y^{d-d_i} f_i(\mathbf{x}, y) = p(y) = 0$ and $\displaystyle\sum_{i=1}^{s} v_i (y-1)^{d-d_i} f_i(\mathbf{x}, y) = q(y) = 0$

or $\sum_{i=1}^{s} u_i a_i(\mathbf{x}) = 0$ and $\sum_{i=1}^{s} v_i a_i(\mathbf{x}) = 0$. In the latter case $a_1(\mathbf{x}) = \cdots = a_s(\mathbf{x}) = 0$. If the latter case does not occur, there is an i with $a_i(\mathbf{x}) \neq 0$. Therefore, $\sum_{i=1}^{s} u_i a_i(\mathbf{x}) \neq 0$ and $\sum_{i=1}^{s} v_i a_i(\mathbf{x}) \neq 0$. Hence, by (6), $y \in \widetilde{K(\mathbf{x},\mathbf{u})} \cap \widetilde{K(\mathbf{x},\mathbf{v})} = \widetilde{K(\mathbf{x})}$. Since $y \neq 0$ or $y \neq 1$, (6) implies $f_1(\mathbf{x}, y) = \cdots = f_s(\mathbf{x}, y) = 0$.

Conversely, suppose (4) holds. Then $\sum_{j=1}^{r} g_j(\mathbf{x}) m_j(\mathbf{u}, \mathbf{v}) = g(\mathbf{u}, \mathbf{v}, \mathbf{x}) = 0$. Hence, $g_1(\mathbf{x}) = \cdots = g_r(\mathbf{x}) = 0$. $\square$

Remark 19.5.4: If $r = 0$, then (4a) holds for each $\mathbf{x} \in \mathbb{A}^n$ for which not all leading coefficients of $f_1, \ldots, f_s$ vanish. Thus, if $r = 0$ and $x_1, \ldots, x_n$ are algebraically independent over K, then $f_1(\mathbf{x}, Y), \ldots, f_s(\mathbf{x}, Y)$ have a common factor $d(\mathbf{x}, Y)$ in $K(\mathbf{x})[Y]$ of positive degree. Without loss (Gauss' lemma) we may take $d(\mathbf{x}, Y)$ to be a primitive nontrivial common factor of $f_1(\mathbf{x}, Y), \ldots, f_s(\mathbf{x}, Y)$ in $K[\mathbf{x}, Y]$. $\square$

Remark 19.5.5: *Universal bounds on degrees.* Suppose K in Lemma 19.5.3 is an arbitrary field and $f_1, \ldots, f_s$ are nonzero polynomials in $K[\mathbf{X}, Y]$. The proof of Lemma 19.5.3 then proves the existence of the resultant system $g_1, \ldots, g_r \in K[\mathbf{X}]$ of $f_1, \ldots, f_s$. Moreover, the proof gives a bound on r and $\deg(g_1), \ldots, \deg(g_r)$ which depends on $\deg(f_1), \ldots, \deg(f_s)$ but neither on

19.5 Constructive Algebraic Geometry

K nor on the coefficients of $f_1, \ldots, f_s$. (We will henceforth say that r and $\deg(g_1), \ldots, \deg(g_s)$ are **bounded**.)

Indeed, let p and q be as in (5). Put $k = \deg(p)$ and $l = \deg(q)$. The resultant $g(\mathbf{u}, \mathbf{v}, \mathbf{X})$ of the polynomials (5) is a $(k+l) \times (k+l)$ determinant in the coefficients of $f_1(\mathbf{X}, Y), \ldots, f_s(\mathbf{X}, Y)$ viewed as polynomials in Y. Therefore, $\deg(g(\mathbf{u}, \mathbf{v}, \mathbf{X}))$ is bounded. By definition, $g(\mathbf{u}, \mathbf{v}, \mathbf{X}) = \sum_{j=1}^{r} g_j(\mathbf{X}) m_j(\mathbf{u}, \mathbf{v})$. Consequently, r and $\deg(g_1), \ldots, \deg(g_r)$ are bounded. □

PROPOSITION 19.5.6: *Let $f_1, \ldots, f_s$ be explicitly given nonzero polynomials in $K[X_1, \ldots, X_n]$. Then we can effectively decompose the K-closed set $A = V(f_1, \ldots, f_s)$ into a union $A = V_1 \cup \cdots \cup V_m$ of K-varieties. Furthermore, we can explicitly give a generic point and the dimension of V_i, $i = 1, \ldots, m$.*

Proof: Adjoin n elements $t_1, \ldots, t_n$, algebraically independent over the universal domain containing K, to K. We construct polynomials over $K(\mathbf{t})$ in $X_1, \ldots, X_n$ and then eliminate $X_1, \ldots, X_n$, one by one. The components V_i are defined by polynomials that appear in Part D and Part E of this proof. As in the proof of Lemma 19.5.2 we may assume that K is infinite.

PART A: *Elimination of X_n.* Apply a linear transformation to $X_1, \ldots, X_n$ to assume (Lemma 19.5.1) that the coefficient of X_n in f_1 is 1. Consider this collection of polynomials in $K[\mathbf{t}, \mathbf{X}, Z]$:

(7a) $$e(\mathbf{t}, \mathbf{X}, Z) = Z - t_1 X_1 - \cdots - t_n X_n; \quad \text{and}$$
(7b) $$f_1(\mathbf{X}), \ldots, f_s(\mathbf{X}).$$

Compute (Lemma 19.5.3) a resultant system of (7) with respect to X_n:

(8) $$g_{1,j}(\mathbf{t}, X_1, \ldots, X_{n-1}, Z), \quad j = 1, \ldots, r_1.$$

Since the polynomials of (7) have no common factor in $K[\mathbf{t}, \mathbf{X}, Z]$, Remark 19.5.4 implies $r_1 \geq 1$. If $r_1 = 1$, let $h_1 = g_{1,1}$ and $h_2 = \cdots = h_n = 1$. Then the elimination procedure stops here and we skip to the second paragraph of Part D.

Otherwise, $r_1 \geq 2$. Apply (3c) of Section 19.1 and Lemma 19.1.3 in $K[\mathbf{t}, X_1, \ldots, X_{n-1}, Z]$ to find $h_1(\mathbf{t}, X_1, \ldots, X_{n-1}, Z) = \gcd(g_{1,1}, \ldots, g_{1,r_1})$ and $e_{1,1}, \ldots, e_{1,r_1} \in K[\mathbf{t}, X_1, \ldots, X_{n-1}, Z]$ such that $g_{1,j} = h_1 e_{1,j}$, $j = 1, \ldots, r_1$. If Z occurs in $e_{1,j}$ for some j, say in $e_{1,1}$, then, with another linear change of variables (not affecting X_n), we may assume that the leading coefficient of Z in $e_{1,1}$ is 1. Compute a resultant system of $e_{1,1}, \ldots, e_{1,r_1}$ with respect to Z:

(9) $$d_{1j}(\mathbf{t}, X_1, \ldots, X_{n-1}), \quad j = 1, \ldots, m_1.$$

Consider the collection (9) as polynomials in $\mathbf{t}$ and display their nonzero coefficients (in $K[X_1, \ldots, X_{n-1}]$):

(10) $$f_{1j}(X_1, \ldots, X_{n-1}), \quad j = 1, \ldots, s_1.$$

418   Chapter 19. Effective Field Theory and Algebraic Geometry

PART B: *Properties of the collection (10):*
(11a) If $(\mathbf{x}, z)$ is a common zero of the system (7), then $(\mathbf{x}, z)$ is a zero of h_1, or of all $e_{1,1}, \ldots, e_{1,r_1}, f_{1,1}, \ldots, f_{1,s_1}$ (Lemma 19.5.3);
(11b) If $(x_1, \ldots, x_{n-1}, z)$ is a presented zero of h_1 over $K(\mathbf{t})$, then we can present an algebraic zero x_n of $f_1(x_1, \ldots, x_{n-1}, X_n)$ over $K(x_1, \ldots, x_{n-1})$, such that $\mathbf{x} \in A$ and $z = \sum_{i=1}^{n} t_i x_i$.
(11c) If $(x_1, \ldots, x_{n-1}, z)$ is a presented common zero of all $e_{1,j}$ and $f_{1,j}$, then we can present an algebraic zero x_n of $f_1(x_1, \ldots, x_{n-1}, X_n)$ over $K(x_1, \ldots, x_{n-1})$, such that $\mathbf{x} \in A$ and $z = \sum_{i=1}^{n} t_i x_i$.
(11d) The collection (10) is nonempty.

Apply Remark 19.5.4 to see (11c). Otherwise, $e_{1,1}, \ldots, e_{1,r_1}$ have a common linear factor in $K[\mathbf{t}, X_1, \ldots, X_{n-1}, Z]$, contrary to their definition.

PART C: *Elimination of X_{n-1}.* Apply Part A to

$$\{e_{1,1}, \ldots, e_{1,r_1}; f_{1,1}, \ldots, f_{1,s_1}\}$$

as it is applied to (7). That is, make a linear transformation so that the leading coefficient of X_{n-1} in $f_{1,1}$ is 1. Then display a resultant system of $\{e_{1,1}, \ldots, e_{1,r_1}; f_{1,1}, \ldots, f_{1,s_1}\}$ with respect to X_{n-1}:

$$g_{2j}(\mathbf{t}, X_1, \ldots, X_{n-2}, Z), \quad j = 1, \ldots, r_2.$$

Since $e_{1,1}, \ldots, e_{1,r_1}$ have no common factor, $r_2 \geq 1$. If $r_2 = 1$, let $h_2 = g_{2,1}$, $h_3 = \cdots = h_n = 1$, and skip to Part D. Otherwise, find

$$h_2(\mathbf{t}, X_1, \ldots, X_{n-2}, Z) = \gcd(g_{2,1}, \ldots, g_{2,r_2})$$

and $e_{2,j}(\mathbf{t}, X_1, \ldots, X_{n-2}, Z)$ such that $g_{2,j} = h_2 e_{2,j}$, $j = 1, \ldots, r_2$. After a suitable linear transformation which makes the leading coefficient of Z in $e_{2,1}$ equal 1, construct a resultant system,

$$d_{2,j}(\mathbf{t}, X_1, \ldots, X_{n-2}), \quad j = 1, \ldots, m_2,$$

of $e_{2,1}, \ldots, e_{2,r_2}$ with respect to Z. Display the nonzero coefficients of $d_{2,j}$, $j = 1, \ldots, m_2$, as

(12) $f_{2j}(X_1, \ldots, X_{n-2}), \quad j = 1, \ldots, s_2$ (again $s_2 \geq 1$).

Then the following holds:
(13a) If $(\mathbf{x}, z)$ is a common zero of the system (7), then $(\mathbf{x}, z)$ is a zero of h_1, or of h_2, or of all e_{2j} and f_{2j}.
(13b) If $(x_1, \ldots, x_{n-2}, z)$ is a presented zero of h_2 over $K(\mathbf{t})$, then we can present algebraic elements x_{n-1}, x_n over $K(x_1, \ldots, x_{n-2})$, such that $\mathbf{x} \in A$ and $z = \sum_{i=1}^{n} t_i x_i$.
(13c) If $(x_1, \ldots, x_{n-2}, z)$ is a presented common zero of all $e_{1,j}$ and $f_{1,j}$, then we can present algebraic elements x_{n-1}, x_n over $K(x_1, \ldots, x_{n-2})$ such that $\mathbf{x} \in A$ and $z = \sum_{i=1}^{n} t_i x_i$.

19.5 Constructive Algebraic Geometry

PART D: *The induction step.* Proceed by induction to find m between 1 and n and for each i between m and n polynomials $h_i(\mathbf{t}, X_1, \ldots, X_{n-i}, Z)$, $e_{ij}(\mathbf{t}, X_1, \ldots, X_{n-i})$, $j = 1, \ldots, r_i$, and $f_{ij}(X_1, \ldots, X_{n-i})$, $j = 1, \ldots, s_i$, as in Parts A, B, and C such that $r_n, \ldots, r_{m+1} \geq 2$ and $r_m = 1$. Let $h_1 = \ldots = h_{m-1} = 1$. These polynomials have the following properties:

(14a) If $(\mathbf{x}, z)$ is a common zero of the system (7), then $(\mathbf{x}, z)$ is a zero of one of the h_i's. (If $m = n$, then f_{n1} is a nonzero constant, so $f_{n1}(\mathbf{x}) \neq 0$.)

(14b) Let $1 \leq i \leq n$. If $(x_1, \ldots, x_{n-i}, z)$ is a presented zero of h_i over $K(\mathbf{t})$, then we can present algebraic elements $x_{n-i+1}, \ldots, x_n$ over $K(x_1, \ldots, x_{n-i})$, such that $\mathbf{x} \in A$ and $z = \sum_{i=1}^n t_i x_i$.

(14c) If $(x_1, \ldots, x_{n-i}, z)$ is a presented common zero of all $e_{i,j}$ and $f_{i,j}$, then we can present algebraic elements $x_{n-i+1}, \ldots, x_n$ over $K(x_1, \ldots, x_{n-i})$ such that $\mathbf{x} \in A$ and $z = \sum_{i=1}^n t_i x_i$.

Decompose h_i into irreducible factors in $K[\mathbf{t}, X_1, \ldots, X_{n-i}, Z]$;

$$(15) \qquad h_i(\mathbf{t}, X_1, \ldots, X_{n-i}, Z) = \prod_{j=1}^{b_i} h_{ij}(\mathbf{t}, X_1, \ldots, X_{n-i}, Z)^{a_j},$$

$i = 1, \ldots, n$. For each i and j, display the coefficients of

$$h_{ij}(\mathbf{t}, X_1, \ldots, X_{n-i}, t_1 X_1 + \ldots + t_n X_n)$$

(as a polynomial in $t_1, \ldots, t_n$):

$$(16) \qquad h_{ijk}(X_1, \ldots, X_n), \qquad k = 1, \ldots, l_{ij}.$$

Now consider the K-closed set set $V_{ij} = V(h_{ij,1}, \ldots, h_{ij,l_{ij}})$. If h_i happens to be 1, then V_{ij} is empty for all j.

PART E: $A = \bigcup_{i=1}^n \bigcup_{j=1}^{b_i} V_{ij}$. Indeed, let $\mathbf{x} \in A$ and $z = t_1 x_1 + \cdots + t_n x_n$. Then $(\mathbf{x}, z)$ is a zero of (7). By (14a) there is an i with $h_i(\mathbf{t}, \mathbf{x}, z) = 0$. By (15) there exists a j with $h_{ij}(\mathbf{t}, x_1, \ldots, x_{n-i}, z) = 0$. Thus, $\mathbf{x}$ is a common zero of the h_{ijk}, $k = 1, \ldots, l_{ij}$, and $\mathbf{x} \in V_{ij}$.

Conversely, let $(x_1, \ldots, x_n) \in V_{ij}$. Then $h_i(\mathbf{t}, x_1, \ldots, x_{n-i}, z) = 0$, where $z = t_1 x_1 + \cdots + t_n x_n$. By (14b), there exist $x'_{n-i+1}, \ldots, x'_n$ algebraic over $K(x_1, \ldots, x_{n-i})$ with

$$(x_1, \ldots, x_{n-i}, x'_{n-i+1}, \ldots, x'_n) \in A \text{ and}$$
$$z = t_1 x_1 + \cdots + t_{n-i} x_{n-i} + t_{n-i+1} x'_{n-i+1} + \cdots + t_n x'_n.$$

But then $x'_j = x_j$, $j = n - i + 1, \ldots, n$. Hence, $\mathbf{x} \in A$.

PART F: V_{ij} is either empty or it is a K-variety of dimension $n - i$. Suppose h_{ij} is nonconstant. Then Z actually occurs in h_{ij}. Otherwise, we may start with a zero $(x_1, \ldots, x_{n-i})$ of $h_{ij}(\mathbf{t}, X_1, \ldots, X_{n-i})$ and conclude for two distinct elements z and z' the existence of $x_{n-i+1}, \ldots, x_n$, algebraic over $K(x_1, \ldots, x_{n-i})$ such that both $(\mathbf{x}, z)$ and $(\mathbf{x}, z')$ are zeros of (7a). Then $z = z'$, a contradiction. From this we produce a generic point of V_{ij}: Starting from elements $x_1, \ldots, x_{n-i}$, algebraically independent over K, choose z with $h_{ij}(\mathbf{t}, x_1, \ldots, x_{n-i}, z) = 0$. Then $h_i(\mathbf{t}, x_1, \ldots, x_{n-i}, z) = 0$. By (14b), there are $x_{n-i+1}, \ldots, x_n$ in the algebraic closure of $K(x_1, \ldots, x_{i-1})$ such that $z = \sum_{i=1}^{n} t_i x_i$. It follows that $\mathbf{x}$ belongs to V_{ij} and has maximal transcendence degree over K. Therefore, $\dim(V_{ij}) = n - i$.

Moreover, $\mathbf{x}$ is a generic point of V_{ij}, so V_{ij} is a K-variety. Indeed, $(\mathbf{t}, x_1, \ldots, x_{n-i}, z)$ is a generic point of the K-variety $V(h_{ij})$. Consider $\mathbf{x}' \in V_{ij}$. Let $z' = t_1 x_1' + \cdots + t_n x_n'$. Then $h_{ij}(\mathbf{t}, x_1', \ldots, x_{n-i}', z') = 0$. Hence, $(\mathbf{t}, x_1', \ldots, x_{n-i}', z')$ is a K-specialization of $(\mathbf{t}, x_1, \ldots, x_{n-i}, z)$. As in Part E, this specialization extends to a K-specialization

$$(\mathbf{t}, \mathbf{x}, z) \to (\mathbf{t}, x_1', \ldots, x_{n-i}', x_{n-i+1}'', \ldots, x_n'', z')$$

with

$$z' = t_1 x_1' + \cdots + t_{n-i} x_{n-i}' + t_{n-i+1} x_{n-i+1}'' + \cdots + t_n x_n''.$$

Hence, $x_i' = x_i''$, $i = n-i+1, \ldots, n$. It follows that $\mathbf{x}'$ is a K-specialization of $\mathbf{x}$. This proves our claim and concludes the proof of the proposition. $\square$

Remark 19.5.7: (a) A K-closed set A in $\mathbb{A}^n$ is said to be **presented** if polynomials $f_1, \ldots, f_s \in K[X_1, \ldots, X_n]$ are given explicitly such that $A = V(f_1, \ldots, f_s)$.

(b) If A and B are two given K-closed sets in $\mathbb{A}^n$, then Proposition 19.5.6 allows us to effectively decompose A into a union of K-varieties and to compute generic points of each of them. Substituting these points in the polynomials that define B, we can decide if $A \subseteq B$.

(c) The decomposition of A described in Proposition 19.5.6 may not be into K-irreducible components, because there might be inclusions among the V_{ij}'s. This, however, can be checked using (b). Throw away the unnecessary K-varieties from the union to effectively obtain a decomposition of A into K-irreducible components. $\square$

Remark 19.5.8: *Universal bounds on degrees.* Suppose K in Proposition 19.5.6 is an arbitrary field and $f_1, \ldots, f_s$ are nonzero polynomials in $K[\mathbf{X}]$. Then the proof of Proposition 19.5.6 proves the existence of a decomposition of $A = V(f_1, \ldots, f_s)$ into a union $V_1 \cup \cdots \cup V_m$ of K-varieties. Moreover, it gives polynomials in $K[\mathbf{X}]$ which define V_j with degrees bounded by a number e. Both m and e depend only on the degrees of $f_1, \ldots, f_s$ but neither on K nor on the coefficients of $f_1, \ldots, f_s$:

First, by Remark 19.5.5, $g_{1,j}$ in (8) have bounded degrees. Then both h_1 and $e_{1,j}$, as factors of $g_{1,j}$, have bounded degrees. Again, by Remark 19.5.5,

19.5 Constructive Algebraic Geometry

the polynomials d_{1j} in (9) have bounded degrees. Hence, the degrees of their coefficients f_{1j} are bounded. The proof of Proposition 19.5.6 proceeds in Part C and Part D by induction on n. After at most n steps it produces polynomials $h_i \in K(t_1, \ldots, t_n, X_1, \ldots, X_{n-i}, Z)$, $i = 1, \ldots, n$, with bounded degrees. Let h_{ij} be the irreducible factors of h_i, as in (15). Then $h_{ijk}(X_1, \ldots, X_n)$, $k = 1, \ldots, l_{ij}$, are the coefficients of $h_{ij}(\mathbf{t}, X_1, \ldots, X_{n-i}, t_1 X_1 + \cdots + t_n X_n)$. Both the degrees of h_{ij} and their number are bounded. Moreover, $A = \bigcup_{i=1}^{n} \bigcup_{j=1}^{b_i} V_{ij}$, where V_{ij} is defined by h_{ijk}, $k = 1, \ldots, l_{ij}$. Part F of the proof of Proposition 19.5.6 then proves that each of the V_{ij} is either empty or it is a K-variety. This completes the proof of our claim.

Note that each V_{ij} can be defined by a bounded number of polynomials. Indeed, suppose $V_{ij} = V(g_1, \ldots, g_q)$ where each g_k is a polynomial in $K[X_1, \ldots, X_n]$ of degree at most e. There are only finitely many monomials in $X_1, \ldots, X_n$, say r, of degree at most e (indeed, $r = \binom{n+e}{e}$ — see Exercise 3). Hence, the dimension of the space of all polynomials in $K[X_1, \ldots, X_n]$ of degree at most e is r. Renumbering $g_1, \ldots, g_q$, we may assume $g_1, \ldots, g_{r'}$, with $r' \leq \min(q, r)$, generate the K-vector space $\sum_{i=1}^{q} K g_i$. Then, $V_{ij} = V(g_1, \ldots, g_{r'})$, as claimed. □

As a corollary of Remark 19.5.8 we prove that the concept of an absolutely irreducible variety is elementary. We have been able to circumvent this result in this book and instead use the easier result that absolute irreducibility of polynomials is elementary. For example, we have proved that pseudo algebraic closeness of a field can be checked by absolutely irreducible polynomials and deduce that "K is PAC" is an elementary statement on K (Section 11.3). However, there are occasions where one has to use the elementary nature of absolute irreducibility of varieties directly. We therefore put this result here as a convenient reference.

Let I be the set of all n-tuples $(i_1, \ldots, i_n)$ of nonnegative integers with $i_1 + \cdots + i_n \leq d$. Put $r = |I|$. Choose a bijective map $j \colon I \to \{1, \ldots, r\}$. Then the general polynomial in $X_1, \ldots, X_n$ of degree d can be written as $f(\mathbf{T}, X_1, \ldots, X_n) = \sum_{\mathbf{i} \in I} T_{j(\mathbf{i})} X_1^{i_1} \cdots X_n^{i_n}$, with $\mathbf{T} = (T_1, \ldots, T_r)$. Every polynomial in $K[X_1, \ldots, X_n]$ of degree at most d can then be written as $f(\mathbf{a}, X_1, \ldots, X_n)$ with $\mathbf{a} \in K^r$.

PROPOSITION 19.5.9: *For all positive integers d, m, n there is a formula $\theta(\mathbf{T}_1, \ldots, \mathbf{T}_m)$ in $\mathcal{L}(\text{ring})$ satisfying this: Let K be a field and $f_1, \ldots, f_m$ be polynomials in $K[X_1, \ldots, X_n]$ of degree at most d with vectors of coefficients $\mathbf{a}_1, \ldots, \mathbf{a}_m$, respectively. Then the Zariski K-closed subset of $\mathbb{A}^n$ defined by the system of equations $f(\mathbf{a}_i, X_1, \ldots, X_n) = 0$, $i = 1, \ldots, m$, is absolutely irreducible if and only if $\theta(\mathbf{a}_1, \ldots, \mathbf{a}_m)$ holds in K.*

Proof: Let K be a field and V a Zariski closed subset of $\mathbb{A}^n$ defined by polynomials in $K[X_1, \ldots, X_n]$ of degrees at most d. Remark 19.5.8 gives positive integers e, s, and k depending only on n and d such that "V is irreducible over $\tilde{K}$" is equivalent to the following statement:

(17) There exist no polynomials $g_{ij} \in \tilde{K}[X_1, \ldots, X_n]$, $i = 1, \ldots, k$, $j = 1, \ldots, s$, of degree at most e and with $k \geq 2$ such that

$$V = \bigcup_{i=1}^{k} V(g_{i1}, \ldots, g_{is})$$

and no $V(g_{i1}, \ldots, g_{is})$ is contained in another.

For all distinct i, i' the statement $V(g_{i1}, \ldots, g_{is}) \not\subseteq V(g_{i'1}, \ldots, g_{i's})$ is equivalent over $\tilde{K}$ to "There exist $\mathbf{x}$ with $g_{i1}(\mathbf{x}) = \cdots = g_{ir}(\mathbf{x}) = 0$ and $g_{i'j}(\mathbf{x}) \neq 0$ for at least one j." Statement (17) is therefore equivalent to a formula $\tilde{\theta}(\mathbf{T}_1, \ldots, \mathbf{T}_m)$ of $\mathcal{L}(\text{ring})$ with the following property:

(18) Let $\tilde{K}$ be an algebraically closed field and $\mathbf{a}_1, \ldots, \mathbf{a}_m \in \tilde{K}^r$. Then $V(f(\mathbf{a}_1, \mathbf{X}), \ldots, f(\mathbf{a}_m, \mathbf{X}))$ is irreducible over $\tilde{K}$ if and only if $\tilde{\theta}(\mathbf{a}_1, \ldots, \mathbf{a}_m)$ is true in $\tilde{K}$.

Elimination of quantifiers (Theorem 9.2.1) gives a quantifier free formula $\theta(\mathbf{T}_1, \ldots, \mathbf{T}_m)$ which is equivalent to $\tilde{\theta}(\mathbf{T}_1, \ldots, \mathbf{T}_m)$ over every algebraically closed field. Observe that a quantifier free sentence with parameters in K is true in K if and only if it is true in $\tilde{K}$. It follows that for every field K and all $\mathbf{a}_1, \ldots, \mathbf{a}_m \in K^r$ the following chain of equivalencies holds:

$$\theta(\mathbf{a}_1, \ldots, \mathbf{a}_m) \text{ is true in } K$$
$$\iff \theta(\mathbf{a}_1, \ldots, \mathbf{a}_m) \text{ is true in } \tilde{K}$$
$$\iff \tilde{\theta}(\mathbf{a}_1, \ldots, \mathbf{a}_m) \text{ is true in } \tilde{K}$$
$$\iff V(f(\mathbf{a}_1, \mathbf{X}), \ldots, f(\mathbf{a}_m, \mathbf{X})) \text{ is irreducible over } \tilde{K}$$
$$\iff V(f(\mathbf{a}_1, \mathbf{X}), \ldots, f(\mathbf{a}_m, \mathbf{X})) \text{ is absolutely irreducible.}$$

This completes the proof of the proposition. □

19.6 Presented Rings and Constructible Sets

Let R be an integral domain presented in its quotient field K. Assume K has elimination theory. The finitely generated integral domains over K are exactly the set of coordinate rings of K-varieties. Later we use the following discussion to consider reduction of these varieties modulo prime ideals of R.

Let $S = R[\mathbf{x}]$, where $\mathbf{x} = (x_1, \ldots, x_n)$ is a presented n-tuple over K. Assume without loss that $x_1, \ldots, x_r$ is a transcendence base for $K(\mathbf{x})/K$ (Lemma 19.2.9). For each $r < i \leq n$ write $\text{irr}(x_i, K(x_1, \ldots, x_{i-1}))$ in the form $\frac{f_i(x_1, \ldots, x_{i-1}, X_i)}{h(x_1, \ldots, x_r)}$ with $f_i \in R[X_1, \ldots, X_i]$ and $h \in R[X_1, \ldots, X_r]$. Each element of $S' = R[x_1, \ldots, x_n, h(x_1, \ldots, x_r)^{-1}]$ has a unique presentation as

(1) $$\sum \frac{g_i(x_1, \ldots, x_r)}{h(x_1, \ldots, x_r)^{d(i)}} \cdot x_{r+1}^{i_{r+1}} \cdots x_n^{i_n},$$

19.6 Presented Rings and Constructible Sets

where the sum ranges over values of $i_{r+1}, \ldots, i_n$ from 0 to $\deg_{X_{r+1}}(f_{r+1}) - 1, \ldots, \deg_{X_n}(f_n) - 1$, respectively, and for each $i = (i_{r+1}, \ldots, i_n)$ the polynomial $g_i(x_1, \ldots, x_r)$ of $R[x_1, \ldots, x_r]$ is not divisible in $K[x_1, \ldots, x_r]$ by $h(x_1, \ldots, x_r)$.

An arbitrary element y of $K(\mathbf{x})$ may be presented in the form

$$(2) \quad \sum \frac{g_i(x_1, \ldots, x_r)}{h_i(x_1, \ldots, x_r)} \cdot x_{r+1}^{i_{r+1}} \cdots x_n^{i_n},$$

where the exponents $i_{r+1}, \ldots, i_n$ satisfy the same condition as above and g_i, h_i are relatively prime in $K[x_1, \ldots, x_r]$. A necessary condition for y to belong to S' is that each $h_i(x_1, \ldots, x_r)$ divides a power of $h(x_1, \ldots, x_r)$ in $K[x_1, \ldots, x_r]$. In this case, rewrite the ith coefficient as $\frac{f_i(x_1, \ldots, x_r)}{c_i \cdot h(x_1, \ldots, x_r)^{d(i)}}$ with $f_i \in R[x_1, \ldots, x_r]$ and $c_i \in R$. Then y belongs to S' if and only if $\frac{1}{c_i} f_i \in R[x_1, \ldots, x_r]$ for each i. We can check this effectively. We sum up this discussion:

LEMMA 19.6.1: *In the notation above, S' is also a primitive recursive subset of $K(\mathbf{x})$.*

DEFINITION 19.6.2: Let $h_1, \ldots, h_m \in K[X_1, \ldots, X_n]$ and P a Boolean polynomial in the symbols $\cup, \cap,$ and $'$ (i.e. taking complements). Call $A = P(V(h_1), \ldots, V(h_m))$ a **K-constructible set** in $\mathbb{A}^n$. The **dimension** of A is $\max\left(\operatorname{trans.deg}(K(\mathbf{x})/K) \mid \mathbf{x} \in A\right)$.

If L is a field containing K, denote the set of points of A with coordinates in L by $A(L)$. Call A **presented** if $h_1, \ldots, h_m$ and P are presented. Note that A can be presented in many ways. In particular, if K is the quotient field of a presented integral domain R, then we may assume that $h_1, \ldots, h_m$ have coefficients in R. □

LEMMA 19.6.3 (Chevalley): *Let $f_1, \ldots, f_m \in K[X_1, \ldots, X_n]$. Define a K-morphism $\varphi \colon \mathbb{A}^n \to \mathbb{A}^m$ by $\varphi(\mathbf{x}) = (f_1(\mathbf{x}), \ldots, f_m(\mathbf{x}))$. If $A \subseteq \mathbb{A}^n$ is a presented K-constructible set, then we can effectively describe $B = \varphi(A)$ as a presented K-constructible set.*

Proof: Let Ω be a universal domain containing K. Then

$$B = \{(y_1, \ldots, y_m) \in \Omega^m \mid \theta(y_1, \ldots, y_m) \text{ is true in } \Omega\},$$

where $\theta(Y_1, \ldots, Y_m)$ is the formula

$$(\exists X_1) \cdots (\exists X_n)[f_1(\mathbf{X}) = Y_1 \wedge \cdots \wedge f_m(\mathbf{X}) = Y_m \wedge \mathbf{X} \in A].$$

Theorem 9.3.1(a) effectively gives a quantifier-free formula $\theta'(Y_1, \ldots, Y_m)$ (in the language $\mathcal{L}(\text{ring}, K)$) equivalent to $\theta(Y_1, \ldots, Y_m)$ over Ω. This formula gives B as a constructible subset of $\mathbb{A}^m$. □

Call a K-constructible subset A of $\mathbb{A}^n$ K-**basic** if $A = V \smallsetminus V(g)$, where $V = V(f_1, \ldots, f_m)$ with $f_1, \ldots, f_m \in K[X_1, \ldots, X_n]$ is a K-variety and $g \in K[X_1, \ldots, X_n]$ does not vanish on V. If $\mathbf{x}$ is a generic point of V, then call $K[A] = K[\mathbf{x}, g(\mathbf{x})^{-1}]$ the **coordinate ring** of A and $K(A) = K(\mathbf{x})$ the **function field** of A. Then the dimension of A is the transcendence degree of $K(A)/K$. Furthermore, the K-basic set A is **normal** if $K[A]$ is an integrally closed domain, and A is **presented** if the polynomials $f_1, \ldots, f_m, g$ and the ring $K[A]$ are presented.

Remark 19.6.4: Let $A = V \smallsetminus V(g)$ be a K-basic set, $\mathbf{x}$ a generic point of V, and $u \in K(\mathbf{x})$. By Lemma 2.4.4, u is in $K[A] = K[\mathbf{x}, g(\mathbf{x})^{-1}]$ if and only if for each $\mathbf{a} \in A$ there are $h_1, h_2 \in K[\mathbf{X}]$ with and $h_2(\mathbf{a}) \neq 0$ and $u = \frac{h_1(\mathbf{x})}{h_2(\mathbf{x})}$. In particular, consider a K-basic subset $B = V' \smallsetminus V(h)$ of A, with $\mathbf{x}'$ a generic point of V'. Then $g(\mathbf{b}) \neq 0$ for each $\mathbf{b} \in B$. Thus, $K[\mathbf{x}', g(\mathbf{x}')^{-1}] \subseteq K[B]$. In other words, the K-specialization extends to a K-homomorphism of $K[A]$ into $K[B]$. $\square$

Definition 19.6.5: Let P be a property of constructible sets (e.g. basic, normal, nonsingular, etc.). A P-**stratification** of a constructible set A is a finite collection $\{A_i \mid i \in I\}$ of disjoint constructible sets having property P such that $A = \bigcup_{i \in I} A_i$. Refer to A_i as a P-set, $i \in I$. $\square$

LEMMA 19.6.6 (The Stratification Lemma): *Let P be a property of constructible sets. Suppose for each presented nonempty K-basic set A we can effectively compute a nonempty K-basic P-set B, K-open in A. Then we can effectively produce a P-stratification of each presented nonempty constructible set.*

Proof: Start with a presented nonempty K-constructible set A in $\mathbb{A}^n$ and do an induction on $\dim(A)$.

Without loss assume $A = \bigcup_{i=1}^{r} \bigcap_{j=1}^{s_i} [V(f_{ij}) \smallsetminus V(g_{ij})]$, with $f_{ij}, g_{ij} \in K[X_1, \ldots, X_n]$. Use the identity

$$\bigcap_{j=1}^{s_i} [V(f_{ij}) \smallsetminus V(g_{ij})] = V(f_{i1}, \ldots, f_{i,s_i}) \smallsetminus V(g_{i1} \cdots g_{i,s_i})$$

to rewrite A in the form

$$(3) \qquad A = \bigcup_{i=1}^{r} (V_i \smallsetminus U_i),$$

where V_i and U_i are presented Zariski K-closed sets. Apply Proposition 19.5.6 to write each V_i as a union of K-irreducible components and thus assume that each V_i is K-irreducible.

If $V_i = V_j$, replace $(V_i \smallsetminus U_i) \cup (V_j \smallsetminus U_j)$ by $V_i \smallsetminus U_i \cap U_j$. Thus, assume the V_i's are distinct. Replace $V_i \smallsetminus U_i$ by $V_i \smallsetminus V_i \cap U_i$ to assume $U_i \subseteq V_i$ for each

i. Then assume $U_i \subset V_i$. Finally, reorder the indices such that $\dim(U_r) \leq \dim(U_i)$ for $i = 1, \ldots, r-1$ and rewrite A as a disjoint union: $A_1 \cup A_2 \cup A_3$ with $A_1 = V_r \smallsetminus U_r$, $A_2 = \bigcup_{i=1}^{r-1}[V_i \smallsetminus (V_r \cup U_i)]$ and $A_3 = \bigcup_{i=1}^{r-1}[(V_i \cap U_r) \smallsetminus U_i]$. Hence, $\dim(A_3) \leq \dim(U_r) < \dim(V_r) \leq \dim(A)$. Apply the induction hypothesis to effectively produce a P-stratification of A_3. The A_2 term has less than r components. Use induction on r to assume that the union is disjoint. Thus, assume that

(4) $\qquad A = V \smallsetminus U \quad \text{with} \quad U = V(g_1, \ldots, g_k)$.

Since $A \neq \emptyset$, there is an i with $V(g_i) \cap V \subset V$. Then $A = [V \smallsetminus V(g_i)] \cup [V \cap V(g_i) \smallsetminus U]$ and the second term in the union has dimension less than $\dim(A)$. By the induction hypothesis, it suffices to stratify the first term. Hence, we may assume that A is K-basic.

By assumption, we can effectively find a nonempty P-set B, K-open in A with $A = B \cup (A \smallsetminus B)$. Since $\dim(A \smallsetminus B) < \dim(A)$, we can effectively P-stratify A. □

19.7 Basic Normal Stratification

We first give a constructive version of the normalization procedure. The next lemma circumvents difficulties that arise when K is imperfect:

LEMMA 19.7.1: Let $(x_1, \ldots, x_n)$ be a presented n-tuple over K. Then we can effectively compute a finite purely inseparable extension K' of K with $K'(\mathbf{x})/K'$ separable. Moreover, if $\mathrm{char}(K) \neq 0$, we can effectively find a basis $w_1, \ldots, w_l$ of K'/K and a power q of $\mathrm{char}(K)$ with $w_1 = 1$ and $w^q \in K$ for all $w \in K'$.

Proof: Without loss $\mathrm{char}(K) = p > 0$. Assume $x_1, \ldots, x_r$ is a transcendence base for $K(\mathbf{x})/K$. If $r = n$, take $K' = K$. Otherwise, compute a nonzero polynomial $f \in K[X_1, \ldots, X_{r+1}]$ of minimal degree with

(1) $\qquad f(x_1, \ldots, x_{r+1}) = \sum_{i \in I} a_i x_1^{i_1} \cdots x_{r+1}^{i_{r+1}} = 0, \quad \text{with} \quad a_i \in K$.

Let q be the maximal power of p that divides all exponents of (1). Put $K_1 = K(a_i^{1/q} \mid i \in I)$. Then $x_1, \ldots, x_{r+1}$ satisfy the following irreducible relation over K_1:

$$\sum_{i \in I} a_i^{1/q} x_1^{i_1/q} \cdots x_{r+1}^{i_{r+1}/q} = 0.$$

One of the variables, say x_1, appears in this relation in a monomial which is not a pth power. Hence, x_1 is separably algebraic over $K_1(x_2, \ldots, x_n)$. Now use induction on n to compute a finite purely inseparable extension K' of K_1 such that $K'(x_2, \ldots, x_n)/K'$ is separable. Then $K'(\mathbf{x})/K'$ is also separable. In addition, find a basis $w_1'', \ldots, w_{l''}''$ for K'/K_1.

Now order the a_i's in a sequence $a_1, \ldots, a_r$. Examine which of the $a_j^{1/q}$ K-linearly depends on $a_1^{1/q}, \ldots, a_{j-1}^{1/q}$ to find a basis $w_1', \ldots, w_{l'}'$ for K_1/K. Then $\{w_i' w_j'' \mid 1 \leq i \leq l', \ 1 \leq j \leq l''\}$ is a basis for K'/K which can be ordered as $w_1, \ldots, w_l$. Finally, replace w_j, if necessary, by $\frac{w_j}{w_1}$ to assure $w_1 = 1$. □

LEMMA 19.7.2: *Let $(x_1, \ldots, x_n, z)$ be a presented $(n+1)$-tuple over K with z separably algebraic over $K(\mathbf{x})$. Then we can effectively find a polynomial $g \in K[\mathbf{X}]$ with these properties:*

(2) *$g(\mathbf{x}) \neq 0$, and $K[\mathbf{x}, g(\mathbf{x})^{-1}]$ is integrally closed and presented in its function field.*

(3) *The ring $K[\mathbf{x}, g(\mathbf{x})^{-1}, z]$ is a presented cover of the ring $K[\mathbf{x}, g(\mathbf{x})^{-1}]$ with z a primitive element.*

Proof: Suppose we have found $g \in K[\mathbf{X}]$ satisfying (2). Present $\mathrm{irr}(z, K(\mathbf{x}))$ as a quotient of a polynomial in $K[\mathbf{x}, Z]$ by a polynomial in $K[\mathbf{x}]$. Then multiply g by the product of the denominator and the discriminant of $\mathrm{irr}(z, K(\mathbf{x}))$. By the remarks leading up to Lemma 19.6.1, the ring $K[\mathbf{x}, g(\mathbf{x})^{-1}, z]$ is presented over K. Also (Definition 6.1.3), z is a primitive element for the cover $K[\mathbf{x}, g(\mathbf{x})^{-1}, z]/K[\mathbf{x}, g(\mathbf{x})^{-1}]$.

The argument that finds g satisfying (2) breaks into two cases.

CASE A: *$K(\mathbf{x})/K$ is separable.* Reorder $x_1, \ldots, x_n$ according to Lemma 19.2.4 to assume x_i is separable over $K(x_1, \ldots, x_{i-1}), i = 1, \ldots, n$.

Use induction on n to effectively find a polynomial $g_1 \in K[X_1, \ldots, X_{n-1}]$ with $g_1(\mathbf{x}) \neq 0$ and $K[x_1, \ldots, x_{n-1}, g_1(\mathbf{x})^{-1}]$ integrally closed. If x_n is transcendental over $K(x_1, \ldots, x_{n-1})$, then $K[x_1, \ldots, x_n, g_1(\mathbf{x})^{-1}]$ is also integrally closed [Zariski-Samuel2, p. 85 or p. 126]. Otherwise, x_n is separable algebraic over $K(x_1, \ldots, x_{n-1})$ and from the above, with $z = x_n$, we can effectively find a multiple $g \in K[X_1, \ldots, X_{n-1}]$ of g_1 such that $K[x_1, \ldots, x_n, g(\mathbf{x})^{-1}]$ is a ring cover of $K[x_1, \ldots, x_{n-1}, g(\mathbf{x})^{-1}]$. In particular $K[\mathbf{x}, g(\mathbf{x})^{-1}]$ is integrally closed.

CASE B: *$K(\mathbf{x})/K$ is general.* Assume without loss $\mathrm{char}(K) \neq 0$. Apply Lemma 19.7.1 to find a finite purely inseparable extension K' of K with $K'(\mathbf{x})/K'$ separable. Lemma 19.7.1 gives a linear basis $w_1, \ldots, w_l$ with $w_1 = 1$ for K'/K and a power q of $\mathrm{char}(K)$ with $w^q \in K$ for each $w \in K'$. The w_i's need not be linearly independent over $K(\mathbf{x})$. Find among them a basis, say $w_1, \ldots, w_k$, for $K'(\mathbf{x})/K(\mathbf{x})$ and compute a polynomial $a \in K[\mathbf{X}]$ with $a(\mathbf{x}) \neq 0$ and $w_i \in K[\mathbf{x}, a(\mathbf{x})^{-1}, w_1, \ldots, w_k], i = 1, \ldots, l$. Note: $K(\mathbf{x}) \cap K'[\mathbf{x}] \subseteq K[\mathbf{x}, a(\mathbf{x})^{-1}]$.

Use Case A to find $h \in K'[\mathbf{X}]$ with $h(\mathbf{x}) \neq 0$ and $K'[\mathbf{x}, h(\mathbf{x})^{-1}]$ integrally closed. Now let $g = ah^q$. Then $K(\mathbf{x}) \cap K'[\mathbf{x}, h(\mathbf{x})^{-1}]$ is contained in the ring $K[\mathbf{x}, g(\mathbf{x})^{-1}]$, which is therefore integrally closed. □

Combine Lemma 19.7.2 with the stratification lemma:

Exercises

PROPOSITION 19.7.3: *There is an effective procedure for producing a basic normal stratification of a given constructible set.*

Exercises

1. (a) Use the inequality $p_{n+1} \leq (p_1 p_2 \cdots p_n) - 1$ (Euclid) to show for each $n \in \mathbb{N}$ that $p_1 p_2 \cdots p_n \leq 2^{2^n}$, where $p_1 < p_2 < \cdots$ is the sequence of primes.

   (b) Use (a) to show for each $r \in \mathbb{N}$ that $\{p_1^{j_1} \cdots p_n^{j_n} \mid n \in \mathbb{N}, 1 \leq j_1, \ldots, j_n \leq r\}$ is a primitive recursive subset of $\mathbb{N}$.

2. *The racetrack problem.* Consider the real points $C_1(\mathbb{R})$ in the ellipse $X^2 + 2Y^2 = 1$ as the inside rail of a racetrack. Let $C_2(\mathbb{R})$ be the locus of points (U, V) traced out by the points on the outside of the ellipse that are 1 unit distance from the ellipse along a perpendicular to the ellipse. Let $W \subseteq \mathbb{A}^4$ with coordinates (X, Y, U, V) be given by the equations

$$f_1 = X^2 + 2Y^2 - 1 = 0,$$
$$f_2 = XV - 2YU + XY = 0,$$
$$f_3 = (U - X)^2 + (V - Y)^2 - 1 = 0.$$

Let $\pi \colon \mathbb{A}^4 \to \mathbb{A}^2$ with $\pi(X, Y, U, V) = (U, V)$ be the projection onto the last coordinates.

   (a) Show that $C_2(\mathbb{R}) = \pi(W)(\mathbb{R})$.

   (b) Show that $\pi(W) = V(h)$, where h is a polynomial of degree 8 ($= \deg(f_1)\deg(f_2)\deg(f_3)$) that generates the ideal $\mathbb{Q}[U, V] \cap I(f_1, f_2, f_3)$ of $\mathbb{Q}[U, V]$:

   (b1) First eliminate X to show that Y, U, V satisfy:

$$g_1 = (1 - 2Y^2)(V + Y)^2 - 4(YU)^2 = 0,$$
$$g_2 = (UV - YU)^2 + ((V - Y)^2 - 1)(V + Y)^2 = 0.$$

Notice that both g_1 and g_2 are polynomials of degree 4 in Y with coefficients in $\mathbb{Z}[U, V]$ and with leading coefficients -2 and 1, respectively.

   (b2) Now apply the proof of Theorem 9.2.1 (expression (1)) to g_1 and g_2 to eliminate Y. Alternatively, apply the resultant to g_1 and g_2.

3. Prove that the number of monomials in $X_1, \ldots, X_n$ of degree at most d is $\binom{n+d}{d}$. Hint: Given integers $i_1, \ldots, i_n$ between 0 and d define $j_1 = i_1 + 1$, $j_2 = i_1 + i_2 + 2$, ..., $j_n = i_1 + i_2 + \cdots + i_n + n$. Prove that the map $(i_1, \ldots, i_n) \mapsto (j_1, \ldots, j_n)$ defines a bijective map between the set of all monomials in $X_1, \ldots, X_n$ of degree at most d and the set of all subsets with n elements of the set $\{1, 2, \ldots, n + d\}$.

Notes

This Chapter is an elaboration of [Fried-Haran-Jarden, §2]. Most of it is due to Kronecker. The common feature is the use of the "method of indeterminates." Our sources include [Waerden3] for the proofs of Lemmas 19.1.3, 19.2.2, and 19.3.2; [Waerden1] for the proof of Lemma 19.5.3; a model for the proof of Lemma 19.2.4 from [Lang4]; [Zariski-Samuel2] for the proofs of Lemmas 19.3.1 and 19.5.1; and an elaboration of [Waerden2] for the proof of Proposition 19.5.6.

Kronecker's algorithm is no longer the state of the art for a deterministic procedure for factoring polynomials with coefficients in $\mathbb{Z}$ (Lemma 19.1.3). Indeed, [Lenstra-Lenstra-Lovász] gives a bound on the time of factorization which is a polynomial in the height of the polynomial and [Lenstra] extends this to polynomials in several variables. No one has yet tested the effect of these procedures on such practical applications as the algorithm of Proposition 19.5.6. But this would be a key ingredient in producing a computer program that actually accomplishes the Galois stratification procedure and its corollaries of Chapters 30 and 31.

L. v. d. Dries and K. Schmidt [v.d.Dries-Schmidt] use nonstandard approach and the notion of faithful flatness to prove a stronger result than that achieved in Remark 19.5.8: Given positive integers d and n, there is a positive integer e such that for every field K and all polynomials $f_1, \ldots, f_m \in K[X_1, \ldots, X_n]$ of degree at most d, the ideal I generated by $f_1, \ldots, f_m$ is prime if and only if for all $g, h \in K[X_1, \ldots, X_n]$ of degree at most e, the relation $gh \in I$ implies $g \in I$ or $h \in I$. The idea of checking absolute irreducibility over the algebraic closure of a field K first and then using elimination of quantifiers to descend to K was communicated to the authors by v. d. Dries.

Chapter 20.
The Elementary Theory of e-Free PAC Fields

This chapter presents one of the highlights of this book, the study of the elementary theory of e-free PAC fields. We apply the elementary equivalence theorem for arbitrary PAC fields (Theorem 20.3.3) to the theory of perfect e-free PAC fields containing a fixed countable base field K. If K is finite and $e = 1$ or K is Hilbertian and $e \geq 1$, then this theory coincides with the theory of all sentences with coefficients in K that are true in $\tilde{K}(\sigma)$, for almost all $\sigma \in G(K)^e$ (Section 20.5). In particular, if K is explicitly given with elimination theory, then this theory is recursively decidable. In the special case where K is a global field and $e = 1$ we prove a transfer theorem (Theorem 20.9.3): A sentence θ of $\mathcal{L}(\text{ring}, O_K)$ is true among the fields $\tilde{K}(\sigma)$ with probability equal to the probability that θ is true among the residue fields of K.

Finally, we prove that the elementary theory of finite fields is recursively decidable (Theorem 20.10.6).

20.1 $\aleph_1$-Saturated PAC Fields

We start with a result that strengthens the PAC property of a field which is also $\aleph_1$-saturated (Section 7.7):

LEMMA 20.1.1: Let K be an $\aleph_1$-saturated PAC field and R an integral domain which is countably generated over K.
(a) Suppose R is contained in a field F regular over K. Then there exists a K-homomorphism $\varphi \colon R \to K$.
(b) Suppose in addition, $\text{char}(K) = p > 0$. Let S be a subset of R, p-independent over F^p with $|S| \leq \min(\aleph_0, [K : K^p])$. Then φ can be chosen such that $\varphi(S)$ is p-independent over K^p.

Proof: By assumption, $R = K[x_1, x_2, x_3, \ldots]$. Denote the ideal of all polynomials in $K[X_1, \ldots, X_n]$ that vanish at $(x_1, \ldots, x_n)$ by I_n. Let $f_{n,1}, \ldots, f_{n,r_n}$ be a system of generators for I_n. Then $I_1 \subseteq I_2 \subseteq I_3 \cdots$. Since

$$K[X_1, \ldots, X_n]/I_n \cong_K K[x_1, \ldots, x_n]$$

and $K(x_1, \ldots, x_n)$, as a subfield of F, is regular over K, $V(I_n)$ is a variety defined over K (Corollary 10.2.2(a)). Since K is PAC, there exist elements $a_1, \ldots, a_n \in K$ with $f_{nj}(a_1, \ldots, a_n) = 0$, $j = 1, \ldots, r_n$. Each f_{ij} with $1 \leq i \leq n$ and $1 \leq j \leq r_i$ is a linear combination of $f_{n1}, \ldots, f_{nr_n}$. Hence,

(1) $$\bigwedge_{i=1}^{n} \bigwedge_{j=1}^{r_i} f_{ij}(a_1, \ldots, a_i) = 0.$$

The saturation property of K gives $b_1, b_2, b_3, \ldots$ in K with $f_{nj}(b_1, \ldots, b_n) = 0$ for all n and j. The map $x_i \mapsto b_i$, $i = 1, 2, 3, \ldots$ extends to a K-homomorphism $\varphi \colon R \to K$.

Now assume $\operatorname{char}(F) = p > 0$ and S is as in (b). The essential case occurs when $[K : K^p] \geq \aleph_0$ and $S = \{s_1, s_2, s_3, \ldots\}$ is infinite. Write $s_n = g_n(x_1, \ldots, x_{k_n})$ with $g_n \in K[X_1, \ldots, X_{k_n}]$. Without loss assume that $n \leq k_n \leq k_{n+1}$ for all n. For each n Proposition 11.4.1 gives $a_1, \ldots, a_{k_n}$ in K such that, in addition to (1), $g_i(a_1, \ldots, a_{k_i})$, $i = 1, \ldots, n$, are p-independent over K^p. Then proceed as before. □

20.2 The Elementary Equivalence Theorem of $\aleph_1$-Saturated PAC Fields

Conditions for the elementary equivalence of two PAC fields E and F are central to the model theory of PAC fields. If E and F are elementarily equivalent, then E and F have the same characteristic and thus the same prime field K. Since a polynomial $f \in K[X]$ has a zero in E if and only if it has a zero in F, it follows that $E \cap \tilde{K} \cong F \cap \tilde{K}$ (Lemma 20.6.3). Assume therefore that $L = E \cap \tilde{K}$ also equals $F \cap \tilde{K}$. More generally (Section 23.4), if L' is a finite Galois extension of L and E' is a finite Galois extension of E containing L', then there exists a finite Galois extension F' of F, containing L', and a commutative diagram

(1)
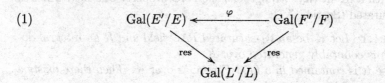

with φ an isomorphism. If E and F are countable, then an inverse limit gives a similar diagram with absolute Galois groups replacing corresponding relative Galois groups.

The next lemma shows this last condition to be essentially sufficient for the elementary equivalence of E and F. The basic element is Lemma 20.2.2 whose main ingredients, the field crossing argument and the use of existence of rational points on varieties, have already appeared in the proof of the Chebotarev density theorem and will appear again in Proposition 24.1.1.

Let L, M be fields and $\Phi \colon L_s \to M_s$ be an embedding with $\Phi(L) \subseteq M$. Then Φ **induces** a homomorphism $\varphi \colon \operatorname{Gal}(M) \to \operatorname{Gal}(L)$ with $\varphi(\sigma)x = \Phi^{-1}(\sigma \Phi(x))$ for all $\sigma \in \operatorname{Gal}(M)$ and $x \in L_s$. If Φ is an isomorphism with $\Phi(L) = M$, then φ is an isomorphism.

LEMMA 20.2.1: *Let E/L be a regular extension of fields. Then E_s/L_s is regular.*

Proof: By assumption, E is linearly disjoint from $\tilde{L}$ over L, so EL_s is linearly disjoint from $\tilde{L}$ over L_s. Since E_s/EL_s is separable and $E\tilde{L}/EL_s$ is purely inseparable, E_s is linearly disjoint from $E\tilde{L}$ over EL_s. It follows from the tower property of linear disjointness that E_s is linearly disjoint from $\tilde{L}$ over L_s. In other words, E_s/L_s is regular. $\square$

LEMMA 20.2.2 (Embedding Lemma [Jarden-Kiehne, p. 279]): *Let E/L and F/M be separable field extensions satisfying: E is countable and F is PAC and $\aleph_1$-saturated. In addition, suppose there are an isomorphism $\Phi_0: L_s \to M_s$ with $\Phi_0(L) = M$ and a commutative diagram*

(2)
$$\begin{array}{ccc} \mathrm{Gal}(E) & \xleftarrow{\varphi} & \mathrm{Gal}(F) \\ {\scriptstyle \mathrm{res}}\downarrow & & \downarrow {\scriptstyle \mathrm{res}} \\ \mathrm{Gal}(L) & \xleftarrow{\varphi_0} & \mathrm{Gal}(M) \end{array}$$

where φ_0 is the isomorphism induced by Φ_0 and φ is a homomorphism. If $\mathrm{char}(L) = p > 0$, add the assumption that $[E : E^p] \leq [F : F^p]$.

Then there exists an extension of Φ_0 to an embedding $\Phi: E_s \to F_s$ which induces φ with $F/\Phi(E)$ separable.

Proof: Since $E \cap \tilde{L}/L$ is a separable extension, $E \cap \tilde{L} = E \cap L_s$. Similarly, $F \cap \tilde{M} = F \cap M_s$. In addition, $\mathrm{res}_{L_s}(\mathrm{Gal}(E)) = \mathrm{Gal}(E \cap \tilde{L})$ and $\mathrm{res}_{M_s}(\mathrm{Gal}(F)) = \mathrm{Gal}(F \cap \tilde{M})$. The hypotheses thus give $\Phi_0(E \cap \tilde{L}) = F \cap \tilde{M}$. Therefore, replace L and M, respectively, by $E \cap \tilde{L}$ and $F \cap \tilde{M}$ to assume that E/L and F/M are regular extensions. By Lemma 20.2.1, E_s/L_s and F_s/M_s are regular. Without loss assume $L = M$, Φ_0 and φ_0 are the identity isomorphisms, and E is algebraically independent from F over L. It follows from Corollary 2.6.8 that

(3) EF is a separable (and regular) extension of both E and F.

Also, by Lemma 2.6.7, E_s is linearly disjoint from F_s over L_s. We divide the rest of the proof into three parts to separate out the use of the PAC property.

PART A: *The field crossing argument.* From (2), $\varphi(\sigma)x = \sigma x$ for each $\sigma \in \mathrm{Gal}(F)$ and $x \in L_s$. Thus, by Lemma 2.5.5, each $\sigma \in \mathrm{Gal}(F)$ extends uniquely to a $\tilde{\sigma} \in \mathrm{Gal}(E_s F_s/EF)$ with

$$\tilde{\sigma}x = \begin{cases} \varphi(\sigma)x & \text{if } x \in E_s \\ \sigma x & \text{if } x \in F_s. \end{cases}$$

The map $\sigma \mapsto \tilde{\sigma}$ embeds $\mathrm{Gal}(F)$ into $\mathrm{Gal}(E_s F_s/EF)$ and $\mathrm{res}_{F_s E_s/F_s}\tilde{\sigma} = \sigma$. Let D be the fixed field of the image of $\mathrm{Gal}(F)$. Then res: $\mathrm{Gal}(E_s F_s/D) \to \mathrm{Gal}(F)$ is an isomorphism, so $D \cap F_s = F$ and $DF_s = E_s F_s$. Since D/EF is separable, (3) implies that D is a separable extension of both E and F. Hence, by Lemma 2.6.4,

(4) D is a regular extension of F.

PART B: *Use of the PAC property.* Note that $E_s F_s$ is an algebraic extension of D. Hence, $E_s \subseteq E_s F_s = D[F_s] = F_s[D]$. For each $x \in E_s$ choose $y_i \in F_s$ and $d_i \in D$, where i ranges over a finite set I_x such that $x = \sum_{i \in I_x} y_i d_i$. Let $D_0 = E \cup \{d_i \mid x \in E_s, i \in I_x\}$. Then, $E_s \subseteq F_s[D_0]$ and since E is countable, so is D_0. We illustrate the relations among all rings and fields mentioned so far in the following diagram.

(5)
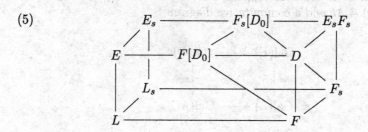

If $p = \text{char}(L) > 0$, let S be a p-basis of E over E^p. Since $F(D_0)/E$ is separable, S is p-independent over $F(D_0)^p$. Also, $|S| \leq [E : E^p] \leq [F : F^p]$. Applying Lemma 20.1.1 using the hypothesis and (4), we conclude that there exists an F-homomorphism $\Psi \colon F[D_0] \to F$ with $\Psi(S)$ p-independent over F^p (if $p > 0$). Since E is a field, Ψ is injective on E, so $\Psi(S)$ is a p-basis of $\Psi(E)$ over $\Psi(E)^p$. Therefore, $F/\Psi(E)$ is separable. Since D is linearly disjoint from F_s over F, Ψ extends to an F_s-homomorphism $\tilde{\Psi} \colon F_s[D_0] \to F_s$ (Lemma 2.5.5).

PART C: *Conclusion of the proof.* Check that

(6) $$\tilde{\Psi}(\tilde{\sigma} x) = \sigma \tilde{\Psi}(x) \text{ for each } \sigma \in \text{Gal}(F)$$

and each $x \in F_s \cup D_0$. Thus, (6) holds for each $x \in F_s[D_0]$. Since $E_s \subseteq F_s[D_0]$, (6) holds for each $x \in E_s$. Denote the restriction of $\tilde{\Psi}$ to E_s by Φ. Then Φ is an L_s-embedding of E_s into F_s which satisfies the conclusion of the Lemma. $\square$

Recall that if K is a field, then $\mathcal{L}(\text{ring}, K)$ denotes the first order language of the theory of rings augmented with constant symbols for the elements of K (Example 7.3.1). We apply Skolem-Löwenheim (Proposition 7.4.2) and the Cantor back and forth argument to improve Lemma 20.2.2:

LEMMA 20.2.3: *Let E/L and F/M be separable field extensions with both L and M countable and containing a given field K. Assume E and F have the same imperfect degree, they are PAC and $\aleph_1$-saturated, and there exists a K-isomorphism $\Phi_0 \colon L_s \to M_s$ with $\Phi_0(L) = M$. Assume in addition that (2) is a commutative diagram with φ an isomorphism. Then E is K-elementarily equivalent to F.*

Proof: Skolem-Löwenheim gives a countable K-elementary subfield M_1 of F that contains M. Since M_1/M is separable, we may apply Lemma 20.2.2

20.3 Elementary Equivalence of PAC Fields

to the diagram

$$\begin{array}{ccc} \mathrm{Gal}(E) & \xrightarrow{\mathrm{res}\circ\varphi^{-1}} & \mathrm{Gal}(M_1) \\ \downarrow\mathrm{res} & & \downarrow\mathrm{res} \\ \mathrm{Gal}(L) & \xrightarrow{\varphi_0^{-1}} & \mathrm{Gal}(M) \end{array}$$

to conclude that there is an extension of Φ_0^{-1} to an embedding $\Psi_1\colon M_{1,s} \to E_s$ with these properties: $E/\Psi_1(M_1)$ is separable and the diagram

$$\begin{array}{ccc} \mathrm{Gal}(E) & \xrightarrow{\mathrm{res}\circ\varphi^{-1}} & \mathrm{Gal}(M_1) \\ \downarrow\mathrm{res} & & \downarrow\mathrm{res} \\ \mathrm{Gal}(M_1') & \xrightarrow{\psi_1} & \mathrm{Gal}(M_1) \end{array}$$

is commutative, where $M_1' = \Psi_1(M_1)$ and ψ_1 is the isomorphism induced by Ψ_1.

Now reverse the roles of E and F to find a countable K-elementary subfield L_1 of E that contains M_1' and an embedding $\Phi_1\colon L_{1,s} \to F_s$ that extends Ψ_1^{-1} appropriately.

Proceed by induction to construct two towers of countable fields

$$L \subseteq L_1 \subseteq L_2 \subseteq \cdots \subseteq E \quad \text{and} \quad M \subseteq M_1 \subseteq M_2 \subseteq \cdots \subseteq F$$

and embeddings $\Phi_i\colon L_i \to M_{i+1}$, and $\Psi_i\colon M_i \to L_i$ satisfying: L_i (resp. M_i) is a K-elementary subfield of E (resp. F), the map Φ_i extends Ψ_i^{-1}, and Ψ_i extends Φ_{i-1}^{-1}. Let $L_\infty = \bigcup_{i=1}^\infty L_i$ and $M_\infty = \bigcup_{i=1}^\infty M_i$. By Lemma 7.4.1(b), L_∞ and M_∞ are K-elementary subfields of E and F, respectively. Moreover, the Φ_i combine to give a K-isomorphism $\Phi_\infty\colon L_\infty \to M_\infty$. Consequently, $E \equiv_K F$. □

Remark 20.2.4: Suppose E and F have the same cardinality, say m and they are m^+-saturated. Then with m^+-saturated replacing $\aleph_1$-saturated in Lemma 20.2.3, we may conclude that $E \cong_K F$. The introduction of the m^+-saturated concept in this approach usually forces one to use the continuum hypothesis $2^{\aleph_0} = \aleph_1$. Of course, a general principle of set theory asserts that arithmetical theorems proved using the continuum hypothesis hold even without assuming it. Our exposition achieves the same result directly. □

20.3 Elementary Equivalence of PAC Fields

We wish to remove the restriction of saturation put on the fields E and F from Lemma 20.2.2. This requires, however, a strengthening of (2) of Section 20.2 to an ultraproduct statement:

LEMMA 20.3.1: Let $\{E_i \mid i \in I\}$ and $\{F_i \mid i \in I\}$ be families of fields. For each $i \in I$ let $\varphi_i \colon \mathrm{Gal}(F_i) \to \mathrm{Gal}(E_i)$ be an isomorphism. Let $\mathcal{D}$ be an ultrafilter of I. Put $E^* = \prod E_i/\mathcal{D}$, $F^* = \prod F_i/\mathcal{D}$, and $\varphi^* = \prod \varphi_i/\mathcal{D}$. Then:
(a) The isomorphism $\varphi^* \colon \prod \mathrm{Gal}(F_i)/\mathcal{D} \to \prod \mathrm{Gal}(E_i)/\mathcal{D}$ induces an isomorphism $\varphi \colon \mathrm{Gal}(F^*) \to \mathrm{Gal}(E^*)$.
(b) Suppose in addition, there exists a field K which is contained in all E_i and F_i. For each i, let L_i be a finite Galois extension of K such that the diagram

(1)
$$\begin{array}{ccc} \mathrm{Gal}(E_i) & \xleftarrow{\varphi_i} & \mathrm{Gal}(F_i) \\ & \searrow{\mathrm{res}} \quad \swarrow{\mathrm{res}} & \\ & \mathrm{Gal}(L_i/K) & \end{array}$$

is commutative. Assume $\{i \in I \mid L \subseteq L_i\} \in \mathcal{D}$ for each finite extension L of K. Then the diagram

(2)
$$\begin{array}{ccc} \mathrm{Gal}(E^*) & \xleftarrow{\varphi} & \mathrm{Gal}(F^*) \\ & \searrow{\mathrm{res}} \quad \swarrow{\mathrm{res}} & \\ & \mathrm{Gal}(K) & \end{array}$$

is commutative.

Proof of (a): Let N be a Galois extension of F^* of degree n. Then there exists $y \in \prod F_{i,s}/\mathcal{D}$ such that $N = F^*(y)$. Choose a representative $\{y_i \mid i \in I\}$ of y. Then there exists $D \in \mathcal{D}$ such that $N_i = F_i(y_i)$ is a Galois extension of F_i of degree n for all $i \in D$, and $N = \prod N_i/\mathcal{D}$. Denote the fixed field in $E_{i,s}$ of $\varphi_i(\mathrm{Gal}(N_i))$ by M_i. Let $\varphi_{N,i}$ be the isomorphism of $\mathrm{Gal}(N_i/F_i)$ onto $\mathrm{Gal}(M_i/E_i)$ induced by φ_i. Then $M = \prod M_i/\mathcal{D}$ is a Galois extension of E^* and $\varphi_N = \prod \varphi_{N,i}/\mathcal{D}$ is an isomorphism of $\mathrm{Gal}(N/F^*)$ onto $\mathrm{Gal}(M/E^*)$ which is induced by φ^*.

If N' is a finite Galois extension of F^* that contains N, then the extension M' of E^* corresponding to N' contains M and $\varphi_{N'}$ induces φ_N on the group $\mathrm{Gal}(N/F^*)$. Thus, the inverse limit of the φ_N's defines an isomorphism $\varphi \colon \mathrm{Gal}(F^*) \to \mathrm{Gal}(E^*)$ which is induced by φ^*.

Proof of (b): Consider $y \in K_s$. Let L be a finite Galois extension of K which contains y. By assumption, $D = \{i \in I \mid L \subseteq L_i\} \in \mathcal{D}$. For each $i \in I$ let $N_i = LF_i$ and $N = \prod N_i/\mathcal{D}$. For each $i \in D$ and each $\sigma_i \in \mathrm{Gal}(N_i/F_i)$ the commutativity of (1) implies $\varphi_{N,i}(\sigma_i)(y) = \sigma_i(y)$. Hence, for each $\sigma \in \mathrm{Gal}(N/F^*)$, $\varphi_N(\sigma)(y) = \sigma(y)$. Therefore, $\varphi(\sigma)(y) = \sigma(y)$ for each $\sigma \in \mathrm{Gal}(F^*)$. Thus, (2) is commutative. $\square$

In the special case where $E_i = E$, $F_i = F$ and $\varphi_i = \varphi_0$ for each $i \in I$, E and F are canonically embedded in E^* and F^*, respectively, and φ^* induces φ_0 on $\mathrm{Gal}(F)$:

20.3 Elementary Equivalence of PAC Fields

COROLLARY 20.3.2: *Let E and F be fields, $\varphi\colon \mathrm{Gal}(F) \to \mathrm{Gal}(E)$ an isomorphism, and $\mathcal{D}$ an ultrafilter of I. Put $E^* = E^I/\mathcal{D}$ and $F^* = F^I/\mathcal{D}$. Then there exists a commutative diagram*

(3)
$$\begin{array}{ccc} \mathrm{Gal}(E^*) & \xleftarrow{\varphi^*} & \mathrm{Gal}(F^*) \\ {\scriptstyle \mathrm{res}}\downarrow & & \downarrow{\scriptstyle \mathrm{res}} \\ \mathrm{Gal}(E) & \xleftarrow{\varphi} & \mathrm{Gal}(F) \end{array}$$

where φ^* is an isomorphism.

THEOREM 20.3.3 (Elementarily Equivalence Theorem): *Let E/L and F/M be separable field extensions with both L and M containing a field K. Suppose E and F are PAC fields having the same imperfect degree. In addition, suppose there exist a K-isomorphism $\Phi_0\colon L_s \to M_s$ with $\Phi_0(L) = M$ and a commutative diagram*

(4)
$$\begin{array}{ccc} \mathrm{Gal}(E) & \xleftarrow{\varphi} & \mathrm{Gal}(F) \\ {\scriptstyle \mathrm{res}}\downarrow & & \downarrow{\scriptstyle \mathrm{res}} \\ \mathrm{Gal}(L) & \xleftarrow{\varphi_0} & \mathrm{Gal}(M), \end{array}$$

where φ is an isomorphism and φ_0 is the isomorphism induced by Φ_0. Then E is K-elementarily equivalent to F.

Proof: Assume first L is countable. Choose a countable set I and a nonprincipal ultrafilter $\mathcal{D}$ of I. Then, in the notation of Corollary 20.3.2, combine the diagrams (3) and (4) to get a commutative diagram

(5)
$$\begin{array}{ccc} \mathrm{Gal}(E^*) & \xleftarrow{\varphi^*} & \mathrm{Gal}(F^*) \\ {\scriptstyle \mathrm{res}}\downarrow & & \downarrow{\scriptstyle \mathrm{res}} \\ \mathrm{Gal}(L) & \xleftarrow{\varphi_0} & \mathrm{Gal}(M) \end{array}$$

where φ^* is an isomorphism. The fields E^* and F^* are PAC (Corollary 11.3.3) and $\aleph_1$-saturated (Lemma 7.7.4). By Lemma 20.2.3, $E^* \equiv_K F^*$. Therefore, $E \equiv_K F$.

For the general case, let θ be a sentence of $\mathcal{L}(\mathrm{ring}, K)$ which is true in E. There are only finitely many elements of K, say $x_1, \ldots, x_n$, that occur in θ. Let K_0 be a countable subfield of K that contains $x_1, \ldots, x_n$. By Skolem-Löwenheim (Proposition 7.4.2), L contains an elementary countable subfield L_0 that contains K_0. Let $M_0 = \Phi_0(L_0)$. Then L/L_0 and M/M_0

are separable extensions and θ is a sentence of $\mathcal{L}(\mathrm{ring}, K_0)$. Also, there is a commutative diagram

$$\begin{array}{ccc} \mathrm{Gal}(L) & \xleftarrow{\varphi_0} & \mathrm{Gal}(M) \\ \downarrow & & \downarrow \\ \mathrm{Gal}(L_0) & \xleftarrow{\varphi_0} & \mathrm{Gal}(M_0) \end{array}$$

where both horizontal arrows are the isomorphisms induced by Φ_0. The first part of the proof gives $E \equiv_{K_0} F$. Hence, θ is true in F. $\square$

A special case of Theorem 20.3.3 is useful in investigating model completeness of PAC fields:

COROLLARY 20.3.4: *Let F/K be a separable extension of PAC fields such that F and K have the same imperfect degree. Suppose* res: $\mathrm{Gal}(F) \to \mathrm{Gal}(K)$ *is an isomorphism. Then F is an elementary extension of K.*

The elementary equivalence theorem for separably closed fields is another special case of Theorem 20.3.3.

COROLLARY 20.3.5 ([Ershov1, Prop.]): *Let E and F be separably closed fields of characteristic $p > 0$ with the same imperfect degree. Then E and F are elementarily equivalent.*

Proof: Set $L = M = \tilde{\mathbb{F}}_p$ in Theorem 20.3.3. $\square$

20.4 On e-Free PAC Fields

In Theorem 20.3.3 we may assume without loss that E/L and F/M are regular extensions (e.g. as in the proof of Lemma 20.2.2). It is the relation between the groups $\mathrm{Gal}(E)$ (resp. $\mathrm{Gal}(F)$) and $\mathrm{Gal}(L)$ (resp. $\mathrm{Gal}(M)$) via restriction that complicates applications. When $\mathrm{Gal}(E)$ and $\mathrm{Gal}(F)$ are isomorphic to the free profinite group, $\hat{F}_e$, on e generators, Gaschütz' lemma (Lemma 17.7.2) comes to our aid:

PROPOSITION 20.4.1: *Let E and F be e-free PAC fields with the same imperfect degree and with a common subfield K. Suppose $E/\tilde{K} \cap E$ and $F/\tilde{K} \cap F$ are separable extensions and $\tilde{K} \cap E \cong_K \tilde{K} \cap F$. Then $E \equiv_K F$.*

Proof: Put $L = \tilde{K} \cap E$. By Proposition 17.7.3, there exists an isomorphism $\varphi \colon \mathrm{Gal}(F) \to \mathrm{Gal}(E)$ satisfying $\mathrm{res}_{E_s/L_s} \circ \varphi = \mathrm{res}_{F_s/L_s}$. Hence, by Theorem 20.3.3, $E \equiv_K F$. $\square$

If E is perfect, then $\tilde{K} \cap E$ is also a perfect field and therefore $E/\tilde{K} \cap E$ is a separable extension. This simplifies Proposition 20.4.1:

COROLLARY 20.4.2: *Let K be a subfield of perfect e-free PAC fields E and F. If $\tilde{K} \cap E \cong_K \tilde{K} \cap F$, then $E \equiv_K F$.*

20.4 On e-Free PAC Fields

COROLLARY 20.4.3: *Let E and F be e-free PAC fields with the same imperfect degree. If F is a regular extension of E, then F is an elementary extension of E.*

We axiomatize the concept of e-free PAC fields for model-theoretic applications:

PROPOSITION 20.4.4: *Let K be a field and e a positive integer. Then there exists a set $\mathrm{Ax}(K,e)$ of axioms in the language $\mathcal{L}(\mathrm{ring}, K)$ such that a field extension F of K satisfies the axioms if and only if it is perfect, PAC and e-free.* The axioms are sentences that interpret the field axioms, perfectness axioms, $[p \neq 0] \vee (\forall X)(\exists Y)[Y^p = X]$, as p ranges over primes, the positive diagram of K (Example 7.3.1) and the following axioms:
(a) *PAC axioms*: Every absolutely irreducible polynomial $f(X,Y)$ of degree d has a zero, $d = 1, 2, \ldots$.
(b) *e-free axioms*: The finite groups which appear as Galois groups over F are exactly the groups of rank bounded by e.

Proof: Section 11.3 translates the PAC axioms into elementary statements. Thus, it suffices to translate the e-free axioms into elementary statements. For this, consider a polynomial

$$f(X) = X^n + u_1 X^{n-1} + \cdots + u_n$$

with indeterminate coefficients $u_1, \ldots, u_n$ and a subgroup G of S_n which is given by its action on $\{1, 2, \ldots, n\}$. Suppose the following assertion is an elementary statement on $u_1, \ldots, u_n$: "The polynomial f is irreducible, separable, normal and has G as a Galois group." Then consider all subgroups $G_1, \ldots, G_r$ of S_n which are generated by e elements. Restate axiom (b): "For each n and for every irreducible, separable, and normal polynomial f of degree $\leq n$, the Galois group of f is isomorphic to one of the groups $G_1, \ldots, G_r$; and for each group G of order $\leq n$ and of rank $\leq e$, there exists an irreducible, separable, normal polynomial of degree at most n with Galois group isomorphic to G."

The normality condition on f means that a root z of $f(X)$ (in the algebraic closure), gives all other roots as polynomials in z of degree at most $n-1$ with coefficients in F. To eliminate the reference to $\tilde{F}$, use congruences modulo $f(X)$ as follows:

There exist polynomials $p_1(Z) = Z, p_2(Z), \ldots, p_n(Z)$ of degree at most $n-1$ with

(1) $$f(X) \equiv \prod_{i=1}^{n}(X - p_i(Z)) \mod f(Z).$$

Of course, (1) is actually n congruence conditions on the coefficients of the powers of X on both sides. For example, equating the free coefficients on both sides gives the condition

$$u_n \equiv (-1)^n p_1(Z) \cdots p_n(Z) \mod f(Z),$$

which is equivalent to the existence of a polynomial g of degree at most $n(n-1)$ with $u_n = (-1)^n p_1(Z) \cdots p_n(Z) + g(Z)f(Z)$. Thus, the normality condition on f is elementary.

The condition "$\text{Gal}(f, F)$ is isomorphic to G as a permutation group" may be shown to be elementary by considering the action on the roots of f. Then eliminate the reference to $\tilde{F}$ as before. Indeed, suppose f is monic, irreducible, separable, normal, and $p_1(z) = z = z_1$, $p_2(z) = z_2$, ..., $p_n(z) = z_n$ are the roots of f with $p_i \in F[X]$ a polynomial of degree at most $|\text{Gal}(f, F)| - 1$. Suppose $\sigma \in \text{Gal}(f, F)$. Then, for each i, $\sigma z_i = p_i(\sigma z)$. Hence, $z_{\sigma(i)} = p_i(z_{\sigma(1)})$. Hence, $p_{\sigma(i)}(z) = p_i(z_{\sigma(1)})$. Conversely, suppose $\sigma \in S_n$ satisfies $p_{\sigma(i)}(z) = p_i(z_{\sigma(1)})$ for $i = 1, \ldots, n$. Let τ be the unique element of $\text{Gal}(f, F)$ with $\tau(1) = \sigma(1)$. Then $p_{\sigma(i)}(z) = p_i(z_{\sigma(1)}) = p_i(z_{\tau(1)}) = p_{\tau(i)}(z)$. Hence, $\sigma(i) = \tau(i)$ for each i. Therefore, $\sigma = \tau \in \text{Gal}(f, F)$. Consequently, "$\text{Gal}(f, F) \cong G$" is equivalent to

$$\bigwedge_{\sigma \in G} \bigwedge_{i=1}^{n} [p_{\sigma(i)}(z) = p_i(z_{\sigma(1)})] \wedge \bigwedge_{\sigma \in S_n \setminus G} \bigvee_{i=1}^{n} [p_{\sigma(i)}(z) \neq p_i(z_{\sigma(1)})]. \qquad \square$$

Remark 20.4.5: (a) If, in an application, K is the quotient field of a distinguished subring R, we may replace the positive diagram of K by the positive diagram of R.

(b) When desired, axioms indicating that the imperfect exponent of F is m ($0 \leq m \leq \infty$) may replace the perfect axioms.

(c) If K is presented with elimination theory, then $\text{Ax}(K, e)$ can be effectively presented.

(d) Let K be a field and G a group of order n. Then "G occurs as a Galois group over K" is an elementary statement on K. Indeed, the proof of Proposition 20.4.4 presents the equivalent statement "there is a monic Galois polynomial $f \in K[X]$ with $\text{Gal}(f, K) \cong G$" as an elementary statement. $\square$

PROPOSITION 20.4.6 ([Klingen1]): *Let K and L be elementarily equivalent fields. Suppose $\text{Gal}(K)$ is a small profinite group. Then $\text{Gal}(K) \cong \text{Gal}(L)$.*

Proof: By Remark 20.4.5, a finite group G occurs as a Galois group over K if and only if G occurs as a Galois group over L. Thus, $\text{Gal}(K)$ and $\text{Gal}(L)$ have the same finite quotients. It follows from Proposition 16.10.7 that $\text{Gal}(K) \cong \text{Gal}(L)$. $\square$

20.5 The Elementary Theory of Perfect e-Free PAC Fields

We interpret the elementary theory of perfect e-free PAC fields that contain a field K in the following cases:

(1a) K is finite and $e = 1$.

20.5 The Elementary Theory of Perfect e-Free PAC Fields

(1b) K is countable and Hilbertian, and $e \geq 1$.

In each of these cases (K, e) is called a **Hilbertian pair**.

For $\sigma = (\sigma_1, \ldots, \sigma_e)$ in $\mathrm{Gal}(K)^e$, let $\tilde{K}(\sigma) = K_s(\sigma)_{\mathrm{ins}}$ be the maximal purely inseparable extension of $K_s(\sigma)$. It is the fixed field in $\tilde{K}$ of the unique extension of σ to automorphisms of $\tilde{K}$. The fields $K_s(\sigma)$ and $\tilde{K}(\sigma)$ have the same absolute Galois group. If $K_s(\sigma)$ is PAC, so is $\tilde{K}(\sigma)$. But $\tilde{K}(\sigma)$ is a perfect field. Apply Corollary 18.5.9 and Proposition 18.6.4 to case (1a) and Theorems 18.5.6 and 18.6.1 to case (1b):

THEOREM 20.5.1: *Suppose (K, e) is a Hilbertian pair. Then $\tilde{K}(\sigma)$ is a perfect e-free PAC field for almost all $\sigma \in \mathrm{Gal}(K)^e$.*

Recall the regular ultrafilters of Section 7.6 when the index set S is $\mathrm{Gal}(K)^e$ and "small sets" are the subsets of $\mathrm{Gal}(K)^e$ of measure zero. In particular, a regular ultrafilter of $\mathrm{Gal}(K)^e$ contains all subsets of $\mathrm{Gal}(K)^e$ of measure 1. We will compare an arbitrary e-free PAC field with a regular ultraproduct of the fields $\tilde{K}(\sigma)$.

Denote the theory of all sentences of $\mathcal{L}(\mathrm{ring}, K)$ which are true in $\tilde{K}(\sigma)$ for almost all $\sigma \in \mathrm{Gal}(K)^e$ by $\mathrm{Almost}(K, e)$.

LEMMA 20.5.2: *Let (K, e) be a Hilbertian pair.*
(a) *A field F is a model of $\mathrm{Almost}(K, e)$ if and only if it is K-elementarily equivalent to a regular ultraproduct of the fields $\tilde{K}(\sigma)$.*
(b) *Every regular ultraproduct of the fields $\tilde{K}(\sigma)$ is a perfect e-free PAC field.*

Proof: Statement (a) is a special case of Proposition 7.8.1(b). Statement (b) follows from Proposition 20.4.4 and Theorem 20.5.1. □

We define the **corank** of a field K as the rank of $\mathrm{Gal}(K)$.

LEMMA 20.5.3: *Let K be a field and e a positive integer. Then, for every perfect field F of corank at most e that contains K, there exists a regular ultraproduct E of the $\tilde{K}(\sigma)$'s with $\tilde{K} \cap E \cong_K \tilde{K} \cap F$.*

Proof: Let $\tau_1, \ldots, \tau_e$ be generators of $\mathrm{Gal}(F)$. For each finite Galois extension L of K the set $S(L) = \{\sigma \in \mathrm{Gal}(K)^e \mid \mathrm{res}_L(\sigma) = \mathrm{res}_L(\tau)\}$ has a positive Haar measure, so $S(L)$ is not small. If L is contained in a larger finite Galois extension L' of K, then $S(L') \subseteq S(L)$. By Lemma 7.6.1 there exists a regular ultrafilter $\mathcal{D}$ of $\mathrm{Gal}(K)^e$ which contains the sets $S(L)$ as L runs over all finite Galois extensions of K.

Let $E = \prod \tilde{K}(\sigma)/\mathcal{D}$. Then $L \cap E = L \cap F$ for every finite Galois extension L of K. Therefore, $K_s \cap E = K_s \cap F$. Since $\tilde{K} \cap E$ and $\tilde{K} \cap F$ are perfect, they are equal. □

THEOREM 20.5.4: *If (K, e) is a Hilbertian pair, then $\mathrm{Ax}(K, e)$ (Proposition 20.4.4) is a set of axioms for $\mathrm{Almost}(K, e)$. Specifically, a field F is perfect,* e-free, PAC, and contains K if and only if it satisfies each sentence θ which is true in $\tilde{K}(\sigma)$ for almost all $\sigma \in \text{Gal}(K)^e$.

Proof: Suppose that $F \models \text{Ax}(K, e)$. By Lemma 20.5.3 there exists a regular ultraproduct E of the $\tilde{K}(\sigma)$'s such that $\tilde{K} \cap E \cong_K \tilde{K} \cap F$. By Lemma 20.5.2(b), $E \models \text{Ax}(K, e)$. Corollary 20.4.2 now gives $E \equiv_K F$. From Lemma 20.5.2(a), $F \models \text{Almost}(K, e)$.

Theorem 20.5.1 gives the converse. $\square$

20.6 The Probable Truth of a Sentence

Let K be a field and let e be a positive integer. For a sentence θ of $\mathcal{L}(\text{ring}, K)$, consider the **truth set** of θ:

(1) $$S(K, e, \theta) = \{\sigma \in G(K)^e \mid \tilde{K}(\sigma) \models \theta\}.$$

Refer to the case where K has elimination theory (Definition 19.2.8) and e and θ are explicitly given, as the **explicit case**. Regard the measure of $S(K, e, \theta)$ (if it exists) as the probability that θ is true among the $\tilde{K}(\sigma)$'s.

Call a sentence λ of the form

(2) $$P((\exists X)[f_1(X) = 0], \ldots, (\exists X)[f_m(X) = 0])$$

with $f_1, \ldots, f_m \in K[X]$ separable polynomials and P a Boolean polynomial (Section 7.6), a **test sentence**. In this case it is fairly easy to describe the set $S(K, e, \lambda)$. Indeed, the splitting field L of the polynomial $f_1 \cdots f_m$ is a finite Galois extension of K. Denote the set of all $\tau \in \text{Gal}(L/K)^e$ with $L(\tau) \models \lambda$ by S_0. Then

(3) $$S(K, e, \lambda) = \{\sigma \in G(K)^e \mid \text{res}_L(\sigma) \in S_0\}.$$

Indeed, if λ has the form $(\exists X)[f_1(X) = 0]$, then S_0 consists exactly of the $\tau \in \text{Gal}(L/K)^e$ for which $L(\tau)$ contains at least one root of $f_1(X)$. Since L contains all roots of $f_1(X)$, this gives (3). An induction on the structure of P gives (3) in general. From (3)

(4) $$\mu(S(K, e, \lambda)) = \frac{|S_0|}{[L:K]^e}.$$

In the explicit case the right hand side of (4) can be computed effectively from Lemma 19.3.2.

LEMMA 20.6.1: *Let K be a field, e be a positive integer, and λ a test sentence. Then $\mu(S(K, e, \lambda))$ is a rational number which, in the explicit case, can be effectively computed.*

The reduction of arbitrary sentences to test sentences depends on a general result of field theory:

20.6 The Probable Truth of a Sentence

LEMMA 20.6.2: *Let $K \subseteq L \subseteq L'$ be a tower of fields. Suppose L'/K is algebraic and $L \cong_K L'$. Then $L = L'$.*

Proof: Consider $x \in L'$. Let $f = \mathrm{irr}(x, K)$. Denote the set of all zeros of f in L (resp. L') by Z (resp. Z'). Then $Z \subseteq Z'$ and $|Z| = |Z'|$, so $Z = Z'$. Therefore, $x \in L$. □

LEMMA 20.6.3: *Let E and F be fields having a common subfield K.*
(a) *Suppose each irreducible polynomial $f \in K[X]$ which has a root in E has a root in F. Then there exists a K-embedding of $\tilde{K} \cap E$ into $\tilde{K} \cap F$.*
(b) *Suppose an irreducible polynomial $f \in K[X]$ has a root in E if and only if it has a root in F. Then $\tilde{K} \cap E \cong_K \tilde{K} \cap F$.*

Proof: Statement (b) follows from (a) by Lemma 20.6.2. Thus, it suffices to prove (a). Assume without loss E and F are algebraic over K. If K is finite and $x \in E$, then $K(x)$ is a Galois extension of K. By assumption, $\mathrm{irr}(x, K)$ has a root in F. Hence, $K(x) \subseteq F$. Therefore, $E \subseteq F$.

Now assume K is infinite. Let L be a finite extension of K in E. Choose a finite normal extension $\hat{L}$ of K which contains L. Put $L' = \hat{L} \cap F$. List the K-isomorphisms of L' into $\tilde{K}$ as $\sigma_1, \ldots, \sigma_n$. Put $L_i = \sigma_i(L')$, $i = 1, \ldots, n$. By assumption, for each $x \in L$ the polynomial $f = \mathrm{irr}(x, K)$ has a root $x' \in L'$. Extend the map $x \mapsto x'$ to a K-automorphism τ of $\hat{L}$. Then $\tau^{-1}|L' = \sigma_i$ for some i, so $x = \tau^{-1}(x') = \sigma_i(x') \in L_i$. It follows that $L \subseteq L_1 \cup \cdots \cup L_n$. Now consider each of the fields $L, L_1, \ldots, L_n$ as a subspace of the finite dimensional vector space $\hat{L}$ over K. Since K is infinite, there exists j with $L \subseteq L_j$. Therefore, $\sigma_j^{-1}(L) \subseteq L' \subseteq F$.

Denote the finite nonempty set of all K-embeddings of L into F by $I(L)$. If L is contained in another finite extension L_1 of K, contained in E, then restriction defines a canonical map of $I(L_1)$ into $I(L)$. Take the inverse limit of the $I(L)$'s to establish the existence of a K-embedding of E into F (Corollary 1.1.4), as desired. □

COROLLARY 20.6.4: *Let E and F be perfect fields with a common subfield K.*
(a) *Suppose each separable irreducible polynomial $f \in K[X]$ which has a root in E has a root in F. Then $\tilde{K} \cap E$ can be K-embedded into $\tilde{K} \cap F$.*
(b) *Suppose a separable irreducible polynomial $f \in K[X]$ has a root in E if and only if it has a root in F. Then $\tilde{K} \cap E \cong_K \tilde{K} \cap F$.*

Proof: Lemma 20.6.3 covers the case where $\mathrm{char}(K) = 0$, so we assume $\mathrm{char}(K) = p > 0$. Let $f \in K[X]$ be an irreducible polynomial. Then there exists a separable irreducible polynomial $g \in K[X]$ and a power q of p such that $f(X) = g(X^q)$. Since E (resp. F) is perfect, f has a root in E (resp. F) if and only if g has one, as well. Therefore, we may apply Lemma 20.6.3 to conclude both (a) and (b). □

We combine Corollary 20.4.2 with Corollary 20.6.4:

LEMMA 20.6.5: *Let (K, e) be a Hilbertian pair and E and F models of Almost(K, e). Then $E \equiv_K F$ is and only if E and F satisfy exactly the same test sentences.*

Proof: Suppose E and F satisfy the same test sentences. Then a separable polynomial $f \in K[X]$ has a root in E if and only if f has a root in F. Since E and F are models of Almost(K, e), they are perfect. Hence, by Corollary 20.6.4, $\tilde{K} \cap E \cong_K \tilde{K} \cap F$. Thus, by Corollary 20.4.2, $E \equiv_K F$. ☐

PROPOSITION 20.6.6: *Let (K, e) be a Hilbertian pair. For each sentence θ of $\mathcal{L}(\text{ring}, K)$ there exists a test sentence λ satisfying:*
(a) *The sets $S(K, e, \theta)$ and $S(K, e, \lambda)$ differ only by a zero set; the sentence $\theta \leftrightarrow \lambda$ belongs to Almost(K, e).*
(b) *There is a formal proof $(\delta_1, \ldots, \delta_n)$ of $\theta \leftrightarrow \lambda$ from the set of axioms Ax(K, e); both λ and $(\delta_1, \ldots, \delta_n)$ can be found in the explicit case in a recursive way by checking all proofs from Ax(K, e).*

Proof: Proposition 7.8.2 gives a test sentence λ satisfying (a). By Theorem 20.5.4, Ax(K, e) is a set of axioms for Almost(K, e). Therefore, by Corollary 8.2.6, Ax$(K, e) \vdash \theta \leftrightarrow \lambda$, which is the first part of (b). The second part of (b) follows from Proposition 8.7.2. ☐

We use Proposition 20.6.6 to generalize Lemma 20.6.1 to arbitrary sentences:

THEOREM 20.6.7 ([Jarden-Kiehne]): *Let (K, e) be a Hilbertian pair and θ a sentence of $\mathcal{L}(\text{ring}, K)$. Then $\mu(S(K, e, \theta))$ is a rational number which, in the explicit case, can be recursively computed. In particular, the theory Almost(K, e) is recursively decidable.*

Proof: Indeed, $\mu(S(K, e, \theta)) = 1$ if and only if the sentence θ belongs to Almost(K, e). ☐

Remark 20.6.8: Chapter 31 uses algebraic geometry to prove that Almost(K, e) is primitive recursive. ☐

20.7 Change of Base Field

Suppose θ is a sentence of $\mathcal{L}(\text{ring}, K)$ and K' is a field containing K. Then θ is also a sentence of $\mathcal{L}(\text{ring}, K')$. It is therefore natural, to relate $S(K, e, \theta)$ and $S(K', e, \theta)$. More generally, we relate $S(K, e, \theta)$ and $S(K', e, \theta)$ when θ is an "infinite sentence". To define the latter concept, we adjoin the symbol $\bigvee_{i=1}^{\infty}$ to the language and define the set of **infinite sentences** to be the smallest set of strings that satisfy the following rules:
(1a) Every sentence of $\mathcal{L}(\text{ring}, K)$ is an infinite sentence.
(1b) If θ is an infinite sentence, then so is $\neg \theta$.
(1c) For $\theta_1, \theta_2, \theta_3, \ldots$ a sequence of infinite sentences, $\bigvee_{i=1}^{\infty} \theta_i$ is an infinite sentence.

20.7 Change of Base Field

The interpretation of infinite sentences is given by the following rule: For F an extension field of K, $F \models \bigvee_{i=1}^{\infty} \theta_i$ if $F \models \theta_i$ for some integer i.

If θ is an infinite sentence, use (1) of Section 20.6 to define its truth set, $S(K, e, \theta)$. Observe that the operator $S(K, e, *)$ commutes with infinite disjunctions. Also, if fields E and F containing K are K-elementarily equivalent, then the same infinite sentences are true in both of them.

Consider a regular extension K' of K. Let $\rho\colon \mathrm{Gal}(K')^e \to \mathrm{Gal}(K)^e$ be the restriction map. Then $\tilde{K} \cap \tilde{K}'(\sigma) = \tilde{K}(\rho(\sigma)) = \tilde{K}(\sigma)$ for each $\sigma \in \mathrm{Gal}(K')^e$. Denote the measure of $\mathrm{Gal}(K')^e$ by μ' and use the rule $\mu'(\rho^{-1}(A)) = \mu(A)$ for each measurable subset A of $\mathrm{Gal}(K)^e$ (Proposition 18.2.2).

THEOREM 20.7.1 ([Jacobson-Jarden1, Thm. 1.1]): *Let (K, e) and (K', e) be Hilbertian pairs such that K' is a regular extension of K. Then:*
(a) *For almost all $\sigma \in \mathrm{Gal}(K')^e$, $\tilde{K}(\sigma) \prec \tilde{K}'(\sigma)$.*
(b) *For each infinite sentence θ of $\mathcal{L}(\mathrm{ring}, K)$, $S(K', e, \theta)$ and $\rho^{-1}(S(K, e, \theta))$ differ only by a zero set.*
(c) $\mu'(S(K', e, \theta)) = \mu(S(K, e, \theta))$.

Proof: Denote the set of all σ in $\mathrm{Gal}(K)^e$ (resp. in $\mathrm{Gal}(K')^e$) with $\tilde{K}(\sigma)$ e-free and PAC by S (resp. S'). By Theorem 20.5.1, $\mu(S) = 1$ and $\mu'(S') = 1$. Hence, $\mu'(\rho^{-1}(S) \cap S') = 1$. By Corollary 20.4.3, $\tilde{K}(\sigma)$ is an elementary subfield of $\tilde{K}'(\sigma)$ for every $\sigma \in \rho^{-1}(S) \cap S'$. This completes the proof of (a). Statement (b) follows from (a); and (c) follows from (b). □

Consider the special case where $K' = K(t_1, \ldots, t_r)$ is the field of rational functions over K in the variables $t_1, \ldots, t_r$. Let $R = K[t_1, \ldots, t_r]$ be the corresponding ring of polynomials. Regard a sentence of $\mathcal{L}(\mathrm{ring}, R)$ as a formula $\theta(t_1, \ldots, t_r)$ of $\mathcal{L}(\mathrm{ring}, K)$ involving the variables $t_1, \ldots, t_r$. If $a_1, \ldots, a_r \in K$, then $\theta(\mathbf{a})$ is a sentence of $\mathcal{L}(\mathrm{ring}, K)$.

The next theorem generalizes Hilbert's irreducibility theorem (Exercise 4):

THEOREM 20.7.2: *Let K be a countable Hilbertian field, e a positive integer, $\theta(t_1, \ldots, t_r)$ a sentence of $\mathcal{L}(\mathrm{ring}, R)$. Then there exists a separable Hilbert subset H of K^r such that*

$$(2) \qquad \mu'(S(K', e, \theta(\mathbf{t}))) = \mu(S(K, e, \theta(\mathbf{a})))$$

for each $\mathbf{a} \in H$.

Proof: Using test sentences divides the proof into two parts.

PART A: *Reduction to test sentences.* Proposition 20.6.6 gives a test sentence $\lambda(\mathbf{t})$ for $\theta(\mathbf{t})$ of the form $P((\exists X)[f_1(\mathbf{t}, X) = 0], \ldots, (\exists X)[f_m(\mathbf{t}, X) = 0])$ with $f_1, \ldots, f_m \in K(\mathbf{t})[X]$ separable polynomials and P a boolean polynomial. Moreover, there exists a formal proof $(\delta_1(\mathbf{t}), \ldots, \delta_n(\mathbf{t}))$ of $\theta(\mathbf{t}) \leftrightarrow \lambda(\mathbf{t})$ from $\mathrm{Ax}(K', e)$. The axioms of the positive diagram of K' (Example 7.3.1) involved in this proof have the form $r_1(\mathbf{t}) + r_2(\mathbf{t}) = r_3(\mathbf{t})$ or $r_1(\mathbf{t}) \cdot r_2(\mathbf{t}) = r_3(\mathbf{t})$

where $r_1, r_2, r_3 \in K(\mathbf{t})$. Define U to be the K-Zariski open set of all $\mathbf{a} \in \mathbb{A}^r$ at which none of the denominators of the f_i's and the r_j's vanishes. Thus, $\tilde{K}(\boldsymbol{\sigma}) \models \delta_i(\mathbf{a})$ for each $\mathbf{a} \in U(K)$, $i = 1, \ldots, n$. Therefore, for each $\mathbf{a} \in U(K)$ and for each $\boldsymbol{\sigma} \in \text{Gal}(K)^e$ with $\tilde{K}(\boldsymbol{\sigma})$ e-free and PAC, $\tilde{K}(\boldsymbol{\sigma}) \models \theta(\mathbf{a}) \leftrightarrow \lambda(\mathbf{a})$. Now apply Theorem 20.5.1 to both K and K' to obtain

(3a) $\qquad \mu'(S(K', e, \theta(\mathbf{t}))) = \mu'(S(K', e, \lambda(\mathbf{t}))$ and
(3b) $\qquad \mu(S(K, e, \theta(\mathbf{a}))) = \mu(S(K, e, \lambda(\mathbf{a})))$ for each $\mathbf{a} \in U(K)$.

PART B: *Test sentences.* Let L' be the splitting field of $f_1(\mathbf{t}, X) \cdots f_m(\mathbf{t}, X)$ over K'. Take a primitive element z for L'/K' which is integral over R and let $g(\mathbf{t}, X) = \text{irr}(z, K')$. Put $h = g f_1 \cdots g_m$. Denote the set of all $\mathbf{a} \in U(K)$ such that the discriminants (and therefore also the leading coefficients) of $f_1, \ldots, f_m, g$ remain nonzero under the specialization $\mathbf{t} \to \mathbf{a}$ by H'. Make H' smaller, if necessary, to assume that the specialization $\mathbf{t} \to \mathbf{a}$ induces an isomorphism of $\text{Gal}(L'/K')$ onto $\text{Gal}(L/K)$, where L is the splitting field of $h(\mathbf{a}, X)$ over K, that preserves the operation on the roots of $f_1, \ldots, f_n$ (Lemma 13.1.1(a)). For $\mathbf{a} \in H'$, the number of $\sigma' \in \text{Gal}(L'/K')^e$ with $L'(\sigma') \models \lambda(\mathbf{t})$ is equal to the number of $\boldsymbol{\sigma} \in \text{Gal}(L/K)^e$ such that $L(\boldsymbol{\sigma}) \models \lambda(\mathbf{a})$. Hence $\mu'(S(K', e, \lambda(\mathbf{t}))) = \mu(S(K, e, \lambda(\mathbf{a})))$. Therefore, (2) follows from (3).

By Lemma 13.1.1, H' contains a Hilbert subset of K^r. The theorem follows. $\square$

Remark: With K fixed, Section 30.6 analyzes the effect of a change in e on $\mu(S(K, e, \theta))$. $\square$

20.8 The Fields $K_s(\sigma_1, \ldots, \sigma_e)$

The free generators theorem (Theorem 18.5.6) and the PAC Nullstellensatz (Theorem 18.6.1) establish properties satisfied by almost all fields $K_s(\boldsymbol{\sigma})$. As applications, however, the last sections developed a theory of properties shared by almost all fields $\tilde{K}(\boldsymbol{\sigma})$. The next section explains this shift. Comparison of the theory of these fields to the theory of finite (and therefore perfect) fields forces us to consider the perfect fields $\tilde{K}(\boldsymbol{\sigma})$ rather than the imperfect fields $K_s(\boldsymbol{\sigma})$.

Nevertheless, if we replace $\tilde{K}(\boldsymbol{\sigma})$ by $K_s(\boldsymbol{\sigma})$ and make some obvious changes, the results in Sections 20.4, 20.5, and 20.6 as well as their proofs remain valid. First, the analog of Corollary 20.4.2:

PROPOSITION 20.8.1: *Let E and F be separable extensions of a field K with the same imperfect degree. Suppose E and F are e-free, PAC, $K_s \cap E \cong_K K_s \cap F$. Then $E \equiv_K F$.*

Now replace $\text{Ax}(K, e)$ by a set of axioms $\text{Ax}'(K, e)$. A field extension F of K satisfies $\text{Ax}'(K, e)$ if and only if F is PAC, e-free, $[F : F^p] = [K : K^p]$,

20.8 The Fields $K_s(\sigma_1, \ldots, \sigma_e)$

and F/K is separable. To express the separability of F/K by sentences of $\mathcal{L}(\text{ring}, K)$, choose a p-basis B of K over K^p. Then "F/K is separable" if and only if "B_0 is p-independent over F^p" for all finite subsets B_0 of B.

Having done this, we write the analog of Theorem 20.5.1:

THEOREM 20.8.2: Suppose (K, e) is an Hilbertian pair. Then, for almost all $\sigma \in \text{Gal}(K)^e$, the field $K_s(\sigma)$ is e-free, PAC, separable over K, and $[K_s(\sigma) : K_s(\sigma)^p] = [K : K^p]$.

We denote the theory of all sentences of $\mathcal{L}(\text{ring}, K)$ which are true in $K_s(\sigma)$ for almost all $\sigma \in \text{Gal}(K)^e$ by Almost$'(K, e)$. Then the analog of Lemma 20.5.2 holds:

LEMMA 20.8.3: Let (K, e) be an Hilbertian pair.
(a) A field F is a model of Almost$'(K, E)$ if and only if it is K-elementarily equivalent to a regular ultraproduct of the fields $K_s(\sigma)$.
(b) Every regular ultraproduct of the fields $K_s(\sigma)$ is e-free and PAC, has the same imperfect degree as K, and is separable over K.

Now we present the analog of Lemma 20.5.3:

LEMMA 20.8.4: Let K be a field, e a positive integer, and F a field of corank at most e which is separable over K. Then there exists a regular ultraproduct E of the $K_s(\sigma)$'s with $K_s \cap E \cong_K K_s \cap F$.

This gives the analog of Theorem 20.5.4:

THEOREM 20.8.5: Let (K, e) be an Hilbertian pair. Then $\text{Ax}'(K, e)$ is a set of axioms for Almost$'(K, e)$.

Finally, for a sentence θ of $\mathcal{L}(\text{ring}, K)$ let

$$S'(K, e, \theta) = \{\sigma \in \text{Gal}(K)^e \mid K_s(\sigma) \models \theta\}.$$

Then the following analog of Theorem 20.6.7 holds:

THEOREM 20.8.6: Let (K, e) be an Hilbertian pair and θ a sentence of $\mathcal{L}(\text{ring}, K)$. Then $\mu(S'(K, e, \theta))$ is a rational number which, in the explicit case, can be recursively computed. In particular, the theory Almost$'(K, e)$ is recursively decidable.

The results of Section 20.7 are in general false for the fields $K_s(\sigma)$. Suppose for example, $K = \mathbb{F}_p(t)$ and $K' = \mathbb{F}_p(t, u)$ with t, u algebraically independent elements over $\mathbb{F}_p$. Then the separable extensions of K have imperfect exponent 1 while the separable extensions of K' have imperfect exponent 2. Therefore, no $K_s'(\sigma)$ is elementarily equivalent to $K_s(\sigma)$.

20.9 The Transfer Theorem

This section connects the elementary theory of finite fields with the elementary theory of the fields $\tilde{K}(\sigma)$.

Let K be a global field and O_K the ring of integers of K. Denote the set of nonzero prime ideals of O_K by $P(K)$. It is equipped with the Dirichlet density δ (Section 6.3). We consider models of the language $\mathcal{L}(\text{ring}, O_K)$ that are field extensions either of K or of one of the residue fields $\bar{K}_\mathfrak{p}$, for $\mathfrak{p} \in P(K)$.

By Proposition 7.8.1, a sentence θ of $\mathcal{L}(\text{ring}, O_K)$ is true in $\bar{K}_\mathfrak{p}$ for almost all $\mathfrak{p} \in P(K)$ if and only if θ is true in every nonprincipal ultraproduct of the $\bar{K}_\mathfrak{p}$'s. If $F = \prod \bar{K}_\mathfrak{p}/\mathcal{D}$ is one of these ultraproducts, then, by Proposition 7.9.1 and Corollary 11.3.4, F is a perfect, 1-free, PAC field that contains K. Since K is a global field, it is Hilbertian (Theorem 13.4.2). Hence, by Theorem 20.5.4, F is a model of Almost$(K, 1)$. We have therefore proved:

LEMMA 20.9.1: *If a sentence θ of $\mathcal{L}(\text{ring}, O_K)$ is true in $\tilde{K}(\sigma)$ for almost all $\sigma \in \text{Gal}(K)$, then θ is true in $\bar{K}_\mathfrak{p}$ for almost all $\mathfrak{p} \in P(K)$.*

For a given sentence θ of $\mathcal{L}(\text{ring}, O_K)$ we compare the sets

$$S(\theta) = S(K, 1, \theta) = \{\sigma \in \text{Gal}(K) \mid \tilde{K}(\sigma) \models \theta\} \quad \text{and}$$
$$A(\theta) = A(K, \theta) = \{\mathfrak{p} \in P(K) \mid \bar{K}_\mathfrak{p} \models \theta\},$$

using the Dirichlet density δ of $P(K)$ and the Haar measure μ of $\text{Gal}(K)$.

LEMMA 20.9.2: *Let λ be the test sentence*

(1) $$p((\exists X)[f_1(X) = 0], \ldots, (\exists X)[f_m(X) = 0]),$$

where $f_1, \ldots, f_m \in K[X]$ are separable polynomials and p is a Boolean polynomial. Let B be the set of all $\mathfrak{p} \in P(K)$ such that all coefficients of f_i are $\mathfrak{p}$-integral and the leading coefficient and the discriminant of the f_i's are $\mathfrak{p}$-units, $i = 1, \ldots, m$. Denote the splitting field of $f_1 \cdots f_m$ over K by L. Then:

(a) *For each $\mathfrak{p} \in B$, every $\mathfrak{P} \in P(L)$ over $\mathfrak{p}$, every $\sigma \in D_\mathfrak{P}$, and every field extension F of $\bar{K}_\mathfrak{p}$ satisfying $\bar{L}_\mathfrak{P} \cap F = \bar{L}_\mathfrak{P}(\bar{\sigma})$ (where $\bar{\sigma}$ is the image of σ under the map $D_\mathfrak{P} \to \text{Gal}(\bar{L}_\mathfrak{P}/\bar{K}_\mathfrak{p})$ induced by $\mathfrak{P}$) we have $L(\sigma) \models \lambda \iff F \models \lambda$.*

(b) *Let $\bar{S}(\lambda) = \{\sigma \in \text{Gal}(L/K) \mid L(\sigma) \models \lambda\}$. Then $\delta(A(\lambda)) = \frac{|\bar{S}(\lambda)|}{[L:K]}$.*

Proof of (a): First note that B is a cofinite set, because each f_i is separable. Suppose $\mathfrak{p} \in B$. Then $\mathfrak{p}$ is unramified in L (Section 6.2). If λ is $(\exists X)[f_i(X) = 0]$, statement (a) is a reinterpretation of Lemma 6.1.8(a). The general case follows by induction on the structure of λ.

Proof of (b): Use (a) and the Chebotarev density theorem (Theorem 6.3.1). □

20.9 The Transfer Theorem

THEOREM 20.9.3 (The Transfer Theorem): *Let O_K be the ring of integers of a global field K and θ a sentence of $\mathcal{L}(\mathrm{ring}, O_K)$. Then $S(\theta)$ is measurable, $A(\theta)$ has a Dirichlet density, and $\mu(S(\theta)) = \delta(A(\theta))$.*

Proof: Proposition 20.6.6 provides a test sentence λ of the form (1) with $\theta \leftrightarrow \lambda$ true in $\tilde{K}(\sigma)$ for almost all $\sigma \in \mathrm{Gal}(K)$. Without loss assume the coefficients of the polynomials $f_1, \ldots, f_m$ belong to O_K. Thus (Lemma 20.9.1), $\theta \leftrightarrow \lambda$ is true in $\bar{K}_\mathfrak{p}$, for almost all $\mathfrak{p} \in P(K)$. Hence, $S(\theta) \approx S(\lambda)$ (i.e. $S(\theta)$ and $S(\lambda)$ differ by a set of measure zero) and $A(\theta) \approx A(\lambda)$ (i.e. $A(\theta)$ and $A(\lambda)$ differ by a finite set). Therefore, it suffices to prove the theorem for λ, rather than for θ.

Let L be the splitting field of the polynomial $f_1 \ldots f_m$. Then L is a finite Galois extension of K, $\bar{S}(\lambda) = \{\tau \in \mathrm{Gal}(L/K) \mid L(\tau) \models \lambda\}$ is a union of conjugacy classes of $\mathrm{Gal}(L/K)$, and $S(\lambda) = \{\sigma \in \mathrm{Gal}(K) \mid \mathrm{res}_L \sigma \in \bar{S}(\lambda)\}$. Hence, $\mu(S(\lambda)) = \frac{|\bar{S}(\lambda)|}{[L:K]}$. By Lemma 20.9.2,

$$(2) \qquad A(\lambda) \approx \{\mathfrak{p} \in P(K) \mid \left(\frac{L/K}{\mathfrak{p}}\right) \subseteq \bar{S}(\lambda)\}$$

and $\delta(A(\lambda)) = \frac{|\bar{S}(\lambda)|}{[L:K]}$. Consequently, $\mu(S(\lambda)) = \delta(A(\lambda))$. □

If $\delta(A(\lambda)) = 0$, then $\mu(S(\lambda)) = 0$ and $\bar{S}(\lambda)$ is empty. Therefore, by (2), $A(\lambda)$, hence $A(\theta)$, are finite sets. Therefore, Theorem 20.6.7 gives the following supplement to Theorem 20.9.3:

THEOREM 20.9.4 ([Ax1, p. 161, Cor.]): *Let θ be a sentence of $\mathcal{L}(\mathrm{ring}, O_K)$. Then $\delta(A(\theta))$ is a rational number which is positive if $A(\theta)$ is infinite. In the explicit case $\delta(A(\theta))$ can be recursively computed.*

The special case $K = \mathbb{Q}$ gives the decidability of the theory of sentences which are true in $\mathbb{F}_p$ for almost all p. The next result represents this theory in two more ways.

Consider the set $\mathcal{Q}$ of all powers of prime numbers. Call a subset B of $\mathcal{Q}$ **small** if only finitely many prime numbers divide the elements of B.

PROPOSITION 20.9.5: *The following three statements about a sentence θ of $\mathcal{L}(\mathrm{ring})$ are equivalent:*
(a) $\tilde{\mathbb{Q}}(\sigma) \models \theta$ *for almost all* $\sigma \in \mathrm{Gal}(\mathbb{Q})$.
(b) $\mathbb{F}_p \models \theta$ *for almost all* $p \in P(\mathbb{Q})$.
(c) $\mathbb{F}_q \models \theta$ *for almost all* $q \in \mathcal{Q}$ *(i.e. for all but a small set).*

Proof: The equivalence "(a) $\Longleftrightarrow$ (b)" is a special case of the transfer theorem. The implication "(c) $\Longrightarrow$ (b)" follows from the definitions. Finally, the implication "(a) $\Longrightarrow$ (c)" follows from Lemma 20.5.2(a), because every regular ultraproduct $\prod \mathbb{F}_q / \mathcal{D}$, where $\mathcal{D}$ contains all complements of small sets of $\mathcal{Q}$, is a 1-free PAC field of characteristic zero (Corollary 11.3.4). □

448   Chapter 20. The Elementary Theory of e-Free PAC Fields

COROLLARY 20.9.6 ([Ax2, p. 265]): *Let K be a given global field. Then the theory of all sentences in $\mathcal{L}(\mathrm{ring}, O_K)$ which are true in $\tilde{K}_\mathfrak{p}$ for almost all $\mathfrak{p} \in P(K)$ is recursively decidable.*

Proof: By Theorem 20.9.4, the set of all sentences θ of $\mathcal{L}(\mathrm{ring}, K)$ with $\delta(A(\theta)) = 1$ is recursive. □

If $\delta(A(\theta)) = 1$, then there are only finitely many $\mathfrak{p} \in P(K)$ for which θ may be false in $\tilde{K}_\mathfrak{p}$. The proof of the next theorem gives a recursive procedure for displaying these primes:

THEOREM 20.9.7 ([Ax2, p. 264]): *Let K be a given global field. Then the theory of all sentences true in $\tilde{K}_\mathfrak{p}$ for all $\mathfrak{p} \in P(K)$ is recursively decidable.*

Proof: We follow the pattern of proof of Theorem 20.9.3.

PART A: *Finding a test sentence.* Let $\theta \in \mathcal{L}(\mathrm{ring}, O_K)$. Proposition 20.6.4 recursively gives a test sentence λ of the form (2) of Section 20.6 and a formal proof $(\delta_1, \ldots, \delta_n)$ from $\mathrm{Ax}(K, 1)$ (in the language $\mathcal{L}(\mathrm{ring}, K)$) of $\theta \leftrightarrow \lambda$. Let A_0 be the set of all $\mathfrak{p} \in P(K)$ that divide one of the denominators of the elements of K involved in $\delta_1, \ldots, \delta_n$. If a field F of characteristic not in A_0 contains a homomorphic image of O_K and if the axioms among the δ_i are true in F, then $(\delta_1, \ldots, \delta_n)$ is a valid proof in F. In particular, $\theta \leftrightarrow \lambda$ is true in F.

PART B: *Reduction modulo $\mathfrak{p}$ of the PAC axioms.* Let $1 \leq i \leq n$. If δ_i is the axiom "each absolutely irreducible polynomial $f(X, Y)$ of degree d has a zero", define A_i to be the set of all $\mathfrak{p} \in P(K)$ with $|\bar{K}_\mathfrak{p}| \leq (d-1)^4$. By Corollary 5.4.2, if $\mathfrak{p} \in P(K) \smallsetminus A_i$, then $\tilde{K}_\mathfrak{p} \models \delta_i$. Otherwise, let $A_i = A_0$. Let $B_1 = A_0 \cup A_1 \cup \cdots \cup A_n$. Since $\tilde{K}_\mathfrak{p}$ is perfect and 1-free, $\tilde{K}_\mathfrak{p} \models \theta \leftrightarrow \lambda$ for each $\mathfrak{p} \in P(K) \smallsetminus B_1$.

PART C: *Exceptional primes for λ.* Construct the splitting field L of the product of the polynomials $f_1, \ldots, f_m$ appearing in λ and check if there exists $\tau \in \mathrm{Gal}(L/K)$ with $L(\tau) \not\models \lambda$. In this case, λ (and therefore θ) is false in $\tilde{K}_\mathfrak{p}$ for almost all $\mathfrak{p} \in P(K)$ with $\tau \in \left(\frac{L/K}{\mathfrak{p}}\right)$ (Lemma 20.9.2). By Chebotarev, infinitely many primes $\mathfrak{p}$ satisfy the latter condition. Assume therefore that $L(\tau) \models \lambda$ for each $\tau \in \mathrm{Gal}(L/K)$. Compute a finite subset B_2 of $P(K)$ including all $\mathfrak{p}$ which are either ramified in L or divide one of the discriminants (or the leading coefficients) of $f_1, \ldots, f_m$. Let $B = B_1 \cup B_2$. Then $\tilde{K}_\mathfrak{p} \models \lambda$ (Lemma 20.9.2), so $\tilde{K}_\mathfrak{p} \models \theta$ for each $\mathfrak{p} \in P(K) \smallsetminus B$.

Complete the procedure by checking whether $\tilde{K}_\mathfrak{p} \models \theta$ for each of the finitely many exceptional primes $\mathfrak{p} \in B$. □

20.10 The Elementary Theory of Finite Fields

We conclude with a discussion of the theory of finite fields and its decidability.

Call a field F **pseudo finite** if F is perfect, $\mathrm{Gal}(F) \cong \hat{\mathbb{Z}}$ (i.e. 1-free), and PAC.

20.10 The Elementary Theory of Finite Fields

LEMMA 20.10.1: *Every nonprincipal ultraproduct of distinct finite fields is pseudo finite.*

Proof: Let $F_1, F_2, F_3, \ldots$ be distinct finite fields. Consider a nonprincipal ultrafilter $\mathcal{D}$ of $\mathbb{N}$ and let $F = \prod F_i/\mathcal{D}$. Then F is perfect, 1-free, and PAC (Proposition 20.4.4). Thus, F is pseudo finite. □

The following result is a special case of Corollary 20.4.2 in the case $e = 1$:

PROPOSITION 20.10.2: *Let E and F be pseudo finite fields.*
(a) *Suppose E and F contain a common field K and $E \cap \tilde{K} \cong_K F \cap \tilde{K}$. Then $E \equiv_K F$.*
(b) *Suppose F is a regular extension of E. Then $E \prec F$.*

LEMMA 20.10.3: *Let K be a finite field and L an algebraic extension of K. For each positive integer n denote the unique extension of K of degree n by K_n. Then there exists a nonprincipal ultraproduct $D = \prod K_n/\mathcal{D}$ such that $D \cap \tilde{K} = L$.*

Proof: For each positive integer d the set $A_d = \{n \in \mathbb{N} \mid K_n \cap K_d = L \cap K_d\}$ is infinite. If $d|d'$, then $A_{d'} \subseteq A_d$. Thus, given $d_1, \ldots, d_r$, we put $d = \text{lcm}(d_1, \ldots, d_r)$ and conclude from the relation $A_d \subseteq A_{d_1} \cap \cdots \cap A_{d_r}$ that the latter intersection is an infinite set. By Lemma 7.5.4, there exists a nonprincipal ultrafilter $\mathcal{D}$ on $\mathbb{N}$ which contains each A_d. Let $D = \prod K_n/\mathcal{D}$ be the corresponding ultraproduct.

We prove that $D \cap \tilde{K} = L$: Consider a positive integer $d \in D$, let x be a primitive element for K_d over K, and put $f = \text{irr}(x, K)$. Then f is Galois over K. If $K_d \subseteq L$, then $K_d \subseteq K_n$ for each $n \in A_d$, so f has a root in K_n for each $n \in A_d$. By Łoś (Proposition 7.7.1), f has a root in D, so $D \subseteq L$. If $K_d \not\subseteq L$, then $L \cap K_d \subset K_d$, so f has no root in K_n for each $n \in A_d$. By Łoś, f has no root in D, so $K_d \not\subseteq D$. It follows that $D \cap \tilde{K} = L$. □

PROPOSITION 20.10.4: *Let θ be a sentence of $\mathcal{L}(\text{ring})$. Then θ is true in almost all (i.e. all but finitely many) finite fields if and only if θ is true in every pseudo finite field.*

Proof: First suppose θ is true in almost all finite fields. Consider a pseudo finite field F. By Lemma 20.10.3, there exists a nonprincipal ultraproduct $D = \prod \mathbb{F}_{p^n}/\mathcal{D}$ such that $F \cap \tilde{\mathbb{F}}_p = D \cap \tilde{\mathbb{F}}_p$. By Lemma 20.10.1, D is pseudo finite. Hence, by Proposition 20.10.2, $F \equiv_K D$. By Łoś (Proposition 7.7.1), θ is true in D. Therefore, θ is true in F.

Conversely, suppose θ is false in a sequence $F_1, F_2, F_3, \ldots$ of distinct finite fields. Choose a nonprincipal ultrapower $\mathcal{D}$ of $\mathbb{N}$. Put $F = \prod F_i/\mathcal{D}$. By Łoś, θ is false in F. By Lemma 20.10.1, F is pseudo finite. □

COROLLARY 20.10.5 ([Ax2, p. 240]): *A field F is pseudo finite if and only if F is an infinite model of the theory of finite fields.*

Proof: First suppose F is pseudo finite. Then, by Proposition 20.10.4, each sentence θ which holds in every finite field holds in F.

Conversely, suppose F is an infinite model of the theory of finite fields. Then F is perfect and $\mathrm{Gal}(F) \cong \hat{\mathbb{Z}}$ (Proposition 20.4.4). For each positive integer d, let θ_d be the sentence "there are at most $(d-1)^4$ distinct elements or every absolutely irreducible polynomial in the variables X, Y of degree at most d has a zero". By Corollary 5.4.2(b), θ_d holds in each finite field. Hence, θ_d holds in F. But F is infinite. Hence, every absolutely irreducible polynomial of degree d has a zero in F. Consequently, F is PAC and therefore pseudo finite. $\square$

Like the theory of all residue fields of a given global field, the theory of finite fields is decidable. This was a problem raised by Tarski and solved by Ax:

THEOREM 20.10.6 ([Ax2, p. 264]): *The theory of all sentences of $\mathcal{L}(\mathrm{ring})$ which are true in every finite field is recursively decidable.*

Proof: Consider $\theta \in \mathcal{L}(\mathrm{ring})$. Follow the proof of Theorem 20.9.7 in the case $K = \mathbb{Q}$ to check if θ is true in $\mathbb{F}_p$ for all $p \in P(\mathbb{Q})$. In the affirmative case, choose an integer n greater than $(d-1)^4$ for all d's that appear in Part B and greater than of the primes belonging to Part C of that proof. Then conclude from Lemma 20.9.2 at the end of Part C of Theorem 20.9.7 that $\mathbb{F}_{p^i} \models \theta$ for each $p \geq n$ and each i.

We therefore only need to check if, for a given prime p, the sentence θ is true in $\mathbb{F}_{p^i}$ for all $i \in \mathbb{N}$. This is equivalent to checking if θ is true in $\overline{\mathbb{F}_p(t)}_\mathfrak{p}$ for each $\mathfrak{p} \in P(\mathbb{F}_p(t))$. This, again is a special case of Theorem 20.9.7. Now, the proof is complete. $\square$

Example 20.10.7: Pseudo finite fields.

(a) If a sentence θ of $\mathcal{L}(\mathrm{ring})$ holds in infinitely many finite fields, then it holds in every nonprincipal ultraproduct of those fields. Hence, by Lemma 20.10.1, θ holds in every pseudo finite field.

Similarly, let K be a global field and θ a sentence of $\mathcal{L}(\mathrm{ring}, O_K)$. Suppose θ holds in $\bar{K}_\mathfrak{p}$ with $\mathfrak{p}$ ranging over an infinite set A of prime divisors of K. Then θ holds in each nonprincipal ultraproduct F of the $\bar{K}_\mathfrak{p}$'s. Reduction modulo $\mathfrak{p}$ embeds O_K in F. Thus, F extends K.

(b) Let K be a global field. Then $\tilde{K}(\sigma)$ is pseudo finite for almost all $\sigma \in \mathrm{Gal}(K)$ (Theorem 20.5.1 for $e = 1$). Consider a sentence θ of $\mathcal{L}(\mathrm{ring}, O_K)$ which holds in $\bar{K}_\mathfrak{p}$ for infinitely many prime $\mathfrak{p}$ of K. By Theorem 20.9.4, $\delta(A(\theta)) > 0$. Hence, by the Transfer Theorem, $\mu(S(\theta)) > 0$. In particular, there is a $\sigma \in \mathrm{Gal}(K)$ such that $\tilde{K}(\sigma)$ is pseudo finite and θ holds in $\tilde{K}(\sigma)$.

(c) Let F be an infinite algebraic extension of $\mathbb{F}_p$. Then F is perfect and PAC (Corollary 11.2.4). Moreover, $\mathrm{Gal}(F) \cong \prod_{l \in S} \mathbb{Z}_l$ for some set S of prime numbers (Exercise 8 of Chapter 1). Suppose F has an extension of degree l for every prime number l. Then $\mathrm{Gal}(F) \cong \hat{\mathbb{Z}}$. Hence, F is a pseudo finite field. $\square$

Exercises

THEOREM 20.10.8: *Let K be a global field. Denote the set of all nonprincipal ultraproducts $\prod \bar{K}_{\mathfrak{p}}/\mathcal{D}$, where $\mathfrak{p}$ ranges over $P(K)$, by $\mathcal{F}$. Let T be the set of all sentences $\theta \in \mathcal{L}(\mathrm{ring}, O_K)$ which hold in almost all residue fields $\bar{K}_{\mathfrak{p}}$. Then:*

(a) *Each $F \in \mathcal{F}$ is pseudo finite.*
(b) *A sentence θ of $\mathcal{L}(\mathrm{ring}, O_K)$ belongs to T if and only if θ holds in every $F \in \mathcal{F}$.*
(c) *Every model of T is elementarily equivalent in $\mathcal{L}(\mathrm{ring}, O_K)$ to some $F \in \mathcal{F}$.*
(d) *For every $\sigma \in \mathrm{Gal}(K)$ there is an $F \in \mathcal{F}$ with $F \cap \tilde{K} \cong_K \tilde{K}(\sigma)$.*

Proof: Statement (a) is a special case of Lemma 20.10.1. Statement (b) is a special case of Proposition 7.8.1(a). Statement (c) is a special case of Proposition 7.8.1(b). Statement (d) can be proved directly from the Chebotarev density theorem, but we deduce it here from Lemma 20.5.3. By that lemma, there is a regular ultraproduct E of the $\tilde{K}(\tau)$'s with $E \cap \tilde{K} \cong_K \tilde{K}(\sigma)$. By Lemma 20.5.2, E is a pseudo finite field. By (c), there is an $F \in \mathcal{F}$ which is elementarily equivalent in $\mathcal{L}(\mathrm{ring}, O_K)$ to E. It follows from 20.6.3, that $E \cap \tilde{K} \cong_K F \cap \tilde{K}$. Consequently, $F \cap \tilde{K} \cong_K \tilde{K}(\sigma)$. $\square$

Exercises

1. Give an example showing that the hypothesis of Corollary 20.3.4 can hold nontrivially, by taking K to be a PAC field for which $\mathrm{Gal}(K)$ is finitely generated. Then, let F be a nonprincipal ultraproduct of countably many copies of K.

2. Let λ be the test sentence (Section 20.6) $(\exists X)[f_1(X) = 0 \vee f_2(X) = 0 \vee f_3(X) = 0]$ with $f_i(X) = X^2 - a_i$ and $a_i \in \mathbb{Z}$ for $i = 1, 2, 3$, nonzero integers. What is the exact condition on a_1, a_2, a_3 such that $\mu(S(K, 1, \lambda)) = 1$? Now answer the same question for $\mu(S(K, 2, \lambda)) = 1$.

3. [Jarden8, p. 149] It is a consequence of Proposition 20.6.6(a) that if (K, e) is a Hilbertian pair and θ is a sentence of $\mathcal{L}(\mathrm{ring}, K)$, then $S(K, e, \theta)$ is a measurable set. We outline an alternative proof, valid for every countable field K and every positive integer e:

Let $\varphi(X_1, \ldots, X_n)$ be a formula of $\mathcal{L}(\mathrm{ring}, K)$. For $x_1, \ldots, x_n \in \tilde{K}$ define

$$S(K, e, \varphi(\mathbf{x})) = \{\boldsymbol{\sigma} \in \mathrm{Gal}(K)^e \mid x_1, \ldots, x_n \in \tilde{K}(\boldsymbol{\sigma}) \text{ and } \tilde{K}(\boldsymbol{\sigma}) \models \varphi(\mathbf{x})\}$$

(a) Use induction on structure to show that $S(K, e, \varphi(\mathbf{x}))$ is a Borel set. Hint: Use the identity $S(K, e, (\exists Y)\varphi(\mathbf{x}, Y)) = \bigcup_{y \in \tilde{K}} S(K, e, \varphi(\mathbf{x}, y))$.

(b) Deduce that $S(K, e, \theta)$ is a Borel set for each infinite sentence of $\mathcal{L}(\mathrm{ring}, K)$.

4. Prove the converse of Theorem 20.7.2: Let K be a field and t an indeterminate. Consider the ring $R = K[t]$ and the field $K' = K(t)$. Suppose for each positive integer e and each sentence $\theta(t)$ of $\mathcal{L}(\mathrm{ring}, R)$ there exists a nonempty subset H of K with $\mu'(S(K', e, \theta(t))) = \mu(S(K, e, \theta(a)))$ for each $a \in H$. Prove that K is Hilbertian.

Hint: Let $f \in K[X]$ be separable. Then f is irreducible over K if and only if for all large e there exists $S \subseteq \mathrm{Gal}(K)^e$ of positive measure such that f is irreducible over $\tilde{K}(\boldsymbol{\sigma})$ for all $\boldsymbol{\sigma} \in S$.

5. ([Ax2, p. 260]) Let K be a global field and $\tau \in \mathrm{Gal}(K)$. Prove that there exists a nonprincipal ultraproduct $F = \prod \bar{K}_{\mathfrak{p}}/\mathcal{D}$ with $\tilde{K} \cap F \cong \tilde{K}(\tau)$.

Hint: Either reproduce Ax's direct application of the Chebotarev density theorem, or combine the Transfer theorem, Lemma 20.5.3 and Proposition 7.8.1.

6. Give an example of an infinite sentence of $\mathcal{L}(\mathrm{ring})$ which is true in each field $\tilde{\mathbb{Q}}(\sigma)$ but is false in each field $\mathbb{F}_p$. Thus, the Transfer theorem does not generalize to infinite sentences. Indeed, the Haar measure is σ-additive while the Dirichlet density is only finitely-additive.

7. Verify the Transfer theorem over $\mathbb{Q}$ for the test sentence $(\exists X)[X^2 = a]$, where a is an integer, by using quadratic reciprocity and the Dirichlet density theorem (Corollary 6.3.2) for primes in arithmetic progressions.

8. Let K be a global field and $f \in K[X]$ be separable and irreducible with $\deg(f) = n > 1$. Denote the set of prime numbers p such that f has no zero mod p by A. Put $G = \mathrm{Gal}(f, K)$ and let S be the set of all $\sigma \in G$ which fix no zero of f.

   (a) Apply the Transfer theorem to prove: $\delta(A) = \frac{|S|}{|G|}$. Alternatively, use the Chebotarev density theorem to prove the equality directly.

   (b) Suppose $\mathrm{Gal}(f, K) \cong S_n$. Prove: $\delta(A) = \frac{1}{2!} - \frac{1}{3!} + \cdots + (-1)^n \frac{1}{n!}$.

9. Let $f \in \mathbb{Z}[X, Y]$ be an absolutely irreducible polynomial. Prove that for almost all primes p there exists $a \in \mathbb{F}_p$ such that $f(a, Y)$ decomposes into linear factors in $\mathbb{F}_p[Y]$.

Hint: Let $F = \mathbb{Q}(x, y)$ be the function field of $V(f)$. Choose a stabilizing element t for the extension $F/\mathbb{Q}$ (Theorem 18.9.3). Let $\hat{F}$ be the Galois hull of $F/\mathbb{Q}(t)$. One way to complete the proof is to specialize t to elements of $\mathbb{Q}$ infinitely often so that x, y are integral over the corresponding local ring and such that the residue fields of $\hat{F}$ form a linearly disjoint sequence of Galois extensions of $\mathbb{Q}$ (as in the proof of Theorem 18.10.3). Use Borel-Cantelli (Lemma 18.3.5) to prove that for almost all $\sigma \in \mathrm{Gal}(K)$ there exists $a \in \tilde{\mathbb{Q}}(\sigma)$ such that $f(a, Y)$ decomposes into linear factors. Now use the transfer principle.

Alternatively, let z be a primitive element for the extension $\hat{F}/\mathbb{Q}(t)$. Let $g \in \mathbb{Z}[T, Z]$ be an absolutely irreducible polynomial with $g(t, z) = 0$. Now find an $h \in \mathbb{Z}[X]$, $h \neq 0$ such that for almost all p and for each $b \in \mathbb{F}_p$

with $h(b) \neq 0$ and $g(b,T)$ has zeros in $\mathbb{F}_p$ the specialization $t \to b$ extends to a specilization $(t,x) \to (b,a)$ and the polynomial $f(a,Y)$ decomposes into linear factors. Thus, the result follows directly from Corollary 5.4.2.

10. Show that the small sets of $\mathcal{Q}$ in Section 20.9 cannot be replaced by finite sets, by giving a sentence $\theta \in \mathcal{L}(\text{ring})$ for which θ is true for all $p \in P(\mathbb{Q})$ but the $q \in \mathcal{Q}$ for which $\mathbb{F}_q \not\models \theta$ is infinite.

11. Let K be a number field. Denote the set of all prime ideals of O_K by $P(K)$. For each finite extension L of K let $\text{Splt}(L/K)$ be the set of all $\mathfrak{p} \in P(K)$ which split completely in L.
(a) Prove that there exists an ultrafilter $\mathcal{L}$ of $P(K)$ which contains $\text{Splt}(L/K)$ for each finite extension L of K and each subset of $P(K)$ of Dirichlet density 1.
(b) Let $\mathcal{L}$ be as in (a). Put $F = \prod_{\mathfrak{p} \in P(K)} \tilde{K}_\mathfrak{p}/\mathcal{L}$. Prove that $\tilde{\mathbb{Q}} \subseteq F$.
(c) Let $\mathcal{L}'$ be another ultrafilter of $P(K)$ satisfying the conditions of (a). Put $F' = \prod \tilde{K}_\mathfrak{p}/\mathcal{L}'$. Prove that $F' \equiv F$.

Notes

Ax initiated the investigation of free PAC fields in [Ax1] and [Ax2] by connecting the theory of finite fields with the theory of pseudo finite fields. This connection is based on the special case of the embedding lemma (Lemma 20.2.2) where E and F are 1-free and $\text{Gal}(L)$ is procyclic (Ax's proof [Ax2, p. 248] is very complicated) and on Exercise 5. The transfer theorem [Jarden2] strengthens this connection.

The treatment of perfect e-free PAC fields can be found in [Jarden-Kiehne]. [Cherlin-v.d.Dries-Macintyre] introduces the treatment of imperfect fields that are e-free and PAC. Finally, we follow [Jacobson-Jarden1] and include finite fields, with $e = 1$ as base fields, in addition to the countable Hilbertian fields.

In [Ax1, Lemma 5], Ax proves Lemma 20.6.3(b) essentially for the case where both $\tilde{K} \cap E/K$ and $\tilde{K} \cap F/K$ are separable extensions. His reduction of the general case to the separable case is vague. The argument which appears in our proof and which uses vector spaces is due independently to B. Poizat [Poizat] and to H. W. Lenstra.

Chapter 21.
Problems of Arithmetical Geometry

We apply the model theory - measure theory technique of Chapter 20 to concrete field arithmetic problems. The transfer principles between properties of finite fields and properties of the fields $\tilde{K}(\sigma)$ can often be accomplished through direct application of the Chebotarev density theorem. Usually, however, application of the transfer theorem avoids a repetition of arguments.

This chapter includes the theory of C_i-fields, Kronecker conjugacy of global field extensions, Davenport's problem on value sets of polynomials over finite fields, a solution of Schur's conjecture on permutation polynomials, and a solution of the generalized Carlitz's conjecture on the degree of a permutation polynomial in characteristic p. Each of these concrete problems focuses our attention on rich historically motivated concepts that could be overlooked in an abstract model theoretic viewpoint.

21.1 The Decomposition-Intersection Procedure

The classical diophantine concern is the description of the $\mathbb{Q}$-rational (resp. $\mathbb{Z}$-rational) points of a Zariski $\mathbb{Q}$-closed set A. The decomposition-intersection procedure reduces this concern to the study of a union A^* of subvarieties (Section 10.2) of A defined over $\mathbb{Q}$.

We start with an arbitrary base field K. To each nonempty Zariski K-closed set A in affine space, $\mathbb{A}^n$, or in the projective space, $\mathbb{P}^n$, there corresponds a canonical K-closed subset A' defined as follows: First, decompose A into its K-components, $A = \bigcup_i V_i$. Then decompose each V_i into its $\tilde{K}$-components, $V_i = \bigcup_j W_{ij}$. For a fixed i, $\{W_{ij}\}_j$ is a complete set of conjugate varieties, defined over $\tilde{K}$ (Section 10.2). Thus, the intersection $U_i = \bigcap_j W_{ij}$ is invariant under the action of $\mathrm{Gal}(K)$: U_i is a K-closed set. Denote $\bigcup_i U_i$ by A'.

Continue the procedure to obtain a descending sequence $A \supseteq A^{(1)} \supseteq A^{(2)} \supseteq \cdots \supseteq A^{(m)} \supseteq A^{(m+1)} \supseteq \cdots$ of K-closed sets, with $A^{(m+1)} = (A^{(m)})'$. Hilbert's basis theorem (Lemma 10.1.1) gives an integer m with $A^{(r)} = A^{(m)}$ for all $r \geq m$. Denote $A^{(m)}$ by A^*.

Let $W_1, \ldots, W_s$ be a list of all $\tilde{K}$-components of $A, A^{(1)}, \ldots, A^{(m)}$. The compositum, L', of the fields of definition of $W_1, \ldots, W_s$ is a finite normal extension of K. The maximal separable extension L of K in L' is the smallest Galois extension of K such that each of $W_1, \ldots, W_s$ is L-closed. Call L the **Galois splitting field** of A over K.

LEMMA 21.1.1: *Let M be a field extension of K with $L \cap M = K$. Then:*
(a) *A^* contains all M-closed subvarieties of A. In particular, $A^*(M) = A(M)$.*

(b) A^* is nonempty if and only if A contains an M-closed nonempty subvariety.

Proof of (a): Let W be a nonempty M-closed subvariety of A. Choose be a generic point $\mathbf{x}$ of W over M. Then $\mathbf{x}$ is a generic point of W over LM. Also, there exist i and j with $\mathbf{x} \in W_{ij}$. Hence, $W \subseteq W_{ij}$. Each automorphism of L/K extends to an automorphism of LM/M. Since, for a fixed i, the W_{ij}'s are conjugate under the action of $\mathrm{Gal}(L/K)$, we have $W \subseteq \bigcap_j W_{ij} = U_i \subseteq A'$. Proceed by induction to show that $W \subseteq A^*$.

In particular, each M-rational point of A belongs to A^*. Therefore, $A(M) = A^*(M)$.

Proof of (b): Suppose first A^* is nonempty. Assume without loss that $A = A^*$. Then $A = A'$ and A is nonempty. For each i we have $U_i \subseteq V_i$. Conversely, let $\mathbf{x}$ be a generic point of V_i over K. Then, there exists an i' such that $\mathbf{x} \in U_{i'}$. Hence, $\mathbf{x} \in V_{i'}$, so $V_i = V_{i'}$. Thus, $i = i'$ and $V_i \subseteq U_i$. It follows that $U_i = V_i$. This implies that V_i is absolutely irreducible. Otherwise there would be at least two W_{ij}'s and hence we would have the contradiction $\dim(U_i) = \dim(\bigcap_j W_{ij}) < \dim(V_i)$ (Lemma 10.1.2). Thus, V_i is a variety which is K-closed, hence also M-closed.

The converse follows from (a). □

21.2 C_i-Fields and Weakly C_i-Fields

Recall that a form $f(X_0, \ldots, X_n)$ with coefficients in a field K defines a projective Zariski K-closed set in $\mathbb{P}^n$ (Section 10.7). If $\mathbf{x} = (x_0, \ldots, x_n) \neq (0, \ldots, 0)$ is a nontrivial zero of f, then the $(n+1)$-tuple $(ax_0, \ldots, ax_n)$ defines the same point of $\mathbb{P}^n$ for each $a \in K^\times$. Since projective hypersurfaces are the simplest projective sets, they occupy a special place in diophantine investigations. We follow the lead of a problem due to Artin:

Does each form $f \in \mathbb{F}_q[X_0, \ldots, X_d]$ of degree d have a nontrivial zero?

Chevalley's affirmative solution [Chevalley1] motivated Lang [Lang1] to explore the concept of a C_i-field:

Definition 21.2.1: The field K is called $C_{i,d}$ if each form $f \in K[X_0, \ldots, X_n]$ of degree d with $d^i \leq n$ has a nontrivial zero in K^{n+1}. Call it C_i if it is $C_{i,d}$ for each $d \in \mathbb{N}$. In this case, K is C_j for each $j \geq i$. □

Example 21.2.2: Every algebraically closed field K is C_0. Indeed, every form $f \in K[X_0, \ldots, X_n]$ of positive degree with $n \geq 1$ has a nontrivial zero in K^{n+1}. □

LEMMA 21.2.3 (Chevalley-Warning): Let $f_1, \ldots, f_m \in \mathbb{F}_q[X_1, \ldots, X_n]$ be polynomials with $\deg(f_1 \cdots f_m) < n$. Put, $A = V(f_1, \ldots, f_m)$. Then $|A(\mathbb{F}_q)| \equiv 0 \bmod p$.

Proof: Consider $g(\mathbf{X}) = \prod_{j=1}^m (1 - f_j(\mathbf{X})^{q-1})$. If $\mathbf{x} \in A(\mathbb{F}_q)$, then $f_j(\mathbf{x}) = 0$, $j = 1, \ldots, m$. Hence, $g(\mathbf{x}) = 1$. If $\mathbf{x} \in \mathbb{F}_q^n \smallsetminus A(\mathbb{F}_q)$, then $f_j(\mathbf{x}) \neq 0$ for some

j. Therefore, $f_j(\mathbf{x})^{q-1} = 1$ and $g(\mathbf{x}) = 0$. This gives

(1) $$|A(\mathbb{F}_q)| + p\mathbb{Z} = \sum_{\mathbf{x} \in \mathbb{F}_q^n} g(\mathbf{x}).$$

Rewrite $g(\mathbf{X})$ as $\sum c_\mathbf{i} X_1^{i_1} \cdots X_n^{i_n}$ where I is a finite set and $c_\mathbf{i} \in \mathbb{F}_q^\times$ for each $\mathbf{i} \in I$. Then

(2) $$\sum_{\mathbf{x} \in \mathbb{F}_q^n} g(\mathbf{x}) = \sum_{\mathbf{i} \in I} c_\mathbf{i} \Big(\sum_{x_1 \in \mathbb{F}_q} x_1^{i_1} \Big) \cdots \Big(\sum_{x_n \in \mathbb{F}_q} x_n^{i_n} \Big).$$

Let $\mathbf{i} \in I$. If $\mathbf{i} = (0, \ldots, 0)$, then $\sum_{x_1 \in \mathbb{F}_q} x_1^{i_1} = 0 + p\mathbb{Z}$. Otherwise, $i_1 + \cdots + i_n \leq \deg(g) \leq \sum_{j=1}^m \deg(f_j)(q-1) < n(q-1)$, so there is a j with $1 \leq i_j < q-1$. Put $k = i_j$. Since the polynomial $X^k - 1$ has at most k roots in $\mathbb{F}_q^\times$, there is an $a \in \mathbb{F}_q^\times$ with $a^k \neq 1$. Hence,

$$\sum_{x \in \mathbb{F}_q} x^k = \sum_{x \in \mathbb{F}_q} (ax)^k = a^k \sum_{x \in \mathbb{F}_q} x^k,$$

so $\sum_{x \in \mathbb{F}_q} x^k = 0$. Thus, the right hand side of (2) is 0. By (1), $|A(\mathbb{F}_q)| \equiv 0 \bmod p$. □

If $f_1, \ldots, f_m$ are forms, $(0, \ldots, 0) \in A(\mathbb{F}_q)$. Hence, by Lemma 21.2.3, $V(\mathbb{F}_q)$ has at least p points. An application of this argument to the case when $m = 1$ gives Chevalley's result:

PROPOSITION 21.2.4: *Every finite field is C_1.*

Ax applied ultraproducts to deduce from Chevalley's result that each perfect PAC field with Abelian absolute Galois group is C_1 (Theorem 22.9.6). He left unsolved this question:

PROBLEM 21.2.5 (Ax): *Is every perfect PAC field C_1?*

If Ax's problem has an affirmative solution, then each form of degree d over $\mathbb{Q}$ in $d+1$ variables has a nonempty $\mathbb{Q}$-closed subvariety (Lemma 21.3.1). This motivates the introduction of the weak C_i condition. We use it to prove some results about PAC fields which are C_i.

Definition 21.2.6: A field K is called **weakly $C_{i,d}$**, if for each form $f \in K[X_0, \ldots, X_n]$ of degree d, with $d^i \leq n$, the Zariski K-closed set $V(f)$ of $\mathbb{P}^n$ contains a subvariety W which is Zariski K-closed. If K is weakly $C_{i,d}$ for each $d \in \mathbb{N}$, we say that K is **weakly C_i**. □

Remark 21.2.7: By definition, every $C_{i,d}$ field is also weakly $C_{i,d}$. In addition, every perfect PAC field K which is weakly $C_{i,d}$ is also $C_{i,d}$. Perfectness here guarantees that a Zariski K-closed variety is defined over K (Lemma 10.2.3). □

21.2 C_i-Fields and Weakly C_i-Fields

Example 21.2.8: Separably closed fields. Let K be a field of positive characteristic p. Suppose K has an infinite sequence $a_1, a_2, a_3, \ldots$ of p-independent elements over K^p. Then, $a_1, a_2, a_3, \ldots$ are p-independent over K_s^p (Lemma 2.7.3). Hence, for each i and for all $x_0, \ldots, x_{q^i} \in K_s$ the relation $\sum_{j=0}^{q^i} a_j x_j^p = 0$ implies $x_0, \ldots, x_{q^i} = 0$. Therefore, K_s is not C_i although K_s is PAC (Section 11.1). On the other hand, each form $f \in K[X_0, X_1]$ has a nontrivial zero $(x_0, x_1) \in \tilde{K}^2$. Hence, K_s is weakly C_0.

For example, take $K = \mathbb{F}_p(t_1, t_2, t_3, \ldots)$ with $t_1, t_2, t_3, \ldots$ algebraically independent over $\mathbb{F}_p$. Then $t_1, t_2, t_3, \ldots$ are p-independent over K^p (proof of Lemma 2.7.2). Thus, K_s is weakly C_i for no positive integer i. □

We explore the behavior of the weakly C_i property under field extensions. Recall: An extension F of K is primary if $K_s \cap F = K$ (Lemma 2.6.13).

LEMMA 21.2.9: *A field K is weakly $C_{i,d}$ if and only if every form $f \in K[X_0, \ldots, X_n]$ of degree d with $d^i \leq n$ has a nontrivial zero $\mathbf{x}$ such that $K(\mathbf{x})$ is a primary extension of K.*

Proof: Suppose first K is weakly $C_{i,d}$. Let f be as stated in the lemma. Then $V(f)$ contains a K-closed variety W. It is defined over a purely inseparable extension K' of K. Let $\mathbf{p}$ be a generic point of W over K'. Then $\mathbf{p}$ is represented by a zero $(x_0, \ldots, x_n)$ of f where x_0, say, is transcendental over $K'(\mathbf{p}) = K'(\frac{x_1}{x_0}, \ldots, \frac{x_n}{x_0})$ and $K'(\mathbf{p})/K'$ is regular. Therefore, $K_s \cap K(x_0, \ldots, x_n) = K$.

Conversely, suppose that $f(\mathbf{x}) = 0$ and that $K(\mathbf{x})/K$ is a primary extension. Denote the K-closed subset of $\mathbb{P}^n$ generated by $\mathbf{x}$ by W. Since $K(\mathbf{x})$ is linearly disjoint from K_s over K, the set W is K_s-irreducible (lemma 10.2.1). It follows that W is a variety (Lemma 10.2.4). □

The form that appears in the next lemma might aptly be called a **weakly normic form**.

LEMMA 21.2.10: *Assume the field K is not separably closed. Let e_0 be an integer. Then there exist an integer $e > e_0$ and a form $h \in K[Y_1, \ldots, Y_e]$ of degree e satisfying this: Suppose $K(y_1, \ldots, y_e)$ is a primary extension of K. Then*

(3) $$h(y_1, \ldots, y_e) = 0 \implies y_1 = \cdots = y_e = 0.$$

Proof: By assumption, there exists a nontrivial Galois extension L/K. Denote its degree by l and its Galois group by G. Let $w_1, \ldots, w_l$ be a basis for L/K and consider the form

$$g(\mathbf{X}) = \prod_{\sigma \in G}(w_1^\sigma X_1 + \cdots + w_l^\sigma X_l),$$

of degree l and with coefficients in K. Suppose $K(\mathbf{y})/K$ is a primary extension and $g(\mathbf{y}) = 0$. Then there exists $\sigma \in G$ with $w_1^\sigma y_1 + \cdots + w_l^\sigma y_l = 0$. Moreover, $K(\mathbf{y})$ is linearly disjoint from L over K. Hence, $y_1 = \cdots = y_l = 0$.

Next consider the form
$$g_2 = g(g(X_1,\ldots,X_l), g(X_{l+1},\ldots,X_{2l}),\ldots, g(X_{(l-1)l+1},\ldots,X_{l^2})).$$
It is of degree l^2 and it has property (3) for $e = l^2$. Iterate this procedure to obtain, finally, a form h of degree exceeding e_0. □

The next results treat C_i-fields and weakly C_i-fields simultaneously:

LEMMA 21.2.11: Let K be a C_i-field (resp. weakly C_i-field) and $f_1, f_2, \ldots, f_r$ forms over K of degree d in n variables with $n > rd^i$. Then $f_1, f_2, \ldots, f_r$ have a common nontrivial zero in K (resp. some primary extension of K).

Proof: First suppose K is separably closed. Then K is C_i for no i but is weakly C_0 (Example 21.2.8). By the projective dimension theorem (Section 10.7), $V(f_1,\ldots,f_r)$ is a nonempty Zariski closed subset of $\mathbb{P}^{n-1}$. Thus, $f_1,\ldots,f_r$ have a nontrivial zero in $\tilde{K}$, which is a primary extension of K.

Assume now K is not separably closed. Then Lemma 21.2.10 gives a weakly normic form $h \in K[Y_1,\ldots,Y_e]$ of degree $e > r$ satisfying

(4) $$n\left(\frac{n}{r}-1\right)\left[\frac{e}{r}\right] \geq n+1.$$

For each positive integer j, let $\mathbf{X}_j = (\mathbf{X}_{j1},\ldots,\mathbf{X}_{jn})$ be a vector of variables. Let
$$h_1(\mathbf{X}_1,\ldots,\mathbf{X}_q) = h(f_1(\mathbf{X}_1),\ldots,f_r(\mathbf{X}_1),\ldots,f_1(\mathbf{X}_q),\ldots,f_r(\mathbf{X}_q),0,\ldots,0)$$
with $q = \left[\frac{e}{r}\right]$. Then h_1 is a form of degree $d_1 = de$ in $n_1 = n\left[\frac{e}{r}\right]$ variables. If $(\mathbf{x}_1,\ldots,\mathbf{x}_q)$ is a nontrivial zero with coordinates in K (resp. a primary extension of K), then there exists j between 1 and q with $\mathbf{x}_j \neq 0$ and $f_1(\mathbf{x}_j) = \cdots = f_r(\mathbf{x}_j) = 0$.

Replace h by h_1 in the above definition and define a form h_2 of degree $d_2 = d^2 e$ in $n_2 = n\left[\frac{n_1}{r}\right]$ variables. Continue by induction to define, for every positive integer k, a form h_k of degree $d_k = d^k e$ in $n_k = n\left[\frac{n_{k-1}}{r}\right]$ variables. Every nontrivial zero of h_k in K (resp. in a primary extension E of K) defines a common nontrivial zero of $f_1,\ldots,f_r$ in K (resp. in E). By assumption, a zero of that type exists if k satisfies $n_k > d_k^i$. Thus, it suffices to prove that $n_k > d_k^i$ for all large k.

The lemma is done, then, if we choose k such that $n_k > d_k^i$. Indeed,

(5) $$n_{k+1} = n\left[\frac{n_k}{r}\right] > \frac{n}{r}n_k - n.$$

Use (5) for n_k rather than for n_{k+1} and substitute in the right hand side of (5) to obtain $n_{k+1} > (\frac{n}{r})^2 n_{k-1} - n(\frac{n}{r}+1)$. Continue in this manner, inductively, to obtain the inequality

(6) $$n_{k+1} > \left(\frac{n}{r}\right)^k n_1 - n\left(\left(\frac{n}{r}\right)^{k-1} + \left(\frac{n}{r}\right)^{k-2} + \cdots + 1\right)$$
$$= \frac{1}{b}\left(\left(nb\left[\frac{e}{r}\right] - n\right)\left(\frac{n}{r}\right)^k + n\right),$$

21.2 C_i-Fields and Weakly C_i-Fields

with $b = \frac{n}{r} - 1$. Use (4) to see that the right hand side of (6) bounds $\frac{1}{b}((\frac{n}{r})^k + n)$. Therefore, $\frac{n_{k+1}}{d_{k+1}^i} > \frac{1}{d^i e^i b}((\frac{n}{rd^i})^k + \frac{n}{d^{ik}})$. Since $n > rd^i$, the right side tends to infinity with k. Consequently, $n_k > d_k^i$ for all large k, as claimed. □

PROPOSITION 21.2.12: *Let K be a C_i-field (resp. weakly C_i-field) and E is an extension of K of transcendence degree j. Then E is C_{i+j} (resp. weakly C_{i+j}).*

Proof: It suffices to prove the proposition for simple transcendental extensions and then for finite separable extensions.

PART A: $E = K(t)$, t is transcendental over K. Let $f \in K(t)[X_0, \ldots, X_n]$ be a form of degree d with $n \geq d^{i+1}$. Assume without loss the coefficients of f lie in $K[t]$. Denote the maximum of the degrees of the coefficients of f by r. Let s be a positive integer with $(n - d^{i+1} + 1)s > d^i(r+1) - (n+1)$. Consider the $(n+1)(s+1)$ variables Y_{jk}, $j = 0, \ldots, n$ and $k = 0, \ldots, s$. There exist forms $f_0(\mathbf{Y}), \ldots, f_{ds+r}(\mathbf{Y})$ over K of degree d with

$$f(\sum_{k=0}^{s} Y_{0k}t^k, \ldots, \sum_{k=0}^{s} Y_{nk}t^k) = \sum_{l=0}^{ds+r} f_l(\mathbf{Y})t^l.$$

Since $(n+1)(s+1) > d^i(ds+r+1)$, Lemma 21.2.11 gives a nonzero $\mathbf{y}$ in $K^{(n+1)(s+1)}$ (resp. $K(\mathbf{y})/K$ is a primary extension) with $f_0(\mathbf{y}) = \cdots = f_{ds+r}(\mathbf{y}) = 0$. Put $x_j = \sum_{k=0}^{s} y_{jk}t^k$, $j = 0, \ldots, n$. These elements are not all zero and $f(\mathbf{x}) = 0$.

When K is C_i, $\mathbf{x}$ is $K(t)$-rational. Therefore, $K(t)$ is C_{i+1}.

When K is weakly C_i, we may choose $\mathbf{y}$ with $K(\mathbf{y})$ algebraically independent from $K(t)$ over K. By Lemma 2.6.15(a), $K(\mathbf{y}, t)$ is a primary extension of $K(t)$. Hence, $K(\mathbf{x}, t)/K(t)$ is primary. It follows from Lemma 21.2.9 that $K(t)$ is weakly C_{i+1}.

PART B: E/K is finite. Let $f \in E[X_0, \ldots, X_n]$ be a form of degree d, with $n \geq d^i$. With $w_1, \ldots, w_e$ a basis for E/K, introduce new variables Z_{jk}, $j = 0, \ldots, n$ and $k = 1, \ldots, e$. Then there exist forms $f_1(\mathbf{Z}), \ldots, f_e(\mathbf{Z})$ over K of degree d with

$$f(\sum_{k=1}^{e} Z_{0k}w_k, \ldots, \sum_{k=1}^{e} Z_{nk}w_k) = \sum_{k=1}^{e} f_k(\mathbf{Z})w_k.$$

Since $(n+1)e > ed^i$, Lemma 21.2.11 again gives a nonzero $\mathbf{z}$ such that $f_1(\mathbf{z}) = \cdots = f_e(\mathbf{z}) = 0$ and $K(\mathbf{z}) = K$ (resp. $K(\mathbf{z})/K$ is primary). The elements $x_j = \sum_{k=1}^{e} z_{jk}w_k$, $j = 0, \ldots, n$, satisfy $f(\mathbf{x}) = 0$ and not all of them are zero. If K is C_i, then $E(\mathbf{x}) = E$. Therefore, E is C_i.

If K is weakly C_i, then $E(\mathbf{z})/E$ is primary (Lemma 2.6.15(a)). Hence, $E(\mathbf{x})/E$ is primary. Consequently, E is weakly C_i. □

Weakly C_i-fields have a large class of subfields that are also weakly C_i. Note, however, that this result does not hold for C_i fields (Exercise 5).

PROPOSITION 21.2.13: *If L is a primary extension of a field K and L is weakly C_i, then K is also weakly C_i.*

Proof: Since a primary extension $L(\mathbf{x})$ of L is, under the hypotheses, a primary extension of K, this is immediate from Lemma 21.2.9. □

21.3 Perfect PAC Fields which are C_i

First we note a relation between weakly C_i-fields, C_i-fields and the fields $\tilde{K}(\sigma)$:

LEMMA 21.3.1: *Let K be a countable Hilbertian field. Then the following are equivalent:*
(a) *K is weakly C_i.*
(b) *For each $e \in \mathbb{N}$ and for almost all $\sigma \in \mathrm{Gal}(K)^e$, $\tilde{K}(\sigma)$ is C_i.*

Proof: If (a) holds, then (Proposition 21.2.12) every algebraic extension of K is weakly C_i.

By Theorem 18.6.1, $\tilde{K}(\sigma)$ is a perfect PAC field for almost all $\sigma \in \mathrm{Gal}(K)^e$. Since a weakly C_i PAC field is C_i (Remark 21.2.7), this gives (b).

Assume now (b) holds. Let $f \in K[X_0, \ldots, X_n]$ be a form of degree d with $d^i \leq n$. Apply the decomposition-intersection procedure to the subset $A = V(f)$ of $\mathbb{P}^n$ (Section 21.1). Let L be the Galois splitting field of A over K. Choose generators $\sigma_1', \ldots, \sigma_e'$ for $\mathrm{Gal}(L/K)$. Then there exist $\sigma_1, \ldots, \sigma_e \in \mathrm{Gal}(K)$ that respectively extend $\sigma_1', \ldots, \sigma_e'$, and for which $\tilde{K}(\sigma)$ is C_i. In particular, $L \cap \tilde{K}(\sigma) = K$ and A has a $\tilde{K}(\sigma)$-rational point. By Lemma 21.1.1, A^* is nonempty. With K replacing M in Lemma 21.1.1 conclude that A contains a nonempty subvariety which is K-closed. Hence, K is weakly C_i. □

We do not expect a general field extension of a C_i-field to be C_i (Exercise 6). But, under simple conditions this is true for weakly C_i fields:

PROPOSITION 21.3.2: *The following conditions on a countable Hilbertian field K are equivalent:*
(a) *K is weakly C_i.*
(b) *Every field extension F of K is weakly C_i.*
(c) *Every perfect PAC field extension F of K is C_i.*

Proof of "(a) $\Longrightarrow$ (b)": Assume without loss F is also countable. If F/K is algebraic, then F is weakly C_i (Proposition 21.2.12). Suppose F/K is transcendental. Let K' be a purely transcendental extension of K with F algebraic over K'. By Theorem 13.2.1, K' is Hilbertian. By Lemma 21.3.1, the field $\tilde{K}(\sigma)$ is C_i for each $e \in \mathbb{N}$ and almost all $\sigma \in \mathrm{Gal}(K)^e$. Hence, by Theorem 20.7.1, $\tilde{K}'(\sigma)$ is C_i for each $e \in \mathbb{N}$ and almost all $\sigma \in \mathrm{Gal}(K')^e$. It

21.3 Perfect PAC Fields which are C_i

follows from Lemma 21.3.1 that K' is weakly C_i. Consequently (Proposition 21.2.12), F is weakly C_i.

Proof of "(b) $\Longrightarrow$ (c)": Use Remark 21.2.7.

Proof of "(c) $\Longrightarrow$ (a)": By Theorem 18.6.1 and assumption (c), $\tilde{K}(\sigma)$ is C_i for each $e \in \mathbb{N}$ and for almost all $\sigma \in \text{Gal}(K)^e$. Hence, by Lemma 21.3.1, K is weakly C_i. $\square$

Proposition 21.2.4 says that every finite field is C_1. Hence, by Proposition 21.2.12, every algebraic extension of a finite field is C_1. Every other field contains either $\mathbb{Q}$ or $\mathbb{F}_p(t)$ for some p and transcendental element t. Each of the latter fields is countable and Hilbertian (Theorem 13.4.2). An application of Proposition 21.3.2 to $i = 1$ therefore gives this reformulation of Ax's problem 21.2.5:

COROLLARY 21.3.3: *The following conditions are equivalent:*
(a) *Every field is weakly C_1.*
(b) *Each of the fields $\mathbb{Q}$ and $\mathbb{F}_p(t)$ is weakly C_1.*
(c) *Every perfect PAC field is C_1.*

LEMMA 21.3.4: *Let K be a weakly C_i-field and F an extension of K. Then F is weakly C_{i+1}. If, in addition, F is perfect and PAC, then F is C_{i+1}.*

Proof: The second statement follows from the first by remark 21.2.7. To prove the first statement, suppose first F/K is algebraic. By Proposition 21.2.12, F is weakly C_i. Hence, F is weakly C_{i+1}.

Now suppose F/K is transcendental. Use Proposition 21.2.13 and replace K by a smaller field to assume K is countable. Choose $t \in F$ transcendental over K. Then $K(t)$ is Hilbertian (Theorem 13.4.2) and weakly C_{i+1} (Proposition 21.2.12). It follows from Proposition 21.3.2 that F is weakly C_{i+1}. $\square$

LEMMA 21.3.5: *Let F be a field.*
(a) *Suppose F contains an algebraically closed field. Then F is weakly C_1.*
(b) *Suppose F has a positive characteristic. Then F is weakly C_2.*
(c) *Suppose $\text{Gal}(F)$ is procyclic. Then F is weakly C_1.*
(d) *Suppose F contains $\tilde{\mathbb{Q}}(\sigma)$ with $\sigma \in \text{Gal}(\mathbb{Q})$. Then F is weakly C_2.*

Proof of (a): By Example 21.2.2, each algebraically closed field is C_0. Hence, by Lemma 21.3.4, F is weakly C_1.

Proof of (b): By assumption, F contains a finite field. The latter is C_1 (Proposition 21.2.4). Hence, by Lemma 21.3.4, F is weakly C_2.

Proof of (c): Let K_0 be the prime field of F. Consider a form $f \in F[X_0, \ldots, X_d]$ of degree d. The coefficients of f are algebraic over a finitely generated extension K of K_0. Also, $\text{Gal}(\tilde{K} \cap F)$ is procyclic. Suppose $V(f)$ has a subvariety W which is $\tilde{K} \cap F$-closed. Then W is also F-closed. So, assume without loss, F is algebraic over K. Finally, replace F by F_{ins}, if necessary, to assume F is perfect (Proposition 21.2.13).

Now K is either a finite field, or algebraic over $\mathbb{Q}$, or transcendental over K_0. In all cases $(K, 1)$ is a Hilbertian pair in the sense of Section 20.5 (Theorem 13.4.2). By Lemma 20.5.3, there is a regular ultraproduct E of the fields $\tilde{K}(\sigma)$ with $\tilde{K} \cap E \cong_K F$. By Lemma 20.5.2(b), E is pseudo finite.

For each positive integer d the $C_{1,d}$ property of a field is elementary. Since every finite field has this property (Proposition 21.2.4), so does every pseudo finite field (Proposition 20.10.2). Therefore, E is C_1. Consequently, by Proposition 21.2.13, F is weakly C_1.

Proof of (d): Use (c) and Lemma 21.3.4. □

Suppose F in Lemma 2.3.5 is PAC and perfect. Then we may replace "weakly C_i" by "C_i" in each of the statement of that lemma. This gives the main result of this section:

THEOREM 21.3.6: *Let F be a perfect PAC field.*
(a) *Suppose F contains an algebraically closed field $\tilde{K}$. Then F is C_1.*
(b) *Suppose F has a positive characteristic. Then F is C_2.*
(c) *Suppose $\text{Gal}(F)$ is procyclic. Then F is C_1.*
(d) *Suppose F contains a field $\tilde{\mathbb{Q}}(\sigma)$, where $\sigma \in \text{Gal}(\mathbb{Q})$. Then F is C_2.*

Remark 21.3.7: Theorem 22.9.6 strengthens part (c) of Theorem 21.3.6 relaxing the condition "Gal(F) is procyclic" to "Gal(F) is Abelian."

János Kollár proves in [Kollár3, Thm. 1] that every PAC field of characterisitc 0 is C_1. This settles Problem 21.2.5 in characterisitic 0. □

21.4 The Existential Theory of PAC Fields

In Section 28.10 it is shown that the theory of PAC fields is undecidable. It is therefore of interest to observe that the decomposition-intersection procedure gives a decision procedure for the existential theory of PAC fields:

Definition 21.4.1: Call a sentence of $\mathcal{L}(\text{ring})$ **existential** if it has the form

(1) $$(\exists X_1) \cdots (\exists X_n) \left[\bigvee_i \bigwedge_j [f_{ij}(\mathbf{X}) = 0 \wedge g_i(\mathbf{X}) \neq 0] \right]$$

with $f_{ij}, g_i \in \mathbb{Z}[\mathbf{X}]$, and i, j range over finite sets. □

Replace each inequality $g_i(\mathbf{X}) \neq 0$ in (1) by the equivalent formula $(\exists Y_i)[Y_i g_i(\mathbf{X}) - 1 = 0]$ to assume the sentence has the form

(2) $$(\exists X_1) \cdots (\exists X_n) \left[\bigvee_i \bigwedge_j f_{ij}(\mathbf{X}) = 0 \right].$$

The bracketed expression in (2) defines a Zariski $\mathbb{Q}$-closed subset A of $\mathbb{A}^n$. Rewrite (2) as

(3) $\qquad\qquad\qquad (\exists X_1)\cdots(\exists X_n)[\mathbf{X} \in A].$

To test if (3) is true in every PAC field of characteristic 0, apply the decomposition-intersection procedure (Section 21.1, effective by Proposition 19.5.6). Let L be the Galois splitting field of A over $\mathbb{Q}$ and $\tau_1,\ldots,\tau_e$ generators for $\text{Gal}(L/\mathbb{Q})$.

The set of $\boldsymbol{\sigma} \in \text{Gal}(\mathbb{Q})^e$ whose restriction to L is $\boldsymbol{\tau}$ has positive measure ($= 1/[L:\mathbb{Q}]^e$). Hence, by Theorem 18.6.1, there is a $\boldsymbol{\sigma}$ such that $M = \tilde{\mathbb{Q}}(\boldsymbol{\sigma})$ is PAC and $L \cap M = \mathbb{Q}$.

If A^* is empty, then A has no M-rational point (Lemma 21.1.1). Thus, there is a PAC field of characteristic 0 for which (3) is false.

If A^* is nonempty, then (Lemma 21.1.1) each $\mathbb{Q}$-component W of A^* is a subvariety of A defined over $\mathbb{Q}$ (effectively computable). That variety has an M-rational point in each PAC field M of characteristic 0. In this case, (3) is true in every PAC field of characteristic 0.

Now we check the PAC fields of positive characteristic. Proposition 10.4.2 gives us a finite set of primes S, such that for each $p \notin S$ the variety W (as above) is well defined and remains a variety when considered as a Zariski $\mathbb{F}_p$-closed set. For each $p \in S$, repeat the decomposition-intersection procedure for A over $\mathbb{F}_p$. If A^* is nonempty in each of these cases, then (3) is true in every PAC field. If, however, there exists $p \in S$ for which A^* is empty over $\mathbb{F}_p$, then, as above, there exists a PAC field M such that $A(M)$ is empty. In this case (3) fails in some PAC field.

We summarize:

THEOREM 21.4.2: *Both the existential theory of PAC fields of a given characteristic and the existential theory of all PAC fields are primitive recursively decidable.*

21.5 Kronecker Classes of Number Fields

Our next example discusses the "Kronecker conjugacy" of polynomials. We place it in the framework of classical algebraic number theory and mention some results and open problems, with partial reformulation in terms of group theory.

It is customary to credit Kronecker (in 1880) for the impetus to investigate the decomposition of primes in number field extensions [Jehne]. Throughout this section fix a global field K.

Definition 21.5.1: Let L be a finite separable extension of K. Denote the set of all $\mathfrak{p} \in P(K)$ that have a prime divisor $\mathfrak{P} \in P(L)$ of relative degree 1 (i.e. $\bar{L}_\mathfrak{P} = \bar{K}\mathfrak{p}$) by $V(L/K)$. Call a finite separable extension M of K **Kronecker conjugate** to L (over K) if $V(L/K)$ and $V(M/K)$ differ by only finitely many elements. The set of all finite separable extensions M of K which are Kronecker conjugate to L is called the **Kronecker class** of L/K. We denote it by $\mathcal{K}(L/K)$. □

LEMMA 21.5.2 (Dedekind, Kummer): *Let R be a Dedekind ring with quotient field K. Consider a finite separable extension $L = K(x)$ of K. Suppose x is integral over R and let $f = \mathrm{irr}(x, K)$. If $R_\mathfrak{p}[x]$ is the integral closure of $R_\mathfrak{p}$ in L (by lemma 6.1.2, this holds for almost all $\mathfrak{p}$) and*

$$f(X) \equiv f_1(X)^{e_1} \cdots f_r(X)^{e_r} \mod \mathfrak{p}$$

is the factorization of $f(X)$ as a product of powers of distinct monic irreducible polynomials modulo $\mathfrak{p}$, then

$$\mathfrak{p}S = \mathfrak{P}_1^{e_1} \cdots \mathfrak{P}_r^{e_r}$$

where $\mathfrak{P}_1, \ldots, \mathfrak{P}_r$ are distinct prime ideals of S with $f(\mathfrak{P}_i/\mathfrak{p}) = \deg(f_i)$, $i = 1, \ldots, r$.

Proof: See [Lang5, p. 27] or [Janusz, p. 32]. □

The arithmetic condition for Kronecker conjugacy is equivalent to a Galois theoretic condition:

LEMMA 21.5.3: *Let L_1 and L_2 be finite separable extensions of K and L a finite Galois extension of K which contains both L_1 and L_2. Then L_1 and L_2 are Kronecker conjugate over K if and only if, with $G = \mathrm{Gal}(L/K)$,*

(1) $$\bigcup_{\sigma \in G} \mathrm{Gal}(L/L_1)^\sigma = \bigcup_{\sigma \in G} \mathrm{Gal}(L/L_2)^\sigma.$$

Proof: For $i = 1, 2$ let $x_i \in O_{L_i}$ be a primitive element of the extension L_i/K, and let $f_i = \mathrm{irr}(x_i, K)$. By Lemma 21.5.2, $V(L_i/K)$ differs from the set of primes $\mathfrak{p}$ for which $(\exists X)[f_i(X) = 0]$ is true in $\tilde{K}_\mathfrak{p}$ by only finitely many primes. Let θ be the sentence

$$(\exists X)[f_1(X) = 0] \leftrightarrow (\exists X)[f_2(X) = 0].$$

Then, Kronecker conjugacy of L_1 and L_2 over K is equivalent to the truth of θ in almost all $\tilde{K}_\mathfrak{p}$. By the transfer theorem (Theorem 20.9.3) applied to test sentences, this is equivalent to the truth of θ in $L(\sigma)$, for all $\sigma \in \mathrm{Gal}(L/K)$. But this is just a restatement of (1). □

If L_i/K is Galois, $i = 1, 2$, and L_1 and L_2 are Kronecker conjugate over K, Lemma 21.5.3 implies $L_1 = L_2$. This is Bauer's theorem (see also Exercise 5 of Chapter 6). But, in general, Kronecker conjugacy does not imply conjugacy of the fields L_1 and L_2 over K. This is demonstrated in the following example:

21.5 Kronecker Classes of Number Fields

Example 21.5.4: *Kronecker conjugate extensions of $\mathbb{Q}$ of different degrees [Schinzel, Lemma 4].* Let $L = \mathbb{Q}(\cos\frac{2\pi}{7})$ and $M = \mathbb{Q}(\sqrt{2\cos\frac{2\pi}{7}})$. Then $L/\mathbb{Q}$ is a cyclic extension of degree 3, M/L is a quadratic extension, M is not Galois over $\mathbb{Q}$, but M is Kronecker conjugate to L over $\mathbb{Q}$. We give proofs of these statements.

PART A: $\mathrm{irr}(2\cos\frac{2\pi}{7}, \mathbb{Q}) = X^3 + X^2 - 2X - 1$. Let $\zeta = e^{2\pi i/7}$ and $\eta = 2\cos\frac{2\pi}{7} = \zeta + \zeta^{-1}$. Then $\mathbb{Q}(\zeta)/\mathbb{Q}$ is a cyclic extension of degree 6 and η is fixed by complex conjugation. Hence, $\mathbb{Q}(\eta)/\mathbb{Q}$ is a cyclic extension of degree 3. The conjugates of η over $\mathbb{Q}$ are $\eta_i = \zeta^i + \zeta^{-i}$, $i = 1, 2, 3$ and they satisfy

(2) $\quad \eta_1\eta_2\eta_3 = 1, \quad \eta_1\eta_2 + \eta_1\eta_3 + \eta_2\eta_3 = -2, \quad \eta_1 + \eta_2 + \eta_3 = -1$

Hence, $\mathrm{irr}(\eta, \mathbb{Q}) = X^3 + X^2 - 2X - 1$.

PART B: $\sqrt{\eta} \notin L$. Assume $\eta = \theta^2$ with θ in L. Then $\mathrm{irr}(\theta, \mathbb{Q})$ is a polynomial of degree 3 with integral coefficients. The element θ is a zero of $f(X) = X^6 + X^4 - 2X^2 - 1$. Therefore, $\mathrm{irr}(\theta, \mathbb{Q})$ divides $f(X)$:

$$X^6 + X^4 - 2X^2 - 1 = (X^3 + aX^2 + bX + 1)(X^3 + cX^2 + dX - 1)$$

with $a, b, c, d \in \mathbb{Z}$. Thus, $a + c = 0$, $d - b = 0$, $ad + bc = 0$, $ac + b + d = 1$ and $c - a + bd = -2$. Eliminate b and c to come down to the equations $2d = a^2 + 1$ and $2a = d^2 + 2$. Then $d \geq a$ from the first and $a > d$ from the second, a contradiction.

PART C: *The structure of the Galois group* $G = \mathrm{Gal}(\hat{M}/\mathbb{Q})$. By Part B, $M = \mathbb{Q}(\sqrt{\eta_1})$ is a quadratic extension of L. Similarly, $\mathbb{Q}(\sqrt{\eta_2})$ is a quadratic extension of L which we claim to be different from $\mathbb{Q}(\sqrt{\eta_1})$. Otherwise, there exists $x \in L$ such that $\eta_2 = x^2\eta_1$. By (2), $\eta_3^{-1} = (x\eta_1)^2$, a contradiction to Part B. Also $\sqrt{\eta_3} = \sqrt{\eta_1^{-1}\eta_2^{-1}}$ implies that $\hat{M} = L(\sqrt{\eta_1}, \sqrt{\eta_2})$ is an extension of L of degree 4 with $\mathrm{Gal}(\hat{M}/L) \cong \mathbb{Z}/2\mathbb{Z} \times \mathbb{Z}/2\mathbb{Z}$. The generator of the cyclic group $\mathrm{Gal}(L/\mathbb{Q})$ acts on $\mathrm{Gal}(\hat{M}/L)$ by permuting the three subgroups of order 2 in a cyclic way. So, G is the semidirect product of $\mathbb{Z}/3\mathbb{Z}$ with $\mathbb{Z}/2\mathbb{Z} \times \mathbb{Z}/2\mathbb{Z}$.

PART D: *The Kronecker conjugacy of L and M.* By part C we have

$$\bigcup_{\sigma \in G} \mathrm{Gal}(\hat{M}/M)^\sigma = \bigcup_{i=1}^{3} \mathrm{Gal}(\hat{M}/\mathbb{Q}(\sqrt{\eta_i})) = \mathrm{Gal}(\hat{M}/L) = \bigcup_{\sigma \in G} \mathrm{Gal}(\hat{M}/L)^\sigma.$$

Hence, by Lemma 21.5.3, M and L are Kronecker conjugate. □

Lemma 21.5.3 immediately gives $L_1 \subseteq L_2$ if $L_1/\mathbb{Q}$ is Galois and L_1 and L_2 are Kronecker conjugate — just as we saw in Schinzel's example. Call a field M of a Kronecker class $\mathcal{K}(L/K)$ **minimal** if it contains no proper subfields of the same class.

LEMMA 21.5.5: *Let L_1 and L_2 be minimal fields of a Kronecker class over K. Then their Galois closures $\hat{L}_1$ and $\hat{L}_2$ over K are equal.*

Proof: Let N be a finite Galois extension of K that contains $\hat{L}_1$ and $\hat{L}_2$. By Lemma 21.5.3

$$\bigcup_{\sigma \in G} \operatorname{Gal}(N/L_1)^\sigma = \bigcup_{\sigma \in G} \operatorname{Gal}(N/L_2)^\sigma,$$

with $G = \operatorname{Gal}(N/K)$. Restrict both sides to $\hat{L}_2$ to conclude that

$$\bigcup_{\sigma \in G_2} \operatorname{Gal}(\hat{L}_2/L_1 \cap \hat{L}_2)^\sigma = \bigcup_{\sigma \in G_2} \operatorname{Gal}(\hat{L}_2/L_2)^\sigma,$$

with $G_2 = \operatorname{Gal}(\hat{L}_2/K)$. By Lemma 21.5.3, $L_1 \cap \hat{L}_2$ is Kronecker conjugate to L_2 and therefore to L_1. Minimality of L_1 implies that $L_1 = L_1 \cap \hat{L}_2 \subseteq \hat{L}_2$. Therefore, $\hat{L}_1 \subseteq \hat{L}_2$; and, by symmetry, $\hat{L}_1 = \hat{L}_2$. □

COROLLARY 21.5.6: *Each Kronecker class contains only finitely many minimal fields.*

Remark 21.5.7: On the size of Kronecker classes. Fix a number field K.

(a) Call the number of nonconjugate minimal fields of $\mathcal{K}(L/K)$ the **width** and denote it by $\omega(L/K)$. [Jehne, §4] gives L with $\omega(L/K)$ arbitrarily large. The proof incorporates Exercise 7 and the theorem of Scholz-Shafarevich that each p-group is a Galois group over K.

(b) On the other hand, [Jehne, Theorem 3] gives examples in the number field case where $\mathcal{K}(L/K)$ is infinite. Indeed, this is the case if $\operatorname{Aut}(L/K)$ contains either a nontrivial automorphism of odd order, or a cyclic group of order 8, or a quaternion group of order 8.

(c) For a quadratic extension L/K [Jehne, §6] conjectures that $\mathcal{K}(L/K)$ consists only of L. This conjecture has been proved by Saxl.

To prove the conjecture, assume M is another field in $\mathcal{K}(L/K)$. By Lemma 21.5.3, $L \subset M$. Replace M by a smaller field to assume M is a minimal extension of L. Let $\hat{M}$ be the Galois closure of M/K, $G = \operatorname{Gal}(\hat{M}/K)$, $H = \operatorname{Gal}(\hat{M}/L)$, and $U = \operatorname{Gal}(\hat{M}/M)$. Then $(G : H) = 2$, U is a maximal subgroup of H, and $H = \bigcup_{g \in G} U^g$. By [Jehne, Thm. 5], H has a unique minimal normal subgroup N which is non-Abelian and simple. Moreover, $UN = H$. So, $V = U \cap N$ is a proper subgroup of N. The minimality of N implies $N \triangleleft G$. Choose $\sigma \in G \smallsetminus H$. View σ as an automorphism of N acting as conjugation. Then, $N = \bigcup_{h \in H} V^h \cap \bigcup_{h \in H} V^{\sigma h}$. But this contradicts the main result of [Saxl]:

Let N be a finite simple group and V a proper subgroup. Then there exists $a \in N$ lying in no $\operatorname{Aut}(N)$-orbit of V.

Saxl's proof uses the classification of finite simple groups, proving the result case by case. The case where $N = \operatorname{PSL}(2, p^r)$ with $p \neq 2$ appears already in [Jehne, Thm. 5]. The case $N = A_n$ with $n \geq 5$ is due to [Klingen3].

(d) Saxl's solution of Jehne's conjecture has an interesting arithmetic consequence: Let d be a square free integer and $f \in \mathbb{Z}[X]$ an irreducible polynomial without a root in $\mathbb{Q}(\sqrt{d})$. Then there are infinitely many prime numbers p such that $\left(\frac{d}{p}\right) = 1$ but $f(X) \equiv 0$ mod p has no solution.

Otherwise, let x be a root of $f(X)$ in $\tilde{\mathbb{Q}}$, $M = \mathbb{Q}(\sqrt{d}, x)$, $\hat{M}$ the Galois closure of $M/\mathbb{Q}$, and $G = \text{Gal}(\hat{M}/\mathbb{Q})$. Then, for almost p with $\left(\frac{d}{p}\right) = 1$ the equation $f(X) \equiv 0$ mod p has a solution. By the transfer principle,

$$\text{Gal}(\hat{M}/\mathbb{Q}(\sqrt{d})) = \bigcup_{\sigma \in G} \text{Gal}(\hat{M}/M)^\sigma.$$

Thus, $M \in \mathcal{K}(\mathbb{Q}(\sqrt{d})/\mathbb{Q})$, in contrast to Jehne-Saxl. □

Another problem of Jehne is still open (see the notes at the end of the chapter for a profinite version):

PROBLEM 21.5.8 ([Jehne, §7]): *Are there fields $K \subset L \subset M$, with K global, L/K finite separable, M/K infinite separable, such that $M_0 \in \mathcal{K}(L/K)$ for every intermediate field $L \subset M_0 \subset M$ of finite degree over K?*

21.6 Davenport's Problem

Let K be a global field of characteristic p and let $f, g \in K[Y]$ be nonconstant polynomials which are not p-powers. For each prime $\mathfrak{p} \in P(K)$ for which f is defined modulo $\mathfrak{p}$, consider the **value set** of f:

$$V_\mathfrak{p}(f) = \{f(y) \mid y \in \bar{K}_\mathfrak{p}\}.$$

If $g(Y) = f(aY + b)$ with $a, b \in K$ and $a \neq 0$, then g and f are said to be **strictly linearly related**. In this case,
(1) $V_\mathfrak{p}(f) = V_\mathfrak{p}(g)$ for almost all $\mathfrak{p} \in P(K)$.

Call the pair (f, g) **exceptional** if f and g satisfy (1) but they are not strictly linearly related.

Remark 21.6.1: $V_p(X^8) = V_p(16X^8)$ for every prime number p. To prove this statement, verify the identity

(2) $\quad X^8 - 16 = (X^2 - 2)(X^2 + 2)((X - 1)^2 + 1)((X + 1)^2 + 1).$

By the multiplicativity of the Legendre symbol, $\left(\frac{2}{p}\right)\left(\frac{-2}{p}\right)\left(\frac{-1}{p}\right) = \left(\frac{4}{p}\right) = 1$ for each odd prime number p. Hence, at least one of the first three factors on the right hand side of (2) is divisible by p for some value of X. It follows that 16 is an 8th power modulo every p. Thus, for each $x \in \mathbb{F}_p$ there is a $y \in \mathbb{F}_p$ with $x^8 = 16y^8$, as claimed.

Consider now a polynomial $h(X) = c_n X^n + c_{n-1} X^{n-1} + \cdots + c_0$ in $\mathbb{Q}[X]$ with $n \geq 1$ and $c_n \neq 0$. Then $V_p(h(X^8)) = V_p(h(16X^8))$ for each p which divides no denominator of $c_0, \ldots, c_{n-1}, c_n$. Assume there were $a, b \in \mathbb{Q}$ with $a \neq 0$ and $h(16X^8) = h((aX+b)^8)$. Comparing the coefficients of X^{8n} on both sides this would give $16 = a^8$, which is a contradiction. It follows, $(h(X^8), h(16X^8))$ is an exceptional pair. $\square$

PROBLEM 21.6.2:
(a) Are all exceptional pairs in $\mathbb{Q}[X]$ of the form $(h(X^8), h(16X^8))$ with $h \in \mathbb{Q}[X]$ (Müller)?
(b) For a given global field K find all exceptional pairs (f, g) with $f, g \in K[X]$ (Davenport).

A complete solution of this problem is still unavailable. Transfer principles, however, allow us to reduce (1) to a condition on a finite Galois extension of $K(t)$, in complete analogy with the Galois theoretic version of Kronecker conjugacy of global field extensions. From this point we derive certain relations between f and g (e.g. $\deg(f) = \deg(g)$ if the degrees are relatively prime to $\mathrm{char}(K)$). In the special case where $K = \mathbb{Q}$ and $\deg(f)$ is a prime, we prove there is no $g \in \mathbb{Q}[X]$ such that (f, g) is exceptional. But we mention the existence of exceptional pairs (f, g) with $\deg(f)$ prime over other number fields.

Clearly (1) is equivalent to the truth of the sentence

(3) $$(\forall T)\big[(\exists X)[f(X) = T] \leftrightarrow (\exists Y)[g(Y) = T]\big]$$

in almost all fields $\tilde{K}_\mathfrak{p}$. By Theorem 20.9.3, this is equivalent to the truth of (3) in $\tilde{K}(\sigma)$, for almost all $\sigma \in \mathrm{Gal}(K)$. First we eliminate $(\forall T)$ to reduce the problem to a test sentence over $K' = K(t)$. To this end we choose $x, y \in \widetilde{K(t)}$ with $f(x) = t$ and $g(y) = t$ and put $E = K(t, x)$ and $F = K(t, y)$. Assume that $E/K(t)$ and $F/K(t)$ are separable extensions. When $p = \mathrm{char}(K) > 0$ this means, $f(X)$ and $g(X)$ are not a pth powers in $\tilde{K}[X]$.

LEMMA 21.6.3: Let N a finite Galois extension of $K(t)$ containing E and F. Put $G = \mathrm{Gal}(N/K(t))$. Then (1) is equivalent to

(4) $$\bigcup_{\nu \in G} \mathrm{Gal}(N/E)^\nu = \bigcup_{\nu \in G} \mathrm{Gal}(N/F)^\nu.$$

Proof: Put $K' = K(t)$. If (3) holds in $\tilde{K}(\sigma)$ for almost all $\sigma \in \mathrm{Gal}(K)$, then (3) holds in $\tilde{K}'(\sigma)$ for almost all $\sigma \in \mathrm{Gal}(K')$ (Theorem 20.7.2). In particular, formula in the outer brackets of (3) hold for t. That is, the sentence

(5) $$(\exists X)[f(X) = t] \leftrightarrow (\exists Y)[g(Y) = t]$$

is true in $\tilde{K}'(\sigma)$ for almost all $\sigma \in \mathrm{Gal}(K')$. Therefore, (5) holds in $N(\sigma)$ for all $\sigma \in G$. Consequently, (4) is true.

Conversely, suppose that the elements $\tau \in G$ that fix a root of $f(X) - t$ are exactly the elements that fix a root of $g(Y) - t$.

21.6 Davenport's Problem

We prove (3) holds in $\tilde{K}(\sigma)$ for all $\sigma \in \text{Gal}(K)$. Indeed, let $\sigma \in \text{Gal}(K)$ and $a \in \tilde{K}(\sigma)$. Extend the K-specialization $t \to a$ to a homomorphism φ of the integral closure R of $K[t]$ in N into $\tilde{K}$. The residue field $\bar{N} = \varphi(R)$ is a normal extension of $K(a)$. Denote the decomposition group $\{\tau \in G \mid \tau(\text{Ker}(\varphi)) = \text{Ker}(\varphi)\}$ of φ by H. Then φ induces an epimorphism φ_* of H onto $\text{Aut}(\bar{N}/K(a))$ through the formula (Section 6.1)

(6) $$\varphi_*(\tau)(\varphi(z)) = \varphi(\tau(z)) \text{ for each } \tau \in H \text{ and } z \in R.$$

Observe that the roots of $f(X) = t$ and $g(X) - t$ are in R. Hence, φ maps the roots of $f(X) - t$ (resp. $g(Y) - t$) onto the roots of $f(X) - a$ (resp. $g(Y) - a$). Therefore, if $b \in \tilde{K}(\sigma)$ satisfies $f(b) = a$, then there exists a zero x of $f(X) = t$ with $\varphi(x) = b$. Note that $H \cap \text{Gal}(N/K(x))$ is the decomposition group of φ over $K(x)$. Hence, by Lemma 6.1.1(a), the restriction of φ_* to $H \cap \text{Gal}(N/K(x))$ maps that subgroup onto $\text{Aut}(\bar{N}/K(b))$. In particular, there exists $\tau \in H \cap \text{Gal}(N/K(x))$ with

(7) $$\varphi_*(\tau) = \text{res}_{\bar{N}}(\sigma).$$

Since $\tau(x) = x$, the assumptions give a zero y of $g(Y) = t$ with $\tau(y) = y$. Thus, $c = \varphi(y)$ is a zero of $g(Y) = a$ for which (6) and (7) give $\sigma(c) = c$. □

Remark 21.6.4: The switch from the surjectivity of φ_* on H to its surjectivity on $H \cap \text{Gal}(N/K(x))$ is a subtlety in the proof of Lemma 21.6.3. From the surjectivity of φ_* on H alone we could choose $\tau \in H$ such that (7) holds. It follows that there exist zeros x and x' of $f(X) = t$ with $\tau(x) = x'$ and $\varphi(x) = \varphi(x') = b$. If $f(X) - a$ is separable, then this implies that $x = x'$, and conclude the proof easily. Nothing, however, in the assumptions guaranties separability of $f(X) - a$. This same subtlety occurs in the stratification procedure of Chapter 30. □

Following Section 21.5 we say that fields E and F (and the polynomials f and g) satisfying (4) are **Kronecker conjugate** over $K(t)$. This is independent of N. If E and F are minimal extensions of $K(t)$ with this property, then as in Lemma 21.5.5, they have the same Galois closure over $K(t)$. The next results, however, allow us the same conclusion under a less restrictive assumption than minimality. Throughout let $\deg(f) = m$, $\deg(g) = n$.

COROLLARY 21.6.5: *Suppose E and F are Kronecker conjugate over $K(t)$ and $[E : K(t)]$ and $[F : K(t)]$ are not divisible by $\text{char}(K)$. Let $\hat{E}$ (resp. $\hat{F}$) be the Galois closure of $E/K(t)$ (resp. $F/K(t)$). Then $[E : K(t)] = [F : K(t)]$ and $\hat{E} = \hat{F}$.*

Proof: By assumption, $\mathfrak{p}_\infty$ is tamely ramified in all fields conjugate to E or to F. Hence, $\mathfrak{p}_\infty$ is tamely ramified in the composite N of all these fields [Cassels-Fröhlich, p. 31]. Let $\mathfrak{P}_\infty$ be a prime divisor of N that lies over $\mathfrak{p}_\infty$.

Then the decomposition group of $\mathfrak{P}_\infty$ is cyclic [Cassels-Fröhlich, p. 30]. Let τ be a generator of it. Then $\mathfrak{p}_\infty$ is unramified in $N(\tau)$. Since $\mathfrak{p}_\infty$ is totally ramified in E (Example 2.3.11), the two fields are linearly disjoint over $K(t)$ (Lemma 2.5.8). In particular, the polynomial $f(X) - t$ (a root of which generates E over $K(t)$) remains irreducible over $N(\tau)$. Thus, the group $\langle\tau\rangle$ operates transitively on the roots $x_1, \ldots, x_m$ of $f(X) - t$. Therefore, τ is an m-cycle on $x_1, \ldots, x_m$.

By symmetry, τ is an n-cycle on $y_1, \ldots, y_n$, the roots of $g(X) - t$. If $m < n$, then τ^m would fix each of $x_1, \ldots, x_m$ but would move each of $y_1, \ldots, y_n$. This contradicts (4). Thus, $m \geq n$; and by symmetry, $m = n$.

Next restrict relation (4) to $\hat{F}$ to conclude that $E \cap \hat{F}$ and F are Kronecker conjugate over $K(t)$. By the first part of the corollary, $[E \cap \hat{F} : K(t)] = [F : K(t)] = [E : K(t)]$. Therefore, $E = E \cap \hat{F} \subseteq \hat{F}$. Hence, $\hat{E} \subseteq \hat{F}$. By symmetry $\hat{E} = \hat{F}$. $\square$

LEMMA 21.6.6: *Assume the hypotheses of Corollary 21.6.5. Then the polynomial $f(X) - g(Y)$ is reducible in $K[X, Y]$; and there are constants $a, b \in K$ and distinct roots $x_1, \ldots, x_k$ of $f(X) - t$ with $1 \leq k < \deg(f)$ and*

(8) $$ay + b = x_1 + \cdots + x_k.$$

Proof: Assume $f(X) - g(Y)$ is irreducible in $K[X, Y]$. Then, since y is transcendental over K, $f(X) - t = f(X) - g(y)$ is irreducible in $F[X]$. Lemma 13.3.2 gives $\sigma \in \text{Gal}(N/F)$ that moves each root of $f(X) - t$. This is a contradiction, since $\sigma(y) = y$. Thus, $f(X) - g(Y)$ is reducible.

Now consider a nontrivial factorization $f(X) - g(Y) = p(X, Y)q(X, Y)$ in $K[X, Y]$. By Corollary 21.6.5,

$$\deg_X(p) + \deg_X(q) = \deg_X(f(X) - g(Y))$$
$$= \deg(f(X) - g(Y)) = \deg(p) + \deg(q).$$

Since $\deg_X(p) \leq \deg(p)$ and $\deg_X(q) \leq \deg(q)$, it follows that $\deg_X(p) = \deg(p) = k$, with $1 \leq k < \deg(f)$. Write $p(X, Y)$ as

(9) $$c_{00}X^k + (c_{10}Y + c_{11})X^{k-1} + p_2(Y)X^{k-2} + \cdots + p_k(Y),$$

with $c_{00}, c_{10}, c_{11} \in K$, $c_{00} \neq 0$, and $p_2, \ldots, p_k \in K[Y]$. Since, $f(X) - t = p(X, y)q(X, y)$, the k roots of $p(X, y)$, say $x_1, \ldots, x_k$, are roots of $f(X) - t$. Since $f(X) - t$ is irreducible and separable over $K(t)$, $x_1, \ldots, x_k$ are distinct. By (9), $ay + b = x_1 + \cdots + x_k$, where $a = -c_{00}^{-1}c_{10}$ and $b = -c_{00}^{-1}c_{11}$, as claimed. $\square$

We summarize:

PROPOSITION 21.6.7: *Let K be a global field and $f, g \in K[X]$ nonconstant polynomials of degree not divisible by $\text{char}(K)$. Suppose $V_\mathfrak{p}(f) = V_\mathfrak{p}(g)$ for almost all $\mathfrak{p} \in P(K)$. Then:*

21.6 Davenport's Problem

(a) $\deg(f) = \deg(g)$.
(b) $f(X)-t$ and $g(X)-t$ are Kronecker conjugate over the rational function field $K(t)$.
(c) The splitting fields of $f(X) - t$ and $g(X) - t$ over $K(t)$ coincide.
(d) $f(X) - g(Y)$ is reducible in $K[X, Y]$.
(e) There exist constants $a, b \in K$ and distinct roots $x_1, \ldots, x_k$ of $f(X) - t$ with $1 \leq k < \deg(f)$ and $ay + b = x_1 + \cdots + x_k$.

Note that conditions (a)-(e) are all necessary for f and g to be strictly linearly related. Here is a case when they are also sufficient. Let K be a field. Call a polynomial $f \in K[X]$ **decomposable** over K if $f(X) = f_1(f_2(X))$ with $f_1, f_2 \in K[X]$, $\deg(f_1) > 1$, and $\deg(f_2) > 1$. (Then, $\deg(f) = \deg(f_1)\deg(f_2)$.) Otherwise, f is **indecomposable** over K (e.g. if f is of prime degree).

THEOREM 21.6.8 ([Fried4]): *Let $f, g \in \mathbb{Q}[X]$. Suppose $V_p(f) = V_p(g)$ for almost all prime number p and f is indecomposable. Then g and f are strictly linearly related.*

Proof: We give an elementary proof in the case that $\deg(f) = l$ is a prime. The proof in the general case depends on deeper group theoretic assertions, specifically the theory of doubly transitive permutation groups [Fried4].

Suppose without loss f is monic and let c be the leading coefficient of g. By Proposition 21.6.7(d)

(10) $$f(X) - g(Y) = r(X, Y)s(X, Y)$$

with $r, s \in \mathbb{Q}[X, Y]$ and $m = \deg(r) \geq 1$, $n = \deg(s) \geq 1$. Write $r = \sum_{i=0}^{m} r_i$ and $s = \sum_{i=0}^{n} s_i$ with $r_i, s_i \in \mathbb{Q}[X, Y]$ homogeneous of degree i. By (10),

(11) $$X^l - cY^l = r_m(X, Y)s_n(X, Y).$$

Since $\frac{Z^l-1}{Z-1}$ is irreducible in $\mathbb{Q}[X]$ [Lang7, p. 184], one of the factors r_m or s_m must be linear. Thus, either $m = 1$ or $n = 1$. Suppose for example $m = 1$. Then $r(X, Y) = a_1 X + a_2 Y + a_3$ with $a_1, a_2, a_3 \in \mathbb{Q}$ and, say, $a_1 \neq 0$. Let $a = -\frac{a_2}{a_1}$ and $b = -\frac{a_3}{a_1}$. Substituting X in (10) by $aY + b$ gives $f(aY + b) = g(Y)$, as claimed. $\square$

Let n be one of the integers 7, 11, 13, 15, 21, or 31. [Fried8] and [Fried11] give indecomposable polynomials f and g of degree n and a number field K such that (f, g) is an exceptional pair over K. For indecomposable polynomials, these numbers are the only exceptional degrees: a consequence of the classification of finite simple groups.

21.7 On Permutation Groups

We gather in this section various definitions and results about permutation groups which enter the proof of the Schur Conjecture in the next section.

Remark 21.7.1: *The affine linear group.*

(a) Let G be a group. Suppose G acts on a set X from the left. For each $x \in X$ put $G_x = \{\sigma \in G \mid \sigma x = x\}$. Call G **transitive** if for all $x, y \in X$ there is a $\sigma \in G$ with $\sigma x = y$. Call G **regular** if $G_x = 1$ for each $x \in X$. Note that if G is both transitive and regular, then $|G| = |X|$. Finally, call G **faithful** if $\bigcap_{x \in X} G_x = 1$.

(b) Let G be a permutation group of a set X and H a subgroup of G. Suppose H is regular and transitive. Then $H \cap G_x = 1$ and $G = HG_x = G_x H$ for each $x \in X$.

Indeed, if $\eta \in H \cap G_x$, then $\eta x = \eta$, so $\eta = 1$. For each $\sigma \in G$ there is an $\eta \in H$ with $\eta x = \sigma x$. Hence, $\sigma = \eta \cdot \eta^{-1} \sigma \in HG_x$. Thus, $G = HG_x$. Applying the same argument to σ^{-1} we find that $G = G_x H$.

If in addition H is normal in G, then $G = G_x \ltimes H$.

(c) Denote the cyclic group of order n by C_n and the **dihedral group** of order $2n$ by D_n. The latter group is generated by elements σ, τ with defining relations $\tau^2 = 1$, $\sigma^n = 1$, and $\sigma^\tau = \sigma^{-1}$.

(d) Let l be a prime number. Denote the group of all affine maps of $\mathbb{F}_l$ by $\mathrm{AGL}(1, \mathbb{F}_l)$. It consists of all maps $\sigma_{a,b} \colon \mathbb{F}_l \to \mathbb{F}_l$ with $a \in \mathbb{F}_l^\times$ and $b \in \mathbb{F}_l$ defined by $\sigma_{a,b}(x) = ax + b$. Thus, $|\mathrm{AGL}(1, \mathbb{F}_l)| = (l-1)l$. The subgroup $L = \{\sigma_{1,b} \mid b \in \mathbb{F}_l\}$ is the unique l-Sylow subgroup of $\mathrm{AGL}(1, \mathbb{F}_l)$. Thus, $L \cong C_l$ and L is normal. Its action on $\mathbb{F}_l$ is transitive and regular. Hence, by (b), $\mathrm{AGL}(1, \mathbb{F}_l) = \mathrm{AGL}(1, \mathbb{F}_l)_x \ltimes L$. Moreover, $\mathrm{AGL}(1, \mathbb{F}_l)_x \cong \mathbb{F}_l^\times$. Thus, $\mathrm{AGL}(1, \mathbb{F}_l)$ is a metacyclic group. It particular, $\mathrm{AGL}(1, \mathbb{F}_l)$ is solvable.

Claim: Each $\sigma \in \mathrm{AGL}(1, \mathbb{F}_l) \smallsetminus L$ fixes exactly one $x \in \mathbb{F}_l$. Indeed, $\sigma = \sigma_{a,b}$ with $a \neq 1$. Hence, $\sigma x = x$ if and only if $x = \frac{b}{1-a}$. □

The following lemma establishes a converse to Remark 21.7.1(d):

LEMMA 21.7.2: *Let l be an odd prime number and G a nontrivial finite solvable group acting faithfully and transitively on a set X of l elements. Let L be a minimal normal subgroup of G. Then:*

(a) *L is transitive and regular.*
(b) *$L \cong C_l$ and $G = G_x \ltimes L$ for each $x \in X$.*
(c) *L is its own centralizer in G.*
(d) *G/L is a cyclic group of order dividing $l - 1$ and $G_x \cong G/L$ for each $x \in X$.*
(e) *L is the unique l-Sylow subgroup of G.*
(f) *G is isomorphic as a permutation group to a subgroup of $\mathrm{AGL}(1, \mathbb{F}_l)$.*
(g) *Each $\rho \in G \smallsetminus L$ belongs to exactly one G_x.*
(h) *Let I be a subgroup of G which contains L and P a p-Sylow subgroup of I with $p \neq l$. Suppose $P \triangleleft I$. Then $P = 1$.*

21.7 On Permutation Groups

(i) Suppose $(G:L) = 2$ and let $x \in X$. Then G is the dihedral group D_l generated by the involution τ of G_x and a generator λ of L with the relation $\lambda^\tau = \lambda^{-1}$.

Proof of (a) and (b): Since G is solvable, L is Abelian.

Choose a system Y of representatives of the L-orbits of X. Then $X = \bigcup_{y \in Y} Ly$ and $l = \sum_{y \in Y} |Ly|$. Given $y, y' \in Y$, choose $\sigma \in G$ with $\sigma y = y'$. Then $\sigma(Ly) = \sigma L \sigma^{-1} \sigma y = Ly'$. Thus, all L-orbits have the same length n. Therefore, $l = |Y|n$.

By definition, $L \neq 1$. Choose $\lambda \in L$, $\lambda \neq 1$. Then there is an $x \in X$ with $\lambda x \neq x$, because G is faithful. Hence, $L_x < L$. Thus, $n = |Lx| = (L : L_x) > 1$. Since l is prime, $n = l$ and $|Y| = 1$. Therefore, L is transitive.

Next consider $\lambda \in L$ and $x \in X$ with $\lambda x = x$. For each $x' \in X$ choose $\lambda' \in L$ with $\lambda' x = x'$. Then $\lambda x' = \lambda \lambda' x = \lambda' \lambda x = \lambda' x = x'$, so $\lambda = 1$. Thus, $L_x = 1$ and L is regular. Hence, by Remark 21.7.1(b), $G = G_x \ltimes L$. In addition, $|L| = (L : L_x) = l$. Therefore, $L \cong C_l$.

Proof of (c): Assume $\sigma \in C_G(L)$. Fix $x \in X$ and write $\sigma = \lambda \rho$ with $\lambda \in L$ and $\rho \in G_x$. Since L is Abelian, $\rho \in C_G(L)$. For each $x' \in X$ choose $\lambda' \in L$ with $\lambda' x = x'$. Then $\rho x' = \rho \lambda' x = \lambda' \rho x = \lambda' x = x'$. Therefore, $\rho = 1$ and $\sigma \in L$, as asserted.

Proof of (d): Conjugation of L by elements of G embeds G/L into $\mathrm{Aut}(L)$ (by (c)). The latter is isomorphic to $\mathbb{F}_l^\times$. Hence, G/L is a cyclic group of order dividing $l - 1$. By (b), $G_x \cong G/L$.

Proof of (e): By (d), $l^2 \nmid |G|$, so L is an l-Sylow subgroup of G. Since L is normal, it is the unique l-Sylow subgroup of G.

Proof of (f) and (g): Let λ be a generator of L. Choose $x_0 \in X$. By (a) and (b), $X = \{x_0, \lambda x_0, \ldots, \lambda^{l-1} x_0\}$. The bijection $X \to \mathbb{F}_l$ mapping $\lambda^b x_0$ onto b for $b = 0, \ldots, l-1$ identifies L as a permutation group of $\mathbb{F}_l$ such that $\lambda^b 0 = b$ for all $b \in \mathbb{F}_l$. For an arbitrary $x \in \mathbb{F}_l$ we have $\lambda^b x = \lambda^b \lambda^x 0 = \lambda^{b+x} 0 = x + b$.

Let $\sigma \in G$ and $a \in \mathbb{F}_l$ with $\sigma \lambda \sigma^{-1} = \lambda^a$. Suppose first $\sigma 0 = 0$. Then $\sigma x = \sigma \lambda^x 0 = \sigma \lambda^x \sigma^{-1} 0 = \lambda^{ax} 0 = ax$. In the general case there is a $b \in \mathbb{F}_l$ with $\sigma 0 = \lambda^b 0$. Then $(\lambda^{-b} \sigma) \lambda (\lambda^{-b} \sigma)^{-1} = \lambda^a$, so by the preceding case, $\lambda^{-b} \sigma x = ax$. Therefore, $\sigma x = ax + b$.

This gives an embedding of G into $\mathrm{AGL}(1, \mathbb{F}_l)$ as permutation groups. In particular, each $\sigma \in G \smallsetminus L$ belongs to exactly one G_x (Remark 21.7.1(d)).

Proof of (h): Under the assumptions of (h), $P \cap L = 1$ and $P, L \triangleleft I$. Hence, $P \leq C_G(L) = L$. Therefore, $P = 1$.

Proof of (i): Suppose $(G : L) = 2$. Choose a generator λ of L and a generator τ of G_x. Then $\tau^2 = 1$, $\lambda^l = 1$, and $\lambda^\tau = \lambda^t$ for some $t \in \mathbb{F}_l$ satisfying $t^2 = 1$. By (c), $\tau \notin L = C_G(L)$. Hence, $t = -1$. Consequently, $G \cong D_l$. $\square$

DEFINITION 21.7.3: *Primitive permutation groups.* Let G be a transitive permutation group of a nonempty set X. A **block** of G is a subset Y of X with the following property: For each $\sigma \in G$ either $\sigma Y = Y$ or $Y \cap \sigma Y = \emptyset$. Let $G_Y = \{\sigma \in G \mid \sigma Y = Y\}$. Choose a system of representatives R for the left cosets of G_Y in G. Then $X = \bigcup_{\rho \in R} \rho Y$ and G acts transitively on the set $\{\rho Y \mid \rho \in R\} = \{\sigma Y \mid \sigma \in G\}$.

The set X itself, the empty set, and each $\{x\}$ with $x \in X$ are **trivial blocks**. Call G **imprimitive** if it has a nontrivial block, otherwise G is **primitive**. □

LEMMA 21.7.4: *Let G be a transitive permutation group of a set X and let $x \in X$. Then G is primitive if and only if G_x is maximal in G.*

Proof: Suppose first G is imprimitive. Then G has a nontrivial block Y. Since G is transitive, we may assume $x \in Y$. Then $G_x \leq G_Y$. Since $Y \subset X$ and G is transitive, $G_Y < G$. Now choose $y \in Y$, $y \neq x$, and $\sigma \in G$ with $\sigma x = y$. Then $\sigma Y = Y$ but $\sigma x \neq x$, so $\sigma \in G_Y \smallsetminus G_x$. Therefore, G_x is not maximal.

Conversely, suppose G has a subgroup H with $G_x < H < G$. Put $Y = Hx$. Then $H = G_Y$. Moreover, Y is a G-block.

Indeed, let $\sigma \in G$ and $y, y' \in Y$ with $\sigma y = y'$. Then $\sigma \eta x = \eta' x$ for some $\eta, \eta' \in H$. Hence, $(\eta')^{-1} \sigma \eta \in G_x < H$, so $\sigma \in H$. Therefore, $\sigma Y = Y$.

Further, $\sigma Y \neq Y$ for each $\sigma \in G \smallsetminus H$. Hence, $Y \subset X$.

By definition, $x \in Y$. In addition, $\sigma x \neq x$ for each $\sigma \in H \smallsetminus G_x$. Hence, Y contains more than one element. Thus, Y is a nontrivial G-block. Consequently, G is imprimitive. □

LEMMA-DEFINITION 21.7.5: *Let G be a permutation group of a set A. We say G is **doubly transitive** if one of the following equivalent conditions hold:*
(a) *G is transitive on A and there exists $a \in A$ such that G_a is transitive on $A \smallsetminus \{a\}$.*
(b) *G is transitive on A and for each $a \in A$, G_a is transitive on $A \smallsetminus \{a\}$.*
(c) *Let $a_1, a_2, b_1, b_2 \in A$ with $a_1 \neq a_2$ and $b_1 \neq b_2$. Then there is a $\tau \in G$ with $\tau a_1 = b_1$ and $\tau a_2 = b_2$.*

Proof: Statements (a) and (b) are equivalent because G is transitive. Thus, it suffices to prove that (b) and (c) are equivalent:

Suppose (b) holds. Let a_1, a_2, b_1, b_2 be as in (b). Choose $a \in A$ and $\alpha, \beta \in G$ with $\alpha a_1 = a$ and $\beta b_1 = a$. Then $\alpha a_2 \neq a$ and $\beta b_2 \neq a$. Hence, there is a $\sigma \in G_a$ with $\sigma \alpha a_2 = \beta b_2$. Put $\tau = \beta^{-1} \sigma \alpha$. Then $\tau a_1 = b_1$ and $\tau a_2 = b_2$.

Conversely, suppose (c) holds. Let $a \in A$ and $b, b' \in A \smallsetminus \{a\}$. Then there is a $\sigma \in G$ with $\sigma a = a$ and $\sigma b = b'$. □

REMARK 21.7.6: *Doubly transitive subgroups of* $\mathrm{AGL}(1, \mathbb{F}_p)$. If x_1, x_2, y_1, y_2 are elements in $\mathbb{F}_p$ with $x_1 \neq x_2$ and $y_1 \neq y_2$, then there are unique $a \in \mathbb{F}_p^\times$ and $b \in \mathbb{F}_p$ such that $ax_1 + b = y_1$ and $ax_2 + b = y_2$. Thus, $\mathrm{AGL}(1, \mathbb{F}_p)$ is doubly transitive.

Conversely, let G be a doubly transitive subgroup of $\mathrm{AGL}(1, \mathbb{F}_p)$. Then for each $(a, b) \in (\mathbb{F}_p^\times, \mathbb{F}_p)$ there is a $\sigma \in G$ with $\sigma(1) = a + b$ and $\sigma(0) = b$. Then $\sigma(x) = ax + b$ for all $x \in \mathbb{F}_p$. It follows that $G = \mathrm{AGL}(1, \mathbb{F}_p)$. □

One of the main tools in the proof of Schur's Proposition below is the group ring $\mathbb{Z}[H]$ of a finite Abelian group H (Section 16.6). Each $\rho \in \mathbb{Z}[H]$ has a unique presentation as a sum $\sum_{\eta \in H} c_\eta \eta$ with **coefficients** $c_\eta \in \mathbb{Z}$. Define the **support** of ρ to be the set $\mathrm{Supp}(\rho) = \{\eta \in H \mid c_\eta \neq 0\}$. Then $\langle \mathrm{Supp}(\rho) \rangle = \langle \eta \in H \mid c_\eta \neq 0 \rangle$.

Let $\rho' = \sum_{\eta \in H} c'_\eta \eta$ be another element of $\mathbb{Z}[H]$. Suppose $c_\eta, c'_\eta \geq 0$ for all η. Then $\rho \rho' = \sum_{\tau \in H} \left(\sum_{\eta \eta' = \tau} c_\eta c'_{\eta'} \right) \tau$ and there are no cancellations among the coefficients of $\rho \rho'$. Hence, $\mathrm{Supp}(\rho \rho') = \mathrm{Supp}(\rho) \mathrm{Supp}(\rho')$.

Consider a prime number p. We write $\rho \equiv \rho' \bmod p$ if $c_\eta \equiv c'_\eta \bmod p$ for all $\eta \in H$. Since $\mathbb{Z}[H]$ is a commutative ring, $\rho^p \equiv \sum_{\eta \in H} c_\eta \eta^p \bmod p$.

PROPOSITION 21.7.7 (Schur): *Let G be a primitive permutation group of $A = \{1, 2, \ldots, n\}$. Suppose G contains an n-cycle ν. Then G is doubly transitive or n is a prime number.*

Proof: We assume G is not doubly transitive, n is a composite number, and draw a contradiction.

Denote the identity map of A by ε. Let $H = \langle \nu \rangle$.

CLAIM A: $G = HG_1 = G_1 H$ and $H \cap G_1 = \{\varepsilon\}$. Indeed, H is transitive. Hence, for each $\sigma \in G$ there is an $\eta \in H$ with $\eta(1) = \sigma(1)$. Therefore, $\sigma = \eta \cdot \eta^{-1} \sigma \in HG_1$. Consequently, $G = HG_1$. Likewise, $G = G_1 H$.

It follows that each $\sigma \in G$ can be uniquely written as $\sigma = \eta \sigma_1$ with $\eta \in H$ and $\sigma_1 \in G_1$. Likewise, there are unique $\eta' \in H$ and $\sigma'_1 \in G_1$ with $\sigma = \sigma'_1 \eta'$

CLAIM B: Consider $\mathbb{Z}[H]$ as a subring of $\mathbb{Z}[G]$. Let $\gamma = \sum_{\sigma \in G_1} \sigma$ and $R = \{\rho \in \mathbb{Z}[H] \mid \gamma \rho = \rho \gamma\}$. Then there is a partition $H = \bigcup_{j=1}^r H_j$ with $H_1 = \{\varepsilon\}$ and $r \geq 3$ such that $R = \sum_{j=1}^r \mathbb{Z} \alpha_j$ where $\alpha_j = \sum_{\eta \in H_j} \eta$.

Let $A_1, \ldots, A_r$ with $A_1 = \{1\}$ be the G_1-orbits of A. Since G is not doubly transitive, G_1 is not transitive on $\{2, \ldots, n\}$ (Lemma-Definition 21.7.5). Hence, $r \geq 3$.

Next define a map $f \colon H \to A$ by $f(\eta) = \eta(1)$. Check that f is bijective and $G_1 \eta G_1 \cap H = f^{-1}(G_1 \cdot \eta(1))$ for each $\eta \in H$. For each j choose $\eta_j \in H$ with $\eta_j(1) \in A_j$. Let $H_j = G_1 \eta_j G_1 \cap H = f^{-1}(G_1 \cdot \eta_j(1)) = f^{-1}(A_j)$. Then $H = \bigcup_{j=1}^r H_j$ and $H_1 = \{\varepsilon\}$.

Now consider an element $\rho = \sum_{\eta \in H} c_\eta \eta$ in $\mathbb{Z}[H]$ with $c_\eta \in \mathbb{Z}$ for each $\eta \in H$. Then $\rho \in R$ if and only if

(1) $$\sum_{\sigma \in G_1} \sum_{\eta \in H} c_\eta \sigma \eta = \sum_{\sigma \in G_1} \sum_{\eta \in H} c_\eta \eta \sigma.$$

By Claim A, the $\sigma \eta$ (resp. $\eta \sigma$) on the left (resp. right) hand side of (1) are distinct. Hence, (1) holds if and only if for all $\sigma, \sigma' \in G_1$ and $\eta, \eta' \in H$ the equality $\sigma\eta = \eta'\sigma'$ implies $c_\eta = c_{\eta'}$. In other words, c_η is constant as η ranges on $G_1\eta G_1 \cap H$. In particular, c_η is constant on each H_j. Thus, there are $c_1,\ldots,c_r \in \mathbb{Z}$ with $\rho = \sum_{j=1}^r c_j \sum_{\eta \in H_j} \eta = \sum_{j=1}^r c_j \alpha_j$. This concludes the proof of Claim B.

CLAIM C: *Let $\rho \in R$. Put $K = \langle\mathrm{Supp}(\rho)\rangle$. Then $K = \{\varepsilon\}$ or $K = H$.*

Write $\rho = \sum_{\eta \in H} c_\eta \eta$. By Claim B, c_η are constant on H_j, $j = 1,\ldots,r$. Put $\kappa = \sum_{\eta \in \mathrm{Supp}(\rho)} \eta$. Then $\mathrm{Supp}(\kappa) = \mathrm{Supp}(\rho)$ and the coefficients of κ (which are 0 or 1) are also constant on each H_j. Hence, by Claim B, $\kappa \in R$. This means $\kappa\gamma = \gamma\kappa$. Since the coefficients of both κ and γ are nonnegative, $\mathrm{Supp}(\kappa)\mathrm{Supp}(\gamma) = \mathrm{Supp}(\gamma)\mathrm{Supp}(\kappa)$. Therefore, $KG_1 = G_1K$. This implies that KG_1 is a subgroup of G which contains G_1.

Since G is primitive, G_1 is a maximal subgroup of G (Lemma 21.7.4). Hence, $KG_1 = G_1$ or $KG_1 = G$. Thus, $|K| \cdot |G_1| = |G_1|$ or $|K| \cdot |G_1| = |G| = |H| \cdot |G_1|$ (by Claim A). Therefore, $K = \{\varepsilon\}$ or $K = H$, as claimed.

Now choose a prime divisor p of n. Denote the unique subgroup of H of order p by P. Since n is composite, P is a proper subgroup of H.

CLAIM D: *For each j, $H_j \smallsetminus P$ is a union of cosets of P.* Indeed, $P = \{\eta \in H \mid \eta^p = 1\}$. Hence, for each $\eta_0 \in H$ the set $\{\eta \in H \mid \eta^p = \eta_0\}$ is either empty or a coset of P. In the latter case, it consists of p elements. If we prove for each $\eta_0 \in H$ with $\eta_0 \neq \varepsilon$ that

(2) $\quad p \mid \#\{\eta \in H_j \mid \eta^p = \eta_0\}$,

then $\{\eta \in H_j \mid \eta^p = \eta_0\}$ is either empty or coincides with $\{\eta \in H \mid \eta^p = \eta_0\}$. In the latter case it is a coset of P. This will prove Claim D.

To prove (2), consider an element $\rho = \sum_{\eta \in H} c_\eta \eta$ in $\mathbb{Z}[H]$. For each $\eta \in H$ let c'_η be the unique integer between 0 and $p - 1$ satisfying $c'_\eta \equiv c_\eta$ mod p. Put $\rho' = \sum_{\eta \in H} c'_\eta \eta$. If c_η is constant on each H_j, then so is c'_η. So, by Claim B, $\rho \in R$ implies $\rho' \in R$. By definition, $\mathrm{Supp}(\rho') \subseteq \mathrm{Supp}(\rho)$. Also, $\rho_1 \equiv \rho_2$ mod p, with $\rho_1, \rho_2 \in \mathbb{Z}[H]$, is equivalent to $\rho'_1 = \rho'_2$.

Specifically, let $\rho = \sum_{\eta \in H_j} \eta^p$. Then $\mathrm{Supp}(\rho') \subseteq \mathrm{Supp}(\rho) \subseteq \{\tau^p \mid \tau \in H\}$. The right hand side is a proper subgroup of H (because $p|n$). Hence,

(3) $\quad\quad\quad\quad\quad\quad\quad \langle\mathrm{Supp}(\rho')\rangle < H.$

In addition, $\rho \equiv \left(\sum_{\eta \in H_j} \eta\right)^p \equiv \alpha_j^p$ mod p. Hence, $\rho' = (\alpha_j^p)'$. By definition, R is a subring of $\mathbb{Z}[H]$. As α_j is in R (Claim B), so is α_j^p. Therefore, $\rho' \in R$. It follows from (3) and Claim C that $\mathrm{Supp}(\rho') \subseteq \langle\mathrm{Supp}(\rho')\rangle \leq \{\varepsilon\}$. In particular, the coefficient of η_0 in ρ' is 0. In other words, the coefficient of η_0 in ρ is divisible by p. But the latter coefficient is exactly the right hand side of (2). Consequently, (2) is true.

CLAIM E: *Suppose $1 < j \leq r$ and $|H_j \cap P| \leq \frac{p-1}{2}$. Then $\alpha_j \eta = \alpha_j$ for all $\eta \in P$.*

Put $\alpha = \alpha_j$, $\pi = \sum_{\eta \in P} \eta$, and $\mu = \sum_{\eta \in H_j \cap P} \eta$. By Claim D, $H_j \smallsetminus P = \bigcup_{i=1}^m \lambda_i P$ with $\lambda_1,\ldots,\lambda_m \in H$. Put $\lambda = \sum_{i=1}^m \lambda_i$. Then

21.7 On Permutation Groups

(4) $\lambda \in \mathbb{Z}[H]$ and $\alpha = \sum_{\eta \in H_j} \eta = \sum_{\eta \in H_j \smallsetminus P} \eta + \sum_{\eta \in H_j \cap P} \eta = \lambda \pi + \mu$

For each $\eta \in P$, multiplication from the right by η yields a bijection of P onto itself. Hence, $\pi \eta = \pi$. Thus, in order to prove that $\alpha \eta = \eta$, it suffices (by (4)) to prove that $\mu = 0$.

To this end observe that $\pi^2 = \sum_{\eta \in P} \pi \eta = \sum_{\eta \in P} \pi = p\pi$. Similarly, $\pi \mu = k\pi$ with $k = |H_j \cap P|$. Since $\mathbb{Z}[H]$ is a commutative ring, this implies $\alpha^2 - 2k\alpha = p\lambda^2 \pi + \mu^2 - 2k\mu$. Hence, $(\mu^2 - 2k\mu)' = (\alpha^2 - 2k\alpha)'$. By Claim B, α is in R. Therefore, by the proof of Claim D, $(\alpha^2 - 2k\alpha)' \in R$. Thus, $(\mu^2 - 2k\mu)' \in R$. In addition, $\mathrm{Supp}(\mu^2 - 2k\mu)' \subseteq \mathrm{Supp}(\mu^2 - 2k\mu) \subseteq P$. Since $p \neq n$, P is a proper subgroup of H. Therefore, by Claim C, $\mathrm{Supp}((\mu^2 - 2k\mu)') \subseteq \{\varepsilon\}$. It follows that

(5) $$\mu^2 - 2k\mu = p\rho + l\varepsilon$$

with $\rho \in \mathbb{Z}[H]$, $\varepsilon \notin \mathrm{Supp}(\rho)$, and $l \in \mathbb{Z}$. By definition, $\mu = \sum_{i=1}^{k} \eta_i$, where $\eta_1, \ldots, \eta_k$ are distinct elements of H. Each row of the matrix $B = (\eta_i \eta_j)_{1 \leq i,j \leq k}$ consists of distinct elements of H. Hence, each element of H appears at most k times as an entry of B and therefore as a summand in $\sum_{i,j=1}^{k} \eta_i \eta_j = \mu^2$. Thus, the coefficients of $\mu^2 - 2k\mu$ lie between $-2k$ and k. Since $2k < p$, this implies that ρ in (5) is 0.

It follows that (5) may be rewritten as $\mu^2 = 2k\mu + l\varepsilon$. Thus, $\eta_i \eta_j \in \{\eta_1, \ldots, \eta_k, \varepsilon\}$ for all i, j. This means that $\mathrm{Supp}(\mu) \cup \{\varepsilon\} = \{\eta_1, \ldots, \eta_k, \varepsilon\}$ is a subgroup of P. But $k + 1 < p$. Hence, $\mathrm{Supp}(\mu) \cup \{\varepsilon\} = \{\varepsilon\}$. As $j > 1$ $\varepsilon \notin \mathrm{Supp}(\mu)$ (Claim B). Thus, $\mathrm{Supp}(\mu) = \emptyset$. Consequently, $\mu = 0$, as claimed.

END OF PROOF: By Claim B, $\bigcup_{j=2}^{r} H_j \cap P = P \smallsetminus \{\varepsilon\}$. Hence, $\sum_{j=2}^{r} |H_j \cap P| = p - 1$. Since $r \geq 3$, there is a $j \geq 2$ with $|H_j \cap P| \leq \frac{p-1}{2}$. Hence, by Claim E, P is contained in the subgroup $L = \{\sigma \in G \mid \gamma \alpha_j \sigma = \gamma \alpha_j\}$ of G. By Claim B, $\gamma \alpha_j = \alpha_j \gamma$. By definition of γ, $\gamma \sigma = \gamma$ for each $\sigma \in G_1$. Therefore, $G_1 \leq L$ and $PG_1 \leq L$. But $P \cap G_1 = 1$. Hence, $G_1 < L$. By the primitivity of G, $L = G$. In particular, $(\gamma \alpha_j)^{-1} \in L$. Therefore, $\gamma \alpha_j = \varepsilon$. It follows that $\varepsilon \sigma = \varepsilon$ for each $\sigma \in G$, that is G is trivial. This contradiction concludes the proof of the proposition. □

The following result completes Schur's Proposition:

PROPOSITION 21.7.8 (Burnside): *Let G be a transitive permutation group of a set A of p elements for some prime number p. Then G is doubly transitive or isomorphic (as a permutation group) to a subgroup of* $\mathrm{AGL}(1, \mathbb{F}_p)$.

Proof: For each $a \in A$ we have $(G : G_a) = |A| = p$. Hence, p divides $|G|$. By Cauchy's theorem, G has an element π of order p. Decompose π as a product of disjoint cycles. The order of π, namely p, is the least common multiple of the lengths of these cycles. Thus, π must be a cycle of length p. We may therefore identify A with $\mathbb{F}_p$ such that $\pi(a) = a + 1$ for each $a \in \mathbb{F}_p$.

Denote the vector space of all functions from $\mathbb{F}_p$ to $\mathbb{F}_p$ by V. The operations in V are defined as follows: $(f + g)(x) = f(x) + g(x)$ and $(af)(x) =$

$af(x)$. For each $f \in V$ there is a unique polynomial $h(X) = \sum_{i=0}^{p-1} a_i X^i$ in $\mathbb{F}_p[X]$ with $\sum_{i=0}^{p-1} a_i x^i = f(x)$ for all $x \in \mathbb{F}_p$. Indeed, the latter condition gives a Vandermonde system of equations for $a_0, \ldots, a_{p-1}$ which determines them uniquely. Define the **degree** of f to be $\deg(h)$. Then let V_k be the subspace of all $f \in V$ with $\deg(f) \leq k$.

The action of G on $\mathbb{F}_p$ naturally gives an action of G on V: $(\sigma f)(x) = f(\sigma^{-1} x)$. The rest of the proof is divided into four parts:

PART A: *A basis of V_k.* Let k be an integer between 0 and $p-1$. Then $1, X, \ldots, X^k$ give a basis of V_k. Thus, $\dim(V_k) = k+1$. Conversely, suppose $f_0, \ldots, f_k$ are elements of V with $\deg(f_j) = j$, then they are linearly independent, so they form a basis of V_k.

Suppose now $1 \leq k \leq p-1$. Consider $f \in V$ of degree k given by $f(x) = a_k x^k + a_{k-1} x^{k-1} + \cdots + a_0$ with $a_i \in \mathbb{F}_p$ and $a_k \neq 0$. Then

$$(\pi f)(x) = a_k (\pi^{-1} x)^k + a_{k-1} (\pi^{-1} x)^{k-1} + \cdots + a_0$$
$$= a_k (x-1)^k + a_{k-1}(x-1)^{k-1} + \cdots + a_0$$
$$= a_k x^k + (-k a_k + a_{k-1}) x^{k-1} + \text{lower terms}.$$

Hence, $f(x) - (\pi f)(x) = k a_k x^{k-1} + \text{lower terms}$. Thus, $\deg(\pi f) = k$ and $\deg(f - \pi f) = k - 1$.

Inductively define $f_k = f$ and $f_{j-1} = f_j - \pi f_j$ for $j = k, k-1, \ldots, 1$. Then $\deg(f_j) = j$, so $f_0, \ldots, f_k$ form a basis of V. Since $f_0, \ldots, f_k$ are linear combinations of $f, \pi f, \ldots, \pi^k f$, the latter k functions also form a basis of V all of its elements have degree k.

PART B: *The fixed space of G_0.* Assume from now on G is not doubly transitive. Then $\mathbb{F}_p$ has at least three G_0-orbits, A_0, A_1, A_2. Let $g_i \colon \mathbb{F}_p \to \mathbb{F}_p$ be the characteristic function of A_i, $i = 0, 1, 2$. Then $\sigma g_i = g_i$ for each $\sigma \in G_0$ and g_0, g_1, g_2 are linearly independent. Thus, at most one of the functions g_0, g_1, g_2 is of degree 0 and each of them belongs to the subspace $U = \{f \in V \mid \sigma f = f \text{ for each } \sigma \in G_0\}$ of V.

We prove that U contains a function f with $1 \leq \deg(f) \leq p - 2$. If $1 \leq \deg(g_i) \leq p - 2$ for some i between 0 and 2, take $f = g_i$. Otherwise, either $\deg(g_0) = \deg(g_1) = \deg(g_2) = p-1$ or, say, $\deg(g_0) = 0$ and $\deg(g_1) = \deg(g_2) = p-1$. In the former case there are $b_1, b_2, c_1, c_2 \in \mathbb{F}_p^\times$ with $\deg(b_1 g_0 + b_2 g_1) \leq p - 2$ and $\deg(c_1 g_0 + c_2 g_2) \leq p - 2$. Since the functions $b_1 g_0 + b_2 g_1$ and $c_1 g_0 + c_2 g_2$ are linearly independent, at least one of them is not of degree 0 (because $\dim(V_0) = 1$), so has degree between 1 and $p - 2$. In the latter case there are $b_1, b_2 \in \mathbb{F}_p^\times$ with $1 \leq \deg(b_1 g_1 + b_2 g_2) \leq p-2$. Let $k = \deg(f)$.

PART C: *The subspaces V_k and V_{k-1} are G-invariant.* By part A, $f, \pi f, \ldots, \pi^k f$ form a basis of V_k. Hence, $\langle \pi \rangle$ leaves V_k invariant. In addition, $\langle \pi \rangle$ is transitive on $\mathbb{F}_p$. Hence, $G = \langle \pi \rangle G_0 = G_0 \langle \pi \rangle$. Moreover, $\sigma f = f$ for each $\sigma \in G_0$. Therefore, $G_0 V_k = V_k$. Consequently, $G V_k = V_k$.

Next consider the G-invariant subspace

$$W = \Big\{ \sum_{\sigma \in G} c_\sigma \cdot \sigma f \,\Big|\, \sum_{\sigma \in G} c_\sigma = 0 \Big\}$$

of V. We claim that $W \subseteq V_{k-1}$. Indeed, write each $\sigma \in G$ as $\sigma = \sigma_0 \pi^i$ with $\sigma_0 \in G_0$. Then, in the notation of Part A, $(\sigma f)(x) = (\pi^i f)(x) = f(\pi^{-i} x) = f(x - i) = a_k x^k +$ lower terms. Hence, if $\sum_{\sigma \in G} c_\sigma = 0$, then $\deg(\sum_{\sigma \in G} c_\sigma \cdot \sigma f) \leq k-1$, as claimed. In particular, $\dim(W) \leq \dim(V_{k-1}) = k$.

Finally, note that $f - \pi f, \pi f - \pi^2 f, \ldots, \pi^{k-1} f - \pi^k f$ are in W and they are linearly independent. Hence, $\dim(W) = k$. Therefore, $W = V_{k-1}$. It follows that V_{k-1} is G-invariant.

PART D: *End of proof.* Define the product gh of two functions $g, h \in V$ by $(gh)(x) = g(x)h(x)$. Then, $\sigma(gh) = \sigma(g)\sigma(h)$ for each $\sigma \in G$ and $V_1 = \{g \in V \mid gh \in V_k$ for all $h \in V_{k-1}\}$. Since both V_{k-1} and V_k are G-invariant, so is V_1. In particular, applying σ^{-1} to the identity map of $\mathbb{F}_p$, we find $a \in \mathbb{F}_p^\times$ and $b \in \mathbb{F}_p$ with $\sigma(x) = ax + b$ for all $x \in \mathbb{F}_p$. This gives the desired embedding of G into $\mathrm{AGL}(1, \mathbb{F}_p)$. $\square$

The combination of Schur's result and Burnside's result gives the following characterization of transitive proper subgroups of $\mathrm{AGL}(1, \mathbb{F}_p)$:

COROLLARY 21.7.9: *Let G be a transitive permutation group of a set A of n elements. Suppose G contains a cycle of length n but G is not doubly transitive. Then n is a prime number p and G is isomorphic (as a permutation group) to a subgroup of $\mathrm{AGL}(1, \mathbb{F}_p)$. In particular, G is solvable.*

Proof: By Schur (Proposition 21.7.7), $n = p$ is a prime number. By Burnside (Proposition 21.7.8), G is isomorphic to a subgroup of $\mathrm{AGL}(1, \mathbb{F}_p)$. It follows from Remark 21.7.1(d) that G is solvable. $\square$

21.8 Schur's Conjecture

Let K be a field and $f \in K[X]$. We say that f **permutes** K, f is a **permutation polynomial** on K, or f is **bijective** on K if the map $x \mapsto f(x)$ is a bijection of K onto itself. Let R be a ring, $f \in R[X]$, and P a maximal ideal of R. We say f is a **permutation polynomial** modulo P, if the reduction of f modulo P is a permutation polynomial on R/P.

In this section we consider a global field K and a polynomial $f \in O_K[X]$ with $\mathrm{char}(K) \nmid \deg(f)$ which is a permutation polynomial modulo infinitely many prime ideals of O_K. We prove a conjecture of Schur: f is composed of linear polynomials and Dickson polynomials $D_n(a, X)$. The latter are defined in $\mathbb{Z}[a, X]$ by induction on n: $D_0(a, X) = 2$, $D_1(a, X) = X$, and

(1) $\quad D_n(a, X) = X D_{n-1}(a, X) - a D_{n-2}(a, X)$ for $n \geq 2$.

Thus, $D_2(a,X) = X^2 - 2a$, $D_3(a,X) = X^3 - 3aX$, and $D_4(a,X) = X^4 - 4aX^2 + 2a^2$. Call $D_n(a,X)$ the **Dickson polynomial** of degree n with parameter a.

LEMMA 21.8.1: *Let K be a field, $a \in K$, and n a nonnegative integer. Use X and Z for variables. Then:*
(a) $D_n(a, Z + aZ^{-1}) = Z^n + a^n Z^{-n}$.
(b) *Suppose $f \in K(Z)$ satisfies $f(Z + aZ^{-1}) = Z^n + a^n Z^{-n}$. Then $f(X) = D_n(a,X)$.*
(c) $b^n D_n(a,X) = D_n(b^2 a, bX)$.
(d) *For $n \geq 3$ we have $D_n(a,X) = X^n - naX^{n-2} + g(X)$ with $g \in \mathbb{Z}[a,X]$ of degree at most $n-3$.*

Proof of (a): Check (a) for $n = 0$ and $n = 1$. Next assume (a) by induction for $n - 2$ and $n - 1$. Then

$$D_n(a, Z + aZ^{-1}) = (Z + aZ^{-1})D_{n-1}(a, Z + aZ^{-1}) - aD_{n-2}(a, Z + aZ^{-1})$$
$$= (Z + aZ^{-1})(Z^{n-1} + a^{n-1}Z^{1-n}) - a(Z^{n-2} + a^{n-2}Z^{2-n})$$
$$= Z^n + a^n Z^{-n}$$

Proof of (b): Put $X' = Z + aZ^{-1}$. Then $f(X') = D_n(a, X')$ (by (a)) and $[K(Z) : K(X')] \leq 2$. Hence, X' is a variable. Therefore, $f(X) = D_n(a,X)$.

Proof of (c) and (d): Carry out induction on n. □

Definition 21.8.2: Linear relation of polynomials. Let K be a field and $f, g \in K[X]$ polynomials. We say f and g are **linearly related** over K if there exist $a, b \in K^\times$ and $c, d \in K$ with $f(X) = ag(bX + c) + d$. □

LEMMA 21.8.3: *Let K be a field, $f \in K[X]$, K' an extension of K, and n a positive integer. Suppose $\mathrm{char}(K) \nmid n$ and $f(X)$ is linearly related over K' to $D_n(a', X)$ for some $a' \in K'$. Then $f(X)$ is linearly related over K to $D_n(a, X)$ for some $a \in K$.*

Proof: We prove the lemma only for $n \geq 3$. The case $n \leq 2$ may be checked directly.

By assumption, there are $\alpha', \beta' \in (K')^\times$ and $a', \gamma', \delta' \in K'$ with $f(X) = \alpha' D_n(a', \beta'X + \gamma') + \delta'$. Put $a = (\beta')^{-2}a'$, $\alpha = \alpha'(\beta')^n$, $\gamma = (\beta')^{-1}\gamma'$, and $\delta = \delta'$. By Lemma 21.8.1(c), $f(X) = \alpha D_n(a, X + \gamma) + \delta$. By Lemma 21.8.1(d),

$$f(X) = \alpha(X + \gamma)^n - n\alpha a(X+\gamma)^{n-2} + \text{lower terms}$$
$$= \alpha X^n + n\alpha\gamma X^{n-1} + \alpha(\tbinom{n}{2}\gamma^2 - na)X^{n-2} + \text{lower terms}$$

Hence, $\alpha \in K$, $n\alpha\gamma \in K$, and $\alpha(\binom{n}{2}\gamma^2 - na) \in K$. By assumption, $n \neq 0$ in K. Therefore, $\gamma \in K$ and $a \in K$. Finally, $\delta = f(0) - \alpha D_n(a, \gamma) \in K$. Consequently, $f(X)$ is linearly related to $D_n(a, X)$ over K. □

21.8 Schur's Conjecture

Special cases of Dickson polynomials give rise to well known families of polynomials. For $a = 0$ and $n \geq 1$ induction or Lemma 21.8.1(a) give $D_n(0, X) = X^n$. The special case $a = 1$ gives the **Chebyshev polynomials**: $T_n(X) = D_n(1, X)$. We extract properties of Chebyshev polynomials from Lemma 21.8.1:

LEMMA 21.8.4: Let K be a field and n a nonnegative integer. Then:
(a) $T_0(X) = 2$, $T_1(X) = X$, and $T_n(X) = XT_{n-1}(X) - T_{n-2}(X)$ for $n \geq 2$.
(b) $T_n(Z + Z^{-1}) = Z^n + Z^{-n}$.
(c) Suppose $f \in K(Z)$ satisfies $f(Z + Z^{-1}) = Z^n + Z^{-n}$. Then $f = T_n$.
(d) For each $n \geq 3$ there is a $g \in \mathbb{Z}[X]$ of degree at most $n - 3$ satisfying $T_n(X) = X^n - nX^{n-2} + g(X)$.
(e) Suppose $\mathrm{char}(K) \nmid n$ and $n > 2$. Then, for each $a \in \tilde{K}$ there are distinct $x_1, x_2 \in \tilde{K}$ with $T_n(x_1) = T_n(x_2) = a$.

Proof of (e): Choose $y \in \tilde{K}$ satisfying $y^2 - ay + 1 = 0$. Then $y + y^{-1} = a$. Since $\mathrm{char}(K) \nmid n$, there are distinct $z_1, \ldots, z_n \in \tilde{K}$ with $z_i^n = y$, $i = 1, \ldots, n$. Since $n \geq 3$, there are i, j such that $x_i = z_i + z_i^{-1}$ and $x_j = z_j + z_j^{-1}$ are distinct. They satisfy $T_n(x_i) = T_n(x_j) = a$. □

Remark 21.8.5: *Automorphisms of $K(z)$.* Let K be a field and z an indeterminate. Recall that $\mathrm{PGL}(2, K)$ is the quotient of $\mathrm{GL}(2, K)$ by the group of scalar matrices. Denote the image of a matrix $\begin{pmatrix} a & b \\ c & d \end{pmatrix} \in \mathrm{GL}(2, K)$ in $\mathrm{PGL}(2, K)$ by $\begin{bmatrix} a & b \\ c & d \end{bmatrix}$. Thus, $\begin{bmatrix} a & b \\ c & d \end{bmatrix} = \begin{bmatrix} a' & b' \\ c' & d' \end{bmatrix}$ means that there is an $e \in K^{\times}$ with $\begin{pmatrix} a' & b' \\ c' & d' \end{pmatrix} = \begin{pmatrix} ea & eb \\ ec & ed \end{pmatrix}$. Remark 11.7.4 identifies $\mathrm{PGL}(2, K)$ with $\mathrm{Aut}(K(z)/K)$, where the action of $\begin{bmatrix} a & b \\ c & d \end{bmatrix}$ on $K(z)$ is defined by the formula: $\begin{bmatrix} a & b \\ c & d \end{bmatrix} z = \frac{az+b}{cz+d}$. □

LEMMA 21.8.6: Let K be an algebraically closed field and $n \geq 3$ an integer with $\mathrm{char}(K) \nmid n$. Consider the elements $\sigma = \begin{bmatrix} 1 & 0 \\ 0 & \zeta_n \end{bmatrix}$ and $\tau = \begin{bmatrix} 0 & 1 \\ 1 & 0 \end{bmatrix}$ of $\mathrm{PGL}(2, K)$. Put $\Delta = \langle \sigma, \tau \rangle$. Then:
(a) $\sigma^n = \tau^2 = 1$, $\sigma^{\tau} = \sigma^{-1}$, and $\Delta \cong D_n$.
(b) Let $\Delta_1 = \langle \sigma_1, \tau_1 \rangle$ be a subgroup of $\mathrm{PGL}(2, K)$ with $\sigma_1^n = \tau_1^2 = 1$ and $\sigma_1^{\tau_1} = \sigma_1^{-1}$ (so, $\Delta_1 \cong D_n$). Then there is a $\lambda \in \mathrm{PGL}(2, K)$ with $\langle \sigma_1 \rangle^{\lambda} = \langle \sigma \rangle$, $\tau_1^{\lambda} = \tau$, and $\Delta_1^{\lambda} = \Delta$.
(c) Let z be an indeterminate. Then the fixed field of τ (resp. Δ) in $K(z)$ is $K(z + z^{-1})$ (resp. $K(z^n + z^{-n})$).

Proof of (a): Put $\zeta = \zeta_n$. All we need to do is to verify the relation $\sigma^{\tau} = \sigma^{-1}$:

$$\begin{bmatrix} 0 & 1 \\ 1 & 0 \end{bmatrix} \begin{bmatrix} 1 & 0 \\ 0 & \zeta \end{bmatrix} \begin{bmatrix} 0 & 1 \\ 1 & 0 \end{bmatrix} = \begin{bmatrix} \zeta & 0 \\ 0 & 1 \end{bmatrix} = \begin{bmatrix} 1 & 0 \\ 0 & \zeta^{-1} \end{bmatrix}.$$

Proof of (b): Denote the image of $\sigma \in \mathrm{GL}(2, K)$ in $\mathrm{PGL}(2, K)$ by $[\sigma]$. Since K is algebraically closed, there are $\tilde{\sigma}_1, \tilde{\tau}_1 \in \mathrm{GL}(2, K)$ with $[\tilde{\sigma}_1] = \sigma_1$, $[\tilde{\tau}_1] = \tau_1$, and $\tilde{\sigma}_1^n = \tilde{\tau}_1^2 = 1$. The minimal polynomial of $\tilde{\sigma}_1$ divides $X^n - 1$. Since $n \neq \mathrm{char}(K)$, $X^n - 1$ has n distinct roots in K. Hence, $\tilde{\sigma}_1$ is conjugate to a diagonal matrix $\begin{pmatrix} a & 0 \\ 0 & d \end{pmatrix}$ with $a^n = d^n = 1$. Therefore, $[\tilde{\sigma}_1]$ is conjugate to $\begin{bmatrix} 1 & 0 \\ 0 & a^{-1}d \end{bmatrix}$ with $a^{-1}d = \zeta^k$ and $\gcd(k,n) = 1$. Replace ζ by ζ^k, if necessary, to assume $[\tilde{\sigma}_1] = \sigma$.

Let $\tau_1 = \begin{bmatrix} w & x \\ y & z \end{bmatrix}$ with $\begin{pmatrix} w & x \\ y & z \end{pmatrix} \in \text{GL}(2, K)$. Since $\tau_1\sigma = \sigma^{-1}\tau_1$, there is an $a \in K^\times$ with

$$\begin{pmatrix} w & x \\ y & z \end{pmatrix}\begin{pmatrix} 1 & 0 \\ 0 & \zeta \end{pmatrix} = \begin{pmatrix} a & 0 \\ 0 & a\zeta^{-1} \end{pmatrix}\begin{pmatrix} w & x \\ y & z \end{pmatrix}.$$

Therefore, $w = aw$, $x\zeta = ax$, $y = a\zeta^{-1}y$, and $z\zeta = a\zeta^{-1}z$. Assume $w \neq 0$. Then $a = 1$, $x = 0$, $y = 0$, and $z = 0$ (use $n \neq 2$). This contradiction proves that $w = 0$. Hence, $x \neq 0$ and $y \neq 0$. Therefore, $a = \zeta$ and $z = 0$. Consequently, $\tau_1 = \begin{bmatrix} 0 & x \\ y & 0 \end{bmatrix}$.

Put $\rho = \begin{bmatrix} \sqrt{x} & 0 \\ 0 & \sqrt{y} \end{bmatrix}$. Then, $\rho^{-1}\tau_1\rho = \tau$ and $\rho^{-1}\sigma\rho = \sigma$. So, there exists $\lambda \in \text{PGL}(2, K)$ as claimed.

Proof of (c): Let $x = z + z^{-1}$ and $t = z^n + z^{-n}$. Denote the fixed field of τ (resp. Δ) in $K(z)$ by E_1 (resp. E_2). Then $[K(z) : E_1] = 2$ and $[K(z) : E_2] = 2n$.

Deduce from $\tau z = z^{-1}$ and $\tau z^{-1} = z$ that $\tau x = x$. Hence, $K(x)$ is contained in E_1. On the other hand, z is a root of $X^2 - xX + 1$. Therefore, $[K(z) : K(x)] \leq 2$. Combined with the preceding paragraph, this implies $K(x) = E_1$.

Similarly, $K(t) \subseteq E_2$ and z is a root of the equation $X^{2n} - tX^n + 1 = 0$. Hence, $K(t) = E_2$. □

PROPOSITION 21.8.7 (Müller): *Let K be an algebraically closed field of characteristic p, $f \in K[X]$ a polynomial of prime degree $l \neq p$, and t an indeterminate. Suppose $G = \text{Gal}(f(X) - t, K(t))$ is solvable. Then either $G \cong C_l$ and f is linearly related to X^l or $G \cong D_l$ and f is linearly related to the Chebyshev polynomial T_l.*

Proof: By assumption, $f(X) - t$ is a separable polynomial of degree l which is irreducible. List its zeros in $K(t)_s$ as $x_1, \ldots, x_l$. Then $F = K(x_1, \ldots, x_l)$ is the Galois closure of $K(x_j)/K(t)$, $G = \text{Gal}(F/K(t))$ acts faithfully and transitively on $\{x_1, \ldots, x_l\}$. When $l = 2$, $G \cong C_2$ and $f(X)$ is linearly related to X^2 (use that $2 \neq p$). So, assume from now on $l > 2$. Then, we may apply Lemma 21.7.2 with $\{x_1, \ldots, x_l\}$ replacing X and $H_j = \text{Gal}(F/K(x_j))$ replacing G_{x_j}, $j = 1, \ldots, l$. In particular, let L be a minimal normal subgroup of G. Then, by Lemma 27.7.2(b,e), $L \cong C_l$ and L is the unique l-Sylow subgroup of G.

Let $m = [F : K(x_j)]$ and $g = \text{genus}(F/K)$. In the remaining parts of the proof we apply Riemann-Hurwitz to prove that $g = 0$ and $m \leq 2$. We then show: If $m = 1$, then $G \cong C_l$ and f is linearly related to X^l. If $m = 2$, then $G \cong D_l$ and f is linearly related to T_l.

21.8 Schur's Conjecture

PART A: *The Riemann-Hurwitz formula.* Denote the set of all prime divisors of F/K which ramify over $K(t)$ by R. It includes all prime divisors of F/K which ramify over $K(x_j)$. Hence, $\text{Diff}(F/K(t)) = \sum_{\mathfrak{q}\in R} d_\mathfrak{q} \mathfrak{q}$ and $\text{Diff}(F/K(x_j)) = \sum_{\mathfrak{q}\in R} d_{\mathfrak{q},j} \mathfrak{q}$ (Section 3.6). Here $d_\mathfrak{q}$ (resp. $d_{\mathfrak{q},j}$) are the different exponents of $\mathfrak{q}$ over $K(t)$ (resp. $K(x_j)$). Since K is algebraically closed, $\deg(\mathfrak{q}) = 1$ for each $\mathfrak{q} \in R$. Therefore, Riemann-Hurwitz formula for $F/K(t)$ and $F/K(x_j)$ becomes:

(2a) $$2g - 2 = -2ml + \sum_{\mathfrak{q}\in R} d_\mathfrak{q}$$

(2b) $$2g - 2 = -2m + \sum_{\mathfrak{q}\in R} d_{\mathfrak{q},j}, \quad j = 1,\ldots,l.$$

Subtract the sum of the l equations (2b) from (2a):

(3) $$2(g-1)(1-l) = \sum_{\mathfrak{q}\in R} \left(d_\mathfrak{q} - \sum_{j=1}^{l} d_{\mathfrak{q},j}\right).$$

Now consider $\mathfrak{q} \in R$. Denote the inertia group of $\mathfrak{q}$ over $K(t)$ by $I_\mathfrak{q}$. Then the inertia group of $\mathfrak{q}$ over $K(x_j)$ is $H_j \cap I_\mathfrak{q}$. To compute the $\mathfrak{q}$th term in (3) we distinguish between two cases.

CASE A1: $I_\mathfrak{q}$ *acts intransitively on* $\{x_1,\ldots,x_l\}$. By Lemma 21.7.2(a), $L \not\leq I_\mathfrak{q}$. Since $L \cong C_l$, we have $L \cap I_\mathfrak{q} = 1$. Thus, $I_\mathfrak{q} \cong LI_\mathfrak{q}/L \leq G/L$. Since G/L is cyclic, $I_\mathfrak{q}$ is cyclic. Choose a generator τ of $I_\mathfrak{q}$. Then $\tau \notin L$. Hence, by Lemma 21.7.2(g), there is a k between 1 and l with $\tau \in H_k$ and therefore $I_\mathfrak{q} \leq H_k$. This implies that $\mathfrak{q}|_{K(x_k)}$ is unramified over $K(t)$. Therefore, $d_\mathfrak{q} = d_{\mathfrak{q},k}$ (Lemma 3.5.8). For $j \neq k$ we have $I_\mathfrak{q} \cap H_j = 1$ (Lemma 21.7.2(g)). Hence, $\mathfrak{q}$ is unramified over $K(x_j)$ and $d_{\mathfrak{q},j} = 0$ (Remark 3.5.7(b)). Consequently, the $\mathfrak{q}$th term in the right hand side of (3) is 0.

CASE A2: $I_\mathfrak{q}$ *acts transitively on* $\{x_1,\ldots,x_l\}$. Then for each j we have $(I_\mathfrak{q} : I_\mathfrak{q} \cap H_j) = l$. Hence, l divides $|I_\mathfrak{q}|$. Therefore, by Lemma 21.7.2(e), $L \leq I_\mathfrak{q}$.

Let P be the trivial group if $p = 0$ and a p-Sylow subgroup of $I_\mathfrak{q}$ if $p \neq 0$. Then P is a normal subgroup of $I_\mathfrak{q}$ [Cassels-Fröhlich, p. 29, Thm. 1(ii)]. Hence, by Lemma 21.7.2(h), P is trivial. Thus, $I_\mathfrak{q}$ is cyclic [Cassels-Fröhlich, p. 31, Cor. 1]. In particular, each Sylow subgroup of $I_\mathfrak{q}$ other than L is normal, hence trivial (Lemma 21.7.2(h)). Therefore, $I_\mathfrak{q} = L$.

Denote the fixed field of L, hence of $I_\mathfrak{q}$, in F by E. By Lemma 21.7.2(b), $E/K(t)$ is Galois of degree m, $K(x_j)E = F$, and $K(x_j) \cap E = K(t)$. By the preceding paragraph, $\mathfrak{q}$ is totally ramified over E and $\mathfrak{q}|_E$ is unramified over $K(t)$. Hence, $\mathfrak{q}$ is unramified over $K(x_j)$ and $\mathfrak{q}|_{K(x_j)}$ is totally and tamely ramified over $K(t)$. Therefore, $d_\mathfrak{q} = l - 1$ and $d_{\mathfrak{q},j} = 0$ (Remark 3.5.7).

There are exactly m prime divisors of F/K lying over $\mathfrak{q}|_{K(t)}$. Hence, they contribute $m(l-1)$ to the right hand side of (3).

PART B: *Computation of g.* Consider a prime divisor $\mathfrak{p}$ of $K(t)/K$. Let $\mathfrak{q}$ be a prime divisor of F/K over $\mathfrak{p}$. Suppose $I_\mathfrak{q}$ acts transitively on $\{x_1,\ldots,x_l\}$. By Case A2, $\mathfrak{p}$ totally ramifies in $K(x_1)$. Hence, there are only finitely many such $\mathfrak{p}$. Denote their number by r.

Conversely, suppose $I_\mathfrak{q}$ acts intransitively on $\{x_1,\ldots,x_l\}$. Case A1 gives k with $\mathfrak{q}|_{K(x_k)}$ unramified over $K(t)$. Choose $\rho \in G$ with $\rho x_k = x_1$. Then $\rho\mathfrak{q}|_{K(x_1)}$ is unramified over $K(t)$. Hence, $\mathfrak{p}$ is not totally ramified in $K(x_1)$.

Thus, by Case A1, only the r prime divisors $\mathfrak{p}$ with $I_\mathfrak{q}$ transitive contribute to the right hand side of (3). The contribution of each of them is, by Case A2, $m(l-1)$. Thus, (3) simplifies to $2(l-1)(1-g) = rm(l-1)$ and furthermore to

(4) $$2(1-g) = rm.$$

By Example 2.3.11, the infinite prime divisor $\mathfrak{p}_\infty$ of $K(t)/K$ totally ramifies in $K(x_1)$. Hence, $r \geq 1$. Therefore, by (4), $g = 0$ and $rm = 2$. It follows that, either $m = 1$ and $r = 2$ or $m = 2$ and $r = 1$.

PART C: *Suppose $m = 1$ and $r = 2$.* Then $K(x_1) = F$ is a Galois extension of degree l of $K(t)$. Hence, $G \cong C_l$.

Since $r = 2$, $K(t)$ has a finite place $\mathfrak{p}_c$, mapping t to $c \in K$, which totally ramifies in F. In particular, $\mathfrak{p}_c$ has only one prime divisor in F. Hence, $f(X) - c$ has only one root in K. It follows that there are $a \in K^\times$ and $b \in K$ with $f(X) - c = (aX + b)^l$. Thus, $f(X)$ is linearly related to X^l.

PART D: *Suppose $m = 2$ and $r = 1$.* By Lemma 21.7.2(i), G is the dihedral group D_l generated by the involution τ of H_1 and a generator σ of L with the relation $\sigma^\tau = \sigma^{-1}$. By part B, $g = 0$. Since K is algebraically closed, each prime divisor of F/K has degree 1. Hence, $F = K(z)$ for some z (Example 3.2.4). Lemma 21.8.6(b) gives an automorphism λ of $\mathrm{Aut}(F/K)$ with $\sigma' = \lambda\sigma\lambda^{-1} = \begin{bmatrix} 1 & 0 \\ 0 & \zeta \end{bmatrix}$ (and $\zeta = \zeta_l$) and $\tau' = \lambda\tau\lambda^{-1} = \begin{bmatrix} 0 & 1 \\ 1 & 0 \end{bmatrix}$. Put $z' = \lambda z$, $x' = \lambda x_1$, and $t' = \lambda t$. Then, $K(x')$ is the fixed field in $K(z')$ of $\begin{bmatrix} 0 & 1 \\ 1 & 0 \end{bmatrix}$ and $K(t')$ is the fixed field in $K(z')$ of $\left\langle \begin{bmatrix} 1 & 0 \\ 0 & \zeta \end{bmatrix}, \begin{bmatrix} 0 & 1 \\ 1 & 0 \end{bmatrix} \right\rangle$. By Lemma 21.8.6(c), the former field is $K(z' + (z')^{-1})$ and the latter field is $K((z')^l + (z')^{-l})$. Therefore, there are $\kappa_1 \in \mathrm{Aut}(K(x')/K)$ and $\kappa_2 \in \mathrm{Aut}(K(t')/K)$ with $x' = \kappa_1(z' + (z')^{-1})$ and $t' = \kappa_2((z')^l + (z')^{-l})$.

Identify $\lambda, \kappa_1, \kappa_2$ with Möbius transformations having coefficients in K. Put $\mu = \lambda^{-1}\kappa_1$ and $\nu = \lambda^{-1}\kappa_2$. Then $x_1 = \mu(z' + (z')^{-1})$ and $t = \nu((z')^l + (z')^{-l})$. Hence, $\nu^{-1} \circ f \circ \mu(z' + (z')^{-1}) = (z')^l + (z')^{-l}$. Since z' is transcendental over K, Lemma 21.8.4(c) implies $\nu^{-1} \circ f \circ \mu = T_l$.

PART E: *Conclusion of the proof.* It remains to prove that μ and ν are linear polynomials.

Assume first ν is not a polynomial. Then ν^{-1} is not a polynomial. Thus, $\nu^{-1}(X) = \frac{aX+b}{cX+d}$ with $a, b, c, d \in K$ and $c \neq 0$. It follows that $\nu^{-1}(\infty) = \frac{a}{c}$. Use Lemma 21.8.4(e) to find distinct $e_1, e_2 \in K$ with $T_l(e_i) = \frac{a}{c}$. Then

21.8 Schur's Conjecture

$f(\mu(e_i)) = \nu(T_l(e_i)) = \nu(\frac{a}{c}) = \infty$. Since f is polynomial, $\mu(e_i) = \infty$, $i = 1, 2$. This contradiction to the injectivity of μ on $K \cup \{\infty\}$ proves that ν is a linear polynomial.

Finally, assume that μ is not a polynomial. Then $e = \mu(\infty) \in K$. Hence, by the preceding paragraph, $T_l(\infty) = \nu^{-1}(f(\mu(\infty))) = \nu^{-1}(f(e)) \in K$, which is a contradiction. Consequently, μ is a linear polynomial. □

LEMMA 21.8.8: *Let K be an algebraically closed field, $f \in K[X]$ a polynomial of degree n, x a transcendental element over K, and $t = f(x)$. Suppose $\mathrm{char}(K) \nmid n$. Then $\mathrm{Gal}(f(X) - t, K(t))$ contains an n-cycle.*

Proof: By assumption, $f(X) - t$ has n distinct roots $x_1, \ldots, x_n$ in $K(t)_s$. Put $x = x_1$ and $F = K(x_1, \ldots, x_n)$. Let v_∞ be the valuation of $K(t)/K$ with $v_\infty(t) = -1$. By Example 2.3.11, v_∞ is totally ramified in $K(x)$. Thus, v_∞ has a unique extension w to $K(x)$.

Denote the completion of $E = K(t)$ at v_∞ by $\hat{E}$ (By Example 3.5.1, $\hat{E} \cong K((t^{-1}))$.) By Lemma 3.5.3, $f(X) - t$ is irreducible over $\hat{E}$ and $\hat{E}(x)/\hat{E}$ is a totally ramified extension of degree n. Since $\mathrm{char}(K) \nmid n$, $\hat{E}(x)/\hat{E}$ is tamely ramified. Put $\hat{F} = F\hat{E}$. Then $\hat{F} = \hat{E}(x_1)\cdots\hat{E}(x_n)$ is a compositum of tamely ramified extensions of $\hat{E}$. Hence, $\hat{F}/\hat{E}$ is tamely ramified [Cassels-Fröhlich, p. 31, Cor. 2]. Since K is algebraically closed, the residue degree of $\hat{F}/\hat{E}$ is 1. Therefore, $\hat{F}/\hat{E}$ is totally and tamely ramified. Therefore, $\mathrm{Gal}(\hat{F}/\hat{E})$ is cyclic [Cassels-Fröhlich, p. 31, Cor. 1]. It follows that $\hat{E}(x)/\hat{E}$ is a Galois extension. Hence, $\hat{F} = \hat{E}(x)$ and $\mathrm{Gal}(f(X) - t, \hat{E}) \cong \mathrm{Gal}(\hat{F}/\hat{E})$ is cyclic of order n. It follows that $\mathrm{Gal}(f(X) - t, \hat{E})$ is transitive.

Choose a generator σ of $\mathrm{Gal}(f(X) - t, \hat{E})$. It decomposes into disjoint cycles whose lengths sum up to n. Since $\langle\sigma\rangle$ is transitive, there is only one cycle. Thus, σ is an n-cycle. Finally, note that $\sigma \in \mathrm{Gal}(f(X) - t, K(t))$. □

Let K be a field and $f \in K[X]$ a polynomial with $\deg(f) > 1$. Then $f = f_1 \circ f_2 \circ \cdots \circ f_r$ with $f_i \in K[X]$ indecomposable and $\deg(f_i) > 1$. Call each f_i a **decomposition factor** of f over K.

LEMMA 21.8.9: *Let K be a field, t an indeterminate, and $f \in K[X]$ an indecomposable polynomial over K of degree n with $\mathrm{char}(K) \nmid n$. Then $\mathrm{Gal}(f(X) - t, K(t))$ is a primitive group.*

Proof: Let x be a root of $f(X) - t$ in $K(t)_s$ and F the Galois closure of $K(x)/K(t)$. By Lemma 21.7.4, it suffices to prove that $\mathrm{Gal}(F/K(x))$ is a maximal subgroup of $\mathrm{Gal}(F/K(t))$. In other words, we have to prove that $K(x)/K(t)$ is a minimal extension.

Assume there is a field E which lies strictly between $K(t)$ and $K(x)$. Let $d = [K(x) : E]$ and $m = [E : K(t)]$. Then $d > 1$ and $m > 1$. Let v_∞ be the valuation of $K(t)/K$ with $v_\infty(t) = -1$. Then v_∞ is totally ramified in $K(x)$. Thus, v_∞ has a unique extension w' to $K(x)$, and w' satisfies $w'(x) = -1$ and has the same residue field as v_∞, namely K (Example 2.3.11). Let w be the restriction of w' to E. Then the residue field of E at w is K. By Lüroth

(Remark 3.6.2(a)), E is a rational function field over K. Therefore, there is a $y \in E$ with $w(y) = -1$ and $v(y) \geq 0$ for all other valuations v of E/K. Therefore, $w'(y) = -d$ and $v(y) \geq 0$ for all other valuations v of $K(x)/K$. It follows that $y = h(x)$ for some $h \in K[X]$.

The degree of the pole divisor of y as an element of $K(x)$ is d. Therefore, $[K(x) : K(y)] = d$. Since $K(y) \subseteq E$, this implies $E = K(y)$. Therefore, $\deg(h) = d$.

Repeat the above arguments for t, y replacing y, x to find a polynomial $g \in K[X]$ of degree m with $t = g(y)$. This gives $f(x) = t = g(y) = g(h(x))$. Hence, $f(X) = g(h(X))$. This contradiction to the indecomposability of f proves that $K(x)/K(t)$ is minimal. □

Let K be a field and $f \in K[X]$ a polynomial of positive degree. Put

$$f^*(X, Y) = \frac{f(X) - f(Y)}{X - Y}.$$

PROPOSITION 21.8.10: *Let K be an algebraically closed field and $f \in K[X]$ a polynomial of degree l. Suppose $\mathrm{char}(K) \nmid l$, f is indecomposable over K, and $f^*(X, Y)$ is reducible. Then l is prime and $f(X)$ is linearly related to X^l or to $T_l(X)$.*

Proof: Let t be an indeterminate. Since $\mathrm{char}(K) \nmid l$, $f(X) - t$ has l distinct roots $x_1, \ldots, x_l$ in $K(t)_s$. Hence, $F = K(x_1, \ldots, x_l)$ is a Galois extension of $K(t)$. Put $G = \mathrm{Gal}(F/K(t))$ and $H_j = \mathrm{Gal}(F/K(x_j))$, $j = 1, \ldots, l$.

View G as a permutation group of $\{x_1, \ldots, x_l\}$. Since $f(X) - t$ is irreducible in $K(t)[X]$, G is transitive. Since f is indecomposable, G is primitive (Lemma 21.8.9). Further, by Lemma 21.8.8, G contains an l-cycle.

Since $f^*(X, Y)$ is reducible, $l - 1 \geq 2$. By Gauss, $f^*(x_1, Y)$ is reducible in $K(x_1)[Y]$ of degree $l - 1$. Therefore, H_1 is intransitive on the roots $x_2, \ldots, x_l$ of $f^*(x_1, Y)$. This means that G is not doubly transitive (Lemma-Definition 21.7.5). It follows from Corollary 21.7.9 that G is solvable. Consequently, by Proposition 21.8.7, f is linearly related to X^l or to $T_l(X)$. □

LEMMA 21.8.11 ([Fried-MacCrae, Thm. 3.5]): *Let K be a field, $f \in K[X]$ a polynomial, and L a field extension of K. Suppose $\mathrm{char}(K) \nmid \deg(f)$ and f is decomposable over L. Then f is decomposable over K.*

Proof: There are $g(X) = \sum_{i=0}^m a_i X^i$ and $h(X) = \sum_{j=0}^n b_j X^j$ in $L[X]$ with $f(X) = g(h(X))$ and $\deg(g), \deg(h) < \deg(f)$. Assume without loss L is algebraically closed. Divide g and f by a_m, if necessary, to assume $a_m = 1$. Then replace X by $b_n^{-1/n} X$, if necessary, to assume $b_n = 1$. It follows that f, g, and h are monic. Finally, choose $\beta \in L$ with $h(\beta) = 0$. Then replace X by $X + \beta$ to assume $b_0 = 0$. We prove that $g, h \in K[X]$. Indeed,

$$f(X) = (X^n + b_{n-1} X^{n-1} + \cdots + b_1 X)^m$$

(5)
$$+ \sum_{i=0}^{m-1} a_i (X^n + b_{n-1} X^{n-1} + \cdots + b_1 X)^i.$$

21.8 Schur's Conjecture

The highest power of X involved in the sum on the right hand side of (5) is $X^{(m-1)n}$. Hence, for each k between 1 and $n-1$, only the first term on the right hand side of (5) contributes to the coefficient of X^{mn-k}. Therefore, this coefficient is $mb_{n-k} + p_k(b_{n-k+1}, \ldots, b_{n-1}, 1)$ with $p_k \in \mathbb{Z}[X_{n-k+1}, \ldots, X_n]$. Observe that $\deg(f) = mn$, so $\mathrm{char}(K) \nmid m$. Induction on k proves that $b_{n-k} \in K$. Thus, $h \in K[X]$.

Finally, suppose $a_{m-1}, \ldots, a_{m-r+1}$ have been proved to be in K. Then $a_{m-r}h(X)^{m-r} + a_{m-r-1}h(X)^{m-r-1} + \cdots + a_0$ is a polynomial in $K[X]$. Its leading coefficient is a_{m-r}. Hence, $a_{m-r} \in K$. Consequently, $g \in K[X]$. $\square$

Let K be a field and $f \in K[X]$. We say f is **injective** (resp. **surjective**) on K if the map $x \mapsto f(x)$ of K into itself is injective (resp. surjective).

PROPOSITION 21.8.12: *Let K be a perfect field and $f \in K[X]$ a polynomial of degree l with $\mathrm{char}(K) \nmid l$.*
(a) *Suppose K is PAC and f is injective on K and indecomposable over K. Then l is a prime number and f is linearly related over K to a Dickson polynomial of degree l.*
(b) *Suppose K is pseudo finite and f is injective or surjective on K. Then each composition factor of f is linearly related over K to a Dickson polynomial of a prime degree.*

Proof of (a): By assumption, there are no $x, y \in K$ with $f(x) = f(y)$ and $x \neq y$. Hence, there are no $x, y \in K$ with $f^*(x, y) = 0$ and $x \neq y$. By Proposition 11.1.1, $f^*(X, Y)$ is reducible over $\tilde{K}$. Therefore, by Proposition 21.8.10, l is prime and f is linearly related over $\tilde{K}$ to X^l or to $T_l(X)$. Both X^l and $T_l(X)$ are Dickson polynomials. It follows from Lemma 21.8.3 that f is linearly related over K to a Dickson polynomial of degree l.

Proof of (b): Every injective (resp. surjective) polynomial on a finite field F of degree n is bijective. Since this is an elementary statement, it holds for every pseudo finite field (Proposition 20.10.4). In particular, f is bijective on K.

Suppose $f(X) = g(h(X))$ with $g, h \in K[X]$. Then g is surjective on K and h is injective. By the preceding paragraph both g and h are bijective on K. It follows that each composition factor q of f over K is bijective on K. By (a), each composition factor of f is linearly related to a Dickson polynomial of a prime degree. $\square$

THEOREM 21.8.13 (Schur's Conjecture): *Let K be a global field and $f \in K[X]$ a polynomial with $\mathrm{char}(K) \nmid \deg(f)$. Suppose f is a permutation polynomial modulo $\mathfrak{p}$ for infinitely many prime divisors $\mathfrak{p}$ of K. Then each composition factor of f is linearly related over K to a Dickson polynomial of a prime degree.*

Proof: The statement "f is a permutation polynomial" is elementary. By Example 20.10.7(a) there is a pseudo finite field F containing K such that f is a permutation polynomial on F. By Proposition 21.8.12(b), each composition factor of f is linearly related over F, hence over K (Lemma 21.8.3) to a Dickson polynomial of a prime degree. □

THEOREM 21.8.14: *Let K be a finite field and $f \in K[X]$ a polynomial with* $\text{char}(K) \nmid \deg(f)$. *Suppose f permutes infinitely many finite extensions L of K. Then each composition factor of f is linearly related over K to a Dickson polynomial of a prime degree.*

Proof: Take a nonprincipal ultraproduct F of those finite extensions of K on which f is a permutation polynomial. Then f is a permutation polynomial on F. By Example 20.10.7(a), F is pseudo finite and $K \subseteq F$. By Proposition 21.8.12(b), each composition factor of f is linearly related over F, hence over K, to a Dickson polynomial of a prime degree. □

We end this section with examples which give converses to Proposition 21.8.12 and Theorem 21.8.14:

Example 21.8.15: Cyclic polynomials. Let n be a positive integer and F a finite field of order q. Then $\mathbb{F}_q^\times$ is a cyclic group of order $q-1$. Hence, the map $x \mapsto x^n$ is bijective on F if and only if $\gcd(n, q-1) = 1$.

Suppose $n \geq 3$ is odd. Let $q_1, \ldots, q_r$ be the prime divisors of n. Then both 1 and 2 are relatively prime to q_i, $i = 1, \ldots, r$. Choose an integer a with $a \equiv 1 \mod q_i$, $i = 1, \ldots, r$. Then both a and $a+1$ are relatively prime to n. Dirichlet's theorem (Corollary 6.3.2) gives infinitely many prime numbers $p \equiv a+1 \mod n$. By the preceding paragraph, X^n is a permutation polynomial modulo p for each of these p. □

The next lemma will be used to give more examples of Dickson polynomials which are permutation polynomials modulo infinitely many primes:

LEMMA 21.8.16: *Let n be a positive integer, q a power of a prime number, and $a \in \mathbb{F}_q$. Suppose $\gcd(n, q^2 - 1) = 1$. Then $D_n(a, X)$ permutes $\mathbb{F}_q$.*

Proof: Put $K = \mathbb{F}_q$ and $L = \mathbb{F}_{q^2}$. Then $L^\times$ is a cyclic group of order $q^2 - 1$. Hence, the map $z \mapsto z^n$ is bijective on L.

Now consider $x_1, x_2 \in K$ with $D_n(a, x_1) = D_n(a, x_2)$. Choose $z_1, z_2 \in L$ with $z_i + a z_i^{-1} = x_i$, $i = 1, 2$. Then $D_n(a, z_1 + a z_1^{-1}) = D_n(a, z_2 + a z_2^{-1})$. By Lemma 21.8.1(a), $z_1^n + a^n z_1^{-n} = z_2^n + a^n z_2^{-n}$. Denote the common value of the two sides of the latter equality by b. The product of the two solutions of the equation $X + a^n X^{-1} = b$ is a^n. Both z_1^n and z_2^n are solutions. Hence, $z_1^n = z_2^n$ or $z_1^n = a^n z_2^{-n}$. By the preceding paragraph, $z_1 = z_2$ or $z_1 = a z_2^{-1}$. In both cases, $x_1 = x_2$. Consequently, $D_n(a, X)$ is a permutation polynomial on K. □

Example 21.8.17: Permutation polynomials modulo infinitely many primes.

(a) Let n be a positive integer, K a number field, and $a \in O_K$. Suppose $\gcd(6, n) = 1$ and $K \cap \mathbb{Q}(\zeta_n) = \mathbb{Q}$. Then $D_n(a, X)$ is a permutation polynomial modulo infinitely many prime ideals of O_K.

Indeed, list the prime divisors of n as $q_1, \ldots, q_r$. Choose an integer b with $b \equiv 2 \bmod q_i$, $i = 1, \ldots, r$. Then $\gcd(b, n) = 1$ and $\gcd(b^2 - 1, n) = 1$. Denote the Galois closure of $K(\zeta_n)/\mathbb{Q}$ by L. Then choose $\sigma \in \text{Gal}(L/K)$ with $\sigma\zeta_n = \zeta_n^b$. Chebotarev density theorem (Theorem 6.3.1) gives infinitely many prime ideals $\mathfrak{P}$ of O_L with $[\frac{L/\mathbb{Q}}{\mathfrak{P}}] = \sigma$. Let $\mathfrak{P}$ be one of them. Then the decomposition group of $\mathfrak{P}$ over $\mathbb{Q}$ is in $\text{Gal}(L/K)$. Put $\mathfrak{p} = \mathfrak{P}|_K$. Then $p = N\mathfrak{p}$ is a prime number. By the proof of Corollary 6.3.2, $p \equiv b \bmod n$. Hence, by the choice of b, $\gcd(|\bar{K}_\mathfrak{p}|^2 - 1, n) = 1$. Consequently, by Lemma 21.8.16, $D_n(a, X)$ is a permutation polynomial modulo $\mathfrak{p}$.

(b) Let n be a positive integer, p a prime number, q a power of p, K a finite separable extension of $\mathbb{F}_q(t)$ which is regular over $\mathbb{F}_q$, and $a \in O_K$. Suppose $\gcd(p(q^2 - 1), n) = 1$. Then $D_n(a, X)$ is a permutation polynomial modulo infinitely many prime divisors of $K/\mathbb{F}_q$.

Indeed, let L be the Galois closure of $K/\mathbb{F}_q(t)$, $\mathbb{F}_{q^m}$ the algebraic closure of $\mathbb{F}_q$ in L, and $\varphi(n)$ the Euler totient function. Then $q^{\varphi(n)} \equiv 1 \bmod n$. Consider a large positive integer k. Choose τ in $\text{Gal}(L/K)$ whose restriction to $\mathbb{F}_{q^m}$ coincides with the restriction of $\text{Frob}_q^{k\varphi(n)+1}$. Denote the conjugacy class of τ in $\text{Gal}(L/\mathbb{F}_q(t))$ by $\text{Con}(\tau)$. Proposition 6.4.8 gives a prime divisor $\mathfrak{p}$ of $\mathbb{F}_q(t)$ with $\deg(\mathfrak{p}) = k\varphi(n) + 1$ and $(\frac{L/\mathbb{F}_q(t)}{\mathfrak{p}}) = \text{Con}(\tau)$. Hence, there is a prime divisor $\mathfrak{Q}$ of L with $[\frac{L/\mathbb{F}_q(t)}{\mathfrak{Q}}] = \tau$. Denote the restriction of $\mathfrak{Q}$ to K by $\mathfrak{q}$. Then $\deg(\mathfrak{q}) = \deg(\mathfrak{p}) = k\varphi(n) + 1$. Hence, $|\bar{K}_\mathfrak{q}|^2 - 1 = q^{2(k\varphi(n)+1)} - 1 \equiv q^2 - 1 \bmod n$. Consequently, by Lemma 21.8.16, $D_n(a, X)$ is a permutation polynomial modulo $\mathfrak{q}$. $\square$

21.9 Generalized Carlitz's Conjecture

Let q be a power of a prime number p and $f \in \mathbb{F}_q[X]$ a polynomial of degree n. Suppose f is a permutation polynomial of $\mathbb{F}_{q^k}$ for infinitely many k but f is not a p-power in $\mathbb{F}_q[X]$. Theorem 21.8.14 describes the composition factors of f when $p \nmid n$. In the general case the generalized Carlitz's Conjecture states that $\gcd(n, q-1) = 1$. In particular, for $p \neq 2$, the conjecture predicts that n is odd (Carlitz's Conjecture).

The paper [Fried-Guralnick-Saxl] gives more precise information about permutation polynomials than Carlitz's Conjecture does. Not only that it proves that conjecture for $p > 3$, but it gives information about $\text{Gal}(f(X) - t, \tilde{\mathbb{F}}_q(t))$. Suppose without loss that f is indecomposable over $\mathbb{F}_q$. Then three are three cases:
(a) $p \nmid n$ and G is cyclic or isomorphic to the dihedral group D_n. This is essentially contained in Section 21.8.
(b) $n = p^m$ and $G \cong H \ltimes \mathbb{F}_p^m$, where H is a subgroup of $\text{GL}(m, \mathbb{F}_p)$ acting linearly on $\mathbb{F}_p^m$.
(c) $p \in \{2, 3\}$, $n = \frac{p^a(p^a - 1)}{2}$ with $a \geq 3$ odd, and G normalizes the simple group $\text{PSL}(2, \mathbb{F}_{p^a})$.

The proof that no other cases arise uses the classification of finite simple groups and is beyond the scope of this book.

One can find an elementary proof of Carlitz's Conjecture in [Lenstra]. That proof argues with decomposition and inertia groups over $\mathbb{F}_q(t)$. Here we follow [Cohen-Fried] which, following a suggestion of [Lenstra, Remark 1], takes place over $\mathbb{F}_q((t))$.

The first step in the proof translates the assumption "f permutes infinitely many finite extensions of $\mathbb{F}_q$" into a statement of a general nature:

LEMMA 21.9.1: *Let $f \in \mathbb{F}_q[X]$ be a polynomial which permutes infinitely many finite extensions of $\mathbb{F}_q$. Then $f^*(X, Y) = \frac{f(X) - f(Y)}{X - Y}$ has no absolutely irreducible factor which belongs to $\mathbb{F}_q[X, Y]$.*

Proof: Assume $h \in \mathbb{F}_q[X, Y]$ is an absolutely irreducible factor of $f^*(X, Y)$. By Corollary 5.4.2 there is an n_0 such that for every integer $n > n_0$ there are distinct $x, y \in \mathbb{F}_{q^n}$ with $h(x, y) = 0$. Then $f(x) = f(y)$. Hence, f does not permutes $\mathbb{F}_{q^n}$, in contradiction to our assumption. □

The converse of Lemma 21.9.1 is also true (see Exercise 14). However, it is not used in proof of the generalized Carlitz Conjecture.

LEMMA 21.9.2: *Let N/M be a finite cyclic extension, σ an element of $\mathrm{Gal}(M)$ whose restriction to N generates $\mathrm{Gal}(N/M)$, and $h \in M[X]$ a separable irreducible polynomial which becomes reducible over N. Then $\sigma x \neq x$ for every root x of h.*

Proof: Assume h has a root x with $\sigma x = x$. By assumption, $N \cap M_s(\sigma) = M$. Hence, $N \cap M(x) = M$, so N and $M(x)$ are linearly disjoint over M. Therefore, $[N(x) : N] = [M(x) : M]$. It follows that $h = \mathrm{irr}(x, M)$ is irreducible over N, a contradiction. □

LEMMA 21.9.3: *Let L/K be a finite cyclic extension, $f \in K[X]$ a polynomial of degree at least 2 with a nonzero derivative f', and z an indeterminate. Suppose each irreducible factor of $f^*(X, Y) = \frac{f(X) - f(Y)}{X - Y}$ in $K[X, Y]$ is reducible over L. Then $f(X) - z$ is separable and each $\rho \in \mathrm{Gal}(K(z))$ whose restriction to L generates $\mathrm{Gal}(L/K)$ fixes exactly one root of $f(X) - z$.*

Proof: By assumption, $(f(X) - z)' = f'(X) \neq 0$. All roots of $f(X) - z$ are transcendental over K while all roots of $f'(X)$ are algebraic over K. Hence, $f(X) - z$ is relatively prime in $K(z)[X]$ to its derivative. Therefore, $f(X) - z$ is separable. Let $\hat{F}$ be the finite Galois extension of $E = K(z)$ generated by $L(z)$ and all roots of $f(X) - z$.

Choose a generator ρ_0 of L/K and let $G^* = \{\rho \in \mathrm{Gal}(\hat{F}/E) \mid \rho|_L = \rho_0\}$. For each root x of $f(X) - z$ let $G_x^* = \{\rho \in G^* \mid \rho x = x\}$. We have to prove that $G^* = \bigcup_x G_x^*$, where x ranges over all roots of $f(X) - z$. We divide the rest of the proof into two parts.

21.9 Generalized Carlitz's Conjecture

PART A: *Disjointness.* Let x and y be distinct roots of $f(X)-z$ in $\hat{F}$. Then $f(x) = z$ and $f(y) = z$, so x, y are transcendental over K. By the above, $f^*(X,y) = \frac{f(X)-f(y)}{X-y} = \frac{f(X)-z}{X-y} \in K(y)[X]$ is a separable polynomial and x is a root of $f^*(X,y)$. Let $h \in K[y, X]$ be an irreducible factor of $f^*(X,y)$ with $h(x) = 0$. In particular, h is primitive. By assumption, $h(X)$ is reducible in $L[y, X]$. Hence, $h(X)$ is reducible in $L(y)[X]$.

Now consider $\rho \in G_y^*$. Then $\rho \in \text{Gal}(\hat{F}/K(y))$ and $\rho|_{L(y)}$ generates the cyclic group $\text{Gal}(L(y)/K(y))$. By Lemma 21.9.2, ρ fixes no root of h. In particular, $\rho x \ne x$. Thus, $G_x^* \cap G_y^* = \emptyset$.

PART B: Every $\rho \in G^*$ fixes a root of $f(X) - z$. Indeed, extend ρ_0 in the unique possible way to a generator ρ_0^* of $\text{Gal}(L(z)/K(z))$. Then $G^* = \{\rho \in \text{Gal}(\hat{F}/E) \mid \rho|_{L(z)} = \rho_0^*\}$. Hence, $|G^*| = [\hat{F} : L(z)]$.

Next let x be a root of $f(X)-z$ in $\hat{F}$ and extend ρ_0^* in the unique possible way to a generator ρ_1^* of $\text{Gal}(L(x)/K(x))$. Then

$$G_x^* = \{\rho \in \text{Gal}(\hat{F}/K(x)) \mid \rho|_{L(x)} = \rho_1^*\}.$$

Hence, $|G_x^*| = [\hat{F} : L(x)]$.

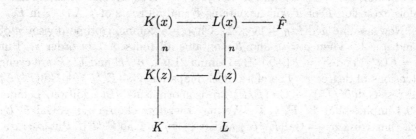

Finally, let $x_1, \ldots, x_n$ be the distinct roots of $f(X) - z$. Then $n = \deg(f) = [K(x_i) : K(z)] = [L(x_i) : L(z)]$. By Part A, $|\bigcup_{i=1}^n G_{x_i}^*| = \sum_{i=1}^n |G_{x_i}^*| = n[\hat{F} : L(x_i)] = [L(x_i) : L(z)][\hat{F} : L(x_i)] = [\hat{F} : L(z)] = |G^*|$. Hence, $G^* = \bigcup_{i=1}^n G_{x_i}^*$. Thus, each $\rho \in G^*$ fixes some x_i. □

LEMMA 21.9.4: *Let K be a field, $f \in K[X]$ a monic polynomial of degree $n > 1$, r a divisor of n with $\text{char}(K) \nmid r$, and z an indeterminate. Put $E = K((z^{-1}))$ and let x be a root of $f(X) - z$ in $\tilde{E}$. Then $E(x)$ contains an rth root $\sqrt[r]{z}$ and $[E(\sqrt[r]{z}) : E] = r$.*

Proof: Let v be the unique discrete complete valuation of E/K such that $v(z^{-1}) = 1$ (Example 3.5.1). Extend v to $E(x)$ in the unique possible way. By Lemma 3.5.2, both E and $E(x)$ satisfy Hensel's lemma.

Now write $f(X) = X^n + a_{n-1}X^{n-1} + \cdots + a_0$. Put $g(Y) = 1 + a_{n-1}Y + \cdots + a_1 Y^{n-1} + a_0 Y^n$ and $y = x^{-1}$. Then $x^n g(y) = f(x)$. Consider the polynomial $h(T) = T^r - g(y) \in E(x)[T]$. Then $h(1) = -a_{n-1}y - \cdots - a_1 y^{n-1} - a_0 y^n$ and $h'(1) = r \ne 0$. By assumption, $x^n + a_{n-1}x^{n-1} + \cdots + a_0 = z$. Hence,

$v(x) < 0$, so $v(y) > 0$. It follows that, $v(h(1)) > 0$ and $v(h'(1)) = 0$. Hensel's lemma gives $t \in E(x)$ with $t^r = g(y)$.

Finally, let $w = x^{n/r}t$. Then $w \in E(x)$ and $w^r = x^n t^r = x^n g(y) = f(x)$. Moreover, $E(w)/E$ is a totally ramified extension of degree r (Example 2.3.8). □

THEOREM 21.9.5 (Lenstra): *Let p be a prime number, q a power of p, and $f \in \mathbb{F}_q[X]$ a polynomial of degree $n > 1$. Suppose $f \ne f_1^p$ for all $f_1 \in \mathbb{F}_q[X]$ and f permutes infinitely many finite extensions of $\mathbb{F}_q$. Then $\gcd(n, q-1) = 1$.*

Proof: Put $K = \mathbb{F}_q$. Since f is not a pth power of a polynomial in $K[X]$ and K is perfect, the derivative of f is nonzero. Put $f^*(X,Y) = \frac{f(X)-f(Y)}{X-Y}$. By Lemma 21.9.1, no absolutely irreducible factor of $f^*(X,Y)$ belongs to $K[X,Y]$. Choose a finite, necessarily cyclic, extension L such that all absolutely irreducible factors of $f^*(X,Y)$ belong to $L[X,Y]$. Then every irreducible factor of $f^*(X,Y)$ over K is reducible over L. Let z be an indeterminate. Put $E = K((z^{-1}))$ and $F = L((z^{-1}))$. Then E/K is a regular extension and $EL = F$ (Example 3.5.1). Hence, res: $\text{Gal}(F/E) \to \text{Gal}(L/K)$ is an isomorphism. By Lemma 21.9.3, $f(X) - z$ is separable. Choose a finite Galois extension $\hat{F}$ of E which contains F and all roots of $f(X) - z$ in E_s.

Now assume $\gcd(n, q-1) > 1$. Choose a common prime divisor r of n and $q - 1$. Then $p \ne r$ and K contains all roots of 1 of order r. Put $E' = E(\sqrt[r]{z})$ and $F' = F(\sqrt[r]{z})$. By Lemma 21.9.4, E'/E and F'/F are cyclic extensions of degree r. Therefore, both maps res: $\text{Gal}(F'/F) \to \text{Gal}(E'/E)$ and res: $\text{Gal}(F'/E') \to \text{Gal}(F/E)$ are isomorphisms. In addition, Lemma 21.9.4 implies that $E', F' \subseteq \hat{F}$. We may therefore choose $\sigma, \tau \in \text{Gal}(\hat{F}/E)$ such that $\langle \text{res}_{F'}\sigma \rangle = \text{Gal}(F'/E')$ and $\langle \text{res}_{F'}\tau \rangle = \text{Gal}(F'/F)$. Put $\rho = \sigma\tau$. Then $\text{res}_L\rho$ generates $\text{Gal}(L/K)$.

$$\begin{array}{ccc} E(x) & \!\!\!\!-\!\!\!\!- F(x) -\!\!\!\!- & \hat{F} \\ | & | & \\ E' & \!\!\!\!-\!\!\!\!- F' & \\ | & | & \\ E & \!\!\!\!-\!\!\!\!- F & \\ | & | & \\ K & \!\!\!\!-\!\!\!\!- L & \end{array}$$

By Lemma 21.9.3, ρ fixes a root x of $f(X) - z$ in $\hat{F}$. By Lemma 21.9.4, $E(x)$ contains a root of z of order r, so $E' \subseteq E(x)$. Hence, $\text{res}_{E'}\rho = \text{id}$. On the other hand, $\text{res}_{E'}\rho = \text{res}_{E'}\tau \ne \text{id}$. This contradiction proves that $\gcd(n, q-1) = 1$. □

Exercises

1. Let K be a perfect field and V a Zariski K-closed set with $V(K)$ infinite. Use the decomposition-intersection procedure and Corollary 10.5.3 to prove that V contains a curve defined over K.

2. For each positive integer i, give an example of a PAC field K which is C_{i+1} but not C_i. Hint: Take K to be imperfect, follow Example 21.2.8 and use Propositions 21.2.4 and 21.2.12.

3. Let $f \in \mathbb{Q}[X_0, \ldots, X_l]$ be an irreducible form of prime degree l. Prove that $V(f)$ contains a variety defined over $\mathbb{Q}$. Hint: The intersection of l hyperplanes in $\mathbb{P}^l$ is nonempty.

4. Prove that almost all $\sigma \in \text{Gal}(\mathbb{Q})$ satisfy this: Let $f \in \tilde{\mathbb{Q}}(\sigma)[X_0, \ldots, X_n]$ be a polynomial of degree $d \leq n$ with a $\tilde{\mathbb{Q}}(\sigma)$-zero. Then $V(f)$ contains a curve defined over $\tilde{\mathbb{Q}}(\sigma)$. Hint: Use Chevalley-Warning, the transfer theorem and Exercise 1.

5. Give an example of fields K_0 and K such that K_0 is algebraically closed in K, K is C_1 but K_0 is not. Suggestion: Take K_0 to be $\tilde{\mathbb{Q}}(\tau)$ where τ is the conjugation on $\mathbb{C}$. Then use Lemmas 20.5.3 and 21.3.5.

6. Show that $\mathbb{F}_p(t)$, with t a transcendental, is not C_1. Suggestion: Choose a nonsquare $a \in \mathbb{F}_p$ and prove that $X^2 + aY^2 + tZ^2$ has no nontrivial zero in $\mathbb{F}_p(t)$.

7. Let M and L be finite separable extensions of a global field K. Suppose that for almost all $\mathfrak{p} \in P(K)$ the number of primes $\mathfrak{P} \in P(L)$ lying over $\mathfrak{p}$ and having a relative degree 1 is equal to the number of primes $\mathfrak{P} \in P(M)$ lying over $\mathfrak{p}$ and having relative degree 1. Prove that $[L : K] = [M : K]$ and the Galois closure $\hat{L}$ of L/K is equal to the Galois closure $\hat{M}$ of M/K [Kronecker]. Hint: Follow these steps.

   (a) Choose primitive elements x and y for L/K and M/K, respectively, integral over O_K. Put $f = \text{irr}(x, K)$ and $g = \text{irr}(y, K)$. Use Lemma 21.5.2 to prove that for almost all $\mathfrak{p} \in P(K)$ the numbers of roots of f and g in $\bar{K}_\mathfrak{p}$ are equal.

   (b) Let N be a finite Galois extension of K that contains both L and M. Use the transfer theorem (or alternatively the Chebotarev density theorem) to prove that every $\sigma \in \text{Gal}(N/K)$ fixes the same numbers of roots of f and of g.

   (c) Apply (b) to the elements $\sigma \in \text{Gal}(N/\hat{L})$ and the elements $\tau \in \text{Gal}(N/\hat{M})$ and prove that $\deg(f) = \deg(g)$.

   (d) Observe: For $\sigma \in \text{Gal}(N/\hat{L})$, all roots of g belong to $N(\sigma)$, so $\hat{M} \subseteq \hat{L}$.

8. Let G be a finite group and H, I, N subgroups. Suppose $G = H \ltimes N$ and $\bigcup_{x \in G} I^x = \bigcup_{x \in G} H^x$. Prove that $G = I \ltimes N$. Deduce that H is not a proper subgroup of I. Hint: Observe that $\bigcup_{x \in G}(IN)^x = G$.

9. (a) Let $A \triangleleft B \leq G$ be a tower of finite groups. Prove that $|\bigcup_{x \in G} A^x| \leq (G:B)|A|$.

(b) Let $K \subseteq L$ be a finite separable extension of a global field. Observe: The Dirichlet density δ of the set $V(L/K)$ (Section 20.5) is equal to the Haar measure of $\bigcup_{\sigma \in \text{Gal}(K)} \text{Gal}(L)^\sigma$.

(c) Use (a), (b) and Lemma 21.5.3 to prove that if M is a finite Galois extension of L which is Kronecker conjugate to L over K, then $[M:L] \leq \delta^{-1}$ [Klingen2].

10. Exercise 1(c) of Chapter 13 gives a pair of polynomials $f(X) = X^4 + 2X^2$, $g(X) = -4X^4 - 4X^2 - 1$ for which $f(X) - g(Y)$ is reducible. Show however, that $f(X) - t$ and $g(Y) - t$ are not Kronecker conjugate over $\mathbb{Q}(t)$. Hint: Use Corollary 21.6.5.

11. Prove for nonzero integers $a, b_1, \ldots, b_n$ that the following two statements are equivalent.

(a) There exist $\varepsilon_1, \ldots, \varepsilon_n \in \{0, 1\}$ and $c \in \mathbb{Z}$ satisfying $a = b_1^{\varepsilon_1} \ldots b_n^{\varepsilon_n} c^2$.

(b) For almost all primes p, if $b_1, \ldots, b_n$ are quadratic residues modulo p, then so is a.

Hint: Combine the transfer theorem (or directly the Chebotarev density theorem) with Kummer theory for quadratic extensions over $\mathbb{Q}$.

12. Let n be a positive integer, q a power of a prime number p, and $a \in \mathbb{F}_q$. Suppose $\gcd(n, q-1) > 1$. Find distinct $x, y \in \mathbb{F}_q$ with $D_n(a, x) = D_n(a, y)$. Combine this with Theorem 21.8.13 to supply a proof of the generalized Carlitz's Conjecture (Theorem 21.9.5) in the case where $p \nmid \deg(f)$.

Hint: Choose a common prime divisor l of n and $q - 1$. Then choose $x = z + az^{-1}$ and $y = \zeta z + a^{-1}\zeta z$ with $\zeta, z \in \mathbb{F}_1$, $\zeta^l = 1$, and $\zeta \neq 1$.

13. Let p, l be distinct prime numbers, q a power of p, and $a \in \mathbb{F}_q^\times$. Suppose $l|p^2 - 1$. Then $D_l(a, X)$ permutes only finitely many fields $\mathbb{F}_{q^k}$.

Hint: Use the hint of Exercise 12 to reduce to the case $l|q+1$ and $l \neq 2$. Then

(1) $$\frac{D_l(a, X) - D_l(a, Y)}{X - Y} = \prod_{i=1}^{(l-1)/2} (X^2 - \alpha_i XY + Y^2 + \beta_i^2 a),$$

where $\alpha_i = \zeta_l^i + \zeta_l^{-i}$ and $\beta_i = \zeta_l^i - \zeta_l^{-i}$. [Schinzel2, p. 52]. Observe that $\zeta_l^q = \zeta_l^{-1}$, so $\alpha_i, \beta_i^2 \in \mathbb{F}_q$. Next note that each of the factors on the right hand side of (1) is absolutely irreducible. Therefore, it has $\mathbb{F}_{q^k}$-zeros off the diagonal if k is large.

14. Prove the following generalization of a theorem of MacCluer (See also [Fried6, Thm. 1]): Let K be a pseudo finite field, $f \in K[X]$ a separable polynomial of degree $n > 0$, x an indeterminate. Put $t = f(x)$. Suppose no K-irreducible factor of $f^*(X, Y) = \frac{f(X) - f(Y)}{X - Y}$ is absolutely irreducible. Then f permutes K.

Hint: Let $x_1, \ldots, x_n$ be the n distinct roots of $f(X) = t$. Put $F = K(x_1, \ldots, x_n)$. Let $\hat{K}$ be the algebraic closure of K in F, $T = \{\tau \in \mathrm{Gal}(F/K(t)) \mid \mathrm{res}_{\hat{K}} \tau = \mathrm{res}_{\hat{K}} \sigma\}$, and $T_i = T \cap \mathrm{Gal}(F/K(x_i))$, $i = 1, \ldots, n$. Prove: $|T| = \sum_{i=1}^n |T_i|$ and $T_i \cap T_j = \emptyset$ for $i \neq j$. Conclude that $T = \bigcup_{i=1}^n T_i$. Now consider $a \in K$. Find a $\hat{K}$-place $\psi \colon F \to \tilde{K} \cup \{\infty\}$ with $\psi(t) = a$. Then choose σ in the decomposition group of ψ which belongs to T. There is an i with $\sigma x_i = x_i$. Hence, $b = \psi(x_i)$ is in K and satisfies $f(b) = a$. Consequently, f is surjective on K. Since K is pseudo finite, f is also injective.

Notes

The decomposition-intersection procedure has been used by several authors. For example, [Greenleaf] uses it to prove that for each $d > 0$ almost all fields $\mathbb{Q}_p$ are $C_{2,d}$. Also, it appears in the first version [Fried-Sacerdote], of the stratification procedure.

[Klingen4] gives a comprehensive survey on Kronecker classes of number fields.

Problem 21.5.8 has a negative solution if the following statement holds for each infinite profinite group G of rank $\leq \aleph_0$: For each closed subgroup H of G of infinite index the set $\bigcup_{g \in G} H^g$ does not contain an open neighborhood of 1. There is an attempt in [Klingen2, Thm. 6] to give a negative answer to Problem 21.5.8. Unfortunately there is an error in the proof: A closed subgroup H of a profinite group G of countable rank may have uncountably many, rather that countably many (as claimed in [Klingen2]), conjugates in G. For example, there are uncountably many involutions in $G(\mathbb{Q})$, all conjugate. Thus, Problem 21.5.8 is still open.

Davenport's original problem [Fried-Jarden3, Problem 19.26] asked if $V_p(f) = V_p(g)$ for nonconstants $f, g \in \mathbb{Q}[X]$ and almost all p implies f and g are linearly related. Remark 21.6.1 supplies counter examples to that problem. Davenport's present problem 21.6.2 is a modification of the older one.

Lemma 21.7.2 overlaps with a result attributed to Galois [Huppert, p. 163, Satz 3.6].

More on Dickson's polynomials can be found in [Schinzel2, §1.4-1.5] and [Lidl-Mullen-Turnwald].

The original proof of Proposition 21.7.7 appears in [Schur2]. Our proof is an elaboration of [Lidl-Mullen-Turnwald, p. 126]. Likewise, the original proof of Proposition 21.7.8 appears in [Burnside1]. We have elaborated [Lidl-Mullen-Turnwald p. 127].

[Schur1] proves that every polynomial $f \in \mathbb{Z}[X]$ of prime degree which is a permutation polynomial modulo infinitely many prime numbers p is a composition of polynomials which are linearly related over $\tilde{\mathbb{Q}}$ to Cyclic polynomials or Chebyshev polynomials. Schur conjectured that his result holds for an arbitrary number field K. [Fried1] uses the theory of Riemann surfaces to prove Schur's conjecture. [Turnwald, Thm. 2] uses algebraic methods to generalize Fried's result to arbitrary global field (See also [Lidl-Mullen-Turnwald]). Our proof of Schur's conjecture (Theorem 21.8.13) is based on a result of Müller (Proposition 21.8.7).

Our version of the proof of Cohen-Fried proof of the generalized Carlitz's Conjecture (Theorem 21.9.5) uses improvements of Bensimhoun and Haran.

Partially building on ideas of [Lenstra], [Guralnick-Müller, §8] further generalizes the generalized Conjecture to nonconstant separable morphisms between smooth projective curves over $\mathbb{F}_q$.

Chapter 22.
Projective Groups and Frattini Covers

Every profinite group is a Galois group of some Galois extension (Leptin, Proposition 2.6.12). However, not every profinite group is an absolute Galois group. For example, the only finite groups that appear as absolute Galois groups are the trivial group and $\mathbb{Z}/2\mathbb{Z}$ (Artin [Lang7, p. 299, Cor. 9.3]). This raises the question of characterizing the absolute Galois groups among all profinite groups. This question is still wide open. A more restrictive question finds a complete solution in projective groups: By Theorem 11.6.2, the absolute Galois group of a PAC field is projective. Conversely, for each projective group G there exists a PAC field K such that $\mathrm{Gal}(K) \cong G$ (Theorem 23.1.2). Thus, a profinite group G is projective if and only if it is isomorphic to the absolute Galois group of a PAC field.

In this chapter we define a projective group as a profinite group G for which every embedding problem is weakly solvable. By Gruenberg's theorem (Lemma 22.3.2), it suffices to weakly solve only finite embedding problems. This leads to the second characterization of projective groups as those profinite groups which are isomorphic to closed subgroups of free profinite groups (Corollary 22.4.6).

Projective groups also appear as the universal Frattini covers of profinite groups (Proposition 22.6.1). Thus, as a preparation for decidability and undecidability results about families of PAC fields, we introduce in this chapter the basis properties of Frattini covers.

22.1 The Frattini Group of a Profinite Group

Consider a profinite group G. Denote the intersection of all maximal open subgroups of G by $\Phi(G)$. Here we call a subgroup M **maximal** if there is no subgroup M' of G such that $M < M' < G$. The characteristic (and therefore normal) closed subgroup $\Phi(G)$ is called the **Frattini group** of G. We characterize the elements of $\Phi(G)$ as "dispensable generators" of G:

LEMMA 22.1.1: *Let G be a profinite group. An element $g \in G$ belongs to $\Phi(G)$ if and only if there is no proper closed subgroup H of G for which $\langle H, g \rangle = G$. Also, if H is a closed subgroup for which $H \cdot \Phi(G) = G$, then $H = G$.*

Proof: Let $g \in \Phi(G)$ and let H be a closed subgroup of G for which $\langle H, g \rangle = G$. If $H \neq G$, then H is contained in a maximal open subgroup M of G. Therefore, $G = \langle H, g \rangle \leq M < G$, a contradiction.

Conversely, suppose an element $g \in G$ satisfies the above condition. Let M be a maximal open subgroup of G. If $g \notin M$, then $G = \langle M, g \rangle$.

Hence, $M = G$, a contradiction. Therefore, $g \in M$, and since M is arbitrary, $g \in \Phi(G)$.

Finally, let H be a closed subgroup of G such that $H \cdot \Phi(G) = G$. If $H < G$, then H is contained in a maximal open subgroup M of G and $M = M \cdot \Phi(G) = G$, a contradiction. Thus, $H = G$. $\square$

LEMMA 22.1.2: *The Frattini group of a profinite group G is pronilpotent.*

Proof: Assume first G is finite. Let P be a p-Sylow group of $\Phi(G)$. For $g \in G$ the group P^g is also a p-Sylow subgroup of $\Phi(G)$. Hence, there exists $a \in \Phi(G)$ such that $P^g = P^a$. Therefore, ga^{-1} is in the normalizer, $N_G(P)$, of P in G. Thus, $G = N_G(P)\Phi(G)$. By Lemma 22.1.1, $G = N_G(P)$. Hence, P is normal in G, and therefore in $\Phi(G)$. It follows that $\Phi(G)$ is the direct product of its p-Sylow groups. That is, G is nilpotent [Huppert, p. 260].

In the general case observe that for any open normal subgroup N of G, $\Phi(G)/N \cap \Phi(G) \cong N\Phi(G)/N \leq \Phi(G/N)$. Since $\Phi(G/N)$ is a nilpotent finite group, so is $\Phi(G)/N \cap \Phi(G)$. Therefore, $\Phi(G) = \varprojlim \Phi(G)/N \cap \Phi(G)$ is a pronilpotent group. $\square$

LEMMA 22.1.3: *Let A be a minimal normal subgroup of a finite group G. If $A \leq \Phi(G)$, then A is p-elementary Abelian for some prime number p.*

Proof: By Lemma 22.1.2, A is a nilpotent group. Hence, the center $Z(A)$ of A is nontrivial [Huppert, p. 260] and normal in G. Thus, $Z(A) = A$ and A is Abelian. As such, A is a direct product of its Sylow subgroups and each of them is normal in G. Therefore, A is an Abelian p-group for some prime number p. The subgroup of A consisting of all elements x with $x^p = 1$ is normal in G. Consequently, $A = (\mathbb{Z}/p\mathbb{Z})^m$ for some $m \geq 0$. $\square$

As an operation, taking the Frattini subgroup of a group has functorial properties:

LEMMA 22.1.4:
(a) *Let $\theta \colon G \to H$ be an epimorphism of profinite groups. Then $\theta(\Phi(G)) \leq \Phi(H)$. If $\mathrm{Ker}(\theta) \leq \Phi(G)$, then $\theta(\Phi(G)) = \Phi(H)$.*
(b) *Suppose U is a closed subgroup of G and N is a closed subgroup of $\Phi(U)$, normal in G. Then $N \leq \Phi(G)$.*
(c) *Let N be a closed normal subgroup of G. Then $\Phi(N) \leq \Phi(G)$.*
(d) *Let $G = \prod_{i \in I} G_i$ be a direct product of profinite groups. Then $\Phi(G) = \prod_{i \in I} \Phi(G_i)$.*

Proof of (a): The inverse image of each maximal open subgroup of H is a maximal open subgroup of G. Hence, $\theta(\Phi(G)) \leq \Phi(H)$.

Proof of (b): Assume $N \not\leq \Phi(G)$. Then G has a maximal open subgroup M with $N \not\leq M$. Thus, $MN = G$. Hence, $U = U \cap MN = (U \cap M)N$. Since $N \leq \Phi(U)$, Lemma 22.1.1 implies $U = U \cap M$. Therefore, $N \leq M$, contrary to the assumption.

22.2 Cartesian Squares

Proof of (c): Apply (b) to N and $\Phi(N)$ instead of to U and N.

Proof of (d): Each G_i is a closed normal subgroup of G. Therefore, (c) gives $\Phi(G_i) \leq \Phi(G)$ and $\prod_{i \in I} \Phi(G_i) \leq \Phi(G)$.

Conversely, let M be a maximal open subgroup of G_j. Then $M \times \prod_{i \neq j} G_i$ is a maximal open subgroup of G. Thus $\Phi(G) \leq M \times \prod_{i \neq j} G_i$. Running over all M, this gives $\Phi(G) \leq \Phi(G_j) \times \prod_{i \neq j} G_i$. Since this is true for each $j \in I$, we conclude that $\Phi(G) \leq \prod_{i \in I} \Phi(G_i)$. □

Remark 22.1.5: The map $\theta \colon \Phi(G) \to \Phi(H)$ in Lemma 22.1.4(a) need not be surjective. Let $G = \langle b \rangle \ltimes \langle a \rangle$ be the semidirect product of a cyclic group $\langle a \rangle$ of order 5 with a cyclic group $H = \langle b \rangle$ of order 4 with the relation $a^b = a^2$. Let $\theta \colon G \to H$ be the quotient map. Both $\langle a \rangle$ and $\langle b \rangle$ are maximal subgroups of G, so $\Phi(G) = 1$. On the other hand, H has order 4, so $\Phi(H)$ is of order 2. Thus, $\theta(\Phi(G)) < \Phi(H)$. □

22.2 Cartesian Squares

The usual direct product of group theory has a useful generalization:

Let $\alpha \colon B \to A$ and $\gamma \colon C \to A$ be homomorphisms of profinite groups. Let

$$B \times_A C = \{(b, c) \in B \times C \mid \alpha(b) = \gamma(c)\}.$$

Since A is Hausdorff, $B \times_A C$ is a closed subgroup of $B \times C$. It is therefore a profinite group called the **fiber product** of B and C over A. There are natural projection maps $\mathrm{pr}_B \colon B \times_A C \to B$ and $\mathrm{pr}_C \colon B \times_A C \to C$ defined by $\mathrm{pr}_B(b, c) = b$ and $\mathrm{pr}_C(b, c) = c$, respectively.

It is possible to change A, B, and C such that the projection maps will be surjective: Put $A_0 = \{a \in A \mid \exists (b, c) \in B \times C \colon \alpha(b) = \gamma(c) = a\}$, $B_0 = \alpha^{-1}(A_0)$, and $C_0 = \gamma^{-1}(A_0)$. Then A_0 (resp. B_0, C_0) is a closed subgroup of A (resp. B, C), $B_0 \times_{A_0} C_0 = B \times_A C$, and the projections maps of $B_0 \times_{A_0} C_0$ are surjective.

PROPOSITION 22.2.1: Consider a commutative diagram of profinite groups

(1)
$$\begin{array}{ccc} D & \xrightarrow{\delta} & C \\ {\scriptstyle\beta}\downarrow & & \downarrow{\scriptstyle\gamma} \\ B & \xrightarrow{\alpha} & A \end{array}$$

where β, α, γ, and δ are homomorphisms. The following statements are equivalent:
(a) There exists an isomorphism $\theta \colon D \to B \times_A C$ with $\beta \circ \theta^{-1} = \mathrm{pr}_B$ and $\delta \circ \theta^{-1} = \mathrm{pr}_C$.
(b) Let G be a profinite group and $\varphi \colon G \to B$ and $\psi \colon G \to C$ homomorphisms. Suppose $\alpha \circ \varphi = \gamma \circ \psi$. Then there exists a unique homomorphism

$\pi\colon G \to D$ which makes the following diagram commutative:

(2)

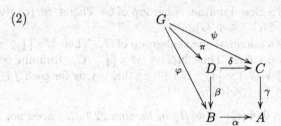

Proof of "(a)$\Longrightarrow$(b)": Without loss assume $D = B \times_A C$, $\beta = \mathrm{pr}_B$, and $\delta = \mathrm{pr}_C$. Define a map $\pi\colon G \to D$ by $\pi(g) = (\varphi(g), \psi(g))$. Then π is the unique homomorphism which makes (2) commutative.

Proof of "(b)$\Longrightarrow$(a)": Let $G = B \times_A C$, $\varphi = \mathrm{pr}_B$, and $\psi = \mathrm{pr}_C$. Then (b) gives a unique π that makes (2) commutative.

Define $\theta\colon D \to B \times_A C$ by $\theta(d) = (\beta(d), \delta(d))$ for $d \in D$. Apply the uniqueness property to maps from D to D and from G to G. It implies the map $\pi \circ \theta$ (resp. $\theta \circ \pi$) is the identity, and θ satisfies (a). $\square$

Definition 22.2.2: A commutative diagram (1) satisfying the conditions of Proposition 22.2.1 is said to be a **cartesian square**. $\square$

Lemma 22.2.3: Let (1) be a cartesian square of homomorphisms of profinite groups.
(a) If $b \in B$ and $c \in C$ satisfy $\alpha(b) = \gamma(c)$, then there exists a unique $d \in D$ with $\beta(d) = b$ and $\delta(d) = c$.
(b) If α (resp. γ) is surjective, then δ (resp. β) is surjective.

Proof of (a): Assume without loss that $D = B \times_A C$. If $b \in B$ and $c \in C$ satisfy $\alpha(b) = \gamma(c)$, then $d = (b, c)$ is the unique element of D satisfying $\beta(d) = b$ and $\delta(d) = c$.

Proof of (b): Consider $c \in C$. By assumption, there is a $b \in B$ with $\alpha(b) = \gamma(c)$. By (a), there is a $d \in D$ with $\delta(d) = c$. Thus, δ is surjective. $\square$

LEMMA 22.2.4: Let (1) be a commutative square of epimorphisms of profinite groups. Then (1) is cartesian if and only if $\mathrm{Ker}(\alpha \circ \beta) = \mathrm{Ker}(\beta) \times \mathrm{Ker}(\delta)$.

Proof: First suppose (1) is a cartesian square. Clearly $\mathrm{Ker}(\beta) \cdot \mathrm{Ker}(\delta) \leq \mathrm{Ker}(\alpha \circ \beta)$.

Assume without loss, $D = B \times_A C$, $\beta = \mathrm{pr}_B$, and $\delta = \mathrm{pr}_C$. Let $(b, c) \in \mathrm{Ker}(\alpha \circ \beta)$. Then $\alpha(b) = 1 = \gamma(c)$. Hence, $(1, c)$ and $(b, 1)$ belong to D. Moreover, $(b, c) = (1, c) \cdot (b, 1) \in \mathrm{Ker}(\beta) \cdot \mathrm{Ker}(\delta)$. Finally, $\mathrm{Ker}(\beta) \cap \mathrm{Ker}(\delta) = 1$. Consequently, $\mathrm{Ker}(\alpha \circ \beta) = \mathrm{Ker}(\beta) \times \mathrm{Ker}(\delta)$.

Now suppose $\mathrm{Ker}(\alpha \circ \beta) = \mathrm{Ker}(\beta) \times \mathrm{Ker}(\delta)$. Define a homomorphism $\theta\colon D \to B \times_A C$ by $\theta(d) = (\beta(d), \delta(d))$. If $\theta(d) = (1, 1)$, then $d \in \mathrm{Ker}(\beta) \cap \mathrm{Ker}(\delta) = 1$. Thus, θ is injective.

22.2 Cartesian Squares

To prove that θ is surjective, consider $(b,c) \in B \times_A C$. Then $\alpha(b) = \gamma(c)$. There exist $d_1, d_2 \in D$ with $\beta(d_1) = b$ and $\delta(d_2) = c$. Then $\alpha(\beta(d_1 d_2^{-1})) = \alpha(\beta(d_1)) \cdot \gamma(\delta(d_2))^{-1} = \alpha(b)\gamma(c)^{-1} = 1$. By assumption, $d_1 d_2^{-1} = d_3^{-1} d_4$ for some $d_3 \in \mathrm{Ker}(\beta)$ and $d_4 \in \mathrm{Ker}(\delta)$. Let $d = d_3 d_1 = d_4 d_2$. Then $\beta(d) = b$ and $\delta(d) = c$. Therefore, $\theta(d) = (b,c)$. $\square$

LEMMA 22.2.5: *Let (1) be a cartesian square of epimorphisms of profinite groups. Then β maps $\mathrm{Ker}(\delta)$ isomorphically onto $\mathrm{Ker}(\alpha)$.*

Proof: Apply Lemma 22.2.4:

$$\mathrm{Ker}(\alpha) = \beta(\mathrm{Ker}(\alpha \circ \beta)) = \beta(\mathrm{Ker}(\beta) \times \mathrm{Ker}(\delta)) = \beta(\mathrm{Ker}(\delta)).$$

In addition, $\mathrm{Ker}(\delta) \cap \mathrm{Ker}(\beta) = 1$. Hence, $\beta\colon \mathrm{Ker}(\delta) \to \mathrm{Ker}(\alpha)$ is an isomorphism. $\square$

LEMMA 22.2.6: *Let (1) and (2) be commutative diagrams of homomorphisms of profinite groups. Suppose (1) is cartesian and all maps except possibly π are surjective. Then:*
(a) *If π is surjective, then $\mathrm{Ker}(\alpha \circ \varphi) = \mathrm{Ker}(\varphi)\mathrm{Ker}(\psi)$.*
(b) *If $\mathrm{Ker}(\alpha \circ \varphi) \le \mathrm{Ker}(\varphi)\mathrm{Ker}(\psi)$, then π is surjective.*

Proof of (a): The condition $\alpha \circ \varphi = \gamma \circ \psi$ implies $\mathrm{Ker}(\varphi)\mathrm{Ker}(\psi) \le \mathrm{Ker}(\alpha \circ \varphi)$.

Now suppose π is surjective. Let $g \in \mathrm{Ker}(\alpha \circ \varphi)$. Put $b = \varphi(g)$. Then $\alpha(b) = 1$. Hence, there is a $d \in D$ with $\beta(d) = b$ and $\delta(d) = 1$. Choose an $h \in G$ with $\pi(h) = d$. Then $\psi(h) = \delta(\pi(h)) = \delta(d) = 1$ and $\varphi(gh^{-1}) = b\beta(\pi(h))^{-1} = b\beta(d)^{-1} = 1$. Therefore, $g = gh^{-1} \cdot h \in \mathrm{Ker}(\varphi)\mathrm{Ker}(\psi)$.

Proof of (b): Suppose $\mathrm{Ker}(\alpha \circ \varphi) \le \mathrm{Ker}(\varphi)\mathrm{Ker}(\psi)$. Consider $d \in D$. Put $b = \beta(d)$ and $c = \delta(d)$. Choose $g, h \in G$ with $\varphi(g) = b$ and $\psi(h) = c$. Then $\alpha(\varphi(gh^{-1})) = \alpha(\varphi(g))\gamma(\psi(h))^{-1} = \alpha(b)\gamma(c)^{-1} = \alpha(\beta(d))\gamma(\delta(d))^{-1} = 1$. Thus, there are $g_1 \in \mathrm{Ker}(\varphi)$ and $h_1 \in \mathrm{Ker}(\psi)$ with $gh^{-1} = g_1 h_1$. The element $g_1^{-1} g = h_1 h$ of G satisfies $\varphi(g_1^{-1} g) = b$ and $\psi(h_1 h) = c$. Therefore, by Lemma 22.2.3(a), $\pi(g_1^{-1} g) = d$. $\square$

Example 22.2.7: Fiber products. Here are four examples where fiber products naturally appear. The verification that the occurring commutative squares are cartesian follows from Lemma 22.2.4:

(a) Let M and M' be Galois extensions of a field K. Put $L = M \cap M'$ and $N = MM'$. Then $\mathrm{Gal}(N/K)$ is the fiber product of $\mathrm{Gal}(M/K)$ and $\mathrm{Gal}(M'/K)$ over $\mathrm{Gal}(L/K)$ with respect to the restriction maps.

(b) Similarly, let G be a profinite group and K, L, M, N closed normal subgroups. Suppose $K \cap L = N$ and $KL = M$. Then $G/N = G/K \times_{G/M} G/L$ with respect to the quotient maps.

(c) Let $\varphi\colon B \to A$ be an epimorphism of profinite groups, $C = \mathrm{Ker}(\varphi)$, and B_0 a closed normal subgroup of B. Suppose $B_0 \cap C = 1$. Put $\bar{B} = B/B_0$, $A_0 = \varphi(B_0)$, $\bar{A} = A/A_0$, $\alpha\colon A \to \bar{A}$ and $\beta\colon B \to \bar{B}$ the canonical maps, and $\bar{\varphi}\colon \bar{B} \to \bar{A}$ the map induced from φ. Then $B \cong \bar{B} \times_{\bar{A}} A$.

Indeed, if $b \in B$ satisfies $\alpha(\varphi(b)) = 1$, then $\varphi(b) \in A_0$. Hence, there exists $b_0 \in B_0$ with $\varphi(b_0) = \varphi(b)$. Therefore, $b = b_0 \cdot b_0^{-1} b \in \mathrm{Ker}(\beta) \mathrm{Ker}(\varphi)$. It follows from Lemma 22.2.4 that $B \cong \bar{B} \times_{\bar{A}} A$.

Suppose now $B_0 \cap C$ is not necessarily trivial. Then by the preceding paragraph, $B/B_0 \cap C \cong \bar{B} \times_{\bar{A}} A$. Hence, for each $a \in A$ and $b \in B$ satisfying $\bar{\varphi}(\beta(b)) = \alpha(a)$ there is a $b' \in B$ (not necessarily unique) with $\beta(b') = \beta(b)$ and $\varphi(b') = a$.

(d) Condition (1b) of Section 17.3 on a family $\mathcal{C}$ of finite groups is equivalent to "$\mathcal{C}$ is closed under fiber products". Thus, a formation is a family of finite groups which is closed under taking quotients and fiber products. □

The homomorphism π of diagram (2) need not be surjective even if all other maps are surjective. Here is a condition for this to happen:

LEMMA 22.2.8: *Let $\varphi\colon G \to B$ and $\psi\colon G \to C$ be epimorphisms of profinite groups. Then there is a commutative diagram (2), unique up to an isomorphism such that (1) is a cartesian square with $\beta, \alpha, \gamma, \delta, \pi$ epimorphisms.*

Proof: Let $M = \mathrm{Ker}(\varphi)\mathrm{Ker}(\psi)$, $N = \mathrm{Ker}(\varphi) \cap \mathrm{Ker}(\psi)$, $A = G/M$, $D = G/N$, and $\pi\colon G \to G/N$ the quotient map. Now find $\beta, \alpha, \gamma, \delta$ that make (1) cartesian and (2) commutative. □

The following lemma gives a useful criterion for δ in the cartesian diagram (1) to have a group theoretic section:

LEMMA 22.2.9: *Let (1) be a cartesian diagram of homomorphisms of profinite groups. Suppose there exists a homomorphism $\varphi\colon C \to B$ with $\alpha \circ \varphi = \gamma$. Then there exists a monomorphism $\delta'\colon C \to D$ such that $\delta \circ \delta' = \mathrm{id}_C$.*

Proof: Let $\psi = \mathrm{id}_C$. Then $\alpha \circ \varphi = \gamma \circ \psi$. Hence, by Proposition 22.2.1(b), there exists a homomorphism $\delta'\colon C \to D$ such that $\delta \circ \delta' = \mathrm{id}_C$ (and $\beta \circ \delta' = \varphi$), as claimed. Note that the existence of φ implies that δ is surjective. □

22.3 On $\mathcal{C}$-Projective Groups

Embedding problems and projective groups have already appeared, for example, in Sections 16.5 and 17.7. Now we consider the subject in detail:

Definition 22.3.1: Embedding problems. An **embedding problem** for a profinite group G is a pair

(1) $\qquad\qquad (\varphi\colon G \to A, \; \alpha\colon B \to A)$

in which φ and α are epimorphisms of profinite groups. We call $\mathrm{Ker}(\alpha)$ the **kernel of the problem**. We call the problem **finite** if B is finite. We say (1) **splits** if α has a **group theoretic section**. That is, there is a homomorphism $\alpha'\colon A \to B$ with $\alpha \circ \alpha' = \mathrm{id}_A$.

22.3 On $\mathcal{C}$-Projective Groups

Embedding problem (1) is said to be **solvable** (resp. **weakly solvable**) if there exists an epimorphism (resp. homomorphism) $\gamma\colon G \to B$ with $\alpha\circ\gamma = \varphi$. The map γ is a **solution** (resp. **weak solution**) to (1).

Suppose $\mathcal{C}$ is a Melnikov formation of finite groups (Section 17.3). Let G be a pro-$\mathcal{C}$ group. Then call (1) a $\mathcal{C}$-**embedding problem** (resp. **pro-$\mathcal{C}$ embedding problem**), if B is a $\mathcal{C}$-group (resp. pro-$\mathcal{C}$).

Call G $\mathcal{C}$-**projective** if every pro-$\mathcal{C}$ embedding problem (1) for G is weakly solvable. □

LEMMA 22.3.2 (Gruenberg): *Let $\mathcal{C}$ be a Melnikov formation of finite groups and G be a pro-$\mathcal{C}$ group. Suppose every finite $\mathcal{C}$-embedding problem for G is weakly solvable. Then G is $\mathcal{C}$-projective.*

Proof: Consider embedding problem (1) for G with B a pro-$\mathcal{C}$ group. Let $C = \mathrm{Ker}(\alpha)$. The transition from the finite case to the general case in Part C of the proof forces us to consider pairs (1) in which φ is a homomorphism which is not necessarily surjective.

PART A: *Suppose $\varphi\colon G \to A$ is a homomorphism, $B \in \mathcal{C}$, and $\alpha\colon B \to A$ is an epimorphism. Then there is a homomorphism $\beta\colon G \to B$ with $\alpha\circ\beta = \varphi$.*

Indeed, B and $A_0 = \varphi(G)$ are in $\mathcal{C}$. Hence, so is C and therefore also $B_0 = \alpha^{-1}(A_0)$. By assumption, there is a homomorphism $\beta\colon G \to B_0$ with $\alpha\circ\beta = \varphi$.

PART B: *C is finite.* Choose an open normal subgroup U of B with $C\cap U = 1$. Since $B/U \in \mathcal{C}$, Part A gives a homomorphism $\bar\beta\colon G \to B/U$ for which

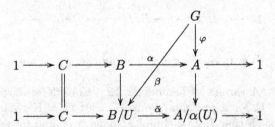

is a commutative diagram. The right square is cartesian (Example 22.2.7(c)). Therefore (Proposition 22.2.1(b)), there is a homomorphism $\gamma\colon G \to B$ with $\alpha\circ\gamma = \varphi$.

PART C: *The general case.* Apply Zorn's lemma to Part B. Let Λ be the set of pairs (L, λ) where L is a closed normal subgroup of B contained in C and $\lambda\colon G \to B/L$ is a homomorphism which makes the following diagram commutative:

(2)
$$1 \longrightarrow C/L \longrightarrow B/L \xrightarrow{\alpha_L} A \longrightarrow 1.$$

Here α_L is the epimorphism induced by α. The pair $(C, \alpha_C^{-1} \circ \varphi)$ belongs to Λ, so Λ is nonempty. Partially order Λ by $(L', \lambda') \leq (L, \lambda)$ if $L' \leq L$ and the following triangle is commutative:

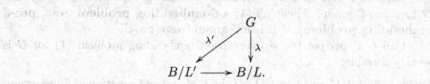

Suppose $\Lambda_0 = \{(L_i, \lambda_i) \mid i \in I\}$ is a descending chain in Λ. Then $\varprojlim B/L_i = B/L$ with $L = \bigcap_{i \in I} L_i$. The λ_i's define a homomorphism $\lambda \colon G \to B/L$ with (2) commutative. Therefore, (L, λ) is a lower bound for Λ_0.

By Zorn's Lemma, Λ has a minimal element (L, λ). It suffices to prove that $L = 1$.

Assume $L \neq 1$. Then B has an open normal subgroup N with $L \not\leq N$. Therefore, $L' = N \cap L$ is a proper open subgroup of L which is normal in B. Part B gives a homomorphism $\lambda' \colon G \to B/L'$ which makes the following diagram commutative:

(3)
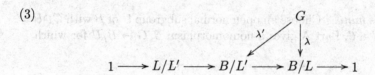

Therefore, (L', λ') is an element of Λ less than (L, λ), a contradiction. □

Remark 22.3.3: *A variant of Lemma 22.3.2.* Consider an arbitrary embedding problem (1) for a profinite group G. Put $C = \mathrm{Ker}(\alpha)$. Suppose for all $L, L' \leq C$ such that $L, L' \triangleleft B$ and L' open in L and for each homomorphism $\lambda \colon G \to B/L$ there is a homomorphism $\lambda' \colon G \to B/L'$ making (3) commutative. The proof of Lemma 22.3.2 proves that (1) is weakly solvable. □

LEMMA 22.3.4: *Let $\mathcal{C}$ be a Melnikov formation of finite groups and G a pro-$\mathcal{C}$ group. Suppose every $\mathcal{C}$-embedding problem (1) in which $\mathrm{Ker}(\alpha)$ is a minimal normal subgroup of B is weakly solvable. Then G is $\mathcal{C}$-projective.*

Proof: By Lemma 22.3.2, it suffices to weakly solve each $\mathcal{C}$-embedding problem (1). Put $C = \mathrm{Ker}(\alpha)$.

If C is minimal normal in B, then (1) is solvable by assumption. Suppose C is not minimal normal in B and proceed by induction on $|C|$.

By assumption, there is a nontrivial normal subgroup L of B which is

22.3 On $\mathcal{C}$-Projective Groups

properly contained in C. Consider the diagram

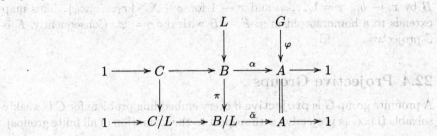

in which $\bar{\alpha}$ is induced by α. Since $|C/L| < |C|$, the induction hypothesis gives a homomorphism $\beta \colon G \to B/L$ with $\bar{\alpha} \circ \beta = \varphi$. Also $|L| < |C|$. Using the induction hypothesis again, there is a homomorphism $\gamma \colon G \to B$ with $\pi \circ \gamma = \beta$. Therefore, $\alpha \circ \gamma = \varphi$. □

PROPOSITION 22.3.5: *Let $\mathcal{C}$ be a full formation of finite groups and G a pro-$\mathcal{C}$ group. Suppose each $\mathcal{C}$-embedding problem (1) for G, where $\mathrm{Ker}(\alpha)$ is a minimal Abelian p-elementary normal subgroup of B, is weakly solvable. Then G is $\mathcal{C}$-projective.*

Proof: Let $C = \mathrm{Ker}(\alpha)$. By Lemma 22.3.4, it suffices to prove that every embedding problem (1) for G, where $B \in \mathcal{C}$ and C is a minimal normal subgroup of B, is weakly solvable. We distinguish between two cases:

CASE A: Suppose $C \not\leq \Phi(B)$. Then B has a maximal subgroup B_1 with $C \not\leq B_1$. Let α_1 be the restriction of α to B_1. Then $B_1 C = B$ and $\alpha_1(B_1) = \alpha(CB_1) = \alpha(B) = A$. Consider the commutative diagram

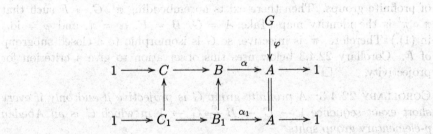

with $C_1 = C \cap B_1$. Since $\mathcal{C}$ is full, $B_1 \in \mathcal{C}$. An induction hypothesis on $|B|$ gives a homomorphism $\gamma \colon G \to B_1$ with $\alpha_1 \circ \gamma = \varphi$, hence $\alpha \circ \gamma = \varphi$.

CASE B: $C \leq \Phi(B)$. By Lemma 22.1.3, C is a minimal Abelian p-elementary normal subgroup of B. Hence, by assumption, (1) is solvable. □

The following lemma gives the basic example for projective $\mathcal{C}$-groups:

LEMMA 22.3.6: *Let $\mathcal{C}$ be a Melnikov formation of finite groups. Then each free pro-$\mathcal{C}$ group F is $\mathcal{C}$-projective.*

Proof: Choose a basis X for F. Consider a $\mathcal{C}$-embedding problem $(\varphi \colon F \to A, \alpha \colon B \to A)$. Choose $x_1, \ldots, x_n \in X$ such that $\varphi(x_1), \ldots, \varphi(x_n)$ generate

A. Then choose $b_1, \ldots, b_n \in B$ with $\alpha(b_i) = \varphi(x_i)$, $i = 1, \ldots, n$. Map X into B by $x_i \mapsto b_i$, $i = 1, \ldots, n$ and $x \mapsto 1$ for $x \in X \smallsetminus \{x_1, \ldots, x_n\}$. This map extends to a homomorphism $\gamma \colon F \to B$ with $\alpha \circ \gamma = \varphi$. Consequently, F is $\mathcal{C}$-projective. $\square$

22.4 Projective Groups

A profinite group G is **projective** if every embedding problem for G is weakly solvable (i.e. G is projective with respect to the formation of all finite groups).

LEMMA 22.4.1: *If $\mathcal{C}$ is a full formation of finite groups and G is $\mathcal{C}$-projective, then G is projective.*

Proof: By Proposition 22.3.5, it suffices to prove that every finite embedding problem

(1) $\qquad\qquad\qquad (\varphi \colon G \to A, \ \alpha \colon B \to A)$

with $\mathrm{Ker}(\alpha) = (\mathbb{Z}/p\mathbb{Z})^m$ and p a prime number is weakly solvable.

If $p \nmid |A|$, then Schur-Zassenhaus [Huppert, p. 126] gives a homomorphism $\alpha' \colon A \to B$ with $\alpha \circ \alpha' = \mathrm{id}_A$. Thus, $\alpha' \circ \varphi$ is a weak solution of (1).

Now suppose p divides $|A|$. Then A contains an isomorphic copy of $\mathbb{Z}/p\mathbb{Z}$ (Cauchy). Since A is in $\mathcal{C}$, so is $\mathbb{Z}/p\mathbb{Z}$. Hence, $(\mathbb{Z}/p\mathbb{Z})^m \in \mathcal{C}$. Therefore, $B \in \mathcal{C}$. By assumption, (1) has a weak solution. $\square$

Remark 22.4.2: Suppose G is projective and $\pi \colon F \to G$ is an epimorphism of profinite groups. Then there exists an embedding $\pi' \colon G \to F$ such that $\pi \circ \pi'$ is the identity map (Take $A = G$, $B = F$, $\alpha = \pi$, and $\varphi = \mathrm{id}_G$ in (1).) Therefore, π' is injective, so G is isomorphic to a closed subgroup of F. Corollary 22.4.3 below uses this observation to give a criterion for projectivity. $\square$

COROLLARY 22.4.3: *A profinite group G is projective if and only if every short exact sequence $1 \to C \to B \xrightarrow{\alpha} G \to 1$ in which C is an Abelian p-elementary group splits.*

Proof: By Remark 22.4.2, we have only to prove sufficiency. By Lemma 22.3.5 it suffices to give a weak solution to each finite embedding problem $(\varphi \colon G \to A, \ \alpha \colon B \to A)$ with $\mathrm{Ker}(\alpha) = (\mathbb{Z}/p\mathbb{Z})^m$ for some prime number p and a positive integer m.

Complete the embedding problem to a commutative diagram

$$\begin{array}{ccccccccc} 1 & \to & (\mathbb{Z}/p\mathbb{Z})^m & \to & B \times_A G & \to & G & \to & 1 \\ & & \| & & \downarrow & & \downarrow \varphi & & \\ 1 & \to & (\mathbb{Z}/p\mathbb{Z})^m & \to & B & \xrightarrow{\alpha} & A & \to & 1 \end{array}.$$

22.4 Projective Groups

By assumption, the upper row splits. Hence, there is a homomorphism $\gamma\colon G \to B$ with $\alpha \circ \gamma = \varphi$. □

Remark 22.4.4: Cohomological interpretation of projectivity. In cohomological terms Corollary 22.4.3 signifies that a profinite group G is projective if and only if its cohomological dimension is bounded by 1 [Ribes, p. 211]. □

COROLLARY 22.4.5: *Let C be a full formation of finite groups and F a free pro-C-group. Then F is projective.*

Proof: By Lemma 22.3.6, F is C-projective. Hence, by Lemma 22.4.1, F is projective. □

COROLLARY 22.4.6: *A profinite group G is projective if and only if it is isomorphic to a closed subgroup of a free profinite group.*

Proof: By Corollary 17.4.8, G is a quotient of a free profinite group F. So, by Remark 22.4.2, G is isomorphic to a closed subgroup of F.

Conversely, let H be a closed subgroup of a free profinite group F. Let $\varphi\colon H \to A$ and $\alpha\colon B \to A$ be epimorphisms, with B finite. Lemma 1.2.5(c) gives an open subgroup H' of F containing H and an epimorphism $\varphi'\colon H' \to A$ extending φ. By Proposition 17.6.2, H' is free. Hence, by Corollary 22.4.5, H' is projective. Therefore, there exists a homomorphism $\gamma'\colon H' \to B$ with $\alpha \circ \gamma' = \varphi'$. Denote the restriction of γ' to H by γ. Then $\alpha \circ \gamma = \varphi$. It follows that H is projective. □

PROPOSITION 22.4.7: *Let G be a projective group and H a closed subgroup. Then H is projective. Moreover, H is either trivial or infinite. In particular, G is torsion free.*

Proof: By Corollary 22.4.6, G is isomorphic to a closed subgroup of a free profinite group F. Hence, H is also isomorphic to a closed subgroup of F. A second application of Corollary 22.4.6 now proves that H is projective.

Assume G has a nontrivial finite subgroup H. By Cauchy's theorem, H contains an isomorphic copy of $\mathbb{Z}/p\mathbb{Z}$ for some p. By the preceding paragraph, $\mathbb{Z}/p\mathbb{Z}$ is projective. It follows from Corollary 22.4.3 that the natural map $\mathbb{Z}/p^2\mathbb{Z} \to \mathbb{Z}/p\mathbb{Z}$ has a group theoretic section. Hence, $\mathbb{Z}/p^2\mathbb{Z} \cong \mathbb{Z}/p\mathbb{Z} \times \mathbb{Z}/p\mathbb{Z}$. This is a contradiction. □

PROPOSITION 22.4.8: *Let G be a projective group and C a full formation of finite groups. Then the maximal pro-C quotient of G is projective. In particular, for each p the maximal pro-p quotient of G is projective.*

Proof: Let $\pi\colon G \to \bar{G}$ be the quotient map of G onto its maximal pro-C quotient (Definition 17.3.2). Consider a C-embedding problem ($\varphi\colon \bar{G} \to A$, $\alpha\colon B \to A$) for $\bar{G}$. Since G is projective, there is a homomorphism $\gamma\colon G \to B$ such that $\alpha \circ \gamma = \varphi \circ \pi$. Since $B \in C$ and C is full, $\gamma(G) \in C$. Hence, γ factors through a homomorphism $\bar{\gamma}\colon \bar{G} \to B$; that is $\bar{\gamma} \circ \pi = \gamma$. It follows that $\alpha \circ \bar{\gamma} = \varphi$. By Lemma 22.4.1, $\bar{G}$ is projective. □

LEMMA 22.4.9: *For profinite groups G and H there exists a unique profinite group $G * H$ with the following properties:*
(a) *G and H are closed subgroups of $G * H$, $G \cap H = 1$, $\langle G, H \rangle = G * H$.*
(b) *Each pair $\alpha: G \to C$ and $\beta: H \to C$ of homomorphisms of profinite groups uniquely extends to a homomorphism $\gamma: G * H \to C$.*

We call $G * H$ the **free product** of G and H.

Proof: Let F be the free product of G and H in the category of abstract groups. It is the unique abstract group that contains G and H such that $G \cap H = 1$, $\langle G, H \rangle = F$, and every pair $\alpha: G \to C$ and $\beta: H \to C$ of homomorphisms of abstract groups uniquely extends to a homomorphism $\gamma: F \to C$ [Kurosh, §33].

Denote the collection of normal subgroups N of F of finite index such that $N \cap G$ and $N \cap H$ are, respectively, open in G and H, by $\mathcal{N}$. Then $\mathcal{N}$ is directed (Section 17.2). Apply Lemma 17.2.1 to the profinite completion $\hat{F}$ of F with respect to $\mathcal{N}$. For an open normal subgroup N_0 of G the maps $G \to G/N_0$ and $H \to 1$ extend to a homomorphism $F \to G/N_0$ whose kernel N satisfies $N \cap G = N_0$ and $N \cap H = H$, so $N \in \mathcal{N}$. It follows that the canonical map $g \mapsto (gN)_{N \in \mathcal{N}}$ embeds G into $\hat{F}$. Similarly H naturally embeds into $\hat{F}$. Since G and H generate F, they also generate $\hat{F}$.

Let $\alpha: G \to C$ and $\beta: H \to C$ be homomorphisms of profinite groups. Extend them to a homomorphism $\gamma: F \to C$ of abstract groups. Then $\gamma^{-1}(C_0) \in \mathcal{N}$ for each open normal subgroup C_0 of C. By Lemma 17.2.2, γ uniquely defines a homomorphism $\hat{\gamma}: \hat{F} \to C$ of profinite groups which extends both α and β.

In particular, the maps $\mathrm{id}_G: G \to G$ and $H \to 1$ extend to a homomorphism $\hat{F} \to G$. Hence, $G \cap H = 1$. □

PROPOSITION 22.4.10: *The free product $G * H$ of projective groups is a projective group.*

Proof: Let $(\varphi: G * H \to A, \alpha: B \to A)$ be an embedding problem for $G * H$. Then there are homomorphisms $\gamma_G: G \to B$ and $\gamma_H: H \to B$ with $\alpha \circ \gamma_G = \varphi|_G$ and $\alpha \circ \gamma_H = \varphi|_H$. The extension $\gamma: G * H \to B$ of γ_G and γ_H is a weak solution of the embedding problem. □

22.5 Frattini Covers

Frattini covers allow us to form profinite groups that inherit properties from one of their finite quotients:

Definition 22.5.1: Frattini covers. A homomorphism $\varphi: H \to G$ of profinite groups is called a **Frattini cover** if it satisfies one, hence all, of the following equivalent conditions (Lemma 22.1.1):
(1a) φ is surjective and $\mathrm{Ker}(\varphi) \leq \Phi(H)$.
(1b) A closed subgroup H_0 of H is equal to H if and only if $\varphi(H_0) = G$.
(1c) A subset S of H generates H if and only if $\varphi(S)$ generates G.

22.5 Frattini Covers

In particular, φ maps the set of all open maximal open subgroups of H onto the set of all maximal open subgroups of G, so $\varphi(\Phi(H)) = \Phi(G)$. □

LEMMA 22.5.2: *Let G be a profinite group and X a subset of G which contains 1 and converges to 1. Then the following holds:*
(a) *X is closed.*
(b) *Every closed subset X_0 of X which does not contain 1 is finite.*
(c) *Let $\varphi\colon H \to G$ be an epimorphism of profinite groups. Then H has a closed subset X' which contains 1 and converges to 1 such that φ maps X' homeomorphically onto X.*

Proof of (a): Let $g \in G \smallsetminus X$. Then $g \neq 1$. Hence, G has an open normal subgroup N_0 with $g \notin N_0$, so $xN_0 \neq gN_0$ for each $x \in X \cap N_0$. In addition, $X \smallsetminus N_0$ is finite and $g \notin X \smallsetminus N_0$. Therefore, there exists an open normal subgroup N_1 of G with $N_1 \leq N_0$ and $xN_1 \neq gN_1$ for each $x \in X \smallsetminus N_1$. By the above, $xN_1 \neq gN_1$ for all $x \in X \cap N_1$. Thus, the set gN_1 is an open neighborhood of g which is disjoint from X. Consequently, X is closed.

Proof of (b): By (a), $G \smallsetminus X_0$ is an open neighborhood of 1. Hence, it contains an open normal subgroup N. By assumption, $X \smallsetminus N$ is finite. Therefore, X_0 is finite.

Proof of (c): By Lemma 1.2.7, there exists a continuous map $\varphi'\colon G \to H$ such that $\varphi \circ \varphi' = \mathrm{id}_G$. Put $G' = \varphi'(G)$ and $X' = \varphi'(X)$. Then G' and X' are closed subsets of H and φ maps G' (resp. X') homeomorphically onto G (resp. X).

Now consider an open normal subgroup N of H. Then $X' \smallsetminus N$ is a closed subset of X' which does not contains 1. Hence, $\varphi(X' \smallsetminus N)$ is a closed subset of X which does not contain 1. By (b), $\varphi(X' \smallsetminus N)$ is finite. Hence, $X' \smallsetminus N$ is finite. Consequently, X' converges to 1.

Alternatively, we could use Exercise 1 of Chapter 17. □

COROLLARY 22.5.3: *If $\varphi\colon H \to G$ is a Frattini cover, then $\mathrm{rank}(H) = \mathrm{rank}(G)$.*

Proof: Choose a system of generators X for G in the following way: If $\mathrm{rank}(G) < \infty$, then $|X| = \mathrm{rank}(G)$. If $\mathrm{rank}(G) = \infty$, then X converges to 1 (Proposition 17.1.1) and $1 \in X$. In the former case choose a subset X' of H which φ maps bijectively onto X. In the latter case Lemma 22.5.2 gives a subset X' of H which converges to 1 and which φ maps homeomorphically onto X. By Definition 22.5.1(1c), X' generates H. Therefore, in both cases, $\mathrm{rank}(H) = |X'| = |X| = \mathrm{rank}(G)$. □

The following rules follow directly from Definition 22.5.1:

LEMMA 22.5.4: *Let $H \xrightarrow{\psi} G \xrightarrow{\varphi} A$ be homomorphisms of profinite groups.*
(a) *If φ and ψ are Frattini covers, then $\varphi \circ \psi$ is a Frattini cover.*
(b) *If φ is a Frattini cover and $\varphi \circ \psi$ is surjective, then ψ is surjective.*

(c) If $\varphi \circ \psi$ is a Frattini cover and ψ is surjective, then both φ and ψ are Frattini covers.

(d) If both φ and $\varphi \circ \psi$ are Frattini covers, then so is ψ.

LEMMA 22.5.5: *Consider a cartesian square of epimorphisms (Definition 22.2.2):*

$$\begin{array}{ccc} D & \xrightarrow{\delta} & C \\ \beta \downarrow & & \downarrow \gamma \\ B & \xrightarrow{\alpha} & A \end{array}$$

If δ is a Frattini cover, then so is α.

Proof: By Lemma 22.2.5, $\beta(\text{Ker}(\delta)) = \text{Ker}(\alpha)$. Since δ is a Frattini cover, $\text{Ker}(\delta) \leq \Phi(D)$. By Lemma 22.1.4(a), $\beta(\Phi(D)) \leq \Phi(B)$. Hence, $\text{Ker}(\alpha) \leq \Phi(B)$ and α is a Frattini cover. □

For each profinite group epimorphism, restriction of the domain gives a Frattini cover:

LEMMA 22.5.6: *Let $\varphi: H \to G$ be an epimorphism of profinite groups. Then H has a closed subgroup H' such that $\varphi|_{H'}: H' \to G$ is a Frattini cover.*

Proof: Let $\{H_i \mid i \in I\}$ be a decreasing chain of closed subgroups of H with $\varphi(H_i) = G$ for each $i \in I$. By Lemma 1.2.2(c), $\varphi(\bigcap_{i \in I} H_i) = G$. Applying Zorn's Lemma, we conclude that H has a minimal closed subgroup H_0 with $\varphi(H_0) = G$. The restriction of φ to H_0 is a Frattini cover of G. □

We combine the fiber product construction with Lemma 22.5.6:

LEMMA 22.5.7: *Let $\alpha: B \to A$ and $\gamma: C \to A$ be Frattini covers. Then there is a commutative diagram*

$$\begin{array}{ccc} D & \xrightarrow{\delta} & C \\ \beta \downarrow & \searrow^{\varphi} & \downarrow \gamma \\ B & \xrightarrow{\alpha} & A \end{array}$$

in which β, δ, and φ are Frattini covers.

Proof: Let $D_1 = B \times_A C$ (Section 22.2), $\alpha_1 = \text{pr}_B$, $\delta_1 = \text{pr}_C$, and $\varphi_1 = \gamma \circ \delta_1 = \alpha \circ \beta_1$. This gives a cartesian square, but the maps in it may not be Frattini covers. Apply Lemma 22.5.6 to find a closed subgroup D of D_1 such that $\varphi = \varphi_1|_D$ is a Frattini cover. Lemma 22.5.4(d) implies that $\beta = \beta_1|_D$ and $\delta = \delta_1|_D$ are Frattini covers. □

Let

(2) $\qquad\qquad (\varphi: G \to A,\ \alpha: B \to A)$

22.5 Frattini Covers

be an embedding problem for a profinite group. Call (2) a **Frattini embedding problem** if α is a Frattini cover. The following proposition shows that in order to solve an arbitrary embedding problem for G, it suffices to solve Frattini embedding problem followed by a split embedding problem:

PROPOSITION 22.5.8: *Let (2) be an embedding problem for a profinite group G. Suppose the following two conditions are satisfied:*
(a) *Every Frattini embedding problem ($\varphi\colon G \to A$, $\alpha_0\colon B_0 \to A$) where B_0 is a subgroup of B is solvable.*
(b) *Every split embedding problem ($\gamma_0\colon G \to B_0$, $\alpha'\colon B' \to B_0$) for G with $\mathrm{Ker}(\alpha') \cong \mathrm{Ker}(\alpha)$ is solvable.*

Then (2) is solvable.

Proof: Choose a closed subgroup B_0 of B such that $\alpha_0 = \alpha|_{B_0}\colon B_0 \to A$ is a Frattini cover (Lemma 22.5.6). Thus,

(3) $\qquad\qquad\qquad (\varphi\colon G \to A,\ \alpha_0\colon B_0 \to A)$

is a Frattini embedding problem. By (a), there is an epimorphism $\gamma_0\colon G \to B_0$ with $\alpha_0 \circ \gamma_0 = \varphi$.

Put $C = \mathrm{Ker}(\alpha)$ and let $B' = B_0 \ltimes C$ with B_0 acting on C by conjugation. Then $\pi\colon B' \to B$ given by $\pi(b_0, c) = b_0 c$ with $b_0 \in B_0$ and $c \in C$ is an epimorphism [Nobusawa]. Let $\alpha'\colon B' \to B_0$ be the projection on B_0. Then $\alpha \circ \pi = \alpha_0 \circ \alpha'$ and

(4) $\qquad\qquad\qquad (\gamma_0\colon G \to B_0,\ \alpha'\colon B' \to B_0)$

is a split embedding problem for G with $\mathrm{Ker}(\alpha') = C$. Condition (b) gives an epimorphism $\gamma'\colon G \to B'$ with $\alpha' \circ \gamma' = \gamma_0$:

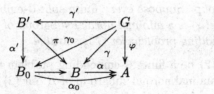

Then $\gamma = \pi \circ \gamma'$ solves embedding problem (2). $\qquad\square$

Proposition 22.5.8 is useful when G is projective:

PROPOSITION 22.5.9: *Let G be a projective group. Then:*
(a) *Every embedding problem (2) in which α is a Frattini cover is solvable.*
(b) *If every (finite) split embedding problem for G is solvable, then every (finite) embedding problem for G is solvable.*
(c) *If $\psi\colon \tilde G \to G$ is a Frattini cover, then ψ is an isomorphism.*

Proof of (a): Since G is projective, there exists a homomorphism $\gamma\colon G \to B$ with $\alpha \circ \gamma = \varphi$. Then $\alpha(\gamma(G)) = \varphi(G) = A$. By (1b), $\gamma(G) = B$. Thus, γ is a solution of (2).

Proof of (b): Use (a) and Proposition 22.5.8.

Proof of (c): Take $B = A = G$, $\varphi = \mathrm{id}_G$, and $\alpha = \psi$ to see that there exists an epimorphism $\gamma\colon G \to G$ with $\psi \circ \gamma = \mathrm{id}_G$. Thus, ψ is an isomorphism. □

Many properties of projective groups are determined by their quotients modulo their Frattini groups:

COROLLARY 22.5.10: *Let G and H be profinite groups with H projective.*
(a) *Each epimorphism $\theta_0\colon H/\Phi(H) \to G/\Phi(G)$ has a lift to an epimorphism $\theta\colon H \to G$.*
(b) *If in addition G is projective, then each isomorphism $\theta_0\colon H/\Phi(H) \to G/\Phi(G)$ has a lift to an isomorphism $\theta\colon H \to G$.*

Proof of (a): Let $\pi_H\colon H \to H/\Phi(H)$ and $\pi_G\colon G \to G/\Phi(G)$ be the quotient maps. Then apply Proposition 22.5.9 to find an epimorphism θ with $\pi_G \circ \theta = \theta_0 \circ \pi_H$.

Proof of (b): Suppose θ_0 is an isomorphism. Note that both π_G and π_H are Frattini covers. Hence, by Lemma 22.5.4, θ is a Frattini cover. Since G is projective, a symmetrical argument gives a Frattini cover $\theta'\colon G \to H$. Since $\theta' \circ \theta\colon H \to H$ is also a Frattini cover, Proposition 22.5.9(c) implies that $\theta' \circ \theta$ is an isomorphism. Consequently, θ is an isomorphism. □

Let $\mathcal{C}$ be an arbitrary formation of finite groups. The proof of Proposition 22.5.9(b) does not extend to $\mathcal{C}$-projective groups, because the proof of Proposition 22.5.8 involves a subgroups B_0 of a $\mathcal{C}$-group B which need not be a $\mathcal{C}$-group. Nevertheless, the result itself is true:

PROPOSITION 22.5.11: *Let $\mathcal{C}$ be a formation of finite groups and G a $\mathcal{C}$-projective group. Suppose every finite split $\mathcal{C}$-embedding problem (2) for G such that $\mathrm{Ker}(\alpha)$ is a minimal normal subgroup of B is solvable. Then every finite $\mathcal{C}$-embedding problem for G is solvable.*

Proof: Let (2) be a finite $\mathcal{C}$-embedding problem for G. First suppose $C = \mathrm{Ker}(\alpha)$ is a minimal normal subgroup of B but (2) does not necessarily split. Since G is $\mathcal{C}$-projective, there is a homomorphism $\gamma\colon G \to B$ with $\alpha \circ \gamma = \varphi$. Put $\bar{G} = G/\mathrm{Ker}(\gamma)$. Let $\pi\colon G \to \bar{G}$ be the quotient map and $\bar{\gamma}\colon \bar{G} \to B$ and $\bar{\varphi}\colon \bar{G} \to A$ the homomorphisms induced by γ and φ, respectively. Then $\alpha \circ \bar{\gamma} = \bar{\varphi}$. This leads to a commutative diagram

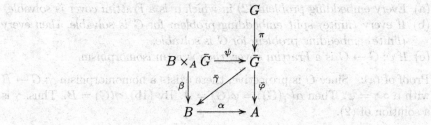

in which ψ and β are the projections on the coordinates. By Lemma 22.2.5, β maps $\mathrm{Ker}(\psi)$ isomorphically onto C. Hence, $\mathrm{Ker}(\psi)$ is a minimal normal subgroup of $B \times_A \bar{G}$. Lemma 22.2.9 gives a group theoretic section to ψ. Therefore, by assumption, there is an epimorphism $\zeta: G \to B \times_A \bar{G}$. The epimorphism $\beta \circ \zeta: G \to B$ solves (2).

Finally suppose C is not a minimal normal subgroup of B. Then C has a proper nontrivial subgroup C_0 which is normal in B. Put $\bar{B} = B/C_0$, $\bar{\beta}: B \to \bar{B}$ the quotient map, and $\bar{\alpha}: \bar{B} \to A$ the epimorphism induced by α. This gives a $\mathcal{C}$-embedding problem $(\varphi: G \to A, \bar{\alpha}: \bar{B} \to A)$ whose kernel C/C_0 has a smaller order than $|C|$. An induction hypothesis gives an epimorphism $\bar{\varphi}: G \to \bar{B}$ with $\bar{\alpha} \circ \bar{\varphi} = \varphi$. This gives an embedding problem $(\bar{\varphi}: G \to \bar{B}, \bar{\beta}: B \to \bar{B})$ whose kernel C_0 also has a smaller order than $|C|$. Again, an induction hypothesis gives an epimorphism $\gamma: G \to B$ with $\bar{\beta} \circ \gamma = \bar{\varphi}$. The epimorphism φ solves (2). □

22.6 The Universal Frattini Cover

The study of projective groups and the study of Frattini covers have a common subject: "universal Frattini covers":

Starting from a profinite group G, we partially order the epimorphisms of profinite groups onto G (also called **covers** of G). Let $\theta_i: H_i \to G, i = 1, 2$, be covers. We write $\theta_2 \geq \theta_1$ and say that θ_2 is **larger** than θ_1 if there is an epimorphism $\theta: H_2 \to H_1$ with $\theta_1 \circ \theta = \theta_2$. If θ is an isomorphism, then θ_1 is said to be **isomorphic** to θ_2.

An epimorphism $\varphi: \tilde{G} \to G$ is called a **projective cover** if $\tilde{G}$ is a projective group. In this case, $\tilde{G}$ is called a **projective cover** of G.

PROPOSITION 22.6.1: *Each profinite group G has a cover $\tilde{\varphi}: \tilde{G} \to G$, unique up to an isomorphism, called the **universal Frattini cover** and satisfying the following equivalent conditions:*
(a) *$\tilde{\varphi}$ is a projective Frattini cover of G.*
(b) *$\tilde{\varphi}$ is the largest Frattini cover of G.*
(c) *$\tilde{\varphi}$ is the smallest projective cover of G.*

In particular, each projective group is its own universal Frattini cover.

Proof: By Corollary 17.4.8, there exists an epimorphism $\varphi: F \to G$ with F a free profinite group. Every closed subgroup of F is projective (Corollary 22.4.6). Apply Lemma 22.5.6 to produce a closed subgroup $\tilde{G}$ of F such that $\tilde{\varphi} = \mathrm{res}_{\tilde{G}}(\varphi)$ is a Frattini (and projective) cover of G. The cover $\tilde{\varphi}: \tilde{G} \to G$ appears throughout this proof.

Proof of (a) $\Longrightarrow$ (b): Let $\theta: G_1 \to G$ be a Frattini cover. Proposition 22.5.9 gives an epimorphism $\gamma: \tilde{G} \to G_1$ with $\theta \circ \gamma = \tilde{\varphi}$. Thus, $\theta \leq \tilde{\varphi}$.

Proof of (b) $\Longrightarrow$ (a): Let $\varphi': G' \to G$ be a Frattini cover that is larger than any Frattini cover. Thus, there exists an epimorphism $\tilde{\theta}: G' \to \tilde{G}$ such that $\tilde{\varphi} \circ \tilde{\theta} = \varphi'$. By Lemma 22.5.4(c), $\tilde{\theta}$ is a Frattini cover. By "(a) $\Longrightarrow$ (b)",

$\tilde{\varphi}$ is also a maximal Frattini cover of G. Thus, there exists a Frattini cover $\theta'\colon \tilde{G} \to G'$ with $\varphi' \circ \theta' = \tilde{\varphi}$. It follows that $\tilde{\theta} \circ \theta'\colon \tilde{G} \to \tilde{G}$ is a projective Frattini cover. By Proposition 22.5.9(c), $\tilde{\theta} \circ \theta'$ is an isomorphism. Hence, $\theta'\colon \tilde{G} \to G'$ is an isomorphism. Consequently, G' is projective and (a) holds.

Proof of (a) $\Longrightarrow$ (c): Let $\varphi\colon P \to G$ be a projective cover. Proposition 22.5.9 gives an epimorphism $\gamma\colon P \to \tilde{G}$ with $\tilde{\varphi} \circ \gamma = \varphi$. Thus, $\tilde{\varphi} \le \varphi$.

Proof of (c) $\Longrightarrow$ (a): Let $\varphi'\colon G' \to G$ be a projective cover which is smaller than any other projective cover of G. Since $\tilde{G}$ is projective, there is an epimorphism $\tilde{\theta}\colon \tilde{G} \to G'$ with $\varphi' \circ \tilde{\theta} = \tilde{\varphi}$. By Lemma 22.5.4(c), φ' is a Frattini cover. □

LEMMA 22.6.2: *Let $\mathcal{C}$ be a full formation of finite groups and G a pro-$\mathcal{C}$-group. Then the smallest projective cover $\tilde{G}$ of G is a pro-$\mathcal{C}$-group and* $\mathrm{rank}(\tilde{G}) = \mathrm{rank}(G)$.

Proof: Let $m = \mathrm{rank}(G)$. By Corollary 22.4.5, $\hat{F}_m(\mathcal{C})$ is a projective cover of G. Hence, $\tilde{G}$ is a quotient of $\hat{F}_m(\mathcal{C})$. The equality of the ranks is a special case of Corollary 22.5.3. □

The next lemma characterizes the quotients of the universal Frattini cover of a profinite group:

LEMMA 22.6.3: *Let $\tilde{\varphi}\colon \tilde{G} \to G$ be the universal Frattini cover of a profinite group G. Then a profinite group H is a quotient of $\tilde{G}$ if and only if H is a Frattini cover of a quotient of G.*

Proof: Suppose $\psi\colon \tilde{G} \to H$ is an epimorphism. Lemma 22.2.8 gives a commutative diagram of epimorphisms

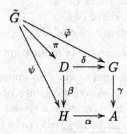

where the square is cartesian. By Lemma 22.5.4(c), δ is a Frattini cover. Thus (Lemma 22.5.5), so is α.

Conversely, suppose $\alpha\colon H \to A$ is a Frattini cover and $\gamma\colon G \to A$ is an epimorphism. Apply the projectivity of $\tilde{G}$ and Proposition 22.5.9 to find an epimorphism $\psi\colon \tilde{G} \to H$. □

22.7 Projective Pro-p-Groups

Fix a prime number p for the whole section. By Example 17.3.3, the family of finite p-groups is full. Thus (Corollary 22.4.5), each free pro-p-group is projective. We prove here a converse to this statement:

LEMMA 22.7.1: *Let m be a cardinal number and $V = \mathbb{F}_p^m$ be the direct product of m copies of the additive group of $\mathbb{F}_p$. Then* rank$(V) = m$.

Proof: Suppose first m is finite. Consider V as a vector space over $\mathbb{F}_p$. Then rank$(V) = \dim(V) = m$.

Now suppose m is infinite. Choose a set I of cardinality m and for each $i \in I$ an isomorphic copy F_i of $\mathbb{F}_p$. Each open subgroup H of $\prod_{i \in I} F_i$ contains $\prod_{i \in J} F_i$ for some cofinite subset J of I. The cardinality of the set of cofinite subsets of I is m. For each such J only finitely many open subgroups of $\prod_{i \in I} F_i$ contain $\prod_{i \in J} F_i$. Hence, the cardinality of all open subsets of $\prod_{i \in I} F_i$ is m, as claimed. □

LEMMA 22.7.2: *Let I be a set, $G = (\mathbb{F}_p)^I$, and N a closed subgroup of G. Then $G = N \times N'$ for some closed subgroup N' of G.*

Proof: Let $\varphi \colon G \to G/N$ be the quotient map. Lemma 22.5.6 gives a closed subgroup N' of G such that $\varphi' = \varphi|_{N'} \colon N' \to G/N$ is a Frattini cover. Thus, Ker$(\varphi') \leq \Phi(N') \leq \Phi(G) = \Phi(\mathbb{F}_p)^I = 1$ (Lemma 22.1.4(c,d)). Consequently, $G = N \times N'$. □

LEMMA 22.7.3: *Let m be a cardinal number, $G = \mathbb{F}_p^m$, and H a closed subgroup of G. Then $H \cong \mathbb{F}_p^k$ for some cardinal number $k \leq m$.*

Proof: Choose a set J of cardinality $|G|$. Let $\mathcal{U}$ be the set of all triples (U, I, φ) where $U \leq H$, $I \subseteq J$, and $\varphi \colon H/U \to \mathbb{F}_p^I$ is an isomorphism. Define a partial ordering on $\mathcal{U}$ by the following rule: $(U', I', \varphi') \leq (U, I, \varphi)$ if $U' \leq U$, $I \subseteq I'$ and the following diagram is commutative:

(1)
$$\begin{array}{ccc} H/U' & \xrightarrow{\pi_{U',U}} & H/U \\ \varphi' \downarrow & & \downarrow \varphi \\ \mathbb{F}_p^{I'} & \xrightarrow{\rho_{I',I}} & \mathbb{F}_p^I \end{array}$$

Here $\pi_{U',U}$ is the quotient map and $\rho_{I',I}$ is the projection.

The triple $(H, \emptyset, \mathrm{id})$ belongs to $\mathcal{U}$. Suppose $\{(U_\alpha, I_\alpha, \varphi_\alpha) \mid \alpha \in A\}$ is a descending chain in $\mathcal{U}$. Let $U_0 = \bigcap_{\alpha \in A} U_\alpha$, $I_0 = \bigcup_{\alpha \in A} I_\alpha$, and $\varphi_0 = \varprojlim \varphi_\alpha$ (we assume here $0 \notin A$). Then (U_0, I_0, φ_0) is an element of $\mathcal{U}$ which is smaller or equal to each $(U_\alpha, I_\alpha, \varphi_\alpha)$.

Zorn's lemma gives a minimal element (U, I, φ) in $\mathcal{U}$. Assume U is not trivial. Lemma 22.7.2 gives a closed subgroup V of G with $V \times U = G$. Then V is a proper subgroup of G. Hence, V is contained in an open subgroup

G' of G of index p. Put $H' = G' \cap H$ and $U' = G' \cap U$. Then $H/H' \cong G/G' \cong \mathbb{F}_p$ and $H/U' \cong H/U \times H/H'$. Since $|G| \geq |H/U| = 2^{|I|} > |I|$, there is a $j \in J \smallsetminus I$. Put $I' = I \cup \{j\}$. Then it is possible to lift φ to an isomorphism $\varphi' \colon H/U' \to \mathbb{F}_p^{I'}$ such that (1) is commutative. This contradicts the minimality of (U, I, φ) and proves that U is trivial.

It follows that $H \cong (\mathbb{F}_p)^I$. By Lemma 22.7.1 and Corollary 17.1.5, $|I| = \operatorname{rank}(H) \leq \operatorname{rank}(G) = m$, as claimed. $\square$

For a profinite group G let

$$G^p = \langle g^p \mid g \in G \rangle \quad \text{and} \quad [G, G] = \langle [g_1, g_2] \mid g_1, g_2 \in G \rangle.$$

Both closed subgroups of G are characteristic.

LEMMA 22.7.4: *Let G be a pro-p-group of rank m. Then $\Phi(G) = G^p[G, G]$ and $G/\Phi(G)$ is isomorphic to the vector space $\mathbb{F}_p^m$.*

Proof: Maximal subgroups of finite p-groups are normal subgroups of index p. Therefore, maximal open subgroups of G are open normal subgroups of index p. This gives a canonical embedding $G/\Phi(G) \to \prod G/N$, where N ranges over all open normal subgroups of G of index p. By Lemma 22.7.3, $G/\Phi(G) \cong \mathbb{F}_p^m$ with $m = \operatorname{rank}(G/\Phi(G)) = \operatorname{rank}(G)$ (Corollary 22.5.3).

Now let $G_0 = G^p[G, G]$. Since $G/\Phi(G) \cong \mathbb{F}_p^m$, the canonical map $G \to G/\Phi(G)$ maps each $g^p[a, b]$ to 1. Hence, $G_0 \leq \Phi(G)$. On the other hand let U be an open normal subgroup of G that contains G_0. Then G/U is an Abelian elementary p-group. In particular, U is the intersection of all open normal subgroups of index p that contain U. Hence, $\Phi(G) \leq U$. Since $G_0 \triangleleft G$, the intersection of all such U is G_0. Thus, $\Phi(G) \leq G_0$. Consequently, $\Phi(G) = G_0$. $\square$

Remark 22.7.5: *Subgroups of finite index.* Let m be an infinite cardinal number. By Lemma 22.7.1, the cardinality of the set of open subgroups of $\mathbb{F}_p^m$ of index p is m. On the other hand, $|\mathbb{F}_p^m| = 2^m$. Hence, any basis B of $\mathbb{F}_p^m$, where $\mathbb{F}_p^m$ is considered as a vector space over $\mathbb{F}_p$, has cardinality 2^m. For each $v \in B$ the subgroup generated by $B \smallsetminus \{v\}$ has index p. All these subgroups are distinct. Therefore, $\mathbb{F}_p^m$ has subgroups of index p which are not open.

In contrast, Serre proved that every subgroup of finite index of a finitely generated pro-p group G is open [Serre11, p. 32, Exercises 5 and 6] (see also [Ribes-Zalesski, Thm. 4.2.8]) and wrote he did not know if that statement holds for arbitrary finitely generated profinite groups.

Recently [Nikolov-Segal] gave an affirmative answer to Serre's question:
(2) Every subgroup N of a finite index of a finitely generated profinite group G is open.

The proof of (2) depends on properties of certain "verbal subgroups": Consider a group theoretic word $w(X_1, \ldots, X_m)$ in the variables $X_1, \ldots, X_m$. For each group G let $w(G)$ be the subgroup generated (in the sense of abstract

22.7 Projective Pro-p-Groups

groups) by $w(\mathbf{x})$ with $\mathbf{x} \in G^m$. Given a positive integer d, we say w is d-**locally finite** if every group H which is generated (in the abstract sense) by d elements and satisfies $w(H) = 1$ is finite. The existence of d-locally finite words is proved in the introduction of [Nikolov-Segal]:

(3) For each finite group A which is generated by d elements there exists a d-locally finite word w with $w(A) = 1$.

To prove (3) consider the free group F on $x_1, \ldots, x_d$. Let $N = \bigcap \mathrm{Ker}(\theta)$, where θ ranges over all homomorphisms from F to A. Then N is a normal subgroup of a finite index. By Corollary 17.5.8, N is finitely generated. Let $y_1, \ldots, y_k$ be generators of N. For each i write $y_i = w_i(\mathbf{x})$, where w_i is a word in $x_1, \ldots, x_d$. Then consider the word $w(\mathbf{X}) = w_1(\mathbf{X}_1) \cdots w_k(\mathbf{X}_k)$, where $\mathbf{X}_i = (X_{i1}, \ldots, X_{id})$ and $\mathbf{X} = (\mathbf{X}_1, \ldots, \mathbf{X}_k)$.

Each of the generators $y_i = w_i(\mathbf{x})$ of N belongs to $w(F)$, so $N \leq w(F)$. Conversely, let $x'_1, \ldots, x'_d \in F$. Then the map $x_i \mapsto x'_i$, $i = 1, \ldots, d$, extends to a homomorphism $\kappa \colon F \to F$. Let $\theta \colon F \to A$ be an arbitrary homomorphism. Then $\theta \circ \kappa$ is also a homomorphism from F to A. Hence, $\theta(w_i(\mathbf{x}')) = \theta(w_i(\kappa(\mathbf{x}))) = \theta \circ \kappa(w_i(\mathbf{x})) = \theta \circ \kappa(y_i) = 1$ for $i = 1, \ldots, k$. It follows that $w(F) \leq N$. Consequently, $w(F) = N$.

Now let H be a group with d generators and $w(H) = 1$. Let $\pi \colon F \to H$ be an epimorphism. Then $\pi(w(F)) = w(H) = 1$. Hence, $|H| \leq (F : w(F)) < \infty$. Consequently, w is d-locally finite, as desired.

The key result in the proof of (2) bounds the number of factors in the elements of $w(H)$:

(4) Let d be a positive integer and $w(X_1, \ldots, X_m)$ a d-locally finite word. Then there exists a positive integer r such that for each finite group A generated by d elements and for each $a \in w(A)$ there are $\mathbf{b}_1, \ldots, \mathbf{b}_r \in A^m$ and $\beta_1, \ldots, \beta_r \in \{\pm 1\}$ such that $a = w(\mathbf{b}_1)^{\beta_1} \cdots w(\mathbf{b}_r)^{\beta_r}$ [Nikolov-Segal, Thm. 2.1].

The proof of (4) uses the classification of finite simple groups. Compactness arguments generalize (4) to profinite groups:

(5) Let d be a positive integer and $w(X_1, \ldots, X_m)$ a d-locally finite word. Then there exists a positive integer r such that for each profinite group G generated (in the profinite sense) by d elements and for each $g \in w(G)$ there are $\mathbf{b}_1, \ldots, \mathbf{b}_r \in G^m$ and $\beta_1, \ldots, \beta_r \in \{\pm 1\}$ such that $g = w(\mathbf{b}_1)^{\beta_1} \cdots w(\mathbf{b}_r)^{\beta_r}$. In particular, $w(G)$ is closed.

Now consider a profinite group G generated (in the profinite sense) by elements $x_1, \ldots, x_d$ and let N be a subgroup of finite index. Without loss assume that N is normal in G. Then G/N is finite. By (3), there exists a d-locally finite word $w(X_1, \ldots, X_m)$ such that $w(G/N) = 1$. It follows that $w(G) \leq N$.

Let G_0 be the abstract subgroup of G generated by $x_1, \ldots, x_d$. Then $w(G_0) \triangleleft G_0$, $w(G_0/w(G_0)) = 1$, and $G_0/w(G_0)$ is generated (in the sense of abstract groups) by d elements. Thus, $G_0/w(G_0)$ is finite and $(G_0 : w(G_0)) < \infty$. Therefore, $(G_0 w(G) : w(G)) = (G_0 : G_0 \cap w(G)) \leq (G_0 : w(G_0)) < \infty$.

By (5), $w(G)$ is closed, so $G_0 w(G)$ is closed. Since G_0 is dense in G, we have $G_0 w(G) = G$. It follows that $w(G)$ has a finite index in G. Consequently, $w(G)$ is open, so also N is open, as claimed. □

PROPOSITION 22.7.6 (Tate): *A pro-p-group G is projective if and only if it is pro-p free.*

Proof: By Corollary 22.4.5, it suffices to show that G projective implies G is pro-p free. Let $m = \operatorname{rank}(G)$. By Lemma 22.7.4,

$$G/\Phi(G) \cong \mathbb{F}_p^m \cong \hat{F}_m(p)/\Phi(\hat{F}_m(p)).$$

Therefore, by Corollary 22.5.10(b), $G \cong \hat{F}_m(p)$. □

Corollary 22.4.6 immediately gives an analog of Schreier's theorem (Proposition 17.5.6) for discrete free groups:

COROLLARY 22.7.7 (Tate): *A closed subgroup of a free pro-p-group is pro-p free.*

Proof: Let G be a closed subgroup of a free pro-p group F. By Proposition 22.7.6, F is projective. By Proposition 22.4.7, G is projective. Therefore, by Proposition 22.7.6, G is free pro-p. □

COROLLARY 22.7.8: *Let G be a pro-p group of rank m. Then the universal Frattini cover of G is $\hat{F}_m(p)$.*

Proof: Let $\tilde{G}$ be the universal Frattini cover of G. By Lemma 22.6.2, $\tilde{G}$ is a pro-p group of rank m. In addition, $\tilde{G}$ is projective. Hence, by Proposition 22.7.6, $G \cong \hat{F}_m(p)$. □

Here is a generalization of Lemma 22.1.4(c) for pro-p groups:

LEMMA 22.7.9: *Let G be a pro-p group and H a closed subgroup. Then $\Phi(H) \leq \Phi(G)$.*

Proof: By Lemma 22.7.4, $\Phi(H) = H^p[H, H] \leq G^p[G, G] = \Phi(G)$. □

PROPOSITION 22.7.10: *Every pro-p group G which is small is finitely generated.*

Proof: By Lemma 22.7.4, $\Phi(G)$ is the intersection of all open subgroups of G of index p. By assumption, there are only finitely many of them. Hence, $G/\Phi(G)$ is a finite group. In particular, $G/\Phi(G)$ is finitely generated. It follows from Lemma 22.1.1 that G is also finitely generated. □

COROLLARY 22.7.11: *Let A be an Abelian projective pro-p group. Then either A is trivial or $A \cong \mathbb{Z}_p$.*

Proof: By Proposition 22.7.6, A is pro-p free. Since A is Abelian, $\operatorname{rank}(A) \leq 1$. Otherwise, A would have a non-Abelian finite quotients of rank 2. So, either $A = 1$ or $A \cong \mathbb{Z}_p$. □

The following result surveys the structure of an arbitrary Abelian pro-p group:

22.7 Projective Pro-p-Groups

PROPOSITION 22.7.12: *Let A be an additive Abelian pro-p group. Then:*
(a) *A is a $\mathbb{Z}_p$-module.*
(b) *Suppose A is finitely generated. Then $A \cong \mathbb{Z}/p^{k_1}\mathbb{Z} \oplus \cdots \oplus \mathbb{Z}/p^{k_m}\mathbb{Z} \oplus \mathbb{Z}_p^r$, with $k_1 \geq \cdots \geq k_m \geq 1$ and r is a nonnegative integer. The sequence $k_1, \ldots, k_m, r$ are uniquely determined by A.*
(c) *Suppose A is finitely generated and torsion free. Let B be a subgroup. Then A and B are free $\mathbb{Z}_p$ modules, A has a $\mathbb{Z}_p$-basis $a_1, \ldots, a_n$ and there are positive integer $\alpha_1, \ldots, \alpha_m$ with $0 \leq m \leq n$ satisfying this: $\alpha_1 a_1, \ldots, \alpha_m a_m$ form a $\mathbb{Z}_p$-basis of B and $\alpha_i | \alpha_{i+1}$, $i = 1, \ldots, m-1$.*
(d) *In general, $A \cong \mathbb{Z}_p^m \times A'$ where m is a cardinal number and A' is the intersection of all kernels of the epimorphisms of A onto $\mathbb{Z}_p$.*
(e) *If A is torsionfree, then $A \cong \mathbb{Z}_p^m$ for some cardinal number m.*

Proof of (a): Let B be an open subgroup of A. Then, $(A : B) = p^n$ with n a nonnegative integer. Hence, $p^n A \leq B$. Thus, $\bigcap_{n=1}^{\infty} p^n A = 0$.

The rule $(z + p^n \mathbb{Z})(a + p^n A) = za + p^n A$ defines a continuous action of $\mathbb{Z}/p^n \mathbb{Z}$ on $A/p^n A$. Going to the limit, the first paragraph gives an action of $\mathbb{Z}_p$ on A.

Proof of (b): Statement (b) is a special case of the main theorem on finitely generated modules over principal ideal domains [Lang7, §III, Thms. 7.3 and 7.5].

Proof of (c): See [Lang7, §III, Thm. 7.8].

Proof of (d): See [Geyer-Jensen1, p. 337, (4)].

Proof of (e): See [Ribes-Zalesskii, Thm. 4.3.3]. □

Example 22.7.13: *The rank of local Galois groups.* Let K be a local field, that is, K is a finite extension of $\mathbb{Q}_p$ or of $\mathbb{F}_p((t))$. Denote the maximal tamely ramified extension of K by K_{tr}. Let $G = \text{Gal}(K)$ and $P = \text{Gal}(K_{\text{tr}})$. Then P is a closed normal pro-p group of G. By Lemma 22.1.4(c), $\Phi(P) \leq \Phi(G)$. By Lemma 22.7.4, $\Phi(P) = P^p[P, P]$. Hence, $\text{rank}(G) = \text{rank}(G/[P, P])$ (Corollary 22.5.3).

This observation, is implicitly used in [Jannsen, Section 3] to determine the number of generators of G. Indeed, if K is a finite extension of $\mathbb{Q}_p$, then by results of Iwasawa and local class field theory, $\text{rank}(G/[P, P]) = [K : \mathbb{Q}_p] + 3$. Hence, $\text{rank}(G) = [K : \mathbb{Q}_p] + 3$. In particular, $\text{rank}(\text{Gal}(\mathbb{Q}_p)) = 4$. If $\text{char}(K) = p$, then the rank of $G/[P, P]$ and therefore also of $\text{Gal}(K)$ is infinite.

Denote the fixed field of $\Phi(G)$ in $\tilde{K}$ by F. It is proved in [Jarden-Ritter] that F is the compositum of all Galois extensions N of K containing K_{tr} such that $\text{Gal}(N/K_{\text{tr}}) \cong \mathbb{Z}/p\mathbb{Z}$. □

22.8 Supernatural Numbers

The next definition generalizes the index and the order of finite groups to profinite groups:

Definition 22.8.1: Supernatural numbers. A **supernatural number** is a formal product $n = \prod p^{n_p}$, where n_p is either a nonnegative integer or ∞, and p ranges over all primes.

Let m and n be supernatural numbers. We say that m **divides** n if $m_p \leq n_p$ for every prime p. Define the **product** of supernatural numbers n_i, for $i \in I$, by the formula

$$\prod_{i \in I} \prod_p p^{n_{i,p}} = \prod_p p^{\sum_{i \in I} n_{i,p}}.$$

Here $\sum_{i \in I} n_{i,p} = \infty$ if $n_{i,p} > 0$ for infinitely many $i \in I$ or $n_{i,p} = \infty$ for at least one $i \in I$.

Similarly, there is a **greatest common divisor** and a **least common multiple** of supernatural numbers:

$$\gcd(m,n) = \prod_p p^{\min(m_p, n_p)} \quad \text{and} \quad \operatorname{lcm}(m,n) = \prod_p p^{\max(m_p, n_p)}.$$

Define the **index** of a closed subgroup H in a profinite group G to be

$$(G:H) = \operatorname{lcm}\bigl((G:U) \mid U \text{ is an open subgroup of } G \text{ that contains } H\bigr).$$

Then define the **order** of G to be

$$\#G = (G:1) = \operatorname{lcm}\bigl((G:U) \mid U \text{ is an open subgroup of } G\bigr).$$

Note that the index generalizes the usual index, $(G:H)$, if H is an open subgroup of G. □

Example 22.8.2: Order of three classical Abelian groups.

$$\#\mathbb{Z}_p = p^\infty; \quad \#\hat{\mathbb{Z}} = \prod_p p^\infty; \quad \text{and} \quad \#\prod_p \mathbb{Z}/p\mathbb{Z} = \prod_p p. \quad \square$$

The usual rules for indices of finite groups hold for profinite groups:

LEMMA 22.8.3:
(a) Let $K \leq H \leq G$ be profinite groups. Then $(G:K) = (G:H)(H:K)$.
(b) Let $\{H_i \mid i \in I\}$ be a directed family of closed subgroups of a profinite group G. Put $H = \bigcap_{i \in I} H_i$. Then $(G:H) = \operatorname{lcm}_{i \in I}(G:H_i)$.
(c) Let $\langle G_i, \pi_{ji}\rangle_{j,i \in I}$ be an inverse limit of profinite groups such that π_{ji} is an epimorphism for $j \geq i$. Then the inverse limit $G = \varprojlim G_i$ satisfies $\#G = \operatorname{lcm}_{i \in I}\{\#G_i\}$.

22.8 Supernatural Numbers

(d) $\#\prod_{i\in I} G_i = \prod_{i\in I} \#G_i$.

Proof of (a): Let $\mathcal{N}$ be the set of all open normal subgroups of G. Then $(G : H) = \text{lcm}\,((G : HN)|\ N \in \mathcal{N})$ and $(H : K) = \text{lcm}\,((H : K(H \cap N))|\ N \in \mathcal{N})$. The identities $K(H \cap N) = H \cap KN$ and $(HN : KN) = (H : H \cap KN)$ imply

(1) $\qquad\qquad (H : K) = \text{lcm}\{(HN : KN)|\ N \in \mathcal{N}\}.$

In addition,

(2) $\qquad\qquad (G : KN) = (G : HN)(HN : KN).$

Let p be a prime number. If $p^\infty|(G : K)$, then (1) and (2) imply $p^\infty|(G : H)$ or $p^\infty|(H : K)$. Suppose therefore that $p^\infty \nmid (G : K)$. Let n_1, n_2, and n_3 be the maximal exponents of p-powers that divide $(G : K)$, $(G : H)$, and $(H : K)$, respectively. Then $n_1, n_2, n_3 < \infty$. Also, there exists a group $N \in \mathcal{N}$ such that p^{n_1} (resp. p^{n_2}, p^{n_3}) is the greatest power of p dividing $(G : KN)$ (resp. $(G : HN)$, $(HN : KN)$). By (2), $n_1 = n_2 + n_3$, as desired.

Proof of (b): By (a), $(G : H_i)|(G : H)$ for each $i \in I$. Hence, $\text{lcm}_{i\in I}(G : H_i)|(G : H)$. Conversely, let U be an open subgroup of G that contains H. Then there is an $i \in I$ with $H_i \leq U$ (Lemma 1.2.2(a)). Therefore, $(G : U)|(G : H_i)$, so $(G : U)|\text{lcm}_{i\in I}(G : H_i)$. Thus, $\text{lcm}_U(G : U)|\text{lcm}_{i\in I}(G : H_i)$ with U ranging over all open subgroups containing H. This gives (b).

Proof of (c): Let $\pi_i: G \to G_i$ be the canonical epimorphism and $N_i = \text{Ker}(\pi_i)$. Then $(G : N_i) = \#G_i$ and $\bigcap N_i = 1$. Apply (b): $\#G = \text{lcm}\,\#G_i$.

Proof of (d): Denote the family of finite subsets of I by $\mathcal{F}$. Partially order $\mathcal{F}$ by inclusion. Then $\prod G_i = \varprojlim \prod_{i\in F} G_i$, where F ranges over $\mathcal{F}$. Hence, by (a) and (c),

$$\#\prod_{i\in I} G_i = \text{lcm}_{F\in\mathcal{F}} \#\prod_{i\in F} G_i = \text{lcm}_{F\in\mathcal{F}} \prod_{i\in F} \#G_i = \prod_{i\in I} \#G_i. \qquad\square$$

LEMMA 22.8.4: *Let G be a profinite group and H a closed subgroup. Suppose $p \nmid (G : H)$ for all but finitely many p and $p^\infty \nmid (G : H)$ for all p. Then H is open.*

Proof: For each p let p^{n_p} be the maximal power of p which divides $(G : H)$. By assumption, $(G : H) = \prod p^{n_p} < \infty$, so H is open. $\square$

LEMMA 22.8.5: *Let A be a finitely generated Abelian group and B a closed subgroup of infinite index. Then B is contained in an open subgroup of index p for infinitely many p or B is contained in a closed subgroup C with $A/C \cong \mathbb{Z}_p$ for some p.*

Proof: Divide out by B, if necessary, to assume $B = 0$. We have to prove: $\mathbb{Z}/p\mathbb{Z}$ is a factor of A for infinitely many p or $\mathbb{Z}_p$ is a factor of A for some p.

Proposition 22.7.12(b) presents A as

(3) $$A = \bigoplus_p \mathbb{Z}/p^{k_{p,1}}\mathbb{Z} \oplus \cdots \oplus \mathbb{Z}/p^{k_{p,m(p)}}\mathbb{Z} \oplus \mathbb{Z}_p^{k_p},$$

with nonnegative integers $k_p, m(p)$ and positive integers $k_{p,i}$. Each of the direct summands in (3) appears also as a factor group of A. In particular, if $m(p) > 0$, then $\mathbb{Z}/p^{k_{p,1}}\mathbb{Z}$ is a factor of A. Hence, so is $\mathbb{Z}/p\mathbb{Z}$.

By Lemma 22.8.4, $m(p) > 0$ for infinitely many p or there is a p with $k_p > 0$. This together with the preceding arguments concludes the proof of the lemma. □

Remark 22.8.6: Degree of algebraic extensions. Let L/K be an algebraic extension. Define $[L : K]$ to be the least common multiple of all degrees $[E : K]$ with E ranging over all finite subextensions of L/K. Call $[L : K]$ the **degree** of L/K. When L/K is separable, Galois correspondence implies that $[L : K] = (\mathrm{Gal}(K) : \mathrm{Gal}(L))$. Hence, we may rephrase Lemma 22.8.3 in field theoretic terms:

(4a) Let $K \subseteq L \subseteq M$ be fields with M/K separable algebraic. Then $[M : K] = [M : L][L : K]$.
(4b) Let $\{L_i \mid i \in I\}$ be a family of separable algebraic extensions of K. Put $L = \prod_{i \in I} L_i$. Then $[L : K] = \mathrm{lcm}_{i \in I}[L_i : K]$.
(4c) Suppose in (4b) the L_i are linearly disjoint over K. Then $[L : K] = \prod_{i \in I}[L_i : K]$.

Now suppose L/K is an arbitrary algebraic extension. Let E be a finite subextension. Denote the maximal separable extension of K in E (resp. L) by E_0 (resp. L_0). Then E is linearly disjoint from L_0 over E_0, so $[E : E_0] = [L_0 E : L_0]$. Thus, $[L : K] = [L : L_0][L_0 : K]$. This and similar considerations show the separability assumption in (4a) is redundant. □

22.9 The Sylow Theorems

The concept of Sylow groups carries over to profinite groups:

Definition 22.9.1: Let p be a prime number. A closed subgroup P of a profinite group G is said to be a p-**Sylow group** of G if P is a pro-p-group and $p \nmid (G : P)$. □

PROPOSITION 22.9.2: *Let G be a profinite group and p a prime number. Then:*
(a) *G has a p-Sylow group and every pro-p subgroup of G is contained in a p-Sylow group of G.*
(b) *The p-Sylow groups of G are conjugate.*
(c) *If $\theta \colon G \to H$ is an epimorphism of profinite groups and P is a p-Sylow group of G, then $\theta(P)$ is a p-Sylow group of H.*
(d) *If N is a closed normal subgroup of G and P a p-Sylow group of G, then $N \cap P$ is a Sylow group of N.*

22.9 The Sylow Theorems

(e) $\#G = \prod_p \#G_p$ where G_p is a p-Sylow group of G and p ranges over all prime numbers.

Proof of (a): Let $\mathcal{N}$ be the set of all open normal subgroups of G and H a pro-p-subgroup of G (H may be trivial). For each $N \in \mathcal{N}$ denote the set of all p-Sylow groups of G/N which contain HN/N by $\mathcal{P}(N)$. By Sylow's theorem for finite groups, $\mathcal{P}(N)$ is finite and nonempty. If $M \in \mathcal{N}$ and $M \leq N$, then the quotient map $G/M \to G/N$ maps p-Sylow groups of G/M onto p-Sylow groups of G/N. It therefore defines a canonical map of $\mathcal{P}(M)$ into $\mathcal{P}(N)$. By Lemma 1.1.3, the inverse limit of the sets $\mathcal{P}(N)$ is nonempty. Thus, there is a $(P_N)_{N \in \mathcal{N}}$ such that $P_N \in \mathcal{P}(N)$ and P_M is mapped onto P_N for all $M, N \in \mathcal{N}$ with $M \leq N$. The inverse limit $P = \varprojlim P_N$ is a pro-p-subgroup of $G = \varprojlim G/N$ that contains H. If $N \in \mathcal{N}$, then $PN/N = P_N$. Hence, $p \nmid (G : PN)$. Therefore, $p \nmid (G : P)$. It follows that P is a p-Sylow group of G that contains H.

Proof of (b): Let P and P' be p-Sylow groups of G. If $N \in \mathcal{N}$, then PN/N and $P'N/N$ are p-Sylow groups of the finite group G/N. Hence, they are conjugate. It follows from Lemma 1.2.2(e) that P and P' are conjugate.

Proof of (c): The group $Q = \theta(P)$ is a pro-p-group and $(H : Q) = (G : \theta^{-1}(Q))$ divides $(G : P)$, which is relatively prime to p. Hence, Q is a p-Sylow group of H.

Proof of (d): Use (a) to choose a p-Sylow group Q of N and a p-Sylow group P' of G which contains Q. Then $(N \cap P' : Q)$ divides $(P' : Q)$ (Lemma 22.8.3(a)), hence $\#P'$, so $(N \cap P' : Q)$ is a power of p. On the other hand, $(N \cap P' : Q)$ divides $(N : Q)$, so $(N \cap P' : Q)$ is relatively prime to p. It follows that $(N \cap P' : Q) = 1$ and $N \cap P' = Q$.

By (b), there exists $g \in G$ with $P^g = P'$. Since $N \triangleleft G$, we have $(N \cap P)^g = N \cap P' = Q$, so $N \cap P = Q^{g^{-1}}$ is a p-Sylow group of N.

Proof of (e): The formula $\#G = \prod_l \#G_l$ follows from the formula $\#G = (G : G_l) \cdot \#G_l$ for each prime l (Lemma 22.8.3(a)). □

Finite nilpotent groups are direct products of their unique p-Sylow groups as p runs over all prime numbers. The next proposition is an immediate consequence of this and Lemma 22.1.2:

PROPOSITION 22.9.3: *Each pronilpotent group N is the direct product of its p-Sylow groups. In particular, if G is a profinite group, $\Phi(G)$ is a direct product of its p-Sylow groups and each of them is normal in G.*

Proof: For each prime number p the group N has a unique p-Sylow group, N_p, and it is characteristic in N, since this is the case for the finite quotients of N. The groups N_p generate N and the intersection of N_p with the group generated by all other Sylow groups of N is trivial. It follows that $N = \prod N_p$.

Since each p-Sylow group of $\Phi(G)$ is characteristic, Lemma 22.1.2 gives the second part of the proposition. □

PROPOSITION 22.9.4: *Let A be an Abelian projective group. Then there is a set S of prime numbers with $A \cong \prod_{p \in S} \mathbb{Z}_p$.*

Proof: By Proposition 22.9.3, $A \cong \prod_p A_p$, where A_p is the p-Sylow group of A. By Proposition 22.4.7, A_p is projective. Hence, by Corollary 22.7.11, A_p is either trivial or is isomorphic to $\mathbb{Z}_p$. □

We conclude this section with some applications to fields:

COROLLARY 22.9.5: *Let K be an algebraic extension of a finite field. Then there exists a set S of prime numbers with $\mathrm{Gal}(K) \cong \prod_{p \in S} \mathbb{Z}_p$.*

Proof: By Section 1.5, $\mathrm{Gal}(K)$ is a subgroup of $\hat{\mathbb{Z}}$. Hence, $\mathrm{Gal}(K)$ is projective (Corollary 22.4.6). It follows from Proposition 22.9.4 that $\mathrm{Gal}(K) \cong \prod_{p \in S} \mathbb{Z}_p$ for some set S of prime numbers. □

THEOREM 22.9.6 ([Ax2, Thm. D]): *Let K be a perfect PAC field with $\mathrm{Gal}(K)$ Abelian. Then K is C_1.*

Proof: By Theorem 11.6.2, $\mathrm{Gal}(K)$ is projective. Hence, by Proposition 22.9.4, $\mathrm{Gal}(K)$ is procyclic. It follows from Theorem 21.3.6(c) that K is C_1. □

And here is an application to Hilbertian fields:

THEOREM 22.9.7 ([Uchida, Thm. 3(ii)]): *Let $K \subseteq L \subseteq N$ be fields. Suppose K is Hilbertian, N is a pronilpotent extension of K, and $[L : K]$ is divisible by two distinct prime numbers. Then L is Hilbertian.*

Proof: By Proposition 22.9.3, $\mathrm{Gal}(N/K)$ is the direct product of its Sylow subgroups G_p. Choose a prime divisor q of $[L : K]$, let N_q be the fixed field in N of G_q, and N_q' the fixed field in N of $\prod_{p \neq q} G_p$. Then $N_q N_q' = N$, $L \not\subseteq N_q$, and $L \not\subseteq N_q'$. By the diamond theorem (Theorem 13.8.3), L is Hilbertian. □

22.10 On Complements of Normal Subgroups

Let N be a closed normal subgroup of a profinite group G. A closed subgroup H of G is called a **complement** to N in G if $N \cap H = 1$ and $NH = G$. The Schur-Zassenhaus theorem, whose finite version appears in the proof of Lemma 22.4.1, gives the existence of a complement, unique up to conjugation in G, under a simple index condition.

Now suppose $N, M \triangleleft G$, $H \leq G$, and $M \leq N \cap H$. We say, H is a **complement** to N in G/M if H/M is a complement to N/M in G/M. In other words, $N \cap H = M$ and $NH = G$.

LEMMA 22.10.1 (Schur-Zassenhaus): *Let N be a closed normal subgroup of a profinite group G. Suppose $\gcd(\#N, (G : N)) = 1$. Then N has a complement in G. Furthermore, all complements to N in G are conjugate.*

Proof: First we prove the uniqueness part of the lemma: Let H and H' be two complements to N in G. Let M be an open normal subgroup of G. Then

22.10 On Complements of Normal Subgroups

$(HM : M) = (G : NM)|(G : N)$ and $(NM : M)|\#N$. Hence, $(HM : M)$ and $(NM : M)$ are relatively prime. Therefore, $NM \cap HM = M$. It follows that HM/M is a complement to NM/M in G/M. Similarly, $H'M/M$ is a complement to NM/M in G/M. By the uniqueness part of the finite group version of Schur-Zassenhaus [Huppert, p. 128], HM/M and $H'M/M$ are conjugate. Consequently, by Lemma 1.2.2(e), H and H' are conjugate.

The proof of the existence of a complement is divided into two parts:

PART A: N is finite. Then G has an open normal subgroup M with $N \cap M = 1$. Apply Schur-Zassenhaus [Huppert, p. 126] to the finite group G/M to find a complement H to NM in G/M. Then H is a complement to N in G.

PART B: N is infinite. With $U_0 = G$ let $\{U_\alpha \mid \alpha < \lambda\}$ be a well ordering of the open normal subgroups of G. For each $\beta \leq \lambda$ let $M_\beta = N \cap \bigcap_{\alpha < \beta} U_\alpha$. We apply a transfinite induction to find a decreasing sequence $\{H_\beta \mid \beta \leq \lambda\}$ of closed subgroups of G such that H_β is a complement to N in G/M_β.

Indeed, let $H_0 = G$. Assume the statement is true for β. Use Part A and that $M_{\beta+1}$ is open in M_β to find a complement $H_{\beta+1}$ to M_β in $H_\beta/M_{\beta+1}$. Then $H_{\beta+1}$ is a complement to N in $G/M_{\beta+1}$.

Now assume $\gamma \leq \lambda$ is a limit ordinal and H_β has been defined for each $\beta < \gamma$. Put $H_\gamma = \bigcap_{\beta < \gamma} H_\beta$. Then $N \cap H_\gamma = M_\gamma$. By Lemma 1.2.2(b), $G = \bigcap_{\beta < \gamma} NH_\beta = NH_\gamma$.

We conclude from $M_\lambda = 1$ that H_λ is a complement to N in G. □

LEMMA 22.10.2 (Gaschütz [Huppert, p. 121]): *Let $A \leq H \leq G$ be a tower of profinite groups. Suppose $\gcd(\#A, (G : H)) = 1$. In addition suppose A is finite, Abelian, normal in G, and has a complement in H. Then A has a complement in G.*

Proof: Choose a subgroup K of G with

(1) $$AK = H \quad \text{and} \quad A \cap K = 1.$$

The rest of the proof splits into two parts as G is finite or not:

PART A: G is finite [Brandis, §2]. For $x, y \in G$ with $Hx = Hy$ there is a unique $k(x, y) \in K$ with $xy^{-1}A = k(x,y)A$. The function $k(x,y)$ satisfies the following rules (use normality of A in G):

(2a) $k(x,x) = 1$, $k(y,x) = k(x,y)^{-1}$, and $k(x,y)k(y,z) = k(x,z)$.
(2b) $x^{-1}k(x,y)y \in A$ and $k(xg, yg) = k(x,y)$ for each $g \in G$.
(2c) $k(xa, y) = k(x, y)$ for each $a \in A$.

Let $\mathcal{S}$ be a set of representative systems of right cosets of G modulo H. Thus, for all $\Phi \in \mathcal{S}$ and $x \in G$ we have $|\Phi \cap Hx| = 1$. Define a convolution product between $\Phi_1, \Phi_2 \in \mathcal{S}$:

(3) $\Phi_1^{-1} * \Phi_2 = \prod r_1^{-1} k(r_1, r_2) r_2$, where the product ranges over all pairs (r_1, r_2) with $r_1 \in \Phi_1$, $r_2 \in \Phi_2$, and $r_1 \in Hr_2$.

By (2b), each of the factors of the product belongs to the Abelian group A, so that the product is well defined. Use (2) to check the following rules for $\Phi_1, \Phi_2, \Phi_3 \in \mathcal{S}$:

(4a) $\Phi_1^{-1} * \Phi_1 = 1$.
(4b) $(\Phi_1^{-1} * \Phi_2) \cdot (\Phi_2^{-1} * \Phi_3) = \Phi_1^{-1} * \Phi_3$.
(4c) $(\Phi_1^{-1} * \Phi_2)^{-1} = \Phi_2^{-1} * \Phi_1$.
(4d) $(\Phi_1 g)^{-1} * (\Phi_2 g) = (\Phi_1^{-1} * \Phi_2)^g$ for each $g \in G$.
(4e) $(\Phi_1 a)^{-1} * \Phi_1 = a^{-(G:H)}$ for each $a \in A$.

Each $\Phi \in \mathcal{S}$ defines a map $\varphi : G \to A$ by $\varphi(g) = (\Phi g)^{-1} * \Phi$. By (4b) and (4d), $\varphi(g_1 g_2) = \varphi(g_1)^{g_2} \varphi(g_2)$. That is, φ is a **crossed homomorphism**. In particular, $\text{Ker}(\varphi) = \{g \in G \mid \varphi(g) = 1\}$ is a subgroup of G (which is not necessarily normal). Since $\varphi(a) = a^{-(G:H)}$ for each $a \in A$, the restriction of φ to A is an automorphism. In particular, $\text{Ker}(\varphi) \cap A = 1$. Also, since A is an Abelian group, $\varphi(ga) = \varphi(g)^a \varphi(a) = \varphi(g)\varphi(a)$, for $a \in A$. Moreover, for each $g \in G$ there is an $a \in A$ with $\varphi(a) = \varphi(g)^{-1}$. Thus, $g = (ga) \cdot a^{-1} \in \text{Ker}(\varphi) A$. Consequently, $\text{Ker}(\varphi)$ is a complement to A in G.

PART B: *G is infinite.* Since A is finite, K is open in H. Therefore, there exists an open normal subgroup N of G with $H \cap N \leq K$. Consider the finite quotient group G/N. Observe that $AN \cap KN = N$: if $x = an_1 = kn_2$, with $a \in A$, $k \in K$, and $n_1, n_2 \in N$, then $k^{-1}a = n_2 n_1^{-1} \in H \cap N$. Hence, $a \in A \cap K = 1$ and $x = n_1 \in N$. By Part A, there exists an open subgroup L of G with $ANL = G$ and $AN \cap L = N$. Also, $A \cap L = (A \cap AN) \cap L = A \cap N \leq A \cap K = 1$. Thus, L is a complement to A in G. □

Remark 22.10.3: Uniqueness. Assume in Lemma 22.10.2 that all complements to A in H are conjugate. If G is finite, then Gaschütz' theorem [Huppert, p 121] asserts that all complements to A in G are conjugate. As in the proof of Lemma 22.10.1, one can deduce for arbitrary G that A has a unique (up to conjugacy) complement in G. □

We apply Lemma 22.10.2 to prove another criterion for projectivity:

PROPOSITION 22.10.4: *A profinite group G is projective if and only if for each p all p-Sylow groups of G are free pro-p groups.*

Proof: If G is projective and P is a p-Sylow subgroup of G, then P is projective (Proposition 22.4.7). Thus, P is a free pro-p group (Proposition 22.7.6).

Conversely, suppose all p-Sylow groups of G are free pro-p groups. By Corollary 22.4.3, it suffices to prove that each exact sequence $1 \to A \to E \to G \to 1$, in which A is a finite Abelian p-group, splits. From Proposition 22.9.2, E has a p-Sylow group P that contains A. Then P/A, which is isomorphic to a p-Sylow group of G, is by assumption a free pro-p group. In particular, P/A is projective (Corollary 22.4.5). Hence, A has a complement in P (Remark 22.4.2). By definition, the index $(E : P)$ is relatively prime to p. Consequently, by Lemma 22.10.2, A has a complement in E. □

22.10 On Complements of Normal Subgroups

We apply these results to Frattini covers:

PROPOSITION 22.10.5: *Let H be a Frattini cover of a profinite group G. Then $\#G$ and $\#H$ have the same prime divisors.*

Proof: G is a homomorphic image of H. Hence, each prime divisor p of $\#G$ divides $\#H$.

Conversely, assume p divides $\#H$ but not $\#G$. By assumption, there is a Frattini cover $\varphi\colon H \to G$. Thus, $\text{Ker}(\varphi) \leq \Phi(H)$. Hence, $p \nmid (H : \Phi(H))$, so $p|\#\Phi(H)$. Therefore, $\Phi(H)$ has a nontrivial p-Sylow subgroup P. By Proposition 22.9.3, P is normal in H. Thus, Schur-Zassenhaus (Lemma 22.10.1) gives a closed subgroup B of H with $P \cap B = 1$ and $PB = H$. By Lemma 22.1.1, $B = H$. Consequently, $P = 1$, a contradiction. □

COROLLARY 22.10.6: *Let $G_1, G_2, G_3, \ldots$ be a sequence of profinite groups of mutually relatively prime orders, and let $G = \prod_{i=1}^{\infty} G_i$.*
(a) If each G_i is projective, then so is G.
(b) The universal Frattini covers satisfy $\tilde{G} \cong \prod_{i=1}^{\infty} \tilde{G}_i$.

Proof: A p-Sylow group of G_i is a p-Sylow group of G. Thus, (a) is a consequence of Proposition 22.10.4.

By Proposition 22.10.5, the groups $\tilde{G}_i$ have mutually relatively prime orders. Hence, by (a), $\prod_{i=1}^{\infty} \tilde{G}_i$ is a projective group. In addition, if $\tilde{\varphi}_i\colon \tilde{G}_i \to G_i$ is the universal Frattini cover, $i = 1, 2, \ldots$, and $\tilde{\varphi} = \prod_{i=1}^{\infty} \tilde{\varphi}_i$, then

$$\text{Ker}(\tilde{\varphi}) = \prod_{i=1}^{\infty} \text{Ker}(\tilde{\varphi}_i) \leq \prod_{i=1}^{\infty} \Phi(\tilde{G}_i) = \Phi(\prod_{i=1}^{\infty} \tilde{G}_i)$$

(Lemma 22.1.4(d)). Thus, $\prod_{i=1}^{\infty} \tilde{G}_i$ is a Frattini cover of G. It follows from Proposition 22.6.1 that $\prod_{i=1}^{\infty} \tilde{G}_i$ is the universal Frattini cover of G. □

For finite groups the next result improves Proposition 22.10.5:

PROPOSITION 22.10.7: *Let $\tilde{G}$ be the universal Frattini cover of a finite group G. Then $\#G$ and $\#\Phi(\tilde{G})$ have the same prime divisors.*

Proof: If p divides $\#\Phi(\tilde{G})$, then $p|\#\tilde{G}$. Therefore, p divides $\#G$ (Proposition 22.10.5).

Conversely, if p divides $\#G$, then $\tilde{G}$ has a nontrivial p-Sylow group P. Since $\tilde{G}$ is projective, so is P. Therefore, P is infinite (Proposition 22.4.7). On the other hand, $\tilde{G}/\Phi(\tilde{G})$, a quotient of G, is finite. Therefore, $\Phi(\tilde{G}) \cap P$ is nontrivial. Consequently, p divides $\#\Phi(\tilde{G})$. □

Remark 22.10.8: The finiteness assumption in Proposition 22.10.7 is essential. The Frattini group of $\hat{F}_e$, for $e \geq 2$, is trivial (e.g. Corollary 24.10.4(d)). Since each p divides $\#\hat{F}_e$ and $\hat{F}_e$ is its own universal Frattini cover (Proposition 22.6.1), Proposition 22.10.7 fails for arbitrary profinite groups. □

A Corollary of Proposition 22.10.7 is that the condition on $\mathcal{C}$ in Lemma 22.4.1 to be full is necessary. More precisely, suppose a Melnikov formation $\mathcal{C}$

contains no simple Abelian groups. Then each free pro-$\mathcal{C}$ group is $\mathcal{C}$-projective (Lemma 22.3.6) but not projective. This follows from the following result:

PROPOSITION 22.10.9: *Let $\mathcal{C}$ be a Melnikov formation of finite groups and G a pro-$\mathcal{C}$ group. Suppose G is projective. Then $C_p \in \mathcal{C}$ for each p dividing the order of G.*

Proof: Let p be a prime divisor of $\#G$. Then there exists an epimorphism $\varphi\colon G \to A$ onto a finite group A having an order divisible by p. Let $\alpha\colon \tilde{A} \to A$ be the universal Frattini cover of A. Since G is projective, there is an epimorphism $\gamma\colon G \to \tilde{A}$ with $\alpha \circ \gamma = \varphi$ (Proposition 22.5.9(a)). In particular, $\tilde{A}$ is a pro-$\mathcal{C}$ group. It follows that $\Phi(\tilde{A})$ is also a pro-$\mathcal{C}$ group. In addition, $\Phi(\tilde{A})$ is pronilpotent (Lemma 22.1.2) and its p-Sylow subgroup $\Phi(\tilde{A})_p$ is nontrivial (Proposition 22.10.7). Thus, C_p is a quotient of $\tilde{A}$, hence belongs to $\mathcal{C}$. □

22.11 The Universal Frattini p-Cover

Fix a prime number p for the whole section.

Call a profinite group G **p-projective** if each p-Sylow group of G is projective, hence a free pro-p group (Proposition 22.7.6). In cohomological terms this means the pth cohomological dimension of G is at most 1. The following result, a p analog of Remark 22.4.2, justifies the name p-projective:

PROPOSITION 22.11.1: *Let $1 \to P \to H \xrightarrow{\pi} G \to 1$ be a short exact sequence of profinite groups. Suppose G is p-projective and P is a pro-p group. Then the short exact sequence splits.*

Proof: Choose a p-Sylow group H_p of H which contains P (Proposition 22.9.2(a)). Then $G_p = \pi(H_p)$ is a p-Sylow group of G (Proposition 22.9.2(c)). By assumption, G_p is projective. Hence, the short exact sequence $1 \to P \to H_p \xrightarrow{\pi} G_p \to 1$ splits. This means that H_p has a subgroup G'_p which π maps isomorphically onto G_p. In other words, $P \cap G'_p = 1$ and $PG'_p = H_p$. The rest of the proof breaks up into two parts.

PART A: *P is finite.* First suppose P is Abelian. By definition, $\gcd(\#P, (H : H_p)) = 1$. Hence, by Gaschütz (Lemma 22.10.2), P has a complement in H.

Now suppose P is non-Abelian. Then, its center Z is a nontrivial Abelian normal subgroup of H. Let $\bar{H} = H/Z$, $\bar{P} = P/Z$, and $\bar{\pi}\colon \bar{H} \to G$ be the reduction of π modulo Z. Then $\#\bar{P} < \#P$. An induction hypothesis gives a splitting of $\bar{\pi}$. Thus, H has a subgroup H_0 with $P \cap H_0 = Z$ and $PH_0 = H$. Hence, $1 \to Z \to H_0 \xrightarrow{\pi} G \to 1$ is a short exact sequence. The preceding paragraph gives a subgroup H_1 of H_0 with $H_1 \cap Z = 1$ and $H_1 Z = H_0$. This subgroup is a complement to P in H.

PART B: *P is infinite.* We repeat the arguments of Part B of the proof of Lemma 22.10.1. We find a decreasing transfinite sequence $\{P_\beta \mid \beta \leq m\}$

of closed normal subgroups of H with $P_0 = P$, $P_\beta = \bigcap_{\alpha<\beta} P_\alpha$ for each limit ordinal $\beta \leq m$, and $P_m = 1$. By transfinite induction we construct a decreasing sequence $\{H_\beta \mid \beta \leq m\}$ of closed subgroups of H with $P \cap H_\beta = P_\beta$ and $PH_\beta = H$.

Let $\beta < m$. Suppose P has a complement H_β in H/P_β. Since $P_\beta/P_{\beta+1}$ is finite and $H_\beta/P_\beta \cong H/P \cong G$, Part A gives a complement $H_{\beta+1}$ to P_β in $H_\beta/P_{\beta+1}$. Then $H_{\beta+1}$ is a complement to P in $P_{\beta+1}$.

Now consider a limit ordinal $\gamma \leq m$. Suppose H_β has been constructed for each $\beta < \gamma$. Put $H_\gamma = \bigcap_{\beta<\gamma} H_\beta$. Then H_γ is a complement to P in H/P_γ (use Lemma 1.2.2(b)).

Transfinite induction gives a complement to P in H/P_m, that is, in H. □

An embedding problem with a pro-p kernel is said to be a **p-embedding problem**.

COROLLARY 22.11.2: *Let G be a p-projective profinite group. Then every p-embedding problem for G is weakly solvable.*

Proof: Let $(\varphi: G \to A,\ \alpha: B \to A)$ be an embedding problem for G. Let $H = B \times_A G$ and let $\pi: H \to G$ be the projection on the second coordinate. Then, $\text{Ker}(\pi) \cong \text{Ker}(\alpha)$ (Lemma 22.2.5) is a pro-p group. Proposition 22.11.1 gives a section $\pi': G \to H$ of π. Let $\beta: H \to B$ be the projection onto the first coordinate. Then $\beta \circ \pi'$ is a week solution of the given embedding problem. □

LEMMA 22.11.3: *Every closed subgroup of a p-projective group is p-projective.*

Proof: Let G be a p-projective group and H a closed subgroup. Choose a p-Sylow subgroup H_p of H and a p-Sylow subgroup G_p of G which contains H_p (Proposition 22.9.2(a)). Since G_p is projective, so is H_p (Proposition 22.4.7). Hence, H is p-projective. □

Certain quotients of p-projective groups are p-projective:

LEMMA 22.11.4: *Let G be a p-projective group and N a closed normal subgroup. Suppose $p \nmid \#N$. Then G/N is p-projective.*

Proof: Let G_p be a p-Sylow subgroup of G. By assumption, $G_p \cap N = 1$, hence $G_pN/N \cong G_p$, so G_pN/N is projective. In addition, G_pN/N is a p-Sylow subgroup of G/N (Proposition 22.9.2(c)). Consequently, G/N is p-projective. □

A **p-cover** of a profinite group G is an epimorphism $\varphi: H \to G$ with a pro-p kernel. Call φ a **Frattini p-cover** if in addition φ is Frattini.

Construction 22.11.5: p-projective Frattini p-cover. Let G be a profinite group and $\tilde\varphi: \tilde G \to G$ its universal Frattini cover. Put $K = \text{Ker}(\tilde\varphi)$. Then, $K \leq \Phi(\tilde G)$. Hence, K is pronilpotent (Lemma 22.1.2). Thus, by Proposition

22.9.3, $K = \prod K_l$, where l ranges over all prime numbers and K_l is the unique l-Sylow subgroup of K. In particular, K_l is normal in $\tilde{G}$. Since $\tilde{G}$ is projective (Proposition 22.6.1), each K_l is projective, hence pro-l free.

Put $K'_p = \prod_{l \neq p} K_l$. Then K'_p is a closed normal subgroup of $\tilde{G}$. Put $\hat{G} = \tilde{G}/K'_p$, $\hat{K} = K/K'_p$, and $\hat{\varphi}\colon \hat{G} \to G$ the epimorphism induced by $\tilde{\varphi}$. Then $\hat{K} = \mathrm{Ker}(\hat{\varphi})$ and $\hat{K} \cong K_p$ is pro-p free. In addition, $\hat{K} \leq \Phi(\hat{G})$. Therefore, $\hat{\varphi}\colon \hat{G} \to G$ is a Frattini p-cover. Finally, by Lemma 22.11.4, $\hat{G}$ is p-projective. □

We apply the existence of a p-projective Frattini p-cover of a profinite group to prove a converse to Corollary 22.11.2:

LEMMA 22.11.6: *Let G be a profinite group. Suppose each finite p-embedding problem for G is weakly solvable. Then G is p-projective.*

Proof: Let

(1) $\qquad\qquad\qquad (\varphi\colon G \to A,\ \alpha\colon B \to A)$

be an embedding problem for G with $C = \mathrm{Ker}(\alpha)$ pro-p.

First suppose C is finite. Choose an open normal subgroup N of B with $N \cap C = 1$. Let $\bar{A} = B/NC$, $\bar{B} = B/N$, and $\bar{C} = NC/N$. Let $\bar{\alpha}\colon \bar{B} \to \bar{A}$, $\pi_B\colon B \to \bar{B}$, and $\pi_A\colon A \to \bar{A}$ be the quotient maps. Then $\mathrm{Ker}(\bar{\alpha}) = \bar{C} \cong C$ and the square

$$\begin{array}{ccc} B & \xrightarrow{\alpha} & A \\ \pi_B \downarrow & & \downarrow \pi_A \\ \bar{B} & \xrightarrow{\bar{\alpha}} & \bar{A} \end{array}$$

is cartesian (Example 22.2.7(b)). By assumption, there exists a homomorphism $\bar{\gamma}\colon G \to \bar{B}$ with $\bar{\alpha} \circ \bar{\gamma} = \pi_A \circ \varphi$. Hence, there exists a homomorphism $\gamma\colon G \to B$ such that $\alpha \circ \gamma = \varphi$.

Now consider the general case. Suppose L and L' are closed normal subgroups of B, $L' \leq L \leq C$, and L' is open in L. Then L/L' is a finite p-group. By the first case, each embedding problem for G with kernel L/L' is weakly solvable, so by Remark 22.3.3, (1) is weakly solvable.

It follows that each cover of G with a pro-p kernel has a section. In particular, this holds for the universal Frattini p-cover $\hat{\varphi}\colon \hat{G} \to G$ (Construction 22.11.5). Thus, G is isomorphic to a subgroup of $\hat{G}$. Since $\hat{G}$ is p-projective, so is G (Lemma 22.11.3). □

LEMMA 22.11.7: *Let G be a p-projective group. Then every p-embedding problem*

(2) $\qquad\qquad\qquad (\varphi\colon G \to H,\ \alpha\colon B \to A)$

22.12 Examples of Universal Frattini p-covers

in which α is a Frattini p-cover is solvable. Furthermore, each Frattini p-cover $\psi\colon H \to G$ is an isomorphism.

Proof: By Corollary 22.11.2, there is a homomorphism $\gamma\colon G \to B$ with $\alpha \circ \gamma = \varphi$. Then $\alpha(\gamma(G)) = \varphi(G) = B$, so $\gamma(G) = B$. Thus, γ solves embedding problem (2).

In particular, there is an epimorphism $\gamma\colon G \to H$ with $\psi \circ \gamma = \mathrm{id}$. Therefore, ψ is an isomorphism. □

PROPOSITION 22.11.8: *Let $\varphi'\colon G' \to G$ be a p-cover of a profinite groups. The following conditions are equivalent.*
(a) *φ' is a p-projective Frattini p-cover.*
(b) *φ' is a maximal Frattini p-cover.*
(c) *φ' is a minimal p-projective p-cover.*

Proof of "(a) $\Longrightarrow$ (b)": Suppose $\eta\colon H \to G$ is a Frattini p-cover. Lemma 22.11.7 gives an epimorphism $\gamma\colon G' \to H$ with $\eta \circ \gamma = \varphi'$. Hence, φ' is a maximal Frattini p-cover.

Proof of "(a) $\Longrightarrow$ (c)": Suppose $\eta\colon H \to G$ is a p-projective p-cover. Lemma 22.11.7 gives an epimorphism $\gamma'\colon H \to G'$ with $\varphi' \circ \gamma' = \eta$. Hence, φ' is a minimal p-projective p-cover.

Proof of "(b) $\Longrightarrow$ (a)": Let $\hat{\varphi}\colon \hat{G} \to G$ be the p-projective Frattini p-cover that Construction 22.11.5 gives. The maximality of φ' gives an epimorphism $\gamma\colon G' \to \hat{G}$ with $\hat{\varphi} \circ \gamma = \varphi'$. Since both φ' and $\hat{\varphi}$ are Frattini, so is γ (Lemma 22.5.4(c)).

Lemma 22.11.7 gives an epimorphism $\gamma'\colon \hat{G} \to G'$ with $\varphi' \circ \gamma' = \hat{\varphi}$. Again, by Lemma 22.5.4(c), γ' is Frattini.

By Lemma 22.5.4(a), $\gamma' \circ \gamma$ is Frattini. By construction, $\mathrm{Ker}(\gamma' \circ \gamma) \leq \mathrm{Ker}(\varphi')$. Hence, $\mathrm{Ker}(\gamma' \circ \gamma)$ is a pro-p group. Therefore, by Lemma 22.11.7, $\gamma' \circ \gamma$ is an isomorphism, so γ and γ' are isomorphisms. In particular, G' is p-projective.

Proof of "(c) $\Longrightarrow$ (a)": Consider again the p-projective Frattini p-cover $\hat{\varphi}\colon \hat{G} \to G$ that Construction 22.11.5 gives. The minimality of φ' gives an epimorphism $\gamma'\colon \hat{G} \to G'$ with $\varphi' \circ \gamma' = \hat{\varphi}$. By Lemma 22.5.4(c), φ' is Frattini. □

Remark 22.11.9: Universal Frattini p-cover. We call each cover $\varphi'\colon G' \to G$ satisfying the equivalent conditions of Proposition 22.11.8 a **universal Frattini p-cover** of G. Construction 22.11.5 establishes the existence of a universal Frattini p-cover of G. The proof of Proposition 22.11.8 shows it is unique. That is, given another universal Frattini p-cover $\varphi''\colon G'' \to G$ there is an isomorphism $\gamma\colon G' \to G''$ with $\varphi'' \circ \gamma = \varphi'$.

Occasionally we refer also to the group $\hat{G}$ of Construction 22.11.5 as the universal Frattini p-cover of G. □

22.12 Examples of Universal Frattini p-covers

The universal Frattini cover of a pro-p group P of rank e is $\hat{F}_e(p)$ (Corollary 22.7.8). It coincides with the universal Frattini p-cover of P. We generalize this observation and prove that if a finitely generated profinite group A with $p \nmid \#A$ acts on P, then this action lifts to an action on $\hat{F}_e(p)$ and $A \ltimes \hat{F}_e(p)$ is the universal Frattini p-cover of $A \ltimes P$.

Although $\mathrm{Aut}(P)$ is a profinite group (Remark 16.10.5), it is not necessarily pro-p. For example, $\mathrm{Aut}(\mathbb{Z}/p\mathbb{Z}) = \mathbb{Z}/(p-1)\mathbb{Z}$. However, $\mathrm{Aut}(P)$ contains an open normal pro-p subgroup. This follows from the following result:

LEMMA 22.12.1: *Let P be a finitely generated pro-p group. Then*

$$K = \{\kappa \in \mathrm{Aut}(P) \mid \kappa(g)\Phi(P) = g\Phi(P) \; \forall g \in P\}$$

is a pro-p group.

Proof: Since $\Phi(P)$ is an open characteristic subgroup of P, each $\kappa \in \mathrm{Aut}(P)$ defines an automorphism $\bar{\kappa} \in \mathrm{Aut}(P/\Phi(P))$ by $\bar{\kappa}(g\Phi(P)) = \kappa(g)\Phi(P)$. The group K is the kernel of the homomorphism $\kappa \mapsto \bar{\kappa}$ of $\mathrm{Aut}(P)$ into $\mathrm{Aut}(P/\Phi(P))$. The rest of the proof naturally breaks up into two parts:

PART A: *P is finite.* Choose generators $x_1, \ldots, x_e$ of P. Let

$$T = \{(x_1 f_1, \ldots, x_e f_e) \mid f_1, \ldots, f_e \in \Phi(P)\}.$$

For each $\kappa \in K$ and $(x_1 f_1, \ldots, x_e f_e) \in T$ there are $f'_1, \ldots, f'_e \in \Phi(P)$ with $\kappa(x_i f_i) = x_i f'_i$, $i = 1, \ldots, e$. By Lemma 22.1.1, $x_1 f_1, \ldots, x_e f_e$ generate P, so if $\kappa(x_1 f_1, \ldots, x_e f_e) = (x_1 f_1, \ldots, x_e f_e)$, then $\kappa = 1$. This means, K acts regularly on T. Thus, each K-orbit of T has the same number of elements as K. Denote the number of these orbits by k. Then $|T| = k|K|$. On the other hand, $|T| = |\Phi(P)|^e$ is a p-power. Hence, K is a p-group, as claimed.

PART B: *The general case.* Consider an open characteristic subgroup N of P. Denote reduction modulo N by a bar. Let

$$K_N = \{\kappa \in \mathrm{Aut}(\bar{P}) \mid \kappa(\bar{g})\Phi(\bar{P}) = \bar{g}\Phi(\bar{P}) \; \forall \bar{g} \in \bar{P}\}.$$

By Part A, K_N is a finite p-group.

Each $\kappa \in \mathrm{Aut}(P)$ defines $\kappa_N \in \mathrm{Aut}(\bar{P})$ by $\kappa_N(\bar{g}) = \overline{\kappa(g)}$. Suppose $\kappa \in K$. Then $\kappa_N(\bar{g})\overline{\Phi(P)} = \bar{g}\overline{\Phi(P)}$. By Lemma 22.1.4(c), $\overline{\Phi(P)} \leq \Phi(\bar{P})$. Hence, $\kappa_N(\bar{g})\Phi(\bar{P}) = \bar{g}\Phi(\bar{P})$. Therefore, $\kappa_N \in K_N$.

By Remark 16.10.5, the intersection of all N as above is trivial. Thus, the maps $\kappa \mapsto \kappa_N$ define an embedding of K into $\varprojlim K_N$. The latter group is pro-p. Consequently, K is a pro-p group. □

22.12 Examples of Universal Frattini p-covers

PROPOSITION 22.12.2: *Let P be a pro-p group of finite rank e and A a finitely generated profinite group. Suppose A acts on P and $p \nmid \#A$. Put $G = A \ltimes P$. Then the action of A on P lifts to an action of A on $\hat{F}_e(p)$ and $\hat{G} = A \ltimes \hat{F}_e(p)$ is the universal Frattini p-cover of G.*

Proof: The action of A on P is defined by a homomorphism $\psi: A \to \mathrm{Aut}(P)$ with $y^{\psi(a)} = y^a$ for each $y \in P$. Let $B = \psi(A)$. Choose generators $y_1, \ldots, y_e$ for P and generators $x_1, \ldots, x_e$ for $\hat{F}_e(p)$. Extend the map $x_i \mapsto y_i$, $i = 1, \ldots, e$, to an epimorphism $\varphi_1: \hat{F}_e(p) \to P$. Consider $b \in B$. For each i choose $x_i' \in \hat{F}_e(p)$ with $\varphi_1(x_i') = y_i^b$. Then lift b to a homomorphism $\hat{b}: \hat{F}_e(p) \to \hat{F}_e(p)$ satisfying $x_i^{\hat{b}} = x_i'$. By definition, $\varphi_1(\hat{F}_e(p)^{\hat{b}}) = P$. Since φ_1 is a Frattini cover, $\hat{F}_e(p)^{\hat{b}} = \hat{F}_e(p)$. By Proposition 16.10.6(a), $\hat{b}$ is an automorphism of $\hat{F}_e(p)$. It satisfies, $\varphi_1(x^{\hat{b}}) = \varphi_1(x)^b$ for each $x \in \hat{F}_e(p)$. In particular, $\hat{b}$ leaves $\mathrm{Ker}(\varphi_1)$ invariant.

By Remark 16.10.5, $\mathrm{Aut}(\hat{F}_e(p))$ is a profinite group. Let $\hat{B}$ be the closed subgroup of $\mathrm{Aut}(\hat{F}_e(p))$ generated by all $\hat{b}$. Then each β in $\hat{B}$ leaves $\mathrm{Ker}(\varphi_1)$ invariant. Therefore, for each $\beta \in \hat{B}$ there exists a unique $\bar{\beta} \in \mathrm{Aut}(P)$ with $\varphi_1 \circ \beta = \bar{\beta} \circ \varphi_1$. Let $\varphi_2(\beta) = \bar{\beta}$. If $\beta = \hat{b}$ for some $b \in B$, then $\bar{\beta} = b$. Thus, $\varphi_2: \hat{B} \to B$ defined by $\varphi_2(\beta) = \bar{\beta}$ is an epimorphism satisfying $\varphi_2(\hat{b}) = b$ for each $b \in B$.

Consider $\beta \in \mathrm{Ker}(\varphi_2)$ and $x \in \hat{F}_e(p)$. Then, $\varphi_1(x^\beta) = \varphi_1(x)^{\varphi_2(\beta)} = \varphi_1(x)$, so $x^\beta \mathrm{Ker}(\varphi_1) = x\mathrm{Ker}(\varphi_1)$. We have already noticed that $\varphi_1: \hat{F}_e(p) \to P$ is a Frattini cover. Thus, $\mathrm{Ker}(\varphi_1) \leq \Phi(\hat{F}_e(p))$, so $x^\beta \Phi(\hat{F}_e(p)) = x\Phi(\hat{F}_e(p))$. It follows that $\mathrm{Ker}(\varphi_2)$ is contained in the pro-p group K of Lemma 22.12.1, so $\mathrm{Ker}(\varphi_2)$ is a pro-p group. Since $p \nmid \#A$ and therefore $p \nmid \#B$, Schur-Zassenhaus (Lemma 22.10.1) gives a section φ_2' to φ_2.

The homomorphism $\varphi_2' \circ \psi: A \to \mathrm{Aut}(\hat{F}_e(p))$ defines an action of A on $\hat{F}_e(p)$ compatible with its action on P through φ_1. Construct the semidirect product $\hat{G} = A \ltimes \hat{F}_e(p)$ and extend φ_1 to an epimorphism $\varphi: \hat{G} \to G$ with $\varphi(\hat{F}_e(p)) = P$ and $\varphi(a) = a$ for each $a \in A$.

Since $p \nmid \#A$, the group $\hat{F}_e(p)$ is the p-Sylow group of $\hat{G}$. Thus, $\hat{G}$ is p-projective. In addition, $\mathrm{Ker}(\varphi) = \mathrm{Ker}(\varphi_1) \leq \Phi(\hat{F}_e(p)) \leq \Phi(\hat{G})$ (Lemma 22.1.4(c)). Hence, φ is Frattini. Consequently, $\varphi: \hat{G} \to G$ is a universal Frattini p-cover. □

Examples 22.12.3:

(a) Let p be an odd prime number. The dihedral group D_{p^n} of order $2p^n$ is generated by two elements σ, τ with the defining relations $\sigma^{p^n} = 1$, $\tau^2 = 1$, and $\sigma^\tau = \sigma^{-1}$ (Remark 21.7.1). Thus, $D_{p^n} = \pm 1 \ltimes C_{p^n}$, where C_{p^n} is a multiplicative copy of $\mathbb{Z}/p^n\mathbb{Z}$. By Proposition 22.12.2, the universal Frattini p-cover of D_{p^n} is $D_{p^\infty} = \pm 1 \ltimes C$, where C is a multiplicative copy of $\mathbb{Z}_p$. We denote the generator of ± 1 by t and the image of $1 \in \mathbb{Z}_p$ in C by s. Then $s^t = s^{-1}$.

(b) Proposition 22.12.2 has a global analog. Let $G = A \ltimes B$ be a semidirect product of profinite groups with $\gcd(\#A, \#B) = 1$. Then, the universal

Frattini cover of G is $\tilde{G} = \tilde{A} \ltimes \tilde{B}$, where $\tilde{A}$ and $\tilde{B}$ are the universal Frattini covers of A and B, respectively [Ribes2, p. 392, Thm. 3.2]. □

22.13 The Special Linear Group $\mathrm{SL}(2, \mathbb{Z}_p)$

Algebraic groups over $\mathbb{Z}_p$ are a source of interesting examples of Frattini p-covers. We describe one example in detail and refer the reader to the general result:

Remark 22.13.1: The group $\mathrm{SL}(2, \mathbb{Z}_p)$. Consider the group $\mathrm{SL}(2, \mathbb{Z}_p)$ of all 2×2 matrices with entries in $\mathbb{Z}_p$ and determinant 1. For each positive integer n, reduction modulo p^n defines an epimorphism $\pi_n \colon \mathrm{SL}(2, \mathbb{Z}_p) \to \mathrm{SL}(2, \mathbb{Z}/p^n\mathbb{Z})$. Let $\Gamma_n = \mathrm{Ker}(\pi_n) = \{s \in \mathrm{SL}(2, \mathbb{Z}_p) \mid s \equiv 1 \bmod p^n\}$. Since $\bigcap_{n=1}^{\infty} \Gamma_n = 1$, we have $\mathrm{SL}(2, \mathbb{Z}_p) = \varprojlim \mathrm{SL}(2, \mathbb{Z}/p^n\mathbb{Z})$. Thus, $\mathrm{SL}(2, \mathbb{Z}_p)$ is a profinite group.

Each element s of Γ_n has the form $1 + p^n u$ with $u = \begin{pmatrix} a & b \\ c & d \end{pmatrix}$ in $M(2, \mathbb{Z}_p)$ (the additive group of 2×2 matrices with entries in $\mathbb{Z}_p$). Then $1 = \det(s) = 1 + p^n(a + d) + p^{2n}(ad - bc)$, so $\mathrm{trace}(u) = a + d \equiv 0 \bmod p$. Hence, the map $\lambda \colon 1 + p^n u \mapsto u \bmod p$ is a homomorphism of Γ_n onto the hypersurface

$$\mathfrak{sl}(2, \mathbb{F}_p) = \{\bar{u} \in M(2, \mathbb{F}_p) \mid \mathrm{trace}(\bar{u}) = 0\}$$

of $M(2, \mathbb{F}_p)$ whose kernel is Γ_{n+1}. Note that $\dim(\mathfrak{sl}(2, \mathbb{F}_p)) = 3$. Thus, to prove that λ is surjective, it suffices to show that its image contains three linearly independent elements. Consider the following three elements: $u_1 = \begin{pmatrix} 0 & 1 \\ 0 & 0 \end{pmatrix}$, $u_2 = \begin{pmatrix} 0 & 0 \\ 1 & 0 \end{pmatrix}$, and $u_3 = \begin{pmatrix} 1 & 0 \\ 0 & -1 + b_{n+1}p + b_{n+2}p^2 + \cdots \end{pmatrix}$ where $b_{n+1}, b_{n+2}, \ldots$ are integers between 0 and $p - 1$ such that $(1 + p^n)^{-1} = 1 - p^n + b_{n+1}p^{n+1} + b_{n+2}p^{n+2} + \cdots$. Then

$$\lambda(1 + p^n u_1) = \begin{pmatrix} 0 & 1 \\ 0 & 0 \end{pmatrix}, \quad \lambda(1 + p^n u_2) = \begin{pmatrix} 0 & 0 \\ 1 & 0 \end{pmatrix}, \quad \lambda(1 + p^n u_3) = \begin{pmatrix} 1 & 0 \\ 0 & -1 \end{pmatrix}$$

are linearly independent, as desired. It follows that $\Gamma_n/\Gamma_{n+1} \cong (\mathbb{Z}/p\mathbb{Z})^3$. Thus, $(\Gamma_n : \Gamma_{n+1}) = p^3$ and $(\Gamma_1 : \Gamma_n) = p^{3(n-1)}$. Therefore, $\Gamma_1 = \varprojlim \Gamma_1/\Gamma_n$ is a pro-p group. Consequently, $\pi_1 \colon \mathrm{SL}(2, \mathbb{Z}_p) \to \mathrm{SL}(2, \mathbb{F}_p)$ is a p-cover. □

PROPOSITION 22.13.2: *Suppose* $p \geq 5$. *Then the map* $\pi_1 \colon \mathrm{SL}(2, \mathbb{Z}_p) \to \mathrm{SL}(2, \mathbb{F}_p)$ *is a Frattini p-cover.*

Proof: Consider a closed subgroup H of $\mathrm{SL}(2, \mathbb{Z}_p)$ which π_1 maps onto $\mathrm{SL}(2, \mathbb{F}_p)$. By Remark 22.13.1, it suffices to prove that $H = \mathrm{SL}(2, \mathbb{Z}_p)$. For this it suffices to prove that $\pi_n(H) = \mathrm{SL}(2, \mathbb{Z}/p^n\mathbb{Z})$ for all n. Inductively assuming this holds for n, it suffices to consider $s \equiv 1 \bmod p^n$ in $\mathrm{SL}(2, \mathbb{Z}_p)$ and to find $h \in H$ with $s \equiv h \bmod p^{n+1}$.

Write $s = 1 + p^n u$ with $u \in M(2, \mathbb{Z}_p)$. By Remark 22.13.1, $\mathrm{trace}(u) \equiv 0 \bmod p$.

22.13 The Special Linear Group $SL(2, \mathbb{Z}_p)$

CLAIM: *Each $v \in sl(2, \mathbb{F}_p)$ is the sum of matrices with zero squares.* Indeed, the square of each of the matrices

$$v_1 = \begin{pmatrix} 0 & 1 \\ 0 & 0 \end{pmatrix} \quad v_2 = \begin{pmatrix} 0 & 0 \\ 1 & 0 \end{pmatrix} \quad v_3 = \begin{pmatrix} 1 & -1 \\ 1 & -1 \end{pmatrix}$$

is zero. Moreover, v_1, v_2, v_3 are linearly independent, so they form a basis of $sl(2, \mathbb{F}_p)$. Thus, each $v \in sl(2, \mathbb{F}_p)$ is a sum of the v_i's, each taken several times.

Using the claim, we may write $u = \sum_{i=1}^{k} u_i$ with $u_i^2 \equiv 0 \mod p$, $i = 1, \ldots, k$. Suppose we have found $h_i \in H$ with $1 + p^n u_i \equiv h_i \mod p^{n+1}$, $i = 1, \ldots, k$. Then $h = \prod_{i=1}^{k} h_i \in H$ and $1 + p^n u \equiv \prod_{i=1}^{k}(1 + p^n u_i) \equiv h \mod p^{n+1}$. Thus, we may assume $u^2 \equiv 0 \mod p$.

The induction hypothesis, applied to $1 + p^{n-1}u$, gives $h_0 \in H$ and $v \in M(2, \mathbb{Z}_p)$ with $h_0 = 1 + p^{n-1}u + p^n v$. Let $h = h_0^p$. Then

$$h = 1 + p(p^{n-1}u + p^n v) + \binom{p}{2}(p^{n-1}u + p^n v)^2 + \cdots + (p^{n-1}u + p^n v)^p$$

$$= 1 + p^n(u + pv) + p^{2(n-1)}\binom{p}{2}(u + pv)^2 + \cdots + p^{p(n-1)}(u + pv)^p.$$

First suppose $n = 1$. Then, using $u^2 \equiv 0 \mod p$, we find that $(u + pv)^2 \equiv p(uv + vu) \mod p^2$. Hence, $(u+pv)^p \equiv p^2(uv+vu)^2(u+pv)^{p-4} \equiv 0 \mod p^2$, because $p \geq 5$. Thus, $h \equiv 1 + pu \mod p^2$.

If $n \geq 2$, then $1 + 2(n-1) \geq n + 1$. Therefore, $h \equiv 1 + p^n u \mod p^{n+1}$. $\square$

COROLLARY 22.13.3: *Suppose $p \geq 5$. Put*

$$u = \begin{pmatrix} 1 & 1 \\ 0 & 1 \end{pmatrix} \quad \text{and} \quad u' = \begin{pmatrix} 1 & 0 \\ 1 & 1 \end{pmatrix}.$$

(a) *Considered as elements of $SL(2, \mathbb{F}_p)$, the matrices u and u' have order p. Moreover, $SL(2, \mathbb{F}_p) = \langle u, u' \rangle$.*

(b) *Considered as elements of $SL(2, \mathbb{Z}_p)$, the matrices u and u' generate $SL(2, \mathbb{Z}_p)$.*

Proof: By Proposition 22.13.2, it suffices to prove (a). Indeed, $u^k = \begin{pmatrix} 1 & k \\ 0 & 1 \end{pmatrix}$ and $(u')^k = \begin{pmatrix} 1 & 0 \\ k & 1 \end{pmatrix}$ for every positive integer k. Thus, both u and u' are of order p.

Multiplying a matrix $s = \begin{pmatrix} a & b \\ c & d \end{pmatrix}$ in $SL(2, \mathbb{F}_p)$ on the left with u^k (resp. $(u')^k$) amounts to adding the second line (resp. first line) multiplied by k to the first line (resp. second line). Assuming $c \neq 0$, three successive row operations of this type map $\begin{pmatrix} a & b \\ c & d \end{pmatrix}$ to $\begin{pmatrix} 1 & b' \\ c & d \end{pmatrix}$, then to $\begin{pmatrix} 1 & b' \\ 0 & 1 \end{pmatrix}$, and finally to $\begin{pmatrix} 1 & 0 \\ 0 & 1 \end{pmatrix}$. Thus, $s \in \langle u, u' \rangle$.

If $c = 0$, then $a \neq 0$. Adding the first line to the second one, we make $c \neq 0$. Then we proceed as before. □

Dividing out $\mathrm{SL}(2, \mathbb{F}_p)$ by its center (which consists of the matrices ± 1), produces the **projective special linear group** $\mathrm{PSL}(2, \mathbb{F}_p)$. When $p \geq 5$, this group is simple [Lang7, p. 539, Thm. 8.4]. Likewise, $\mathrm{PSL}(2, \mathbb{Z}_p) = \mathrm{SL}(2, \mathbb{Z}_p)/\pm 1$. These groups fit into the following commutative diagram:

(1)
$$\begin{array}{ccc} \mathrm{SL}(2, \mathbb{Z}_p) & \xrightarrow{\pi_1} & \mathrm{SL}(2, \mathbb{F}_p) \\ {\scriptstyle \tilde{\gamma}}\downarrow & & \downarrow{\scriptstyle \gamma} \\ \mathrm{PSL}(2, \mathbb{Z}_p) & \xrightarrow{\bar{\pi}_1} & \mathrm{PSL}(2, \mathbb{F}_p), \end{array}$$

where both γ and $\tilde{\gamma}$ are the quotient maps with respect to ± 1 and $\bar{\pi}_1$ is induced by π_1.

COROLLARY 22.13.4: *Suppose $p \geq 5$. Then each of the maps in diagram (1) is a Frattini cover.*

Proof: First note that γ is Frattini. Indeed, suppose H is a subgroup of $\mathrm{SL}(2, \mathbb{F}_p)$ which is mapped onto $\mathrm{PSL}(2, \mathbb{F}_p)$. Then $(\mathrm{SL}(2, \mathbb{F}_p) : H) \leq 2$. Each of the generators u and u' of $\mathrm{SL}(2, \mathbb{F}_p)$ given by Corollary 22.13.3 has order p. Hence, $u, u' \in H$. Thus, $H = \mathrm{SL}(2, \mathbb{F}_p)$, which proves our claim.

By Proposition 22.13.2, π_1 is a Frattini cover. Hence, $\gamma \circ \pi_1$ is Frattini (Lemma 22.5.4(a)). Therefore, by Lemma 22.5.4(c), both $\bar{\pi}_1$ and $\tilde{\gamma}$ are Frattini. □

Remark 22.13.5: Group schemes. Let G be a simple simply-connected affine group scheme. Then reduction modulo p gives a Frattini cover $G(\mathbb{Z}_p) \to G(\mathbb{F}_p)$ for all $p \geq 3$ except for $\mathrm{SL}(2, \mathbb{Z}_3)$. When $p = 2$, the map is Frattini in all but at most ten cases [Weigel4, Thm. B]. □

Example 22.13.6: A $\mathcal{C}$-projective nonprojective group. Let $\mathcal{C}$ be a formation of finite groups. If $\mathcal{C}$ is full, then each $\mathcal{C}$-projective group is projective (Lemma 22.4.1). If $\mathcal{C}$ is not full, this is not necessarily the case, as the following example shows:

Let $\mathcal{C}$ be the Melnikov formation consisting of all finite groups whose composition groups are only A_5. By [Huppert, p. 183, Satz 6.14(3)], $\mathrm{PSL}(2, \mathbb{F}_5) \cong A_5$. By Lemma 22.3.6, the free pro-$\mathcal{C}$ group $F = \hat{F}_2(\mathcal{C})$ on two generators is $\mathcal{C}$-projective. By Lemma 22.13.3, $\mathrm{PSL}(2, \mathbb{F}_5)$ is generated by two elements. Let $\varphi \colon F \to \mathrm{PSL}(2, \mathbb{F}_5)$ be a epimorphism. By Corollary 22.13.4, the quotient map $\gamma \colon \mathrm{SL}(2, \mathbb{F}_5) \to \mathrm{PSL}(2, \mathbb{F}_5)$ is a Frattini cover.

Assume F is projective. Then there exists an epimorphism $\psi \colon F \to \mathrm{SL}(2, \mathbb{F}_5)$ with $\gamma \circ \psi = \varphi$ (Proposition 22.5.9(a)). Hence, $\mathrm{SL}(2, \mathbb{F}_5)$ is a $\mathcal{C}$-group, so $\mathbb{Z}/2\mathbb{Z} \cong \mathrm{Ker}(\gamma)$ is a $\mathcal{C}$-group. This means that $\mathbb{Z}/2\mathbb{Z} \cong A_5$. We conclude from this contradiction that F is not projective. □

22.14 The General Linear Group $\mathrm{GL}(2,\mathbb{Z}_p)$

Although the p-cover $\pi_1\colon \mathrm{SL}(2,\mathbb{Z}_p) \to \mathrm{SL}(2,\mathbb{F}_p)$ is Frattini if $p \geq 5$ (Proposition 22.13.2), it is far from being a universal Frattini p-cover. If it were, $\Gamma_1 = \mathrm{Ker}(\pi_1)$ would be a free pro-p group of finite rank (indeed 3). By Nielsen-Schreier (Proposition 17.5.7), the rank of the open subgroups of Γ_1 would tend to infinity with the index. This would contradict the result below saying the rank is bounded:

Notation 22.14.1: Let n be a positive integer and $G = \mathrm{GL}(n,\mathbb{Z}_p)$. For each positive integer k let $G_k = \{g \in G \mid g \equiv 1 \bmod p^k\}$. Then G_k is an open normal subgroup of G, $G_{k+1} \leq G_k$, and $\bigcap_{k=1}^\infty G_k = 1$. □

LEMMA 22.14.2: *Let p be an prime number and k a positive integer. Suppose $k \geq 2$ if $p = 2$. Then:*
(a) Every element of G_{k+1} is a pth power of an element of G_k.
(b) The map $\lambda_k\colon G_k/G_{k+1} \to G_{k+1}/G_{k+2}$ given by $gG_{k+1} \mapsto g^p G_{k+2}$ for $g \in G_k$ is an isomorphism.
(c) The map $(1+p^k a)G_{k+1} \to \bar{a}$, where $\bar{a}$ is the residue of a modulo p, is an isomorphism $\mu_k\colon G_k/G_{k+1} \cong M(n,\mathbb{F}_p)$. Thus, $G_{k+1} = G_k^p$ and G_k/G_{k+1} may be viewed as an $\mathbb{F}_p$-vector space of dimension n^2.
(d) G_k is a pro-p group, $\Phi(G_k) = G_{k+1}$, and $\mathrm{rank}(G_k) = n^2$.

Proof of (a): Consider $a \in M(n,\mathbb{Z}_p)$. We find $x \in M(n,\mathbb{Z}_p)$ with $(1+p^k x)^p = 1 + p^{k+1} a$.

To this end put $x_1 = a$. Then $(1+p^k x_1)^p \equiv 1 + p^{k+1}a \bmod p^{k+2}$. (Note the use of $k \geq 2$ when $p = 2$.) Let $r \geq 1$. By induction assume we have found $x_r \in M(n,\mathbb{Z}_p)$ which commutes with a such that $(1+p^k x_r)^p \equiv 1 + p^{k+1}a \bmod p^{k+r+1}$. Thus, there is a $b \in M(n,\mathbb{Z}_p)$ with $(1+p^k x_r)^p = 1 + p^{k+1}a + p^{k+r+1}b$. Then b commutes with a. Put $x_{r+1} = x_r - p^r(1+p^k x_r)^{-p+1}b$. Then x_{r+1} commutes with a and

$$(1+p^k x_{r+1})^p = (1+p^k x_r - p^{k+r}(1+p^k x_r)^{-p+1}b)^p$$
$$\equiv (1+p^k x_r)^p - p^{k+r+1}(1+p^k x_r)^{p-1}(1+p^k x_r)^{-p+1}b$$
$$\equiv (1+p^k x_r)^p - p^{k+r+1}b \equiv 1 + p^{k+1}a \bmod p^{k+r+2}.$$

This completes the induction.

The sequence x_r converges to an element $x \in M(n,\mathbb{Z}_p)$ which commutes with a and satisfies $(1+p^k x)^p = 1 + p^{k+1}a$, as needed.

Proof of (b): Consider an element $g = 1 + p^k a$ of G_k with $a \in M(n,\mathbb{Z}_p)$. Then $g^p \equiv 1 + p^{k+1}a \bmod p^{k+2}$ (again, we use $k \geq 2$ if $p = 2$), so there is a $b \in M(n,\mathbb{Z}_p)$ with

$$g^p = 1 + p^{k+1}a + p^{k+2}b = (1+p^{k+1}a)(1+p^{k+2}(1+p^{k+1}a)^{-1}b) \in G_{k+1}.$$

If in addition, $a \equiv 0 \bmod p$, then $g^p \in G_{k+2}$, so λ is well defined.

If $\lambda_k(gG_{k+1}) = 1$, then $a \equiv 0 \mod p$, so $g \in G_{k+1}$. This means that λ_k is injective. Finally, by (a), λ_k is surjective.

Proof of (c): The only nontrivial point is proving that μ_k is an isomorphism, is the multiplicativity of μ_k. This follows as in the proof of (b): $(1+p^k a)(1+p^k b) = 1 + p^k(a+b) + p^{2k}ab \in (1 + p^k(a+b))G_{k+1}$.

It follows from the isomorphism $G_k/G_{k+1} \cong M(n, \mathbb{F}_p)$ that $G_{k+1} \leq G_k^p$. Combining this with (a), we get $G_{k+1} = G_k^p$.

Proof of (d): For each $l \geq k$, (c) implies $G_l/G_{l+1} \cong (\mathbb{Z}/p\mathbb{Z})^{n^2}$, so G_k/G_l is a finite p-group. Thus, $G_k = \varprojlim G_k/G_l$ is a pro-p group. Also, $[G_k, G_k] \leq G_{k+1}$. Hence, $\Phi(G_k) = G_k^p[G_k, G_k] = G_{k+1}[G_k, G_k] = G_{k+1}$ (use (c) and Lemma 22.7.4). Consequently,

$$\mathrm{rank}(G_k) = \mathrm{rank}(G_k/G_{k+1}) = \dim_{\mathbb{F}_p}(G_k/G_{k+1}) = n^2$$

(Corollary 22.5.3). □

LEMMA 22.14.3: *Let P be a pro-p group and e a positive integer. Inductively define an ascending sequence of closed normal subgroups P_i of P by $P_1 = P$ and $P_{i+1} = P_i^p$. For each $i \geq 1$ suppose:*
(1a) $P_i/P_{i+1} \cong \mathbb{F}_p^e$.
(1b) *The map $xP_{i+1} \mapsto x^p P_{i+2}$ is an isomorphism $P_i/P_{i+1} \cong P_{i+1}/P_{i+2}$.*

Then $\mathrm{rank}(Q) \leq e$ for every closed subgroup Q of P.

Proof: Let N be an open normal subgroup of P. Then $(P : N) = p^l$ for some positive integer l. Hence, $\bigcap_{i=1}^\infty P_i \leq P^{p^l} \leq N$. It follows that $\bigcap_{i=1}^\infty P_i = 1$.

Let Q be a closed subgroup of P. Using compactness arguments, it suffices to prove $\mathrm{rank}(QP_n/P_n) \leq e$ for each n. Dividing out P by P^n, we may assume that P is finite, (1a) holds for $i = 1, \ldots, n-1$, and that (1b) holds for $i = 1, \ldots, n-2$.

Put $K = Q \cap P_2$ and $k = \mathrm{rank}(K)$. Then $Q/K \cong QP_2/P_2$ is a vector space over $\mathbb{F}_p$ (by (1a)) of dimension, say, d. Choose $x_1, \ldots, x_d \in Q$ with $Q = \langle x_1, \ldots, x_d \rangle K$. Then $QP_2 = \langle x_1, \ldots, x_d \rangle P_2$.

Since P_2 satisfies the same assumptions as P and has a smaller order, an induction hypothesis gives $k \leq e$. Since $P_3 = P_2^p$ and P_2/P_3 is Abelian, $P_3 = \Phi(P_2)$ (use Lemma 22.7.4). Hence, by Lemma 22.1.4, $\Phi(K) \leq P_3$.

Let $\pi \colon P/P_2 \to P_2/P_3$ be the isomorphism given by $\pi(xP_2) = x^p P_3$ (use (1b)). Note that $x_i^p \in Q \cap P_2 = K$. In addition, the elements $x_1, \ldots, x_d$ are linearly independent modulo K, so they are linearly independent modulo P_2. Hence, $x_1^p, \ldots, x_d^p$ are linearly independent modulo P_3. Thus, $x_1^p, \ldots, x_d^p$ are linearly independent modulo $\Phi(K)$. Therefore,

$$k = \mathrm{rank}(K) = \dim_{\mathbb{F}_p}(K/\Phi(K)) \geq \dim_{\mathbb{F}_p}(\langle x_1^p, \ldots, x_d^p \rangle \Phi(K)/\Phi(K))$$
$$\geq \dim_{\mathbb{F}_p}(\langle x_1^p, \ldots, x_d^p \rangle P_3/P_3)$$
$$= \dim_{\mathbb{F}_p}(\pi(\langle x_1, \ldots, x_d \rangle P_2/P_2))$$
$$= \dim_{\mathbb{F}_p}(QP_2/P_2) = \dim_{\mathbb{F}_p}(Q/K) = d.$$

Also, there are $y_1, \ldots, y_{k-d} \in K$ with $K = \langle x_1^p, \ldots, x_d^p, y_1, \ldots, y_{k-d}\rangle \Phi(K)$. Then $K = \langle x_1^p, \ldots, x_d^p, y_1, \ldots, y_{k-d}\rangle$, so

$$Q = \langle x_1, \ldots, x_d\rangle K = \langle x_1, \ldots, x_d, y_1, \ldots, y_{k-d}\rangle.$$

Consequently, $\mathrm{rank}(Q) \leq k \leq e$, as claimed. □

We are now in a position to prove that the rank of closed subgroups of $\mathrm{GL}(n, \mathbb{Z}_p)$ is bounded:

PROPOSITION 22.14.4: *Let m be maximum $\mathrm{rank}(A)$ with A ranging on all subgroups of $\mathrm{GL}(n, \mathbb{F}_p)$. Then $\mathrm{rank}(H) \leq n^2 + m$ for every closed subgroup H of $\mathrm{GL}(n, \mathbb{Z}_p)$.*

Proof: Use Notation 22.14.1. By Lemmas 22.14.2 and 22.14.3, $\mathrm{rank}(H \cap G_1) \leq n^2$. Using $\mathrm{GL}(n, \mathbb{Z}_p)/G_1 \cong \mathrm{GL}(n, \mathbb{F}_p)$, we find that $\mathrm{rank}(H) \leq \mathrm{rank}(H \cap G_1) + \mathrm{rank}(H/H \cap G_1) \leq n^2 + \mathrm{rank}(HG_1/G_1) \leq n^2 + m$, as claimed. □

Remark 22.14.5: *Characterization of p-adically linear profinite groups.* The converse of Proposition 22.14.4 is also true: Each pro-p group P with a bound on the rank of its closed subgroups is isomorphic to a subgroup of $\mathrm{GL}(n, \mathbb{Z}_p)$ for some n [Dixon-du.Sautoy-Mann-Segal, p. 155, Thm. 7.19]. Moreover, suppose a profinite group G has an open pro-p subgroup P whose closed subgroups have a bounded rank. Then the induced representation from P to G will embed G into $\mathrm{GL}(n', \mathbb{Z}_p)$ with n' possibly larger than n. See also [Lubotzky2, Thm. 1]. □

The group $\mathrm{SL}(2, \mathbb{Z}_p)$ is a Frattini cover of $\mathrm{PSL}(2, \mathbb{F}_p)$ (Corollary 22.13.4), but as noted at the beginning of this section, it is not the universal Frattini cover of $\mathrm{PSL}(2, \mathbb{F}_p)$.

PROBLEM 22.14.6: *Describe the universal Frattini p-cover of $\mathrm{PSL}(2, \mathbb{Z}_p)$ (hence of $\mathrm{PSL}(2, \mathbb{F}_p)$).*

Exercises

1. Use fiber products and Proposition 22.3.5 to prove the following: Let $\mathcal{C}$ be a full formation of finite groups and G a pro-$\mathcal{C}$-group. Suppose every short exact pro-$\mathcal{C}$ sequence $1 \to K \to H \to G \to 1$ in which K is a minimal Abelian p-elementary subgroup of H splits. Then G is $\mathcal{C}$-projective.

2. Let $\theta\colon H \to G$ be a Frattini cover of profinite groups. Prove that $\theta^{-1}(\Phi(G)) = \Phi(H)$. Deduce that if $\Phi(G) = 1$, then $G \cong H/\Phi(H)$.

3. Let G and H be closed subgroups of a free profinite group. Prove that $G/\Phi(G) \cong H/\Phi(H)$ implies $G \cong H$.

4. Let F be a profinite group and G, H closed subgroups. Suppose for each pair $\alpha\colon G \to C$, $\beta\colon H \to C$ of homomorphisms of profinite groups there is a unique homomorphism $\gamma\colon F \to C$ which extends both α and β.

Prove that $G \cap H = 1$, $\langle G, H\rangle = F$, and there is an isomorphism $F \to G * H$ whose restriction to both G and H is the identity map.

Hint: To prove that $F = \langle G, H\rangle$, extend the inclusion maps of G and H into $\langle G, H\rangle$ to a homomorphism $\gamma\colon F \to \langle G, H\rangle$. Then compose each of the three maps with the inclusion $\langle G, H\rangle \to F$.

5. Let n be a positive integer relatively prime to the order of a profinite group G. Prove that the map φ_n that takes g to g^n for $g \in G$ is surjective (although it may not be a homomorphism). Hint: φ_n induces a surjective map $G/N \to G/N$ for each open normal subgroup N of G.

6. Let $\mathcal{C}$ be a full family of finite groups. Suppose a prime p divides the order of a pro-$\mathcal{C}$-group. Use the Sylow theorems to prove that every p-group belongs to $\mathcal{C}$.

7. A celebrated theorem of Artin says: If the absolute Galois group $\mathrm{Gal}(K)$ of a field K is finite, then $|\mathrm{Gal}(K)| = 1$ or 2 [Lang7, p. 299, Cor. 9.3]. Use the Sylow theorems to generalize this as follows:

Let K be a field and $\#\mathrm{Gal}(K) = \prod p^{n_p}$. Prove the following statements:
(a) $n_2 = 0, 1$, or ∞.
(b) $n_2 = 1$ only if $\mathrm{char}(K) = 0$.
(c) If $p \neq 2$, then $n_p = 0$ or ∞.

8. [Ershov2] suggests the following alternative proof to Douady's theorem (Proposition 17.1.1):

(a) Let P be a pro-p-group and m a cardinal number with $P/\Phi(P) \cong \mathbb{F}_p^m$ (Lemma 22.7.1). Apply Corollary 22.5.10 to prove that there is an epimorphism $\pi\colon \hat{F}_m(p) \to P$. The image of a basis of $\hat{F}_m(p)$ is a set of generators of P that converges to 1.

(b) Let G be a profinite group. For each prime number p choose a p-Sylow group G_p of G. Prove that the collection of G_p generates G.

(c) For each p use (a) to choose a set of generators S_p converging to 1. Prove that $S = \bigcup S_p$ is a set of generators of G which converges to 1.

9. (Haran) Let $\varphi\colon G \to B$ and $\psi\colon G \to C$ be epimorphisms of profinite groups. Suppose that $\Phi(B) = 1$ and that ψ is a Frattini cover. Apply Lemma 22.5.5 to show there exists a unique epimorphism $\pi\colon C \to B$ such that $\pi \circ \psi = \varphi$.

10. [Lim] With A and B subsets of a profinite group G, denote the closed subgroup generated by all commutators $[a, b] = a^{-1}b^{-1}ab$ with $a \in A$ and $b \in B$ by $[A, B]$. The **lower central series** of G is the descending sequence of closed characteristic groups $G = G_1 \geq G_2 \geq G_3 \geq \ldots$ defined inductively by $G_{i+1} = [G_i, G]$, $i = 1, 2, \ldots$. Prove the equivalence of the following three conditions (Proposition 22.9.3):

(a) G is pronilpotent.
(b) G is the direct product of its p-Sylow groups.
(c) $\bigcap_{i=1}^{\infty} G_i = 1$.

Hint: Show that if $\varphi \colon G \to N$ is an epimorphism, then $\varphi([A,B]) = [\varphi(A), \varphi(B)]$ for all subgroups A and B of G. Then use the equivalence of the conditions for G finite.

11. [Ershov2, p. 361]) Call a Melnikov formation $\mathcal{C}$ of finite groups **admissible** if $\mathcal{C}$ is closed under taking finite Frattini covers. Prove:
(a) Every full family $\mathcal{D}$ of finite groups is admissible.
(b) For each $e \in \mathbb{N}$, the family of all $D \in \mathcal{D}$ of rank at most e is admissible.
Hint: Use $\hat{F}_e(\mathcal{D})$ to obtain D with $\mathrm{rank}(D) \leq e$.

12. Let g be an element of a profinite group G and α an element of $\hat{\mathbb{Z}}$ with p-component $\alpha(p) = 0$ for each prime p that divides $\#g = \#\langle g \rangle$. Prove that $g^\alpha = 1$ (Exercise 12 of Chapter 1).

13. Let $G = \prod_{i=1}^{\infty} G_i$ be a direct product of profinite groups of relatively prime orders. Let N be a closed normal subgroup of G. Prove that $N = \prod_{i=1}^{\infty}(N \cap G_i)$ and $G/N \cong \prod_{i=1}^{\infty} G_i/(N \cap G_i)$ are again direct products of profinite groups of relatively prime orders.

Hint: If N is open, there exists a positive integer k with $\prod_{i=k+1}^{\infty} G_i \leq N$. Therefore, each $n \in N$ can be presented as a product $n = g_1 \ldots g_k g$ with $g_i \in N \cap G_i$, $i = 1, \ldots, k$, and $g \in \prod_{i=k+1}^{\infty} G_i$. For each i with $1 \leq i \leq k$ consider the element $\alpha_i \in \hat{\mathbb{Z}}$ with component $\alpha_i(p) = 1$ if $p | \#G_i$ and $\alpha_i(p) = 0$ otherwise. Then $g_i = n^{\alpha_i}$ (Exercise 12).

14. Let S be a non-Abelian simple group and let $\tilde{S}$ be its universal Frattini cover.
(a) Observe that a nontrivial profinite group H is a quotient of $\tilde{S}$ if and only if there exists a Frattini cover $\pi \colon H \to S$ such that $\mathrm{Ker}(\pi) = \Phi(H)$ (Lemma 22.6.3).
(b) Deduce from (a) that each closed normal subgroup N of H, with $N \neq H$, is contained in $\mathrm{Ker}(\pi)$. In particular, $H = [H, H]$ is its own commutator.

15. Let G be a profinite group and H a closed subgroup. Prove that the universal Frattini cover of H is isomorphic to a closed subgroup of the universal Frattini cover of G. The same applies to the universal Frattini p-covers of G and H. Hint: Let $\tilde{\varphi} \colon \tilde{G} \to G$ be the universal Frattini cover of G. Use Lemma 22.5.6 to find a closed subgroup $\tilde{H}$ of $\tilde{\varphi}^{-1}(H)$ such that $\mathrm{res}_{\tilde{H}} \colon \tilde{H} \to H$ is a Frattini cover. Then $\tilde{H}$ is isomorphic to the universal Frattini cover of H.

16. (Efrat-Völklein) Let G be an infinite profinite group. Suppose each nontrivial closed subgroup of G is open. Follow the following steps to prove that $G \cong \mathbb{Z}_p$ for some prime number p:
(a) Use the Sylow theorems to prove that G is a pro-p-group for some p.

(b) Reduce to the case where G has a normal open subgroup N isomorphic to $\mathbb{Z}_p$.
(c) Choose $g \in G \smallsetminus N$ and prove that g acts trivially on N.
(d) Then prove $G = \langle g \rangle \cong \mathbb{Z}_p$.

Notes

Our earliest source material for projective groups is [Gruenberg]. Investigation of PAC fields from 1980 on brought refinement to the study of projective groups.

A family $\mathcal{C}$ of finite groups is **almost full** if $\mathcal{C}$ is closed under taking quotients, direct products, and subgroups but not necessarily under forming extensions. For example, the families of finite Abelian groups and finite nilpotent groups are almost full, but neither full nor Melnikov. Nevertheless, the theory of almost full families, as appears in [Fried-Jarden3], is not very exciting. In contrast, the theory of Melnikov formation is rather rich. We return to the latter theory in Chapter 25.

There is a vast literature on free products of profinite groups. We mention here only a few results:

Let $\mathcal{C}$ be a full formation of profinite groups. Denote the free product of finitely many pro-$\mathcal{C}$ groups G_i, $i \in I$, by $\coprod_{i \in I}^{\mathcal{C}} G_i$. This is the unique pro-$\mathcal{C}$ group G containing isomorphic copies of G_i satisfying: Given homomorphisms α_i of G_i into a pro-$\mathcal{C}$ group C, there is a unique homomorphism $\alpha \colon G \to C$ extending each α_i. Then $G_i^g \cap G_j \neq 1$ for $g \in G$ implies $i = j$ and $g \in G_i$ [Herfort-Ribes, Thm. B']. More on free products of finitely many pro-$\mathcal{C}$ groups can be found in [Ribes-Zalesskii, §9.1].

[Binz-Neukirch-Wenzel] defines a free pro-$\mathcal{C}$ product $G = \coprod_{i \in I} G_i$ of groups converging to 1 for full families $\mathcal{C}$ and proves that each open subgroup of G is again a free pro-$\mathcal{C}$ product of groups converging to 1.

Finally, [Haran3] and [Melnikov4] define free products of groups indexed by **profinite spaces**, that is inverse limits of discrete finite spaces.

Proposition 22.5.9(b) is called "Jarden's lemma" in [Matzat2, p. 231]. The proof that appears in [Matzat2], is a simplified version of the proof of Proposition 22.5.11. We deduce Proposition 22.5.9(b) from Proposition 22.5.8. The latter appears in a less explicit form in [Malle-Matzat, p. 270, Thm. 1.8], where it is attributed to Ikeda and Nobusawa.

Several authors have independently developed the theory of Frattini covers. We mention here [Cossey-Kegel-Kovacs], [Ershov-Fried], [Ershov2], [Haran-Lubotzky], and [Cherlin-v.d.Dries-Macintyre].

A cohomological proof that a closed subgroup of a projective group is projective (Corollary 22.4.7) can be found in [Ribes, p. 204]. [Cossey-Kegel-Kovacs] use wreath products to prove this. PAC fields provide still another proof (Exercise 1 of Chapter 23).

Likewise, [Ribes, p. 235, Thm. 6.5 and p. 211, Prop. 3.1] gives cohomological proofs to Propositions 22.10.4 and 22.11.1, respectively. Our proof is based on Brandis' non-cohomological proof of Gaschütz theorem (Lemma 22.10.2).

It is traditional to call Lemma 22.10.1 after Schur and Zassenhaus although they proved it only in the finite case where either H or G/H is solvable. For an arbitrary finite group the proof requires the Feit-Thompson theorem that every finite group of odd order is solvable.

Proposition 22.11.1 is applied in [Kuhlmann-Pank-Roquette] to split short exact sequences of Galois groups over Henselian fields: Let (K,v) be a Henselian field. Let $G_u = \{\sigma \in \mathrm{Gal}(K) \mid v(\sigma x - x) \geq 0 \text{ for each } x \in K_s \text{ with } v(x) \geq 0\}$ be the **inertia group** and $G_r = \{\sigma \in \mathrm{Gal}(K) \mid v\left(\frac{\sigma x}{x} - 1\right) > 0 \text{ for all } x \in K_s^\times\}$ the **ramification group** of $\mathrm{Gal}(K)$. Denote the fixed field of G_u (resp. G_r) in K_s by K_u (resp. K_r). Then the following two exact sequences split:

$$1 \longrightarrow \mathrm{Gal}(K_r/K_u) \longrightarrow \mathrm{Gal}(K_r/K) \longrightarrow \mathrm{Gal}(K_u/K) \longrightarrow 1$$
$$1 \longrightarrow \mathrm{Gal}(K_r) \longrightarrow \mathrm{Gal}(K) \longrightarrow \mathrm{Gal}(K_r/K) \longrightarrow 1$$

Frattini p-covers are discussed in [Fried13, Part II, p. 126]. Proposition 22.12.2 with outline of the proof is mentioned on [Fried13, p. 134].

Corollary 22.13.4 has special significance when $p = 5$. Then $\mathrm{PSL}(2, \mathbb{F}_5) \cong A_5$ [Huppert, p. 183, Satz 6.14]. Thus, $\mathrm{PSL}(2, \mathbb{Z}_5)$ is a Frattini 5-cover of A_5. The universal Frattini 5-cover of A_5 is discussed in [Fried13, pp. 129–136].

Chapter 23. PAC Fields and Projective Absolute Galois Groups

The absolute Galois group of a PAC field is projective (Theorem 11.6.2). But not every field with projective absolute Galois group is PAC (Example 23.1.4). Nevertheless, for each projective absolute Galois group G there is a PAC field K with $\mathrm{Gal}(K) \cong G$ (Corollary 23.1.2). Moreover, if $\mathrm{rank}(G) \leq \aleph_0$, we may choose K to be algebraic over $\mathbb{Q}$ (Theorem 23.2.3). It follows that the elementary theory of PAC fields of characteristic 0 is determined by the PAC fields which are algebraic over $\mathbb{Q}$ (Corollary 23.2.6).

Sections 23.3 -23.7 treat the theory of perfect PAC fields of corank bounded by a positive integer e that contain a fixed countable Hilbertian field K. Although the models of this theory of the form $\tilde{K}(\sigma_1, \ldots, \sigma_e)$ which are not e-free are indexed by a set $\boldsymbol{\sigma} \in G(K)^e$ of measure zero, they nevertheless determine the theory (Theorem 23.6.3). In analogy to the theory of perfect e-free PAC fields containing K (Section 20.5), each sentence is equivalent to a basic sentence modulo the theory of perfect PAC fields of corank $\leq e$ (as in Proposition 20.6.6). If K has elimination theory, this leads to a recursive decision procedure for the theory. Frattini covers play a decisive role in the proofs.

23.1 Projective Groups as Absolute Galois Groups

We characterize absolute Galois groups of PAC fields among all profinite groups as projective groups:

THEOREM 23.1.1: *Let L/K be a Galois extension, G a projective group, and $\alpha\colon G \to \mathrm{Gal}(L/K)$ an epimorphism. Then K has an extension E which is perfect, PAC, and linearly disjoint from L, and there exists an isomorphism $\gamma\colon \mathrm{Gal}(E) \to G$ such that $\alpha \circ \gamma = \mathrm{res}_L$.*

Proof: Proposition 1.3.3 gives a Galois extension F_1/E_1 with the following properties: E_1 is an extension of K, linearly disjoint from L, F_1 is an extension of L, and there is an isomorphism $\gamma_1\colon \mathrm{Gal}(F_1/E_1) \to G$ with $\alpha \circ \gamma_1 = \mathrm{res}_{F_1/L}$. Let E_2 be a regular extension of E_1 which is PAC (Proposition 13.4.6). Then $\mathrm{res}_{F_1 E_2/F_1}\colon \mathrm{Gal}(F_1 E_2/E_2) \to \mathrm{Gal}(F_1/E_1)$ is an isomorphism. In particular, $\mathrm{Gal}(F_1 E_2/E_2)$ is a projective group. Remark 22.4.2 gives an embedding $\theta\colon \mathrm{Gal}(F_1 E_2/E_2) \to \mathrm{Gal}(E_2)$ such that $\mathrm{res}_{E_{2,s}/F_1 E_2} \circ \theta$ is the identity on $\mathrm{Gal}(F_1 E_2/E_2)$. Let E be the maximal purely inseparable extension of the fixed field in $E_{2,s}$ of $\theta(\mathrm{Gal}(F_1 E_2/E_2))$. Put $\gamma = \gamma_1 \circ \mathrm{res}_{\tilde{E}/F_1}$. Then $\gamma\colon \mathrm{Gal}(E) \to G$ is an isomorphism and $\alpha \circ \gamma = \mathrm{res}_{\tilde{E}/L}$. In particular, $\mathrm{res}_{\tilde{E}/L}\colon \mathrm{Gal}(E) \to \mathrm{Gal}(L/K)$ is an epimorphism, so E and L are linearly disjoint over K. Since E is an algebraic extension of E_2, it is PAC (Corollary 11.2.5). $\square$

23.1 Projective Groups as Absolute Galois Groups

The special case $L = K$ gives a converse to Ax' theorem 11.6.2:

COROLLARY 23.1.2 ([Lubotzky-v.d.Dries, p. 44]): *Given a projective group G and a field K, there is an extension F of K which is perfect and PAC with $\mathrm{Gal}(F) \cong G$.*

The combination of Theorem 11.6.2 and Corollary 23.1.2 leads to the already announced characterization of absolute Galois groups of PAC fields among all profinite group:

COROLLARY 23.1.3: *A profinite G is projective if and only if G is isomorphic to the absolute Galois group of a PAC field K.*

Example 23.1.4: Projectivity of the Galois group does not imply PAC. For each prime number p, $\mathrm{Gal}(\mathbb{F}_p) \cong \hat{\mathbb{Z}}$ is a projective group (Corollary 22.4.5), but $\mathbb{F}_p$ is not a PAC field (Proposition 11.1.1). □

Remark 23.1.5: On the characterization of absolute Galois groups. Let G be a profinite group. By Leptin (Corollary 1.3.4), G is always isomorphic to a Galois group $\mathrm{Gal}(L/K)$ for some Galois extension L/K. However, G is not always an absolute Galois group. For example, by Artin-Schreier, if G is finite, then G is an absolute Galois group if and only if G is trivial, or $G \cong \mathbb{Z}/2\mathbb{Z}$ [Lang7, p. 299, Cor. 9.3]. Corollary 23.1.3 is another example of characterizing a class of projective groups as a class of absolute Galois groups of fields with a given arithmetic property (namely PAC). More examples of this nature are given in [Haran-Jarden2] (PRC versus real projectivity) and in [Haran-Jarden4] (PpC versus p-projectivity) and various generalizations thereof. However, the question of characterizing absolute Galois groups by topological and group theoretic means is still wide open. □

Here is a model theoretic application of Theorem 23.1.1:

THEOREM 23.1.6: *Let θ be a sentence of $\mathcal{L}(\mathrm{ring})$ and e a positive integer. Then:*
(a) *θ is true in every perfect PAC field of infinite corank if and only if θ is true in every perfect PAC field.*
(b) *θ is true in every perfect PAC field of finite corank at least e if and only if θ is true in every perfect PAC field of finite corank.*

Proof of (a): Suppose that θ is true in every perfect PAC field of infinite corank and let K be a perfect PAC field of finite corank. We prove that θ is true in K.

By Theorem 11.6.2, $\mathrm{Gal}(K)$ is projective. By Corollary 22.4.5, $\hat{F}_\omega(p)$ is projective for each p. Therefore, by Proposition 22.4.10, the free product $\mathrm{Gal}(K) * \hat{F}_\omega(p)$ is projective. Since $\mathrm{Gal}(K)$ is a quotient of $\mathrm{Gal}(K) * \hat{F}_\omega(p)$, Theorem 23.1.1 produces a perfect PAC field L_p, regular over K, with $\mathrm{Gal}(L_p) \cong \mathrm{Gal}(K) * \hat{F}_\omega(p)$.

Let $L = \prod L_p/\mathcal{D}$ be a nonprincipal ultraproduct of the fields L_p. Then L is a regular perfect PAC extension of K. Moreover, every finite quotient H of

Gal(K) is a Galois group over each of the fields L_p, so H is a Galois group over L (Remark 20.4.5(d)). Conversely, if H is a finite quotient of Gal(L), then H is a finite quotient of Gal(L_p)) for infinitely many p (Corollary 7.7.2). Choose p larger than $|H|$ to find an epimorphism Gal(K)$*\hat{F}_\omega(p) \to H$ with the factor $\hat{F}_\omega(p)$ in the kernel. Thus, H is a quotient of Gal(K). Therefore, Gal(K) and Gal(L) have the same finite quotients. Since Gal(K) is finitely generated, Gal(K) $\cong$ Gal(L) (Proposition 16.10.7(b)). Since L is a regular extension of K, the restriction map of Gal(L) is an epimorphism onto Gal(K). Hence, by Proposition 16.10.6(a), it is an isomorphism. It follows from Corollary 20.3.4 that L is an elementary extension of K.

By assumption, θ is true in each of the fields L_p. Hence, θ is true in L. Therefore, θ is true in K.

Proof of (b): Same as the proof of (a) except, choose L_p with Gal(L_p) $\cong$ Gal(K) $* \hat{F}_e(p)$. □

Remark 23.1.7: In contrast to Theorem 23.1.6, the theory of PAC fields does not coincide with the theory of PAC fields of finite corank. Remark 28.10.3 gives an example of a sentence θ which holds in every perfect PAC field of finite corank, but does not hold in every perfect PAC field. □

23.2 Countably Generated Projective Groups

For our model theoretic applications it suffices to consider only projective groups of rank $\leq \aleph_0$. This section realizes each such group as the absolute Galois group of a perfect PAC field which are algebraic over a given countable Hilbertian field K. We conclude that the perfect PAC fields algebraic over K determine the elementary theory of all perfect PAC fields over K.

Every finite embedding problem over K has a solution over a finite extension of K. This is the content of the following result:

LEMMA 23.2.1 ([Kuyk1, Thm. 3]): *Let K be a Hilbertian field, L a finite Galois extension, G a finite group, and $\alpha: G \to$ Gal(L/K) an epimorphism. Then there is a tower $K' \subseteq L' \subseteq M'$ of Galois extensions with the following properties: K'/K is a finite separable extension, $L \cap K' = K$, $LK' = L'$, and there is an isomorphism $\beta:$ Gal(M'/K') $\to G$ with $\alpha \circ \beta = \text{res}_{M'/L}$.*

Proof: Embed G in S_n for some positive integer n. Lemma 16.2.7 gives a Galois extension M_1/K with Gal(M_1/K) $= S_n$ and $M_1 \cap L = K$. Put $M' = LM_1$. Then, Gal(M'/K) $\cong$ Gal(L/K) $\times S_n$. Let G^* be the subgroup of Gal(M'/K) consisting of the pairs $(\sigma, \alpha(\sigma))$ for $\sigma \in G$. The projection $(\sigma, \alpha(\sigma)) \mapsto \sigma$, for $\sigma \in G$, gives an isomorphism $\beta: G^* \to G$. Let K' be the fixed field of G^* in M' and $L' = LK'$. Then $G^* =$ Gal(M'/K') and restriction of elements of G^* to L is given by projection of $(\sigma, \alpha(\sigma)) \in G^*$ onto the second factor. Consequently, $\alpha \circ \beta = \text{res}_{M'/L}$. □

23.2 Countably Generated Projective Groups

PROPOSITION 23.2.2: *Let K be a countable Hilbertian field, L a finite Galois extension of K, G a profinite group of rank at most $\aleph_0$, and $\alpha\colon G \to \mathrm{Gal}(L/K)$ a homomorphism. Then there are a PAC field E, separably algebraic over K, a Galois extension F of E, and an isomorphism $\beta\colon \mathrm{Gal}(F/E) \to G$ with a commutative diagram*

$$\begin{array}{ccc} & & \mathrm{Gal}(F/E) \\ & \beta\swarrow & \downarrow \mathrm{res} \\ G & \xrightarrow{\alpha} & \mathrm{Gal}(L/K). \end{array}$$

Proof: Example 17.1.7(a) gives a sequence

$$\cdots \xrightarrow{\alpha_2} G_2 \xrightarrow{\alpha_1} G_1 \xrightarrow{\alpha_0} G_0$$

of finite groups and epimorphisms with $G_0 = \alpha(G)$, $G = \varprojlim G_i$ and $\alpha = \varprojlim \alpha_i$. Order the nonconstant absolutely irreducible polynomials of $K[T, X]$ which are separable in X in a sequence $f_1, f_2, f_3, \ldots$. Apply Lemma 23.2.1 to inductively construct a sequence of finite Galois extensions, L_n/K_n, with the following properties:
(1a) K_0 is the fixed field in L of $\alpha(G)$ and $L_0 = L$.
(1b) K_n is a finite separable extension of K with a zero of f_n.
(1c) $K_n \subseteq K_{n+1}$, $L_n \subseteq L_{n+1}$, $L_n \cap K_{n+1} = K_n$.
(1d) There exists a commutative diagram

$$\begin{array}{ccc} \mathrm{Gal}(L_{n+1}/K_{n+1}) & \xrightarrow{\mathrm{res}} & \mathrm{Gal}(L_n/K_n) \\ \beta_{n+1}\downarrow & & \downarrow \beta_n \\ G_{n+1} & \xrightarrow{\alpha_n} & G_n \end{array}$$

in which β_n and β_{n+1} are isomorphisms.

The union $E = \bigcup_{n=0}^\infty K_n$ is a separable algebraic extension of K. By Theorem 11.2.5 it is a PAC field. Also, $F = \bigcup_{n=0}^\infty L_n E$ is a Galois extension of E and $L_n \cap E = K_n$, $n = 0, 1, 2, \ldots$. Thus, the diagram of (1d) induces a commutative diagram

$$\begin{array}{ccc} \mathrm{Gal}(L_{n+1}E/E) & \xrightarrow{\mathrm{res}} & \mathrm{Gal}(L_nE/E) \\ \tilde{\beta}_{n+1}\downarrow & & \downarrow \tilde{\beta}_n \\ G_{n+1} & \xrightarrow{\alpha_n} & G_n \end{array}$$

in which $\tilde{\beta}_n$ and $\tilde{\beta}_{n+1}$ are isomorphisms. Take β to be the inverse limit of $\{\tilde{\beta}_n\}$. $\square$

If G is projective, we may apply Remark 22.4.2 to the epimorphism res: $\mathrm{Gal}(E) \to \mathrm{Gal}(F/E)$ and then replace E by the maximal purely inseparable extension of the fixed field of the image of $\mathrm{Gal}(F/E)$ in E_s to get the following result:

THEOREM 23.2.3: *Let K be a countable Hilbertian field, L a finite Galois extension of K, G a profinite group of rank at most $\aleph_0$, and $\alpha\colon G \to \mathrm{Gal}(L/K)$ a homomorphism. Then there exists a perfect PAC field E, algebraic over K, and an isomorphism $\beta\colon \mathrm{Gal}(E) \to G$ with $\alpha \circ \beta = \mathrm{res}_{\tilde{E}/L}$.*

Now, for the model theoretic applications:

PROPOSITION 23.2.4: *Let K be a countable Hilbertian field and F a countable perfect PAC extension of K. Then F is K-elementarily equivalent to an ultraproduct $\prod_{n=1}^{\infty} E_n/\mathcal{D}$ of perfect PAC fields with E_n algebraic over K and $\mathrm{Gal}(E_n) \cong \mathrm{Gal}(F)$ for each $n \in \mathbb{N}$.*

Proof: Let $L_1 \subseteq L_2 \subseteq L_3 \subseteq \cdots$ be an ascending sequence of finite Galois extensions of K whose union is K_s. For each n, Theorem 23.2.3 gives a perfect PAC field E_n, algebraic over K, and an isomorphism β_n which makes the following diagram commutative, $n = 1, 2, \ldots$:

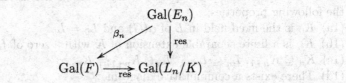

Let $\mathcal{D}$ be a nonprincipal ultraproduct of $\mathbb{N}$. Put $E^* = \prod E_n/\mathcal{D}$ and $F^* = F^{\mathbb{N}}/\mathcal{D}$. Lemma 20.3.1 gives an isomorphism β which makes the following diagram commutative:

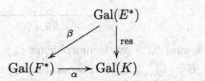

Since E^* and F^* are perfect PAC fields, the elementary equivalence theorem (Theorem 20.3.3) implies $E^* \equiv_K F^*$. Consequently, $E^* \equiv_K F$. $\square$

THEOREM 23.2.5: *Let K be a countable Hilbertian field and $\mathcal{P}$ a class of projective groups with this property: If E and F are elementary equivalent PAC fields and $\mathrm{Gal}(F) \in \mathcal{P}$, then $\mathrm{Gal}(E) \in \mathcal{P}$. Then a sentence θ of $\mathcal{L}(\mathrm{ring}, K)$ is true in all perfect PAC fields F with $K \subseteq F$ and $\mathrm{Gal}(F) \in \mathcal{P}$ if and only if θ is true in all perfect PAC fields E, algebraic over K, with $\mathrm{Gal}(E) \in \mathcal{P}$.*

Proof: Let F be a perfect PAC field containing K with $\mathrm{Gal}(F) \in \mathcal{P}$. By Skolem-Löwenheim (Proposition 7.4.2), F has a countable elementary subfield F_0 that contains K. By Proposition 23.2.4, $F_0 \equiv_K \prod E_i/\mathcal{D}$ with E_i a perfect PAC field, algebraic over K, and $\text{Gal}(E_i) \cong \text{Gal}(F_0)$, for each i. By assumption, $\text{Gal}(E_i) \in \mathcal{P}$. Hence, θ is true in E_i for each i. Therefore, θ is true in F. □

COROLLARY 23.2.6: Let K be a countable Hilbertian field. Then a sentence θ of $\mathcal{L}(\text{ring}, K)$ is true in each perfect PAC field which contains K if and only if θ is true in each perfect PAC field which is algebraic over K.

23.3 Perfect PAC Fields of Bounded Corank

Throughout the rest of this chapter, K is a fixed countable Hilbertian field and e is a fixed positive integer. We introduce four classes of fields:
PAC(K, $\leq$ e), the class of perfect PAC fields of corank $\leq e$ that contain K;
PAC(K, e), the class of perfect PAC fields of corank e that contain K;
$\text{NF}(K, e) = \{E \in \text{PAC}(K, e) \mid \text{Gal}(E) \neq \hat{F}_e\}$, and
$\text{FMP}(K, e) = \{E \in \text{PAC}(K, e) \mid \text{only finitely many primes divide } \#\text{Gal}(E)\}$.

Since each projective group is the absolute Galois group of a perfect PAC field that contains K (Corollary 23.1.2),

(1) $\quad \text{FMP}(K, e) \subset \text{NF}(K, e) \subset \text{PAC}(K, e) \subset \text{PAC}(K, \leq e)$.

The first three classes of fields in (1) have no characterization by axioms of $\mathcal{L}(\text{ring}, K)$ (i.e. they are **nonelementary**). Indeed, for each prime number p consider the projective group $\hat{F}_e(p)$ (Corollary 22.4.5) and choose a perfect PAC field E_p that contains K with $\text{Gal}(E_p) \cong \hat{F}_e(p)$ (Corollary 23.1.2). Let $E = \prod E_p / \mathcal{D}$. Then $E_p \in \text{FMP}(K, e)$ and $\text{Gal}(E)$ is trivial, so $E \notin \text{PAC}(K, e)$. By Corollary 7.7.2, a sentence of $\mathcal{L}(\text{ring}, K)$ which holds in a sequence of fields holds in each ultraproduct of those fields. It follows that none of the classes $\text{FMP}(K, e)$, or $\text{NF}(K, e)$, or $\text{PAC}(K, e)$ is elementary.

Nevertheless, PAC(K, $\leq$ e) is elementary. Thus, there is a set of axioms $\Delta(K, \leq e)$ of $\mathcal{L}(\text{ring}, K)$ such that for each structure E of $\mathcal{L}(\text{ring}, K)$, $E \models \Delta(K, \leq e)$ if and if $E \in \text{PAC}(K, \leq e)$. Indeed, by Proposition 20.4.4(a), the PAC axioms are elementary. Next, for each positive integer n list the finite groups of order at most $n!$ of rank at most e as $G_{n,1}, \ldots, G_{n,r(n)}$. Then include in $\Delta(K, \leq e)$ all statements saying that for each separable polynomial $f \in K[X]$ of degree n the Galois group $\text{Gal}(f, K)$ is isomorphic to $G_{n,i}$ for some i between 1 and $r(n)$. By Remark 20.4.5(d) each of these statements is elementary. If K is an explicitly given field with elimination theory (Section 19.2), then $\Delta(K, \leq e)$ can be explicitly given.

As previously, if $\mathcal{M}$ is a class of structures for a given language $\mathcal{L}$, denote the theory of $\mathcal{M}$ in $\mathcal{L}$ by $\text{Th}(\mathcal{M})$. This is the set of sentences of $\mathcal{L}$ which are true in every structure in $\mathcal{M}$. Applying this functor to the tower (1) we get

(2) $\text{Th}(\text{PAC}(K, \leq e)) \subseteq \text{Th}(\text{PAC}(K, e)) \subseteq \text{Th}(\text{NF}(K, e)) \subseteq \text{Th}(\text{FMP}(K, e))$.

Theorem 23.6.3 shows all four sets in (2) coincide. Moreover, if K is an explicitly given field with elimination theory, $\text{Th}(\text{PAC}(K, \leq e))$, and therefore each of these theories, is decidable.

23.4 Basic Elementary Statements

Let K be a countable Hilbertian field, e a positive integer, and E, F perfect PAC field extensions of K of corank at most e with $E \cap \tilde{K} = F \cap \tilde{K}$. A sufficient condition for E and F to be K-equivalent is the existence an isomorphism $\varphi: \text{Gal}(F) \to \text{Gal}(E)$ such that $\text{res}_{\tilde{F}/\tilde{K}} = \text{res}_{\tilde{E}/\tilde{K}} \circ \varphi$ (Theorem 20.3.3). To establish the existence of φ, it suffices to prove that a finite group A of rank at most e is a Galois group over E if and only if A is a Galois group over F. The proof of the second condition on φ is much more difficult, one has to prove that the same embedding problems over finite extensions of K for $\text{Gal}(E)$ and $\text{Gal}(F)$ are solvable.

We show that the solvability of embedding problems can be expressed by elementary statements and establish a procedure to check the solvability of embedding problems over PAC fields.

Let L be a finite separable extension of K and $h \in L[X]$ a monic irreducible separable Galois polynomial of degree m. Let A be a subgroup of $\text{Gal}(h, L)$ (again, considered as a permutation group) and $\alpha: B \to A$ an epimorphism of finite groups. Let $\theta_L(h, B \xrightarrow{\alpha} A)$ be the following statement about a field E that contains K:

(1a) There is a K-embedding of L in E (whose image we identify with L);
(1b) $\text{Gal}(h, E) = A$;
(1c) there is a monic irreducible normal separable polynomial $g \in E[X]$ of degree $|B|$ whose splitting field over E contains the splitting field of h; and
(1d) there is an isomorphism $\beta: \text{Gal}(g, E) \to B$ such that the following diagram is commutative

(2)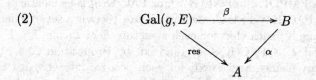

Remark 23.4.1: *Interpretation of $\theta_L(h, B \xrightarrow{\alpha} A)$ as an elementary statement.* Condition (1a) which implies that $\text{Gal}(h, E)$ is a subgroup of $\text{Gal}(h, L)$, gives sense to (1b). Let L_1 be the splitting field of h over L, and let L_0 be the fixed field of A (regarded as a subgroup of $\text{Gal}(L_1/L)$) in L_1. Then (1b) implies $L_1 \cap E = L_0$.

We produce a sentence θ in the language $\mathcal{L}(\text{ring}, K)$ such that
(3) $E \models \theta$ if and only if $\theta_L(h, B \xrightarrow{\alpha} A)$ is true in E,

for each field extension E of K.

Suppose $L = K(x_0)$ and $p(X) = \text{irr}(x_0, K)$. Then (1a) is equivalent to "$p(X)$ has a root x in E." Identify x with x_0. Working in $L[X]$ modulo $p(X)$, we may interpret $\mathcal{L}(\text{ring}, L)$ in $\mathcal{L}(\text{ring}, K)$. Thus, it suffices to produce θ in $\mathcal{L}(\text{ring}, L)$.

23.4 Basic Elementary Statements

For (1b), view A as a subgroup of S_m and use the proof of Proposition 20.4.4 to rewrite "$\text{Gal}(h, E) \cong A$" as an elementary statement $\varphi(x)$ about E, where $\varphi(X)$ is a formula in $\mathcal{L}(\text{ring}, L)$ with the free variable X. The coefficients of h are expressed as polynomials in x of degree less than $\deg(p)$.

In order to interpret (1c) in $\mathcal{L}(\text{ring}, L)$, consider B as a subgroup of S_n with $n = |B|$. Then the conjunction of (1c) and (1d) is equivalent to the following statement: "there exists a monic irreducible separable Galois polynomial $g(X)$ of degree n in $E[X]$ with Galois group B and roots $z_1, \ldots, z_n$ in K_s satisfying this: if $y_1, \ldots, y_m$ are the roots of h, then there exists a polynomial $q \in E[X]$ of degree at most $n - 1$ with $y_1 = q(z_1)$ and $y_{\alpha(\tau)(1)} = q(z_{\tau(1)})$ for each $\tau \in B$." Finally, use congruences modulo g and h, respectively, to eliminate reference to the elements $z_1, \ldots, z_n$ and $y_1, \ldots, y_m$ in the resulting elementary statement. $\square$

Call $\theta_L(h, B \xrightarrow{\alpha} A)$ a **basic statement**. If $\text{rank}(B) \le e$, call it an e-**basic statement**. Denote the Boolean algebra in $\mathcal{L}(\text{ring}, K)$ generated by all e-basic statements by $\mathcal{B}_e$.

Remark 23.4.2: Interpretation of embedding problems as elementary statements. Let $h \in K[X]$ be a monic irreducible separable normal polynomial, B a finite subgroup and

(4) $$\alpha \colon B \to \text{Gal}(h, K)$$

an epimorphism. Remark 23.4.1 interprets $\theta_K(h, B \xrightarrow{\alpha} \text{Gal}(h, K))$ as an elementary statement in $\mathcal{L}(\text{ring})$ with parameters in K, namely the coefficients of h. By (1c) and (1d), $\theta_K(h, B \xrightarrow{\alpha} \text{Gal}(h, K))$ is true in K if and only if the embedding problem (4) for $\text{Gal}(K)$ is solvable. Consequently, solvability of embedding problems over K is elementary. $\square$

The next lemma interprets the truth of special statements belonging to $\mathcal{B}_e$ in some perfect PAC field in terms of Frattini covers. Section 23.5 shows that any Boolean combination of basic statements is equivalent modulo $\text{Th}(\text{PAC}(K, \le e))$ to one of these special statements (given by (5)).

LEMMA 23.4.3: Let

$$\theta_L(h, B \xrightarrow{\alpha} A) \quad \text{and} \quad \theta_L(h, B_i \xrightarrow{\alpha'_i} A), \quad i = 1, \ldots, r,$$

be e-basic statements and θ the statement

(5) $$\theta_L(h, B \xrightarrow{\alpha} A) \wedge \bigwedge_{i=1}^{r} \neg \theta_L(h, B_i \xrightarrow{\alpha'_i} A).$$

Consider also the following assertion:

(C) For no i, $1 \leq i \leq r$, does there exist a commutative diagram

(6)
$$\begin{array}{ccccc} B & \xrightarrow{\varphi_i} & \bar{B}_i & \xleftarrow{\varphi'_i} & B'_i \\ & \searrow^{\alpha} & \downarrow^{\psi} & \swarrow^{\alpha'_i} & \\ & & A & & \end{array}$$

in which φ'_i is a Frattini cover and φ_i and ψ are epimorphisms.
Then:
(a) If there exists a perfect PAC field E, with $K \subseteq E$, in which θ is true, then (C) holds.
(b) If (C) holds, then there exists $E \in \mathrm{FMP}(K, e)$ in which θ is true.

Proof of (a): Suppose there is a perfect PAC field E which contains K in which θ is true. Let M be the splitting field of h over E. Then $A = \mathrm{Gal}(h, E) = \mathrm{Gal}(M/E)$. By (1), E has a finite Galois extension F that contains M and there exists a commutative diagram

(7)
$$\begin{array}{ccc} \mathrm{Gal}(F/E) & \xrightarrow{\beta} & B \\ & \searrow_{\mathrm{res}} \quad \swarrow_{\alpha} & \\ & A & \end{array}$$

Assume for some i, there exists a commutative diagram (6) in which φ'_i is a Frattini cover and φ_i and ψ are epimorphisms. Combine (6) with (7) to obtain the following commutative diagram

(8)
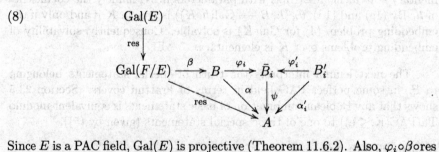

Since E is a PAC field, $\mathrm{Gal}(E)$ is projective (Theorem 11.6.2). Also, $\varphi_i \circ \beta \circ \mathrm{res}$ is an epimorphism. Since φ'_i is a Frattini cover, there is an epimorphism $\rho \colon \mathrm{Gal}(E) \to B'_i$ that completes (8) to a commutative diagram (Proposition 22.5.9(a)). The fixed field, F'_i, of $\mathrm{Ker}(\rho)$ is therefore a finite Galois extension of E that contains M, and the isomorphism $\beta'_i \colon \mathrm{Gal}(F'_i/E) \to B'_i$ induced by ρ makes the following diagram commutative

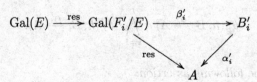

23.4 Basic Elementary Statements

That is, $\theta_L(h, B'_i \xrightarrow{\alpha} A)$ is true in E. This contradicts the assumption that θ is true in E.

Proof of (b): Conversely, suppose (C) holds. Let p be a prime number which divides none of the orders of $B, B'_1, \ldots, B'_r$. Consider the group $H = B \times (\mathbb{Z}/p\mathbb{Z})^e$. Let $b_1, \ldots, b_e$ be generators of B. For $i = 1, \ldots, e$, let v_i be the element of $(\mathbb{Z}/p\mathbb{Z})^e$ with 1 in the ith coordinate and 0 in all other coordinates. Since the orders of b_i and v_i are relatively prime, $(b_1, v_1), \ldots, (b_e, v_e)$ generate H. Since $(\mathbb{Z}/p\mathbb{Z})^e$ is of rank e, so is H. Let $\tilde{\gamma} \colon \tilde{H} \to H$ be the universal Frattini cover of H. Then $\tilde{H}$ is a projective group of rank e (Proposition 22.6.1). Moreover, the order of $\tilde{H}$ is divisible exactly by the prime numbers dividing the order of the finite group H (Proposition 22.10.5). Let M_0 be the splitting field of h over L and let L_0 be the fixed field of A in M_0. Theorem 23.2.3 gives a perfect PAC field E, algebraic over L, and a commutative diagram

(9)
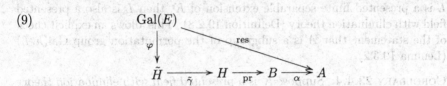

in which φ is an isomorphism. In particular, $E \in \mathrm{FMP}(K, e)$. We prove that θ is true in E.

Let M be the fixed field of $\mathrm{Ker}(\alpha \circ \mathrm{pr} \circ \tilde{\gamma} \circ \varphi) = \mathrm{Ker}(\mathrm{res})$ in $\tilde{E}$ and F the fixed field of $\mathrm{Ker}(\mathrm{pr} \circ \tilde{\gamma} \circ \varphi)$ in K_s. Then M is the splitting field of h over E and (9) induces the commutative diagram (7) in which β is an isomorphism. Thus, $\theta_L(h, B \xrightarrow{\alpha} A)$ is true in E.

Assume there is an i with $\theta_L(h, B_i \xrightarrow{\alpha} A)$ true in E. In particular, E has a Galois extension F' containing M that gives rise to a commutative diagram

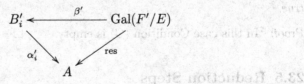

in which β' is an isomorphism. Then, with $\bar{F} = F \cap F'$, $\bar{B}_i = \mathrm{Gal}(\bar{F}/E)$, $\psi = \mathrm{res}_{\bar{F}/M}$, $\varphi_i = \mathrm{res}_{F/\bar{F}} \circ \beta^{-1}$, $\varphi'_i = \mathrm{res}_{F'/\bar{F}} \circ \beta'^{-1}$, we obtain a commutative diagram (6) in which φ_i, φ'_i, and ψ are epimorphisms. If φ'_i is shown to be a Frattini cover, then (C) is false, and this contradiction will conclude the lemma.

By Corollary 22.7.8 and Corollary 22.10.6, $\mathrm{Gal}(E) \cong \tilde{H} \cong \tilde{B} \times \hat{F}_e(p)$, so $\hat{F}_e(p)$ is the p-Sylow group of $\mathrm{Gal}(E)$ (Proposition 22.10.5). Denote the fixed field of $\hat{F}_e(p)$ in $\tilde{E}$ by N. Then $\mathrm{Gal}(N/E) \cong \tilde{B}$. Since the orders of $B = \mathrm{Gal}(F/E)$ and $B_i = \mathrm{Gal}(F'/E)$ are relatively prime to p, $FF' \subseteq N$. Hence, res: $\text{Gal}(N/E) \to \text{Gal}(F/E)$ is a Frattini cover. Hence, res: $\text{Gal}(FF'/E) \to \text{Gal}(F/E)$ is also Frattini (Lemma 22.5.4(c)). Since $F \cap F' = \bar{F}$,

$$\begin{array}{ccc}
\text{Gal}(FF'/E) & \xrightarrow{\text{res}} & \text{Gal}(F'/E) \\
{\scriptstyle\text{res}}\downarrow & & \downarrow{\scriptstyle\text{res}} \\
\text{Gal}(F/E) & \xrightarrow[\text{res}]{} & \text{Gal}(\bar{F}/E)
\end{array}$$

is a cartesian square (Example 22.2.7(a)). By Lemma 22.5.5, res: $\text{Gal}(F'/E) \to \text{Gal}(\bar{F}/E)$ (and hence φ_i) is Frattini. Consequently, θ is true in E. $\square$

Since $B, B_1, \ldots, B_r$ are finite groups, there is an effective check for condition (C). If, in addition, K is a presented field with elimination theory and L is a presented finite separable extension of K, then L is also a presented field with elimination theory (Definition 19.2.8). This allows an explicit check of the statement that A is a subgroup of the permutation group $\text{Gal}(h, L)$ (Lemma 19.3.2).

COROLLARY 23.4.4: *Suppose K is a presented field with elimination theory and θ is given by (5). Then there is a primitive recursive procedure for deciding the existence of a field $E \in \text{PAC}(K, \le e)$ in which θ is true.*

Proof: By Lemma 23.4.3, the truth of (C) is equivalent to the existence of a field $E \in \text{PAC}(K, \le e)$ in which θ is true. $\square$

COROLLARY 23.4.5: *If there exists a perfect PAC field E that contains K in which θ is true, then there exists a field $E' \in FMP(K, e)$ in which θ is true.*

COROLLARY 23.4.6: *Suppose A is a subgroup of $\text{Gal}(h, L)$ of rank at most e and $r = 0$ in (5). Then there exists a field $E \in FMP(K, e)$ in which θ is true.*

Proof: In this case Condition (C) is empty. $\square$

23.5 Reduction Steps

This section shows that any Boolean combination of e-basic statements is equivalent modulo $\text{Th}(\text{PAC}(K, \le e))$ to an elementary statement given by (3) of Section 23.4. As in that section, K is a countable Hilbertian field.

LEMMA 23.5.1: *Let L be a finite separable extension of K, $\theta_L(h, B \xrightarrow{\alpha} A)$ an e-basic statement, and $q \in L[X]$ a separable polynomial. Suppose h divides q. Then there are epimorphisms $\alpha_i \colon B_i \to A_i$, of finite groups with $\text{rank}(B_i) \le e$, $i = 1, \ldots, r$, such that $\theta_L(h, B \xrightarrow{\alpha} A)$ is equivalent modulo $\text{Th}(\text{PAC}(K, \le e))$ to the disjunction $\bigvee_{i=1}^{r} \theta_L(q, B_i \xrightarrow{\alpha_i} A_i)$.*

23.5 Reduction Steps

Proof: Let M and N be the splitting fields of h and q over L, respectively. Denote the fixed field of A in M by M_0. Then $M \subseteq N$. Let $N_1, \ldots, N_r$ be the fields satisfying

(1) $\qquad N_i \subseteq N \quad \text{and} \quad M_0 = M \cap N_i, \qquad i = 1, \ldots, r.$

For each i, let $A_i = \mathrm{Gal}(N/N_i)$ (regarded as a subgroup of $\mathrm{Gal}(q, L)$). Find all subgroups B_{ij} of $B \times A_i$ of rank $\leq e$ with these properties: The restriction, α_{ij}, of the projection onto A_i and the restriction of pr_B to B_{ij} are surjective and the diagram

(2)
$$\begin{array}{ccccc} B_{ij} & \xrightarrow{\alpha_{ij}} & A_i & = & \mathrm{Gal}(N/N_i) \\ {\scriptstyle \mathrm{pr}_B} \downarrow & & \downarrow {\scriptstyle \mathrm{res}} & & \downarrow {\scriptstyle \mathrm{res}} \\ B & \xrightarrow{\alpha} & A & = & \mathrm{Gal}(M/M_0) \end{array}$$

is commutative. We prove $\theta_L(h, B \xrightarrow{\alpha} A)$ is equivalent modulo $\mathrm{Th}(\mathrm{PAC}(K, \leq e))$ to $\bigvee_{i,j} \theta_L(q, B_{ij} \xrightarrow{\alpha_{ij}} A_i)$.

Indeed, let $E \in \mathrm{PAC}(K, \leq e)$ be a field in which $\theta_L(h, B \xrightarrow{\alpha} A)$ is true. Then $M_0 = M \cap E$ and there exists a commutative diagram

(3)
$$\begin{array}{ccc} \mathrm{Gal}(F/E) & \xrightarrow{\beta} & B \\ {\scriptstyle \mathrm{res}} \downarrow & & \downarrow {\scriptstyle \alpha} \\ \mathrm{Gal}(ME/E) & = & A \end{array}$$

in which F is a Galois extension of E and β is an isomorphism. Thus, $N_i = N \cap E$ is one of the fields for which (1) holds. Also, $F_i = NF$ is a finite Galois extension of E.

$$\begin{array}{ccccc} N & \text{---} & NE & \text{---} & F_i \\ | & & | & & | \\ M & \text{---} & MN_i & \text{---} & ME & \text{---} & F \\ | & & | & & | \\ M_0 & \text{---} & N_i & \text{---} & E \\ | \\ L \end{array}$$

In particular, $\mathrm{rank}(\mathrm{Gal}(F_i/E)) \leq \mathrm{rank}(\mathrm{Gal}(E)) \leq e$. Hence, $\mathrm{Gal}(F_i/E)$ is isomorphic to one of the groups B_{ij} satisfying (2). Moreover, there is a commutative diagram

(4)
$$\begin{array}{ccc} \mathrm{Gal}(F_i/E) & \xrightarrow{\beta_{ij}} & B_{ij} \\ {\scriptstyle \mathrm{res}} \downarrow & & \downarrow {\scriptstyle \alpha_{ij}} \\ \mathrm{Gal}(NE/E) & = & A_i \end{array}$$

in which β_{ij} is an isomorphism and α_{ij} is an epimorphism. This means that $E \models \theta_L(q, B_{ij} \xrightarrow{\alpha_{ij}} A_i)$.

Conversely, suppose $E \in \text{PAC}(K, \leq e)$ and $E \models \theta_L(q, B_{ij} \xrightarrow{\alpha_{ij}} A_j)$ for one of the groups B_{ij} satisfying (2). Then there exists a Galois extension F_i of E and an isomorphism β_{ij}: $\text{Gal}(F_i/E) \to B_{ij}$ with (4) commutative. Denote the fixed field in F_i of the kernel of the projection map $B_{ij} \to B$ by F. This gives a commutative diagram

$$\begin{array}{ccc} \text{Gal}(F/E) & \xrightarrow{\beta} & B \\ \text{res} \downarrow & & \downarrow \alpha \\ \text{Gal}(M/M_0) & = & A \end{array}$$

in which β is an isomorphism. Consequently, $E \models \theta_L(h, B \xrightarrow{\alpha} A)$. □

LEMMA 23.5.2: Let $\theta_L(h, B_i \xrightarrow{\alpha_i} A)$, be an e-basic statement, $i = 1, \ldots, r$. Then there are e-basic statements $\theta_L(h, C_j \xrightarrow{\rho_j} A)$, $j = 1, \ldots, s$, such that $\bigwedge_{i=1}^{r} \theta_L(h, B_i \xrightarrow{\alpha_i} A)$ is equivalent modulo $\text{Th}(\text{PAC}(K, \leq e))$ to the disjunction $\bigvee_{j=1}^{s} \theta_L(h, C_j \xrightarrow{\rho_j} A)$.

Proof: Consider the subgroups C of $B_1 \times \cdots \times B_r$ of rank at most e, with these properties: The projection pr_i of $B_1 \times \cdots \times B_r$ onto B_i maps C surjectively onto B_i, $i = 1, \ldots, r$; and $\alpha_i \circ \text{pr}_i = \alpha_{i'} \circ \text{pr}_{i'}$ on C, for all i and i' between 1 and r. List these groups as $C_1, \ldots, C_s$ and define ρ_j to be $\alpha_i \circ \text{pr}_i$ restricted to C_j, $j = 1, \ldots, s$ (this is independent of i). Then (C_j, ρ_j), $j = 1, \ldots, s$, satisfy the conclusion of the lemma. Note: If no such group C exists, then the conjunction has no model in $\text{PAC}(K, \leq e)$. In this case, regard the empty disjunction as a false sentence modulo $\text{Th}(\text{PAC}(K, \leq e))$. □

PROPOSITION 23.5.3: Let θ_i be an e-basic statement, $i = 1, \ldots, r$, and P a Boolean polynomial in r variables. Put $\theta = P(\theta_1, \ldots, \theta_r)$. The following holds:

(a) If θ has a model $E \in \text{PAC}(K, \leq e)$, then it has a model $E' \in \text{FMP}(K, e)$; and

(b) If K is a presented field with elimination theory, then it is possible to decide (primitive recursively) if θ has a model $E \in \text{PAC}(K, \leq e)$.

Proof: Write θ in normal form as

$$\bigvee_i [\bigwedge_j \theta_{L_{ij}}(h_{ij}, B_{ij} \xrightarrow{\alpha_{ij}} A_{ij}) \wedge \bigwedge_k \neg \theta_{L_{ik}}(h_{ik}, B_{ik} \xrightarrow{\alpha_{ik}} A_{ik})]$$

where i, j and k range over finite sets. The theorem follows from Corollaries 23.4.4, 23.4.5 and 23.4.6 if we can reduce to the case that θ is of the form given by (5) of Section 23.4.

23.5 Reduction Steps

First observe that if (a) and (b) are true for each of the disjuncts, then they are true for θ. We may therefore assume that θ is

$$\bigwedge_{i\in I} \theta_{L_i}(h_i, B_i \xrightarrow{\alpha_i} A_i) \wedge \bigwedge_{j\in J} \neg \theta_{L_j}(h_j, B_j \xrightarrow{\alpha_j} A_j),$$

where I and J are disjoint finite sets. The reduction of θ to the form (5) of Section 22.2 is divided into five parts:

PART A: *Reduction to the case that $\alpha_i = $ id. for some $i \in I$.* Let N be a finite Galois extension of K that contains L_i and the splitting field of h_i over L_i, for each $i \in I$; and L_j and the splitting field of h_j over L_j, for each $j \in J$. Choose an irreducible polynomial $h \in K[X]$ with a root that generates N over K. Let $\{N_r \mid r \in R\}$ be the set of all fields between K and N with $A_r = \mathrm{Gal}(N/N_r)$ generated by e elements. If $E \in \mathrm{PAC}(K, \leq e)$, then $N \cap E = N_r$ for some $r \in R$. The latter equality is equivalent to $E \models \theta_{N_r}(h, A_r \xrightarrow{\mathrm{id}} A_r)$. Therefore, θ is equivalent modulo $\mathrm{Th}(\mathrm{PAC}(K, \leq e))$ to the nonempty disjunction

$$\bigvee_{r\in R} [\theta_{N_r}(h, A_r \xrightarrow{\mathrm{id}} A_r) \wedge \bigwedge_{i\in I} \theta_{L_i}(h_i, B_i \xrightarrow{\alpha_i} A_i) \wedge \bigwedge_{j\in J} \neg \theta_{L_j}(h_j, B_j \xrightarrow{\alpha_j} A_j)]$$

Again, it suffices to consider each of the disjuncts separately. Hence, assume θ is

$$\theta_{N_r}(h, A_r \xrightarrow{\mathrm{id}} A_r) \wedge \bigwedge_{i\in I} \theta_{L_i}(h_i, B_i \xrightarrow{\alpha_i} A_i) \wedge \bigwedge_{j\in J} \neg \theta_{L_j}(h_j, B_j \xrightarrow{\alpha_j} A_j)$$

PART B: *Reduction to the case with $N_r = L_i = L_j$ for all $i \in I$ and $j \in J$.* Suppose θ has a model E in $\mathrm{PAC}(K, \leq e)$ and let $i \in I$. Denote the splitting field of h_i over L_i by $\hat{L}_i$. Let L'_i be the fixed field of A_i in $\hat{L}_i$. Then $L'_i = \hat{L}_i \cap E = \hat{L}_i \cap N \cap E = \hat{L}_i \cap N_r \subseteq N_r$ for each $i \in I$ where N is the field of Part A. We may thus replace L_i by N_r for each $i \in I$. Similarly define $\hat{L}_j$ and L'_j for each $j \in J$. Assume $L'_j \not\subseteq N_r$ for some j. Then θ and the conjunction without $\neg \theta_{L_j}(h_j, B_j \xrightarrow{\alpha_j} A_j)$ have the same models in $\mathrm{PAC}(K, \leq e)$. Hence, we may drop $\neg \theta_{L_j}(h_j, B_j \xrightarrow{\alpha_j} A_j)$ from the conjunction. Therefore, we may replace L_j by N_r for each $j \in J$.

Thus, without loss, assume θ is

$$\bigwedge_{i\in I} \theta_L(h_i, B_i \xrightarrow{\alpha_i} A_i) \wedge \bigwedge_{j\in J} \neg \theta_L(h_j, B_j \xrightarrow{\alpha_j} A_j),$$

where I is nonempty.

PART C: *Reduction to the case where all h_i and h_j are the same.* Let $h = \text{lcm}(h_i, h_j \mid i \in I, j \in J)$. By Lemma 23.5.1, $\theta_L(h_i, B_i \xrightarrow{\alpha_i} A_i)$ is equivalent modulo the theory $\text{PAC}(K, \le \text{e})$ to a disjunction of the form $\bigvee_k \theta_L(h, D_k \xrightarrow{\alpha_k} C_k)$. If this disjunction is empty, then θ has no model. Otherwise, consider each of the disjuncts separately to assume that θ is

$$\bigwedge_{i \in I} \theta_L(h, B_i \xrightarrow{\alpha_i} A_i) \wedge \bigwedge_{j \in J} \neg \theta_L(h_j, B_j \xrightarrow{\alpha_j} A_j).$$

Similarly $\theta_L(h_j, B_j \xrightarrow{\alpha_j} A_j)$ is equivalent to a disjunction as above. Thus, we may assume θ is

$$\bigwedge_{i \in I} \theta_L(h, B_i \xrightarrow{\alpha_i} A_i) \wedge \bigwedge_{j \in J} \neg \theta_L(h, B_j \xrightarrow{\alpha_j} A_j),$$

where I is nonempty.

PART D: *Reduction to the case that all A_i's and A_j's are the same.* Suppose there are $i_1, i_2 \in I$ with $A_{i_1} \ne A_{i_2}$. Since it is impossible for a field $E \in \text{PAC}(L, \le \text{e})$ to exist with $\text{Gal}(h, E) = A_{i_1}$ and $\text{Gal}(h, E) = A_{i_2}$, θ has no model. We may therefore assume that $A_i = A$ for all $i \in I$. In particular, if θ holds in $E \in \text{PAC}(L, \le \text{e})$, then $\text{Gal}(h, E) = A$, because $I \ne \emptyset$. If there is a $j \in J$ with $A_j \ne A$, then $E \models \neg \theta_L(h, B_j \xrightarrow{\alpha_j} A_j)$ and the corresponding conjunct can be dropped from the conjunction. Thus, we may assume $A_j = A$ for each $j \in J$ and θ is

$$\bigwedge_{i \in I} \theta_L(h, B_i \xrightarrow{\alpha_i} A) \wedge \bigwedge_{j \in J} \neg \theta_L(h, B_j \xrightarrow{\alpha_j} A).$$

PART E: *Conclusion.* The conjunction $\bigwedge_{i \in I} \theta_L(h, B_i \xrightarrow{\alpha_i} A)$ is equivalent modulo $\text{Th}(\text{PAC}(K, \le \text{e}))$ to a disjunction of the form $\bigvee_{s \in S} \theta_L(h, C_s \xrightarrow{\rho_s} A)$ (Lemma 23.5.2). If S is empty, then θ has no model. Otherwise, consider each of the disjuncts separately to assume θ is

$$\theta_L(h, C \xrightarrow{\rho} A) \wedge \bigwedge_{j \in J} \neg \theta_L(h, B_j \xrightarrow{\alpha_j} A).$$

This concludes the proof of the proposition. □

23.6 Application of Ultraproducts

This section applies ultraproducts to show that each sentence of $\mathcal{L}(\text{ring}, K)$ is equivalent modulo $\text{Th}(\text{PAC}(K, \le \text{e}))$ to a boolean combination of basic elementary statements. With this the results of Section 23.4 give the proof of the main theorem:

23.6 Application of Ultraproducts

LEMMA 23.6.1: *Let E and F be models of* $\mathrm{Th}(\mathrm{PAC}(K, \leq e))$. *Suppose*

$$E \models \theta_L(h, B \xrightarrow{\pi} A) \iff F \models \theta_L(h, B \xrightarrow{\pi} A)$$

for each e-basic statement $\theta_L(h, B \xrightarrow{\pi} A)$. Then E and F are K-elementarily equivalent.

Proof: Our assumption implies, in particular, that $K_s \cap E \cong_K K_s \cap F$. Since E and F are perfect fields, $\tilde{K} \cap E \cong_K \tilde{K} \cap F$. Identify $\tilde{K} \cap E$ with $\tilde{K} \cap F$ and denote it by M. For each positive integer n, let the composite of all extensions of E (resp. M and F) of degrees at most n be E_n (resp. M_n and F_n). Since $\mathrm{Gal}(E)$ and $\mathrm{Gal}(F)$ are finitely generated, E_n, M_n and F_n are finite Galois extensions of E, M and F, respectively. Moreover, $M_n \subseteq E_n$ and $M_n \subseteq F_n$. Use the primitive element theorem to prove that there exists a finite separable extension L of K, with L contained in M, and a separable polynomial $h \in L[X]$ with splitting field L_n such that $L_n \cap M = L$ and $L_n M = M_n$. Then $\mathrm{Gal}(h, L) = \mathrm{Gal}(h, M) = \mathrm{Gal}(h, E) = \mathrm{Gal}(h, F)$. The e-basic statement $\theta_L(h, \mathrm{Gal}(E_n/E) \xrightarrow{\mathrm{res}} \mathrm{Gal}(h, L))$ is true in E. Hence, it is also true in F. Thus, F has a finite Galois extension F'_n with a commutative diagram

(1)
$$\mathrm{Gal}(F'_n/F) \xrightarrow{\beta} \mathrm{Gal}(E_n/E)$$
$$\searrow_{\mathrm{res}} \quad \swarrow_{\mathrm{res}}$$
$$\mathrm{Gal}(M_n/M)$$

in which β is an isomorphism. In particular, from the definition of E_n, F_n is the composite of extensions of F of degree at most n. Thus, $F'_n \subseteq F_n$, and $[E_n : E] = [F'_n : F] \leq [F_n : F]$. Interchange the roles of E and F to get the inequality $[F_n : F] \leq [E_n : E]$. Therefore, $[F_n : F] = [E_n : E]$, $F'_n = F_n$ and (1) turns out to be the diagram

(2)
$$\mathrm{Gal}(F_n/F) \xrightarrow{\beta} \mathrm{Gal}(E_n/E)$$
$$\searrow_{\mathrm{res}} \quad \swarrow_{\mathrm{res}}$$
$$\mathrm{Gal}(M_n/M)$$

This proves that the finite set B_n of all isomorphisms β for which the diagram (2) is commutative, is nonempty. The groups $\mathrm{Gal}(F_{n+1}/F_n)$ and $\mathrm{Gal}(E_{n+1}/E_n)$ are the intersections of all subgroups of $\mathrm{Gal}(F_{n+1}/F)$ and $\mathrm{Gal}(E_{n+1}/E)$, respectively, of index at most n. Hence, if $\gamma \in B_{n+1}$, then $\gamma(\mathrm{Gal}(F_{n+1}/F_n)) = \mathrm{Gal}(E_{n+1}/E_n)$. Therefore, γ induces an element of B_n. Thus, the collection $\{B_n\}_{n=1}^\infty$ forms an inverse system of finite nonempty sets.

By Lemma 1.1.3, $\varprojlim B_n$ is nonempty. Each element in the inverse limit defines an isomorphism φ such that the following diagram is commutative

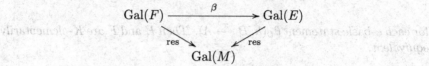

By Theorem 20.3.3, $E \equiv_K F$. □

Since the language $\mathcal{L}(\mathrm{ring}, K)$ is countable, each model of $\mathrm{Th}(\mathrm{PAC}(K, \leq e))$ has a countable submodel (Skolem-Löwenheim). Thus, the equivalence classes of $\mathrm{Th}(\mathrm{PAC}(K, \leq e))$ have a complete system of representatives $\{E_i \mid i \in I\}$ with $|I| \leq 2^{\aleph_0}$ and E_i is countable for each $i \in I$.

For a sentence θ of $\mathcal{L}(\mathrm{ring}, K)$, let $S(\theta) = \{i \in I \mid E_i \models \theta\}$. Let $\mathcal{S}$ be the Boolean algebra of I which is generated by all sets of the form $S(\theta_L(h, B \xrightarrow{\pi} A))$ with $\theta_L(h, B \xrightarrow{\pi} A)$ an e-basic statement.

PROPOSITION 23.6.2: *For each* $\theta \in \mathcal{L}(\mathrm{ring}, K)$, $S(\theta) \in \mathcal{S}$.

Proof: Assume $S(\theta) \notin \mathcal{S}$. Then, there are ultrafilters $\mathcal{E}$ and $\mathcal{F}$ of I with $\mathcal{E} \cap \mathcal{S} = \mathcal{F} \cap \mathcal{S}$, $S(\theta) \in \mathcal{E}$, but $S(\theta) \notin \mathcal{F}$ (Lemma 7.6.2). Let $E = \prod E_i/\mathcal{E}$ and $F = \prod E_i/\mathcal{F}$. Then the same e-basic statements hold in E and in F. By Lemma 23.6.1, $E \equiv_K F$. Since θ is true in E, but false in F, this is a contradiction. □

For the main result, use the notation of Section 23.1:

THEOREM 23.6.3: *Let K be a countable Hilbertian field. Then:*
(a) $\mathrm{Th}(\mathrm{FMP}(K, e)) = \mathrm{Th}(\mathrm{PAC}(K, \leq e))$.
(b) *If K is a presented field with elimination theory, then* $\mathrm{Th}(\mathrm{PAC}(K, \leq e))$ *is (recursively) decidable [Cherlin-v.d.Dries-Macintyre].*

Proof of (a): By (2) of Section 23.3, $\mathrm{Th}(\mathrm{PAC}(K, \leq e)) \subseteq \mathrm{Th}(\mathrm{FMP}(K, e))$. Conversely, let $\eta \in \mathrm{Th}(\mathrm{FMP}(K, e))$. Assume $\eta \notin \mathrm{Th}(\mathrm{PAC}(K, \leq e))$. Then there exists a model $E \in \mathrm{PAC}(K, \leq e)$ with $E \models \neg\eta$. By Proposition 23.6.2, η is equivalent modulo $\mathrm{Th}(\mathrm{PAC}(K, \leq e))$ to a Boolean combination θ of e-basic statements. In particular, $E \models \neg\theta$. By Proposition 23.5.3, $\neg\theta$ has a model $E' \in \mathrm{FMP}(K, e)$. It follows that $E' \models \neg\theta$. This contradicts the assumption that $\eta \in \mathrm{Th}(\mathrm{FMP}(K, e))$ and proves that $\eta \in \mathrm{Th}(\mathrm{PAC}(K, \leq e))$.

Proof of (b): Let K be a presented field with elimination theory. Consider a sentence $\eta \in \mathcal{L}(\mathrm{ring}, K)$. By Proposition 23.6.2, $\mathrm{Th}(\mathrm{PAC}(K, \leq e)) \models \eta \leftrightarrow \theta$ for some Boolean combination θ of e-basic statements. By Gödel's completeness theorem (Corollary 8.2.6), $\mathrm{Th}(\mathrm{PAC}(K, \leq e)) \vdash \eta \leftrightarrow \theta$. Use the system of axioms $\Delta(K, \leq e)$ for the class $\mathrm{PAC}(K, \leq e)$ (Section 23.3) and to find θ in finitely many steps (Lemma 8.7.1). Then apply Corollary 23.4.4 to decide if θ belongs to $\mathrm{Th}(\mathrm{PAC}(K, \leq e))$. Whatever the conclusion is, it applies to η. This gives (b). □

Exercises

1. Combine Corollary 23.1.2, Theorem 11.6.2, Artin's theorem about fields with finite absolute Galois groups, and Corollary 11.2.5 to give an alternative proof to the second part of Corollary 22.4.6: Every closed subgroup of a projective group G is projective and G is torsion free.

2. [Wheeler, Thm 2.3] Let K be a perfect field. Denote the class of all regular extensions of K by $\text{Reg}(K)$. Apply Proposition 11.3.5 and Exercise 13 of Chapter 2 to prove that the following conditions on $E \in \text{Reg}(K)$ are equivalent:

   (a) E is **existentially complete** in $\text{Reg}(K)$ (i.e. if $E \subseteq F$ and $F \in \text{Reg}(K)$, then E is existentially closed in F); and

   (b) E is perfect and PAC and the map res: $\text{Gal}(E) \to \text{Gal}(K)$ is the universal Frattini cover of $\text{Gal}(K)$.

3. In the notation of Exercise 2, apply Theorem 20.3.3 to prove that fields E and E' that are existentially complete fields in $\text{Reg}(K)$ are K-elementarily equivalent.

4. Let f and g be polynomials in $\mathbb{Q}[X]$.

   (a) Find a Boolean combination of basic statements, $\theta_{\mathbb{Q}}(h, B \xrightarrow{\alpha} A)$, which is equivalent modulo $\text{Th}(\text{PAC}(\mathbb{Q}, \leq e))$ to the sentence $[(\exists X)f(X) = 0] \leftrightarrow [(\exists X)g(X) = 0]$.

   (b) How can you decide if the above sentence is true for all field $E \in \text{PAC}(\mathbb{Q}, \leq e)$?

5. Let $e = 1, 2$, or 3. Does there exists an irreducible polynomial $f \in \mathbb{Q}[X]$ of degree 4 such that for all $E \in \text{PAC}(\mathbb{Q}, \leq e)$, $\sqrt{2} \in E$ if and only if f has a root in E?

   Hint: Consider first the case where $e = 1$. Note that $\text{Gal}(f, \mathbb{Q})$ is a transitive subgroup of S_4 and distinguish between the different cases. Compare also with Section 21.5. Deduce the cases where $e = 2$ or 3 from the first case.

6. Use Example 21.5.4 to show that the sentence

$$(\exists X)[X^3 + X^2 - 2X - 1 = 0] \longleftrightarrow (\exists X)[X^6 + X^4 - 2X^2 - 1 = 0]$$

holds for all $E \in \text{PAC}(\mathbb{Q}, \leq 1)$ but that it does not belong to $\text{Th}(\text{PAC}(\mathbb{Q}, \leq 2))$.

Notes

[Kuyk1] uses wreath products to prove Lemma 23.2.1, whereas [Jarden2] applies the free generators theorem (Theorem 18.5.6) to achieve the same result. The proof we give here is based on the field crossing argument.

Chapter 24. Frobenius Fields

The embedding property (Proposition 17.7.3) for free profinite groups is essential to the primitive recursive procedure for perfect PAC fields with free absolute Galois groups (Theorem 30.6.2). Since this is only a special case of a general result, we focus attention here on PAC fields whose absolute Galois groups have the embedding property: the **Frobenius fields**. The "field crossing argument" (e.g., the proofs of Lemma 6.4.8, Part G in Section 6.5, Part C of Proposition 16.8.6, Lemma 20.2.2, and Lemma 23.2.1) applies to give an analog, for Frobenius fields, of the Chebotarev density theorem (Proposition 24.1.4). The remainder of the Chapter concentrates on the embedding property. For example, if a profinite group has the embedding property, so does its universal Frattini cover (Proposition 24.3.5). We further show that every profinite group G has a smallest cover $E(G)$ with the embedding property (Proposition 24.4.5). In particular, for G finite, $E(G)$ is finite and unique (Proposition 24.4.6). The construction of $E(G)$ leads to a decision procedure for projective groups with the embedding property (Corollary 24.5.3).

24.1 The Field Crossing Argument

The decidability algorithms of Chapters 20 and 23 are based on the elementary equivalence theorem for PAC fields (Theorem 20.3.3). In addition to the PAC assumption, this theorem also requires certain infinite embedding problems for the absolute Galois groups of the fields involved to be solvable. The resulting decision procedure are recursive. Establishing a primitive recursive decision procedures, requires a finite version of the embedding properties of the absolute Galois groups. We achieve this in this section by using the field crossing argument:

Let K be a field, S/R be a ring cover (Definition 6.1.3) with $K \subseteq R$, and F/E the corresponding field cover. We say S/R is **regular** (resp. **finitely generated, Galois**) over K if E/K is regular (resp. R is finitely generated over K, F/E is Galois).

Let $\varphi \colon S \to K_s$ be a K-homomorphism. The decomposition group of φ is (Remark 6.1.6)

$$D_\varphi = \{\sigma \in \mathrm{Gal}(F/E) \mid \varphi(x) = 0 \iff \varphi(\sigma x) = 0, \text{ for all } x \in S\}.$$

LEMMA 24.1.1: *Let K be a PAC field, S/R a regular finitely generated Galois ring cover over K, and F/E the corresponding Galois field cover. Denote the algebraic closure of K in F by L. Let E' be a field between E and F with* res: $\mathrm{Gal}(F/E') \to \mathrm{Gal}(L/K)$ *surjective. Suppose there is an epimorphism* $\tilde\gamma \colon \mathrm{Gal}(K) \to \mathrm{Gal}(F/E')$ *with* $\mathrm{res}_{F/L} \circ \tilde\gamma = \mathrm{res}_{K_s/L}$. *Denote the fixed field*

24.1 The Field Crossing Argument

of $\mathrm{Ker}(\tilde{\gamma})$ in K_s by M. Then there exists a K-epimorphism $\varphi\colon S \to M$ satisfying:

(1a) $\varphi(R) = K$ and $D_\varphi = \mathrm{Gal}(F/E')$.

(1b) If $\mathrm{char}(K) = p$ and $y_1,\ldots,y_m \in R$ are p-independent over E^p with m bounded by $[K:K^p]$, then φ can be chosen with $\varphi(y_1),\ldots,\varphi(y_m)$ being p-independent over K^p.

Proof: The field M is a Galois extension of K which contains L and $\tilde{\gamma}$ induces an isomorphism $\gamma\colon \mathrm{Gal}(M/K) \to \mathrm{Gal}(F/E')$ with $\mathrm{res}_{F/L}\circ\gamma = \mathrm{res}_{M/L}$. By assumption, F is a regular extension of L, so $E'M \cap F = E'L$ and res: $\mathrm{Gal}(E'M/E') \to \mathrm{Gal}(M/K)$ is an isomorphism. Put $F' = FM$ and consider $\mathrm{Gal}(F'/E')$ as a subgroup of $\mathrm{Gal}(F/E') \times \mathrm{Gal}(M/K)$. Then

$$\mathrm{Gal}(F'/E') = \{(\sigma_1,\sigma_2) \in \mathrm{Gal}(F/E') \times \mathrm{Gal}(M/K) \mid \mathrm{res}_L(\sigma_1) = \mathrm{res}_L(\sigma_2)\}.$$

Let $\Delta = \{(\gamma(\sigma),\sigma) \mid \sigma \in \mathrm{Gal}(M/K)\}$. Denote the fixed field of Δ in F' by D. Then $\Delta \subseteq \mathrm{Gal}(F'/E')$ and $\Delta \cap \mathrm{Gal}(M/K) = \Delta \cap \mathrm{Gal}(F/E') = 1$. In addition, for each $(\sigma_1,\sigma_2) \in \mathrm{Gal}(F'/E')$,

$$(\sigma_1,\sigma_2) = (\gamma(\sigma_2),\sigma_2)\cdot(\gamma(\sigma_2)^{-1}\sigma_1,1) \in \Delta\cdot\mathrm{Gal}(F/E'),$$

and with $\sigma_1 = \gamma(\tau_1)$, $\tau_1 \in \mathrm{Gal}(M/K)$

$$(\sigma_1,\sigma_2) = (\gamma(\tau_1),\tau_1))(1,\tau_1^{-1}\sigma_2) \in \Delta\cdot\mathrm{Gal}(M/K).$$

Thus, $\Delta\cdot\mathrm{Gal}(M/K) = \Delta\cdot\mathrm{Gal}(F/E') = \mathrm{Gal}(F'/E')$. It follows that $DM = DF = F'$, $D \cap F = E'$, and $D \cap M = K$. In particular, D is a regular extension of K.

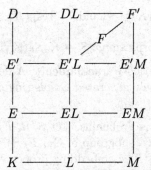

The integral closure U of R in D is finitely generated over R [Lang4, p. 120], hence over K. Since K is PAC, there exists a K-epimorphism $\psi\colon U \to K$. By Lemma 2.5.10, the integral closure V of U in F' is $UM = U\otimes_K M$. In particular, $S \subseteq V$. Thus, ψ extends to an M-epimorphism $\psi'\colon V \to M$. Since $[F':D] = [M:K]$, the decomposition group $D_{\psi'}$ is Δ (Lemma 6.1.4). Let φ be the restriction of ψ' to S. Then $\mathrm{Gal}(F/E') = \mathrm{res}_F D_{\psi'} \leq D_\varphi \leq \mathrm{Gal}(F/E')$. Hence, $D_\varphi = \mathrm{Gal}(F/E')$. This proves (1a).

Moreover, since $\varphi(R) = K$, $\varphi(S) = M_0$ is a Galois extension of K in M with $\mathrm{Gal}(M_0/K) \cong D_\varphi = \mathrm{Gal}(F/E') \cong \mathrm{Gal}(M/K)$ (Remark 6.1.4). Therefore, $\varphi(S) = M$.

If $\mathrm{char}(K) = p$ and $y_1, \ldots, y_m$ are as in (1b), then, since D/E is separable, they are p-independent over D^p. By Proposition 10.11, we may choose ψ so that $\psi(y_1), \ldots, \psi(y_m)$ are p-independent over K^p. □

Definition 24.1.2: A profinite group G has the **embedding property** if each embedding problem

(2) $\qquad (\zeta: G \to A, \ \alpha: B \to A)$

where ζ and α are epimorphisms and $B \in \mathrm{Im}(G)$ (i.e. B is a finite quotient of G) is solvable. That is, there exists an epimorphism $\gamma: G \to B$ with $\alpha \circ \gamma = \zeta$.

Note that each quotient of a group B in $\mathrm{Im}(G)$ belongs to $\mathrm{Im}(G)$. Hence, by induction on the order of $\mathrm{Ker}(\alpha)$, it suffices to consider only embedding problems (2) where $B \in \mathrm{Im}(G)$ and $\mathrm{Ker}(\alpha)$ is a minimal normal subgroup of B. □

In addition to the groups $\hat{F}_e(\mathcal{C})$ with $\mathcal{C}$ a formation of finite groups, Section 24.6 contains many examples of groups with the embedding property.

Definition 24.1.3: (a) A field K is called a **Frobenius field** if K is PAC and $\mathrm{Gal}(K)$ has the embedding property.

(b) Let S/R be a regular finitely generated Galois ring cover over K and F/E the corresponding field cover. Denote the algebraic closure of K in F by L. We say that S/R satisfies the **decomposition group condition** if for each subgroup H of $\mathrm{Gal}(F/E)$ with

(3) $\qquad H \in \mathrm{Im}(\mathrm{Gal}(K)) \quad \text{and} \quad \mathrm{res}_L(H) = \mathrm{Gal}(L/K)$

there exists a K-homomorphism $\varphi: S \to K_s$ satisfying (1). □

PROPOSITION 24.1.4 ([Fried-Haran-Jarden]): *A field K is Frobenius if and only if every regular finitely generated Galois ring cover over K satisfies the decomposition group condition.*

Proof: Suppose first K is Frobenius. Let S/R, F/E, and L be as in Definition 24.3(b). Let H be a subgroup of $\mathrm{Gal}(F/E)$ satisfying (3). Let E' be the fixed field of H in F. The embedding property of $\mathrm{Gal}(K)$ gives an epimorphism $\tilde{\gamma}: \mathrm{Gal}(K) \to \mathrm{Gal}(F/E')$ with $\mathrm{res}_{F/L} \circ \tilde{\gamma} = \mathrm{res}_{K_s/L}$. Lemma 24.1.1 then gives a homomorphism $\varphi: S \to K_s$ satisfying (1). Thus S/R satisfies the decomposition condition.

Conversely, suppose every regular finitely generated Galois ring cover over K satisfies the decomposition group condition. Consider a variety V over K. Let F be its function field. Choose an integrally closed domain R, finitely generated over K, containing the coordinate ring of V and having F as its quotient field. Take $S = R$, $E = F$, and $H = 1$ in the decomposition group condition to assure the existence of a K-homomorphism $\varphi \colon R \to K$. Thus $V(K)$ is nonempty. This proves K is PAC.

Next we prove $\mathrm{Gal}(K)$ has the embedding property. Let L be a finite Galois extension of K and $\alpha \colon G \to \mathrm{Gal}(L/K)$ an epimorphism, with $G \in \mathrm{Im}(\mathrm{Gal}(K))$. Consider a set $X = \{x^\sigma \mid \sigma \in G\}$ of $|G|$ algebraically independent elements over L. Let G act on X from the right in the obvious manner, and on L by $a^\sigma = a^{\alpha(\sigma)}$. This defines a faithful action of G on the rational function field $F = L(X)$. Denote the fixed field of G in F by E. Then $E \cap L = K$. Since F/L is regular, so is EL/L. Therefore, E/K is a finitely generated regular extension (Lemma 10.5.1). Let S/R be a finitely generated regular ring cover over K with F/E the corresponding field cover (Definition 6.1.3).

The decomposition group condition gives a K-epimorphism $\varphi \colon S \to M$, where M is a Galois extension of K containing L, $\varphi(R) = K$, and $D_\varphi = \mathrm{Gal}(F/E)$. Assume without loss $\mathrm{res}_L(\varphi)$ is the identity map of L. Let $\varphi^* \colon \mathrm{Gal}(M/K) \to \mathrm{Gal}(F/E)$ be the isomorphism induced by φ (Lemma 6.1.4). Identify $\mathrm{Gal}(F/E)$ with G and $\mathrm{res}_{F/L}$ with α to find

$$\varphi^* \circ \mathrm{res}_{K_s/M} \colon \mathrm{Gal}(K) \to G$$

is an epimorphism and $\alpha \circ \varphi^* \circ \mathrm{res}_{K_s/M} = \mathrm{res}_L$. □

For general profinite groups neither the projective property nor the embedding property implies the other (Examples 24.6.1 and 24.6.7). We call a profinite group G **superprojective** if it is both projective and has the embedding property.

PROPOSITION 24.1.5: *Let F be a Frobenius field. Then $\mathrm{Gal}(F)$ is superprojective.*

Proof: By definition, F is PAC. Hence, by Theorem 11.6.2, $\mathrm{Gal}(F)$ is projective. By definition, $\mathrm{Gal}(F)$ has the embedding property. Consequently, $\mathrm{Gal}(F)$ is superprojective. □

A special case of Theorem 23.1.1 yields a converse to Proposition 24.1.5:

PROPOSITION 24.1.6: *Let L/K be a Galois extension, L_0 a field between K and L and G is a superprojective group. Suppose $\mathrm{Gal}(L/L_0)$ is a quotient of G. Then there exists a field extension F of K which is perfect and Frobenius such that $L \cap F = L_0$ and $\mathrm{Gal}(F) \cong G$.*

24.2 The Beckmann-Black Problem

Let L/K be a finite Galois extension of fields with Galois group G and t an indeterminate. The **Beckmann-Black Problem** for K, L, and G asks whether $K(t)$ has a Galois extension F with the following properties:
(1a) $\mathrm{Gal}(F/K(t)) \cong G$.
(1b) F/K is a regular extension.

(1c) There is a prime divisor $\mathfrak{p}$ of F/K with decomposition field $K(t)$ and residue field L.

The following observation shows that the Beckmann-Black problem is apparently stronger than the regular inverse Galois problem (Section 16.2):

LEMMA 24.2.1 ([Dèbes, Prop. 1.2]): *Let K_0 be a field and G a finite group. Suppose each field extension K of K_0 has the following property: If G is realizable over K, then G is regular over K. Then G is regular over K_0 (although G need not be realizable over K_0).*

Proof: Let $L = K_0(x^\sigma \mid \sigma \in G)$ with x^σ algebraically independent over K_0. Define an action of G on L by $(x^\sigma)^\tau = x^{\sigma\tau}$ and $a^\sigma = a$ for $a \in K_0$. Let K be the fixed field of G in L. Then $\mathrm{Gal}(L/K) = G$. By assumption, $K(t)$ has a Galois extension F with Galois group G such that F/K is regular. Then $L(t) = K_0(t, x^\sigma \mid \sigma \in G)$ is a purely transcendental extension of K_0, $\mathrm{Gal}(LF/L(t)) \cong G$, and LF/L is a regular extension. Thus, LF/K_0 is a regular extension. □

The Beckmann-Black problem has an affirmative solution over PAC fields, even in a stronger form:

THEOREM 24.2.2 ([Dèbes, Thm. 3.2]): *Let K be a PAC field, t an indeterminate, and G a finite group. For $i = 1, \ldots, n$ let G_i be a subgroup of G and L_i a Galois extension of K with Galois group G_i. Then $K(t)$ has a Galois extension F satisfying this:*

(2a) $\mathrm{Gal}(F/K(t)) \cong G$.
(2b) F/K is a regular extension.
(2c) *For each i there exists a prime divisor $\mathfrak{p}_i$ of F/K with decomposition group over $K(t)$ equal to G_i and with residue field L_i. Moreover, $\mathfrak{p}_1, \ldots, \mathfrak{p}_n$ are distinct.*

Proof: Proposition 16.12.2 gives F with properties (2a) and (2b). Put $R = K[t]$ and $E = K(t)$. Choose a finitely generated ring cover S/R over K with corresponding field cover F/E. For each i let E_i be the fixed field of G_i in F. Then apply Lemma 24.1.1 with E', L, M replaced by E_i, K, L_i. This gives a K-epimorphism $\varphi_i \colon S \to L_i$ with $\varphi_i(R) = K$ and $D_{\varphi_i} = G_i$. By Proposition 3.3.4, S is a Dedekind domain. In particular, the local ring of S at $\mathrm{Ker}(\varphi_i)$ is a valuation ring. Hence, φ_i defines a prime divisor $\mathfrak{p}_i$ of F/K with the same decomposition group.

Finally, to ensure that $\mathfrak{p}_i$ is different from $\mathfrak{p}_1, \ldots, \mathfrak{p}_{i-1}$, choose $y_j \in E \smallsetminus O_{\mathfrak{p}_j}$, $j = 1, \ldots, i-1$, and replace R by $R_i = R[y_1^{-1}, \ldots, y_{i-1}^{-1}]$. None of the places $\varphi_1, \ldots, \varphi_{i-1}$ will be finite on R_i. Therefore, $\mathfrak{p}_i \neq \mathfrak{p}_j$ for all $j < i$. □

PROBLEM 24.2.3: *Prove or disprove: Every field K with an affirmative solution for the regular inverse Galois problem has an affirmative solution for the Beckmann-Black problem.*

24.3 The Embedding Property and Maximal Frattini Covers

We prove that the universal Frattini cover $\tilde{G}$ of a profinite group G with the embedding property has the embedding property. Thus, $\tilde{G}$ is superprojective.

LEMMA 24.3.1: Let G be a group with the embedding property and $\mathcal{N}$ a family of open normal subgroups of G with this property: If N' is an open normal subgroup of G and $G/N' \cong G/N$ for some $N \in \mathcal{N}$ then $N' \in \mathcal{N}$. Denote the intersection of all $N \in \mathcal{N}$ by N_0. Then G/N_0 has the embedding property.

Proof: Let $\varphi\colon G/N_0 \to A$ and $\alpha\colon B \to A$ be epimorphisms with $B \in \mathrm{Im}(G/N_0)$. That is, G has an open normal subgroup M that contains N_0 and there is an isomorphism $\beta\colon G/M \to B$. Choose $N_1, \ldots, N_n \in \mathcal{N}$ with $J = N_1 \cap \cdots \cap N_n \leq M$. Let $\gamma\colon G/J \to G/M$ and $\pi\colon G \to G/N_0$ be the quotient maps.

The embedding property for G gives an epimorphism $\theta_1\colon G \to G/J$ with $\alpha\circ\beta\circ\gamma\circ\theta_1 = \varphi\circ\pi$. Put $N'_i = \theta_1^{-1}(N_i/J)$, $i = 1,\ldots,n$. Then, $\mathrm{Ker}(\theta_1) = N'_1 \cap \cdots \cap N'_n$ and $G/N_i \cong G/N'_i$, $i = 1,\ldots,n$. By assumption, $N'_1,\ldots,N'_n \in \mathcal{N}$, so $N_0 \leq \mathrm{Ker}(\theta_1)$. Thus, θ_1 induces an epimorphism $\theta_0\colon G/N_0 \to G/J$ with $\theta_0\circ\pi = \theta_1$. The map $\theta = \beta\circ\gamma\circ\theta_0$ is an epimorphism of G/N_0 onto B with $\alpha\circ\theta = \varphi$.

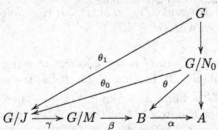

It follows that G/N_0 has the embedding property. □

LEMMA 24.3.2:
(a) If a profinite group G has the embedding property, then so does $G/\Phi(G)$.
(b) Let G be a profinite group and let $\varphi\colon \tilde{G} \to G$ be the universal Frattini cover. Suppose $\Phi(G) = 1$. Then $\mathrm{Ker}(\varphi) = \Phi(\tilde{G})$ and $\tilde{G}/\Phi(\tilde{G}) \cong G$.
(c) Suppose $\Phi(G) = 1$ and $\tilde{G}$ has the embedding property. Then G has the embedding property.

Proof of (a): For M a maximal open subgroup of G, the intersection $N(M)$ of all conjugates of M in G is an open normal subgroup of G. The collection $\mathcal{N} = \{N(M)\mid M \text{ ranges over maximal open subgroups of } G\}$ consists of all maximal open normal subgroups of G. It satisfies the hypotheses of Lemma 24.3.1. Furthermore $\Phi(G) = \bigcap N(M)$. Thus, (a) is a special case of Lemma 24.3.1.

Proof of (b): Since φ is a Frattini cover, $\mathrm{Ker}(\varphi) \le \Phi(\tilde{G})$. Thus, $\varphi(\Phi(\tilde{G})) \le \Phi(G) = 1$ (Lemma 22.1.4(a)), and this implies $\Phi(\tilde{G}) \le \mathrm{Ker}(\varphi)$. Consequently, $\mathrm{Ker}(\varphi) = \Phi(\tilde{G})$.

Proof of (c): Combine (a) and (b). □

Here is an important collection of groups with the embedding property:

LEMMA 24.3.3: *Let $\mathcal{C}$ be a formation of finite groups and F is a free pro-$\mathcal{C}$-group. Then F has the embedding property.*

Proof: Proposition 17.7.3 gives the result when F has finite rank. Assume F has infinite rank. Let $\varphi\colon F \to A$ and $\alpha\colon B \to A$ be epimorphisms with $B \in \mathrm{Im}(F)$. In particular, B belongs to $\mathcal{C}$. Let X be a basis of F. Then $X_2 = X \smallsetminus \mathrm{Ker}(\varphi)$ is a finite set. For each $x \in X_2$ choose $\gamma_2(x) \in B$ with $\alpha(\gamma_2(x)) = \varphi(x)$. Next choose a finite subset X_1 of $X \cap \mathrm{Ker}(\varphi)$ with $|X_1| = |\mathrm{Ker}(\alpha)|$ and a bijective map $\gamma_1\colon X_1 \to \mathrm{Ker}(\alpha)$. Define $\gamma\colon X \to B$ as follows: $\gamma(x) = \gamma_1(x)$ if $x \in X_1$, $\gamma(x) = \gamma_2(x)$ if $x \in X_2$, and $\gamma(x) = 1$ otherwise. Then γ extends to an epimorphism $\gamma\colon F \to B$ with $\alpha \circ \gamma = \varphi$. □

COROLLARY 24.3.4: *Each profinite group G has a cover H with the embedding property such that $\mathrm{rank}(G) = \mathrm{rank}(H)$. If G is finite and $\mathrm{rank}(G) = e$, then H can be chosen with $|H| \le |G|^{|G|^e}$.*

Proof: By Corollary 17.4.8, G is the quotient of a free profinite group F with $\mathrm{rank}(F) = \mathrm{rank}(G)$. By Lemma 24.3.3, F has the embedding property.

Now suppose G is finite and $\mathrm{rank}(G) = e$. Then there are at most $|G|^e$ epimorphisms of F onto G. List the kernels of these epimorphisms as $N_1, \ldots, N_k$, with $k \le |G|^e$. Put $H = F/(N_1 \cap \cdots \cap N_k)$. Then H covers G, and Lemma 24.3.1 shows H has the embedding property. Finally, $|H| \le \prod_{i=1}^k (F : N_i) = |G|^k$. □

Finally, we give a stronger version of the converse of Lemma 24.3.2(c):

PROPOSITION 24.3.5 ([Haran-Lubotzky]): *If a profinite group G has the embedding property, then so does its universal Frattini cover $\tilde{G}$. That is, $\tilde{G}$ is superprojective.*

Proof: Let $\varphi\colon \tilde{G} \to A$ and $\alpha\colon B \to A$ be epimorphisms with $B \in \mathrm{Im}(\tilde{G})$, and let $\tilde{\varphi}\colon \tilde{G} \to G$ be the universal Frattini cover of G. We prove the existence of an epimorphism $\theta\colon \tilde{G} \to B$ with $\alpha \circ \theta = \varphi$ in cases:

CASE A: $B \in \mathrm{Im}(G)$. By Lemma 22.2.8, φ and $\tilde{\varphi}$ are part of a commutative

24.4 The Smallest Embedding Cover of a Profinite Group

diagram of epimorphisms

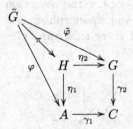

in which the square from H to C is Cartesian. Since G has the embedding property and $A \in \mathrm{Im}(G)$, there is an epimorphism $\zeta\colon G \to A$ with $\gamma_1 \circ \zeta = \gamma_2$. As the square is Cartesian, there is a homomorphism $\eta\colon G \to H$ with $\eta_2 \circ \eta = \mathrm{Id}_G$ and $\eta_1 \circ \eta = \zeta$. Applying Lemma 22.5.4(c) to $\tilde\varphi = \eta_2 \circ \pi$ we conclude that η_2 is a Frattini cover. Since $\eta_2(\eta(G)) = G$, this implies $\eta(G) = H$. Therefore, $\eta \circ \eta_2 = \mathrm{Id}_H$.

The embedding property of G gives an epimorphism $\beta\colon G \to B$ with $\alpha \circ \beta = \eta_1 \circ \eta$. Thus, $\theta = \beta \circ \tilde\varphi$ is an epimorphism of $\tilde G$ onto B with $\alpha \circ \theta = \alpha \circ \beta \circ \tilde\varphi = \eta_1 \circ \eta \circ \eta_2 \circ \pi = \eta_1 \circ \pi = \varphi$, as needed.

PART B: $B \in \mathrm{Im}(\tilde G)$. Lemma 22.6.5 gives a Frattini cover $\beta\colon B \to B'$ with $B' \in \mathrm{Im}(G)$. The maps α and β define a commutative diagram of epimorphisms

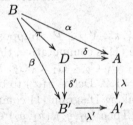

where the square from D to A' is Cartesian (Lemma 22.2.8). Part A gives an epimorphism $\theta'\colon \tilde G \to B'$ with $\lambda' \circ \theta' = \lambda \circ \varphi$. Hence, there exists a homomorphism $\zeta\colon \tilde G \to D$ with $\delta' \circ \zeta = \theta'$ and $\delta \circ \zeta = \varphi$. Applying Lemma 22.5.4(c) to the Frattini cover $\beta = \pi \circ \delta'$ we may conclude that both π and δ' are Frattini covers. Since $\delta'(\zeta(\tilde G)) = \theta'(\tilde G) = B'$, ζ is surjective. Since $\tilde G$ is projective (Proposition 22.6.1) and π is a Frattini cover, there is an epimorphism $\theta\colon \tilde G \to B$ with $\pi \circ \theta = \zeta$ (Proposition 22.5.9(a)). Therefore, $\alpha \circ \theta = \delta \circ \pi \circ \theta = \delta \circ \zeta = \varphi$. It follows that $\tilde G$ has the embedding property. $\square$

24.4 The Smallest Embedding Cover of a Profinite Group

Every profinite group G has a smallest cover $\tilde G$ with the embedding property (Proposition 24.4.5). By Chatzidakis, $\tilde G$ is unique. We prove the uniqueness only under the assumption that $\mathrm{rank}(G) \leq \aleph_0$ (Lemma 24.4.8).

The construction of $\tilde{G}$ uses properties of double covers: Let G_1 and G_2 be profinite groups. Consider the collection $\mathcal{P} = \mathcal{P}(G_1, G_2)$ of triples (π_1, π_2, A) with $\pi_i\colon G_i \to A$ an epimorphism, $i = 1, 2$. Partially order $\mathcal{P}$ by $(\pi_1, \pi_2, A) \leq (\pi'_1, \pi'_2, A')$ if there exists an epimorphism $\pi\colon A' \to A$ which makes the diagram

(1)
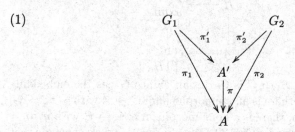

commutative. Write $(\pi_1, \pi_2, A) \sim (\pi'_1, \pi'_2, A')$ if π is an isomorphism. Then $\leq$ induces a partial ordering on the quotient of $\mathcal{P}$ by $\sim$.

Let $\mathrm{pr}_i\colon G_1 \times G_2 \to G_i$ be projection onto the ith coordinate, $i = 1, 2$. Define $\mathcal{H} = \mathcal{H}(G_1, G_2)$, dual to $\mathcal{P}$, to be the collection of closed subgroups H of $G_1 \times G_2$ with $\mathrm{pr}_i(H) = G_i$, $i = 1, 2$. Partially order $\mathcal{H}$ by inclusion.

The collections $\mathcal{P}$ and $\mathcal{H}$ are related by a map $T\colon \mathcal{P} \to \mathcal{H}$ defined by $T(\pi_1, \pi_2, A) = G_1 \times_A G_2$ (Section 22.2).

LEMMA 24.4.1: *The map $T\colon \mathcal{P} \to \mathcal{H}$ induces an order reversing bijection between $\mathcal{P}/\sim$ and $\mathcal{H}$.*

Proof: Let $(\pi_1, \pi_2, A), (\pi'_1, \pi'_2, A') \in \mathcal{P}$. Suppose there exists an epimorphism π such that Diagram (1) is commutative, then $G_1 \times_{A'} G_2 \leq G_1 \times_A G_2$. Conversely, assume that $G_1 \times_{A'} G_2 \leq G_1 \times_A G_2$. Each $a' \in A'$ is of the form $\pi'_i(g_i)$ for some $g_i \in G_i$, $i = 1, 2$. Define $\pi(a')$ to be $\pi_1(g_1)$. This gives a well defined map $\pi\colon A' \to A$ making Diagram (1) commutative. Indeed, suppose $h_1 \in G_1$ and $\pi'_1(h_1) = a'$. Then $(g_1, g_2), (h_1, g_2) \in G_1 \times_{A'} G_2 \leq G_1 \times_A G_2$, so $\pi_1(g_1) = \pi_2(g_2) = \pi_1(h_1)$.

Thus, T induces an order reversing injection from $\mathcal{P}/\sim$ into $\mathcal{H}$. To see that T is surjective let $H \in \mathcal{H}$ and $\eta_i = \mathrm{pr}_i|_H$, $i = 1, 2$. Lemma 22.2.8 produces a commutative diagram of epimorphisms

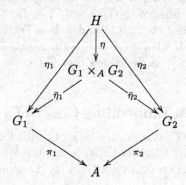

24.4 The Smallest Embedding Cover of a Profinite Group

with $\bar{\eta}_i$ the restriction of pr_i to $G_1 \times_A G_2$, $i = 1, 2$. For $(g_1, g_2) \in H$ we have $\bar{\eta}_i(\eta(g_1, g_2)) = \eta_i(g_1, g_2) = g_i$, $i = 1, 2$. Therefore, $\eta(g_1, g_2) = (g_1, g_2)$ and $H = G_1 \times_A G_2 = T(\pi_1, \pi_2, A)$. □

LEMMA 24.4.2:
(a) Each $H \in \mathcal{H}$ contains a minimal element of $\mathcal{H}$.
(b) For each $(\pi_1, \pi_2, A) \in \mathcal{P}$ there exists a maximal element $(\pi'_1, \pi'_2, A') \in \mathcal{P}$ with $(\pi_1, \pi_2, A) \leq (\pi'_1, \pi'_2, A')$.

Proof: Let $\{H_\alpha\}$ be a descending chain in $\mathcal{H}$. By Lemma 1.2.2(c), $\text{pr}_i(\bigcap H_\alpha) = \bigcap \text{pr}_i(H_\alpha) = G_i$, $i = 1, 2$, so $\bigcap H_\alpha \in \mathcal{H}$. Thus, statement (a) follows from Zorn's lemma.
  Statement (b) follows from (a) by Lemma 24.4.1. □

The universal Frattini cover of a profinite group G is the smallest projective cover of G (Proposition 22.6.1). Thus, every Frattini cover of G is a quotient of each projective cover of G. The next definition introduces two types of covers, the "embedding covers" and "I-covers" which relate to each other as projective covers relate to Frattini covers:

Definition 24.4.3: Call an epimorphism $\varphi \colon H \to G$ of profinite groups an **embedding cover** of G if H has the embedding property. In this case we also say that H is an **embedding cover** of G.
  Call an epimorphism $\varepsilon \colon E \to G$ of profinite groups an **I-cover** if for each embedding cover $\varphi \colon H \to G$ there exists an epimorphism $\gamma \colon H \to E$ with $\varepsilon \circ \gamma = \varphi$.
  Note that If $\varepsilon \colon E \to G$ and $\zeta \colon F \to E$ are I-covers, then so is $\varepsilon \circ \zeta$. Conversely, if ε and ζ are epimorphisms and $\varepsilon \circ \zeta$ is an I-cover, then so is ε.
  Finally, an epimorphism $\varepsilon \colon E \to G$ which is both an embedding cover and an I-cover is a **smallest embedding cover**. □

LEMMA 24.4.4: *If a profinite group G does not have the embedding property, then G has an I-cover $\varepsilon \colon E \to G$ with a nontrivial finite kernel.*

Proof: Suppose each I-cover of G with a finite kernel is an isomorphism. We show that G has the embedding property.
  Indeed, let $\alpha \colon B \to A$ and $\alpha' \colon G \to A$ be epimorphisms with $B \in \text{Im}(G)$. Lemma 24.4.2 gives a commutative diagram of epimorphisms

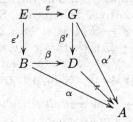

with (β, β', D) a maximal element of $\mathcal{P}(B, G)$, and $E = B \times_D G$. We prove ε is an I-cover.

Let $\varphi\colon H \to G$ be an embedding cover. Then $B \in \mathrm{Im}(H)$. Therefore, there exists an epimorphism $\varphi'\colon H \to B$ with $\beta \circ \varphi' = \beta' \circ \varphi$. Since the square from D to E is Cartesian, there exists a homomorphism $\eta\colon H \to E$ with $\varepsilon' \circ \eta = \varphi'$ and $\varepsilon \circ \eta = \varphi$. In particular, $\eta(H) \in \mathcal{H}(B,G)$. It follows from the minimality of E in $\mathcal{H}(B,G)$ (Lemma 24.4.1) that $\eta(H) = E$. Therefore, $\varepsilon\colon E \to G$ is an I-cover.

By hypothesis ε is an isomorphism. Hence, β is an isomorphism and $\beta^{-1} \circ \beta'$ is an epimorphism from G to B satisfying $\alpha \circ \beta^{-1} \circ \beta' = \alpha'$. Thus, G has the embedding property. □

PROPOSITION 24.4.5: *Every profinite group G has a smallest embedding cover.*

Proof: Choose a free profinite group F and a cover $\varphi\colon F \to G$ (Proposition 17.4.8). Let $K = \mathrm{Ker}(\varphi)$ and m be a cardinal number greater than $|F|$. By transfinite induction construct a descending sequence $\{L_\kappa\}_{\kappa \leq m}$ of closed normal subgroups of F in the following way. First choose $L_0 = K$. Next suppose L_κ has been chosen such that the quotient map $F/L_\kappa \to F/K$ is an I-cover. If L_κ has a proper open subgroup M which is normal in F and the quotient map $F/M \to F/L_\kappa$ is an I-cover, choose $L_{\kappa+1} = M$, otherwise let $L_{\kappa+1} = L_\kappa$. In both cases the quotient map $F/L_{\kappa+1} \to F/K$ is an I-cover. If $\lambda \leq m$ is a limit ordinal, let $L_\lambda = \bigcap_{\kappa < \lambda} L_\kappa$. Again, the canonical map $F/L_\lambda \to F/K$ is an I-cover. Indeed, suppose $\varphi\colon H \to F/K$ is an embedding cover. By transfinite induction construct epimorphisms $\varphi_\kappa\colon H \to F/L_\kappa$, for $\kappa < \lambda$, which commute with the canonical maps. Their limit $\varphi_\lambda = \varprojlim \varphi_\kappa$ is an epimorphism φ_λ which gives φ when composed with the quotient map $F/L_\lambda \to F/K$.

Each open subgroup of L_κ is the intersection of an open subgroup of F with L_κ. Thus, in the definition of $L_{\kappa+1}$, the case that $L_{\kappa+1} \neq L_\kappa$ cannot occur more than $|F|$ times (Proposition 17.1.2). Since $|F| < m$, there is a $\kappa_0 < m$ with $L_\kappa = L_{\kappa+1}$ for each $\kappa \geq \kappa_0$. In particular, $L = L_m$ is a normal closed subgroup of F contained in K, the quotient map $\pi\colon F/L \to F/K$ is an I-cover and there is no open normal subgroup M of F properly contained in L such that the quotient map $F/M \to F/L$ is an I-cover.

To prove that π is a smallest embedding cover, it suffices to verify that F/L has the embedding property. If not, then Lemma 24.4.4, gives an I-cover $\varepsilon\colon E \to F/L$ with a nontrivial finite kernel. Since F has the embedding property (Lemma 24.3.3), $F \to F/L$ is an embedding cover. Hence, there is an epimorphism $\theta\colon F \to E$ such that $\varepsilon \circ \theta$ is the quotient map. Let $M = \mathrm{Ker}(\theta)$. Then $M \triangleleft F$, $M < L$, and the quotient map $F/M \to F/K$ is an I-cover. This contradicts the preceding paragraph. □

LEMMA 24.4.6: *Let $\varepsilon\colon E \to G$ be a smallest embedding cover of a profinite group G and $\mathcal{C}$ a formation of finite groups.*
(a) *If G is a pro-$\mathcal{C}$-group, then so is E.*
(b) $\mathrm{rank}(E) = \mathrm{rank}(G)$.

24.4 The Smallest Embedding Cover of a Profinite Group

(c) If $\varepsilon'\colon E' \to G$ is another smallest embedding cover of G, then there exist epimorphisms $\theta\colon E \to E'$ and $\theta'\colon E' \to E$ with $\varepsilon' \circ \theta = \varepsilon$ and $\varepsilon \circ \theta' = \varepsilon'$. In particular, $\mathrm{Im}(E) = \mathrm{Im}(E')$.

(d) If G is small, then θ and θ' of (c) are isomorphisms. Thus, in this case $\varepsilon\colon E \to G$ is, up to isomorphism, uniquely determined. We denote E by $E(G)$.

(e) If G is a finite group, then there is a primitive recursive procedure to construct $E(G)$ and ε from G.

Proof: To prove (a) and (b), choose a cover $\varphi\colon \hat{F}_m(\mathcal{C}) \to G$ with $m = \mathrm{rank}(G)$ (Proposition 17.4.8). Since $\hat{F}_m(\mathcal{C})$ has the embedding property (Lemma 24.3.3), there is an epimorphism $\theta\colon \hat{F}_m(\mathcal{C}) \to E$. Hence, E is a pro-$\mathcal{C}$-group and $\mathrm{rank}(E) = \mathrm{rank}(G)$.

Assertion (c) follows from the definition of a smallest e-embedding cover. Assertion (d) is a consequence of (c) and Proposition 16.10.6(b).

Finally, let G be a finite group of rank e. Then G has an embedding cover H of order at most $|G|^{|G|^e}$ (Corollary 24.3.4). Note that $E(G)$ is the finite cover of G of a smallest order with the embedding property. Thus, we may find $E(G)$ by checking the finitely many groups of order bounded by $|G|^{|G|^e}$ that cover G. □

Small profinite groups are characterized by their finite quotients (Proposition 16.10.7(b)). Thus, the smallest embedding cover of such groups is unique (Lemma 24.4.6(d)). A back and forth argument partially generalizes this to profinite groups of countable rank:

LEMMA 24.4.7: *Let G and G' be profinite groups of at most countable rank, A a finite group, and $\varphi\colon G \to A$, $\varphi'\colon G' \to A$ epimorphisms. Suppose $\mathrm{Im}(G) = \mathrm{Im}(G')$ and both G and G' have the embedding property. Then there exists an isomorphism $\theta\colon G \to G'$ with $\varphi' \circ \theta = \varphi$.*

Proof: Choose descending sequences $\mathrm{Ker}(\varphi) \geq K_1 \geq K_2 \geq K_3 \geq \cdots$ and $\mathrm{Ker}(\varphi') \geq K_1' \geq K_2' \geq K_3' \geq \cdots$ of open normal subgroups of G and G', respectively, with $\bigcap_{i=1}^{\infty} K_i = 1$ and $\bigcap_{i=1}^{\infty} K_i' = 1$. Inductively define descending sequences $\mathrm{Ker}(\varphi) \geq L_1 \geq M_1 \geq L_2 \geq M_2 \geq \cdots$, and $\mathrm{Ker}(\varphi') \geq L_1' \geq M_1' \geq L_2' \geq M_2' \geq \cdots$ of open normal subgroups of G and G', respectively, and isomorphisms $\theta_i'\colon G'/L_i' \to G/L_i$, $\theta_i\colon G/M_i \to G'/M_i'$, $i = 1, 2, \ldots$ satisfying this:

(2a) $L_1 = K_1$, $L_i = K_i \cap M_{i-1}$, $i = 2, 3, \ldots$.

(2b) $M_i' = K_i' \cap L_i'$, $i = 1, 2, \ldots$.

(2c) $\bar{\varphi} \circ \theta_1' = \bar{\varphi}'$, where $\bar{\varphi}\colon G/L_1 \to A$ is induced by φ and $\bar{\varphi}'\colon G'/L_1' \to A$ is induced by φ'.

(2d) For each $n \geq 2$ the following diagram, where the vertical arrows are defined as the quotient maps, is commutative.

(3)
$$\begin{array}{ccc} G' & & G \\ \pi'_n \downarrow & & \downarrow \pi_n \\ G'/L'_n & \xrightarrow{\theta'_n} & G/L_n \\ \lambda'_n \downarrow & & \downarrow \lambda_n \\ G'/M'_{n-1} & \xleftarrow{\theta_{n-1}} & G/M_{n-1} \\ \mu'_n \downarrow & & \downarrow \mu_n \\ G'/L'_{n-1} & \xrightarrow{\theta'_{n-1}} & G/L_{n-1} \end{array}$$

Indeed, $G/L_1 \in \text{Im}(G) = \text{Im}(G')$. Hence, there exists an epimorphism $\zeta'_1 \colon G' \to G/L_1$ with $\bar{\varphi} \circ \zeta'_1 = \varphi'$. Put $L'_1 = \text{Ker}(\zeta'_1)$ and let $\theta'_1 \colon G'/L'_1 \to G/L_1$ be the isomorphism induced by ζ'_1. Then $\bar{\varphi} \circ \theta'_1 = \bar{\varphi}'$.

Let now $n \geq 2$. Inductively suppose the bottom line of (3) has already been constructed. Following (2b), put $M'_{n-1} = K'_{n-1} \cap L'_{n-1}$. Then $G'/M'_{n-1} \in \text{Im}(G') = \text{Im}(G)$. Hence, there is an epimorphism $\zeta_{n-1} \colon G \to G'/M_{n-1}$ such that $\theta'_{n-1} \circ \mu'_n \circ \zeta_{n-1} \colon G \to G/L_{n-1}$ is the quotient map. Put $M_{n-1} = \text{Ker}(\zeta_{n-1})$ and let $\theta_{n-1} \colon G/M_{n-1} \to G'/M'_{n-1}$ be the isomorphism which ζ_{n-1} induces. Then the lower square of (3) is commutative.

Similarly one constructs an isomorphism $\theta'_n \colon G'/L'_n \to G/L_n$ making the upper square of (3) commutative. This concludes the induction

The isomorphisms θ_i, $i = 1, 2, \ldots$, define an isomorphism $\theta \colon G \to G'$ with $\varphi' \circ \theta = \varphi$. □

COROLLARY 24.4.8: *Let G be a profinite group of rank at most $\aleph_0$. Suppose $\varepsilon \colon E \to G$ and $\varepsilon' \colon E' \to G'$ are smallest embedding covers of G. Then $E \cong E'$.*

Proof: By Lemma 24.4.6(b), $\text{rank}(E) = \text{rank}(G) = \text{rank}(E')$. Hence, both $\text{rank}(E)$ and $\text{rank}(E')$ are at most $\aleph_0$. Also, $\text{Im}(E) = \text{Im}(E')$ (Lemma 24.4.6(c)). Since both E and E' have the embedding property, Lemma 24.4.7 applied to $A = 1$ implies $E \cong E'$. □

Actually, the uniqueness of the smallest embedding cover of a small group (Lemma 24.4.6(d)) and Corollary 24.4.8 are special cases of a result of Chatzidakis asserting the uniqueness of the smallest embedding cover of each profinite group (see Notes to this chapter).

24.5 A Decision Procedure

Let $\mathcal{A}$ and $\mathcal{B}$ be finite collections of finite groups. This section establishes a decision procedure for the existence of a superprojective group that has the elements of $\mathcal{A}$ among its quotients but none of the elements of $\mathcal{B}$. This is a key ingredient in a decision procedure for Frobenius fields (Chapter 30).

24.5 A Decision Procedure

LEMMA 24.5.1:
(a) Let A and A' be finite quotients of a group Γ with the embedding property. Then each minimal element H of $\mathcal{H}(A, A')$ is a quotient of Γ.
(b) Let $A_1, \ldots, A_m$ be finite quotients of a group Γ with the embedding property and e a positive integer. Suppose $\operatorname{rank}(A_i) \leq e$, $i = 1, \ldots, m$. Then $A_1 \times \cdots \times A_m$ contains a finite quotient H of Γ of rank bounded by e with $\operatorname{pr}_i(H) = A_i$, $i = 1, \ldots, m$.

Proof of (a): Let H be a minimal element of $\mathcal{H}(A, A')$. Then the projections $\eta\colon H \to A$ and $\eta'\colon H \to A'$ are surjective. By Lemma 24.4.1, there exist a group B and epimorphisms $\alpha\colon A \to B$ and $\alpha'\colon A' \to B$ such that $H = A \times_B A'$. Let $\varphi\colon \Gamma \to A$ be an epimorphism. Since Γ has the embedding property and since $A' \in \operatorname{Im}(\Gamma)$ there is an epimorphism $\varphi'\colon \Gamma \to A'$ with $\alpha \circ \varphi = \alpha' \circ \varphi'$. So, there is a homomorphism $\gamma\colon \Gamma \to H$ with $\eta \circ \gamma = \varphi$ and $\eta' \circ \gamma = \varphi'$. Since the image $\gamma(\Gamma)$ is contained in $\mathcal{H}(A, A')$, the minimality of H implies γ is surjective.

Proof of (b): First consider the case $m = 2$. Choose a minimal element H of $\mathcal{H}(A_1, A_2)$. By (a), $H \in \operatorname{Im}(\Gamma)$. Also, $A_i \in \operatorname{Im}(\hat{F}_e)$, $i = 1, 2$, and $\hat{F}_e$ has the embedding property (Lemma 24.3.3). Therefore, by (a), $H \in \operatorname{Im}(\hat{F}_e)$. Thus, $\operatorname{rank}(H) \leq e$. An induction on m proves (b). □

PROPOSITION 24.5.2 ([Haran-Lubotzky, p. 200]): *Let $\mathcal{C}$ be a full formation of finite groups. Consider groups $A_1, \ldots, A_m, B_1, \ldots, B_n$ in $\mathcal{C}$ of rank at most e with $m \geq 1$ and $n \geq 0$. Let $\mathcal{A}$ be the collection of all subgroups A of $A_1 \times \cdots \times A_m$ with $\operatorname{rank}(A) \leq e$ and $\operatorname{pr}_i(A) = A_i$, $i = 1, \ldots, m$. Then the following conditions are equivalent:*
(a) *There exists a superprojective pro-$\mathcal{C}$-group Γ with $A_1, \ldots, A_m \in \operatorname{Im}(\Gamma)$ and $B_1, \ldots, B_n \notin \operatorname{Im}(\Gamma)$.*
(b) *There exists a superprojective pro-$\mathcal{C}$-group Γ of rank at most e with $A_1, \ldots, A_m \in \operatorname{Im}(\Gamma)$ $B_1, \ldots, B_n \notin \operatorname{Im}(\Gamma)$.*
(c) *There exists $A \in \mathcal{A}$ such that none of the groups $B_1, \ldots, B_n$ is a Frattini cover of a quotient group of $E(A)$.*

Proof of "(a) $\Longrightarrow$ (c)": Lemma 24.5.1 gives $A \in \mathcal{A} \cap \operatorname{Im}(\Gamma)$. Since Γ is an embedding cover of A, the smallest embedding cover $E = E(A)$ of A is a quotient of Γ. The universal Frattini cover $\tilde{E}$ of E is the smallest projective cover of E (Proposition 22.6.1). Therefore, as Γ is a projective cover of E, $\tilde{E}$ is a quotient of Γ. Thus, none of the groups $B_1, \ldots, B_n$ is a quotient of $\tilde{E}$. It follows from Lemma 22.6.5 that no $B_1, \ldots, B_n$ is a Frattini cover of a quotient of E.

Proof of "(c) $\Longrightarrow$ (b)": By Proposition 24.4.6, $E = E(A)$ is a pro-$\mathcal{C}$-group and $\operatorname{rank}(E) \leq e$. Lemma 22.6.4 implies $\tilde{E}$ is a pro-$\mathcal{C}$-group and $\operatorname{rank}(\tilde{E}) \leq e$. Also, $\tilde{E}$ is superprojective (Proposition 24.3.5). Clearly $A_1, \ldots, A_m \in \operatorname{Im}(\tilde{E})$ and (by Lemma 22.6.5), none of the groups $B_1, \ldots, B_m$ is a quotient of $\tilde{E}$. □

COROLLARY 24.5.3: *Let C be a primitive recursive full formation of finite groups and e a positive integer. There is a primitive recursive procedure to decide whether given groups $A_1, \ldots, A_m, B_1, \ldots, B_n \in C$ of rank at most e satisfy the conditions of Proposition 24.5.2.*

Proof: There is a primitive recursive construction of the collection $\mathcal{A}$ of Lemma 24.5.2 corresponding to $A_1, \ldots, A_m$. Also, there is a primitive recursive construction of $E(A)$ for each $A \in \mathcal{A}$ (Proposition 24.4.6(e)). Thus, there is a primitive recursive check for the validity of (c) of Proposition 24.5.2. □

24.6 Examples

This section presents examples that, in particular, show the independence of the concepts introduced in Sections 24.1 - 24.5.

Example 24.6.1: Finite simple groups. Each finite simple groups G has the embedding property (but is not projective). The universal Frattini cover, $\tilde{G}$, of G is superprojective (Proposition 24.3.5). □

LEMMA 24.6.2: *Let G be a profinite group of at most countable rank. Then G has the embedding property if and only if for each pair of epimorphisms $\varphi: G \to A$ and $\psi: G \to A$ onto a finite group A, there exists an epimorphism $\theta: G \to G$ with $\psi \circ \theta = \varphi$.*

Proof: Necessity is given by Lemma 24.4.7. Conversely, suppose G has the property stated in the lemma relative to each finite group A. Let $\varphi: G \to A$ and $\alpha: B \to A$ be epimorphisms. Assume that $B \in \text{Im}(G)$. Then there exists an epimorphism $\pi: G \to B$. By assumption, there exists an epimorphism θ of G onto G with $\alpha \circ \pi \circ \theta = \varphi$. The epimorphism $\gamma = \pi \circ \theta$ of G onto B satisfies $\alpha \circ \gamma = \varphi$. Thus, G has the embedding property. □

Example 24.6.3: [Ershov3, p. 511]. Lemma 24.4.7 requires A to be a finite group. Suppose $G = \hat{F}_\omega = A$ and $\varphi = \text{Id}$. Let $\{x_1, x_2, \ldots\}$ be a basis of $\hat{F}_\omega$ and define the epimorphism $\psi: \hat{F}_\omega \to \hat{F}_\omega$ by $\psi(x_1) = 1$ and $\psi(x_{i+1}) = x_i$, $i = 1, 2, \ldots$. Then there exists no epimorphism $\theta: \hat{F}_\omega \to \hat{F}_\omega$ with $\psi \circ \theta = \text{id}$ because any $x \in \hat{F}_\omega$ with $\theta(x) = x_1$ would give $x = \psi(x_1) = 1$, so $x_1 = 1$, a contradiction. □

Example 24.6.4: $\mathbb{Z}/p^m\mathbb{Z}$ modules with and without the embedding property. Consider the group $G = (\mathbb{Z}/p^{m_1}\mathbb{Z}) \times \cdots \times (\mathbb{Z}/p^{m_r}\mathbb{Z})$ with $1 \leq m_1 \leq \cdots \leq m_r = m$. Thus, G is a finitely generated $\mathbb{Z}/p^m\mathbb{Z}$ module. We show G has the embedding property if and only if it is a free $\mathbb{Z}/p^m\mathbb{Z}$ module (i.e. $m_1 = m$).

First suppose $m_1 < m$. Let $\varphi: G \to \mathbb{Z}/p^{m_1}\mathbb{Z}$ be the projection onto the first factor. Let $\psi: G \to \mathbb{Z}/p^{m_1}\mathbb{Z}$ be the projection onto the rth factor of G multiplied by p^{m-m_1} (We identify $\mathbb{Z}/p^{m_1}\mathbb{Z}$ with the group of all elements of $\mathbb{Z}/p^m\mathbb{Z}$ divisible by p^{m-m_1}.) Assume G has the embedding property. Then there exists an epimorphism $\theta: G \to G$ with $\psi \circ \theta = \varphi$. Let $\theta(1, 0, \ldots, 0) =$

$(a_1, a_2, \ldots, a_r)$. Then $\psi \circ \theta(1,0,\ldots,0) = p^{m-m_1}a_r$. Since $(1,0,\ldots,0)$ is of order p^{m_1}, we have $p^{m_1}a_r = 0$. As $m_1 < m$, this implies $p^{m_1-1}\psi \circ \theta(1,0,\ldots,0) = p^{m-1}a_r = 0$. On the other hand $\varphi(1,0,\ldots,0) = 1 + p^{m_1}\mathbb{Z}$ has order p^{m_1}, so $p^{m_1-1}(p^{m-m_1}a_r) \neq 0$. This contradiction proves that G does not have the embedding property.

Now, suppose $m_1 = \cdots = m_r = m$. Choose a basis $x_1, \ldots, x_r$ for G as a $\mathbb{Z}/p^m\mathbb{Z}$ module. Let $\varphi \colon G \to A$ and $\psi \colon G \to A$ be epimorphisms. Then there are generators $y_1, \ldots, y_r$ of G with $\psi(y_i) = \varphi(x_i)$, $i = 1, \ldots, r$ (e.g. by Gaschütz's lemma, Lemma 17.7.2). Extend the map $(x_1, \ldots, x_r) \to (y_1, \ldots, y_r)$ to an automorphism θ of G with $\psi \circ \theta = \varphi$. □

Example 24.6.5: *Necessity of the condition $\Phi(G) = 1$ in Lemma 24.3.2(c).* Consider $H = \mathbb{Z}/p^2\mathbb{Z} \times \mathbb{Z}/p^2\mathbb{Z}$ and $G = \mathbb{Z}/p\mathbb{Z} \times \mathbb{Z}/p^2\mathbb{Z}$. Then the epimorphism $\varepsilon \colon H \to G$ which is multiplication by p on the first coordinate and the identity on the second coordinate has kernel isomorphic to $\mathbb{Z}/p\mathbb{Z}$. By Example 24.6.4, H is the smallest embedding cover of G. By Corollary 22.7.8, $\tilde{G} = \tilde{H} = \hat{F}_2(p)$ and $\tilde{G}$ is a superprojective group (Lemma 24.3.3). Note that $\Phi(G) = \Phi(\mathbb{Z}/p^2\mathbb{Z}) = \mathbb{Z}/p\mathbb{Z} \neq 1$ (Lemma 22.1.4(d)). Thus, the condition $\Phi(G) = 1$ is necessary for the conclusion of Corollary 24.3.2(c) to hold. □

LEMMA 24.6.6:
(a) Let G be a small profinite group with the embedding property. Let N and N' be open normal subgroups with $G/N' \cong G/N$. Then $N \cong N'$.
(b) Let A and B be finitely generated profinite groups. Suppose B acts nontrivially on A. Choose an isomorphic copy B' of B. Then $B' \times (B \ltimes A)$ does not have the embedding property.

Proof of (a): Let $\pi \colon G \to G/N$ and $\pi' \colon G \to G/N'$ be the quotient maps and $\alpha \colon G/N' \to G/N$ an isomorphism. Lemma 24.6.2 gives an epimorphism $\theta \colon G \to G$ with $\alpha \circ \pi' \circ \theta = \pi$. Since, G is finitely generated, θ is an automorphism (Corollary 16.10.8). Hence, $\theta(N) = N'$ and $N \cong N'$.

Proof of (b): Set $N = B \ltimes A$, $N' = B' \times A$ and $G = B' \times (B \ltimes A)$ in (a). □

Example 24.6.7: *A PAC field which is not a Frobenius field.* The dihedral group D_n is the semidirect product of $\mathbb{Z}/n\mathbb{Z}$ and $\{\pm 1\}$, where -1 acts on $\mathbb{Z}/n\mathbb{Z}$ by the rule $x \mapsto -x$. For $n > 2$, this action is nontrivial. Hence, by Lemma 24.6.6(b), $G = \mathbb{Z}/2\mathbb{Z} \times D_n$ does not have the embedding property. Suppose in addition, $n = p$ is a prime. Then both ± 1 and $\mathbb{Z}/p\mathbb{Z}$ are maximal subgroups of D_p. So, $\Phi(G) = \Phi(D_p) = 1$. Apply Lemma 24.3.2(c) to conclude that $\tilde{G}$ does not have the embedding property. Thus, $\tilde{G}$ is a projective group which is not superprojective. Corollary 23.1.2, gives a PAC field F with $G(F) \cong \tilde{G}$. Thus, F is not a Frobenius field. □

Each algebraic extension of a PAC field is a PAC field (Corollary 11.2.5). The remainder of this section concludes with examples showing the analogous result for Frobenius fields is false.

Example 24.6.8: Products of symmetric groups without the embedding property and without nontrivial nilpotent normal subgroups. Let m and n be distinct positive integers exceeding 4, and let $G = S_m \times S_n$. The subgroups $N_1 = A_m \times S_n$ and $N_2 = S_m \times A_n$ are normal, and $G/N_1 \cong \mathbb{Z}/2\mathbb{Z} \cong G/N_2$. If we prove $N_1 \not\cong N_2$, we can deduce from Lemma 24.6.6 that G does not have the embedding property (Lemma 24.25(a)).

Assume, there is an isomorphism $\varphi \colon N_1 \to N_2$. Then $\mathrm{pr}_2(\varphi(A_m \times 1)) \triangleleft A_n$, where pr_2 is the projection on the second factor of $S_m \times A_n$. Since A_m and A_n are simple nonisomorphic groups, $\mathrm{pr}_2(\varphi(A_m \times 1)) = 1$. Thus, $\varphi(A_m \times 1) = A_m \times 1$. Hence, $A_m \times 1$ is of index 2 in $\varphi^{-1}(S_m \times 1)$. Therefore, $\mathrm{pr}_2(\varphi^{-1}(S_m \times 1))$ is a normal subgroup of S_n of order at most 2. But S_n has no normal subgroup of order 2. It follows that $\mathrm{pr}_2(\varphi^{-1}(S_m \times 1)) = 1$ and $\varphi^{-1}(S_m \times 1) \leq A_m \times 1$. This is a clear contradiction.

Note in addition that if N is a normal nilpotent subgroup of G, then so are its projections onto S_m and S_n. This implies $N = 1$. □

LEMMA 24.6.9:
(a) Suppose $G = \prod_{i=1}^\infty G_i$ is a direct product of profinite groups with the embedding property and $G_1, G_2, \ldots$ are of pairwise relatively prime orders. Then G has the embedding property.
(b) Each projective pronilpotent group N has the embedding property.

Proof of (a): Let $\varphi \colon G \to A$ and $\alpha \colon B \to A$ be epimorphisms with $B \in \mathrm{Im}(G)$. Then $A = \prod_{i=1}^\infty A_i$, $B = \prod_{i=1}^\infty B_i$ and φ and α are the direct product of epimorphisms $\varphi_i \colon G_i \to A_i$ and $\alpha_i \colon G_i \to A_i$ with $B_i \in \mathrm{Im}(G_i)$, $i = 1, 2, \ldots$ (Exercise 13 of Chapter 22). Since G_i has the embedding property, the result follows.

Proof of (b): By Proposition 22.9.3, N is the direct product of its p-Sylow groups. Each p-Sylow group is p-free (Proposition 22.10.4). Therefore, each p-Sylow group has the embedding property (Lemma 24.3.3). Thus, N satisfies the hypotheses of (a), and it too has the embedding property. □

LEMMA 24.6.10: *Let G be a small profinite group with the embedding property and N a closed characteristic subgroup of G. Then G/N has the embedding property.*

Proof: Let φ and ψ be epimorphisms of G/N onto a finite group A. Let $\pi \colon G \to G/N$ be the quotient map. Lemma 24.6.2 gives an epimorphism $\theta \colon G \to G$ with $\psi \circ \pi \circ \theta = \varphi \circ \pi$. By Proposition 16.10.6, θ is an automorphism. Since N is characteristic, $\theta(N) = N$. Hence, θ induces an automorphism $\bar{\theta}$ of G/N with $\psi \circ \bar{\theta} = \varphi$. If follows from Lemma 24.6.2 that G/N has the embedding property. □

Example 24.6.11: [Haran-Lubotzky, p. 199] *Neither projective subgroups nor overgroups of finite index of superprojective groups must be superprojective.* We give a sequence of projective groups, $G_1 < G_2 < G_3$, with G_1 open and normal in G_3, G_1, and G_3 superprojective, and G_2 not superprojective.

Consider a finite group G without the embedding property, having no nontrivial nilpotent normal subgroups. For example $G = S_m \times S_n$ with m, n distinct integers larger than 4 (Example 24.6.8). Embed G into a simple group S. For example first embed G in S_k, with $k \geq 3$. Then embed S_k into A_{k+2} by $f(\pi) = \pi$ for $\pi \in S_k$ even and $f(\pi) = (k+1 \ k+2)\pi$ for π odd.

Let $\varphi \colon \tilde{S} \to S$ be the universal Frattini cover of S. Put $G_3 = \tilde{S}$, $G_2 = \varphi^{-1}(G)$ and $G_1 = \Phi(\tilde{S})$. Since G_3 is projective (Proposition 22.6.1), so are G_1 and G_2 (Proposition 22.4.7). Furthermore, G_1 is pronilpotent (Lemma 22.1.2). It therefore has the embedding property (Lemma 24.6.9(b)). In addition, $\tilde{S}$ has the embedding property (Example 24.6.1 and Proposition 24.3.5). We show G_2 does hot have the embedding property.

First note that since $\Phi(S) = 1$, we have $\operatorname{Ker}(\varphi) = G_1$. Thus, G_1 is open in G_3. If N is an open normal nilpotent subgroup of G_2, then $\varphi(N)$ is a normal nilpotent subgroup of G. Hence, it is trivial, and N is contained in G_1. Thus, G_1 is a maximal open normal nilpotent subgroup of G_2. In particular, G_1 is an open characteristic subgroup of G_2. By Corollary 22.5.3, $\tilde{S}$ is finitely generated. Hence, by Corollary 17.6.3, G_2 is also finitely generated. Since, however, G does not have the embedding property, Lemma 24.6.10 implies that G_2 also does not have it. □

COROLLARY 24.6.12: *There exists a tower of PAC fields, $K \subset L \subset M$, satisfying this: M is a finite Galois extension of K, both K and M are Frobenius fields, but L is not.*

Proof: Let $G_1 < G_2 < G_3$ be as in Example 24.6.11. Corollary 23.1.2 gives a perfect PAC field K with $\operatorname{Gal}(K) \cong G_3$. Choose $L = \tilde{K}(G_2)$ and $M = \tilde{K}(G_1)$. □

24.7 Non-projective Smallest Embedding Cover

Although "projective cover" and "embedding cover" play analogous roles in the theory of profinite groups, their roles are not completely symmetric. Whereas the smallest projective cover of a profinite group with the embedding property has the embedding property (Proposition 24.3.5), there are projective groups whose smallest embedding covers are not projective (Lemma 24.7.3):

LEMMA 24.7.1: *Let N be a non trivial closed normal subgroup of a profinite group G. For each prime l choose an l-Sylow group G_l of G. Suppose $N \neq 1$. Then there is an l with $N \cap G_l \neq 1$.*

Proof: Choose an l which divides the order of N. Then its l-Sylow group $N \cap G_l$ (Proposition 22.9(d)) is nontrivial. □

LEMMA 24.7.2: *Let G be a profinite group and $\varepsilon \colon E \to G$ the smallest embedding cover of G. Then the orders of G and E are divisible by the same prime numbers.*

Proof: Denote the set of all prime numbers that divide $\#G$ by S. Let $\mathcal{C}$ be the formation of all finite groups whose orders are divisible only by prime numbers belonging to S. Then $\mathcal{C}$ is full. Lemma 17.4.8 gives an epimorphism $\varphi\colon F \to G$, where F is a free pro-$\mathcal{C}$ group. By Lemma 24.3.3, F has the embedding property. Hence, there is an epimorphism $\gamma\colon F \to E$ such that $\varepsilon \circ \gamma = \varphi$. It follows that $\#G$ and $\#E$ are divisible by the same prime numbers. $\square$

PROPOSITION 24.7.3 ([Chatzidakis3, Prop. 1.2]): *Let p be an odd prime number and let $H = H_2 \ltimes H_p$ be the profinite group defined by the following rules:*
$H_2 = \langle a, b \rangle$ *is the free pro-2-group on* a, b,
$H_p = \langle c \rangle \cong \mathbb{Z}_p$, *and*
$c^a = c^{-1}$, $c^b = c$.
Then H is projective but its smallest embedding cover is not projective.

Proof: For each prime number l, every l-Sylow group of H is l-free. Hence, H is projective (Proposition 22.10.4). We prove in four parts that the smallest embedding cover $\pi\colon E \to H$ is not projective.

PART A: *H is generated by two elements, namely a and bc.* Indeed, choose a generator u for $\mathbb{Z}_p$ and a generator v for $\mathbb{Z}_2$. Since $bc = cb$, the map $(u, v) \mapsto (b, c)$ extends to an epimorphism $\mathbb{Z}_2 \times \mathbb{Z}_p \to \langle b, c \rangle$. Since uv generates $\mathbb{Z}_2 \times \mathbb{Z}_p$, bc generates $\langle b, c \rangle$. Hence, $H = \langle a, b, c \rangle = \langle a, bc \rangle$, as claimed.

PART B: *H does not have the embedding property.* Indeed, consider the Klein group $A_2 = \langle a_0, b_0 \rangle$ of order 4 defined by the relations $a_0^2 = b_0^2 = 1$ and $a_0 b_0 = b_0 a_0$. The group A_2 acts on the cyclic group $A_p = \langle c_0 \rangle$ of order p by $c_0^{a_0} = c_0^{-1}$ and $c_0^{b_0} = c_0$. The semidirect product $A = A_2 \ltimes A_0$ is a quotient of H via the map $(a, b, c) \to (a_0, b_0, c_0)$. Consider the epimorphism $\alpha\colon A \to A_2$ defined by $\alpha(a_0) = b_0$, $\alpha(b_0) = a_0$, and $\alpha(c_0) = 1$. Its kernel is $\langle c_0 \rangle$. Consider the epimorphism $\eta\colon H \to A_2$ defined by $\eta(a) = a_0$, $\eta(b) = b_0$, and $\eta(c) = 1$.

Assume there is an epimorphism $\theta\colon H \to A$ with $\alpha \circ \theta = \eta$. Then η maps the p-Sylow group $\langle c \rangle$ of H onto the p-Sylow group $\langle c_0 \rangle$ of A, so $\theta(c) = c_0^i$, where i is relatively prime to p. Also, $\theta(a) = c_0^j b_0$. Apply θ to the relation $a^{-1}ca = c^{-1}$ to get $c_0^i = c_0^{-i}$. Hence, $p | 2i$, a contradiction. Therefore, θ does not exist and H does not have the embedding property, as claimed.

PART C: *E_p is Abelian.* Indeed, let F be the free profinite group on two generators x, y. Use Part A to define an epimorphism $\varphi\colon F \to H$ by $\varphi(x) = a$ and $\varphi(y) = bc$. Let $U = \varphi^{-1}(\langle c \rangle)$. Then $F/U \cong H/\langle c \rangle \cong H_2$ is the free pro-2 group of rank 2, $N = \mathrm{Ker}(\varphi) \leq U$, and $U/N \cong \langle c \rangle \cong \mathbb{Z}_p$.

Suppose U_0 is a closed normal subgroup of F, $U_0 \leq U$, and F/U_0 is a pro-2 group. Then $\mathrm{rank}(F/U_0) = 2$, so the quotient map $F/U_0 \to F/U$ is an isomorphism (Lemma 17.4.11). Hence, $U_0 = U$. Thus, U is the smallest closed normal subgroup of F such that F/U is a pro-2 group. As such, U is a characteristic subgroup of F.

24.8 A Theorem of Iwasawa

Let V be the smallest closed normal subgroup of U such that U/V is an Abelian pro-p group. Then V is characteristic in U, hence in F. Since $U/N \cong \langle c \rangle \cong \mathbb{Z}_p$, the group N contains V. Let $\varphi' \colon F/V \to H$ be the epimorphism which φ induces.

Since F has the embedding property, so does F/V (Lemma 24.6.10). Hence, there exists an epimorphism $\gamma \colon F/V \to E$ with $\pi \circ \gamma = \varphi'$. Note that U/V is a p-Sylow group of F/V. Thus, γ maps U/V onto a p-Sylow group E_p of E. Since U/V is Abelian, so is E_p.

PART D: *Conclusion of the proof.* Since N/V is contained in U/V and the latter group is pro-p, the intersection of N/V with $(F/V)_2$ is trivial. Thus, φ' is injective on $(F/V)_2$. Hence, π is injective on $E_2 = \gamma((F/V)_2)$. The only prime numbers which divide the order of F/V are 2 and p. By Part B, π is not injective (otherwise $E \cong H$ and H would have the embedding property, in contradiction to Part A). Thus, $\mathrm{Ker}(\pi) \neq 1$. By Lemma 24.7.2, 2 and p are the only prime numbers dividing $\#E$. Hence, by Lemma 24.7.1, $\mathrm{Ker}(\pi) \cap E_p \neq 1$, so π is not injective on E_p. Since $\pi(E_p) = H_p \cong \mathbb{Z}_p$ and E_p is Abelian, this implies that E_p is not pro-p free. Consequently, E itself is not projective. □

Remark 24.7.4: Solution of two open problems of [Fried-Jarden3]. The example, $E \to H$, which Proposition 24.7.3 gives solves [Fried-Jarden3, Problem 23.16(a)] negatively. Since H is projective, $H = \tilde{H}$ is its own smallest projective cover. On the other hand, as $E = E(H)$ is not projective, so the smallest projective cover of E is different from E. Thus, Proposition 24.7.3 also gives a negative answer to [Fried-Jarden3, Problem 23.16(b)]. □

24.8 A Theorem of Iwasawa

The back and fourth argument of Lemma 24.4.7 led Iwasawa to characterize $\hat{F}_\omega$ as a profinite group of at most countable rank for which every finite embedding problem is solvable [Iwasawa, p. 567]. Here are some generalizations of that characterization:

THEOREM 24.8.1: *Let $\mathcal{C}$ be a formation of finite groups and F a pro-$\mathcal{C}$ group of at most countable rank. Then $F \cong \hat{F}_\omega(\mathcal{C})$ if and only if $\mathrm{Im}(F) = \mathcal{C}$ and F has the embedding property.*

Proof: By Lemma 24.3.3, $\hat{F}_\omega(\mathcal{C})$ has the embedding property. In addition, $\mathrm{Im}(\hat{F}_\omega(\mathcal{C})) = \mathcal{C}$. Thus, our theorem is a special case of Lemma 24.4.7. □

COROLLARY 24.8.2: *Let $\mathcal{C}$ be a formation of finite groups and F a pro-$\mathcal{C}$ group of at most countable rank. Each of the following condition suffices (and also necessary) for F to be isomorphic to $\hat{F}_\omega(\mathcal{C})$:*
(a) *Every $\mathcal{C}$-embedding problem for F is solvable.*
(b) *F is $\mathcal{C}$-projective and every split $\mathcal{C}$-embedding problem for F is solvable.*

Proof: For each $B \in \mathcal{C}$ the embedding problem $(F \to 1, B \to 1)$ is finite and splits, hence solvable. Thus, $B \in \mathrm{Im}(F)$.

If either every $\mathcal{C}$-embedding problem for F is solvable or F is $\mathcal{C}$-projective and each split-embedding problem for F is solvable, then F has the embedding property (Proposition 22.5.9). Therefore, by Theorem 24.8.1, $F \cong \hat{F}_\omega(\mathcal{C})$. □

The special case where $\mathcal{C}$ is the formation of all finite groups gives Iwasawa's theorem:

COROLLARY 24.8.3 ([Iwasawa, p. 567]): *Let F be a profinite group of at most countable rank. Suppose every finite embedding problem for F is solvable. Then $F \cong \hat{F}_\omega$.*

Denote the maximal prosolvable extension of a field K by K_{solv}.

COROLLARY 24.8.4: *Let K be a Hilbertian field. Suppose $\text{Gal}(K)$ is projective and of rank at most $\aleph_0$ (e.g. K is countable). Then $\text{Gal}(K_{\text{solv}}/K) \cong \hat{F}_\omega(\text{solv})$.*

Proof: Put $G = \text{Gal}(K_{\text{solv}}/K)$. Consider a finite embedding problem

(1) $\qquad\qquad (\varphi: G \to A, \alpha: B \to A)$

where B is solvable and $C = \text{Ker}(\alpha)$ is a minimal normal subgroup of B. Then C is Abelian. If (1) splits, it has a solution, by Proposition 16.4.5. Hence, by Proposition 22.5.9, (1) has a solution even if (1) does not split. It follows from Corollary 24.8.2 that $G \cong \hat{F}_\omega(\text{solv})$. □

Example 24.8.5: *Countable Hilbertian fields with projective absolute Galois groups.*

(a) Consider the maximal Abelian extension $\mathbb{Q}_{\text{ab}}$ of $\mathbb{Q}$. By Theorem 16.11.3, $\mathbb{Q}_{\text{ab}}$ is Hilbertian. Class field theory implies $\text{Gal}(\mathbb{Q}_{\text{ab}})$ is projective [Ribes, p. 302]. Hence, by Corollary 24.8.4, $\text{Gal}(\mathbb{Q}_{\text{solv}}/\mathbb{Q}_{\text{ab}}) \cong \hat{F}_\omega(\text{solv})$. A long standing conjecture of Iwasawa predicts that $\text{Gal}(\mathbb{Q}_{\text{solv}}) \cong \hat{F}_\omega$. Suppose we could prove that every non-Abelian finite simple group is GAR over $\mathbb{Q}_{\text{ab}}$ (Definition 16.8.1). Then, by Matzat (Proposition 16.9.1), every finite embedding problem for $\text{Gal}(\mathbb{Q}_{\text{ab}})$ would be solvable. Therefore, by Corollary 24.8.3, $\text{Gal}(\mathbb{Q}_{\text{ab}}) \cong \hat{F}_\omega$.

(b) Let K be a countable PAC Hilbertian field. Then $\text{Gal}(K)$ is projective (Theorem 11.6.2). Thus, by Corollary 24.8.3, $\text{Gal}(K_{\text{solv}}/K) \cong \hat{F}_\omega(\text{solv})$. This is also a consequence of the much stronger result

(2) $\qquad\qquad \text{Gal}(K) \cong \hat{F}_\omega$

which was stated as Problem 24.41 of [Fried-Jarden3]. This was proved for the first time when $\text{char}(K) = 0$ in [Fried-Völklein2]. In the general case one considers a finite split embedding problem

(3) $\qquad\qquad (\varphi: \text{Gal}(K) \to A, \alpha: B \to A)$.

and the associated embedding problem

(4) $\qquad (\varphi \circ \mathrm{res}\colon \mathrm{Gal}(K(t)) \to A, \alpha\colon B \to A).$

When K is PAC (or even ample) (3) is solvable ([Pop1, Main Theorem A] or [Haran-Jarden6, Thm. B]).

When K is Hilbertian, one uses Lemma 13.1.1 to specialize the solution of (4) to a solution of (3). It follows from Corollary 24.8.2 that $\mathrm{Gal}(K) \cong \hat{F}_\omega$. $\square$

Remark 24.8.5(b) gives evidence to the following generalization of Iwasawa's conjecture:

CONJECTURE 24.8.6 ([Fried-Völklein2, p. 470]): *Let K be a Hilbertian field with $\mathrm{Gal}(K)$ projective. Then each finite embedding problem for $\mathrm{Gal}(K)$ is solvable.*

24.9 Free Profinite Groups of at most Countable Rank

We prove for a Melnikov formation $\mathcal{C}$ that if an open subgroup M of a closed normal subgroup of a free pro-$\mathcal{C}$ group is pro-$\mathcal{C}$, then M has the embedding property (Corollary 24.9.4). If in addition, $\mathrm{rank}(M) \leq \aleph_0$ and $\mathrm{Im}(M) = \mathcal{C}$, then $M \cong \hat{F}_\omega(\mathcal{C})$. This is an analog of Weissauer's theorem (Theorem 13.9.1) about extensions of Hilbertian fields.

Let G be a profinite group and S a finite simple group. Denote the intersection of all open normal subgroups H of G with $G/H \cong S$ by $M_G(S)$. First suppose, S is the cyclic group C_p of a prime order p. Let N_p be the intersection of all closed normal subgroup H of G with G/H a p-group. Then G/N_p is the maximal pro-p quotient of G (Definition 17.3.2), $M_G(S)/N_p = \Phi(G/N_p)$, and $G/M_G(S) \cong C_p^m$ for some cardinal number m (Lemma 22.7.1). Now suppose S is non-Abelian and G has m open normal subgroups N with $G/N \cong S$. By Lemma 18.3.11, $G/M_G(S) \cong S^m$. In both cases denote m by $r_G(S)$.

For a fixed G, r_G is a function from the set of all finite simple groups to the set of all cardinal numbers at most $\max(\aleph_0, \mathrm{rank}(G))$. We call r_G the **S-rank function** of G. We write $r_G \leq r_H$ for profinite groups G, H if $r_G(S) \leq r_H(S)$ for each finite simple group S.

Recall that open normal subgroups $N_1, \ldots, N_r$ of G are μ-independent when $G/\bigcap_{i=1}^r N_i \cong \prod_{i=1}^r G/N_i$ (Lemma 18.3.7). For $r = 2$, this is equivalent to $N_1 N_2 = G$.

LEMMA 24.9.1: *Let G be a profinite group, S a finite simple group, and $N_1, \ldots, N_r$ μ-independent open normal subgroups with $G/N_i \cong S$, $i = 1, \ldots, r$.*
(a) *Let $N_1', \ldots, N_{r+s}'$ be μ-independent open normal subgroups with $G/N_j' \cong S$, $j = 1, \ldots, r+s$. Then there are distinct $j_1, \ldots, j_s$ in $\{1, \ldots, r+s\}$ such that $N_1, \ldots, N_r, N_{j_1}', \ldots, N_{j_s}'$ are μ-independent.*

(b) Let $L_1, \ldots, L_s$ be open normal subgroups of G with $G/L_j \cong S$, $j = 1, \ldots, s$. Put $L = \bigcap_{j=1}^{s} L_j$ and $N = \bigcap_{i=1}^{r} N_i$. Let M be an open subgroup of N with $M \triangleleft G$ and $r = r_{G/M}(S)$. Suppose L and N are μ-independent. Then L and M are μ-independent.

Proof of (a): Put $N = \bigcap_{i=1}^{r} N_i$ and $N' = \bigcap_{j=1}^{r+s} N'_j$. Suppose first $S \cong C_p$. Then $G/N \cap N' \cong \mathbb{F}_p^t$ for some $t \geq r+s$. Now use Steinitz's Replacement Theorem: Let $v_1, \ldots, v_r$ and $v'_1, \ldots, v'_{r+s}$ be two sets of linearly independent vectors in a vector space V. Then there are $j_1, \ldots, j_s$ in $\{1, \ldots, r+s\}$ such that $v_1, \ldots, v_r, v'_{j_1}, \ldots, v'_{j_s}$ are linearly independent.

Now suppose S is non-Abelian. Choose $j_1, \ldots, j_s$ with $N_1, \ldots, N_r$, $N'_{j_1}, \ldots, N'_{j_s}$ distinct. By Example 18.3.11, these subgroups are μ-independent.

Proof of (b): By assumption, $NL = G$ and $N/M = M_{G/M}(S)$. Also, G/ML, being a quotient of G/L, is a direct product of isomorphic copies of S. Since $r_{G/M}(S) = r_{G/N}(S)$, we have $N \leq ML$. Thus, $ML = G$, so L and M are μ-independent. $\square$

The following result gives two simple properties of $r_G(S)$:

LEMMA 24.9.2:
(a) If $\bar{G}$ is a homomorphic image of G, then $r_{\bar{G}} \leq r_G$.
(b) Let G_1, G_2 be finite groups. Then $r_{G_1 \times G_2} = r_{G_1} + r_{G_2}$.

Proof of (a): Let $\varphi \colon G \to \bar{G}$ be an epimorphism. Then

$$M_G(S) \leq \varphi^{-1}(M_{\bar{G}}(S)).$$

Hence, $r_{\bar{G}}(S) \leq r_G(S)$ for each finite simple group S.

Proof of (b): We have to prove that $r_{G_1 \times G_2}(S) = r_{G_1}(S) + r_{G_2}(S)$ for each finite simple group S. For $S = C_p$ use that the dimension of a direct sum of vector spaces is the sum of their dimensions. For S non-Abelian use Lemma 18.3.10. $\square$

Example 24.6.11 shows, by illustration with finite groups, it is rare that each open subgroup of a profinite group with the embedding property has the embedding property. However:

LEMMA 24.9.3: Suppose each open normal subgroup of a profinite group G has the embedding property. Then each closed normal subgroup of G has the embedding property.

Proof: Let N be a closed normal subgroup of G and

(1) $\qquad\qquad (\varphi \colon N \to A, \ \alpha \colon B \to A)$

a finite embedding problem for G with $B \in \text{Im}(N)$ and $C = \text{Ker}(\alpha)$ minimal normal in G (Definition 24.1.2). Let $N_1 = \text{Ker}(\varphi)$ and let M be an open normal subgroup of N with $N/M \cong B$. By Lemma 1.2.5, G has an open normal subgroup G_0 such that $N \cap G_0 \leq N_1 \cap M$. Then $K = NG_0$ is an open normal subgroup of G, φ extends to a homomorphism $\lambda \colon K \to A$ by $\lambda(ng_0) = \varphi(n)$ for $g_0 \in G_0$ and $n \in N$, and $K/MG_0 \cong N/M \cong B$.

By assumption, K has the embedding property. Thus, there exists an epimorphism $\theta \colon K \to B$ with $\alpha \circ \theta = \lambda$. Put $K_1 = \mathrm{Ker}(\lambda)$ and $K_2 = \mathrm{Ker}(\theta)$. Then $N \cap K_1 = N_1$, $NK_1 = K$, $K/K_2 \cong B$, and $K_1/K_2 \cong C$. In particular, N_1 and K_2 are normal in K. Hence, $N_1 K_2/K_2$ is a normal subgroup of K/K_2 which is contained in K_1/K_2. The latter group is minimal in K/K_2, so either $N_1 K_2 = K_1$ or $N_1 \leq K_2$.

CASE 1: $N_1 K_2 = K_1$. Then $NK_2 = K$. Hence, $\theta(N) = \theta(K) = B$ and $\theta|_N \colon N \to B$ solves embedding problem (2).

CASE 2: $N_1 \leq K_2$. Then $L = NK_2$ is normal in K and $L/K_2 \cong N/N_1 \cong A$. Hence, $L \cap K_1 = K_2$. Thus,

$$(2) \qquad B \cong K/K_2 \cong L/K_2 \times K_1/K_2 \cong A \times C.$$

Since C is a minimal normal subgroup of B, it is isomorphic to a direct product $\prod_{i=1}^{s} S_i$ of isomorphic copies of a single finite simple group S (Remark 16.8.4). Let $r = r_A(S)$ and $E_1, \ldots, E_r$ μ-independent open normal subgroups of N which contain N_1 and with $N/E_i \cong S$, $i = 1, \ldots, r$. Since $B \in \mathrm{Im}(N)$, Lemma 24.9.2 implies, $r_N(S) \geq r_B(S) = r + s$. Thus, N has μ-independent open normal subgroups $F_1, \ldots, F_{r+s}$ with $N/F_j \cong S$, $j = 1, \ldots, r+s$. Lemma 24.9.1(a) gives distinct $j_1, \ldots, j_s$ in $\{1, \ldots, r+s\}$ such that $E_1, \ldots, E_r, F_{j_1}, \ldots, F_{j_s}$ are μ-independent. Put $F = F_{j_1} \cap \cdots \cap F_{j_s}$. Then $N/F \cong S^s \cong C$, and N_1 and F are μ-independent (Lemma 24.9.1(b)). Thus, $N/N_1 \cap F \cong N/N_1 \times N/F \cong A \times C = B$. We conclude that embedding problem (1) is solvable. $\square$

COROLLARY 24.9.4: *Let C be a Melnikov formation of finite groups, F a free pro-C-group, N a closed normal subgroup of G, and M an open subgroup of N. Suppose M is a pro-C group. Then*
(a) *M has the embedding property.*
(b) *If, in addition, $\mathrm{rank}(F) \leq \aleph_0$ and $\mathrm{Im}(M) = C$, then $M \cong \hat{F}_\omega(C)$.*

Proof of (a): Choose an open normal subgroup D of F with $D \cap N \leq M$ (Lemma 1.2.5). Then E is an open subgroup of F and $M \triangleleft E$. Moreover, D, M, and N are pro-C groups, so E is also a pro-C group. By Proposition 17.6.2, E is pro-C free. Hence, each open normal subgroup of E is also pro-C free, so has the embedding property (Lemma 24.3.3). It follows from Lemma 24.9.3 that M has the embedding property.

Proof of (b): Use (a) and Theorem 24.8.1. $\square$

REMARK 24.9.5: We prove in Proposition 24.10.3 that if M in Corollary 24.9.4(b) is properly contained in N, then the condition $\mathrm{Im}(M) = C$ is satisfied and $M \cong \hat{F}_\omega(C)$. $\square$

EXAMPLE 24.9.6: *Free profinite groups have closed subgroups without the embedding property.* Indeed, if G is a projective, but not superprojective group (Example 24.6.7), then G is isomorphic to a closed subgroup of a free profinite group (Corollary 22.4.6). But G does not have the embedding property. ◻

24.10 Application of the Nielsen-Schreier Formula

Let $\mathcal{C}$ be a Melnikov formation of finite groups and F a free pro-$\mathcal{C}$ group with $2 \le \mathrm{rank}(F) \le \aleph_0$. Then every closed normal subgroup N of F is a pro-$\mathcal{C}$ group with the embedding property (Corollary 24.9.4). Yet, N need not be pro-$\mathcal{C}$ free, because $\mathrm{Im}(N)$ may be a proper subset of $\mathcal{C}$. For example, if $\mathcal{C}$ is the formation of all finite groups and N is the minimal closed subgroup such that F/N is a pro-p group, then C_p is not a quotient of N, so N is not free.

We establish sufficient conditions for $\mathrm{Im}(N) = \mathcal{C}$ to hold, and therefore for N to be pro-$\mathcal{C}$ free. They are based on the Nielsen-Schreier's rank formula (Proposition 17.6.2) for an open subgroup E of F which is pro-$\mathcal{C}$:

$$\mathrm{rank}(E) = 1 + (F:E)(\mathrm{rank}(F) - 1).$$

LEMMA 24.10.1 ([Lubotzky-v.d.Dries, p. 35]): *Let F be a free pro-$\mathcal{C}$-group with $2 \le \mathrm{rank}(F) \le \aleph_0$ and N a closed normal subgroup F of infinite index. Suppose N is contained in an open subgroup H of F with*

(1) $$\mathrm{rank}(H/N) < 1 + (F:H)(e - 1).$$

Then $N \cong \hat{F}_\omega(\mathcal{C})$.

Proof: By Corollary 24.9.4(b) it suffices to prove that each group $G \in \mathcal{C}$ is a quotient of N.

Indeed, there exists an open subgroup E of H with $N \le E \triangleleft F$ and $(H:E) \ge \mathrm{rank}(G)$. Applying Corollary 17.6.3 to the group H/N and its open subgroup E/N we conclude that $\mathrm{rank}(E/N) \le 1 + (H:E)(\mathrm{rank}(H/N) - 1)$. By (1), $\mathrm{rank}(H/N) \le (F:H)(e-1)$. Hence, $\mathrm{rank}(E/N) \le 1 + (F:E)(e-1) - (H:E)$. By Proposition 17.6.2, E is $\mathcal{C}$-free. So, Proposition 17.7.4 gives a commutative diagram of epimorphisms

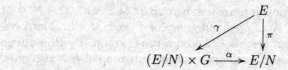

where π is the quotient map and $\alpha(\pi(x), g) = \pi(x)$ for each $x \in E$. In particular, for each $g \in G$ there is an $x \in E$ with $\gamma(x) = (1, g)$. Then $\pi(x) = \alpha(\gamma(x)) = 1$, so $x \in N$. Thus, $\gamma(N) = 1 \times G$ and G is a quotient of N. ◻

The following result is an analog of Kuyk's result (Theorem 16.11.3) on Abelian extensions of Hilbertian fields:

24.10 Application of the Nielsen-Schreier Formula

PROPOSITION 24.10.2 ([Jarden-Lubotzky1, Lemma 1.4]): *Let F be a free pro-$\mathcal{C}$ group of rank m and N a closed normal subgroup of F. Suppose $2 \leq m \leq \aleph_0$ and F/N is Abelian and infinite. Then $N \cong \hat{F}_\omega(\mathcal{C})$.*

Proof: First suppose $m < \aleph_0$. Put $A = F/N$ and let $\pi \colon F \to A$ be the quotient map. Choose a prime number p dividing the order of A. The map $a \mapsto a^p$ maps A onto its subgroup A^p. Put $H = \pi^{-1}(A^p)$. Then $(A : A^p) > 1$ and $\mathrm{rank}(H/N) = \mathrm{rank}(A^p) \leq \mathrm{rank}(A) \leq m < 1 + (A : A^p)(m-1) = 1 + (F : H)(m-1)$. Hence, by Lemma 24.10.1, $N \cong \hat{F}_\omega(\mathcal{C})$.

Now consider the case where $m = \aleph_0$. By Corollary 24.9.4(b), it suffices to prove that each $\mathcal{C}$-group G is a quotient of N. Indeed, choose an epimorphism $\varphi \colon F \to \hat{F}_e(\mathcal{C})$ with $e \geq \mathrm{rank}(G)$. Put $\bar{F} = \hat{F}_e(\mathcal{C})$ and $\bar{N} = \varphi(N)$. Then $\bar{F}/\bar{N}$ is Abelian. If $(\bar{F} : \bar{N}) < \infty$, then $\bar{N}$ is free of rank at least e (Proposition 17.6.2). If $(\bar{F} : \bar{N}) = \infty$, then $\bar{N} \cong \hat{F}(\mathcal{C})$, by the preceding paragraph. In both cases G is a quotient of $\bar{N}$, hence also of N. □

The condition on M in the next proposition is the group theoretic version of the condition on the field M in Weissauer's extension theorem of Hilbertian fields (Theorem 13.9.1(b)).

PROPOSITION 24.10.3 (Lubotzky-Melnikov-v.d.Dries): *Let F be a free pro-$\mathcal{C}$-group with $2 \leq \mathrm{rank}(F) \leq \aleph_0$, N a closed normal subgroup of infinite index, and M a proper open subgroup of N. Suppose M is pro-$\mathcal{C}$. Then $M \cong \hat{F}_\omega(\mathcal{C})$.*

Proof: The proof splits into two parts according to the rank of F.

PART A: $\mathrm{rank}(F) = e$ is finite. Choose an open normal subgroup H of F with $N \cap H \leq M$ (Lemma 1.2.5(a)) and put $E = MH$. Then H and M are pro-$\mathcal{C}$, hence so is also E. Moreover, E is open in F and $M \triangleleft E$.

Then E is $\mathcal{C}$-free and M is a closed normal subgroup of E. By Proposition 17.6.2, $(F : NE)(e-1) = \mathrm{rank}(NE) - 1$. Hence,

$$\begin{aligned}\mathrm{rank}(E/M) = \mathrm{rank}(NE/N) &\leq 1 + (F : NE)(e-1) \\ &< 1 + (N : M)(F : NE)(e-1) \\ &\leq 1 + (NE : E)(\mathrm{rank}(NE) - 1).\end{aligned}$$

It follows from Lemma 24.10.1, with NE replacing F and E replacing H, that $M \cong \hat{F}_\omega(\mathcal{C})$.

PART B: $\mathrm{rank}(F) = \aleph_0$. By Corollary 24.9.4(b) it suffices to prove that each group $G \in \mathcal{C}$ is a quotient of M. Indeed, there is an open subgroup E of F with $N \cap E = M$. Let E_0 be a normal subgroup of finite index in F which is contained in E_0, and let X be a basis of F. Choose a finite subset Y of X with $|Y| > \mathrm{rank}(G)$ and $X \smallsetminus Y \subseteq E_0$. Then, the smallest closed normal subgroup L of F which contains $X \smallsetminus Y$ is contained in E. By Lemma

17.4.9(a), F/L is isomorphic to the free pro-$\mathcal{C}$-group with basis Y. Also, ML is an open proper subgroup of the closed normal subgroup NL of F.

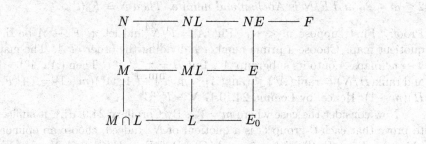

There are two possibilities: if $(F : NL) < \infty$, then ML/L is a free pro-$\mathcal{C}$-group of rank exceeding rank(G) (Proposition 17.6.2); if $(F : NL) = \infty$, then Part A, with F/L replacing F, gives $ML/L \cong \hat{F}_\omega(\mathcal{C})$. In both cases $G \in \text{Im}(ML/L) = \text{Im}(M/M \cap L) \subseteq \text{Im}(M)$.   □

Theorem 25.4.7(a) generalizes Proposition 24.10.3 to the case where $m = \text{rank}(F) \geq \aleph_0$. Proposition 24.10.3 has consequences analogous to Theorem 16.11.6:

PROPOSITION 24.10.4: *Let F be a free pro-$\mathcal{C}$-group with $2 \leq \text{rank}(F) \leq \aleph_0$ and N a nontrivial closed normal subgroup of F of infinite index. Then*
(a) *N has infinite rank.*
(b) *N is non-Abelian (in particular the center of F is trivial).*
(c) *If $\mathcal{C}$ contains a nonsolvable group, then N is not prosolvable.*
(d) *If $\mathcal{C}$ does not consist of p-groups for some prime p, then N is not pronilpotent. In particular, the Frattini subgroup of F is trivial.*
(e) *If $\mathcal{C}$ is the formation of all finite p-groups, then $N \cong \hat{F}_\omega(p)$.*

Proof: Since N is nontrivial, it has an open normal proper subgroup M. In particular, M is a pro-$\mathcal{C}$ group. By Proposition 24.10.3, M is isomorphic to $\hat{F}_\omega(\mathcal{C})$. So, rank$(M) = \aleph_0$. Therefore, rank$(N) = \aleph_0$. This proves (a).

Proof of (b): Each Melnikov formation is nonempty and closed under extensions. In particular, $\mathcal{C}$ contains non-Abelian groups. Thus, M is non-Abelian. Therefore, also N is non-Abelian.

Proof of (c): If $\mathcal{C}$ contains nonsolvable group, M is not prosolvable, so also N is not prosolvable.

Proof of (d): Suppose there are distinct prime numbers p and q dividing the orders of groups in $\mathcal{C}$. By (c), we may assume $\mathcal{C}$ contains only solvable groups. Hence, C_p and C_q occur as composition factors of groups in $\mathcal{C}$, so $C_p, C_q \in \mathcal{C}$. Let C_p operate on $(C_q)^p$ by permuting the factors of the as the cycle $(12 \cdots p)$. Denote the corresponding semidirect product by E. Then, $E \in \mathcal{C}$ and has C_p as a nonnormal p-Sylow subgroup. Hence, E is nonnilpotent. Since E is a quotient of M, M is not pronilpotent. Therefore, also N is not pronilpotent. By Lemma 22.1.2, $\Phi(F)$ is pronilpotent, so $\varphi(F)$ must be trivial.

24.10 Application of the Nielsen-Schreier Formula

Proof of (e): By Corollary 22.7.7, N is a free pro-p group. Since $\text{rank}(N)$ is infinite, $N \cong \hat{F}_\omega(p)$. □

Theorem 25.4.7 generalizes Proposition 24.10.4 to the case where $\text{rank}(F)$ is an arbitrary infinite cardinal number.

PROPOSITION 24.10.5 ([Jarden-Lubotzky1, Thm. 1.9]): *Let F be a free pro-$\mathcal{C}$ group with $2 \leq \text{rank}(F) \leq \aleph_0$. Let N_1 and N_2 be normal subgroups of F neither of which contains the other. Suppose $N = N_1 \cap N_2$ has infinite index. Then $N \cong \hat{F}_\omega(\mathcal{C})$.*

Proof: The proof has three parts.

PART A: *Reduction steps.* By assumption, N is a proper normal subgroup of both N_1 and N_2. Hence, if $(N_1 : N) < \infty$ or $(N_2 : N) < \infty$, then $N \cong \hat{F}_\omega(\mathcal{C})$ (Proposition 24.10.3). We may therefore assume $(N_1 : N) = \infty$ and $(N_2 : N) = \infty$.

For $i = 1, 2$ choose proper open normal subgroup N_i' of N_i which contain N. By Proposition 24.10.3, $N_i' \cong \hat{F}_\omega(\mathcal{C})$. Also, $N_1' \not\leq N_2$ (otherwise $N = N_1'$ has a finite index in N_1). Similarly, $N_2' \not\leq N_1'$. Next observe: $N_1 N_2/N_1' N_2' \cong N_1/N_1' \times N_2/N_2'$. Hence, by Proposition 24.10.3, $N_1' N_2'$ is pro-$\mathcal{C}$ free. Further, both N_1' and N_2' are normal in $N_1' N_2'$ and $N_1' \cap N_2' = N$. Replacing F, N_1, N_2 respectively by $N_1' N_2', N_1', N_2'$ we may assume $N_1 \cong N_2 \cong \hat{F}_\omega(\mathcal{C})$ and $N_1 N_2 = F$.

PART B: *Suppose $\text{rank}(F) < \infty$.* For $i = 1, 2$ choose an open normal subgroup M_i of N_i of large index which contains N such that

(2) $$2 + ((N_1 : M_1) + (N_2 : M_2))(\text{rank}(F/N) - 1)$$
$$< 1 + (N_1 : M_1)(N_2 : M_2)(\text{rank}(F/N) - 1).$$

By Corollary 17.6.3, $\text{rank}(M_i/N) \leq 1 + (N_i : M_i)(\text{rank}(N_i) - 1)$, $i = 1, 2$. The assumptions made in Part A imply $F/N \cong N_1/N \times N_2/N$. Hence, N_i/N is a quotient of F/N and $(F : M_1 M_2) = (N_1 : M_1)(N_2 : M_2)$.

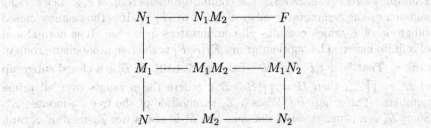

Therefore,

$$\begin{aligned}\mathrm{rank}(M_1M_2/N) &\le \mathrm{rank}(M_1/N) + \mathrm{rank}(M_2/N)\\ &\le 2 + (N_1:M_1)(\mathrm{rank}(N_1)-1) + (N_2:M_2)(\mathrm{rank}(N_2)-1)\\ &\le 2 + ((N_1:M_1)+(N_2:M_2))(\mathrm{rank}(F/N)-1)\\ &< 1 + (N_1:M_1)(N_2:M_2)(\mathrm{rank}(F/N)-1)\\ &= 1 + (F:M_1M_2)(\mathrm{rank}(F/N)-1).\end{aligned}$$

By Lemma 24.10.1, $N \cong \hat{F}_\omega(\mathcal{C})$.

PART C: *Suppose* $\mathrm{rank}(F) = \aleph_0$. By Corollary 24.9.4, it suffices to prove each $G \in \mathcal{C}$ is a quotient of N. So, choose an epimorphism $\varphi\colon F \to \hat{F}_e(\mathcal{C})$ with $e \ge \max(2, \mathrm{rank}(G))$ and such that none of the groups $\bar{N}_1 = \varphi(N_1)$ and $\bar{N}_2 = \varphi(N_2)$ contains the other. Denote the image of elements and subgroups of F under φ with a bar. By Parts A and B, $\bar{N}_1 \cap \bar{N}_2$ is pro-$\mathcal{C}$ free with rank at least e.

Each element of N_1 commutes modulo N with all elements of N_2. Hence, each element of $\bar{N}_1 \cap \bar{N}_2$ commutes modulo $\bar{N}$ with each element of $\bar{N}_1$ and $\bar{N}_2$. Thus, $\bar{N}_1 \cap \bar{N}_2/\bar{N}$ lies in the center of $\bar{F}/\bar{N}$. In particular, $\bar{N}_1 \cap \bar{N}_2/\bar{N}$ is Abelian. Hence, by Propositions 17.6.2 and 24.10.2, $\bar{N}$ is pro-$\mathcal{C}$ free with rank at least e. Therefore, G is a quotient of $\bar{N}$, hence also of N. □

Corollary 25.4.5 generalizes Proposition 24.10.5 to the case where m is an arbitrary cardinal number at least 2.

COROLLARY 24.10.6 ([Jarden-Lubotzky1, Cor. 1.10]): *Let F be a free pro-$\mathcal{C}$ group with $2 \le \mathrm{rank}(F) \le \aleph_0$ and let N be a closed normal subgroup of F. Suppose G/N is pronilpotent of infinite index and two distinct prime numbers divide $(G:N)$. Then $N \cong \hat{F}_\omega(\mathcal{C})$.*

Proof: By Proposition 22.9.3, G/N is the direct product of its p-Sylow groups and each of them is normal. Hence, G has closed normal subgroups N_1 and N_2 with $N_1 \cap N_2 = N$. It follows from Proposition 24.10.5 that $N \cong \hat{F}_\omega(\mathcal{C})$. □

Corollary 25.4.6 generalizes Corollary 24.10.6 to the case where m is an arbitrary cardinal number greater or equal to 2.

Example 24.10.7 (v. d. Dries): *The commutator subgroup of $\hat{F}_e$.* Let $F = \hat{F}_e$ and denote the commutator subgroup of F by $[F, F]$. It is the smallest closed subgroup of F which contains all commutators $a^{-1}b^{-1}ab$. It is normal and of infinite index. The quotient group $F/[F, F]$ is the free proabelian group of rank e. That is, $F/[F,F] \cong \hat{\mathbb{Z}}^e$ (Lemma 17.4.10). If B is a closed subgroup of $\hat{\mathbb{Z}}^e = \prod \mathbb{Z}_p^e$, then $B = \prod(B \cap \mathbb{Z}_p^e)$, where the p ranges over all prime numbers. The group $B \cap \mathbb{Z}_p^e$ is a $\mathbb{Z}_p$-submodule of the free $\mathbb{Z}_p$-module $\mathbb{Z}_p^e$. Since $\mathbb{Z}_p$ is a principal ideal domain, $B \cap \mathbb{Z}_p^e$ is also a free $\mathbb{Z}_p$-module of rank at most e [Lang7, p. 146, Thm. 7.1]. Therefore, $\mathrm{rank}(B) \leq e$. Assuming $e \geq 2$, choose B as $H/[F,F]$ with H an open subgroup of F of index at least 2. Then $\mathrm{rank}(B) < 1 + (F:H)(e-1)$. It follows from Lemma 24.10.1 that $[F,F] \cong \hat{F}_\omega$ if $e \geq 2$. □

Example 24.10.8: *Necessity of hypothesis (1) in Lemma 24.10.1.* Let $\mathcal{B} \subset \mathcal{C}$ be nonempty Melnikov formations of finite groups (e.g. $\mathcal{B}$ is the family of p-groups and $\mathcal{C}$ is the family of all finite groups). Consider the free pro-$\mathcal{C}$-group $F = \hat{F}_e(\mathcal{C})$. Suppose there is a $B \in \mathcal{B}$ with $\mathrm{rank}(B) \leq e$. Let N be the intersection of all open normal subgroups E of F with $F/E \in \mathcal{B}$. Then N is a closed normal subgroup of F and $F/N \cong \hat{F}_e(\mathcal{B})$ (Lemma 17.4.10). By Proposition 17.6.2, $\mathrm{rank}(E) = \mathrm{rank}(E/N)$ for each open subgroup E of F that contains N By Corollary 17.6.5, $\hat{F}_e(\mathcal{B})$ is infinite, so $(F:N) = \infty$. On the other hand, none of the groups $G \in \mathcal{C} \smallsetminus \mathcal{B}$ is a quotient of N. Thus, $N \not\cong \hat{F}_\omega(\mathcal{C})$. Hence, neither hypothesis (1) nor the conclusion of Lemma 24.10.1 hold. □

Example 24.10.9 ([Lubotzky-v.d.Dries, p. 34]): *A nonfree prosolvable profinite group which satisfies the Nielsen-Schreier index formula for open subgroups.* Let $p_0 < p_1 < p_2 < \cdots$ be a sequence of prime numbers. With $e > 2$, inductively construct a descending sequence $\hat{F}_e = N_0 > N_1 > N_2 > \ldots$ of closed characteristic subgroups of $\hat{F}_e$. Given $N_n \cong \hat{F}_{e_n}$ with $e_n = 1 + (\hat{F}_e : N_n)(e-1)$, let N_{n+1} be the unique open normal subgroup of N_n with $N_n/N_{n+1} \cong (\mathbb{Z}/p_n\mathbb{Z})^{e_n}$. Let $N = \bigcap_{i=0}^\infty N_n$. Then $\mathrm{rank}(N_n/N) = e_n$ for each n. Hence, $\mathrm{rank}(H/N) = 1 + (F:H)(e-1)$ for each open subgroup H of $\hat{F}_e$ that contains N (Exercise 13). Clearly $G = \hat{F}_e/N$ is a prosolvable group. But, since $\#G = \prod_{n=1}^\infty p_n^{e_n}$ is a supernatural product of finite prime powers, G is certainly nonfree. □

Exercises

1. Let G be a profinite group. Suppose every finite embedding problem $(\varphi: G \to A, \alpha: B \to A)$ where $\mathrm{Ker}(\alpha)$ is a minimal normal subgroup of B is solvable. Prove that G has the embedding property.

2. Prove that each inverse limit of profinite groups with the embedding property has the embedding property.

3. Let $\alpha: B \to A$, $\beta: C \to B$ be epimorphisms of finite groups. Prove that If $\alpha \circ \beta$ is an I-cover, then so is β.

4. Let $\beta': B \to D$ be an I-cover of finite groups and $\beta: G \to D$ an epimorphism of profinite groups. Combine the construction appearing in the proof of Lemma 24.4.4 with Exercise 3 to construct an I-cover $\varepsilon: E \to G$ and an epimorphism $\varepsilon': E \to B$ such that $\beta' \circ \varepsilon' = \beta \circ \varepsilon$.

5. Let $\alpha, \alpha': C \to A$ be epimorphisms of finite groups. Construct I-covers $\varepsilon, \varepsilon': E \to C$ with E finite and $\alpha \circ \varepsilon = \alpha' \circ \varepsilon'$.

6. Let $\varphi\colon H \to A$ be an embedding cover of a finite group A and let $\alpha\colon B \to A$ be an epimorphism. Suppose B is a quotient of H and $\mathrm{rank}(B) \leq \aleph_0$. Construct an epimorphism $\gamma\colon H \to B$ with $\alpha \circ \gamma = \varphi$.

7. Let G be a profinite group and F a smallest embedding cover of G (Proposition 24.4.5). Prove that $E(\tilde{G})$ is a quotient of $\tilde{F}$.

8. Let G be a group of even order $2k$ which is generated by elements of order 2. Suppose that 4 divides k. Show that the right regular representation of G maps G into a subgroup of A_k. Deduce that for $m, n > 2$, the group $S_m \times S_n$ can be embedded into the simple group $A_{m!n!}$. Hint: Factor the image in S_k of each involution of G into a product of disjoint 2-cycles.

9. For a profinite group G, define the descending Frattini sequence inductively: $\Phi_0(G) = G$ and $\Phi_{n+1}(G) = \Phi(\Phi_n(G))$. Prove that if G is pro-p then $\bigcap_{n=1}^{\infty} \Phi_n(G) = 1$.

Hint: If N is an open subgroup of G, then there is a normal sequence, $N = N_n \triangleleft \cdots \triangleleft N_1 \triangleleft N_0 = G$, of groups such that $(N_i : N_i + 1) = p$, $i = 0, \ldots, n-1$. Show that $N_i \leq \Phi_i(G)$.

10. Let N be a closed normal subgroup of a finitely generated pro-p-group G such that $N \not\leq \Phi(G)$. Apply Lemma 22.7.4 to conclude that $\mathrm{rank}(G/N) < \mathrm{rank}(G)$.

11. Let G be a pro-p-group of rank e such that $\mathrm{rank}(H) = 1 + (G : H)(e-1)$ for each open subgroup H. Prove that $G \cong \hat{F}_e(p)$. [Lubotzky1, Thm. 4.2]

Hint: Present G as a quotient $\hat{F}_e(p)/N$, where N is a closed normal subgroup of $\hat{F}_e(p)$. If $N \neq 1$, apply Exercise 2 to find an integer n such that $N \leq \Phi_n(\hat{F}_e(p))$ but $Nn \leq \Phi_{n+1}(\hat{F}_e(p))$. Now apply Exercise 10.

12. Give an example of a profinite group G of rank at most α_0 such that $\mathrm{Im}(G)$ consists of all finite groups but G is not free.

13. Let G be a profinite group of rank e. Suppose that G has a descending sequence $G = G_0 > G_1 > G_1 > \ldots$ of open subgroups, with trivial intersection, such that $\mathrm{rank}(G_n) - 1 = (G : G_n)(e - 1)$ for each n. Use Corollary 17.5.2 to prove that $\mathrm{rank}(H) - 1 = (G : H)(e-1)$ for each open subgroup H of G.

Notes

Lemma 24.1.1 gives a homomorphism φ whose decomposition group $D(\varphi)$ is a given subgroup $\mathrm{Gal}(F/E')$ of $\mathrm{Gal}(F/E)$. Even if $\mathrm{Gal}(F/E')$ is a cyclic group with a generator σ, there is no distinguished element in $D(\varphi)$ which is made equal to σ. Thus, Lemma 24.1.1 is an analog to Frobenius density theorem, which preceded the Chebotarev density theorem. This explains our choice of the name Frobenius field. An appropriate analog to the Chebotarev density theorem can be found in [Jarden9].

The proof of Lemma 24.2.2 is a simplified version of the proof of [Dèbes, Prop. 1.2]. A proof of the Beckman-Black problem in characteristic 0 appears in [Colliot-Thélène] and in general in [Moret-Bailly] and [Haran-Jarden7, Thm. 2.2]. See [Dèbes] for a history of the problem and results over non-ample fields.

[Haran-Lubotzky] proves the existence and uniqueness of the smallest embedding covers of each finitely generated profinite group. We generalize their methods and prove the same result for arbitrary profinite groups (Proposition 24.4.5). Chatzidakis translated the conditions entering in the definition of the smallest embedding cover of a profinite group G into a countably many sorted first order language (see [Chatzadakis1] and its reproduction [Chatzidakis4].) This, together with the translation of the data about the finite quotients of $E(G)$ gives an "ω-stable" theory T. An application of the existence (Morley) and uniqueness (Shelah) of the prime model of T allows Chatzidakis to prove the existence and the uniqueness of the smallest embedding cover of G. Proposition 24.4.5 gives a pure profinite group proof of existence. It is desirable to prove the uniqueness along the same lines.

The decision problem solved by Proposition 24.4.6 is raised in [Fried-Haran-Jarden, Section 1]. The problem of finding a PAC field which is not Frobenius is raised in [Fried-Haran-Jarden, Problem 1.9] and is first answered in [Ershov-Fried] by proving the universal Frattini cover of $S_2 \times S_3$ is projective but not superprojective. This is a special case of our Lemma 24.6.6(b). Corollary 24.6.12 solves Problem 1.8 of [Fried-Haran-Jarden].

[Haran-Lubotzky, Cor. 2.12] states without proof that $\widetilde{E(H)} = E(\tilde{H})$ for every finitely generated profinite group H. [Fried-Jarden3, Problem 23.16] states this and a related question as an open problem. Following [Chatzidakis3], Section 24.7 solves the open problem and refutes [Haran-Lubotzky, Cor. 2.12].

Chapter 25.
Free Profinite Groups of Infinite Rank

This chapter studies free pro-$\mathcal{C}$ groups of infinite rank and their closed subgroups, where $\mathcal{C}$ is a Melnikov formation of finite groups.

The data needed to characterize a pro-$\mathcal{C}$ group F of rank m as $\hat{F}_m(\mathcal{C})$ becomes more complicated with increasing m. We distinguish between three cases. If $m < \infty$, it suffices to know that each $\mathcal{C}$-group of rank at most m is an image of F (Lemma 17.7.1). If $m = \aleph_0$, each finite embedding problem for F has to be solvable (Corollary 24.8.2). When $m > \aleph_0$, each $\mathcal{C}$-embedding problem for F must have m distinct solutions (Lemma 25.1.2 and Theorem 25.1.7).

As a first application of this result we prove that a closed subgroup H of $\hat{F}_m(\mathcal{C})$ which is pro-$\mathcal{C}$ and does not lie too deep is isomorphic to $\hat{F}_m(\mathcal{C})$ (Proposition 25.2.2). By "not lying too deep" we mean that there are less than m open subgroups of $\hat{F}_m(\mathcal{C})$ containing H. When $m = \aleph_0$, this means that H is open. In the general case, H may well be of infinite index.

This basic result then leads to a group theoretic analog of Haran's diamond theorem for Hilbertian field (Theorem 13.8.3): Let N_1, N_2, M be closed subgroups of $\hat{F}_m(\mathcal{C})$. Suppose m is infinite, N_1, N_2 are normal, M is pro-$\mathcal{C}$, $N_1 \cap N_2 \leq M$, but neither N_1 nor N_2 are contained in M. Then $M \cong \hat{F}_m(\mathcal{C})$ (Theorem 25.4.3).

As a result, several types of closed subgroups of $\hat{F}_m(\mathcal{C})$, with m infinite, are proved to be free pro-$\mathcal{C}$ of rank m. For example, let M be a proper open subgroup of a normal closed subgroup N of $\hat{F}_m(\mathcal{C})$. Suppose M is pro-$\mathcal{C}$. Then $M \cong \hat{F}_m(\mathcal{C})$ (Theorem 25.4.7).

Each closed normal subgroup N of $\hat{F}_e(\mathcal{C})$ of infinite index, with $2 \leq e < \infty$, is contained in a closed normal subgroup E which is isomorphic to $\hat{F}_\omega(\mathcal{C})$ (Proposition 25.8.3). So, if M is a pro-$\mathcal{C}$ proper open subgroup of N, then $M \cong \hat{F}_\omega(\mathcal{C})$ (Proposition 25.8.4).

Not every closed normal subgroup of $\hat{F}_m(\mathcal{C})$ is free pro-$\mathcal{C}$. For example, for each non-Abelian group S in $\mathcal{C}$ there is a closed normal subgroup N of $\hat{F}_m(\mathcal{C})$ such that S is the only simple quotient of N. More generally, let f be a function from the set of finite simple groups to the set of cardinal numbers at most m. Suppose $f(S) = 0$ for each $S \notin \mathcal{C}$ and $f(C_p)$ is either 0 or m. Then there exists a closed normal subgroup N of $\hat{F}_m(\mathcal{C})$ with $r_N = f$ (Proposition 25.7.7). Here r_N is the S-rank function of N introduced in Section 24.9. Conversely, r_N satisfies the same conditions as f for each closed normal subgroup N of $\hat{F}_m(\mathcal{C})$ (Proposition 25.7.4). Moreover, the function r_N characterizes the normal closed subgroup N up to an isomorphism (Theorem 25.7.3). For example, $N \cong \hat{F}_m(\mathcal{C})$ if and only if $r_N(S) = m$ for all simple S in $\mathcal{C}$.

A class of subgroups of $\hat{F}_m(\mathcal{C})$ that lie more deeply than closed normal subgroups is that of accessible subgroups. They are intersections of normal transfinite sequences of closed subgroups of $\hat{F}_m(\mathcal{C})$. Like closed normal subgroups, accessible subgroups of pro-$\mathcal{C}$ groups are pro-$\mathcal{C}$ (Lemma 25.9.1). By Lemma 25.9.3, each accessible subgroup N of a pro-$\mathcal{C}$ group is the intersection of a normal sequence of closed subgroups of at most countable rank.

Accessible subgroups of $\hat{F}_m(\mathcal{C})$ are characterized as homogeneous pro-$\mathcal{C}$ groups (Theorem 25.9.11). Here we call a pro-$\mathcal{C}$ group G of infinite rank **homogeneous** if every pro-$\mathcal{C}$ embedding problem $(\varphi\colon G \to A,\ \alpha\colon B \to A)$ satisfying
(1a) $\mathrm{Ker}(\alpha)$ is contained in every open normal subgroup of B, and
(1b) $\mathrm{rank}(G) = \aleph_0$ and B finite, or $\mathrm{rank}(A) < \mathrm{rank}(G)$ and $\mathrm{rank}(B) \le \mathrm{rank}(G)$
is solvable.

Like closed normal subgroups, accessible subgroups are uniquely determined up to an isomorphism by their S-rank functions (Proposition 25.7.2). However, in contrast to closed normal subgroups, the only restriction on these functions is that their values at $S \notin \mathcal{C}$ are 0 (Proposition 25.9.10).

25.1 Characterization of Free Profinite Groups by Embedding Problems

In order to characterize profinite groups of rank exceeding $\aleph_0$ by their finite quotients, in addition to the embedding property, we must add an hypothesis concerning the cardinality of the set of solutions of finite embedding problems.

LEMMA 25.1.1: *Let F be a profinite group and*

(1) $$(\varphi\colon F \to A,\ \alpha\colon B \to A)$$

a finite embedding problem for F. If $\mathrm{rank}(F) < \infty$, then the number of solutions to (1) is at most $|B|^{\mathrm{rank}(F)}$. If $\mathrm{rank}(F) = \infty$, then this number is at most $\mathrm{rank}(F)$.

Proof: The finite case is clear. Suppose $\mathrm{rank}(F) = \infty$.

To each solution, $\gamma\colon F \to B$, of (1), associate the open normal subgroup $\mathrm{Ker}(\gamma)$ of F. If $x_1,\ldots,x_n$ generate F modulo $\mathrm{Ker}(\gamma)$, then γ is determined by its values on $x_1,\ldots,x_n$. There are at most $|B|^n$ possibilities for these values, so the map $\gamma \mapsto \mathrm{Ker}(\gamma)$ has finite fibers. By Proposition 17.1.2, F has $\mathrm{rank}(F)$ open normal subgroups. Thus, there are at most $\mathrm{rank}(F)$ possible epimorphisms γ that solve embedding problem (1). □

If α in (1) is an isomorphism, then (1) has a unique solution, $\alpha^{-1} \circ \varphi$. The next result gives the number of solutions of (1) when $\mathrm{Ker}(\alpha) \ne 1$ and F is free of infinite rank.

LEMMA 25.1.2: *Let $\mathcal{C}$ be a formation of finite groups and F a free pro-$\mathcal{C}$ group with basis X. Suppose $B \in \mathcal{C}$ and $\mathrm{Ker}(\alpha) \neq 1$ in embedding problem (1). If $|B| < |X| < \infty$, then (1) has at least $2^{|X|-|B|}$ solutions. If X is infinite, then (1) has exactly $|X|$ solutions.*

Proof: Choose a set theoretic section $\alpha' \colon A \to B$ to $\alpha \colon B \to A$ with $\alpha'(1) = 1$. Thus, $\alpha \circ \alpha' = \mathrm{id}_A$. The proof splits up into two parts.

PART A: $|B| < |X| < \infty$. List the elements of $\varphi(X) \smallsetminus \{1\}$ as $a_1, \ldots, a_e$ and the elements of $\mathrm{Ker}(\alpha)$ as $b_{e+1}, \ldots, b_n$. Put $b_i = \alpha'(a_i)$, $i = 1, \ldots, e$. Then $n \leq |A| - 1 + |\mathrm{Ker}(\alpha)| \leq |A| \cdot |\mathrm{Ker}(\alpha)| = |B|$ and $B = \langle b_1, \ldots, b_n \rangle$. Now choose distinct elements $x_1, \ldots, x_n$ of X with $\varphi(x_1) = a_1, \ldots, \varphi(x_e) = a_e$. Define a map $\gamma_0 \colon X \to B$ by $\gamma_0(x_i) = \alpha'(\varphi(x_i))$ for $i = 1, \ldots, e$, $\gamma_0(x_j) = b_j \alpha'(\varphi(x_j))$ for $j = e+1, \ldots, n$, and $\gamma_0(x) = b \alpha'(\varphi(x))$ for $x \in X \smallsetminus \{x_1, \ldots, x_n\}$ with an arbitrary $b \in \mathrm{Ker}(\alpha)$. The number of possible γ_0 is $|\mathrm{Ker}(\alpha)|^{|X|-n}$, which is at least $2^{|X|-|B|}$. Each γ_0 extends to a homomorphism $\gamma \colon F \to B$.

By definition, $\alpha(\gamma(x)) = \varphi(x)$ for each $x \in X$, so $\alpha \circ \gamma = \varphi$ and $\alpha(\gamma(F)) = \varphi(F) = A$. Moreover, for each j between $e+1$ and n either $\varphi(x_j) = 1$ or there is an i between 1 and e with $\varphi(x_j) = \varphi(x_i)$. In the former case $b_j = \gamma(x_j) \in \langle \gamma(F) \rangle$. In the latter case $b_j = \gamma_0(x_j) \alpha'(\varphi(x_j))^{-1} = \gamma(x_j) \alpha'(\varphi(x_i))^{-1} = \gamma(x_j)\gamma(x_i)^{-1} \in \langle \gamma(F) \rangle$. Thus, $\mathrm{Ker}(\alpha) \leq \langle \gamma(F) \rangle$. Consequently, $\langle \gamma(F) \rangle = B$, as desired.

PART B: X is infinite. Let $X_0 = X \cap \mathrm{Ker}(\varphi)$ and $X_1 = X \smallsetminus X_0$. Since $\mathrm{Ker}(\alpha) \neq 1$ and $|X| = |X_0|$, the number of surjective maps $\gamma_0 \colon X_0 \to \mathrm{Ker}(\alpha)$ that map all but finitely many elements of X_0 to 1 is equal to $|X|$. Extend each γ_0 to a map $\gamma_1 \colon X \to B$ by $\gamma_1(x) = \gamma_0(x)$ if $x \in X_0$ and $\gamma_1(x) = \alpha'(\varphi(x))$ if $x \in X_1$. Then γ_1 maps almost every element of X to 1 and satisfies $\alpha(\gamma_1(x)) = \varphi(x)$ for each $x \in X$. Hence, γ_1 uniquely extends to an epimorphism $\gamma \colon F \to B$ with $\alpha \circ \gamma = \varphi$. Distinct γ_0's induce distinct γ's. It follows that embedding problem (1) has at least $|X|$ solutions. Combining this with Lemma 25.1.1, we conclude that the number of solutions is exactly $|X|$. □

If $\gamma \colon F \to B$ is a solution of embedding problem (1), we call $\mathrm{Ker}(\gamma)$ a **solution group** of (1).

LEMMA 25.1.3: *Let $\mathcal{C}$ be a formation of finite groups and F a free pro-$\mathcal{C}$ group of rank m. Suppose $B \in \mathcal{C}$ and $\mathrm{Ker}(\alpha) \neq 1$. Then the number of solution groups of (1) lies between $\frac{2^{m-|B|}}{|\mathrm{Aut}(B)|}$ and $\frac{|B|^m}{|\mathrm{Aut}(B)|}$ if $|B| < m < \infty$ and is exactly m if m is infinite.*

Proof: Denote the set of all solutions of (1) by Γ and the set of all solution groups of (1) by $\mathcal{N}$. Then $\mathrm{Ker}(\gamma) \in \mathcal{N}$ for each $\gamma \in \Gamma$. Two solutions $\gamma, \gamma' \in \Gamma$ have the same kernel if and only if there is a $\theta \in \mathrm{Aut}(B)$ with $\gamma' = \theta \circ \gamma$. So, the fibers of the map $\gamma \mapsto \mathrm{Ker}(\gamma)$ of Γ onto $\mathcal{N}$ have cardinality $|\mathrm{Aut}(B)|$. Now apply Lemmas 25.1.1 and 25.1.2 to verify the assertion of the lemma. □

25.1 Characterization of Free Profinite Groups by Embedding Problems

LEMMA 25.1.4: *Let C be a formation of finite groups, F a pro-C group, and m a cardinal number. Suppose every C-embedding problem (1) with $B \neq 1$, where $\mathrm{Ker}(\alpha)$ is a minimal normal subgroup of B, has at least m solutions. Then every C-embedding problem (1) with $\mathrm{Ker}(\alpha)$ and nontrivial has at least m solutions.*

Proof: Put $C = \mathrm{Ker}(\alpha)$. By assumption (1) has m solutions if C is a finite minimal normal subgroup of B. Otherwise, choose a non trivial proper subgroup C_0 of C which is normal in B. Then $(\varphi\colon F \to A, \bar\alpha\colon B/C_0 \to A)$, with $\bar\alpha$ induced by α, is a pro-C-embedding problem for F with B/C_0 nontrivial. Since $\mathrm{Ker}(\bar\alpha) = C/C_0$ is nontrivial and has a smaller order than C, induction gives m epimorphisms $\beta_i\colon F \to B/C_0$ with $\bar\alpha \circ \beta_i = \varphi$.

For each i consider the pro-C-embedding problem

$$(\beta_i\colon F \to B/C_0,\ \pi\colon B \to B/C_0),$$

where π is the quotient map. Its kernel C_0 is nontrivial and has a smaller order than C. Induction gives an epimorphism $\gamma_i\colon F \to B$ with $\pi \circ \gamma_i = \beta_i$. Each of the γ_i solves (1) and they are all distinct. □

LEMMA 25.1.5: *Let C be a formation of finite groups and F a pro-C group of infinite rank. Suppose every finite C-embedding problem for F with a nontrivial kernel has $\mathrm{rank}(F)$ solutions. Then every pro-C-embedding problem*

(2) $$(\varphi\colon F \to G,\ \alpha\colon H \to G)$$

in which $\mathrm{Ker}(\alpha)$ is finite and $\mathrm{rank}(G) < \mathrm{rank}(F)$ is solvable.

Proof: Assume without loss that $\mathrm{Ker}(\alpha) \neq 1$. As in the proof of Lemma 25.1.4, an induction on $|\mathrm{Ker}(\alpha)|$ reduces the lemma to the case where $\mathrm{Ker}(\alpha)$ is a minimal nontrivial closed normal subgroup of H. Choose an open normal subgroup H_1 of H with $H_1 \cap \mathrm{Ker}(\alpha) = 1$. This gives a commutative diagram

(3)
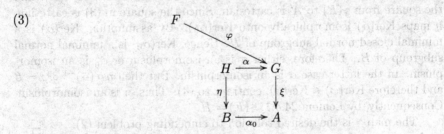

in which $B = H/H_1$ and $A = G/\alpha(H_1)$ are finite groups and the square from H to A is cartesian. In particular, $\mathrm{Ker}(\alpha_0) \cong \mathrm{Ker}(\alpha)$ is nontrivial.

Let $\mathcal{B}$ be the set of all epimorphisms $\beta\colon F \to B$ with $\alpha_0 \circ \beta = \xi \circ \varphi$. Put $\mathcal{B}_0 = \{\beta \in \mathcal{B}\mid \mathrm{Ker}(\varphi) \leq \mathrm{Ker}(\beta)\}$. Denote the set of all epimorphisms $\zeta\colon G \to B$ with $\alpha_0 \circ \zeta = \xi$ by $\mathcal{G}$. For each $\beta \in \mathcal{B}_0$ there is a unique $\zeta \in \mathcal{G}$ with $\zeta \circ \varphi = \beta$. Moreover, the map $\mathcal{B}_0 \to \mathcal{G}$ given by $\beta \mapsto \zeta$ is bijective. By Lemma

25.1.1, either $\mathcal{G}$ is finite or $\mathcal{G}$ is infinite and $|\mathcal{G}| \leq \text{rank}(G)$. By assumption, $\text{rank}(G) < \text{rank}(F) = |\mathcal{B}|$, so $|\mathcal{B}_0| < |\mathcal{B}|$. This gives an epimorphism $\beta\colon F \to B$ with

(4) $\qquad\qquad \alpha_0 \circ \beta = \xi \circ \varphi$ and $\text{Ker}(\varphi) \not\leq \text{Ker}(\beta)$.

Since the square from H to A in (3) is cartesian, there is a homomorphism $\gamma\colon F \to H$ which makes the following diagram commutative.

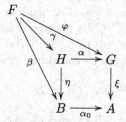

Thus, $\alpha(\gamma(F)) = \varphi(F) = G$ and $\eta(\gamma(F)) = \beta(F) = B$.

CLAIM: γ is surjective. Indeed, since $\gamma(F) \leq H$, Lemma 24.4.1 gives a commutative diagram of epimorphisms

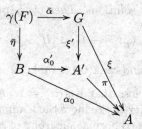

where $\bar{\alpha}$ and $\bar{\eta}$ are, respectively, the restrictions of α and η to $\gamma(F)$, and the square from $\gamma(F)$ to A' is cartesian. Since the square in (3) is cartesian, η maps $\text{Ker}(\alpha)$ isomorphically onto $\text{Ker}(\alpha_0)$. By assumption, $\text{Ker}(\alpha)$ is a minimal closed normal subgroup of H. Hence, $\text{Ker}(\alpha_0)$ is a minimal normal subgroup of B. Therefore, either π is an isomorphism or α'_0 is an isomorphism. In the latter case $\bar{\alpha}$ is an isomorphism. But then $\bar{\eta} \circ (\bar{\alpha})^{-1} \circ \varphi = \beta$ and therefore $\text{Ker}(\varphi) \leq \text{Ker}(\beta)$, contrary to (4). Thus, π is an isomorphism. Consequently, by Lemma 24.4.1, $\gamma(F) = H$.

The map γ is the desired solution to embedding problem (2). $\qquad\square$

PROPOSITION 25.1.6 ([Melnikov3], [Chatzidakis4, §3.2, Cor.]): *Let $\mathcal{C}$ be a formation of finite groups and F and F' pro-$\mathcal{C}$ groups. Suppose $\text{rank}(F) = \text{rank}(F')$ are infinite and every finite $\mathcal{C}$-embedding problem for F and for F' with a nontrivial kernel has $\text{rank}(F)$ solutions. Then $F \cong F'$.*

Proof: Let $\text{rank}(F) = m$. Order all open normal subgroups of F and F' in transfinite sequences $\{L_\alpha \mid \alpha < m\}$ and $\{L'_\alpha \mid \alpha < m\}$, respectively,

25.1 Characterization of Free Profinite Groups by Embedding Problems

with $L_0 = F$ and $L'_0 = F'$. Apply transfinite induction to construct two descending sequences $\{N_\beta \mid \beta < m\}$ and $\{N'_\beta \mid \beta < m\}$ of closed normal subgroups of F and F', respectively, satisfying:

(5a) For all $\beta < m$ there are isomorphisms $\theta_\beta \colon F/N_\beta \to F'/N'_\beta$ with commutative diagrams

$$\begin{array}{ccc} F/N_\gamma & \xrightarrow{\theta_\gamma} & F'/N'_\gamma \\ \downarrow & & \downarrow \\ F/N_\beta & \xrightarrow[\theta_\beta]{} & F'/N'_\beta \end{array}$$

for $\beta < \gamma < m$, where the vertical arrows are the quotient maps.

(5b) $\mathrm{rank}(F/N_\beta) < m$ for each $\beta < m$.

(5c) $N_\beta \leq L_\alpha$ and $N'_\beta \leq L'_\alpha$ for all $\alpha < \beta < m$.

Consider $\delta < m$. Suppose N_γ, N'_γ, and θ_γ have been defined for each $\gamma < \delta$ such that (5) holds for each $\beta < \gamma$. There are two cases to consider:

CASE A: $\delta = \gamma + 1$. Let $M_\delta = L_\gamma \cap N_\gamma$. Lemma 25.1.5 gives an epimorphism θ' making the following diagram commutative:

$$\begin{array}{c} F' \\ {\theta'}\swarrow \quad \downarrow \\ \quad\quad F'/N'_\gamma \\ \quad\quad \downarrow {\theta_\gamma^{-1}} \\ 1 \longrightarrow N_\gamma/M_\delta \longrightarrow F/M_\delta \longrightarrow F/N_\gamma \longrightarrow 1. \end{array}$$

For $M'_\delta = \mathrm{Ker}(\theta')$, this induces a commutative diagram

$$\begin{array}{ccc} F/M_\delta & \xleftarrow{\bar\theta'} & F'/M'_\delta \\ \downarrow & & \downarrow \\ F/N_\gamma & \xrightarrow[\theta_\gamma]{} & F'/N'_\gamma \end{array}$$

in which $\bar\theta'$ is an isomorphism. Define N'_δ to be $L'_\gamma \cap M'_\delta$. Now apply Lemma 25.1.5 to F to obtain a closed normal subgroup N_δ of F which is contained in M_δ and for which

$$\begin{array}{ccc} F/N_\delta & \xrightarrow{\theta_\delta} & F'/N'_\delta \\ \downarrow & & \downarrow \\ F/M_\delta & \xrightarrow[(\bar\theta')^{-1}]{} & F'/M'_\delta \end{array}$$

is commutative with θ_δ an isomorphism. Then $\operatorname{rank}(F/N_\delta) = \operatorname{rank}(F'/N'_\delta) < m$.

CASE B: δ is a limit ordinal. In this case let $N_\delta = \bigcap_{\gamma<\delta} N_\gamma$ and $N'_\delta = \bigcap_{\gamma<\delta} N'_\gamma$. The collection of isomorphisms θ_γ defines an isomorphism θ_δ which satisfies (5).

Finally, note by (5c) that $\bigcap_{\beta<m} N_\beta = 1$ and $\bigcap_{\beta<m} N'_\beta = 1$. Hence, the isomorphisms θ_β define the desired isomorphism $\theta\colon F \to F'$. □

THEOREM 25.1.7: Let $\mathcal{C}$ be a formation of finite groups, m an infinite cardinal, and F a pro-$\mathcal{C}$ group. Suppose $\operatorname{rank}(F) \leq m$ and each finite pro-$\mathcal{C}$ embedding problem

(6) $\qquad (\varphi\colon F \to A,\ \alpha\colon B \to A)$

where $B \neq 1$ and $\operatorname{Ker}(\alpha)$ is a minimal normal subgroup of B has m solutions. Then $F \cong \hat{F}_m(\mathcal{C})$.

Proof: Choose a simple group S in $\mathcal{C}$. Then $(F \to 1, S \to 1)$ is a $\mathcal{C}$-embedding problem with a minimal normal kernel. By assumption, it has m solutions. Only finitely many solutions have the same kernel. Hence, F has at least m open normal subgroups N with $F/N \cong S$, so $\operatorname{rank}(F) \geq m$. Consequently, $\operatorname{rank}(F) = m$.

By Lemma 25.1.4, each $\mathcal{C}$-embedding problem (6) for F with $\operatorname{Ker}(\alpha) \neq 1$ has m solutions. By Lemma 25.1.2, each $\mathcal{C}$-embedding problem for $\hat{F}_m(\mathcal{C})$ with a nontrivial kernel has m solutions. Therefore, by Proposition 25.1.6, $F \cong \hat{F}_m(\mathcal{C})$. □

Here is a criterion for freeness that applies to projective groups:

LEMMA 25.1.8: Let $\mathcal{C}$ be a formation of finite groups, m an infinite cardinal, and M a $\mathcal{C}$-projective group of rank at most m (equivalently, M is a closed subgroup of $\hat{F}_m(\mathcal{C})$). Suppose every split $\mathcal{C}$-embedding problem for M with a nontrivial kernel has m solutions. Then $M \cong \hat{F}_m(\mathcal{C})$.

Proof: Consider a $\mathcal{C}$-embedding problem

(7) $\qquad (\varphi\colon M \to A,\ \alpha\colon B \to A)$

with $\operatorname{Ker}(\alpha) \neq 1$. By Theorem 25.1.7, it suffices to construct m solutions to (7). The construction follows that of Proposition 22.5.8.

Since M is $\mathcal{C}$-projective, there is a homomorphism $\gamma\colon M \to B$ with $\alpha \circ \gamma = \varphi$. Put $\bar{M} = M/\operatorname{Ker}(\gamma)$. Let $\pi\colon M \to \bar{M}$ be the quotient map and $\bar{\gamma}\colon \bar{M} \to B$ and $\bar{\varphi}\colon \bar{M} \to A$ the homomorphisms induced by γ and φ, respectively. Construct the fiber product $B \times_A \bar{M}$ with the projections $\psi\colon B \times_A \bar{M} \to \bar{M}$ and $\beta\colon B \times_A \bar{M} \to B$. Lemma 22.2.9 gives a section to ψ. Thus,

(8) $\qquad (\pi\colon M \to \bar{M},\ \psi\colon B \times_A \bar{M} \to \bar{M})$

is a finite split $\mathcal{C}$-embedding problem for M with $\operatorname{Ker}(\psi) \cong \operatorname{Ker}(\alpha) \neq 1$. By assumption, (8) has distinct solutions γ_i with i ranging on a set I of cardinality m.

For each $i \in I$, $\beta \circ \gamma_i$ is a solution of (7). If $i \neq j$, there is an $x \in M$ with $\gamma_i(x) \neq \gamma_j(x)$. Also, $\psi(\gamma_i(x)) = \pi(x) = \psi(\gamma_j(x))$. Hence, $\beta(\gamma_i(x)) \neq \beta(\gamma_j(x))$. Consequently, $\beta \circ \gamma_i \neq \beta \circ \gamma_j$. □

25.2 Applications of Theorem 25.1.7

The characterization of $\hat{F}_m$ with $m \geq \aleph_0$ by the m-fold solvability of each finite embedding problem with a nontrivial kernel has many applications. For example, the absolute Galois group of an ultraproduct of fields with free absolute Galois groups is free (Theorem 25.2.3). Secondly, the completion of the free abstract group of rank m with respect to the family of all normal subgroups of finite index is $\hat{F}_{2^m}$ (Theorem 25.3.4). But most important is the generalization of Proposition 17.6.2 from open subgroups of $\hat{F}_m$ to closed subgroups which do not lie too deep (Proposition 25.2.2).

To be more precise, recall that the rank of a profinite group G is the smallest cardinality of a set of generators of G which converges to 1. When $\text{rank}(G)$ is infinite, it is also the cardinality of the set of all open normal subgroups of G (Proposition 17.1.2). In particular, if N is a closed normal subgroup of G and G/N is not finitely generated, then $\text{rank}(G/N)$ is the cardinality of the set of all open normal subgroups of G which contain N. We generalize the latter observation to closed subgroups of G which are not necessarily normal.

Let H be a closed subgroup of G. Denote the set of all open subgroups of G which contains H by $\text{OpenSubgr}(G/H)$. Define the **weight** of G/H to be 1 if H is open and $|\text{OpenSubgr}(G/H)|$ if $(G : H) = \infty$. Here are some properties of the weight function which we use in the sequel:

LEMMA 25.2.1: *Let G be a profinite group, H a closed subgroup, and m an infinite cardinal number.*
(a) *Suppose $H \triangleleft G$ and $\text{rank}(G/H) = \infty$. Then $\text{weight}(G/H) = \text{rank}(G/H)$.*
(b) *Suppose $H = \bigcap_{i \in I} H_i$, $|I| < m$, and H_i is an open subgroup of G for each $i \in I$. Then $\text{weight}(G/H) < m$.*
(c) *Let N be the largest closed normal subgroup of G which is contained in H. Suppose $\text{weight}(G/H) < m$. Then $\text{rank}(G/N) < m$.*
(d) *Suppose $H_2 \leq H_1 \leq G$ are closed subgroups with $\text{weight}(G/H_1) < m$ and $\text{weight}(H_1/H_2) < m$. Then $\text{weight}(G/H_2) < m$.*

Proof of (a): The intersection of all conjugates of an open subgroup of G which contains H is open, normal, and contains H. Conversely, every open normal subgroup of G is contained in only finitely many open subgroups of G. Hence, $\text{weight}(G/H) = \text{rank}(G/H)$.

Proof of (b): If I is finite, then H is open in G and $\text{weight}(G/H) = 1 < m$. Suppose I is infinite. Then $\aleph_0 \leq |I| < m$. Since the cardinality of the set of all finite subsets of I is $|I|$, we may replace $\{H_i \mid i \in I\}$ by the collection $\{\bigcap_{i \in I_0} H_i \mid I_0 \text{ is a finite subset of } I\}$, if necessary, to assume that the set $\{H_i \mid i \in I\}$ is closed under finite intersections. By

Lemma 1.2.2(a), $\mathrm{OpenSubgr}(G/H) = \bigcup_{i \in I} \mathrm{OpenSubgr}(G/H_i)$. It follows that, $\mathrm{weight}(G/H) = |\bigcup_{i \in I} \mathrm{OpenSubgr}(G/H_i)| \leq \aleph_0 \cdot |I| < m$.

Proof of (c): Let H_i, $i \in I$, be the open subgroups of G which contain H. For each $i \in I$ let N_i be the largest open normal subgroup of G which is contained in H_i. Then, $N = \bigcap_{g \in G} H^g = \bigcap_{g \in G} \bigcap_{i \in I} H_i^g = \bigcap_{i \in I} \bigcap_{g \in G} H_i^g = \bigcap_{i \in I} N_i$. Therefore, by (a) and (b), $\mathrm{rank}(G/N) < m$.

Proof of (d): Denote the set of open subgroups of G which contain H_1 by $\mathrm{OpenSubgr}(G/H_1)$. By assumption, $|\mathrm{OpenSubgr}(G/H_1)| < m$, so each $E \in \mathrm{OpenSubgr}(H_1/H_2)$ is the intersection of less than m open subgroups of G. By (b), $\mathrm{weight}(G/E) < m$. The intersection of all $E \in \mathrm{OpenSubgr}(H_1/H_2)$ is H_2. There are less than m such E's. Therefore, by (b), $\mathrm{weight}(G/H_2) < m$. □

PROPOSITION 25.2.2: *Let $\mathcal{C}$ be a Melnikov formation of finite groups, F a free pro-$\mathcal{C}$ group of infinite rank m, and N a closed pro-$\mathcal{C}$ subgroup of F with $\mathrm{weight}(F/N) < m$. Then $N \cong \hat{F}_m(\mathcal{C})$.*

Proof: Consider a $\mathcal{C}$-embedding problem for N

(1) $\quad\quad\quad\quad\quad\quad (\varphi \colon N \to A,\ \pi \colon B \to A)$

where $B \neq 1$ and $C = \mathrm{Ker}(\pi)$ is a minimal normal subgroup of B. By Theorem 25.1.7, it suffices to construct m solutions to (1).

Indeed, $N' = \mathrm{Ker}(\varphi)$ is an open normal subgroup of N. Choose an open normal subgroup E_0 of F with $N_0 = N \cap E_0 \leq N'$. Put $E' = N'E_0$ and $E = NE_0$. Then E is an open subgroup of F and $E/E_0 \cong N/N_0$ is in $\mathcal{C}$. Since E_0 is normal in F, it is a pro-$\mathcal{C}$ group. Hence, E is a pro-$\mathcal{C}$ group (because $\mathcal{C}$ is Melnikov). By Proposition 17.6.2, $E \cong \hat{F}_m(\mathcal{C})$. Also, $N \cap E' = N'$ and $NE' = E$. Extend φ to an epimorphism $\tilde{\varphi} \colon E \to A$ by $\tilde{\varphi}(ne') = \varphi(n)$ for $n \in N$ and $e' \in E'$. Then $\mathrm{Ker}(\tilde{\varphi}) = E'$.

Now let β be an ordinal number less than m. Inductively suppose (1) has distinct solutions ψ_α, $\alpha < \beta$. Then $\mathrm{Ker}(\psi_\alpha)$ is an open normal subgroup of N and $M = \bigcap_{\alpha < \beta} \mathrm{Ker}(\psi_\alpha)$ is a closed normal subgroup of N with $M \leq N'$ and $\mathrm{rank}(N/M) < m$ (Lemma 25.2.1(b)). Combined with $\mathrm{weight}(E/N) < m$ and Lemma 25.2.1(d), this gives $\mathrm{weight}(E/M) < m$. Let M_0 be the maximal closed normal subgroup of E which is contained in M. By Lemma 25.2.1(c), $\mathrm{rank}(E/M_0) < m$.

Lemma 25.1.2 supplies m solutions to the embedding problem ($\tilde{\varphi} \colon E \to A$, $\pi \colon B \to A$). One of them is an epimorphism $\tilde{\psi} \colon E \to B$ such that $\pi \circ \tilde{\psi} = \tilde{\varphi}$ and $E_\beta = \mathrm{Ker}(\tilde{\psi})$ does not contain M_0. Thus, $\tilde{\psi}(E') = C$. It follows that $\tilde{\psi}(M_0)$ is a nontrivial normal subgroup of B which is contained in C. Since C is minimal, $\tilde{\psi}(M_0) = C$, so $C \leq \tilde{\psi}(N)$. Combined with $\pi(\tilde{\psi}(N)) = \varphi(N) = A$, this gives $\tilde{\psi}(N) = B$. Put $\psi_\beta = \tilde{\psi}|_N$. Then ψ_β solves (1). In addition, $\mathrm{Ker}(\psi_\beta) = N \cap E_\beta$ does not contain M. Therefore, $\mathrm{Ker}(\psi_\beta) \neq \mathrm{Ker}(\psi_\alpha)$, hence $\psi_\beta \neq \psi_\alpha$ for all $\alpha < \beta$. This concludes the induction step. □

25.2 Applications of Theorem 25.1.7

THEOREM 25.2.3 ([Chatzidakis4, Thm. 3.4]): *Let $\{K_i \mid i \in I\}$ be a set of fields with free absolute Galois groups and $\mathcal{D}$ an ultrafilter on I. Put $K = \prod K_i/\mathcal{D}$ and $e_i = \text{rank}(\text{Gal}(K_i))$. Then:*
(a) $\text{Gal}(K)$ *is a free profinite group.*
(b) *If for each $e \in \mathbb{N}$ the set $\{i \in I \mid e < e_i < \infty\}$ belongs to $\mathcal{D}$, then* $\text{rank}(\text{Gal}(K)) = |\prod 2^{e_i}/\mathcal{D}|$;
(c) *otherwise* $\text{rank}(\text{Gal}(K)) = |\prod e_i/\mathcal{D}|$.

Proof: First suppose there is a positive integer e with $\{i \in I \mid e_i = e\} \in \mathcal{D}$ for a set of i's in $\mathcal{D}$. Then $\text{Gal}(K) \cong \hat{F}_e$ (Lemma 17.7.1 and Remark 20.4.5(d)) and $\text{rank}(\text{Gal}(K)) = e = |\prod e_i/\mathcal{D}|$. This concludes the case that $\text{Gal}(K)$ has finite rank.

Now suppose there does not exist e as above. Then $\{i \in I \mid e < e_i\} \in \mathcal{D}$ for each e, so $\text{rank}(\text{Gal}(K)) \geq \aleph_0$. Let U_n (resp. $U_{i,n}$) be the set of all Galois extensions of K (resp. K_i) of degree at most n. Then $U_n = \prod U_{i,n}/\mathcal{D}$.

Consider $i \in I$ with $e_i < \infty$ and a finite group A. The number of finite Galois extensions of K_i with Galois group isomorphic to A is at most $|A|^{e_i}$. Denote the number of finite groups of order at most n by $g(n)$. Then $|U_{i,n}| \leq g(n) n^{e_i} \leq 2^{e_i k(n)}$, where $k(n)$ is the smallest integer greater than $\frac{\log n + \log g(n)}{\log 2}$. On the other hand, if $e_i \geq 2$, then K_i has exactly 2^{e_i} quadratic extensions.

Proof of (b): Suppose for each $e \in \mathbb{N}$ the set $I_e = \{i \in I \mid e < e_i < \infty\}$ belongs to $\mathcal{D}$. Then $\prod 2^{e_i}/\mathcal{D}$ is infinite, so for $n \geq 2$,

$$|\prod 2^{e_i}/\mathcal{D}| \leq |\prod U_{i,n}/\mathcal{D}| \leq |\prod 2^{e_i}/\mathcal{D}|^{k(n)} = |\prod 2^{e_i}/\mathcal{D}|.$$

Hence, $|U_n| = |\prod 2^{e_i}/\mathcal{D}|$. By Proposition 17.1.2, $\text{rank}(\text{Gal}(K)) = |\prod 2^{e_i}/\mathcal{D}|$.

Proof of (c): Suppose $I_\infty = \{i \in I \mid e_i \text{ is infinite}\}$ belongs to $\mathcal{D}$. For each $i \in I_\infty$ let X_i be a basis of $\text{Gal}(K_i)$. Then, for each $n \geq 2$, $|U_{i,n}|$ is equal to the number of finite subsets of X_i of cardinality at most n. That is, $|U_{i,n}| = e_i$. Thus, $|U_n| = |\prod e_i/\mathcal{D}|$, so $\text{rank}(\text{Gal}(K)) = |\prod e_i/\mathcal{D}|$.

Proof of (a): Consider a finite Galois extension L/K, a finite group B, and an epimorphism $\alpha \colon B \to \text{Gal}(L/K)$ with a nontrivial kernel. Let $\mathcal{M}$ be the set of all finite Galois extensions M of K for which M contains L and there exists a commutative diagram

(2)
$$\begin{array}{ccc} & & \text{Gal}(M/K) \\ & \beta \swarrow & \downarrow \text{res} \\ B & \xrightarrow{\alpha} & \text{Gal}(L/K) \end{array}$$

in which β is an isomorphism.

Write L in the form $L = \prod L_i/\mathcal{D}$ where L_i/K_i is a Galois extension and $D = \{i \in I \mid \text{Gal}(L_i/K_i) \cong \text{Gal}(L/K)\}$ is in $\mathcal{D}$. For each $i \in D$ define a set $\mathcal{M}_i$ with respect to the extension L_i/K_i as $\mathcal{M}$ is defined with respect to L/K. Since the existence of the commutative diagram (2) is an elementary statement on fields (Section 23.4), $\mathcal{M} = \prod \mathcal{M}_i/\mathcal{D}$.

As mentioned in the proof of Lemma 25.1.3, $|\text{Aut}(B)| \cdot |\mathcal{M}|$ (resp. $|\text{Aut}(B)| \cdot |\mathcal{M}_i|$) is the number of solutions of the embedding problem (res: $\text{Gal}(K) \to \text{Gal}(L/K)$, $\alpha\colon B \to \text{Gal}(L/K)$) (resp. (res: $\text{Gal}(K_i) \to \text{Gal}(L_i/K_i)$, $\alpha_i\colon B \to \text{Gal}(L_i/K_i)$), where α_i are epimorphisms such that $\prod \alpha_i/\mathcal{D} = \alpha$). By Theorem 25.1.7, it suffices to prove that $|\text{Aut}(B)| \cdot |\mathcal{M}| = \text{rank}(\text{Gal}(K))$.

Distinguish between two cases. First suppose $I_e \in \mathcal{D}$ for each e. Let $i \in I_e$. Then $e_i < \infty$. By Lemma 25.1.3, $2^{e_i - |B|} \leq |\text{Aut}(B)| \cdot |\mathcal{M}_i| \leq |B|^{e_i}$. Hence, $|B|^{e_i} \leq 2^{(e_i - |B|)l}$ for all large e_i, where $l = \left\lceil \frac{\log |B|}{\log 2} \right\rceil + 1$. In addition, $\prod 2^{e_i - |B|}/\mathcal{D}$ is an infinite cardinal number, so

$$\left|\prod 2^{e_i - |B|}/\mathcal{D}\right| \leq \left|\prod |\text{Aut}(B)||\mathcal{M}_i|/\mathcal{D}\right| \leq \left|\prod 2^{e_i - |B|}/\mathcal{D}\right|^l = \left|\prod 2^{e_i - |B|}/\mathcal{D}\right|.$$

As $|\mathcal{M}| = |\prod \mathcal{M}_i/\mathcal{D}|$ and $|\prod 2^{e_i - |B|}/\mathcal{D}| = |2^{|B|} \prod 2^{e_i - |B|}/\mathcal{D}| = |\prod 2^{e_i}/\mathcal{D}|$, we deduce from (b) that $\text{rank}(\text{Gal}(K)) = |\text{Aut}(B)| \cdot |\mathcal{M}|$, as desired.

Now suppose $I_\infty \in \mathcal{D}$. Then, $\{i \in I \mid |\text{Aut}(B)| \cdot |\mathcal{M}_i| = e_i\} \in \mathcal{D}$ (Lemma 25.1.2). Hence, by (c), $|\text{Aut}(B)| \cdot |\mathcal{M}| = |\prod |\text{Aut}(B)| \cdot |\mathcal{M}_i|/\mathcal{D}| = |\prod e_i/\mathcal{D}| = \text{rank}(\text{Gal}(K))$. Our proof is complete in both cases. $\square$

25.3 The Pro-$\mathcal{C}$ Completion of a Free Discrete Group

Let $\mathcal{C}$ be a formation of finite groups and F a free abstract group with basis X'. Suppose there is a $C \in \mathcal{C}$ with $\text{rank}(C) \leq |X'|$. Consider the completion $\hat{F} = \hat{F}(\mathcal{C}) = \varprojlim F/N$ with N ranging over all normal subgroups of F satisfying $F/N \in \mathcal{C}$. Denote the canonical map from F into $\hat{F}(\mathcal{C})$ by θ. Put $X = \theta(X')$. Then X generates $\hat{F}(\mathcal{C})$.

LEMMA 25.3.1: *Let G be a pro-$\mathcal{C}$ group. Then:*
(a) *For each map $\varphi_0'\colon X' \to G$ with $\langle \varphi_0'(X') \rangle = G$ there is a unique epimorphism $\varphi\colon \hat{F} \to G$ with $\varphi \circ \theta|_{X'} = \varphi_0'$.*
(b) *θ maps X' bijectively onto X.*
(c) *Each map φ_0 of X into a pro-$\mathcal{C}$ group G with $\langle \varphi_0(X) \rangle = G$ uniquely extends to an epimorphism $\varphi\colon \hat{F}(\mathcal{C}) \to G$.*
(d) *If X is infinite, then $\text{rank}(\hat{F}(\mathcal{C})) = 2^{|X|}$.*

Proof: We repeat the arguments of the proof of Proposition 17.4.2.

Proof of (a): Consider an open normal subgroup M of G. Let $\pi\colon G \to G/M$ be the quotient map. Then $\pi \circ \varphi_0'\colon X' \to G/M$ extends to an epimorphism

25.3 The Pro-$\mathcal{C}$ Completion of a Free Discrete Group

$\varphi'_M \colon F \to G/M$. Put $M' = \operatorname{Ker}(\varphi'_M)$ and let $\bar{\varphi}_M \colon F/M' \to G/M$ be the induced isomorphism. Thus, $F/M' \cong G/M \in \mathcal{C}$. So, θ induces an isomorphism $\theta_M \colon F/M' \to \hat{F}/\hat{M}$, where $\hat{M} = \langle \theta(M') \rangle$. Let $\hat{\pi} \colon \hat{F} \to \hat{F}/\hat{M}$ be the quotient map and $\varphi_M = \bar{\varphi}_M \circ \theta_M^{-1} \circ \hat{\pi} \colon \hat{F} \to G/M$.

The collection of all φ_M defines an epimorphism $\varphi \colon \hat{F} \to G$ with $\varphi \circ \theta|_{X'} = \varphi'_0$.

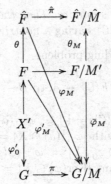

Since $\langle \theta(X') \rangle = F$, the map φ is unique.

Proof of (b): Consider distinct $x', y' \in X'$. Choose nontrivial $C \in \mathcal{C}$ with $r = \operatorname{rank}(C) \le |X'|$. Let $c_1, \dots, c_r$ be generators of C. Define a map $\varphi'_0 \colon X' \to C$ as follows. First suppose $r = 1$. Put $\varphi'_0(x') = c_1$ and $\varphi'_0(z') = 1$ for all $z' \in X' \smallsetminus \{x'\}$. Now suppose $r \ge 2$. Choose distinct elements $x'_1, \dots, x'_r \in X'$ with $x'_1 = x'$ and $x'_r = y'$. Then put $\varphi'_0(x'_i) = c_i$, $i = 1, \dots, r$. In both cases $\varphi'_0(x') \ne \varphi'_0(y')$.

Part (a) gives an epimorphism $\varphi \colon \hat{F} \to C$ with $\varphi \circ \theta|_{X'} = \varphi'_0$. In particular, $\varphi(\theta(x')) \ne \varphi(\theta(y'))$. Thus, $\theta(x') \ne \theta(y')$.

Proof of (c): Use (a).

Proof of (d): The rank of $\hat{F}$ is the number of all open normal subgroups of $\hat{F}$ (Proposition 17.1.2). The latter is the number of normal subgroups N of F with $F/N \in \mathcal{C}$. By (c), this number is $2^{|X|}$. □

Following Lemma 25.3.1, we identify X' with X via θ and consider X as a subset of both F and $\hat{F}(\mathcal{C})$. In its latter role we call X a **free set of generators** for $\hat{F}(\mathcal{C})$. Note, however, that unlike a basis of a free profinite group (Definition 17.4.1), X need not converge to 1. This is reflected by the exponentially large rank, $2^{|X|}$, of $\hat{F}(\mathcal{C})$.

LEMMA 25.3.2: *Let $x_0 \in X$. Put $Y = \{x_0\} \cup \{x_0^{-1} x \mid x \in X \text{ and } x \ne x_0\}$. Then Y is a basis of F and a free set of generators of $\hat{F}(\mathcal{C})$.*

Proof: The map $x \mapsto x_0^{-1} x$ for $x \ne x_0$ and $x_0 \mapsto x_0$ of X onto Y is bijective and Y generates both F and $\hat{F}(\mathcal{C})$. So, it extends to an epimorphism $\varphi \colon \hat{F}(\mathcal{C}) \to \hat{F}(\mathcal{C})$ (Lemma 25.3.1).

Note that $w(\varphi(x_1), \ldots, \varphi(x_n))$ is a nonempty reduced word in Y whenever $w(x_1, \ldots, x_n)$ is a nonempty reduced word in X, so Y is a basis of F. The map $x_0^{-1}x \mapsto x$ for $x \neq x_0$ and $x_0 \mapsto x_0$ therefore extends to an epimorphism $\varphi' \colon \hat{F}(\mathcal{C}) \to \hat{F}(\mathcal{C})$ which is the inverse of φ (use the uniqueness in Lemma 25.3.1). Consequently, φ is an isomorphism and Y is a free set of generators of $\hat{F}(\mathcal{C})$. □

LEMMA 25.3.3: *Suppose X is infinite. Then the number of solutions of each $\mathcal{C}$-embedding problem for $\hat{F}(\mathcal{C})$ having a nontrivial kernel is $2^{|X|}$.*

Proof: Consider a $\mathcal{C}$-embedding problem

(1) $$(\varphi \colon \hat{F}(\mathcal{C}) \to A, \ \alpha \colon B \to A)$$

with $\mathrm{Ker}(\alpha) \neq 1$. Then $\hat{F}(\mathcal{C}) = \bigcup_{a \in A} \varphi^{-1}(a)$. Since A is finite and X infinite, there is an $a \in A$ with $|\varphi^{-1}(a) \cap X| = |X|$. Choose $x_0 \in X$ with $\varphi(x_0) = a$. Then $|\mathrm{Ker}(\varphi) \cap x_0^{-1}X| = |\varphi^{-1}(\varphi(x_0)) \cap X| = |X|$. Applying Lemma 25.3.2, we may assume $|\mathrm{Ker}(\varphi) \cap X| = |X|$. Put $X_0 = \mathrm{Ker}(\varphi) \cap X$ and $X_1 = X \smallsetminus X_0$.

Now choose a set theoretic section $\alpha' \colon A \to B$ to α. Since $\mathrm{Ker}(\alpha) \neq 1$, the number of surjective maps $\gamma_0 \colon X_0 \to \mathrm{Ker}(\alpha)$ is $2^{|X|}$. Each γ_0 extends to a map $\gamma_1 \colon X \to B$ by setting $\gamma_1(x) = \alpha'(\varphi(x))$ for $x \in X_1$. Then γ_1 further extends to a solution $\gamma \colon \hat{F}(\mathcal{C}) \to B$ of (1). This gives at least $2^{|X|}$ distinct solutions.

On the other hand, Lemma 25.1.1 bounds the number of these solutions by the number of open normal subgroups of $\hat{F}(\mathcal{C})$, which is at most $2^{|X|}$ because $\mathrm{rank}(\hat{F}(\mathcal{C})) = 2^{|X|}$ (Lemma 25.3.1(d)). Thus, (1) has exactly $2^{|X|}$ solutions. □

THEOREM 25.3.4 ([Melnikov1]): *Let F be the free abstract group of an infinite rank m. Then $\hat{F}(\mathcal{C}) \cong \hat{F}_{2^m}(\mathcal{C})$.*

Proof: By Lemma 25.3.1, $\mathrm{rank}(\hat{F}(\mathcal{C})) = 2^m$. By Lemma 25.3.1(d), each finite $\mathcal{C}$-embedding problem has 2^m solutions. Therefore, by Theorem 25.1.7, $\hat{F}(\mathcal{C}) \cong \hat{F}_{2^m}(\mathcal{C})$. □

25.4 The Group Theoretic Diamond Theorem

The diamond theorem for Hilbertian fields (Theorem 13.8.3) has an analog for free profinite groups of infinite rank. We start with a technical result which gives sufficient conditions for a closed subgroup M of $\hat{F}_m$ to be isomorphic to $\hat{F}_m$ in terms of twisted wreath products:

PROPOSITION 25.4.1: *Let m be an infinite cardinal, $F = \hat{F}_m$, and M a closed subgroup of F. Suppose for each open subgroup E of F containing M and each open normal subgroup D of E there exist*
(1a) *an open subgroup F_0 of E with $M \leq F_0$,*

25.4 The Group Theoretic Diamond Theorem

(1b) an open normal subgroup L of F with $L \leq F_0 \cap D$, and
(1c) a closed normal subgroup N of F with $N \leq M \cap L$

such that no finite embedding problem

(2) $\qquad (\bar{\varphi}\colon F/N \to F/L,\; \bar{\alpha}\colon \bar{A}\,\mathrm{wr}_{F_0/L} F/L \to F/L)$

where $\bar{A} \neq 1$, and $\bar{\alpha}$, $\bar{\varphi}$ are the quotient maps is solvable.
Then $M \cong \hat{F}_m$.

Proof: The proof naturally breaks up into several parts:

PART A: *Preliminaries.* Consider a finite nontrivial group A and a finite group G_1 acting on A. Let $\alpha_1\colon G_1 \ltimes A \to G_1$ be the projection map and $\mu\colon M \to G_1$ an epimorphism. In order to prove that $M \cong \hat{F}_m$, it suffices to construct m distinct solutions to the embedding problem

(3) $\qquad (\mu\colon M \to G_1,\; \alpha_1\colon G_1 \ltimes A \to G_1)$

(Lemma 25.1.8).

To this end choose an open normal subgroup D of F with $M \cap D \leq \mathrm{Ker}(\mu)$ (Lemma 1.2.5). Let $E = MD$. Then let F_0, L, and N be as in (1). Then $M \cdot (F_0 \cap D) = F_0$. Also, let $\varphi\colon F \to F/L = G$ and $\varphi_0\colon F_0 \to F_0/L = G_0$ be the quotient maps:

$$
\begin{array}{ccccc}
M & \text{---} & F_0 & \text{---} & E \text{---} F \\
| & & | & & | \\
\mathrm{Ker}(\mu) & & & & \\
| & & | & & | \\
M \cap D & \text{---} & F_0 \cap D & \text{---} & D \\
| & & | & & \\
N \text{---} & M \cap L & \text{---} & L &
\end{array}
$$

Extend μ to an epimorphism $\varphi_1\colon F_0 \to G_1$ by setting $\varphi_1(ad) = \mu(a)$ for $a \in M$ and $d \in F_0 \cap D$. Since $L \leq F_0 \cap D \leq \mathrm{Ker}(\varphi_1)$, the epimorphism φ_1 decomposes as $\varphi_1 = \bar{\varphi}_1 \circ \varphi_0$ in the following commutative diagram:

(4)
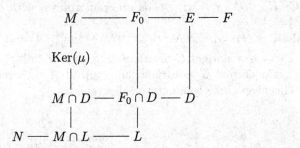

Here G_0 acts on A via $\bar{\varphi}_1$, α_0 is the quotient map, $\rho|_{G_0} = \bar{\varphi}_1$, and $\rho|_A = \mathrm{id}_A$.

PART B: *Epimorphisms onto a wreath product.* Let $\alpha\colon A\operatorname{wr}_{G_0} G \to G$ be the wreath product. Fix a set I of cardinality m. For each $i \in I$ we construct an epimorphism $\psi_i\colon F \to A\operatorname{wr}_{G_0} G$ with $\alpha \circ \psi_i = \varphi$.

To this end choose a basis X of F of cardinality m. Put $X_0 = X \smallsetminus L$ and $X_1 = X \cap L$. Then X_0 is finite and $|X_1| = m$. Since $|\operatorname{Ind}_{G_0}^G(A) \times I| = m$, there is a bijection $\operatorname{Ind}_{G_0}^G(A) \times I \to X_1$ which we write as $(f, i) \mapsto x_{f,i}$. Thus, $X = X_0 \cup \{x_{f,i} \mid f \in \operatorname{Ind}_{G_0}^G(A),\ i \in I\}$.

Define $\psi_i\colon X \to A\operatorname{wr}_{G_0} G$ by $\psi_i(x_{f,k}) = f$ if $k = i$, $\psi_i(x_{f,k}) = 1$ if $k \neq i$, and $\psi_i(x_0) = \varphi(x_0)$ for $x_0 \in X_0$ (here we identify G with a subgroup of $A\operatorname{wr}_{G_0} G$ via α). Then ψ_i converges to 1 and $\alpha \circ \psi_i(x) = \varphi(x)$ for each $x \in X$. In particular, $\langle \alpha(\psi_i(X)) \rangle = G$. In addition, $\psi_i(X_1) = \operatorname{Ind}_{G_0}^G(A) = \operatorname{Ker}(\alpha)$. Therefore, ψ_i extends to an epimorphism $\psi_i\colon F \to A\operatorname{wr}_{G_0} G$ with

(5) $$\alpha \circ \psi_i = \varphi.$$

Let $\pi\colon \operatorname{Ind}_{G_0}^G(A) \to A$ be the epimorphism given by $f \mapsto f(1)$. By definition, $\pi(f)^{\sigma_0} = f(1)^{\sigma_0} = f(\sigma_0) = f^{\sigma_0}(1)$ for $f \in \operatorname{Ind}_{G_0}^G(A)$ and $\sigma_0 \in G_0$. Hence, π extends to an epimorphism

$$\pi\colon G_0 \ltimes \operatorname{Ind}_{G_0}^G(A) \to G_0 \ltimes A$$

which is the identity map on G_0.

PART C: If $\psi\colon F \to A\operatorname{wr}_{G_0} G$ is an epimorphism with $\alpha \circ \psi = \varphi$, then $\pi(\psi(N)) = A$.

Indeed, $\varphi(N) = 1$ and $N \triangleleft F$, so $\psi(N) \leq \operatorname{Ind}_{G_0}^G(A)$ and $\psi(N) \triangleleft A\operatorname{wr}_{G_0} G$. Hence, $\psi(N)$ is a normal G-invariant subgroup of $\operatorname{Ind}_{G_0}^G(A)$. Thus, $A_1 = \pi(\psi(N))$ is a normal G_0-invariant subgroup of A. Therefore, G_0 acts on A/A_1. This gives three commutative diagrams:

(6)
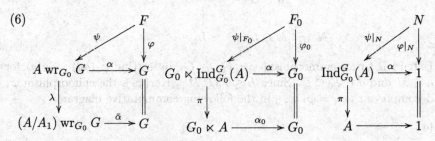

in which λ is the epimorphism induced by the quotient map $A \to A/A_1$.

Now, $\psi(N) \leq \pi^{-1}(A_1) = \{f \in \operatorname{Ind}_{G_0}^G(A) \mid f(1) \in A_1\}$ and $\psi(N)$ is a G-invariant subgroup of $\operatorname{Ind}_{G_0}^G(A)$. Hence,

$$\psi(N) \leq \bigcap_{\sigma \in G} \{f \in \operatorname{Ind}_{G_0}^G(A) \mid f(1) \in A_1\}^{\sigma}$$
$$= \bigcap_{\sigma \in G} \{f \in \operatorname{Ind}_{G_0}^G(A) \mid f(\sigma) \in A_1\} = \operatorname{Ker}(\lambda).$$

25.4 The Group Theoretic Diamond Theorem

It follows that $\lambda \circ \psi$ induces an epimorphism $\bar{\psi}\colon F/N \to (A/A_1)\,\mathrm{wr}_{G_0} G$ that solves (2) with $\bar{A} = A/A_1$. By assumption, this cannot happen, unless $A_1 = A$, as claimed.

PART D: *Solutions of (3).* Fix $i \in I$. By (5) and the middle diagram of (6) with ψ_i replacing ψ, $\rho \circ \pi \circ \psi_i$ maps F_0 into $G_1 \ltimes A$ and $\alpha_0 \circ \pi \circ \psi_i|_{F_0} = \varphi_0$. By (4), $\alpha_1 \circ \rho \circ \pi \circ \psi_i|_{F_0} = \bar{\varphi}_1 \circ \alpha_0 \circ \pi \circ \psi_i|_{F_0} = \bar{\varphi}_1 \circ \varphi_0 = \varphi_1$, so $\rho \circ \pi \circ \psi_i|_M$ is a homomorphism of M into $G_1 \ltimes A$ satisfying $\alpha_1 \circ (\rho \circ \pi \circ \psi_i|_M) = \varphi_1|_M$.

By Part C, $\pi \circ \psi_i(N) = A$. Therefore,

$$\rho \circ \pi \circ \psi_i(M) \geq \rho \circ \pi \circ \psi_i(N) = \rho(A) = A = \mathrm{Ker}(\alpha_1).$$

It follows that $\rho \circ \pi \circ \psi_i|_M\colon M \to G_1 \ltimes A$ is surjective. Hence, it solves (3).

PART E: *The solutions are distinct.* Consider distinct i,j in I. We have to prove $\rho \circ \pi \circ \psi_i|_M \neq \rho \circ \pi \circ \psi_j|_M$.

To this end put $\hat{A} = A \times A$. Let $\pi_1\colon \hat{A} \to A$ and $\pi_2\colon \hat{A} \to A$ be the projections on the coordinates. Then G_0 acts on $\hat{A}$ coordinatewise. This gives a wreath product $\hat{\alpha}\colon \hat{A}\,\mathrm{wr}_{G_0} G \to G$ and a commutative diagram

(7)
$$\begin{array}{ccc} \hat{A}\,\mathrm{wr}_{G_0} G & \xrightarrow{\pi_2^*} & A\,\mathrm{wr}_{G_0} G \\ {\scriptstyle \pi_1^*}\downarrow & {\scriptstyle \hat{\alpha}}\searrow & \downarrow{\scriptstyle \alpha} \\ A\,\mathrm{wr}_{G_0} G & \xrightarrow{\alpha} & G \end{array}$$

in which π_r^* is the identity on G and $\pi_r^*(\hat{f}) = \pi_r \circ \hat{f}$ for each $\hat{f} \in \mathrm{Ind}_{G_0}^G(\hat{A})$, $r = 1, 2$.

Use the basis X of F from Part B to define a map $\hat{\psi}\colon X \to \hat{A}\,\mathrm{wr}_{G_0} G$ by

$$\hat{\psi}(x_{f,k}) = \begin{cases} (f, 1) & k = i \\ (1, f) & k = j \\ 1 & k \neq i, j \end{cases}$$

and $\hat{\psi}(x_0) = \varphi(x_0)$ for each $x_0 \in X_0$ (again, we identify G with a subgroup of $\hat{A}\,\mathrm{wr}_{G_0} G$ via $\hat{\alpha}$). Then $\hat{\psi}$ extends to an epimorphism $\hat{\psi}\colon F \to \hat{A}\,\mathrm{wr}_{G_0} G$ with $\hat{\alpha} \circ \hat{\psi} = \varphi$. Furthermore, $\pi_1^* \circ \hat{\psi} = \psi_i$ and $\pi_2^* \circ \hat{\psi} = \psi_j$.

By Part C (with $\hat{A}$ replacing A), $\hat{\pi} \circ \hat{\psi}(N) = \hat{A}$, where $\hat{\pi}\colon \mathrm{Ind}_{G_0}^G(\hat{A}) \to \hat{A}$ is the map given by $\hat{f} \mapsto \hat{f}(1)$. Thus, there is an $x \in N$ with $\hat{\pi} \circ \hat{\psi}(x) = (a, 1)$, with $1 \neq a \in A$. Clearly, $\pi_1 \circ \hat{\pi} = \pi \circ \pi_1^*$ and $\pi_2 \circ \hat{\pi} = \pi \circ \pi_2^*$. Hence, $\pi \circ \psi_i(x) = \pi \circ \pi_1^* \circ \hat{\psi}(x) = \pi_1 \circ \hat{\pi} \circ \hat{\psi}(x) = \pi_1(a, 1) = a \neq 1$ while $\pi \circ \psi_j(x) = 1$. Since ρ is the identity on A, $\rho \circ \pi \circ \psi_i(x) \neq \rho \circ \pi \circ \psi_j(x)$. Consequently, $\rho \circ \pi \circ \psi_i \neq \rho \circ \pi \circ \psi_j$. □

Proposition 25.4.1 gives a new proof of Proposition 17.6.2 in the case where $\mathcal{C}$ is the formation of all finite groups and $\mathrm{rank}(\hat{F}) \geq \aleph_0$. This proof is devoid of combinatorial constructions for discrete free groups.

PROPOSITION 25.4.2: *Let m be an infinite cardinal and M an open subgroup of $\hat{F}_m$. Then $M \cong \hat{F}_m$.*

Proof: Put $F = \hat{F}_m$. Let E be an open subgroup of F containing M and D an open normal subgroup of E. Choose an open normal subgroup L of F in $M \cap D$. Put $F_0 = M$ and $N = L$. Then consider an embedding problem $(\bar{\varphi}\colon F/N \to F/L, \bar{\alpha}\colon \bar{A}\operatorname{wr}_{F_0/L} F/L \to F/L)$ with $\bar{A}$ a nontrivial finite group on which F_0/L acts and $\bar{\alpha}$ the projection onto F/L. Since $\bar{\varphi}$ is an isomorphism while $\bar{\alpha}$ is not, the embedding problem has no solution. We conclude from Proposition 25.4.1 that $M \cong \hat{F}_m$. □

THEOREM 25.4.3 (Group theoretic diamond dheorem [Haran5, Thm. 3.2]):
Let m be an infinite cardinal, M_1, M_2 closed normal subgroups of $\hat{F}_m$, and M a closed subgroup of $\hat{F}_m$. Suppose $M_1 \cap M_2 \leq M$, $M_1 \not\leq M$ and $M_2 \not\leq M$. Then $M \cong \hat{F}_m$.

Proof: We put $F = \hat{F}_m$ and use Proposition 25.4.2 to assume $(F : M) = \infty$. Then we first prove the theorem under an additional assumption:
(8) Either $M_1 M_2 = F$ or $(M_1 M : M) > 2$.

The proof of the theorem in this case utilizes Proposition 25.4.1. It has two parts.

PART A: *Construction of L, F_0, and N.* Let $D \triangleleft E$ be open subgroups of F with $M \leq E$. Choose an open normal subgroup L of F in D and put $F_0 = ML$. Let $G = F/L$ and $\varphi\colon F \to G$ the quotient map. Put $G_0 = \varphi(M) = F_0/L$, $G_1 = \varphi(M_1)$, and $G_2 = \varphi(M_2)$. Then
(9a) $G_1, G_2 \triangleleft G$.

Moreover, choosing L sufficiently small, the following holds:
(9b) $G_1, G_2 \not\leq G_0$ (use $M_1, M_2 \not\leq M$).
(9c) $(G : G_0) > 2$ (use $(F : M) = \infty$).
(9d) $G_1 G_2 = G$ or $(G_1 G_0 : G_0) > 2$ (use (8)).

This implies:
(10) $G_2 \not\leq G_1 G_0$ or $(G_1 G_0 : G_0) > 2$.

Indeed, suppose both $G_2 \leq G_1 G_0$ and $G_1 G_2 = G$. Then $G = G_1 G_0$, so by (9c), $(G_1 G_0 : G_0) > 2$.

Now let $N = L \cap M_1 \cap M_2$. Then $N \leq M$.

PART B: *An embedding problem.* Suppose G_0 acts on a nontrivial finite group $\bar{A}$. Put $H = \bar{A}\operatorname{wr}_{G_0} G$. Consider the embedding problem

(11) $\qquad\qquad (\varphi\colon F \to G, \pi\colon H \to G)$

where π is the quotient map. We have to prove that (11) has no solution which factors through F/N.

Assume $\psi\colon F \to H$ is an epimorphism with $\pi \circ \psi = \varphi$ and $\psi(N) = 1$. For $i = 1, 2$ put $H_i = \psi(M_i)$. Then $H_i \triangleleft H$ and $\pi(H_i) = \varphi(M_i) = G_i$.

We use (10) to find $h_1 \in H_1$ and $h_2 \in H_2$ with $\pi(h_1) = 1$ and $[h_1, h_2] \neq 1$. Suppose first $G_2 \not\leq G_1 G_0$. Then there is an $h_2 \in H_2$ with $\pi(h_2) \notin G_1 G_0$. By (9b), $G_1 \not\leq G_0$, so Lemma 13.7.4(b) provides the required $h_1 \in H_1$. Now suppose $(G_1 G_0 : G_0) > 2$. We use (9b) to find $h_2 \in H_2$ with $\pi(h_2) \notin G_0$. Lemma 13.7.4(a) gives the required $h_1 \in H_1$.

Having chosen h_i choose $\gamma_i \in M_i$ with $\psi(\gamma_i) = h_i$. Then $\varphi(\gamma_1) = \pi(h_1) = 1$. So, $\gamma_1 \in L$. Then $[\gamma_1, \gamma_2] \in [L, M_2] \cap [M_1, M_2] \leq L \cap (M_1 \cap M_2) = N$. So, $[h_1, h_2] = [\psi(\gamma_1), \psi(\gamma_2)] \in \psi(N) = 1$. This contradiction proves that ψ as above does not exist.

CONCLUSION OF THE PROOF: In the general case we use $M_1 \not\leq M$ to conclude that $(M_1 M : M) \geq 2$. The case $(M_1 M : M) > 2$ is covered by the special case proved above. Suppose $(M_1 M : M) = 2$. Choose an open subgroup K_2 of F containing M but not $M_1 M$. Thus, $K_2 \cap M_1 M = M$. Put $K = K_2 M_1 M$. Then $(K : K_2) = (M_1 M : M) = 2$, hence $K_2 \triangleleft K$. Observe: $M_1 K_2 = K$ and $K_2 \cap M_1 \leq K_2 \cap M_1 M = M \leq K$. Furthermore, $K_2 \not\leq M$, because $(K_2 : M) = \infty$.

By Proposition 25.4.2, $K \cong \hat{F}_m$, so the first alternative of (8) applies with K replacing F and K_2 replacing M_2. Consequently, $M \cong \hat{F}_m$. □

Theorem 25.4.3 gives a diamond theorem for the category of pro-$\mathcal{C}$ groups, when $\mathcal{C}$ is a Melnikov formation:

THEOREM 25.4.4: *Let $\mathcal{C}$ be a Melnikov formation of finite groups, m an infinite cardinal number, and M_1, M_2, M closed subgroups of $F = \hat{F}_m(\mathcal{C})$. Suppose M is pro-$\mathcal{C}$, $M_1, M_2 \triangleleft F$, $M_1 \cap M_2 \leq M$, but $M_1 \not\leq M$ and $M_2 \not\leq M$. Then $M \cong \hat{F}_m(\mathcal{C})$.*

Proof: Let $\hat{F} = \hat{F}_m$ and $\hat{N} = M_{\hat{F}}(\mathcal{C})$ the intersection of all open normal subgroups K of $\hat{F}$ with $\hat{F}/K \in \mathcal{C}$. Lemma 17.4.10 gives an epimorphism $\varphi \colon \hat{F} \to F$ with $\text{Ker}(\varphi) = \hat{N}$. Put $\hat{M}_1 = \varphi^{-1}(M_1)$, $\hat{M}_2 = \varphi^{-1}(M_2)$, and $\hat{M} = \varphi^{-1}(M)$. Then $\hat{M}_1, \hat{M}_2 \triangleleft \hat{F}$, $\hat{N} \leq \hat{M}_1 \cap \hat{M}_2 \leq \hat{M}$ but $\hat{M}_1 \not\leq \hat{M}$ and $\hat{M}_2 \not\leq \hat{M}$. By Theorem 25.4.3, $\hat{M} \cong \hat{F}_m$.

Suppose L is an open normal subgroup of $\hat{M}$ with $\hat{M}/L \in \mathcal{C}$. Then $\hat{N}/L \cap \hat{N} \cong L\hat{N}/L$ and $L\hat{N}/L \triangleleft \hat{M}/L$, so $\hat{N}/L \cap \hat{N} \in \mathcal{C}$. By Lemma 17.4.10, $L \cap \hat{N} = \hat{N}$, hence $\hat{N} \leq L$. Thus, $\hat{N} \leq M_{\hat{M}}(\mathcal{C})$. On the other hand, $\hat{M}/\hat{N} \cong M$ is pro-$\mathcal{C}$. Hence, $M_{\hat{M}}(\mathcal{C}) \leq \hat{N}$. It follows that $\hat{N} = M_{\hat{M}}(\mathcal{C})$. By the first paragraph, $\hat{M} \cong \hat{F}_m$. Therefore, by Lemma 17.4.10, $M \cong \hat{M}/M_{\hat{M}}(\mathcal{C}) \cong \hat{F}_m(\mathcal{C})$, as claimed. □

A special case of the diamond theorem generalizes Proposition 24.10.5:

COROLLARY 25.4.5 ([Jarden-Lubotzky1, Thm. 1.9]): *Let $\mathcal{C}$ be a Melnikov formation of finite groups, $m \geq 2$ a cardinal number, and N_1, N_2 closed normal subgroups of $\hat{F}_m(\mathcal{C})$ neither of which contains the other. Suppose $N = N_1 \cap N_2$ has infinite index in $\hat{F}_m(\mathcal{C})$. Then $N \cong \hat{F}_m(\mathcal{C})$.*

Proof: The case where $m \leq \aleph_0$ is covered by Proposition 24.10.5. The case where $m \geq \aleph_0$ is a special case of Theorem 25.4.4. □

COROLLARY 25.4.6 ([Jarden-Lubotzky1, Cor. 1.10]): *Let $\mathcal{C}$ be a Melnikov formation of finite groups, $m \geq 2$ a cardinal number, and N a closed normal subgroup of $\hat{F}_m(\mathcal{C})$. Suppose $\hat{F}_m(\mathcal{C})/N$ is pronilpotent of infinite index and two distinct prime numbers divide $(\hat{F}_m(\mathcal{C}) : N)$. Then $N \cong \hat{F}_m(\mathcal{C})$.*

Proof: By Proposition 22.9.3, $\hat{F}_m(\mathcal{C})/N$ is the direct product of its p-Sylow groups and each of them is normal. Hence, $\hat{F}_m(\mathcal{C})$ has closed normal subgroups N_1 and N_2 neither of which contains the other such that $N_1 \cap N_2 = N$. By Proposition 25.4.5, $N \cong \hat{F}_m(\mathcal{C})$. □

We use the diamond theorem to generalize Proposition 24.10.4 to closed subgroups of free pro-$\mathcal{C}$ groups of arbitrary infinite rank:

THEOREM 25.4.7: *Let $\mathcal{C}$ be a Melnikov formation of finite groups, m an infinite cardinal number, and N a closed normal nontrivial subgroup of $F = \hat{F}_m(\mathcal{C})$.*

(a) *Let M be a proper open subgroup of N. Suppose M is pro-$\mathcal{C}$. Then $M \cong \hat{F}_m(\mathcal{C})$.*
(b) *rank$(N) = m$.*
(c) *N is non-Abelian. In particular, the center of F is trivial.*
(d) *If $\mathcal{C}$ contains a nonsolvable group, then N is not prosolvable.*
(e) *If $\mathcal{C}$ does not consist of p-groups for some p, then N is not pronilpotent. In particular, the Frattini subgroup of F is trivial.*
(f) *Suppose $\mathcal{C}$ is the formation of all finite p-groups. Then $N \cong \hat{F}_m(p)$.*

Proof of (a): The case where $(F : M) < \infty$ is covered by Proposition 17.6.2. Suppose $(F : M) = \infty$. Then choose an open normal subgroup L of F with $N \cap L \leq M$. Our assumptions imply $N \not\leq M$ and $L \not\leq M$. By Theorem 25.4.4, $M \cong \hat{F}_m(\mathcal{C})$.

Proof of (b): Since N is nontrivial, it has a proper open normal subgroup M. By (a), $M \cong \hat{F}_m(\mathcal{C})$, so rank$(M) = m$. It follows that rank$(N) = m$.

Proof of (c), (d), (e), (f): Repeat the proof of Proposition 24.10.4(b),(c),(d),(e), respectively. □

Our first application of Theorem 25.4.7 generalizes Proposition 24.10.2 to free profinite groups of arbitrary rank:

COROLLARY 25.4.8: *Let $\mathcal{C}$ be a Melnikov formation of finite groups, $m \geq 2$ a cardinal number, and N a closed normal subgroup of $F = \hat{F}_m(\mathcal{C})$. Suppose F/N is infinite and Abelian. Then $N \cong \hat{F}_m(\mathcal{C})$.*

Proof: The case where m is finite is covered by Proposition 24.10.2. Assume $m \geq \aleph_0$.

Choose $x \in F \smallsetminus N$. Put $M = \langle N, x \rangle$. Then M is a normal subgroup of F and M/N is a nontrivial procyclic group. Choose a proper open subgroup M_0 of M which contains N. By Theorem 25.4.7(a), $M_0 \cong \hat{F}_m(\mathcal{C})$. Moreover, M_0/N is procyclic. By Proposition 25.2.2, $N \cong \hat{F}_m(\mathcal{C})$. □

We do not know if the diamond theorem holds for $\hat{F}_e$ with $2 \leq e < \infty$:

25.5 The Melnikov Group of a Profinite Group

PROBLEM 25.4.9: Let $\mathcal{C}$ be a Melnikov formation of finite groups, $e \geq 2$ an integer, and M, M_1, M_2 closed subgroups of $\hat{F}_e(\mathcal{C})$. Suppose M is pro-$\mathcal{C}$ of infinite index, $M_1, M_2 \triangleleft F$, $M_1 \cap M_2 \leq M$, but $M_1, M_2 \not\leq M$. Is $M \cong \hat{F}_\omega(\mathcal{C})$?

Remark 25.4.10: Lior Bary-Soroker adjusted the method of proof of Theorem 25.4.3 and gave an affirmative answer to Problem 25.4.9 [Bary-Soroker1]. □

25.5 The Melnikov Group of a Profinite Group

The Frattini group $\Phi(G)$ of a profinite group G has been defined in Section 22.1 as the intersection of all maximal open subgroups of G. It is a characteristic closed subgroup of G characterized by the following property: If H is a closed subgroup of G and $H \cdot \Phi(G) = G$, then $H = G$. Numerous applications of the Frattini group in this book prove this concept to be indispensable for the study of closed subgroups of profinite groups.

Likewise the **Melnikov group** of G is vital for the study of closed normal subgroups of G. Define $M(G)$ to be the intersection of all maximal open normal subgroups of G. Again, $M(G)$ is a closed characteristic subgroup of G.

If N is a maximal open normal subgroup of G, then G/N is a finite simple group. Thus, $M(G) = \bigcap_S M_G(S)$, where S ranges over all finite simple groups and where $M_G(S)$ is the intersection of all N with $G/N \cong S$. Therefore, $G/M(G) \cong \prod_S S^{r_G(S)}$ (Section 24.9). In particular, $M(G) = 1$ if and only if G is a direct product of finite simple groups.

Each maximal open normal subgroup N of G is contained in a maximal subgroup M of G. Indeed, N is the intersection of all conjugates of M, so $\Phi(G) \leq N$. Therefore, $\Phi(G) \leq M(G)$. If G is a pro-p group, then every maximal subgroup is normal of index p, so $\Phi(G) = M(G)$.

Here are some basic properties of $M(G)$:

LEMMA 25.5.1: Let G and H be profinite groups.
(a) Let $\varphi: G \to H$ be an epimorphism. Then $\varphi(M(G)) \leq M(H)$.
(b) If $H \triangleleft G$ and $H \cdot M(G) = G$, then $H = G$.
(c) If $H \triangleleft G$, then $M(H) \leq M(G)$.

Proof of (a): If H_1 is a maximal open normal subgroup of H, then $\varphi^{-1}(H_1)$ is a maximal open normal subgroup of G. Thus, $M(G) \leq \bigcap \varphi^{-1}(H_1) = \varphi^{-1}(\bigcap H_1) = \varphi^{-1}(M(H))$. Consequently, $\varphi(M(G)) \leq M(H)$.

Proof of (b): Assume $H \neq G$. Then H is contained in a maximal open normal subgroup N of G. Hence, $H \cdot M(G) \leq N < G$, a contradiction.

Proof of (c): Assume $M(H) \not\leq M(G)$. Then G has a maximal open normal subgroup N which does not contain $M(H)$, so $N \cdot M(H) = G$. Hence, $(N \cap H)M(H) = H$. By (b), $N \cap H = H$. Thus, $M(H) \leq H \leq N$, a contradiction. □

Example 25.5.2: The Frattini group can be a proper subgroup of the Melnikov group. For example, A_3 is the unique nontrivial proper normal subgroup of S_3. Hence, $M(S_3) = A_3$. On the other hand, both A_3 and $\{1, (1\,2)\}$ are maximal subgroups of S_3, so $\Phi(S_3) = 1$. □

LEMMA 25.5.3: *Let $G = \prod_{i \in I} S_i$ be a direct product of finite simple groups and N a closed normal subgroup. Then:*
(a) *Put $I_a = \{i \in I \mid S_i \text{ is Abelian}\}$, $I_n = \{i \in I \mid S_i \text{ is non-Abelian}\}$, $G_a = \prod_{i \in I_a} S_i$, and $G_n = \prod_{i \in I_n} S_i$. Then $G = G_a \times G_n$ and $N = (G_a \cap N) \times (G_n \cap N)$.*
(b) *N is a direct product of finite simple groups and*
(c) *$G = K \times N$ for some closed normal subgroup K.*
(d) *Let $\varphi \colon G \to H$ be an epimorphism. Then G has a normal subgroup K which φ maps isomorphically onto H. Thus, H is a direct product of finite simple groups.*

Proof of (a): The statement $G = G_a \times G_n$ is clear. Consider $x \in N$ and write it as $x = yz$ with $y \in G_a$ and $z \in G_n$. For each $i \in I$ let $\pi_i \colon G \to S_i$ be the ith projection. Let $j \in I_n$ with $S_j \not\leq N$. By Lemma 18.3.9(a), $\pi_j(N) = 1$, so $\pi_j(z) = \pi_j(x) = 1$. Therefore, $z \in N$, hence $y \in N$. Consequently, $N = (G_a \cap N) \times (G_n \cap N)$.

Proof of (b) and (c): Put $N_n = G_n \cap N$ and $N_a = G_a \cap N$. For each prime number p let G_p (resp. N_p) be the unique p-Sylow group of G (resp. N). Then $N_a = \prod_p N_p$, where p ranges over all prime numbers.

By (a), $G = G_a \times G_n$ and $N = N_a \times N_n$. By Lemma 18.3.9, N_n is a direct product of finite simple groups and G_n has a closed normal subgroup K_n with $G_n = K_n \times N_n$.

Lemma 22.7.2 gives for each p a closed subgroup K_p of G_p with $G_p = K_p \times N_p$. Put $K_a = \prod_p K_p$. By Lemma 22.7.3, N_p is a direct product of copies of C_p. Put $K = K_a \times K_n$. Then N is a direct product of finite simple groups, K is a closed normal subgroup of G, and $G = K \times N$.

Proof of (d): Let $N = \mathrm{Ker}(\varphi)$ and K the normal complement that (c) gives. Then φ maps K isomorphically onto H. By (b), K is a direct product of finite simple groups. Hence, so is H. □

As a corollary, we get a property of the Melnikov group which the Frattini group does not have (Remark 22.1.5):

LEMMA 25.5.4: *Let $\varphi \colon G \to H$ be an epimorphism of profinite groups. Then $\varphi(M(G)) = M(H)$.*

Proof: Since $H/\varphi(M(G))$ is a quotient of $G/M(G)$, it is a direct product of finite simple groups (Lemma 25.5.3(d)). Therefore, $\varphi(M(G))$ is the intersection of maximal open normal subgroups, so $M(H) \leq \varphi(M(G))$. On the other hand, $\varphi(M(G)) \leq M(H)$ (Lemma 25.5.1(a)). Consequently, $\varphi(M(G)) = M(H)$. □

25.6 Homogeneous Pro-$\mathcal{C}$ Groups

A Frattini cover is an epimorphism $\varphi\colon G \to H$ of profinite groups such that $\operatorname{Ker}(\varphi) \leq \Phi(G)$. Likewise, we call φ a **Melnikov cover** if $\operatorname{Ker}(\varphi) \leq M(G)$. Alternatively, φ has the following property: $G_1 \triangleleft G$ and $\varphi(G_1) = H$ implies $G_1 = G$. The latter definition immediately gives:

LEMMA 25.5.5: *Let $\alpha\colon B \to A$ and $\beta\colon C \to B$ be epimorphisms of profinite groups. Then $\alpha \circ \beta$ is a Melnikov cover if and only if both α and β are.*

LEMMA 25.5.6: *Let $\varphi\colon G \to H$ be a Melnikov cover of profinite groups. Then the quotient map $\bar{\varphi}\colon G/M(G) \to H/M(H)$ is an isomorphism.*

Proof: By Lemma 25.5.4, $\varphi(M(G)) = M(H)$, so $\bar{\varphi}$ is well defined. By definition, $\operatorname{Ker}(\varphi) \leq M(G)$. Therefore, $\bar{\varphi}$ is an isomorphism. □

25.6 Homogeneous Pro-$\mathcal{C}$ Groups

One of the central concepts of Chapter 24 is the embedding property. Among others each closed normal subgroup of a free profinite group F has the embedding property (Lemma 24.9.4). Closed normal subgroups of infinite index of F have an additional property called homogeneity. Its definition uses Melnikov covers:

Fix a Melnikov formation $\mathcal{C}$ of finite groups for the whole section. Let G be a pro-$\mathcal{C}$ group. Call an embedding problem

(1) $\qquad\qquad\qquad (\varphi\colon G \to A,\ \alpha\colon B \to A)$

for G **Melnikov** if α is a Melnikov cover. Now suppose $m = \operatorname{rank}(G) \geq \aleph_0$. Call G $\mathcal{C}$**-homogeneous** if each Melnikov pro-$\mathcal{C}$ embedding problem (1) which satisfies one of the following conditions is solvable:

(2a) $m = \aleph_0$ and B is finite.
(2b) $m > \aleph_0$, $\operatorname{rank}(A) < m$, and $\operatorname{rank}(B) \leq m$.

Lemma 25.6.4 replaces "$\operatorname{rank}(A) < m$" in (2b) by the weaker condition "$\operatorname{rank}(M(A)) < m$". This leads to the characterization of $\mathcal{C}$-homogeneous group G by its rank and the quotient $G/M(G)$ (Proposition 25.7.2). By lemma 25.6.3, each nontrivial closed normal subgroup N of infinite index of $\hat{F}_m(\mathcal{C})$ with $m \geq 2$ is $\mathcal{C}$-homogeneous of rank $\max(\aleph_0, m)$. Thus, N is uniquely determined among all closed normal subgroups of G, up to isomorphism, by the quotient $N/M(N)$.

LEMMA 25.6.1: *Let G be a profinite group of an uncountable rank m, H a closed subgroup, and N a closed normal subgroup of H. Suppose $\operatorname{rank}(H/N) < m$. Then G has a closed normal subgroup M such that $\operatorname{rank}(G/M) < m$ and $M \cap H \leq N$.*

Proof: Let N_i with i ranging on a set I be all open normal subgroups of H which contain N. By Lemma 17.1.2, $|I| \leq \max(\aleph_0, \operatorname{rank}(H/N)) < m$. For each $i \in I$ choose an open normal subgroup M_i of G with $M_i \cap H \leq N_i$ (Lemma 1.2.5(a)). By Lemma 25.2.1(b), $M = \bigcap_{i \in I} M_i$ satisfies the requirements. □

PROPOSITION 25.6.2: *Let F be a free pro-$\mathcal{C}$ group of infinite rank m and*

(3) $\qquad\qquad\qquad (\varphi\colon F \to A, \alpha\colon B \to A)$

a pro-$\mathcal{C}$ embedding problem with $\mathrm{rank}(A) < m$ and $\mathrm{rank}(B) \leq m$. Then (3) is solvable. In particular, F is $\mathcal{C}$-homogeneous.

Proof: The case where B is finite is covered by Lemma 25.1.2. Assume B is infinite. Let X be a basis of F. Then, $\varphi(X) \smallsetminus \{1\} = \bigcup_N (\varphi(X) \smallsetminus N)$ where N ranges over the set $\mathcal{N}$ of all open normal subgroups of A. For $N \in \mathcal{N}$, the group $\varphi^{-1}(N)$ is open and normal in N, hence $X \smallsetminus \varphi^{-1}(N)$ is finite, so $\varphi(X) \smallsetminus N \subseteq \varphi(X - \varphi^{-1}(N))$ is also finite. By Proposition 17.1.2, $|\mathcal{N}| = \mathrm{rank}(A) < m$. It follows that, $|\varphi(X)| < m$. Choose a subset X_0 of X with $\varphi(X_0) = \varphi(X)$ and $|X_0| < m$. Put $X_1 = X \smallsetminus X_0$. Then $|X_1| = m$.

Lemma 1.2.7 gives a continuous set theoretic section to α, that is, a continuous map $\alpha'\colon A \to B$ with $\alpha \circ \alpha' = \mathrm{id}_A$. Then, the set $Z_0 = \alpha'(\varphi(X_0))$ converges to 1. If $\mathrm{rank}(\mathrm{Ker}(\alpha)) < \infty$, choose a finite set of generators Z_1 for $\mathrm{Ker}(\alpha)$. If $\mathrm{rank}(\mathrm{Ker}(\alpha)) = \infty$, choose a set of generators Z_1 of $\mathrm{Ker}(\alpha)$ converging to 1. In the latter case, $|Z_1| = \mathrm{rank}(\mathrm{rank}(\alpha)) \leq \mathrm{rank}(B) \leq m$ (Lemma 17.1.5). Thus, in each case, $Z_0 \cup Z_1$ is a set of generators of B which converges to 1 of cardinality at most m.

Finally, choose a surjective map $\gamma_1\colon X_1 \to Z_1$. Extend γ_1 to a map $\gamma\colon X \to Z_0 \cup Z_1$ by setting $\gamma(x) = \alpha'(\varphi(x))$ for $x \in X_0$. By construction, γ converges to 1. Since X is a basis of F, γ extends to an epimorphism $\gamma\colon F \to B$. In addition, $\alpha(\gamma(x)) = \varphi(x)$ for each x in X_0 or in X_1. Hence, $\alpha \circ \gamma = \varphi$ and γ is a solution of (3). □

LEMMA 25.6.3: *Let F be a free pro-$\mathcal{C}$ group of rank $m \geq 2$, and N a nontrivial closed normal subgroup of infinite index. Then N is $\mathcal{C}$-homogeneous with $\mathrm{rank}(N) = \max(\aleph_0, m)$.*

Proof: If $m \leq \aleph_0$, then $\mathrm{rank}(N) = \aleph_0$ (Proposition 24.10.4(a)). If $m \geq \aleph_0$, then $\mathrm{rank}(N) = m$ (Theorem 25.4.7(b)).

Now consider a pro-$\mathcal{C}$ Melnikov embedding problem

(4) $\qquad\qquad\qquad (\varphi\colon N \to A,\ \alpha\colon B \to A)$

satisfying this:
(5a) either $\mathrm{rank}(N) = \aleph_0$ and B is finite,
(5b) or $\mathrm{rank}(N) > \aleph_0$, $\mathrm{rank}(A) < m$, and $\mathrm{rank}(B) \leq m$.

Put $N_1 = \mathrm{Ker}(\varphi)$ and distinguish between three cases. In each case construct a closed normal subgroup L_1 of F with $N \cap L_1 \leq N_1$ and put $L = NL_1$. Then $L \triangleleft F$, so L is a pro-$\mathcal{C}$ group. These subgroups should satisfy additional properties:

CASE A: $\mathrm{rank}(N) = \aleph_0$, $m < \aleph_0$ *and B is finite.* Then $(N : N_1) = |A| < \infty$. Use Proposition 17.6.2 to choose L_1 open with $\mathrm{rank}(L) \geq \mathrm{rank}(B)$. Then L is pro-$\mathcal{C}$ free of finite rank.

25.6 Homogeneous Pro-$\mathcal{C}$ Groups

CASE B: $\mathrm{rank}(N) = \aleph_0$, $m = \aleph_0$ and B is finite. Again, $(N : N_1) < \infty$. Choose L_1 open. Then $L \cong \hat{F}_\omega(\mathcal{C})$ (Proposition 17.6.2).

CASE C: $\mathrm{rank}(N) = m > \aleph_0$, $\mathrm{rank}(A) < m$, and $\mathrm{rank}(B) \leq m$. Then $\mathrm{rank}(N/N_1) = \mathrm{rank}(A) < m$. Use Lemma 25.6.1 to choose L_1 with $\mathrm{rank}(F/L_1) < m$. Then $\mathrm{rank}(F/L) < m$. By Proposition 25.2.2, $L \cong \hat{F}_m(\mathcal{C})$.

In each case extend φ to an epimorphism $\tilde{\varphi}\colon L \to A$ by $\tilde{\varphi}(nl_1) = \varphi(n)$ for $n \in N$ and $l_1 \in L_1$. In Case A, Proposition 17.7.3 gives an epimorphism $\gamma\colon L \to B$ with $\alpha \circ \gamma = \tilde{\varphi}$. In Cases B and C, Proposition 25.6.2 supplies γ. Thus, $\gamma(N)$ is a normal subgroup of B with $\alpha(\gamma(N)) = A$. Since $\alpha\colon B \to A$ is Melnikov, $\gamma(N) = B$. Consequently, $\gamma|_N$ is a solution of (4). □

The next lemma relaxes the assumption "A is finite" (resp. $\mathrm{rank}(A) < m$) in (5) to "$M(A)$ is finite" (resp. $\mathrm{rank}(M(A)) < m$):

LEMMA 25.6.4: *Let G be a $\mathcal{C}$-homogeneous pro-$\mathcal{C}$ group of infinite rank m and*

(6) $\qquad\qquad\qquad (\varphi\colon G \to A,\ \alpha\colon B \to A)$

a Melnikov pro-$\mathcal{C}$ embedding problem. Suppose
(7a) *either $m = \aleph_0$, $M(A)$ is finite, and $\mathrm{rank}(B) \leq \aleph_0$,*
(7b) *or $m > \aleph_0$, $\mathrm{rank}(M(A)) < m$, and $\mathrm{rank}(B) \leq m$.*
Then (6) is solvable.

Proof: Put $C = \mathrm{Ker}(\alpha)$ and break up the proof into two parts:

PART A: *C is finite.* By assumption, $C \leq M(B)$. Hence, $M(B)/C = M(A)$ (Lemma 25.5.4). Thus,
(8a) either $m = \aleph_0$, $M(B)$ is finite, and $\mathrm{rank}(B) \leq \aleph_0$,
(8b) or $m > \aleph_0$, $\mathrm{rank}(M(B)) < m$, and $\mathrm{rank}(B) \leq m$.

If (8a) holds, choose an open normal subgroup L of B with $M(B) \cap L = 1$. If (8b) holds, choose a closed normal subgroup L of B with $M(B) \cap L = 1$ and $\mathrm{rank}(B/L) < m$ (Lemma 25.6.1).

Put $K = \alpha(L)$. Let $\pi_A\colon A \to A/K$ and $\pi_B\colon B \to B/L$ be the quotient maps and let $\bar{\varphi} = \pi_A \circ \varphi$. Let $\bar{\alpha}\colon B/L \to A/K$ be the homomorphism induced by α. Then $\mathrm{Ker}(\bar{\alpha}) = CL/L \leq M(B)L/L = M(B/L)$ (Lemma 25.5.4), so $\bar{\alpha}$ is a Melnikov cover. If (8a) holds, B/L is finite. If (8b) holds, then $\mathrm{rank}(A/K) \leq \mathrm{rank}(B/L) < m$. The homogeneity of G gives an epimorphism ξ which makes the following diagram commutative:

(9)

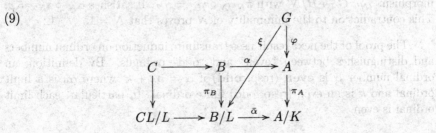

Since $\operatorname{Ker}(\alpha) \cap \operatorname{Ker}(\pi_B) = C \cap L = 1$, the right square in (9) is cartesian. Hence, there is a homomorphism $\gamma: G \to B$ with $\pi_B \circ \gamma = \xi$ and $\alpha \circ \gamma = \varphi$. We have to prove that γ is surjective.

Indeed, α induces an isomorphism $\alpha': B/M(B) \to A/M(A)$ (Lemma 25.5.6). Let $\mu_A: A \to A/M(A)$ be the quotient map. Put $\psi = (\alpha')^{-1} \circ \mu_A \circ \varphi$. Then consider the diagram

(10)
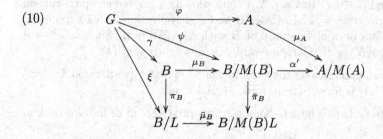

where μ_B, π_B, and $\bar{\mu}_B$ are the quotient maps and $\xi = \pi_B \circ \gamma$. It is commutative and all maps except possibly γ are surjective. Since $M(B) \cap L = 1$, the square in (10) is cartesian (Example 22.2.7(b)). Since ψ is surjective, $\psi(M(G)) = M(B/M(B)) = 1$ (Lemma 25.5.4), so $M(G) \le \operatorname{Ker}(\psi)$. Similarly, $\xi(M(G)) = M(B/L)$, so $\xi^{-1}(M(B/L)) = M(G)\operatorname{Ker}(\xi)$. Thus,

$$\operatorname{Ker}(\bar{\mu}_B \circ \xi) = \xi^{-1}(\operatorname{Ker}(\bar{\mu}_B)) = \xi^{-1}(M(B)L/L)$$
$$= \xi^{-1}(M(B/L)) = M(G)\operatorname{Ker}(\xi) \le \operatorname{Ker}(\psi)\operatorname{Ker}(\xi).$$

We conclude from Lemma 22.2.6(b) that γ is surjective.

PART B: *C is arbitrary.* Use Zorn's lemma to find a minimal closed normal subgroup N of B contained in C with an epimorphism $\gamma_N: G \to B/N$ such that $\alpha_N \circ \gamma_N = \varphi$. Here $\alpha_N: B/N \to A$ is the epimorphism induced by α. We have to prove that $N = 1$.

Assume N is nontrivial. Then N has an open subgroup N' which is normal in B. Let $\pi_{N'/N}: B/N' \to B/N$ be the quotient map. Then $\operatorname{Ker}(\pi_{N'/N}) = N/N'$ is finite. By Lemma 25.5.5, $\pi_{N',N}$ and α_N are Melnikov, so $M(B/N) \cong M(A)$ (Lemma 25.5.6). Thus, either $m = \aleph_0$ and $M(B/N)$ is finite or $m > \aleph_0$ and $\operatorname{rank}(M(B/N)) < m$. Part A gives an epimorphism $\gamma_{N'}: G \to B/N'$ with $\pi_{N'/N} \circ \gamma_{N'} = \gamma_N$. It satisfies $\alpha_{N'} \circ \gamma_{N'} = \varphi$. This contradiction to the minimality of N proves that $N = 1$. $\square$

The proof of the next result uses transfinite induction on ordinal numbers and distinguishes between "even" and "odd" ordinals. By definition, an ordinal number α is **even** (resp. **odd**) if $\alpha = \alpha_0 + k$, where α_0 is a limit ordinal and k is an even (resp. odd) finite ordinal. In particular, each limit ordinal is even.

25.6 Homogeneous Pro-$\mathcal{C}$ Groups

PROPOSITION 25.6.5: *Let $\pi_i\colon G_i \to A$ be a Melnikov cover of pro-$\mathcal{C}$ groups, $i = 1, 2$. Suppose G_1 and G_2 are $\mathcal{C}$-homogeneous of the same infinite rank m. In addition suppose either $m = \aleph_0$ and $M(A)$ is finite or $m > \aleph_0$ and $\mathrm{rank}(M(A)) < m$. Then there is an isomorphism $\varphi\colon G_1 \to G_2$ with $\pi_2 \circ \varphi = \pi_1$.*

Proof: Put $N_{i,0} = \mathrm{Ker}(\pi_i)$, $i = 1, 2$. Since π_i is Melnikov, $N_{i,0} \leq M(G_i)$. Also, $M(G_i)/N_{i,0} \cong M(A)$ (Lemma 25.5.4). Thus, either $m = \aleph_0$ and $M(G_i)/N_{i,0}$ is finite or $m > \aleph_0$ and $\mathrm{rank}(M(G_i)/N_{i,0}) < m$. Let

$$\varphi_0\colon G_1/N_{1,0} \to G_2/N_{2,0}$$

be the isomorphism induced by π_1 and π_2. We have to lift φ_0 to an isomorphism $\varphi\colon G_1 \to G_2$.

To do so list the open normal subgroups of G_1 as $K_{1,\alpha}$ with α ranging over all even ordinals at most m. List the open normal subgroups of G_2 as $K_{2,\alpha}$ with α ranging over all odd ordinal numbers less than m.

By transfinite induction construct for each $\alpha \leq m$ a closed normal subgroup $N_{i,\alpha}$ of G_i which is contained in $M(G_i)$. For each even ordinal number $\alpha \leq m$ construct an isomorphism $\varphi_\alpha\colon G_1/N_{1,\alpha} \to G_2/N_{2,\alpha}$. Finally, for each odd ordinal $\alpha < m$ construct an isomorphism $\psi_\alpha\colon G_2/N_{2,\alpha} \to G_1/N_{1,\alpha}$. This data should satisfy the following conditions:

(11a) $N_{i,\alpha} \geq N_{i,\beta}$ for $\alpha < \beta \leq m$ and $N_{i,0} \cap \bigcap_{\alpha < \beta} K_{i,\alpha} \leq N_{i,\beta}$ for $\beta \leq m$.
(11b) For each $\alpha < m$ either $m = \aleph_0$ and $M(G_i)/N_{i,\alpha}$ is finite or $m > \aleph_0$ and $\mathrm{rank}(M(G_i)/N_{i,\alpha}) < m$.
(11c) φ_β lifts φ_α for $\alpha < \beta \leq m$.
(11d) Let α be an even ordinal. Then $\psi_{\alpha+1}$ lifts φ_α^{-1} and $\varphi_{\alpha+2}$ lifts $\psi_{\alpha+1}^{-1}$.

To carry out the induction step consider $\beta < m$ and assume all objects have been defined for $\alpha < \beta$. Suppose first $\beta = \alpha + 1$ is a successor ordinal.

Assume α is even. Put $N_{1,\beta} = K_{1,\alpha} \cap N_{1,\alpha}$. Let $\pi_{2,\alpha}\colon G_2 \to G_2/N_{2,\alpha}$ and $\pi_{1,\beta,\alpha}\colon G_1/N_{1,\beta} \to G_1/N_{1,\alpha}$ be the quotient maps. Then

(12) $\qquad (\pi_{2,\alpha}\colon G_2 \to G_2/N_{2,\alpha},\ \varphi_\alpha \circ \pi_{1,\beta,\alpha}\colon G_1/N_{1,\beta} \to G_2/N_{2,\alpha})$

is a pro-$\mathcal{C}$ embedding problem for G_2. Its kernel is $N_{1,\alpha}/N_{1,\beta} \leq M(G_1/N_{1,\beta})$, so (12) is Melnikov. Taking (11b) into account, Lemma 25.6.4 gives an epimorphism $\tilde{\psi}_\beta\colon G_2 \to G_1/N_{1,\beta}$ which solves (12). Put $N_{2,\beta} = \mathrm{Ker}(\tilde{\psi}_\beta)$. Let $\psi_\beta\colon G_2/N_{2,\beta} \to G_1/N_{1,\beta}$ be the isomorphism induced by $\tilde{\psi}_\beta$. Then ψ_β lifts φ_α^{-1}. If $m = \aleph_0$, then $M(G_1)/N_{1,\beta}$ is finite whereas if $m > \aleph_0$, then $\mathrm{rank}(M(G_1)/N_{1,\beta}) < m$ (use (11b)). Thus, (11b) holds for $i = 2$ and β replacing α.

The case with α odd is treated in the same way as the case with α even. Consider the case where β is a limit ordinal. Let $\varphi_\beta\colon G_1/N_{1,\beta} \to G_2/N_{2,\beta}$ be the limit of the isomorphisms φ_α, with $\alpha < \beta$. Then φ_β is an isomorphism which lifts all φ_α and ψ_α^{-1} with $\alpha < \beta$. Put $N_{i,\beta} = \bigcap_{\alpha < \beta} N_{i,\alpha}$. Then

$N_{i,0} \cap \bigcap_{\alpha<\beta} K_{i,\alpha} \leq N_{i,\beta}$. If $\beta < m$, then $m > \aleph_0$ and the set of all ordinals $\alpha < \beta$ has cardinality less than m. Hence, by (11b) and Lemma 25.2.1(a,b), $\text{rank}(M(G_i)/N_{i,\beta}) < m$. If $\beta = m$, then $N_{i,m} \leq \bigcap_{\alpha<m} K_{i,\alpha} = 1$, so $N_{i,m} = 1$. The isomorphism $\varphi = \varphi_m$ is the desired lifting of φ_0. □

25.7 The S-rank of Closed Normal Subgroups

Let $\mathcal{C}$ be a Melnikov formation of finite groups and F a free pro-$\mathcal{C}$ group of infinite rank m. We classify the closed normal subgroups N of F up to isomorphism by their S-rank functions r_N (Theorem 25.7.3). Moreover, let f be a function from the set of all simple finite groups to the set of cardinal numbers at most m. Then there is a closed normal subgroup N of F with $r_N = f$ if and only if this holds: $f(S) = 0$ if $S \notin \mathcal{C}$, $0 \leq f(S) \leq m$ if S is non-Abelian, and $f(C_p)$ is either 0 or m (Propositions 25.7.4 and 25.7.7).

LEMMA 25.7.1: *Let F be a free pro-$\mathcal{C}$ group of infinite rank m and S a finite simple group. Then $r_F(S) = 0$ if $S \notin \mathcal{C}$ and $r_F(S) = m$ if $S \in \mathcal{C}$.*

Proof: Suppose $S \in \mathcal{C}$. An application of Lemma 25.1.2 to the embedding problem $(F \to 1, S \to 1)$ gives $r_S(F) \geq m$, so $r_S(F) = m$. □

PROPOSITION 25.7.2 ([Melnikov2, Thm. 3.2]): *Let G_1 and G_2 be $\mathcal{C}$-homogeneous pro-$\mathcal{C}$ groups of infinite rank m. Suppose $r_{G_1} = r_{G_2}$. Then:*
(a) *There is an isomorphism $\varphi_0: G_1/M(G_1) \to G_2/M(G_2)$.*
(b) *Every isomorphism as in (a) can be lifted to an isomorphism $\varphi: G_1 \to G_2$.*
(c) *Let G be a $\mathcal{C}$-homogeneous pro-$\mathcal{C}$ group of infinite rank m. Suppose the following holds for each finite simple group S: $r_G(S) = m$ if $S \in \mathcal{C}$ and $r_G(S) = 0$ if $S \notin \mathcal{C}$. Then $G \cong \hat{F}_m(\mathcal{C})$.*

Proof: Statement (a) follows from the definitions:

$$G_1/M(G_1) \cong \prod_S S^{r_{G_1}(S)} = \prod_S S^{r_{G_2}(S)} \cong G_2/M(G_2)$$

where S ranges over all finite simple groups (Section 25.5).

Now consider an isomorphism φ_0 as in (a). Put $A = G_2/M(G_2)$, $\pi_2: G_2 \to A$ the quotient map, and $\pi_1: G_1 \to A$ the compositum of the quotient map $G_1 \to G_1/M(G_1)$ with φ_0. Then both π_1 and π_2 are Melnikov covers and $M(A) = M(G_2/M(G_2)) = 1$. Proposition 25.6.5 gives φ as in (b).

Statement (c) follows from Lemma 25.7.1 by (b). □

THEOREM 25.7.3 ([Melnikov1, Thm. 3.1]): *Let $\mathcal{C}$ be a Melnikov formation of finite groups, $m \geq 2$ a cardinal number, and N, N' closed normal subgroups of $\hat{F}_m(\mathcal{C})$. Assume N, N' are of infinite index if $m < \aleph_0$.*
(a) *Suppose $r_N = r_{N'}$. Then $N \cong N'$.*
(b) *Suppose $r_N(S) = m$ for each simple group $S \in \mathcal{C}$. Then*

$$N \cong \hat{F}_{\max(m,\aleph_0)}(\mathcal{C}).$$

25.7 The S-rank of Closed Normal Subgroups

Proof of (a): If N is trivial, then $r_{N'} = r_N = 0$, so N' is also trivial. Assume therefore both N and N' are nontrivial.

Put $m^* = \max(m, \aleph_0)$. By Proposition 25.6.3, both N and N' are $\mathcal{C}$-homogeneous of rank m^*. Hence, by Proposition 25.7.2, $N \cong N'$.

Proof of (b): Apply Proposition 25.7.2(c) to N. □

PROPOSITION 25.7.4: *Let $\mathcal{C}$ be a Melnikov formation of finite groups and F a free pro-$\mathcal{C}$ group of rank $m \geq 2$. Let N be a closed normal subgroup of infinite index of F and S a finite simple group. Then,*
(a) $r_N(S) = 0$ if $S \notin \mathcal{C}$,
(b) $0 \leq r_N(S) \leq \max(m, \aleph_0)$ if $S \in \mathcal{C}$ is non-Abelian, and
(c) *either $r_N(S) = 0$ or $r_N(S) = \max(m, \aleph_0)$ if $S = C_p \in \mathcal{C}$.*

Proof: Statements (a) and (b) are clear. To prove (c), assume $S = C_p \in \mathcal{C}$ and N has an open normal subgroup M with $N/M \cong C_p$. We have to prove $r_N(S) = \max(m, \aleph_0)$.

Indeed, F has an open normal subgroup D with $N \cap D \leq M$. Put $E = ND$. Then E is open in F, so $E \cong \hat{F}_{m'}(\mathcal{C})$ with $m \leq m' < \infty$ if m is finite and $m' = m$ if m is infinite (Proposition 17.6.2). Also, $E/MD \cong N/M \cong C_p$, hence $M_E(C_p) \leq MD$ (Section 25).

Lemma 17.4.10 gives an epimorphism $\varphi \colon E \to \hat{F}_{m'}(p)$ with $\text{Ker}(\varphi) \leq MD$. Then $N \cap \text{Ker}(\varphi) \leq N \cap MD = M$, so $\varphi(N)/\varphi(M) \cong N/M \cong C_p$. In particular $\varphi(N)$ is a closed normal nontrivial subgroup of $\hat{F}_{m'}(p)$. By Corollary 22.7.7, $\varphi(N)$ is a free pro-p group.

Distinguish between two cases. First suppose m is finite. Fix a positive integer e. Since $(F : N) = \infty$, we may choose D with $(F : E)$ large enough such that $m' \geq e$ (Proposition 17.6.2). If $(\hat{F}_{m'}(p) : \varphi(N)) = \infty$, then, by Theorem 24.10.4, $\text{rank}(\varphi(N)) = \aleph_0$. If $(\hat{F}_{m'}(p) : \varphi(N)) < \infty$, then $\text{rank}(\varphi(N)) \geq m' \geq e$ (Proposition 17.6.2). It follows that in both cases, $\text{rank}(\varphi(N)) = \aleph_0$ and $r_N(C_p) \geq \text{rank}(\varphi(N)) = \aleph_0$. Therefore, $r_N(C_p) = \aleph_0$.

Now assume m is infinite. Then $\text{rank}(\varphi(N)) = m$ (Theorem 25.4.7(b)). Hence, $m \geq r_N(C_p) \geq r_{\varphi(N)}(C_p) = m$ (Lemma 25.7.1). Consequently, $r_N(C_p) = m$. □

THEOREM 25.7.5: *Let $\mathcal{C}$ be a Melnikov formation of finite groups, $m \geq 2$ a cardinal number, and N a closed normal subgroup of $\hat{F}_m(\mathcal{C})$ which is of infinite index if $m < \aleph_0$. Suppose:*
(1a) *C_p is a quotient of N for every prime number p.*
(1b) *$S^{\max(m,\aleph_0)}$ is a quotient of N for every non-Abelian simple group $S \in \mathcal{C}$.*

Then $N \cong \hat{F}_{\max(m,\aleph_0)}(\mathcal{C})$.

Proof: Let $F = \hat{F}_{\max(m,\aleph_0)}(\mathcal{C})$. By Lemma 25.7.1, by (1), and by Proposition 25.7.4, $r_N = r_F$. Hence, by Theorem 25.7.2(c) $N \cong F$.

COROLLARY 25.7.6: *Let C be a Melnikov formation of finite groups, $2 \leq m \leq \aleph_0$ a cardinal number, and N a closed normal subgroup of $\hat{F}_m(C)$ which is of infinite index if $m < \aleph_0$. Suppose:*

(2a) *C_p is a quotient of N for every prime number p.*

(2b) *S^q is a quotient of N for every non-Abelian simple group $S \in C$ and every positive integer q.*

Then $N \cong \hat{F}_\omega(C)$.

Proof: By (2b), $r_N(S) = \aleph_0$ for every non-Abelian simple group $S \in C$, so we may apply Theorem 25.7.5. □

PROPOSITION 25.7.7: *[Melnikov1, Thm. 3.3] Let C be a Melnikov formation of finite groups and F a free pro-C group of infinite rank m. Consider a function f from the set of all finite simple groups to the set of all cardinal numbers at most m. Suppose each finite simple group S satisfies the following condition:*

(3a) *$f(S) = 0$ if $S \notin C$, and*

(3b) *either $f(S) = 0$ or $f(S) = m$ if $S = C_p \in C$.*

Then F has a closed normal subgroup N with $r_N = f$.

Proof: Denote the set of all simple groups in C by $\mathcal{S}$. Consider $S \in \mathcal{S}$. If $S = C_p$ put $L(S) = F$. If S is non-Abelian, use $F/M_F(S) \cong S^m$ (Lemma 25.7.1) to choose a closed normal subgroup $L(S)$ of F containing $M_F(S)$ with $L(S)/M_F(S) \cong S^{f(S)}$. The rest of the proof naturally divides into three parts.

PART A: *Let $K = \bigcap_{S \in \mathcal{S}} L(S)$. Then $K \cdot M_F(S) = L(S)$ for all $S \in \mathcal{S}$.*

Consider $S \in \mathcal{S}$. List the elements of $\mathcal{S}$ in a sequence $S_0, S_1, S_2, \ldots$ with $S_0 = S$. Write $K_0 = L(S)$, $K_n = \bigcap_{i=0}^n L(S_i)$ for $n = 1, 2, 3, \ldots$, and $K = \bigcap_{i=0}^\infty K_i = \bigcap_{S' \in \mathcal{S}} L(S')$. For each $n \geq 1$, $K_{n-1}/K_n \cong K_{n-1} \cdot L(S_n)/L(S_n) \leq F/L(S_n)$, so the only composition factor of K_{n-1}/K_n is S_n. It follows that S is not a composition factor of $L(S)/K$, so S is not a composition factor of $L(S)/K \cdot M_F(S)$. On the other hand, the only possible composition factor of $L(S)/K \cdot M_F(S)$ is S. Therefore, $K \cdot M_F(S) = L(S)$.

PART B: *Construction of N.* Let $\mathcal{L}$ be the set of all closed normal subgroups L of F with $L \cdot M_F(S) = L(S)$ for each $S \in \mathcal{S}$. By Part A, $\mathcal{L}$ is nonempty. Suppose $\{L_i \mid i \in I\}$ is a descending chain in $\mathcal{L}$. Put $L = \bigcap_{i \in I} L_i$. By Lemma 1.2.2(b), $L \cdot M_F(S) = \bigcap_{i \in I} L_i \cdot M_F(S) = L(S)$, so $L \in \mathcal{L}$.

Zorn's Lemma gives a minimal element N in $\mathcal{L}$.

PART C: *$r_N = f$.* Let S be a finite simple group. First suppose, $S \notin \mathcal{S}$. Since N is a pro-C group, S is not a factor of N. Hence, $r_N(S) = 0$.

Next suppose $S = C_p \in \mathcal{S}$. Then $N \cdot M_F(S) = F$ and $F/M_F(S)$ is a quotient of N. Hence, $r_N(S) \geq m$ (Lemma 25.7.1). On the other hand $r_N(S) \leq \text{rank}(N) \leq m$. Therefore, $r_N(S) = m$.

Finally, suppose S is in $\mathcal{S}$ but non-Abelian. Then $N/N \cap M_F(S) \cong L(S)/M_F(S) \cong S^{f(S)}$, so $M_N(S) \leq N \cap M_F(S)$. Lemma 18.3.9 applied to the closed normal subgroup $N \cap M_F(S)/M_N(S)$ of $N/M_N(S)$ gives a closed normal subgroup N_0 of F with $N_0 \cap (N \cap M_F(S)) = M_N(S)$ and $N_0 \cdot (M \cap M_F(S)) = N$. Thus, $N_0 \cdot M_F(S) = L(S)$. The minimality of N implies $N_0 = N$, so $M_N(S) = N \cap M_F(S)$. Thus, $N/M_N(S) \cong L(S)/M_F(S)$. Consequently, $r_N(S) = f(S)$. □

As a corollary, we classify the closed normal subgroups of a free pro-$\mathcal{C}$ groups as $\mathcal{C}$-homogeneous groups with special S-rank functions:

THEOREM 25.7.8: *Let $\mathcal{C}$ be a Melnikov formation of finite groups and G a pro-$\mathcal{C}$ group of infinite rank m. Then G is isomorphic to a closed normal subgroup of $\hat{F}_m(\mathcal{C})$ if and only if G is $\mathcal{C}$-homogeneous and for each prime number p either $r_G(C_p) = 0$ or $r_G(C_p) = m$.*

Proof: First suppose G is isomorphic to a closed normal subgroup N of $\hat{F}_m(\mathcal{C})$. By Lemma 25.6.3, N is $\mathcal{C}$-homogeneous. Moreover, for each p, either $r_G(C_p) = 0$ or $r_G(C_p) = m$ (Proposition 25.7.4).

Conversely, suppose G is $\mathcal{C}$-homogeneous and for each p either $r_G(C_p) = 0$ or $r_G(C_p) = m$. In addition, for each finite simple group S, $r_G(S) \leq$ rank$(G) \leq m$ and $r_G(S) = 0$ if $S \notin \mathcal{C}$. By Proposition 25.7.7, $\hat{F}_m(\mathcal{C})$ has a closed normal subgroup N with $r_N = r_G$. It follows from Proposition 25.7.2 that $G \cong N$. □

25.8 Closed Normal Subgroups with a Basis Element

Let $\mathcal{C}$ be a Melnikov formation of finite groups and F_0 a free abstract group. Then the canonical map of F_0 into its pro-$\mathcal{C}$ completion F is injective (Proposition 17.5.11). It is rare that a closed normal subgroup N of F of infinite index intersects F_0 nontrivially. But, when it happens, N is a free pro-$\mathcal{C}$ group (Proposition 25.8.1). This implies that every closed normal subgroup N of $\hat{F}_e(\mathcal{C})$ of infinite index with $2 \leq e < \aleph_0$ is contained in a closed normal subgroup N_0 which is isomorphic to $\hat{F}_\omega(\mathcal{C})$ (Proposition 25.8.3). As a consequence we prove an analog of Theorem 25.4.7(a) for free profinite groups of finite rank.

Our first result is a substantial generalization of the fact $(\hat{F}_e)' \cong \hat{F}_\omega$ proved for $e \geq 2$ in Example 24.10.7:

PROPOSITION 25.8.1 ([Melnikov1, p. 10]): *Let $\mathcal{C}$ be a Melnikov formation of finite groups, F a free pro-$\mathcal{C}$-group of rank $m \geq 2$, X a basis of F, F_0 the abstract subgroup of F generated by X, and N a closed normal subgroup of F of infinite index. Suppose $F_0 \cap N \neq 1$. Then $N \cong \hat{F}_{\max(m,\aleph_0)}(\mathcal{C})$.*

Proof: By Remark 17.6.1, F_0 is free on X. Put $N_0 = N \cap F_0$. The rest of the proof splits up into two parts according to the rank of F:

PART A: *m is finite.* Lemma 17.6.4(b) gives an open normal subgroup E of F containing N such that rank$(E/N) < 1 + (F : E)(m - 1)$. By Lemma 24.10.1, $N \cong \hat{F}_\omega(\mathcal{C})$.

PART B: m is infinite. By Theorem 25.7.3(b), it suffices to prove $r_N(S) = m$ for each finite simple group $S \in \mathcal{C}$. Indeed, Lemma 17.6.4(a) gives an open normal subgroup E of F which contains N and a basis Y of E with an element $y \in Y \cap N$. Distinguish now between two cases:

CASE B1: S is non-Abelian. Choose a set I of cardinality m, put $G = S^I$ and choose $g \in G$ with $g_i \ne 1$ for each $i \in I$. Since S is simple, the smallest normal subgroup of S which contains g_i is S itself. Hence, by Lemma 18.3.9, the smallest closed normal subgroup of G which contains g is G itself.

By Corollary 17.6.1(b), $\operatorname{rank}(G) = m$. Choose a set G_0 of generators of G of cardinality m which converges to 1 (Proposition 17.1.2). Construct a surjective map $\varphi_0 \colon Y \to G_0 \cup \{g\}$ with $\varphi_0(y) = g$. Then φ_0 extends to an epimorphism $\varphi \colon E \to G$ (Definition 17.4.1). In particular, $\varphi(N)$ is a closed normal subgroup of G which contains g. By the choice of g, $\varphi(N) = G$, so $r_S(N) \ge r_S(G) = m$. Therefore, $r_S(N) = m$.

CASE B2: $S = C_p$. Choose a map $\varphi_0 \colon Y \to C_p$ which maps y onto a generator of C_p and all other elements of Y onto 1. Extend φ_0 to an epimorphism $\varphi \colon E \to C_p$. Then, $\varphi(N) = C_p$. Hence, by Proposition 25.7.4(c), $r_N(C_p) = m$. □

LEMMA 25.8.2: *The intersection of finitely many nontrivial normal subgroups of a free abstract group is nontrivial.*

Proof: It suffices to consider two nontrivial normal subgroups A, B of a free abstract group F and to prove that $A \cap B \ne 1$. Assume $A \cap B = 1$. Choose $a \in A$ and $b \in B$ with $a, b \ne 1$. Then $\langle a \rangle \cong \mathbb{Z}$, $\langle b \rangle \cong \mathbb{Z}$, and $ab = ba$. Hence, $C = \langle a, b \rangle \cong \mathbb{Z} \times \mathbb{Z}$. On the other hand, C, as a subgroup of F, must be free (Proposition 17.5.6). It follows from this contradiction that $A \cap B \ne 1$. □

PROPOSITION 25.8.3 ([Melnikov1, Thm. 4.2]): *Let $\mathcal{C}$ be a Melnikov formation of finite groups, $e \ge 2$ an integer, and N a closed normal subgroup of $\hat{F}_e(\mathcal{C})$ of infinite index. Then N is contained in a closed normal subgroup M of $\hat{F}_e(\mathcal{C})$ which is isomorphic to $\hat{F}_\omega(\mathcal{C})$.*

Proof: Put $F = \hat{F}_e(\mathcal{C})$. If $r_N(S) = \aleph_0$ for each simple group $S \in \mathcal{C}$, then $N \cong \hat{F}_\omega(\mathcal{C})$ (Theorem 25.7.3(b)). Suppose there is a simple group $S \in \mathcal{C}$ with $r = r_N(S)$ finite. Choose a basis $x_1, \ldots, x_e$ of F. Let F_0 be the abstract subgroup of F generated by $x_1, \ldots, x_e$. By Remark 17.6.1, F_0 is free and $x_1, \ldots, x_e$ form a basis for F_0.

Let E be an open normal subgroup of F which contains N. Then E is a free pro-$\mathcal{C}$ group with $f = \operatorname{rank}(E) = 1 + (F : E)(e - 1)$ (Proposition 17.6.2). Choose $(F : E)$ large enough to imply $f - 1 \ge \operatorname{rank}(S^{r+1})$. Then S^{r+1} is a quotient of $\hat{F}_{f-1}(\mathcal{C})$, so $r_{\hat{F}_{f-1}(\mathcal{C})}(S) \ge r + 1$.

Put $E_0 = F_0 \cap E$. By Lemma 17.2.1, $(F_0 : E_0) = (F : E)$ and E is the closure of E_0 in F. By Proposition 17.5.7, E_0 is a free abstract group with a finite basis $y_1, \ldots, y_f$. Hence, E is the free pro-$\mathcal{C}$ group with basis $y_1, \ldots, y_f$ (Lemma 17.4.6).

Let K be the closed normal subgroup of E generated by y_1. Assume KN is open in E. Let $\varphi \colon E \to \hat{F}_{f-1}(S)$ be an epimorphism with $\varphi(y_1) = 1$. Then $\varphi(N) = \varphi(KN)$ is an open subgroup of $\hat{F}_{f-1}(S)$. Hence, $\varphi(N)$ is a free pro-S group of rank at least $f-1$. Therefore, $r_N(S) \geq r_{\varphi(N)}(S) \geq r_{\hat{F}_{f-1}(\mathcal{C})}(S) > r_N(S)$. It follows from this contradiction that $(E : KN) = \infty$.

Let $z_1, \ldots, z_d$ be representatives for the left cosets of E_0 in F_0. Then $z_1, \ldots, z_d$ are representatives for the left cosets of E in F. Therefore, $L = \bigcap_{j=1}^d K^{z_j}$ is a closed normal subgroup of F. Hence, so is LN.

Let K_0 be the normal abstract subgroup of E_0 generated by y_1. Put $L_0 = \bigcap_{j=1}^d K_0^{z_j}$.

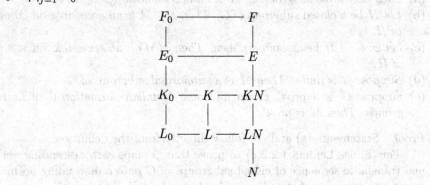

By Lemma 25.8.2, $L_0 \neq 1$. It follows from Proposition 25.8.1 that $LN \cong \hat{F}_\omega(\mathcal{C})$. □

Proposition 25.8.3 may be used to reduce questions about closed normal subgroups of $\hat{F}_e(\mathcal{C})$ to closed normal subgroups of $\hat{F}_\omega(\mathcal{C})$. For example, we prove a result which would also follow from an affirmative answer to Problem 25.4.9:

PROPOSITION 25.8.4: Let $\mathcal{C}$ be a Melnikov formation of finite groups, $e \geq 2$ an integer, $F = \hat{F}_e(\mathcal{C})$, N a closed normal subgroup of F, and M a proper open subgroup of N. Suppose $(F : N) = \infty$ and M is pro-$\mathcal{C}$. Then $M \cong \hat{F}_\omega(\mathcal{C})$.

Proof: Choose a closed normal subgroup E of F which contains N and is isomorphic to $\hat{F}_\omega(\mathcal{C})$ (Proposition 25.8.3). By Theorem 25.4.7(a), with E replacing F, we have $M \cong \hat{F}_\omega(\mathcal{C})$. □

25.9 Accessible Subgroups

Projective groups have been defined as profinite groups for which every finite embedding problem is solvable. They are characterized as closed subgroups of free profinite groups. Likewise, homogeneous groups have been defined as profinite groups of infinite rank for which every Melnikov embedding problem of a certain type is solvable. Here we characterize homogeneous groups as "accessible subgroups" of free profinite groups.

We call a closed subgroup M of a profinite group G **accessible** if M is the intersection of a descending transfinite normal sequence of closed subgroups of G. Thus, $M = \bigcap_{\alpha < \mu} N_\alpha$, where μ is an ordinal number, α ranges over all ordinals less than μ, $N_0 = G$, $N_{\alpha+1} \triangleleft N_\alpha$, and $N_\beta = \bigcap_{\alpha < \beta} N_\alpha$ for each limit ordinal $\beta < \mu$. When μ is finite, M is just a **subnormal subgroup** of G.

Here are some basic properties of accessible subgroups:

LEMMA 25.9.1: *Let M be an accessible subgroup of a profinite group G. Then:*
(a) *Each accessible subgroup of M is an accessible subgroup of G.*
(b) *Let H be a closed subgroup of G. Then $H \cap M$ is an accessible subgroup of H.*
(c) *Let $\varphi: G \to H$ be an epimorphism. Then $\varphi(M)$ is an accessible subgroup of H.*
(d) *Suppose G is finite. Then M is a subnormal subgroup of G.*
(e) *Suppose G is a pro-$\mathcal{C}$ group for some Melnikov formation $\mathcal{C}$ of finite groups. Then M is pro-$\mathcal{C}$.*

Proof: Statements (a) and (b) follow directly from the definition.

For (c) use Lemma 1.2.2(c) to prove that φ maps each descending normal transfinite sequence of closed subgroups of G onto a descending normal transfinite sequence of closed normal subgroups of H. In particular, φ maps the intersection of the former sequence onto the intersection of the latter sequence.

Now suppose G is finite. Let m be the maximal length of a normal sequence of subgroups of G. Then each normal transfinite sequence of subgroups of G collapses into a normal sequence of at most m subgroups. In particular, M is the smallest subgroup of that sequence. Thus, M is subnormal in G. This proves (d).

Finally, let G and $\mathcal{C}$ be as in (e). If G is finite, induction on the length of a normal sequence from G to M proves that M is a $\mathcal{C}$-group.

In the general case consider an open normal subgroup N of M. Choose an open normal subgroup H of G with $H \cap M \leq N$. Put $\bar{G} = G/H$ and let $\varphi: G \to \bar{G}$ be the quotient map. Then $\varphi(M)/\varphi(N) \cong M/N$. By (c) and (d), $\varphi(M)$ is subnormal, so $\varphi(M) \in \mathcal{C}$. Hence, $M/N \in \mathcal{C}$. Consequently, M is pro-$\mathcal{C}$. $\square$

It is somewhat surprising that each accessible subgroup of a profinite group G is the intersection of a countable normal sequence of subgroups of G. In order to see this we introduce notation which will also be useful in the next chapter:

Let S be a subset of a profinite group G. Denote the smallest closed normal subgroup of G containing S by $[S]_G$. Thus, $[S]_G$ is the intersection

25.9 Accessible Subgroups

of all closed normal subgroups of G which contain S. Alternatively, $[S]_G = \langle s^g \mid s \in S, g \in G \rangle$. When $S = \{s_1, \ldots, s_n\}$, write $[s_1, \ldots, s_n]_G$ instead of $[S]_G$. For example, let K be a field and $\sigma_1, \ldots, \sigma_e \in \text{Gal}(K)$. Then, in the notation of Section 18.10, $\text{Gal}(K_s[\sigma_1, \ldots, \sigma_e]) = [\sigma_1, \ldots, \sigma_e]_{\text{Gal}(K)}$.

Here are some obvious rules for this notation:

LEMMA 25.9.2: *Let G be a profinite group and S a subset of G.*
(a) *Suppose $H \triangleleft G$ and $S \subseteq H$. Then $[S]_G \leq H$.*
(b) *Suppose $H \leq G$ and $S \subseteq H$. Then $[S]_H \leq [S]_G$.*
(c) *Suppose $S \subseteq S' \subseteq G$. Then $[S]_G \leq [S']_G$.*
(d) *Suppose $\varphi \colon G \to H$ is an epimorphism. Then $\varphi([S]_G) = [\varphi(S)]_H$.*

Again, let S be a subset of a profinite group G. Inductively define $[S]_G^{(0)} = G$, $[S]_G^{(1)} = [S]_G$, and $[S]_G^{(i+1)} = [S]_{[S]_G^{(i)}}$. This gives a normal descending sequence of closed subgroups of G: $G \triangleright [S]_G \triangleright [S]_G^{(2)} \triangleright [S]_G^{(3)} \triangleright \cdots \supseteq S$. All rules of Lemma 25.9.2 inductively extend to the generalized operation. In particular, if $\varphi \colon G \to H$ is an epimorphism, then $\varphi([S]_G^{(i)}) = [\varphi(S)]_H^{(i)}$.

LEMMA 25.9.3: *The following holds for a closed subgroup M of a profinite group G.*
(a) *If M is accessible, then $\bigcap_{i=1}^{\infty} [M]_G^{(i)} = M$.*
(b) *Suppose for each epimorphism $\varphi \colon G \to H$ with H finite, $\varphi(M)$ is subnormal in H. Then M is accessible.*
(c) *Let $\varphi \colon G \to H$ be a Melnikov cover. Suppose M is accessible and $\varphi(M) = H$. Then $M = G$.*

Proof: Put $M_i = [M]_G^{(i)}$, $i = 1, 2, 3, \ldots$, and $M_\omega = \bigcap_{i=1}^{\infty} M_i$. Then $M \leq M_\omega$.

Proof of (a): First suppose G is finite. By Lemma 25.9.1(d), M is subnormal. Thus, there is a finite normal sequence $G \triangleright H_1 \triangleright H_2 \triangleright \cdots \triangleright H_r = M$. It follows that $M_1 \leq H_1$. Hence, $M_2 = [M]_{M_1} \leq M_1 \cap H_2 \leq H_2$. Repeating this argument r times gives $M_r \leq H_r = M$. Therefore, $M_\omega = M$.

In the general case let $\varphi \colon G \to \bar{G}$ be an epimorphism with $\bar{G}$ finite. Denote images under φ by a bar. Using induction on i, it follows from Lemma 25.9.2(d), that $\overline{M_i} = \bar{M}_i$, $i = 1, 2, \ldots$, hence $\overline{M_\omega} = \bar{M}_\omega$ (Lemma 1.2.2(c)). By Lemma 25.9.1(c), $\bar{M}$ is an accessible subgroup of $\bar{G}$. Hence, by the finite case, $\bar{M} = \bar{M}_\omega = \overline{M_\omega}$. Since φ is arbitrary and both M and M_ω are closed, $M = M_\omega$.

Proof of (b): Let $\varphi \colon G \to \bar{G}$ be an epimorphism with $\bar{G}$ finite. Using the bar notation as in the proof of (a), our assumption is that $\bar{M}$ is subnormal. Hence, by (a) and its proof, $\overline{M_\omega} = \bar{M}_\omega = \bar{M}$. It follows that, $\bar{M}_\omega = M$. Therefore, M is accessible.

Proof of (c): By induction assume $M_i = G$. Then $M_{i+1} \triangleleft G$ and $\varphi(M_{i+1}) = H$, so $M_{i+1} = G$. This completes the induction.

Now we may use (a) to conclude that $M = \bigcap_{i=1}^{\infty} M_i = G$. □

LEMMA 25.9.4: *Intersection of accessible subgroups of a profinite group is accessible.*

Proof: Let G be a profinite group and M_i, $i \in I$, a set of accessible subgroups. We have to prove $N = \bigcap_{i \in I} M_i$ is accessible.

Put $N_j = [N]_G^{(j)}$ and $M_{ij} = [M_i]_G^{(j)}$. By induction suppose $N_{j-1} \leq M_{i,j-1}$. Then $N_j = [N]_{N_{j-1}} \leq [M_i]_{M_{i,j-1}} = M_{ij}$ (Lemma 25.9.2), so this inclusion holds for all j. By Lemma 25.9.3(a), $\bigcap_{j=1}^{\infty} M_{i,j} = M_i$. Hence, $\bigcap_{j=1}^{\infty} N_j \leq \bigcap_{i \in I} \bigcap_{j=1}^{\infty} M_{i,j} = \bigcap_{i \in I} M_i = N$. Therefore $\bigcap_{j=1}^{\infty} N_j = N$ and N is accessible. □

Given an epimorphism $\varphi \colon G \to A$ of profinite groups, there is a closed subgroup B of G such that $\varphi|_B \colon G \to A$ is a Frattini cover (Lemma 22.5.6). Accessible subgroups give rise to an analog of that result for Melnikov covers:

PROPOSITION 25.9.5: *Let $\varphi \colon G \to A$ be an epimorphism of profinite groups. Then G has an accessible subgroup B such that $\varphi|_B \colon B \to A$ is a Melnikov cover.*

Proof: Let N_i, $i \in I$, be a descending chain of accessible subgroups of G with $\varphi(N_i) = A$. Then $N = \bigcap_{i \in I} N_i$ is an accessible subgroup (Lemma 25.9.4) and $\varphi(N) = A$ (Lemma 1.2.2(c)). Thus, by Zorn's Lemma, G has a minimal accessible subgroup B with $\varphi(B) = A$. Each proper closed normal subgroup B_0 of B is accessible. By minimality, $\varphi(B_0) < A$. Consequently, $\varphi|_B$ is a Melnikov cover of A. □

LEMMA 25.9.6 ([Ribes-Zalesskii, Thm. 8.4.2]): *Let $\mathcal{C}$ be a Melnikov formation of finite groups, F a free pro-$\mathcal{C}$ group of infinite rank m, and N an accessible subgroup with weight$(F/N) < m$. Then $N \cong \hat{F}_m(\mathcal{C})$.*

Proof: By Lemma 25.9.1(e), N is a pro-$\mathcal{C}$ group. Hence, by Proposition 25.2.2, $N \cong \hat{F}_m(\mathcal{C})$. □

The computation of the rank of an accessible subgroup of a free profinite group in Proposition 25.9.9 and Theorem 25.9.12 uses special profinite groups which we construct in the following two lemmas:

LEMMA 25.9.7: *Let T be a finite simple non-Abelian group, p a prime number, and W a finite dimensional faithful $\mathbb{F}_p[T]$-module. Then W has a faithful composition factor.*

Proof: We prove the lemma by induction on $n = \dim_{\mathbb{F}_p}(W)$. If W is irreducible, W is the desired composition factor. Otherwise, W has a submodule V of dimension m with $0 < m < n$. Choose an $\mathbb{F}_p$-basis of V and extend it to an $\mathbb{F}_p$-basis of W. The action of T on W defines an embedding ρ of T into the subgroup H of $\mathrm{GL}_n(\mathbb{F}_p)$ consisting of all matrices $\begin{pmatrix} A & B \\ 0 & C \end{pmatrix}$ with $A \in \mathrm{GL}_m(\mathbb{F}_p)$, $B \in M_{m \times (n-m)}(\mathbb{F}_p)$, and $C \in \mathrm{GL}_{n-m}(\mathbb{F}_p)$. Let N be the normal subgroup of H consisting of all matrices $\begin{pmatrix} I_m & B \\ 0 & I_{n-m} \end{pmatrix}$ with I_m and

25.9 Accessible Subgroups

I_{n-m} being the unit matrices. Then N is nilpotent, so $\rho(T) \cap N = 1$. Thus, ρ induces an embedding $\bar{\rho}\colon T \to H/N$. But H/N is isomorphic to the group of all matrices $\left(\begin{smallmatrix} A & 0 \\ 0 & B \end{smallmatrix}\right)$ with $A \in \mathrm{GL}_m(\mathbb{F}_p)$ and $B \in \mathrm{GL}_{n-m}(\mathbb{F}_p)$, that is, to $\mathrm{GL}_m(\mathbb{F}_p) \times \mathrm{GL}_{n-m}(\mathbb{F}_p)$. Take $t \in T$ with $(A, B) = \bar{\rho}(t) \neq 1$. Assume without loss $A \neq 1$. Let $\alpha\colon \mathrm{GL}_m(\mathbb{F}_p) \times \mathrm{GL}_{n-m}(\mathbb{F}_p) \to \mathrm{GL}_m(\mathbb{F}_p)$ be the projection on the first coordinate. Then $\alpha \circ \bar{\rho}(t) \neq 1$. This means $\mathrm{Ker}(\alpha \circ \bar{\rho}) < T$. Since T is simple, $\alpha \circ \bar{\rho}$ is injective. It follows that V is a faithful $\mathbb{F}_p[T]$-module. Induction gives a faithful composition factor of V, hence of W. □

LEMMA 25.9.8: *Let I be a set and S, T finite simple groups. Suppose $S \neq T$ if both S and T are Abelian. Then there exists a pro-$\{S, T\}$ group B with open normal subgroups L, M satisfying this:*
(a) *$B/M \cong T$ and M is the unique maximal normal subgroup of B (thus, $M = M(B)$).*
(b) *$L \leq M$ and $L \cong \prod_{i \in I} L_i$ where each L_i is isomorphic to a direct product of finitely many isomorphic copies of S.*
(c) *If I is infinite, $\mathrm{rank}(B) = |I|$.*
(d) *If S or T are non-Abelian, then $L = M$.*

Proof: We construct a finite group $\hat{T}$, an epimorphism $\pi\colon \hat{T} \to T$, a finite group V, and an action of $\hat{T}$ on V satisfying this:

(2a) $M(\hat{T}) = \mathrm{Ker}(\pi)$.
(2b) The trivial group and V itself are the only subgroups of V which are $\hat{T}$-invariants.
(2c) V is the direct product of at least two and at most finitely many isomorphic copies of S.
(2d) If S or T are non-Abelian, then $\hat{T} = T$ and π is the identity map.

Having constructed $\hat{T}$, π, and V, we choose for each $i \in I$ an isomorphic copy L_i of V and let $L = \prod_{i \in I} L_i$. The action of $\hat{T}$ on each L_i gives an action of $\hat{T}$ on L. Let $B = \hat{T} \ltimes L$. Then, each L_i is a minimal normal subgroup of B.

Now let M be the kernel of the combined epimorphism $B \to \hat{T} \xrightarrow{\pi} T$. Thus, $L \leq M$ and $B/M \cong T$. If I is infinite, then $\mathrm{rank}(B) = \mathrm{rank}(L) = |I|$ (Corollary 17.1.6(b)).

To prove $M(B) = M$, we have to show that M is the unique maximal open normal subgroup of B. So, let K be a maximal open normal subgroup of B. Assume $M \not\leq K$. Since $M/L = M(B/L)$ (by (2a)), this assumption implies $L \not\leq K$, so there is a $j \in I$ with $L_j \not\leq K$. Thus, $K \cap L_j$ is a proper subgroup of L_j which is normal in B. The minimality of L_j implies $K \cap L_j = 1$. Since K is maximal normal, $KL_j = B$. Therefore, $B/K \cong L_j$. This is a contradiction, because B/K is simple while L_j is not (by (2c)). Finally, (2d) implies (d).

The construction of $\hat{T}$, π, and V splits up into three cases.

CASE A: *S is non-Abelian.* Put $\hat{T} = T$ and let $\pi\colon \hat{T} \to T$ be the identity map. Let $V = S^T$ be the group of all functions $f\colon T \to S$ with multiplication rule $(fg)(t) = f(t)g(t)$. Thus, $V = \prod_{t \in T} S_t$ where $S_t = \{f \in V \mid f(T \smallsetminus \{t\}) = 1\}$ is isomorphic to S (Remark 13.7.5). Define the action of T on V by $f^t(t') = f(tt')$. Then $S_t^{t'} = S_{(t')^{-1}t}$ for all $t, t' \in T$. Thus, T acts transitively on the set $\{S_t \mid t \in T\}$. Since S is non-Abelian, the only normal subgroups of V are $\prod_{t \in T_0} S_t$ with $T_0 \subset T$ (Lemma 18.3.9). Hence, the only normal subgroups of V which are invariant under the action of T are the trivial group and V itself.

CASE B: $S = C_p$ and T is non-Abelian. Again, put $\hat{T} = T$ and let $\pi \colon \hat{T} \to T$ be the identity map. Then T has a faithful $\mathbb{F}_p[T]$-module W of finite dimension n over $\mathbb{F}_p$ with $n > 1$. For example, $\mathbb{F}_p[T]$ itself is such a module. Lemma 25.9.7 gives a faithful composition factor V of W. Thus, V is an irreducible $\mathbb{F}_p[T]$-module. Its dimension is greater than 1, otherwise T is a subgroup of $\mathbb{F}_p^\times$, so T is Abelian, which is a contradiction. The module V will be a finite group with a T-action satisfying (2b).

CASE C: $S = C_p$ and $T = C_q$ with $p \ne q$. Choose a positive integer k with $p \not\equiv 1 \bmod q^k$, so that $\zeta_{q^k} \notin \mathbb{F}_p$. Let $\hat{T} = C_{q^k}$ and $\pi \colon \hat{T} \to T$ an epimorphism. Then $\mathrm{Ker}(\pi) = M(\hat{T})$.

Now put $V = \mathbb{F}_p[\zeta_{q^k}]$. Multiplication by ζ_{q^k} makes V an $\mathbb{F}_p[C_{q^k}]$-module with $\dim_{\mathbb{F}_p}(V) > 1$. Moreover, V is an irreducible module. Indeed, $f = \mathrm{irr}(\zeta_{q^k}, \mathbb{F}_p)$ is the characteristic polynomial of ζ_{q^k} as an automorphism of V. By assumption, $\zeta_{q^k} \notin \mathbb{F}_p$, hence $\deg(f) > 1$. Any proper nontrivial submodule of V leads to a factorization of f over $\mathbb{F}_p$. Hence, there is no such submodule.

This concludes the construction of $\hat{T}$ and V and the proof of the lemma. ◻

Every subgroup of a finite p-group is subnormal [Huppert, p. 308], so every closed subgroup N of a pro-p group F is accessible (Lemma 25.9.3(b)). If F is pro-p free, then so is N. But there are no constraints on the rank of N. This explains the exclusion of the formation of p-groups from the next result, which partially generalizes Lemma 25.6.3.

PROPOSITION 25.9.9 ([Melnikov2, Thm. 3.4] and [Ribes-Zalesskii, Prop. 8.5.10]): *Let $\mathcal{C}$ be a Melnikov formation of finite groups, F a free pro-$\mathcal{C}$ group of rank $m \geq 2$, and N a nontrivial accessible subgroup of infinite index. Suppose $\mathcal{C}$ is not the formation of p-groups for a single prime number p. Then N is $\mathcal{C}$-homogeneous with $\mathrm{rank}(N) = \max(\aleph_0, m)$. If in addition, $r_N(S) = m^{\max(\aleph_0, m)}$ for each finite simple group $S \in \mathcal{C}$, then $N \cong \hat{F}_{\max(\aleph_0, m)}(\mathcal{C})$.*

Proof: In contrast to the situation of Lemma 25.6.3 we do not know $\mathrm{rank}(N)$ in advance, so we first solve enough embedding problems of F and then compute $\mathrm{rank}(N)$:

25.9 Accessible Subgroups

PART A: *An embedding problem.* Consider a pro-$\mathcal{C}$ Melnikov embedding problem

$$(3) \qquad (\varphi\colon N \to A,\ \alpha\colon B \to A)$$

for N satisfying one of the following conditions:
(4a) either $m < \aleph_0$ and B is finite,
(4b) or $m = \aleph_0$, $\mathrm{rank}(B) \leq \aleph_0$, and A is finite,
(4c) or $m > \aleph_0$, $\mathrm{rank}(B) \leq m$, and $\mathrm{rank}(A) < m$.

Put $N_1 = \mathrm{Ker}(\varphi)$. In each of the above three cases we construct a closed normal subgroup L_1 of F with $N \cap L_1 \leq N_1$ and put $L = NL_1$. Then L is accessible in F, so L is a pro-$\mathcal{C}$ group (Lemma 25.9.1(e)). These subgroups should satisfy additional properties:

If (4a) holds, use Proposition 17.6.2 to choose L_1 open with $\mathrm{rank}(L) \geq \mathrm{rank}(B)$. Then L is pro-$\mathcal{C}$ free.

If (4b) holds, choose L_1 open. Then $L \cong \hat{F}_\omega(\mathcal{C})$ (Proposition 17.6.2).

If (4c) holds, use Lemma 25.6.1 to choose L_1 with $\mathrm{rank}(F/L_1) < m$. Then $\mathrm{rank}(F/L) < m$. By Proposition 25.2.2, $L \cong \hat{F}_m(\mathcal{C})$.

In each case extend φ to an epimorphism $\tilde{\varphi}\colon L \to A$ by $\tilde{\varphi}(nl_1) = \varphi(n)$ for $n \in N$ and $l_1 \in L_1$. In Case (4a), Proposition 17.7.3 gives an epimorphism $\gamma\colon L \to B$ with $\alpha \circ \gamma = \varphi$. In Cases (4b) and (4c) Lemma 25.6.2 supplies such γ. Thus, $\gamma(N)$ is a subnormal subgroup of B with $\alpha(\gamma(N)) = A$. Since $\alpha\colon B \to A$ is Melnikov, $\gamma(N) = B$ (Lemma 25.9.3(c)). Therefore, $\gamma|_N$ is a solution of (3).

PART B: *Computation of* $\mathrm{rank}(N)$. By assumption, N is nontrivial. Hence, there is a finite simple group T and an epimorphism $\varphi\colon N \to T$. We choose a finite simple group S in $\mathcal{C}$ in the following way. If T is non-Abelian, choose $S = T$. If T is Abelian, then $T \cong C_q$ for some prime number q. By assumption, $\mathcal{C}$ is not the formation of finite q-groups, so we may choose $S \neq T$ in $\mathcal{C}$. Let I be a set of cardinality $\max(\aleph_0, m)$, B be the profinite group which Lemma 25.9.8 supplies, $M = M(B)$, and $\alpha\colon B \to B/M \cong T$ the quotient map. Then B is a pro-$\mathcal{C}$ group of rank $\max(\aleph_0, m)$ and α is a Melnikov cover. Part A gives an epimorphism $\gamma\colon N \to B$ with $\alpha \circ \gamma = \varphi$. Thus, $\mathrm{rank}(N) \geq \mathrm{rank}(B)$. Consequently, $\mathrm{rank}(N) = \max(\aleph_0, m)$.

PART C: *The final statement.* Suppose now $r_N(S) = \max(\aleph_0, m)$ for each finite simple group $S \in \mathcal{C}$. Then $N \cong \hat{F}_{\max(\aleph_0, m)}(\mathcal{C})$ (Proposition 25.7.2(c)). $\square$

The next result says that, in contrast to closed normal subgroups of $\hat{F}_m(\mathcal{C})$, there are no constraints on the S-rank of an accessible subgroup except the obvious ones:

PROPOSITION 25.9.10 ([Ribes-Zalesskii Thm. 8.5.3]): *Let $\mathcal{C}$ be a Melnikov formation of finite groups, $m \geq 2$ a cardinal number, and f a function from the set of finite simple groups to the set of cardinal numbers at most*

$\max(\aleph_0, m)$. Suppose $f(S) = 0$ if $S \notin \mathcal{C}$. Then $\hat{F}_m(\mathcal{C})$ has an accessible subgroup N with $r_N = f$.

Proof: Put $F = \hat{F}_m(\mathcal{C})$. If $m < \aleph_0$ replace F by an accessible subgroup H which is isomorphic to $\hat{F}_\omega(\mathcal{C})$. For example, choose H to be a proper open normal subgroup of a closed normal subgroup of infinite index of F (Proposition 24.10.3). Thus, assume without loss, $m \geq \aleph_0$.

Denote the set of all simple groups in $\mathcal{C}$ by $\mathcal{S}$. Consider the direct product $G = \prod_{S \in \mathcal{S}} S^{f(S)}$. By Corollary 17.1.6(b), $\mathrm{rank}(G) \leq m$. Thus, there is an epimorphism $\varphi \colon F \to G$ (Proposition 17.4.8). Proposition 25.9.5 gives an accessible subgroup N of F such that $\varphi|_N \colon N \to G$ is a Melnikov cover. By Lemma 25.5.6, $N/M(N) \cong G/M(G) \cong G$. It follow that $r_N = r_G = f$. □

The results accumulated in the last sections culminate with the classification of $\mathcal{C}$-homogeneous groups:

THEOREM 25.9.11 ([Melnikov2, Thm. 3.2]): *Let $\mathcal{C}$ be a Melnikov formation of finite groups and G a pro-$\mathcal{C}$ group of infinite rank m. Suppose $\mathcal{C}$ is not the formation of all finite p-groups for a single prime number p. Then, G is $\mathcal{C}$-homogeneous if and only if G is isomorphic to a nontrivial accessible subgroup of $\hat{F}_m(\mathcal{C})$.*

Proof: By Proposition 25.9.9, each accessible subgroup of $\hat{F}_m(\mathcal{C})$ is $\mathcal{C}$-homogeneous. Conversely, suppose G is $\mathcal{C}$-homogeneous. Then $r_G(S) \leq m$ for each finite simple group S, so Proposition 25.9.10 gives an accessible subgroup N of $\hat{F}_m(\mathcal{C})$ with $r_N = r_G$. In other words, $N/M(N) \cong G/M(G)$. If N is open in $\hat{F}_m(\mathcal{C})$, then N is $\mathcal{C}$-free (Proposition 17.6.2), hence $\mathcal{C}$-homogeneous (Proportion 25.6.2). If the index of N in $\hat{F}_m(\mathcal{C})$ is infinite, then N is $\mathcal{C}$-homogeneous of rank m (Proposition 25.9.9). In both cases we conclude from Proposition 25.7.2 that $G \cong N$. □

Let $\mathcal{C}$ be a formation of finite groups and G a pro-$\mathcal{C}$ group. We call G **virtually free pro-$\mathcal{C}$** if G has an open subgroup which is free pro-$\mathcal{C}$.

THEOREM 25.9.12 ([Ribes-Zalesskii, Cor. 8.5.15]): *Let $\mathcal{C}$ be a Melnikov formation of finite groups, $m \geq 2$ a cardinal number, N a nontrivial accessible subgroup of $\hat{F}_m(\mathcal{C})$ of infinite index. Suppose $\mathcal{C}$ contains a non-Abelian simple group. Then N is virtually free pro-$\mathcal{C}$.*

Proof: Put $m^* = \max(\aleph_0, m)$.

CLAIM: *Suppose M is an open normal subgroup of N with $T = N/M$ a simple group. Let S be a finite simple group belonging to $\mathcal{C}$. Suppose S or T are non-Abelian. Then $r_M(S) = m^*$. If T is non-Abelian, then $M \cong \hat{F}_{m^*}(\mathcal{C})$.*

Indeed, M is accessible, so M is $\mathcal{C}$-homogeneous and $\mathrm{rank}(M) = m^*$ (Proposition 25.9.9).

Lemma 25.9.8 gives a Melnikov cover $\alpha \colon B \to N/M$ where $\mathrm{Ker}(\alpha)$ is a direct product of m^* isomorphic copies of S. Let $\varphi \colon N \to N/M$ be the quotient map. Since N is $\mathcal{C}$-homogeneous (Theorem 25.9.11), there is an epimorphism $\gamma\colon N \to B$ with $\alpha\circ\gamma = \varphi$. It maps M onto $\mathrm{Ker}(\alpha)$, so $r_M(S) = m^*$.

If T is non-Abelian, then the latter conclusion holds for every finite simple group $S \in \mathcal{C}$. Hence, by Proposition 25.7.2(c), $M \cong \hat{F}_{m^*}(\mathcal{C})$. This concludes the proof of the claim.

Now choose a maximal open normal subgroup M of N. Thus, N/M is simple. If N/M is non-Abelian, $M \cong \hat{F}_{m^*}(\mathcal{C})$, by the Claim. Otherwise, choose a non-Abelian simple group S in $\mathcal{C}$. By the Claim, $r_M(S) = m^*$. In particular, M has an open normal subgroup L with $M/L \cong S$. An application of the Claim with M, L replacing N, M gives $L \cong \hat{F}_{m^*}(\mathcal{C})$. □

Remark 25.9.13: *$\mathcal{C}$-Homogeneous Melnikov covers.* Let $\mathcal{C}$ be a Melnikov formation of finite groups. We call an epimorphism $\varphi\colon H \to G$ of pro-$\mathcal{C}$ groups a **$\mathcal{C}$-homogeneous cover** if H is $\mathcal{C}$-homogeneous and refer to $\mathrm{rank}(H)$ as the **rank of the cover**.

Suppose $\mathcal{C}$ contains a non-Abelian simple group. Let G be a nontrivial $\mathcal{C}$-group and $m \geq \max(\aleph_0, \mathrm{rank}(G))$ a cardinal number. Then G has a pro-$\mathcal{C}$ homogeneous Melnikov cover of rank m.

Indeed, choose an epimorphism $\tilde{\varphi}\colon \hat{F}_m(\mathcal{C}) \to G$. Proposition 25.9.5 gives an accessible subgroup H of $\hat{F}_m(\mathcal{C})$ such that $\varphi = \tilde{\varphi}|_H\colon H \to G$ is a Melnikov cover. By Proposition 25.9.9, H is $\mathcal{C}$-homogeneous of rank m.

Let $\varphi'\colon H' \to G$ be another $\mathcal{C}$-homogeneous Melnikov cover of rank m. If $m = \aleph_0$ and G is finite or $\mathrm{rank}(M(G)) < m$, there is an isomorphism $\theta\colon H \to H'$ satisfying $\varphi' \circ \theta = \varphi$ (Proposition 25.6.5). In other words, in this case, the $\mathcal{C}$-homogeneous Melnikov cover $\varphi\colon H \to G$ of G is unique.

In the general case, $H/M(H) \cong G/M(G) \cong H'/M(H')$ (Lemma 25.5.6). Hence, $r_{H_1} = r_{H_2}$. So, by Proposition 25.7.2, $H \cong H'$. However, it is not clear whether there exists an isomorphism $\theta\colon H \to H'$ with $\varphi' \circ \theta = \varphi$.

For example, let $\varphi\colon \hat{F}_\omega \to \hat{F}_\omega$ and $\varphi'\colon \hat{F}_\omega \to \hat{F}_\omega$ be Melnikov covers. Does there exist $\theta \in \mathrm{Aut}(\hat{F}_\omega)$ with $\varphi' \circ \theta = \varphi$? □

Notes

The basic source for this Chapter is [Melnikov1]. Other sources are [Melnikov2], [Chatzidakis1], [Jarden-Lubotzky1], and [Haran5]. [Ribes-Zalesskii, Chap8] gives an excellent presentation to a great part of the Chapter.

[Melnikov1, Prop. 2.1] proves Proposition 25.2.2 under the assumption $N \triangleleft F$. Likewise [Ribes-Zalesskii, Thm. 8.7.1] proves Theorem 25.4.7 under the weaker assumption that M is a proper open normal subgroup of N.

The notation $M(G)$ for the Melnikov group, albeit not the name, appears in [Melnikov2]. See also [Ribes-Zalesskii, Sec. 8.5]. Likewise, the notion of Homogeneous pro-$\mathcal{C}$ groups is due to [Melnikov2].

The definition for a $\mathcal{C}$-homogeneous group G that appears in [Ribes-Zalesskii, p. 322] replaces Condition (2a) of Section 25.6 by the seemingly stronger condition:

(1) $m = \aleph_0$, rank$(B) \leq \aleph_0$, and A is finite.

Nevertheless, one can break up the epimorphism $\alpha\colon B \to A$ into a sequence of epimorphisms $B/L_{n+1} \to B/L_n$ with L_n open normal in B, $L_0 = \mathrm{Ker}(\alpha)$, $L_{n+1} \leq L_n$, and $\bigcap_{n=0}^{\infty} L_n = 1$. An application of (2a) of Section 25.6 and induction gives a compatible sequence of epimorphisms $\gamma_n\colon G \to B/L_n$. This gives a solution of embedding problem (1) of Section 25.6. Thus, our definition is equivalent to that of [Ribes-Zalesskii].

The notion of "accessible group" appears in [Melnikov2].

Lemma 25.9.8 is a workout of [Ribes-Zalesskii, Lemma 8.5.8]. Both results consider a set I and finite simple groups S, T which are distinct if both of them are Abelian. Then [Ribes-Zalesskii, Lemma 8.5.8] constructs a profinite group B with an open normal subgroup L such that $B/L \cong T$, $M(B) = L$, and $L \cong \prod_{i \in I} L_i$ with L_i isomorphic to a direct product of finitely many copies of S. When $S \cong C_p$, the proof of [Ribes-Zalesskii, Lemma 8.5.8] chooses each L_i as an irreducible $\mathbb{F}_p[T]$-module of dimension > 1. Case B of the proof of Lemma 25.9.8 supplies the missing construction when T is non-Abelian. However, if $T \cong C_q$ and q divides $p - 1$ all irreducible $\mathbb{F}_p[T]$-modules are of dimension 1. To overcome this difficulty, Lemma 25.9.8 modifies the construction of B in this case. We then use the modified B in the proof of Proposition 25.9.9 in the same way as [Zalesskii-Ribes] do with the original B.

Theorem 25.9.12 assumes $\mathcal{C}$ contains a non-Abelian simple group. In contrast, [Ribes-Zalesskii, Cor. 8.5.15] assumes $\mathcal{C}$ is not the formation of p-groups with a single p. However, the proof of [Ribes-Zalesskii, Cor. 8.5.15(b)] uses the existence of a non-Abelian simple group in $\mathcal{C}$.

Chapter 26.
Random Elements in Profinite Groups

Let n be a positive integer and $F = \hat{F}_n$ the free profinite group of rank n. For each e-tuple $(x_1, \ldots, x_e)$ in F^e we consider the closed (resp. normal closed) subgroup $\langle \mathbf{x} \rangle$ (resp. $[\mathbf{x}]$) generated by $x_1, \ldots, x_n$ in F. We investigate the probability that $\langle \mathbf{x} \rangle$ (resp. $[\mathbf{x}]$) is an open subgroup of F. Having done so, we strive to prove that with probability 1 each of the groups $\langle \mathbf{x} \rangle$ and $[\mathbf{x}]$ are $\mathcal{C}$-free of specific ranks. We mention here only the most striking results:

For almost all $\mathbf{x} \in F^e$ the group $\langle \mathbf{x} \rangle$ has an infinite index in F and is isomorphic to $\hat{F}_e$ (Lemma 26.1.7). This settles problem 16.16 of [Fried-Jarden3].

For almost all $\mathbf{x} \in (\hat{\mathbb{Z}}^n)^e$ the group $\langle \mathbf{x} \rangle$ is of infinite index if $e \leq n$ and is open in $\hat{\mathbb{Z}}^n$ if $e > n$ (Theorem 26.3.5). In addition, for almost all $\mathbf{x} \in (\hat{\mathbb{Z}}^n)^e$ the group $\langle \mathbf{x} \rangle$ is isomorphic to $\hat{\mathbb{Z}}^e$ if $e \leq n$ and to $\hat{\mathbb{Z}}^n$ if $e > n$ (Theorem 26.3.6). This generalizes Lemma 18.5.8.

Suppose $n \geq 2$. Then F^e has a subset C of positive measure such that $[\mathbf{x}]$ has infinite index for each $\mathbf{x} \in C$ (Theorem 26.4.5(b)). The proof of this result uses the Golod-Shafarevich inequality. If $e > n$, then F^e has a subset B of positive measure such that $[\mathbf{x}] = \hat{F}_n$ (Theorem 26.4.5(a)). Here we must use the classification of finite simple groups. Finally, we prove that for almost all $\mathbf{x} \in C$ the group $[\mathbf{x}]$ is isomorphic to $\hat{F}_\omega$ (Theorem 26.5.6). The proof uses Melnikov's criterion for a closed normal subgroup of $\hat{F}_n$ to be free of countable rank (Corollary 25.7.6).

26.1 Random Elements in a Free Profinite Group

This section generalizes Lemma 18.5.8 from $\hat{\mathbb{Z}}$ to an arbitrary free profinite group F of rank at least 2: The closed subgroup of F generated by an e-tuple $(x_1, \ldots, x_e)$ chosen at random is of infinite index and isomorphic to $\hat{F}_n$.

Notation 26.1.1: For a finite group G and a positive integer e let

$$d_e(G) = \max(m \in \mathbb{N} \mid G^m \text{ is generated by } e \text{ elements})$$
$$= \max(m \in \mathbb{N} \mid G^m \text{ is a homomorphic image of } \hat{F}_e)$$
$$D_e(G) = \{(x_1, \ldots, x_e) \in G^e \mid \langle x_1, \ldots, x_e \rangle = G\}. \qquad \square$$

The next Lemma implies that $d_e(S)$ is a positive integer if e is large enough:

LEMMA 26.1.2 (P. Hall): Let S be a finite simple non-Abelian group, $s = \operatorname{rank}(S)$, and e a positive integer. Then:
(a) $d_e(S)$ is the number of open normal subgroups N of $\hat{F}_e$ with $\hat{F}_e/N \cong S$.

(b) $d_e(S) = r_{\hat{F}_e}(S) = r_{\hat{F}_e(S)}(S)$ (Section 24.9).
(c) $d_e(S) = \frac{|D_e(S)|}{|\text{Aut}(S)|}$.
(d) $d_e(S) \leq |S|^{e-1}$ and $|S|^{e-2\text{rank}(S)} \leq d_e(S)$ if $\text{rank}(S) \leq e$.
(e) $d_e(S) < d_{e+1}(S)$.

Proof of (a) and (b): By the opening discussion of Section 24.9, $d_e(S) = r_{\hat{F}_e}(S)$. By Lemma 16.8.3, $d_e(S)$ is then the number of normal subgroups N of $\hat{F}_e$ with $\hat{F}_e/N \cong S$.

Similar statements hold for the free pro-S group $\hat{F}_e(S)$ or rank e.

Proof of (c): Put $d = \frac{|D_e(S)|}{|\text{Aut}(S)|}$. Fix a basis $z_1, \ldots, z_e$ of $\hat{F}_e$. The map

$$\psi \mapsto (\psi(z_1), \ldots, \psi(z_e))$$

is a bijection between the set of all epimorphisms $\psi \colon \hat{F}_e \to S$ and $D_e(S)$. Two epimorphisms ψ and ψ' have the same kernel if and only if there is an $\alpha \in \text{Aut}(S)$ with $\alpha \circ \psi = \psi'$. Hence, $\hat{F}_e$ has exactly d normal subgroups N with $\hat{F}_e/N \cong S$. By (a), $d = d_e(S)$.

Proof of (d): Since S is simple and non-Abelian, mapping each $x \in S$ onto the corresponding inner automorphism embeds S into $\text{Aut}(S)$. Hence, $|S| \leq |\text{Aut}(S)|$. By definition, $|D_e(S)| \leq |S|^e$. Hence, by (b), $d_e(S) \leq |S|^{e-1}$.

Now suppose $s = \text{rank}(G) \leq e$. Let $x_1, \ldots, x_s$ be generators of S. Then $x_1, \ldots, x_s, x_{s+1}, \ldots, x_e$ generate S for all $x_{s+1}, \ldots, x_e \in S$. Hence, $|D_n(S)| \geq |S|^{e-s}$. Also, each $\alpha \in \text{Aut}(S)$ is uniquely determined by $(\alpha(x_1), \ldots, \alpha(x_s))$. Therefore, $|\text{Aut}(S)| \leq |S|^s$. It follows from (c) that $d_e(S) \geq |S|^{e-2s}$. □

Consider the alternative group A_k on k letters. The probability that a pair $(x, y) \in A_k$ generates A_k approaches 1 as $k \to \infty$. In other words,

(1) $$\frac{|D_2(A_k)|}{(k!)^2/4} \xrightarrow[k\to\infty]{} 1$$

[Dixon]. The inequality stated in the next lemma is not good enough to prove (1). Nevertheless, it suffices for the application in the proof of Lemma 26.1.4:

LEMMA 26.1.3: *Let $k \geq 7$ be an odd integer. Then*

$$|D_2(A_k)| \geq (k-3)!(k-7)!.$$

Proof: Let $\gamma = (1\,2\,3)$ and let $\rho = (b_3\,b_4\,\cdots\,b_k)$ be a cyclic permutation of the set $B = \{3, 4, \ldots, k\}$. We claim:

(2) $$A_k = \langle \gamma, \rho \rangle.$$

To prove (2), it suffices to prove that $H = \langle \gamma, \rho \rangle$ contains each 3-cycle. Assume without loss that $b_3 = 3$. Since k is odd,

$$\sigma = \rho^\gamma = (1\,b_4\,\cdots\,b_k) \in H \quad \text{and} \quad \tau = \rho^{\gamma^2} = (2\,b_4\,\cdots\,b_k) \in H.$$

26.1 Random Elements in a Free Profinite Group

Hence, for each $j \geq 3$

$$(b_j \ 2 \ 3) = \gamma^{\sigma^{j-3}}, \quad (1 \ b_j \ 3) = \gamma^{\tau^{j-3}}, \quad \text{and} \quad (1 \ 2 \ b_j) = \gamma^{\rho^{j-3}}$$

belong to H. Finally, let b, c, d be distinct elements of B. Then $(2 \ c \ b) = (1 \ 2 \ b)^{(1 \ 2 \ c)}$, $(c \ b \ 1) = (1 \ b \ 3)^{(1 \ c \ 3)}$, $(b \ 3 \ c) = (b \ 2 \ 3)^{(c \ 2 \ 3)}$, and $(d \ b \ c) = (2 \ b \ c)^{(2 \ d \ 3)}$ belong to H. Thus, every 3-cycle of $\{1, 2, \ldots, k\}$ belongs to H. Hence, $H = A_k$, as asserted.

Next check the residues modulo 6 to find a positive integer m with $k - 6 \leq m \leq k - 3$ which is prime to 6. Each cyclic permutation $\alpha = (a_1 \ a_2 \ \cdots \ a_m)$ of m integers in $A = \{4, 5, \ldots, k\}$ belongs to A_k. Moreover, $(\alpha\gamma)^m = \alpha^m \gamma^m = \gamma^m = \gamma^{\pm 1}$. Hence, by (2), $A_k = \langle \alpha\gamma, \rho \rangle$. There are $(k-3)(k-4)\cdots(k-2-m)/m$ permutations α and $(k-3)!$ permutations ρ. The former number is $\geq (m-1)!$. Hence, $|D_2(A_k)| \geq (k-7)!(k-3)!$, as asserted. □

LEMMA 26.1.4: *For each odd integer* $n \geq 15$, $L_k = A_k^{\frac{(k-3)!(k-7)!}{k!}}$ *is generated by 2 elements.*

Proof: By [Huppert, p. 175], $\mathrm{Aut}(A_k) \cong S_k$. So, $|\mathrm{Aut}(A_k)| = k!$. It follows from Lemmas 26.1.2 and 26.1.3 that

$$d_2(A_k) = \frac{|D_2(A_k)|}{k!} \geq \frac{(k-3)!(k-7)!}{k!} \geq 1.$$

To prove the latter inequality, verify that $7! \geq 14 \cdot 13 \cdot 12$ and then prove inductively that $(n - 7)! \geq n(n - 1)(n - 2)$ for each $n \geq 14$. Thus, L_k is generated by 2 elements. □

LEMMA 26.1.5: *The probability that an e-tuple of elements of L_k generates L_k approaches 0 as k approaches infinity over the odd positive integers.*

Proof: In order for an e-tuple of elements of L_k to generate L_k its projection on each of the factors must generate A_k. The probability of the last event is at most $1 - \frac{1}{k^e}$, because an e-tuple of elements which belong to the subgroup A_{k-1} of index k, does not generate A_k. Hence, the probability that an e-tuple of elements of L_k generates L_k is at most

$$(3) \quad \left(1 - \frac{1}{k^e}\right)^{\frac{(k-3)!(k-7)!}{k!}} = \left\{\left(1 - \frac{1}{k^e}\right)^{k^e}\right\}^{\frac{(k-3)!(k-7)!}{k!k^e}}$$

The expression in the braces approaches the inverse of the basis of the natural logarithms (which is usually denoted by the same letter e which we are using here for another purpose). The exponent of the braces approaches infinity as k approaches infinity. Consequently, the right hand side of (3) approaches 0 as $k \to \infty$. □

LEMMA 26.1.6: *For $n \geq 2$ and $e \geq 1$, the probability for an e-tuple of elements of $\hat{F}_n$ to generate $\hat{F}_n$ is 0.*

Proof: Let $k \geq 15$ be an odd integer. By Lemma 26.1.4, there is an epimorphism $\psi \colon \hat{F}_n \to L_k$. If $(x_1, \ldots, x_e) \in (\hat{F}_n)^e$ generates $\hat{F}_n$, then its image under ψ generates L_k. Hence, the probability for an e-tuple of elements of $\hat{F}_n$ to generate $\hat{F}_n$ is at most the probability for an e-tuple of elements of L_k to generate L_k. By Lemma 26.1.5, the latter probability approaches 0 as $k \to \infty$. Therefore, the former probability is 0. □

PROPOSITION 26.1.7: *Let F be a free profinite group of rank at least 2 and e a positive integer. Then for almost all $(x_1, \ldots, x_e) \in F^e$*
(a) *$\langle x_1, \ldots, x_e \rangle$ has infinite index [Kantor-Lubotzky], and*
(b) *$\langle x_1, \ldots, x_e \rangle \cong \hat{F}_e$ [Lubotzky3].*

Proof of (a): If $\operatorname{rank}(F)$ is infinite, then so is the rank of each open subgroup. Hence, we may assume $F = \hat{F}_n$ with $2 \leq n < \infty$.

By Proposition 17.6.2, each open subgroup of $\hat{F}_n$ is isomorphic to $\hat{F}_s$ for some s. For each s, the group $\hat{F}_n$ has only finitely many open subgroups of index at most s (Lemma 16.10.2). Applying Lemma 26.1.6 to each of these subgroups, we conclude that the probability of an e-tuple to generate an open subgroup of F is 0. This proves our claim.

Proof of (b): First note that there is an epimorphism $\psi \colon F \to \hat{F}_n$ with $2 \leq n < \infty$. If $x_1, \ldots, x_e \in F$ and $\langle \psi(x_1), \ldots, \psi(x_e) \rangle \cong \hat{F}_e$, then $\langle x_1, \ldots, x_e \rangle \cong \hat{F}_e$ (Lemma 17.14.11). Thus, if (b) holds for the quotient, it holds for F. Therefore, we may assume that $F = \hat{F}_n$ with $2 \leq n < \infty$. There are two cases to consider:

CASE A: $n \geq e + 3$. To prove $G = \langle x_1, \ldots, x_e \rangle$ is isomorphic to $\hat{F}_e$, it suffices to show that each finite group B which is generated by e elements is a quotient of G (Lemma 17.7.1). Since there are only countably many finite groups, it suffices to fix a finite group B with $\operatorname{rank}(B) \leq e$ and to prove that for almost all $(x_1, \ldots, x_e) \in F^e$ the group B is a quotient of $\langle x_1, \ldots, x_e \rangle$.

Indeed, Let $l = |B|$. Then B can be embedded into the symmetric group S_l. Consider the cycle $\kappa = (l+1\ l+2)$ of S_{l+2}. Define an embedding f of S_l into A_{l+2} by the following rule: $f(\pi) = \pi$ if $\pi \in A_l$ and $f(\pi) = \pi\kappa$ if $\pi \notin A_l$. Let $k(B) = \max(7, l+2)$. Then, we can view B as a subgroup of A_k for each $k \geq \max(7, l+2)$.

Let $k \geq 7$ be an odd integer. Since A_k is generated by two elements (Lemma 26.1.3), $|D_n(A_k)| \geq |A_k|^{n-2}$ (Proof of Part (c) of Lemma 26.1.2). Also, $|\operatorname{Aut}(A_k)| = |S_k| = 2|A_k|$ [Suzuki, p. 299, Statement 2.17]. Hence, by Lemma 26.1.2,

(4) $$d_n(A_k) = \frac{|D_n(A_k)|}{|\operatorname{Aut}(A_k)|} \geq \frac{1}{2}|A_k|^{n-3} \geq \frac{1}{2}|A_k|^e.$$

By definition, $A_k^{d_n(A_k)}$ is generated by n elements. Hence, $A_k^{d_n(A_k)}$ is a quotient of F, with kernel N. Since A_k is simple non-Abelian, F has exactly

26.1 Random Elements in a Free Profinite Group

$d_n(A_k)$ open normal subgroups N_i which contain N with $F/N_i \cong A_k$ (Lemma 26.1.2(a)).

For each i between 1 and $d_n(A_k)$ let $\varphi_i \colon F \to A_k$ be an epimorphism with kernel N_i and
$$B_{k,i} = \{(x_1, \ldots, x_e) \in F^e \mid \langle \varphi_i(x_1) \cdots, \varphi_i(x_e) \rangle \cong B\}.$$
Denote the probability that e elements of B generate B by $p_e(B)$. Then

(5) $$\mu(B_{k,i}) = \left(\frac{|B|}{|A_k|}\right)^e p_e(B).$$

The sets $B_{k,i}$ with $k \geq k(B)$ and $i = 1, \ldots, d_n(A_k)$ are μ-independent (Example 18.3.11). By (4) and (5)

$$\sum_{k \geq k(B)}^{d_n(A_k)} \sum_{i=1} \mu(B_{k,i}) = \sum_{k \geq k(B)} d_n(A_k) p_e(B) \left(\frac{|B|}{|A_k|}\right)^e$$

$$\geq \sum_{k \geq k(B)} \frac{1}{2} p_e(B) |B|^e = \infty,$$

because all terms are constant and $p_e(B) \neq 0$, since B is generated by e elements.

By Lemma 18.3.4, $\mu\bigl(\bigcup_{k,i} B_{k,i}\bigr) = 1$. Each e-tuple in the union generates a closed subgroup of F which has B as a quotient.

CASE B: *The general case.* By (a), almost all $(x_1, \ldots, x_e)$ generate a closed subgroup G of F of infinite index, so G is contained in an open subgroup H of F of index at least $e + 2$. Since $k \geq 2$, the group H is free of rank at least $e + 3$ (Proposition 17.6.2). Hence, by Case A, the probability for G to be contained in H and not to be free is zero. Since F has only countably many open subgroups, the probability for G not to be free is zero. This concludes the proof of (b). □

Remark 26.1.8: Proposition 26.1.7 solves Problem 16.16 of [Fried-Jarden3]. □

Remark 26.1.9: Generalization of (1). [Liebeck-Shalev] completes earlier works of [Dixon] and [Kantor-Lubotzky] and prove: The probability that a pair (x, y) of elements of a finite non-Abelian simple group S generates S approaches 1 as $|S|$ approaches infinity. Thus, in Notation 26.1.1,

$$\lim_{|S| \to \infty} \frac{|D_2(S)|}{|S|^2} = 1,$$

where S ranges over the finite non-Abelian simple groups. The proof of this result uses the classification of finite simple groups. □

Corollary 17.6.5 says that the free pro-$\mathcal{C}$ group is infinite if $\mathcal{C}$ is a Melnikov formation. The following example shows this does not hold for an arbitrary formations:

Example 26.1.10: Finite free pro-$\mathcal{C}$ groups. Let S be a finite non-Abelian simple group. Put $\mathcal{C} = \{S^d \mid d = 0, 1, 2, \ldots\}$.

CLAIM A: *$\mathcal{C}$ is a formation of finite groups (Section 17.3).*

Indeed, for each normal subgroup N of S^d there exists another normal subgroup N' with $S^d = N \times N'$. Moreover, $N' \cong S^k$ for some k (Lemma 18.3.9). Thus, $\mathcal{C}$ is closed under quotients. That $\mathcal{C}$ is closed under fiber products is a consequence of Lemma 18.3.11. Consequently, $\mathcal{C}$ is a formation.

CLAIM B: *Let $n \geq 2$ be an integer and $d = d_n(S)$ (Notation 26.1.1). Then S^d is the free pro-$\mathcal{C}$ group of rank n.*

Indeed, let F be the free abstract group with basis $x_1, \ldots, x_n$. By definition, F has a normal subgroup N with $F/N \cong S^d$. Moreover, d is the maximal number m such that S^m is a quotient of F.

Let $\pi \colon F \to S^d$ be an epimorphism with $\mathrm{Ker}(\pi) = N$. Put $y_i = \pi(x_i)$, $i = 1, \ldots, n$. We claim that $y_1, \ldots, y_n$ are free generators of S^d in the category of pro-$\mathcal{C}$ groups. Indeed, suppose $z_1, \ldots, z_n$ are generators of S^m for some m. Then there is an epimorphism $\varphi \colon F \to S^m$. If $N \not\leq \mathrm{Ker}(\varphi)$, then by Claim A, $F/(\mathrm{Ker}(\varphi) \times N) \cong S^e$ for some $e > d$, in contradiction the to maximality of d. It follows that $N \leq \mathrm{Ker}(\varphi)$, so there exists an epimorphism $\bar{\varphi} \colon S^d \to S^m$ with $\pi \circ \bar{\varphi} = \varphi$. In particular, $\bar{\varphi}(y_i) = z_i$, $i = 1, \ldots, n$, as needed. $\square$

Here is an application of Proposition 26.1.7 to fields:

Example 26.1.11: A non-Hilbertian field K with $K_s(\boldsymbol{\sigma})$ PAC and e-free for all e and almost all $\boldsymbol{\sigma} \in \mathrm{Gal}(K)^e$.

Let K be a PAC field with $\mathrm{Gal}(K) \cong \hat{F}_2$. For example, starting from a countable Hilbertian field E, almost all fields $E_s(\sigma_1, \sigma_2)$ are 2-free PAC (Theorem 20.5.1). By Proposition 26.1.7(b), for each positive integer e and for almost all $\boldsymbol{\sigma} \in \mathrm{Gal}(K)^e$ the field $K_s(\boldsymbol{\sigma})$ is e-free. By Ax-Roquette (Corollary 11.2.5), each of the fields $K_s(\boldsymbol{\sigma})$ is PAC. However, by Lemma 16.12.5, K is not Hilbertian because $\mathrm{Gal}(K)$ is finitely generated. $\square$

26.2 Random Elements in Free pro-p Groups

Theorem 26.1.7 gets a new form if we consider free pro-p-groups instead of free profinite groups:

LEMMA 26.2.1: *Let $f \in \mathbb{Z}_p[X_1, \ldots, X_n]$ be a nonzero polynomial. Then the set $A = \{(x_1, \ldots, x_n) \in \mathbb{Z}_p^n \mid f(x_1, \ldots, x_n) = 0\}$ has measure 0.*

Proof: View f as a polynomial in X_n with coefficients in $\mathbb{Z}_p[X_1, \ldots, X_{n-1}]$. Choose a nonzero coefficient $g \in \mathbb{Z}_p[X_1, \ldots, X_{n-1}]$. An induction hypothesis on n implies that $B = \{(x_1, \ldots, x_{n-1}) \in \mathbb{Z}_p^{n-1} \mid g(x_1, \ldots, x_{n-1}) = 0\}$ has measure 0. For each $(x_1, \ldots, x_{n-1}) \in \mathbb{Z}_p^{n-1} \smallsetminus B$ the set $A_{(x_1, \ldots, x_{n-1})} = \{x_n \in \mathbb{Z}_p \mid (x_1, \ldots, x_{n-1}, x_n) \in A\}$ is finite, hence has measure 0. It follows from Fubini [Halmos, p. 147, Thm. A] that $\mu(A) = 0$. $\square$

26.2 Random Elements in Free pro-p Groups

PROPOSITION 26.2.2 ([Lubotzky3, Prop. 7]): *Let $F = \hat{F}_n(p)$ be the free pro-p-group of rank n and let e be a positive integer. Let*

$$A_e = \{(x_1,\ldots,x_e) \in F^e \mid \langle x_1,\ldots,x_e\rangle \text{ is open in } F\}$$
$$B_e = \{(x_1,\ldots,x_e) \in F^e \mid \langle x_1,\ldots,x_e\rangle \cong \hat{F}_e(p)\}.$$

Then:
(a) *If $e < n$, then $\mu(A_e) = 0$ and $\mu(B_e) = 1$.*
(b) *$0 < \mu(A_n) < 1$ and $\mu(B_n) = 1$.*
(c) *If $e > n$, then $0 < \mu(A_e) < 1$ and $0 < \mu(B_e) < 1$.*

Proof of (b): By Nielsen-Schreier (Proposition 17.6.2), the rank of each proper open subgroup of F is greater than n. Hence,

(1) $$A_n = \{(x_1,\ldots,x_n) \in F^n \mid \langle x_1,\ldots,x_n\rangle = F\}.$$

Let $V = \mathbb{F}_p^n \cong F/\Phi(F)$ (Lemma 22.7.4). By definition, $x_1,\ldots,x_n$ generate F if and only if their reductions $v_1,\ldots,v_n$ modulo $\Phi(F)$ generate V. The latter happens exactly if $v_1,\ldots,v_n$ are linearly independent. Hence, $\mu(A_n)$ is the probability in V^n that $v_1,\ldots,v_n$ are linearly independent. Thus

(2) $$\mu(A_n) = \frac{p^n - 1}{p^n} \cdot \frac{p^n - p}{p^n} \cdots \frac{p^n - p^{n-1}}{p^n}$$
$$= \left(1 - \frac{1}{p^n}\right)\left(1 - \frac{1}{p^{n-1}}\right) \cdots \left(1 - \frac{1}{p}\right).$$

Hence, $0 < \mu(A_n) < 1$.

To compute $\mu(B_n)$ let $Z = \mathbb{Z}_p^n$ and choose an epimorphism $\pi \colon F \to Z$. Consider each element of Z as a column with r entries in $\mathbb{Z}_p$. In this notation $(z_1 \cdots z_n)$ denotes an $n \times n$ matrix with entries in $\mathbb{Z}_p$. Then

$$\bar{B}_n = \{(z_1,\ldots,z_n) \in Z^n \mid \langle z_1,\ldots,z_n\rangle \cong Z\}$$
$$= \{(z_1,\ldots,z_n) \in Z^n \mid \text{rank}\langle z_1,\ldots,z_n\rangle = n\}$$
$$= \{(z_1,\ldots,z_n) \in \mathbb{Z}_p^{n^2} \mid \text{rank}(z_1 \cdots z_n) = n\}$$
$$= \{(z_1,\ldots,z_n) \in \mathbb{Z}_p^{n^2} \mid \det(z_1 \cdots z_n) \neq 0\}.$$

By Lemma 26.2.1, $\mu(\bar{B}_n) = 1$.

If $x_1,\ldots,x_n \in F$ and $(\pi(x_1),\ldots,\pi(x_n)) \in \bar{B}_n$, then $\text{rank}\langle x_1,\ldots,x_n\rangle = n$. Since each closed subgroup of F is a free pro-p-group (Corollary 22.7.7), this implies $\langle x_1,\ldots,x_n\rangle \cong F$. Hence, $(x_1,\ldots,x_n) \in B_n$. Thus, $\pi^{-1}(\bar{B}_n) \subseteq B_n$. It follows from the preceding paragraph that $\mu(B_n) = 1$.

Proof of (a): By Nielsen-Schreier, the rank of each open subgroup of F is at least e. Hence, in case (a), $A_e = \emptyset$, so $\mu(A_e) = 0$.

To compute $\mu(B_e)$ consider the projection $\tau\colon F^n \to F^e$ on the first e coordinates. Suppose $(x_1,\ldots,x_n) \in B_n$. Then, $\mathrm{rank}\langle x_1,\ldots,x_e\rangle = e$, so $\langle x_1,\ldots,x_e\rangle \cong \hat{F}_e(p)$. Therefore, $(x_1,\ldots,x_e) \in B_e$. Thus, $B_n \subseteq \tau^{-1}(B_e)$. By (b) and Proposition 18.2.2, $\mu(B_e) = 1$.

Proof of (c): Let $\rho\colon F^e \to F^n$ be the projection on the first n coordinates. Let $(x_1,\ldots,x_e) \in \rho^{-1}(A_n)$. Then, $(x_1,\ldots,x_n) \in A_n$. By (1), $\langle x_1,\ldots,x_n\rangle = F$ and therefore $\langle x_1,\ldots,x_e\rangle = F$. Thus, $\rho^{-1}(A_n) \subseteq A_e$. Hence, by (b), $0 < \mu(A_n) \le \mu(A_e)$. Also, since $F \not\cong \hat{F}_e(p)$, we have, $\rho^{-1}(A_n) \subseteq F^e \smallsetminus B_e$. Therefore, $\mu(B_e) < 1$.

Next use Nielsen-Schreier formula to choose an open subgroup U of F with $l = \mathrm{rank}(U) > e$. The rank of each open subgroup of U is also greater than e. Hence, $U^e \cap A_e = \emptyset$. Since $\mu(U^e) > 0$, this implies that $\mu(A_e) < 1$.

Finally, let $\lambda\colon F^l \to F^e$ be the projection on the first e coordinates. Then $B_l \subseteq \lambda^{-1}(B_e)$. Hence, $\mu(B_l) \le \mu(B_e)$. Applying (b) to U and l instead of to F and e, we conclude that $\mu(B_l) > 0$, so $\mu(B_e) > 0$. This concludes the proof of (c) and the proposition. $\square$

It will be interesting to compute the measure of A_e and B_e in the case where F is the free prosolvable group on n generators. The proof of Proposition 26.2.2 does not apply to this case.

26.3 Random e-tuples in $\hat{\mathbb{Z}}^n$

By Lemma 18.5.8, $\langle \mathbf{a}\rangle \cong \hat{\mathbb{Z}}$ for almost all $\mathbf{a} \in \hat{\mathbb{Z}}^e$. Moreover, $\langle \mathbf{a}\rangle$ is open in $\hat{\mathbb{Z}}$ if $e \ge 2$ and of infinite index if $e = 1$. Theorems 26.3.5 and 26.3.6 below generalize this result to $\hat{\mathbb{Z}}^n$.

We fix positive integers e,n and use l to denote a prime number.

LEMMA 26.3.1: *Suppose $e \ge n$. Then the number of e-tuples $(\mathbf{a}_1,\ldots,\mathbf{a}_e) \in (\mathbb{F}_l^n)^e$ that generate $\mathbb{F}_l^n$ is $\prod_{i=0}^{n-1}(l^e - l^i)$. If $e < n$, there are none.*

Proof: Consider the $\mathbb{F}_l$-vector spaces $\mathbb{F}_l^n$ and $\mathbb{F}_l^e$. Then $\dim(\mathbb{F}_l^n) = n$. If $e < n$, no e-tuple $(\mathbf{a}_1,\ldots,\mathbf{a}_e) \in (\mathbb{F}_l^n)^e$ generates $\mathbb{F}_l^n$.

Suppose $e \ge n$. Put

$$R = \{(\mathbf{a}_1,\ldots,\mathbf{a}_e) \in (\mathbb{F}_l^n)^e \mid \mathbf{a}_1,\ldots,\mathbf{a}_e \text{ generate } \mathbb{F}_l^n\}$$

$$C = \{(\mathbf{a}^{(1)},\ldots,\mathbf{a}^{(n)}) \in (\mathbb{F}_l^e)^n \mid \mathbf{a}^{(1)},\ldots,\mathbf{a}^{(n)} \text{ are linearly independent}\}$$

Define a map $\alpha\colon (\mathbb{F}_l^n)^e \to (\mathbb{F}_l^e)^n$ as follows. With each $(\mathbf{a}_1,\ldots,\mathbf{a}_e) \in (\mathbb{F}_l^n)^e$ associate the matrix $A = (a_{ij})_{1 \le i \le e,\, 1 \le j \le n}$. Then let

$$\alpha(\mathbf{a}_1,\ldots,\mathbf{a}_e) = (\mathbf{a}^{(1)},\ldots,\mathbf{a}^{(n)})$$

with $\mathbf{a}^{(j)}$ being the jth column of A, $j = 1,\ldots,n$. This map is bijective.

26.3 Random e-tuples in $\hat{\mathbb{Z}}^n$

Suppose $(\mathbf{a}_1, \ldots, \mathbf{a}_e) \in R$. Then $\mathbf{a}_1, \ldots, \mathbf{a}_e$ generate $\mathbb{F}_l^n$, so the row rank of A is n. Hence, the column rank of A is also n. Since A has exactly n columns, they are linearly independent. Conversely, if the columns of A are linearly independent, then the rows generate $\mathbb{F}_l^n$. Consequently, α maps R bijectively onto C.

An n-tuple $(\mathbf{a}^{(1)}, \ldots, \mathbf{a}^{(n)})$ of $(\mathbb{F}_l^e)^n$ belongs to C if and only if the vectors $\mathbf{a}^{(1)}, \ldots, \mathbf{a}^{(i)}$ are linearly independent for $i = 1, \ldots, n$. Equivalently, $\mathbf{a}^{(i+1)} \in \mathbb{F}_l^e \smallsetminus \langle \mathbf{a}^{(1)}, \ldots, \mathbf{a}^{(i)} \rangle$ for $i = 0, \ldots, n-1$. If $\mathbf{a}^{(1)}, \ldots, \mathbf{a}^{(i)}$ are linearly independent, then $|\mathbb{F}_l^e \smallsetminus \langle \mathbf{a}^{(1)}, \ldots, \mathbf{a}^{(i)} \rangle| = l^e - l^i$. Therefore, $|R| = |C| = \prod_{i=0}^{n-1}(l^e - l^i)$. □

LEMMA 26.3.2: *Let n be a positive integer, Z one of the groups $\mathbb{Z}^n$, $\mathbb{Z}_l^n$, or $\hat{\mathbb{Z}}^n$, and Y a subgroup of Z of finite index. Then $Y \cong Z$.*

Proof: Let $m = (Z : Y)$. Then mZ is isomorphic to Z and $mZ \leq Y$.

First suppose $Z = \mathbb{Z}^n$. Then Z is a free $\mathbb{Z}$-module of rank n. Hence, Y is a free $\mathbb{Z}$-module of rank at most n [Lang7, p. 146, Thm. 7.1]. Similarly, $n = \text{rank}(Z) = \text{rank}(mZ) \leq \text{rank}(Y)$. Therefore, $\text{rank}(Y) = n$ and $Y \cong \mathbb{Z}^n$.

Next suppose Z is $\hat{\mathbb{Z}}^n$ or $\mathbb{Z}_l^n$. Then mZ is an open subgroup of Z and Y is a union of cosets of mZ, so Y is open.

When $Z = \hat{\mathbb{Z}}^n$ let $\mathcal{N}$ be the set of all subgroups of $\mathbb{Z}^n$ of finite index. When $Z = \mathbb{Z}_l^n$ add the assumption that the index is a power of l. In both cases Z is the completion of $\mathbb{Z}$ with respect to $\mathcal{N}$. Hence, Y is the completion of a subgroup Y_0 of $\mathbb{Z}^n$ of finite index (Lemma 17.2.1). The case $Z = \mathbb{Z}^n$ implies $Y_0 \cong \mathbb{Z}^n$. Therefore, Y is isomorphic to $\hat{\mathbb{Z}}^n$ in the former case and to $\mathbb{Z}_l^n$ in the latter case. □

Denote the number of open subgroups of index n of a profinite group G by $a_n(G)$. Lemma 16.10.2 says $a_n(G)$ is finite if G is finitely generated.

LEMMA 26.3.3: $a_{l^k}(\mathbb{Z}_l^n) \leq l^{nk}$.

Proof: Use induction on k and start with $k = 1$. Each open subgroup of $\mathbb{Z}_l^n$ of index l contains $l\mathbb{Z}_l^n$, so $a_l(\mathbb{Z}_l^n) = a_l(\mathbb{F}_l^n)$. By duality, $a_l(\mathbb{F}_l^n)$ is equal to the number of subspaces of $\mathbb{F}_l^n$ of dimension 1. The latter is $\frac{l^n - 1}{l - 1}$ which is smaller than l^n.

Suppose $k \geq 2$. Consider an open subgroup H of $\mathbb{Z}_l^n$ of index l^k. Then H is contained in an open subgroup M of $\mathbb{Z}_l^n$ of index l^{k-1}. By Lemma 26.3.2, $M \cong \mathbb{Z}_l^n$. By the first paragraph, M has at most l^n open subgroups of index l. By induction, there are at most $l^{n(k-1)}$ possibilities for M. Therefore, there are at most l^{nk} possibilities for H. □

LEMMA 26.3.4: *Suppose $e > n$. Then almost all e-tuples $(\mathbf{a}_1, \ldots, \mathbf{a}_e)$ in $(\mathbb{Z}_l^n)^e$ generate an open subgroup of $\mathbb{Z}_l^n$.*

Proof: For each positive integer k list the open subgroups of $\mathbb{Z}_l^n$ of index l^k

as $H_{k1},\ldots,H_{k,r(k)}$. By Lemma 26.3.3, $r(k) = a_{l^k}(\mathbb{Z}_l^n) \le l^{nk}$. Hence,

$$\sum_{k=1}^\infty \sum_{i=1}^{r(k)} \mu(H_{ki}) = \sum_{k=1}^\infty \frac{r(k)}{l^{ek}} \le \sum_{k=1}^\infty \frac{1}{l^{(e-n)k}} < \infty.$$

Denote the set of all $\mathbf{a} = (\mathbf{a}_1,\ldots,\mathbf{a}_e) \in (\mathbb{Z}_l^n)^e$ such that $\langle \mathbf{a}_1,\ldots,\mathbf{a}_e \rangle \le H_{ki}$ for infinitely many pairs (k,i) by A. By Borel-Cantelli (Lemma 18.3.5(a)), $\mu(A) = 0$. If $\mathbf{a} \in (\mathbb{Z}_l^n)^e \smallsetminus A$, then $\langle \mathbf{a}_1,\ldots,\mathbf{a}_e \rangle$ is contained in only finitely many open subgroups of $\mathbb{Z}_l^n$, so $\langle \mathbf{a}_1,\ldots,\mathbf{a}_e \rangle$ is open. □

THEOREM 26.3.5 ([Kantor-Lubotzky, Prop. 12]): *Given positive integers e and n the following statements hold for an e-tuple $(\mathbf{a}_1,\ldots,\mathbf{a}_e) \in (\hat{\mathbb{Z}}^n)^e$:*
(a) *The probability that $\mathbf{a}_1,\ldots,\mathbf{a}_e$ generate $\hat{\mathbb{Z}}^n$ is $\prod_{i=e-n+1}^e \zeta(i)^{-1}$ (which is positive) if $e > n$ and 0 if $e \le n$. Here $\zeta(s)$ is the Riemann zeta function.*
(b) *The probability that $\mathbf{a}_1,\ldots,\mathbf{a}_e$ generate an open subgroup of $\hat{\mathbb{Z}}^n$ is 1 if $e > n$ and 0 if $e \le n$.*
(c) *If $e < n$, then $\langle \mathbf{a}_1,\ldots,\mathbf{a}_e \rangle$ is of infinite index in $\hat{\mathbb{Z}}^n$.*

Proof of (a): Suppose $e \ge n$. For each l let π_l be the epimorphism of $\hat{\mathbb{Z}}^n$ onto $\mathbb{F}_l^n$ with kernel $N_l = l\hat{\mathbb{Z}}^n$. Put

$$A_l = \{(\mathbf{a}_1,\ldots,\mathbf{a}_e) \in (\hat{\mathbb{Z}}^n)^e \mid \pi_l(\mathbf{a}_1),\ldots,\pi_l(\mathbf{a}_e) \text{ generate } \mathbb{F}_l^n\}.$$

Then, by Lemma 26.3.1,

$$\mu(A_l) = \frac{1}{l^{ne}} \prod_{i=0}^{n-1}(l^e - l^i) = \prod_{i=e-n+1}^{e}\left(1 - \frac{1}{l^i}\right).$$

Put $A = \bigcap_l A_l$. If $\mathbf{a} = (\mathbf{a}_1,\ldots,\mathbf{a}_e) \in A$, then for each l we have $\langle \mathbf{a}, N_l \rangle = \hat{\mathbb{Z}}^n$, so $\langle \mathbf{a} \rangle$ is contained in no open subgroup of index l. Hence, $\langle \mathbf{a} \rangle$ is contained in no proper open subgroup of $\hat{\mathbb{Z}}^n$. Therefore, $\langle \mathbf{a} \rangle = \hat{\mathbb{Z}}^n$. Conversely, the latter condition implies $\mathbf{a} \in A$.

To compute the measure of A suppose first $e > n$. As l ranges on all prime numbers, the indices $l^{en} = (\hat{\mathbb{Z}} : N_l)$ are relatively prime in pairs. Hence, the A_l are μ-independent (Example 18.3.8). Therefore,

$$(1) \qquad \mu(A) = \prod_{i=e-n+1}^{e} \prod_l \left(1 - \frac{1}{l^i}\right) = \prod_{i=e-n+1}^{n} \zeta(i)^{-1}.$$

In particular, since each i appearing in (1) is at least 2, the ith factor is positive.

Suppose $e = n$. Then, in the above notation, $\mu(A) = \prod_{i=1}^e \prod_l (1 - \frac{1}{l^i})$. Since $\prod_l (1 - \frac{1}{l}) = 0$ [LeVeque, p. 99, Thm. 6-11], $\mu(A) = 0$.

26.3 Random e-tuples in $\hat{\mathbb{Z}}^n$

Finally, if $e < n$, then each A_l is empty, so $\mu(A) = 0$.

Proof of (b): First suppose $e > n$. Let $A' = \bigcup_{m=1}^{\infty} \bigcap_{l \geq m} A_l$ be the set of all $(\mathbf{a}_1, \ldots, \mathbf{a}_e) \in (\hat{\mathbb{Z}}^n)^e$ which belong to all but finitely many A_l's. By the proof of (a), $\mu(\bigcap_l A_l) \neq 0$. Hence, by Borel-Cantelli (Lemma 18.3.5(c)), $\mu(A') = 1$.

Next denote the projection of $\hat{\mathbb{Z}}^n$ onto $\hat{\mathbb{Z}}_l^n$ by $\hat{\pi}_l$. Let A_l'' be the set of all $(\mathbf{a}_1, \ldots, \mathbf{a}_e) \in (\hat{\mathbb{Z}}^n)^e$ with $\langle \hat{\pi}_l(\mathbf{a}_1), \ldots, \hat{\pi}_l(\mathbf{a}_e) \rangle$ open in $\hat{\mathbb{Z}}_l^n$. Put $A'' = \bigcap_l A_l''$. By Lemma 26.3.4, $\mu(A_l'') = 1$ for each l, so $\mu(A'') = 1$. Consequently, $\mu(A' \cap A'') = 1$.

Consider $\mathbf{a} = (\mathbf{a}_1, \ldots, \mathbf{a}_e) \in A' \cap A''$. Put $H = \langle \mathbf{a}_1, \ldots, \mathbf{a}_e \rangle$. Assume $(\hat{\mathbb{Z}}^n : H) = \infty$. By Lemma 22.8.4, H is contained in an open subgroup of $\hat{\mathbb{Z}}^n$ of index l for infinitely many l or there is an l such that H is contained in a closed subgroup N of $\hat{\mathbb{Z}}^n$ with $\hat{\mathbb{Z}}^n/N \cong \mathbb{Z}_l$. The former case contradicts $\mathbf{a} \in A'$. The latter contradicts $\mathbf{a} \in A''$. It follows that $(\hat{\mathbb{Z}}^n : H) < \infty$.

Finally, suppose $e \leq n$. Then $\prod_l \mu(A_l) = 0$ (proof of (a)). Since the A_l's are μ-independent, almost all $\mathbf{a} \in (\hat{\mathbb{Z}}^n)^e$ belong to infinitely many $\hat{\mathbb{Z}}^n \smallsetminus A_l$ (Lemma 18.3.5(d)). Suppose $\mathbf{a}$ is one of them. For each l with $\mathbf{a} \in \hat{\mathbb{Z}} \smallsetminus A_l$ the group $\langle \mathbf{a} \rangle$ is contained in an open subgroup of index l. Since there are infinitely many such l's, $(\hat{\mathbb{Z}}^n : \langle \mathbf{a} \rangle) = \infty$. This completes the proof of (b).

Proof of (c): Suppose $e < n$. Then, for each l, $\langle \mathbf{a}_1, \ldots, \mathbf{a}_e \rangle$ is contained in all open subgroups of $\hat{\mathbb{Z}}^n$ of index l. Hence, $\langle \mathbf{a}_1, \ldots, \mathbf{a}_e \rangle$ has an infinite index in $\hat{\mathbb{Z}}^n$. □

THEOREM 26.3.6 ([Jarden-Lubotzky2, Thm. 3.1]): *Let e, n be positive integers. Then $\langle \mathbf{a}_1, \ldots, \mathbf{a}_e \rangle \cong \hat{\mathbb{Z}}^{\min(e,n)}$ for almost all $(\mathbf{a}_1, \ldots, \mathbf{a}_e) \in (\hat{\mathbb{Z}}^n)^e$.*

Proof: We put $A = \hat{\mathbb{Z}}^n$ and divide the proof into three parts.

PART A: $e > n$. By Theorem 26.3.5(b), for almost all $\mathbf{a} \in A^e$ the group $\langle \mathbf{a} \rangle$ is open in A. By Lemma 26.3.2, $\langle \mathbf{a} \rangle \cong \hat{\mathbb{Z}}^n$.

PART B: $e = n$. For each prime number p consider the quotient group $A_p = \mathbb{Z}_p^n$ of A.

CLAIM: *For almost all $\mathbf{a} \in A_p^n$ we have $\langle \mathbf{a} \rangle \cong \mathbb{Z}_p^n$.* Consider each $\mathbf{a} \in A_p^n$ as a column of height n whose ith entry is a row $\mathbf{a}_i = (a_{i1}, a_{i2}, \ldots, a_{in})$ of elements of $\mathbb{Z}_p$. In this way identify $\mathbf{a}$ with an $n \times n$ matrix with entries in $\mathbb{Z}_p$. Then $\langle \mathbf{a} \rangle$ is a free $\mathbb{Z}_p$-module of rank which is equal to the rank of the matrix $\mathbf{a}$. Thus, $\langle \mathbf{a} \rangle \cong \mathbb{Z}_p^n$ if and only if $\det(\mathbf{a}) \neq 0$. By Lemma 26.2.1, the latter condition holds for a subset of $M_n(\mathbb{Z}_p)$ of measure 1. Therefore, $\langle \mathbf{a} \rangle \cong \mathbb{Z}_p^n$ for almost all $\mathbf{a} \in A_p^n$.

By the Claim, for almost all $\mathbf{a} \in A^n$ each Abelian group which is generated by n elements is a quotient of $\langle \mathbf{a} \rangle$. Since $\langle \mathbf{a} \rangle$ is generated by n elements, Lemma 17.7.1 implies $\langle \mathbf{a} \rangle \cong \hat{\mathbb{Z}}^n$.

PART C: $e < n$. Consider now each $\mathbf{a} = (\mathbf{a}_1, \ldots, \mathbf{a}_n) \in A^n$ as a pair $\mathbf{a} = (\mathbf{b}, \mathbf{c})$, where $\mathbf{b} = (\mathbf{a}_1, \ldots, \mathbf{a}_e) \in A^e$ and $\mathbf{c} = (\mathbf{a}_{e+1}, \ldots, \mathbf{a}_n) \in A^{n-e}$.

By Part B,

(2) $$\langle \mathbf{a} \rangle \cong \hat{\mathbb{Z}}^n,$$

holds for almost all $\mathbf{a} \in A^n$. Hence, by Fubini's theorem [Halmos, p. 147, Thm. A], for almost all $\mathbf{b} \in A^e$ the set of $\mathbf{c} \in A^{n-e}$ such that (2) holds for $\mathbf{a} = (\mathbf{b}, \mathbf{c})$ has measure 1.

Choose $\mathbf{c} \in A^{n-e}$ for which (2) holds. For each Abelian profinite group B and each $(b'_1, \ldots, b'_e) \in B^e$, we may extend the map $(\mathbf{b}_1, \ldots, \mathbf{b}_e) \mapsto (b'_1, \ldots, b'_e)$ to a map of $\mathbf{a}$ into B (say $\mathbf{a}_i \mapsto 0$, $i = e+1, \ldots, n$), hence to a homomorphism $h \colon \langle \mathbf{a} \rangle \to B$. The restriction of h to $\langle \mathbf{b} \rangle$ is a homomorphism into B. Consequently, $\langle \mathbf{b} \rangle \cong \hat{\mathbb{Z}}^e$. $\square$

Remark 26.3.7: Positively generated groups. Let G be a profinite group and k a positive integer. Put $A_k = \{(x_1, \ldots, x_k) \in G^k \mid \langle x_1, \ldots, x_k \rangle = G\}$. Note that a k-tuple $(x_1, \ldots, x_k) \in G^k$ belongs to A_k if and only if $\langle x_1, \ldots, x_k \rangle N = G$ for each open normal subgroup N of G, so A_k is closed. We say that G is **positively finitely generated** if there is a positive integer k with $\mu(A_k) > 0$. In this case G is finitely generated.

Free profinite groups of rank at least 2 are not positively finitely generated (Proposition 26.1.7(a)). In contrast, $\hat{F}_e(p)$ is positively finitely generated. Indeed, (2) of Section 26.2 gives an explicit positive value for $\mu(A_e)$. By Theorem 26.3.5(a), $\hat{\mathbb{Z}}^n$ is positively finitely generated. More generally, every finitely generated profinite group G which contains an open prosolvable group is positively finitely generated [Mann, Thm. 10]. In particular, every closed subgroup G of $\mathrm{GL}(n, \mathbb{Z}_p)$ is positively finitely generated. Indeed, G is finitely generated (Proposition 22.14.4) and $\{g \in G \mid g \equiv 1 \bmod p\}$ is an open pro-p (hence prosolvable) subgroup of G (Lemma 22.14.2(d)).

Finally, for each profinite group and a positive integer n let $m_n(G)$ be the number of maximal open subgroups of G of index n. By [Lubotzky-Segal, Thm. 11.1], G is positively finitely generated if and only if there is a $c > 0$ and a positive integer r with $m_n(G) \leq cn^r$ for all n. $\square$

26.4 On the Index of Normal Subgroups Generated by Random Elements

Let G be a profinite group. For each $\boldsymbol{\sigma} = (\sigma_1, \ldots, \sigma_e) \in G^e$ let $[\boldsymbol{\sigma}] = [\boldsymbol{\sigma}]_G$ be the closed normal subgroup of G generated by $\sigma_1, \ldots, \sigma_e$ (Section 25.9). In other words, $[\boldsymbol{\sigma}]$ is the intersection of all closed normal subgroups of G which contain $\langle \boldsymbol{\sigma} \rangle$. Observe that if H is a closed normal subgroup of G and $\boldsymbol{\sigma} \in H^e$, then $[\boldsymbol{\sigma}]_H \leq [\boldsymbol{\sigma}]_G \leq H$ but it may happen that $[\boldsymbol{\sigma}]_H < [\boldsymbol{\sigma}]_G$. In addition note that if $\varphi \colon G \to A$ is an epimorphism, then $\varphi([\boldsymbol{\sigma}]) = [\varphi(\boldsymbol{\sigma})]$. Finally, if G is Abelian, then $[\boldsymbol{\sigma}] = \langle \boldsymbol{\sigma} \rangle$.

26.4 On the Index of Normal Subgroups Generated by Random Elements

The purpose of this section is to study the index of $[\boldsymbol{\sigma}]$ in $\hat{F}_n$ when $\boldsymbol{\sigma}$ is taken at random in $\hat{F}_n^e$.

For a profinite group G and a closed normal subgroup H we set:

$$B_e(G,H) = \{\boldsymbol{\sigma} \in G^e \mid [\boldsymbol{\sigma}]_G = H\},$$
$$B_e(G) = B_e(G,G) = \{\boldsymbol{\sigma} \in G^e \mid [\boldsymbol{\sigma}]_G = G\}$$
$$C_e(G) = \{\boldsymbol{\sigma} \in G^e \mid (G : [\boldsymbol{\sigma}]) = \infty\}$$

LEMMA 26.4.1: *Let G be a profinite group and let H be a closed normal subgroup.*
(a) *$B_e(G, H)$ is closed subset of G^e.*
(b) *Suppose $\operatorname{rank}(G) \leq \aleph_0$. Then $C_e(G)$ is a measurable subset of G^e.*

Proof of (a): For each open normal subgroup N of G, $\{\boldsymbol{\sigma} \in G^e \mid [\boldsymbol{\sigma}]_G N = HN\}$ is an open-closed subset of G. The intersection of all these sets is $B_e(G,H)$. Hence, $B_e(G,H)$ is closed.

Proof of (b): We have $C_e(G) = G^e \smallsetminus \bigcup_H B_e(G,H)$, where H ranges over all open normal subgroup of G. By Proposition 17.1.2, G has at most $\aleph_0$ open subgroups. By (a), each $B_e(G,H)$ is closed. Hence, C_e is measurable. □

The **relation rank** of a finite p-group G of rank d is defined as the minimal number r for which there exists a short exact sequence $1 \to [y_1,\ldots,y_r] \to \hat{F}_d(p) \to G \to 1$. The relation rank and the rank of G are related by the **Golod-Shafarevich inequality**: $r > \frac{1}{4}d^2$ [Huppert, p. 395, Satz 18.1]. This ensures that $C_e(G)$ has positive measure (Theorem 25.4.5):

LEMMA 26.4.2: *Let $n \geq 2$ and $e \geq 1$ be integers. Put $P = \hat{F}_n(p)$. Then, $\mu(C_e(P)) > 0$.*

Proof: Assume $\mu(C_e(P)) = 0$. Take an open normal subgroup Q of P of index m such that

(1) $$em \leq \frac{1}{4}(m(n-1)+1)^2.$$

Then $\Phi(Q)$ is open in P, so $\mu(\Phi(Q)^e) > 0$. Hence, we may choose $\boldsymbol{\sigma} \in \Phi(Q)^e \smallsetminus C_e(P)$. Then $[\boldsymbol{\sigma}]_P$ is an open normal subgroup of P, hence also of Q. Therefore, $G = Q/[\boldsymbol{\sigma}]_P$ is a finite p-group.

By Nielsen-Schreier (Proposition 17.6.2), $\operatorname{rank}(Q) = m(n-1)+1$. Since $[\boldsymbol{\sigma}]_P \leq \Phi(Q)$, the quotient map $Q \to G$ is a Frattini cover, so $\operatorname{rank}(G) = \operatorname{rank}(Q) = m(n-1)+1$ (Corollary 22.5.3).

On the other hand, let $\tau_1,\ldots,\tau_m$ be representatives for the left cosets of P modulo Q. Then $[\boldsymbol{\sigma}]_P = [\sigma_i^{\tau_j} \mid i = 1,\ldots,e,\ j = 1,\ldots,m]_Q$, so relation.$\operatorname{rank}(G) \leq em$. By Golod-Shafarevich,

$$\text{relation.rank}(G) > \frac{1}{4}\operatorname{rank}(G)^2,$$

so $em > \frac{1}{4}(m(n-1)+1)^2$. This contradiction to (1) proves that $\mu(C_e(P)) > 0$.
□

The following result overlaps with Lemma 18.3.11:

LEMMA 26.4.3: *Let G be a profinite group, e a positive integer, and M_i, N_j, $i,j = 1,2,3,\ldots$ distinct open normal subgroups. Suppose G/M_i is a simple Abelian group, $i = 1,2,3,\ldots$, and G/N_j is a simple non-Abelian group, $j = 1,2,3,\ldots$. Then $\bigcup_{i=1}^{\infty} M_i^e, N_1^e, N_2^e, N_2^e, \ldots$ are μ-independent.*

Proof: Let $i_1 < \cdots < i_k$ and n be positive integers. Put $M = M_{i_1}^e \cap \cdots \cap M_{i_k}^e$ and $N = N_1^e \cap \cdots \cap N_n^e$. Then no composition factor of G^e/N is Abelian. Hence, $(G^e : M \cap N) = (G^e : M)(G^e : N)$.

Next let m be a positive integer. Apply the inclusion-exclusion principle (Lemma 18.3.1(a)) and the preceding paragraph to the sets $M_1^e \cap N^e, \ldots, M_m^e \cap N^e$:

$$\mu(\bigcup_{i=1}^{m} M_i^e \cap N^e) = \sum_{k=1}^{m}(-1)^{k-1} \sum_{1 \leq i_1 < \cdots < i_k \leq m} \mu(M_{i_1}^e \cap \cdots \cap M_{i_k}^e \cap N^e)$$

$$= \sum_{k=1}^{m}(-1)^{k-1} \sum_{1 \leq i_1 < \cdots < i_k \leq m} \mu(M_{i_1}^e \cap \cdots \cap M_{i_k}^e)\mu(N^e)$$

$$= \mu(\bigcup_{i=1}^{m} M_i^e)\mu(N^e).$$

Taking the limit on m, we conclude that $\mu(\bigcup_{i=1}^{\infty} M_i^e \cap N) = \mu(\bigcup_{i=1}^{\infty} M_i^e)\mu(N)$. Finally, by Example 18.3.10, $N_1^e, \ldots, N_n^e$ are μ-independent, so

$$\mu(\bigcup_{i=1}^{\infty} M_i^e \cap N_1^e \cap \cdots \cap N_n^e) = \mu(\bigcup_{i=1}^{\infty} M_i^e)\mu(N_1^e) \cap \cdots \cap \mu(N_n)^e.$$

This implies the statement of the Lemma. □

LEMMA 26.4.4: *Let n be a positive integer and $\varphi \colon \hat{F}_n \to \hat{\mathbb{Z}}^n$ an epimorphism. Then every open normal subgroup of $\hat{F}_n$ of prime index contains $\mathrm{Ker}(\varphi)$.*

Proof: Let N be an open normal subgroup of $\hat{F}_n$ of a prime index p. Choose an epimorphism $\pi \colon \hat{\mathbb{Z}}^n \to \mathbb{F}_p^n$ and let $\psi = \pi \circ \varphi$. It suffices to prove that $\mathrm{Ker}(\psi) \subseteq N$.

Assume $\mathrm{Ker}(\psi)$ is not contained in N. Then $\hat{F}_n/(\mathrm{Ker}(\psi) \cap N) \cong \mathbb{F}_p^{n+1}$, which is impossible, because the rank of each quotient of $\hat{F}_n$ is at most n. Thus, $\mathrm{Ker}(\psi) \subseteq N$. □

THEOREM 26.4.5 ([Jarden-Lubotzky2, Thm. 1.4]): *Let $n \geq 2$ be a positive integer. Put $F = \hat{F}_n$.*

26.4 On the Index of Normal Subgroups Generated by Random Elements

(a1) If $e \leq n$, then $\mu(B_e(F)) = 0$. In fact, if $e < n$, then $B_e(F) = \emptyset$ while if $e = n$, then $B_e(F) \neq \emptyset$.
(a2) If $e > n$, then $0 < \mu(B_e(F)) < 1$.
(b1) If $e \leq n$, then $\mu(C_e(F)) = 1$. In fact, if $e < n$, then $C_e(F) = F^e$ while if $e = n$, then $C_e(F) \neq F^e$.
(b2) If $e > n$, then $0 < \mu(C_e(F)) < 1$.

Proof: The proof has five parts:

PART A: In each case $\mu(C_e(F)) > 0$. Choose a prime number p. Let $P = \hat{F}_n(p)$ be the free pro-p group on n generators and $\pi\colon F \to P$ be an epimorphism. If $\boldsymbol{\sigma} \in F^e$, then $\pi([\boldsymbol{\sigma}]) = [\pi(\boldsymbol{\sigma})]$. Hence, if $[\pi(\boldsymbol{\sigma})]$ has infinite index in P, then $[\boldsymbol{\sigma}]$ has infinite index in F. Thus, $\pi^{-1}(C_e(P)) \subseteq C_e(F)$. By Lemma 26.4.2, $\mu(C_e(P)) > 0$. Hence, $\mu(C_e(F)) > 0$.

PART B: If $e < n$, then $(F : [\boldsymbol{\sigma}]) = \infty$ for each $\boldsymbol{\sigma} \in F^e$. Again, choose a prime number p and let $\varphi\colon F \to \mathbb{Z}_p^n$ be an epimorphism. Each open subgroup of $\mathbb{Z}_p^n$ is isomorphic to $\mathbb{Z}_p^n$ (Lemma 26.3.2), hence can not be generated by less than n elements. Let $\boldsymbol{\sigma}$ be an arbitrary e-tuple of F. Since $\varphi([\boldsymbol{\sigma}]) = [\varphi(\boldsymbol{\sigma})] = \langle\varphi(\boldsymbol{\sigma})\rangle$, the index of $\varphi([\boldsymbol{\sigma}])$ in $\mathbb{Z}_p^n$ is infinite. Consequently, $(F : [\boldsymbol{\sigma}]) = \infty$.

PART C: If $e = n$, then $\mu(C_e(F)) = 1$. Choose an epimorphism $\varphi\colon F \to \hat{\mathbb{Z}}^n$. By Theorem 26.3.5(b), $[\boldsymbol{\sigma}] = \langle\boldsymbol{\sigma}\rangle$ has an infinite index in $\hat{\mathbb{Z}}^n$ for almost all $\boldsymbol{\sigma} \in (\hat{\mathbb{Z}}^n)^n$. Hence, $(F : [\boldsymbol{\sigma}]) = \infty$ for almost all $\boldsymbol{\sigma} \in F^n$.

PART D: If $e > n$, then $\mu(B_e(F)) > 0$. We have to prove that

$$\mu(F^e \smallsetminus B_e(F)) < 1.$$

Indeed, $\boldsymbol{\sigma} \in F^e \smallsetminus B_e(F)$ if and only if $\sigma_1, \ldots, \sigma_e$ belong to a normal subgroup N of G with G/N simple. Thus,

$$F^e \smallsetminus B_e(F) = \bigcup_{p} \bigcup_{F/N \cong \mathbb{Z}/p\mathbb{Z}} N^e \cup \bigcup_{S \in \text{SNA}} \bigcup_{F/N \cong S} N^e.$$

Here SNA is the set of all non-Abelian finite simple groups.

Recall that if A and B are independent subsets of a probability space, then $1 - \mu(A \cup B) = (1 - \mu(A))(1 - \mu(B))$. Hence, $\mu(A \cup B) < 1$ if and only if $1 - \mu(A) > 0$ and $1 - \mu(B) > 0$. Also, if B is a union of a sequence $B_1, B_2, B_3, \ldots$ of independent sets, then $1 - \mu(B) = \prod_{j=1}^{\infty}(1 - \mu(B_j))$. Thus, $\mu(B) < 1$ if and only if $\mu(B_j) < 1$ for each j and $\sum_{j=1}^{\infty} \mu(B_j) < \infty$ (Lemma 18.3.2).

Let in our case $A = \bigcup_p \bigcup_{F/N \cong \mathbb{Z}/p\mathbb{Z}} N^e$ and let B_j range over all N^e with $F/N \in \text{SNA}$. By Lemma 26.4.3, $A, B_1, B_2, B_3, \ldots$ are μ-independent. Moreover, as in Part C, let $\varphi\colon F \to \hat{\mathbb{Z}}^n$ be an epimorphism. By Lemma 26.4.4, each open normal subgroup N of F with $F/N \cong \mathbb{Z}/p\mathbb{Z}$ contains

Ker(φ). Therefore, $\mu(A)$ is equal to the measure of all $\sigma \in (\hat{\mathbb{Z}}^n)^e$ which are contained in a maximal subgroup of a prime index. The latter measure is equal to the measure of all $\sigma \in (\hat{\mathbb{Z}}^n)^e$ which do not generate $\hat{\mathbb{Z}}^n$. Hence, by Theorem 26.3.5(a), $1 - \mu(A) = \prod_{e-n < i \le e} \zeta(i)^{-1} > 0$, where ζ is the Riemann zeta function.

To prove the condition on the B_j's, let $S \in SNA$. By Example 18.3.11, the sets N^e, with N ranging over SNA are μ-independent. For each $S \in $ SNA, $d_n(S)$ is the number of open normal subgroups N of F with $F/N \cong S$ (Lemma 26.1.2(a)). By Lemma 26.1.2(c), $d_n(S) \le |S|^{n-1}$. Hence,

$$(2) \qquad \sum_{S \in SNA} \sum_{F/N \cong S} \mu(N^e) = \sum_{S \in SNA} \frac{d_n(S)}{|S|^e}$$

$$\le \sum_{S \in SNA} \frac{1}{|S|^{e+1-n}} \le \sum_{S \in SNA} \frac{1}{|S|^2} < \infty.$$

The last inequality holds because for each positive integer n there are at most two simple groups of order n ([Kimmerle-Lyons-Sandling-Teaque, Thm. 5.1] proves this result by using the classification of finite simple groups). Consequently, $\mu(C) > 0$.

PART E: *Conclusion of the proof.* If $e < n$, then F is not generated by e elements. Hence, $B_e(F) = \emptyset$. If $e = n$, then F is generated by e elements and therefore $B_e(F) \neq \emptyset$. However, since $B_e(F) \subseteq F^e \smallsetminus C_e(F)$, Part C implies that $\mu(B_e(F)) = 0$. This concludes the proof of (a1).

Part D of the proof takes care of the left inequality of (a2). In order to prove also the right inequality of (a2) we take a proper open normal subgroup E of F. Then $E^e \subseteq F^e \smallsetminus B_e(F)$ and $\mu(E^e) > 0$. Therefore, $\mu(B_e(F)) < 1$.

Parts B and C give (b1).

Finally, if $e > n$, then, by Part D, $\mu(F^e \smallsetminus C_e(F)) \ge \mu(B_e(F)) > 0$. Together with Part A we get $0 < \mu(C_e(F)) < 1$. □

Remark 26.4.6: Note that the only application of the classification of simple groups in the proof of Theorem 26.4.5 occurs in the proof of Part D, and therefore in the proof of the inequalities $\mu(B_e(F)) > 0$ and $\mu(C_e(F)) < 1$. In addition, we apply the classification of finite simple groups in Proposition 26.4.7(b) below. □

If $e > n$, then $[\sigma]$ is, with a positive probability, of infinite index. Nevertheless, as the next result shows, the quotient $F/[\sigma]$ is small compared to F.

PROPOSITION 26.4.7: *Let $e > n$. Then almost all $\sigma \in F^e$ have the following properties:*
(a) *The maximal Abelian quotient of $F/[\sigma]$ is finite.*
(b) *There are only finitely many open maximal normal subgroups N of F which contain $[\sigma]$ such that F/N is simple and non-Abelian.*

Proof of (a): Choose an epimorphism $\varphi\colon F \to \hat{\mathbb{Z}}^n$. Then, for each $\boldsymbol{\sigma} \in f^e$, the maximal Abelian quotient of $F/[\boldsymbol{\sigma}]$ is $\hat{\mathbb{Z}}^n/\langle\varphi(\boldsymbol{\sigma})\rangle$. By Lemma 26.3.5(b), $\langle\boldsymbol{\tau}\rangle$ has a finite index in $\hat{\mathbb{Z}}^n$ for almost all $\boldsymbol{\tau} \in (\hat{\mathbb{Z}}^n)^e$. Hence, for almost all $\boldsymbol{\sigma} \in F^e$, the maximal Abelian quotient of $F/[\boldsymbol{\sigma}]$ is finite.

Proof of (b): By (2) and Borel-Cantelli, almost all $\boldsymbol{\sigma} \in F^e$ belong to only finitely many open normal subgroups S with $S/N \in \text{SNA}$. This proves (b). □

26.5 Freeness of Normal Subgroups Generated by Random Elements

Let $e \geq 1$ and $n \geq 2$ be integers. Put $F = \hat{F}_n$. We prove here that $[\boldsymbol{\sigma}]$ is a free profinite group for almost all $\boldsymbol{\sigma} \in F^e$. The case where the index is finite is well known. Therefore, it suffices to prove that $[\boldsymbol{\sigma}] \cong \hat{F}_\omega$ for almost all $\boldsymbol{\sigma} \in F^e$ which satisfy $(F : [\boldsymbol{\sigma}]) = \infty$. A basic tool is the following special case of Corollary 25.7.6:

LEMMA 26.5.1: *Let N be a closed normal subgroup of F satisfying:*
(1a) $\mathbb{Z}/p\mathbb{Z}$ *is a quotient of N for every p.*
(1b) S^q *is a quotient of N for every finite simple group S and every positive integer q.*
Then $N \cong \hat{F}_\omega$.

LEMMA 26.5.2: *For almost all $\boldsymbol{\sigma} \in F^e$ and for each prime number p the group $\mathbb{Z}/p\mathbb{Z}$ is a quotient of $[\boldsymbol{\sigma}]$.*

Proof: For each p let $\varphi_p\colon F \to \hat{F}_n(p)$ be an epimorphism. Then $\text{Ker}(\varphi_p)$ has an infinite index in F, hence measure 0. It follows that $U = \bigcup_p \text{Ker}(\varphi_p)^e$ is a zero subset of F^e. If $\boldsymbol{\sigma} \notin U$, then for each p, $\varphi_p[\boldsymbol{\sigma}]$ is a nontrivial pro-p-group. Therefore, $\mathbb{Z}/p\mathbb{Z}$ is a quotient of $\varphi_p[\boldsymbol{\sigma}]$, hence of $[\boldsymbol{\sigma}]$. □

Lemma 26.5.2 settles Condition (1a) of Lemma 26.5.1. The proof of Condition (1b) uses the notion the "S-rank" of a profinite group (Section 24.9).

For a profinite group G and a closed subgroup H let $\mathcal{N}(G, H)$ be the family of all closed normal subgroups M of G satisfying $HM = G$. Denote the intersection of all $M \in \mathcal{N}(G, H)$ by $\mathbf{N}(G, H)$.

The following rule is a group theoretic analog of the tower property of linear disjointness of fields (Lemma 2.5.3):

RULE 26.5.3: *Let G be a profinite group, H an open subgroup of G, and M and N closed normal subgroups of G such that $M \leq N$. Then $HM = G$ if and only if $HN = G$ and $(H \cap N)M = N$.*

LEMMA 26.5.4: *Let G be a profinite group, S a simple non-Abelian group, k a positive integer, and H an open normal subgroup of G with $G/H \cong S^k$. Then, $H \cdot \mathbf{N}(G, H) = G$.*

Proof: By Lemma 1.2.2(b), it suffices to prove $\mathcal{N}(G,H)$ is closed under finite intersections. So, let M and N be open normal subgroups of G with $HM = HN = G$. By induction on k we prove $H(M \cap N) = G$.

Consider first the case $k = 1$. By assumption, $HM/H = G/H \cong S$ is a non-Abelian simple group. Hence, there exist $m, m' \in M$ with $[m, m'] \notin H$. Since $HN = G$ there exist $h \in H$ and $n \in N$ with $m' = hn$. Using the identity $[m, m'] = [m, hn] = [m, n][m, h]^n$ and the relation $[m, h]^n \in H$, we conclude that $[m, n] \notin H$. On the other hand, $[m, n] \in M \cap N$. Hence, $M \cap N \not\leq H$. It follows that $H(M \cap N) = G$, because $F/H \cong S$ is simple.

Now suppose $k > 1$ and the statement holds for $k - 1$. Then G has an open normal subgroup E containing H such that $G/E \cong S$ and $E/H \cong S^{k-1}$. It satisfies, $EM = EN = G$. By Rule 26.5.3, $H(E \cap M) = E$ and $H(E \cap N) = E$. By the induction hypothesis, applied to E instead of G, we have $H \cdot (E \cap M \cap N) = E$. By the case $k = 1$, $E(M \cap N) = G$. Hence, by Rule 26.5.3, $H(M \cap N) = G$. □

Let S be a finite simple group. Recall (Remark 17.4.7) that a pro-S group is a profinite group whose only composition factor is S. For an arbitrary profinite group G, we denote the intersection of all open normal subgroups M of G with $G/M \cong S$ by $M_G(S)$. It satisfies $G/M_G(S) \cong S^{r_G(S)}$ (Section 24.9).

Let F be a free pro-S group and N a nontrivial closed subgroup of F of infinite index. If $S = \mathbb{Z}/p\mathbb{Z}$, then $\text{rank}(N) = \infty$ (Proposition 24.10.4(a)), so $r_N(S) = \infty$. It is quite surprising that for non-Abelian S there exists N with $r_N(S) < \infty$. This will follow from Lemma 26.5.5(a). Nevertheless, there are only countably many such N (Lemma 26.5.5(d)).

LEMMA 26.5.5: *Let G be a profinite group, S a simple non-Abelian group, and k a nonnegative integer.*

(a) *Let $H' \leq G'$ be open normal subgroups of G with $G'/H' \cong S^k$. Then $\mathbf{N}(G', H') \triangleleft G$ and $r_{\mathbf{N}(G',H')}(S) = k$.*

(b) *Suppose N is a closed normal subgroup of G with $r_N(S) = k$. Then there exist open normal subgroups $H' \leq G'$ of G with $G'/H' \cong S^k$, $H'N = G'$, and $H' \cap N = M_N(S)$.*

(c) *Suppose G is a pro-S-group. Let N be a closed normal subgroup of G with $r_N(S) = k$. Then there exists open normal subgroups $H' \leq G'$ of G with $N = \mathbf{N}(G', H')$ and $G'/H' \cong S^k$.*

(d) *Suppose G is a pro-S-group. The number of closed normal subgroups of G with $r_N(S) < \infty$ is bounded by $\max(\aleph_0, \text{rank}(G))$. In particular, if $\text{rank}(G) \leq \aleph_0$, then there are at most countably many closed normal subgroups N of G with $r_N(S) < \infty$.*

Proof of (a): Let $N = \mathbf{N}(G', H')$ and $M = H' \cap N$. The family $\mathcal{N}(G', H')$ is closed under conjugation by elements of G. Hence, $N \triangleleft G$.

By Lemma 26.5.4, $H'N = G'$. Hence, $N/M \cong G'/H' \cong S^k$. Therefore, $r_N(S) \geq k$. If $r_N(S) > k$, then N has an open normal subgroup N_0 such

26.5 Freeness of Normal Subgroups Generated by Random Elements

that $N/N_0 \cong S$ and $M \not\leq N_0$. Since S is simple, this implies $MN_0 = N$. The group N_0 need not be normal in G'. So, consider $N_1 = \mathbf{N}(N, M)$. Then $N_1 \leq N_0$. By the preceding paragraph, $N_1 \triangleleft G'$. By Lemma 26.5.4, $MN_1 = N$ and therefore $H'N_1 = G'$. By the definition of N, this implies that $N \leq N_1 \leq N_0 < N$. We conclude from this contradiction that $r_N(S) = k$.

Proof of (b): Since $N \triangleleft G$, the set of all open $N' \triangleleft N$ such that $N/N' \cong S$ is closed under conjugation by elements of G. Hence, $M_N(S)$ is normal in G and open in N. Choose an open normal subgroup H' of G with $H' \cap N = M_N(S)$. Put $G' = H'N$. Then $G'/H' \cong N/M_N(S) \cong S^k$, as required.

Proof of (c): Put $M = M_N(S)$. By assumption, $N/M \cong S^k$. Let G' and H' be as in (b). Then, $N \in \mathcal{N}(G', H')$. Hence, $N_0 = \mathbf{N}(G', H') \leq N$. In particular, N_0 is a closed normal subgroup of N with $H'N_0 = G'$ (Lemma 26.5.4), hence $MN_0 = N$. Since G is a pro-S group, so is N/N_0. If $N_0 < N$, then N contains an open normal subgroup N_1 which contains N_0 with $N/N_1 \cong S$. In particular, $MN_1 = N$. On the other hand, by definition of M, we have $M \leq N_1$, hence $N_1 = N$. This contradiction implies $N_0 = N$.

Proof of (d): The cardinality of the set of all open normal subgroups of G is at most $\max(\aleph_0, \mathrm{rank}(G))$ (Proposition 17.1.2). Now use (c). $\square$

Let F be a profinite group. Then $C_e(F) = \{\sigma \in F^e \mid (F : [\sigma]) = \infty\}$.

THEOREM 26.5.6 ([Jarden-Lubotzky2, Thm. 2.7]): *Let $e \geq 1$ and $n \geq 2$ be integers. Put $F = \hat{F}_n$. Then $[\sigma] \cong \hat{F}_\omega$ for almost all $\sigma \in C_e(F)$.*

Proof: By Lemmas 26.5.1 and 26.5.2 it suffices to consider a non-Abelian simple group S and to prove that $r_{[\sigma]}(S) = \infty$ for almost all $\sigma \in C_e(F)$.

To this end denote the set of all open normal subgroups of F by $\mathcal{E}$. Each $E \in \mathcal{E}$ is a free profinite group and $\mathrm{rank}(E) = 1 + (F : E)(n-1)$ (Proposition 17.6.2). Hence, there is an epimorphism h_E of E onto the free pro-S-group with $\mathrm{rank}(E)$ generators. Denote the latter group by $\bar{E}$. If $\sigma \in E^e$, then $[\sigma] = [\sigma]_F$ is a normal subgroup of E. Hence, $h_E([\sigma])$ is a normal subgroup of $\bar{E}$. (Nevertheless, since $[\sigma]$ may properly contain $[\sigma]_E$, we may have $[h_E(\sigma)]_{\bar{E}} < h_E([\sigma])$.) Let

$$C_e(F, E) = \{\sigma \in C_e(F) \cap E^e \mid (\bar{E} : h_E([\sigma])) = \infty\} \qquad C = \bigcup_{E \in \mathcal{E}} C_e(F, E).$$

The proof will be concluded, once we prove Claims A and B below:

CLAIM A: $r_{[\sigma]}(S) = \aleph_0$ *for each $\sigma \in C_e(F) \smallsetminus C$.* Consider $\sigma \in C_e(F) \smallsetminus C$. Then $(F : [\sigma]) = \infty$ and $h_E([\sigma])$ is open normal in $\bar{E}$ for each open normal subgroup E of F that contains $[\sigma]$. By Proposition 17.6.2, $h_E([\sigma])$ is a free pro-S-group. Moreover,

$$\bar{r} = \mathrm{rank}(h_E([\sigma])) \geq \mathrm{rank}(\bar{E}) = \mathrm{rank}(E) = 1 + (F : E)(n-1).$$

Let $m = \text{rank}(S)$. By the remarks preceding Lemma 26.5.5 and by Lemma 26.1.2(b),
$$r_{[\sigma]}(S) \geq r_{h_E([\sigma])}(S) = d_{\bar{r}}(S) \geq |S|^{\bar{r}-2m}.$$
Since $(F:E)$ is unbounded and $n > 1$, we find that $r_{[\sigma]}(S) = \aleph_0$.

CLAIM B: $r_{[\sigma]}(S) = \aleph_0$ for almost all $\sigma \in C$. For each $E \in \mathcal{E}$ let $\mathcal{B}(\bar{E})$ be the set of all pairs (G', H') such that $H' \leq G'$ are open normal subgroups of $\bar{E}$ and $(\bar{E}:\mathbf{N}(G', H')) = \infty$. Let $B(\bar{E})$ be the union of all the sets $\mathbf{N}(G', H')^e$ with $(G', H') \in \mathcal{B}(\bar{E})$. By the index assumption, each set $\mathbf{N}(G', H')^e$ with $(G', H') \in \mathcal{B}(\bar{E})$ has measure zero. Since $\mathcal{B}(\bar{E})$ is countable, $B(\bar{E})$ is a zero set in $\bar{E}^e$. It follows that $B = \bigcup_{E \in \mathcal{E}} h_E^{-1}(B(\bar{E}))$ is a zero set in F^e.

If $\sigma \in C \smallsetminus B$, then there exists $E \in \mathcal{E}$ such that $\sigma \in C_e(F, E)$ and $h_E(\sigma) \notin B(\bar{E})$. Thus, $h_E([\sigma])$ is a closed normal subgroup of $\bar{E}$ of an infinite index. If $r_{h_E([\sigma])}(S) < \aleph_0$, there exists $(G', H') \in \mathcal{B}(\bar{E})$ with $h_E([\sigma]) = \mathbf{N}(G', H')$ (Lemma 26.5.5(c)). Since $h_E(\sigma) \in h_E([\sigma])^e$, we have that $\sigma \in B(\bar{E})$. This contradiction proves $r_{[\sigma]}(S) \geq r_{h_E([\sigma])}(S) = \aleph_0$.

This concludes the proof of Claim B. □

COROLLARY 26.5.7: Let $F = \hat{F}_n$ with $n \geq 2$.
(a) If $e \leq n$ then $[\sigma] \cong \hat{F}_\omega$ for almost all $\sigma \in F^e$.
(b) If $e > n$, then $[\sigma] \cong \hat{F}_\omega$ for a set of $\sigma \in F^e$ of a positive measure (but less than 1).

Proof: By Theorem 26.4.5(b), $\mu(C_e(F)) = 1$ if $e \leq n$ and $0 < \mu(C_e(F)) < 1$ if $e > n$. Now apply Theorem 26.5.6. □

COROLLARY 26.5.8 ([Jarden-Lubotzky2, Cor. 2.9]): For each $e \geq 1$ and for almost all $\sigma \in F^e$, $[\sigma]$ is a free profinite group.

Proof: If $\sigma \in F^e \smallsetminus C_e(F)$, then $[\sigma]$ is open in F and is therefore a free profinite group (Proposition 17.6.2). By Theorem 26.5.6, $[\sigma]$ is free for almost all $\sigma \in C_e(F)$. Hence, $[\sigma]$ is free for almost all $\sigma \in F^e$. □

Example 26.5.9: Exceptional σ. Let $F = \hat{F}_n$ with $n \geq 2$. Denote the intersection of all open normal subgroups N of F such that F/N is a solvable group by F^{solv}. The index of F^{solv} in F is infinite. By Lemma 17.4.10, each simple quotient of F^{solv} is non-Abelian. Hence, $F^{\text{solv}} \neq \hat{F}_\omega$.

Let $\mathcal{N}$ be the set of all open normal subgroups N of F^{solv} with F^{solv}/N simple. Denote the intersection of all $N \in \mathcal{N}$ by M. Then F^{solv}/M is the direct product of simple non-Abelian groups. Choose $\sigma \in F^{\text{solv}}$ with $\sigma \notin N$ for all $N \in \mathcal{N}$. Then, $[\sigma] = F^{\text{solv}}$. Thus, the conclusion of Corollary 26.5.8 does not hold for all $\sigma \in F^e$. □

Notes

Proposition 26.1.7(a) appears in [Kantor-Lubotzky, Prop. 11]. Our proof follows [Lubotzky3, Thm. 1(a)].

Chapter 27. Omega-free PAC Fields

Let K be a countable Hilbertian field. Then each finite extension of K is Hilbertian (Proposition 12.3.3). Thus, we could be tempted to think of Hilbertian algebraic extensions of K as "small". On the other hand, $K_s(\boldsymbol{\sigma})$ is PAC for almost all $\boldsymbol{\sigma} \in \text{Gal}(K)^e$ (Theorem 18.6.1), so PAC algebraic extensions of K could be thought of as "large".

In this chapter we construct an abundance of algebraic extensions N of K which are both small and large in that sense, thus they are both Hilbertian and PAC (Theorem 27.4.8). Indeed, we construct N as an ω-free PAC field and conclude that N is Hilbertian (Corollary 27.3.3). Moreover, we prove that ω-free and PAC is equivalent to the following stronger form of Hilbertianity: For every variety V over N with a generic point $\mathbf{x}$ and for every polynomial $f \in N[\mathbf{x}, Y]$ which is irreducible over $N(\mathbf{x})$ there exists $\mathbf{a} \in V(N)$ such that $f(\mathbf{a}, Y)$ is irreducible over N (a version of Lemma 27.2.1 which holds when $\text{char}(K) = 0$).

These algebraic results have model theoretic interpretations: We extend $\mathcal{L}(\text{ring})$ by adding a sequence of predicate to the language and denote the extended language by $\mathcal{L}_R(\text{ring})$. Every field can be naturally considered as a structure for $\mathcal{L}(\text{ring})$. A field extension L/K is an extension of structures for $\mathcal{L}_R(\text{ring})$ if and only if K is algebraically closed in L. We prove that if a field K of characteristic 0 is existentially closed within the language $\mathcal{L}_R(\text{ring})$ in every field L in which K is algebraically closed, then K is PAC, and ω-free (Proposition 27.2.2). It follows that ω-free PAC fields of characteristic 0 are the models of a model companion of a theory T_R of fields in the language $\mathcal{L}_R(\text{ring})$ (Theorem 27.2.3).

27.1 Model Companions

The absolute Galois group of a countable field is isomorphic to $\hat{F}_\omega$ if and only if each finite embedding problem for $\text{Gal}(K)$ is solvable (a special case of Corollary 24.8.2). An arbitrary field which has this property is said to be ω-**free**. This section prepares an augmentation of the theory of fields in Section 27.2 to a theory in which ω-free PAC fields play an existential completeness role analogous to the role played by algebraically closed fields for the ordinary theory of fields.

Definition 27.1.1: Let T be a theory in a language $\mathcal{L}$. A theory $\tilde{T}$ in $\mathcal{L}$ is called the **model companion** of T if the following holds:
(1a) Each model of $\tilde{T}$ is a model of T.
(1b) Each model of T can be embedded in a model of $\tilde{T}$.
(1c) $\tilde{T}$ is model complete; that is, $\mathcal{A} \prec \mathcal{B}$ whenever $\mathcal{A}$ and $\mathcal{B}$ are models of $\tilde{T}$ with $\mathcal{A} \subseteq \mathcal{B}$ (Section 9.1).

A theory T is said to have the **amalgamation property**, if whenever two models $\mathcal{B}$ and $\mathcal{C}$ of T contain a common model $\mathcal{A}$ there are embeddings $f\colon \mathcal{B} \to \mathcal{D}$ and $g\colon \mathcal{C} \to \mathcal{D}$ into a common model $\mathcal{D}$, such that f and g coincide on $\mathcal{A}$.

Call the theory $\tilde{T}$ a **model completion** of T if in addition to (1)
(2) T has the amalgamation property.

Example 27.1.2: If T is the theory of fields, then the theory $\tilde{T}$ of algebraically closed fields is the model completion of T (Corollary 9.3.2).

Examples of model companions of theories which are not model completions appear in the next section. □

First we characterize the models of $\tilde{T}$ among the models of T:

Definition 27.1.3: Let $\mathcal{A} \subseteq \mathcal{B}$ be structures for a language $\mathcal{L}$ with domains $A \subseteq B$. Denote the language $\mathcal{L}$ augmented with new constant symbols for the elements of A by $\mathcal{L}(A)$. Then $\mathcal{A}$ is **existentially closed** in $\mathcal{B}$ if each existential sentence θ of $\mathcal{L}(A)$ (Definition 21.4.1) which is true in $\mathcal{B}$ is also true in $\mathcal{A}$. Note that this definition is equivalent to the one we gave in Section 7.3. □

LEMMA 27.1.4: *Let $\mathcal{A} \subseteq \mathcal{B}$ be structures for a language $\mathcal{L}$. Then $\mathcal{A}$ is existentially closed in $\mathcal{B}$ if and only if $\mathcal{B}$ can be embedded in a structure $\mathcal{A}^*$ for $\mathcal{L}$ which is an elementary extension of $\mathcal{A}$ (Section 7.3).*

Proof: The sufficiency of the condition is obvious. Now assume that $\mathcal{A}$ is existentially closed in $\mathcal{B}$.

Consider the set T of all sentences $\varphi(b_1, \ldots, b_n)$ of the language $\mathcal{L}(B)$ where $\varphi(X_1, \ldots, X_n)$ is a quantifier free formula of $\mathcal{L}$ and $b_1, \ldots, b_n$ are elements of B such that $\mathcal{B} \models \varphi(b_1, \ldots, b_n)$. We show that the union of T with the theory of $\mathcal{A}$ in the language $\mathcal{L}(A)$ has a model $\mathcal{A}^*$. Then each $b \in B$ corresponds to a distinguished element b^* of $\mathcal{A}^*$. The correspondence $b \mapsto b^*$ is an embedding of $\mathcal{B}$ into $\mathcal{A}^*$ and $\mathcal{A} \prec \mathcal{A}^*$ as models of $\mathcal{L}$.

The construction of $\mathcal{A}^*$ applies the compactness theorem (Proposition 7.7.6). By that theorem, it suffices to construct a model for $T_0 \cup \text{Th}(\mathcal{A})$ for each finite subset T_0 of T. The conjunction of all sentences in T_0 has the form $\varphi(b_1, \ldots, b_n)$ as above. Since $\mathcal{A}$ is existentially closed in $\mathcal{B}$, there exists $a_1, \ldots, a_n \in A$ such that $\mathcal{A} \models \varphi(a_1, \ldots, a_n)$. Interpret $b_1, \ldots, b_n$ in A, respectively, as $a_1, \ldots, a_n$, and interpret all other elements of B in A arbitrarily. This structure of $\mathcal{L}(B)$ gives the desired model for $T_0 \cup \text{Th}(\mathcal{A})$. □

Definition 27.1.5: If $\mathcal{A} \subseteq \mathcal{B}$ are structures for a language $\mathcal{L}$, we call $\mathcal{B}$ an **extension** of $\mathcal{A}$ (Section 7.3). If in addition $\mathcal{B}$ is a model of a theory T, we call $\mathcal{B}$ a T-**extension** of $\mathcal{A}$.

A theory T is said to be **inductive** if the union of each ascending transfinite sequence of models of T is also a model of T. For example, the theories

27.1 Model Companions

of groups, rings and fields are inductive. But the theory of finite fields is not inductive. □

LEMMA 27.1.6: *Every model complete theory is inductive.*

Proof: Let T be a model complete theory and let $\{\mathcal{A}_\alpha \mid \alpha < \gamma\}$ be an ascending transfinite sequence of models of T. Then $\mathcal{A}_\alpha \prec \mathcal{A}_\beta$ for each $\alpha \leq \beta < \gamma$. Hence, the union $\mathcal{A}$ of this transfinite sequence is an elementary extension of each of its members (Lemma 7.4.1(a)). In particular, $\mathcal{A}$ is a model of T. □

By definition, each model companion $\tilde{T}$ of a theory T is model complete. Hence, by Lemma 27.1.6, $\tilde{T}$ is an inductive theory.

A sentence θ of a language $\mathcal{L}$ is said to be of **type** $\forall\exists$ if θ is of the form $(\forall X_1)\cdots(\forall X_m)(\exists Y_1)\cdots(\exists Y_n)\varphi(\mathbf{X},\mathbf{Y})$, (abbreviate to $(\forall \mathbf{X})(\exists \mathbf{Y})\varphi(\mathbf{X},\mathbf{Y})$) where $\varphi(\mathbf{X},\mathbf{Y})$ is a quantifier free formula of $\mathcal{L}$.

The use of the compactness theorem in the next lemma parallels its use in Lemma 27.1.4. For a theory T of a language $\mathcal{L}$, the collection $T_{\forall\exists}$ of sentences of $\mathcal{L}$ of type $\forall\exists$ valid in $\mathrm{Mod}(T)$ is called the **set of logical consequences** of T of type $\forall\exists$. In particular, each model of T is a model of $T_{\forall\exists}$.

LEMMA 27.1.7: *Let T be a theory of a language $\mathcal{L}$ and let $\mathcal{A}$ be a structure for $\mathcal{L}$. Then $\mathcal{A} \in \mathrm{Mod}(T_{\forall\exists})$ if and only if $\mathcal{A}$ is existentially closed in some T-extension $\mathcal{B}$.*

Proof: Suppose first that $\mathcal{A}$ is existentially closed in some T-extension $\mathcal{B}$ and let $(\forall \mathbf{X})(\exists \mathbf{Y})\varphi(\mathbf{X},\mathbf{Y})$ be a sentence in $T_{\forall\exists}$ where $\mathbf{X} = (X_1,\ldots,X_m)$ and $\mathbf{Y} = (Y_1,\ldots,Y_n)$. Then $\mathcal{B} \models (\forall \mathbf{X})(\exists \mathbf{Y})\varphi(\mathbf{X},\mathbf{Y})$. Denote the domain of $\mathcal{A}$ by A. For each $\mathbf{x} \in A^m$, $\mathcal{B} \models (\exists \mathbf{Y})\varphi(\mathbf{x},\mathbf{Y})$. Since $\mathcal{A}$ is existentially closed in $\mathcal{B}$, we have $\mathcal{A} \models (\exists \mathbf{Y})\varphi(\mathbf{x},\mathbf{Y})$. It follows that $\mathcal{A} \in \mathrm{Mod}(T_{\forall\exists})$.

Conversely, suppose that $\mathcal{A} \in \mathrm{Mod}(T_{\forall\exists})$. Let S be the set of all sentences $(\forall \mathbf{Y})\varphi(\mathbf{a},\mathbf{Y})$ of the augmented language $\mathcal{L}(A)$, where $\varphi(\mathbf{X},\mathbf{Y})$ is a quantifier free formula of $\mathcal{L}$ and $\mathbf{a} \in A^m$ such that $\mathcal{A} \models (\forall \mathbf{Y})\varphi(\mathbf{a},\mathbf{Y})$.

We use the compactness theorem (Proposition 7.7.6) to show that $S \cup T$ has a model: Since S is closed under finite conjunctions (up to logical equivalence), it suffices to prove that $T \cup \{(\forall \mathbf{Y})\varphi(\mathbf{a},\mathbf{Y})\}$ has a model, where $(\forall \mathbf{Y})\varphi(\mathbf{a},\mathbf{Y})$ is as in the last paragraph.

If not, then
$$\mathrm{Mod}(T) \models \neg(\exists \mathbf{X})(\forall \mathbf{Y})\varphi(\mathbf{X},\mathbf{Y}).$$

Hence, $\mathrm{Mod}(T) \models (\forall \mathbf{X})(\exists \mathbf{Y})\neg\varphi(\mathbf{X},\mathbf{Y})$. The latter, however, is a $\forall\exists$-sentence of $\mathcal{L}$. It, therefore belongs to $T_{\forall\exists}$. Thus, $\mathcal{A} \models (\forall \mathbf{X})(\exists \mathbf{Y})\neg\varphi(\mathbf{X},\mathbf{Y})$, contrary to $\mathcal{A} \models (\forall \mathbf{Y})\varphi(\mathbf{a},\mathbf{Y})$.

Therefore, let $\mathcal{B}$ be a model of $S \cup T$; in particular $\mathcal{B}$ is a structure for $\mathcal{L}(A)$. Let $(\exists \mathbf{Y})\psi(\mathbf{a},\mathbf{Y})$ be an existential sentence of $\mathcal{L}(A)$ which is true in $\mathcal{B}$, where $\psi(\mathbf{X},\mathbf{Y})$ is a quantifier free formula of $\mathcal{L}$ and $\mathbf{a} \in A^m$. If $\mathcal{A} \not\models (\exists \mathbf{Y})\psi(\mathbf{a},\mathbf{Y})$, then $\mathcal{A} \models (\forall \mathbf{Y})\neg\psi(\mathbf{a},\mathbf{Y})$. Consequently, $(\forall \mathbf{Y})\neg\psi(\mathbf{a},\mathbf{Y})$

belongs to S and therefore is true in $\mathcal{B}$, a contradiction. Thus, $\mathcal{B}$ extends $\mathcal{A}$ and $\mathcal{A}$ is existentially closed in $\mathcal{B}$. □

LEMMA 27.1.8: *If T is an inductive theory in a language $\mathcal{L}$, then* $\mathrm{Mod}(T) = \mathrm{Mod}(T_{\forall\exists})$.

Proof: We must prove that each model $\mathcal{A}$ of $T_{\forall\exists}$ is also a model of T.

For this, we inductively construct an ascending sequence $\mathcal{A} = \mathcal{A}_0 \subseteq \mathcal{B}_0 \subseteq \mathcal{A}_1 \subseteq \mathcal{B}_1 \subseteq \cdots$ of structures for $\mathcal{L}$ such that $\mathcal{B}_n \in \mathrm{Mod}(T)$, $\mathcal{A}_n$ is existentially closed in $\mathcal{B}_n$ and $\mathcal{A}_n \prec \mathcal{A}_{n+1}$ for each $n \geq 0$. Suppose the sequence has been constructed up to $\mathcal{B}_n$. Apply Lemma 27.1.4 to find an extension $\mathcal{A}_{n+1}$ of $\mathcal{B}_n$ with $\mathcal{A}_n \prec \mathcal{A}_{n+1}$. Thus, $\mathcal{A} \prec \mathcal{A}_{n+1}$, so $\mathcal{A}_{n+1} \in \mathrm{Mod}(T_{\forall\exists})$. By Lemma 27.1.7, $\mathcal{A}_{n+1}$ has a T-extension $\mathcal{B}_{n+1}$ in which $\mathcal{A}_{n+1}$ is existentially closed. Since T is inductive, $\mathcal{B}_\omega$, the union of the sequence, is in $\mathrm{Mod}(T)$. But it is also an elementary extension of $\mathcal{A}$ (Lemma 7.4.1(a)). It follows that $\mathcal{A} \in \mathrm{Mod}(T)$. □

A model $\mathcal{A}$ of a theory T is said to be **T-existentially closed** if $\mathcal{A}$ is existentially closed in each of its T-extensions. For example, a field K is algebraically closed if and only if K is T-existentially closed, where T is the theory of fields in $\mathcal{L}(\mathrm{ring})$. The next result generalizes this observation to arbitrary theories:

PROPOSITION 27.1.9: *Let $\tilde{T}$ be a model companion of a theory T of a language $\mathcal{L}$. Then a model $\mathcal{A}$ of T is also a model of $\tilde{T}$ if and only if $\mathcal{A}$ is T-existentially closed.*

Proof: First suppose $\mathcal{A} \in \mathrm{Mod}(\tilde{T})$. Let $\mathcal{B}$ be a T-extension of $\mathcal{A}$. Then $\mathcal{B}$ can be extended to a model $\tilde{\mathcal{B}}$ of $\tilde{T}$. Since $\tilde{T}$ is model complete, $\tilde{\mathcal{B}}$ is an elementary extension of $\mathcal{A}$. Hence, $\mathcal{A}$ is existentially closed in $\mathcal{B}$.

Conversely, suppose that $\mathcal{A}$ is T-existentially closed. By (1b), $\mathcal{A}$ has a $\tilde{T}$-extension $\tilde{\mathcal{A}}$ and $\mathcal{A}$ is existentially closed in $\tilde{\mathcal{A}}$. Lemma 27.1.7 implies that $\mathcal{A} \in \mathrm{Mod}(\tilde{T}_{\forall\exists})$. Since $\tilde{T}$ is model complete, it is inductive (Lemma 27.1.6). By Lemma 27.1.8, $\mathrm{Mod}(\tilde{T}) = \mathrm{Mod}(\tilde{T}_{\forall\exists})$. Consequently, $\mathcal{A} \in \mathrm{Mod}(\tilde{T})$. □

The remainder of this section develops a criterion for the existence of a model companion of an inductive theory:

LEMMA 27.1.10: *If T is an inductive theory of a language $\mathcal{L}$, then each model $\mathcal{A}$ of T has a T-existentially closed extension $\mathcal{A}'$.*

Proof: Let m be the maximum of the cardinal numbers $|A|$, $|\mathcal{L}|$, and $\aleph_0$, where A is the domain of $\mathcal{A}$. Let $\{\varphi_\alpha \mid \alpha < m\}$ be a wellordering of all existential sentences of the augmented language $\mathcal{L}(A)$. Define an ascending transfinite sequence $\{\mathcal{A}_\beta \mid \beta < m\}$ of structures for $\mathcal{L}$: $\mathcal{A}_0 = \mathcal{A}$; if $\beta = \alpha + 1$ is a successor ordinal and φ_α is true in some T-extension $\mathcal{B}$ of $\mathcal{A}_\alpha$, then $\mathcal{A}_\beta = \mathcal{B}$, otherwise let $\mathcal{A}_\beta = \mathcal{A}_\alpha$; and if β is a limit ordinal, $\mathcal{A}_\beta = \bigcup_{\alpha<\beta} \mathcal{A}_\alpha$ ($\in \mathrm{Mod}(T)$, since T is inductive).

Now let $\mathcal{A}^{(1)} = \bigcup_{\alpha<m} \mathcal{A}_\alpha$. Then $\mathcal{A}^{(1)} \in \text{Mod}(T)$, and each existential sentence of $\mathcal{L}(A)$ which is true in some T-extension of $\mathcal{A}^{(1)}$ is already true in $\mathcal{A}^{(1)}$.

Apply the same construction to $\mathcal{A}^{(1)}$ to obtain a model $\mathcal{A}^{(2)}$ of T with domain $A^{(2)}$ such that each existential sentence of $\mathcal{L}(A^{(1)})$ which is true in some T-extension is already true in $\mathcal{A}^{(2)}$. Iterate this process countably many times to find an ascending sequence $\mathcal{A} \subseteq \mathcal{A}^{(1)} \subseteq \mathcal{A}^{(2)} \subseteq \cdots$ for which $\mathcal{A}' = \bigcup_{n=1}^{\infty} \mathcal{A}^{(n)}$ is a T-extension of $\mathcal{A}$ which is T-existentially closed. $\square$

LEMMA 27.1.11 (Robinson's Test for Model Completeness): *A theory T in a language $\mathcal{L}$ is model complete if and only if each model of T is T-existentially closed.*

Proof: It suffices to prove that the condition is sufficient.

Let $\mathcal{A} \subseteq \mathcal{B}$ be two models of T. Apply Lemma 27.1.4 to find an extension $\mathcal{A}_2$ of $\mathcal{B}$ with $\mathcal{A} \prec \mathcal{A}_2$. Continue inductively to find a sequence $\mathcal{A}_1 \subseteq \mathcal{B}_1 \subseteq \mathcal{A}_2 \subseteq \mathcal{B}_2 \subseteq \mathcal{A}_3 \subseteq \mathcal{B}_3 \subseteq \cdots$ of models of T, where $\mathcal{A}_1 = \mathcal{A}$, $\mathcal{B}_1 = \mathcal{B}$ such that $\mathcal{A}_n \prec \mathcal{A}_{n+1}$ and $\mathcal{B}_n \prec \mathcal{B}_{n+1}$ for each positive integer n. By Lemma 7.4.1(a), the union of these sequences is an elementary extension of both $\mathcal{A}$ and $\mathcal{B}$. Hence, $\mathcal{A} \prec \mathcal{B}$. Thus, T is model complete. $\square$

PROPOSITION 27.1.12: *Let T be an inductive theory in a language $\mathcal{L}$. Denote the class of all T-existentially closed structures by $\text{Exis}(T)$. Then T has a model companion if and only if there exists a theory $\tilde{T}$ of $\mathcal{L}$ such that $\text{Exis}(T) = \text{Mod}(\tilde{T})$. In this case $\tilde{T}$ is a model companion of T.*

Proof: First suppose that $\tilde{T}$ is a model companion of T. Then, by Proposition 27.1.9, $\text{Exis}(T) = \text{Mod}(\tilde{T})$. Conversely, let $\tilde{T}$ be a theory of $\mathcal{L}$ with $\text{Exis}(T) = \text{Mod}(\tilde{T})$. We prove that $\tilde{T}$ is a model companion of T. Let $\tilde{\mathcal{B}} \subseteq \tilde{\mathcal{C}}$ be models of $\tilde{T}$. Then $\tilde{\mathcal{B}}, \tilde{\mathcal{C}} \in \text{Exis}(T)$, so $\tilde{\mathcal{B}}$ is existentially closed in $\tilde{\mathcal{C}}$. Thus, $\tilde{\mathcal{B}}$ is $\tilde{T}$-existentially closed. By Lemma 27.1.11, $\tilde{T}$ is model complete. Since T is inductive, each model $\mathcal{A}$ of T has a T-existentially closed extension $\tilde{\mathcal{A}}$, which is then a model of $\tilde{T}$. Consequently, $\tilde{T}$ is a model companion of T. $\square$

27.2 The Model Companion in an Augmented Theory of Fields

We augment the language $\mathcal{L}(\text{ring})$ to a language $\mathcal{L}_R(\text{ring})$ by adding an n-ary relation symbol R_n for each positive integer n. Let T_R be the theory of fields together with the axioms

(1) $\qquad R_n(X_1, \ldots, X_n) \leftrightarrow (\exists Z)[Z^n + X_1 Z^{n-1} + \cdots + X_n = 0]$,

In a field K, interpret R_n as a relation on K, $n = 1, 2, 3, \ldots$, in the unique way such that (1) holds. Then consider K also as a model of T_R. If

K and L are fields, regarded as models of T_R, then K is a substructure of L if and only if $K \subseteq L$ and K is algebraically closed in L.

Clearly T_R is an inductive theory. We construct a theory $\tilde{T}_R$ such that $\mathrm{Exis}(T_R) = \mathrm{Mod}(\tilde{T}_R)$. The models of $\tilde{T}_R$ are 1-perfect ω-free PAC fields. By Proposition 27.1.12, $\tilde{T}_R$ will be a model companion for T_R.

LEMMA 27.2.1: *The following conditions on a field K are equivalent:*
(a) *K is ω-free and PAC.*
(b) *Let V be a variety defined over K with a generic point $\mathbf{x}$, $f(\mathbf{x}, Y) \in K[\mathbf{x}, Y]$ an irreducible separable polynomial in Y over $K(\mathbf{x})$, and $0 \neq h(\mathbf{x}) \in K[\mathbf{x}]$. Then there exists $\mathbf{a} \in V(K)$ such that $h(\mathbf{a}) \neq 0$ and $f(\mathbf{a}, Y)$ is irreducible in $K[Y]$.*
(c) *Let V be a variety defined over K with a generic point $\mathbf{x}$, $f_1(\mathbf{x}, Y), \ldots, f_k(\mathbf{x}, Y) \in K[\mathbf{x}, Y]$ irreducible separable polynomials in Y over $K(\mathbf{x})$, and $0 \neq g(\mathbf{x}) \in K[\mathbf{x}]$. Then there is an $\mathbf{a} \in V(K)$ such that*

$$f_1(\mathbf{a}, Y), \ldots, f_k(\mathbf{a}, Y)$$

are irreducible over K and $g(\mathbf{a}) \neq 0$.

Furthermore, suppose $\mathrm{char}(K) = p$, the imperfect exponent of K is at least m, and $h_1(\mathbf{x}), \ldots, h_m(\mathbf{x}) \in K[\mathbf{x}]$ are p-independent over $K(\mathbf{x})^p$. Then (a) implies that $\mathbf{a}$ can be chosen in (b) and (c) such that $h_1(\mathbf{a}), \ldots, h_m(\mathbf{a})$ are p-independent over K^p.

Proof of "(a) $\Longrightarrow$ (b)": By assumption, $\mathrm{Im}(\mathrm{Gal}(K))$ consists of all finite groups and $\mathrm{Gal}(K)$ has the embedding property (Definition 24.1.2). Since K is also PAC, it is a Frobenius field (Definition 24.1.3). By Lemma 19.7.2 there exists $g_0(\mathbf{x}) \in K[\mathbf{x}]$, $g_0(\mathbf{x}) \neq 0$, such that $K[\mathbf{x}, g_0(x)^{-1}]$ is integrally closed. Let F be the splitting field of $f(\mathbf{x}, Y)$ over $K(\mathbf{x})$. Choose a root y of $f(\mathbf{x}, Y)$ in F. Let $f_0(\mathbf{x})$ be the coefficient of the highest power of Y in $f(\mathbf{x}, Y)$ and let $d_0(\mathbf{x})$ be the discriminant of y over $K(\mathbf{x})$. Now choose a primitive element z for $F/K(\mathbf{x})$, integral over $K[\mathbf{x}]$ and let $d_1(\mathbf{x})$ be its discriminant over $K(\mathbf{x})$. Finally, let $g(\mathbf{x}) = g_0(\mathbf{x}) f_0(\mathbf{x}) d_0(\mathbf{x}) d_1(\mathbf{x}) h(\mathbf{x})$. Then consider the ring $R = K[\mathbf{x}, g(\mathbf{x})^{-1}]$. By Lemma 6.1.2, $R[y]$ and $R[z]$ are the respective integral closures of R in $K(\mathbf{x}, y)$ and F. Let L be the algebraic closure of K in F. By Corollary 10.2.2(a), $K(\mathbf{x})/K$ is a regular extension, so $\mathrm{res}_L \mathrm{Gal}(F/K(\mathbf{x})) = \mathrm{Gal}(L/K)$. Since K is ω-free, $\mathrm{Gal}(F/K(\mathbf{x})) \in \mathrm{Im}(\mathrm{Gal}(K))$. Thus, $R[z]/R$ satisfies the decomposition group condition (Definition 24.1.3). Since K is Frobenius, Proposition 24.1.4 gives a K-homomorphism $\varphi \colon R[z] \to K_s$ with decomposition group $\mathrm{Gal}(F/K(\mathbf{x}))$ such that $\varphi(R) = K$. In particular, $\mathbf{a} = \varphi(\mathbf{x})$ is a point of $V(K)$ and for $c = \varphi(z)$,

(2) $$[K(c) : K] = [F : K(\mathbf{x})].$$

If $\mathrm{char}(K) = p$, $h_1(\mathbf{x}), \ldots, h_m(\mathbf{x}) \in K[\mathbf{x}]$ are p-independent over $K(\mathbf{x})^p$ and m is at most the imperfect degree of K, then Proposition 24.1.4 allows us to choose $\mathbf{a}$ such that $h_1(\mathbf{a}), \ldots, h_m(\mathbf{a})$ are p-independent in K.

27.2 The Model Companion in an Augmented Theory of Fields

Let $b = \varphi(y)$. Then, $[K(b) : K] \leq [K(\mathbf{x}, y) : K(\mathbf{x})]$ and $[K(c) : K(b)] \leq [F : K(\mathbf{x}, y)]$. By (2), $[K(b) : K] = [K(\mathbf{x}, y) : K(\mathbf{x})]$. Hence, $f(\mathbf{a}, Y)$ is irreducible in $K[Y]$.

Proof of "(b) $\Longrightarrow$ (a)": Condition (b) implies that K is a PAC field. We need only show that every finite embedding problem for $\mathrm{Gal}(K)$ is solvable.

Let L be a finite Galois extension of K, B a finite groups, and $\alpha \colon B \to \mathrm{Gal}(L/K)$ an epimorphism. We have to produce an homomorphism $\gamma \colon \mathrm{Gal}(K) \to B$ satisfying $\alpha \circ \gamma = \mathrm{res}_{K_s/L}$.

Lemma 11.6.1 gives a finitely generated regular extension $E = K(\mathbf{x})$ of K and a finite Galois extension F of E that contains L with $B = \mathrm{Gal}(F/E)$ and $\alpha = \mathrm{res}_{F/L}$. Let S/R be a ring cover for F/E with $R = K[\mathbf{x}, h(\mathbf{x})^{-1}]$ and $0 \neq h(\mathbf{x}) \in K[\mathbf{x}]$ and let z be a primitive element for S/R (Definition 6.1.3). If $f \in K[\mathbf{x}, Z]$ is an irreducible polynomial in Z over $K(\mathbf{x})$ such that $f(\mathbf{x}, z) = 0$, then (b) gives a K-rational specialization $\mathbf{x} \to \mathbf{a}$ such that $f(\mathbf{a}, Z)$ is irreducible in $K[Z]$ and $h(\mathbf{a}) \neq 0$. This extends to an L-homomorphism φ of S onto a field $N = K(\varphi(z))$ which is Galois over K and contains L. By Lemma 6.1.4, φ induces an embedding φ^* of $\mathrm{Gal}(N/K)$ into $\mathrm{Gal}(F/E)$ such that $\mathrm{res}_{F/L} \circ \varphi^* = \mathrm{res}_{N/L}$. Since $f(\mathbf{a}, Z)$ is irreducible over K, $[N : K] = [F : E]$ and φ^* is bijective. Thus, $\gamma = \varphi^* \circ \mathrm{res}_N \colon \mathrm{Gal}(K) \to B$ is an epimorphism which solves our embedding problem.

Proof of "(b) $\Longleftrightarrow$ (c)": Use the primitive element theorem. $\square$

Recall that a field K is 1-imperfect if either $\mathrm{char}(K) = 0$ or $\mathrm{char}(K) = p$ and $[K : K^p] = p$ (Section 2.7).

Let T_R again be the theory of fields together with the axioms of (1) as at the beginning of this section.

PROPOSITION 27.2.2: *A field K is an existentially closed model of T_R if and only if K is 1-imperfect, ω-free, and PAC.*

Proof: Let $p = \mathrm{char}(K)$. Each of the two parts of the proof proves one direction of the proposition:

PART A: *Suppose K is an existentially closed model of T_R.*

PART A1: *Proof that K is 1-imperfect.* We need only treat the case where $p > 0$. If t is transcendental over K, then K is existentially closed in $K(t)$. Lemma 27.1.4 produces an elementary extension K^* of K containing $K(t)$ such that $K(t)$ is algebraically closed in K^*. In particular, $t^{1/p} \notin K^*$. Hence, K^* is imperfect. Therefore, K is imperfect.

On the other hand assume the imperfect exponent of K exceeds 1. Then there exist p-independent elements a, b in K. Let x, y be transcendental elements over K such that $ax^p + by^p = 1$. By Lemma 2.7.4, K is algebraically closed in $K(x, y)$. Hence, K is existentially closed in $K(x, y)$, so there are $c, d \in K$ with $ac^p + bd^p = 1$, contrary to the p-independence of a and b in K. It follows that K is 1-imperfect.

PART A2: *Proof that K is ω-free and PAC.* It suffices to prove that K satisfies Condition (b) of Lemma 27.2.1. Let $V = V(h_1, \ldots, h_r)$ be a variety defined over K with generic point $\mathbf{x}$, $f(\mathbf{x}, Y) \in K[\mathbf{x}, Y]$ an irreducible separable polynomial over $K(\mathbf{x})$, and $0 \neq g(\mathbf{x}) \in K[\mathbf{x}]$. By Corollary 10.2.2, $K(\mathbf{x})/K$ is regular. Hence, by Lemma 27.1.4, $K(\mathbf{x})$ is algebraically closed in an elementary extension K^* of K. Therefore, $f(\mathbf{x}, Y)$ is irreducible over K^*. In addition, $h_1(\mathbf{x}) = \cdots = h_r(\mathbf{x}) = 0$. Since irreducibility of a polynomial over a field is an elementary statement, there exists $\mathbf{a} \in V(K)$ such that $f(\mathbf{a}, Y)$ is irreducible in $K[Y]$ and $g(\mathbf{a}) \neq 0$.

PART B: *Suppose K is a 1-imperfect ω-free PAC field.* We have to prove that if K is algebraically closed in an extension E, then K is existentially closed in E. Indeed, let $\varphi(X_1, \ldots, X_n)$ be a quantifier free formula of $\mathcal{L}_R$(ring) and $x_1, \ldots, x_n$ elements in E such that $E \models \varphi(\mathbf{x})$. Then K is algebraically closed in $K(\mathbf{x})$ and $K(\mathbf{x}) \models \varphi(\mathbf{x})$. Thus, we may assume that $E = K(\mathbf{x})$.

By Lemma 27.1.4, it suffices to prove that E is algebraically closed in an elementary extension K^* of K. Since K is 1-perfect, $K(\mathbf{x})$ is regular over K (Lemma 2.7.5). Thus, $(\mathbf{x})$ generates a variety V over K (Corollary 10.2.2).

PART B1: *Data for being algebraically closed.* Extend the language $\mathcal{L}$(ring, K) to a language $\mathcal{L}$(ring, K, $\mathbf{x}$) by adding n constant symbols corresponding to $x_1, \ldots, x_n$. Call a system,

$$(3) \quad (f_1(\mathbf{x}, Y), \ldots, f_k(\mathbf{x}, Y); g_1(\mathbf{x}), \ldots, g_l(\mathbf{x}); h_1(\mathbf{x}), \ldots, h_m(\mathbf{x})),$$

of polynomials with coefficients in K **data for being algebraically closed**, if the following hold:
(4a) $f_1(\mathbf{x}, Y), \ldots, f_k(\mathbf{x}, Y)$ are separable and irreducible in $E[Y]$.
(4b) $g_1(\mathbf{x}), \ldots, g_l(\mathbf{x}) \neq 0$.
(4c) $h_1(\mathbf{x}), \ldots, h_m(\mathbf{x}) \in K[\mathbf{x}] \smallsetminus E^p$.

Suppose that for each choice of the data (3) for being algebraically closed there exists $\mathbf{a} \in V(K)$ such that
(5a) $f_1(\mathbf{a}, Y), \ldots, f_k(\mathbf{a}, Y)$ are separable and irreducible in $K[Y]$;
(5b) $g_1(\mathbf{a}), \ldots, g_l(\mathbf{a}) \neq 0$; and
(5c) $h_1(\mathbf{a}), \ldots, h_m(\mathbf{a}) \in K \smallsetminus K^p$.

Then $(K, \mathbf{a})$ is a structure for $\mathcal{L}$(ring, K, $\mathbf{x}$) in which the sentences of (4) are true. By the compactness theorem (Proposition 7.7.6) there is a structure (K^*, x^*) for $\mathcal{L}$(ring, K, $\mathbf{x}$) such that K^* is an elementary extension of K, $\mathbf{x}^* \in V(K^*)$, $f(\mathbf{x}^*, Y)$ is separable and irreducible in $K^*[Y]$ for every separable and irreducible polynomial $f(\mathbf{x}, Y)$ in $E[Y]$, $g(\mathbf{x}^*) \neq 0$ for every nonzero $g(\mathbf{x}) \in K[\mathbf{x}]$; and $h(\mathbf{x}^*) \in K^* \smallsetminus (K^*)^p$ for every $h(\mathbf{x}) \in K[\mathbf{x}] \smallsetminus E^p$. It follows that the K-specialization $\mathbf{x} \to \mathbf{x}^*$ maps E isomorphically onto a subfield E^* of K^* such that E^* is algebraically closed in K^*.

If $p = 0$, we remove all occurrences of h_i from Part B1.

27.2 The Model Companion in an Augmented Theory of Fields

PART B2: *Existence of the specialization satisfying (5).* The proof of the Proposition is complete if, given data (3) for being algebraically closed, we demonstrate the existence of $\mathbf{a} \in V(K)$ that satisfies (5).

Indeed, let $u_1, \ldots, u_r$ be a p-basis for E over E^p with $p = \text{char}(K)$. If $r \geq 2$, choose $2(r-1)$ elements $z_2, w_2, \ldots, z_r, w_r$ such that $z_2, \ldots, z_r$ are algebraically independent elements over E and

(6) $$u_1 z_i^p + u_i w_i^p = 1, \qquad i = 2, \ldots, r.$$

Consider the field extension $F = K(\mathbf{x}, \mathbf{z}, \mathbf{w})$ of E, where $F = E$ if $r = 1$ or if $\text{char}(K) = 0$. By Lemma 2.7.4, E is algebraically closed in F. Therefore, so is K. By Lemma 2.7.5, F is a regular extension of K.

For each i between 1 and m, $E(h_i(\mathbf{x})^{1/p}) \subseteq E(u_1^{1/p}, \ldots, u_r^{1/p})$. Since E is algebraically closed in F, $h_i(\mathbf{x})^{1/p} \notin F$. Hence, by (6),

$$F \subset F(h_i(\mathbf{x})^{1/p}) \subseteq F(u_1^{1/p}, \ldots, u_r^{1/p}) = F(u_1^{1/p}).$$

It follows that $F(h_i(\mathbf{x})^{1/p}) = F(u_1^{1/p})$. Thus, there exist relations

(7) $$s_i(\mathbf{x}, \mathbf{z}, \mathbf{w}) u_1^{1/p} = \sum_{j=0}^{p-1} t_{ij}(\mathbf{x}, \mathbf{z}, \mathbf{w}) h_i(\mathbf{x})^{j/p}, \qquad i = 1, \ldots, m,$$

with $s_i, t_{ij} \in K[\mathbf{x}, \mathbf{z}, \mathbf{w}]$ and $s_i(\mathbf{x}, \mathbf{z}, \mathbf{w}) \neq 0$.

Since F is regular over K, $(\mathbf{x}, \mathbf{u}, \mathbf{z}, \mathbf{w})$ generates a variety over K. Hence, by Lemma 27.2.1, there exists a K-rational specialization $(\mathbf{x}, \mathbf{u}, \mathbf{z}, \mathbf{w}) \to (\mathbf{a}, \mathbf{b}, \mathbf{c}, \mathbf{d})$ such that (5a) and (5b) hold, $b_1 \notin K^p$ and $s_i(\mathbf{a}, \mathbf{c}, \mathbf{d}) \neq 0$, $i = 1, \ldots, m$. By (7),

$$s_i(\mathbf{a}, \mathbf{c}, \mathbf{d}) b_1^{1/p} = \sum_{j=0}^{p-1} t_{ij}(\mathbf{a}, \mathbf{c}, \mathbf{d}) h_i(\mathbf{a})^{j/p}, \qquad i = 1, \ldots, m.$$

Hence, (5c) is also satisfied. Thus, $\mathbf{a}$ belongs to $V(K)$ and satisfies (5). □

The results of this section prepare all necessary assumptions for an application of Proposition 27.1.12 to T_R:

THEOREM 27.2.3: *The theory T_R has a model companion $\tilde{T}_R$. A field K is a model of $\tilde{T}_R$ if and only if K is 1-imperfect, ω-free and PAC.*

Proof: We show there exists a set of sentences $\tilde{T}_R$ in $\mathcal{L}(\text{ring})$ such that a field K is a model of $\tilde{T}_R$ if and only if K is 1-imperfect, ω-free and PAC. Indeed, axiomatize 1-imperfect with the sentences

$$p = 0 \to (\exists X)(\forall Y)[Y^p \neq X \land (\forall Z)(\exists U_0, \ldots, \exists U_{p-1})[Z = \sum_{i=0}^{p-1} U_i^p X^i]],$$

one for each prime p. That is, if $p = 0$ in K there exists an element in $K \smallsetminus K^p$ and any two elements of K are p-dependent over K^p. Proposition 11.3.2 axiomatizes the PAC property. Finally, by Remark 23.4.2, the statement "each finite embedding problem over K is solvable" is elementary.

It follows from Proposition 27.2.2 that $\text{Mod}(\tilde{T}_R) = \text{Exis}(T)$. Consequently, by Proposition 27.1.12, $\tilde{T}_R$ is a model companion of T_R. □

The following example proves that T_R does not have the amalgamation property. Thus, although T_R has a model companion, it does not have a model completion.

Example 27.2.4: Let t, u, v be algebraically independent elements over $\mathbb{F}_p$ and put $w = u - v$. Then $t, u, v^{1/p}$ are algebraically independent over $\mathbb{F}_p$. Therefore, $\mathbb{F}_p(t, u)$ is algebraically closed in $\mathbb{F}_p(t, u, v^{1/p})$. Similarly $\mathbb{F}_p(t, u)$ is algebraically closed in $\mathbb{F}_p(t, u, w^{1/p})$. If, however, L is a T_R-extension of both $\mathbb{F}_p(t, u, v^{1/p})$ and $\mathbb{F}_p(t, u, w^{1/p})$, then they are both algebraically closed in L. In particular, $v^{1/p}$, which is algebraic over $\mathbb{F}_p(t, u, v) = \mathbb{F}_p(t, u, w)$, belongs to $\mathbb{F}_p(t, u, w^{1/p})$. This means that $v^{1/p} f(t, u, w^{1/p}) = g(t, u, w^{1/p})$, for two polynomials f, g with coefficients in $\mathbb{F}_p$, where $f \neq 0$. Raise this expression to the pth power and substitute $t = 0$ to obtain $(u - w) f(0, u^p, w) = g(0, u^p, w)$. The exponents of the powers of u on the right side are multiples of p, whereas they are congruent to 1 modulo p on the left hand side, a contradiction. □

27.3 New Non-Classical Hilbertian Fields

We extend the theory T_R of Section 27.2 to a theory $T_{R,Q}$ in an extended language $\mathcal{L}_{R,Q}(\text{ring})$. We prove that $T_{R,Q}$ has a model completion $\tilde{T}_{R,Q}$ whose models are the ω-imperfect ω-free PAC fields.

Augment $\mathcal{L}_R(\text{ring})$ to a language $\mathcal{L}_{R,Q}(\text{ring})$ by adding n-ary relation symbols, $Q_{p,n}$, one for each prime p and each positive integer n. Let $T_{R,Q}$ be the theory of $\mathcal{L}_{R,Q}(\text{ring})$ consisting of T_R (Section 27.2) together with the axioms

(1) $Q_{p,n}(X_1, \ldots, X_n) \leftrightarrow p = 0 \wedge (\exists U_i)[\sum_i U_i^p X_1^{i_1} \cdots X_n^{i_n} = 0 \wedge \bigvee_i U_i \neq 0]$,

one for each (p, n) where $\mathbf{i}$ ranges over all n-tuples $(i_1, \ldots, i_n)$ of integers between 0 and $p - 1$. Given a field K, we may uniquely regard K as a model of $T_{R,Q}$. Thus, if $p = \text{char}(K)$ and $x_1, \ldots, x_n \in K$, then $x_1, \ldots, x_n$ are p-dependent if and only if $Q_{p,n}(x_1, \ldots, x_n)$ holds in K. Therefore, if K is a substructure of L (as models of $T_{R,Q}$), then L is linearly disjoint from $K^{1/p}$ over K. By Lemma 2.6.1, L is separable over K. Since L is a T_R-extension of K, K is algebraically closed in L, so L is regular over K (Lemma 2.6.4). If K is a substructure of both L and M, replace M by a K-isomorphic copy, if necessary, to assume that M is algebraically independent from L

27.3 New Non-Classical Hilbertian Fields

over K and both are contained in a common field. By Corollary 2.6.8(a), LM is a regular extension of both L and M. It follows that $T_{R,Q}$ has the amalgamation property.

Conversely, if L is a regular field extension of K, then L/K is an extension of models of $T_{R,Q}$.

THEOREM 27.3.1: *The theory $T_{R,Q}$ has a model completion whose models are the ω-imperfect ω-free PAC fields.*

Proof (Sketch): This is similar to the proof of Theorem 27.2.3 (and its main ingredient Proposition 27.2.2). We elaborate on two points that differ from the arguments of the result for T_R:

First: If a field K of characteristic p is $T_{R,Q}$-existentially closed and $x_1, \ldots, x_n$ are n algebraically independent elements over K, then $x_1, \ldots, x_n$ are p-independent in $E = K(\mathbf{x})$. Thus, the existential sentence

$$(\exists X_1, \ldots, \exists X_n)[\neg Q_{p,n}(\mathbf{X})]$$

is true in E (take X_i to be x_i, $i = 1, \ldots, n$). Hence, it holds in K. Therefore, K is ω-imperfect.

Second: If K is an ω-imperfect ω-free and PAC field of characteristic p, and if $E = K(\mathbf{x})$ is a finitely generated regular extension of K, apply Condition (c) of Lemma 27.2.1 to find a K-specialization of $\mathbf{x}$ into K such that a given p-independent elements $u_1, \ldots, u_m$ of E are mapped into p-independent elements of K. ☐

Remark 27.3.2: There are two obvious generalizations of Theorems 27.2.3 and 27.3.1. One may augment the theory T_R with an axiomatization of a Melnikov formation of finite groups $\mathcal{C}$ in the language of fields. This may give a theory with a model companion (respectively, model completion) whose countable models are the 1-imperfect (resp. ω-imperfect) PAC fields K with $\mathrm{Gal}(K) \cong \hat{F}_\omega(\mathcal{C})$. Alternatively, one may augment T_R so as to bound the imperfect exponent of the fields by m, for some given positive integer m. ☐

We apply Lemma 27.2.1(b) in the case that V runs over the affine spaces $\mathbb{A}^n$, $n = 1, 2, \ldots$, to rephrase the condition that K is Hilbertian:

COROLLARY 27.3.3 ([Roquette1]): *Every ω-free PAC field is Hilbertian.*

Recall that a field K is Hilbertian if and only if $\mathbb{A}^1(K)$ is a nonthin set (Proposition 13.5.3). If K is a number field, then K is Hilbertian but $C(K)$ is a thin set for each curve of genus at least 1 (Remark 13.5.4). We show that ω-free PAC fields are not only Hilbertian but $V(K)$ is nonthin for every variety V over K. Thus, ω-free PAC fields are Hilbertian in a stronger sense.

PROPOSITION 27.3.4: *A field K is ω-free and PAC if and only if $V(K)$ is nonthin for every variety V over K.*

Proof: Suppose first K is ω-free and PAC. Let V be a variety defined over K and $\mathbf{x}$ a generic point $\mathbf{x}$ of V over K. For $i = 1, \ldots, m$ consider an element

$y_i \in K(\mathbf{x})_s$ which is integral over $K[\mathbf{x}]$ such that $K(\mathbf{x}, y_i)/K$ is regular and $[K(\mathbf{x}, y_i) : K(\mathbf{x})] \geq 2$. Let $h_i \in K[\mathbf{X}, Y]$ be a polynomial which is monic in Y such that $h_i(\mathbf{x}, Y) = \text{irr}(y_i, K(\mathbf{x}))$. Let $g \in K[\mathbf{X}]$ with $g(\mathbf{x}) \neq 0$. By Lemma 27.2.1(c), there is an $\mathbf{a} \in V(K)$ such that $h_1(\mathbf{a}, Y), \ldots, h_m(\mathbf{a}, Y)$ are irreducible over K and $g(\mathbf{a}) \neq 0$. In particular, $h_i(\mathbf{a}, Y)$ has no zero in K, $i = 1, \ldots, m$. By Remark 13.5.2(g), $V(K)$ is nonthin.

Now suppose $V(K)$ is nonthin for every variety V defined over K. We prove that K satisfies Condition (b) of Lemma 27.2.1.

Let V be a variety over K with a generic point $\mathbf{x}$, $f(\mathbf{x}, Y) \in K[\mathbf{x}, Y]$ an irreducible separable polynomial in Y over $K(\mathbf{x})$, and $h \in K[\mathbf{X}]$ with $h(\mathbf{x}) \neq 0$. We use a simplified version of Proof B of Lemma 13.1.2 to find $\mathbf{a} \in V(K)$ such that $h(\mathbf{a}) \neq 0$ and $f(\mathbf{a}, Y)$ is irreducible over K.

Assume without loss, $f(\mathbf{x}, Y)$ is monic. Write $f(\mathbf{x}, Y) = \prod_{i=1}^n (Y - y_i)$ with $y_1, \ldots, y_n \in K(\mathbf{x})_s$. Denote the collection of all nonempty subsets of $\{1, \ldots, n\}$ by $\mathcal{I}$. For each $I \in \mathcal{I}$ let $f_I(Y) = \prod_{i \in I}(Y - y_i)$. Since $f(\mathbf{x}, Y)$ is irreducible over $K(\mathbf{x})$, there is a coefficient z_I of f_I which does not belong to $K(\mathbf{x})$. Thus, there is a polynomial $g_I \in K[\mathbf{X}, Z]$ monic and of degree at least 2 in Z such that $g_I(\mathbf{x}, Z) = \text{irr}(z_I, K(\mathbf{x}))$.

Since $V(K)$ is nonthin, there is an $\mathbf{a} \in V(K)$ such that $h(\mathbf{a}) \neq 0$, $f(\mathbf{a}, Y)$ is separable, and $g_I(\mathbf{a}, c) \neq 0$ for all $c \in K$ (Remark 13.5.2(g)). Assume $f(\mathbf{a}, Y) = p(Y)q(Y)$ is a nontrivial factorization in $K[Y]$ with monic p and q. Extend the K-specialization $\mathbf{x} \to \mathbf{a}$ to a K-specialization $(\mathbf{x}, \mathbf{y}) \to (\mathbf{a}, \mathbf{b})$, with $\mathbf{b} = (b_1, \ldots, b_n) \in K_s^n$. Then $f(\mathbf{a}, Y) = \prod_{i=1}^n (Y - b_i)$. Thus, there is an $I \in \mathcal{I}$ such that $p(Y) = \prod_{i \in I}(Y - b_i)$, the polynomial $f_I(Y)$ maps onto $p(Y)$, and z_I maps onto a coefficient c of $p(Y)$. Then c lies in K and satisfies $g_I(\mathbf{a}, c) = 0$. This contradiction to the choice of $\mathbf{a}$ proves that $f(\mathbf{a}, Y)$ is irreducible over K. $\square$

Our previous examples of Hilbertian fields include number fields, transcendental finitely generated extensions of fields, fields of power series $K_0((X_1, \ldots, X_r))$ with $r \geq 2$, and composition of pairs of Galois extensions of Hilbertian fields. Corollary 27.3.3 provides new examples of Hilbertian fields. They are large algebraic extensions of given countable Hilbertian fields:

Example 27.3.5: Algebraic ω-free PAC fields. Let K be a countable Hilbertian field. Take $G = \hat{F}_\omega$ in Theorem 23.2.3 to conclude the existence of an ω-free PAC field E algebraic over K. Alternatively take $(\sigma_1, \sigma_2) \in \text{Gal}(K)^2$ such that $K_s(\boldsymbol{\sigma})$ is 2-free PAC field (Theorem 20.8.2). Then use the results of Section 24.10 to provide an algebraic extension E of $K_s(\boldsymbol{\sigma})$ such that $\text{Gal}(E) \cong \hat{F}_\omega$ (e.g. take E as the maximal Abelian extension of $K_s(\boldsymbol{\sigma})$ as in Example 24.10.7). Then E is an ω-free PAC field. By Corollary 27.3.3, E is Hilbertian.

Note that $E = K_s(\boldsymbol{\sigma})_{\text{ab}}$ (with $\boldsymbol{\sigma} = (\sigma_1, \sigma_2)$ taken at random in $\text{Gal}(K)^2$) can not be proved to be Hilbertian by the diamond theorem (Theorem 13.8.3) over K. Indeed, assume L and M are Galois extensions of K such that $K_s(\sigma_{\text{ab}}) \not\subseteq L$, $K_s(\boldsymbol{\sigma})_{\text{ab}} \not\subseteq M$, by $K_s(\boldsymbol{\sigma})_{\text{ab}} \subseteq LM$. Then $K_s(\boldsymbol{\sigma}) \subseteq L$

or $K_s(\boldsymbol{\sigma}) \subseteq M$, otherwise $K_s(\boldsymbol{\sigma})$ is Hilbertian (by the diamond theorem), so $K_s(\boldsymbol{\sigma})$ has a linearly disjoint sequence of quadratic extension (Corollary 16.2.7), in contradiction to Lemma 16.11.2. Suppose for example that $K_s(\boldsymbol{\sigma}) \subseteq L$. By Proposition 16.12.6, $L = K_s$. Hence, $K_s(\boldsymbol{\sigma})_{\mathrm{ab}} \subseteq L$, in contradiction to our assumption. □

We have mentioned in Example 24.8.5(b) that the converse of Corollary 27.3.3 is also true: Every Hilbertian PAC field is ω-free.

27.4 An Abundance of ω-Free PAC Fields

Let K be a countable Hilbertian field. By Theorem 18.10.2, $K_s[\boldsymbol{\sigma}]$ is PAC for almost all $\boldsymbol{\sigma} \in \mathrm{Gal}(K)^e$. Here we prove that $K_s[\boldsymbol{\sigma}]$ is ω-free for almost all $\boldsymbol{\sigma} \in \mathrm{Gal}(K)^e$. This will imply that $K_s[\boldsymbol{\sigma}]$ is Hilbertian.

This result presents a climax of Field Arithmetic. It uses several major results proved in this book and two results whose full proofs are unfortunately beyond the framework of the book. The latter are the regularity of finite groups over complete discrete fields (Proposition 16.12.1) and the stability of fields (Theorem 18.9.3). The former include Borel-Cantelli Lemma, the theorem of Ax-Roquette that every algebraic extension of a PAC field is PAC (Theorem 11.2.5), Weissauer's theorem about infinite extensions of Hilbert fields (Theorem 13.9.1) Roquette's theorem that every ω-free PAC field is Hilbertian (Theorem 27.3.3), and Melnikov's characterization of closed normal subgroups of $\hat{F}_\omega$ by finite simple quotients (Corollary 25.7.6).

The strategy of the proof is for a $\boldsymbol{\sigma}$ taken at random in $\mathrm{Gal}(K)^e$ to embed $[\boldsymbol{\sigma}]$ as a closed normal subgroup of $\hat{F}_\omega$, to prove that each of the groups $\mathbb{Z}/p\mathbb{Z}$ is a quotient $[\boldsymbol{\sigma}]$, and every finite non-Abelian simple is a quotient of $[\boldsymbol{\sigma}]$ infinitely often. By Corollary 25.7.6, this will prove that $[\boldsymbol{\sigma}] \cong \hat{F}_\omega$.

We begin by enhancing some notation introduced in Section 18.10: Let N/K be a Galois extension, L/K a Galois subextension, and $\boldsymbol{\sigma} = (\sigma_1, \ldots, \sigma_e) \in \mathrm{Gal}(N/K)^e$. Then $L(\boldsymbol{\sigma})$ is the fixed field in L of $\sigma_1, \ldots, \sigma_e$, $L[\boldsymbol{\sigma}]$ is the maximal Galois extension of K which is contained in $L(\boldsymbol{\sigma})$. It follows that $L[\boldsymbol{\sigma}] = L \cap N[\boldsymbol{\sigma}]$.

LEMMA 27.4.1: *Let K be a Hilbertian field. Then for almost all $(\boldsymbol{\sigma}, \tau) \in \mathrm{Gal}(K)^{e+1}$, the field $K_s[\boldsymbol{\sigma}, \tau]$ is properly contained in $K_s[\boldsymbol{\sigma}]$.*

Proof: Corollary 16.2.7 gives a linearly disjoint sequence $L_1, L_2, L_3, \ldots$ of Galois extensions of K with $\mathrm{Gal}(L_i/K) \cong (\mathbb{Z}/2\mathbb{Z})^{e+1}$, $i = 1, 2, 3, \ldots$. For each i let $\sigma_{i1}, \ldots, \sigma_{ie}, \tau_i$ be a system of generators for $\mathrm{Gal}(L_i/K)$. For almost all $(\boldsymbol{\sigma}, \tau) \in \mathrm{Gal}(K)^{e+1}$ there is an i with $\mathrm{res}_{L_i}(\boldsymbol{\sigma}, \tau) = (\boldsymbol{\sigma}_i, \tau_i)$ (Lemma 18.5.3(b)). Then $K = L_i(\boldsymbol{\sigma}, \tau) = L_i[\boldsymbol{\sigma}, \tau]$. Since $\mathrm{Gal}(L_i/K)$ is not generated by e elements, $K \subset L_i(\boldsymbol{\sigma}_i)$. Since L_i/K is Abelian, $L_i(\boldsymbol{\sigma}_i) = L_i[\boldsymbol{\sigma}_i]$. Consequently, $K_s[\boldsymbol{\sigma}, \tau] \subset K_s[\boldsymbol{\sigma}]$. □

LEMMA 27.4.2: *Let K be a countable Hilbertian field. Then for almost all $\boldsymbol{\sigma} \in \mathrm{Gal}(K)^e$, the field $K_s[\boldsymbol{\sigma}]$ is a Galois extension of an ω-free PAC field.*

Proof: Let S be the set of all $(\boldsymbol{\sigma}, \tau) \in \text{Gal}(K)^{e+1}$ with the following properties:
(1a) $K_s[\boldsymbol{\sigma}]$ is PAC.
(1b) $K_s[\boldsymbol{\sigma}]$ properly contains $K_s[\boldsymbol{\sigma}, \tau]$.

By Theorem 18.10.2 and by Lemma 27.4.1, $\mu(S) = 1$. Hence, by Fubini's theorem, for almost all $\boldsymbol{\sigma} \in \text{Gal}(K)^e$ the set of all $\tau \in \text{Gal}(K)$ with $(\boldsymbol{\sigma}, \tau) \in S$ has measure 1 [Halmos, p. 147, Thm. A]. Thus, for almost all $\boldsymbol{\sigma} \in \text{Gal}(K)^e$ there is a $\tau \in \text{Gal}(K)$ satisfying (1).

Let $(\boldsymbol{\sigma}, \tau) \in S$. Choose a proper finite extension M of $K_s[\boldsymbol{\sigma}, \tau]$ in $K_s[\boldsymbol{\sigma}]$. By Weissauer (Theorem 13.9.1), M is Hilbertian. By Ax-Roquette (Corollary 11.2.5), M is PAC. Hence, by Example 24.8.5(b), M is ω-free. $\square$

PROPOSITION 27.4.3: *Let K be a field and G a finite group. Then there exists a variety V over K with the following property: If L is a field extension of K and $V(L) \neq \emptyset$, then G is regular over L.*

Proof: Consider the formal power series $E = K((t))$ in the indeterminate t. By Proposition 16.12.1, G is regular over E. Let $f \in E[Y, Z]$ be a Z-stable polynomial which is Galois with respect to Z such that $\text{Gal}(f(Y, Z), E(Y)) \cong G$ (Proposition 16.2.8). Let $x_1, \ldots, x_n$ be the elements of E which appear as coefficients of f. Thus, $f(Y, Z) = h(\mathbf{x}, Y, Z)$ for some $h \in K[\mathbf{X}, Y, Z]$. By Example 3.5.1, E is a regular extension of K. By Lemma 10.2.2(c), $\mathbf{x}$ generates a variety W over K in $\mathbb{A}^n$. Lemma 16.1.4 gives a nonempty Zariski K-open subset W_0 of W such that $h(\mathbf{a}, Y, Z)$ is absolutely irreducible, Galois with respect to Z, and $\text{Gal}(h(\mathbf{a}, Y, Z), L(Y)) \cong G$ for every field extension L of K and every $\mathbf{a} \in W_0(L)$. Thus, G is regular over L (Remark 16.2.2).

Now choose a polynomial $g \in K[\mathbf{X}]$ which vanishes on $W \smallsetminus W_0$ but not on W. Let V be the variety generated in $\mathbb{A}^{n+1}$ over K by $(\mathbf{x}, g(\mathbf{x})^{-1})$. If L is a field extension of K and $(\mathbf{a}, b) \in V(L)$, then $\mathbf{a} \in W_0(L)$. Hence, by the preceding paragraph, G is regular over L. $\square$

LEMMA 27.4.4: *Let K be a Hilbertian field and G a finite group. Then there exists a positive integer n and a linearly disjoint sequence $L_1, L_2, L_3, \ldots$ of Galois extensions of K such that for each i, $\text{Gal}(L_i/K) \cong S_n$ and L_i has a linearly disjoint sequence $L_{i1}, L_{i2}, L_{i3}, \ldots$ of Galois extensions with $\text{Gal}(L_{ij}/L_i) \cong G$ for each j.*

Proof: Let V be the variety given by Proposition 27.4.3. Choose a generic point $\mathbf{x}$ of V over K and put $F = K(\mathbf{x})$. Then F is a regular extension of K (Corollary 10.2.2) of transcendence degree, say, r. Theorem 18.9.3 gives a separating transcendence base $t_1, \ldots, t_r$ for F/K such that the Galois closure $\hat{F}$ of $F/K(\mathbf{t})$ is a regular extension of K and $\text{Gal}(\hat{F}/K) \cong S_n$ for some positive integer n.

Since K is Hilbertian, Lemma 16.2.6 gives a linearly disjoint sequence $L_1, L_2, L_3, \ldots$ of Galois extensions of K with $\text{Gal}(L_i/K) \cong S_n$ and $V(L_i) \neq \emptyset$ for each i.

27.4 An Abundance of ω-Free PAC Fields

Let i be a positive integer. By Lemma 27.4.3, G is regular over L_i. Hence, by Lemma 16.2.6, L_i has a linearly disjoint sequence $L_{i1}, L_{i2}, L_{i3}, \ldots$ with $\mathrm{Gal}(L_{ij}/L_i) \cong G$ for each i. $\square$

LEMMA 27.4.5: *Let K be a Hilbertian field and G a finite group. Suppose G is the normal subgroup of itself generated by e elements. Then, for almost all $\boldsymbol{\sigma} \in \mathrm{Gal}(K)^e$, the group G is realizable over $K_s[\boldsymbol{\sigma}]$.*

Proof: Let L_i and L_{ij} be the fields which Lemma 27.4.4 gives. For each i and j choose $\boldsymbol{\sigma}_{ij} \in \mathrm{Gal}(L_{ij}/L_i)^e$ with $L_{ij}[\boldsymbol{\sigma}_{ij}] = L_i$. Then put $A_{ij} = \{\boldsymbol{\sigma} \in \mathrm{Gal}(L_i) \mid \mathrm{res}_{L_{ij}} \boldsymbol{\sigma} = \boldsymbol{\sigma}_i\}$. Since the L_{ij} are linearly disjoint over L_i with fixed degree, $\mu(\bigcup_{j=1}^{\infty} A_{ij}) = \mu(\mathrm{Gal}(L_i))$ (Lemma 18.5.3(b)). Similarly, $\mu(\bigcup_{i=1}^{\infty} \mathrm{Gal}(L_i)) = 1$. Hence, $\mu(\bigcup_{i,j} A_{ij}) = 1$.

Consider $\boldsymbol{\sigma} \in A_{ij}$. Then, $L_{ij} \cap K_s[\boldsymbol{\sigma}] = L_i$. Hence,

$$\mathrm{Gal}(L_{ij} K_s[\boldsymbol{\sigma}]/K_s[\boldsymbol{\sigma}]) \cong \mathrm{Gal}(L_{ij}/L_i) \cong G,$$

as claimed. $\square$

LEMMA 27.4.6: *Let S be a finite non-Abelian simple group and n a positive integer. Then, for almost all $\boldsymbol{\sigma} \in \mathrm{Gal}(K)^e$, the group S^n occurs as a Galois group over $K_s[\boldsymbol{\sigma}]$.*

Proof: Rewrite S^n as $\prod_{i=1}^n S_i$ with $S_i \cong S$. Choose $s = (s_1, \ldots, s_n) \in S^n$ with $s_i \neq 1$ for all i. Let N be the normal subgroup of G generated by s. Then the projection of N on the ith coordinate is the whole group S_i. By Lemma 18.3.9, $N = S^n$. It follows from Lemma 27.4.5 that S^n occurs as a Galois group over $K_s[\boldsymbol{\sigma}]$ for almost all $\boldsymbol{\sigma} \in \mathrm{Gal}(K)^e$. $\square$

LEMMA 27.4.7: *Let p be a prime number. Then, for almost all $\boldsymbol{\sigma} \in \mathrm{Gal}(K)^e$, the group $\mathbb{Z}/p\mathbb{Z}$ occurs as a Galois group over $K_s[\boldsymbol{\sigma}]$.*

Proof: Lemma 16.3.6 gives a linearly disjoint sequence $L_1, L_2, L_3, \ldots$ of Galois extensions of K with $\mathrm{Gal}(L_i/K) \cong \mathbb{Z}/p\mathbb{Z}$ for each i. Choose a generator $\bar{\sigma}_i$ of $\mathrm{Gal}(L_i/K)$. For almost all $\boldsymbol{\sigma} \in \mathrm{Gal}(K)^e$ there is an i with $\mathrm{res}_{L_i}(\sigma_1, \ldots, \sigma_e) = (\bar{\sigma}_i, \ldots, \bar{\sigma}_i)$ (Lemma 18.5.3(b)). Hence, $\mathbb{Z}/p\mathbb{Z}$ is realizable over $K_s[\boldsymbol{\sigma}]$. $\square$

We sum up and prove our main result:

THEOREM 27.4.8 ([Jarden16, Thm. 2.7]): *Let K be a countable Hilbertian field. Then, for almost all $\boldsymbol{\sigma} \in \mathrm{Gal}(K)^e$, the field $K_s[\boldsymbol{\sigma}]$ is ω-free and PAC. In particular, $K_s[\boldsymbol{\sigma}]$ is Hilbertian.*

Proof: Let S be the set of all $\boldsymbol{\sigma} \in \mathrm{Gal}(K)^e$ with the following properties:
(2a) $K_s[\boldsymbol{\sigma}]$ is PAC.
(2b) K has an algebraic extension M in $K_s[\boldsymbol{\sigma}]$ which is ω-free.
(2c) For each finite non-Abelian group S and for every positive integer n, the group S^n occurs over $K_s[\boldsymbol{\sigma}]$ as a Galois group.
(2d) $\mathbb{Z}/p\mathbb{Z}$ occurs over $K_s[\boldsymbol{\sigma}]$ as a Galois group for every prime number p.

By Theorem 18.10.2, Lemma 27.4.2, Lemma 27.4.6, and Lemma 27.4.7, $\mu(S) = 1$. Let $\sigma \in S$ and $G = \text{Gal}(K_s[\sigma])$. Then G is a closed normal subgroup of $\text{Gal}(M)$ and $\text{Gal}(M) \cong \hat{F}_\omega$. By (2c), (2d), and Corollary 25.7.6, $G \cong \hat{F}_\omega$.

It follows from (2a) and Corollary 27.3.3 that $K_s[\sigma]$ is Hilbertian. □

Notes

The notions of model completion and model companion are due to A. Robinson. More in this direction can be found in [Hirschfeld-Wheeler].

The predicate symbols R_n of Section 24.6 appear for the first time in [Adler-Kiefe, p. 306] in order to prove that the theory of 1-free perfect PAC fields is the model completion of the theory of perfect fields of corank at most 1.

One can find the predicates $Q_{p,n}$ in [K. Schmidt, p. 96]. Schmidt uses an ultraproduct criterion to prove (in the notation of Section 24.7) that the class of $T_{R,Q}$-existentially closed models is axiomatizable. Hence, $T_{R,Q}$ has a model companion $\tilde{T}_{R,Q}$ [K. Schmidt1, p. 107]. He also observes that models of $\tilde{T}_{R,Q}$ are PAC and Hilbertian [K. Schmidt1, p. 97]. Theorems 27.2.3 and 27.3.1 are results of a correspondence with K. Schmidt. They also appear in [K. Schmidt2, p. 92]. One can find in [Ershov3, p. 512] results without proof on model completions of various theories of fields in the spirit of Remark 27.3.2.

Corollary 27.3.3 is proved in [Jarden9, p. 145] by model theoretic methods.

Chapter 28. Undecidability

In contrast to the theories considered so far (e.g. the theory of finite fields, the theory of almost all fields $\tilde{\mathbb{Q}}(\sigma_1, \ldots, \sigma_e)$ for fixed e, and the theory of perfect PAC fields of bounded corank), the theory of perfect PAC fields is undecidable. This is the main result of this chapter (Corollary 28.10.2). An application of Cantor's diagonalization process to Turing machines shows that certain families of Turing machines are nonrecursive. An interpretation of these machines in the theory of graphs shows the latter theory to be undecidable. Finally, Frattini covers interpret the theory of graphs in the theory of fields. This applies to demonstrate the undecidability of the theory of perfect PAC fields.

28.1 Turing Machines

A Turing machine may be considered as an abstract primitive computer. As with the "machine languages" in use in the 1950's, description and interpretation of the machine causes no difficulties, but programming the machine requires immense work.

Definition: A **Turing machine** is a system $M = \langle Q, R, L, S, \text{pr}, \text{mv}, \text{md} \rangle$, consisting of these components: a finite set $Q = \{q_1, \ldots, q_e\}$ of **operational modes**; the **move commands** R ("move right"), L ("move left") and S ("stay"); and its **working functions**, pr: $Q \times \{0,1\} \to \{0,1\}$ ("print"), mv: $Q \times \{0,1\} \to \{R, L, S\}$ ("move") and md: $Q \times \{0,1\} \to Q$ ("next mode").

A **tape** $T = \boxed{t_1 | t_2 | t_3 | t_4 | \cdots}$ is an infinite strip (sequence) with $t_i \in \{0, 1\}$, $i = 1, 2, \ldots$. The sequence of squares without reference to the t_i's is called the **ribbon** of M. Each Turing machine M operates on a tape T, the **initial tape**, as a mechanical device whose operational modes correspond to the elements of Q. Intuitively the initial tape is "input data" and the functions pr, mv, and md are the "program." We use the word **instant** to refer to the time interval required for the sequence of machine operations called a **beat** to be executed. We denote the value of t_y at the instant x by $\tau(x, y)$, and let $\kappa(x)$ be the mode of the machine at instant x. If M is at initial mode q_1, in instant 1 the machine goes through beat 1: it scans square 1; then it prints $\tau(2, 1) = \text{pr}(q_1, t_1)$ in square 1; moves one square to the right if $\text{mv}(q_1, t_1) = R$, and it stays at square 1 if $\text{mv}(q_1, t_1) = S$; and then it changes the mode to $\text{md}(q_1, t_1)$. At the xth instant with the machine in mode $\kappa(x)$ and at the yth square it goes through beat x: it reads the value $\tau(x, y)$ in square y; then it prints $\tau(x + 1, y) = \text{pr}(\kappa(x), \tau(x, y))$ in the yth square; it then moves one square to the right if $\text{mv}(\kappa(x), \tau(x, y)) = R$, it stays at the yth square if $\text{mv}(\kappa(x), \tau(x, y)) = S$, and it moves one square to the left if $\text{mv}(\kappa(x), \tau(x, y)) = L$ (unless $y = 1$, in which case the machine stops); and then it changes the mode to $\kappa(x + 1) = \text{md}(\kappa(x), \tau(x, y))$.

If the machine does not stop at the end of the xth instant, it proceeds to perform the $(x+1)$th beat. At the end of the xth beat the machine has changed the original tape $T = T_1$ to a tape T_x. The action of M is said to be **eventually stationary** if there exists an instant b when the machine is at square y, $\text{mv}(\kappa(b), \tau(b,y)) = S$, and $\tau(x,y) = \tau(b,y)$ and $\kappa(x) = \kappa(b)$ for each $x \geq b$ (i.e. each successive beat after b consists of the same operation; and $T_b = T_{b+1} = \cdots$). If M stops at instant c, then we call $T' = T_{b+1}$ the **final tape**.

The machine M is uniquely determined by its working functions pr, mv, and md. Let $p_1 < p_2 < \cdots$ be the sequence of primes. Encode M as a positive integer as follows: For $t \in \{0,1\}$ and $q_i \in Q$, let

$$\gamma(q_i, t) = \begin{cases} 5 & \text{if } \text{pr}(q_i, t) = 0 \\ 7 & \text{if } \text{pr}(q_i, t) = 1, \end{cases}$$

$$\delta(q_i, t) = \begin{cases} 11 & \text{if } \text{mv}(q_i, t) = R \\ 13 & \text{if } \text{mv}(q_i, t) = L \\ 17 & \text{if } \text{mv}(q_i, t) = S, \text{ and} \end{cases}$$

$$\varepsilon(q_i, t) = p_{10+j} \quad \text{if} \quad \text{md}(q_i, t) = q_j.$$

Now define $\text{code}(M)$ to be $2^{a(M)} 3^{b(M)}$ with

$$a(M) = \prod_{i=1}^{e} p_{10+i}^{19^{\gamma(q_i,0)} 23^{\delta(q_i,0)} 29^{\varepsilon(q_i,0)}} \quad \text{and} \quad b(M) = \prod_{i=1}^{e} p_{10+i}^{19^{\gamma(q_i,1)} 23^{\delta(q_i,1)} 29^{\varepsilon(q_i,1)}}.$$

The map from M to $\text{code}(M)$ is injective, and the set of codes of all Turing machines is primitive recursive (Exercise 2). Moreover, one may reconstruct $\gamma, \delta, \varepsilon$ and therefore M from its code.

28.2 Computation of Functions by Turing Machines

Denote the set of nonnegative integers by N. For a positive integer n, partition the ribbon R of the Turing machine M into n subribbons, $R = R_1 \cup \cdots \cup R_n$, where R_k consists of all squares y with $y \equiv k \bmod n$. Call each such subribbon an **n-track** (or just **track**). In each track we may **store** an integer $z_0 \in N$ by printing 1 in the first z_0 squares of the track and 0 in the rest of the squares. Let $f \colon N^m \to N$ be a function and let $k_1, \ldots, k_{m+1} \leq n$ be distinct positive integers. We say that M **computes f (from the n-tracks $k_1, \ldots, k_m$ to the k_{m+1}th n-track)** if when M operates on a tape T with $x_1, \ldots, x_m$ stored in the n-tracks $k_1, \ldots, k_m$, respectively, and 0 stored in the remaining tracks, then M eventually stops and the final tape T' is identical with T except that M has stored $f(x_1, \ldots, x_m)$ in the k_{m+1}th n-track.

PROPOSITION 28.2.1: *For each recursive function $f(x_1, \ldots, x_m)$ there exists a Turing machine M which computes f.*

28.2 Computation of Functions by Turing Machines

Proof: A recursive function arises from a list of basic functions and basic operations (Section 8.5), and the parts of the proof treat each of these separately. The first step in each part is the choice of an integer n and integers $k_1, \ldots, k_{m+1}$ with $1 < k_1 < \cdots < k_{m+1} \leq n$ such that M computes f from the n-tracks $k_1, \ldots, k_m$ to the k_{m+1}th n-track without changing the n-tracks $k_1, \ldots, k_m$ (where the arguments for f are stored). In each case we store 1 in the first track and we leave this track unchanged through the whole operation of M. Also, in each case to each mode of operation q_i there corresponds a number l_i, with $1 \leq l_i \leq n$, such that whenever M is in mode q_i it scans a square in the l_ith n-track. In particular, the first track throughout the action of M corresponds only to the initial mode q_1 and the final mode q_e. Thus, if M is in mode q_e and it reads 1, then it scans square 1. If $e \neq 1$, we define $\mathrm{mv}(q_1, 1) = R$ and $\mathrm{mv}(q_e, 1) = L$. This gives complete control on the action of M at the last instant, where M is at the mode q_e and it reads 1. We bother to define pr, mv and md of M only for those arguments which actually occur in their computation. Thus, our definitions appear as commands for a computer (with all commands, that would never be executed, left unstated). Throughout the proof the variable t ranges over $\{0, 1\}$.

PART A: *The zero function* $f(x) = 0$. With $n = 3$, $k_1 = 2$, $k_2 = 3$ and $e = 1$, the machine M computes f from the 2nd track to the 3rd. It does this in one beat:

$$\mathrm{pr}(q_1, 1) = 1, \quad \mathrm{mv}(q_1, 1) = L, \quad \mathrm{md}(q_1, 1) = q_1.$$

For example, here is the starting (and final) tape for $x = 4$

| 1 | 1 | 0 | 0 | 1 | 0 | 0 | 1 | 0 | 0 | 1 | 0 | 0 | 0 | 0 | $\cdots$ |

PART B: *The successor function* $f(x) = x + 1$. With $n = 3$, $k_1 = 2$, $k_2 = 3$ and $e = 6$, M computes f from the 2nd track to the 3rd. In outline: M moves to the right and scans the squares in the second track printing 1 in the 3rd track square to the right of the second track square. The last time it does this is when it has scanned 0 in the second track. Then, after printing 1 in the next square to the right, it moves only left until it stops:

$$\begin{aligned}
&\mathrm{pr}(q_1, t) = t, &&\mathrm{mv}(q_1, t) = R, &&\mathrm{md}(q_1, t) = q_2; \\
&\mathrm{pr}(q_2, 1) = 1, &&\mathrm{mv}(q_2, 1) = R, &&\mathrm{md}(q_2, 1) = q_3; \\
&\mathrm{pr}(q_3, 0) = 1, &&\mathrm{mv}(q_3, 0) = R, &&\mathrm{md}(q_3, t) = q_1; \\
&\mathrm{pr}(q_2, 0) = 0, &&\mathrm{mv}(q_2, 0) = R, &&\mathrm{md}(q_2, 0) = q_4; \\
&\mathrm{pr}(q_4, t) = 1, &&\mathrm{mv}(q_4, t) = L, &&\mathrm{md}(q_4, t) = q_5; \\
&\mathrm{pr}(q_5, t) = t, &&\mathrm{mv}(q_5, t) = L, &&\mathrm{md}(q_5, t) = q_6; \\
&\mathrm{pr}(q_6, t) = t, &&\mathrm{mv}(q_6, t) = L, &&\mathrm{md}(q_6, t) = q_4.
\end{aligned}$$

Example: Computation of $f(3)$.

inst\sqr	1	2	3	4	5	6	7	8	9	10	11	12	mode
1	**1**	1	0	0	1	0	0	1	0	0	0		1
2	1	**1**	0	0	1	0	0	1					2
3	1	1	**0**	0	1	0	0	1					3
4	1	1	1	**0**	1	0	0	1					1
5	1	1	1	0	**1**	0	0	1					2
6	1	1	1	0	1	**0**	0	1					3
7	1	1	1	0	1	1	**0**	1					1
8	1	1	1	0	1	1	0	**1**					2
9	1	1	1	0	1	1	0	1	**0**				3
10	1	1	1	0	1	1	0	1	1	**0**			1
11	1	1	1	0	1	1	0	1	1	0	**0**		2
12	1	1	1	0	1	1	0	1	1	0	0	**0**	4
13	1	1	1	0	1	1	0	1	1	0	**0**	1	5
14	1	1	1	0	1	1	0	1	1	**0**	0	1	6
15	1	1	1	0	1	1	0	1	**1**	0	0	1	4
16	1	1	1	0	1	1	0	**1**	1	0	0	1	5
17	1	1	1	0	1	1	**0**	1	1	0	0	1	6
18	1	1	1	0	1	**1**	0	1	1	0	0	1	4
19	1	1	1	0	**1**	1	0	1	1	0	0	1	5
20	1	1	1	**0**	1	1	0	1	1	0	0	1	6
21	1	1	**1**	0	1	1	0	1	1	0	0	1	4
22	1	**1**	1	0	1	1	0	1	1	0	0	1	5
23	**1**	1	1	0	1	1	0	1	1	0	0	1	6

STOP

The xth line indicates the tape and the mode of operation at the beginning of the xth instant. The square with the bold figure is that being scanned.

PART C: *The coordinate projection function $f(x_1, \ldots, x_m) = x_l$, $1 \leq l \leq m$.* With $n = m+2$ and $e = 2n+1$, M computes f from the n-tracks $2, \ldots, m+1$ to the $(m+2)$th n-track. In outline: as M moves to the right and scans 1 in the $(l+1)$th n-track it prints 1 in the next square of the $(m+2)$th n-track. When, however, M scans 0 in the $(l+1)$th n-track, it starts to move to the left until it stops:

$$\mathrm{pr}(q_i, t) = t, \quad \mathrm{mv}(q_i, t) = R, \quad \mathrm{md}(q_i, t) = q_{i+1}, \ i = 1, \ldots, l,$$
$$l+2, \ldots, n-1;$$

$$\mathrm{pr}(q_{l+1}, 1) = 1, \quad \mathrm{mv}(q_{l+1}, 1) = R, \quad \mathrm{md}(q_{l+1}, 1) = q_{l+2};$$
$$\mathrm{pr}(q_n, 0) = 1, \quad \mathrm{mv}(q_n, 0) = R, \quad \mathrm{md}(q_n, t) = q_1;$$
$$\mathrm{pr}(q_{l+1}, 0) = 0, \quad \mathrm{mv}(q_{l+1}, 0) = L, \quad \mathrm{md}(q_{l+1}, 0) = q_{n+l};$$
$$\mathrm{pr}(q_i, t) = t, \quad \mathrm{mv}(q_i, t) = L, \quad \mathrm{md}(q_i, t) = q_{i-1}, \ i = n+3, \ldots, 2n+1;$$
$$\mathrm{pr}(q_{n+2}, t) = t, \quad \mathrm{mv}(q_{n+2}, t) = L, \quad \mathrm{md}(q_{n+2}, t) = q_{2n+1}.$$

28.2 Computation of Functions by Turing Machines

PART D: *Iteration.* There are two ingredients. First: Given a machine M that computes a function $f(x_1, \ldots, x_m)$ from the n-tracks $k_1, \ldots, k_m$ to the k_{m+1}th n-track we may, for any $r > n$, replace M by a machine M' that computes $f(x_1, \ldots, x_m)$ from the r-tracks $l_1, \ldots, l_m$ to the l_{m+1}th r-track.

Second: Given M_1 and M_2 that, respectively, compute the functions $f_1(x_1, \ldots, x_m)$ and $f_2(x_1, \ldots, x_m)$, we may construct a machine M that first computes f_1 and then computes f_2. Indeed, adjust the computation tracks of f_1 and f_2, if necessary, so that the corresponding tracks in M will be disjoint. Let $q_1, \ldots, q_e$ denote the modes of M_1 and $q_{e+1}, \ldots, q_{e+e'}$ the modes of M_2. Then in the program for M_1 replace the values of pr, mv, and md at $(q_e, 1)$ by

$$\text{pr}(q_e, 1) = 1, \quad \text{mv}(q_e, 1) = S, \quad \text{md}(q_e, 1) = q_{e+1}.$$

With the commands for M_2 now listed below those for M_1 we have M.

Each of the remaining parts of the proof considers one of the recursion operations:

PART E: *Composition,*

$$f(x_1, \ldots, x_m) = g(h_1(x_1, \ldots, x_m), \ldots, h_l(x_1, \ldots, x_m)).$$

Suppose that for each i, $1 \leq i \leq l$, there exists a machine M_i that computes $h_i(x_1, \ldots, x_m)$ from the n_i-tracks $k_{i,1}, \ldots, k_{i,m}$ to the $k_{i,m+1}$th n_i-track and that there exists M_0 that computes $g(y_1, \ldots, y_l)$ from the n_0-tracks $k_{0,1}, \ldots, k_{0,l}$ to the $k_{0,l+1}$th n_0-track. Choose n suitably large and use Part D to construct M that successively computes $y_i = h_i(x_1, \ldots, x_m)$ from the n-tracks $2, \ldots, m+1$ to the $(m+1+i)$th n-tracks, $i = 1, \ldots, l$, and then computes $g(y_1, \ldots, y_l)$ from the n-tracks $m+2, \ldots, m+1+l$ to the $(m+2+l)$th n-track. Altogether M computes $f(x_1, \ldots, x_m)$ from the n-tracks $2, \ldots, m+1$ to the $(m+l+2)$nd n-track.

PART F: *Induction,*

$$f(x_1, \ldots, x_m, 0) = f_0(x_1, \ldots, x_m)$$
$$f(x_1, \ldots, x_m, y+1) = g(x_1, \ldots, x_m, y, f(x_1, \ldots, x_m, y)).$$

Suppose f_0 and g can be computed by Turing machines. Apply Part D to the argument below, with n chosen suitably large. First, M copies y from the $(m+2)$nd n-track to the $(m+4)$th n-track. Second, M computes $f_0(x_1, \ldots, x_m)$ from the n-tracks $2, \ldots, m+1$ to the $(m+3)$rd track. Third, M successively computes

$$g(x_1, \ldots, x_m, z, f(x_1, \ldots, x_m, z)),$$

for $z = 0, \ldots, y-1$ from the n-tracks $2, \ldots, m+1, m+5, m+3$ to the $(m+3)$rd n-track; it subtracts 1 from the $(m+4)$th n-track and adds 1 to the $(m+5)$th n-track. The computation ends when 0 has been stored in the

$(m+4)$th n-track, and M has computed $f(x_1,\ldots,x_m,y)$ from the n-tracks $2,\ldots,m+2$ to the $(m+3)$rd n-track.

Note the loop in the computation. The number of times the machine has to repeat the loop is known in advance; it is $y+1$.

PART G: *The minimalization function,*

$$f(x_1,\ldots,x_m) = \min\{y \in N \mid g(x_1,\ldots,x_m,y) = 0\}.$$

Assume that g is a computable function for which, given $x_1,\ldots,x_m$, there exists $y \in N$ such that $g(x_1,\ldots,x_m,y) = 0$.

A loop appears in M whose number of repetitions is not known in advance. In the yth repetition M computes $z = g(x_1,\ldots,x_m,y)$ from the tracks $2,\ldots,m+2$ to the track $m+3$. If $z=0$ the machine stops. Otherwise, the value of y in the track $m+2$ is increased by 1 and the loop is repeated once more. The computation begins with $y=0$. The value of y in the track $m+2$ at the end of the computation is $f(x_1,\ldots,x_m)$. □

Remark 28.2.2: Each of the Turing machines constructed in Proposition 28.2.1 uses a different number of tracks for the computation. Thus, if M_1 and M_2 compute functions of one variable $f_1(x)$ and $f_2(x)$ using, respectively, n_1 and n_2 tracks with $n_1 \neq n_2$, then the argument x has been stored differently in the initial tapes of M_1 and M_2. In order to unify the storage of the arguments and the value of a function (dependent only on the number of variables) let $1 < k_1,\ldots,k_m \leq n+1$ be distinct integers and construct a Turing machine M that first changes a tape with the numbers $1, x_1,\ldots,x_m$ stored in the $(m+1)$-tracks $1, 2,\ldots,m+1$ to a tape with the same numbers now stored in the $(n+1)$-tracks $1, k_1,\ldots,k_m$. Since the actual construction of M is lengthy, we do not carry it out here. This allows us to follow the proof of Proposition 28.2.1 and to construct, for each recursive function $f(x_1,\ldots,x_m)$, a machine M that changes a tape on which $1, x_1,\ldots,x_m$ are stored in the $(m+1)$-tracks $1, 2,\ldots,m+1$ to a tape on which 1 and $f(x_1,\ldots,x_m)$ are stored in the 1st and 2nd 2-tracks. □

28.3 Recursive Inseparability of Sets of Turing Machines

Recursive sets and recursive functions (Section 8.6) are technical tools that interpret decidability of theories. Undecidable theories correspond to nonrecursive sets. In order to prove that a subset A of N ($=$ the set of nonnegative integers) is nonrecursive, one may prove a stronger property, namely that A is "recursively inseparable" from a certain subset B of N. The advantage of the stronger concept lies in Lemma 28.3.2. That lemma gives a convenient tool to deduce the recursive inseparability of sets A' and B' from the recursive inseparability of A and B.

Definition 28.3.1: Recursively inseparable sets. We call disjoint subsets A and B of N **recursively separable** if there exists a recursive set C such that

28.3 Recursive Inseparability of Sets of Turing Machines

$A \subseteq C$ and $B \cap C = \emptyset$. Otherwise, A and B are **recursively inseparable**. In the latter case both A and B are nonrecursive. □

LEMMA 28.3.2: *Let A and B be recursively inseparable subsets of N. Suppose that A' and B' are disjoint subsets of N and that there exists a recursive function $f\colon N \to N$ such that $f(A) \subseteq A'$ and $f(B) \subseteq B'$. Then A' and B' are recursively inseparable.*

Proof: Assume that there exists a recursive set C such that $A' \subseteq C$ and $B' \cap C = \emptyset$. Then $A \subseteq f^{-1}(C)$ and $B \cap f^{-1}(C) = \emptyset$. Since the characteristic function of $f^{-1}(C)$ satisfies $\chi_{f^{-1}(C)}(x) = \chi_C(f(x))$, the set $f^{-1}(C)$ is recursive. Thus, contrary to assumption, A and B are recursively separable. □

Definition 28.3.3: Recursively inseparable. Two collections of Turing machines, A and B, are said to be **recursively inseparable** if $\mathrm{code}(A)$ and $\mathrm{code}(B)$ are recursively inseparable subsets of N. □

We construct recursively inseparable sets of machines by considering the action of each machine on its own code.

Definition 28.3.4: The tapes T_n. For n a nonnegative integer, denote the tape on which 1 and n, respectively, are stored in the 1st and 2nd 2-tracks by T_n. Let $\mathrm{Hlt}(n)$ (resp. $\mathrm{Stat}(n)$) be the collection of Turing machines which eventually stop (resp. become stationary) when they work on T_n. Finally, denote the collection of Turing machines M which belong to $\mathrm{Hlt}(\mathrm{code}(M))$ (resp. $\mathrm{Stat}(\mathrm{code}(M))$) by Hlt (resp. Stat). □

LEMMA 28.3.5: *The sets of Turing machines Hlt and Stat are recursively inseparable.*

Proof: Assume that there exists a recursive set C such that

(1) $\qquad \mathrm{code}(\mathrm{Stat}) \subseteq C \quad \text{and} \quad \mathrm{code}(\mathrm{Hlt}) \cap C = \emptyset.$

By definition, χ_C is a recursive function. By Remark 28.2.2 there exists a machine M_0 which eventually stops, when applied to T_x, to produce the final tape $T_{\chi_C(x)}$. Let q_e be the final operation mode and, with no loss, assume that $e > 1$. Then, from the opening statements of the proof of Proposition 28.2.1,

(2) $\qquad \mathrm{pr}(q_e, 1) = 1, \quad \mathrm{mv}(q_e, 1) = L, \quad \mathrm{md}(q_e, 1) = q_d,$

for some d, $1 \le d \le e$. Add modes q_{e+1}, q_{e+2}, and q_{e+3} and change (2) to

(3) $\qquad \mathrm{pr}(q_e, 1) = 1, \quad \mathrm{mv}(q_e, 1) = R, \quad \mathrm{md}(q_e, 1) = q_{e+1},$

with the following addition to the operations of M_0 (and (3)):

$$\mathrm{pr}(q_{e+1}, 1) = 1, \quad \mathrm{mv}(q_{e+1}, 1) = L, \quad \mathrm{md}(q_{e+1}, 1) = q_{e+2};$$
$$\mathrm{pr}(q_{e+2}, 1) = 1, \quad \mathrm{mv}(q_{e+2}, 1) = L, \quad \mathrm{md}(q_{e+2}, 1) = q_{e+2};$$
$$\mathrm{pr}(q_{e+1}, 0) = 0, \quad \mathrm{mv}(q_{e+1}, 0) = L, \quad \mathrm{md}(q_{e+1}, 0) = q_{e+3};$$
$$\mathrm{pr}(q_{e+3}, 1) = 1, \quad \mathrm{mv}(q_{e+3}, 1) = S, \quad \mathrm{md}(q_{e+3}, 1) = q_{e+3}.$$

This produces a new machine M_1 for which one checks that

(4) $\quad x \in C$ implies $M_1 \in \text{Hlt}(x)$, otherwise $M_1 \in \text{Stat}(x)$.

Consider $\text{code}(M_1)$: If $\text{code}(M_1) \in C$, then $M_1 \in \text{Hlt}(\text{code}(M_1))$. Hence, $M_1 \in \text{Hlt}$ and (by (1)) $\text{code}(M_1) \notin C$. If $\text{code}(M_1) \notin C$, then $M_1 \in \text{Stat}(\text{code}(M_1))$. Hence, $M_1 \in \text{Stat}$ and $\text{code}(M_1) \in C$. These contradictions conclude the lemma. $\square$

LEMMA 28.3.6: *The collections* $\text{Stat}(0)$ *and* $\text{Hlt}(0)$ *are recursively inseparable.*

Proof: To each Turing machine M we associate a new machine, M', such that

(5a) $\qquad\qquad M \in \text{Stat} \Longrightarrow M' \in \text{Stat}(0),$
(5b) $\qquad\qquad M \in \text{Hlt} \Longrightarrow M' \in \text{Hlt}(0).$

In outline, define M' so that applied to T_0, it first changes T_0 to T_m with $m = \text{code}(M)$, and then it continues as M. Thus, if $M \in \text{Stat}$, then M' eventually becomes stationary, and if $M \in \text{Hlt}$, then M' eventually stops.

In more detail, if M has modes $q_1, \ldots, q_e$, then M' has $4m + e$ modes $q'_1, \ldots, q'_{4m+e}$. The definitions involving the first $4m$ modes allow M' to change T_0 to T_m:

$\text{pr}(q'_1, 1) = 1, \quad \text{mv}(q'_1, 1) = R, \quad \text{md}(q'_1, 1) = q'_2,$
$\text{pr}(q'_{2i}, 0) = 1, \quad \text{mv}(q'_{2i}, 0) = R, \quad \text{md}(q'_{2i}, 0) = q'_{2i+1}, \ i = 1, \ldots, m,$
$\text{pr}(q'_{2i+1}, 0) = 0, \quad \text{mv}(q'_{2i+1}, 0) = R, \quad \text{md}(q'_{2i+1}, 0) = q'_{2i+2}, \ i = 1, \ldots, m-1,$
$\text{pr}(q'_{2m+i}, t) = t, \quad \text{mv}(q'_{2m+i}, t) = L, \quad \text{md}(q'_{2m+i}, t) = q'_{2m+i+1}, \ i = 1, \ldots, 2m.$

The definitions involving the last e modes of M' essentially give the operations of M:

$\text{pr}(q'_{4m+i}, t) = \text{pr}(q_i, t), \quad \text{mv}(q'_{4m+i}, t) = \text{mv}(q_i, t), \quad \text{md}(q'_{4m+i}, t) = \text{md}(q_i, t),$

$i = 1, \ldots, e$. In the notation of Section 28.1, $\text{code}(M') = 2^{a(M')} 3^{b(M')}$ with

$$a(M') = \prod_{i=1}^{4m+e} p_{10+i}^{19^{\gamma(q'_i, 0)} 23^{\delta(q'_i, 0)} 29^{\varepsilon(q'_i, 0)}}, \quad b(M') = \prod_{i=1}^{4m+e} p_{10+i}^{19^{\gamma(q'_i, 1)} 23^{\delta(q'_i, 1)} 29^{\varepsilon(q'_i, 1)}}.$$

By Exercise 2, the set of all codes of Turing machines is primitive recursive. Therefore, functions $\text{code}(M) \mapsto a(M')$ and $\text{code}(M) \mapsto b(M')$ are primitive recursive.

Thus, the function $f \colon N \to N$ defined by $f(\text{code}(M)) = \text{code}(M')$ for each Turing machine M, and $f(n) = 0$ if n is not a code of a machine, is primitive recursive. By (5), $f(\text{code}(\text{Stat})) \subseteq \text{code}(\text{Stat}(0))$ and $f(\text{code}(\text{Hlt})) \subseteq \text{code}(\text{Hlt}(0))$. From Lemmas 28.3.2 and 28.3.5, the sets $\text{code}(\text{Stat}(0))$ and $\text{code}(\text{Hlt}(0))$ are therefore recursively inseparable. $\square$

28.4 The Predicate Calculus

For each positive integer n denote the first order language with countably many m-ary predicate symbols for each $m \leq n$ which does not contain the equality symbol by $\mathcal{L}_n$(predicate). This section interprets Turing machines in $\mathcal{L}_2$(predicate) and gives two recursively inseparable sets of sentences arising from Stat(0) and Hlt(0) (Proposition 28.4.4):

Definition 28.4.1: Stationary relations. For each positive integer m define $\pi_m \colon \mathbb{N} \to \{1,\ldots,m\}$ by $\pi(x) = x$ for $x = 1,\ldots,m$ and $\pi(x) = m$ for $x \geq m$. A k-ary relation R of $\mathbb{N}$ is said to be m-**stationary** if

$$(x_1,\ldots,x_i,\ldots,x_k) \in R \iff (x_1,\ldots,\pi_m(x_i),\ldots,x_k) \in R$$

for all $x_1,\ldots,x_k \in \mathbb{N}$ and each i, $1 \leq i \leq k$. In other words, if R is m-stationary, then $(x_1,\ldots,x_{i-1},j,x_{i+1},\ldots,x_k) \in R$ for some $j \geq m$ if and only if the k-tuple is in R for all $j \geq m$. Thus, if R is m-stationary, then R is also n-stationary for each $n \geq m$.

Let $\mathcal{L}'$ be a language that extends a language $\mathcal{L}$ and let $\mathcal{A} = \langle A, \ldots \rangle$ be a structure for $\mathcal{L}$. Call a sentence θ' of $\mathcal{L}'$ **satisfiable** in $\mathcal{A}$ if there exist interpretations of the relation symbols of $\mathcal{L}' \smallsetminus \mathcal{L}$ as relations of A such that θ' is true in the corresponding structure $\mathcal{A}'$ for $\mathcal{L}'$. Consider the structure $\langle \mathbb{N}, 1,' \rangle$ with $'$ as the successor function and let $\mathcal{L}$ be the corresponding language. A sentence θ in an extended language $\mathcal{L}'$ is said to be **stationarily satisfiable** in $\langle \mathbb{N}, 1,' \rangle$ if there exists a positive integer m such that the relations (excluding the successor function) that interpret the additional relation symbols are m-stationary. □

LEMMA 28.4.2: *To each Turing machine M there effectively corresponds a quantifier-free formula $\lambda(U,X,V,Y)$ in $\mathcal{L}_2$(predicate) such that:*
(a) $M \notin \mathrm{Hlt}(0)$ *if and only if* $(\forall X,Y)\lambda(1,X,X',Y)$ *is satisfiable in* $\langle \mathbb{N}, 1,' \rangle$.
(b) $M \in \mathrm{Stat}(0)$ *if and only if* $(\forall X,Y)\lambda(1,X,X',Y)$ *is stationarily satisfiable in* $\langle \mathbb{N}, 1,' \rangle$.

Proof: The proof divides into parts consisting of the construction of λ and the proofs of (a) and (b)

PART A: *Construction of λ.* Let M be a Turing machine with the operational modes $q_1,\ldots,q_e$ and the working functions pr, mv, and md. Define functions r, v, and d from $\{1, 2, \ldots, e\} \times \{0, 1\}$ to $\{0, 1\}$, $\{-1, 0, 1\}$, and $\{1, 2, \ldots, e\}$, respectively, as follows: $r(i, t) = \mathrm{pr}(q_i, t)$; $v(i, t) = -1, 0,$ or 1 as $\mathrm{mv}(q_i, t) = L, S,$ or R, respectively; and $\mathrm{md}(q_i, t) = q_{d(i,t)}$. Then $\lambda(1, X, X', Y)$ (whose terms will be explained in Parts B and C) consists of a conjunction of ten formulas:
(1a) $\bigvee_{i=1}^{e} Q_i(X) \wedge \bigvee_{i \neq j} \neg[Q_i(X) \wedge Q_j(X)]$.
(1b) $[T_0(X,Y) \vee T_1(X,Y)] \wedge \neg[T_0(X,Y) \wedge T_1(X,Y)]$.
(1c) $\bigvee_{j=-1}^{1} [V_j(X) \wedge \bigwedge_{i \neq j} \neg[V_i(X) \wedge V_j(X)]]$.
(1d) $Q_1(1) \wedge T_1(1,1) \wedge T_0(1,Y') \wedge \mathrm{Sc}(1,1) \wedge \neg \mathrm{Sc}(1,Y')$.

(1e) $\bigwedge_{i=1}^{e} \bigwedge_{t=0}^{1} [\mathrm{Sc}(X,Y) \wedge Q_i(X) \wedge T_t(X,Y)$
$\qquad \to T_{r(i,t)}(X',Y) \wedge V_{v(i,t)}(X) \wedge Q_{d(i,t)}(X')].$
(1f) $\neg \mathrm{Sc}(X,Y) \to [T_1(X',Y) \leftrightarrow T_1(X,Y)].$
(1g) $V_1(X) \to [\mathrm{Sc}(X,Y) \leftrightarrow \mathrm{Sc}(X',Y')] \wedge \neg \mathrm{Sc}(X',1).$
(1h) $V_0(X) \to [\mathrm{Sc}(X,Y) \leftrightarrow \mathrm{Sc}(X',Y)].$
(1i) $V_{-1}(X) \to [\mathrm{Sc}(X,Y') \leftrightarrow \mathrm{Sc}(X',Y)].$
(1j) $\neg[\mathrm{Sc}(X,1) \wedge V_{-1}(X)].$

Replace each occurrence of 1 in $\lambda(1, X, X', Y)$ by U and each occurrence of X' by V to obtain $\lambda(U, X, V, Y)$.

PART B: *Interpretation of λ and proof of (a)*. The formula λ derives from the effect of M as applied to the tape T_0. We interpret the relation symbols of λ as relations on $\mathbb{N}$ as follows:

(2a) $Q_i(x)$ holds $\iff$ the mode of M at instant x is q_i.
(2b) $V_j(x)$ holds $\iff$ the move command at instant x is L, S, R, respectively, if $j = -1, 0, 1$.
(2c) $T_t(x,y)$ holds $\iff$ the tape symbol in the yth square at instant x is t.
(2d) $\mathrm{Sc}(x,y)$ holds $\iff$ M scans the yth square at instant x.

For example, (1a) interprets that M is in one and only one of the modes $q_1, \ldots, q_e$ at any instant and (1b) interprets that the tape symbol in the yth square at instant x is either 0 or 1. Continuing in this way, (1d) interprets that M works on T_0 and (1j) interprets that if M scans square 1 at instant x, then the next move command is not L; in other words, M never stops. Thus, if $M \notin \mathrm{Hlt}(0)$, then $\lambda(1, x, x', y)$ is true for each $(x, y) \in \mathbb{N} \times \mathbb{N}$.

Conversely, suppose that $(\forall X, Y)\lambda(1, X, X', Y)$ is satisfiable in $\langle \mathbb{N}, 1, ' \rangle$. Then there are relations Q_i, V_j, T_t, and Sc on $\mathbb{N}$ such that $\lambda(1, x, x', y)$ is true in the corresponding extension of $\langle \mathbb{N}, 1, ' \rangle$ for each $x, y \in \mathbb{N}$. We must show that these relations then correspond to the operations of M as given by (2). With this conclude from (1j) that $M \notin \mathrm{Hlt}(0)$.

Indeed, if M starts to work at the instant $x = 1$ on the tape T_0, then from the validity of (1c), (1d), and (1e) for $x = 1$ it follows that (2) is true for $x = 1$ and for every y. For example, since the mode of operation of M at the instant $x = 1$ is q_1 and since $Q_1(1)$ is true (by (1d)), (2a) is true for $x = 1$. Next it follows from (1d) that $Q_1(1) \wedge T_1(1,1) \wedge \mathrm{Sc}(1,1)$ is true. Hence, $V_{v(1,1)}$ is true (by (1e)). Thus, by the definition of v, a move command L, S, or R at $x = 1$ implies that $V_{-1}(1)$, $V_0(1)$, or $V_1(1)$, respectively, is true. The converse follows from this and from $\bigwedge_{i \neq j} \neg(V_i(1) \wedge V_j(1))$ (by (1c)). Similarly verify (2c) and (2d) for $x = 1$.

Now assume by induction that (2) is true for x and for each y. It follows from (1a) and (1e) that (2a) is true for x'; from (1g)-(1i) that (2d) is true for x' and for each y; from (1c) and (1e) that (2b) is true for x'; and from (1b), (1e), and (1f) that (2c) is true for x', and for each y.

PART C: *Proof of (b)*. Suppose that $M \in \mathrm{Stat}(0)$. Define Q_i, V_j, T_t, and Sc by (2). Then $M \notin \mathrm{Hlt}(0)$ and, therefore (Part B), $\lambda(1, x, x', y)$ is true for all

28.4 The Predicate Calculus

$(x, y) \in \mathbb{N} \times \mathbb{N}$. By definition, there exist $a, b \in \mathbb{N}$ such that M scans square a at each instant $x \geq b$ and the tape and mode remain unchanged. Since M moves at most one square to the right at each instant, none of the squares greater than b has been scanned until instant b. Thus, $a \leq b$. Therefore, the relations Q_i, V_j, T_t, and Sc are $(b+1)$-stationary.

Conversely, suppose Q_i, V_j, T_t, and Sc are m-stationary relations and $\lambda(1, x, x', y)$ is true for all $(x, y) \in \mathbb{N} \times \mathbb{N}$. Apply M to the tape T_0 at instant $x = 1$. Then, as in Part B, Q_i, V_j, T_j, and Sc satisfy (2). Thus, at the latest, M is stationary from the mth instant on some square $a < m$. Consequently, $M \in \text{Stat}(0)$. □

The next lemma translates stationary satisfiability into satisfiability in finite models:

LEMMA 28.4.3: *Let $\lambda(U, X, V, Y)$ be a quantifier-free formula of $\mathcal{L}_n$(predicate) without an occurrence of* '. *Denote the sentence*

$$(\exists U)(\forall X)(\exists V)(\forall Y)\lambda(U, X, V, Y)$$

by θ and the sentence $(\forall X, Y)\lambda(1, X, X', Y)$ by θ'.
(a) *If θ has a model, then θ' is satisfiable in $\langle \mathbb{N}, 1, ' \rangle$.*
(b) *If θ' is stationarily satisfiable in $\langle \mathbb{N}, 1, ' \rangle$, then θ is true in infinitely many finite models.*

Proof of (a): Let $\mathcal{A} = \langle A, R_1, \ldots, R_k \rangle$ be a model of θ. Then there exists $a_1 \in A$ and for each $x \in A$ there exists $s(x) \in A$ such that $\mathcal{A} \models \lambda(a_1, x, s(x), y)$ for all $y \in A$. Inductively define a function $\pi: \mathbb{N} \to A$ by $\pi(1) = a_1$ and $\pi(n') = s(\pi(n))$. With $a_n = \pi(n)$, $n = 1, 2, 3, \ldots$ consider the set $A_0 = \{a_n \mid n \in \mathbb{N}\}$. Restriction of the relations of $\mathcal{A}$ to A_0 gives a model, $\mathcal{A}_0 = \langle A_0, a_1, s, R_{01}, \ldots, R_{0k} \rangle$, of the sentence $(\forall X, Y)\lambda(a_1, X, s(X), Y)$.

For each integer r, if R_i is an r-ary relation on A, define the r-ary relation $\bar{R}_i$ on $\mathbb{N}$ by

$$\bar{R}_i = \{(x_1, \ldots, x_r) \in \mathbb{N}^r \mid (\pi(x_1), \ldots, \pi(x_r)) \in R_{0i}\}, \ i = 1, \ldots, k.$$

Induction on structure shows that, with $\mathcal{N} = \langle \mathbb{N}, 1, ', \bar{R}_1, \ldots, \bar{R}_k \rangle$,

$$\mathcal{N} \models \varphi(x_1, \ldots, x_n) \iff \mathcal{A}_0 \models \varphi(\pi(x_1), \ldots, \pi(x_n))$$

for each quantifier-free formula $\varphi(X_1, \ldots, X_n)$ and for each $(x_1, \ldots, x_n) \in \mathbb{N}^n$ (Note, however, that since π is not necessarily an injection, this would not hold if φ would contain the equality symbol.) In particular, $\mathcal{N} \models \lambda(1, x, x', y)$ for each $(x, y) \in \mathbb{N} \times \mathbb{N}$.

Proof of (b): Suppose that θ' is satisfiable in $\langle \mathbb{N}, 1, ' \rangle$ such that the interpretation of each relation symbol of $\lambda(U, X, V, Y)$ in $\mathbb{N}$ is m-stationary. Define a function s from the set $A = \{1, 2, \ldots, m\}$ into itself by $s(x) = x'$ for $x = 1, 2, \ldots, m-1$ and $s(m) = m$. Let $\pi: \mathbb{N} \to A$ be the function $\pi(x) = x$

for $x = 1,\ldots,m$ and $\pi(x) = m$ for $x \geq m$. Restrict the relations of λ to A. This gives a finite model $\mathcal{A}$ in which $(\forall X, Y)\lambda(1, X, s(X), Y)$ is true. Drop 1 and s from $\mathcal{A}$ to get a finite model in which θ is true.

Since the relations of λ are also n-stationary for each $n \geq m$, this gives infinitely many finite models for θ. □

PROPOSITION 28.4.4 (Büchi): *The set of sentences of $\mathcal{L}_2$(predicate) which are true in infinitely many finite models and the set of sentences of $\mathcal{L}_2$(predicate) which have no models are recursively inseparable.*

Proof: For each Turing machine M let $\lambda(U, X, V, Y)$ be the quantifier-free formula of $\mathcal{L}_2$(predicate) that Lemma 28.4.2 attaches to M. Let θ and θ' be as in Lemma 28.4.3. If $M \in \text{Stat}(0)$, then θ' is stationarily satisfiable in $\langle \mathbb{N}, 1,' \rangle$ (Lemma 28.4.2(b)). Hence, θ is true in infinitely many finite models (Lemma 28.4.3(b)). If $M \in \text{Hlt}(0)$, then θ' is not satisfiable in $\langle \mathbb{N}, 1,' \rangle$ (Lemma 28.4.2(a)). Hence, θ has no model (Lemma 28.4.3(a)). Moreover, the map $f \colon \text{code}(M) \mapsto \text{code}(\theta)$ (Section 8.6) is recursive. In addition, $f(\text{Stat}(0))$ (resp. $f(\text{Hlt}(0))$) is contained in the set of codes of sentences true in infinitely many finite models (resp. with no models). By Lemma 28.3.6, Stat(0) and Hlt(0) are recursively inseparable. Therefore, by Lemma 28.3.2, the Proposition holds. □

28.5 Undecidability in the Theory of Graphs

We interpret the language $\mathcal{L}_2$(predicate) in the language of graphs and use Büchi's theorem to construct recursively inseparable theories in this language.

A **graph** in this chapter is a structure $\Gamma = \langle A, R \rangle$ with A a set and R a binary symmetric nonreflexive relation on A. Denote the language of the theory of graphs without equality by $\mathcal{L}$(graph). It contains exactly one binary predicate symbol P. Thus, a graph is a structure for $\mathcal{L}$(graph) which satisfies the axiom

$$(\forall X, Y)[\neg P(X, X) \wedge [P(X, Y) \leftrightarrow P(Y, X)]]$$

Visually we present the relation R of a graph Γ by a diagram which shows the points of A and an edge between each pair of points that belongs to R. For example, the diagram

expresses that (x, y) and (y, z) belongs to R but (x, z) does not. A broken line through points $x_1, x_2, \ldots, x_n$ of A

$$\begin{array}{ccccc} \bullet\,\text{-}\,\text{-}\,\text{-} & \bullet\,\text{-}\,\text{-}\,\text{-} & \bullet\,\text{-}\,\text{-}\,\text{-} & \cdots\,\text{-}\,\text{-} & \bullet \\ x_1 & x_2 & x_3 & & x_n \end{array}$$

signifies that each pair (x_i, x_j), with $i \neq j$, belongs to R.

28.5 Undecidability in the Theory of Graphs

PROPOSITION 28.5.1 (Lavrov [Ershov-Lavrov-Taimanov-Taitslin, p. 79]): *The set of sentences of $\mathcal{L}$(graph) which are true in infinitely many finite graphs and the set of sentences of $\mathcal{L}$(graph) which have no models are recursively inseparable.*

Proof: The proof divides into parts consisting of the construction and the verification of properties of a recursive map f from the set of sentences of $\mathcal{L}_2$(predicate) into the set of sentences of $\mathcal{L}$(graph) for which the following holds. If θ belongs to the set of sentences S_1 of $\mathcal{L}_2$(predicate) which are true in infinitely many finite models, (resp. sentences S_2 which have no model), then $f(\theta)$ is true in infinitely many finite graphs (resp. $f(\theta)$ has no model). Proposition 28.4.4 implies that S_1 and S_2 are recursively inseparable. Thus, Lemma 28.3.2 with S_1 and S_2, respectively, replacing A and B, implies the proposition.

PART A: *Definition of f.* Let θ be a sentence of $\mathcal{L}_2$(predicate) which involves only unary relation symbols $U_1, \ldots, U_m$ and binary relation symbols $B_1, \ldots, B_n$. Denote the following formula of $\mathcal{L}$(graph) by $\mu(X)$:

$$(\exists Y_1) \cdots (\exists Y_{m+n+3}) \Big[\bigwedge_{i=1}^{m+n+3} P(X, Y_i) \wedge \bigwedge_{i \neq j} P(Y_i, Y_j) \Big].$$

Let $\nu_i(X)$ be the formula

$$(\exists T_1) \cdots (T_{i+1}) \Big[\bigwedge_{j=1}^{i+1} \neg \mu(T_j) \wedge P(X, T_j) \wedge \bigwedge_{k \neq j} P(T_k, T_j) \wedge$$

$$\neg (\exists T)[\neg \mu(T) \wedge P(X, T) \wedge \bigwedge_{j=1}^{i+1} P(T, T_j)] \Big]$$

$i = 1, \ldots, m$. Finally, let $\beta_j(X, Y)$ be the formula

$$(\exists U)(\exists W)(\exists Z_1) \cdots (\exists Z_i) \Big[\bigwedge_{j=1}^{i} P(W, Z_j) \wedge P(X, U) \wedge P(U, W) \wedge P(Y, W)$$

$$\wedge \neg \mu(U) \wedge \neg \mu(W) \wedge \bigwedge_{j=1}^{i} \neg \mu(Z_j) \wedge \bigwedge_{j \neq k} P(Z_j, Z_k)$$

$$\wedge \neg (\exists Z) \Big[\neg \mu(Z) \wedge P(Z, W) \wedge \bigwedge_{j=1}^{i} P(Z, Z_j) \Big] \Big],$$

$i = 1, \ldots, n$.

Now define a recursive map $\varphi \mapsto \varphi^*$ from formulas of $\mathcal{L}_2$(predicate) to formulas of $\mathcal{L}$(graph) by a structure induction: $U_i^*(X)$ is $\nu_i(X)$, $i = 1, \ldots, m$, and $B_j^*(X,Y)$ is $\beta_j(X,Y)$, $i = 1, \ldots, n$. Next make the map commute with negation and conjunction. If $\varphi(\mathbf{X},Y)$ is $(\exists Y)\psi(\mathbf{X},Y)$ and $\psi^*(\mathbf{X},Y)$ has already been defined, then $\varphi^*(\mathbf{X},Y)$ is $(\exists Y)[\mu(Y) \wedge \psi^*(\mathbf{X},Y)]$. Finally, define $f(\theta)$ for a sentence θ of $\mathcal{L}_2$(predicate) as $\theta^* \wedge (\exists X)\mu(X)$.

PART B: *Interpretation of finite models for $\mathcal{L}_2$(predicate) as finite graphs.*
Suppose that θ is true in infinitely many finite models. Let

$$\mathcal{M} = \langle M, U_1, \ldots, U_m, B_1, \ldots, B_n \rangle$$

be one of these. Without loss assume that $|M| > m+n+3$. We construct a finite graph $\Gamma = \langle A, R \rangle$, with $|A| \geq |M|$, in which $f(\theta)$ is true.

For each i, $1 \leq i \leq m$, and for each $x \in M$ which belongs to U_i, add $i+1$ new elements, $t(x,i,1), \ldots, t(x,i,i+1)$, to M. For each j, $1 \leq j \leq n$, and for each pair (x,y) of elements of M which belongs to B_j, add $j+2$ new elements, $u(x,y,j)$, $w(x,y,j)$; $z(x,y,j,1), \ldots, z(x,y,j,j)$. Denote the resulting set by A.

Define a symmetric nonreflexive relation R on A. Two elements x,y of A relate to each other exactly when they are connected by an edge in one of the following diagrams:

(1a)

$$\overset{x}{\bullet} \longrightarrow \overset{y}{\bullet}$$

for $x \neq y$ and $x, y \in M$;

(1b)

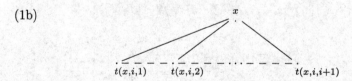

for $x \in M$, $x \in U_i$ and $i = 1, \ldots, m$; and

(1c)

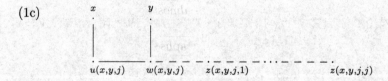

for $x, y \in M$, $(x, y) \in B_j$ and $j = 1, \ldots, n$. Let $\Gamma = \langle A, R \rangle$ be the resulting graph. In particular, no point of M is connected to itself. Thus, if $P(x,y)$ holds for $x, y \in M$, then $x \neq y$. It follows that the following statements hold:
(2a) $x \in M \iff \Gamma \models \mu(x)$ (indeed an element of A relates to $m+n+3$ elements if and only if it is in M).
(2b) If $x \in M$, then $x \in U_i \iff \Gamma \models \nu_i(x)$, $i = 1, \ldots, m$.
(2c) If $x, y \in M$, then $(x,y) \in B_j \iff \Gamma \models \beta_j(x,y)$, $j = 1, \ldots, n$.

28.5 Undecidability in the Theory of Graphs

It follows by induction on structure that for each formula $\varphi(X_1,\ldots,X_r)$ of $\mathcal{L}_2$(predicate) and for all $x_1,\ldots,x_r \in M$, $M \models \varphi(\mathbf{x})$ if and only if $\Gamma \models \varphi^*(\mathbf{x})$. In particular, $\Gamma \models f(\theta)$. Consequently, $f(\theta)$ is true in infinitely many finite graphs.

PART C: *Extracting a model for θ from a graph.* Conversely, let $\Gamma = \langle A, R \rangle$ be a graph in which $f(\theta)$ is true. Then the set $M = \{x \in A \mid \Gamma \models \mu(x)\}$ is nonempty. Define relations U_i and B_j on M:

$$U_i = \{x \in M \mid \Gamma \models \nu_i(x)\}, i = 1,\ldots,m: \quad \text{and}$$
$$B_j = \{(x,y) \in M \times M \mid \Gamma \models \beta_j(x,y)\}, \; j = 1,\ldots,n.$$

The structure $\mathcal{M} = \langle M, U_1,\ldots,U_m, B_1,\ldots,B_n \rangle$ is a model of θ. □

Lemma 28.5.2: *Let θ be a sentence of $\mathcal{L}$(graph). Suppose θ holds in each infinite graph. Then θ is true in almost all finite graphs.*

Proof: For each positive integer m there are, up to isomorphism, only finitely many graphs $\langle A, R \rangle$ with $|A| \leq m$. Assume $\neg\theta$ holds in infinitely many finite graphs. Then, each nonprincipal ultraproduct of those graphs is an infinite graph in which $\neg\theta$ is true. We conclude from this contradiction, that θ is true in almost all finite graphs. □

By Definition 28.3.1, if $A \subseteq B$ are subsets of N, then A and $N \smallsetminus B$ are recursively inseparable if and only if there exists no recursive subset C of N with $A \subseteq C \subseteq B$. We use this observation in the proof of the following result:

COROLLARY 28.5.3: *Consider the following lattice of sets of sentences of $\mathcal{L}$(graph).*

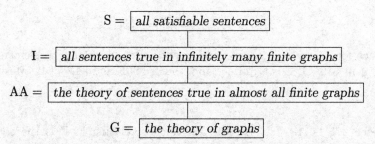

(a) *There is no recursive set between I and S.*
(b) *There is no recursive set between G and AA.*

In particular, none of the four sets is decidable.

Proof of (a): The set of satisfiable sentences is the complement of the set of sentences which are false in each graph. Thus, (a) reformulates Proposition 28.5.1.

Proof of (b): The map $\theta \mapsto \neg\theta$ is an injective recursive map from the set of sentences of $\mathcal{L}$(graph) into itself. It maps the set of sentences with no model onto G, and it maps I onto the set of sentences, I', false in infinitely many finite models. By Lemma 28.3.2 and Proposition 28.5.1, I' and G are recursively inseparable. Since AA is the complement of I', (b) follows.

If one of the sets in the statement of the corollary is recursive, then it is a recursive set between either I and S, or between G and AA. Since this contradicts either (a) or (b), we are done. □

PROPOSITION 28.5.4: *The four theories of Corollary 28.5.3 are distinct.*

Proof (Haran): The parts of the proof consist of introducing the equality relation to $\mathcal{L}(\text{graph})$ and the proof that G, AA, I, and S are distinct.

PART A: *Introducing the equality relation.* Let $\varepsilon(X, Y)$ be the formula

$$(\forall Z)[P(X, Z) \leftrightarrow P(Y, Z)].$$

For each graph $\Gamma = \langle A, R \rangle$ define an equivalence relation on A:

$$x \sim y \iff \Gamma \models \varepsilon(x, y).$$

That is, x is equivalent to y if and only if y is related to exactly the same elements of A as is x. Denote the equivalence class of an element $x \in A$ by $\bar{x}$ and denote the set of all equivalence classes by $\bar{A}$. Then $\bar{R} = \{(\bar{x}, \bar{y}) \in \bar{A} \times \bar{A} \mid (x, y) \in R\}$ is a well defined symmetric nonreflexive relation on A. Therefore, $\bar{\Gamma} = \langle \bar{A}, \bar{R} \rangle$ is a graph and

$$\Gamma \models P(x, y) \iff \bar{\Gamma} \models P(\bar{x}, \bar{y})$$

for each $x, y \in A$. An induction on structure shows that

$$\Gamma \models \varphi(x_1, \ldots, x_n) \iff \bar{\Gamma} \models \varphi(\bar{x}_1, \ldots, \bar{x}_n)$$

for each formula $\varphi(X_1, \ldots, X_n)$ of $\mathcal{L}(\text{graph})$. In particular, Γ is elementarily equivalent to $\bar{\Gamma}$. If Γ is a finite graph, then so is $\bar{\Gamma}$ and $|\bar{A}| \leq |A|$. Moreover

$$\bar{\Gamma} \models \varepsilon(a, b) \iff a = b$$

for each $a, b \in \bar{A}$. That is, the equivalence relation in $\bar{\Gamma}$ is the equality relation.

PART B: $G \neq AA$. Consider the sentence θ

$$(\exists X_1)[(\exists X_2)P(X_1, X_2) \wedge (\forall Y)[P(X_1, Y) \leftrightarrow \varepsilon(X_2, Y)]$$
$$\wedge (\forall Z_2)[\neg \varepsilon(X_1, Z_2) \to (\exists Z_1, Z_3)[P(Z_1, Z_2) \wedge P(Z_2, Z_3) \wedge \neg \varepsilon(Z_1, Z_3)$$
$$\wedge (\forall Y)[P(Z_2, Y) \to \varepsilon(Y, Z_1) \vee \varepsilon(Y, Z_3)]]]].$$

That is, θ interprets the existence of an element x_1 which relates (up to equivalence) to exactly one element, and each element z_2 not equivalent to x_1

relates (up to equivalence) to exactly two nonequivalent elements. Replace Γ by $\bar{\Gamma}$ to assume that each equivalence class of Γ contains exactly one element. Suppose that Γ is a model for θ. We do an induction to show, for each integer $n \geq 0$, that A has $n+1$ distinct elements $x_1, \ldots, x_{n+1}$ which relate according to the following diagram

(3)
$$\underset{\cdot}{x_1} \underset{\cdot}{\quad} \underset{\cdot}{x_2} \ldots \underset{\cdot}{x_{n-1}} \underset{\cdot}{\quad} \underset{\cdot}{x_n} \underset{\cdot}{\quad} \underset{\cdot}{x_{n+1}}$$

Let $x_1 \in A$ be related to exactly one element. Apply the induction assumption to assume that diagram (3) exists without the point x_{n+1}. But, since x_i is related to x_{i-1} and to x_{i+1}, for $i = 2, \ldots, n-1$, none of $x_1, \ldots, x_{n-2}$ are related to x_n. Hence, x_n is related to a new element x_{n+1}, which gives (3). In particular, A is an infinite set.

It follows that $\neg\theta$ is true in all finite graphs. Also, θ is true in the graph $\langle \mathbb{N}, R \rangle$, where $R = \{(n, n+1) \mid n \in \mathbb{N}\}$. Consequently, $G \neq AA$.

PART C: $AA \neq I$. The sentence $(\forall X, Y)[\neg P(X, Y)]$ interprets the triviality of the graph (i.e. no two points are related). There are infinitely many trivial finite graphs and there are infinitely many nontrivial finite graphs. Hence, $AA \neq I$.

PART D: $I \neq S$. The sentence θ of Part B is satisfiable but false in each finite graph. □

Remark 28.5.5: Let IG be the theory of infinite graphs and FG the theory of finite graphs. The arguments of Part A of the proof of Proposition 28.5.4 show that $IG = G$. Indeed to each point x of a finite graph Γ we can add infinitely many points, each of which is equivalent to x. The new graph so obtained is infinite and elementarily equivalent to Γ. Similarly, each finite graph is elementarily equivalent to an arbitrarily large finite graph. Thus $AA = FG$. In particular, by Corollary 28.5.3, both IG and FG are undecidable theories.

Finally, each finite graph Γ has a recursive theory $\text{Th}(\Gamma)$. The arguments of the preceeding paragraph show that $FG = AA \subset \text{Th}(\Gamma) \subset I$. This settles Problem 33 of [Fried-Jarden5] (see also [Fried-Jarden5, Remark 28.5.5]). □

28.6 Assigning Graphs to Profinite Groups

In the following sections we interpret the theory of graphs in the theory of perfect PAC fields. With this we prove the undecidability of the latter theory. The interpretation goes through several stages. This section uses a pair of finite groups, (D, W), as an auxiliary parameter, to assign to each profinite group G a graph Γ_G. Sections 28.7 and 28.8 develop conditions that guarantee that a given graph Γ equals Γ_G for some profinite group G. Section 28.9 shows that under these conditions a given graph Γ is $\Gamma_{\text{Gal}(K)}$ for some perfect PAC field K. And, finally, Section 28.10 concludes the undecidability of the theory of perfect PAC fields as an application of Proposition 28.8.3.

Fix a pair (D, W) of finite groups with D nontrivial. For each profinite group G define the graph $\Gamma_G = (A_G, R_G)$ as follows: A_G is the set of open normal subgroups N of G with $G/N \cong D$; and R_G is the set of pairs $(N_1, N_2) \in A_G \times A_G$ such that $N_1 N_2 = G$ and there exists an open normal subgroup M of G with $M \leq N_1 \cap N_2$ and $G/M \cong W$.

Note that if R_G is nonempty there must be a sequence
$$G/M \to G/N_1 \cap N_2 \to G/N_1 \times G/N_2$$
that induces an epimorphism $W \to D \times D$. Clearly R_G is a symmetric nonreflexive relation on A_G. Hence, if A_G is nonempty, then $\Gamma_G = \langle A_G, R_G \rangle$ is a graph. This notation suppresses the dependence of Γ_G on (D, W).

LEMMA 28.6.1: *Suppose the respective Frattini subgroups, $\Phi(D)$ and $\Phi(W)$, of D and W are trivial. Assume also that $\pi \colon H \to G$ is a Frattini cover of profinite groups. If Γ_G is nonempty, then Γ_H is nonempty, and $\Gamma_G \cong \Gamma_H$. In particular, if $\tilde{G}$ is the universal Frattini cover of G, then $\Gamma_{\tilde{G}} \cong \Gamma_G$.*

Proof: From the isomorphisms $H/\pi^{-1}(N) \cong G/N$ for open normal subgroups N of G, the map $N \mapsto \pi^{-1}(N)$ maps A_G injectively into A_H. If $(N_1, N_2) \in R_G$, then $(\pi^{-1}(N_1), \pi^{-1}(N_2)) \in R_H$.

Conversely, if M is an open normal subgroup of H and either $H/M \cong D$ or $H/M \cong W$, then $\mathrm{Ker}(\pi) \leq \Phi(H) \leq M$ (Lemma 22.1.4(a)). Hence, $M = \pi^{-1}(\pi(M))$. Thus, the map $A_G \to A_H$ induces an isomorphism of graphs. □

28.7 The Graph Conditions

We develop sufficient conditions on the finite groups D and W that guarantee the surjectivity of the map $G \mapsto \Gamma_G$ of profinite groups to graphs.

For D a finite group and I any set regard the collection of all functions from I to D as a profinite group, denoted D^I (Lemma 1.2.6). Each $\mathbf{d} \in D^I$ represents a vector whose ith coordinate is d_i, and multiplication is defined componentwise. For each $i \in I$ let $\pi_i \colon D^I \to D$ be projection on the ith coordinate (i.e. $\pi_i(\mathbf{d}) = d_i$). The kernel of π_i is $D'_i = \{\mathbf{d} \in D^I \mid d_i = 1\}$, and with $D_i = \{\mathbf{d} \in D^I \mid d_j = 1 \text{ for each } j \neq i\}$, $D^I = D_i \times D'_i$. In particular, for $I = \{1, 2\}$, $D \times D = D^I = D_1 \times D_2$.

Recall that every finite group G has a **composition series**: a sequence of subgroups $G = G_0 \triangleright G_1 \triangleright \cdots \triangleright G_r = 1$ such that the quotient G_i/G_{i+1} is a simple group, $i = 0, \ldots, r-1$. The Jordan-Hölder theorem states that the length and the set of quotients of a composition series are invariants of G [Huppert, p. 63]. The quotient groups are the composition factors of G.

Consider two finite groups D and U such that $D \neq 1$ and $D \times D$ acts on U as a group of automorphism. Let $W = (D \times D) \ltimes U$ be the semidirect product of U and $D \times D$ (Definition 13.7.1) and the associated short split exact sequence:

(1) $\qquad 1 \longrightarrow U \longrightarrow W \overset{\lambda}{\underset{\theta}{\rightleftarrows}} D \times D \longrightarrow 1,$

28.7 The Graph Conditions

We assume the following properties for D and U:
(2a) D and U have no composition factor in common.
(2b) For each finite set I and each epimorphism $\pi\colon D^I \to D$ there exists $i \in I$ such that $\operatorname{Ker}(\pi) = \operatorname{Ker}(\pi_i)$.
(2c) $\Phi(W) = \Phi(D) = 1$.
(2d) For each embedding θ' of $D \times D$ into W as a **semidirect complement** of U (i.e. $U \cdot \theta'(D \times D) = W$ and $U \cap \theta'(D \times D) = 1$) and for each nontrivial normal subgroup N of U none of the factors D_1 and D_2 acts trivially (via conjugation) on N.

Remark 28.7.1: If I is an arbitrary set and $\pi\colon D^I \to D$ is a continuous epimorphism, then there exists a finite subset J of I and an epimorphism $\bar{\pi}\colon D^J \to D$ such that $\pi = \bar{\pi} \circ \pi_J$, where $\pi_J\colon D^I \to D^J$ is the suitable coordinate projection. By (2b) there exists $i \in J$ such that $\operatorname{Ker}(\pi) = \operatorname{Ker}(\pi_i)$. □

Definition 28.7.2: A pair of finite groups (D, W) is said to satisfy the **graph conditions** if W can be factored as a semidirect product $(D \times D) \ltimes U$ such that Condition (2) holds. □

The following two examples provide the reader with some practice in the mechanics of the graph conditions.

Example 28.7.3: [Ershov4]. Let D be a simple non-Abelian group and let H be a finite group whose composition factors are non-Abelian and distinct from D. The action of $D \times D$ on $U = H^{D \times D}$ is given by permutation of the coordinates:

(3) $\qquad (h_\delta)^{\delta'} = h_{\delta'\delta}$ for $\mathbf{h} \in U$; $\quad \delta, \delta' \in D \times D.$

The semidirect product $W = U \rtimes (D \times D)$ is the wreath product of $D \times D$ with H (Remark 13.7.7). Condition (2a) is immediate. Condition (2b) follows from Lemma 16.8.3(a).

Proof of (2c): Since D is simple, $\Phi(D) = 1$. In the notation of (1), Lemma 22.1.4 implies that $\lambda(\Phi(W)) \leq \Phi(D \times D) = \Phi(D) \times \Phi(D) = 1$. Hence, $\Phi(W) \triangleleft U$. By Lemma 22.1.2, however, $\Phi(W)$ is nilpotent. If $\Phi(W)$ were nontrivial, U (and therefore H) would have a cyclic composition factor, contrary to the assumptions.

Proof of (2d): Suppose that $D \times D$ is a subgroup of W in some way. Let N be a nontrivial normal subgroup of U. With no loss assume that N is a minimal normal subgroup of U. We show that if $\delta \in D \times D$ normalizes N, then $\delta = 1$. This will imply (2d).

By Lemma 18.3.9, N is equal to one of the groups

$$H_{\delta'} = \{h \in U \mid h_{\delta''} = 1 \text{ for all } \delta'' \neq \delta'\}.$$

with $\delta' \in D \times D$. By assumption, $H_{\delta'} = H_{\delta'}^\delta = H_{\delta'\delta^{-1}}$. Hence, $\delta' = \delta'\delta^{-1}$ and $\delta = 1$, as claimed. □

Example 28.7.4: *[Cherlin-v.d.Dries-Macintyre].* Let p and q be distinct odd primes. Consider the dihedral group $D = D_p$ of order $2p$ generated by β of order 2 and γ of order p with the relation $\beta\gamma\beta = \gamma^{-1}$. Then $\alpha = \gamma\beta$ is of order 2. Let τ be the unique epimorphism of D (with $\text{Ker}(\tau)$ generated by γ) onto the multiplicative group $\{\pm 1\}$ so that $\tau(\beta) = -1$. Let U be the multiplicative cyclic group of order q and let $D \times D$ operate on U by the rule

(4) $$u^{(x,y)} = u^{\tau(x)\tau(y)}, \quad \text{for} \quad x, y \in D \quad \text{and} \quad u \in U.$$

The composition factors of D_p, namely $\mathbb{Z}/2\mathbb{Z}$ and $\mathbb{Z}/p\mathbb{Z}$, are distinct from $\mathbb{Z}/q\mathbb{Z}$, the unique composition factor of U. This gives (2a).

Proof of (2b): Let I be a finite set and let $\pi \colon D^I \to D$ be an epimorphism. Denote the elements of the ith factor D_i of D^I that correspond to α, β, and γ by α_i, β_i, and $\gamma_i = \alpha_i \beta_i$. If there exists $j \in I$ such that $D_j \leq \text{Ker}(\pi)$, then consider π as an epimorphism from $D^{I \smallsetminus \{j\}}$ onto D and apply an induction to prove the existence of $i \in I \smallsetminus \{j\}$ such that $\text{Ker}(\pi) = \text{Ker}(\pi_i)$. Otherwise, for each $i \in I$, $\text{Ker}(\pi) \cap D_i$, as a proper normal subgroup of D_i is either 1 or $\langle \gamma_i \rangle$. If $\text{Ker}(\pi) \cap D_i = \langle \gamma_i \rangle$ for all $i \in I$, then D is generated by pairwise commuting copies of $\mathbb{Z}/2\mathbb{Z}$. Hence, D is Abelian, a contradiction. If $\text{Ker}(\pi) \cap D_i = 1$ for some $i \in I$, say for $i = 1$, then $\pi(D_1) = D$. Hence, for each $i \in I \smallsetminus \{1\}$, each element of $\pi(D_i)$ commutes with each element of D. That is, the center of D is nontrivial, a contradiction unless $I = \{1\}$. This proves (2b).

Proof of (2c): Both $\langle \alpha \rangle$ and $\langle \beta \rangle$ have index p in D. Hence, $\Phi(D) \leq \langle \alpha \rangle \cap \langle \beta \rangle = 1$, so $\Phi(D) = 1$. Therefore, $\Phi(W) \leq U$. By (1), $\theta(D \times D)$ has index q in W. Thus, $\Phi(W) \leq U \cap \theta(D \times D) = 1$.

Proof of (2d): Let θ' be an embedding of $D \times D$ into W as a semidirect complement of U. Since the orders of U and $D \times D$ are relatively prime, there is a $u \in U$ with $\theta'(D \times D) = \theta(D \times D)^u$ (Schur-Zassenhaus - Lemma 22.10.1). By (2b), $D \times D$ has a unique factorization as a direct product of two copies of D. Thus, $\theta'(D \times D) = \theta(D)^u \times \theta(D)^u$. Take $x \in D$ such that $\tau(x) = -1$ and $v \in U$, $v \neq 1$. Since q is odd,

$$v^{\theta(x)^u} = v^{u^{-1}\theta(x)u} = v^{\tau(x)u} = (v^{-1})^u = v^{-1} \neq v.$$

Hence, none of the direct factors of $\theta'(D \times D)$ acts trivially on U. □

28.8 Assigning Profinite Groups to Graphs

Suppose (D, W) is a pair of finite groups that satisfy the graph conditions. In particular, there is a split short exact sequence

(1) $$1 \longrightarrow U \longrightarrow W \mathrel{\mathop{\rightleftarrows}^{\lambda}_{\theta}} D \times D \longrightarrow 1.$$

28.8 Assigning Profinite Groups to Graphs

Let $\Gamma = \langle A, R \rangle$. We construct a projective group G such that $\Gamma_G \cong \Gamma$, with $\Gamma_G = \langle A_G, R_G \rangle$ defined in Section 28.6.

To that end consider the profinite group $D^A \times W^R$ whose elements are pairs $(\mathbf{d}, \mathbf{w})$ with $\mathbf{d} \in D^A$ and $\mathbf{w} \in W^R$. Consider, also the following coordinate projections

$$\pi_A \colon D^A \times W^R \to D^A, \quad \pi_A(\mathbf{d}, \mathbf{w}) = \mathbf{d};$$
$$\pi_a \colon D^A \times W^R \to D, \quad \pi_a(\mathbf{d}, \mathbf{w}) = d_a, \quad \text{for} \quad a \in A; \quad \text{and}$$
$$\pi_r \colon D^A \times W^R \to W, \quad \pi_r(\mathbf{d}, \mathbf{w}) = w_r, \quad \text{for} \quad r \in R.$$

The subgroup

$$G = \{(\mathbf{d}, \mathbf{w}) \in D^A \times W^R \mid r = (a, b) \in R \Longrightarrow \lambda(w_r) = (d_a, d_b)\},$$

of $D^A \times W^R$ is closed, and therefore profinite. Moreover, if Γ is a finite graph, then G is a finite group.

LEMMA 28.8.1: $\pi_A(G) = D^A$ and $\operatorname{Ker}(\pi_A) \cap G = U^R$.

Proof: Indeed, if $\mathbf{d} \in D^A$ and $r = (a, b) \in R$, choose $w_r \in W$ such that $\lambda(w_r) = (d_a, d_b)$. Then $(\mathbf{d}, \mathbf{w})$ defined by this procedure belongs to G and $\pi_A(\mathbf{d}, \mathbf{w}) = \mathbf{d}$. Thus, $\pi_A(G) = D^A$.

Suppose that $\mathbf{u} \in U^R$. Then $\lambda(u_r) = (1, 1)$ for each $r \in R$. Hence, $(1, \mathbf{u}) \in \operatorname{Ker}(\pi_A) \cap G$. Conversely, if $(\mathbf{d}, \mathbf{w}) \in \operatorname{Ker}(\pi_A) \cap G$, then $\mathbf{d} = 1$ and $\lambda(w_r) = (1, 1)$ for each $r \in R$. Thus, $\operatorname{Ker}(\pi_A \cap G) = U^R$. □

Lemma 28.8.1 shows that the sequence

(2) $$1 \to U^R \to G \xrightarrow{\pi'_A} D^A \to 1,$$

with $\pi'_A = \pi_A|_G$, is exact.

LEMMA 28.8.2: There exists a continuous homomorphism $\eta \colon D^A \to G$ such that $\pi'_A \circ \eta = \operatorname{Id}$. (i.e. (2) splits). Also, if $a' \in A$, $r = (a, b) \in R$ and $a' \notin \{a, b\}$, then $D_{a'}$, acts trivially on U_r through η.

Proof: If $\mathbf{d} \in D^A$, define η by $\eta(\mathbf{d}) = (\mathbf{d}, \mathbf{w})$, where $w_r = \theta(d_a, d_b)$ for each $r = (a, b) \in R$. Now assume that a' and $r = (a, b)$ satisfy the hypotheses of the lemma. If $\mathbf{d} \in D_{a'}$, then $d_a = d_b = 1$. Hence, $w_r = \theta(d_a, d_b) = 1$. Thus, $\eta(\mathbf{d}) = (\mathbf{d}, \mathbf{w})$ commutes with each element of U_r. □

PROPOSITION 28.8.3: The graphs $\langle A, R \rangle$ and $\langle A_G, R_G \rangle$ are isomorphic.

Proof: We present the construction and properties of a map from $\langle A, R \rangle$ into $\langle A_G, R_G \rangle$ in three parts:

PART A: *Map of A onto A_G.* For each $a \in A$ put $\pi'_a = \pi_a|_G$ and $\mathcal{N}(a) = \text{Ker}(\pi'_a)$. Then the following diagram commutes:

(3)
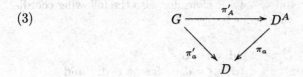

By (2), π'_a is surjective. Hence, $G/\mathcal{N}(a) \cong D$ and $\mathcal{N}(a) \in A_G$. We regard $\mathcal{N}$ as a map from A into A_G. By the commutativity of (3), $\pi'_A(\text{Ker}(\pi'_a)) = \text{Ker}(\pi_a)$. Since $\text{Ker}(\pi_a) = \text{Ker}(\pi_b)$ implies that $a = b$, $\mathcal{N}$ is injective. Now we show surjectivity of $\mathcal{N}$.

Consider an open normal subgroup N of G with $G/N \cong D$ and let $\pi': G \to D$ be the corresponding epimorphism. Since the composition factors of $\pi'(U^R)$ are among the compositions factors of U, graph condition (2a) of Section 28.7 implies that $U^R \leq \text{Ker}(\pi') = N$. By (2) there exists an epimorphism $\pi: D^A \to D$ such that $\pi' = \pi \circ \pi'_A$. By Remark 28.7.1, there is an $a \in A$ with $\text{Ker}(\pi) = \text{Ker}(\pi_a)$. Therefore, $N = \text{Ker}(\pi'_a) = \mathcal{N}(a)$.

PART B: *Map of R into R_G.* Let $r = (a,b) \in R$. Then $a \neq b$. Hence, by (3) and by (2),

$$\mathcal{N}(a)\mathcal{N}(b) = \text{Ker}(\pi'_a)\text{Ker}(\pi'_b)$$
$$= (\pi'_A)^{-1}(\text{Ker}(\pi_a))(\pi'_A)^{-1}(\text{Ker}(\pi_b)) = (\pi'_A)^{-1}(D^A) = G.$$

In addition, with $\pi'_r = \pi_r|_G$, the diagram

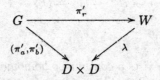

commutes. Thus, $\text{Ker}(\pi'_r) \leq \mathcal{N}(a) \cap \mathcal{N}(b)$. Therefore, if we show that π'_r is surjective, then $G/\text{Ker}(\pi'_r) \cong W$ and $(\mathcal{N}(a), \mathcal{N}(b)) \in R_G$. It will follow that the map $r \mapsto (\mathcal{N}(a), \mathcal{N}(b))$ from R into R_G is well defined.

Indeed, let $w_r \in W$ and let $\lambda(w_r) = (d_a, d_b)$. For each $a' \in A$ distinct from both a and b, choose any element $d_{a'} \in D$. Also, for $r' = (a', b') \in R$, $r' \neq r$, choose any element $w_{r'} \in W$ such that $\lambda(w_{r'}) = (d_{a'}, d_{b'})$. This defines $(\mathbf{d}, \mathbf{w})$ in G for which $\pi_r(\mathbf{d}, \mathbf{w}) = w_r$. Thus, π'_r is surjective.

If $r' = (a', b')$ is an element of R with $(\mathcal{N}(a), \mathcal{N}(b)) = (\mathcal{N}(a'), \mathcal{N}(b'))$, then $\mathcal{N}(a) = \mathcal{N}(a')$ and $\mathcal{N}(b) = \mathcal{N}(b')$. By Part A, $a = a'$ and $b = b'$. Hence, R maps injectively into R_G.

28.9 Assigning Fields to Graphs

PART C: *R maps surjectively onto R_G.* Let $a, b \in A$ with $(\mathcal{N}(a), \mathcal{N}(b)) \in R_G$. Then $\mathcal{N}(a)\mathcal{N}(b) = G$ and there exists an open normal subgroup M of G such that $M \le \mathcal{N}(a) \cap \mathcal{N}(b)$ with $G/M \cong W$. This gives a commutative diagram of epimorphisms

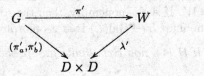

The remaining parts of the proof show that (a, b) is in R:

PART C1: $\operatorname{Ker}(\lambda') = U$ *and* $\pi'(U^R) = U$. Indeed, the composition factors of $\lambda'(U)$ are among the composition factors of U. But, since $\lambda'(U)$ is a subgroup of $D \times D$, these are among the composition factors of D. By graph condition (2a) of Section 28.7, $\lambda'(U) = 1$. Hence, $\operatorname{Ker}(\lambda) \le \operatorname{Ker}(\lambda')$. In addition, $|\operatorname{Ker}(\lambda)| = |W|/|D \times D| = |\operatorname{Ker}(\lambda')|$. Therefore, $\operatorname{Ker}(\lambda') = U$.

The same argument shows that $\lambda'(\pi'(U^R)) = 1$. Hence, $\pi'(U^R) \le \operatorname{Ker}(\lambda') = U$. Therefore, $U/\pi'(U^R)$ is a quotient of $(\pi')^{-1}(U)/U^R$, hence of D^A (by (2)). Thus, the decomposition factors of $U/\pi'(U^R)$ are among those of both U and D. By the graph condition (2a) of Section 28.7, $U = \pi'(U^R)$.

PART C2: *An element $r \in R$.* Part C1 gives $r \in R$ such that the normal subgroup $U' = \pi'(U_r)$ of U is nontrivial. In addition, Part C1 gives a commutative diagram

(4)
$$\begin{array}{ccccccccc} 1 & \longrightarrow & U^R & \longrightarrow & G & \xrightarrow{\pi'_A} & D^A & \longrightarrow & 1 \\ & & \pi' \downarrow & & \pi' \downarrow & (\pi'_a,\pi'_b)\searrow & \downarrow (\pi_a,\pi_b) & & \\ 1 & \longrightarrow & U & \longrightarrow & W & \xrightarrow{\lambda'} & D \times D & \longrightarrow & 1 \end{array}$$

Note that (π_a, π_b) maps $D_a \times D_b$ isomorphically onto $D \times D$. Hence, the section $\eta \colon D^A \to G$ of π'_A (Lemma 28.8.2) induces a section $\theta' \colon D \times D \to W$ of λ'. In particular, θ' embeds $D \times D$ into W as a semidirect complement of U.

PART C3: $(a, b) \in R$. Write r as (a', b'). If $a \notin \{a', b'\}$, then Lemma 28.8.2 shows that D_a acts trivially on U_r. By (4), the first factor D_1 of $D \times D$ acts (via θ') trivially on U'. This contradicts graph condition (2d) of Section 28.7. Therefore, $a \in \{a', b'\}$. Similarly $b \in \{a', b'\}$, so $r = (a, b)$ or $r = (b, a)$. In either case $(a, b) \in R$. □

28.9 Assigning Fields to Graphs

Again, let (D, W) be a pair of finite groups with D nontrivial and let K be a field. We define A_K to be the set of all Galois extensions L/K with $\mathrm{Gal}(L/K) \cong D$ and R_K to be the set of pairs $(L_1, L_2) \in A_K \times A_K$ with $L_1 \cap L_2 = K$ for which there exists a Galois extension N/K with $L_1 L_2 \subseteq N$ and $\mathrm{Gal}(N/K) \cong W$. If A_K is nonempty, then the structure $\Gamma_K = \langle A_K, R_K \rangle$ is a graph, and the map $L \mapsto \mathrm{Gal}(L)$ is an isomorphism of Γ_K onto $\Gamma_{\mathrm{Gal}(K)}$.

LEMMA 28.9.1: *If H is a nontrivial finite group, then*

$$\mathrm{rank}(H^n) \geq \frac{\log n}{\log |H|}.$$

Proof: The number of epimorphisms of H^n onto H is at most $|H|^{\mathrm{rank}(H^n)}$. On the other hand, each coordinate projection of H^n is an epimorphism. Hence, $n \leq |H|^{\mathrm{rank}(H^n)}$. Apply log to both sides for the result. □

Note that the proof of Lemma 28.8.3 does not use graph condition (2c) of Section 28.7. It is this condition however that assures that a perfect PAC field can be assigned to a graph.

PROPOSITION 28.9.2: *Let (D, W) be a pair of finite groups with $D \neq 1$ that satisfies the graph conditions. Then, for each graph $\Gamma = \langle A, R \rangle$ there exists a perfect PAC field K such that $\Gamma \cong \Gamma_K$. If Γ is finite, then $\mathrm{corank}(K) < \infty$. Furthermore $\mathrm{rank}(\mathrm{Gal}(K)) \geq \mathrm{rank}(D^A) \geq \log|A|/\log|D|$.*

Proof: Lemma 28.8.3 constructs a profinite group G such that $\Gamma \cong \Gamma_G$. With $\tilde{G}$ the universal Frattini cover of G, Lemma 28.6.1 shows that $\Gamma_{\tilde{G}} \cong \Gamma_G$. Apply Corollary 23.1.2 to produce a perfect PAC field K with $\mathrm{Gal}(K) \cong \tilde{G}$. Then, $\Gamma \cong \Gamma_K$.

If Γ is a finite graph, then G above is finite. By Corollary 22.5.3, $\mathrm{rank}(\mathrm{Gal}(K)) = \mathrm{rank}(G) < \infty$. By (2) of Section 28.8, $\mathrm{rank}(G) \geq \mathrm{rank}(D^A)$. Now Lemma 28.9.1 gives the result. □

Remark 28.9.3: The field K in Proposition 28.9.2 may be chosen to contain any given field K_0. Thus, the undecidability results of Section 28.10 generalize to the appropriate theories. □

28.10 Interpretation of the Theory of Graphs in the Theory of Fields

Let (D, W) be a pair of finite groups with $|D| = m$ and $|W| = n$. We attach a formula φ' of $\mathcal{L}(\mathrm{ring})$ to each formula φ of $\mathcal{L}(\mathrm{graph})$.

For l an integer let $f_{\mathbf{X}}(T) = T^l + X_1 T^{l-1} + \cdots + X_l$. If K is a field and $\mathbf{a} \in K^l$, denote the splitting field of $f_{\mathbf{a}}(T)$ over K by $K_{\mathbf{a}}$. Let H be a finite group. Use Remark 20.4.5(d) to effectively construct a formula $\alpha_{l,H}(\mathbf{X})$ of $\mathcal{L}(\mathrm{ring})$ such that for each field K and each $\mathbf{x} \in K^l$, $\alpha_{l,H}(\mathbf{x})$ holds in

28.10 Interpretation of the Theory of Graphs in the Theory of Fields

K if and only if $f_{\mathbf{x}}(T)$ is separable and $\mathrm{Gal}(f_{\mathbf{x}}(T), K) \cong H$. Use α for $l = m, n$ to construct a formula $\pi(\mathbf{X}, \mathbf{Y}, \mathbf{Z})$ of $\mathcal{L}(\mathrm{ring})$ such that for $(\mathbf{x}, \mathbf{y}, \mathbf{z}) \in K^m \times K^m \times K^n$, $\pi(\mathbf{x}, \mathbf{y}, \mathbf{z})$ is true in K if and only if $K_{\mathbf{x}} \cap K_{\mathbf{y}} = K$, $\mathrm{Gal}(K_{\mathbf{x}}/K) \cong \mathrm{Gal}(K_{\mathbf{y}}/K) \cong D$, $\mathrm{Gal}(K_{\mathbf{z}}/K) \cong W$, and $K_{\mathbf{x}}, K_{\mathbf{y}} \subseteq K_{\mathbf{z}}$.

Now define φ' by induction on the structure of φ:

$$R(X,Y)' = (\exists \mathbf{Z})\pi(\mathbf{X}, \mathbf{Y}, \mathbf{Z}), \qquad [\neg \varphi]' = \neg[\varphi'],$$
$$[\varphi_1 \vee \varphi_2]' = \varphi_1' \vee \varphi_2' \quad \text{and} \quad [(\exists X)\varphi]' = (\exists \mathbf{X})[\alpha_{m,D}(\mathbf{X}) \wedge \varphi'],$$

where $\mathbf{X} = (X_1, \ldots, X_m)$, $\mathbf{Y} = (Y_1, \ldots, Y_m)$, and $\mathbf{Z} = (Z_1, \ldots, Z_n)$. It follows that for each formula $\varphi(X_1, \ldots, X_k)$ in $\mathcal{L}(\mathrm{graph})$, for each field K with $A_K \neq \emptyset$, and for all $\mathbf{a}_1, \ldots, \mathbf{a}_k \in K^m$ we have: $K \models \varphi'(\mathbf{a}_1, \ldots, \mathbf{a}_k)$ if and only if $K_{\mathbf{a}_1}, \ldots, K_{\mathbf{a}_k} \in A_K$ and $\Gamma_K \models \varphi(K_{\mathbf{a}_1}, \ldots, K_{\mathbf{a}_k})$. In particular, if θ is a sentence of $\mathcal{L}(\mathrm{graph})$:

(1) $$\Gamma_K \models \theta \iff K \models \theta'.$$

Finally, denote the sentence $(\exists \mathbf{X})\alpha_{m,D}(\mathbf{X}) \wedge \theta'$ by θ''. Then, for each field K:

(2) $$A_K \neq \emptyset \text{ and } \Gamma_K \models \theta \iff K \models \theta''.$$

THEOREM 28.10.1: *Let Q be the set of sentences θ of $\mathcal{L}(\mathrm{ring})$ for which θ is true in at least one perfect PAC field of finite corank exceeding e for each positive integer e. Let P be the set of sentences θ of $\mathcal{L}(\mathrm{ring})$ such that θ is false in each field. Then P and Q are recursively inseparable.*

Proof: The map $\theta \mapsto \theta''$ of sentences of $\mathcal{L}(\mathrm{graph})$ to sentences of $\mathcal{L}(\mathrm{ring})$ is primitive recursive. Suppose θ is true in infinitely many finite graphs $\Gamma_1, \Gamma_2, \ldots$. Then $\lim_{i \to \infty} |\Gamma_i| = \infty$. Let K_i be the perfect PAC of finite corank that corresponds to Γ_i by Proposition 28.9.2, $i = 1, 2, \ldots$. Apply (2) to see that θ'' is true in each of the fields $K_1, K_2, \ldots$.

Suppose on the other hand, that θ is false in each graph. If there exists a field K in which θ'' is true, then, by (2), θ is true in Γ_K, a contradiction. Thus, θ'' is false in each field.

The theorem now follows from Lemma 28.3.2 and Lavrov's theorem (Proposition 28.5.1). □

COROLLARY 28.10.2: *Consider the following diagram of sets of sentences of* $\mathcal{L}(\text{ring})$:

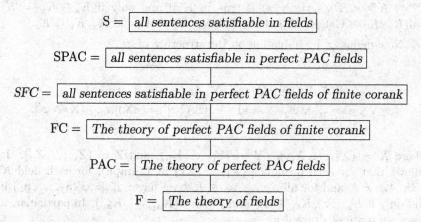

(a) *There is no recursive set between SFC and S.*
(b) *There is no recursive set between F and FC.*

In particular, none of the six sets is decidable.

Proof: Let θ be a sentence of $\mathcal{L}(\text{ring})$ which is true in a perfect PAC field K of finite corank. Then, for each integer e, θ is true in a perfect PAC field of finite corank exceeding e. Otherwise, there exists e_0 such that $\neg\theta$ is true in each perfect PAC field of finite corank exceeding e_0. But Theorem 23.1.6(b) shows that $\neg\theta$ is true in each perfect PAC field of finite corank, a contradiction. Thus, SFC is equal to the set Q of Theorem 28.10.1. Since S is the complement of the set of sentences of $\mathcal{L}(\text{ring})$ which are false in each field, (a) follows from Theorem 28.10.1. Statement (b) follows from (a) by taking negations and complements. □

Remark 28.10.3: The six sets of Corollary 28.10.2 are distinct. Indeed, by Proposition 28.5.4, there exists a sentence θ of $\mathcal{L}(\text{graph})$ which is true in all finite graphs but is false in an infinite graph Γ'. If K is a perfect PAC field of finite corank, then Lemma 16.10.2 implies that K has only finitely many Galois extensions L with $\text{Gal}(L/K) \cong D$. It follows that A_K is a finite set. If A_K is empty, then $K \models \neg(\exists \mathbf{X})\alpha_{m,D}(\mathbf{X})$. If however $A_K \neq \emptyset$, then, by (1), $K \models \theta'$. In both cases $K \models [(\exists X)\alpha_{m,D}(\mathbf{X})] \to \theta'$.

On the other hand let K' be a perfect PAC field with $\Gamma' \cong \Gamma_{K'}$. Then $K' \models (\exists \mathbf{X})\alpha_{m,D}(\mathbf{X})$. But, by (1), $K' \not\models \theta'$, so $K' \not\models [(\exists \mathbf{X})\alpha_{m,D}(\mathbf{X})] \to \theta'$.

This shows that PAC and FC are distinct. We leave to the reader the remainder of the remark (Exercise 7). □

Remark 28.10.4: The theory of algebraically closed fields of characteristic 0 is recursive and it lies between the sets FC and SFC of Corollary 28.10.2. □

Exercises

1. Construct a Turing machine M such that when M is applied to a 2-track tape T with 1 and n, respectively, stored in the 2-tracks 1 and 2, it ends with a 3-track tape T' with 1, n, and 0, respectively, stored in tracks 1, 2, and 3.

2. Use Exercise 1 of Chapter 19 to show that the codes of Turing machines (Section 28.1) form a primitive recursive subset of $\mathbb{N}$.

3. (a) Draw an inst-mode diagram, similar to that of Part B of the proof of Proposition 28.2.1 for the coordinate projection function $f(x_1, x_2, x_3) = x_2$ as applied to $(2,3,4)$.

   (b) Draw an inst-mode diagram for the computation of $f(x_1, x_2, x_3) = x_2 + 1$ as applied to $(2,3,4)$ using Part E of the proof of Proposition 28.2.1.

4. Draw an inst-mode diagram to compute the function $f(x,y) = xy$ at $x = y = 3$ by induction from $f_0(x) = 2x$ and $g(x,y,z) = y+z$ as in Part F of the proof of Proposition 28.2.1.

5. Let $D = \mathbb{Z}/p\mathbb{Z}$, $W = \mathbb{Z}/p\mathbb{Z} \times \mathbb{Z}/p\mathbb{Z}$, and $G = \mathbb{Z}_p^3$, where p is a prime. Describe the graph Γ_G of Section 28.6.

6. In this problem we consider graph condition (2b) of Section 28.7:

   (a) Suppose that D is generated by nontrivial groups, $A_1, \ldots, A_e$ with $e > 1$, having these properties: A_i is a quotient of D, $i = 1, \ldots, e$; and A_i centralizes A_j (as subgroups of D) for $1 \leq i \neq j \leq e$. With $I = \{1, 2, \ldots, e\}$, find an epimorphism $\pi \colon D^I \to D$ such that $\mathrm{Ker}(\pi) \neq \mathrm{Ker}(\pi_i)$, $i = 1, \ldots, e$.

   (b) Use the argument of Examples 28.7.3 and 28.7.4 to show that D satisfies graph condition (2b) if and only if there exist no groups $A_1, \ldots, A_e$, with $e > 1$, satisfying the conditions of (a).

7. Finish the proof of Remark 28.10.3 that the 6 sets of Corollary 28.10.2 are distinct.

Notes

Sections 28.1-28.5 are based on [Ershov-Lavrov-Taimanov-Taitslin].

The interpretation of the theory of graphs in the theory of perfect PAC fields is due independently to [Cherlin-v.d.Dries-Macintyre] and to [Ershov4]. Both sources use Frattini covers.

Corollary 28.5.3 of [Fried-Jarden5] erroneously claims that the theories G, FG, IG, and AA of Section 28.5 are distinct. Indeed, Part C of that corollary incorrectly claims that the sentence $(\exists X, Y)[P(X, Y)]$ is true in all infinite graphs, while it is false in each graph with no edges. The correct relations are $G = IG \subset FG = AA$, as indicated in Remark 28.5.5 of the present edition. We are indebted to Eric Rosen for pointing out this error.

Chapter 29.
Algebraically Closed Fields with Distinguished Automorphisms

Let K be an explicitly given countable Hilbertian field and e a positive integer. Denote the theory of all sentences θ of the language $\mathcal{L}(\text{ring}, K)$ which hold in $\tilde{K}(\sigma)$ for almost all $\sigma \in \text{Gal}(K)^e$ by Almost(K, e). By Theorem 20.6.7, Almost(K, e) is decidable. Moreover, for each sentence θ of $\mathcal{L}(\text{ring}, K)$ let $\text{Truth}(\theta) = \{\sigma \in \tilde{K}(\sigma) \mid \tilde{K}(\sigma) \models \theta\}$ and let $\text{Prob}(\theta)$ be the Haar measure of $\text{Truth}(\theta)$. Then $\text{Prob}(\theta)$ is a rational number between 0 and 1 which can be computed if θ is explicitly given.

In this chapter we add e unary function symbols $\Sigma_1, \ldots, \Sigma_e$ to $\mathcal{L}(\text{ring}, K)$ and denote the resulting language by $\mathcal{L}(\text{ring}, K, \Sigma_1, \ldots, \Sigma_e)$. We prove that, in contrast to the decidability result of the preceding paragraph, the theory of all sentences θ of $\mathcal{L}(\text{ring}, K, \Sigma_1, \ldots, \Sigma_e)$ which hold in $\langle \tilde{K}, \sigma_1, \ldots, \sigma_e \rangle$ for almost all $\sigma \in \text{Gal}(K)^e$ is undecidable (Proposition 29.2.3). Moreover, as θ ranges over all sentences of $\mathcal{L}(\text{ring}, K, \Sigma_1, \ldots, \Sigma_e)$, the probability $\text{Prob}(\theta)$ that θ holds in $\langle \tilde{K}, \sigma_1, \ldots, \sigma_e \rangle$ ranges over all definable real numbers between 0 and 1 (Theorems 29.4.5 and 29.4.6). In particular, $\text{Prob}(\theta)$ takes also transcendental values.

29.1 The Base Field K

Throughout this chapter we work over a fixed infinite base field K, finitely generated over its prime field. By Theorem 13.4.2, K is Hilbertian.

We summarize some properties of almost all fields $\tilde{K}(\sigma)$ for $e \geq 2$ that eventually lead to our undecidability results. We denote the group of roots of unity of a field F by $U(F)$.

PROPOSITION 29.1.1: *The following statements hold for every integer $e \geq 2$ and almost all $\sigma \in \text{Gal}(K)^e$:*
(a) $\tilde{K}(\sigma)$ *is PAC.*
(b) $\text{Gal}(\tilde{K}(\sigma)) \cong \hat{F}_e$.
(c) $U(\tilde{K}(\sigma))$ *is a finite group.*
(d) $U(\tilde{K}(\sigma)) = \{z \in \tilde{K} \mid \bigwedge_{i=1}^{e} \sigma_i z = z \land (\exists \alpha \in \tilde{K})[\alpha \neq 0 \land \sigma_1 \alpha = z\alpha]\}$.

In addition, for every positive integer n the measure of the set of all $\sigma \in \text{Gal}(K)^e$ with $|U(\tilde{K}(\sigma))| \geq n$ is positive.

Proof (a) and (b): Statements (a) and (b) are special cases of Theorem 20.5.1. Statement (c) repeats Theorem 18.11.7.

To prove (d), let $L = \tilde{K}(\sigma)$. Consider an element z of the right hand side of (d). Let α be as above and denote the degree of the Galois closure of $L(\alpha)$ over L by n. Then $\sigma_1 \alpha = z\alpha$. Hence, $\alpha = \sigma_1^n \alpha = z^n \alpha$. Therefore, $z^n = 1$.

29.1 The Base Field K

For the converse, we may assume that (b) and (c) hold. Let z be an element of $U(L)$. Denote the order of z by n. Since $\text{Gal}(L)$ is free, L has a cyclic extension N of degree n. By Kummer theory, N is generated over L by a nonzero element α satisfying $\alpha^n \in L$. Then $\sigma_1\alpha = \zeta_n\alpha$ for some primitive root of unity $\zeta_n \in L$ and $z = \zeta_n^i$ for some i. Hence, $\sigma_1\alpha^i = z\alpha^i$. Thus, z belongs to the right hand side of (d).

The last part of the proposition follows from the fact that $\text{Gal}(K(\zeta_n))^e$ has positive measure. $\square$

LEMMA 29.1.2 ([Duret, 4.3 and 5.2]): *Let $a_1, \ldots, a_k, b_1, \ldots, b_l$ be distinct elements of a field L.*
(a) *Let n be a positive integer satisfying $\text{char}(L) \nmid n$ and let c be a nonzero element of L. Then, the algebraic subset V of $\mathbb{A}^{1+k+l}$ defined by the system of equations*

$$X + a_i = Y_i^n, \quad i = 1, \ldots, k; \qquad X + b_j = cZ_j^n, \quad j = 1, \ldots, l$$

is an absolutely irreducible curve which is defined over L.
(b) *Suppose $\text{char}(L) = p > 0$ and $a_1, \ldots, a_k, b_1, \ldots, b_l$ are linearly independent over $\mathbb{F}_p$. Let $c \in L$. Then the algebraic subset V of $\mathbb{A}^{1+k+l}$ defined by the system of equations*

$$a_i X = Y_i^p - Y_i, \quad i = 1, \ldots, k$$
$$b_j X + c = Y_j^p - Y_i, \quad j = 1, \ldots, l$$

is an absolutely irreducible curve which is defined over L.

Proof of (a): Let x be an indeterminate. Choose algebraic elements y_i, z_j over $L(x)$ with $x + a_i = y_i^n$ and $x + b_j = cz_j^n$ for all i and j. Since $x + a_1, \ldots, x + a_k, x + b_1, \ldots, x + b_l$ are distinct prime elements of $\tilde{L}[x]$, they are multiplicatively linearly independent modulo $(\tilde{L}(x)^\times)^n$. Thus, by Kummer theory, $\text{Gal}(\tilde{L}(x, \mathbf{y}, \mathbf{z})/\tilde{L}(x)) \cong (\mathbb{Z}/n\mathbb{Z})^{k+l}$ [Lang7, p. 295, Thm. 8.2]. Hence, $Y_i^n - (x + a_i)$ is an irreducible polynomial over $\tilde{L}(x, y_1, \ldots, y_{i-1})$ for $i = 1, \ldots, k$ and $cZ_j^n - (x + b_j)$ is an irreducible polynomial over $\tilde{L}(x, \mathbf{y}, z_1, \ldots, z_{j-1})$ for $j = 1, \ldots, l$. Therefore, if $(\xi, \boldsymbol{\eta}, \boldsymbol{\zeta}) \in V(\tilde{L})$, then the $\tilde{L}$-specialization $x \to \xi$ can be successively extended to an $\tilde{L}$-specialization $(x, \mathbf{y}, \mathbf{z}) \to (\xi, \boldsymbol{\eta}, \boldsymbol{\zeta})$. It follows that V is an absolutely irreducible curve with generic point $(x, \mathbf{y}, \mathbf{z})$. Finally, note that $n^{k+1} = [\tilde{L}(x, \mathbf{y}, \mathbf{z}) : \tilde{L}(x)] \leq [L(x, \mathbf{y}, \mathbf{z}) : L(x)] \leq n^{k+l}$. Hence, $L(x, \mathbf{y}, \mathbf{z})$ and $\tilde{L}(x)$ are linearly disjoint over $L(x)$, so $L(x, \mathbf{y}, \mathbf{z})$ is a regular extension of L. Consequently, V is defined over L (Corollary 10.2.2).

Proof of (b): Replace Kummer theory by Artin-Schreier theory [Lang7, p. 296, Thm. 8.3]. Note that the assumption about the linear independence over $\mathbb{F}_p$ of $a_1, \ldots, a_k, b_1, \ldots, b_l$ implies that the additive group $\wp(\tilde{L}(x)) = \{u^p - u \mid u \in \tilde{L}(x)\}$ has index p^{k+l} in the additive subgroup of $\tilde{L}(x)$ generated by $a_1 x, \ldots, a_k x, b_1 x + c, \ldots, b_l x + c$ and $\wp(\tilde{L}(x))$. $\square$

29.2 Coding in PAC Fields with Monadic Quantifiers

Every first order language $\mathcal{L}$ naturally extends to a language $\mathcal{L}_n$, the **language of n-adic quantifiers**. It is the simplest extension of $\mathcal{L}$ which allows for each $m \leq n$ quantification over certain m-ary relations on the underlying sets of structures of $\mathcal{L}$. To obtain $\mathcal{L}_n$ from $\mathcal{L}$ adjoin for each $m \leq n$ a sequence of **m-ary variable symbols** $X_{m1}, X_{m2}, X_{m3}, \ldots$. The variable symbols of $\mathcal{L}$ are taken here as $x_1, x_2, x_3, \ldots$. An **atomic formula** of $\mathcal{L}_n$ is either an atomic formula of $\mathcal{L}$ or a formula $(x_{i_1}, \ldots, x_{i_m}) \in X_{mj}$, where $m \leq n$ and $i_1, \ldots, i_m, j$ are positive integers. As usual we close the set of formulas of $\mathcal{L}_n$ under negation, disjunction, conjunction, and quantification on variables. A **structure** for $\mathcal{L}_n$ (or an n-adic structure for $\mathcal{L}$) is a system $\langle A, \mathcal{Q}_1, \ldots, \mathcal{Q}_n \rangle$, where A is a structure for $\mathcal{L}$ and, for each $m \leq n$, $\mathcal{Q}_m$ is a nonempty collection of m-ary relations on the underlying set of A (which we also denote by A). The structure is **weak** if for each m, all relations in $\mathcal{Q}_m$ are finite. We interpret the variables x_i as elements of A and the variables X_{mj} as elements of $\mathcal{Q}_j$. Thus, "$(x_1, \ldots, x_m) \in X_{mj}$" means "$(x_1, \ldots, x_m)$ belongs to X_{mj}", "$\exists x_i$" means "there exists an element x_i in A", and "$\exists X_{mj}$" means "there exists an element X_{mj} in $\mathcal{Q}_m$".

Theories of $\mathcal{L}_n$, also called **n-adic theories**, are often undecidable. Thus, whenever we "interpret" such a theory in another theory (e.g. a theory of PAC fields), the latter also turns out to be undecidable.

To be more precise let T ad T^* be theories of languages $\mathcal{L}$ and $\mathcal{L}^*$, respectively. An **interpretation** of T in T^* is a recursive map $\theta \mapsto \theta^*$ of sentences of $\mathcal{L}$ onto sentences of $\mathcal{L}^*$ such that $T \models \theta$ if and only if $T^* \models \theta^*$. Obviously, if T is undecidable, then so is T^*.

We are mainly interested in the case where $\mathcal{L} = \mathcal{L}(\mathrm{ring}, K)$ is the language of rings enriched by constant symbols for each element of K. For integers $q \geq 2$ and p, and for a field F we say that **hypothesis $H(p, q)$ holds in** F if at least one of the following conditions holds:
(1a) $\mathrm{char}(F) = p$, $p \nmid q$, $\zeta_q \in F$, and $(F^\times)^q \neq F^\times$.
(1b) $\mathrm{char}(F) = p$, $p | q$, and $\wp(F) = \{u^p - u \mid u \in F\} \neq F$.

Similarly we say that a class, $\mathcal{F}$, of n-adic structures over fields satisfies **hypothesis $H(p,q)$** if for each structure $\langle F, \mathcal{Q}_1, \ldots, \mathcal{Q}_n \rangle$ in $\mathcal{F}$, F is a field that satisfies hypothesis $H(p,q)$.

For the next lemma consider a class $\mathcal{F}$ of **weak monadic structures** (i.e. weak 1-adic structures) over PAC fields that contain K and satisfy condition $H(p,q)$ for some p and q. To each $\langle F, \mathcal{Q} \rangle$ in $\mathcal{F}$ we associate another monadic structure $\langle F, \mathcal{Q}' \rangle$ and denote the class of all $\langle F, \mathcal{Q}' \rangle$'s by $\mathcal{F}'$. The definition of $\langle F, \mathcal{Q}' \rangle$ is divided into two cases:

CASE A: $p \nmid q$. $\mathcal{Q}'$ is the collection of all sets

$$D(A, x) = \{a \in A \mid (\exists y \in F)[y \neq 0 \wedge a + x = y^q]\}$$

with $A \in \mathcal{Q}$ and $x \in F$.

29.2 Coding in PAC Fields with Monadic Quantifiers

CASE B: $p|q$. $\mathcal{Q}'$ is the collection of all sets

$$E(A, u, x) = \left\{a \in A \mid (\exists y \in F)\left[\frac{x}{u+a} = y^p - y\right]\right\}$$

with $A \in \mathcal{L}$ and $u, x \in F$. In both cases each $A' \in \mathcal{Q}'$ is contained in some $A \in \mathcal{Q}$.

LEMMA 29.2.1:
(a) For each structure $\langle F, \mathcal{Q}\rangle$ in $\mathcal{F}$ the collection $\mathcal{Q}'$ consists of all subsets of the sets $A \in \mathcal{Q}$.
(b) The monadic theory $\mathrm{Th}(\mathcal{F}')$ is interpretable in $\mathrm{Th}(\mathcal{F})$.

Proof: We treat each of the above cases separately.

CASE A: Let $\langle F, \mathcal{Q}\rangle$ be a structure in $\mathcal{F}$. Choose an element $c \in F \smallsetminus F^q$. Let $A \in \mathcal{Q}$ and let A' be a subset of A. By assumption, A is finite and F is PAC. Hence, by Proposition 29.1.2(a), there exist $x \in F$ and $y_a \in F^\times$ for each $a \in A$ such that $a + x = y_a^q$ for all $a \in A'$ and $a + x = cy_a^q$ for all $a \in A \smallsetminus A'$. Then $A' = D(A, x)$, because $(F^\times)^q \cap c(F^\times)^q = \emptyset$. This proves (a).

Now define a map $\varphi \mapsto \varphi^*$ from formulas of $\mathcal{L}_1$ onto formulas of $\mathcal{L}_1$ by induction on the structure of φ. If φ is an atomic formula of $\mathcal{L}$, let $\varphi^* = \varphi$. If φ is the formula $a \in X$, define φ^* to be the formula

$$a \in A_X \wedge (\exists y_X)[y_X \neq 0 \wedge a + x_X = y_X^q]$$

where x_X, y_X are variable symbols on elements and A_X is a variable symbol on sets attached to the variable X. Next let the star operation commute with negation, disjunction, conjunction, and quantification on elements. Finally, if ψ^* has been defined for a formula ψ and φ is the formula $(\exists X)\psi$, then define φ^* to be $(\exists A_X)(\exists x_X)\psi^*$.

One verifies by induction on the structure of a formula $\varphi(\mathbf{z}, X_1, \ldots, X_n)$ that for each monadic structure $\langle F, \mathcal{Q}\rangle$ in $\mathcal{F}$, for $A_1, \ldots, A_n \in \mathcal{Q}$, and $x_1, \ldots, x_n \in F$ we have

(2) $$\langle F, \mathcal{Q}\rangle \models \varphi^*(\mathbf{z}, A_1, x_1, \ldots, A_n, x_n) \iff$$
$$\langle F, \mathcal{Q}'\rangle \models \varphi(\mathbf{z}, D(A_1, x_1), \ldots, D(A_n, x_n)).$$

In particular, if θ is a sentence of $\mathcal{L}_1$, then θ is true in $\langle F, \mathcal{L}'\rangle$ if and only if θ^* is true in $\langle F, \mathcal{L}\rangle$.

CASE B: Again, let $\langle F, \mathcal{Q}\rangle$ be a structure in $\mathcal{F}$. Choose an element $c \in F \smallsetminus \wp(F)$, let $A \in \mathcal{Q}$, and let A' be a subset of A. Since F is an infinite field, there exists $u \in F$ such that $\sum_{a \in A} \frac{\alpha(a)}{u+a} \neq 0$ for every function $\alpha: A \to \mathbb{F}_p$ which is not identically zero. In other words, the elements $\frac{1}{u+a}$ with a ranging on A are linearly independent over $\mathbb{F}_p$. Now apply Proposition 29.1.2(b) to find $x \in F$ and for each $a \in A$ an element $y_a \in F$ such that $\frac{x}{u+a} = y_a^p - y_a$ for each $a \in A'$ and $\frac{x}{u+a} = y_a^p - y_a + c$ for each $a \in A \smallsetminus A'$. Thus, $E(A, u, x) = A'$. This proves (a). The proof of (b) is carried out as in Case A. □

Our next construction allows us to replace monadic structures by certain n-adic structures. As before we start from a class $\mathcal{F}$ of weak monadic structures over PAC fields that satisfies hypotheses $H(p, q)$. For each structure $\langle F, \mathcal{Q} \rangle \in \mathcal{F}$ and every $m \leq n$ let $\mathcal{Q}_m$ be the collection of all subsets of $A_1 \times \cdots \times A_m$, where $A_1, \ldots, A_m \in \mathcal{Q}$. Denote the class of n-adic structures $\langle F, \mathcal{Q}_1, \ldots, \mathcal{Q}_n \rangle$ obtained in this way by $\mathcal{F}_n$.

LEMMA 29.2.2: $\mathrm{Th}(\mathcal{F}_n)$ is interpretable in $\mathrm{Th}(\mathcal{F})$.

Proof: The interpretation of $\mathrm{Th}(\mathcal{F}_n)$ in $\mathrm{Th}(\mathcal{F})$ goes through two auxiliary theories. Specifically, in addition to $\mathcal{F}'$ and $\mathcal{F}_n$, we associate with $\mathcal{F}$ one more auxiliary class $\tilde{\mathcal{F}}_n$, and make the following interpretations:

$$\mathrm{Th}(\mathcal{F}_n) \rightsquigarrow \mathrm{Th}(\tilde{\mathcal{F}}_n) \rightsquigarrow \mathrm{Th}(\mathcal{F}') \rightsquigarrow \mathrm{Th}(\mathcal{F}).$$

The interpretation of $\mathrm{Th}(\mathcal{F}')$ in $\mathrm{Th}(\mathcal{F})$ is done in Lemma 29.2.1(b). It remains to perform the two first interpretations:

PART A: *Interpretation of* $\mathrm{Th}(\mathcal{F}_n)$ *in* $\mathrm{Th}(\tilde{\mathcal{F}}_n)$. Our first step is to blow up m-ary relations, with $m \leq n$, to n-ary relations. We omit each m-ary variable symbol X_{mj} with $m < n$ from $\mathcal{L}_n$ and rename X_{nj} as Y_j. Denote the language obtained in this way by $\tilde{\mathcal{L}}_n$. In addition to the atomic formulas of $\mathcal{L}$ the only atomic formulas of $\tilde{\mathcal{L}}_n$ are $(x_{i_1}, \ldots, x_{i_n}) \in Y_j$ where $i_1, \ldots, i_n, j$ are positive integers.

We define a map $\varphi \mapsto \tilde{\varphi}$ from formulas of $\mathcal{L}_n$ onto formulas $\tilde{\mathcal{L}}_n$ by induction on the structure of φ. If φ is atomic formula of $\mathcal{L}$, let $\tilde{\varphi} = \varphi$. If φ is the formula $(x_{i_1}, \ldots, x_{i_m}) \in X_{mj}$, with $m \leq n$, define $\tilde{\varphi}$ to be the formula $(x_{i_1}, \ldots, x_{i_m}, x_{i_m}, \ldots, x_{i_m}) \in Y_{m+nj}$. Next let the tilde operation commute with negation, disjunction, and quantification on elements. Finally, if $\tilde{\psi}$ has been defined for a formula ψ and φ is the formula $(\exists X_{mj})\psi$, we define $\tilde{\varphi}$ to be the formula $(\exists Y_{m+nj})\tilde{\psi}$.

Let $\tilde{\mathcal{F}}_n$ be the class of all structures $\langle F, \mathcal{Q}_n \rangle$ of $\tilde{\mathcal{L}}_n$ with $\langle F, \mathcal{Q} \rangle \in \mathcal{F}$. For each $\langle F, \mathcal{Q} \rangle \in \mathcal{F}$ and each substitution f of the variables of $\mathcal{L}_n$ we define a substitution $\tilde{f}$ of the variables of $\tilde{\mathcal{L}}_n$ as follows: If $f(X_{mj}) = R$, then $\tilde{f}(Y_{m+nj}) = \{(a_1, \ldots, a_m, a_m, \ldots, a_m) \mid (a_1, \ldots, a_m) \in R\}$. The value of the symbol variables on elements remains unchanged. It follows by induction on the structure of a formula $\varphi(\mathbf{x}, X_{m_1 j_1}, \ldots, X_{m_r j_r})$ of $\mathcal{L}_n$ that

$$\langle F, \mathcal{Q}_1, \ldots, \mathcal{Q}_n \rangle \models \varphi(f(\mathbf{x}), f(X_{m_1 j_1}), \ldots, f(X_{m_r j_r}))$$

if and only if

$$\langle F, \mathcal{Q}_n \rangle \models \tilde{\varphi}(\tilde{f}(\mathbf{x}), \tilde{f}(Y_{m_1+nj_1}), \ldots, \tilde{f}(Y_{m_r+nj_r})).$$

In particular, this holds for each sentence θ of $\mathcal{L}_n$. Thus, $\mathrm{Th}(\mathcal{F}_n)$ is interpretable in $\mathrm{Th}(\tilde{\mathcal{F}}_n)$.

29.2 Coding in PAC Fields with Monadic Quantifiers

PART B: *Interpretation of* $\mathrm{Th}(\tilde{\mathcal{F}}_n)$ *in* $\mathrm{Th}(\mathcal{F}')$. For each $\langle F, \mathcal{Q} \rangle$ in $\mathcal{F}$ consider the bilinear map $\pi \colon F^n \times F^n \to F$ defined by $\pi(\mathbf{c}, \mathbf{x}) = \sum_{i=1}^n c_i x_i$. For each $\mathbf{c} \in F^n$, $A_1, \ldots, A_n \in \mathcal{Q}$, and $B \subseteq \pi(\mathbf{c}, A_1 \times \cdots \times A_n)$, the set

$$S(\mathbf{c}, A_1, \ldots, A_n, B) = \{(x_1, \ldots, x_n) \in A_1 \times \cdots \times A_n \mid \pi(\mathbf{c}, \mathbf{x}) \in B\}$$

belongs to $\mathcal{Q}_n$. Conversely, let $A_1, \ldots, A_n \in \mathcal{Q}$. Since F is infinite and $A_1, \ldots, A_n$ are finite, there exists $\mathbf{c} \in F^n$ with $\sum_{i=1}^n (x_i - x_i') c_i \neq 0$ for all distinct $\mathbf{x}, \mathbf{x}' \in A_1 \times \cdots \times A_n$. Then the map $\mathbf{x} \mapsto \pi(\mathbf{c}, \mathbf{x})$ from $A_1 \times \cdots \times A_n$ into F is injective. Hence, if we start from a subset R of $A_1 \times \cdots \times A_n$ and define $B = \{\pi(\mathbf{c}, \mathbf{x}) \mid \mathbf{x} \in R\}$, then $R = S(\mathbf{c}, A_1, \ldots, A_n, B)$. This representation of $\mathcal{Q}_n$ allows us to interpret $\mathrm{Th}(\tilde{\mathcal{F}}_n)$ in $\mathrm{Th}(\mathcal{F}')$:

To each formula φ of $\tilde{\mathcal{L}}_n$ we associate a formula φ' of $\mathcal{L}_1$. The definition proceeds by induction on the structure of φ. If φ is the formula $(x_1, \ldots, x_n) \in Y$ where $Y = Y_j$ for some j, φ' is the formula

$$\bigwedge_{i=1}^n x_i \in A_{Y,i} \wedge \pi(\mathbf{c}_Y, *) \text{ is injective on } A_{Y,1} \times \cdots \times A_{Y,n} \wedge \pi(\mathbf{c}_Y, \mathbf{x}) \in B_Y,$$

where $\mathbf{c}_Y = (c_{Y,1}, \ldots, c_{Y,n})$ are variables symbols on elements, and $A_{Y,1}, \ldots, A_{Y,n}, B_Y$ are 1-ary variable symbols on sets attached to Y. As before, make the prime operation commute with negation, disjunction, and quantification on elements. Finally, if φ is the formula $(\exists Y)\psi$, where Y is as above and ψ is a formula for which ψ' has been defined, then φ' is the formula

(3)
$$(\exists A_{Y,1}) \cdots (\exists A_{Y,n})(\exists B_Y)(\exists c_{Y,1}) \cdots (\exists c_{Y,n})$$
$$[\pi(\mathbf{c}_Y, *) \text{ is injective on } A_{Y,1} \times \cdots \times A_{Y,n} \wedge \pi(\mathbf{c}_Y, \mathbf{x}) \in B_Y \wedge \psi^*].$$

For all formulas $\varphi(\mathbf{x}, Y_1, \ldots, Y_s)$ of $\mathcal{L}_n$ where $\mathbf{x} = (x_{i_1}, \ldots, x_{i_r})$, for all $\langle F, \mathcal{Q} \rangle \in \mathcal{F}$, for all $A_{i1}, \ldots, A_{in} \in \mathcal{Q}$, for all $c_{i1}, \ldots, c_{in} \in F$ such that the map $\pi(\mathbf{c}_i, *) \colon A_{i1} \times \cdots \times A_{in} \to F$ is injective, and for all subsets B_i of $\pi(\mathbf{c}_i, A_{i1} \times \cdots \times A_{in})$, $i = 1, \ldots, s$, put $S_i = S(\mathbf{c}_i, A_{i1}, \ldots, A_{in}, B_i)$. Let $b_1, \ldots, b_r \in F$. An induction on the structure of φ proves that

$$\langle F, \mathcal{Q}_n \rangle \models \varphi(\mathbf{b}, S_1, \ldots, S_s)$$

if and only if

$$\langle F, \mathcal{Q}' \rangle \models \varphi'(\mathbf{b}, \mathbf{c}_1, \mathbf{A}_1, B_1 \ldots, \mathbf{c}_s, \mathbf{A}_s, B_s),$$

where $\mathbf{c}_i = (c_{i1}, \ldots, c_{in})$ and $\mathbf{A}_i = (A_{i1}, \ldots, A_{in})$. In particular, if θ is a sentence of $\mathcal{L}_n$, then $\theta \in \mathrm{Th}(\tilde{\mathcal{F}}_n)$ if and only if $\theta' \in \mathrm{Th}(\mathcal{F}')$, as desired.
□

PROPOSITION 29.2.3: *Let $\mathcal{F}$ be a class of weak monadic structures over PAC fields that satisfies hypotheses $H(p,q)$ and the following assumption:*
(4) *For each positive integer n there exists $\langle F, \mathcal{Q} \rangle \in \mathcal{F}$ and there exists $A \in \mathcal{Q}$ of cardinality at least n.*
Then $\mathrm{Th}(\mathcal{F})$ is undecidable.

Proof: Applying Lemma 29.2.2 to $n = 2$, it suffices to prove that $\mathrm{Th}(\mathcal{F}_2)$ is undecidable. By Remark 28.5.5, the theory of finite graphs is undecidable, so it suffices to interpret that theory in $\mathrm{Th}(\mathcal{F}_2)$.

To each sentence θ of the language $\mathcal{L}(\mathrm{graph})$ of graphs recursively associate the following sentence θ^* of $\mathcal{L}_2$:

$$(\forall A \in \mathcal{Q}_1)((\forall R \in \mathcal{Q}_2)[R \subseteq A \times A, \text{ and}$$
$$R \text{ is symmetric and nonreflexive} \to (A, R) \models \theta].$$

By (3), θ is true in each finite symmetric graph if and only if θ^* is true in each $\langle F, \mathcal{Q} \rangle \in \mathcal{F}$. Thus, the map $\theta \mapsto \theta^*$ is an interpretation of the theory of finite graphs into $\mathrm{Th}(\mathcal{F}_2)$, as desired. $\square$

29.3 The Theory of Almost all $\langle \tilde{K}, \sigma_1, \ldots, \sigma_e \rangle$'s

We combine the methods developed in Section 29.2 with the algebraic background of Section 29.1 to obtain undecidability results for theories over PAC fields.

Recall that we are working over a fixed infinite base field K, finitely generated over its prime field. For each $e \geq 1$ we extend the languages $\mathcal{L}(\mathrm{ring}, K)$ and $\mathcal{L}(\mathrm{ring}, \tilde{K})$ to languages

$$\mathcal{L} = \mathcal{L}(\mathrm{ring}, K, \Sigma_1, \ldots, \Sigma_e) \quad \text{and} \quad \tilde{\mathcal{L}} = \mathcal{L}(\mathrm{ring}, \tilde{K}, \Sigma_1, \ldots, \Sigma_e)$$

by adding e unary function symbols $\Sigma_1, \ldots, \Sigma_e$. Every e-tuple $(\sigma_1, \ldots, \sigma_e)$ of automorphisms of K_s over K uniquely extends to an e-tuple of automorphisms of $\tilde{K}$, also denoted by $\sigma_1, \ldots, \sigma_e$. Thus, $\langle \tilde{K}, \boldsymbol{\sigma} \rangle$ is a structure for $\tilde{\mathcal{L}}$. Let $\mathrm{Almost}(\tilde{K}, \Sigma_1, \ldots, \Sigma_e)$ be the set of all sentences θ of $\tilde{\mathcal{L}}$ which are true in $\langle \tilde{K}, \boldsymbol{\sigma} \rangle$ for almost all $\boldsymbol{\sigma} \in \mathrm{Gal}(K)^e$.

Each elementary statement on $\tilde{K}(\boldsymbol{\sigma})$ in $\mathcal{L}$ has a natural interpretation as a statement on $\langle \tilde{K}, \boldsymbol{\sigma} \rangle$ in $\tilde{\mathcal{L}}$. More precisely, we associate with each formula $\varphi(x_1, \ldots, x_n)$ of $\mathcal{L}$ a formula $\varphi^*(x_1, \ldots, x_n)$ of $\tilde{\mathcal{L}}$ such that for all $\boldsymbol{\sigma} \in \mathrm{Gal}(K)^e$ and $\mathbf{a} \in \tilde{K}(\boldsymbol{\sigma})^e$

$$\tilde{K}(\boldsymbol{\sigma}) \models \varphi(\mathbf{a}) \iff \langle \tilde{K}, \boldsymbol{\sigma} \rangle \models \varphi^*(\mathbf{a}).$$

In particular, if θ is a sentence of $\mathcal{L}$, then $\theta \in \mathrm{Almost}(K, e)$ if and only if $\theta^* \in \mathrm{Almost}(\tilde{K}, \Sigma_1, \ldots, \Sigma_e)$. The star operation leaves each atomic formula of $\mathcal{L}$ unchanged and commutes with disjunction and negation. If φ is a formula $(\exists x)\psi$ and ψ^* has been defined, then φ^* is the formula $(\exists x)[\bigwedge_{i=1}^{e} \sigma_i x = x \wedge \psi^*]$.

29.3 The Theory of Almost all $\langle \tilde{K}, \sigma_1, \ldots, \sigma_e \rangle$'s

The **truth set** of a sentence θ of $\tilde{\mathcal{L}}$ is defined to be

$$\text{Truth}(\theta) = \{\sigma \in \text{Gal}(K)^e \mid \langle \tilde{K}, \sigma \rangle \models \theta\}.$$

It is a measurable set. Indeed, if $\varphi(x_1, \ldots, x_n)$ is a quantifier free formula and $a_1, \ldots, a_n \in \tilde{K}$, then the truth of $\varphi(a_1, \ldots, a_n)$ in $\langle \tilde{K}, \sigma \rangle$ depends only on the restriction of σ to the normal closure N of $K(a_1, \ldots, a_n)/K$, hence on the restriction of σ to the maximal Galois extension N_0 of K in N. Therefore, $\text{Truth}(\varphi(a_1, \ldots, a_n))$ is an open-closed set. For an arbitrary formula $\varphi(\mathbf{x}, y)$ we have

$$\text{Truth}((\exists y)\varphi(\mathbf{x}, y)) = \bigcup_{y \in \tilde{K}} \text{Truth}(\varphi(\mathbf{x}, y)).$$

We conclude by induction on the structure of φ that $\text{Truth}(\theta)$ is even a Borel subset of $\text{Gal}(K)^e$.

The measure of $\text{Truth}(\theta)$ may be considered as the probability of θ to be true among the $\langle \tilde{K}, \sigma \rangle$'s. We write $\text{Prob}(\theta) = \mu(\text{Truth}(\theta))$.

THEOREM 29.3.1: *For $e \geq 2$, $\text{Almost}(\tilde{K}, \Sigma_1, \ldots, \Sigma_e)$ is an undecidable theory.*

Proof: Let

$$S = \bigcap_{\theta \in \text{Almost}(\tilde{K}, \Sigma_1, \ldots, \Sigma_e)} \text{Truth}(\theta) \cap \{\sigma \in \text{Gal}(K)^e \mid U(\tilde{K}(\sigma)) \text{ is finite}\}.$$

By definition, $\mu(\text{Truth}(\theta)) = 1$ for each $\theta \in \text{Almost}(\tilde{K}, \Sigma_1, \ldots, \Sigma_e)$. By Proposition 29.1.1, $U(\tilde{K}(\sigma))$ is finite for almost all $\sigma \in \tilde{K}(\sigma)$. Since $\text{Almost}(\tilde{K}, \Sigma_1, \ldots, \Sigma_e)$ is countable, $\mu(S) = 1$. By Proposition 29.1.1, each $\sigma \in S$ has these properties:
(1a) $\tilde{K}(\sigma)$ is PAC.
(1b) $\text{Gal}(\tilde{K}(\sigma)) = \hat{F}_e$.
(1c) $U(\tilde{K}(\sigma))$ is a finite group.
(1d) $U(\tilde{K}(\sigma)) = \{z \in \tilde{K} \mid \bigwedge_{i=1}^{e} \sigma_i z = z \land (\exists \alpha \in \tilde{K})[\alpha \neq 0 \land \sigma_1 \alpha = z\alpha]\}$.

By (1b), $F = \tilde{K}(\sigma)$ has an extension F' of degree 2. If $\text{char}(K) \neq 2$, then $F' = F(\sqrt{u})$ with $u \in F^\times \smallsetminus (F^\times)^2$. If $\text{char}(K) = 2$, then $F' = F(x)$ where $x^2 - x = a$ and $a \in F \smallsetminus \wp(F)$. In each case F satisfies hypothesis $H(\text{char}(K), 2)$.

Let $\mathcal{Q}_\sigma = \{U(\tilde{K}(\sigma))\}$ and $\mathcal{F} = \{\langle \tilde{K}(\sigma), \mathcal{Q}_\sigma \rangle \mid \sigma \in S\}$. Then $\mathcal{F}$ is a set of weak monadic structures with $|U(\tilde{K}(\sigma))|$ unbounded (Proposition 29.1.1). By Proposition 29.2.3, $\mathcal{F}$ is undecidable.

Condition (1d) suggests an interpretation of $\text{Th}(\mathcal{F})$ in $\text{Almost}(\tilde{K}, \Sigma_1, \ldots, \Sigma_e)$: replace $z \in X$ by

$$\bigwedge_{i=1}^{e} \sigma_i z = z \land (\exists a)[a \neq 0 \land \sigma_1 z = za].$$

If φ^* is an interpretation of a formula φ of $\mathcal{L}_1$, then φ^* is also the interpretation of $(\exists X)\varphi$. An induction on the structure of formulas in $\mathcal{L}_1$ proves the following statement: Let $\varphi(z_1,\ldots,z_m,X_1,\ldots,X_n)$ be a formula of $\mathcal{L}_1$ and $a_1,\ldots,a_m \in \tilde{K}$. Then $\varphi^*(z_1,\ldots,z_n)$ has its free variables among the $z_1,\ldots,z_m$'s and for all $\sigma \in S$

$$\langle \tilde{K}(\sigma), \mathcal{Q}_\sigma \rangle \models \varphi(\mathbf{a}, U(\tilde{K}(\sigma)),\ldots,U(\tilde{K}(\sigma))) \iff \langle \tilde{K},\sigma \rangle \models \varphi^*(\mathbf{a}).$$

In particular, a sentence θ of $\mathcal{L}_1$ belongs to $\text{Th}(\mathcal{F})$ if and only if θ^* is in Almost$(\tilde{K},\Sigma_1,\ldots,\Sigma_e)$.

It follows that Almost$(\tilde{K},\Sigma_1,\ldots,\Sigma_e)$ is undecidable. $\square$

PROBLEM 29.3.2: *Is $T(K,1)$ undecidable?*

29.4 The Probability of Truth Sentences

As before K is an infinite finitely generated field extension of its prime field. We survey here a stronger undecidability result for Almost$(\Sigma_1,\ldots,\Sigma_e)$ than that proved in Section 29.3. Moreover, we represent the set of all Prob(θ), with θ ranging over all sentences of $\mathcal{L}(\text{ring},K,e)$, as the set of all "definable numbers" between 0 and 1.

The stronger decidability result involves **Arithmetic**. This is the complete theory of the structure $\mathcal{N} = \langle \mathbb{N}, +, \cdot, 1 \rangle$.

LEMMA 29.4.1 ([Cherlin-Jarden, Prop. 2.4]): *Let $\mathcal{F}$ be a class of weak monadic structures over PAC fields that satisfies Hypotheses $H(p,q)$. Suppose for each $\langle F,\mathcal{Q}\rangle \in \mathcal{F}$ the cardinality of the sets $A \in \mathcal{Q}$ is unbounded. Then $\text{Th}(\mathcal{N})$ is interpretable in $\text{Th}(\mathcal{F})$.*

Moreover, there is a recursive map $\varphi(x) \mapsto \varphi^*(X)$ from formulas of arithmetic to formulas of $\mathcal{L}_1$ satisfying this: for all $\langle F,\mathcal{Q}\rangle \in \mathcal{F}$ and $A \in \mathcal{Q}$ we have $\mathcal{N} \models \varphi(|A|)$ if and only if $\langle F,\mathcal{Q}\rangle \models \varphi^*(A)$.

It is well known Arithmetic is undecidable [Ershov-Lavrov-Taimanov-Taitslin, Thm. 3.2.4]. Thus, each $\mathcal{F}$ satisfying the assumptions of Lemma 29.4.1 is undecidable.

The encoding of Arithmetic in Almost$(\Sigma_1,\ldots,\Sigma_e)$ uses some facts about torsion of elliptic curves over the fields $\tilde{K}(\sigma)$. The first of them is already cited in Section 18.11:

PROPOSITION 29.4.2 ([Geyer-Jarden1, Thm. 1.1]): *For $e \geq 2$ and for almost all $\sigma \in \text{Gal}(K)^e$ the set $E_{\text{tor}}(\tilde{K}(\sigma))$ is finite.*

The second one is an easier result:

PROPOSITION 29.4.3 ([Cherlin-Jarden, Prop. 1.4]): *For each $e \geq 1$, for almost all $\sigma \in \text{Gal}(K)^e$, and for every $n \in \mathbb{N}$ there exists an elliptic curve E over K which has a $\tilde{K}(\sigma)$-rational point of order n.*

Finally, we need an analog of Proposition 29.1.1 for elliptic curves:

29.4 The Probability of Truth Sentences

PROPOSITION 29.4.4 ([Cherlin-Jarden, Cor. 1.6]): *For each $e \geq 1$, for almost all $\sigma \in \mathrm{Gal}(K)^e$, and for every elliptic curve defined over $\tilde{K}(\sigma)$,*

$$E_{\mathrm{tor}}(\tilde{K}(\sigma)) = \{\mathbf{z} \in E(\tilde{K}) \mid \bigwedge_{i=1}^{e} \sigma_i \mathbf{z} = \mathbf{z} \wedge (\exists \mathbf{a} \in E(\tilde{K})) \, \sigma_1 \mathbf{a} = \mathbf{a} + \mathbf{z}\}.$$

A real number r is said to be **arithmetically definable** if there exists a formula $\varphi(x,y)$ of $\mathcal{N}$ such that for all $k, m \in \mathbb{N}$, $r > \frac{k}{m}$ if and only if $\mathcal{N} \models \varphi(k,m)$. For example, every rational number is arithmetically definable.

THEOREM 29.4.5 ([Cherlin-Jarden, Thm. 5.3]): *Suppose K is an explicitly given finitely generated extension of its prime field and e is a positive integer. Let θ be a sentence of $\mathcal{L}(\mathrm{ring}, \tilde{K}, \Sigma_1, \ldots, \Sigma_e)$. Then $\mathrm{Prob}(\theta)$ is an arithmetically definable real number.*

The proof of Theorem 29.4.5 uses manipulation of recursive functions and convergent series of real numbers and explicit computation in Galois groups over K (Lemma 19.3.2) but none of the results 29.4.2–cs29.4.4. Those tools are needed for the following converse of Theorem 29.4.5:

THEOREM 29.4.6 ([Cherlin-Jarden, Thms. 6.5 and 7.2]): *Let K be an infinite finitely generated field over its prime field and let $e \geq 2$ be an integer. Then for every definable real number r between 0 and 1 there exists a sentence θ of $\mathcal{L}(\mathrm{ring}, K, \Sigma_1, \ldots, \Sigma_e)$ such that $\mathrm{Prob}(\theta) = r$.*

Remark 29.4.7: Sentences with transcendental probabilities. We have already mentioned in the introduction to this chapter that the probability of a sentence in $\mathcal{L}(\mathrm{ring}, K)$ to be true in $\tilde{K}(\sigma)$ is a rational number. In contrast, the probability of a sentence of $\mathcal{L}(\mathrm{ring}, \tilde{K}, \Sigma_1, \ldots, \Sigma_e)$ to be true in $\langle \tilde{K}, \sigma_1, \ldots, \sigma_e \rangle$ may take a transcendental value.

By Theorem 29.4.6 it suffices to give an example of a definable transcendental number between 0 and 1. The example we give is $\frac{\pi}{4}$ which can be expressed as $\sum_{i=0}^{\infty}(-1)^{i+1}\frac{1}{2i+1}$. This expression follows from the relation $\tan \frac{\pi}{4} = 1$ and the formula $\arctan x = \sum_{i=0}^{\infty}(-1)^{i+1}\frac{x^i}{2i+1}$. For each positive integer n consider the partial sum $s_n = \sum_{i=1}^{n}(-1)^{i+1}\frac{1}{2i+1}$. Then $s_1, s_3, s_5, \ldots$ is a monotonically descending sequence which converges to $\frac{\pi}{4}$. Write $s_{2n+1} = \frac{a_n}{b_n}$ with a_n and b_n relatively prime positive integers. Then each of the sequences $a_1, a_2, a_3, \ldots$ and $b_1, b_2, b_3, \ldots$ is recursive. By a theorem of Gödel, each recursive subset of $\mathbb{N}$ is **representable** [Shoenfield, p. 128]. Thus, there are formulas $\alpha(x)$ and $\beta(x)$ of Arithmetic such that $\mathcal{N} \models \alpha(a)$ if and only if a is one of the a_i's and $\mathcal{N} \models \beta(b)$ if and only if b is one of the b_i's. We may therefore write a formula $\varphi(x,y)$ of Arithmetic such that $\mathcal{N} \models \varphi(a,b)$ if and only if $\frac{a}{b} > s_1$ or there are positive integers m, n satisfying $s_{2m+1} \leq \frac{a}{b} < s_{2n+1}$. Thus, $\mathcal{N} \models \varphi(a,b)$ if and only if $\frac{a}{b} > \frac{\pi}{4}$. Consequently, $\frac{\pi}{4}$ is a definable number. □

Chapter 30. Galois Stratification

Chapter 30 extends the constructive field theory and algebraic geometry of Chapter 19, in contrast to Chapter 20, to give effective decision procedures through elimination of quantifiers. Such an elimination of quantifiers requires formulas outside of the theory $\mathcal{L}$(ring). We call these more general formulas "Galois formulas". These formulas include data for a stratification of the affine space $\mathbb{A}^n$ into K-normal basic sets A. Each coordinate ring $K[A]$ is equipped with a Galois ring cover C and a collection of conjugacy classes Con of subgroups of the Galois group $\text{Gal}(K(C)/K(A))$. Each successive elimination of a quantifier contributes to the subgroups that appear in Con. This leads to the primitive recursiveness of the theory of all Frobenius fields which contain a given field K with elimination theory (Theorem 30.6.1).

30.1 The Artin Symbol

For K, a fixed field, denote the class of all perfect Frobenius fields M (Section 24.1) that contain K by Frob(K). We refer to the situation where K is a presented field with elimination theory (Definition 19.2.8) as the **explicit case**.

The results of this chapter hold without restriction on K; but, in the explicit case they become effective in the sense of Chapter 19. Thus, the existence theorems that are established in the general case become effective in the explicit case.

Let A be a normal K-basic set and C an integral domain extending $K[A]$ (Section 19.6). We call C/A a **(Galois) ring/set cover** over K if $C/K[A]$ is a (Galois) ring cover (Definition 6.1.3). In the explicit case suppose A is a presented K-basic set and F is a presented finite separable extension of $K(A)$. Then Lemma 19.7.2 effectively produces a normal K-basic open subset A' of A and an integral domain C such that $K(C) = F$ and C/A' is a ring/set cover over K.

Let C/A be a Galois ring/set cover over K with

$$K[A] = K[x_1, \ldots, x_n, g(\mathbf{x})^{-1}]$$

and let z be a primitive element for the ring cover $C/K[A]$. Denote the Galois group $\text{Gal}(K(C)/K(A))$ by $\text{Gal}(C/A)$ and consider a field M that contains K. If $(a_1, \ldots, a_n) \in A(M)$, then the K-specialization $\mathbf{x} \to \mathbf{a}$ uniquely extends to a homomorphism φ_0 of $K[A]$ into M. By Lemma 6.1.4, φ_0 extends to a homomorphism φ from C into a Galois extension $N = M(\varphi(z)) = M \cdot \varphi(C)$ of M. Denote the quotient fields of $\varphi(K[A])$ and $\varphi(C)$, respectively, by $\bar{E}$ and $\bar{F}$. Then $\bar{F}/\bar{E}$ is a Galois extension. Let

$$D(\varphi) = \{\sigma \in \text{Gal}(C/A) \mid (\forall u \in C)[\varphi(u) = 0 \implies \varphi(\sigma u) = 0]\}$$

be the decomposition group of φ. Each $\sigma \in D(\varphi)$ induces an element $\bar{\sigma}$ of $\mathrm{Gal}(\bar{F}/\bar{E})$ by the formula $\bar{\sigma}(\varphi(u)) = \varphi(\sigma u)$ for each $u \in C$. By Lemma 6.1.4, the map $\varphi' \colon D(\varphi) \to \mathrm{Gal}(\bar{F}/\bar{E})$ that maps σ to $\bar{\sigma}$ is an isomorphism. Furthermore, φ' maps the subgroup

$$D_M(\varphi) = \{\sigma \in \mathrm{Gal}(C/A) \mid (\forall u \in C)[\varphi(u) \in M \Longrightarrow \varphi(\sigma u) = \varphi(u)]\}$$

of $D(\varphi)$ onto $\mathrm{Gal}(\bar{F}/\bar{F} \cap M)$. If $M = K$, then $D_M(\varphi) = D(\varphi)$. In the general case, the composition of the isomorphism $\mathrm{res}_{\bar{F}} \colon \mathrm{Gal}(N/M) \to \mathrm{Gal}(\bar{F}/\bar{F} \cap M)$ with $(\varphi')^{-1}$ is an isomorphism $\varphi^* \colon \mathrm{Gal}(N/M) \to D_M(\varphi)$, where

$$\varphi(\varphi^*(\sigma)(u)) = \sigma(\varphi(u))$$

for all $\sigma \in \mathrm{Gal}(N/M)$ and $u \in C$. As φ ranges over all possible extensions of φ_0 to C, the group $D_M(\varphi)$ ranges over a conjugacy class of subgroups of $\mathrm{Gal}(C/A)$. We refer to this class as the **Artin symbol** of $\mathbf{a}$ in $\mathrm{Gal}(C/A)$ and denote it by $\mathrm{Ar}(C/A, M, \mathbf{a})$.

If D/A is another Galois ring/set cover such that $C \subseteq D$ and $\mathbf{a} \in A(M)$, then $\mathrm{Ar}(C/A, M, \mathbf{a}) = \mathrm{res}_{K(C)} \mathrm{Ar}(D/A, M, \mathbf{a})$. Indeed, in the notation above let z_1 be a primitive element for $D/K[A]$. Extend φ to a homomorphism φ_1 of D into $N_1 = M(\varphi_1(z_1))$. Then N_1 is a Galois extension of M which contains N and $\mathrm{res}_{K(C)} D_M(\varphi_1) \leq D_M(\varphi)$. Since in the commutative diagram

$$\begin{array}{ccc} \mathrm{Gal}(N_1/M) & \xrightarrow{\varphi_1^*} & D_M(\varphi_1) \\ {\scriptstyle \mathrm{res}_N} \downarrow & & \downarrow {\scriptstyle \mathrm{res}_{K(C)}} \\ \mathrm{Gal}(N/M) & \xrightarrow{\varphi^*} & D_M(\varphi) \end{array}$$

both φ_1^* and φ^* are isomorphisms and res_N is surjective, so is $\mathrm{res}_{K(C)}$. Thus, $\mathrm{res}_{K(C)} D_M(\varphi_1) = D_M(\varphi)$, which proves our claim.

Whenever there is no confusion, we omit reference to the cover from the Artin symbol and write it as $\mathrm{Ar}(A, M, \mathbf{a})$. By definition, $\mathrm{Ar}(A, M, \mathbf{a})$ is a conjugation domain of subgroups of $\mathrm{Gal}(C/A)$. Hence, if $H \in \mathrm{Ar}(A, M, \mathbf{a})$, then $\mathrm{Ar}(A, M, \mathbf{a}) = \{H^\sigma \mid \sigma \in \mathrm{Gal}(C/A)\}$.

If $n = 0$, then $\mathbb{A}^0$ consists of one point, O, the origin. If $A = \mathbb{A}^0$, then $K(A) = K$ and $C = L$ is a finite Galois extension of K. In this case φ_0 is the identity map, φ is an automorphism of L over K and $\mathrm{Ar}(A, M, O) = \{\mathrm{Gal}(L/L \cap M)^\sigma \mid \sigma \in \mathrm{Gal}(L/K)\}$.

In the general definition replacement of A by an open subset A' does not affect the Artin symbol. Indeed, let $h \in K[X_1, \ldots, X_n]$ be a polynomial that does not vanish on A. Put $A' = A \smallsetminus V(h)$ and $C' = C[h(\mathbf{x})^{-1}]$, where $\mathbf{x}$ is a generic point of A. Then C'/A' is also a Galois ring/set cover. If $\mathbf{a} \in A'(M)$, then $\mathrm{Ar}(A', M, \mathbf{a}) = \mathrm{Ar}(A, M, \mathbf{a})$.

More generally, if A' is a K-normal basic set contained in A with a generic point $\mathbf{x}'$, then the specialization $\mathbf{x} \to \mathbf{x}'$ uniquely extends to a K-homomorphism τ_0 of $K[A]$ into $K[A']$ (Remark 19.6.4). With z a primitive element for $C/K[A]$ let $p(Z)$ be the image of $\mathrm{irr}(z, K(A))$ in $K[A'][Z]$ under τ_0. Then any root z' of $p(Z)$ is a primitive element for a Galois ring/set cover C'/A' and τ_0 extends to a homomorphism $\tau\colon C \to C'$ with $\tau(z) = z'$. The cover C'/A' is said to be **induced** by C/A, and τ induces an isomorphism $\tau^*\colon \mathrm{Gal}(C'/A') \to D(\tau)$.

Claim: If $\mathbf{a} \in A'(M)$, then $\tau^*(\mathrm{Ar}(A', M, \mathbf{a})) \subseteq \mathrm{Ar}(A, M, \mathbf{a})$. Indeed, let ψ be a homomorphism of C' into $\tilde{M}$ that extends the specialization $\mathbf{x}' \to \mathbf{a}$, and let $N = M(\psi(z'))$. By definition, $\tau(\tau^*(\sigma)u) = \sigma\tau u$ for all $\sigma \in \mathrm{Gal}(C'/A')$ and $u \in C$. Hence, $\tau^*(D_M(\psi)) \leq D_M(\psi \circ \tau)$. Since both groups are isomorphic to $\mathrm{Gal}(N/M)$, we have $\tau^*(D_M(\psi)) = D_M(\psi \circ \tau)$.

A **conjugacy domain of subgroups** of $\mathrm{Gal}(C/A)$ is a collection of subgroups of $\mathrm{Gal}(C/A)$ which is closed under conjugation by elements of $\mathrm{Gal}(C/A)$.

Consider a family $\mathcal{H}$ of finite groups and let $\mathrm{Con}(A, \mathcal{H})$ be a conjugacy domain of subgroups of $\mathrm{Gal}(C/A)$ belonging to $\mathcal{H}$. Note that if $\mathrm{Ar}(A, M, \mathbf{a}) \cap \mathrm{Con}(A, \mathcal{H}) \neq \emptyset$, then $\mathrm{Ar}(A, M, \mathbf{a}) \subseteq \mathrm{Con}(A, \mathcal{H})$. Assume, as above, that C'/A' is induced by $\tau\colon C \to C'$ from C/A. Define

$$\mathrm{Con}(A', \mathcal{H}) = \{H \leq \mathrm{Gal}(C'/A') \mid H \in \mathcal{H} \text{ and } \tau^*(H) \in \mathrm{Con}(A, \mathcal{H})\}.$$

We say that the conjugacy domain $\mathrm{Con}(A', \mathcal{H})$ of $\mathrm{Gal}(C'/A')$ is **induced** by $\mathrm{Con}(A, \mathcal{H})$. If $\mathbf{a} \in A'(M)$, then $\mathrm{Ar}(A', M, \mathbf{a}) \subseteq \mathrm{Con}(A', \mathcal{H})$ if and only if $\mathrm{Ar}(A, M, \mathbf{a}) \subseteq \mathrm{Con}(A, \mathcal{H})$.

30.2 Conjugacy Domains under Projections

Continue the conventions of Section 30.1. Let $n \geq 0$ and let $\pi\colon \mathbb{A}^{n+1} \to \mathbb{A}^n$ be projection onto the first n coordinates. If $n = 0$, then π maps each point of $\mathbb{A}^1$ onto the only point, O, of $\mathbb{A}^0$. Fix for the whole section a family $\mathcal{H}$ of finite groups.

Let $A \subseteq \mathbb{A}^{n+1}$ and $B \subseteq \mathbb{A}^n$ be normal K-basic sets such that $\pi(A) = B$. Suppose A is equipped with a Galois ring cover and a conjugacy domain $\mathrm{Con}(A, \mathcal{H})$. Suppose B is equipped with a Galois ring cover. We construct a conjugacy domain $\mathrm{Con}(B, \mathcal{H})$ for B for which, with some additional conditions on A and B, the following holds:

(1) For each perfect Frobenius extension M of K and each $\mathbf{b} \in B(M)$, $\mathrm{Ar}(B, M, \mathbf{b}) \subseteq \mathrm{Con}(B, \mathcal{H})$ if and only if there exists $\mathbf{a} \in A(M)$ such that $\pi(\mathbf{a}) = \mathbf{b}$ and $\mathrm{Ar}(A, M, \mathbf{a}) \subseteq \mathrm{Con}(A, \mathcal{H})$.

There are two cases: either $\dim(A) = \dim(B) + 1$ or $\dim(A) = \dim(B)$. Lemmas 30.2.1 and 30.2.3 treat the first case, Lemma 30.2.5 the second.

We start with the first case. Let C/A and D/B be Galois ring/set covers over K with these properties: $A \subseteq \mathbb{A}^{n+1}$; $B = \pi(A)$ (so, $K[B] \subseteq K[A]$); $K(A) = K(B)(y)$ with y transcendental over $K(B)$; and $K(D)$ contains the algebraic closure L of $K(B)$ in $K(C)$. In addition, let z be a primitive element

30.2 Conjugacy Domains under Projections

for the ring cover $C/K[A]$ and let $\mathbf{x}$ be a generic point for B. Suppose $(\mathbf{x}, y)$ is a generic point for A.

LEMMA 30.2.1: In the above notation, there exists a polynomial $h \in K[X_1, \ldots, X_n]$, not vanishing on B, such that for $C' = C[h(\mathbf{x})^{-1}]$, $A' = A \smallsetminus V(h)$, $D' = D[h(\mathbf{x})^{-1}]$, and $B' = B \smallsetminus V(h)$, the pair $(C'/A', D'/B')$ of Galois ring/set covers satisfies these conditions:
(2a) $D' \cap L/K[B']$ is a ring cover.
(2b) $\pi(A') = B'$.
(2c) For each field extension M of K, for each transcendental element y' over M and for each K-homomorphism $\varphi: C' \to \widetilde{M(y')}$ with $\varphi(\mathbf{x}) \in B'(M)$ and $\varphi(y) = y'$ let $N = M[\varphi(D' \cap L)]$ and $F = M(y', \varphi(z))$. Then $[K(C): L(y)] = [F: N(y')]$ and F/N is a regular extension.

Moreover, in the explicit case when A, B, C and D are presented, h can be computed effectively.

(3)
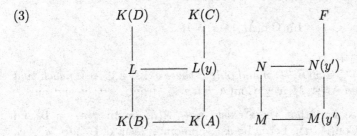

Proof: Let $K[B] = K[\mathbf{x}, g_1(\mathbf{x})^{-1}]$, $K[A] = K[\mathbf{x}, y, g_2(\mathbf{x}, y)^{-1}]$, and $S = L \cap D$. Find a polynomial $f \in S[Y, Z]$, irreducible over L, with $f(y, z) = 0$. Since $L(y, z) = K(C)$ is a regular extension of L, $f(Y, Z)$ is absolutely irreducible. Bertini-Noether (Proposition 10.4.2) produces a nonzero element $u \in S$ with this property: if φ is a homomorphism of C into a field and $\varphi(u) \neq 0$, then the polynomial $f^\varphi = f^\varphi(Y, Z)$ is absolutely irreducible and has the same degree in Z as $f(Y, Z)$. Choose $h \in K[X_1, \ldots, X_n]$ with $h(\mathbf{x}) = g_1(\mathbf{x})^k N_{L/K(B)}(u)$ for some integer $k \geq 0$. Further, a multiplication of h by an appropriate polynomial assures that with D' and B' given in the statement of the lemma, $(D' \cap L)/K[B']$ is a ring cover (Remark 6.1.5).

In order to prove that the pair C'/A' and D'/B' of Galois ring/set covers satisfies (2) we have only to check (2c). Indeed, $h(\varphi(\mathbf{x})) \neq 0$. Hence, $\varphi(u) \neq 0$, and $f^\varphi(y', Z)$ is irreducible over $N(y')$. Therefore $[F: N(y')] = \deg_Z f^\varphi(y', Z) = \deg_Z \varphi(y, Z) = [K(C): L(y)]$. Finally, since f^φ is absolutely irreducible and $F = N(y', \varphi(z))$, the extension F/N is regular. □

We refer to the properties listed in (2) by stating that the pair $(C'/A', D'/B')$ of Galois ring /set covers is **specialization compatible**.

Notation 30.2.2: For a finite group G, denote the family of all subgroups of G by $\mathrm{Subgr}(G)$. Consider a Galois ring/set cover C/A, a subfield E of $K(A)$, and a Galois extension L of E contained in $K(C)$. If $\mathrm{Con}(A, \mathcal{H})$ is a conjugacy domain of subgroups of $\mathrm{Gal}(C/A)$ belonging to $\mathcal{H}$, then $\mathrm{res}_L(\mathrm{Con}(A, \mathcal{H}))$ denotes the collection of groups obtained by restricting elements of $\mathrm{Con}(A, \mathcal{H})$ to L. ☐

LEMMA 30.2.3: *Suppose the pair $(C/A, D/B)$ of Galois ring/set covers of Lemma 30.2.1 is specialization compatible. Let $\mathrm{Con}(A, \mathcal{H})$ be a conjugacy domain of subgroups of $\mathrm{Gal}(C/A)$ belonging to a family $\mathcal{H}$ of finite groups. Define*

$$\mathrm{Con}(B, \mathcal{H}) = \{G \leq \mathrm{Gal}(D/B) \mid G \in \mathcal{H} \text{ and } \mathrm{res}_L G \in \mathrm{res}_L(\mathrm{Con}(A, \mathcal{H}))\}.$$

where L is the algebraic closure of $K(B)$ in $K(C)$. Let

$$\mathcal{I} = \mathrm{Subgr}(\mathrm{Gal}(D/B)) \cup \mathrm{Subgr}(\mathrm{Gal}(C/A)).$$

Then for each Frobenius field M that contains K with

(4) $$\mathrm{Im}(\mathrm{Gal}(M)) \cap \mathcal{I} = \mathcal{H} \cap \mathcal{I},$$

and for each $\mathbf{b} \in B(M)$,
(5) $\mathrm{Ar}(B, M, \mathbf{b}) \subseteq \mathrm{Con}(B, \mathcal{H})$ *if and only if there exists* $\mathbf{a} \in A(M)$ *such that $\pi(\mathbf{a}) = \mathbf{b}$ and $\mathrm{Ar}(A, M, \mathbf{a}) \subseteq \mathrm{Con}(A, \mathcal{H})$.*

Proof: Suppose first that there exists $\mathbf{a} \in A(M)$ with $\pi(\mathbf{a}) = \mathbf{b}$ and $\mathrm{Ar}(A, M, \mathbf{a}) \subseteq \mathrm{Con}(A, \mathcal{H})$. Let φ be a K-homomorphism of C into $\tilde{M}$ such that $\varphi(\mathbf{x}, y) = \mathbf{a}$. By Section 30.1, $\mathrm{res}_{L(y)}(D_M(\varphi)) = D_M(\mathrm{res}_{L(y)}\varphi)$. Also for $S = D \cap L$, $S \cdot K[A]/K[A]$ is a Galois ring cover and $L(y)$ is the quotient field of $S \cdot K[A]$. Put $\psi = \mathrm{res}_{L(y)}\varphi$. By definition, $\mathrm{res}_L D_M(\psi) \subseteq D_M(\mathrm{res}_S \psi)$. Since $M(\varphi(S \cdot K[A])) = M(\varphi(S))$, both groups have the same order, so they coincide. Thus, $\mathrm{res}_L(D_M(\mathrm{res}_{L(y)}\varphi)) = D_M(\mathrm{res}_L \varphi)$. Therefore, $\mathrm{res}_L(D_M(\varphi)) = D_M(\mathrm{res}_L \varphi)$. Since $\varphi(\mathbf{x}) = \mathbf{b}$, this gives $\mathrm{res}_L(\mathrm{Ar}(A, M, \mathbf{a})) \subseteq \mathrm{Ar}(S/B, M, \mathbf{b})$. Since the left hand side of the inclusion is a conjugacy domain and the right hand side is a conjugacy class of subgroups of $\mathrm{Gal}(L/K(B))$ they are equal: $\mathrm{res}_L(\mathrm{Ar}(A, M, \mathbf{a})) = \mathrm{res}_L(\mathrm{Ar}(B, M, \mathbf{b}))$. If $G \in \mathrm{Ar}(B, M, \mathbf{b})$, then $G \in \mathrm{Im}(\mathrm{Gal}(M)) \cap \mathrm{Subgr}(\mathrm{Gal}(C/A))$ and $\mathrm{res}_L G \in \mathrm{res}_L(\mathrm{Ar}(A, M, \mathbf{a})) \subseteq \mathrm{Con}(A, \mathcal{H})$. By (4), $G \in \mathcal{H}$. Hence, $G \in \mathrm{Con}(B, \mathcal{H})$. It follows that $\mathrm{Ar}(B, M, \mathbf{b}) \subseteq \mathrm{Con}(B, \mathcal{H})$.

Suppose now that M is a Frobenius field which contains K and satisfies (4) and $\mathbf{b}$ is a point of $B(M)$ with $\mathrm{Ar}(B, M, \mathbf{b}) \subseteq \mathrm{Con}(B, \mathcal{H})$. The existence of $\mathbf{a} \in A(M)$ with $\pi(\mathbf{a}) = \mathbf{b}$ and $\mathrm{Ar}(A, M, \mathbf{a}) \subseteq \mathrm{Con}(A, \mathcal{H})$ falls into two parts.

PART A: *Specialization of $(\mathbf{x}, y)$ to a point of transcendence degree 1 over M.* Without loss assume $K(D) = L$. Take a transcendental element y' over M and extend the K-specialization $\mathbf{x} \to \mathbf{b}$ to a homomorphism φ of C into the algebraic closure of $M(y')$ such that $\varphi(y) = y'$. Recall that $K[A] =$

30.2 Conjugacy Domains under Projections

$K[\mathbf{x}, y, g_2(\mathbf{x}, y)^{-1}]$ and z is a primitive element for the ring cover $C/K[A]$. Since $\pi(A) = B$, we have $g_2(\mathbf{b}, y') \neq 0$. Let $z' = \varphi(z)$, $N = M \cdot \varphi(D)$, $R = M[y', g_2(\mathbf{b}, y')^{-1}] = M[\varphi(K[A])]$, $E = M(y')$, and $F = E(z')$ (see Diagram (3)). Then $R[z']/R$ is a Galois ring cover over M, with F/E the corresponding field cover. By the specialization compatibility assumption on $(C/A, D/B)$, $[F: N(y')] = [K(C): L(y)]$. Therefore, in the following commutative diagram,

$$
\begin{array}{ccccccccc}
1 & \longrightarrow & \mathrm{Gal}(K(C)/L(y)) & \longrightarrow & \mathrm{Gal}(C/A) & \xrightarrow{\mathrm{res}} & \mathrm{Gal}(D/B) & \longrightarrow & 1 \\
& & \uparrow \varphi^* & & \uparrow \varphi^* & & \uparrow \varphi^* & & \\
1 & \longrightarrow & \mathrm{Gal}(F/N(y')) & \longrightarrow & \mathrm{Gal}(F/E) & \xrightarrow[\mathrm{res}]{} & \mathrm{Gal}(N/M) & \longrightarrow & 1
\end{array}
$$

the left vertical arrow is an isomorphism.

PART B: *Application of the Frobenius property.* By definition,

$$\varphi^*(\mathrm{Gal}(N/M)) \in \mathrm{Ar}(B, M, \mathbf{b}) \in \mathrm{Con}(B, \mathcal{H}),$$

so there is an $H \in \mathrm{Con}(A, \mathcal{H})$ with $\mathrm{res}_{K(D)} H = \varphi^*(\mathrm{Gal}(N/M))$. In particular, $H \in \mathrm{Im}(\mathrm{Gal}(M))$. The commutativity of the diagram in Part A shows that $\mathrm{res}_{K(D)}(\varphi^*(\mathrm{Gal}(F/E))) = \varphi^*(\mathrm{Gal}(N/M))$. Since the left vertical arrow is surjective, $H \leq \varphi^*(\mathrm{Gal}(F/E))$. Hence, there is a subgroup H' of $\mathrm{Gal}(F/E)$ such that $\varphi^*(H') = H$. Since all maps denoted by φ^* are injective, $H' \in \mathrm{Im}(\mathrm{Gal}(M))$ and $\mathrm{res}_N H' = \mathrm{Gal}(N/M)$. By Lemma 30.2.1, N is the algebraic closure of M in F.

Since M is a Frobenius field, Proposition 24.1.4 gives an M-epimorphism ψ of $R[z']$ onto a Galois extension F' of M that contains N such that, in the notation of Section 30.1, $\psi(y') = c \in M$ and $D_M(\psi) = \psi^*(\mathrm{Gal}(F'/M)) = H'$. By definition, $\varphi^*(D_M(\psi)) \leq D_M(\psi \circ \varphi)$. But, since both $D_M(\psi \circ \varphi)$ and $D_M(\psi)$ are isomorphic to $\mathrm{Gal}(F'/M)$, $H = \varphi^*(H') = \varphi^*(D_M(\psi)) = D_M(\psi \circ \varphi)$. The point $\mathbf{a} = (\mathbf{b}, c) = \psi \circ \varphi(\mathbf{x}, y)$ belongs to $A(M)$ and $\pi(\mathbf{a}) = \mathbf{b}$. Hence, $H \in \mathrm{Ar}(A, M, \mathbf{a})$. In addition, $H \in \mathrm{Con}(A, \mathcal{H})$. Consequently, $\mathrm{Ar}(A, M, \mathbf{a}) \subseteq \mathrm{Con}(A, \mathcal{H})$.

This concludes the proof of the lemma. □

Remark 30.2.4: Note for later application that the only use made of the assumption that M is a Frobenius field in the Galois stratification procedure appears in Part B of the proof of Lemma 30.2.3. □

Let C/A and D/B be Galois ring/set covers over K with $K[A]$ integral over $K[B]$ such that $\pi(A) = B$. Let E (resp. F) be the maximal separable extension of $K(B)$ in $K(A)$ (resp. $K(C)$). Both extensions $K(A)/E$ and $K(C)/F$ are purely inseparable. Hence, $K(A)$ and F are linearly disjoint over E and $F \cdot K(A) = K(C)$. If $\mathrm{char}(K) = p \neq 0$, let q be a power of p such that $K(A)^q \subseteq E$ and $K(C)^q \subseteq F$. Then $K(C)^q/K(A)^q$ is a Galois extension and $K(C)^q \cdot E = F$. Therefore, F/E is also a Galois extension and res: $\text{Gal}(C/A) \to \text{Gal}(F/E)$ is an isomorphism. Assume $F \subseteq K(D)$.

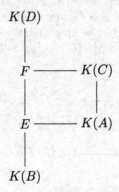

Let $\text{Con}(A, \mathcal{H})$ be a conjugacy domain of subgroups of $\text{Gal}(C/A)$ belonging to $\mathcal{H}$. Define

(6) $$\text{Con}(B, \mathcal{H}) = \{H^\sigma \mid H \leq \text{Gal}(K(D)/E),\ H \in \mathcal{H},\\ \text{res}_F H \in \text{res}_F(\text{Con}(A, \mathcal{H})),\ \sigma \in \text{Gal}(D/B)\}.$$

LEMMA 30.2.5: *In the above notation, for each perfect field M that contains K and satisfies $\text{Im}(\text{Gal}(M)) \cap \text{Subgr}(\text{Gal}(D/B)) \subseteq \mathcal{H}$ and for each $\mathbf{b} \in B(M)$*
(7) *$\text{Ar}(B, M, \mathbf{b}) \subseteq \text{Con}(B, \mathcal{H})$ if and only if there exists $\mathbf{a} \in A(M)$ with $\pi(\mathbf{a}) = \mathbf{b}$ and $\text{Ar}(A, M, \mathbf{a}) \subseteq \text{Con}(A, \mathcal{H})$.*

Proof: Denote the integral closures of $K[B]$ in E and F, respectively, by R and S. Then $R \subseteq K[A]$ and $S \subseteq D \cap C$. Also, let $\mathbf{x}$ and $\mathbf{y}$ be generic points of A and B such that $\pi(\mathbf{x}) = \mathbf{y}$.

Suppose that $\text{Ar}(B, M, \mathbf{b}) \subseteq \text{Con}(B, \mathcal{H})$. Then there exists an $H \leq \text{Gal}(K(D)/E)$ that belongs to $\text{Ar}(B, M, \mathbf{b})$ and there is $G \in \text{Con}(A, \mathcal{H})$ with $\text{res}_F H = \text{res}_F G$. Since $H \in \text{Ar}(b, M, \mathbf{b})$, there is a K-homomorphism φ of D into a Galois extension N of M such that $\varphi(\mathbf{y}) = \mathbf{b}$ and $\varphi^*(\text{Gal}(N/M)) = H$. In particular, $\varphi(R) \subseteq M$. Since M is perfect and $K(C)/F$ is purely inseparable, there is a unique K-homomorphism ψ of C into N with $\psi|_S = \varphi|_S$. Since $K(A)/E$ is purely inseparable, $\mathbf{a} = \psi(\mathbf{x})$ belongs to $A(M)$ and $\pi(\mathbf{a}) = \mathbf{b}$. Also, if we denote $M[\psi(C)]$ by N_0, then $\text{res}_F(\psi^*(\text{Gal}(N_0/M))) = \text{res}_F H$. Since $\text{res}_F: \text{Gal}(C/A) \to \text{Gal}(F/E)$ is an isomorphism, $\psi^*(\text{Gal}(N_0/M)) = G$. Therefore, $G \in \text{Ar}(A, M, \mathbf{a})$ and $\text{Ar}(A, M, \mathbf{a}) \subseteq \text{Con}(A, \mathcal{H})$.

For the converse use the condition $\text{Im}(\text{Gal}(M)) \cap \text{Subgr}(\text{Gal}(D/B)) \subseteq \mathcal{H}$ to reverse the above arguments. □

The condition of specialization compatibility from Lemma 30.2.3 and the special assumptions of Lemma 30.2.5 require us to go through a stratification procedure in order to apply these lemmas to a given set of Galois ring/set covers:

30.3 Normal Stratification

LEMMA 30.2.6: Let $n \geq 0$ and let $\{C_t/A_t \mid t \in T\}$ be a finite collection of Galois ring/set covers over K with $A_t \subseteq \mathbb{A}^{n+1}$, $t \in T$. Let $B \subseteq \mathbb{A}^n$ be a normal K-basic set with $B \subseteq \pi(A_t)$, $t \in T$. Then there exist a nonempty Zariski K-open subset B' of B and a Galois ring/set cover D/B' over K and for each $t \in T$ there exist a subset $I(t)$ of I and Galois ring/set covers C'_{ti}/A'_{ti} over K, $i \in I(t)$, with these properties:

(a) $\pi^{-1}(B') \cap A_t = \bigcup_{i \in I(t)} A'_{ti}$ and $\pi(A'_{ti}) = B'$.
(b) C'_{ti}/A'_{ti} is induced (Section 30.1) by C_t/A_t.
(c) If $\dim(A'_{ti}) = \dim(B')$, then $K[A'_{ti}]$ is integral over $K[B']$ and $K(D)$ contains the maximal separable extension of $K(B')$ in $K(C'_{ti})$.
(d) If $\dim(A'_{ti}) = \dim(B') + 1$, then $K(D)$ contains the algebraic closure of $K(B')$ in $K(C'_{ti})$ and the pair $(C'_{ti}/A'_{ti}, D/B')$ of Galois ring/set covers is specialization compatible.

Moreover, in the explicit case, if C_t/A_t and B are presented, then $I(t)$, C'_{ti}/A'_{ti}, $i \in I(t)$, and D/B' can be effectively computed.

Proof: Let $t \in T$. Apply Proposition 19.7.3 to find a stratification of $A_t \cap \pi^{-1}(B)$ into a disjoint union $\bigcup_{i \in J(t)} A_{ti}$ of normal K-basic sets. In particular, $\pi(A_{ti}) \subseteq B$, for $i \in J(t)$. Let $I(t) = \{j \in J(t) \mid \dim(\pi(A_{tj})) = \dim(B)\}$ and $I'(t) = J(t) \smallsetminus I(t)$. Then

$$B_0 = \bigcup_{t \in T} [\bigcup_{j \in I(t)} (B \smallsetminus \pi(A_{tj})) \cup \bigcup_{j \in I'(t)} \pi(A_{tj})]$$

is of dimension smaller than B. Find a polynomial $f \in K[X_1, \ldots, X_n]$ that vanishes on B_0 but not on B.

Note that for each $i \in I(t)$ either $\dim(A_{ti}) = \dim(B)$ or $\dim(A_{ti}) = \dim(B) + 1$. Let C_{ti} be the ring cover of A_{ti} induced by C_t. If $\dim(A_{ti}) = \dim(B)$, let F_{ti} be the separable closure of $K(B)$ in $K(C_{ti})$. If $\dim(A_{ti}) = \dim(B) + 1$, let F_{ti} be the algebraic closure of $K(B)$ in $K(C_{ti})$. Choose a finite Galois extension P of $K(B)$ containing all F_{ti}, $i \in I(t)$, $t \in T$. Choose a generic point $\mathbf{x}$ for B. Lemmas 19.7.2 and 30.2.1 produce a multiple $g \in K[X_1, \ldots, X_n]$ of f that vanishes on B_0 but not on B such that with $A'_{ti} = A_{ti} \smallsetminus V(g)$ and $C'_{ti} = C_{ti}[g(\mathbf{x})^{-1}]$ the following conditions hold:

(8a) $B' = B \smallsetminus V(g)$ has a Galois ring cover D with $K(D) = P$.
(8b) If $\dim(A_{ti}) = \dim(B)$, then $K[A'_{ti}]$ is integral over $K[B']$.
(8c) If $\dim(A_{ti}) = \dim(B) + 1$, then the pair $(C'_{ti}/A'_{ti}, D/B')$ of Galois set/ring covers is specialization compatible.

These conditions imply the conditions of the lemma. ☐

30.3 Normal Stratification

The condition of specialization compatibility from Lemma 30.2.3 and the special assumptions of Lemma 30.2.5 require us to go through a stratification procedure in order to apply these lemmas to a given set of Galois ring/set covers:

LEMMA 30.3.1: Let $n \geq 0$ and let $\{C_t/A_t \mid t \in T\}$ be a finite collection of Galois ring/set covers over K with $A_t \subseteq \mathbb{A}^{n+1}$, $t \in T$. Let $B \subseteq \mathbb{A}^n$ be a normal K-basic set with $B \subseteq \pi(A_t)$, $t \in T$. Then there exist a nonempty Zariski K-open subset B' of B and a Galois ring/set cover D/B' over K and for each $t \in T$ there exist a subset $I(t)$ of I and Galois ring/set covers C'_{ti}/A'_{ti} over K, $i \in I(t)$, with these properties:

(a) $\pi^{-1}(B') \cap A_t = \bigcup_{i \in I(t)} A'_{ti}$ and $\pi(A'_{ti}) = B'$.
(b) C'_{ti}/A'_{ti} is induced (Section 30.1) by C_t/A_t.
(c) If $\dim(A'_{ti}) = \dim(B')$, then $K[A'_{ti}]$ is integral over $K[B']$ and $K(D)$ contains the maximal separable extension of $K(B')$ in $K(C'_{ti})$.
(d) If $\dim(A'_{ti}) = \dim(B') + 1$, then $K(D)$ contains the algebraic closure of $K(B')$ in $K(C'_{ti})$ and the pair $(C'_{ti}/A'_{ti}, D/B')$ of Galois ring/set covers is specialization compatible.

Moreover, in the explicit case, if C_t/A_t and B are presented, then $I(t)$, C'_{ti}/A'_{ti}, $i \in I(t)$, and D/B' can be effectively computed.

Proof: Let $t \in T$. Apply Proposition 19.7.3 to find a stratification of $A_t \cap \pi^{-1}(B)$ into a disjoint union $\bigcup_{i \in J(t)} A_{ti}$ of normal K-basic sets. In particular, $\pi(A_{ti}) \subseteq B$, for $i \in J(t)$. Let $I(t) = \{j \in J(t) \mid \dim(\pi(A_{tj})) = \dim(B)\}$ and $I'(t) = J(t) \smallsetminus I(t)$. Then

$$B_0 = \bigcup_{t \in T} [\bigcup_{j \in I(t)} (B \smallsetminus \pi(A_{tj})) \cup \bigcup_{j \in I'(t)} \pi(A_{tj})]$$

is of dimension smaller than B. Find a polynomial $f \in K[X_1, \ldots, X_n]$ that vanishes on B_0 but not on B.

Note that for each $i \in I(t)$ either $\dim(A_{ti}) = \dim(B)$ or $\dim(A_{ti}) = \dim(B) + 1$. Let C_{ti} be the ring cover of A_{ti} induced by C_t. If $\dim(A_{ti}) = \dim(B)$, let F_{ti} be the separable closure of $K(B)$ in $K(C_{ti})$. If $\dim(A_{ti}) = \dim(B) + 1$, let F_{ti} be the algebraic closure of $K(B)$ in $K(C_{ti})$. Choose a finite Galois extension P of $K(B)$ containing all F_{ti}, $i \in I(t)$, $t \in T$. Choose a generic point $\mathbf{x}$ for B. Lemmas 19.7.2 and 30.2.1 produce a multiple $g \in K[X_1, \ldots, X_n]$ of f that vanishes on B_0 but not on B such that with $A'_{ti} = A_{ti} \smallsetminus V(g)$ and $C'_{ti} = C_{ti}[g(\mathbf{x})^{-1}]$ the following conditions hold:
(1a) $B' = B \smallsetminus V(g)$ has a Galois ring cover D with $K(D) = P$.
(1b) If $\dim(A_{ti}) = \dim(B)$, then $K[A'_{ti}]$ is integral over $K[B']$.
(1c) If $\dim(A_{ti}) = \dim(B) + 1$, then the pair $(C'_{ti}/A'_{ti}, D/B')$ of Galois set/ring covers is specialization compatible.

These conditions imply the conditions of the lemma. $\square$

Let $n \geq 0$ and let A be a K-constructible set in $\mathbb{A}^n$. A **normal stratification**,

$$\mathcal{A} = \langle A, C_i/A_i \rangle_{i \in I},$$

of A over K is a partition $A = \bigcup_{i \in I} A_i$ of A as a finite union of disjoint normal K-basic sets A_i, each equipped with a Galois ring/set cover C_i. Denote the collections of groups, $\bigcup_{i \in I} \mathrm{Subgr}(\mathrm{Gal}(C_i/A_i))$, by $\mathrm{Subgr}(\mathcal{A})$.

30.4 Elimination of One Variable

LEMMA 30.3.2: *Let* $n \geq 0$ *and let* $\{C_t/A_t \mid t \in T\}$ *be a finite collection of Galois ring/set covers over* K *with* $A_t \subseteq \mathbb{A}^{n+1}$ *for each* $t \in T$. *Let* $B \subseteq \mathbb{A}^n$ *be a normal* K-*basic set such that* $B \subseteq \pi(A_t)$ *for each* $t \in T$. *Then there exist normal stratifications* $\mathcal{B} = \langle B, D_j/B_j \rangle_{j \in J}$ *and* $\mathcal{A} = \langle \pi^{-1}(B), C_{jk}/A_{jk} \rangle_{j \in J, k \in K(j)}$ *satisfying the following conditions:*
(a) $\pi(A_{jk}) = B_j$ *for all* $j \in J$ *and* $k \in K(j)$ *and* $\pi^{-1}(B_j) = \bigcup_{k \in K(j)} A_{jk}$.
(b) *Each* A_{jk} *is contained in a unique* A_t *and* C_{jk}/A_{jk} *is induced by* C_t/A_t.
(c) *If* $\dim(A_{jk}) = \dim(B_j)$, *then* $K[A_{jk}]$ *is integral over* $K[B_j]$ *and* $K(D_j)$ *contains the maximal separable extension of* $K(B_j)$ *in* $K(C_{jk})$.
(d) *If* $\dim(A_{jk}) = \dim(B_j) + 1$, *then* $K(D_j)$ *contains the algebraic closure of* $K(B_j)$ *in* $K(C_{jk})$ *and the pair* $(C_{jk}/A_{jk}, D_j/B_j)$ *is specialization compatible.*

Proof: First suppose B is a normal K-basic set. Let B' be the nonempty open subset of B which Lemma 30.3.1 supplies. Then, in the notation of that lemma, the normal stratifications $\langle \pi^{-1}(B'), C'_{ti}/A'_{ti} \rangle_{t \in T, i \in I(t)}$ and $\langle B', D/B' \rangle$ satisfy Conditions (a)-(d) of our lemma. The general case follows now by the stratification lemma (Lemma 19.6.6). □

30.4 Elimination of One Variable

This section defines Galois stratification and establishes basic lemmas to eliminate quantifiers from the Galois formulas of Section 30.5.

Let $n \geq 0$, A a K-constructible set in $\mathbb{A}^n$, and $\mathcal{A} = \langle A, C_i/A_i \rangle_{i \in I}$ a normal stratification, of A over K. If $\mathcal{H}$ is a family of finite groups, then $\mathcal{A}$ may be augmented to a **Galois stratification** (with respect to $\mathcal{H}$):

$$\mathcal{A}(\mathcal{H}) = \langle A, C_i/A_i, \mathrm{Con}(A_i, \mathcal{H}) \rangle_{i \in I}$$

where, in addition to the above conditions, $\mathrm{Con}(A_i, \mathcal{H})$ is a conjugacy domain of subgroups of $\mathrm{Gal}(C_i/A_i)$ belonging to $\mathcal{H}$, $i \in I$. Then $\mathcal{A}$ is said to be the **underlying normal stratification of** $\mathcal{A}(\mathcal{H})$.

Let M be a field containing K and $\mathbf{a} \in A(M)$. Write $\mathrm{Ar}(\mathcal{A}, M, \mathbf{a}) \subseteq \mathrm{Con}(\mathcal{A}(\mathcal{H}))$ if $\mathrm{Ar}(A_i, M, \mathbf{a}) \subseteq \mathrm{Con}(A_i, \mathcal{H})$ for the unique $i \in I$ with $\mathbf{a} \in A_i$.

If $\mathcal{H}'$ is another family of finite groups with $\mathcal{H} \cap \mathrm{Subgr}(\mathcal{A}) \subseteq \mathcal{H}'$, then we freely rename $\mathrm{Con}(A_i, \mathcal{H})$ to be $\mathrm{Con}(A_i, \mathcal{H}')$, $i \in I$. Then

$$\mathcal{A}(\mathcal{H}') = \langle A, C_i/A_i, \mathrm{Con}(A_i, \mathcal{H}') \rangle_{i \in I}$$

is a Galois stratification with respect to $\mathcal{H}'$. For every field M containing K and each $\mathbf{a} \in A(M)$, clearly $\mathrm{Ar}(\mathcal{A}, M, \mathbf{a}) \subseteq \mathrm{Con}(\mathcal{A}(\mathcal{H}))$ if and only if $\mathrm{Ar}(\mathcal{A}, M, \mathbf{a}) \subseteq \mathrm{Con}(\mathcal{A}(\mathcal{H}'))$.

Suppose $\mathcal{A}' = \langle A, C'_j/A'_j \rangle_{j \in J}$ is another normal stratification of A. Call $\mathcal{A}'$ a **refinement** of $\mathcal{A}$ if for each $j \in J$ there exists a unique $i \in I$ such that $A'_j \subseteq A_i$ and the cover C'_j/A'_j is induced by the cover C_i/A_i. If $\mathcal{A}'(\mathcal{H}) = \langle A, C'_j/A'_j, \mathrm{Con}(A'_j, \mathcal{H}) \rangle_{j \in J}$ is an extension of $\mathcal{A}'$ to a Galois stratification, then $\mathcal{A}'(\mathcal{H})$ is said to be a **refinement** of $\mathcal{A}(\mathcal{H})$ if in addition $\mathrm{Con}(A'_j, \mathcal{H})$ is induced by $\mathrm{Con}(A_i, \mathcal{H})$ (Section 30.1) whenever $A'_j \subseteq A_i$. In this case we have for $\mathbf{a} \in A(M)$ that $\mathrm{Ar}(\mathcal{A}, M, \mathbf{a}) \subseteq \mathrm{Con}(\mathcal{A}(\mathcal{H}))$ if and only if $\mathrm{Ar}(\mathcal{A}', M, \mathbf{a}) \subseteq \mathrm{Con}(\mathcal{A}'(\mathcal{H}))$.

The next two lemmas are based on Lemma 30.2.6. They will allow us in Section 30.5 to eliminate, respectively, one existential or universal quantifier from a given Galois formula.

LEMMA 30.4.1 (Existential Elimination Lemma): Let $n \geq 0$ and let $\mathcal{A}(\mathcal{H}) = \langle \mathbb{A}^{n+1}, C_i/A_i, \mathrm{Con}(A_i, \mathcal{H}) \rangle_{i \in I}$ be a Galois stratification of $\mathbb{A}^{n+1}$ over K with respect to a family $\mathcal{H}$ of finite groups. Then there exists a Galois stratification $\mathcal{B}(\mathcal{H}) = \langle \mathbb{A}^n, D_j/B_j, \mathrm{Con}(B_j, \mathcal{H}) \rangle_{j \in J}$ with the following property: Let $\mathcal{I} = \mathrm{Subgr}(\mathcal{A}) \cup \mathrm{Subgr}(\mathcal{B})$. Suppose M is a perfect Frobenius field that contains K and satisfies

(1) $$\mathrm{Im}(\mathrm{Gal}(M)) \cap \mathcal{I} = \mathcal{H} \cap \mathcal{I}.$$

Then, for each $\mathbf{b} \in \mathbb{A}^n(M)$,

(2) $\mathrm{Ar}(\mathcal{B}, M, \mathbf{b}) \subseteq \mathrm{Con}(\mathcal{B}(\mathcal{H}))$ if and only if there exists $\mathbf{a} \in \mathbb{A}^{n+1}(M)$ such that $\pi(\mathbf{a}) = \mathbf{b}$ and $\mathrm{Ar}(\mathcal{A}, M, \mathbf{a}) \subseteq \mathrm{Con}(\mathcal{A}(\mathcal{H}))$.

Moreover, the underlying normal stratification $\mathcal{B} = \langle \mathbb{A}^n, D_j/B_j \rangle_{j \in J}$ of $\mathcal{B}(\mathcal{H})$ depends only on the normal stratification $\mathcal{A} = \langle \mathbb{A}^{n+1}, C_i/A_i \rangle_{i \in I}$.

In the explicit case, if $\mathcal{H}$ is primitive recursive and $\mathcal{A}(\mathcal{H})$ presented, then $\mathcal{B}(\mathcal{H})$ can be effectively computed.

Proof: The union of the constructible sets $\pi(A_i)$ is equal to $\mathbb{A}^n$. Apply Lemma 19.6.6 to stratify $\mathbb{A}^n$ into a union of disjoint normal K-basic sets U_s, $s \in S$, such that for all $i \in I$ and $s \in S$, either $U_s \subseteq \pi(A_i)$ or $U_s \cap \pi(A_i) = \emptyset$.

Lemma 30.3.2 allows us to stratify U_s and $\pi^{-1}(U_s)$ for each $s \in S$ separately and then combine the separate stratifications into basic normal stratifications $\mathbb{A}^n = \bigcup_{j \in J} B_j$ and $\mathbb{A}^{n+1} = \bigcup_{j \in J} \bigcup_{k \in K(j)} A_{jk}$ with the following properties:

(3a) Each A_{jk} is contained in a unique A_i and has a Galois ring cover C_{jk} which is induced by C_i/A_i.
(3b) $\pi(A_{jk}) = B_j$ for each $j \in J$ and each $k \in K(j)$ and $\pi^{-1}(B_j) = \bigcup_{k \in K(j)} A_{jk}$.
(3c) Each B_j is equipped with a Galois ring cover D_j.
(3d) If $\dim(A_{jk}) = \dim(B_j)$, then $K[A_{jk}]$ is an integral extension of $K[B_j]$ and $K(D_j)$ contains the maximal separable extension of $K(B_j)$ in $K(C_{jk})$.
(3e) If $\dim(A_{jk}) = \dim(B_j) + 1$, then $K(D_j)$ contains the maximal algebraic extension of $K(B_j)$ in $K(C_{jk})$ and the pair $(C_{jk}/A_{jk}, D_j/B_j)$ of Galois ring/set covers is specialization compatible.

Following (3a), let $\mathrm{Con}(A_{jk}, \mathcal{H})$ be the conjugacy domain of subgroups of $\mathrm{Gal}(C_{jk}/A_{jk})$ induced by $\mathrm{Con}(A_i, \mathcal{H})$. Use Lemma 30.2.5 (resp. Lemma

30.5 The Complete Elimination Procedure

30.2.3), to define a conjugacy domain $\text{Con}_k(B_j, \mathcal{H})$ of subgroups of $\text{Gal}(D_j/B_j)$ from $\text{Con}(A_{jk}, \mathcal{H})$ in case (3d) (resp. (3e)).

The stratification $\mathcal{A}'(\mathcal{H}) = \langle \mathbb{A}^{n+1}, C_{jk}/A_{jk}, \text{Con}(A_{jk}, \mathcal{H}) \rangle_{j \in J, k \in K(j)}$ refines $\mathcal{A}(\mathcal{H})$. For each $j \in J$ define $\text{Con}(B_j, \mathcal{H})$ to be $\bigcup_{k \in K(j)} \text{Con}_k(B_j, \mathcal{H})$. Then $\mathcal{B}(\mathcal{H}) = \langle \mathbb{A}^n, D_j/B_j, \text{Con}(B_j, \mathcal{H}) \rangle_{j \in J}$ is a Galois stratification of $\mathbb{A}^n$. Moreover, the conclusion of the lemma follows from Lemmas 30.2.3 and 30.2.5: $\text{Ar}(\mathcal{B}, M, \mathbf{b}) \subseteq \text{Con}(\mathcal{B}(\mathcal{H}))$ if and only if there exists $\mathbf{a} \in \mathbb{A}^{n+1}(M)$ such that $\pi(\mathbf{a}) = \mathbf{b}$ and $\text{Ar}(\mathcal{A}', M, \mathbf{a}) \subseteq \text{Con}(\mathcal{A}'(\mathcal{H}))$ (i.e. $\text{Ar}(\mathcal{A}, M, \mathbf{a}) \subseteq \text{Con}(\mathcal{A}(\mathcal{H}))$). □

LEMMA 30.4.2 (Universal Elimination Lemma): *Let*

$$\mathcal{A}(\mathcal{H}) = \langle \mathbb{A}^{n+1}, C_i/A_i, \text{Con}(A_i, \mathcal{H}) \rangle_{i \in I}$$

be a Galois stratification of $\mathbb{A}^{n+1}$ over K with respect to a family $\mathcal{H}$ of finite groups. Then there exists a Galois stratification

$$\mathcal{B}(\mathcal{H}) = \langle \mathbb{A}^n, D_j/A_j, \text{Con}(B_j, \mathcal{H}) \rangle_{j \in J}$$

such that for each perfect Frobenius field M that contains K and satisfies (1) the following holds: for each $\mathbf{b} \in \mathbb{A}^n(M)$.
(4) $\text{Ar}(\mathcal{B}, M, \mathbf{b}) \subseteq \text{Con}(\mathcal{B}(\mathcal{H}))$ *if and only if* $\text{Ar}(\mathcal{A}, M, \mathbf{a}) \subseteq \text{Con}(\mathcal{A}(\mathcal{H}))$ *for each $\mathbf{a} \in \mathbb{A}^{n+1}(M)$ with $\pi(\mathbf{a}) = \mathbf{b}$.*

Moreover, the underlying normal stratification $\mathcal{B} = \langle \mathbb{A}^n, D_j/A_j \rangle_{j \in J}$ of $\mathcal{B}(\mathcal{H})$ depends only on the normal stratification $\mathcal{A} = \langle \mathbb{A}^{n+1}, C_i/A_i \rangle_{i \in I}$.

In the explicit case, if $\mathcal{H}$ is primitive recursive and $\mathcal{A}(\mathcal{H})$ is presented, then $\mathcal{B}(\mathcal{H})$ can be effectively computed.

Proof: Let $\mathcal{A}^c(\mathcal{H}) = \langle \mathbb{A}^{n+1}, C_i/A_i, \text{Con}^c(A_i, \mathcal{H}) \rangle_{i \in I}$ be the **complementary Galois stratification** to $\mathcal{A}(\mathcal{H})$ of $\mathbb{A}^{n+1}$, where

$$\text{Con}^c(A_i, \mathcal{H}) = \{ H \leq \text{Gal}(C_i/A_i) \mid H \in \mathcal{H} \text{ and } H \notin \text{Con}(A_i, \mathcal{H}) \}.$$

Apply Lemma 30.4.1 to find a Galois stratification

$$\mathcal{B}^c(\mathcal{H}) = \langle \mathbb{A}^n, D_j/B_j, \text{Con}^c(B_j, \mathcal{H}) \rangle_{j \in J}$$

over K such that for each M and $\mathbf{b}$ as in the lemma
$\text{Ar}(\mathcal{B}^c, M, \mathbf{b}) \subseteq \text{Con}(\mathcal{B}^c(\mathcal{H}))$ if and only if there exists $\mathbf{a} \in \mathbb{A}^{n+1}(M)$ such that $\pi(\mathbf{a}) = \mathbf{b}$ and $\text{Ar}(\mathcal{A}^c, M, \mathbf{a}) \subseteq \text{Con}(\mathcal{A}^c(\mathcal{H}))$.

The complementary Galois stratification to $\mathcal{B}^c(\mathcal{H})$ of $\mathbb{A}^n$ satisfies the conclusion of the lemma. □

30.5 The Complete Elimination Procedure

Lemma 30.4.1 and 30.4.2 lead to an elimination of quantifier procedure for "Galois formulas" (Proposition 30.5.2). For a "Galois sentence" θ this procedure produces a finite Galois extension L of K and a conjugation domain $\mathrm{Con}(\mathcal{H})$ of subgroups of $\mathrm{Gal}(L/K)$ such that θ holds in a perfect Frobenius field M with a certain restriction on $\mathrm{Im}(\mathrm{Gal}(M))$ if and only if $\mathrm{Gal}(L/L \cap M) \in \mathrm{Con}(\mathcal{H})$.

Let $m, n \geq 0$ be integers, $Q_1, \ldots, Q_m$ quantifiers, and

$$\mathcal{A}(\mathcal{H}) = \langle \mathbb{A}^{m+n}, C_i/A_i, \mathrm{Con}(A_i, \mathcal{H}) \rangle_{i \in I}$$

a Galois stratification of $\mathbb{A}^{m+n}$ over K with respect to a family $\mathcal{H}$ of finite groups. Then the expression

(1) $\qquad (Q_1 X_1) \cdots (Q_m X_m)[\mathrm{Ar}(\mathbf{X}, \mathbf{Y}) \subseteq \mathrm{Con}(\mathcal{A}(\mathcal{H}))]$

with $\mathbf{X} = (X_1, \ldots, X_m)$ and $\mathbf{Y} = (Y_1, \ldots, Y_n)$ is said to be a **Galois formula** (with respect to K and $\mathcal{A}(\mathcal{H})$) in the free variables $\mathbf{Y}$. Denote it by $\theta = \theta(\mathbf{Y})$. For a field M that contains K and for $b_1, \ldots, b_n \in M$, write $M \models \theta(\mathbf{b})$ if for $Q_1 a_1 \in M, \ldots, Q_m a_m \in M$, $\mathrm{Ar}(\mathcal{A}, M, (\mathbf{a}, \mathbf{b})) \subseteq \mathrm{Con}(\mathcal{A}(\mathcal{H}))$. Here read "$Q_i a_i \in M$" as "there exists an a_i in M" if Q_i is $\exists$, and as "for each a_i in M" if Q_i is $\forall$. In the case that $n = 0$, θ has no free variables and is called a **Galois sentence**.

Remark 30.5.1: Each formula $\varphi(Y_1, \ldots, Y_n)$ in the language $\mathcal{L}(\mathrm{ring}, K)$ can be written (effectively, in the explicit case) in prenex normal form

$$(Q_1 X_1) \cdots (Q_m X_m) \left[\bigvee_{i=1}^{k} \bigwedge_{j=1}^{l} f_{ij}(\mathbf{X}, \mathbf{Y}) = 0 \land g_{ij}(\mathbf{X}, \mathbf{Y}) \neq 0 \right]$$

with $f_{ij}, g_{ij} \in K[\mathbf{X}, \mathbf{Y}]$. The formula in the brackets defines a K-constructible set $A \subseteq \mathbb{A}^{m+n}$. Let $\mathcal{H}$ be the family containing only the trivial group. Construct a K-normal stratification $\mathbb{A}^{m+n} = \bigcup_{i \in I} A_i$ such that for each $i \in I$ either $A_i \subseteq A$ or $A_i \subseteq \mathbb{A}^{m+n} \smallsetminus A$. In the first case let C_i be $K[A_i]$ and define $\mathrm{Con}(A_i, \mathcal{H})$ to be $\mathcal{H}$. In the second case let $C_i = K[A_i]$ and define $\mathrm{Con}(A_i, \mathcal{H})$ to be the empty collection. The corresponding Galois stratification $\mathcal{A}(\mathcal{H})$ defines a Galois formula θ as in (1). If a field M contains K and if $b_1, \ldots, b_n \in M$, then $M \models \theta(\mathbf{b})$ if and only if $M \models \varphi(\mathbf{b})$. Thus, each formula in $\mathcal{L}(\mathrm{ring}, K)$ is equivalent to a Galois formula over K. □

PROPOSITION 30.5.2: Let $\mathcal{C}$ be a family of finite groups. Consider a Galois formula $\theta(Y_1, \ldots, Y_n)$

$$(Q_1 X_1) \cdots (Q_m X_m)[\mathrm{Ar}(\mathbf{X}, \mathbf{Y}) \subseteq \mathrm{Con}(\mathcal{A}(\mathcal{C}))],$$

with respect to a Galois stratification $\mathcal{A}(\mathcal{C})$ of $\mathbb{A}^{m+n}$ over K. Then there exists a finite family $\mathcal{S}$ of finite groups and a normal stratification $\mathcal{B}$ of $\mathbb{A}^n$ over K with $\mathrm{Subgr}(\mathcal{A}) \cup \mathrm{Subgr}(\mathcal{B}) \subseteq \mathcal{S}$. Moreover, for every family $\mathcal{H}$ of finite groups which satisfies $\mathrm{Con}(\mathcal{A}(\mathcal{C})) \subseteq \mathcal{H}$, the normal stratification $\mathcal{B}$ can be expanded to a Galois stratification $\mathcal{B}(\mathcal{H})$ with respect to $\mathcal{H}$ such that for each perfect Frobenius field M that contains K and satisfies $\mathrm{Im}(\mathrm{Gal}(M)) \cap \mathcal{S} = \mathcal{H} \cap \mathcal{S}$ and for each $\mathbf{b} \in \mathbb{A}^n(M)$ the following holds:

(2) $\qquad M \models \theta(\mathbf{b}) \iff \mathrm{Ar}(\mathcal{B}, M, \mathbf{b}) \subseteq \mathrm{Con}(\mathcal{B}(\mathcal{H})).$

In the explicit case, if $\theta(\mathbf{Y})$ is presented, then $\mathcal{S}$ and $\mathcal{B}$ can be effectively computed. In this case, for each primitive recursive $\mathcal{H}$ an effective computation gives $\mathcal{B}(\mathcal{H})$ with the above properties.

Proof: Lemmas 30.4.1 and 30.4.2 give a normal stratification $\mathcal{A}_{m-1}$ of $\mathbb{A}^{m-1+n}$ such that for each family $\mathcal{H}$ of finite groups which satisfies $\mathrm{Con}(\mathcal{A}(\mathcal{C})) \subseteq \mathcal{H}$, the normal stratification $\mathcal{A}_{m-1}$ can be expanded to a Galois stratification $\mathcal{A}_{m-1}(\mathcal{H})$ (depending on Q_m) with the following property: For each perfect Frobenius field M that contains K and satisfies

$\mathrm{Im}(\mathrm{Gal}(M)) \cap (\mathrm{Subgr}(\mathcal{A}) \cup \mathrm{Subgr}(\mathcal{A}_{m-1})) = \mathcal{H} \cap (\mathrm{Subgr}(\mathcal{A}) \cup \mathrm{Subgr}(\mathcal{A}_{m-1}))$

and for each $(a_1, \ldots, a_{m-1}, \mathbf{b}) \in \mathbb{A}^{m-1+n}(M)$,
$\quad \mathrm{Ar}(\mathcal{A}_{m-1}, M, (a_1, \ldots, a_{m-1}, \mathbf{b})) \subseteq \mathrm{Con}(\mathcal{A}_{m-1}(\mathcal{H}))$
if and only if
$\quad Q_m a_m \in M$ such that $\mathrm{Ar}(\mathcal{A}, M, (\mathbf{a}, \mathbf{b})) \subseteq \mathrm{Con}(\mathcal{A}(\mathcal{C}))$.

This eliminates Q_m from θ. Continue to eliminate $Q_{m-1}, \ldots, Q_1$ in order by constructing the corresponding normal stratifications $\mathcal{A}_{m-2}, \ldots, \mathcal{A}_0$. Let $\mathcal{A}_m = \mathcal{A}$, $\mathcal{B} = \mathcal{A}_0$, and $\mathcal{S} = \mathrm{Subgr}(\mathcal{A}_m) \cup \cdots \cup \mathrm{Subgr}(\mathcal{A}_0)$. The proposition follows. $\square$

The case $n = 0$ is of particular interest: θ is a Galois sentence, the normal stratification $\mathcal{B}$ of $\mathbb{A}^0$ is trivial, and $\mathrm{Con}(\mathcal{B}(\mathcal{H}))$ is a conjugacy domain $\mathrm{Con}(\mathcal{H})$ of $\mathrm{Gal}(L/K)$ with L a finite Galois extension of K. The condition $\mathrm{Ar}(\mathcal{B}, M, \mathbf{b}) \subseteq \mathrm{Con}(\mathcal{B}(\mathcal{H}))$ simplifies to $\mathrm{Gal}(L/L \cap M) \in \mathrm{Con}(\mathcal{H})$.

PROPOSITION 30.5.3: *Let $\mathcal{C}$ be the family of finite groups and let θ be a Galois sentence,*

$$(Q_1 X_1) \ldots (Q_m X_m)[\mathrm{Ar}(\mathbf{X}) \subseteq \mathrm{Con}(\mathcal{A}(\mathcal{C}))],$$

with respect to a Galois stratification $\mathcal{A}(\mathcal{C})$ of $\mathbb{A}^m$ over K. Then there exists a finite family $\mathcal{S}$ of finite groups and a finite Galois extension L of K with $\mathrm{Subgr}(\mathcal{A}) \cup \mathrm{Subgr}(\mathrm{Gal}(L/K)) \subseteq \mathcal{S}$. Moreover, for each family $\mathcal{H}$ of finite groups which satisfies $\mathrm{Con}(\mathcal{A}(\mathcal{C})) \subseteq \mathcal{H}$ there exists a conjugacy domain $\mathrm{Con}(\mathcal{H})$ of $\mathrm{Gal}(L/K)$ consisting of groups belonging to $\mathcal{H}$ with the following property:

For each perfect Frobenius field M that contains K and satisfies $\mathrm{Im}(\mathrm{Gal}(M)) \cap \mathcal{S} = \mathcal{H} \cap \mathcal{S}$,

(3) $\qquad M \models \theta \iff \mathrm{Gal}(L/L \cap M) \in \mathrm{Con}(\mathcal{H})$.

In the explicit case, if θ is presented, then an effective computation gives $\mathcal{S}$ and L, and for each primitive recursive $\mathcal{H}$, $\mathrm{Con}(\mathcal{H})$ satisfying (3) can be found effectively.

30.6 Model-Theoretic Applications

The decidability of the theory of Frobenius fields is now a consequence of the elimination of quantifiers procedure of Section 30.5.

Let K be a field and $\mathcal{C}$ a full formation of finite groups. Write $\mathrm{Frob}(K)$ for the class of all perfect Frobenius fields that contain K. Put

$$\mathrm{Frob}(K, \mathrm{pro}\text{-}\mathcal{C}) = \{M \in \mathrm{Frob}(K) \mid \mathrm{Gal}(M) \text{ is a pro-}\mathcal{C}\text{-group}\}.$$

Then let $\mathrm{Th}(\mathrm{Frob}(K, \mathrm{pro}\text{-}\mathcal{C}))$ be the theory of all sentences $\theta \in \mathcal{L}(\mathrm{ring}, K)$ which are true in all fields $M \in \mathrm{Frob}(K, \mathrm{pro}\text{-}\mathcal{C})$.

THEOREM 30.6.1: *Let K be a presented field with elimination theory and $\mathcal{C}$ a primitive recursive full formation of finite groups. Then $\mathrm{Th}(\mathrm{Frob}(K, \mathrm{pro}\text{-}\mathcal{C}))$ is primitive recursive.*

Proof: Let θ_0 be a sentence of $\mathcal{L}(\mathrm{ring}, K)$. Use Remark 30.5.1 to find a Galois stratification $\mathcal{A}$ over K with $\mathrm{Con}(A_i)$ either empty or consisting only of the trivial group for each A_i in the normal stratification underlying $\mathcal{A}$. Then let θ be the associated Galois sentence which is equivalent to θ_0. Proposition 30.5.3 gives a finite family $\mathcal{S}$ of finite groups and a finite Galois extension L of K such that $\mathrm{Subgr}(\mathcal{A}) \cup \mathrm{Subgr}(\mathrm{Gal}(L/K)) \subseteq \mathcal{S}$. List all subfamilies, $\mathcal{H}_1, \ldots, \mathcal{H}_s$, of $\mathcal{C} \cap \mathcal{S}$ which contain $\mathrm{Con}(\mathcal{A})$. For each i, Corollary 24.5.3 gives a check if there exists a superprojective pro-$\mathcal{C}$-group G such that $\mathrm{Im}(G) \cap \mathcal{S} = \mathcal{H}_i$. Find r, $0 \leq r \leq s$, and reorder $\mathcal{H}_1, \ldots, \mathcal{H}_s$ so that $\mathrm{Im}(G) \cap \mathcal{S} = \mathcal{H}_i$ for some superprojective pro-$\mathcal{C}$ group G, if and only if $i \in \{1, 2, \ldots, r\}$. For each i, $0 \leq i \leq r$, Proposition 30.5.3 gives a conjugacy domain $\mathrm{Con}(\mathcal{H}_i)$ of subgroups of $\mathrm{Gal}(L/K)$ belonging to $\mathcal{H}_i$ with the following property: for each perfect Frobenius field M that contains K and satisfies $\mathrm{Im}(\mathrm{Gal}(M)) \cap \mathcal{S} = \mathcal{H}_i$,

(1) $\qquad M \models \theta \iff \mathrm{Gal}(L/L \cap M) \in \mathrm{Con}(\mathcal{H}_i)$.

Then $\mathrm{Con}(\mathcal{H}_i) \subseteq \mathcal{H}_i \cap \mathrm{Subgr}(\mathrm{Gal}(L/K))$ for each i, $1 \leq i \leq r$.
    Suppose first that

(2) $\qquad \mathrm{Con}(\mathcal{H}_i) = \mathcal{H}_i \cap \mathrm{Subgr}(\mathrm{Gal}(L/K)), \quad i = 1, \ldots, r.$

30.6 Model-Theoretic Applications

Then θ is true in each $M \in \mathrm{Frob}(K, \mathrm{pro}\text{-}\mathcal{C})$. Indeed, for each such M, the pro-$\mathcal{C}$-group $\mathrm{Gal}(M)$ is superprojective (Proposition 24.1.5). Hence, there exists an i, $1 \leq i \leq r$ such that $\mathrm{Im}(\mathrm{Gal}(M)) \cap \mathcal{S} = \mathcal{H}_i$. Then

$$\mathrm{Gal}(L/L \cap M) \in \mathrm{Im}(\mathrm{Gal}(M)) \cap \mathcal{S} \cap \mathrm{Subgr}(\mathrm{Gal}(L/K)) = \mathrm{Con}(\mathcal{H}_i).$$

Thus, by (1), $M \models \theta$.

Suppose on the other hand, that there exists i, $1 \leq i \leq r$, and there exists a subgroup H of $\mathrm{Gal}(L/K)$ which belongs to $\mathcal{H}_i$ but not to $\mathrm{Con}(\mathcal{H}_i)$. By assumption, there exists a superprojective pro-$\mathcal{C}$-group G such that $\mathrm{Im}(G) \cap \mathcal{S} = \mathcal{H}_i$. In particular, $H \in \mathrm{Im}(G)$. Lemma 24.1.6 produces a field $M \in \mathrm{Frob}(K)$ such that $\mathrm{Gal}(M) \cong G$ and $\mathrm{Gal}(L/L \cap M) = H$. In particular, $\mathrm{Gal}(L/L \cap M) \notin \mathrm{Con}(\mathcal{H}_i)$ but $\mathrm{Im}(\mathrm{Gal}(M)) \cap \mathcal{S} = \mathcal{H}_i$. We conclude, by (1) that $M \not\models \theta$.

Thus, condition (2) is equivalent to the validity of θ in each field $M \in \mathrm{Frob}(K, \mathrm{pro}\text{-}\mathcal{C})$. A check of (2) therefore decides whether or not θ is true in each $M \in \mathrm{Frob}(K, \mathrm{pro}\text{-}\mathcal{C})$. □

We now specialize the previous theory. Consider a fixed superprojective group G. Let $\mathrm{Frob}(K, G)$ be the class of all fields $M \in \mathrm{Frob}(K)$ with $\mathrm{Im}(G(M)) = \mathrm{Im}(G)$. A simplification of the proof of Theorem 30.6.1 yields a primitive recursive decision procedure for this theory, provided $\mathrm{Im}(G)$ is primitive recursive.

Indeed, let θ be a sentence in $\mathcal{L}(\mathrm{ring}, K)$ or, more generally, a Galois sentence with respect to the family $\mathrm{Im}(G)$. Take $\mathcal{H}$ in Lemmas 30.4.1 and 30.4.2 to be $\mathrm{Im}(G)$. Then Condition (1) of Lemma 30.4.1 is satisfied for each field $M \in \mathrm{Frob}(K, G)$. Use these lemmas to eliminate the quantifiers of θ one by one. The procedure produces in its final stage a finite Galois extension L of K and a conjugacy domain Con of subgroups of $\mathrm{Gal}(L/K)$ which belong to $\mathrm{Im}(G)$ with this property: For each $M \in \mathrm{Frob}(K, G)$,

(3) $$M \models \theta \iff \mathrm{Gal}(L/L \cap M) \in \mathrm{Con}.$$

Therefore, $M \models \theta$ for each such M if and only if

$$\mathrm{Con} = \mathrm{Im}(G) \cap \mathrm{Subgr}(\mathrm{Gal}(L/K)).$$

This is, again, a checkable condition and this establishes the procedure. These arguments prove the following result:

THEOREM 30.6.2: *Let K be a presented field with elimination theory and G a superprojective group such that $\mathrm{Im}(G)$ is primitive recursive. Then there exists a primitive recursive decision procedure for the theory of $\mathrm{Frob}(K, G)$.*

Similarly, a perfect Frobenius field M is determined up to elementarily equivalence by $\mathrm{Im}(\mathrm{Gal}(M))$ and by its algebraic part:

THEOREM 30.6.3: *Let M and M' be perfect Frobenius fields containing a field K. Then M and M' are elementarily equivalent with respect to $\mathcal{L}(\mathrm{ring}, K)$ if and only if $\mathrm{Im}(\mathrm{Gal}(M)) = \mathrm{Im}(\mathrm{Gal}(M'))$ and $K_s \cap M \cong K_s \cap M'$.*

In particular, if M is algebraically closed in M' and $\mathrm{Im}(\mathrm{Gal}(M)) = \mathrm{Im}(\mathrm{Gal}(M'))$, then M is an elementary subfield of M'.

Proof: Suppose $\mathrm{Im}(\mathrm{Gal}(M)) = \mathrm{Im}(\mathrm{Gal}(M'))$ and $K_s \cap M \cong_K K_s \cap M'$. The elementary equivalence of M and M' follows by taking $\mathrm{Gal}(M') = G$ in Theorem 30.6.2: the right hand side of (3) is simultaneously fulfilled or not fulfilled for both M and M'. Hence, so is the left hand side of (3).

Conversely, suppose that $M \equiv M'$. Since the statement "the group G is realizable over M" is elementary (Remark 20.4.5(d)), $\mathrm{Im}(\mathrm{Gal}(M)) = \mathrm{Im}(\mathrm{Gal}(M'))$. Lemma 20.6.3 shows that $K_s \cap M \cong_K K_s \cap M'$. □

Remark 30.6.4: The converse of Remark 30.5.1 also holds: Each Galois sentence over K is equivalent in a certain sense to a sentence of $\mathcal{L}(\mathrm{ring}, K)$. We explain in detail:

Consider a fixed superprojective group G and let θ be a Galois sentence with respect to the family $\mathcal{H} = \mathrm{Im}(G)$. Let L and $\mathrm{Con}(\mathcal{H})$ be as in the discussion preceding Theorem 30.6.2. Choose representatives $H_1, \ldots, H_r$ for the conjugacy classes of $\mathrm{Con}(\mathcal{H})$. For each i, $1 \leq i \leq r$, let $x_{i0}, x_{i1}, \ldots, x_{i,s(i)}$ be elements of L such that $K(x_{i0}) = L(H_i)$ and $K(x_{i1}), \ldots, K(x_{i,s(i)})$ is a list of all proper extensions of $L(H_i)$ in L. For $f_{ij} = \mathrm{irr}(x_{ij}, K)$, consider the sentence λ of $\mathcal{L}(\mathrm{ring}, K)$:

$$(4) \quad \bigvee_{i=1}^{r} [(\exists X)[f_{i0}(X) = 0] \wedge \bigwedge_{j=1}^{s(i)} \neg(\exists X)[f_{ij}(X) = 0]].$$

Obviously, if a field M contains K, then $\mathrm{Gal}(L/L \cap M) \in \mathrm{Con}(\mathcal{H})$ if and only if $M \models \lambda$. From (3), for each $M \in \mathrm{Frob}(K, G)$,

$$(5) \quad M \models \theta \iff M \models \lambda.$$

Note that λ is a test sentence in the sense of (2) of Section 20.6. Thus, (5) generalizes the case $G = \hat{F}_e$ of Proposition 20.6.6(b). When θ is a sentence of $\mathcal{L}(\mathrm{ring}, K)$, note the difference between the methods by which λ is obtained from θ. In Proposition 20.6.6, λ is found by examining all proofs from the axioms (a recursive procedure). In this chapter, however, λ is achieved by an explicit step by step elaboration, using an algebraic-geometric technique (and a primitive recursive procedure).

In the case where $\mathcal{C}$ is the family of all finite groups and θ is a Galois sentence over K with respect to $\mathcal{C}$ we may assert only a weaker statement than that θ is equivalent to one sentence of $\mathcal{L}(\mathrm{ring}, K)$. In the notation of the proof of Theorem 30.6.1, as above we find for each family $\mathcal{H}_i$ a test sentence λ_i such that for each $M \in \mathrm{Frob}(K)$ with $\mathrm{Im}(\mathrm{Gal}(M)) \cap \mathcal{S} = \mathcal{H}_i$,

$$M \models \theta \iff M \models \lambda_i.$$

30.7 A Limit of Theories

Thus, in this case, we say that θ is equivalent to a sequence $\lambda_1, \ldots, \lambda_r$ of sentence of $\mathcal{L}(\text{ring}, K)$ indexed by cases. $\square$

30.7 A Limit of Theories

In Section 20.5, K is a countable Hilbertian field and $\text{Almost}(K, e)$ is the theory of all sentences θ of $\mathcal{L}(\text{ring}, K)$ which are true in $\tilde{K}(\sigma)$, for almost all $\sigma \in \text{Gal}(K)^e$. We now investigate the limit of these theories: namely, the theory of all sentences θ of $\mathcal{L}(\text{ring}, K)$ which belong to $\text{Almost}(K, e)$ for all large e.

Let $\mathcal{C}$ be a family of finite groups. Put

$$\text{Frob}(K, \mathcal{C}) = \{M \in \text{Frob}(K) \mid \text{Im}(\text{Gal}(M)) = \mathcal{C}\}.$$

Denote the theory of all sentences θ in $\mathcal{L}(\text{ring}, K)$ which are true in all $M \in \text{Frob}(K, \mathcal{C})$ by $\text{Th}(\text{Frob}(K, \mathcal{C}))$. For each integer $e \geq 0$ let $\mathcal{C}_e = \{G \in \mathcal{C} \mid \text{rank}(G) \leq e\}$.

We concentrate on the case where $\mathcal{C}$ is a formation of finite groups. In this case if M is a field, then $\text{Im}(\text{Gal}(M)) = \mathcal{C}_e$ if and only if $\text{Gal}(M) \cong \hat{F}_e(\mathcal{C})$ (Lemma 17.7.1). If M is a countable Frobenius field, then $\text{Im}(\text{Gal}(M)) = \mathcal{C}$ if and only if $\text{Gal}(M) \cong \hat{F}_\omega(\mathcal{C})$ (Theorem 24.8.1). Moreover, if $\mathcal{C}$ is full, then $\hat{F}_e(\mathcal{C})$ and $\hat{F}_\omega(\mathcal{C})$ are projective (Corollary 22.4.5). Since, in addition, they have the embedding property (Lemma 24.3.3), they are superprojective. Thus, Lemma 24.1.6 shows that the classes $\text{Frob}(K, \mathcal{C}_e)$ and $\text{Frob}(K, \mathcal{C})$ are nonempty.

The following theorem may be regarded as an interpretation of the statement that $\text{Th}(\text{Frob}(K, \mathcal{C}))$ is a limit of the theories $\text{Th}(\text{Frob}(K, \mathcal{C}_e))$ as e approaches infinity:

THEOREM 30.7.1: *Let $\mathcal{C}$ be a full family of finite groups and θ a sentence of $\mathcal{L}(\text{ring}, K)$. Then $\theta \in \text{Th}(\text{Frob}(K, \mathcal{C}))$ if and only if there exists an integer $e_0 \geq 0$ such that $\theta \in \text{Th}(\text{Frob}(K, \mathcal{C}_e))$ for each $e \geq e_0$.*

In the explicit case, if $\mathcal{C}$ is primitive recursive, then the theories $\text{Th}(\text{Frob}(K, \mathcal{C}_e))$ and $\text{Th}(\text{Frob}(K, \mathcal{C}))$ are primitive recursive. Moreover, the function $e_0(\theta) = \min(e_0' \mid \theta \in \text{Th}(\text{Frob}(K, \mathcal{C}_e)) \text{ for each } e \geq e_0')$ is primitive recursive. Therefore, the intersection $T = \bigcap_{e=1}^{\infty} \text{Th}(\text{Frob}(K, \mathcal{C}_e))$ is also a primitive recursive theory.

Proof: Remark 30.5.1 allows us to assume that θ is a Galois sentence. Proposition 30.5.3 gives a finite family $\mathcal{S}$ of finite groups, a finite Galois extension L of K, and a conjugacy domain Con of $\text{Gal}(L/K)$ consisting of groups of $\mathcal{H} = \mathcal{C} \cap \mathcal{S}$ such that $\text{Subgr}(\text{Gal}(L/K)) \subseteq \mathcal{S}$. Moreover, for each $M \in \text{Frob}(K)$ with $\text{Im}(\text{Gal}(M)) \cap \mathcal{S} = \mathcal{H}$,

(1) $\qquad M \models \theta \iff \text{Gal}(L/L \cap M) \in \text{Con}.$

Let $e_0 = \max\{\text{rank}(G) \mid G \in \mathcal{S}\}$. Then $\mathcal{C}_e \cap \mathcal{S} = \mathcal{H}$ for each $e \geq e_0$.

Suppose $\theta \in \text{Th}(\text{Frob}(K,\mathcal{C}))$, $e \geq e_0$ and $M \in \text{Frob}(K,\mathcal{C}_e)$. Then $\text{Gal}(M) \cong \hat{F}_e(\mathcal{C})$, so $\text{Gal}(L/L\cap M)$ is a quotient of $\hat{F}_\omega(\mathcal{C})$. Lemma 24.1.6 gives a perfect Frobenius field M' which contains K satisfying $L\cap M' = L\cap M$ and $\text{Gal}(M') \cong \hat{F}_\omega(\mathcal{C})$. In particular, $M' \in \text{Frob}(K,\mathcal{C})$. By assumption, $M' \models \theta$. Hence, by (1), $\text{Gal}(L/L\cap M) = \text{Gal}(L/L\cap M') \in \text{Con}$, so $M \models \theta$. Therefore, $\theta \in \text{Th}(\text{Frob}(K,\mathcal{C}_e))$.

Conversely, suppose $\theta \in \text{Th}(\text{Frob}(K,\mathcal{C}_{e_0}))$. Let $M \in \text{Frob}(K,\mathcal{C})$. Then $\text{Gal}(M) \cong \hat{F}_\omega(\mathcal{C})$ and $\text{Gal}(L/L \cap M) \in \mathcal{S}$. Hence, $\text{Gal}(L/L \cap M)$ is a pro-$\mathcal{C}$ group of rank at most e_0. Lemma 24.1.6 gives a field $M' \in \text{Frob}(K,\mathcal{C}_{e_0})$ with $L \cap M' = L \cap M$. By assumption, $M' \models \theta$. Hence, by (1), $\text{Gal}(L/L \cap M) = \text{Gal}(L/L \cap M') \in \text{Con}$, so $M \models \theta$. Consequently, $\theta \in \text{Th}(\text{Frob}(K,\mathcal{C}))$.

Suppose $\theta \in \text{Th}(\text{Frob}(K,\mathcal{C}))$. Then $\theta \in \text{Th}(\text{Frob}(K,\mathcal{C}_{e_0}))$. Apply Theorem 30.6.2 to check if $\theta \in \text{Th}(\text{Frob}(K,\mathcal{C}_e))$ for each $0 \leq e < e_0$. Thus, decide if θ belongs to each of the theories $\text{Th}(\text{Frob}(K,\mathcal{C}_e))$. □

Finally, suppose $\mathcal{C}$ is the family of all finite groups and (K, e) is a Hilbertian pair (Section 20.5). In this case $\text{Frob}(K,\mathcal{C}_e)$ is the class of all perfect e-free PAC fields that contain K. If L is a finite Galois extension of K and H is a subgroup of $\text{Gal}(L/K)$ with $\text{rank}(H) \leq e$, then the set of all $\sigma \in \text{Gal}(K)^e$ such that $\langle \text{res}_L \sigma \rangle = H$ is of positive rational (and effectively computable) measure.

For each sentence θ of $\mathcal{L}(\text{ring}, K)$ let

$$S(K,e,\theta) = \{\sigma \in \text{Gal}(K)^e \mid \tilde{K}(\sigma) \models \theta\}.$$

By theorem 20.6.7, the function $\mu(\theta(K,e,\theta))$ is recursive. The arguments of the preceding paragraph strengthens this result:

THEOREM 30.7.2: *Let (K,e) be a Hilbertian pair. Suppose K has elimination theory. Then the function $\mu(S(K,e,\theta))$ from sentences of $\mathcal{L}(\text{ring}, K)$ to rational numbers is primitive recursive. The theory Almost(K,e) is primitive recursive.*

Similarly, $\text{Frob}(K,\mathcal{C})$ in this case is the family of all ω-free perfect PAC fields which contain K. Its theory is the limit of Almost(K, e) as e approaches infinity.

THEOREM 30.7.3: *Let K be a Hilbertian field with elimination theory. Then the theory of all ω-free perfect PAC fields that contain K is primitive recursive.*

Exercises

1. This exercise traces through the concepts of Section 30.1. Let A be the affine space $\mathbf{A}^n$ over a field K, so that $K[A] = K[\mathbf{x}]$, with $\mathbf{x} = (x_1, \ldots, x_n)$

Exercises

an n-tuple of algebraically independent elements over K. Let $z_1, \ldots, z_n$ be the roots of the general polynomial of degree n

$$f(\mathbf{x}, Z) = \prod_{i=1}^{n}(Z - z_i) = Z^n + \sum_{i=1}^{n}(-1)^i x_i Z^{n-i}$$

so that x_i is the symmetric polynomial of degree n in $z_1, \ldots, z_n$, $i = 1, \ldots, n$. Then $C = K[\mathbf{z}]$ is an integrally closed ring extension of $K[A]$. Use the notation of Section 30.1. Throughout assume that $n > 2$.

(a) Show that the field extension $K(C)/K(A)$ is Galois with group S_n. The discriminant of $f(\mathbf{x}, Z)$ is $d(f) = (-1)^{\frac{n(n-1)}{2}} \prod_{i \neq j}(z_i - z_j)$. For each $c_1, \ldots, c_n \in K$ let $D(\mathbf{c}) = \prod_{i \neq j}(c_i - c_j)$ and $z_\mathbf{c} = \sum_{i=1}^{n} c_i z_i$. Show that if $D(\mathbf{c}) \neq 0$, then $K(C) = K(\mathbf{x}, z_\mathbf{c})$.

(b) Let $A_1 = \mathbb{A}^n \smallsetminus V(d(f))$ and $C_1 = K[\mathbf{z}, d(f)^{-1}]$. Show that if $f_\mathbf{c} = \mathrm{irr}(z_\mathbf{c}, K(A))$ with $D(\mathbf{c}) \neq 0$, then $d(f_\mathbf{c})$ is a unit in $K[\mathbf{x}, z_\mathbf{c}, d(f)^{-1}]$.

(c) Let K be a finite field and let $\mathbf{c} \in \mathbb{A}^n(K)$ such that $D(\mathbf{c}) \neq 0$. With $C_\mathbf{c} = K[\mathbf{x}, z_\mathbf{c}, d(f_\mathbf{c})^{-1}]$ and $A_\mathbf{c} = \mathbb{A}^n \smallsetminus V(d(f_\mathbf{c}))$ consider the Galois ring/set cover $C_\mathbf{c}/A_\mathbf{c}$. Let $\mathbf{a} = (a_1, \ldots, a_n) \in A_\mathbf{c}(K)$. Explicitly describe $\mathrm{Ar}(A_\mathbf{c}, K, \mathbf{a})$ in terms of the degrees of the irreducible factors of $g(\mathbf{a}, Z)$.

(d) Show directly, for each $\sigma \in S_n$, that there exists $\mathbf{a} \in A_\mathbf{c}(K)$ (in (c)) such that $\mathrm{Ar}(A_\mathbf{c}, K, \mathbf{a})$ consists of the conjugates of σ in S_n, if $|K|$ is suitably large.

2. This exercise expands on Exercise 1 to give some practice with the two essential cases that appear in Section 30.2. Here K is a field of characteristic zero.

(a) let A_1 be the Zariski closed subset $V(f)$ of $\mathbb{A}^{n+1}$ where

$$f(X_1, \ldots, X_n, Z) = Z^n + \sum_{i=1}^{n}(-1)^i X_i Z^{n-i}.$$

Let C_1/A_1 be the trivial Galois ring/set cover (i.e. $C_1 = K[A_1]$). Take $\mathcal{H}$ to be the collection of all finite cyclic groups and let $\mathrm{Con}(A_1, \mathcal{H})$ consists of only the trivial group. Consider $B_1 = \pi(A_1)$ where $\pi\colon \mathbb{A}^{n+1} \to \mathbb{A}^n$ is projection onto the first n coordinates. Explicitly find $h \in K[X_1, \ldots, X_n]$ not vanishing on B_1 and a ring D'_1 such that for $C'_1 = C[h^{-1}]$, $A'_1 = A_1 \smallsetminus V(h)$, $B'_1 = B_1 \smallsetminus V(h)$ the pair $(C'_1/A'_1, D'_1/B'_1)$ consists of Galois ring/set covers for which $K(C'_1) \subseteq K(D'_1)$. Also find a conjugacy domain $\mathrm{Con}(B'_1, \mathcal{H})$ consisting of cyclic subgroups of $\mathrm{Gal}(D'_1/B'_1)$ such that condition (7) of Lemma 30.2.5 holds.

(b) With D'_1/B'_1 replacing C_1/A_1 in part (a), $\mathrm{Con}(B'_1, \mathcal{H})$ replacing $\mathrm{Con}(A_1, \mathcal{H})$ and (4) of Lemma 30.2.3 replacing (7) of Lemma 30.2.5, do part (a) applied to the projection of $\mathbb{A}^n$ onto $\mathbb{A}^{n-1}$. Note that, in this case, $\dim(B'_1) = \dim(\pi(B'_1)) + 1$ in contrast to part (a) where $\dim(A_1) = \dim(\pi(A_1))$.

3. As in Lemma 30.2.5 let C/A and D/B be Galois ring/set covers over K with $K[A]$ integral over $K[B]$ such that $\pi(A) = B$. Denote the maximal separable extension of $K(B)$ in $K(A)$ (resp. $K(C)$) by E (resp. F) and assume that $F \subseteq K(D)$. Suppose $G = \mathrm{Gal}(C/A) \cong \mathbb{Z}/3\mathbb{Z}$ and that the group S_3 acts on $(\mathbb{Z}/3\mathbb{Z})^3$ by permutation of the coordinates. The subgroup $U = \{(z_1, z_2, z_3) \in (\mathbb{Z}/3\mathbb{Z})^3 \mid z_1 + z_2 + z_3 = 0\}$ is invariant under this action. Let $H = S_3 \ltimes U$ be the corresponding semidirect product. Suppose there is an isomorphism $\theta \colon \mathrm{Gal}(D/B) \to H$ which maps $\mathrm{Gal}(K(D)/E)$ onto U. Let $\pi_1 \colon U \to \mathbb{Z}/3\mathbb{Z}$ be the projection on the first coordinate. Suppose θ induces an isomorphism $\bar\theta \colon \mathrm{Gal}(F/E) \to \mathbb{Z}/3\mathbb{Z}$ such that $\pi_1(\theta(\sigma)) = \bar\theta(\mathrm{res}_F \sigma)$ for each $\sigma \in \mathrm{Gal}(K(D)/E)$. Now let $\mathcal{H}$ consists of all finite cyclic groups and suppose that $\mathrm{Con}(A, \mathcal{H}) = \{G\}$. Compute $\mathrm{Con}(B, \mathcal{H})$ as in Lemma 30.2.5. Compute also $\mathrm{Con}^c(A, \mathcal{H})$ and $\mathrm{Con}^c(B, \mathcal{H})$ as in the proof of Lemma 30.4.2.

4. This is an analog of Exercise 3 for the case that $\dim(A) = \dim(B)+1$. Use the notation of Lemma 30.2.3. Suppose the conjugacy domain $\mathrm{Con}(A, \mathcal{H})$ of subgroups of $\mathrm{Gal}(C/A)$ is nonempty. For which of the following possibilities for $\mathcal{H}$, and all possible pairs $(C/A, D/B)$ satisfying the above, is $\mathrm{Con}(B, \mathcal{H})$ nonempty:
(a) $\mathcal{H}$ is all finite groups of rank at most 2;
(b) $\mathcal{H}$ is all finite p-groups for some prime p;
(c) $\mathcal{H}$ is all finite p-groups of rank at most 2; or
(d) $\mathcal{H}$ is all Abelian groups.

5. Apply the procedure of Sections 30.4 and 30.5 to the case that $K = \mathbb{Q}$, $\mathcal{H}$ is the collection of finite cyclic groups and θ is the sentence

$$(\forall X)(\exists Y)[T_n(Y) = X],$$

where n is an odd positive integer and T_n is the Chebyshev polynomial in $\mathbb{Q}[Y]$ satisfying $T_n(Z + Z^{-1}) = Z^n + Z^{-n}$ (Section 21.8) by following this outline:

(a) Eliminate Y (after applying Remark 30.5.1) by noting that the splitting field of $T_n(Y) - x$ over $\mathbb{Q}(x)$ is $\mathbb{Q}(\zeta_n, z)$, where x is transcendental over $\mathbb{Q}$, z satisfies $T_n(z + z^{-1}) = x$, and ζ_n is a primitive nth root of 1. Moreover, $\mathrm{Gal}(\mathbb{Q}(\zeta_n, z)/\mathbb{Q}(\zeta_n, x))$ is generated by σ and τ, where $\sigma(z) = \zeta_n z$ and $\tau(z) = z^{-1}$.

(b) Apply (a) and Lemma 30.4.1 to produce

$$\mathcal{B}(\mathcal{H}) = \langle \mathbb{A}^1, D_j/B_j, \mathrm{Con}(B_j, \mathcal{H}) \mid j = 1, 2, 3 \rangle$$

satisfying the conclusion (2) of Lemma 25.6. Indeed, take $B_1 = \{2\}$, $B_2 = \{-2\}$ and $B_3 = \mathbb{A}^1 \smallsetminus \{2, -2\}$. Note that 2 and -2 are the values of X for which $T_n(Y) - X$ has multiples zeros.

(c) Now apply (b) and Lemma 30.4.1 to eliminate X from $(\forall X)[\mathrm{Ar}(X) \subseteq \mathrm{Con}(\mathcal{B}(\mathcal{H}))]$ and finally arrive at a finite Galois extension $L/\mathbb{Q}$ and a set of conjugacy classes Con of $\mathrm{Gal}(L/\mathbb{Q})$ the following holds: Suppose M is a Frobenius field of characteristic 0 and

$$\mathrm{Im}(G(M)) \cap \mathrm{Subgr}(\mathrm{Gal}(\mathbb{Q}(\zeta_n, z))/\mathbb{Q}(x)))$$

consists of exactly the cyclic subgroups of $\mathrm{Gal}(\mathbb{Q}(\zeta_n, z)/\mathbb{Q}(x))$. Then $M \models \theta$ if and only if $\mathrm{Gal}(L/L \cap M) \in \mathrm{Con}$.

(d) Your answer to (c) should be explicit enough that you will be able to conclude that if $\gcd(n, 6) = 1$, then the set of $\sigma \in G(\mathbb{Q})$ such that $\tilde{\mathbb{Q}}(\sigma) \models \theta$ has positive measure.

Notes

The notion of Galois stratification is developed for the first time in [Fried-Sacerdote] in order to establish an explicit decision procedure for the theory of finite fields.

In our opinion the real progress made by introducing Galois stratification is not so much the primitive recursiveness of this algorithm, but the precise algebraic-geometric indication of how to eliminate quantifiers. No notion from recursion theory is required to appreciate its importance.

Chapter 30 is an elaboration of [Fried-Haran-Jarden, Section 3]. Since, however, [Fried-Haran-Jarden] treats only the special case where the Frobenius fields considered have a fixed superprojective group G as their absolute Galois group, this has required additions to [Fried-Haran-Jarden].

A discussion of the theory of Frobenius fields whose absolute Galois groups are pro-$\mathcal{C}$-groups in the case where $\mathcal{C}$ is an arbitrary fixed full family of finite groups appears in [Jarden13].

A recursive variant of Theorem 30.7.3 (and the remark on limits of theories) appears in [Jarden5].

The language $\mathcal{L}(\text{ring})$ is extended in [Ershov5] so that Galois formulas become formulas in the usual sense. Galois stratification then yield elimination of quantifier procedures.

Galois stratification is used in [Jarden15] in order to establish an algebraic dimension for definable sets over Frobenius fields.

[Haran-Lauwers] establishes Galois stratification over e-fold ordered Frobenius fields.

[Denef-Loeser] uses Galois stratification to associate to each reduced and separated scheme of finite type over $\mathbb{Z}$ a canonical rational function which specializes for almost every p to the p-adic Poincaré series of the scheme.

Chapter 31.
Galois Stratification over Finite Fields

Chapter 30 establishes the Galois stratification procedure over a fixed field K with elimination theory. The outcome is an explicit decision procedure for the theory of Frobenius fields that contain K. Section 31.1 modifies this to replace K by localizations $\mathbb{Z}[k^{-1}]$ of $\mathbb{Z}$. This, in particular, extends the above result to establish an explicit decision procedure for the theory of all Frobenius fields.

The next two sections return to the opening subject of the book - finite fields. Although no finite field is Frobenius, as a collection they behave like Frobenius fields with respect to any given Galois sentence over $\mathbb{Z}[k^{-1}]$, provided we forget about finitely many (exceptional) primes. As a consequence this establishes an explicit decision procedure for the theory of finite fields (Theorem 31.2.4).

Section 31.3 applies the elimination of quantifiers procedure for finite fields and a theorem of Artin in representation theory to prove that the zeta function of a Galois formula over a finite field is essentially the radical of a rational function by reducing this to a corollary of Dwork's theorem that the zeta function of a variety over a finite field is a rational function. The explicit version that we use is a result of Bombieri.

The last section of this chapter gives a survey on some generalizations of the above zeta functions to p-adic fields.

31.1 The Elementary Theory of Frobenius Fields

Let K be a presented field with elimination theory. Theorem 30.6.1 states that the theory of all perfect Frobenius fields M that contains the field K is primitive recursive. We now extend this result to ask about the theory of all perfect Frobenius fields in the language $\mathcal{L}(\text{ring})$. This requires a slight change in basic definitions and a minor amendment to the stratification procedure.

Instead of giving Galois statements over $\mathbb{Q}$, we attach to each Galois stratification $\mathcal{A}$ a presented ring of the form $\mathbb{Z}[k^{-1}] = R(\mathcal{A}) = R$ with $k = k(\mathcal{A}) \in \mathbb{N}$ and with this property: each algebraic set that appears in $\mathcal{A}$ is defined by polynomials with coefficients in R, and the coordinate ring of each algebraic set is finitely generated over R. The stratification procedure, in this case, starts with $(\mathcal{A}, R(\mathcal{A}))$ with $\mathcal{A}$ a stratification of $\mathbb{A}^n$. The inductive step produces $(\mathcal{B}, R(\mathcal{B}))$ with $\mathcal{B}$ a Galois stratification of $\mathbb{A}^{n-1}$ and $R(\mathcal{B}) = \mathbb{Z}[k(\mathcal{B})^{-1}]$ where $k(\mathcal{B})$ is a multiple of $k(\mathcal{A})$.

The next lemma is an addition to Lemma 19.7.2 that allows us to construct appropriate normal $\mathbb{Q}$-basic sets in this new context. Recall that a polynomial $h \in \mathbb{Z}[X_1, \ldots, X_r]$ is said to be primitive if the greatest common divisor of its coefficients is 1.

31.1 The Elementary Theory of Frobenius Fields

LEMMA 31.1.1: *Let $(x_1,\ldots,x_n)$ be a presented n-tuple over $\mathbb{Q}$. Then we can effectively find a positive integer k_0 and $h \in \mathbb{Z}[X_1,\ldots,X_n]$ such that $h(\mathbf{x}) \neq 0$ and for each multiple k of k_0, the ring $\mathbb{Z}[k^{-1}, \mathbf{x}, h(\mathbf{x})^{-1}]$ is presented in $\mathbb{Q}(\mathbf{x})$ (Definition 19.1.1). Moreover, if $x_1,\ldots,x_r$ is a transcendence basis for $\mathbb{Q}(\mathbf{x})/\mathbb{Q}$, then h can be chosen in $\mathbb{Z}[X_1,\ldots,X_r]$.*

Proof: Without loss assume that $x_1,\ldots,x_r$ is a transcendence basis for $\mathbb{Q}(\mathbf{x})/\mathbb{Q}$. For each i, $r < i \leq n$ use $\mathbb{Q}(\mathbf{x}) = \mathbb{Q}(x_1,\ldots,x_r)[x_{r+1},\ldots,x_n]$ to write

$$\mathrm{irr}(x_i, \mathbb{Q}(x_1,\ldots,x_{i-1})) = \frac{f_i(x_1,\ldots,x_{i-1}, X_i)}{h(x_1,\ldots,x_r)}$$

with $f_i \in \mathbb{Z}[k_0^{-1}, X_1, \ldots, X_i]$ with $k_0 \in \mathbb{N}$ and $h \in \mathbb{Z}[X_1,\ldots,X_r]$ primitive. Let k be a multiple of k_0. Each element of $S = \mathbb{Z}[k^{-1}, \mathbf{x}, h(\mathbf{x})^{-1}]$ has a unique presentation

$$\sum \frac{g_i(x_1,\ldots,x_r)}{h(x_1,\ldots,x_r)^{d(i)}} \cdot x_{r+1}^{i_{r+1}} \cdots x_n^{i_n},$$

where $i_{r+1},\ldots,i_n$ range from 0 to $\deg_{X_{r+1}}(f_{r+1}) - 1, \ldots, \deg_{X_n}(f_n) - 1$, respectively, and for each $i = (i_{r+1},\ldots,i_n)$ the polynomial $g_i(x_1,\ldots,x_r)$ of $\mathbb{Z}[k^{-1}, x_1,\ldots,x_r]$ is not divisible in $\mathbb{Q}[x_1,\ldots,x_r]$ by $h(x_1,\ldots,x_r)$.

An arbitrary element u of $\mathbb{Q}(\mathbf{x})$ can be presented in the form

(1) $$\sum \frac{g_i(x_1,\ldots,x_r)}{h_i(x_1,\ldots,x_r)} \cdot x_{r+1}^{i_{r+1}} \cdots x_n^{i_n}$$

where the exponents $i_{r+1},\ldots,i_n$ satisfy the same condition as above and $g_i(x_1,\ldots,x_r)$, $h_i(x_1,\ldots,x_r)$ are polynomials in $\mathbb{Z}[x_1,\ldots,x_r]$ which are relatively prime in $\mathbb{Q}[x_1,\ldots,x_r]$. A necessary condition for u to belong to $\mathbb{Q}[\mathbf{x}, h(\mathbf{x})^{-1}]$ is that each $h_i(x_1,\ldots,x_r)$ divides a power of $h(x_1,\ldots,x_r)$ in $\mathbb{Q}[x_1,\ldots,x_r]$. In this case rewrite the ith coefficient in (1) as $\frac{p_i(x_1,\ldots,x_r)}{c_i h(x_1,\ldots,x_r)^{e(i)}}$ with $p_i \in \mathbb{Z}[X_1,\ldots,X_r]$ and $c_i \in \mathbb{N}$. Then u belongs to $\mathbb{Z}[k^{-1}, \mathbf{x}, h(\mathbf{x})^{-1}]$ if and only if $c_i^{-1} p_i \in \mathbb{Z}[k^{-1}, x_1,\ldots,x_r]$ for each i. This check is effective and it concludes a procedure to decide whether u belongs to $\mathbb{Z}[k^{-1}, \mathbf{x}, h(\mathbf{x})^{-1}]$. Thus, $\mathbb{Z}[k^{-1}, \mathbf{x}, h(\mathbf{x})^{-1}]$ is presented. □

LEMMA 31.1.2: *Let $(x_1,\ldots,x_n, z)$ be a presented $(n+1)$-tuple over $\mathbb{Q}$ with z algebraic over $\mathbb{Q}(\mathbf{x})$. Then we can effectively find a polynomial $g \in \mathbb{Q}[\mathbf{X}]$ and a common multiple k_0 of the denominators of the coefficients of g such that for each multiple k of k_0*
(a) *$g(\mathbf{x}) \neq 0$, and the ring $\mathbb{Z}[k^{-1}, \mathbf{x}, g(\mathbf{x})^{-1}]$ is integrally closed and presented in $\mathbb{Q}(\mathbf{x})$; and*
(b) *the ring $\mathbb{Z}[k^{-1}, \mathbf{x}, g(\mathbf{x})^{-1}, z]$ is presented in $\mathbb{Q}(\mathbf{x}, z)$ and is a cover of $\mathbb{Z}[k^{-1}, \mathbf{x}, g(\mathbf{x})^{-1}]$ with primitive element z.*

Proof: We follow the proof of Lemma 19.7.2:

PART A: *(2a) gives rise to (2b).* Suppose g and k_0 are such that for each multiple k of k_0 (2a) holds. Write $\mathrm{irr}(z, \mathbb{Q}(\mathbf{x})) = \frac{f_1(\mathbf{x},Z)}{f_0(\mathbf{x})}$ and $\mathrm{disc}(z) = \frac{d_1(\mathbf{x})}{d_0(\mathbf{x})}$ with f_0, f_1, d_0, d_1 polynomials with coefficients in $\mathbb{Z}$. Lemma 31.1.1 produces a multiple k_2 of k_0 and a multiple $h(\mathbf{x})$ of $d_0(\mathbf{x})d_1(\mathbf{x})f_0(\mathbf{x})g(\mathbf{x})$ such that for each multiple k of k_1 the ring $\mathbb{Z}[k^{-1}, \mathbf{x}, z, g_1(\mathbf{x})^{-1}]$ is presented in $\mathbb{Q}(\mathbf{x}, z)$. By Definition 6.1.3, the latter ring is a cover of $\mathbb{Z}[k^{-1}, \mathbf{x}, g_1(\mathbf{x})^{-1}]$, with primitive element z.

PART B: *Proof of (2a).* The case where $n = 0$ reduces to the fact that $\mathbb{Z}$ is integrally closed.

Suppose $n \geq 1$. An induction on n effectively gives a polynomial $g_0 \in \mathbb{Q}[X_1, \ldots, X_{n-1}]$ and a common multiple k_0 of the denominators of the coefficients of g_0 such that $g_0(x_1, \ldots, x_{n-1}) \neq 0$, and for each multiple k of k_0, the ring $\mathbb{Z}[k^{-1}, x_1, \ldots, x_{n-1}, g_0(x_1, \ldots, x_{n-1})^{-1}]$ is integrally closed and presented in $\mathbb{Q}(x_1, \ldots, x_{n-1})$. If x_n is transcendental over $\mathbb{Q}(x_1, \ldots, x_{n-1})$, then $\mathbb{Z}[k^{-1}, x_1, \ldots, x_n, g_0(x_1, \ldots, x_{n-1})^{-1}]$ is integrally closed [Zariski-Samuel2, p. 85 or p. 126] and presented in $\mathbb{Q}(x_1, \ldots, x_n)$. Otherwise, use Part A with $z = x_n$ to effectively find a multiple $g \in \mathbb{Q}[\mathbf{X}]$ of g_0 and a common multiple k_1 of k_0 and the denominators of the coefficients of g such that $g(\mathbf{x}) \neq 0$ and for each multiple k of k_1 the ring $\mathbb{Z}[k^{-1}, \mathbf{x}, g(\mathbf{x})^{-1}]$ is presented in $\mathbb{Q}(\mathbf{x})$ and is a cover of $\mathbb{Z}[k^{-1}, x_1, \ldots, x_{n-1}, g(\mathbf{x})^{-1}]$. In particular, $\mathbb{Z}[k^{-1}, \mathbf{x}, g(\mathbf{x})^{-1}]$ is integrally closed. $\square$

Let $A = V(f_1, \ldots, f_m) \smallsetminus V(g)$ be a presented normal $\mathbb{Q}$-basic set with a generic point $\mathbf{x} = (x_1, \ldots, x_n)$, and let $\mathbb{Q}[\mathbf{x}, g(\mathbf{x})^{-1}, z]$ be a presented Galois ring cover of A over $\mathbb{Q}$. Apply Lemma 31.1.2 to find a multiple $h \in \mathbb{Q}[\mathbf{X}]$ of g and a positive integer k such that the coefficients of $f_1, \ldots, f_m, g, h$ lie in $\mathbb{Z}[k^{-1}]$, $h(\mathbf{x}) \neq 0$, $\mathbb{Z}[k^{-1}, \mathbf{x}, h(\mathbf{x})^{-1}]$ is an integrally closed ring, presented in $\mathbb{Q}(\mathbf{x})$, and $\mathbb{Z}[k^{-1}, \mathbf{x}, h(\mathbf{x})^{-1}, z]/\mathbb{Z}[k^{-1}, \mathbf{x}, h(\mathbf{x})^{-1}]$ is a ring cover. Then $A_0 = A \smallsetminus V(h)$ is a presented normal $\mathbb{Q}$-basic open subset of A. Call $\mathbb{Z}[k^{-1}, A_0] = \mathbb{Z}[k^{-1}, \mathbf{x}, h(\mathbf{x})^{-1}]$ the **coordinate ring** of A_0 over $\mathbb{Z}[k^{-1}]$. We say that A_0 is **normal** over $\mathbb{Z}[k^{-1}]$. Also, with $C = \mathbb{Z}[k^{-1}, \mathbf{x}, h(\mathbf{x})^{-1}, z]$ call C/A_0 a **Galois ring /set cover over $\mathbb{Z}[k^{-1}]$**.

Let M be a field of characteristic p (which might be 0). Suppose $p \nmid k$. Then reduction modulo p of $\mathbb{Z}$ uniquely extends to $\mathbb{Z}[k^{-1}]$. In this case, with A_0 as above, let

$$A_0(M) = \{\mathbf{a} \in M^n \mid f_1(\mathbf{a}) = \cdots = f_m(\mathbf{a}) = 0 \text{ and } h(\mathbf{a}) \neq 0\}$$

where the coefficients of $f_1, \ldots, f_m$ and h are interpreted as being in $\mathbb{F}_p$ (or in $\mathbb{Q}$ if $p = 0$). For each $\mathbf{a} \in A_0(M)$ there is a unique homomorphism $\varphi_0 \colon \mathbb{Z}[k^{-1}, A_0] \to M$ with $\varphi_0(\mathbf{x}) = \mathbf{a}$. As usual the homomorphism φ_0 extends to a homomorphism φ of C into a finite Galois extension $N = M(\varphi(z))$ of M. We may define the decomposition group of φ with respect to M as in Section 30.1:

$$D_M(\varphi) = \{\sigma \in \mathrm{Gal}(C/A_0) \mid (\forall u \in C)[\varphi(u) \in M \Longrightarrow \varphi(\sigma(u)) = \varphi(u)]\}.$$

31.1 The Elementary Theory of Frobenius Fields

It is a subgroup of $\mathrm{Gal}(C/A_0)$ isomorphic to $\mathrm{Gal}(N/M)$ (Lemma 6.1.4). Then we define $\mathrm{Ar}(C/A_0, M, \mathbf{a})$ (or $\mathrm{Ar}(A_0, M, \mathbf{a})$, if there can be no confusion) to be the conjugacy class of subgroups of $\mathrm{Gal}(C/A_0)$ generated by $D_M(\varphi)$.

Let $n \geq 0$, A a $\mathbb{Q}$-constructible set in $\mathbb{A}^n$, and k be a nonzero integer. A **normal stratification**
$$\mathcal{A} = \langle A, C_i/A_i \rangle_{i \in I}$$
of A over $\mathbb{Z}[k^{-1}]$ is a decomposition $A = \bigcup_{i \in I} A_i$ of A as a finite union of disjoint normal $\mathbb{Z}[k^{-1}]$-basic sets A_i, each equipped with a Galois ring cover C_i over $\mathbb{Z}[k^{-1}]$. Assume also that $A(M) = \bigcup_{i \in I} A_i(M)$ for each field M with $\mathrm{char}(M) = p$ and $p \nmid k$. In this case for each family $\mathcal{H}$ of finite groups, $\mathcal{A}$ can be augmented to a Galois stratification $\mathcal{A}(\mathcal{H})$ with respect to $\mathbb{Z}[k^{-1}]$ by adding conjugacy domains $\mathrm{Con}(A_i, \mathcal{H})$, as in Section 30.3. For each $\mathbf{a} \in A(M)$, write $\mathrm{Ar}(\mathcal{A}, M, \mathbf{a}) \subseteq \mathrm{Con}(\mathcal{A}(\mathcal{H}))$ if $\mathrm{Ar}(A_i, M, \mathbf{a}) \subseteq \mathrm{Con}(A_i, \mathcal{H})$ for the unique $i \in I$ such that $\mathbf{a} \in A_i(M)$.

Here, as above, we interpret the polynomials defining A and A_i as having coefficients in $\mathbb{F}_p$, so that $A(M)$ and $A_i(M)$ are well defined. Note, however, that the reduction of $\mathbb{Z}[k^{-1}, A_i]$ modulo p is not assumed to be integrally closed. Thus, the reduction of $\mathcal{A}$ modulo p need not be a normal stratification of A over $\mathbb{F}_p$.

In case $n = 0$, the affine space $\mathbb{A}^0$ consists of the origin O only. Let $C/\mathbb{Z}[k^{-1}]$ be the corresponding Galois ring cover and let $L = \mathbb{Q}(C)$. Notationally simplify $\mathrm{Ar}(\mathcal{A}, M, O)$ to $\mathrm{Ar}(L/\mathbb{Q}, M)$ and $\mathrm{Con}(\mathcal{A}(\mathcal{H}))$ to $\mathrm{Con}(\mathcal{H})$. If $\mathrm{char}(M) = 0$, then $\mathrm{Ar}(L/\mathbb{Q}, M) \subseteq \mathrm{Con}(\mathcal{H})$ means that $\mathrm{Gal}(L/L \cap M) \in \mathcal{H}$. If $M = \mathbb{F}_p$, and $p \nmid k$, then $\mathrm{Ar}(L/\mathbb{Q}, M)$ is the conjugacy class of subgroups of $\mathrm{Gal}(L/\mathbb{Q})$ generated by the elements of the classical Artin symbol $\left(\frac{L/\mathbb{Q}}{p}\right)$.

Now define Galois formulas and Galois sentences over $\mathbb{Z}[k^{-1}]$ as in Section 30.4. In particular, the stratification lemma (Lemma 19.7.3) and Lemma 31.1.2 imply, as in Remark 30.5.1, that each formula $\varphi(\mathbf{X})$ of $\mathcal{L}(\mathrm{ring})$ is equivalent to a Galois formula $\theta(\mathbf{X})$ over $\mathbb{Z}[k^{-1}]$ for a suitable k which can be effectively computed. Thus, if M is a field with $\mathrm{char}(M) \nmid k$ and $b_1, \ldots, b_n \in M$, then $M \models \theta(\mathbf{b})$ if and only if $M \models \varphi(\mathbf{b})$.

PROPOSITION 31.1.3: *Let θ be a sentence of $\mathcal{L}(\mathrm{ring})$. Then we can effectively find a finite family $\mathcal{S}$ of finite groups, a finite Galois extension L of $\mathbb{Q}$ with $\mathrm{Subgr}(\mathrm{Gal}(L/\mathbb{Q})) \subseteq \mathcal{S}$, and a positive integer k such that the following holds: For each subfamily $\mathcal{H}$ of $\mathcal{S}$ that contains the trivial group we can effectively find a conjugacy class $\mathrm{Con}(\mathcal{H}) \subseteq \mathrm{Subgr}(\mathrm{Gal}(L/K)) \cap \mathcal{H}$ such that for each perfect Frobenius field M with $\mathrm{char}(M) \nmid k$ and $\mathrm{Im}(\mathrm{Gal}(M)) \cap \mathcal{S} = \mathcal{H}$*

(2) $$M \models \theta \iff \mathrm{Ar}(L/\mathbb{Q}, M) \in \mathrm{Con}(\mathcal{H}).$$

Sketch of proof: Write θ in prenex normal form with m quantifier $Q_1, \ldots, Q_m$. Then apply the stratification lemma (Lemma 19.7.3) and Lemma 31.1.2 to find a positive integer k_0 such that θ is defined over $\mathbb{Z}[k_0^{-1}]$, and a Galois stratification $\mathcal{A}_m$ of $\mathbb{A}^m$ over $\mathbb{Z}[k_0^{-1}]$ such that for each basic set A_i in

$\mathcal{A}_m$, $\mathrm{Con}(\mathcal{A}_i)$ is either empty or consists of the trivial group, and the Galois sentence θ',

$$(Q_1 X_1) \cdots (Q_m X_m)[\mathrm{Ar}(\mathbf{X}) \subseteq \mathrm{Con}(\mathcal{A}_m)],$$

is equivalent to θ over each perfect Frobenius field M with $\mathrm{char}(M) \nmid k$. As in the proof of Proposition 30.4.2 apply Lemmas 31.1.2 and 19.7.3 to proceed inductively for i, $0 \leq i \leq m$: find k_i, with k_{i+1} a multiple of k_i, and a normal stratification $\mathcal{A}_{m-i}$ of $\mathbb{A}^{m-i}$ over $\mathbb{Z}[k_i^{-1}]$ which eliminates Q_i from θ' with respect to each perfect Frobenius field M with $\mathrm{char}(M) \nmid k_i$. Now take L as the ring cover of $\mathbb{Q}$ in $\mathcal{A}_0$, $\mathcal{S} = \bigcup_{i=0}^{m} \mathrm{Subgr}(\mathcal{A}_i)$, and the final integer k to be k_m. Then the conclusion of the proposition holds.

For the verification one must replace each occurrence of the phrase "a field M that contains K" in the appropriate lemmas of Chapter 30 that contribute to Propositions 30.5.2 and 30.5.3, by the phrase "a field M of characteristic not dividing k." The stratification procedure is carried out in characteristic 0, over the rings $\mathbb{Z}[k_i^{-1}]$. The decomposition groups with respect to points over $\mathbb{Q}$ behave as in the case of Chapter 30. In particular, they are isomorphic to the Galois groups of the corresponding residue fields. Thus, the conclusion of Propositions 30.5.2 and 30.5.3 hold. □

The next theorem generalizes Theorem 30.5.3 to include all perfect Frobenius fields:

THEOREM 31.1.4: *Let $\mathcal{C}$ be a primitive recursive full family of finite groups.*
(a) *For each given sentence θ of $\mathcal{L}(\mathrm{ring})$ we can effectively find a nonzero integer k with the following property: If θ is true in all fields $M \in \mathrm{Frob}(\mathbb{Q}, \mathrm{pro}\text{-}\mathcal{C})$, then θ is true in all fields $M \in \mathrm{Frob}(\mathbb{F}_p, \mathrm{pro}\text{-}\mathcal{C})$ such that $p \nmid k$.*
(b) *The theory of perfect Frobenius fields M such that $\mathrm{Gal}(M)$ is a pro-$\mathcal{C}$-group is primitive recursive.*

Proof of (a): Find $\mathcal{S}$, L and k as in Proposition 31.1.3. Suppose θ is true in all fields $M \in \mathrm{Frob}(\mathbb{Q}, \mathrm{pro}\text{-}\mathcal{C})$. Let $\mathcal{H}_1, \ldots, \mathcal{H}_r$ be all subfamilies of $\mathcal{C} \cap \mathcal{S}$ such that there exists a superprojective group G_i with $\mathrm{Im}(G_i) \cap \mathcal{S} = \mathcal{H}_i$, $i = 1, \ldots, r$. Consider a group $H \in \mathcal{H}_i \cap \mathrm{Subgr}(\mathrm{Gal}(L/\mathbb{Q}))$. Lemma 24.1.6 produces a Frobenius field M of characteristic 0 such that $\mathrm{Gal}(M) \cong G_i$ and $\mathrm{Gal}(L/L \cap M) = H$. Since $M \models \theta$, Proposition 31.1.3 implies that $H \in \mathrm{Con}(\mathcal{H}_i)$. Therefore,

(3) $\qquad \mathrm{Con}(\mathcal{H}_i) = \mathcal{H}_i \cap \mathrm{Subgr}(\mathrm{Gal}(L/\mathbb{Q})), \quad i = 1, \ldots, r.$

Now consider $M \in \mathrm{Frob}(\mathbb{F}_p, \mathrm{pro}\text{-}\mathcal{C})$ for $p \nmid k$. Then $\mathrm{Im}(\mathrm{Gal}(M)) \cap \mathcal{S} = \mathcal{H}_i$ for some i, $1 \leq i \leq r$. Since $\mathrm{Subgr}(\mathrm{Gal}(L/\mathbb{Q})) \subseteq \mathcal{S}$, (3) gives $\mathrm{Ar}(L/\mathbb{Q}, M) \in \mathrm{Con}(\mathcal{H}_i)$. By Proposition 31.1.3, $M \models \theta$. This proves (a).

Proof of (b): The proof of Theorem 30.6.1 gives an effective check for condition (3). If (3) holds, then Theorem 30.6.1 gives a check to decide for each prime $p | k$ if θ is true in each field $M \in \mathrm{Frob}(\mathbb{F}_p, \mathcal{C})$. This proves (b). □

In particular, we may take $\mathcal{C}$ to be the family of all finite groups.

COROLLARY 31.1.5: *The theory of perfect Frobenius fields is primitive recursive.*

We leave to the reader the formulation and proof of the appropriate analog of Theorem 30.6.3.

31.2 The Elementary Theory of Finite Fields

A special case of Theorem 30.7.2 implies that the theory Almost$(\mathbb{Q}, 1)$ of all sentences θ of $\mathcal{L}(\text{ring})$ which are true in $\tilde{\mathbb{Q}}(\sigma)$ for almost all $\sigma \in \text{Gal}(\mathbb{Q})$, is primitive recursive. By the transfer theorem (Theorem 20.9.3), each such θ is true in $\mathbb{F}_p$, for almost all primes p. Thus, we have a primitive recursive procedure to decide the truth of a given sentence θ of $\mathcal{L}(\text{ring})$ in $\mathbb{F}_p$ for almost all primes p. Still, even if θ is true in $\mathbb{F}_p$ for almost all p, this does not yet include a procedure to compute the finite set of exceptional primes p for which θ is not true in $\mathbb{F}_p$. This section modifies the Galois stratification procedure to fill this gap. The basic idea is to replace the Frobenius fields in the procedure of Section 31.1 by finite fields. Since the procedure depends on the Frobenius property, which does not hold for finite fields, we must show that, relative to a given sentence θ, the finite fields behave like Frobenius fields provided they are sufficiently large. Exactly how large depends on the following quantitative variant of Proposition 6.4.8:

LEMMA 31.2.1 ([Haran-Jarden1, p. 15]): *Let d be a positive integer, M a finite field, $g \in M[Y]$, $f \in M[Y, Z]$ polynomials with $g \neq 0$, y an indeterminate, and $z \in \widetilde{M(y)}$. Put $R = M[y, g(y)^{-1}]$, $C = R[z]$, and $N = \tilde{M} \cap M(y, z)$. Suppose $q = |M| \geq d^4$, $\deg(g) < d$, $f(y, Z) = \text{irr}(z, M(y))$, $\deg(f) \leq d$, and C/R is a Galois ring cover with primitive element z. Consider $\tau \in \text{Gal}(C/R)$ satisfying $\langle \text{res}_N \tau \rangle = \text{Gal}(N/M)$. Then there exists an M-homomorphism $\psi \colon C \to \tilde{M}$ with $\psi(R) = M$ and $D_M(\psi) = \langle \tau \rangle$.*

Proof: Let $E = M(y)$, $F = M(y, z)$ and $P = \tilde{M}(y, z)$. Since $\text{Gal}(M) \cong \hat{\mathbb{Z}}$, $\text{res}_N \tau$ extends to a generator τ_0 of $\text{Gal}(M)$. Then τ extends to an element $\tilde{\tau} \in \text{Gal}(P/E)$ such that $\text{res}_{\tilde{M}} \tilde{\tau} = \tau_0$. The intersection of $D = P(\tilde{\tau})$ with $\tilde{M}$ is M. In particular, D is a regular extension of M and the map $\text{res}_{\tilde{M}} \colon \text{Gal}(P/D) \to \text{Gal}(M)$ is surjective. Since $\text{Gal}(M) \cong \hat{\mathbb{Z}}$, $\text{res}_{\tilde{M}}$ is bijective (Proposition 16.10.6). Hence, $\tilde{M}D = P$, so $[D : E] = [P : \tilde{M}E] \leq [F : E] \leq d$. Thus, D is a function field of one variable over M.

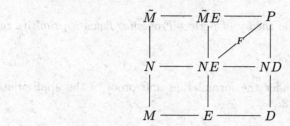

Let n be the number of prime divisors of D/M of degree 1. By the Riemann hypothesis (Theorem 4.5.2)

$$|n - (q+1)| \leq 2g_D\sqrt{q},$$

where $g_D = \text{genus}(D/M)$. Since P is a separable constant field extension of both D and F, $g_D = g_F$ (Proposition 3.4.2). Corollary 5.3.5 now gives

$$g_F \leq \frac{1}{2}(\deg(f) - 1)(\deg(f) - 2) \leq \frac{1}{2}(d-1)^2.$$

By assumption, $\sqrt{q} \geq d^2 \geq (d-1)^2 + 1$. Hence

$$n \geq (q+1) - 2g_F\sqrt{q} \geq (q+1) - (d-1)^2\sqrt{q}$$
$$= 1 + \sqrt{q}(\sqrt{q} - (d-1)^2) \geq 1 + \sqrt{q} \geq 1 + d^2.$$

Thus, D/M has at least $d^2 + 1$ prime divisors of degree 1. Also, E/M has at most $(1 + \deg(g)) \leq d$ prime divisors which are not finite on R. Each of them has at most $[D : E] \leq d$ extensions to D. Using $(1+d^2) - d^2 = 1$, we find an M-rational place $\psi_0 \colon D \to M \cup \{\infty\}$ finite on R such that $\psi_0(R) = M$.

Extend ψ_0 to a place $\tilde{\psi} \colon P \to \tilde{M} \cup \{\infty\}$ such that $\tilde{\psi}|_{\tilde{M}}$ is the identity map. Put $\psi = \tilde{\psi}|_C$. Then $\psi \colon C \to \tilde{M}$ is an M-homomorphism and $\psi(R) = M$.

Let S be the integral closure of R in D. Then $\tilde{M}S$ is the integral closure of R in P (Lemma 2.5.8). From the definitions, $\tilde{\psi}(\tilde{\tau}x) = \tau_0\tilde{\psi}(x)$ for each $x \in \tilde{M}$ and each $x \in S$. Since $C \subseteq \tilde{M}D_0$, this holds also for each $x \in C$. In particular, $\psi(\tau x) = \tau_0\psi(x)$ for each $x \in C$. Let $\psi^* \colon \text{Gal}(\psi(C)/M) \to \text{Gal}(F/E)$ be the embedding induced by ψ. Then $\psi^*(\text{res}_{\psi(C)}\tau_0) = \tau$. Consequently, $D_M(\psi) = \langle \tau \rangle$. □

Finite fields compensate for their being only "approximately" Frobenius fields, by the simplicity of their absolute Galois groups. Using this, there is a simple finite field analog to Proposition 31.1.3 which costs us a few more "exceptional primes."

31.2 The Elementary Theory of Finite Fields

PROPOSITION 31.2.2: *Let k be a positive integer and let $\theta(\mathbf{Y})$ be the Galois formula*

$$(Q_1 X_1) \cdots (Q_m X_m)[\mathrm{Ar}(\mathbf{X}, \mathbf{Y}) \subseteq \mathrm{Con}(\mathcal{A}(\mathcal{C}))]$$

with respect to a Galois stratification $\mathcal{A}(\mathcal{C})$ of $\mathbb{A}^{m+n}$ over $\mathbb{Z}[k^{-1}]$, where $\mathcal{C}$ is the family of all finite cyclic groups. Then we can effectively find a multiple l of k and a Galois stratification $\mathcal{B}(\mathcal{C})$ of $\mathbb{A}^n$ over $\mathbb{Z}[l^{-1}]$ such that the following holds:

(1) *for each finite field M of characteristic not dividing l, and for each $\mathbf{b} \in \mathbb{A}^n(M)$,*

$$M \models \theta(\mathbf{b}) \iff \mathrm{Ar}(\mathcal{B}, M, \mathbf{b}) \subseteq \mathrm{Con}(\mathcal{B}(\mathcal{C})).$$

Sketch of Proof: As in the proof of Proposition 30.5.2, we may use the stratification lemma and Lemma 31.1.2 to find a multiple l_0 of k and to eliminate the quantifiers $Q_m, Q_{m-1}, \ldots, Q_1$ from $\theta(\mathbf{Y})$ with respect to $\mathcal{C}$ over $\mathbb{Z}[l_0^{-1}]$. From this we may successively obtain Galois stratifications $\mathcal{B}_{m-1}(\mathcal{C}), \mathcal{B}_{m-2}(\mathcal{C}), \ldots, \mathcal{B}_0(\mathcal{C})$ of $\mathbb{A}^{m-1+n}, \mathbb{A}^{m-2+n}, \ldots, \mathbb{A}^n$, respectively, over $\mathbb{Z}[l_0^{-1}]$. Denote $\mathcal{A}(\mathcal{C})$ by $\mathcal{B}_m(\mathcal{C})$. Enlarge l_0 if necessary to assume that for each i, $0 \leq i \leq m$, and for each prime $p \nmid l_0$ the reduction of $\mathcal{B}_i$ modulo p gives a stratification of $\mathbb{A}^{i+n}$ in characteristic p (Theorem 9.3.1). Let C/A be a Galois ring/set cover involved in the normal stratification $\mathcal{B}_{i+1}$, $0 \leq i \leq m-1$. Suppose the projection $\pi\colon \mathbb{A}^{i+1+n} \to \mathbb{A}^{i+n}$ maps A onto a basic set B (of $\mathcal{B}_i$) and

(2) $$\dim(A) = \dim(B) + 1.$$

Find a generic point $x = (x_1, \ldots, x_{n+i})$ of B, a transcendental element y over $\mathbb{Q}(B)$ with $\mathbb{Q}(A) = \mathbb{Q}(B)(y)$, a polynomial $g \in \mathbb{Z}[l_0^{-1}, X_1, \ldots, X_{n+i}, Y]$ with $\mathbb{Z}[l_0^{-1}, A] = \mathbb{Z}[l_0^{-1}, \mathbf{x}, y, g(\mathbf{x}, y)^{-1}]$, a primitive element z for the ring cover $C/\mathbb{Z}[l_0^{-1}, A]$ and a polynomial $f \in \mathbb{Z}[k^{-1}, X_1, \ldots, X_{n+i}, Y, Z]$ so that $f(\mathbf{x}, y, z) = 0$ and the coefficient of the highest power of Z in f divides $g(X_1, \ldots, X_{n+i}, Y)$. Define $l(C/A)$ to be the product of l_0 with all primes which do not exceed $\max((\deg_{(Y,Z)} f)^4, (\deg_Y g)^4)$.

Let M be a finite field with $p = \mathrm{char}(M) \nmid l(C/A)$. Then $\mathcal{B}_{i+1}$ and $\mathcal{B}_i$ stratify $\mathbb{A}^{i+1+n}$ and $\mathbb{A}^{i+n}$, respectively. Therefore, π maps the reduction of A modulo p onto the reduction of B modulo p. Let y' be a transcendental element over M. If $\mathbf{b} \in B(M)$ this implies that $g(\mathbf{b}, y') \neq 0$. Hence, $\deg(g(\mathbf{b}, Y)) = \deg_Y g(\mathbf{X}, Y)$ and $\deg(f(\mathbf{b}, Y, Z)) = \deg_{(Y,Z)} f$. Also, the specialization compatibility of the Galois covers of A and B implies that $f(\mathbf{b}, y', Z)$ is irreducible over $M(y')$. Thus, we may replace the use of Proposition 24.1.4 (the Frobenius property) in Part B of the proof of Lemma 30.2.3 by Lemma 31.2.1.

By Remark 30.2.4, Part B of the proof of Lemma 30.2.3 is the only point where the Frobenius property of M is used. Therefore, take l to be the least common multiple of all $l(C/A)$ as i ranges from 0 to $m-1$ and C/A runs over all covers involved in the normal stratification $\mathcal{B}_{i+1}$ for which (2) holds.

With this value of l and with $\mathcal{B}(\mathcal{C}) = \mathcal{B}_0(\mathcal{C})$ the conclusion of the proposition holds. ☐

In the next result we replace $\mathbb{Z}[k^{-1}]$ by a finite field:

PROPOSITION 31.2.3: *Let K be a finite field, $\mathcal{C}$ a family of finite cyclic groups, and $\theta(\mathbf{Y})$ the Galois formula*

$$(Q_1 X_1) \cdots (Q_m X_m)[\mathrm{Ar}(\mathbf{X}, \mathbf{Y}) \subseteq \mathrm{Con}(\mathcal{A}(\mathcal{C}))]$$

with respect to a Galois stratification $\mathcal{A}(\mathcal{C})$ of $\mathbb{A}^{m+n}$ over K. Then we can effectively find a positive integer c and a Galois stratification $\mathcal{B}(\mathcal{C})$ of $\mathbb{A}^n$ over K such that the following holds:
(3) *for each finite field M of cardinality exceeding c that contains K and for each $\mathbf{b} \in \mathbb{A}^n(M)$,*

$$M \models \theta(\mathbf{b}) \iff \mathrm{Ar}(\mathcal{B}, M, b) \subseteq \mathrm{Con}(\mathcal{B}(\mathcal{C})).$$

Proof: Apply the Galois stratification procedure to $\mathcal{A}(\mathcal{C})$ over K. For each Galois ring/set cover C/A that appears in the Galois stratification procedure for which $\dim(A) = \dim(\pi(A)) + 1$ we may assume that $k[A] = K[x_1, \ldots, x_{n+i}, y, g(x_1, \ldots, x_{n+i}, y)^{-1}]$ and $C = K[A][z]$ as in the proof of Proposition 31.2.2. Let $f(\mathbf{x}, y, Z) = \mathrm{irr}(z, K(\mathbf{x}, y))$. Then choose c to exceed $\max((\deg_{(Y,Z)} f)^4, (\deg_Y g)^4)$, for all C/A. Use Lemma 31.2.1 and Remark 30.2.4 to conclude Condition (3). ☐

We are now ready to prove the primitive recursive version of Theorem 20.10.6.

THEOREM 31.2.4 ([Fried-Sacerdote]): *Let θ be a sentence of $\mathcal{L}(\mathrm{ring})$.*
(a) *There is a primitive recursive procedure to decide if θ is true in almost every field $\mathbb{F}_p$ and to compute the set of exceptional primes in case θ is true in almost every field.*
(b) *The theory of sentences θ of $\mathcal{L}(\mathrm{ring})$ which are true in each field $\mathbb{F}_p$ is primitive recursive.*
(c) *The theory of sentences θ of $\mathcal{L}(\mathrm{ring})$ which hold in almost every finite field (equivalently, in every pseudo-finite field, Proposition 20.10.2) is primitive recursive.*
(d) *The theory of finite fields is primitive recursive.*

Proof: Apply Proposition 31.2.2, in the case $n = 0$, to θ. This gives a nonzero integer l, a finite Galois extension L of $\mathbb{Q}$, and a conjugacy class Con of cyclic subgroups of $\mathrm{Gal}(L/\mathbb{Q})$ satisfying this: For each finite field M with $\mathrm{char}(M) \nmid l$,

(4) $$M \models \theta \iff \mathrm{Ar}(L/\mathbb{Q}, M) \subseteq \mathrm{Con}.$$

Suppose there exists $\sigma \in \mathrm{Gal}(L/\mathbb{Q})$ with $\langle \sigma \rangle \notin \mathrm{Con}$. By the Chebotarev density theorem (Theorem 6.3.1), there exist infinitely many primes p not dividing l such that $\left(\frac{L/\mathbb{Q}}{p}\right) = \mathrm{Ar}(L/\mathbb{Q}, \mathbb{F}_p)$ is the conjugacy class of groups generated by $\langle \sigma \rangle$. Thus, $\mathrm{Ar}(L/\mathbb{Q}, \mathbb{F}_p) \not\subseteq \mathrm{Con}$. By (4), θ is false in $\mathbb{F}_p$.

If, on the other hand, Con contains each cyclic subgroup of $\mathrm{Gal}(L/\mathbb{Q})$, then the right hand side of (4), and therefore its left hand side, is true for each finite field M with $\mathrm{char}(M) \nmid l$. In particular, $\mathbb{F}_p \models \theta$ for each prime $p \nmid l$.

Assume that $\mathbb{F}_p \models \theta$ for each $p \nmid l$. Check directly, for each prime $p \mid l$, if $\mathbb{F}_p \models \theta$. Determine in this way the finite set of primes p such that $\mathbb{F}_p \not\models \theta$. This concludes the proofs of (a) and (b).

The argument above actually shows that we can decide the truth of θ in M for each finite field M of characteristic not dividing l. It remains to check the truth of θ in each field of characteristic p where $p|l$. Apply Proposition 31.1.3 to effectively find a positive integer m_p for which we can decide the truth of θ in all finite fields of characteristic p and of cardinality exceeding m_p. If θ is true in all these fields with p running over the divisors of l, then we can check directly the truth of θ in M as M runs over the finite list of finite fields with the property that $\mathrm{char}(M) = p \mid l$ and $|M| \leq m_p$. This proves (d) and (e). □

31.3 Near Rationality of the Zeta Function of a Galois Formula

Let K be the finite field with q elements. Denote the unique extension of K of degree k by K_k. Throughout this section $\mathcal{C}$ is the family of finite cyclic groups. Consider a Galois formula $\theta(Y_1, \ldots, Y_n)$ over K with respect to a Galois stratification $\mathcal{A}(\mathcal{C})$,

$$\mathcal{A}(\mathcal{C}) = \langle \mathbb{A}^{m+n}, C_i/A_i, \mathrm{Con}(A_i, \mathcal{C}) \rangle_{i \in I}.$$

For each positive integer k let $\nu(\theta, k) = \#\{\mathbf{b} \in K_k^n \mid K_k \models \theta(\mathbf{b})\}$. Define the **Poincaré series** of θ to be the formal power series

(1) $$P(\theta, t) = \sum_{k=1}^{\infty} \nu(\theta, k) t^k.$$

The **zeta function**, $Z(\theta, t)$, is then defined by the formulas

(2) $$P(\theta, t) = t \frac{d}{dt}(\log Z(\theta, t)) \quad \text{and} \quad Z(\theta, 0) = 1,$$

where $\frac{d}{dt}$ indicates the operation of taking the formal derivative. Thus,

(3) $$Z(\theta, t) = \exp\left(\sum_{k=1}^{\infty} \frac{\nu(\theta, k)}{k} t^k\right),$$

where $\exp(t) = \sum_{n=0}^{\infty} \frac{t^n}{n!}$.

The remainder of this section consists of the proof that $P(\theta, t)$ is rational in t, that $Z(\theta, t)$ is "nearly" rational in t and that the polynomial expressions involved in these statements are given by a primitive recursive procedure. We give a brief account of the definitions and results of the representation theory of finite groups (see, e.g. [Serre4], for details) as these will be used in the proof.

A **representation** of a finite group G is a homomorphism $\rho\colon G \to \mathrm{GL}(n, \mathbb{C})$. The **character** χ of ρ is the complex valued function on G obtained by composing ρ with the trace: $\chi(\sigma) = \mathrm{trace}(\rho(\sigma))$. If L is a subfield of $\mathbb{C}$ with $\rho(\sigma) \in \mathrm{GL}(n, L)$ for all $\sigma \in G$, then ρ and χ are said to be **rational** over L. In this case $\chi(\sigma) \in L$ for each $\sigma \in G$ (but the converse does not hold [Serre4, Section 12.2]).

Call a character χ of G **irreducible** over L if it is rational over L and it is not the sum $\chi = \chi_1 + \chi_2$ of two rational characters over L. There are only finitely many irreducible characters over L and they are linearly independent. Note (as a property of taking trace) that a character χ of G is constant on conjugacy classes of elements:

(4) $$\chi(\tau^{-1}\sigma\tau) = \chi(\sigma) \text{ for all } \sigma, \tau \in G.$$

Any function $\chi\colon G \to \mathbb{C}$ that satisfies (4) is said to be **central**.

We record here a criterion for a central function to belong to the L-linear space generated by the L-irreducible characters of G [Serre4, Section 12.4] in the case where $L = \mathbb{Q}$.

LEMMA 31.3.1: *Consider a function $\chi\colon G \to \mathbb{Q}$ from a finite group G. The following conditions are equivalent:*
(a) *χ is a linear combination over $\mathbb{Q}$ of the $\mathbb{Q}$-irreducible characters of G over $\mathbb{Q}$.*
(b) *χ is central and $\chi(\sigma) = \chi(\sigma^i)$ for each $\sigma \in G$ and for each $i \in \mathbb{Z}$, with $\gcd(i, \mathrm{ord}(\sigma)) = 1$.*
(c) *$\chi(\sigma) = \chi(\sigma')$ for all $\sigma, \sigma' \in G$ such that $\langle\sigma\rangle$ is conjugate to $\langle\sigma'\rangle$.*

Definition 31.3.2: Refer to a central function $\chi\colon G \to Q$ as $\mathbb{Q}$-**central** if Condition (a) (or (b) or (c)) of Lemma 31.3.1 holds. □

Consider a subgroup H of G. Let $\rho_0\colon H \to \mathrm{GL}(m, \mathbb{C})$ be a representation of H. The group $\mathrm{GL}(m, \mathbb{C})$ acts faithfully on the vector space $W = \mathbb{C}^m$ by multiplication from the left. We may therefore consider ρ_0 as a homomorphism from H into $\mathrm{Aut}(W)$. Let $V = \mathrm{Ind}_H^G W$ be the vector space of all functions $f\colon G \to W$ satisfying $f(\eta\sigma) = \rho_0(\eta)f(\sigma)$ for all $\eta \in H$ and $\sigma \in G$. Define a homomorphism $\rho\colon G \to \mathrm{Aut}(V)$ by

$$(\rho(\tau)f)(\sigma) = f(\sigma\tau), \quad \sigma, \tau \in G.$$

Now choose a basis for V to consider ρ as a representation $\rho\colon G \to \mathrm{GL}(n, \mathbb{C})$, with $n = \dim(V) = (G : H)m$. This is the **induced representation** of ρ_0

31.3 Near Rationality of the Zeta Function of a Galois Formula

from H to G: $\rho = \text{ind}_H^G \rho_0$. The corresponding characters χ_0 and $\chi = \text{ind}_H^G \chi_0$ are related by the formula

$$\chi(\sigma) = \frac{1}{|H|} \sum \chi_0(\tau\sigma\tau^{-1}),$$

where τ ranges over all elements of G satisfying $\tau\sigma\tau^{-1} \in H$.

In particular, let 1_H be the trivial character of H, $1_H(\eta) = 1$ for each $\eta \in H$. Then $1_H^G = \text{ind}_H^G 1_H$ is given by the formula

(5) $$1_H^G(\sigma) = \frac{1}{|H|} \#\{\tau \in G \mid \tau\sigma\tau^{-1} \in H\}.$$

LEMMA 31.3.3: *Let $\chi\colon G \to \mathbb{Q}$ be a $\mathbb{Q}$-central function. Then $\chi = \sum_H r_H 1_H^G$, where H ranges over all cyclic subgroups of G and r_H are rational numbers which can be effectively computed from G and χ.*

Proof: The existence of the r_H's is a well known result of a theorem of Artin [Serre4, p. 103]. Now suppose that the multiplication table of G is known and that χ is presented. For each cyclic subgroup H of G compute the induced character 1_H^G from (5). Then solve the system of linear equations

$$\chi(\sigma) = \sum_H r_H 1_H^G(\sigma), \quad \sigma \in G$$

in the unknowns r_H, where H ranges over all cyclic subgroups of G. $\square$

Now suppose that C/A is a Galois ring/set cover over K with $G = \text{Gal}(C/A)$. Let $\chi\colon G \to \mathbb{Q}$ be a $\mathbb{Q}$-central function. For all $k \in \mathbb{N}$ and $\mathbf{a} \in A(K_k)$ choose an element $\sigma \in G$ with $\langle\sigma\rangle \in \text{Ar}(C/A, K_k, \mathbf{a})$ and define $\chi(k, \mathbf{a})$ to be $\chi(\sigma)$. If σ' is another element of G such that $\langle\sigma'\rangle \in \text{Ar}(C/A, K_k, \mathbf{a})$, then $\langle\sigma\rangle$ is conjugate to $\langle\sigma'\rangle$. Hence, $\chi(\sigma') = \chi(\sigma)$ and consequently $\chi(k, \mathbf{a})$ is well defined. Define the L-series of χ to be

(6) $$L(C/A, \chi, t) = \exp\left(\sum_{k=1}^{\infty} \frac{1}{k} \sum_{\mathbf{a} \in A(K_k)} \chi(k, \mathbf{a}) t^k\right)$$

This definition agrees with other definitions of the L-series (e.g. as in [Serre2, p. 87] or [Dwork1, p. 44]). Indeed, for $\mathbf{a} \in A(K_k)$ let $\deg(\mathbf{a}) = [K(\mathbf{a}) : K]$ be the number of points of A conjugate to $\mathbf{a}$ over K. Let $N(\mathbf{a}) = q^{\deg(\mathbf{a})} = |K(\mathbf{a})|$. Choose a generic point $\mathbf{x}$ of A over K and let $\varphi\colon C \to \bar{K}$ be a homomorphism which extends the specialization $\mathbf{x} \to \mathbf{a}$. Then $\text{Ar}(C/A, K_k, \mathbf{a})$ is the class of subgroups of G conjugate to $D_{K_k}(\varphi)$ (Section 31.1). The group $D_{K(\mathbf{a})}(\varphi)$ is generated by an element σ that corresponds to the automorphism $x \mapsto x^{q^{\deg(\mathbf{a})}}$ of $\varphi(C)$ over $K(\mathbf{a})$ while $D_{K_k}(\varphi)$ is generated by $\sigma^{k/\deg(\mathbf{a})}$. Denote the prime ideal of $K[A]$ corresponding to the K-specialization $\mathbf{x} \to \mathbf{a}$

(and also $x \to \mathbf{a}'$, for each conjugate $\mathbf{a}'$ of $\mathbf{a}$ over K) by $\mathfrak{p}$ and let $N(\mathfrak{p}) = N(\mathbf{a})$. Then, in the notation of [Serre2, p. 87], $\chi(k,\mathbf{a}) = \chi(\mathfrak{p}^{k/\deg(\mathbf{a})})$. Put $t = q^{-s}$ and $m = k/\deg(\mathbf{a})$. Then

$$\sum_{\mathbf{a}'} \frac{\chi(k,\mathbf{a}')t^k}{k} = \frac{\chi(\mathfrak{p}^m)N(\mathfrak{p})^{-ms}}{m},$$

where $\mathbf{a}'$ ranges on the points conjugate to $\mathbf{a}$. Thus

$$L(C/A,\chi,t) = \exp\left(\sum_{\mathfrak{p}} \sum_{m=1}^{\infty} \frac{\chi(\mathfrak{p}^m)N(\mathfrak{p})^{-ms}}{m}\right),$$

where $\mathfrak{p}$ ranges over all primes of A. This is exactly (8) of [Serre2, p. 87].

In the special case where $\chi = 1_G$ the L-series becomes the **zeta function** of A:

(7) $$L(C/A,1_G,t) = Z(A,t) = \exp\left(\sum_{k=1}^{\infty} \frac{1}{k}|A(K_k)|t^k\right).$$

If χ_1 and χ_2 are $\mathbb{Q}$-central functions of G, then (6) implies that

$$L(C/A,\chi_1+\chi_2,t) = L(C/A,\chi_1,t) \cdot L(C/A,\chi_2,t).$$

Hence,

(8) $$L(C/A,r\chi,t) = L(C/A,\chi,t)^r$$

for each $\mathbb{Q}$-central function χ of G and every $r \in \mathbb{Q}$. If H is a subgroup of G let B be a K-variety such that $K[B]$ is the integral closure of $K[A]$ in the fixed field of H in $K(C)$. Then C/B is a Galois ring/set cover and for each character χ of H, rational over $\mathbb{Q}$

(9) $$L(C/B,\chi,t) = L(C/A,\mathrm{ind}_H^G \chi,t)$$

(see [Serre2, p. 88] and also [Cassels-Fröhlich, p. 222] for a proof which is valid in our more general situation).

Definition 31.3.4: The **total degree** of a rational function $r \in \mathbb{C}(t)$ is the sum $\deg(f) + \deg(g)$, where $f,g \in \mathbb{C}[t]$ are relatively prime polynomials with $r = \frac{f}{g}$. Alternatively, the total degree of r is the total number of zeros and poles of r counted with multiplicity. □

In the rest of this section, whenever we speak about a "rational function", we mean "a rational function with coefficients in $\mathbb{C}$".

31.3 Near Rationality of the Zeta Function of a Galois Formula

LEMMA 31.3.5 (Dwork–Bombieri): Let $V = V(f_1, \ldots, f_m)$ be the K-closed subset of $\mathbb{A}^n$ defined by $f_1, \ldots, f_m \in K[X_1, \ldots, X_n]$. Then $Z(V,t)$ is a rational function of total degree at most $(5 + 4\max_{1 \le i \le m}(1 + \deg(f_i)))^{m+n}$.

Proof: The rationality of $Z(V,t)$ is proved in [Dwork1]. In [Dwork2, lemma 14.1, p. 489] Dwork gives a bound on certain vector spaces which leads to a bound on the total degree of $Z(V,t)$. Here we follow [Bombieri2]:

Let $\mathrm{trace}_k \colon K_k \to \mathbb{F}_p$ be the trace from K_k to $\mathbb{F}_p$, $\mathrm{trace}_k(x) = \sum_{i=0}^{ak-1} x^{p^i}$ where $q = p^a$. Then $\psi_k \colon K \to \mathbb{C}$ defined by $\psi_k(x) = e^{2\pi i \cdot \mathrm{trace}_k(x)/p}$ is a non-trivial character of the additive group of K_k. Thus, $\psi_k(x+y) = \psi_k(x)\psi_k(y)$ for all $x, y \in K_k$, $\psi_k(0) = 1$, and there is a $c \in K_k$ with $\psi_k(c) \ne 1$. For $f \in K[X_1, \ldots, X_n]$ use the exponential sum

$$S_k = \sum_{\mathbf{x} \in K_k^n} \psi_k(f(\mathbf{x}))$$

to define an L-series

$$L(\mathbb{A}^n, f, t) = \exp\left(\sum_{k=1}^{\infty} \frac{1}{k} S_k t^k\right).$$

By [Bombierie2, Theorem 2], $L(\mathbb{A}^n, f, t)$ is a rational function and $(5 + 4\deg(f))^n$ is a bound on its total degree. To apply Bombieri's result to $Z(V,t)$ replace n by $n+m$ and take $f(\mathbf{X}, \mathbf{Y}) = \sum_{i=1}^m f_i(\mathbf{X})Y_i$. If $\mathbf{x} \in K_k^n \smallsetminus V(K_k)$, then there is an i between 1 and m with $f_i(\mathbf{x}) \ne 0$. Hence, there exists $\mathbf{b} \in K_k^m$ with $\sum_{i=1}^m f_i(\mathbf{x})b_i = c$. Thus,

$$\sum_{\mathbf{y} \in K_k} \psi_k(f(\mathbf{x}, \mathbf{y})) = \sum_{\mathbf{y} \in K_k} \psi_k(f(\mathbf{x}, \mathbf{y}+\mathbf{b})) = \sum_{\mathbf{y} \in K_k^m} \psi_k\big(f(\mathbf{x},\mathbf{y}) + f(\mathbf{x},\mathbf{b})\big)$$

$$= \sum_{\mathbf{y} \in K_k^m} \psi_k(f(\mathbf{x},\mathbf{y}))\psi_k(c)$$

so $\sum_{\mathbf{y} \in K_k} \psi_k(f(\mathbf{x},\mathbf{y})) = 0$. If $\mathbf{x} \in V(K_k)$, then $f(\mathbf{x}, \mathbf{y}) = 0$. Therefore,

$$S_k = \sum_{\mathbf{x} \in V(K_k)} \sum_{\mathbf{y} \in K_k^m} \psi_k(f(\mathbf{x},\mathbf{y})) = q^{km}|V(K_k)|.$$

Consequently,

$$Z(V, q^m t) = \exp\left(\sum_{k=1}^{\infty} \frac{1}{k}|V(K_k)|(q^m t)^k\right)$$

$$= \exp\left(\sum_{k=1}^{\infty} \frac{1}{k} S_k t^k\right) = L(\mathbb{A}^{n+m}, f, t).$$

Since $\deg(f) \leq \max_{1 \leq i \leq m}(1 + \deg(f_i))$, the expression

$$(5 + 4 \max_{1 \leq i \leq m}(1 + \deg(f_i)))^{m+n}$$

bounds the total degree of $Z(V,t)$. □

The coefficients of a monic polynomial with the roots $x_1, \ldots, x_n$ can be effectively expressed as a polynomials in the sums $\sum_{i=1}^{n} x_i^k$, $k = 1, \ldots, n$, with integral coefficients [Bôcher, p. 241]. To generalize this to rational functions we introduce the following polynomials

$$s_k(\mathbf{X}, \mathbf{Y}) = \sum_{j=1}^{n} Y_j^k - \sum_{i=1}^{m} X_i^k, \quad k = 1, 2, 3, \ldots.$$

LEMMA 31.3.6: *Let F be a field of characteristic 0, $x_1, \ldots, x_m, y_1, \ldots, y_n \in F$, t an indeterminate, and r a positive integer with $m + n \leq r$. Put $g(t) = \prod_{i=1}^{m}(1 - x_i t)$, $h(t) = \prod_{j=1}^{n}(1 - y_j t)$, and $s_k = s_k(\mathbf{x}, \mathbf{y})$, $k = 1, 2, 3, \ldots$. Suppose $(s_1, \ldots, s_r)$ is presented. Then we can effectively compute polynomials $g_1(t), h_1(t) \in \mathbb{Q}(s_1, \ldots, s_r)[t]$ such that $g_1(0) = h_1(0) = 1$ and $\frac{g_1(t)}{h_1(t)} = \frac{g(t)}{h(t)}$.*

Proof: The proof splits into two parts.

PART A: *A system of linear equations for the coefficients of $g(t)$ and $h(t)$.* Let $g(t) = 1 + u_1 t + \cdots + u_m t^m$ and $h(t) = 1 + v_1 t + \cdots + v_n t^n$. A straightforward computation in the field $F((t))$ of formal power series shows that

$$\frac{d(\log(g(t)))}{dt} = -\sum_{k=0}^{\infty} \sum_{i=1}^{m} x_i^{k+1} t^k \quad \text{and} \quad \frac{d(\log(h(t)))}{dt} = -\sum_{k=0}^{\infty} \sum_{j=1}^{n} y_j^{k+1} t^k.$$

Hence,

$$\frac{d}{dt}\left(\log \frac{g(t)}{h(t)}\right) = \sum_{k=0}^{\infty} s_{k+1} t^k.$$

Therefore,

(10) $$\log\left(\frac{g(t)}{h(t)}\right) = \sum_{k=1}^{\infty} \frac{s_k}{k} t^k.$$

Apply exp on both sides of (10):

$$\frac{g(t)}{h(t)} = 1 + \sum_{k=1}^{\infty} \frac{s_k}{k} t^k + \frac{1}{2!}\left(\sum_{k=1}^{\infty} \frac{s_k}{k} t^k\right)^2 + \frac{1}{3!}\left(\sum_{k=1}^{\infty} \frac{s_k}{k} t^k\right)^3 + \cdots.$$

31.3 Near Rationality of the Zeta Function of a Galois Formula

Thus,

$$\text{(11)} \qquad \frac{g(t)}{h(t)} = 1 + \sum_{k=1}^{\infty} d_k t^k,$$

where

$$\text{(12)} \qquad d_k = \frac{1}{k} s_k + h_k(s_1, \ldots, s_{k-1})$$

and h_k is a polynomial with rational coefficients which does not depend on $s_1, \ldots, s_{k-1}$ and which can be effectively computed, $k = 1, 2, 3, \ldots$. Multiply (11) by $h(t)$:

$$\text{(13)} \qquad 1 + u_1 t + \cdots + u_m t^m = (1 + \sum_{k=1}^{\infty} d_k t^k)(1 + v_1 t + \cdots + v_n t^n).$$

Now compare the coefficients of t^k, $k = 1, 2, \ldots, r$ on both sides of (13) to obtain r equalities,

$$\text{(14)} \qquad \sum_{j=1}^{m} e_{kj} u_j + \sum_{j=1}^{n} e_{k,m+j} v_j = d_k, \quad k = 1, 2, \ldots, r,$$

with computable e_{kj} in $\{0, \pm 1, \pm d_1, \ldots, \pm d_{k-1}\}$, $k = 1, 2, \ldots, r$.

PART B: *Computation of $g_1(t)$ and $h_1(t)$*. Put $E = \mathbb{Q}(s_1, \ldots, s_r)$. Consider (14) as a system of linear equations in the variables $u_1, \ldots, u_m, v_1, \ldots, v_n$ with coefficients $e_{kj} \in E$. Since the system has a solution in F^{m+n}, we may effectively compute a solution $(u'_1, \ldots, u'_m, v'_1, \ldots, v'_n) \in E^{m+n}$. Thus, $\sum_{j=1}^{m} e_{kj} u'_j + \sum_{j=m+1}^{n} e_{k,m+j} v'_j = d_k$, $k = 1, \ldots, r$, so

$$\text{(15)} \qquad 1 + u'_1 t + \cdots + u'_m t^m = (1 + \sum_{k=1}^{\infty} d_k t^k)(1 + v'_1 t + \cdots + v'_n t^n).$$

On the other hand let

$$g_1(t) = 1 + u'_1 t + \cdots + u'_m t^m \quad \text{and} \quad h_1(t) = 1 + v'_1 t + \cdots + v'_n t^n.$$

Consider a factorization of g_1 and h_1 in $\tilde{F}[t]$:

$$g_1(t) = \prod_{i=1}^{m}(1 - x'_i t) \quad \text{and} \quad h_1(t) = \prod_{j=1}^{n}(1 - y'_j t),$$

Then put $s'_k = s_k(\mathbf{x}', \mathbf{y}')$, $k = 1, 2, 3, \ldots$. Replacing u_i, v_j, s_k in Part A by u'_i, v'_j, s'_k, respectively, gives

$$1 + \sum_{k=1}^{\infty} \frac{s'_k}{k} t^k + \frac{1}{2!} \left(\sum_{k=1}^{\infty} \frac{s'_k}{k} t^k \right)^2 + \frac{1}{3!} \left(\sum_{k=1}^{\infty} \frac{s'_k}{k} t^k \right)^3 + \cdots = \frac{1 + u'_1 t + \cdots + u'_m t^m}{1 + v'_1 t + \cdots + v'_n t^n},$$

so, by (15), $d_k = \frac{1}{k} s'_k + h_k(s'_1, \ldots, s'_{k-1})$ for $k = 1, \ldots, r$. Hence, by (12), $s'_k = s_k$ for $k = 1, \ldots, r$. Thus, an application of (10) to $\frac{g_1(t)}{h_1(t)}$ gives

$$\log \left(\frac{g(t) h_1(t)}{h(t) g_1(t)} \right) = \sum_{k=r+1}^{\infty} \frac{s_k - s'_k}{k} t^k.$$

Hence,

$$\frac{g(t) h_1(t)}{h(t) g_1(t)} = 1 + \sum_{k=r+1}^{\infty} c_k t^k$$

with $c_k \in \tilde{F}$, $k = r+1, r+2, \ldots$. Therefore,

(16) $$g(t) h_1(t) = h(t) g_1(t) + \sum_{k=r+1}^{\infty} b_k t^k,$$

with $b_k \in \tilde{F}$, $k = r+1, r+2, \ldots$. Since $\deg(gh_1), \deg(hg_1) \leq m + n \leq r$, (16) implies that $g(t) h_1(t) = h(t) g_1(t)$. Consequently, $\frac{g(t)}{h(t)} = \frac{g_1(t)}{h_1(t)}$. □

THEOREM 31.3.7: *Suppose the Galois formula $\theta(Y_1, \ldots, Y_n)$ of the beginning of this section is presented. Then*
(a) *we can effectively compute polynomials $f, g, h \in \mathbb{Q}[t]$ and a positive integer l such that*

(17) $$Z(\theta, t) = \exp\left(f(t) \right) \left(\frac{g(t)}{h(t)} \right)^{1/l}$$

(b) *and we can effectively compute $p, q \in \mathbb{Q}[t]$ such that $P(\theta, t) = \frac{p(t)}{q(t)}$.*

Proof: The assertion about $P(\theta, t)$ is an immediate consequence of (2) and of (a). The proof of (a) applies the Galois stratification procedure. It splits into three parts:

PART A: *Elimination of quantifiers.* By Proposition 31.2.3, we can effectively find a quantifier free Galois formula $\theta'(\mathbf{Y})$ and a positive integer c such that the following holds for each $k > c$ and for each $\mathbf{a} \in (K_k)^n$:

(18) $$K_k \models \theta(\mathbf{a}) \iff K_k \models \theta'(\mathbf{a}).$$

31.3 Near Rationality of the Zeta Function of a Galois Formula

In particular, $\nu(\theta, K) = \nu(\theta', k)$ for $k > c$. Hence

$$Z(\theta, t) = \exp\left(\sum_{k=1}^{c} \frac{\nu(\theta, k) - \nu(\theta', k)}{k} t^k\right) Z(\theta', t).$$

Compute $\nu(\theta, k)$ and $\nu(\theta', k)$, $k = 1, \ldots, c$, to obtain

$$f(t) = \sum_{k=1}^{c} \frac{\nu(\theta, k) - \mu(\theta', k)}{k} t^k.$$

We have therefore only to compute g, h, and l such that $Z(\theta', t) = \left(\frac{g(t)}{h(t)}\right)^{1/l}$.

Suppose we have computed $l, r \in \mathbb{N}$ and proved the existence of polynomials $g(t), h(t) \in \mathbb{C}[t]$ with $g(0) = h(0) = 1$, $\deg(g) + \deg(h) \leq r$, and $Z(\theta', t)^l = \frac{g(t)}{h(t)}$. Then consider a factorization of g and h over $\mathbb{C}$:

$$g(t) = \prod_{i=1}^{m}(1 - x_i t), \qquad h(t) = \prod_{j=1}^{n}(1 - y_j t),$$

where $m = \deg(g)$ and $n = \deg(h)$. Put $s_k = s(\mathbf{x}, \mathbf{y})$, $k = 1, 2, 3, \ldots$. By (10), $\log \frac{g(h)}{h(t)} = \sum_{k=1}^{\infty} \frac{1}{k} s_k t^k$. Now compute $\nu(\theta', k)$ for $k = 1, \ldots, r$. By (3), $\log \frac{g(t)}{h(t)} = \sum_{k=1}^{\infty} \frac{l\nu(\theta', k)}{k} t^k$. Hence, $s_k = l\nu(\theta', k) \in \mathbb{Z}$ for all k. By Lemma 31.3.6 we may effectively compute $g_1, h_1 \in \mathbb{Q}[t]$ such that $Z(\theta', t)^l = \frac{g_1(t)}{h_1(t)}$.

In the remaining two parts of the proof we reduce the computation of l and r to the effective rationality of the zeta function of a variety.

PART B: *The zeta function of a Galois cover.* Let

$$\mathcal{A}(\mathcal{C}) = \langle \mathbb{A}^n, C'_i/A'_i, \mathrm{Con}(A'_i, \mathcal{C})\rangle_{i \in I}$$

be the Galois stratification underlying θ'. For each $i \in I$ and for each positive integer k let

$$\nu(C'_i/A'_i, \mathrm{Con}(A'_i, \mathcal{C}), k) = \#\{\mathbf{b} \in A'_i(K_k) \mid \mathrm{Ar}(C'_i/A'_i, K_k, \mathbf{b}) \subseteq \mathrm{Con}(A'_i, \mathcal{C})\}.$$

Define

$$Z(C'_i/A'_i, \mathrm{Con}(A'_i, \mathcal{C}), t) = \exp\left(\sum_{k=1}^{\infty} \frac{\nu(C'_i/A'_i, \mathrm{Con}(A'_i, \mathcal{C}), k)}{k} t^k\right).$$

Then $\nu(\theta', k) = \sum_{i \in I} \nu(C'_i/A'_i, \mathrm{Con}(A'_i, \mathcal{C}), k)$. Hence,

$$Z(\theta', t) = \prod_{i \in I} Z(C'_i/A'_i, \mathrm{Con}(A'_i, \mathcal{C}), t).$$

It therefore suffices to consider a Galois set/ring cover C/A over K and a conjugacy domain Con of cyclic subgroups of $G = \text{Gal}(C/A)$, and to compute $l(C/A, \text{Con}) \in \mathbb{N}$ such that $Z(C/A, \text{Con}, t)^{l(C/A, \text{Con})}$ is a rational function with an explicit bound on its total degree. Then with $l(\theta') = \text{lcm}_{i \in I}\, l(C'_i/A'_i, \text{Con}(A'_i, C))$, the function $Z(\theta', t)^{l(\theta')}$ is rational with an explicit bound on its total degree.

PART C: *L-series.* Consider the following central function on G:

$$\chi(\sigma) = \begin{cases} 1 & \text{if } \langle \sigma \rangle \in \text{Con} \\ 0 & \text{if } \langle \sigma \rangle \notin \text{Con}. \end{cases}$$

For all $k \in \mathbb{N}$ and $\mathbf{a} \in A(K_k)$ let $\chi(k, \mathbf{a}) = \chi(\sigma)$ if $\langle \sigma \rangle \in \text{Ar}(C/A, K_k, \mathbf{a})$. Then

(19) $$\nu(C/A, \text{Con}, k) = \sum_{\mathbf{a} \in A(K_k)} \chi(k, \mathbf{a}).$$

By (6), $Z(C/A, \text{Con}, t) = L(C/A, \chi, t)$. Compute $r_H \in \mathbb{Q}$ for each cyclic subgroup H of G such that $\chi = \sum_H r_H 1_H^G$ (Lemma 31.3.3). In addition, for each such H compute a normal K-basic set B_H such that $K[B_H]$ is the integral closure of $K[A]$ in the fixed field of H in $K(C)$ (Lemma 19.3.2 and Lemma 19.7.2). By (7), (8), and (9),

$$L(C/A, \chi, t) = \prod_H L(C/A, 1_H^G, t)^{r_H} = \prod_H Z(B_H, t)^{r_H}.$$

Now Lemma 31.3.5 states that each $Z(B_H, t)$ is a rational function and gives a bound on its total degree. Taking $l(C/A, \text{Con})$ to be a common denominator of all r_H's, we conclude that $L(C/A, \chi, t)^{l(C/A, \text{Con})}$ can be presented as a rational function with an effective bound on its total degree. $\square$

Example 31.3.8: *Necessity of the exponential factor in Theorem 31.3.7.* Let $\lambda(Y)$ be the following formula of $\mathcal{L}(\text{ring}, \mathbb{F}_q)$:

$$(\forall X)[X^q = X \wedge Y = 1]$$

Obviously $\nu(\lambda, 1) = 1$ and $\nu(\lambda, k) = 0$ for each $k \geq 2$. Hence, $Z(\lambda, t) = e^t$. Thus, the exponential factor in (17) is necessary. $\square$

Exercises

1. For a positive integer k, let $A_1 = V(1 - kX)$ and $A_2 = \mathbb{A}^1 \smallsetminus A_1$. Show that $\mathbb{A}^1 = A_1 \cup A_2$ is a stratification of $\mathbb{A}^1$ into a disjoint union of normal $\mathbb{Z}$-basic sets.

Exercises

2. Let $A_1 = V(X^2 + 3)$ and $A_2 = \mathbb{A}^1 \smallsetminus A_1$. Prove that 2 is the minimal integer k such that A_i is $\mathbb{Z}[k^{-1}]$-normal, $i = 1, 2$.

3. Let $A_1 = V(X - 1)$ and $A_2 = \mathbb{A}^1 \smallsetminus V(2X - 2)$. Show that $\mathbb{A}^1 = A_1 \cup A_2$ is a stratification of $\mathbb{A}^1$ into disjoint union of normal $\mathbb{Z}$-basic sets. Observe that $A_1(\tilde{\mathbb{F}}_2) = \{1\}$ and $A_2(\tilde{\mathbb{F}}_2) = \emptyset$.

4. (a) Let $A = V(X^2 + Y^3 - pY)$ where p is an odd prime. Show that A is normal over $\mathbb{Z}$, but $V(X^2 + Y^3)$ is not normal over $\mathbb{F}_p$.

   Hint: Let (x, y) be a generic point of A over $\mathbb{F}_p$. Show that $\frac{x}{y}$ is not in $\mathbb{F}_p[x, y]$ but $\left(\frac{x}{y}\right)^2$ is. Also, note that A is normal over $\mathbb{Q}$ because it is nonsingular (Lemma 5.2.3).

   (b) Show that $\mathbb{Z}[A, \frac{1}{2(y-1)}, z]$ with $z^2 = y - 1$, is a Galois ring cover of $\mathbb{Z}[Z, \frac{1}{2(y-1)}]$. Note, however, that $\mathbb{F}_p[A, \frac{1}{2(y-1)}, z]$ is not a ring cover of $\mathbb{F}_p[A, \frac{1}{2(y-1)}]$ because the latter ring is not integrally closed.

5. Let K be a finite field and let K_k be the unique finite extension of K of degree k. For a sentence θ of $\mathcal{L}(\text{ring})$ let $\nu(\theta, k) = 1$ if θ is true in K_k and $\nu(\theta, k) = 0$ otherwise. For each positive integer m let $\chi_m(k) = 1$ if $m|k$ and $\chi_m(k) = 0$ otherwise.

   (a) Suppose $\nu(\theta, k) = \sum_{i=1}^{e} \frac{r_i}{s_i} m_i \chi_{m_i}(k)$, $k = 1, 2, 3, \ldots$, where r_i, s_i are relatively prime integers and m_i is a positive integer, $i = 1, \ldots, e$. Prove that

$$Z(\theta, t) = \sum_{k=1}^{\infty} \frac{\nu(\theta, k)}{k} t^k = \prod_{i=1}^{e} (t^{m_i} - 1)^{s_i/r_i}.$$

   Thus, $l = \text{lcm}(r_1, \ldots, r_e)$ is the minimal integer such that $Z(\theta, t)^l$ is a rational function.

   (b) Let θ be the sentence $(\exists X_1)[f(X_1) = 0] \leftrightarrow (\exists X_2)[g(X_2) = 0]$, where $f = f_1 \ldots f_r$ and $g = g_1 \ldots g_s$ with f_i, g_i irreducible polynomials in $K[X]$. Let $a_i = \deg(f_i)$, $i = 1, \ldots, r$ and $b_j = \deg(g_j)$, $j = 1, \ldots, s$. Prove that $\nu(\theta, k) = 1$ if and only if the following statement holds:

   there exists i such that $a_i \mid k \iff$ there exists j such that $b_j \mid k$.

   Conclude that $\nu(\theta, k)$ is fixed on congruence classes modulo $n = \text{lcm}(a_1, \ldots, a_r, b_1, \ldots, b_s)$.

   (c) Let θ be as in (b) and write $\nu(\theta, k)$ in the form of (a). Hint: Consider the functions

$$f_{\mathbf{a}}(k) = \prod_{i=1}^{r}(1 - \chi_{a_i}(k)) \quad \text{and} \quad f_{\mathbf{b}}(k) = \prod_{j=1}^{s}(1 - \chi_{b_j}(k)).$$

   Then show that $\nu(\theta, k) = (1 - f_{\mathbf{a}}(k))(1 - f_{\mathbf{b}}(k)) + f_{\mathbf{a}}(k) f_{\mathbf{b}}(k)$. Also, note that $\chi_c(k)\chi_d(k) = \chi_{\text{lcm}(c,d)}(k)$.

6. Let $\mathcal{A} = \langle \mathbb{A}^2, C_i/A_i, \mathrm{Con}(A_i, \mathcal{C}) \rangle_{i=1,2}$ be a Galois stratification over $\mathbb{F}_q$ where $A_1 = V(X^q - X - Y)$, $A_2 = \mathbb{A}^2 \smallsetminus A_1$, $C_i = \mathbb{F}_q[A_i]$, $i = 1, 2$, $\mathcal{C}$ is the family of finite cyclic groups, $\mathrm{Con}(A_1)$ consists of the trivial group and $\mathrm{Con}(A_2)$ is empty. Let θ_1 be $(\forall X)[\mathrm{Ar}(X,Y) \subseteq \mathrm{Con}(\mathcal{A})]$ and let θ_2 be $(\exists X)[\mathrm{Ar}(X,Y) \subseteq \mathrm{Con}(\mathcal{A})]$. Thus, in Section 31.3 take $m = 1$ and $n = 1$.

(a) Show that $Z(\theta_1, t) = e^t$ and $Z(\theta_2, t) = (1 - qt)^{-1/q}$.

(b) As in Proposition 31.2.3, compute a minimal positive integer l_i and a Galois stratification $\mathcal{B}_i$ of $\mathbb{A}^1$ such that for each $k \geq l_i$ and for each $b \in \mathbb{F}_{q^k}$, $\mathbb{F}_{q^k} \models \theta_i(b)$ if and only if $\mathrm{Ar}(\mathcal{B}_i, \mathbb{F}_{q^k}, b) \subseteq \mathrm{Con}(\mathcal{B}_i(\mathcal{C}))$, $i = 1, 2$. Let θ'_i be the Galois formula attached to $\mathcal{B}_i$, $i = 1, 2$. Prove that $Z(\theta_1, t) = e^t Z(\theta'_1, t)$ while $Z(\theta_2, t) = Z(\theta'_2, t)$.

7. Let Γ be a smooth curve which is defined over a finite field K. Use the notation of Section 31.3 and define the Zeta function of Γ over K to be $Z_\Gamma(t) = \exp\left(\sum_{k=1}^\infty \frac{N_r}{r} t^r\right)$. Now denote the function field of Γ over K by F and consider the zeta function $Z_{F/K}(t)$ as defined in Section 4.2. Prove that $Z_\Gamma(t) = Z_{F/K}(t)$. Hint: Use the identity $-\sum_{i=1}^{2g} \omega_i^k = N_k - (q^k + 1)$ ((5) of Section 4.5).

Notes

Kiefe proves Part (b) of Theorem 31.3.7 for a formula $\theta(\mathbf{Y})$ of $\mathcal{L}(\mathrm{ring}, \mathbb{F}_q)$ [Kiefe, p. 52]. The elimination of quantifiers step is based on the recursive methods which are introduced in Chapter 19, rather than on Galois stratification. In addition, the representation theory arguments that we use are replaced by combinatorial arguments.

One finds a remark on page 58 of [Kiefe] that the zeta function of a formula is a radical of a rational function. This is inaccurate as Example 31.3.8 shows.

Chapter 32. Problems of Field Arithmetic

32.1 Open Problems of the First Edition

The first edition of "Field Arithmetic" listed 22 open problems. Since the first edition was published fifteen of these problems were solved or partially solved. Here we list those problems and comment on the solutions whenever applicable.

PROBLEM 1:
(a) *Is $\mathbb{Q}_{\text{solv}}$ a PAC field?* ;
(b) *Is there a field K which is neither formally real nor PAC all of whose Henselian hulls are separably closed?*

Comment: A finite field K has no nontrivial valuations. Hence, the condition "all Henselians hulls of K are separably closed" is trivially fulfilled. In addition, K is neither formally real nor PAC (Proposition 11.1.1). Thus, K (trivially) satisfies the conditions of Problem 1(b).

Hrushovski's example (Proposition 11.7.8) supplies a nontrivial example for a field K that satisfies the conditions of Problem 1(b). Indeed, in that example K is a non-PAC field of positive characteristic such that K_{ins} is PAC. In particular, K is neither finite nor formally real. Let v be a valuation of K. Extend v to K_{ins} in the unique possible way. Then $K_{\text{ins}}K_v$ is the Henselian closure of K_{ins} at v. Hence, since K_{ins} is PAC, $K_v K_{\text{ins}}$ is separably closed (Corollary 11.5.5). In fact, $K_{\text{ins}}K_v = \tilde{K}$. Consequently, K_v is separably closed.

Finally, [Geyer-Jarden5, Thm. D and Remark 2.7(a)] constructs for each finite or global field K_0 an infinite regular extension K which is not formally real, each Henselian closure of K is separably closed, and neither K nor K_{ins} are PAC. See also Remark 11.5.11. □

PROBLEM 2: *Let $\sigma \in \text{Gal}(\mathbb{Q})$ for which $\tilde{\mathbb{Q}}(\sigma)$ is neither PAC nor formally real. Does $\tilde{\mathbb{Q}}(\sigma)$ admit a valuation with a non-algebraically closed completion?*

For a field K and $e \in \mathbb{N}$ let $S_e(K) = \{\sigma \in \text{Gal}(K)^e \mid K_s(\sigma) \text{ is PAC }\}$.

PROBLEM 3: *Let $E(t)$ be a rational function field of one variable over an uncountable field E. Is $S_e(E(t))$ nonmeasurable?*

Comment: Proposition 18.8.8(b) gives an affirmative answer to Problem 3 when E is algebraically closed or $E = K_0(T)$, K_0 is an arbitrary field, and T is an uncountable set of indeterminates. □

A field K has the **density property** with respect to a valuation w of $\tilde{K}$ if for each variety V defined over K, $V(K)$ is w-dense in $V(\tilde{K})$.

PROBLEM 4: *Does each PAC field K have the density property with respect to each valuation w of $\tilde{K}$?*

Comment: By Proposition 11.5.3, K is w-dense for every valuation w of $\tilde{K}$. This proves a restricted density property for the affine spaces $\mathbb{A}^n$. □

PROBLEM 5: *Give an example of a PAC field which contains no proper PAC subfield.*

Comment: Example 11.2.6 gives infinite extensions of $\mathbb{F}_p$ which contain no proper PAC subfields, yet these extensions are themselves PAC. □

PROBLEM 6: *Is every perfect PAC field C_1?*

Comment: See Sections 21.2 and 21.3 for discussion on C_i-fields. □

Call a field K ω-**free** if each finite embedding problem over K is solvable.

PROBLEM 7: *Is every PAC Hilbertian field K necessarily ω-free?*

Comment: Problem 7 was one of the main open problems of Field Arithmetic. It was affirmatively solved for the first time by Fried-Völklein in characteristic 0 and then by Pop in general. The general case was also solved by Haran-Jarden. See Example 24.8.5(b) for references. □

PROBLEM 8: *Let L and M be proper Galois extensions of a Hilbertian field K such that $L \cap M = K$. Is LM Hilbertian?*

Comment: By Corollary 13.8.4, if $L \not\subseteq M$ and $M \not\subseteq L$, then LM is Hilbertian. □

PROBLEM 9: *Let K be a Hilbertian field. Prove or disprove: There exist no proper Galois extensions L and M of K such that $L \cap M = K$ and $LM = K_s$.*

Comment: Corollary 13.8.4 affirms the statement of the problem. □

PROBLEM 10: *Let K be a field equipped with an infinite set S of inequivalent discrete valuations. Suppose that for each $a \in K$, $a \neq 0$ the set $\{v \in S \mid v(a) \neq 0\}$ is finite. Is K Hilbertian?*

PROBLEM 11: *Let R be a unique factorization domain with infinitely many primes. Is the quotient field of R Hilbertian?*

Comment: Example 15.5.8, initiated by Corvaja-Zannier, gives a unique factorization domain R with infinitely many primes such that $\text{Quot}(R)$ is non-Hilbertian. This answers both Problem 10 and Problem 11 negatively. □

PROBLEM 12 (Conjecture of Geyer-Jarden): *Let K be a field which is finitely generated over its prime field. Then for almost all $\sigma \in \text{Gal}(K)^e$ each Abelian variety A of dimension at least 2 defined over $\tilde{K}(\sigma)$ has these properties:*
(a) *If $e = 1$, then there exist infinitely many primes l such that $A(\tilde{K}(\sigma))$ has a point of order l.*

32.1 Open Problems of the First Edition

(b) If $e \geq 2$, then for only finitely many primes l, $A(\tilde{K}(\sigma))$ has a point of order l.

(c) If $e = 1$ and l is a prime, then $A(\tilde{K}(\sigma))$ has only finitely many points of an l-power order.

Comment: Problem 12 was stated as Conjecture 16.50 in [Fried-Jarden3]. Part (b) in characteristic 0 and Part (c) in general is proved in [Jacobson-Jarden2]. A weaker version of Part (a) is proved in [Geyer-Jarden6] when K is a number field. See Remark 18.11.3 for more details. □

Extend the language $\mathcal{L}(\text{ring})$ by a unary predicate symbol Σ to a language $\mathcal{L}(\text{ring}, \Sigma)$. For each $\sigma \in \text{Gal}(K)$ let $\langle \tilde{\mathbb{Q}}, \sigma \rangle$ be a system for $\mathcal{L}(\text{ring}, \Sigma)$ with a domain $\tilde{\mathbb{Q}}$ and with σ interpreting Σ.

PROBLEM 13: *Is the theory of all sentences θ of $\mathcal{L}(\text{ring}, \Sigma)$ which are true in $\langle \tilde{\mathbb{Q}}, \sigma \rangle$ for almost all $\sigma \in G(\mathbb{Q})$ decidable?*

Comment: The analogous problem where Σ is replaced by e unary predicate symbols $\Sigma_1, \ldots, \Sigma_e$ with $e \geq 2$ has a negative solution, that is, the corresponding theory is undecidable (Theorem 29.3.1).

PROBLEM 14: *Let $f > e \geq 2$. What are the Haar measures of the following sets:*

(1a) $\qquad \{(x_1, \ldots, x_e) \in (\hat{F}_e)^e \mid \langle x_1, \ldots, x_e \rangle \cong \hat{F}_e\};$ and

(1b) $\qquad \{(x_1, \ldots, x_f) \in (\hat{F}_e)^f \mid (\hat{F}_e : \langle x_1, \ldots, x_f \rangle) < \infty\}$?

Comment: By Proposition 26.1.7, the Haar measure of the set in (1a) is 1 (Lubotzky) and the Haar measure of the set in (1b) is zero (Kantor-Lubotzky). □

PROBLEM 15: *Let G be a profinite group of rank $\leq \aleph_0$. Prove or disprove: For each proper closed subgroup H of G of infinite index the set $\bigcup_{g \in G} H^g$ contains no open neighborhood of 1.*

PROBLEM 16: *Describe the universal Frattini cover of a non-Abelian finite simple group G.*

PROBLEM 17: *Denote the universal Frattini cover of a profinite group G by $\tilde{G}$ and let $E(G)$ be a smallest embedding cover of G.*
(a) *Is $E(G)$ projective whenever G is?*
(b) *Is $E(\tilde{G})$ isomorphic to the universal Frattini cover of $E(G)$?*

Comment: Proposition 24.7.3 gives an Example of Chatzidakis for a projective group G for which $E(G)$ is not projective. This refutes both (a) and (b). □

Algebraic extensions L and L' of a global field K are **Kronecker equivalent** if for almost all primes $\mathfrak{p}$ of K, $\mathfrak{p}$ has a prime divisor of relative degree 1 in L if and only if $\mathfrak{p}$ has a prime divisor of relative degree 1 in L'. Denote the set of all algebraic extensions L' of K which are Kronecker equivalent to L by $\mathcal{K}(L/K)$.

PROBLEM 18 (Conjecture of Jehne): *If L is a quadratic extension of a number field K, then L is the unique element of the class $\mathcal{K}(L/K)$.*

Comment: The conjecture was proved by Saxl. See Remark 21.5.7(c). □

PROBLEM 19 (Problem of Jehne): *Do there exist fields $K \subseteq L \subseteq M$ with K global, L/K finite separable, and M/K infinite separable such that M is Kronecker equivalent to L over K?*

Comment: Problem 19 is related to problem 18. See Notes to Section 21 for details. □

PROBLEM 20 (Problem of Davenport): *Let $f, g \in \mathbb{Z}[X]$ be nonconstant polynomials. Suppose that for almost all primes p*

$$\{f(x) \mid x \in \mathbb{F}_p\} = \{g(x) \mid x \in \mathbb{F}_p\}.$$

Are f and g necessarily strictly linearly related (i.e. there exist $a, b \in \mathbb{Q}$, $a \neq 0$ such that $g(X) = f(aX + b)$)?

Comment: Remark 21.6.1 supplies counter examples to Davenport's problem. □

For the concepts involved in the last two problems see Section 26.4 of [Fried-Jarden3].

PROBLEM 21: *For a formula θ of $\mathcal{L}(\mathrm{ring})$ give an effective computation of the value k_0 such that $H_k(\theta, t)$ is invariant for $k \geq k_0$.*

PROBLEM 22: *Is $H_k^*(\theta, t)$ invariant? If so, give an effective computation of the value k_0 such that $H_k^*(\theta, t)$ is invariant for $k \geq k_0$.*

Comment: Both Problems 21 and 22 were solved affirmatively in [Pas]. □

32.2 Open Problems of the Second Edition

The second edition of "Field Arithmetic" listed 34 open problems. Since that edition was published in 2005 five of these problems were solved or partially solved. Here we list those problems and comment on the solutions whenever applicable.

1. Does there exists a minimal PAC field which is not an algebraic extension of a finite field (Problem 11.2.7)?

2. Prove or disprove: For each non-PAC field K there exists a plane projective curve without K-rational points (Problem 11.2.10).

Comment: The statement is true. Thus, for a field K to be PAC it is necessary and sufficient that each plane projective curve defined over K has a K-rational point (Remark 11.2.10). □

32.2 Open Problems of the Second Edition

3. Let K be a PAC field, w a valuation of K, and V a variety over K. Is $V(K)$ w-dense in $V(\tilde{K})$? (Problem 11.5.4).

Comment: The question has an affirmative answer (Remark 11.5.4). □

4.
(a) Is $\mathbb{Q}_{\text{solv}}$ a PAC field?
(b) Does there exist an infinite field K of a finite transcendence degree over its prime field which is neither finite nor formally real nor PAC all of its Henselian closures are separably closed (Problem 11.5.9)?

5. Does there exist a finitely generated field extension E of $\mathbb{F}_p$ such that E is not PAC but E_{ins} is PAC (Problem 11.7.9)?

6. Let K be an infinite field and H a separable Hilbertian subset of K^r. Does there exist an absolutely irreducible polynomial $f \in K[T_1, \ldots, T_r, X]$ which is monic and separable in X such that $H_K(f) \subseteq H$ (Problem 13.1.5)?

7. Let A be a generalized Krull domain which is not a field.
(a) Is $A[[X]]$ a generalized Krull domain?
(b) Is $\text{Quot}(A[[X]])$ Hilbertian (Problem 15.5.9)?

8. Let K be a finite field and G a finite group. Suppose G is K-regular. Prove or disprove: There is an X-stable polynomial $h \in K[T, X]$ with $\text{Gal}(h(T, X), K(T)) \cong G$ (Problem 16.2.9)?

9. Is every finite p-group regular over $\mathbb{Q}$ (Remark 16.4.6)?

10. Does every finite p-group occurs as a Galois group over every Hilbertian field of characteristic 0 (Remark 16.4.6)?

By Lemmas 13.1.1 and 16.2.1, an affirmative answer to Problem 9 implies an affirmative answer to Problem 10.

11. Prove that A_6 is GAR over $\mathbb{Q}$ (Remark 16.9.5).

12. Prove that A_n, with $n \neq 2, 3, 6$, is GAR over every field K of characteristic 2 (Remark 16.9.5).

13. Following Remark 16.9.5 we ask the following questions:
(a) Is every finite non-Abelian simple group GAR over $\mathbb{Q}$?
(b) Is every finite non-Abelian simple group GAR over $\mathbb{Q}_{\text{solv}}$?
(c) Is every finite non-Abelian simple group GAR over every field K containing $\tilde{\mathbb{Q}}$?

By Definition 16.8.1, if a finite group G is GAR over a field K, then G is GAR over every separable algebraic extension of K. Thus, Problem 13(b) is easier than Problem 13(a).

14. Let K be a number field and S a finite number of prime ideals of O_K. Denote the compositum of all finite Galois extensions of K unramified outside S by K_S. Is the Galois group $\text{Gal}(K_S/K)$ finitely generated (Example 16.10.9(c))?

15. Let $\sigma \in \text{Gal}(\mathbb{Q})$ for which $\tilde{\mathbb{Q}}(\sigma)$ is neither PAC nor formally real. Does $\tilde{\mathbb{Q}}(\sigma)$ admit a valuation with a non-algebraically closed completion? (Problem 18.6.3)

16. Is the following generalization of the bottom theorem true: Let K be a Hilbertian field and e a positive integer. Then for almost all $\sigma \in \text{Gal}(K)^e$, $K_s(\sigma)$ is a finite extension of no proper subfield (Problem 18.7.8).

For a field K and $e \in \mathbb{N}$ let $S_e(K) = \{\sigma \in \text{Gal}(K)^e \mid K_s(\sigma) \text{ is PAC}\}$.

17. Let $E(t)$ be a rational function field of one variable over an uncountable field E. Is $S_e(E(t))$ nonmeasurable (Problem 18.8.10)?

For Problems 18–21 see Conjecture 18.11.2 and Remark 18.11.3:

18. Let K be an infinite finitely generated field and A an Abelian variety. Prove that for almost all $\sigma \in \text{Gal}(K)$ there exist infinitely many prime numbers l such that $A(\tilde{K}(\sigma))$ has a point of order l (Geyer-Jarden).

The following weaker version of Problem 18 is proved when K is a number field in [Geyer-Jarden6]:

19. Let K be an infinite finitely generated field and A an Abelian variety. Then K has a finite Galois extension L with the following property: for almost all $\sigma \in \text{Gal}(L)$ there exist infinitely many prime numbers l such that $A(\tilde{L}(\sigma))$ has a point of order l.

The analog of the following problem for characteristic 0 is proved in [Jacobson-Jarden2]:

20. Let K be an infinite finitely generated field extension of $\mathbb{F}_p$, A an Abelian variety over K, and $e \geq 2$. Then, for almost all $\sigma \in \text{Gal}(K)^e$ there exist only finitely many prime numbers l such that $A(\tilde{K}(\sigma))$ has a point of order l.

21. Is every perfect PAC field C_1 (Problem 21.2.5)?

Comment: The problem has an affirmative solution in characteristic 0 (Remark 21.3.7). □

22. Are there fields $K \subset L \subset M$ with K global, L/K finite separable, and M/K infinite separable such that $M_0 \in \mathcal{K}(L/K)$ for every intermediate field $L \subset M_0 \subset M$ of finite degree over K (Problem 21.5.8).

For the concepts appearing in the following problem see Section 21.6:

23.
(a) Are all exceptional pairs in $\mathbb{Q}[X]$ of the form $(h(X^8), h(16X^8))$ with $h \in \mathbb{Q}[X]$?
(b) For a given global field K find all exceptional pairs (f, g) with $f, g \in K[X]$ (Problem 21.6.2).

32.2 Open Problems of the Second Edition

24. Characterize the class of absolute Galois groups among all profinite groups by means of group theoretic and topological properties (Remark 23.1.5).

25. Describe the universal p-Frattini cover of $\text{PSL}(2, \mathbb{Z}_p)$ (hence of $\text{PSL}(2, \mathbb{F}_p)$) (Problem 22.14.6).

26. Let L/K be a finite Galois extension of fields with Galois group G and t an indeterminate. Does $K(t)$ has a Galois extension F with the following properties:
    (a) $\text{Gal}(F/K(t)) \cong G$.
    (b) F/K is a regular extension.
    (c) There is a prime divisor $\mathfrak{p}$ of F/K with decomposition field $K(t)$ and residue field L (Section 24.2).

27. Prove or disprove: Every field K with an affirmative solution for the regular inverse Galois problem has an affirmative solution for the Beckmann-Black problem (Problem 24.2.3).

28. Prove that $\text{Gal}(\mathbb{Q}_{\text{solv}}) \cong \hat{F}_\omega$ (Example 24.8.5)

29. Let K be a Hilbertian field with $\text{Gal}(K)$ projective. Prove that each finite embedding problem for $\text{Gal}(K)$ is solvable (Conjecture 24.8.6).

30. Give a pure group theoretic proof for the uniqueness of the smallest embedding cover of a profinite group (Notes to Chapter 24).

31. Let $\mathcal{C}$ be a Melnikov formation of finite groups, $e \geq 2$ an integer, and M, M_1, M_2 closed subgroups of $\hat{F}_e(\mathcal{C})$. Suppose M is pro-$\mathcal{C}$ of infinite index, $M_1, M_2 \triangleleft \hat{F}_e(\mathcal{C})$, $M_1 \cap M_2 \leq M$, but $M_1, M_2 \not\leq M$. Is $M \cong \hat{F}_\omega(\mathcal{C})$ (Problem 25.4.9).

Comment: The problem has an affirmative solution (Remark 25.4.10). ☐

32. For $i = 1, 2$ let $\varphi_i \colon \hat{F}_\omega \to \hat{F}_\omega$ be a Melnikov cover. Does there exist an isomorphism $\theta \colon \hat{F}_\omega \to \hat{F}_\omega$ such that $\varphi_2 \circ \theta = \varphi_1$ (Remark 25.9.13)?

33. Let I be the set of all sentences in the language of graphs which hold in infinitely many finite graphs. Let FG be the theory of finite graphs. Does there exist a recursive set of sentences in the language of graphs which lie between FG and I (Remark 28.5.5)?

Comment: Remark 28.5.5 settles Problem 33 by giving a recursive set of sentences in the language of graphs between FG and I. ☐

34. Let $T(\mathbb{Q}, 1)$ be the set of all sentences in $\mathcal{L}(\text{ring}, \mathbb{Q}, \Sigma)$ which hold in $(\tilde{\mathbb{Q}}, \sigma)$ for almost all $\sigma \in \text{Gal}(\mathbb{Q})$. Is $T(\mathbb{Q}, 1)$ decidable (Problem 29.3.2)?

32.3 Open Problems

We list all open problems of the present edition. Some are left over from the second edition, others are reformulation of solved problems of the second edition, and still others are completely new:

1. Does there exists a minimal PAC field which is not an algebraic extension of a finite field (Problem 11.2.7)?

2.
   (a) Is $\mathbb{Q}_{\text{solv}}$ a PAC field?
   (b) Does there exist an infinite field K of a finite transcendence degree over its prime field which is neither finite nor formally real nor PAC all of its Henselian closures are separably closed (Problem 11.5.9)?

3. Does there exist a finitely generated field extension E of $\mathbb{F}_p$ such that E is not PAC but E_{ins} is PAC (Problem 11.7.9)?

4. Let K be an infinite field and H a separable Hilbertian subset of K^r. Does there exist an absolutely irreducible polynomial $f \in K[T_1, \ldots, T_r, X]$ which is monic and separable in X such that $H_K(f) \subseteq H$ (Dèbes and Haran — Problem 13.1.5)?

5. Let A be a generalized Krull domain which is not a field.
   (a) Is $A[[X]]$ a generalized Krull domain?
   (b) Is $\text{Quot}(A[[X]])$ Hilbertian (Problem 15.5.9)?

6. Let K be a finite field and G a finite group. Suppose G is K-regular. Prove or disprove: There is an X-stable polynomial $h \in K[T, X]$ with $\text{Gal}(h(T, X), K(T)) \cong G$ (Problem 16.2.9)?

7. Is every finite p-group regular over $\mathbb{Q}$ (Remark 16.4.6)?

8. Does every finite p-group occurs as a Galois group over every Hilbertian field of characteristic 0 (Remark 16.4.6)?

By Lemmas 13.1.1 and 16.2.1, an affirmative answer to Problem 7 implies an affirmative answer to Problem 8.

9. Prove that A_6 is GAR over $\mathbb{Q}$ (Remark 16.9.5).

10. Prove that A_n, with $n \neq 2, 3, 6$, is GAR over every field K of characteristic 2 (Remark 16.9.5).

11. Following Remark 16.9.5 we ask the following questions:
    (a) Is every finite non-Abelian simple group GAR over $\mathbb{Q}$?
    (b) Is every finite non-Abelian simple group GAR over $\mathbb{Q}_{\text{solv}}$?
    (c) Is every finite non-Abelian simple group GAR over every field K containing $\tilde{\mathbb{Q}}$?

By Definition 16.8.1, if a finite group G is GAR over a field K, then G is GAR over every separable algebraic extension of K. Thus, Problem 11(b) is easier than Problem 11(a).

32.3 Open Problems

12. Let K be a number field and S a finite number of prime ideals of O_K. Denote the compositum of all finite Galois extensions of K unramified outside S by K_S. Is the Galois group $\mathrm{Gal}(K_S/K)$ finitely generated (Shafarevich — Example 16.10.9(c))?

13. Let $\sigma \in \mathrm{Gal}(\mathbb{Q})$ for which $\tilde{\mathbb{Q}}(\sigma)$ is neither PAC nor formally real. Does $\tilde{\mathbb{Q}}(\sigma)$ admit a valuation with a non-algebraically closed completion? (Problem 18.6.3)

14. Is the following generalization of the bottom theorem true: Let K be a Hilbertian field and e a positive integer. Then for almost all $\sigma \in \mathrm{Gal}(K)^e$, $K_s(\sigma)$ is a finite extension of no proper subfield (Problem 18.7.8).

15. Let K be a Hilbertian field. Is K_{alt} PAC (Fried-Völklein — Problem 18.10.6)?

For a field K and $e \in \mathbb{N}$ let $S_e(K) = \{\sigma \in \mathrm{Gal}(K)^e \mid K_s(\sigma) \text{ is PAC}\}$.

16. Let $E(t)$ be a rational function field of one variable over an uncountable field E. Is $S_e(E(t))$ nonmeasurable (Problem 18.8.10)?

For Problems 17–20 see Conjecture 18.11.2 and Remark 18.11.3:

17. Let K be an infinite finitely generated field and A an Abelian variety. Prove that for almost all $\sigma \in \mathrm{Gal}(K)$ there exist infinitely many prime numbers l such that $A(\tilde{K}(\sigma))$ has a point of order l (Geyer-Jarden).

The following weaker version of Problem 17 is proved when K is a number field in [Geyer-Jarden6]:

18. Let K be an infinite finitely generated field and A an Abelian variety. Then K has a finite Galois extension L with the following property: for almost all $\sigma \in \mathrm{Gal}(L)$ there exist infinitely many prime numbers l such that $A(\tilde{L}(\sigma))$ has a point of order l.

The analog of the following problem for characteristic 0 is proved in [Jacobson-Jarden2]:

19. Let K be an infinite finitely generated field extension of $\mathbb{F}_p$, A an Abelian variety over K, and $e \geq 2$. Then, for almost all $\sigma \in \mathrm{Gal}(K)^e$ there exist only finitely many prime numbers l such that $A(\tilde{K}(\sigma))$ has a point of order l.

20. Is every perfect PAC field of positive characteristic C_1 (Ax — Problem 21.2.5)?

21. Are there fields $K \subset L \subset M$ with K global, L/K finite separable, and M/K infinite separable such that $M_0 \in \mathcal{K}(L/K)$ for every intermediate field $L \subset M_0 \subset M$ of finite degree over K (Jehne — Problem 21.5.8).

For the concepts appearing in the following problem see Section 21.6:

22.
(a) Are all exceptional pairs in $\mathbb{Q}[X]$ of the form $(h(X^8), h(16X^8))$ with $h \in \mathbb{Q}[X]$ (Müller)?
(b) For a given global field K find all exceptional pairs (f, g) with $f, g \in K[X]$ (Davenport — Problem 21.6.2).

23. Characterize the class of absolute Galois groups among all profinite groups by means of group theoretic and topological properties (Remark 23.1.5).

24. Describe the universal p-Frattini cover of $\mathrm{PSL}(2, \mathbb{Z}_p)$ (hence of $\mathrm{PSL}(2, \mathbb{F}_p)$) (Problem 22.14.6).

25. Let L/K be a finite Galois extension of fields with Galois group G and t an indeterminate. Does $K(t)$ has a Galois extension F with the following properties:
(a) $\mathrm{Gal}(F/K(t)) \cong G$.
(b) F/K is a regular extension.
(c) There is a prime divisor $\mathfrak{p}$ of F/K with decomposition field $K(t)$ and residue field L (Beckmann-Black — Section 24.2).

26. Prove or disprove: Every field K with an affirmative solution for the regular inverse Galois problem has an affirmative solution for the Beckmann-Black problem (Problem 24.2.3).

27. Prove that $\mathrm{Gal}(\mathbb{Q}_{\mathrm{solv}}) \cong \hat{F}_\omega$ (Iwasawa's Conjecture — Example 24.8.5)

28. Let K be a Hilbertian field with $\mathrm{Gal}(K)$ projective. Prove that each finite embedding problem for $\mathrm{Gal}(K)$ is solvable (Fried-Völklein — Conjecture 24.8.6).

29. Give a pure group theoretic proof for the uniqueness of the smallest embedding cover of a profinite group (Notes to Chapter 24).

30. For $i = 1, 2$ let $\varphi_i \colon \hat{F}_\omega \to \hat{F}_\omega$ be a Melnikov cover. Does there exist an isomorphism $\theta \colon \hat{F}_\omega \to \hat{F}_\omega$ such that $\varphi_2 \circ \theta = \varphi_1$ (Remark 25.9.13)?

31. Let $T(\mathbb{Q}, 1)$ be the set of all sentences in $\mathcal{L}(\mathrm{ring}, \mathbb{Q}, \Sigma)$ which hold in $(\tilde{\mathbb{Q}}, \sigma)$ for almost all $\sigma \in \mathrm{Gal}(\mathbb{Q})$. Is $T(\mathbb{Q}, 1)$ decidable (Problem 29.3.2)?

References

S. S. Abhyankar
1. *Coverings of algebraic curves*, American Journal of Mathematics **79** (1957), 825–856.

A. Adler and C. Kiefe
1. *Pseudofinite fields, procyclic fields and model completion*, Pacific Journal of Mathematics **62** (1976), 305–309.

J. Kr. Arason, B. Fein, M. Schacher, and J. Sonn
1. *cyclic extensions of $K(\sqrt{-1})/K$*, Transactions of the American Mathematical Society **313** (1989), 843-851.

E. Artin
1. *Über eine neue Art von L–Reihen*, Abhandlungen aus dem mathematischen Seminar der Universität Hamburg **3** (1924), 89–108; Collected Papers, Addison–Wesley (1965), 105–124.
2. *Beweis des allgemeinen Reziprozitätsgesetzes*, Abhandlungen aus dem mathematischen Seminar der Universität Hamburg **5** (1927), 353–363; Collected Papers, Addison–Wesley (1965), 131–141.
3. *Algebraic Numbers and Algebraic Functions*, Gordon and Breach, New York, 1967.

M. Aschbacher and R. Guralnick
1. *Some applications of the first cohomology group*, Journal of Algebra **90** (1984), 446-460.

J. Ax
1. *Solving diophantine problems modulo every prime*, Annals of Mathematics **85** (1967), 161–183.
2. *The elementary theory of finite fields*, Annals of Mathematics **88** (1968), 239–271.
3. *Injective endomorphisms of varieties and schemes*, Pacific Journal of Mathematics **31** (1969), 1–7.

J. Ax and S. Kochen
1. *Diophantine problems over local fields I*, American Journal of Mathematics **87** (1965), 605–630.
2. *Diophantine problems over local fields II*, American Journal of Mathematics **87** (1965), 631–648.
3. *Diophantine problems over local fields III*, Annals of Mathematics **83** (1966), 437–456.

L. Bary-Soroker
1. *Diamond theorem for a finitely generated free profinite group*, Mathematische Annalen **336** (2006), 949–961.

J. L. Bell and A. B. Slomson
1. *Models and ultraproducts: an introduction*, North–Holland, Amsterdam, American Elsevier, New York, 1974.

E. Binz, J. Neukirch, and G. H. Wenzel
1. *A subgroup theorem for free products of profinite groups*, Journal of Algebra

19 (1971), 104–109.

M. Bôcher
1. *Introduction to Higher Algebra*, Macmillan, New York, 1947.

E. Bombieri
1. *Counting points on curves over finite fields, (d'apres S. A. Stepanov)*, Seminaire Bourbaki 1972/73, Exp. **430**, Lecture Notes in Mathematics **383**, pp. 234–241, Springer-Verlag, Berlin, 1974.
2. *On exponential sums in finite fields II*, Inventiones mathematicae 47 (1978), 29–39.

Z. I. Borevich and I. R. Shafarevich
1. *Number Theory*, Academic Press, New York, 1966.

N. Bourbaki
1. *Algebra I*, Hermann, Paris, 1974.
2. *Commutative Algebra, Chapters 1–7*, Springer, Berlin, 1989.

A. Brandis
1. *Verschränkte Homomorphismen endlicher Gruppen*, Mathematische Zeitschrift **162** (1978), 205–217.

D. Brink
1. *On alternating and symmetric groups as Galois groups*, Israel Journal of Mathematics (2004).

W. Burnside
1. *On simply transitive groups of prime degree*, Quarterly Journal of Mathematics **37** (1906), 215–221.
2. *Theory of groups of finite order*, 2nd edition, Dover Publications, New York, 1955.

J. W. S. Cassels and A. Fröhlich
1. *Algebraic Number Theory*, Academic Press, London, 1967.

Z. M. Chatzidakis
1. *Model theory of profinite groups*, Dissertation, Yale University, 1984.
2. *Some properties of the absolute Galois group of a Hilbertian field*, Israel Journal of Mathematics **55** (1986), 173–183.
3. *A projective profinite group whose smallest embedding cover is not projective*, Israel Journal of Mathematics **85** (1994), 1–9.
4. *Model theory of profinite groups having the Iwasawa property*, Illinois Journal of Mathematics **42** (1998), 70–96.

G. Cherlin, L. v. d. Dries, and A. Macintyre
1. *The elementary theory of regularly closed fields*, manuscript, 1980.

G. Cherlin and M. Jarden
1. *Undecidability of some elementary theories over PAC fields*, Annals of Pure and Applied Logic **30** (1986), 137–163.

C. Chevalley
1. *Demonstration d'une hypothése de M. Artin*, Abhandlungen aus dem Mathematischen Seminar Hamburg **11** (1936), 73–75.
2. *Introduction to the theory of algebraic functions of one variable*, Mathematical Surveys VI, AMS, Providence, 1951.

S. D. Cohen and M. Fried
   Lenstra's proof of the Carlitz-Wan conjecture on exceptional polynomials: An elementary version. Finite Fields and their Applications **1** (1995), 372–375.

P. M. Cohn
1. *Algebra, Vol. 2*, 2nd Edition, John Wiley & Sons, Chicoster, 1989.

J. -L. Colliot-Thélène
1. *Rational connectedness and Galois covers of the projective line*, Annals of Mathematics **151** (2000), 359-373.

P. Corvaja and U. Zannier
1. *Values of rational functions on non-Hilbertian fields and a question of Weissauer*, Israel Journal of Mathematics **105** (1998), 323–335.

J. Cossey, O. H. Kegel, and L. G. Kovács
1. *Maximal Frattini extensions*, Archiv der Mathematik **35** (1980), 210–217.

P. Dèbes
1. *Galois covers with prescribed fibers: the Beckmann-Black Problem*, Annali della Scuola Normale Superiore di Pisa, Classe di Scienze, Serie IV, **28** (1999), 273–286.

P. Dèbes and D. Haran
1. *Almost hilbertian fields*, Acta Arithmetica **88** (1999), 269–287.

P. Deligne
1. *La conjecture de Weil I*, IHES Publications Mathematiques **43**, (1974), 273–307.

J. Denef, M. Jarden, and D. J. Lewis
1. *On Ax–fields which are C_i*, Quarterly Journal of Mathematics Oxford (2) **34** (1983), 21–36.

J. Denef and F. Loeser
1. *Definable sets, motives and P-adic integrals*, Journal of the American Mathematical Society **14** (2001), 429–469.

M. Deuring
1. *Über den Tchebotareffschen Dichtigkeitssatz*, Mathematische Annalen **110** (1934), 414–415.
2. *Algebren*, Ergebnisse der Mathematik und ihrer Grenzgebiete **4**, Springer, Berlin, 1935.
3. *Lectures on the Theory of Algebraic Functions of One Variable*, Lecture Notes in Mathematics **314**, Springer-Verlag, Berlin, 1973.

J. D. Dixon
1. *The probability of generating the symmetric group*, Mathematische Zeitschrift **110** (1969), 199–205.

J. D. Dixon, M. P. F. du Sautoy, A. Mann, and D. Segal
1. *Analytic pro-p Groups*, Second Edition, Cambridge Studies in Advanced Mathematics **61**, Cambridge University Press, Cambridge, 1999.

J. D. Dixon, L. Pyber, A. Seress, and A. Shalev
1. *Residual properties of free groups and probabilistic methods*, Journal für die reine und angewandte Mathematik **556** (2003) 159-172.

A. Douady
1. *Cohomologie des groupes compacts totalement discontinus*, Séminaire Bourbaki 1959–1960, Exposé 189.

L. v. d. Dries and K. Schmidt
1. *Bounds in the theory of polynomial rings over fields. A nonstandard approach*, Inventiones mathematicae **76** (1984), 77–91.

L. v. d. Dries and R. L. Smith
1. *Decidable regularly closed fields of algebraic numbers*, Journal of Symboloic Logic **50** (1985), 468–475.

V. G. Drinfeld and S. Vlâdut
1. *Number of points on an algebraic curve*, Functional analysis and its applications **17** (1983), 68–69 (Russian).

J.-L. Duret
1. *Les corps faiblement algébraiquement clos non separablement clos ont la propriété d'indépendence*, in Model Theory of Algebra and Arithmetic, Pacholski et. al. editors, Lecture Notes in Mathematics **834** Springer-Verlag, Berlin, 1980.

D. Dwork
1. *On the rationality of the zeta function of an algebraic variety*, American Journal of Mathematics **82** (1960), 631–648.
2. *On the zeta function of a hypersurface III*, Annals of Mathematics **83** (1966), 457–519.

I. Efrat
1. *A Hasse principle for function fields over PAC fields*, Israel Journal of Mathematics **122** (2001), 43-60.

M. Eichler
1. *Zum Hilbertschen Irreduzibilitätssatz*, Mathematische Annalen **116** (1939), 743–748.

Ju. L. Ershov
1. *Fields with a solvable theory*, Soviet Mathematics Doklady **8** (1967), 575–576.
2. *Profinite groups*, Algebra i Logica (Russian) **19** (1980), 552–565; Algebra and Logic (English translation) **19** (1980), 357–366.
3. *Regularly closed fields*, Soviet Mathematics Doklady **31** (1980), 510–512.
4. *Undecidability of regularly closed fields*, Algebra and logic **20** (1981), 257–260.
5. *A talk in a conference in the theory of models*, Oberwolfach, January 1982.
6. *A question of Jarden and Shelah*, Algebra and Logic **28** (1990), 419–420.
7. *Multi-Valued Fields*, Kluwer, 2001.

Ju. L. Ershov and M. Fried
1. *Frattini covers and projective groups without the extension property*, Mathematische Annalen **253** (1980), 233–239.

Ju. L. Ershov, I. A. Lavrov, A.D. Taimanov, and M. A. Taitslin
1. *Elementary theories*, Russian Mathematical Surveys **20** (1965), 35–105.

G. Faltings, G. Wüstholz, et al.
1. *Rational Points*, Seminar Bonn/Wuppertal 1983/84, Vieweg, Braunschweig 1984.

W. Franz
1. *Untersuchungen zum Hilbertschen Irreduzibilitätssatz*, Mathematische Zeitschrift **33** (1931), 275–193.

G. Frey
1. *Pseudo algebraically closed fields with non–archimedian real valuations*, Journal of Algebra **26** (1973), 202–207.

G. Frey and M. Jarden
1. *Approximation theory and the rank of abelian varieties over large algebraic fields*, Proceedings of the London Mathematical Society (3) **28** (1974), 112–128.

M. Fried
1. *On a conjecture of Schur*, Michigan Mathematical Journal **17** (1970), 41-55.
2. *On the Diophantine equation $f(y) - x = 0$*, Acta Arithmetica **19** *(1971)*, 79–87.
3. *A review to Jarden's paper "Elementary statements over algebraic fields"*, Mathematical Reviews **46** (1973), #1975.
4. *The field of definition of function fields and a problem in the reducibility of polynomials in two variables*, Illinois Journal of Mathematics **17** (1973), 128–146.
5. *On Hilbert's irreducibility theorem*, Journal of Number Theory **6** (1974), 211–231.
6. *On a theorem of MacCluer*, Acta Arithmetica **25** (1974), 121–126.
7. *Fields of definition of function fields and Hurwitz families – Groups as Galois groups*, Communications in Algebra **5** (1) (1977), 17–82.
8. *Exposition on arithmetic — group theoretic connection via Riemann's existence theorem*, Proceeding of the Santa Cruz conference on finite groups **37** (1980), AMS, 571–602.
9. *The nonregular analogue of Tchebotarev's theorem*, Pacific Journal of Mathematics **112** (1984), 303–311.
10. *On the Sprindžuk–Weissauer approach to universal Hilbert subsets*, Israel Journal of Mathematics **51** (1985), 347–363.
11. *Variables Separated Polynomials and Moduli Spaces*, Number Theory in Progress, eds. K.Gyory, H.Iwaniec, J.Urbanowicz, proceedings of the Schinzel Festschrift, Summer 1997 Zakopane, Walter de Gruyter, Berlin-New York (Feb. 1999), 169–228.
12. *Toward a general theory of diophantine problems with applications to p–adic fields*, unpublished preprint.
13. *Introduction to modular towers*, Contemporary Mathematics **186** (1995), 111-171.

M. D. Fried, R. Guralnick, and J. Saxl
1. *Schur covers and Carlitz's conjecture*, Israel Journal of Mathematics **82** (1993), 157-225.

M. Fried, D. Haran, and M. Jarden
1. *Galois stratification over Frobenius fields*, Advances in Mathematics **51** (1984), 1–35.

M. Fried and M. Jarden
1. *Stable extensions and fields with the global density property*, Canadian Journal of Mathematics **28** (1976), 774–787.

2. *Diophantine properties of subfields of* $\tilde{\mathbb{Q}}$, American Journal of Mathematics **100** (1978), 653–666.
3. *Field Arithmetic*, Ergebnisse der Mathematik III **11** (First Edition), Springer-Verlag, Heidelberg, 1986.
4. *On Σ-Hilbertian fields*, Pacific Journal of Mathematics **185** (1998), 307–313.
5. M. D. Fried and M. Jarden, *Field Arithmetic, Second edition, revised and enlarged by Moshe Jarden*, Ergebnisse der Mathematik (3) **11**, Springer, Heidelberg, 2005.

M. Fried and R. E. MacCrae
1. *On the invariance of chains of fields*, Illinois Journal of Math. **13** (1969), 165–171.

M. Fried and G. Sacerdote
1. *Solving diophantine problems over all residue class fields of a number field and all finite fields*, Annals of Mathematics **104** (1976), 203–233.

M. D. Fried and H. Völklein
1. *The inverse Galois problem and rational points on moduli spaces*, Mathematische Annalen **290** (1991), 771–800.
2. *The embedding problem over a Hilbertian PAC-field*, Annals of Mathematics **135** (1992), 469–481.

G. Frobenius
1. *Über Beziehungen zwischen den Primidealen eines algebraischen Körpers und den Substitutionen seiner Gruppe*, Sitzbericht preussische Akademie (1896), 689–703.

A. Fröhlich and J. C. Shepherdson
1. *Effective procedures in field theory*, Philosophical Transactions of the Royal Society of London **248** (1956), 407–432.

W. Gaschütz
1. *Zu einem von B. H. Neumann gestellten Problem*, Mathematische Nachrichten **14** (1956) 249–252.

W.-D. Geyer
1. *Galois groups of intersections of local fields*, Israel Journal of Mathematics **30** (1978), 382–396.
2. *Algebraic function fields of one variable over finite fields are stable*, Israel Journal of Mathematics **68** (1989), 102–108.

W.-D. Geyer and C. Jensen
1. *Prodihedral groups as Galois groups over number fields*, Journal of Number Theory **60** (1996), 332–372.
2. *Embeddability of quadratic extensions in cyclic extensions*, Forum Mathematicum **19** (2007), 707–725.

W.-D. Geyer and M. Jarden
1. *Torsion points of elliptic curves over large algebraic extensions of finitely generated fields*, Israel Journal of Mathematics **31** (1978), 257–297.
2. *On the normalizer of finitely generated subgroups of absolute Galois groups of uncountable Hilbertian fields of characteristic 0*, Israel Journal of Mathematics **63** (1988), 323–334.
3. *On stable fields in positive characteristic*, Geometria Dedicatae **29** (1989), 335–375.

4. *Bounded realization of l-groups over global fields*, Nagoya Mathematical Journal **150** (1998), 13–62.
5. *Non PAC fields whose Henselian closures are separably closed*, Mathematical Research Letters **8** (2001), 509–519.
6. *Torsion of Abelian varieties over large algebraic fields*, Finite Field Theory and its Applications, Dedicata **36** (1990), 67–87.

P. C. Gilmore and A. Robinson
1. *Metha–Mathematical considerations on the relative irreducibility of polynomials*, Canadian Journal of Mathematics **7** (1955), 483–489.

K. Gödel
1. *Die Vollständigkeit der Axiome des logischen Funktionenkalküls*, Monatshefte für Mathematik und Physik **37** (1930), 349–360.

L. J. Goldstein
1. *Analytic number theory*, Prentice–Hall, Englewood Cliffs, 1971.

E. S. Golod and I. R. Shafarevich
1. *On class field towers*, American Mathematical Society Translations **48** (1965), 91–102.

V. D. Goppa
1. *Codes on algebraic curves*, Soviet Mathematics Doklady **24** (1981), 170–172.

M. Greenleaf
1. *Irreducible subvarieties and rational points*, American Journal of Mathematics **87** (1965), 25–31.

A. Grothendieck
1. *Revêtement étales et groupe fondamental (SGA 1)*, Lecture Notes in Mathematics **224**, Springer, Berlin, 1971.

K. W. Gruenberg
1. *Projective profinite groups*, Journal of London Mathematical Society **42** (1967), 155–165.

R. M. Guralnick and P. Müller
1. *Exceptional polynomials of affine type*, Journal of Algebra **194** (1997), 429–454.

M. Hall
1. *The Theory of Groups*, Macmillan, New York, 1959.

P. R. Halmos
1. *Measure Theory*, Van Nostrand, Princeton, 1968.

D. Haran
1. *Non–free torsion free profinite groups with open free subgroups*, Israel Journal of Mathematics **50** (1985), 350–352.
2. *The bottom theorem*, Archiv der Mathematik **45** (1985), 229–231.
3. *On closed subgroups of free products of profinite groups*, Proceedings of the London Mathematical Society **55** (1987), 266–289.
4. *Hilbertian fields under separable algebraic extensions*, Inventiones Mathematicae **137** (1999), 113–126.
5. *Free subgroups of free profinite groups*, Journal of Group Theory **2** (1999), 307–317.

D. Haran and M. Jarden
1. *Bounded statements in the theory of algebraically closed fields with distinguished automorphisms*, Journal für die reine und angewandte Mathematik **337** (1982), 1–17.
2. *The absolute Galois group of a pseudo real closed field*, Annali della Scuola Normale Superiore — Pisa, Serie IV, **12** (1985), 449–489.
3. *The absolute Galois group of a pseudo real closed algebraic field*, Pacific Journal of Mathematics **123** (1986), 55–69.
4. *The absolute Galois group of a pseudo p–adically closed field*, Journal für die reine und angewandte Mathematik **383** (1988), 147–206.
5. *Compositum of Galois extensions of Hilbertian fields*, Annales Scientifiques de l'Ecole Normale Superieure (4) **24** (1991), 739–748.
6. *Regular split embedding problems over complete valued fields*, Forum Mathematicum **10** (1998), 329–351.
7. *Regular lifting of covers over ample fields*, manuscript, Tel Aviv, 2000; http://www.tau.ac.il/~haran/Papers/#31.

D. Haran and L. Lauwers
1. *Galois stratification over e-fold ordered Frobenius fields*, Israel Journal of Mathematics **85** (1994), 169–197.

D. Haran and A. Lubotzky
1. *Embedding covers and the theory of Frobenius fields*, Israel Journal of Mathematics **41** (1982), 181–202.

D. Haran and H. Völklein
1. *Galois groups over complete valued fields*, Israel Journal of Mathematics, **93** (1996), 9–27.

D. Harbater
1. *Galois coverings of the arithmetic line*, in Number Theory — New York 1984–85, ed. by D.V. and G.V Chudnovsky, Lecture Notes in Mathematics **1240**, Springer, Berlin, 1987, pp. 165–195.
2. *Abhyankar's conjecture on Galois groups over curves*, Inventiones Mathematicae **117** (1994), 1–25.

R. Hartshorne
1. *Algebraic Geometry*, Springer-Verlag, New York, 1977.

L. Henkin
1. *The completeness of the first order functional calculus*, The Journal of Symbolic Logic **14** (1949), 159–166.

H. Hermes
1. *Enumerability, Decidability, Computability, An Introduction to the Theory of Recursive Functions*, Second revised Edition, Springer-Verlag, New York, 1969.

D. Hilbert
1. *Über die Irreduzibilität ganzer rationaler Funktionen mit ganzzahligen Koeffizienten*, Journal für die reine und angewandte Mathematik **110** (1892), 104–129.

J. Hirschfeld and W. H. Wheeler
1. *Forcing, Arithmetic, Division rings*, Lecture Notes in Mathematics **454**, Springer-Verlag, Berlin, 1975.

J. G. Hocking and G. S. Young
1. *Topology*, Addison-Wesley, Reading, 1961.

E. Hrushovski
1. *A non-PAC field whose maximal purely inseparable extension is PAC*, Israel Journal of Mathematics **85** (1994), 199–202.

B. Huppert
1. *Endliche Gruppen I*, Springer-Verlag, Berlin, 1967.

Y. Ihara
1. *Some remarks on the number of rational points of algebraic curves over finite fields*, Journal of the Faculty of Sciences Tokyo **28** (1981), 721–724.

M. Ikeda
1. *Zur Existenz eigentlicher galoisscher Körper beim Einbettungsproblem für galoissche Algebren*, Abhandlungen aus dem mathematischen Seminar Hamburg **24** (1960), 126–131.

E. Inaba
1. *Über den Hilbertschen Irreduzibilitätssatz*, Japanese Journal of Mathematics **19** (1944), 1–25.

K. Ireland and M. Rosen
1. *A classical introduction to modern number theory*, Graduate texts in Mathematics, Springer-Verlag, New–York, 1972.

K. Iwasawa
1. *On solvable extensions of algebraic number fields*, Annals of Mathematics **58** (1953), 548–572.

N. Jacobson
1. *Finite Dimensional Division Algebras over Fields*, Springer-Verlag, Berlin, 1996.

M. Jacobson and M. Jarden
1. *On torsion of abelian varieties over large algebraic extensions of finitely generated fields*, Mathematika **31** *(1984), 110–116.*
2. *Finiteness theorems for torsion of abelian varieties over large algebraic fields*, Acta Arithmetica, **XCVIII.1** (2001), 15–31.

U. Jannsen
1. *Über Galoisgruppen lokaler Körper*, Inventiones mathematicae **70** (1982), 53–69.

G. J. Janusz
1. *Algebraic Number fields*, Academic Press, New York, 1973.

M. Jarden
1. *Rational points on algebraic varieties over large number fields*, Bulletin of AMS **75** (1969), 603–606.
2. *Elementary statements over large algebraic fields*, Transactions of AMS **164** (1972), 67–91.
3. *Algebraic extensions of finite corank of Hilbertian fields*, Israel Journal of Mathematics **18** (1974), 279–307.
4. *Roots of unity over large algebraic fields*, Mathematische Annalen **213** (1975), 109–127.
5. *The elementary theory of ω–free Ax fields*, Inventiones mathematicae **38**

(1976), 187–206.
6. *An analogue of Artin–Schreier theorem*, Mathematische Annalen **242** (1979), 193–200.
7. *Torsion in linear algebraic groups over large algebraic fields*, Archiv der Mathematik **32** (1979), 445–451.
8. *Transfer principles for finite and p–adic fields*, Nieuw Archief voor Wiskunde, (3) **28** (1980), 139–158.
9. *An analogue of Chebotarev density theorem for fields of finite corank*, Journal of Mathematics of Kyoto University **20** (1980), 141–147.
10. *The Chebotarev density theorem for function fields; an elementary approach*, Mathematische Annalen **261** (1982), 467–475.
11. *Torsion–free profinite groups with open free subgroups*, Archiv der Mathematik **39** (1982), 496–500.
12. *The elementary theory of large e–fold ordered fields*, Acta Mathematica **149** (1982), 239–260.
13. *The elementary theory of normal Frobenius fields*, Michigan Mathematical Journal **30** (1983), 155–163.
14. *Intersection of local algebraic extensions of a Hilbertian field (A. Barlotti et al., eds)*, NATO ASI Series C **333** 343–405, Kluwer, Dordrecht, 1991.
15. *Algebraic dimension over Frobenius fields*, Forum Mathematicum **6** (1994), 43–63.
16. *Large normal extension of Hilbertian fields*, Mathematische Zeitschrift **224** (1997), 555–556.
17. *The projectivity of the fundamental group of an affine line*, The Turkish Journal of Mathematics **23** (1999), 531–547.

M. Jarden and U. Kiehne
1. *The elementary theory of algebraic fields of finite corank*, Inventiones mathematicae **30** (1975), 275–294.

M. Jarden and A. Lubotzky
1. *Hilbertian fields and free profinite groups*, Journal of the London Mathematical Society (2) **46** (1992), 205–227.
2. *Random normal subgroups of free profinite groups*, Journal of Group Theory **2** (1999), 213–224.

M. Jarden and P. Roquette
1. *The Nullstellensatz over p–adicaly closed fields*, Journal of the Mathematical Society Japan **32** (1980), 425–460.

M. Jarden and J. Ritter
1. *On the Frattini subgroup of the absolute Galois group of a local field*, Israel Journal of Mathematics **74** (1991), 81–90.

M. Jarden and S. Shelah
1. *Pseudo algebraically closed fields over rational function fields*, Proceedings of AMS **87**, (1983), 223–228.

W. Jehne
1. *Kronecker classes of algebraic number fields*, Journal of Number Theory **9** (1977), 279–320.

W. M. Kantor and A. Lubotzky
1. *The probability of generating a finite classical group*, Geometriae Dedicata **36**

(1990), 67–87.

W. Kimmerle, R. Lyons, R. Sandling, and D. N. Teaque
1. *Composition factors from the group ring and Artin's theorem on the orders of simple groups*, Proceedings of the London Mathematical Society (3) **60** (1990), 89–122.

N. Klingen
1. *Elementar äquivalente Körper und ihre absolute Galoisgruppe*, Archiv der Mathematik **25** (1974), 604–612.
2. *Zahlkörper mit gleicher Primzerlegung*, Journal für die reine und angewandte Mathematik **299/300** (1978), 342–384.
3. *Atomare Kronecker–Klassen mit speziellen Galoisgruppen*, Abhandlungen aus dem mathematischen Seminar Hamburg **48** (1979), 42–53.
4. *Prime Decomposition and Finite Group Theory*, Oxford University Press, 1998.

U. Kiefe
1. *Sets definable over finite fields: Their zeta–functions*, Transactions of AMS **223** (1976), 45–59.

I. Kiming
1. *Explicit classifications of some 2-extensions of a field of characteristic different from 2*, Canadian Journal of Mathematics **42** (1990), 825–855.

R. Klein
1. *Über Hilbertsche Körper*, Journal für die reine und ungewandte Mathematik **337** (1982), 171–194.

K. Knopp
1. *Infinite series, The Theory and Application of Infinite Series*, Blackie and Son, London, 1928.

J. Kollár
1. *Sharp effective Nullstellensatz*, Journal of the American Mathematical Society **1** (1988), 963–975.
2. *Algebraic varieties over PAC fields*, Israel Journal of Mathematics **161** (2007), 89–101.
3. *A conjecture of Ax and degenerations of Fano varieties*, Israel Journal of Mathematics **162** (2007).

L. Kronecker
1. *Über die Irreduzibilität von Gleichungen*, Werke II, 85–93; Monatsberichte Deutsche Akademie für Wissenschaft (1880), 155–163.

F.-V. Kuhlmann, M. Pank, and P. Roquette
1. *Immediate and purely wild extensions of valued fields*, manuscripta mathematicae **55** (1986), 39–67.

A. G. Kurosh
1. *The theory of groups II*, Chelsea, New York, 1960.

W. Kuyk
1. *Generic approach to the Galois embedding and extension problem*, Journal of Algebra **9** (1968), 393–407.
2. *Extensions de corps hilbertiens*, Journal of Algebra **14** (1970), 112–124.

W. Kuyk and H. W. Lenstra Jr.
1. *Abelian extensions of arbitrary fields*, Mathematische Annalen **216** (1975), 99–104.

J. C. Lagarias, H. L. Montogomery and A. M. Odlyzko
1. *A bound for the least prime ideal in the Chebotarev density theorem*, Inventiones mathematicae **54** (1979), 271–296.

S. Lang
1. *On quasi algebraic closure*, Annals of Mathematics **55** (1952), 373–390.
2. *Abelian varieties*, Interscience Publishers, New York, 1959.
3. *Diophantine Geometry*, Interscience Publishers, New York, 1962.
4. *Introduction to Algebraic Geometry*, Interscience Publishers, New York, 1964.
5. *Algebraic Number Theory*, Addison–Wesley, Reading, 1970.
6. *Introduction to Algebraic and Abelian Functions*, Addison-Wesley, Reading, 1972.
7. *Algebra, Third Edition*, Addison–Wesley, Reading, 1997.

A. Lenstra
1. *Factoring multiinvariant polynomials over finite fields*, Journal of computation and System Science **30** (1985), 235–248.

A. K. Lenstra, H. W. Lenstra Jr., and L. Lovász
1. *Factoring polynomials with rational coefficients*, Mathematische Annalen **261** (1982), 515–534.

H. W. Lenstra Jr.
On the degrees of exceptional polynomials, manuscript.

H. Leptin
1. *Ein Darstellungssatz für kompakte total unzusammenhängende Gruppen*, Archiv der Mathematik **6** (1995), 371–373.

W. J. LeVeque
1. *Topics in number theory I*, Addison–Wesley, Reading, 1958.

R. Lidl, G. L. Mullen, and G. Turnwald
1. *Dickson Polynomials*, Pitman Monographs and Surveys in Pure and Applied Mathematics **65**, Longman, Essex, 1993.

R. Lidl and H. Niederreiter
1. *Finite fields*, Encyclopedia of Mathematics and its applications **20**, Addison–Wesley, Reading, 1983.

M. W. Liebeck and A. Shalev
1. *The probability of generating a finite simple group*, London Mathematical Society Lecture Notes **249**, 163–173, Cambridge University Press, Cambridge, 1998.

C.-K. Lim
1. *Pronilpotent groups*, Nanta Mathematica **6** (1973), 58–59.

Q. Liu
1. *Tout groupe fini est un groupe de Galois sur $\mathbb{Q}_p(T)$*, Proceedings of the 1993 Seattle AMS Summer Conference, "Recent Development in the Inverse Galois Problem."

A. Lubotzky
1. *Combinatorial group theory for pro–p–groups*, Journal of Pure and Applied Algebra **25** (1982), 311–325.
2. *A group theoretic characterization of linear groups*, Journal of Algebra **113** (1988), 207–214.

3. *Random elements of a free profinite group generate a free subgroup*, Illinois Journal of Mathematics **37** (1993), 78–84.

A. Lubotzky and L. v. d. Dries
1. *Subgroups of free profinite groups and large subfields of $\tilde{\mathbb{Q}}$*, Israel Journal of Mathematics **39** (1981), 25–45.

A. Lubotzky and D. Segal
1. *Subgroup Growth*, Progress in Mathematics **212**, Birkhaüser Verlag, Basel, 2003.

R. C. Lyndon and P. E. Schupp
1. *Combinatorial Group Theory*, Ergebnisse der Mathematik und ihrer Grenzgebiete **89**, Springer-Verlag, Berlin, 1997.

A. Macintyre
1. *On definable subsets of p-adic fields*, The Journal of Symbolic Logic **41** (1976), 605–610.

A. Macintyre, K. McKenna, and L. v. d. Dries
1. *Elimination of quantifiers in algebraic structures*, Advances in Mathematics **47** (1983), 74–87.

C. R. MacCluer
1. *A reduction of the Chebotarev density theorem to the cyclic case*, Acta Arithmetica **15** (1968), 45–47.

M. L. Madan and D. J. Madden
1. *On the theory of congruence function fields*, Communications in Algebra **8** (1980), 1687–1697.

G. Malle and B. H. Matzat
1. *Inverse Galois Theory*, Springer, Berlin, 1999.

A. Mann
1. *Positively finitely generated groups*, Forum Mathematicum **8** (1996), 429–459.

H. Matsumura
1. *Commutative ring theory*, Cambridge studies in advanced mathematics **8**, Cambridge University Press, Cambridge, 1994.

B. H. Matzat
1. *Zum Einbettungsproblem der algebraischen Zahlentheorie mit nicht abelschem Kern*, Inventiones mathematicae **80** (1985), 365–374.

O. V. Melnikov
1. *Normal subgroups of free profinite groups*, Mathematics USSR Izvestija **12** (1978), 1–20.
2. *Characterization of accessible subgroups of free profinite groups (Russian)*, Doklady Akademii Nauk BSSR **22** (1978), 677–680.
3. *Projective limits of free profinite groups (Russian, English summary)*, Doklady Akademii Nauk BSSR **24**, *(1980), 968–970, 1051*.
4. *Subgroups and the homology of free products of profinite groups. (Russian)* Izvestiya Akademii Nauk SSSR, Seriya Matematicheskaya 53 (1989), no. 1, 97–120; translation in Mathematics of the USSR-Izvestiya **34** (1990), no. 1, 97–119.

L. Moret-Bailly
1. *Constructions de revêtements de courbe pointées*, Journal of Algebra **240** (2001), 505–534.

P. Müller
1. *A Weil-bound free proof of Schur's conjecture*, Finite fields and their applications **3** (1997), 25–32.

D. Mumford
1. *Abelian varieties*, Tata Institute of Fundamental Research, Bombay, 1974.
2. *The Red Book of Varieties and Schemes*, Lecture Notes in Mathematics **1358**, Springer, Berlin, 1988.

J. Neveu
1. *Mathematical Foundations of the Calculus of Probability*, Holden-Day Series in Probability and Statistics, Holden-Day, 1965, San Francisco.

J. Neukirch, A. Schmidt, and K. Wingberg
1. *Cohomology of Number Fields*, Grundlehren der mathematischen wissenschaften **323**, Springer, 2000, Berlin.

K. Neumann
1. *Every field is stable*, Israel Journal of Mathematics **104** (1998), 221–260.

N. Nikolov and D. Segal
1. *Finite index subgroups in profinite groups*, Comptes Rendus de l'Académie des Sciences **337** (2003), 303–308.

N. Nobusawa
1. *On the embedding problem of fields and Galois algebras*, Abhandlungen aus dem Mathematischen Seminar der Universität Hamburg **26** (1961), 89–92.

J. Pas
1. *Igusa local zeta functions and Meuser's invariant functions*, Journal of Number Theory **38** (1991), 287–299.

B. Poizat
1. *Une preuve d'un théorème de James Ax sur les extension algébrique de corps*, Comptes Rendus de l'Académie des Sciences, Paris **291** (1980), Serie A–245.

L. S. Pontryagin
1. *Topological groups*, Gordon and Breach, New York, 1966.

F. Pop
1. *Embedding problems over large fields*, Annals of Mathematics **144** (1996), 1–34.

A. Prestel
1. *Pseudo real closed fields*, In Set theory and model theory, Lecture Notes in Mathematics **782** (1981), 127–156, Springer-Verlag, Berlin.
2. *Lectures on formally real fields*, Lecture Notes in Mathematics **1093**, Springer-Verlag, Berlin, 1984.

A. Prestel and P. Roquette
1. *Formally p-adic fields*, Lecture Notes in Mathematics 1050, Springer-Verlag, Berlin, 1984.

M. O. Rabin
1. *Computable algebra, General theory and theory of computable fields*, Transactions of AMS **95** (1960), 341–360.

M. Raynaud
1. *Revêtements de la droite affine en caractéristique $p > 0$ et conjecture d'Abhyankar*, Inventiones Mathematicae **116** (1994), 425–462.

H. Reichardt
1. *Der Primdivisorensatz für algebraische Funktionenkörper über einem endlichen Konstantenkörper*, Mathematische Zeitschrift **40** (1936), 713–719.

A. Rényi
1. *Probability theory*, North–Holland/American Elsevier, Amsterdam, 1970.

P. Ribenboim
1. *Théorie des valuations*, Les Presses de l'Université de montréal, Montréal, 1968.

L. Ribes
1. *Introduction to profinite groups and Galois cohomology*, Queen's University, Kingston, 1970.
2. *Frattini covers of profinite groups*, Archiv der Mathematik **44** (1985), 390-396.

L. Ribes and P. Zalesskii
1. *Profinite Groups*, Ergebnisse der Mathematik III **40**, Springer, Berlin, 2000.

J. F. Ritt
1. *On algebraic functions which can be expressed in terms of radicals*, Transactions of AMS **24** (1922), 21–30.

A. Robinson
1. *Nonstandard points on algebraic curves*, Journal of Number Theory **5** (1973), 301–327.

A. Robinson and P. Roquette
1. *On the finiteness theorem of Siegel and Mahler concerning diophantine equations*, Journal of Number Theory **7** (1975), 121–176.

P. Roquette
1. *Applications of nonstandard methods to Hilbert irreducibility theorem and related questions*, Unpublished notes, Heidelberg, 1975.
2. *Nonstandard aspects of Hilbert's irreducibility theorem*, in Model Theory and Algebra, Memorial Tribute to Abraham Robinson, Lecture Notes in Mathematics **498**, Springer-Verlag, Berlin, 1975, pp. 231–275.

M. Rosenlicht
1. *Equivalence relations on algebraic curves*, Annals of Mathematics **56** (1952), 169–191.

G. E. Sacks
1. *Saturated Model Theory*, Benjamin, Reading, 1972.

P. Samuel
1. *Lectures on old and new results on algebraic curves*, Tata Institute, Bombay, 1966.

A. Schinzel
1. *On a theorem of Bauer and some of its applications*, Acta Arithmetica **II** (1968), 333–344.
2. *Polynomials with Special Regard to Reducibility*, Encyclopedia of Mathematics and its Application **77**, Cambridge University Press, Cambridge, 2000.

F. K. Schmidt
1. *Körper, über denen jede Gleichung durch Radikale auflösbar ist*, Sitzungsberichte der Heidelberg Akademie der Wissenschaften **2** (1933), 37–47.

K. Schmidt
1. *Modelltheoretischen Methoden in der algebraischen Geometrie*, Diplomarbeit, Kiel, 1970.
2. *Nonstandard Methods in Field Theory*, Ph.D Thesis, Kiel, 1985.

W. M. Schmidt
1. *Equations over finite fields: An elementary approach*, Lecture Notes in Mathematics **536**, Springer-Verlag, Berlin, 1976.
2. *Zur Methode von Stepanov*, Acta Arithmetica **24** (1973), 347–367.

I. Schur
1. *Über den Zusammenhang zwischen einem Problem der Zahlentheorie und einem Satz über algebraische Funktionen*, Sitzungsberichte der Preussischen Akademie der Wissenschaften, Physikalisch-Mathematische Klasse (1923), 123–134.
2. *Zur Theorie der einfach transitiven Permutationsgruppen*, Sitzungsberichte der Preussischen Akademie der Wissenschaften, Physikalisch-Mathematische Klasse (1933), 598–623.

A. Seidenberg
1. *Elements of the Theory of Algebraic Curves*, Addison–Wesley, Reading, 1968.

J.-P. Serre
1. *Sur la dimension cohomologique des groupes profinis*, Topology **3** (1964–5), 413–420.
2. *Zeta and L–functions*, in Arithmetical Algebraic Geometry (Schilling, ed.), Harper and Row, New York (1965), 82–92.
3. *Abelian l–adic representations and elliptic curves*, Benjamin, New York, 1968.
4. *Corps locaux*, Hermann, Paris, 1968.
5. *Proprietés galoisiennes des points d'ordre fini des courbes elliptiques*, Inventiones mathematicae **15** (1972), 259–331.
6. *Linear representations of finite groups*, Springer-Verlag, Berlin, 1977.
7. *Quelques applications du Théorème de densité de Chebotarev*, Publication Mathématiques, Institut des Hautes Études Scientifiques **54** (1981), 323–401.
8. *Sur le nombre des points rationnels d'une courbe algèbrique sur un corps fini*, Comptes Rendus de l'Académie des Sciences, Paris, sèrie I, 296 (1983), 397–402.
9. *Construction de revêtements étales de la droite affine en charactéristique p*, Comptes Rendus de l'Académie des Sciences, Paris, **311** (1990), 341–346.
10. *Topics in Galois Theory*, Jones and Barlett, Boston 1992.
11. *Galois Cohomology*, Springer, Berlin, 1997.

I. R. Shafarevich
1. *Algebraic number fields (Russian)*, Proceedings of the International Congress of Mathematics 1962, pp. 163–176.

S. Shelah
1. *Every two elementarily equivalent models have isomorphic ultrapowers*, Israel Journal of Mathematics **10** (1971), 224–233.

References

J. R. Shoenfield
1. *Mathematical Logic*, Addison-Wesley, Reading, 1967.

W. Specht
1. *Gruppentheorie*, Springer-Verlag, Berlin, 1956.

V. G. Sprindzuk
1. *Reducibility of polynomials and rational points on algebraic curves*, Soviet Mathematics **21** (1980), 331–334.
2. *Diophantine equations involving unknown primes*, Trudy MIAN SSR **158** (1981), 180–196.

R. Stallings
1. *On torsion free groups with infinitely many ends*, Annals of Mathematics **88** (1968), 312–334.

S. A. Stepanov
1. *The number of points of a hyperelliptic curve over a finite prime field (Russian)*, Izvestiya Akademii Nauk SSSR, Ser. Math **33**, *(1969), 1171–1181*; Mathematics of the USSR-Izvestiya **3** *(1969), 1103–1114*.
2. *An elementary proof of the Hasse–Weil theorem for hyperelliptic curves*, Journal of number theory **4** (1972), 118–143.
3. *An elementary method in the theory of equations over finite fields (Russian)*, Proceeding of the International Congress of Mathematics (Vancouver, B.C. 1974); AMS Translation (2) **109** *(1977), 13–20*.

H. Stichtenoth
1. *Algebraic Function Fields and Codes*, Springer-Verlag, Berlin, 1993.

K. O. Stöhr and J. F. Voloch
1. *Weierstrass points and curves over finite fields*, Proceedings of the London Mathematical Society (3) **52** (1986), 1–19.

M. Suzuki
1. *Group Theory I*, Grundlehren der mathematischen Wissenschaften **247**, Springer-Verlag 1982, New York.

T. Tamagawa
1. *On regularly closed fields*, Contemporary Mathematics **13** (1982), 325–334.

A. Tarski
1. *A decision procedure for elementary algebra and geometry*, Technical report (prepared by MacCkinsy), The Rand Corporation, Santa Monica, 1948.

N. Tschebotarev
1. *Die Bestimmung der Dichtigkeit einer Menge von Primzahlen, welche zu einer gegebenen Substitutionsklasse gehören*, Mathematische Annalen **95** (1926), 191–228.

E. C. Titchmarsh
1. *The theory of the Riemann Zeta-function*, Oxford at the Clarendon Press, 1951.

M. A. Tsfasman, S. G. Vlâdut, and T. Zink
1. *Modular curves, Shimura curves, and Goppa codes better than Warshamov–Gilbert bound*, Mathematische Nachrichten **109** (1982), 21–28.

G. Turnwald
 1. *On Schur's conjecture*, Journal of Australian Mathematical Scociety (series A) **58** (1995), 312-357.

K. Uchida
 1. *Separably Hilbertian fields*, Kodai Mathematical Journal **3** (1980), 83–95.

H. Völklein
 1. *Groups as Galois groups — an Introduction*, Cambridge Studies in Advanced Mathematics **53**, Cambridge University Press, 1996.

J. F. Voloch
 1. Ph.D. Thesis, Cambridge, 1984.

B. L. v. d. Waerden
 1. *Modern Algebra 1, Zweite Auflage*, Springer, 1937.
 2. *Einführung in die algebraische Geometrie*, Springer-Verlag, Berlin, 1973.
 3. *Algebra 1*, Springer-Verlag, New York, 1991.
 4. *Algebra 2*, Springer-Verlag, New York, 1991.

W. C. Waterhouse
 1. *Profinite groups are Galois groups*, Proceedings AMS **42** (1974), 639–640.

T. S. Weigel
 1. *Residual properties of free groups*, Journal of Algebra **160** (1993), 16–41.
 2. *Residual properties of free groups II*, Communications in Algebra **20** (1992), 1395-1425.
 3. *Residual properties of free groups III*, Israel Journal of Mathematics **77** (1992), 65–81.
 4. *On the profinite completion of arithmetic groups of split type*, in: Lois d'algèbres et variétés algébriques (Editor: M. Goze), Collection Travaux en cours **50**, 79–101, Hermann, 1996.

A. Weil
 1. *L'arithmetique sur les courbes algébriques*, Thése, Paris 1928 = Acta Mathematica **52** (1928), 281–315.
 2. *Sur les courbes algebriques et les varietes que s'en deduisent*, actualites Sci. et Ind. **1041**, Hermann, Paris, 1948.
 3. *The field of definition of a variety*, American Journal of Mathematics **78** (1956), 509–524.
 4. *Adeles and Algebraic Groups*, Institute for Advanced Study, Princeton, 1961.
 5. *Foundations of Algebraic Geometry*, American Mathematical Society, Providence, 1962.
 6. *Basic Number Theory*, Springer-Verlag, Berlin, 1967.

R. Weissauer
 1. *Der Hilbertsche Irreduzibilitätssatz*, Journal für die reine und angewandte Mathematik **334** (1982), 203–220.

G. Whaples
 1. *Algebraic extensions of arbitrary fields*, Duke Mathematical Journal **24** (1957), 201–204.

W. H. Wheeler
 1. *Model–complete theories of pseudo–algebraically closed fields*, Annals of Mathematical Logic **17** (1979), 205–226.

References

H. Wielandt
1. *Finite Permutation Groups*, Academic Press, New York, 1968.

E. Witt
1. *Konstruktion von galoisschen Körpern der Charakteristik p zu vorgegebener Gruppe der Ordnung p^f*, Journal für die reine und angewandte Mathematik **174** (1936), 237–245.

U. Zannier
1. *Note on g-hilbertian fields and nonrational curves with the Hilbert property*, manuscripta mathematica **97** (1998), 155–162.

O. Zariski and P. Samuel
1. *Commutative Algebra I*, Springer-Verlag, New York, 1958.
2. *Commutative Algebra II*, Springer-Verlag, New York, 1960.

Index

Abelian extension, 294
Abelian pro-p groups, 308
Abelian profinite, 345
Abhyankar, 70
Abhyankar's conjecture, 70
absolute Galois group, 12
absolute norm (of a prime ideal), 112
absolute value, 238
absolutely converges
   (infinite product), 80, 92
absolutely irreducible (polynomial), 43
absolutely irreducible (variety), 175
abstract variety, 189
accessible, 626
Ackermann, 158
Ackermann function, 158
additive absolute value, 278
adele, 64
admissible (ideal), 123
admissible, 541
Adler-Kiefe, 670
affine n-space $\mathbb{A}^n$, 172
affine plane, 95
affine plane curve, 95
algebraic function field of one variable, 52
algebraic set, 172
algebraically independent (fields), 40
algebraically independent (set), 40
almost all, 146
almost contained (sets), 140
almost equal (sets), 140
almost full, 542
amalgamation property, 656
ample, 335
Arason-Fein-Schacher-Sonn, 307
archimedean (absolute value), 239
Arithmetic, 706
arithmetical primes, 280
arithmetically definable, 707
arithmetic progression, 241
Artin, 41, 64, 76, 130, 494, 537, 738,
Artin reciprocity law, 124
Artin-Schreier extension, 29
Artin symbol, 113, 709
atomic formula (of $\mathcal{L}_n$), 700

atomic formulas, 133
Ax, 141, 148, 162, 191, 200, 207, 209, 217, 218,
   381, 401, 447, 448, 449, 450, 452, 453, 456, 524, 759
Ax-Roquette, 196
Büchi, 682
Bary-Soroker, 613
basic field, 172
basic functions, 402
basic sets, 402
basic statement, 551
basis (of a free abstract group), 346
basis (of a free pro-$\mathcal{C}$ group), 348
Bauer, 130, 236
beat, 671
Beckmann-Black, 760
Beckmann-Black Problem, 565
Bell-Slomson, 145, 148
Bensimhoun, 493
Bertini-Noether, 169, 171, 179
Bezout ring, 276
bijective (polynomial function), 477
Binz-Neukrich-Wenzel, 362, 542
birational equivalence (varieties), 178
birationally equivalent (varieties), 178
block (of a permutation group), 474
Bôcher, 744
Bombieri, 94, 743
Boolean algebra
   generated by a family, 140
Boolean algebra of sets, 140
Boolean polynomials, 140
Borel-Cantelli, 372
Borel field, 362
Borevich-Shafarevich, 61
bottom theorem, 385
bounded (degrees), 417
bounded (occurence of a variable), 133
bounded minimum operator, 157
Bourbaki, 24, 25, 226
branch point, 214
Brandis, 402, 525
Brink, 320, 328

Index   781

Burnside, 206, 477, 495
C_i-field, 453
$C_{i,d}$-field, 453
C-embedding problem, 503
C-homogeneous, 615
C-homogeneous cover, 633
C-projective, 503
canonical class, 55
canonical divisor, 55
Caratheodory, 365
Carlitz's Conjecture, 489
cartesian product, 182
cartesian square, 400
Cassels-Fröhlich, 21, 32, 62, 63, 65, 206
  239, 241, 314, 363, 381, 469, 470
  483, 485, 742
Cauchy sequence, 61
center (of a divisor), 97
center (of a prime divisor), 57
central (function), 740
central simple (K-algebra), 208
Chaevalley, 76
character, 740
character (of a finite abelian group), 122
characteristic (subgroup), 330
characteristic function, 156
Chatzidakis, 401, 580, 593,
  595, 603, 633, 789
Chebotarev, 130
Chebotarev density theorem, 114
Chebyshev polynomials, 479
Chebyshev's inquality, 373
Cherlin, 383
Cherlin-Jarden, 706, 707
Cherlin-v.d.Dries-Macintyre,
  453, 542, 560, 690, 697
Chevalley, 24, 64, 76, 423, 455
Chevalley-Warning, 455
class number, 77
class of divisors, 53
classical Hilbertian fields, 242
coefficients (of an element in $\mathbb{Z}[H]$), 475
cofinite, 143
Cohen-Fried, 490
Cohn, 188
cohomological dimension, 210
Colliot-Thélène, 230, 243, 593
commutator, 309
commutator subgroup, 310

compactness theorem, 144
compatible (maps), 1, 2
complementary Galois stratification, 719
complementary module, 65
complement (of a closed subgroup), 524
complete (measure), 365
complete (theory), 152
complete (valued field), 61
completeness (of projective varieties), 186
completion, 61
components (of a K-closed set), 174
composition factor, 326
composition series, 688
compositum (of places), 23
computes (a function), 672
concurrent, 271
conductor (of a ring extension), 100
cone, 272
congruence test, 132
conjugacy domain of subgroups, 710
conjunction symbol, 133
consequence, 149
conservative
  (function field of one variable), 393
conservative (function field of several
  variables), 392
consistent, 152
constant, 52, 134
constant field extension, 59
constant symbols, 132
constant terms, 153
continuous set theoretic section, 9
contradiction (formula), 152
converge to 1 (subset), 338
convergent to 1 (map), 346
converges (infinite product), 80
convex hull, 280
convex, 274
coordinate ring (of a K-basic set), 424
coordinate ring (of a K-variety), 173
coordinate ring (of a curve), 96
coordinate ring (over $\mathbb{Z}[k^{-1}]$), 732
corank (of a field), 439
Corvaja-Zannier, 266, 289
Cossey-Kegel-Kovács, 542
covers, 513
crossed homomorphism, 526
cusp, 394
cyclotomic (extension), 121

data for being algebraically closed, 662
Davenport (Problem), 753
Davenport, 466
Dèbes, 566, 593
Dèbes-Haran, 236, 758
de-Morgans's laws, 164
decidable theory, 159
decision procedure, 159
decomposable, 471
decomposition factor
   (of a polynomial), 485
decomposition field (of a place), 110
decomposition field
   (of a prime ideal), 107
decomposition group, 25
decomposition group (of a place), 110
decomposition group
   (of a prime ideal), 107
decomposition group condition, 564
Dedekind, 63, 464
Dedekind domain, 32
Dedekind zeta function, 124
deduction theorem, 151
deductively close, 150
defined (absolutely irreducible
   K-variety), 175
defined over K (scheme), 189
degree (of a prime divisor), 52
degree (of a curve), 95
degree (of a divisor), 53
degree (of a field extension), 519
degree (of a polynomial function), 478
degree (of a rational map), 178
degree (of a stabilizing base), 391
Deligne, 94
Denef-Jarden-Lewis, 460
Denef-Loeser, 729
density property, 397, 751
Dentzer, 306
derivation, 48
Deuring, 53, 54, 55, 59, 60, 61, 76, 130, 210
diamond theorem, 260
Dickson polynomial, 480
different, 67
different (of a ring extension), 112
different exponent, 65
differential, 54, 64
dihedral group, 472

dimension (of a constructible set), 423
dimension (of a K-variety), 173
dimension (of a ring), 283
dimension theorem, 174
directed (family), 341
directed (family of subsets), 6
directed partially ordered set, 1
Dirichlet, 114, 130
Dirichlet density, 113
discrete (valuation), 21
discriminant (of an element), 109
discriminant (of a polynomial), 108
discriminant (of a ring extension), 112
distinct (valuations), 20
divides (supernatural number), 520
divisor, 52
divisor (of a differential), 55, 64
divisor of poles, 52
divisor of zeros, 52
Dixon, 636, 639
Dixon-du.Sautoy-Mann-Segal, 539
Dixon-Pyber-Seress-Shalev, 358
domain, 134
domain of definition
   (of a rational map), 178
dominant (rational map), 178
Douady, 338
double implication symbol, 133
double tangents, 392
doubly transitive
   (permutation group), 474
Dries, v. d., 426, 590
Dries-Smith, 401
Drinfeld-Vlădut, 106
Duret, 190, 191, 699
Dwork, 741, 743
Dwork-Bombieri, 743
e-basic statement, 551
e-cycle, 315
e-free (field), 379
effective algorithm, 405
effective computation, 412
Efrat, 207
Efrat-Völklein, 541
Eichler, 266
Eisenbud, 99
Eisenstein's Criterion, 30, 217
elementarily equivalent, 136
elementary equivalence theorem, 435

Index

elementary extension, 137
elementary statement, 136
elementary substructure, 137
elimination (of a quantifier), 164
elimination of quantifiers, 163
elimination theory, 410
embedding (of structures), 137
embedding cover, 571
embedding lemma, 431
embedding problem, 303, 502
embedding property, 564
ends with (word), 351
enlargement, 268
equality symbol, 132
equivalent (absolute values), 239
equivalent (algebras), 208
equivalent (formulas), 150
equivalent ($(n+1)$-tuples), 185
equivalent (places), 20
equivalent (points), 189
equivalent (valuations), 20
Ershov, 196, 362, 402, 436, 540, 541, 542, 538, 539, 576, 670, 689, 697, 729
Ershov-Fried, 542, 593
Ershov-Lavrov-Taimanov-Taitslin, 683, 694, 697, 706
even (ordinal number), 618
eventually stationary, 672
exceptional (pair of polynomials), 467
exchange principle, 45
existential (sentence), 462
existential elimination lemma, 718
existential symbol, 132
existentially closed, 137, 656
existentially complete, 561
explicit case, 440, 708
extension (of function fields), 59
extension (of structures), 136, 656
external (object), 268
faithful (permutation group), 472
Faltings, 248
Faltings-Wüstohlz, 247
fiber product (of groups), 499
fiber product (of schemes), 246
field cover, 109
field crossing argument, 107, 130, 324, 431, 558, 562
field of formal power series, 287
field of p-adic numbers, 17

field with a product formula, 281
filter, 138
final tape, 672
finite (element at a place), 20
finite (embedding problem), 303, 502
finite (K-morphism), 180
finite (place at an element), 20
finite intersection property, 139
finite points (on a plane curve), 95
finitely generated (profinite group), 328
finitely generated (ring cover), 562
first isomorphism theorem
    for compact groups, 5
first order language, 132
formal power series, 62
formal proof, 150
formation, 344
formulas, 133
fractional ideal, 31
Frane, 145
Franz, 266
Frattini cover, 508
Frattini embedding problem, 511
Frattini group, 497
Frattini p-cover, 529
free (abstract group), 346
free (variable), 133
free generators theorem, 379
free occurrence of a variable
    in a formula, 133
free pro-$\mathcal{C}$ group, 346, 349
free pro-$\mathcal{C}$ group of rank m, 348
free product (of profinite groups), 508
free set of generators, 346, 605
free variables (of a formula), 268
Frey, 196, 205, 218, 217
Frey-Geyer, 194
Frey-Jarden, 337, 398
Frey-Prestel, 205
Fried, 130, 206, 244, 264, 266, 290, 471, 492, 494, 540
Fried-Guralnick-Saxl, 489
Fried-Haran-Jarden, 428, 564, 593, 729
Fried-Jarden, 51, 131, 196, 207, 230, 262, 266, 289, 336, 391, 392, 395, 397, 401, 402, 495, 542, 581, 593, 582, 635, 639, 753
Fried-MacCrae, 486
Fried-Sacerdote, 495, 729, 738

Fried-Völklein, 337, 395, 397, 582, 583, 752
Fried-Völklein (Conjecture), 759, 760
Frobenius automorphism, 9, 15
Frobenius automorphism (of a prime ideal), 112
Frobenius, 130
Frobenius density theorem, 130
Frobenius field, 564
Frobenius fields, 559
Fröhlich-Shepherdson, 411
from the n-tracks, 672
full formation, 344
function field, 52
function field (of a curve), 96
function field (of a K-basic set), 424
function field (of a K-variety), 173
function field (of a projective variety), 186
function field of several variables, 391
function symbol, 132
functional prime, 280
$\mathcal{G}$-extesnion, 294
$\mathcal{G}$-extension, 294
g-Hilbertian, 248
GA (realization), 321
Galois, 493
Galois (polynomial), 235
Galois (ring cover), 562
Galois closure, 10
Galois formula, 717
Galois ring/set cover (over $\mathbb{Z}[k^{-1}]$), 732
Galois ring cover, 109
Galois sentence, 720
Galois splitting field (of a Zariski closed set), 454
Galois stratification, 717
GAR (realization), 321
Gaschütz, 362, 525
general polynomial of degree n, 296
generalized Abhyankar's conjecture, 70
generalized Krull domain, 283
generated (formation), 346
generated (σ-algebra), 362
generates (a profinite group), 328
generates (the group $\hat{\mathbb{Z}}$), 14
generic (point), 96
generic point (of a K-variety), 173
genus (of a curve), 96

genus (of a function field), 54
geometric points, 188
Geyer, 51, 176, 211, 243, 288, 337, 390, 393, 402
Geyer-Jarden, 131, 206, 211, 332, 337, 391, 392, 398, 402, 706, 751, 752, 753, 756, 759
Geyer-Jarden (Conjecture), 749
Geyer-Jensen, 307, 336, 519
Gilmore-Robinson, 277, 290
Gilmore-Robinson criterion, 289
global field, 112
Gödel, 162
Gödel completeness theorem, 154
Gödel number, 159
Gödel numbering, 403
Goldstein, 122, 124
Golod-Shafarevich inequality, 647
Goppa, 106
graph conditions, 689
graph, 679
Grauert-Manin, 212
greatest common divisor (of supernatural numbers), 520
Greenleaf, 495
Grothendieck, 70, 187
group of divisor classes, 53
group ring, 310
group theoretic diamond theorem, 610
group theoretic section, 502
Gruenberg, 210, 503, 542
Guralnick-Müller, 487, 496
Guralnick-Thompson, 266
Haar measure, 362
Hall, 252, 635
Halmos, 363, 366, 376, 399, 640, 645, 668
Haran, 130, 207, 385, 402, 496, 540, 542, 610, 633, 686
Haran's diamond theorem, 265
Haran-Jarden, 262, 266, 335, 545, 583, 593, 735, 752
Haran-Lauwers, 729
Haran-Lubotzky, 542, 568, 575, 578, 593
Haran-Völklein, 334
Harbater, 70, 334
Hartshorne, 187, 392
Henkin, 162
Hensel's lemma, 62
Henselian, 203

Index

Henselian closure, 203
Herfort-Ribes, 542
Hermes, 158
Hermite-Minkowski, 331
higher order set, 267
higher order structure, 267
Hilbert, 230, 231, 266, 299, 336
Hilbert's basis theorem, 172
Hilbert's irreducibility theorem, 219
Hilbert's Nullstellensatz, 169
Hilbert set, 219
Hilbert subset, 219
Hilbertian field, 219
Hilbertian (integral domain), 241
Hilbertian pair, 439
Hirschfeld-Wheeler, 670
holomorphy ring, 56
holomorphy ring theorem, 57
homogeneous (pro-$\mathcal{C}$-group), 595
Hrushovski, 215, 751
Huppert, 208, 326, 495, 498, 506,
    526, 536, 543, 630, 637, 647, 688
hyperelliptic, 70
hyperplane, 174
hypothesis $H(p,q)$, 700
I-cover, 568
identity axioms, 151
Ihara, 106
Ikeda, 304, 542
immediate extension, 203
imperfect degree, 45
imperfect exponent, 45
implication symbol, 133
imprimitive (permutation group), 474
Inaba, 191, 230, 266
inclusion-exclusion principle, 370
indecomposable, 471
independent (sets), 369
independent
    (valuations and orderings), 242
index (of a closed subgroup), 520
induced (conjugacy domain), 707
induced (cover), 710
induced (prime divisor), 281
induced representation, 740
induces (homomorphism), 430
induction on structure, 133
inductive, 656
inertia field (of an ideal), 107

inertia field (of a place), 110
inertia group, 25, 543
inertia group (of a place), 110
inertia group (of a prime ideal), 107
inference rule, 149
infinite absolute value, 281
infinite prime, 276
infinite sentences, 442
infinitesimal (element), 274
inflection points, 392
∞-imperfect, 45
initial tape, 671
injective (polynomial function), 487
instant, 671
integral closure, 30
integral over (element), 30
integral over (ring), 31
integrally closed (ring), 30
internal (n-tuples), 269
internal (object), 267
internal function, 270
internal substitutions, 269
interpretation (of an n-adic theory), 700
inverse (of fractional ideal), 31
inverse limit, 1
inverse system, 1
invertible (fractional ideal), 32
Ireland-Rosen, 93
irreducible (character), 740
irreducible (K-closed set), 173, 185
isomorphic (covers), 513
isomorphic (projective curves), 96
isomorphic (structures), 136
isomorphism theorem
    for compact groups, 5
Iwasawa, 516, 578, 581, 582
Iwasawa's Conjecture, 582, 760
Jacobson, 209
Jacobson-Jarden, 337, 398, 443,
    453, 753, 756, 759
Jannsen, 331, 519
Janusz, 69, 123, 124, 130, 464
Jarden, 130, 148, 206, 211, 243, 337, 379,
    380, 381, 384, 398, 401, 402, 451, 453
    561, 592, 669, 670, 729
Jarden-Kiehne, 148, 431, 442, 453
Jarden-Lubotzky, 587, 589, 590, 611,
    612 633, 645, 648, 653, 654
Jarden-Ritter, 516

Jarden-Roquette, 193, 293
Jarden-Shelah, 388, 402
Jarden's lemma, 542
Jehne, 463, 466, 467, 753
Jehne-Saxl, 467
Jehne (Conjecture), 467, 753
Jehne (Problem), 753
Jordan, 394
K-algebraic set (projective), 185
K-basic (constructible set), 424
K-birational (varieties), 178
K-closed subsets, 172
K-components (projective), 185
K-constructible set, 423
K-curve, 173
K-derivation, 48
K-hypersurface, 174
K-isomorphic (curves), 96, 97
K-isomorphism, 180
K-morphism, 180, 186
K-normal (point), 96
K-place, 20
K-rational (point), 178
K-rational map, 178, 186
K-variety, 173
K-variety (projective), 185
Kantor-Lubotzky, 638, 639, 644, 654, 753
kernel of the (embedding) problem, 502
Kiefe, 750
Kiming, 307
Kimmerle-Lyons-Sandling-Teaque, 650
Kollár, 171, 199, 205, 218, 462
Klein, 290
Klingen, 337, 438, 466, 467, 495
Knopp, 92
Kronecker, 411, 428, 461, 493
Kronecker class, 464
Kronecker conjugate, 463, 469
Kronecker equivalent, 753
Kronecker substitution, 199
Kronecker-Weber, 335
Krull domain, 286
Krull topology, 10
Kuhlmann-Pank-Roquette, 543
Kummer, 464
Kurosh, 352, 357, 503, 508
Kurosh subgroup theorem, 361
Kuyk, 333, 546, 558, 561
Kuyk-Lenstra, 336

L-rational point, 188
L-series, 122
Lagarias-Montegomery-Odlyzko, 131
Lang, 11, 12, 18, 22, 23, 24, 30, 37, 38,
  40, 43, 51, 61, 65, 75, 76, 98, 99, 101,
  107, 108, 109, 110, 111, 112, 113, 122,
  123, 124, 171, 172, 173, 174, 175, 176,
  183, 190, 191, 193, 196, 205, 207, 222,
  230, 237, 241, 245, 248, 258, 259, 266,
  282, 287, 296, 299, 300, 301, 307, 308,
  309, 316, 323, 331, 384, 398, 406, 407,
  416, 428, 455, 464, 471, 497, 519, 536,
  540, 545, 563, 591, 643, 699
language, 132
language of n-adic quantifiers, 700
larger (cover), 513
Lavrov, 680
least common multiple
  (of supernatural numbers), 520
Leibniz, 266
length (of a proof), 150
Lenstra, 426, 453, 490, 492, 496
Lenstra-Lenstra-Lovácz, 428
Leptin, 12, 18, 43, 494
letters, 132
LeVeque, 37, 113, 309, 310, 372, 380, 644
Levi, 357
lexicographic order, 19
Lidl-Mullen-Turnwald, 496
Lidl-Niederreiter, 93, 106
Liebeck-Shalev, 639
lies over (prime divisor), 59
lies over v (valuation), 24
Lim, 540
Lindon-Schupp, 361
line at infinity, 95
line, 95
linear fractional transformation, 213
linearly disjoint (fields), 34, 35, 36
linearly equivalent (divisors), 53
linearly independent (over a ring), 23
linearly related, Liu, 334
local parameter, 52
local ring (at a point), 96
local ring (of a K-variety at a point), 173
local ring, 31
localize (a ring at a prime ideal), 107
locally finite (word), 517
logically deducible, 150

Index

logically equivalent (formulas), 163
logically valid formula, 149
long multiplications, 156
long summations, 156
Łoś, 142
lower central series, 540
Lubotzky, 357, 539, 592, 638, 641, 654, 753
Lubotzky-Melnikov-v.d.Dries, 584
Lubotzky-Segal, 646
Lubotzky-v.d.Dries, 337, 545, 586, 591
Lüroth's theorem, 69, 320
lying over (prime ideal), 107
Lyndon-Schupp, 362
m-ary variable symbols, 700
m-stationary, 676
Möbius transformation, 213
Moret-Bailly, 593
Morley, 593
MacCluer, 130, 492
Macintyre, 171
Macintyre-McKenna-v.d.Dries, 171
Madan-Madden, 390, 393, 402
Malle-Matzat, 306, 328, 337, 336, 542
Mann, 646
Matsumura, 287, 288
Matsusaka-Zariski, 180
Matzat, 323, 325, 337, 542
maximal pro-$\mathcal{C}$ quotient, 345
maximal (subgroup), 497
meaningful words, 132
measurable (set), 366, 368
measurable rectangles, 376
Melnikov, 402, 542, 598, 606, 620, 622, 623, 624, 630, 631, 632, 633, 634
Melnikov (embedding problem), 615
Melnikov cover, 615
Melnikov formation, 344
Melnikov group, 613
metric (absolute value), 239
minimal (Kronecker class), 465
minimal normal subgroup, 322
minimal PAC field, 196
minimum operator, 157
Minkowski, 69
model, 135
model companion, 655
model complete, 164
model completion, 656

modified valuation, 280
modus ponens, 150
monotone (collection of sets), 362
Mordell-Weil, 247, 398
Morley, 590
morphism (projective), 186
morphism, 180
move commands, 671
μ-independent, 370
Müller, 266, 468, 482, 496, 760
Mumford, 187, 188, 248
n-adic structure, 700
n-adic theories, 700
n-dimensional projective space, 185
n-imperfect, 45
n-track, 672
negation symbol, 132
Neukirch-Schmidt-Wingberg, 305
Neumann, 391, 394, 402
Neveu, 376
Nielsen-Schreier formula, 355
Nikolov-Segal, 516, 517
Nobusawa, 511, 542
nodes, 392
Noether's normalization theorem, 98, 414
Noether-Grell, 32
Noetherian (integral domain), 31
Noetherian (ring), 172
non-archimedean (absolute value), 239
nonconstant curve, 212
nonelementary (classes), 549
nonsingular homogeneous linear transformation, 97
norm (of a prime divisor), 77
normal (basic set), 424
normal (basic set over $\mathbb{Z}[k^{-1}]$), 732
normal (projective model), 393
normal stratification, 716
normal stratification (over $\mathbb{Z}[k^{-1}]$), 733
number field, 112
objects of type, 267
occurs (a group over a field), 294
odd (ordinal number), 618
ω-free, 652, 752
open-closed partition, 3
operational modes, 671
opposite (algebra), 208
order (of a profinite group), 520
order of magnitude, 275

ordered group, 19
orthogonality relations, 122
overring, 23
p-adic integers, 13
p-adic numbers (field of), 61
p-basis, 44
p-cover, 529
p-embedding problem, 529
p-independent, 44, 45
p-independent, 44, 45
p-projective, 528
P-stratification, 424
p-Sylow group, 522
PAC Nullstellensatz, 380
PAC, 192
partial ordering, 1
partial products, 92
Pas, 754
permutation polynomial, 479
permutation representation, 292
permutes (polynomial), 479
place, 20
Poincaré series, 739
point (in an affince space), 172
points at infinity
  (on a projective curve), 95
Poizat, 453
pole, 52, 278
polefinite, 278
polynomial words, 403
Pontryagin, 18
Pop, 337, 583, 749
positive diagram, 135
positively finitely generated, 646
prenex normal form, 163
presented (basic set), 422
presented (field), 404
presented in its quotient field
  (integral domain), 404
presented (K-basic set), 418
presented (K-constructible set), 421
presented over (n-tuple), 407
presented over (element), 407
presented over (field extension), 405
presented over (rational field), 404
presented (ring), 404
Prestel, 204, 218, 383
primary (field extension), 38
prime (of a global field), 239

prime divisor, 52, 278
prime element, 21
primitive (permutation group), 474
primitive element for the cover, 109
primitive polynomial, 179
primitive recursive (function), 155, 404
primitive recursive (relation), 156
primitive recursive (theory), 159
primitive recursive indexing, 403
principal divisor, 53
principal ultrafilter, 139
pro-C group, 344
pro-C embedding problem, 500
pro-l-extension, 249
pro-p groups, 345
pro-S group, 346
procyclic (group), 16
product formula, 280
product measure, 376
product (of fractional ideals), 31
product (of supernatural numbers), 520
profinite completion
  (of a group), 341
profinite group, 5
profinite ring, 13
profinite space, 3, 539
projective, 506
projective (profinite group), 207
projective completion (of a curve), 95
projective cover, 513
projective limit, 1
projective plane, 95
projective plane curve, 95
projective special linear group, 536
projective system, 1
projective variety, 185
pronilpotent, 205
proper (overring), 23
prosolvable extension, 294
prosolvable group, 205, 345
Prüfer group, 12, 14
pseudo algebraically closed (field), 192
pseudo finite, 448
Q-central (function), 740
quadratic subfield, 70
quantifier axioms, 150
quantifier elimination procedure, 164
quasi-p (group), 70
Rabin, 403

Index

radical, 173
ramification group, 543
ramification index, 24
rank, 327
rank 1 (valuation), 21
rank (of a cover), 633
rank (of a profinite group), 339, 346
rank (of an abstract group), 356
rank
  (of separable algebraic extension), 386
rank (of an infinite separable
  algebraic extension), 387
rational (field extension), 321
rational (representation, character), 737
rational map (projective), 186
rational map, 178
Raynaud, 70
real (valuation), 21
realizable (a group over a field), 294
realization of
  twisted wreath products, 259
realize (twisted wreath product), 257
realize (wreath product), 257
reciprocity map, 123
recognizable (elements in a ring), 403
recursive functions, 157
recursive (theory), 159
recursively inseparable, 677
recursively separable, 676
reduced form, 209
reduced norm, 209
reduced (ring), 187
reduced (word), 351
reducible (K-closed set), 173
refinement (of Galois stratification), 718
refinement (of normal stratification), 717
regular (a group over a field), 294
regular (field extension), 38
regular (permutation group), 472
regular (ring cover), 562
regular (ultrafilter), 140
regular inverse Galois problem, 295
regular solution, 303
regular ultraproduct, 146
regularity (of a measure), 362
regularly solvable, 303
Reichardt, 130
relation rank, 647
relation symbol, 132

relations of type, 267
relativizing (variable), 147
Rényi, 373
repartition, 54
representable (recursive set), 707
representation (of a finite group), 740
residually-$\mathcal{C}$, 356
residually finite, 356
residue (of an element), 20
residue field, 20
resultant system, 413
ribbon, 671
Ribenboim, 203, 381
Ribes, 18, 210, 211, 300, 507, 534, 542, 582
Ribes-Zalesski, 4, 357, 513, 519, 542, 625, 628, 630, 631, 632, 633, 634
Riemann-Hurwitz formula
  for tamely ramified extensions, 69
Riemann-Hurwitz genus formula, 67
Riemann-Roch theorem, 54
Riemann hypotheses
  (for function fields), 85
ring cover, 109
ring of p-adic integers, 13
ring of integers, 33
ring of integers (of a global field), 112
ring/set cover, 708
Ritt, 206
Robinson, 170, 266, 290, 670
Robinson's test
  for model completeness, 659
Robinson-Roquette, 237, 276, 290
Roquette, 184, 277, 290, 337, 361, 665
Rosen, 697
Rosenlicht, 393, 394
rule of generalization, 150
S-rank function, 583
Sacks, 171
Samuel, 106, 212
Sansuc, 243
satisfiable, 679
saturated, 143
Saxl, 466, 467
Schinzel, 465, 492, 495
Schmidt, F. K., 82, 333
Schmidt, K., 426, 670
Schmidt, W. M., 94
Schreier, 355

Schreier basis, 352
Schreier construction, 351
Schreier system, 352
Schur, 475, 495
Schur's Conjecture, 487
Schur-Zassenhaus, 524
Scott, 145
section (group theoretic), 252
Seidenberg, 102, 105
semidirect complement, 689
semidirect product (of groups), 252
semilinear rationality criterion, 322
sentence, 133
sentential axioms, 150
separable (field extension), 38
separable (rational map), 178
separable descent, 183
separable Hilbert subset, 219
separating transcendence basis, 38
Serre, 25, 33, 65, 70, 106, 130, 187,
  210, 24 331, 332, 334, 336, 401,
  516, 740, 741, 742
Serre's question, 516
Serre-Stalling, 401
set of axioms, 135
set of basic test sentences, 146
set of logical consequences, 657
set theoretic section, 9
sets of type, 267
Severi-Brauer (variety), 209
Shafarevich, 332, 759
Shelah, 145, 593
Shoenfield, 707
Siegel's theorem, 266
Siegel-Mahler, 276, 290
Siegel, 237
σ-additivity, 363
σ-algebra, 363
simple (point on an affine curve), 98
simple (point on a projective curve), 99
singular (point on a projective curve), 99
Skolem-Löwenheim, 138, 171
small (profinite group), 329
small (set of powers of primes), 445
small sets, 139
smallest embedding cover, 571
smooth (curve), 104
solution (of an embedding problem),
  303, 503

solution field, 302
solution group, 596
solution (over $K(\mathbf{t})$), 302
solvable (embedding problem), 503
Sonn, 337
specialization, 173
specialization compatible, 711
Specht, 362
splits (algebra), 208
splits (embedding problem), 302, 502
splits (short exact sequence), 252
splitting algorithm, 405
Sprindzuk, 290
stabilizing base, 391
stable (field extension), 391
stable (field), 391
Stalling, 401
standard (object), 268
standard decomposition
  (of a projective variety), 186
standard function, 270
standard prime, 280
star-additive absolute value, 280
starfinite subsets, 274
starfinite summation, 273
starts with (word), 351
stationarily satisfiable, 679
stationary, 679
Stepanov, 94
Stichtenoth, 76
Stöhr-Voloch, 94
store (an integer), 672
stratification lemma, 424
strictly linearly related, 467
string, 132
strong approximation theorem, 56, 238
structure, 134
structure (for $\mathcal{L}_n$), 700
subnormal subgroup, 626
substitution, 134, 267
substructure, 136
supernatural number, 520
superprojective, 565
support, 475
surjective (polynomial function), 487
Suzuki, 327, 638
symmetrically stabilizing base, 391
Tamagawa, 201, 218
tamely ramified

Index

(extension of function fields), 69
tamely ramified (valuation), 25
tape, 671
Tarski, 171
Tate, 218, 518
Tate's theory, 217
tautology, 149
terms, 132
test sentence, 146, 440
T-existentially closed (model), 658
T-extension, 656
theory, 135
thin (subset), 245
Thompson, 305
Titchmarsh, 77
topological group, 4
total degree, 742
totally disconnected, 4
totally positive (element), 122
totally ramifies (valuation), 29
tower property, 35
track, 669
transcendence base, 40
transcendence degree, 40
transfer theorem, 447
transitive (permutation group), 472
translation invariance, 364
trivial (place), 30
trivial (valuation), 21
trivial block, 474
truth (of a formula), 268
truth set, 145, 440, 705
truth value, 134
Tsen, 211
Tsfasman-Vlâdut-Zink, 106
Turing machine, 671
Turnwald, 495
twisted wreath product (of groups), 253
type, 267, 657
Uchida, 228, 230, 336, 524
ultrafilter, 138
ultrametric (absolute value), 239
ultrapower, 143
ultraproduct, 141
underlying normal stratification of, 717
underlying set of simple groups, 346
union (of structures), 137
universal domain, 172
universal elimination lemma, 716

universal Frattini p-cover, 531
universal Frattini cover, 513
universal Hilbert subset property, 289
universal quantifier, 133
unramified
(extension of function fields), 69
unramified (ideal), 32
unramified (prime divisor), 59
unramified (valuation), 25
unramified extension of rings, 129
v.d.Dries-Schmidt, 428
Völklein, 334
valuation, 19
valuation ring, 20
value (of infinite product), 92
value group, 20
value set, 467
valued field, 19
variable symbols, 132
variety, 175
virtually free pro-$\mathcal{C}$, 632
Voloch, 94
Waerden, 48, 69, 318, 392, 428
Waterhouse, 18
weak (n-adic structure), 700
weak approximation theorem, 21
weak monadic structures, 700
weak solution
(of an embedding problem), 503
weakly C_i, 456
weakly $C_{i,d}$, 456
weakly normic form, 457
weakly solvable
(embedding problem), 503
Weigel, 358, 536
weight, 601
Weil, 93, 94, 106, 130, 175, 176, 183,
187, 189, 191, 208, 209, 290, 363
Weil differential, 54
Weissauer, 230, 262, 266, 278, 282,
284, 287, 290, 333
Whaples, 313
Wheeler, 561
width (of a Kronecer class), 464
Wielandt, 394
witness, 152
Witt, 300
word, 350
working functions, 671

wreath product, 257
X-stable, 294
Zannier, 248, 266
Zariski-closed (affine set), 176
Zariski-open (affine set), 176
Zariski K-topology, 172
Zariski K-closed sets (projective), 185
Zariski K-closed subsets, 172
Zariski K-topology, 185
Zariski-open (affine set), 176
Zariski-Samuel, 62, 98, 99, 100, 287, 426, 428, 732
Zariski topology, 176
zero, 52
zero set, 364
zeta function, 79, 371, 739

◎ 编辑手记

　　本书是一本大部头专著,又引自国际著名出版机构,自然价格不菲,所以一定要给出一个理由.为什么要引入?为谁引入的?

　　钱锺书曾经这样评论他的几位老师:"叶公超太懒,吴宓太笨,陈福田太俗",虽然经他的夫人杨绛撰文郑重否认过,却并没有动摇学界的判断:这话就是钱锺书说的,别人不敢说也说不出来,且钱锺书说此话时人在西南联大,杨绛当年却在上海,证伪力度不足.有人评论说:杨绛太卫护夫君了,把一个当代稀有的"魏晋人物",生生弄成无趣的方巾之士,未免大煞风景.

　　借用钱先生的句式有人给出了数学圈的评语:官科太忙,民科太蠢,粉丝太浅.

　　所谓"官科"是指体制内的研究者,弥漫其间的是"大干快上,时不我待"的亢奋与焦虑状态.先说研究生,据笔者了解不论是数学硕士生还是数学博士生都直接跃过了读专著的阶段直接读论文写论文,否则时间就不够了.而老师们则被各种考核打分体系弄得焦头烂额,拼命地写项目书、结题报告、报销经费,无暇他顾,所以能够静下心来读本书的一定不会多.至于后两者就更不用说了,由于门槛过高,他们根本看不懂,更何况大多数人都缺少对大家和高深理论的敬畏之心.这是追求学问和成为学者的基本素养,许多东西绝非普通人想象那般简单.比如,钱锺书有句道尽内心款曲的名言:"大抵学问是荒江野老屋中二三素心人商量培养之事,朝市之显学必成俗学",有人以为该句化自陶渊明《移居》中"闻多素心人,乐与数晨夕",但据考证此句竟是出自古典情色小说《品花宝鉴》,书中有"既然娱悦不在声色,其唯二三知己朝夕素心乎".于是有人慨叹:如此清净淡远的学者隐衷,出处竟深锁春宫,钱氏读书之广可见一斑.据传闻,钱锺书曾信手在纸上向人开列数十本淫秽小说书单.

限于本身的专业特点和笔者的学识水平,这里就不一一介绍内容和作者情况了.据 Moshe Jarden 介绍,本书的主题是学习域的种类的基本性质及其在相关的算法问题中使用代数工具.《算术域》的第一版在 1986 年出版,在第一版的结尾作者给出了 21 个开放性的问题.不过值得注意的是自从第一版出版以后,其中 15 个问题已经被部分或完全地解决了.同时,在许多方面,算术域已经发展成代数学与数论中的独立分支,这些发展中的一部分已经在许多著作中被证明了.《算术域》的第三版与第二版相比较,在两个方面进行了改善.首先,第三版更正了一些打字错误和数学表达上的错误,特别是填补了第二版中关于吉尔摩与罗宾逊、坎特与鲁伯兹凯有关的所有参考文献的空白.其次,第三版报告了 2005 年(第二版出版)以后出现的五个开放性问题.János Kollár 解决了第二个问题及第三个问题,第 31 个问题也被 Lior Bary-Soroker 解决了.最后,Eric Rosen 建议承认第二版中的推论 28.5.3,导致第 33 个问题也被成功地解决了.不幸的是,原版中前四个解决方案的完整描述没有出现在这一版里.能够告诉读者的只有一点,那就是它是一本物有所值的书,值得精读与收藏.光是许多标题就很令数论爱好者遐想:函数域上的黎曼假设,平面曲线,契巴塔廖夫密度定理,代数几何基础,希尔伯特域上的伽罗瓦群,哈尔测度,算术几何问题,弗罗比尼乌斯域,不可判定性,等等.当然引进这只是笔者的一厢情愿,消费者是否认可就是另外一个问题了,这可能就涉及一个较大的话题,即供给侧改革和共识的撕裂.

供给者完全不考虑受众,以自嗨的方式选择生产什么、生产多少.从社会大生产角度而论这是个复杂与敏感的话题,暂且不论,单就文化生产领域它也不是那么简单的,比如喝咖啡为什么一定要 decaf 呢? 查寻了一下 decaf 的来源,根据《韦氏大字典》解释,decaf 是 decaffeinated coffee 的缩写,decaf 最早出现于 1984 年.1970 年,法国哲学家波德里亚便在他的《消费社会》中指出,消费将变成一个"能指"的符号,物的消费将不再仅仅是物的使用价值和交换价值,更重要的是其符号价值,而"消费的过程成为一种意义的建构,比如社会地位、身份标识、文化品位以及美学趣味的彰显等."可见,decaf 是现代社会的产物,并无疑在某种程度上,成为了某种能指的符号,成为了某种"社会地位、身份标识、文化品位以及美学趣味的彰显".喝上一杯 decaf,不是为了让自己保持清醒,而是消费一种咖啡的感觉和氛围.

别的出版机构如何决定引进版权图书的决策机制笔者不了解,但我们工作室就一个因素,即笔者的趣味,这当然极不科学,但这才是形成特色的根本之道.

多年的出版实践使我们认识到:我们永远也不要低估了任何一位潜在读者的阅读趣味和能力.笔者在 2015 年第 2 期《十月》(108 页)上读到一段文字:

> 在 1994 年,深圳打工者李家淳给家人的信中这么写道:最近读了《文化苦旅》才知道散文流变极快,余秋雨老师算是开辟了某种新的散文文体,比之传统,语言风格也有了很大的突破和创新,……散文贵在真实,这种真实是指精神意象的真实,不过据我粗浅的翻读,我觉察出了散文语言太过于追求新颖和变化,也许容易陷于某种"语言虚假",《文化苦旅》中的某些段落,便散发出了这种味道(李家淳:《打工者书信》,《天涯》,2013 年第 6 期).

笔者作为专业出版人几乎同时读到过这本书,但心得比之打工者李家淳差多了.在另一位专业出版人启航著的《无间书道》(新星出版社,2010 年,北京)中有一段关于出版人的论述,笔者深以为是:

> 在《出版人:汤姆·麦奇勒回忆录》的封底印着麦奇勒的一段话,对于正在从事出版工作的人来说,可以把它当成一面镜子——经常有人问我,我如何决定是否应该出版某一本书.要回答这个问题很难,因为做出这样的选择完全是一种个人行为,带有很强的主观性,没有什么规律可循.我只能说,于我而言,我很少出于商业原因来甄选书籍或者作者.要想做好出版,出版人就必须对书籍本身充满热情.对我来说,要想做到这一点,我就必须真正喜欢这本书,而要喜欢这本书,我就必须真正赞赏这本书的品质.这就是我唯一的原则.一旦做出了决定,接下来就开始操作.首先在出版社内部传播这种信念,然后再传播

到外界.

  不论学历如何，不论曾经做过什么，作为一名当下的出版从业者，应该认真问一问自己，我真的热爱图书吗？我清楚自己最喜欢哪一种类型的图书吗？如果对于这两个问题能有坚定的答案，并在行动上忠诚于自己的心，同时充满前行的热情，那么，成为一名优秀的出版者将指日可待. 这就是汤姆·麦奇勒给我的启示.

刘培杰

2017 年 9 月 10 日

于哈工大

哈尔滨工业大学出版社刘培杰数学工作室
已出版(即将出版)图书目录

书　名	出版时间	定　价	编号
新编中学数学解题方法全书(高中版)上卷	2007—09	38.00	7
新编中学数学解题方法全书(高中版)中卷	2007—09	48.00	8
新编中学数学解题方法全书(高中版)下卷(一)	2007—09	42.00	17
新编中学数学解题方法全书(高中版)下卷(二)	2007—09	38.00	18
新编中学数学解题方法全书(高中版)下卷(三)	2010—06	58.00	73
新编中学数学解题方法全书(初中版)上卷	2008—01	28.00	29
新编中学数学解题方法全书(初中版)中卷	2010—07	38.00	75
新编中学数学解题方法全书(高考复习卷)	2010—01	48.00	67
新编中学数学解题方法全书(高考真题卷)	2010—01	38.00	62
新编中学数学解题方法全书(高考精华卷)	2011—03	68.00	118
新编平面解析几何解题方法全书(专题讲座卷)	2010—01	18.00	61
新编中学数学解题方法全书(自主招生卷)	2013—08	88.00	261
数学眼光透视(第2版)	2017—06	78.00	732
数学思想领悟	2008—01	38.00	25
数学应用展观(第2版)	2017—08	68.00	737
数学建模导引	2008—01	28.00	23
数学方法溯源	2008—01	38.00	27
数学史话览胜(第2版)	2017—01	48.00	736
数学思维技术	2013—09	38.00	260
数学解题引论	2017—05	48.00	735
从毕达哥拉斯到怀尔斯	2007—10	48.00	9
从迪利克雷到维斯卡尔迪	2008—01	48.00	21
从哥德巴赫到陈景润	2008—05	98.00	35
从庞加莱到佩雷尔曼	2011—08	138.00	136
数学奥林匹克与数学文化(第一辑)	2006—05	48.00	4
数学奥林匹克与数学文化(第二辑)(竞赛卷)	2008—01	48.00	19
数学奥林匹克与数学文化(第二辑)(文化卷)	2008—07	58.00	36'
数学奥林匹克与数学文化(第三辑)(竞赛卷)	2010—01	48.00	59
数学奥林匹克与数学文化(第四辑)(竞赛卷)	2011—08	58.00	87
数学奥林匹克与数学文化(第五辑)	2015—06	98.00	370

哈尔滨工业大学出版社刘培杰数学工作室
已出版(即将出版)图书目录

书　名	出版时间	定　价	编号
世界著名平面几何经典著作钩沉——几何作图专题卷(上)	2009—06	48.00	49
世界著名平面几何经典著作钩沉——几何作图专题卷(下)	2011—01	88.00	80
世界著名平面几何经典著作钩沉(民国平面几何老课本)	2011—03	38.00	113
世界著名平面几何经典著作钩沉(建国初期平面三角老课本)	2015—08	38.00	507
世界著名解析几何经典著作钩沉——平面解析几何卷	2014—01	38.00	264
世界著名数论经典著作钩沉(算术卷)	2012—01	28.00	125
世界著名数学经典著作钩沉——立体几何卷	2011—02	28.00	88
世界著名三角学经典著作钩沉(平面三角卷Ⅰ)	2010—06	28.00	69
世界著名三角学经典著作钩沉(平面三角卷Ⅱ)	2011—01	38.00	78
世界著名初等数论经典著作钩沉(理论和实用算术卷)	2011—07	38.00	126
发展你的空间想象力	2017—06	38.00	785
走向国际数学奥林匹克的平面几何试题诠释(上、下)(第1版)	2007—01	68.00	11,12
走向国际数学奥林匹克的平面几何试题诠释(上、下)(第2版)	2010—02	98.00	63,64
平面几何证明方法全书	2007—08	35.00	1
平面几何证明方法全书习题解答(第1版)	2005—10	18.00	2
平面几何证明方法全书习题解答(第2版)	2006—12	18.00	10
平面几何天天练上卷·基础篇(直线型)	2013—01	58.00	208
平面几何天天练中卷·基础篇(涉及圆)	2013—01	28.00	234
平面几何天天练下卷·提高篇	2013—01	58.00	237
平面几何专题研究	2013—07	98.00	258
最新世界各国数学奥林匹克中的平面几何试题	2007—09	38.00	14
数学竞赛平面几何典型题及新颖解	2010—07	48.00	74
初等数学复习及研究(平面几何)	2008—09	58.00	38
初等数学复习及研究(立体几何)	2010—06	38.00	71
初等数学复习及研究(平面几何)习题解答	2009—01	48.00	42
几何学教程(平面几何卷)	2011—03	68.00	90
几何学教程(立体几何卷)	2011—07	68.00	130
几何变换与几何证题	2010—06	88.00	70
计算方法与几何证题	2011—06	28.00	129
立体几何技巧与方法	2014—04	88.00	293
几何瑰宝——平面几何500名题暨1000条定理(上、下)	2010—07	138.00	76,77
三角形的解法与应用	2012—07	18.00	183
近代的三角形几何学	2012—07	48.00	184
一般折线几何学	2015—08	48.00	503
三角形的五心	2009—06	28.00	51
三角形的六心及其应用	2015—10	68.00	542
三角形趣谈	2012—08	28.00	212
解三角形	2014—01	28.00	265
三角学专门教程	2014—09	28.00	387
距离几何分析导引	2015—02	68.00	446
图天下几何新题试卷.初中	2017—01	58.00	714

哈尔滨工业大学出版社刘培杰数学工作室
已出版(即将出版)图书目录

书 名	出版时间	定 价	编号
圆锥曲线习题集(上册)	2013—06	68.00	255
圆锥曲线习题集(中册)	2015—01	78.00	434
圆锥曲线习题集(下册·第1卷)	2016—10	78.00	683
论九点圆	2015—05	88.00	645
近代欧氏几何学	2012—03	48.00	162
罗巴切夫斯基几何学及几何基础概要	2012—07	28.00	188
罗巴切夫斯基几何学初步	2015—06	28.00	474
用三角、解析几何、复数、向量计算解数学竞赛几何题	2015—03	48.00	455
美国中学几何教程	2015—04	88.00	458
三线坐标与三角形特征点	2015—04	98.00	460
平面解析几何方法与研究(第1卷)	2015—05	18.00	471
平面解析几何方法与研究(第2卷)	2015—06	18.00	472
平面解析几何方法与研究(第3卷)	2015—07	18.00	473
解析几何研究	2015—01	38.00	425
解析几何学教程.上	2016—01	38.00	574
解析几何学教程.下	2016—01	38.00	575
几何学基础	2016—01	58.00	581
初等几何研究	2015—02	58.00	444
大学几何学	2017—01	78.00	688
关于曲面的一般研究	2016—11	48.00	690
十九和二十世纪欧氏几何学中的片段	2017—01	58.00	696
近世纯粹几何学初论	2017—01	58.00	711
拓扑学与几何学基础讲义	2017—04	58.00	756
物理学中的几何方法	2017—06	88.00	767
平面几何.中考.高考.奥数一本通	2017—07	28.00	820
几何学简史	2017—08	28.00	833
俄罗斯平面几何问题集	2009—08	88.00	55
俄罗斯立体几何问题集	2014—03	58.00	283
俄罗斯几何大师——沙雷金论数学及其他	2014—01	48.00	271
来自俄罗斯的5000道几何习题及解答	2011—03	58.00	89
俄罗斯初等数学问题集	2012—05	38.00	177
俄罗斯函数问题集	2011—03	38.00	103
俄罗斯组合分析问题集	2011—01	48.00	79
俄罗斯初等数学万题选——三角卷	2012—11	38.00	222
俄罗斯初等数学万题选——代数卷	2013—08	68.00	225
俄罗斯初等数学万题选——几何卷	2014—01	68.00	226
463个俄罗斯几何老问题	2012—01	28.00	152
超越吉米多维奇.数列的极限	2009—11	48.00	58
超越普里瓦洛夫.留数卷	2015—01	28.00	437
超越普里瓦洛夫.无穷乘积与它对解析函数的应用卷	2015—05	28.00	477
超越普里瓦洛夫.积分卷	2015—06	18.00	481
超越普里瓦洛夫.基础知识卷	2015—06	28.00	482
超越普里瓦洛夫.数项级数卷	2015—07	38.00	489
初等数论难题集(第一卷)	2009—05	68.00	44
初等数论难题集(第二卷)(上、下)	2011—02	128.00	82,83
数论概貌	2011—03	18.00	93
代数数论(第二版)	2013—08	58.00	94
代数多项式	2014—06	38.00	289
初等数论的知识与问题	2011—02	28.00	95
超越数论基础	2011—03	28.00	96
数论初等教程	2011—03	28.00	97
数论基础	2011—03	18.00	98
数论基础与维诺格拉多夫	2014—03	18.00	292

哈尔滨工业大学出版社刘培杰数学工作室
已出版(即将出版)图书目录

书　　名	出版时间	定价	编号
解析数论基础	2012—08	28.00	216
解析数论基础(第二版)	2014—01	48.00	287
解析数论问题集(第二版)(原版引进)	2014—05	88.00	343
解析数论问题集(第二版)(中译本)	2016—04	88.00	607
解析数论基础(潘承洞,潘承彪著)	2016—07	98.00	673
解析数论导引	2016—07	58.00	674
数论入门	2011—03	38.00	99
代数数论入门	2015—03	38.00	448
数论开篇	2012—07	28.00	194
解析数论引论	2011—03	48.00	100
Barban Davenport Halberstam 均值和	2009—01	40.00	33
基础数论	2011—03	28.00	101
初等数论 100 例	2011—05	18.00	122
初等数论经典例题	2012—07	18.00	204
最新世界各国数学奥林匹克中的初等数论试题(上、下)	2012—01	138.00	144,145
初等数论(Ⅰ)	2012—01	18.00	156
初等数论(Ⅱ)	2012—01	18.00	157
初等数论(Ⅲ)	2012—01	28.00	158
平面几何与数论中未解决的新老问题	2013—01	68.00	229
代数数论简史	2014—11	28.00	408
代数数论	2015—09	88.00	532
代数、数论及分析习题集	2016—11	98.00	695
数论导引提要及习题解答	2016—01	48.00	559
素数定理的初等证明.第2版	2016—09	48.00	686
数论中的模函数与狄利克雷级数(第二版)	2017—11	78.00	837
谈谈素数	2011—03	18.00	91
平方和	2011—03	18.00	92
复变函数引论	2013—10	68.00	269
伸缩变换与抛物旋转	2015—01	38.00	449
无穷分析引论(上)	2013—04	88.00	247
无穷分析引论(下)	2013—04	98.00	245
数学分析	2014—04	28.00	338
数学分析中的一个新方法及其应用	2013—01	38.00	231
数学分析例选:通过范例学技巧	2013—01	88.00	243
高等代数例选:通过范例学技巧	2015—06	88.00	475
三角级数论(上册)(陈建功)	2013—01	38.00	232
三角级数论(下册)(陈建功)	2013—01	48.00	233
三角级数论(哈代)	2013—06	48.00	254
三角级数	2015—07	28.00	263
超越数	2011—03	18.00	109
三角和方法	2011—03	18.00	112
整数论	2011—05	38.00	120
从整数谈起	2015—10	28.00	538
随机过程(Ⅰ)	2014—01	78.00	224
随机过程(Ⅱ)	2014—01	68.00	235
算术探索	2011—12	158.00	148
组合数学	2012—04	28.00	178
组合数学浅谈	2012—03	28.00	159
丢番图方程引论	2012—03	48.00	172
拉普拉斯变换及其应用	2015—02	38.00	447
高等代数.上	2016—01	38.00	548
高等代数.下	2016—01	38.00	549

哈尔滨工业大学出版社刘培杰数学工作室
已出版(即将出版)图书目录

书　名	出版时间	定　价	编号
高等代数教程	2016—01	58.00	579
数学解析教程.上卷.1	2016—01	58.00	546
数学解析教程.上卷.2	2016—01	38.00	553
数学解析教程.下卷.1	2017—04	48.00	781
数学解析教程.下卷.2	2017—06	48.00	782
函数构造论.上	2016—01	38.00	554
函数构造论.中	2017—06	48.00	555
函数构造论.下	2016—09	48.00	680
数与多项式	2016—01	38.00	558
概周期函数	2016—01	48.00	572
变叙的项的极限分布律	2016—01	18.00	573
整函数	2012—08	18.00	161
近代拓扑学研究	2013—04	38.00	239
多项式和无理数	2008—01	68.00	22
模糊数据统计学	2008—03	48.00	31
模糊分析学与特殊泛函空间	2013—01	68.00	241
谈谈不定方程	2011—05	28.00	119
常微分方程	2016—01	58.00	586
平稳随机函数导论	2016—03	48.00	587
量子力学原理·上	2016—01	38.00	588
图与矩阵	2014—08	40.00	644
钢丝绳原理:第二版	2017—01	78.00	745
代数拓扑和微分拓扑简史	2017—06	68.00	791
受控理论与解析不等式	2012—05	78.00	165
解析不等式新论	2009—06	68.00	48
建立不等式的方法	2011—03	98.00	104
数学奥林匹克不等式研究	2009—08	68.00	56
不等式研究(第二辑)	2012—02	68.00	153
不等式的秘密(第一卷)	2012—02	28.00	154
不等式的秘密(第一卷)(第2版)	2014—02	38.00	286
不等式的秘密(第二卷)	2014—01	38.00	268
初等不等式的证明方法	2010—06	38.00	123
初等不等式的证明方法(第二版)	2014—11	38.00	407
不等式·理论·方法(基础卷)	2015—07	38.00	496
不等式·理论·方法(经典不等式卷)	2015—07	38.00	497
不等式·理论·方法(特殊类型不等式卷)	2015—07	48.00	498
不等式的分拆降维降幂方法与可读证明	2016—01	68.00	591
不等式探究	2016—03	38.00	582
不等式探秘	2017—01	88.00	689
四面体不等式	2017—01	68.00	715
数学奥林匹克中常见重要不等式	2017—09	38.00	845
同余理论	2012—05	38.00	163
[x]与{x}	2015—04	48.00	476
极值与最值.上卷	2015—06	28.00	486
极值与最值.中卷	2015—06	38.00	487
极值与最值.下卷	2015—06	28.00	488
整数的性质	2012—11	38.00	192
完全平方数及其应用	2015—08	78.00	506
多项式理论	2015—10	88.00	541

哈尔滨工业大学出版社刘培杰数学工作室
已出版（即将出版）图书目录

书　名	出版时间	定　价	编号
历届美国中学生数学竞赛试题及解答（第一卷）1950—1954	2014—07	18.00	277
历届美国中学生数学竞赛试题及解答（第二卷）1955—1959	2014—04	18.00	278
历届美国中学生数学竞赛试题及解答（第三卷）1960—1964	2014—06	18.00	279
历届美国中学生数学竞赛试题及解答（第四卷）1965—1969	2014—04	28.00	280
历届美国中学生数学竞赛试题及解答（第五卷）1970—1972	2014—06	18.00	281
历届美国中学生数学竞赛试题及解答（第六卷）1973—1980	2017—07	18.00	768
历届美国中学生数学竞赛试题及解答（第七卷）1981—1986	2015—01	18.00	424
历届美国中学生数学竞赛试题及解答（第八卷）1987—1990	2017—05	18.00	769
历届 IMO 试题集（1959—2005）	2006—05	58.00	5
历届 CMO 试题集	2008—09	28.00	40
历届中国数学奥林匹克试题集（第2版）	2017—03	38.00	757
历届加拿大数学奥林匹克试题集	2012—08	38.00	215
历届美国数学奥林匹克试题集：多解推广加强	2012—08	38.00	209
历届美国数学奥林匹克试题集：多解推广加强（第2版）	2016—03	48.00	592
历届波兰数学竞赛试题集. 第1卷,1949～1963	2015—03	18.00	453
历届波兰数学竞赛试题集. 第2卷,1964～1976	2015—03	18.00	454
历届巴尔干数学奥林匹克试题集	2015—05	38.00	466
保加利亚数学奥林匹克	2014—10	38.00	393
圣彼得堡数学奥林匹克试题集	2015—01	38.00	429
匈牙利奥林匹克数学竞赛题解. 第1卷	2016—05	28.00	593
匈牙利奥林匹克数学竞赛题解. 第2卷	2016—05	28.00	594
超越普特南试题：大学数学竞赛中的方法与技巧	2017—04	98.00	758
历届国际大学生数学竞赛试题集（1994—2010）	2012—01	28.00	143
全国大学生数学夏令营数学竞赛试题及解答	2007—03	28.00	15
全国大学生数学竞赛辅导教程	2012—07	28.00	189
全国大学生数学竞赛复习全书（第2版）	2017—05	58.00	787
历届美国大学生数学竞赛试题集	2009—03	88.00	43
前苏联大学生数学奥林匹克竞赛题解（上编）	2012—04	28.00	169
前苏联大学生数学奥林匹克竞赛题解（下编）	2012—04	38.00	170
历届美国数学邀请赛试题集	2014—01	48.00	270
全国高中数学竞赛试题及解答. 第1卷	2014—07	38.00	331
大学生数学竞赛讲义	2014—09	28.00	371
普林斯顿大学数学竞赛	2016—06	38.00	669
亚太地区数学奥林匹克竞赛题	2015—07	18.00	492
日本历届（初级）广中杯数学竞赛试题及解答. 第1卷（2000～2007）	2016—05	28.00	641
日本历届（初级）广中杯数学竞赛试题及解答. 第2卷（2008～2015）	2016—05	38.00	642
360个数学竞赛问题	2016—08	58.00	677
奥数最佳实战题. 上卷	2017—06	38.00	760
奥数最佳实战题. 下卷	2017—05	58.00	761
哈尔滨市早期中学数学竞赛试题汇编	2016—07	28.00	672
全国高中数学联赛试题及解答:1981—2015	2016—08	98.00	676
20世纪50年代全国部分城市数学竞赛试题汇编	2017—07	28.00	797
高考数学临门一脚（含密押三套卷）（理科版）	2017—01	45.00	743
高考数学临门一脚（含密押三套卷）（文科版）	2017—01	45.00	744
新课标高考数学题型全归纳（文科版）	2015—05	72.00	467
新课标高考数学题型全归纳（理科版）	2015—05	82.00	468
洞穿高考数学解答题核心考点（理科版）	2015—11	49.80	550
洞穿高考数学解答题核心考点（文科版）	2015—11	46.80	551
高考数学题型全归纳:文科版. 上	2016—05	53.00	663
高考数学题型全归纳:文科版. 下	2016—05	53.00	664
高考数学题型全归纳:理科版. 上	2016—05	58.00	665
高考数学题型全归纳:理科版. 下	2016—05	58.00	666

哈尔滨工业大学出版社刘培杰数学工作室
已出版(即将出版)图书目录

书 名	出版时间	定价	编号
王连笑教你怎样学数学:高考选择题解题策略与客观题实用训练	2014—01	48.00	262
王连笑教你怎样学数学:高考数学高层次讲座	2015—02	48.00	432
高考数学的理论与实践	2009—08	38.00	53
高考数学核心题型解题方法与技巧	2010—01	28.00	86
高考思维新平台	2014—03	38.00	259
30分钟拿下高考数学选择题、填空题(理科版)	2016—10	39.80	720
30分钟拿下高考数学选择题、填空题(文科版)	2016—10	39.80	721
高考数学压轴题解题诀窍(上)	2012—02	78.00	166
高考数学压轴题解题诀窍(下)	2012—03	28.00	167
北京市五区文科数学三年高考模拟题详解:2013~2015	2015—08	48.00	500
北京市五区理科数学三年高考模拟题详解:2013~2015	2015—09	68.00	505
向量法巧解数学高考题	2009—08	28.00	54
高考数学万能解题法(第2版)	即将出版	38.00	691
高考物理万能解题法(第2版)	即将出版	38.00	692
高考化学万能解题法(第2版)	即将出版	28.00	693
高考生物万能解题法(第2版)	即将出版	28.00	694
高考数学解题金典(第2版)	2017—01	78.00	716
高考物理解题金典(第2版)	即将出版	68.00	717
高考化学解题金典(第2版)	即将出版	58.00	718
我一定要赚分:高中物理	2016—01	38.00	580
数学高考参考	2016—01	78.00	589
2011~2015年全国及各省市高考数学文科精品试题审题要津与解法研究	2015—10	68.00	539
2011~2015年全国及各省市高考数学理科精品试题审题要津与解法研究	2015—10	88.00	540
最新全国及各省市高考数学试卷解法研究及点拨评析	2009—02	38.00	41
2011年全国及各省市高考数学试题审题要津与解法研究	2011—10	48.00	139
2013年全国及各省市高考数学试题解析与点评	2014—01	48.00	282
全国及各省市高考数学试题审题要津与解法研究	2015—02	48.00	450
新课标高考数学——五年试题分章详解(2007~2011)(上、下)	2011—10	78.00	140,141
全国中考数学压轴题审题要津与解法研究	2013—04	78.00	248
新编全国中考数学压轴题审题要津与解法研究	2014—05	58.00	342
全国及各省市5年中考数学压轴题审题要津与解法研究(2015版)	2015—04	58.00	462
中考数学专题总复习	2007—04	28.00	6
中考数学较难题、难题常考题型解题方法与技巧.上	2016—01	48.00	584
中考数学较难题、难题常考题型解题方法与技巧.下	2016—01	58.00	585
中考数学较难题常考题型解题方法与技巧	2016—09	48.00	681
中考数学难题常考题型解题方法与技巧	2016—09	48.00	682
中考数学选择填空压轴好题妙解365	2017—05	38.00	759
中考数学小压轴汇编初讲	2017—07	48.00	788
中考数学大压轴专题微言	2017—09	48.00	846
北京中考数学压轴题解题方法突破(第2版)	2017—03	48.00	753
助你高考成功的数学解题智慧:知识是智慧的基础	2016—01	58.00	596
助你高考成功的数学解题智慧:错误是智慧的试金石	2016—04	58.00	643
助你高考成功的数学解题智慧:方法是智慧的推手	2016—04	68.00	657
高考数学奇思妙解	2016—04	38.00	610
高考数学解题策略	2016—05	48.00	670
数学解题泄天机	2016—06	48.00	668
高考物理压轴题全解	2017—04	48.00	746
高中物理经典问题25讲	2017—05	28.00	764
2016年高考文科数学真题研究	2017—04	58.00	754
2016年高考理科数学真题研究	2017—04	78.00	755
初中数学、高中数学脱节知识补缺教材	2017—06	48.00	766
赢在小题	2017—08	48.00	834
高考数学核心素养解读	2017—09	38.00	839
高考数学客观题解题方法和技巧	2017—10	38.00	847

哈尔滨工业大学出版社刘培杰数学工作室
已出版（即将出版）图书目录

书　名	出版时间	定　价	编号
新编 640 个世界著名数学智力趣题	2014—01	88.00	242
500 个最新世界著名数学智力趣题	2008—06	48.00	3
400 个最新世界著名数学最值问题	2008—09	48.00	36
500 个世界著名数学征解问题	2009—06	48.00	52
400 个中国最佳初等数学征解老问题	2010—01	48.00	60
500 个俄罗斯数学经典老题	2011—01	28.00	81
1000 个国外中学物理好题	2012—04	48.00	174
300 个日本高考数学题	2012—05	38.00	142
700 个早期日本高考数学试题	2017—02	88.00	752
500 个前苏联早期高考数学试题及解答	2012—05	28.00	185
546 个早期俄罗斯大学生数学竞赛题	2014—03	38.00	285
548 个来自美苏的数学好问题	2014—11	28.00	396
20 所苏联著名大学早期入学试题	2015—02	18.00	452
161 道德国工科大学生必做的微分方程习题	2015—05	28.00	469
500 个德国工科大学生必做的高数习题	2015—06	28.00	478
360 个数学竞赛问题	2016—08	58.00	677
德国讲义日本考题.微积分卷	2015—04	48.00	456
德国讲义日本考题.微分方程卷	2015—04	38.00	457
二十世纪中叶中、英、美、日、法、俄高考数学试题精选	2017—06	38.00	783
中国初等数学研究　2009 卷（第 1 辑）	2009—05	20.00	45
中国初等数学研究　2010 卷（第 2 辑）	2010—05	30.00	68
中国初等数学研究　2011 卷（第 3 辑）	2011—07	60.00	127
中国初等数学研究　2012 卷（第 4 辑）	2012—07	48.00	190
中国初等数学研究　2014 卷（第 5 辑）	2014—02	48.00	288
中国初等数学研究　2015 卷（第 6 辑）	2015—06	68.00	493
中国初等数学研究　2016 卷（第 7 辑）	2016—04	68.00	609
中国初等数学研究　2017 卷（第 8 辑）	2017—01	98.00	712
几何变换（Ⅰ）	2014—07	28.00	353
几何变换（Ⅱ）	2015—06	28.00	354
几何变换（Ⅲ）	2015—01	38.00	355
几何变换（Ⅳ）	2015—12	38.00	356
博弈论精粹	2008—03	58.00	30
博弈论精粹.第二版（精装）	2015—01	88.00	461
数学 我爱你	2008—01	28.00	20
精神的圣徒　别样的人生——60 位中国数学家成长的历程	2008—09	48.00	39
数学史概论	2009—06	78.00	50
数学史概论(精装)	2013—03	158.00	272
数学史选讲	2016—01	48.00	544
斐波那契数列	2010—02	28.00	65
数学拼盘和斐波那契魔方	2010—07	38.00	72
斐波那契数列欣赏	2011—01	28.00	160
数学的创造	2011—02	48.00	85
数学美与创造力	2016—01	48.00	595
数海拾贝	2016—01	48.00	590
数学中的美	2011—02	38.00	84
数论中的美学	2014—12	38.00	351
数学王者　科学巨人——高斯	2015—01	28.00	428
振兴祖国数学的圆梦之旅:中国初等数学研究史话	2015—06	98.00	490
二十世纪中国数学史料研究	2015—10	48.00	536
数字谜、数阵图与棋盘覆盖	2016—01	58.00	298
时间的形状	2016—01	38.00	556
数学发现的艺术:数学探索中的合情推理	2016—07	58.00	671
活跃在数学中的参数	2016—07	48.00	675

哈尔滨工业大学出版社刘培杰数学工作室
已出版(即将出版)图书目录

书 名	出版时间	定 价	编号
数学解题——靠数学思想给力(上)	2011—07	38.00	131
数学解题——靠数学思想给力(中)	2011—07	48.00	132
数学解题——靠数学思想给力(下)	2011—07	38.00	133
我怎样解题	2013—01	48.00	227
数学解题中的物理方法	2011—06	28.00	114
数学解题的特殊方法	2011—06	48.00	115
中学数学计算技巧	2012—01	48.00	116
中学数学证明方法	2012—01	58.00	117
数学趣题巧解	2012—03	28.00	128
高中数学教学通鉴	2015—05	58.00	479
和高中生漫谈:数学与哲学的故事	2014—08	28.00	369
算术问题集	2017—03	38.00	789
自主招生考试中的参数方程问题	2015—01	28.00	435
自主招生考试中的极坐标问题	2015—04	28.00	463
近年全国重点大学自主招生数学试题全解及研究.华约卷	2016—02	38.00	441
近年全国重点大学自主招生数学试题全解及研究.北约卷	2016—05	38.00	619
自主招生数学解证宝典	2015—09	48.00	535
格点和面积	2012—07	18.00	191
射影几何趣谈	2012—04	28.00	175
斯潘纳尔引理——从一道加拿大数学奥林匹克试题谈起	2014—01	28.00	228
李普希兹条件——从几道近年高考数学试题谈起	2012—10	18.00	221
拉格朗日中值定理——从一道北京高考试题的解法谈起	2015—10	18.00	197
闵科夫斯基定理——从一道清华大学自主招生试题谈起	2014—01	28.00	198
哈尔测度——从一道冬令营试题的背景谈起	2012—08	28.00	202
切比雪夫逼近问题——从一道中国台北数学奥林匹克试题谈起	2013—04	38.00	238
伯恩斯坦多项式与贝齐尔曲面——从一道全国高中数学联赛试题谈起	2013—03	38.00	236
卡塔兰猜想——从一道普特南竞赛试题谈起	2013—06	18.00	256
麦卡锡函数和阿克曼函数——从一道前南斯拉夫数学奥林匹克试题谈起	2012—08	18.00	201
贝蒂定理与拜姆贝克莫斯尔定理——从一个拣石子游戏谈起	2012—08	18.00	217
皮亚诺曲线和豪斯道夫分球定理——从无限集谈起	2012—08	18.00	211
平面凸图形与凸多面体	2012—10	28.00	218
斯坦因豪斯问题——从一道二十五省市自治区中学数学竞赛试题谈起	2012—07	18.00	196
纽结理论中的亚历山大多项式与琼斯多项式——从一道北京市高一数学竞赛试题谈起	2012—07	28.00	195
原则与策略——从波利亚"解题表"谈起	2013—04	38.00	244
转化与化归——从三大尺规作图不能问题谈起	2012—08	28.00	214
代数几何中的贝祖定理(第一版)——从一道IMO试题的解法谈起	2013—08	18.00	193
成功连贯理论与约当块理论——从一道比利时数学竞赛试题谈起	2012—04	18.00	180
素数判定与大数分解	2014—08	18.00	199
置换多项式及其应用	2012—10	18.00	220
椭圆函数与模函数——从一道美国加州大学洛杉矶分校(UCLA)博士资格考题谈起	2012—10	28.00	219
差分方程的拉格朗日方法——从一道2011年全国高考理科试题的解法谈起	2012—08	28.00	200

哈尔滨工业大学出版社刘培杰数学工作室
已出版(即将出版)图书目录

书 名	出版时间	定 价	编号
力学在几何中的一些应用	2013—01	38.00	240
高斯散度定理、斯托克斯定理和平面格林定理——从一道国际大学生数学竞赛试题谈起	即将出版		
康托洛维奇不等式——从一道全国高中联赛试题谈起	2013—03	28.00	337
西格尔引理——从一道第18届IMO试题的解法谈起	即将出版		
罗斯定理——从一道前苏联数学竞赛试题谈起	即将出版		
拉克斯定理和阿廷定理——从一道IMO试题的解法谈起	2014—01	58.00	246
毕卡大定理——从一道美国大学数学竞赛试题谈起	2014—07	18.00	350
贝齐尔曲线——从一道全国高中联赛试题谈起	即将出版		
拉格朗日乘子定理——从一道2005年全国高中联赛试题的高等数学解法谈起	2015—05	28.00	480
雅可比定理——从一道日本数学奥林匹克试题谈起	2013—04	48.00	249
李天岩—约克定理——从一道波兰数学竞赛试题谈起	2014—06	28.00	349
整系数多项式因式分解的一般方法——从克朗耐克算法谈起	即将出版		
布劳维不动点定理——从一道前苏联数学奥林匹克试题谈起	2014—01	38.00	273
伯恩赛德定理——从一道英国数学奥林匹克试题谈起	即将出版		
布查特—莫斯特定理——从一道上海市初中竞赛试题谈起	即将出版		
数论中的同余数问题——从一道普特南竞赛试题谈起	即将出版		
范·德蒙行列式——从一道美国数学奥林匹克试题谈起	即将出版		
中国剩余定理:总数法构建中国历史年表	2015—01	28.00	430
牛顿程序与方程求根——从一道全国高考试题解法谈起	即将出版		
库默尔定理——从一道IMO预选试题谈起	即将出版		
卢丁定理——从一道冬令营试题的解法谈起	即将出版		
沃斯滕霍姆定理——从一道IMO预选试题谈起	即将出版		
卡尔松不等式——从一道莫斯科数学奥林匹克试题谈起	即将出版		
信息论中的香农熵——从一道近年高考压轴题谈起	即将出版		
约当不等式——从一道希望杯竞赛试题谈起	即将出版		
拉比诺维奇定理	即将出版		
刘维尔定理——从一道《美国数学月刊》征解问题的解法谈起	即将出版		
卡塔兰恒等式与级数求和——从一道IMO试题的解法谈起	即将出版		
勒让德猜想与素数分布——从一道爱尔兰竞赛试题谈起	即将出版		
天平称重与信息——从一道基辅市数学奥林匹克试题谈起	即将出版		
哈密尔顿—凯莱定理:从一道高中数学联赛试题的解法谈起	2014—09	18.00	376
艾思特曼定理——从一道CMO试题的解法谈起	即将出版		
一个爱尔特希问题——从一道西德数学奥林匹克试题谈起	即将出版		
有限群中的爱丁格尔问题——从一道北京市初中二年级数学竞赛试题谈起	即将出版		
贝克码与编码理论——从一道全国高中联赛试题谈起	即将出版		
帕斯卡三角形	2014—03	18.00	294
蒲丰投针问题——从2009年清华大学的一道自主招生试题谈起	2014—01	38.00	295
斯图姆定理——从一道"华约"自主招生试题的解法谈起	2014—01	18.00	296
许瓦兹引理——从一道加利福尼亚大学伯克利分校数学系博士生试题谈起	2014—08	18.00	297
拉姆塞定理——从王诗宬院士的一个问题谈起	2016—04	48.00	299
坐标法	2013—12	28.00	332
数论三角形	2014—04	38.00	341
毕克定理	2014—07	18.00	352
数林掠影	2014—09	48.00	389
我们周围的概率	2014—10	38.00	390
凸函数最值定理:从一道华约自主招生题的解法谈起	2014—10	28.00	391
易学与数学奥林匹克	2014—10	38.00	392

哈尔滨工业大学出版社刘培杰数学工作室
已出版（即将出版）图书目录

书　名	出版时间	定　价	编号
生物数学趣谈	2015—01	18.00	409
反演	2015—01	28.00	420
因式分解与圆锥曲线	2015—01	18.00	426
轨迹	2015—01	28.00	427
面积原理：从常庚哲命的一道CMO试题的积分解法谈起	2015—01	48.00	431
形形色色的不动点定理：从一道28届IMO试题谈起	2015—01	38.00	439
柯西函数方程：从一道上海交大自主招生的试题谈起	2015—02	28.00	440
三角恒等式	2015—02	28.00	442
无理性判定：从一道2014年"北约"自主招生试题谈起	2015—01	38.00	443
数学归纳法	2015—03	18.00	451
极端原理与解题	2015—04	28.00	464
法雷级数	2014—08	18.00	367
摆线族	2015—01	38.00	438
函数方程及其解法	2015—05	38.00	470
含参数的方程和不等式	2012—09	28.00	213
希尔伯特第十问题	2016—01	38.00	543
无穷小量的求和	2016—01	28.00	545
切比雪夫多项式：从一道清华大学金秋营试题谈起	2016—01	38.00	583
泽肯多夫定理	2016—03	38.00	599
代数等式证题法	2016—01	28.00	600
三角等式证题法	2016—01	28.00	601
吴大任教授藏书中的一个因式分解公式：从一道美国数学邀请试题的解法谈起	2016—06	28.00	656
易卦——类万物的数学模型	2017—08	68.00	838
中等数学英语阅读文选	2006—12	38.00	13
统计学专业英语	2007—03	28.00	16
统计学专业英语（第二版）	2012—07	48.00	176
统计学专业英语（第三版）	2015—04	68.00	465
幻方和魔方（第一卷）	2012—05	68.00	173
尘封的经典——初等数学经典文献选读（第一卷）	2012—07	48.00	205
尘封的经典——初等数学经典文献选读（第二卷）	2012—07	38.00	206
代换分析：英文	2015—07	38.00	499
实变函数论	2012—06	78.00	181
复变函数论	2015—08	38.00	504
非光滑优化及其变分分析	2014—01	48.00	230
疏散的马尔科夫链	2014—01	58.00	266
马尔科夫过程论基础	2015—01	28.00	433
初等微分拓扑学	2012—07	18.00	182
方程式论	2011—03	38.00	105
初级方程式论	2011—03	28.00	106
Galois 理论	2011—03	18.00	107
古典数学难题与伽罗瓦理论	2012—11	58.00	223
伽罗华与群论	2014—01	28.00	290
代数方程的根式解及伽罗瓦理论	2011—03	28.00	108
代数方程的根式解及伽罗瓦理论（第二版）	2015—01	28.00	423
线性偏微分方程讲义	2011—03	18.00	110
几类微分方程数值方法的研究	2015—05	38.00	485
N 体问题的周期解	2011—03	28.00	111
代数方程式论	2011—05	18.00	121
线性代数与几何：英文	2016—06	58.00	578
动力系统的不变量与函数方程	2011—07	48.00	137
基于短语评价的翻译知识获取	2012—02	48.00	168
应用随机过程	2012—04	48.00	187
概率论导引	2012—04	18.00	179

哈尔滨工业大学出版社刘培杰数学工作室
已出版（即将出版）图书目录

书　名	出版时间	定　价	编号
矩阵论（上）	2013—06	58.00	250
矩阵论（下）	2013—06	48.00	251
对称锥互补问题的内点法：理论分析与算法实现	2014—08	68.00	368
抽象代数：方法导引	2013—06	38.00	257
集论	2016—01	48.00	576
多项式理论研究综述	2016—01	38.00	577
函数论	2014—11	78.00	395
反问题的计算方法及应用	2011—11	28.00	147
初等数学研究（Ⅰ）	2008—09	68.00	37
初等数学研究（Ⅱ）(上、下)	2009—05	118.00	46,47
数阵及其应用	2012—02	28.00	164
绝对值方程—折边与组合图形的解析研究	2012—07	48.00	186
代数函数论（上）	2015—07	38.00	494
代数函数论（下）	2015—07	38.00	495
偏微分方程论：法文	2015—10	48.00	533
时标动力学方程的指数型二分性与周期解	2016—04	48.00	606
重刚体绕不动点运动方程的积分法	2016—05	68.00	608
水轮机水力稳定性	2016—05	48.00	620
Lévy 噪音驱动的传染病模型的动力学行为	2016—05	48.00	667
铣加工动力学系统稳定性研究的数学方法	2016—11	28.00	710
时滞系统：Lyapunov 泛函和矩阵	2017—05	68.00	784
粒子图像测速仪实用指南：第二版	2017—08	78.00	790
数域的上同调	2017—08	98.00	799
趣味初等方程妙题集锦	2014—09	48.00	388
趣味初等数论选美与欣赏	2015—02	48.00	445
耕读笔记（上卷）：一位农民数学爱好者的初数探索	2015—04	28.00	459
耕读笔记（中卷）：一位农民数学爱好者的初数探索	2015—05	28.00	483
耕读笔记（下卷）：一位农民数学爱好者的初数探索	2015—05	28.00	484
几何不等式研究与欣赏.上卷	2016—01	88.00	547
几何不等式研究与欣赏.下卷	2016—01	48.00	552
初等数列研究与欣赏·上	2016—01	48.00	570
初等数列研究与欣赏·下	2016—01	48.00	571
趣味初等函数研究与欣赏.上	2016—09	48.00	684
趣味初等函数研究与欣赏.下	即将出版		685
火柴游戏	2016—05	38.00	612
智力解谜.第 1 卷	2017—07	38.00	613
智力解谜.第 2 卷	2017—07	38.00	614
故事智力	2016—07	48.00	615
名人们喜欢的智力问题	即将出版		616
数学大师的发现、创造与失误	即将出版		617
异曲同工	即将出版		618
数学的味道	即将出版		798
数贝偶拾——高考数学题研究	2014—04	28.00	274
数贝偶拾——初等数学研究	2014—04	38.00	275
数贝偶拾——奥数题研究	2014—04	48.00	276
集合、函数与方程	2014—01	28.00	300
数列与不等式	2014—01	38.00	301
三角与平面向量	2014—01	28.00	302
平面解析几何	2014—01	38.00	303
立体几何与组合	2014—01	28.00	304
极限与导数、数学归纳法	2014—01	38.00	305
趣味数学	2014—03	28.00	306
教材教法	2014—04	68.00	307
自主招生	2014—05	58.00	308
高考压轴题（上）	2015—01	48.00	309
高考压轴题（下）	2014—10	68.00	310

哈尔滨工业大学出版社刘培杰数学工作室
已出版（即将出版）图书目录

书　名	出版时间	定　价	编号
从费马到怀尔斯——费马大定理的历史	2013—10	198.00	Ⅰ
从庞加莱到佩雷尔曼——庞加莱猜想的历史	2013—10	298.00	Ⅱ
从切比雪夫到爱尔特希（上）——素数定理的初等证明	2013—07	48.00	Ⅲ
从切比雪夫到爱尔特希（下）——素数定理100年	2012—12	98.00	Ⅲ
从高斯到盖尔方特——二次域的高斯猜想	2013—10	198.00	Ⅳ
从库默尔到朗兰兹——朗兰兹猜想的历史	2014—01	98.00	Ⅴ
从比勃巴赫到德布朗斯——比勃巴赫猜想的历史	2014—02	298.00	Ⅵ
从麦比乌斯到陈省身——麦比乌斯变换与麦比乌斯带	2014—02	298.00	Ⅶ
从布尔到豪斯道夫——布尔方程与格论漫谈	2013—10	198.00	Ⅷ
从开普勒到阿诺德——三体问题的历史	2014—05	298.00	Ⅸ
从华林到华罗庚——华林问题的历史	2013—10	298.00	Ⅹ
吴振奎高等数学解题真经（概率统计卷）	2012—01	38.00	149
吴振奎高等数学解题真经（微积分卷）	2012—01	68.00	150
吴振奎高等数学解题真经（线性代数卷）	2012—01	58.00	151
钱昌本教你快乐学数学（上）	2011—12	48.00	155
钱昌本教你快乐学数学（下）	2012—03	58.00	171
高等数学解题全攻略（上卷）	2013—06	58.00	252
高等数学解题全攻略（下卷）	2013—06	58.00	253
高等数学复习纲要	2014—01	18.00	384
三角函数	2014—01	38.00	311
不等式	2014—01	38.00	312
数列	2014—01	38.00	313
方程	2014—01	28.00	314
排列和组合	2014—01	28.00	315
极限与导数	2014—01	28.00	316
向量	2014—09	38.00	317
复数及其应用	2014—08	28.00	318
函数	2014—01	38.00	319
集合	即将出版		320
直线与平面	2014—01	28.00	321
立体几何	2014—04	28.00	322
解三角形	即将出版		323
直线与圆	2014—01	28.00	324
圆锥曲线	2014—01	38.00	325
解题通法（一）	2014—07	38.00	326
解题通法（二）	2014—07	38.00	327
解题通法（三）	2014—05	38.00	328
概率与统计	2014—01	28.00	329
信息迁移与算法	即将出版		330
方程（第2版）	2017—04	38.00	624
三角函数（第2版）	2017—04	38.00	626
向量（第2版）	即将出版		627
立体几何（第2版）	2016—04	38.00	629
直线与圆（第2版）	2016—11	38.00	631
圆锥曲线（第2版）	2016—09	48.00	632
极限与导数（第2版）	2016—04	38.00	635

哈尔滨工业大学出版社刘培杰数学工作室
已出版(即将出版)图书目录

书　名	出版时间	定　价	编号
美国高中数学竞赛五十讲.第1卷(英文)	2014—08	28.00	357
美国高中数学竞赛五十讲.第2卷(英文)	2014—08	28.00	358
美国高中数学竞赛五十讲.第3卷(英文)	2014—09	28.00	359
美国高中数学竞赛五十讲.第4卷(英文)	2014—09	28.00	360
美国高中数学竞赛五十讲.第5卷(英文)	2014—10	28.00	361
美国高中数学竞赛五十讲.第6卷(英文)	2014—11	28.00	362
美国高中数学竞赛五十讲.第7卷(英文)	2014—12	28.00	363
美国高中数学竞赛五十讲.第8卷(英文)	2015—01	28.00	364
美国高中数学竞赛五十讲.第9卷(英文)	2015—01	28.00	365
美国高中数学竞赛五十讲.第10卷(英文)	2015—02	38.00	366
IMO 50 年.第1卷(1959—1963)	2014—11	28.00	377
IMO 50 年.第2卷(1964—1968)	2014—11	28.00	378
IMO 50 年.第3卷(1969—1973)	2014—09	28.00	379
IMO 50 年.第4卷(1974—1978)	2016—04	38.00	380
IMO 50 年.第5卷(1979—1984)	2015—04	38.00	381
IMO 50 年.第6卷(1985—1989)	2015—04	58.00	382
IMO 50 年.第7卷(1990—1994)	2016—01	48.00	383
IMO 50 年.第8卷(1995—1999)	2016—06	38.00	384
IMO 50 年.第9卷(2000—2004)	2015—04	58.00	385
IMO 50 年.第10卷(2005—2009)	2016—01	48.00	386
IMO 50 年.第11卷(2010—2015)	2017—03	48.00	646
历届美国大学生数学竞赛试题集.第一卷(1938—1949)	2015—01	28.00	397
历届美国大学生数学竞赛试题集.第二卷(1950—1959)	2015—01	28.00	398
历届美国大学生数学竞赛试题集.第三卷(1960—1969)	2015—01	28.00	399
历届美国大学生数学竞赛试题集.第四卷(1970—1979)	2015—01	18.00	400
历届美国大学生数学竞赛试题集.第五卷(1980—1989)	2015—01	28.00	401
历届美国大学生数学竞赛试题集.第六卷(1990—1999)	2015—01	28.00	402
历届美国大学生数学竞赛试题集.第七卷(2000—2009)	2015—08	18.00	403
历届美国大学生数学竞赛试题集.第八卷(2010—2012)	2015—01	18.00	404
新课标高考数学创新题解题诀窍:总论	2014—09	28.00	372
新课标高考数学创新题解题诀窍:必修 1～5 分册	2014—08	38.00	373
新课标高考数学创新题解题诀窍:选修 2—1,2—2,1—1,1—2分册	2014—09	38.00	374
新课标高考数学创新题解题诀窍:选修 2—3,4—4,4—5 分册	2014—09	18.00	375
全国重点大学自主招生英文数学试题全攻略:词汇卷	2015—07	48.00	410
全国重点大学自主招生英文数学试题全攻略:概念卷	2015—01	28.00	411
全国重点大学自主招生英文数学试题全攻略:文章选读卷(上)	2016—09	38.00	412
全国重点大学自主招生英文数学试题全攻略:文章选读卷(下)	2017—01	58.00	413
全国重点大学自主招生英文数学试题全攻略:试题卷	2015—07	38.00	414
全国重点大学自主招生英文数学试题全攻略:名著欣赏卷	2017—03	48.00	415
数学物理大百科全书.第1卷	2016—01	418.00	508
数学物理大百科全书.第2卷	2016—01	408.00	509
数学物理大百科全书.第3卷	2016—01	396.00	510
数学物理大百科全书.第4卷	2016—01	408.00	511
数学物理大百科全书.第5卷	2016—01	368.00	512

哈尔滨工业大学出版社刘培杰数学工作室
已出版(即将出版)图书目录

书　名	出版时间	定价	编号
劳埃德数学趣题大全.题目卷.1:英文	2016—01	18.00	516
劳埃德数学趣题大全.题目卷.2:英文	2016—01	18.00	517
劳埃德数学趣题大全.题目卷.3:英文	2016—01	18.00	518
劳埃德数学趣题大全.题目卷.4:英文	2016—01	18.00	519
劳埃德数学趣题大全.题目卷.5:英文	2016—01	18.00	520
劳埃德数学趣题大全.答案卷:英文	2016—01	18.00	521
李成章教练奥数笔记.第1卷	2016—01	48.00	522
李成章教练奥数笔记.第2卷	2016—01	48.00	523
李成章教练奥数笔记.第3卷	2016—01	38.00	524
李成章教练奥数笔记.第4卷	2016—01	38.00	525
李成章教练奥数笔记.第5卷	2016—01	38.00	526
李成章教练奥数笔记.第6卷	2016—01	38.00	527
李成章教练奥数笔记.第7卷	2016—01	38.00	528
李成章教练奥数笔记.第8卷	2016—01	48.00	529
李成章教练奥数笔记.第9卷	2016—01	28.00	530
朱德祥代数与几何讲义.第1卷	2017—01	38.00	697
朱德祥代数与几何讲义.第2卷	2017—01	28.00	698
朱德祥代数与几何讲义.第3卷	2017—01	28.00	699
zeta函数,q-zeta函数,相伴级数与积分	2015—08	88.00	513
微分形式:理论与练习	2015—08	58.00	514
离散与微分包含的逼近和优化	2015—08	58.00	515
艾伦·图灵:他的工作与影响	2016—01	98.00	560
测度理论概率导论,第2版	2016—01	88.00	561
带有潜在故障恢复系统的半马尔柯夫模型控制	2016—01	98.00	562
数学分析原理	2016—01	88.00	563
随机偏微分方程的有效动力学	2016—01	88.00	564
图的谱半径	2016—01	58.00	565
量子机器学习中数据挖掘的量子计算方法	2016—01	98.00	566
量子物理的非常规方法	2016—01	118.00	567
运输过程的统一非局部理论:广义波尔兹曼物理动力学,第2版	2016—01	198.00	568
量子力学与经典力学之间的联系在原子、分子及电动力学系统建模中的应用	2016—01	58.00	569
算术域:第3版	2017—08	158.00	820
第19~23届"希望杯"全国数学邀请赛试题审题要津详细评注(初一版)	2014—03	28.00	333
第19~23届"希望杯"全国数学邀请赛试题审题要津详细评注(初二、初三版)	2014—03	38.00	334
第19~23届"希望杯"全国数学邀请赛试题审题要津详细评注(高一版)	2014—03	28.00	335
第19~23届"希望杯"全国数学邀请赛试题审题要津详细评注(高二版)	2014—03	38.00	336
第19~25届"希望杯"全国数学邀请赛试题审题要津详细评注(初一版)	2015—01	38.00	416
第19~25届"希望杯"全国数学邀请赛试题审题要津详细评注(初二、初三版)	2015—01	58.00	417
第19~25届"希望杯"全国数学邀请赛试题审题要津详细评注(高一版)	2015—01	48.00	418
第19~25届"希望杯"全国数学邀请赛试题审题要津详细评注(高二版)	2015—01	48.00	419
闵嗣鹤文集	2011—03	98.00	102
吴从炘数学活动三十年(1951~1980)	2010—07	99.00	32
吴从炘数学活动又三十年(1981~2010)	2015—07	98.00	491

哈尔滨工业大学出版社刘培杰数学工作室
已出版（即将出版）图书目录

书　名	出版时间	定价	编号
物理奥林匹克竞赛大题典——力学卷	2014—11	48.00	405
物理奥林匹克竞赛大题典——热学卷	2014—04	28.00	339
物理奥林匹克竞赛大题典——电磁学卷	2015—07	48.00	406
物理奥林匹克竞赛大题典——光学与近代物理卷	2014—06	28.00	345
历届中国东南地区数学奥林匹克试题集(2004～2012)	2014—06	18.00	346
历届中国西部地区数学奥林匹克试题集(2001～2012)	2014—07	18.00	347
历届中国女子数学奥林匹克试题集(2002～2012)	2014—08	18.00	348
数学奥林匹克在中国	2014—06	98.00	344
数学奥林匹克问题集	2014—01	38.00	267
数学奥林匹克不等式散论	2010—06	38.00	124
数学奥林匹克不等式欣赏	2011—09	38.00	138
数学奥林匹克超级题库(初中卷上)	2010—01	58.00	66
数学奥林匹克不等式证明方法和技巧(上、下)	2011—08	158.00	134,135
他们学什么：原民主德国中学数学课本	2016—09	38.00	658
他们学什么：英国中学数学课本	2016—09	38.00	659
他们学什么：法国中学数学课本.1	2016—09	38.00	660
他们学什么：法国中学数学课本.2	2016—09	28.00	661
他们学什么：法国中学数学课本.3	2016—09	38.00	662
他们学什么：苏联中学数学课本	2016—09	28.00	679
高中数学题典——集合与简易逻·函数	2016—07	48.00	647
高中数学题典——导数	2016—07	48.00	648
高中数学题典——三角函数·平面向量	2016—07	48.00	649
高中数学题典——数列	2016—07	58.00	650
高中数学题典——不等式·推理与证明	2016—07	38.00	651
高中数学题典——立体几何	2016—07	48.00	652
高中数学题典——平面解析几何	2016—07	78.00	653
高中数学题典——计数原理·统计·概率·复数	2016—07	48.00	654
高中数学题典——算法·平面几何·初等数论·组合数学·其他	2016—07	68.00	655
台湾地区奥林匹克数学竞赛试题.小学一年级	2017—03	38.00	722
台湾地区奥林匹克数学竞赛试题.小学二年级	2017—03	38.00	723
台湾地区奥林匹克数学竞赛试题.小学三年级	2017—03	38.00	724
台湾地区奥林匹克数学竞赛试题.小学四年级	2017—03	38.00	725
台湾地区奥林匹克数学竞赛试题.小学五年级	2017—03	38.00	726
台湾地区奥林匹克数学竞赛试题.小学六年级	2017—03	38.00	727
台湾地区奥林匹克数学竞赛试题.初中一年级	2017—03	38.00	728
台湾地区奥林匹克数学竞赛试题.初中二年级	2017—03	38.00	729
台湾地区奥林匹克数学竞赛试题.初中三年级	2017—03	28.00	730
不等式证题法	2017—04	28.00	747
平面几何培优教程	即将出版		748
奥数鼎级培优教程.高一分册	即将出版		749
奥数鼎级培优教程.高二分册	即将出版		750
高中数学竞赛冲刺宝典	即将出版		751

哈尔滨工业大学出版社刘培杰数学工作室
已出版（即将出版）图书目录

书　名	出版时间	定　价	编号
斯米尔诺夫高等数学.第一卷	2017—02	88.00	770
斯米尔诺夫高等数学.第二卷.第一分册	2017—02	68.00	771
斯米尔诺夫高等数学.第二卷.第二分册	2017—02	68.00	772
斯米尔诺夫高等数学.第二卷.第三分册	2017—02	48.00	773
斯米尔诺夫高等数学.第三卷.第一分册	2017—06	48.00	774
斯米尔诺夫高等数学.第三卷.第二分册	2017—02	58.00	775
斯米尔诺夫高等数学.第三卷.第三分册	2017—02	68.00	776
斯米尔诺夫高等数学.第四卷.第一分册	2017—02	48.00	777
斯米尔诺夫高等数学.第四卷.第二分册	2017—02	88.00	778
斯米尔诺夫高等数学.第五卷.第一分册	2017—04	58.00	779
斯米尔诺夫高等数学.第五卷.第二分册	2017—02	68.00	780
初中尖子生数学超级题典.实数	2017—07	58.00	792
初中尖子生数学超级题典.式、方程与不等式	2017—08	58.00	793
初中尖子生数学超级题典.圆、面积	2017—08	38.00	794
初中尖子生数学超级题典.函数、逻辑推理	2017—08	48.00	795
初中尖子生数学超级题典.角、线段、三角形与多边形	2017—07	58.00	796

联系地址：哈尔滨市南岗区复华四道街 10 号　哈尔滨工业大学出版社刘培杰数学工作室
　　网　　址：http://lpj.hit.edu.cn/
　　邮　　编：150006
联系电话：0451—86281378　　　13904613167
　　E-mail：lpj1378@163.com